KIRK-OTHMER

ENCYCLOPEDIA OF
CHEMICAL
TECHNOLOGY

FOURTH EDITION

VOLUME **23**

SUGAR
TO
THIN FILMS

EXECUTIVE EDITOR
Jacqueline I. Kroschwitz

EDITOR
Mary Howe-Grant

KIRK-OTHMER

ENCYCLOPEDIA OF CHEMICAL TECHNOLOGY

FOURTH EDITION

VOLUME **23**

SUGAR
TO
THIN FILMS

A Wiley-Interscience Publication
JOHN WILEY & SONS

New York • Chichester • Weinheim • Brisbane • Singapore • Toronto

Copyright © 1997 by John Wiley & Sons, Inc.

Library of Congress Cataloging-in-Publication Data

Encyclopedia of chemical technology/executive editor, Jacqueline
  I. Kroschwitz; editor, Mary Howe-Grant.—4th ed.
    p. cm.
  At head of title: Kirk-Othmer.
  "A Wiley-Interscience publication."
  Contents: v. 23, Sugar to thin films
  ISBN 0471-52692-4 (v. 23)
  1. Chemistry, Technical—Encyclopedias.   I. Kirk, Raymond E.
(Raymond Eller), 1890–1957.   II. Othmer, Donald F. (Donald
Frederick), 1904–1995.   III. Kroschwitz, Jacqueline I., 1942– .
IV. Howe-Grant, Mary, 1943– .   V. Title: Kirk-Othmer encyclopedia
of chemical technology.
TP9.E685 1992   91-16789
660′.03—dc20

CONTENTS

EDITORIAL STAFF FOR VOLUME 23

Executive Editor: **Jacqueline I. Kroschwitz**
Editor: **Mary Howe-Grant**
Associate Managing Editor: **Lindy Humphreys**
Copy Editors: **Lawrence Altieri**
              Jonathan Lee
Assistant Managing Editor: **Brendan A. Vilardo**

CONTRIBUTORS TO VOLUME 23

Michael M. Abbott, *Rensselaer Polytechnic Institute, Troy, New York*, Thermodynamics

Douglas A. Balentine, *Lipton, Englewood Cliffs, New Jersey*, Tea

Pamela Banks-Lee, *North Carolina State University, Raleigh*, Testing (under Textiles)

W. D. Betts, *Tar Industries Services, Chesterfield, England*, Tar and pitch

James O. Bledsoe, Jr., *Bush Boake Allen, Inc., Jacksonville, Florida,* Terpenoids

Bob Blumenthal, *Noah Technologies Corporation, San Antonio, Texas,* Thallium and thallium compounds

Barbara H. Bory, *Lever Company, Edgewater, New Jersey,* Surfactants

Timothy A. Calamari, *United States Department of Agriculture, New Orleans, Louisiana,* Finishing (under Textiles)

Michael Capone, *Exxon Engineering, Florham Park, New Jersey,* Sulfur removal and recovery

S. C. Carapella, *Consultant, Tukahoe, New York,* Tellurium and tellurium compounds

Earl Clark, *Phillips Research Center, Bartlesville, Oklahoma,* Sulfolane and sulfones

Margaret A. Clarke, *Sugar Processing Research Institute, Inc., New Orleans, Louisiana,* Cane sugar (under Sugar)

Michael Cleary, *Imperial Holly Corporation, Colorado Springs, Colorado,* Beet sugar (under Sugar)

William J. Colonna, *American Crystal Sugar Company, Moorhead, Minnesota,* Properties of sucrose (under Sugar)

Charles C. Coutant, *Oak Ridge National Laboratory, Oak Ridge, Tennessee,* Thermal pollution

David J. Dixon, *South Dakota School of Mines and Technology, Rapid City,* Supercritical fluids

Michael P. Duncan, *Ferro Corporation, Hammond, Indiana,* Sulfurization and sulfurchlorination

J. Eckert, *H. C. Starck Inc., Newton, Massachusetts,* Tantalum and tantalum compounds

Walter Fabisiak, *Sherwood-Davis and Geck, Connecticut,* Sutures

Mary An Godshall, *Sugar Processing Research Institute, Inc., New Orleans, Louisiana,* Sugar analysis (under Sugar)

R. A. Guest, *James Robinson, Ltd., Yorkshire, England,* Sulfur dyes

Robert J. Harper, Jr., *United States Department of Agriculture, New Orleans, Louisiana,* Finishing (under Textiles)

Ronald E. Hebeda, *CPC International Inc., Summit-Argo, Illinois,* Syrups

James E. Hoffmann, *Jan Reimers and Associates USA Inc., Salt Lake City, Utah,* Tellurium and tellurium compounds

Derk T. A. Huibers, *Union Camp Corporation, Princeton, New Jersey,* Tall oil

Keith P. Johnston, *University of Texas, Austin,* Supercritical fluids

Riaz Khan, *Polytech, Trieste, Italy,* Sugar derivatives (under Sugar)

Edward A. Knaggs, *Consultant, Deerfield, Illinois,* Sulfonation and sulfation

Paul A. Konowicz, *Polytech, Trieste, Italy,* Sugar derivatives (under Sugar)

B. C. Lawes, *E. I. Du Pont de Nemours & Company, Inc., Wilmington, Delaware,* Sulfuric and sulfurous esters

Mary E. Lawson, *SPI Polyols, Inc., New Castle, Delaware,* Sugar alcohols

Thomas D. Lee, *Kraft Foods, Tarrytown, New York,* Sweeteners

O. Griffin Lewis, *Consultant, Norwalk, Connecticut,* Sutures

Jesse L. Lynn, Jr., *Lever Company, Edgewater, New Jersey*, Surfactants

Donald M. Mattox, *Management Plus, Inc., Albuquerque, New Mexico*, Film formation techniques (under Thin films)

Edward F. McCarthy, *Luzenac America, Englewood, Colorado*, Talc

W. B. McCormack, *E. I. Du Pont de Nemours & Company, Inc., Wilmington, Delaware*, Sulfuric and sulfurous esters

Thomas L. Muller, *DuPont Chambers Works, Deepwater, New Jersey*, Sulfuric acid and sulfur trioxide

Philip Myers, *Chevron Research and Technical Company, Orinda, California*, Tanks and pressure vessels

Marshall J. Nepras, *Stepan Company, Northfield, Illinois*, Sulfonation and sulfation

J. E. Oldfield, *Oregon State University, Corvallis*, Tellurium and tellurium compounds

Jan Pegram, *North Carolina State University, Raleigh*, Testing (under Textiles)

R. D. Putnam, *Putnam Environmental Services, Research Triangle Park, North Carolina*, Tellurium and tellurium compounds

Ludwig Rebenfeld, *TRI/Princeton, Princeton, New Jersey*, Survey (under Textiles)

Austin H. Reid, Jr., *E. I. Du Pont de Nemours & Company, New Johnsonville, Tennessee*, Technical service

Upasiri Samaraweera, *American Crystal Sugar Company, Moorhead, Minnesota*, Properties of sucrose (under Sugar)

Stanley R. Sandler, *Elf Atochem North America, Inc., King of Prussia, Pennsylvania*, Sulfur compounds

E. A. Skrabek, *Orbital Sciences Corporation, Germantown, Maryland*, Thermoelectric energy conversion

J. Senior, *James Robinson, Ltd., Yorkshire, England*, Sulfur dyes

Henry E. Sostmann, *Consultant, Albuquerque, New Mexico*, Temperature measurement

James G. Speight, *Western Research Institute, Laramie, Wyoming*, Tar sands

The Sulphur Institute, *Washington, D.C.*, Sulfur

Michio Suzuki, *Nissan Chemical Industries, Ltd., Tokyo, Japan*, Sulfamic acid and sulfamates

Laszlo Toth, *The Western Sugar Company, Denver, Colorado*, Sugar economics (under Sugar)

Terrance B. Tripp, *H. C. Starck Inc., Newton, Massachusetts*, Tantalum and tantalum compounds

Paul S. Tully, *Stepan Company, Northfield, Illinois*, Sulfonic acids

Abraham Ulman, *Polytechnic University, Brooklyn, New York*, Monomolecular layers (under Thin films)

Hendrick C. Van Ness, *Rensselaer Polytechnic Institute, Troy, New York*, Thermodynamics

Edward D. Weil, *Polytechnic University, Brooklyn, New York*, Sulfur compounds

J. S. White, *White Technical Research Group, Argenta, Illinois*, Special sugars (under Sugar)

Rodney Willer, *Gaylord Chemical Corporation, Slidell, Louisiana*, Sulfoxides

W. E. Wood, *James Robinson, Ltd., Yorkshire, England*, Sulfur dyes

Katsumasa Yoshikubo, *Nissan Chemical Industries, Ltd., Tokyo, Japan*, Sulfamic acid and sulfamates

NOTE ON CHEMICAL ABSTRACTS SERVICE REGISTRY NUMBERS AND NOMENCLATURE

Chemical Abstracts Service (CAS) Registry Numbers are unique numerical identifiers assigned to substances recorded in the CAS Registry System. They appear in brackets in the *Chemical Abstracts* (CA) substance and formula indexes following the names of compounds. A single compound may have synonyms in the chemical literature. A simple compound like phenethylamine can be named β-phenylethylamine or, as in *Chemical Abstracts*, benzeneethanamine. The usefulness of the *Encyclopedia* depends on accessibility through the most common correct name of a substance. Because of this diversity in nomenclature careful attention has been given to the problem in order to assist the reader as much as possible, especially in locating the systematic CA index name by means of the Registry Number. For this purpose, the reader may refer to the CAS Registry Handbook—Number Section which lists in numerical order the Registry Number with the *Chemical Abstracts* index name and the molecular formula; eg, **458-88-8**, Piperidine, 2-propyl-, (*S*)-, $C_8H_{17}N$; in the *Encyclopedia* this compound would be found under its common name, coniine [*458-88-8*]. Alternatively, this information can be retrieved electronically from CAS Online. In many cases molecular formulas have also been provided in the *Encyclopedia* text to facilitate electronic searching. The Registry Number is a valuable link for the reader in retrieving additional published information on substances and also as a point of access for on-line data bases.

In all cases, the CAS Registry Numbers have been given for title compounds in articles and for all compounds in the index. All specific substances indexed in *Chemical Abstracts* since 1965 are included in the CAS Registry System as are a large number of substances derived from a variety of reference works. The CAS Registry System identifies a substance on the basis of an unambiguous computer-language description of its molecular structure including stereochemical detail. The Registry Number is a machine-checkable number (like a Social Security number) assigned in sequential order to each substance as it enters the registry system. The value of the number lies in the fact that it is a concise and unique means of substance identification, which is independent of, and therefore bridges, many systems of chemical nomenclature. For polymers, one Registry Number may be used for the entire family; eg, polyoxyethylene (20) sorbitan monolaurate has the same number as all of its polyoxyethylene homologues.

Cross-references are inserted in the index for many common names and for some systematic names. Trademark names appear in the index. Names that are incorrect, misleading, or ambiguous are avoided. Formulas are given very frequently in the text to help in identifying compounds. The spelling and form used, even for industrial names, follow American chemical usage, but not always the usage of *Chemical Abstracts* (eg, *coniine* is used instead of *(S)-2-propylpiperidine*, *aniline* instead of *benzenamine*, and *acrylic acid* instead of *2-propenoic acid*).

There are variations in representation of rings in different disciplines. The dye industry does not designate aromaticity or double bonds in rings. All double bonds and aromaticity are shown in the *Encyclopedia* as a matter of course. For example, tetralin has an aromatic ring and a saturated ring and its structure

appears in the *Encyclopedia* with its common name, Registry Number enclosed in brackets, and parenthetical CA index name, ie, tetralin [*119-64-2*] (1,2,3,4-tetrahydronaphthalene). With names and structural formulas, and especially with CAS Registry Numbers, the aim is to help the reader have a concise means of substance identification.

CONVERSION FACTORS, ABBREVIATIONS, AND UNIT SYMBOLS

SI Units (Adopted 1960)

The International System of Units (abbreviated SI), is being implemented throughout the world. This measurement system is a modernized version of the MKSA (meter, kilogram, second, ampere) system, and its details are published and controlled by an international treaty organization (The International Bureau of Weights and Measures) (1).

SI units are divided into three classes:

BASE UNITS

length	meter[†] (m)
mass	kilogram (kg)
time	second (s)
electric current	ampere (A)
thermodynamic temperature[‡]	kelvin (K)
amount of substance	mole (mol)
luminous intensity	candela (cd)

SUPPLEMENTARY UNITS

plane angle	radian (rad)
solid angle	steradian (sr)

[†]The spellings "metre" and "litre" are preferred by ASTM; however, "-er" is used in the *Encyclopedia*.

[‡]Wide use is made of Celsius temperature (t) defined by

$$t = T - T_0$$

where T is the thermodynamic temperature, expressed in kelvin, and $T_0 = 273.15$ K by definition. A temperature interval may be expressed in degrees Celsius as well as in kelvin.

DERIVED UNITS AND OTHER ACCEPTABLE UNITS

These units are formed by combining base units, supplementary units, and other derived units (2–4). Those derived units having special names and symbols are marked with an asterisk in the list below.

Quantity	Unit	Symbol	Acceptable equivalent
*absorbed dose	gray	Gy	J/kg
acceleration	meter per second squared	m/s^2	
*activity (of a radionuclide)	becquerel	Bq	1/s
area	square kilometer	km^2	
	square hectometer	hm^2	ha (hectare)
	square meter	m^2	
concentration (of amount of substance)	mole per cubic meter	mol/m^3	
current density	ampere per square meter	$A//m^2$	
density, mass density	kilogram per cubic meter	kg/m^3	g/L; mg/cm^3
dipole moment (quantity)	coulomb meter	C·m	
*dose equivalent	sievert	Sv	J/kg
*electric capacitance	farad	F	C/V
*electric charge, quantity of electricity	coulomb	C	A·s
electric charge density	coulomb per cubic meter	C/m^3	
*electric conductance	siemens	S	A/V
electric field strength	volt per meter	V/m	
electric flux density	coulomb per square meter	C/m^2	
*electric potential, potential difference, electromotive force	volt	V	W/A
*electric resistance	ohm	Ω	V/A
*energy, work, quantity of heat	megajoule	MJ	
	kilojoule	kJ	
	joule	J	N·m
	electronvolt[†]	eV[†]	
	kilowatt-hour[†]	kW·h[†]	
energy density	joule per cubic meter	J/m^3	
*force	kilonewton	kN	
	newton	N	$kg·m/s^2$

[†]This non-SI unit is recognized by the CIPM as having to be retained because of practical importance or use in specialized fields (1).

Quantity	Unit	Symbol	Acceptable equivalent
*frequency	megahertz	MHz	
	hertz	Hz	1/s
heat capacity, entropy	joule per kelvin	J/K	
heat capacity (specific), specific entropy	joule per kilogram kelvin	J/(kg·K)	
heat-transfer coefficient	watt per square meter kelvin	W/(m²·K)	
*illuminance	lux	lx	lm/m²
*inductance	henry	H	Wb/A
linear density	kilogram per meter	kg/m	
luminance	candela per square meter	cd/m²	
*luminous flux	lumen	lm	cd·sr
magnetic field strength	ampere per meter	A/m	
*magnetic flux	weber	Wb	V·s
*magnetic flux density	tesla	T	Wb/m²
molar energy	joule per mole	J/mol	
molar entropy, molar heat capacity	joule per mole kelvin	J/(mol·K)	
moment of force, torque	newton meter	N·m	
momentum	kilogram meter per second	kg·m/s	
permeability	henry per meter	H/m	
permittivity	farad per meter	F/m	
*power, heat flow rate, radiant flux	kilowatt	kW	
	watt	W	J/s
power density, heat flux density, irradiance	watt per square meter	W/m²	
*pressure, stress	megapascal	MPa	
	kilopascal	kPa	
	pascal	Pa	N/m²
sound level	decibel	dB	
specific energy	joule per kilogram	J/kg	
specific volume	cubic meter per kilogram	m³/kg	
surface tension	newton per meter	N/m	
thermal conductivity	watt per meter kelvin	W/(m·K)	
velocity	meter per second	m/s	
	kilometer per hour	km/h	
viscosity, dynamic	pascal second	Pa·s	
	millipascal second	mPa·s	
viscosity, kinematic	square meter per second	m²/s	
	square millimeter per second	mm²/s	

Quantity	Unit	Symbol	Acceptable equivalent
volume	cubic meter	m^3	
	cubic diameter	dm^3	L (liter) (5)
	cubic centimeter	cm^3	mL
wave number	1 per meter	m^{-1}	
	1 per centimeter	cm^{-1}	

In addition, there are 16 prefixes used to indicate order of magnitude, as follows:

Multiplication factor	Prefix	Symbol	Note
10^{18}	exa	E	
10^{15}	peta	P	
10^{12}	tera	T	
10^9	giga	G	
10^6	mega	M	
10^3	kilo	k	
10^2	hecto	h[a]	[a]Although hecto, deka, deci, and centi
10	deka	da[a]	are SI prefixes, their use should be
10^{-1}	deci	d[a]	avoided except for SI unit-multiples
10^{-2}	centi	c[a]	for area and volume and nontech-
10^{-3}	milli	m	nical use of centimeter, as for body
10^{-6}	micro	μ	and clothing measurement.
10^{-9}	nano	n	
10^{-12}	pico	p	
10^{-15}	femto	f	
10^{-18}	atto	a	

For a complete description of SI and its use the reader is referred to ASTM E380 (4) and the article UNITS AND CONVERSION FACTORS which appears in Vol. 24.

A representative list of conversion factors from non-SI to SI units is presented herewith. Factors are given to four significant figures. Exact relationships are followed by a dagger. A more complete list is given in the latest editions of ASTM E380 (4) and ANSI Z210.1 (6).

Conversion Factors to SI Units

To convert from	To	Multiply by
acre	square meter (m^2)	4.047×10^3
angstrom	meter (m)	1.0×10^{-10}[†]
are	square meter (m^2)	1.0×10^2[†]

[†]Exact.

To convert from	To	Multiply by
astronomical unit	meter (m)	1.496×10^{11}
atmosphere, standard	pascal (Pa)	1.013×10^5
bar	pascal (Pa)	$1.0 \times 10^{5\dagger}$
barn	square meter (m²)	$1.0 \times 10^{-28\dagger}$
barrel (42 U.S. liquid gallons)	cubic meter (m³)	0.1590
Bohr magneton (μ_B)	J/T	9.274×10^{-24}
Btu (International Table)	joule (J)	1.055×10^3
Btu (mean)	joule (J)	1.056×10^3
Btu (thermochemical)	joule (J)	1.054×10^3
bushel	cubic meter (m³)	3.524×10^{-2}
calorie (International Table)	joule (J)	4.187
calorie (mean)	joule (J)	4.190
calorie (thermochemical)	joule (J)	$4.184^\dagger$
centipoise	pascal second (Pa·s)	$1.0 \times 10^{-3\dagger}$
centistokes	square millimeter per second (mm²/s)	$1.0^\dagger$
cfm (cubic foot per minute)	cubic meter per second (m³/s)	4.72×10^{-4}
cubic inch	cubic meter (m³)	1.639×10^{-5}
cubic foot	cubic meter (m³)	2.832×10^{-2}
cubic yard	cubic meter (m³)	0.7646
curie	becquerel (Bq)	$3.70 \times 10^{10\dagger}$
debye	coulomb meter (C·m)	3.336×10^{-30}
degree (angle)	radian (rad)	1.745×10^{-2}
denier (international)	kilogram per meter (kg/m)	1.111×10^{-7}
	tex‡	0.1111
dram (apothecaries')	kilogram (kg)	3.888×10^{-3}
dram (avoirdupois)	kilogram (kg)	1.772×10^{-3}
dram (U.S. fluid)	cubic meter (m³)	3.697×10^{-6}
dyne	newton (N)	$1.0 \times 10^{-5\dagger}$
dyne/cm	newton per meter (N/m)	$1.0 \times 10^{-3\dagger}$
electronvolt	joule (J)	1.602×10^{-19}
erg	joule (J)	$1.0 \times 10^{-7\dagger}$
fathom	meter (m)	1.829
fluid ounce (U.S.)	cubic meter (m³)	2.957×10^{-5}
foot	meter (m)	$0.3048^\dagger$
footcandle	lux (lx)	10.76
furlong	meter (m)	2.012×10^{-2}
gal	meter per second squared (m/s²)	$1.0 \times 10^{-2\dagger}$
gallon (U.S. dry)	cubic meter (m³)	4.405×10^{-3}
gallon (U.S. liquid)	cubic meter (m³)	3.785×10^{-3}
gallon per minute (gpm)	cubic meter per second (m³/s)	6.309×10^{-5}
	cubic meter per hour (m³/h)	0.2271

†Exact.
‡See footnote on p. xiii.

To convert from	To	Multiply by
gauss	tesla (T)	1.0×10^{-4}
gilbert	ampere (A)	0.7958
gill (U.S.)	cubic meter (m³)	1.183×10^{-4}
grade	radian	1.571×10^{-2}
grain	kilogram (kg)	6.480×10^{-5}
gram force per denier	newton per tex (N/tex)	8.826×10^{-2}
hectare	square meter (m²)	$1.0 \times 10^{4†}$
horsepower (550 ft·lbf/s)	watt (W)	7.457×10^{2}
horsepower (boiler)	watt (W)	9.810×10^{3}
horsepower (electric)	watt (W)	$7.46 \times 10^{2†}$
hundredweight (long)	kilogram (kg)	50.80
hundredweight (short)	kilogram (kg)	45.36
inch	meter (m)	$2.54 \times 10^{-2†}$
inch of mercury (32°F)	pascal (Pa)	3.386×10^{3}
inch of water (39.2°F)	pascal (Pa)	2.491×10^{2}
kilogram-force	newton (N)	9.807
kilowatt hour	megajoule (MJ)	$3.6†$
kip	newton (N)	4.448×10^{3}
knot (international)	meter per second (m/S)	0.5144
lambert	candela per square meter (cd/m³)	3.183×10^{3}
league (British nautical)	meter (m)	5.559×10^{3}
league (statute)	meter (m)	4.828×10^{3}
light year	meter (m)	9.461×10^{15}
liter (for fluids only)	cubic meter (m³)	$1.0 \times 10^{-3†}$
maxwell	weber (Wb)	$1.0 \times 10^{-8†}$
micron	meter (m)	$1.0 \times 10^{-6†}$
mil	meter (m)	$2.54 \times 10^{-5†}$
mile (statute)	meter (m)	1.609×10^{3}
mile (U.S. nautical)	meter (m)	$1.852 \times 10^{3†}$
mile per hour	meter per second (m/s)	0.4470
millibar	pascal (Pa)	1.0×10^{2}
millimeter of mercury (0°C)	pascal (Pa)	$1.333 \times 10^{2†}$
minute (angular)	radian	2.909×10^{-4}
myriagram	kilogram (kg)	10
myriameter	kilometer (km)	10
oersted	ampere per meter (A/m)	79.58
ounce (avoirdupois)	kilogram (kg)	2.835×10^{-2}
ounce (troy)	kilogram (kg)	3.110×10^{-2}
ounce (U.S. fluid)	cubic meter (m³)	2.957×10^{-5}
ounce-force	newton (N)	0.2780
peck (U.S.)	cubic meter (m³)	8.810×10^{-3}
pennyweight	kilogram (kg)	1.555×10^{-3}
pint (U.S. dry)	cubic meter (m³)	5.506×10^{-4}
pint (U.S. liquid)	cubic meter (m³)	4.732×10^{-4}

†Exact.

To convert from	To	Multiply by
poise (absolute viscosity)	pascal second (Pa·s)	$0.10^{\dagger}$
pound (avoirdupois)	kilogram (kg)	0.4536
pound (troy)	kilogram (kg)	0.3732
poundal	newton (N)	0.1383
pound-force	newton (N)	4.448
pound force per square inch (psi)	pascal (Pa)	6.895×10^3
quart (U.S. dry)	cubic meter (m^3)	1.101×10^{-3}
quart (U.S. liquid)	cubic meter (m^3)	9.464×10^{-4}
quintal	kilogram (kg)	$1.0 \times 10^{2\dagger}$
rad	gray (Gy)	$1.0 \times 10^{-2\dagger}$
rod	meter (m)	5.029
roentgen	coulomb per kilogram (C/kg)	2.58×10^{-4}
second (angle)	radian (rad)	$4.848 \times 10^{-6\dagger}$
section	square meter (m^2)	2.590×10^6
slug	kilogram (kg)	14.59
spherical candle power	lumen (lm)	12.57
square inch	square meter (m^2)	6.452×10^{-4}
square foot	square meter (m^2)	9.290×10^{-2}
square mile	square meter (m^2)	2.590×10^6
square yard	square meter (m^2)	0.8361
stere	cubic meter (m^3)	$1.0^{\dagger}$
stokes (kinematic viscosity)	square meter per second (m^2/s)	$1.0 \times 10^{-4\dagger}$
tex	kilogram per meter (kg/m)	$1.0 \times 10^{-6\dagger}$
ton (long, 2240 pounds)	kilogram (kg)	1.016×10^3
ton (metric) (tonne)	kilogram (kg)	$1.0 \times 10^{3\dagger}$
ton (short, 2000 pounds)	kilogram (kg)	9.072×10^2
torr	pascal (Pa)	1.333×10^2
unit pole	weber (Wb)	1.257×10^{-7}
yard	meter (m)	$0.9144^{\dagger}$

†Exact.

Abbreviations and Unit Symbols

Following is a list of common abbreviations and unit symbols used in the *Encyclopedia*. In general they agree with those listed in *American National Standard Abbreviations for Use on Drawings and in Text (ANSI Y1.1)* (6) and *American National Standard Letter Symbols for Units in Science and Technology (ANSI Y10)* (6). Also included is a list of acronyms for a number of private and government organizations as well as common industrial solvents, polymers, and other chemicals.

Rules for Writing Unit Symbols (4):

1. Unit symbols are printed in upright letters (roman) regardless of the type style used in the surrounding text.
2. Unit symbols are unaltered in the plural.
3. Unit symbols are not followed by a period except when used at the end of a sentence.
4. Letter unit symbols are generally printed lower-case (for example, cd for candela) unless the unit name has been derived from a proper name, in which case the first letter of the symbol is capitalized (W, Pa). Prefixes and unit symbols retain their prescribed form regardless of the surrounding typography.
5. In the complete expression for a quantity, a space should be left between the numerical value and the unit symbol. For example, write 2.37 lm, *not* 2.37lm, and 35 mm, *not* 35mm. When the quantity is used in an adjectival sense, a hyphen is often used, for example, 35-mm film. *Exception:* No space is left between the numerical value and the symbols of degree, minute, and second of plane angle, degree Celsius, and the percent sign.
6. No space is used between the prefix and unit symbol (for example, kg).
7. Symbols, not abbreviations, should be used for units. For example, use "A," not "amp," for ampere.
8. When multiplying unit symbols, use a raised dot:

$$\text{N·m} \quad \text{for} \quad \text{newton meter}$$

   In the case of W·h, the dot may be omitted, thus:

$$\text{Wh}$$

   An exception to this practice is made for computer printouts, automatic typewriter work, etc, where the raised dot is not possible, and a dot on the line may be used.
9. When dividing unit symbols, use one of the following forms:

$$\text{m/s} \quad or \quad \text{m·s}^{-1} \quad or \quad \frac{\text{m}}{\text{s}}$$

   In no case should more than one slash be used in the same expression unless parentheses are inserted to avoid ambiguity. For example, write:

$$\text{J/(mol·K)} \quad or \quad \text{J·mol}^{-1}\text{·K}^{-1} \quad or \quad \text{(J/mol)/K}$$

   but *not*

$$\text{J/mol/K}$$

10. Do not mix symbols and unit names in the same expression. Write:

$$\text{joules per kilogram} \quad or \quad \text{J/kg} \quad or \quad \text{J·kg}^{-1}$$

but *not*

$$\text{joules/kilogram} \quad nor \quad \text{joules/kg} \quad nor \quad \text{joules·kg}^{-1}$$

ABBREVIATIONS AND UNITS

A	ampere		AOAC	Association of Official Analytical Chemists
A	anion (eg, HA)			
A	mass number		AOCS	American Oil Chemists' Society
a	atto (prefix for 10^{-18})			
AATCC	American Association of Textile Chemists and Colorists		APHA	American Public Health Association
			API	American Petroleum Institute
ABS	acrylonitrile–butadiene–styrene		aq	aqueous
abs	absolute		Ar	aryl
ac	alternating current, *n.*		*ar-*	aromatic
a-c	alternating current, *adj.*		*as-*	asymmetric(al)
ac-	alicyclic		ASHRAE	American Society of Heating, Refrigerating, and Air Conditioning Engineers
acac	acetylacetonate			
ACGIH	American Conference of Governmental Industrial Hygienists			
			ASM	American Society for Metals
ACS	American Chemical Society		ASME	American Society of Mechanical Engineers
AGA	American Gas Association		ASTM	American Society for Testing and Materials
Ah	ampere hour			
AIChE	American Institute of Chemical Engineers		at no.	atomic number
			at wt	atomic weight
AIME	American Institute of Mining, Metallurgical, and Petroleum Engineers		av(g)	average
			AWS	American Welding Society
			b	bonding orbital
AIP	American Institute of Physics		bbl	barrel
			bcc	body-centered cubic
AISI	American Iron and Steel Institute		BCT	body-centered tetragonal
			Bé	Baumé
alc	alcohol(ic)		BET	Brunauer-Emmett-Teller (adsorption equation)
Alk	alkyl			
alk	alkaline (not alkali)		bid	twice daily
amt	amount		Boc	*t*-butyloxycarbonyl
amu	atomic mass unit		BOD	biochemical (biological) oxygen demand
ANSI	American National Standards Institute			
			bp	boiling point
AO	atomic orbital		Bq	becquerel

C	coulomb
°C	degree Celsius
C-	denoting attachment to carbon
c	centi (prefix for 10^{-2})
c	critical
ca	circa (approximately)
cd	candela; current density; circular dichroism
CFR	Code of Federal Regulations
cgs	centimeter-gram-second
CI	Color Index
cis-	isomer in which substituted groups are on same side of double bond between C atoms
cl	carload
cm	centimeter
cmil	circular mil
cmpd	compound
CNS	central nervous system
CoA	coenzyme A
COD	chemical oxygen demand
coml	commercial(ly)
cp	chemically pure
cph	close-packed hexagonal
CPSC	Consumer Product Safety Commission
cryst	crystalline
cub	cubic
D	debye
D-	denoting configurational relationship
d	differential operator
d	day; deci (prefix for 10^{-1})
d	density
d-	*dextro*-, dextrorotatory
da	deka (prefix for 10^1)
dB	decibel
dc	direct current, *n.*
d-c	direct current, *adj.*
dec	decompose
detd	determined
detn	determination
Di	didymium, a mixture of all lanthanons
dia	diameter
dil	dilute

DIN	Deutsche Industrie Normen
dl-; DL-	racemic
DMA	dimethylacetamide
DMF	dimethylformamide
DMG	dimethyl glyoxime
DMSO	dimethyl sulfoxide
DOD	Department of Defense
DOE	Department of Energy
DOT	Department of Transportation
DP	degree of polymerization
dp	dew point
DPH	diamond pyramid hardness
dstl(d)	distill(ed)
dta	differential thermal analysis
(*E*)-	entgegen; opposed
ϵ	dielectric constant (unitless number)
e	electron
ECU	electrochemical unit
ed.	edited, edition, editor
ED	effective dose
EDTA	ethylenediaminetetraacetic acid
emf	electromotive force
emu	electromagnetic unit
en	ethylene diamine
eng	engineering
EPA	Environmental Protection Agency
epr	electron paramagnetic resonance
eq.	equation
esca	electron spectroscopy for chemical analysis
esp	especially
esr	electron-spin resonance
est(d)	estimate(d)
estn	estimation
esu	electrostatic unit
exp	experiment, experimental
ext(d)	extract(ed)
F	farad (capacitance)
F	faraday (96,487 C)
f	femto (prefix for 10^{-15})

FAO	Food and Agriculture Organization (United Nations)	hyd	hydrated, hydrous
		hyg	hygroscopic
fcc	face-centered cubic	Hz	hertz
FDA	Food and Drug Administration	^{i} (eg, Pr^i)	iso (eg, isopropyl)
		i-	inactive (eg, i-methionine)
FEA	Federal Energy Administration	IACS	International Annealed Copper Standard
FHSA	Federal Hazardous Substances Act	ibp	initial boiling point
		IC	integrated circuit
fob	free on board	ICC	Interstate Commerce Commission
fp	freezing point		
FPC	Federal Power Commission	ICT	International Critical Table
FRB	Federal Reserve Board	ID	inside diameter; infective dose
frz	freezing		
G	giga (prefix for 10^9)	ip	intraperitoneal
G	gravitational constant = 6.67×10^{11} N·m^2/kg^2	IPS	iron pipe size
		ir	infrared
g	gram	IRLG	Interagency Regulatory Liaison Group
(g)	gas, only as in H$_2$O(g)		
g	gravitational acceleration	ISO	International Organization Standardization
gc	gas chromatography		
gem-	geminal	ITS-90	International Temperature Scale (NIST)
glc	gas–liquid chromatography		
		IU	International Unit
g-mol wt; gmw	gram-molecular weight	IUPAC	International Union of Pure and Applied Chemistry
GNP	gross national product	IV	iodine value
gpc	gel-permeation chromatography	iv	intravenous
		J	joule
GRAS	Generally Recognized as Safe	K	kelvin
		k	kilo (prefix for 10^3)
grd	ground	kg	kilogram
Gy	gray	L	denoting configurational relationship
H	henry		
h	hour; hecto (prefix for 10^2)	L	liter (for fluids only) (5)
ha	hectare	l-	$levo$-, levorotatory
HB	Brinell hardness number	(l)	liquid, only as in NH$_3$(l)
Hb	hemoglobin	LC$_{50}$	conc lethal to 50% of the animals tested
hcp	hexagonal close-packed		
hex	hexagonal	LCAO	linear combination of atomic orbitals
HK	Knoop hardness number		
hplc	high performance liquid chromatography	lc	liquid chromatography
		LCD	liquid crystal display
HRC	Rockwell hardness (C scale)	lcl	less than carload lots
		LD$_{50}$	dose lethal to 50% of the animals tested
HV	Vickers hardness number		

LED	light-emitting diode	N-	denoting attachment to nitrogen
liq	liquid		
lm	lumen	n (as n_D^{20})	index of refraction (for 20°C and sodium light)
ln	logarithm (natural)		
LNG	liquefied natural gas	n (as Bu^n),	
log	logarithm (common)	n-	normal (straight-chain structure)
LOI	limiting oxygen index		
LPG	liquefied petroleum gas	n	neutron
ltl	less than truckload lots	n	nano (prefix for 10^9)
lx	lux	na	not available
M	mega (prefix for 10^6); metal (as in MA)	NAS	National Academy of Sciences
M	molar; actual mass	NASA	National Aeronautics and Space Administration
$\overline{M}_w$	weight-average mol wt		
$\overline{M}_n$	number-average mol wt	nat	natural
m	meter; milli (prefix for 10^{-3})	ndt	nondestructive testing
		neg	negative
m	molal	NF	*National Formulary*
m-	meta	NIH	National Institutes of Health
max	maximum		
MCA	Chemical Manufacturers' Association (was Manufacturing Chemists Association)	NIOSH	National Institute of Occupational Safety and Health
		NIST	National Institute of Standards and Technology (formerly National Bureau of Standards)
MEK	methyl ethyl ketone		
meq	milliequivalent		
mfd	manufactured		
mfg	manufacturing		
mfr	manufacturer	nmr	nuclear magnetic resonance
MIBC	methyl isobutyl carbinol		
MIBK	methyl isobutyl ketone	NND	New and Nonofficial Drugs (AMA)
MIC	minimum inhibiting concentration		
		no.	number
min	minute; minimum	NOI-(BN)	not otherwise indexed (by name)
mL	milliliter		
MLD	minimum lethal dose	NOS	not otherwise specified
MO	molecular orbital	nqr	nuclear quadruple resonance
mo	month		
mol	mole	NRC	Nuclear Regulatory Commission; National Research Council
mol wt	molecular weight		
mp	melting point		
MR	molar refraction	NRI	New Ring Index
ms	mass spectrometry	NSF	National Science Foundation
MSDS	material safety data sheet		
mxt	mixture	NTA	nitrilotriacetic acid
μ	micro (prefix for 10^{-6})	NTP	normal temperature and pressure (25°C and 101.3 kPa or 1 atm)
N	newton (force)		
N	normal (concentration); neutron number		

NTSB	National Transportation Safety Board	qv	quod vide (which see)
O-	denoting attachment to oxygen	R	univalent hydrocarbon radical
o-	ortho	(*R*)-	rectus (clockwise configuration)
OD	outside diameter	*r*	precision of data
OPEC	Organization of Petroleum Exporting Countries	rad	radian; radius
o-phen	*o*-phenanthridine	RCRA	Resource Conservation and Recovery Act
OSHA	Occupational Safety and Health Administration	rds	rate-determining step
owf	on weight of fiber	ref.	reference
Ω	ohm	rf	radio frequency, *n*.
P	peta (prefix for 10^{15})	r-f	radio frequency, *adj*.
p	pico (prefix for 10^{-12})	rh	relative humidity
p-	para	RI	Ring Index
p	proton	rms	root-mean square
p.	page	rpm	rotations per minute
Pa	pascal (pressure)	rps	revolutions per second
PEL	personal exposure limit based on an 8-h exposure	RT	room temperature
		RTECS	Registry of Toxic Effects of Chemical Substances
pd	potential difference	ˢ (eg, Buˢ);	
pH	negative logarithm of the effective hydrogen ion concentration	*sec*-	secondary (eg, secondary butyl)
		S	siemens
phr	parts per hundred of resin (rubber)	(*S*)-	sinister (counterclockwise configuration)
p-i-n	positive-intrinsic-negative	*S*-	denoting attachment to sulfur
pmr	proton magnetic resonance	*s*-	symmetric(al)
p-n	positive-negative	s	second
po	per os (oral)	(s)	solid, only as in $H_2O(s)$
POP	polyoxypropylene	SAE	Society of Automotive Engineers
pos	positive		
pp.	pages	SAN	styrene-acrylonitrile
ppb	parts per billion (10^9)	sat(d)	saturate(d)
ppm	parts per million (10^6)	satn	saturation
ppmv	parts per million by volume	SBS	styrene–butadiene–styrene
ppmwt	parts per million by weight	sc	subcutaneous
PPO	poly(phenyl oxide)	SCF	self-consistent field; standard cubic feet
ppt(d)	precipitate(d)		
pptn	precipitation	Sch	Schultz number
Pr (no.)	foreign prototype (number)	sem	scanning electron microscope(y)
pt	point; part		
PVC	poly(vinyl chloride)	SFs	Saybolt Furol seconds
pwd	powder	sl sol	slightly soluble
py	pyridine	sol	soluble

soln	solution	*trans-*	isomer in which
soly	solubility		substituted groups are
sp	specific; species		on opposite sides of
sp gr	specific gravity		double bond between C
sr	steradian		atoms
std	standard	TSCA	Toxic Substances Control
STP	standard temperature and		Act
	pressure (0°C and 101.3	TWA	time-weighted average
	kPa)	Twad	Twaddell
sub	sublime(s)	UL	Underwriters' Laboratory
SUs	Saybolt Universal seconds	USDA	United States Department
syn	synthetic		of Agriculture
^{t} (eg, Bu^t),		USP	*United States*
t-, tert-	tertiary (eg, tertiary		*Pharmacopeia*
	butyl)	uv	ultraviolet
T	tera (prefix for 10^{12}); tesla	V	volt (emf)
	(magnetic flux density)	var	variable
t	metric ton (tonne)	*vic-*	vicinal
t	temperature	vol	volume (not volatile)
TAPPI	Technical Association of	vs	versus
	the Pulp and Paper	v sol	very soluble
	Industry	W	watt
TCC	Tagliabue closed cup	Wb	weber
tex	tex (linear density)	Wh	watt hour
T_g	glass-transition	WHO	World Health
	temperature		Organization (United
tga	thermogravimetric		Nations)
	analysis	wk	week
THF	tetrahydrofuran	yr	year
tlc	thin layer chromatography	(Z)-	zusammen; together;
TLV	threshold limit value		atomic number

Non-SI (Unacceptable and Obsolete) Units		Use
Å	angstrom	nm
at	atmosphere, technical	Pa
atm	atmosphere, standard	Pa
b	barn	cm^2
bar†	bar	Pa
bbl	barrel	m^3
bhp	brake horsepower	W
Btu	British thermal unit	J
bu	bushel	m^3; L
cal	calorie	J
cfm	cubic foot per minute	m^3/s
Ci	curie	Bq
cSt	centistokes	mm^2/s
c/s	cycle per second	Hz

†Do not use bar (10^5 Pa) or millibar (10^2 Pa) because they are not SI units, and are accepted internationally only for a limited time in special fields because of existing usage.

Non-SI (Unacceptable and Obsolete) Units		Use
cu	cubic	exponential form
D	debye	$C \cdot m$
den	denier	tex
dr	dram	kg
dyn	dyne	N
dyn/cm	dyne per centimeter	mN/m
erg	erg	J
eu	entropy unit	J/K
°F	degree Fahrenheit	°C; K
fc	footcandle	lx
fl	footlambert	lx
fl oz	fluid ounce	m^3; L
ft	foot	m
ft·lbf	foot pound-force	J
gf den	gram-force per denier	N/tex
G	gauss	T
Gal	gal	m/s^2
gal	gallon	m^3; L
Gb	gilbert	A
gpm	gallon per minute	(m^3/s); (m^3/h)
gr	grain	kg
hp	horsepower	W
ihp	indicated horsepower	W
in.	inch	m
in. Hg	inch of mercury	Pa
in. H_2O	inch of water	Pa
in.-lbf	inch pound-force	J
kcal	kilo-calorie	J
kgf	kilogram-force	N
kilo	for kilogram	kg
L	lambert	lx
lb	pound	kg
lbf	pound-force	N
mho	mho	S
mi	mile	m
MM	million	M
mm Hg	millimeter of mercury	Pa
$m\mu$	millimicron	nm
mph	miles per hour	km/h
μ	micron	μm
Oe	oersted	A/m
oz	ounce	kg
ozf	ounce-force	N
η	poise	Pa·s
P	poise	Pa·s
ph	phot	lx
psi	pounds-force per square inch	Pa
psia	pounds-force per square inch absolute	Pa
psig	pounds-force per square inch gage	Pa
qt	quart	m^3; L
°R	degree Rankine	K
rd	rad	Gy
sb	stilb	lx
SCF	standard cubic foot	m^3
sq	square	exponential form
thm	therm	J
yd	yard	m

BIBLIOGRAPHY

1. The International Bureau of Weights and Measures, BIPM (Parc de Saint-Cloud, France) is described in Appendix X2 of Ref. 4. This bureau operates under the exclusive supervision of the International Committee for Weights and Measures (CIPM).
2. *Metric Editorial Guide (ANMC-78-1)*, latest ed., American National Metric Council, 5410 Grosvenor Lane, Bethesda, Md. 20814, 1981.
3. *SI Units and Recommendations for the Use of Their Multiples and of Certain Other Units (ISO 1000-1981)*, American National Standards Institute, 1430 Broadway, New York, 10018, 1981.
4. Based on *ASTM E380-89a (Standard Practice for Use of the International System of Units (SI))*, American Society for Testing and Materials, 1916 Race Street, Philadelphia, Pa. 19103, 1989.
5. *Fed. Reg.*, Dec. 10, 1976 (41 FR 36414).
6. For ANSI address, see Ref. 3.

R. P. LUKENS
ASTM Committee E-43 on SI Practice

Continued

SUGAR

PROPERTIES OF SUCROSE

Sucrose [*57-50-1*] (β-D-fructofuranosyl-α-D-glucopyranoside), $C_{12}H_{22}O_{11}$, formula weight 342.3, is a disaccharide composed of glucose and fructose residues joined by an α,β-glycosidic bond (Fig. 1).

The most common sugar in plants, sucrose is formed as a result of photosynthesis and occurs in abundance in sugarcane (*Saccharum officinarum*)

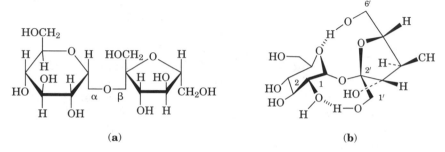

Fig. 1. Structural representations of sucrose: (**a**) Haworth perspective formula, and (**b**) conformational structure of sucrose in solid crystals. Adapted from Ref. 1.

1

and sugarbeets (*Beta vulgaris*) in amounts ranging from 12–15 and 13–20% by weight, respectively (see SUGAR, BEET SUGAR; SUGAR, CANE SUGAR). Commercial quantities are provided only from these two sources. Other sources include sorghum (*Sorghum vulgare*), the sugar maple (*Acer saccharum*), and the date palm (*Phoenix dactylifera*) (2). Sucrose is called maple, cane, beet, and more familiarly, table sugar.

For centuries, sucrose has been valued for its pure, sweet taste. There are indications of sugar production having occurred in what is now New Guinea (the geographic origin of sugarcane) as early as 12,000 BC (3). Cane production spread to other parts of the Pacific and India by 6,000 BC, and to China by ca 800 BC. Sugarcane reached present-day Iran by 500 AD, Egypt and Spain in the eighth century, and Sicily by 950 AD (4). Knowledge of sugar spread throughout Europe during the Crusades (3). In 1493, sugarcane reached the New World with Columbus. The first sugar mill in the Western Hemisphere was built near Santo Domingo in 1508 (2).

The sweetness of sugarbeets was recorded in 1590 AD. Sugar was isolated from beets by Marggraf in 1747 and the first beet sugar factory was established in Silesia by 1802. The industry grew during the Napoleonic wars, and the year 1813 saw the presence of over 300 sugar mills in Germany, France, and Austria–Hungary. The first American beet sugar mill was built in 1838 (3).

In the initial stages of purification, sucrose is recovered in juice form by crushing cane stalks or by extraction of sliced sugarbeets (cossettes) with hot water. The resulting solutions are clarified with lime, then evaporated to thick syrups from which sugar is recovered by crystallization. The final syrup obtained after exhaustive crystallization of sucrose is known as molasses. Enhanced recovery of sucrose from beet molasses is accomplished by ion-exclusion chromatography, a process used in some sugar mills in the United States, Japan, Finland, and Austria.

An *in vitro* enzymatic synthesis of sucrose was carried out in 1944 (5). A successful chemical synthesis was performed by Lemieux and Huber (6) in 1953 from acetylated sugar precursors. However, the economics and chemical complexities of both processes make them unlikely sources of supply.

Production and Consumption Statistics

Sugarcane is cultivated in tropical and semitropical regions, eg, Central and South America, Cuba, India, Australia, Africa, and the Far East. Sugarbeets are grown in more temperate climates such as North America, Europe, and the former Soviet Union. In some nations, eg, the United States, China, and Japan, sucrose is produced from both sources.

Production and consumption statistics for sucrose are shown in Table 1. World production of sucrose during 1993–1994 was ~110 million metric tons, of which ~64% was derived from sugarcane. The largest producer is the European Union (EU), followed closely by India and Brazil. In 1993–1994, the United States ranked fourth in production. World raw sugar prices from 1990–1995 ranged from 20.2¢–32.1¢/kg (10).

Between 1990 and 1995, the leading sugar exporters were the EU, Australia, Brazil, Thailand, Cuba, and the Ukraine, which together provided ~69%

Table 1. Worldwide Production and Consumption Statistics for Beet and Cane Sugar[a]

	World sugar production,[b] 10^6 t			U.S. sugar prdn,[b] 10^6 t	World sugar consumption,[b] 10^6 t
Crop year	Total sugar	Beet sugar	Cane sugar		
1990–1991	114.40	42.00	72.40	6.26	111.92
1991–1992	116.15	38.22	77.93	6.43	113.90
1992–1993	111.70	38.86	72.84	7.07	114.55
1993–1994	110.30	39.39	70.91	6.83	113.72
1994–1995	116.06	35.08	80.98	7.93	114.54

[a]Refs. 7–9.
[b]Raw value.

of total sugar exports. Sucrose demand was highest in the former Soviet Union, the EU, and India, which collectively accounted for ~34% of total consumption. In the United States, the fourth leading consumer, consumption after 1990 remained steady at 29.5 kg/capita (11), roughly two-thirds of consumption levels in the early 1900s. Demand by the food and beverage industry, which uses 70% of refined sucrose, has risen by ~300% since the beginning of the twentieth century (2).

Physical Properties of Sucrose

Physical properties of sucrose are summarized in Table 2. Sucrose is one of the purest substances available in bulk quantities, with purities averaging ~99.96%. Water accounts for about half of the nonsugar impurities (17).

Sucrose crystals are triboluminescent and emit light when they are fractured. Aqueous sucrose solutions rotate polarized light in direct proportion to

Table 2. Physical Properties of Sucrose[a]

Property	Value
density, d_4^{15}, kg/m^3	1587.9
melting point, °C	160–186 dec
specific rotation, degrees	+66.53
solubility in water at 20°C, g/g	2.00
apparent molar volume at 20°C, cm^3/mol	209.5
specific heat, J/mol[b]	
crystalline, at 20°C	415.8
amorphous, at 22°C	90.2
heat of solution, kJ/mol[b]	4.75 ± 0.26
dipole moment, C·m[c]	3.1×10^{-18}
enthalpy of crystallization at 30°C, kJ/mol[b]	10.5
bulk density, kg/m^3	
crystalline	930
powdered	600
normal entropy, J/(mol·K)[b]	360.5
angle of repose, degree	34

[a]Refs. 12–16.
[b]To convert J to cal, divide by 4.184.
[c]To convert C·m to debye, multiply by 3×10^{29}.

sugar concentration. This property is utilized for quantitation, purity evaluations during factory processing, and for setting the selling price of sugar. Sucrose quantitation has also been performed by colorimetric methods. However, in recent years, automated enzymatic analyzers and instrumental methods (eg, ion chromatography and hplc) have become increasingly popular, as they provide greater sensitivity and accuracy. Near infrared (nir) spectroscopy is currently under evaluation as a tool for sucrose quantitation in sugar mills and food processing operations.

The sweet taste of sucrose is its most notable and important physical property and is regarded as the standard against which other sweeteners (qv) are rated. Sweetness is influenced by temperature, pH, sugar concentration, physical properties of the food system, and other factors (18–20). The sweetening powers of sucrose and other sweeteners are compared in Table 3. The sweetness threshold for dissolved sucrose is ~0.2–0.5% and its sweetness intensity is highest at ~32–38°C (19).

Fructose is sweeter than sucrose at low temperatures (~5°C); at higher temperatures, the reverse is true. At 40°C, they have equal sweetness, the result of a temperature-induced shift in the percentages of α- and β-fructose anomers. The taste of sucrose is synergistic with high intensity sweeteners (eg, sucralose and aspartame) and can be enhanced or prolonged by substances like glycerol monostearate, lecithin, and maltol (19).

Sucrose has eight hydroxyl groups capable of hydrogen-bond formation. The glucose and fructose residues in crystalline sucrose are nearly perpendicular to each other and are held in this conformation by hydrogen bonds between the C_1–OH of fructose and the C_2–oxygen of glucose, and C_6–OH of fructose and the ring oxygen of glucose (Fig. 1b). In aqueous sugar solutions, the latter bond is absent; however, the overall conformation is essentially unchanged (1).

Elicitation of sweetness is explained by the AH–B–X Theory, or the Sweetness Triangle (1,23,24), wherein AH and B represent a hydrogen donor and acceptor, respectively, of sucrose. These interact with complementary regions of taste-receptor proteins. An extended hydrophobic region (X) of sucrose docks in a hydrophobic cleft of the receptor, facilitating optimal electrostatic interaction and sensory stimulation. The C_2- and C_3-hydroxyls of glucose (AH and B, respectively) and the back side of the fructose ring (X) are critical to this interaction (1).

The hydroxyl groups of sucrose contribute to its very high water solubility. This is enhanced by the presence of other dissolved solids and diminishes the yield of crystals produced from sugar syrups. Sucrose forms sphenoidic mono-

Table 3. Relative Sweetness of Sucrose and Other Sweet Substances[a]

Sweetener	Relative sweetness	Sweetener	Relative sweetness
fructose	1.2–1.8	saccharin	250–550
sucrose	1.00	aspartame	120–200
glucose	0.60	sucralose	550–750
maltose	~0.5	cyclamate	30–50
lactose	0.15–0.30	acesulfame K	~200
galactose	0.32	alitame	2000

[a]Refs. 18, 20–22.

clinic crystals (Fig. 2), the shapes of which are affected by syrup impurities like raffinose and dextran, which promote growth of the B- and C-axes, respectively, resulting in formation of elongated, needle-like crystals (17,25,26). A secondary consequence is a decrease in bulk density with its accompanying packaging problems (17). Impurities included within sucrose crystals impart color, raise ash levels, and reduce sugar quality. Thus, nonsucrose impurities can seriously impact the recovery and quality of sucrose. Sucrose also has moderate solubility in pyridine, glycerol, methylpyrrolidinone, methylpiperazine, DMSO, and DMF, and is slightly soluble in methanol and ethanol (16,27). The high dielectric constant of sucrose (3.50–3.85) gives it substantial microwave-absorbing capacity and makes it a valuable ingredient in formulating microwavable foods (18).

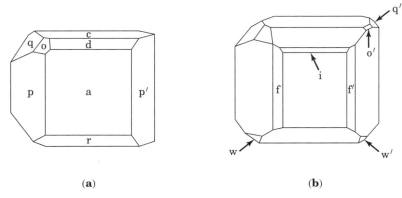

(**a**)   (**b**)

Fig. 2. Normal sucrose crystal showing: (**a**) a combination of the eight most important and frequent forms (aopp'drqo), and (**b**) the 15 simple forms determined with certainty for sugar crystals (acpp'droo'qq'ww'ff'i) (25).

Chemical Properties of Sucrose

The carbonyl groups of fructose and glucose partake in the glycosidic bond of sucrose, making the latter nonreducing. Therefore, sucrose has no anomeric forms, cannot undergo mutarotation or osazone formation, and is inert to alkaline copper reagents. The hydroxyl groups of sucrose are very weakly acidic; the C_2–OH of glucose is the most acidic (1). The pK_as of sucrose at 25°C are $pK_1 = 12.02$, $pK_2 = 12.56$, and $pK_3 = 12.01$ (28). The three primary hydroxyl groups of sucrose are most reactive (6=6' > 1); of the secondary hydroxyls, those on C-2 and C-3' are more reactive (29). The hydroxyls dissociate in alkali to form alcoholates called saccharates, and can be derivatized to produce valuable sucrochemicals.

Oxidation of Sucrose. Sucrose can be oxidized by HNO_3, $KMnO_4$, and peroxide. Under selected conditions using oxygen with palladium or platinum, the 6- or 6'-hydroxyls can be oxidized to form sucronic acid derivatives (29).

Hydrolysis of Sucrose. Sucrose can be enzymatically hydrolyzed to glucose and fructose by invertase. During this reaction, the optical rotation falls to a negative value owing to the large negative specific rotation of fructose. The reversal is called inversion and the resulting glucose–fructose mixture is called invert.

Sugar is destroyed by pH extremes, and inadequate pH control can cause significant sucrose losses in sugar mills. Sucrose is one of the most acid-labile disaccharides known (27), and its hydrolysis to invert is readily catalyzed by heat and low pH; prolonged exposure converts the monosaccharides to hydroxymethyl furfural, which has applications for synthesis of glycols, ethers, polymers, and pharmaceuticals (16,30). The molecular mechanism that occurs during acid hydrolysis operates, albeit slowly, as high as pH 8.5 (18).

Alkaline Degradation.   At high pH, sucrose is relatively stable; however, prolonged exposure to strong alkali and heat converts sucrose to a mixture of organic acids (mainly lactate), ketones, and cyclic condensation products. The mechanism of alkaline degradation is uncertain; however, initial formation of glucose and fructose apparently does not occur (31). In aqueous solutions, sucrose is most stable at ~pH 9.0.

Thermal Degradation.   The heats of formation and combustion of sucrose are -2.26 MJ/mol (540 kcal/mol) and -5.79 MJ/mol (-1384 kcal/mol), respectively (32,33). At high temperatures (160–186°C), sucrose decomposes with charring, emitting an odor of caramel. Thermolysis of crystalline sucrose at 170°C yielded a complex mixture of products (34). The mixture contained several nonreducing trisaccharides, including 6-kestose. Acid-catalyzed thermolysis causes decomposition to glucose and fructofuranosyl cation. The latter reacts with sucrose to form a complex mixture of products, including fructoglucan and several kestoses (35). These substances are examples of fructooligosaccharides (FOS) and are known to promote the growth of beneficial intestinal microorganisms.

Uses for Sucrose

Food Applications.   On the basis of intake, sucrose is the leading food additive (2). Its principal contribution to food is sweetness. However, it provides many other functionalities, eg, body, mouthfeel, texture, and moisture retention. Cereals and baked goods are the leading consumers of sucrose, followed closely by confectionery products (36).

Sucrose helps minimize earthy tastes of vegetables, while enhancing inherent flavors and aromas, and preserving color and texture (37). Addition of sucrose inhibits enzymatic browning of canned and frozen fruits, and prevents loss of color, flavor, and aroma from fruit during processing (38).

In baking, chemicals produced from sucrose by yeast contribute to bread flavor and sponginess of the dough (39); during cooking, reducing sugars formed by sucrose inversion combine with amino acids to form Maillard reaction products called melanoidins that impart brown color to breads, cakes, cookies, and cereals (18,19). Sugar raises the gelatinization temperature of starch, so cakes and breads rise more during baking and become softer and lighter (38). Sucrose also improves moisture retention by baked goods so they remain fresh for longer periods (18).

Subthreshold levels of sucrose enhance meat flavor (19); in cured meats, sucrose is a preservative and improves flavor by reducing the salty taste (38). Similarly, the acidity of condiments is softened by sugar. Sucrose stabilizes protein foams in meringues, and imparts body and smoothness to sauces and puddings. In candy mixes, sucrose raises the cooking temperature and minimizes formation of small sugar crystals during cooling. The reaction of milk proteins with sugars produces the caramel flavor in some candies (18).

Sucrose is often used as a decorative agent to impart a pleasing appearance to baked goods and confections (36). In jams and jellies, sugar raises osmotic pressure and lowers water activity to prevent spoilage (18). Sucrose is a fermentation substrate for lactic acid in cultured buttermilk (40) and lowers the freezing point of ice cream and other frozen desserts to improve product mouthfeel and texture.

Feedstock for Chemical Synthesis. It is estimated that <0.5% of the sucrose produced each year is used for nonfood purposes (41). An alternative application, namely the production of chemicals, is an attractive option as the feedstock is plentiful, renewable, and of consistently high purity. Moreover, the biodegradability of many sucrochemicals makes them environmentally friendly.

Sucrose reacts with fatty acids to produce esters with degrees of esterification (DE) from 1 to 8 and hydrophilic/lipophilic balances that provide them with numerous applications. Primary producers are Japan and the Netherlands, with total production at ~6000 t/yr. Sucrose esters are nontoxic and biodegradable, and are approved for use in the EC, Japan, and the United States.

Sucrose monoesters (SMEs) are used as nonionic surfactants, in detergents and as emulsifiers in foods. Some SMEs have bacteriostatic activity and are used to prevent spoilage in beverages. Semperfresh, an SME produced in the U.K., is used as a coating to extend the shelf life of fruits and vegetables (21,29).

Sucrose esters of mixed degree of esterification improve flour and starch quality, prevent amylose crystallization in baked goods and help maintain crispiness in cereals and other starch-based snacks. They also prevent sugar crystallization in jams, jellies, and ice cream, and ice crystal growth in frozen desserts. In sugar manufacture, sucrose esters improve sucrose recovery from low grade massecuites. Olestra, a lipase-resistant sucrose polyester (SPE) made by Proctor and Gamble (21), has been granted FDA approval and has applications as a low calorie fat mimetic. Other sucrose esters are used as alcohol denaturants, plasticizers, lubricants, bleach boosters, and in lacquer, ink, and cosmetic formulations (16,21,29,41).

Sucrose acrylate derivatives can be converted into polymers and hydrogels that can be used as flocculants, water adsorbents, bioimplantables, and drug delivery devices (42). Sucrose ethers have applications as surfactants and surface coatings, and as feedstocks for synthesis of polyurethane foams and plastics used for insulation, packaging materials, and wood replacements in furniture (21,41,43). Controlled oxidation of sucrose produces carboxylic acid derivatives used as cross-linking agents and detergent additives.

A chlorination process (20,21,44–46) converts sucrose into sucralose [56038-13-2] (4,1',6'-trichloro-4,1',6'-trideoxy-galactosucrose), a heat-stable, noncariogenic, noncaloric, high intensity sweetener. Sucralose is approved for food use in Canada, Australia, and Russia. It is not yet approved for use in the United States.

Several interesting and useful sucrose derivatives are enzymatically synthesized. One of these, called coupling sugar, is a mixture of glucosyl- and maltooligosyl-sucroses formed from starch hydrolysates and sucrose by cyclodextrin-transferase and is used as a noncariogenic sweetener in Japan (47). Similarly, fructosyloligosaccharides (FOS) are mixtures of fructosyl sucroses made from sucrose by fungal invertases (21,47–49) and are used in foods in Japan and Europe. Although less sweet than sucrose, FOSs are noncariogenic and noncaloric, and they promote the growth of beneficial intestinal flora (Bifidobacteria).

To date (ca 1996) many potentially useful sucrose derivatives have been synthesized. However, the economics and complexities of sucrochemical syntheses and the availability of cheaper substitutes have limited their acceptance; hence, only a few of them are in commercial use. A change in the price and availability of petroleum feedstocks could reverse this trend. Additional impetus may come from regioselective, site-specific modifications of sucrose to produce derivatives to facilitate and improve the economics of sucrochemical syntheses. For example, the microbe *Agrobacterium tumifaciens* selectively oxidizes sucrose to a three-keto derivative, a precursor of alkylated sucroses for detergent use (50). Similarly, enzymes have been used for selective synthesis of specific sucrose derivatives (21).

Fermentation Feedstock.   Sucrose, in the form of beet or cane molasses, is a fermentation feedstock for production of a variety of organic compounds, including lactic, glutamic, and citric acids, glycerol, and some antibiotics. Lesser amounts of itaconic, aconitic, and kojic acids, as well as acetone and butanol, are also produced (41,51–53). Rum is made by fermentation of cane molasses. Beet and cane molasses are used for production of baker's and brewer's yeast (qv).

A more abundantly produced substance is ethanol for use in alcoholic beverages, and as a fuel, solvent, and feedstock for organic syntheses. Ethanol (qv) production from sucrose is carried out in Europe (eg, France and the Netherlands), India, Pakistan, China, and on a very large scale in Brazil, where it is used as a motor fuel. A valuable by-product of ethanol fermentation is industrial CO_2 (see CARBON DIOXIDE).

The above chemicals can be obtained by fermentation (qv) of other sugars. However, some compounds require sucrose as a unique feedstock. Examples are the polysaccharides dextran, alternan, and levan, which are produced by specific strains of bacteria (48,54–56). Dextrans are used to make chromatographic separation media, and sulfated dextran derivatives are used as plasma extenders (41). Levans show promise as sweetness potentiators and, along with alternan, have potential as food thickeners and bulking agents in reduced-caloric foods (55,56) (see CARBOHYDRATES).

Sucrose is also a fermentation feedstock for production of palatinose [13718-94-0] (6-O-α-D-glucopyranosylfructose, or isomaltulose) and leucrose [7158-70-5] (5-O-α-D-glucopyranosylfructose). Palatinose is produced by fermentation using *Protaminobacter rubrum*, or enzymatically using α-glucosyltransferase; leucrose is produced by *Leuconostoc mesenteroides*. Both sugars are used as noncariogenic sweeteners in Japan and Europe (47,48,57).

Pharmaceutical Applications.   Sucrose has a long history in the manufacture of pharmaceuticals. It imparts body to syrups and medicinal liquids and masks unpleasant tastes. Sucrose also functions as a diluent to control drug concentrations in medicines, as an ingredient binder for tablets, and to impart chewiness to the latter. Sustained-release medications and protective tablet glazes are prepared using sucrose (41). Sucrose-based sugar pastes are used to promote wound healing (58).

Sucralfate [54182-58-0], an aluminum salt of sucrose octasulfate, is used as an antacid and antiulcer medication (59). Bis- and tris-platinum complexes of sucrose show promise as antitumor agents (60). Sucrose monoesters are used in some pharmaceutical preparations (21). A sucrose polyester is under evaluation as a contrast agent for magnetic resonance imaging (mri) (61). Oral administration of this substance opacifies the gastrointestinal tract and eliminates the need for purging prior to mri.

Nutrition and Health Aspects of Sucrose

For many years, there has been concern by medical professionals and nutritionists over the effects of dietary sugar on human health. Sucrose has been implicated as a cause of juvenile hyperactivity, tooth decay, diabetes mellitus, obesity, atherosclerosis, hypoglycemia, and nutrient deficiencies.

In 1986, the FDA's Sugars Task Force assessed the impact of sugar consumption on human health and nutrition and concluded that sucrose is not an independent risk factor for heart disease, nor does it cause or contribute to the development of diabetes (62). Although diet is important after the onset of diabetes, sucrose can be well tolerated by insulin-dependent diabetics (63–65).

Other studies indicate that sucrose does not cause hyperactivity. Carbohydrate ingestion increases levels of serotonin (5-hydroxytryptamine), a brain neurotransmitter that promotes relaxation and sleep. Dietary sucrose should theoretically have a calming effect and reduce activity, manifestations which have been observed in case studies (63). To date, clinical investigations have failed to show a significant connection between sucrose consumption and aggressive or disruptive behavior (66).

The rising incidence of obesity has not paralleled sucrose consumption. The FDA Task Force concluded that sugars have no unique role in obesity and that dietary fat rather than carbohydrate is a significant contributor to this condition (62,67,68). However, sugar can promote weight gain in individuals with life-styles marked by excess caloric intake and insufficient exercise.

The notion that complex carbohydrates elicit a gradual, steady secretion of insulin while sugars cause a sudden release of this hormone accompanied by a rapid drop in blood glucose has fostered the belief that hypoglycemia is affected by sucrose ingestion. However, research does not support this conclusion (63).

The Sugars Task Force's Select Committee on Nutrition and Human Needs recommended a daily consumption of sugars at 10% of total calories, which approximates current (11%) daily intake levels in the United States. At this level, sucrose does contribute to the development of dental caries; however, no firm evidence exists that it causes dietary imbalances or deficiencies of vitamins (qv), minerals, or trace nutrients (62).

BIBLIOGRAPHY

"Sucrose" under "Sugars (Commercial)" in *ECT* 1st ed., Vol. 13, pp. 247–251, by J. L. Hickson, Sugar Research Foundation, Inc.; "Properties of Sucrose" under "Sugar" in *ECT*

2nd ed., Vol. 19, pp. 151–155, by R. A. McGinnis, Spreckels Sugar Co.; in *ECT* 3rd ed., Vol. 21, pp. 865–870, by R. M. Sequeira, Amstar Corp.

1. F. W. Lichtenthaler and S. Immel, *Int. Sugar J.* **97**(1153), 12–22 (1995).
2. A. H. Ensminger and co-workers, *Foods and Nutrition Encyclopedia*, 2nd ed., CRC Press, Boca Raton, Fla., 1994, pp. 2060–2070.
3. L. Toth and A. B. Rizzuto, in N. L. Pennington and C. W. Baker, eds., *Sugar, A User's Guide to Sucrose*, Van Nostrand Reinhold Publishing Co., New York, 1990, pp. 1–10.
4. J. E. Irvine, in J. C. P. Chen and C.-C. Chou, eds., *Cane Sugar Handbook*, John Wiley & Sons, Inc., New York, 1993, pp. 1–18.
5. W. Z. Hassid, M. Doudoroff, and H. A. Barker, *J. Am. Chem. Soc.* **66**, 1416 (1944).
6. R. V. Lemieux and G. Huber, *J. Am. Chem. Soc.* **75**, 4118 (1953).
7. "World Sugar Statistics 1994/1995," in F. O. Lichts, ed., *World Sugar and Sweetener Yearbook 1995*, Ratzeburg, Germany, 1995, stats 10.
8. *Sugar y Azucar*, 10–12 (Apr. 1995).
9. R. Lord, *Sugar: Background for 1995 Farm Legislation*, Agricultural Economic Report No. 711, Commercial Agriculture Division, Economic Research Service, U.S. Department of Agriculture, Washington, D.C., 1995.
10. *Sugar and Sweetener Situation and Outlook Report/SSSV20N1*, U.S. Department of Agriculture, Washington, D.C., Mar. 1995.
11. *Sugar and Sweetener Situation and Outlook Report/SSSV19N4*, U.S. Department of Agriculture, Washington, D.C., Dec. 1994.
12. R. A. McGinnis, in R. A. McGinnis, ed., *Beet Sugar Technology*, 3rd ed., Beet Sugar Development Foundation, Fort Collins, Colo., 1982, pp. 25–63.
13. N. A. Armstrong and co-workers, in A. Wade and P. J. Weller, eds., *Handbook of Pharmaceutical Excipients*, 2nd ed., American Pharmaceutical Association, Washington, D.C. and The Pharmaceutical Press, Royal Pharmaceutical Society, London, 1994, pp. 500–505.
14. R. L. Knecht, in Ref. 3, pp. 46–65.
15. P. Reiser, G. G. Birch, and M. Mathlouthi, in M. Mathlouthi and P. Reiser, eds., *Sucrose: Properties and Applications*, Chapman and Hall, New York, 1995, pp. 186–222.
16. *The Merck Index*, 9th ed., Merck & Co., Inc., Rahway, N.J., 1976.
17. R. A. McGinnis, in R. A. McGinnis, ed., *Beet Sugar Technology*, 3rd ed., Beet Sugar Development Foundation, Fort Collins, Colo., 1982, pp. 545–566.
18. M. A. Clarke, in Ref. 15, pp. 223–247.
19. M. A. Godshall, in Ref. 15, pp. 248–263.
20. L. L. Hood and L. A. Campbell, *Cereal Foods World* **35**(12), 1171–1182 (1990).
21. R. Khan, in Ref. 15, pp. 264–278.
22. L. O'Brien Nabors and R. C. Gelardi, in L. O'Brien Nabors and R. C. Gelardi, eds., *Alternative Sweeteners*, Marcel Dekker, Inc., New York, 1991, pp. 1–18.
23. R. S. Shallenberger and T. Acree, *Nature* **216**, 480–482 (1967).
24. L. B. Kier, *J. Pharm. Sci.* **61**, 1394–1397 (1972).
25. J. C. P. Chen, in Ref. 4, pp. 343–374.
26. G. Mantovani, *Int. Sugar J.* **93**(1106), 23–24 (1991).
27. K. J. Parker, *La Sucrerie Belge* **93**, 15–27 (1974).
28. C. Francotte, J. Vandegans, and D. Jacqmain, *La Sucrerie Belge* **98**, 61–66 (1979).
29. L. Hough, in F. W. Lichtenthaler, ed., *Carbohydrates as Organic Raw Materials*, VCH Press, New York, 1990, pp. 34–55.
30. G. Descotes, in R. Chandrasekaran, ed., *Frontiers of Carbohydrate Research-2*, Elsevier Applied Science, New York, 1992, pp. 115–127.
31. D. W. Fewkes, K. J. Parker, and A. J. Vlitos, *Sci. Prog. Oxf.* **59**, 25–39 (1971).
32. J. A. Dean, ed., *Lange's Handbook of Chemistry*, 13th ed., McGraw-Hill Book Company, Inc., New York, 1985.

33. R. C. Weast and M. J. Astle, eds., *Handbook of Chemistry and Physics*, 63rd ed., CRC Press, Inc., Boca Raton, Fla., 1982.
34. D. Bollmann and S. Schmidt-Berg, *Z. Zuckerind. Boehm.* **15**, 179–265 (1965).
35. M. Manley-Harris and G. N. Richards, *Carbohydr. Res.* **240**, 183–196 (1993).
36. J. BeMiller, in Y. H. Hui, ed., *Encyclopedia of Food Science and Technology*, Vol. 4, John Wiley & Sons, Inc., New York, 1992, pp. 2441–2442.
37. R. L. Knecht, in Ref. 3, pp. 66–70.
38. N. D. Pintauro, in Ref. 3, pp. 165–181.
39. J. G. Ponte, in Ref. 3, pp. 130–151.
40. M. P. Penfield and A. M. Campbell, *Experimental Food Science*, 3rd ed., Academic Press, Inc., New York, 1990, pp. 485–493.
41. C. B. Broeg, in Ref. 3, pp. 276–287.
42. X. Chen, D. G. Rethwisch, and J. S. Dordick, Abs. No. 72, Div. of Carbohydrate Chemistry, *Abs. Ann. Mfg. Amer. Chem. Soc., Denver, Colo.* (Mar. 28–Apr. 2, 1993).
43. A. R. Meath and L. D. Booth, in J. L. Hickson, ed., *Sucrochemistry*, American Chemical Society, Washington, D.C., 1977, pp. 257–263.
44. U.K. Pat. 1,543,167 (1979), L. S. Hough, S. P. Phadnis, R. Khan, and M. R. Jenner (to Tate & Lyle).
45. U.S. Pat. 4,362,869 (Dec. 7, 1982), M. R. Jenner, D. Waite, and G. Jackson (to Tate & Lyle).
46. U.S. Pat. 4,435,440 (Mar. 6, 1984), L. S. Hough, P. L. Phadnis, and R. Khan (to Tate & Lyle).
47. S. Fujii and M. Komoto, *Zuckerind.* **116**(3), 197–200 (1991).
48. M. A. Clarke, in R. A. Meyers, ed., *Molecular Biology and Biotechnology, a Comprehensive Desk Reference*, VCH Publishers, New York, 1995, pp. 146–149.
49. H. Schiweck and co-workers, *Zuckerind.* **115**(7), 555–565 (1990).
50. M. Pietsch, M. Walter, and K. Buchholz, *Carbohydr. Res.* **254**, 183–194 (1994).
51. J. C. P. Chen, in Ref. 4, pp. 375–431.
52. T. Roukas, *J. Food Sci.* **56**(3), 878–880 (1991).
53. M. Nakagawa and co-workers, *Res. Bull. Obihiro Univ.*, **I**(17), 7–12 (1990).
54. M. J. Beker and co-workers, *Appl. Biochem. Biotechnol.* **24/25**, 265–274 (1990).
55. A. V. Bailey and co-workers, in *Proc. SPRI Res. Conf. San Francisco, Calif.*, 325–321 (May 29–June 1, 1990).
56. G. L. Cote, Abs. No. 291B, *Abs. Ann. Mtg. Amer. Inst. Chem. Eng., Chicago, Ill.* (Nov. 11–16, 1990).
57. Y. Nakajima, *Seito Gijyutso Kenkyu Kaishi* **26**, 55 (1984).
58. K. R. Middleton and D. Seal, *Pharm. J.* **235**, 757–759 (1985).
59. J. BeMiller, in F. W. Lichtenthaler, ed., *Carbohydrates as Raw Materials*, VCH Publishing Co., New York, 1991, pp. 197–206.
60. N. D. Sachinvala and co-workers, *J. Med. Chem.* **36**, 1791 (1993).
61. R. Ballinger and co-workers, *Magnet. Reson. Med.* **19**(1), 199–202 (1991).
62. W. H. Glinsmann, H. Irausquin, and Y. K. Park, *J. Nutr.* **16**(11), S1–S216 (1986).
63. R. M. Black, *Food Technol.* **47**(1), 130–133 (1993).
64. J. M. Steel, D. Mitchell, and R. L. Prescott, *Human Nutr. Appl. Nutr.* **37**, 3–8 (1983).
65. S. Vaaler, K. F. Hanssen, and O. Aagenaes, *Acta Med. Scand.* **207**, 371–373 (1980).
66. J. W. Wade and M. Wolraich, *Am. J. Clin. Nutr.* **62**, 242S–249S (1995).
67. G. H. Anderson, *Am. J. Clin. Nutr.* **62**, 195S–202S (1995).
68. J. O. Hill and A. M. Prentice, *Am. J. Clin. Nutr.* **62**, 264S–274S (1995).

General References

J. C. P. Chen and C.-C. Chou, eds., *Cane Sugar Handbook*, John Wiley & Sons, Inc., New York, 1993; comprehensive reference with emphasis on sucrose extraction and purification.

F. W. Lichtenthaler, ed., *Carbohydrates as Organic Raw Materials*, VCH Press, New York, 1990; detailed treatment of carbohydrate chemistry, with emphasis on synthesis and applications.

M. Mathlouthi and P. Reiser, eds., *Sucrose: Properties and Applications*, Chapman and Hall, New York, 1995; thorough coverage of properties of sucrose and its applications in food systems and chemical syntheses.

R. A. McGinnis, ed., *Beet Sugar Technology*, 3rd ed., Beet Sugar Development Foundation, Fort Collins, Colo., 1982; comprehensive reference with emphasis on sucrose extraction and purification.

N. L. Pennington and C. W. Baker, eds., Sugar, *A User's Guide to Sucrose*, Van Nostrand Reinhold Publishing Co., New York, 1990; good reference on uses of sucrose for food and other applications.

Z. Bubnik, P. Kadlec, D. Urban, and M. Bruhns, eds., *Sugar Technologists' Manual*, 8th ed., Verlag Dr. Albert Bartens, Berlin, Germany, 1995; comprehensive reference for process technology.

Am. J. Clin. Nutr. **62**(1S) (July 1995), workshop on the evaluation of the nutritional and health aspects of sugars.

WILLIAM J. COLONNA
UPASIRI SAMARAWEERA
American Crystal Sugar Company

SUGAR ANALYSIS

Sugar analysis includes the analysis of sucrose and other sugars in sugar processing and commercial trade, and in sugar-containing foodstuffs. In sugar production and sugar trading, sugar analysis also includes the determination of other components present in the sugar.

Since the early 1980s, several important developments have taken place in the field of sugar analysis. (*1*) Worldwide efforts at harmonization of methods have led to greater definition and standardization of the methods, with the subsequent discarding of many obsolete methods; (*2*) polarimetry of sugar solutions has been extended to the higher wavelength range of 880 nm; (*3*) lead abatement has resulted in new clarification agents for polarimetry; (*4*) the 100°*S* point for sucrose has been redefined and is now called the 100°*Z* point; (*5*) chromatographic methods have been improved to the point where they are widely accepted as official methods; and (*6*) near-infrared spectroscopic analysis is developing rapidly as an alternative method for many tests.

Standards and Definitions

The trend toward international standardization and harmonization of methods used in trading has had a significant impact on the methods used for sugar analysis. The Codex Alimentarius Commission was established in 1962 by FAO/WHO of the United Nations to develop an international compilation of food and commodity standards, which includes those pertaining to sugar and many sugar-containing products. In Europe, the European Union (EU) is carrying out a similar function for its member nations. The Nutrition and Labeling Act (NLEA) passed by the U.S. Congress in 1990 has provided some impetus toward developing more accurate analytical methods for sugars.

The International Commission for Uniform Methods of Sugar Analysis (ICUMSA) promulgates official methods of sugar analysis for the cane and beet sugar industry by the standardization and validation of methods through collaborative testing (1) (see SUGAR, CANE SUGAR; SUGAR, BEET SUGAR). The Corn Refiners Association (CRA) establishes methods used in the corn sugar industry (2). The Association of Official Analytical Chemists (AOAC) reviews methods of sugar analysis along with a vast array of other methods, many of which are required by the U.S. Food and Drug Administration (FDA) for setting standards of identity for foodstuffs and for labeling purposes (3).

At least six specifications of standards for granulated sugar quality are applicable in the United States. These include *Codex Alimentarius*, *Food Chemicals Codex* (FCC) (4), *U.S. Pharmacopeia* (USP) and *National Formulary* (NF) (5), National Soft Drink Association (6), National Canners Association, and Military Standard-900 for white sugar. These standards are intended to set limits on various components, including, but not necessarily limited to, polarization, invert or reducing sugar, ash, moisture, color, sulfur dioxide, arsenic, lead, and copper.

Sugar trading is controlled by contractual agreements between buyers and sellers. The contracts set specifications and limits and detail the required tests that must be conducted. In the United States, three parties participate in the analyses of raw sugar: the buyer, the seller, and the New York Sugar Trade Laboratory (NYSTL). The settlement value for polarization is determined as the average of the two closest results; if all three results are equidistant, all three results are averaged.

Physical Methods of Sugar Analysis

The concentration of a pure sugar solution is determined by measurements of polarization (optical rotation), refractive index, and density.

Polarimetry. Polarimetry, or polarization, is defined as the measure of the optical rotation of the plane of polarized light as it passes through a solution. Specific rotation $[\alpha]$ is expressed as $[\alpha] = \alpha/lc$, where α is the direct or observed rotation, l is the length in dm of the tube containing the solution, and c is the concentration in g/mL. Specific rotation depends on temperature and wavelength of measurement, and is a characteristic of each sugar; it may be used for identification (7).

Polarization is the most common method for the determination of sugar in sugar-containing commodities as well as many foodstuffs. Polarimetry is applied in sugar analysis based on the fact that the optical rotation of pure sucrose solutions is a linear function of the sucrose concentration of the solution. Saccharimeters are polarimeters in which the scales have been modified to read directly in percent sucrose based on the normal sugar solution reading 100%.

The normal sugar solution corresponds to 26.000 g of pure sugar dissolved in water at 20.000°C to a final volume of 100.000 mL. The International Sugar Scale is calibrated in °Z. The 100°Z point is the optical rotation of the normal solution of pure sucrose at the wavelength of the green line of the mercury isotope 198 Hg (546.2271 nm *in vacuo*) at 20.00°C in a 200.000-mm tube and is established as 40.777 angular degrees (on the old °S scale, this value was 40.765). For other wavelengths, the Bünnagel formula for the rotatory dispersion of sucrose solutions is used. For quartz wedge instruments, the effective wavelength has been fixed at 587.0000 nm, and the 100°Z point is established as 34.934 (on the old °S scale, this value was 34.924). For the double line of yellow sodium light, the mean effective wavelength has been fixed at 589.4400 nm, and the 100°Z point is established as 34.626 (on the old °S scale, this value was 34.616) (1). Because of the changeover of the sugar scale from °S to °Z in 1988, some saccharimeters are calibrated in both scales, as some contracts continue to call for the older scale.

For turbid or colored sugar solutions, such as raw sugar, clarification of the solution is required. Until recently, lead acetate solution was used, but with the growing prominence of health and environmental concerns, as of the mid-1990s lead is being phased out, and use of other clarification agents, usually based on aluminum salts, is being implemented. An alternative to clarification has been to extend the wavelength of determination into a higher wavelength region, namely 880 nm, where a simple filtration to obtain an optically clear solution suffices to obtain a polarimetric reading. Values obtained using different methods of clarification and different wavelengths are not always the same, and issues of equivalence are still being determined.

Polarimetric determination of the sucrose concentration of a solution is valid when sucrose is the only optically active constituent of the sample. In practice, sugar solutions are almost never pure, but contain other optically active substances, most notably the products of sucrose inversion, fructose and glucose, and sometimes also the microbial polysaccharide dextran, which is dextrorotatory. Corrections can be made for the presence of impurities, such as invert, moisture, and ash. The advantage of polarization is that it is rapid, easy, and very reproducible, having a precision of ±0.001°.

The value obtained as a result of a polarization measurement for sucrose is expressed as "pol" or "polarization" or "degrees pol" and not as percentage of sugar, but it is considered to closely approximate the sugar content.

Double Polarization. The Clerget double polarization method is a procedure that attempts to account for the presence of interfering optically active compounds. Two polarizations are obtained: a direct polarization, followed by acid hydrolysis and a second polarization. The rotation of substances other than sucrose remains constant, and the change in polarization is the result of inversion (hydrolysis) of the sucrose.

Refractive Index.   The refractometric value of sugar solutions is used as a rapid method for the approximate determination of the solids content (also known as dry substance), because it is assumed that the nonsugars present have a similar influence on the refractive index as sucrose. Measurement is usually carried out on a Brix refractometer, which is graduated in percentage of sucrose on a wt/wt basis (g sucrose/100 g solution) according to ICUMSA tables of refractive index at 20.0°C and 589 nm. Tables are available that give mass fraction corrections to refractometric values at temperatures different from 20°C.

ICUMSA (1) has adopted tables showing the relationship between the concentration of aqueous solutions of pure sucrose, glucose, fructose, and invert sugar and refractive index at 20.0°C and 589 nm.

Equations have been developed that determine the relationship of the refractive index of sucrose solutions between 0–85% concentration, 18–40°C, and 546–589 nm.

Density. Measurement of density is widely used in the sugar industry to determine the sugar concentration of syrups, liquors, juices, and molasses. The instrument used is called a hydrometer or a spindle. When it is graduated in sucrose concentration (percent sucrose by weight), it is called a Brix hydrometer or a Brix spindle. Brix is defined as the percent of dry substance by hydrometry, using an instrument or table calibrated in terms of percent sucrose by weight in water solution. Hydrometers are also graduated in °Baumé, still in use in some industries. The relationship between °Baumé and density, d, in g/cm^3, is °Baumé $= 145(1 - 1/d)$.

Although spindles are calibrated for pure sucrose, other components are normally present in sugar that contribute to the density, such as ash and invert. Because the densities of these components are not much different from that of sucrose, the spindle value is considered a measure of total dissolved substances.

Purity. This is a widely used expression in the industry and represents, as a percentage, the proportion between polarization (considered a measure of sucrose) and dry solids (usually obtained by refractometry).

$$\text{purity} = 100 \times \text{sucrose/dry substance}$$

The Determination of Reducing Sugars

Many methods exist that utilize the reduction of copper or other compounds by aldose and ketose sugars, the most important being glucose and fructose. In relatively pure samples, it is assumed that the reducing sugars are present in essentially equal quantities of glucose and fructose, which may not be the case. In less pure samples, such as molasses, it is understood that other reducing substances are present, so the test is a general test for reducing substances.

The most common methods for determining reducing sugars are based on the reduction of the copper(II) complex with tartaric acid in alkaline solutions. The differences among them lie mostly in the composition of the alkaline solution. The choice of the method for reducing sugars depends on the concentration of the reducing sugars as well as the product matrix. Because the reaction is not quite stoichiometric, the reagents and procedures for all copper reduction methods are strictly standardized, and large errors result if deviations from the method occur. The methods for reducing sugars are listed as follows.

Lane and Eynon Constant Volume Procedure. Probably the most common test for reducing sugars, this method is based on the reduction of Fehling's solution, Soxhlet's modification. The constant volume modification, a more recent change to the method, has allowed for greater standardization, increased sensitivity, and the use of a simple formula instead of tables to determine the amount of invert. The method determines reducing sugars in the presence of sucrose, and is used for raw cane sugar, cane processing products, and specialty sugars having low levels of invert. This test forms the basis for some molasses purchasing contracts and is required in several standards, including the *National Formulary* and *Food Chemicals Codex*.

Berlin Institute Method.   This method is for determination of invert sugar in products containing not more than 10% invert in the presence of sucrose. It is a copper reduction method that utilizes Müller's solution, which contains sodium carbonate.

Emmerich Method.   This method is for determination of trace amounts of reducing sugars in pure sucrose and white and refined sugars with reducing sugar content up to 0.15%. The test is carried out in a nitrogen atmosphere and is based on the reduction of 3,6-dinitrophthalic acid.

Knight and Allen.   This is a copper reduction method for reducing sugars in white sugar up to 0.02%. It utilizes EDTA to determine excess unreacted copper. Tests undertaken in 1994 to extend the range of this method were unsuccessful. In spite of poor performance in ring tests, it remains an official ICUMSA method.

Luff Schoorl.   This method is for the determination of total reducing sugars in molasses and refined syrups after hydrolysis. It is a copper-reducing method that forms the basis of some molasses purchasing contracts.

Ofner Method.   This method is for the determination of invert sugar in products with up to 10% invert in the presence of sucrose and is a copper-reduction method that uses Ofner's solution instead of Fehling's. The reduced cuprous oxide is treated with excess standardized iodine, which is black-titrated with thiosulfate using starch indicator.

Other Methods

Colorimetric Methods.   Numerous colorimetric methods exist for the quantitative determination of carbohydrates as a group (8). Among the most popular of these is the phenol–sulfuric acid method of Dubois (9), which relies on the color formed when a carbohydrate reacts with phenol in the presence of hot sulfuric acid. The test is sensitive for virtually all classes of carbohydrates. Colorimetric methods are usually employed when a very small concentration of carbohydrate is present, and are often used in clinical situations. The Somogyi method, of which there are many variations, relies on the reduction of cupric sulfate to cuprous oxide and is applicable to reducing sugars.

Enzymatic Methods.   Since their earliest use to determine blood glucose, applications of enzyme methods have expanded to include sugar analysis in foodstuffs, beverages, and sugar processing (10). Commercial enzyme analyzers are based on immobilized enzymes embedded in membranes. When the membrane or biosensor contacts a solution of the material to be analyzed, glucose is oxidized by glucose oxidase, releasing hydrogen peroxide, which is then measured electronically, giving an estimation of the amount of glucose present:

$$\beta\text{-D-glucose} + O_2 \xrightarrow{\text{Glc O}_x} \text{glucono-}\delta\text{-lactone} + H_2O_2$$

Three enzymes are required to determine sucrose: invertase to hydrolyze sucrose and produce α-glucose, mutarotase to produce β-glucose, and glucose oxidase for the standard reaction. Enzyme methods have the advantage of being rapid and simple, requiring little sample pretreatment except for solubilization and dilution. The methods require frequent calibration. Enzyme membranes have variable lifetimes and may need to be replaced frequently. Enzyme analyzers are used for quality control in sugar processing, for monitoring wastewater, and in determining sugar in animal feed.

Chromatographic Methods.   These methods are ideally suited for the identification and measurement of individual sugars in many matrices. Chromatographic methods have their widest application in research and in commercial laboratories dealing with food analysis, where both gas liquid chromatography (glc) and high performance liquid chromatography (hplc) are in use. Among the older techniques, paper chromatography is obsolete. Thin-layer chromatography is mostly used for qualitative identification, as quantitation of spots by densitometry has not been widely applied.

Although chromatography offers a more accurate measurement of individual sugars in a sample, it has not supplanted polarization as the method of commerce in the sugar industry. Chromatography lacks the precision of polarimetry, being in the range of 0.5–1.0% for sucrose, about a magnitude higher than polarimetry, and it has longer analysis times. As recently as the early 1980s, the precision of chromatography was in the range of 3–5%. The large improvement in precision is the result of incremental advancements in column technology, flow control, and detection systems. Chromatographic methods have been used to show sources of interference that contribute to inaccuracies in polarimetric measurements (11).

The glc analysis of sugars requires chemical derivatization to produce a volatile molecule. Many derivatization methods for sugars exist, but the simplest and most rapid for routine analysis is silylation to produce trimethylsilyl derivatives. Excess water in the sample interferes and must be carefully controlled. Direct silylation produces single peaks for enantiomerically pure nonreducing sugar, such as sucrose, and multiple peaks for reducing sugars, representing the equilibrium of conformations and anomeric configurations. Reactions are available that produce single peaks, but their use adds to the complexity and time of the analysis. Separation is generally done on methyl silicone or phenyl–methyl silicone phases with detection by flame ionization. Glc analysis is limited to lower molecular weight carbohydrates, and does not usually exceed the level of tri- or tetrasaccharides.

Sugar analysis by hplc has advanced greatly as a result of the development of columns specifically designed for carbohydrate separation. These columns fall into several categories. (1) Aminopropyl-bonded silica used in reverse-phase mode with acetonitrile–water as the eluent. (2) Ion-moderated cation-exchange resins using water as the eluent. Efficiency of these columns is enhanced at elevated temperature, ca 80–90°C. Calcium is the usual counterion for carbohydrate analysis, but lead, silver, hydrogen, sodium, and potassium are used to confer specific selectivities for mono-, di-, and oligosaccharides. (3) Size exclusion columns packed with sulfonated polystyrene–divinylbenzene copolymer using water alone as eluent. These columns are designed primarily for analysis of corn syrups containing oligomeric materials. Larger molecules elute before smaller ones. (4) Pellicular anion-exchange columns using dilute sodium hydroxide eluent having pulsed amperometric detection. This latter technique is known variously as high performance anion-exchange chromatography, ion chromatography, hpaec, hpic, or ic.

Near-Infrared Spectroscopy. The relatively new technique of near-infrared spectroscopy (nir) is increasingly applied in the sugar industry for several types of analyses (12). Current (ca 1996) feasible applications include sucrose, pol, Brix, and purity. The technique has the advantage of requiring little or no sample preparation, having a great saving in time resulting from multiple determinations being done simultaneously, and the elimination of chemical usage. It is a secondary technique, depending on results from the primary techniques for calibration, hence it can only be as precise and accurate as the primary method used for calibration. It is expected to eventually have wide usage in process control (qv) and has already been accepted as a commercial method for payment purposes in at least one country.

Determination of Other Components

In the sugar industry, where the goal is to determine the exact amount of sucrose present, the analysis of other components is essential to determine purity. The most important of these, besides reducing sugars discussed, are moisture, ash, and color. Also relevant are methods used to determine particle-size distribution and insoluble matter.

Moisture. In relatively pure sugar solutions, moisture is determined as the difference between 100 and Brix. In crystalline products, it is usually determined by loss-on-drying under specified conditions in an oven or by commercial moisture analyzers that have built-in balances. Moisture in molasses and heavy syrups is determined by a special loss-on-drying technique, which involves coating the sample onto sand to provide a greater surface area for oven drying. The result of this test is usually considered dry substance rather than moisture.

Small amounts of moisture (up to about 0.5%) in crystalline sugars can be determined chemically by titration with Karl Fisher reagent. A volumetric Karl Fisher titration procedure for moisture in molasses is accepted by AOAC. Automatic Karl Fisher titrators are available, and as acceptance of pyridine-free reagents increases, their use may increase.

Ash and Inorganic Constituents. Ash may be measured gravimetrically by incineration in the presence of sulfuric acid or, more conveniently, by conductivity measurement. The gravimetric result is called the sulfated ash. The older carbonate ash method is no longer in use. Ash content of sugar and sugar products is approximated by solution conductivity measurements using standardized procedures and conversion factors.

Tests for elements such as arsenic, lead, and copper are specified in the relevant standards. The methods specified are usually of the colorimetric or atomic absorption types.

Color. The visual color, from white to dark brown, of sugar and sugar products is used as a general indication of quality and degree of refinement. Standard methods are described for the spectrophotometric determination of sugar color that specify solution concentration, pH, filtration procedure, and wavelength of determination. Color or visual appearance may also be assessed by reflectance measurements.

Particle-Size Distribution. Particle-size specifications for sugar are not usually a part of the legislated standards, but they are of concern to commercial users and suppliers and are often specified in contracts. Grain-size distribution is determined by using a series of sieves, either hand-sieved or machine-sieved (13).

Insoluble Matter. Insoluble matter in sugar is determined as the dry weight of material left on a filter or membrane after passage of a sugar solution. This may include bits of sand, filtration medium, plant material, and polymeric material.

BIBLIOGRAPHY

"Sugar Analysis" in *ECT* 1st ed., Vol. 13, pp. 192–203, by E. J. McDonald, National Bureau of Standards; in *ECT* 2nd ed., Vol. 19, pp. 155–166, by E. J. McDonald, U.S. Department of Agriculture; "Sugar Analysis" under "Sugar" in *ECT* 3rd ed., Vol. 21, pp. 871–877, by F. G. Carpenter, U.S. Dept. of Agriculture.

1. *ICUMSA Methods Book*, International Commission for Uniform Methods of Sugar Analysis, Norwich, U.K., 1994.
2. *Standard Analytical Methods*, Corn Industries Research Foundation, Washington, D.C., binder is constantly updated with new and revised methods.
3. K. Helrich, ed., *Official Methods of Analysis of the Association of Official Analytical Chemists*, 15th ed., Arlington, Va., 1990.
4. *Food Chemicals Codex*, 3rd ed., National Academy Press, Washington, D.C., 1981; *Sucrose Monograph* 3rd ed., 3rd Suppl., Washington, D.C., 1992, pp. 149–151.
5. *The United States Pharmacopeia*, 22nd rev. 1990, and *The National Formulary*, 17th ed., Rockville, Md., 1990. These are published together as one volume.
6. *Quality Specifications and Test Procedures for Bottlers' Granulated and Liquid Sugar*, National Soft Drink Association, Washington, D.C., 1975.
7. F. J. Bates and co-workers, *Polarimetry, Saccharimetry and the Sugars*, Circular C440, National Bureau of Standards U.S. Government Printing Office, Washington, D.C., 1942.
8. R. L. Whistler and M. L. Wolfrom, ed., *Methods in Carbohydrate Chemistry*, Vol. 1, *Analysis and Preparation of Sugars*, Academic Press, New York, 1962.
9. M. Dubois and co-workers, *Anal. Chem.* **28**, 350 (1956).
10. M. A. Clarke, ed., *Proceedings of 1992 SPRI Workshop on Sugar Analysis*, Sugar Processing Research Institute, Inc., New Orleans, La., 1993.
11. S. V. Vercellotti and M. A. Clarke, *Int. Sugar J.* **96**, 437 (1994).
12. M. A. Clarke, E. R. Arias, and C. McDonald-Lewis, *Sugar y Azucar*, 45–49 (June 1992).
13. *Proceedings of International Commission for Uniform Methods of Sugar Analysis, 21st Session*, ICUMSA Publications, Norwich, U.K., 1994.

MARY AN GODSHALL
Sugar Processing Research Institute, Inc.

CANE SUGAR

The term sugar describes the chemical class of carbohydrates (qv) of the general formula $C_n(H_2O)_{n-1}$ or $(CH_2O)_n$ for monosaccharides. Colloquially, sugar is the common name for sucrose, the solid crystalline sweetener for foods and beverages. Sucrose, a disaccharide, is found in most plants, but is in sufficient concentrations for commercial recovery only in sugarcane and sugarbeet plants. Cane sugar is the sugar extracted from sugarcane.

Sugarcane is a large perennial tropical grass belonging to the tribe *Andropogoneae* of the family *Gramineae* and the genus *Saccharum*. The genera *Saccharum*, *Erianthus* (sect. *Ripidiam*), *Sclerostachya*, and *Narenga*, most cited in regard to the origin of sugarcane, constitute an interbreeding group that, along with three species of *Saccharum* (*S. officinarum* L., *S. barberi* Jeswiet, and *S. sinense* Roxb), were historically used for commercial sugar production. *Saccharum officinarum* is a progenitor of all modern sugarcane varieties. However, the presence of the interbreeding *Saccharum* complex of the three sugar species as well as its wild relatives, *S. spontaneum* L. and *S. robustum* Brandes and Jeswiet ex Grassl, has provided a genetic pool of unparalleled diversity, allowing for the development of thousands of varieties that are adapted to the areas where sugarcane is grown. Most varieties of sugarcane are interspecific hybrids of two or more of the five *Saccharum* species (1).

The sucrose in cane sugar is identical to that in beet sugar; both white refined products are 99.9% sucrose, with water as the principal nonsucrose component. Trace components from the plant indicate the origin of the sugar.

In 1994–1995, 778×10^6 t of sugarcane were grown on 12.6×10^6 ha in the tropical and subtropical areas of the world. Some 73×10^6 t of cane sugar are produced annually (1994) in sugarcane factories, from harvested sugarcane. Of this, about 26×10^6 t is raw sugar (96–99% sucrose basis) for further refining, and the remainder is direct white, also known as plantation white or mill white sugar, of lower purity (sucrose content) than refined sugar, for consumption in the local area of production. Raw cane sugar is shipped to cane sugar refineries, traditionally in areas of high consumption outside the tropics (North America, northern Europe) but increasingly, in recent years, in the tropics where energy is cheap and the market is increasing. Refined white and brown cane sugars, and liquid sugar products, are sold industrially and directly to the consumer around the world. Sugar is one of the purest food substances made. Food-grade sucrose [57-50-1] also ranks as a very pure organic chemical. Food-grade sugar constitutes the world's largest supply of a high purity, naturally occurring chemical compound. However, more than 99% of all crystalline sugar produced is used as food.

History

Sugarcane, a sweet reed or grass in its earliest forms, probably originated in New Guinea. It was found throughout Southeast Asia, China, the South Pacific, the Indian subcontinent (where some claim it originated), and the Middle East by the fourth century BC. The soldiers of Alexander the Great (356–323 BC) brought from India to Macedonia a plant that produced "honey without bees," thereby bringing sugarcane to the European continent. Arabic travelers spread sugarcane throughout the Mediterranean area. By the twelfth century, sugarcane had reached Europe, and Venice was the center of sugar trade and refining. Marco Polo reported advanced sugar refining in China toward the end of the thirteenth century. Columbus brought sugarcane to the new world on his second voyage. It spread throughout the Western Hemisphere in the next 200 years, and by about 1750 sugarcane had been introduced throughout the world.

The process for extracting juice from the cane is very old. In antiquity, the canes were undoubtedly sucked or chewed for their sweet taste. Also, in the ancient past in various places the canes were cut and crushed by heavy weights, ground with circular stones or by a heavy roller running on a flat surface, pounded in a mortar with a pestle, or soaked in water to better extract the sweet juice. The term grinding survives to this day as the name of the process for extracting the cane juice, even though the process no longer involves a true grinding. Parallel rolls, which are used in the 1990s, were first used in 1449 in Sicily.

The ancient process for obtaining sugar consisted of boiling the juice until solids formed as the syrup cooled. The product looked like gravel and the Sanskrit word for sugar, *shakkara*, has that alternative meaning. Pliny, who traveled widely in the Roman Empire, wrote in 77 AD that sugar was "white and granular." He noted that Indian sugar was more esteemed than Arabian, and that both were used in medicine. By the fourth century AD, the Egyptians were using lime as a purifying agent and carrying out recrystallization, which is still the main step in refining.

Until a few hundred years ago, sugar was strictly a luxury item. Queen Elizabeth I is credited with putting it on the table in the now familiar sugarbowl, but it was so expensive that it was used only on the tables of royalty. Sugar production reached large volume at a reasonable price only by the eighteenth century.

The development of the sugar industry from the sixteenth century onward is closely associated with slavery, which supplied the large amount of labor used at the time. Owing to the low cost of labor and the high price for sugar, many fortunes were made. The abolition of slavery at various times in different countries between 1761 and 1865 profoundly affected the sugar industry. Upon freeing of the slaves, sugar production fell drastically in many producing areas.

The first use of steam power as a replacement for the animal or human power that drove the cane mills occurred in Jamaica in 1768. This first attempt worked only a short time, but steam drive was used successfully a few years later in Cuba. Steam drive for the mills soon spread throughout the world. The use of steam instead of direct firing was soon applied to the evaporating of the cane juice.

Probably the most essential piece of equipment in the modern process is the vacuum pan, invented by Howard (U.K.) in 1813. This accomplishes the evaporation of water at a low temperature and lessens the thermal destruction of sucrose. The bone-char process for decolorization dates from 1820. The other essential piece of equipment is the centrifuge, which was developed by Weston in 1852 and applied to sugar in 1867 in Hawaii. This machine reduces the time for draining the molasses from the sugar crystals from weeks to minutes by applying a force of several orders of magnitude greater than gravity. It is to the everlasting credit of the cane-sugar industry that the greatest energy saver of all time was developed in this industry: the multiple-effect evaporator invented by Norbert Rillieux of Louisiana. The 1846 patents of Rillieux describe every detail of the process. This system is used universally by every industry that has to evaporate water.

The manufacture of sugar was early understood to be an energy-intensive process. Cuba was essentially deforested to obtain the wood that fueled the evaporation of water from the cane juice. When the forests were gone, the bagasse burner was developed to use the dry cane pulp, called bagasse, for fuel. Bagasse was no longer a waste product; its minimal value is the cost of its replacement as fuel.

The principal analytical methods were developed in the mid-nineteenth century: the polariscope by Ventzke in 1842, the Brix hydrometer in 1854, Fehling's method for reducing sugars, and Clerget's method for sucrose in 1846.

Sugar loaves were for centuries the traditional form in which sugar reached the market. These were formed by pouring the mixture of crystals and syrup into a mold. The molds were kept in the hot room to facilitate the draining of the syrup, either through the porous mold or through a hole in the bottom. With a little cooling or drying, the crystals stuck together, forming a convenient marketable loaf of sugar. The sugar loaves required no packaging and were broken up by the user as needed. Only a very small amount of sugar now reaches the market in this form. A few loaves are made in Europe for advertising purposes.

Physical and Chemical Properties

Cane sugar is generally available in one of two forms: crystalline solid or aqueous solution, and occasionally in an amorphous or microcrystalline glassy form. Microcrystalline is here defined as crystals too small to show structure on x-ray diffraction. The melting point of sucrose (anhydrous) is usually stated as 186°C, although, because this property depends on the purity of the sucrose crystal, values up to 192°C have been reported. Sucrose crystallizes as an anhydrous, monoclinic crystal, belonging to space group $P2_1$ (2).

The specific rotation of sucrose in water is $[\alpha]_D^{20} = +66.529°$ (26 g pure sucrose made to 100 cm^3 with water). This property is the basis for measurement of sucrose concentration in aqueous solution by polarimetry. 100°Z indicates 100% sucrose on solids.

Among physical properties of cane sugar that are most important for its use in foods are bulk density, dielectric constant, osmotic pressure, solubility, vapor pressure, and other colligative properties, and viscosity (3). Bulk density, important for cane sugar as an ingredient in dry mixes, is listed in Table 1, as typical values for several types of sugars. Solubility of sucrose with other common sugars is shown in Figure 1 and in Table 2. Colligative properties vary with concentration of sucrose in solution. The strong effect of cane sugar on freezing point depression is widely used in frozen desserts; the reduction in vapor pressure and increase in boiling point are essential for manufacture of hard candy and other confectionery (2,4). The high osmotic pressure generated

Table 1. Bulk Density of Sugars

Sugar type	Typical values[a]	
	kg/m^3	lb/ft^3
confectioners AA	833–881	52–55
sanding	801–833	50–52
manufacturer's or fine granulated	785–833	49–52
bottler's or standard granulated	769–817	48–51
baker's special	785–849	49–53
powdered		
sifted	384–481	24–30
compacted	609–721	38–45
agglomerated	320–384	20–24
soft (brown) compacted	833–993	52–63

[a]Maximum value of bulk density for granulated sugar occurs at grain size 0.2 mm, and is 930 kg/m^3 (no conglomerates).

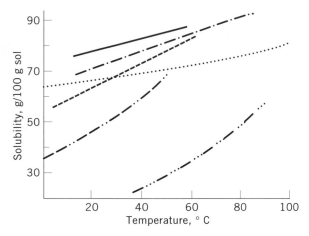

Fig. 1. Solubility of some sugars, where (——) represents fructose; (– · –), sorbitol; (– – –), xylitol; (· · · ·), sucrose; (– · · –), glucose; (– · · · –), lactose.

Table 2. Solubility of Sucrose in Pure Water Under Normal Pressure

t, °C	S,[a] wt %	L_t[b]
0	64.45	1.8127
10	65.43	1.8926
20	66.72	2.0047
30	68.29	2.1535
40	70.10	2.3450
50	72.12	2.5863
60	74.26	2.8856
70	76.48	3.2515
80	78.68	3.6899
90	80.77	4.2004
100	82.65	4.7634

[a]As calculated from the equation $S = 64.447 + 8.222 \cdot 10^{-2} t + 1.6169 \cdot 10^{-3} t^2 - 1.558 \cdot 10^{-6} t^3 - 4.63 \cdot 10^{-8} t^4$.
[b]L_t = gram sucrose per gram water.

by sucrose in solution (Table 3) (2) reduces the water activity and therefore the equilibrium relative humidity, so that insufficient moisture remains to sustain microorganisms, as in jams and preserves. Most common microorganisms require at least 80% equilibrium relative humidity to grow; both crystalline sugar, and concentrated solutions such as jams and preserves, are well under 70%. Dielectric constant values for sucrose in solution are shown in Table 4 (3); the high values make sugar an important ingredient for quick heating in microwaveable foods. The viscosity of cane sugar solutions varies greatly with degree of purity of the sugar; tables for sucrose are readily available (2–7).

Among chemical properties of cane sugar that affect daily use are color, flavor, sweetness, antioxidant properties, and reactions in aqueous solution (3). The purity of cane sugar is generally assessed by its color; lowest color sugars are highest purity sucrose with the lowest content of color and flavor molecules, and other organic and inorganic components. Table 5 shows composition of cane sugar, beet sugar (qv), and other cane sugar products. Brown sugars and golden syrup are generally made from cane sugar, for reasons of flavor.

Sucrose, traditionally cane sugar, is the standard for sweetness, and other sweeteners are ranked against sucrose as 100%, as listed in Table 6 (3). Reactions of cane sugar in aqueous solution are important both in manufacturing

Table 3. Osmotic Pressure of Aqueous Sucrose Solutions at 25°C[a]

Sucrose, g/100 g of water	Osmotic pressure, 10^5 Pa[b]
3	2.17
6	4.56
9	6.95
12	9.33
15	11.72
18	14.11
21	16.49
24	18.89
27	21.27
30	23.66
33	26.05
36	28.43

[a]Ref. 2.
[b]To convert Pa to psi, multiply by 1.45×10^{-4}.

Table 4. Dielectric Constants of Aqueous Sucrose Solutions

Sucrose wt %	Temperature, °C		
	20	25	30
0	80.38	78.54	70.76
10	78.04	76.19	74.43
20	75.45	73.65	71.90
30	72.64	70.86	69.13
40	69.45	67.72	66.05
50	65.88	64.20	62.57
60	61.80	60.19	58.64

Table 5. Composition of Sugars and Syrups[a]

Material	Sucrose, %	Glucose, %	Fructose, %
cane sugar, white	>99.9	<0.01	<0.01
beet sugar, white	>99.9	<0.01	<0.01
brown sugar	90–96	2.5	3–6
golden syrup	32	23–25	22–24
crystalline fructose			<99
palm (date) sugar	72–78	4–5	4–5
molasses			
treacle	32–36	18.22	16–18
fancy, hi-test	22–27	23–28	25–30
medium invert syrup	38–43	28–30	30–32
glucose syrup		20–95	
high fructose syrup (isoglucose)		55–43	42–55
maltose syrup (35% maltose)	4–5		4–5

[a]Dry basis.

Table 6. Sweetness and Flavor of Selected Carbohydrates in Solution

Carbohydrate	Sweetness[a]	Flavor character
Monosaccharides		
glucose	61, 70	sweet, bitter side taste
fructose	130–180	pure sweet, fruity
Disaccharides		
sucrose	100	pure sweet
maltose (malt sugar, maltobiose)	43, 50	sweet, syrupy
lactose (milk sugar)	40, 26, 15–30	faint sweet, fruity
palatinose (isomaltulose, lylose)	50	pure sweet, masks bitter
leucrose (glucose-1,5-fructose)	50	pure sweet
Polyols (sugar alcohols, hydrogenated sugars)		
xylitol	100, 85–120	sweet, cooling
sorbitol (hydrogenated glucose) syrupy	50, 63, 70	sweet, cooling
maltitol (hydrogenated maltose)	68	sweet
mannitol	40, 65	sweet
lactitol (hydrogenated lactose)	30–40	clean, sweet
Mixtures and syrups		
lycasin 80/55 (hydrogenated glucose syrup)	75	sweet
palatinit (Isomalt; 1:1 mix of glu-sorbital + glu-mannitol)	45	pure sweet
high fructose corn syrup (HFCS)	100–160	sweet
invert syrup	105	sweet
maltodextrin (DE < 20)	0+	bland to faintly sweet
neosugar (fructo-oligosaccharides)	46–60	sweet

[a]Relative to sucrose.

(process is almost entirely in solution) and in use as a food and in food processing (qv). Hydrolysis of sucrose, called inversion, forms an equimolar mixture of glucose and fructose, called invert sugar or invert, because of the change in the polarimetric measurement, or inversion from positive to negative, upon hydrolysis. Hydrolysis is the initial step for most reactions of cane sugar in food chemistry. It is depicted in Figure 2 (3). It occurs up to about pH 8; above that, nucleophilic displacement of a proton is the initial reaction in sucrose decomposition. Reactions after initial hydrolysis (inversion) include the following. (1) Reactions in acidic medium which lead to formation of 5-hydroxymethyl furfural (HMF). HMF rapidly decomposes into dark-colored compounds, with off-flavors (2,3,8). (2) Reactions in alkaline medium, including lactic acid formation by chemical means (rather than by fermentation), and the rearrangement of glucose to a mixture of mannose and fructose, which is often responsible for the reported presence of fructose and mannose in products that in actuality contain only glucose. An alkaline environment present during extraction or hydrolysis procedures can cause the transformation of glucose to a mixture of mannose and fructose by this mechanism (2,3,8). (3) Maillard reactions, ie, the reaction of a reducing sugar with an α-amino group to form a condensation product that can subsequently polymerize into dark-colored compounds. This is the basis of the browning reaction observed during baking and cooking processes. Several alternative pathways of color, or melanoidin, formation are possible after the initial Maillard reaction (2,3). (4) Thermal degradation of sucrose and caramel formation. The thermal decomposition of solid sucrose may be the exception to the rule that the common decomposition-related reactions occur in water solution; however, moisture absorption by sucrose as it is heated can account for some thermal degradation along pathways of solution reactions. Multiple reactions, some anhydrous, some involving water, are involved in the formation of the complex mixture known as caramel (2,3).

Color of cane sugar depends on its nonsucrose content; sucrose, glucose, and fructose are white crystalline materials. Colorant compounds are in two classes:

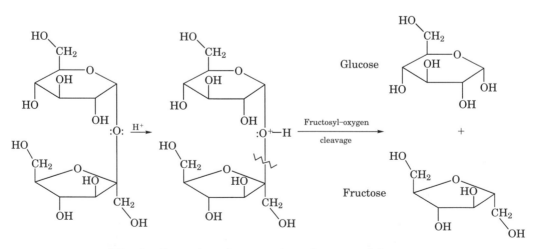

Fig. 2.   Inversion of sucrose into glucose and fructose.

one from the cane plant, including flavonoid and polyphenolic compounds, and one from process-developed colorant, based on sucrose degradation products. These degradation reactions occur in aqueous solution, in process, and in a relatively slow manner in the syrup layer surrounding the sugar crystal. Reactions in solution, included in those described above, that are responsible for color formation include thermal degradation of sugars, with condensation at low pH and caramel formation; alkaline degradation of fructose, with subsequent condensation; and Maillard reactions with primary amines and subsequent melanoidin formation. Many of the colorant compounds are also responsible for the caramel, butterscotch, and toasty flavors of brown cane sugar.

Structure of the Sucrose Molecule. The structure of the sucrose molecule, β-D-fructofuranosyl-α-D-glucopyranoside, in its crystalline state and in aqueous solution has been studied by many groups. The crystalline conformation can be represented as shown in Figure 3, although the inter- and intramolecular hydrogen bonding is still under debate (2). Solution conformations, as measured by nmr and optical spectroscopic methods, are known to be multiple, to interact with water structure, and to be the result of hydrogen bonding and van der Waals forces, but although several structures and theories are each supported by some physical findings, there is no single consistent explanation (2).

Cane sugar is metabolized rapidly, after initial enzyme hydrolysis to glucose and fructose. As a carbohydrate, it yields 16.5 kJ/g (3.94 kcal/g).

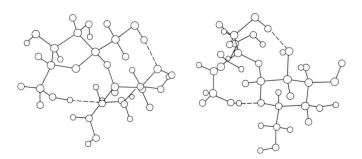

Fig. 3. Representations of sucrose in its crystalline conformation; intramolecular hydrogen bonds are shown as dashed lines.

Cultivation, Harvesting, and Processing of Sugarcane

Cane sugar production is accomplished in one or two stages. At sugarcane factories, located in cane-growing areas, harvested sugarcane is brought in, sugar-containing juice is extracted, and sugar crystallized from the concentrated juice. In the single-stage process, the juice is purified and bleached for the manufacture of plantation white (mill white, direct white) sugar, usually for local consumption. In the two-stage process, partially purified, unbleached juice is crystallized into yellow to brown-colored raw sugar; this is shipped in bulk to the countries of principal cane sugar consumption in North America and northern

Europe, where it is refined into white and colored products for industrial and home use. Sugarcane, once cut (harvested), immediately begins to lose sucrose to deterioration by enzyme, or chemical inversion. The two-stage production system arose because sugarcane cannot be stored. Plantation white sugar, while quite suitable for use within a few weeks after manufacture, cannot be stored for long periods (ie, shipping times) because it contains more water and invert than does refined sugar, and discolors and becomes hardened and lumpy. There is a trend since the late 1970s to increased refining capacity at factories, near the cane production areas, because (1) energy costs are low and sugarcane residual fiber (bagasse) is burned as fuel in the factory; and (2) an increase in consumption is most rapid in the tropical and semitropical countries, especially in processed foods and drinks. As disposable income rises, sweet foods and carbonated beverages are among the first products to show an increase in market strength.

Cultivation.   Sugarcane variety breeding programs are essential for production, from seed crossings and vegetative reproduction, of healthy new varieties with appropriate disease resistance, weather tolerance, and high sugar content, along with agronomic characteristics for each area. The great variation in geography and weather among areas has led to many different varieties and programs. The short growing season in Louisiana has led to development of cold-tolerant varieties, whereas plans for cogeneration of electricity from incineration of bagasse in Florida have placed emphasis on development of high fiber varieties.

Sugarcane requires at least 60 cm moisture each year, whether from rainfall or irrigation. It is propagated vegetatively, from cuttings; each cutting of seed cane must contain at least one bud. Pollinated sugarcane does not breed true because of somaclonal variation (cane is a polyploid hybrid). Most of the world's cane is planted by hand, and some 60% is still harvested by hand in the tropics where labor is low cost and high agricultural employment levels are government policy.

Diseases and Pests.   Sugarcane is subject to a number of bacterial, fungal, and viral diseases, in part because sucrose is such a desirable substrate. At any one time in any given location, there are usually three or four prevalent diseases of concern. The severity of infestations increases and decreases in various parts of the world depending on the varieties grown and control measures. The most recent diseases to appear in the Western Hemisphere are smut, caused by the fungus *Ustilago scitaminea* Sydow, which arrived in the United States (Florida) in 1978, and rust, caused by the fungus *Puccinia melanocephala* H. & P. Sydow, which arrived in 1979. Other important diseases include sugarcane mosaic, a viral disease which caused severe losses throughout the world in the earlier part of the century, and ratoon stunting disease, caused by the bacterium *Clavibacterium xyli*.

Pests include rats, a severe problem in some areas, wild animals, nematodes, and a number of insects. The most severe insect pests are the various types of borers, ie, the sugarcane borer, *Diatrea saccharalis* (F.) and the eldana borer, *Eldana saccharina*, which cause damage first by boring into the cane stalk, then by providing entry points for other diseases, and finally by reducing cane and juice quality.

Weeds cause problems in sugarcane culture by competing for nutrients and crowding or overgrowing the young plants. Perennial grasses are the most serious weeds, harboring insects and diseases. Preemergent herbicides are commonly used for control.

Chemical treatment of diseases is not common, because of legislative controls and costs caused by the difficulty of application through the leaf canopy. Breeding of resistant varieties is the main weapon for disease control. Some diseases, chiefly ratoon stunting disease, are controlled by hot water treatment of cane (6,8).

Sugarcane is the most efficient collector of solar energy in the plant kingdom, converting 2% of available solar energy into chemical bonds of stored compounds, chief among them sucrose (3). Yield in metric tons of cane per hectare varies from 55–60 t in poor growing areas to more than 200 t for cane grown for 18–24 months in optimum areas, eg, Hawaii. The quantity of sugar produced per hectare varies from ca 5.0 (Ethiopia) to ca 26.0 (Campos, Brazil). Sugar recovery averages 10–12% on cane (6).

Harvest season is generally during the cooler, drier part of the year, varying from three months (October–December) in Louisiana, to the first half of the year in most Northern Hemisphere tropics and the second half in most Southern Hemisphere tropics, to year-round in Hawaii, Colombia, and Peru. Generally, replanting is not necessary after each harvest; buds on the plant base and roots remaining sprout again to produce another crop, called ratoon or stubble; this ratooning is repeated until the yearly decline in yield (successive ratoons yield lower cane tonnage) is no longer economical. Ratoon crops vary from none in Hawaii, where pushrake harvesters can harvest roots with the stalks, to eight to ten in optimum regions; two to six ratoon crops are customary. The use of chemical ripeners, or senescence enhancers, to speed up maturation and increase sugar content of cane, is becoming widespread, but requires careful time control.

Harvesting. In hand cutting practice, cane knives range from long machetes to shorter-handled Australian and Brazilian knives with hand guards. Cane leaves and tops (known as trash), which contain little sugar, add weight to transport, hinder cane cutters, and wear down mill rolls, are removed first by burning the cane field and then by hand or mechanical harvesters. Cane stalks are sufficiently high in moisture so that controlled and rapid burns (fire in a 50-ha field is complete in 3 min) incinerate only the leaves, tops, and trash. In Australia, Hawaii, and the Dominican Republic, cane is harvested without burning, to provide more fiber as fuel (for electricity cogeneration at the factory) and for environmental protection. A trash blanket is left on the field to encourage regrowth and discourage disease and pests. The harvesting of green cane is becoming more widespread, for environmental reasons, and as mechanical harvesting progresses. Important factors in cutting are to produce clean, undamaged cane, free of trash, and to leave viable root stock in the field. Mechanical harvesting is found in Australia, the United States, some Caribbean and Latin America countries, and new developing cane areas in Southeast Asia, and is gradually being introduced almost everywhere. Most common are combine harvesters, or chopper harvesters, developed in Australia, which cut cane stalks at the base, cut the stalk into billets, 28–38 cm long, blow excess leaves and trash off the billets, and drop the billets into a cane cart pulled alongside the combine harvester. In

Louisiana, or where tonnage is light, soldier harvesters cut and top erect cane, leaving rows of whole stalks in the field, which are burned after harvest because the canopy is too light to support a burn on standing cane. Other whole-stalk harvesters in Hawaii, where cane tonnage is heaviest, are the V-cutter, which cuts cane at base but not at top, and the push-rake, used on hilly areas, which pushes cane, including the roots, out of the ground, necessitating replanting. Under good conditions, 0.5 t of cane per hour can be cut by hand and 30 t/h of cane by a combine harvester, with other mechanical systems between 15 and 30 t/h. Mechanical cutting is generally more expensive than hand cutting and yields lower quality, more damaged cane, but is increasing for sociological reasons.

Transportation.   Cane loading in the field is accomplished by hand, grab loaders, or continuous belt loaders, into small bins or wagons, which collect at transloader stations for transfer to larger transport containers. In some areas, eg, India, Pakistan, Southeast Asia, and Africa, cane is still transported in small bullock carts. Transport by rail, the cheapest method, continues to be used in Australia and the Philippines; by water, in China, Southeast Asia, and Guyana; and by road, elsewhere. Chopper-harvested cane must be shipped directly to the mill and be processed on arrival, not stored in the millyard, to prevent serious deterioration and loss of sugar; delivery time of less than 24 h is recommended. Harvesting and shipping schedules are decided between grower and processor to ensure a constant supply of cane for the mill, and a fair distribution of maturity and quality.

Cane is usually sampled at the factory gate for payment. Cane payment is generally based on weight, with a deduction for trash, and on sugar content, measured by polarimetric measurement of juice. Where payment for cane quality, ie, weight, sucrose content, fiber, or sugar yield, has been introduced, eg, in the United States, South Africa, Australia, Brazil, Colombia, and the Philippines, the quality and efficiency of the industry have greatly improved (9). The usual split of revenue from sugar is 60–70% to the grower and 30–40% of value to the factory or processor.

Processing.   Sugarcane processing to raw cane sugar is outlined in Figure 4, with equipment and concentrations labeled. Because cane deterioration is a direct function of time delay between harvest and milling, cane is stored in as small amounts and as short a time as possible in the mill yard. Factories run around the clock in most countries, closing for weekends in areas with long seasons or strong labor unions, but cane delivery is usually limited to daylight hours. All factories stop for cleaning of evaporators (unless a spare set is available) and other equipment, every 8–20 days.

After weighing, in very muddy areas sugarcane is washed on the cane table before entering the mill, eg, in Hawaii, Louisiana, and some Central American countries. Cane is then cut into chips by one or two sets of revolving knives, and nowadays often further broken up by a shredder. Shredded cane then moves through a series of mills, usually four to seven mills with four rolls each. Mills were originally three rolls, but a fourth, pressure-feed roll is now usual. Imbibition water, or water of maceration, is run countercurrent to the cane, from the last mills back, to increase extraction of sugar from fiber. Juices from the first mill, ie, the crusher, and other mills are combined, and the mixed juice is pumped to the heaters and to the clarification station. Bagasse comes off the

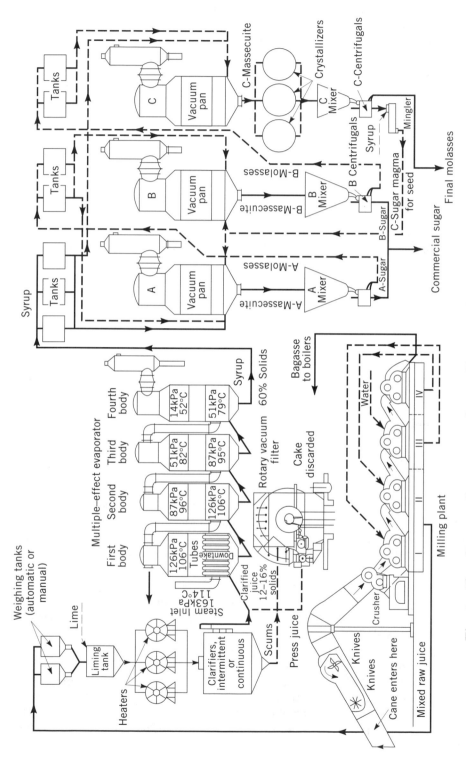

Fig. 4. Flow diagram of a raw sugar factory. To convert kPa to psia, multiply by 0.145.

31

mills at about 50% moisture and goes directly to factory boilers as fuel. To heated (98–105°C) juice is added lime (milk of lime, usually in sugar solution) to pH 7, and flocculation aids, usually polyacrylamides. Cold liming is also employed. Solids are allowed to settle out of juice in juice clarifiers, large settling tanks, with various arrangements of baffles. Heat and lime stop enzyme action in juice and raise pH to minimize inversion. Control of pH is important throughout sugar manufacture because sucrose inverts, or hydrolyzes, to its components, glucose and fructose, at acid pH <7, and all three sugars decompose quickly at high pH (>11.5). Clear juice flows off the upper part of the clarifier; muds are withdrawn below. The settling separation is known as defecation. Muds are pumped to rotary vacuum filters, and residual sucrose is washed out with water spray on the rotating filter. Clear (clarified) juice is pumped to a series of multiple-effect evaporators (4,6,7), where steam from one effect heats the next effect. Nonsugars deposit on the walls and tubes of the evaporator, creating scale and reducing heat transfer; it is removal (boiling out) of this scale that most often causes a routine shutdown of factory operation. Mixed juice (11–16% sucrose) yields clarified juice of 10–15% sucrose, which is concentrated to evaporator syrup of 55–59% sucrose and 60–65 Brix (wt % total solids). Evaporator syrup is sent to vacuum pans, where syrup is heated, under vacuum, to supersaturation: fine seed crystals are added, and the sugar mother liquor yields about 50% by weight crystalline sugar. This is a serial process. The first crystallization of A-sugar or A-strike yields a residual mother liquor (A-molasses) that is concentrated to yield a B-strike. Many schemes of blending and cutting various streams have been developed, leading to open crystallizers stirring lowest grade massecuite (a mixture of crystals and mother liquor) to yield C-sugar and final molasses (blackstrap) from which no more sugar can economically be removed (6,8).

Continuous vacuum pans have been successfully developed for raw sugar crystallization, and are widely applied in South Africa, Australia, South America, and the United States. Continuous crystallizers, developed for beet sugar manufacture, are being adapted for use in cane sugar factories.

After crystallization, crystals and mother liquor are separated in basket-type centrifuges; continuous machines are used for C- and sometimes B-sugars, but batch machines are still best for A-sugars because of crystal breakage in continuous machines. Mother liquor is spun off the crystals, and a fine jet of water is sprayed on the wall of sugar against the centrifugal basket to reduce the syrup coating on each crystal. Raw sugar is dumped onto moving belts, on which it dries as it is moved to storage. In modern factories, washing is increasingly extensive to produce high pol raws, a development of the 1970s that changed the raw sugar market by tailoring a raw material for refineries. Composition of raw cane sugar is shown in Table 7. This is the raw sugar traded on the futures market.

A cane factory generates its own requirements for energy, from burning bagasse to produce electricity; one tonne of mill run bagasse (50% moisture) is equivalent in fuel value, at 3,700 kJ/kg (884 kcal/kg), to one barrel (159 L) of fuel oil. An efficient raw sugar or plantation white factory will use 70–80% of the bagasse available from its cane, and the remainder can be used for cogeneration of electricity for sale to the local grid, as in Hawaii, Mauritius, and elsewhere. The excess power can be used to run a distillery or to run a year-round refinery

Table 7. Composition of Sugars

Component	Raw cane sugar	White refined cane sugar	Mill white	Blanco Directo	Brown cane sugars
sucrose, %	96–99	99.3	99.6	99.9	92.96
glucose, %	0.2–0.3	0.007	0.07	0.02	1–2
fructose, %	0.2–0.3	0.006	0.06	0.03	2–3
color, ICU	900–8,000	35	100–200	40–80	2000–9000
ash, %	0.3–0.6	0.012	0.15	0.05	1–2
moisture, %	0.3–0.7	0.015	0.15	0.03	1–2
organic non-sugars, %	0.3–0.8	0.014	0.40	0.03	1–2
SO_2, mg/kg			20–50	1–5	

to refine raw sugar products from a group of raw sugar factories. This is an increasingly frequent occurrence in Australia, the Far East, and Central and South America, and is developing in Florida.

Diffusion. The alternative to extraction by milling is extraction by diffusion. The sugarcane diffusion process has been developed from sugarbeet diffusion. Here, cane from the shredder must be prepared further in a fiberizer, or extended shredder, for best extraction. Because of this finer preparation, diffusion gives a higher degree of extraction (93–98%) than milling (85–95%); therefore, further cane preparation is increasingly used in mill trains also. Finely prepared cane enters a multicell, countercurrent diffuser of linear, diagonal, or circular design. In the diffusers, shredded cane moves countercurrent to hot water (75°C). This system is for cane diffusion. There are various combinations of sets of mills with a diffuser, for diffusion of partially milled cane; these systems are called bagasse diffusers. Bagasse emerging from the diffuser must be dewatered to reach the approximately 50% moisture of mill-run bagasse; at this moisture bagasse can be fed as fuel to factory boilers. Diffusers tend to be cheaper than mills with a lower energy requirement, but do not handle poor cane and high trash and mud well and are subject to infection.

Direct Consumption Sugar. This sugar (plantation white, mill white, crystal, superior) is the regular table and industrial product in most cane-growing countries outside the United States. This white but not sparkling white crystalline cane sugar product is produced from sugarcane juice by the raw sugar production process (see Fig. 4), with the addition of sulfur dioxide gas, SO_2, generally produced by burning sulfur in air. The SO_2 is injected into juice where it bleaches colorant (by reduction process, or formation of sulfite addition compounds) and is itself oxidized to sulfate. Sulfate reacts with dissolved lime to form calcium sulfate, which precipitates as scale in evaporators and pans. Sulfitation factories operate at rather lower pH than raw cane-sugar factories, and so suffer higher losses. Sulfate is a major anion in sulfitation sugars, often equaling or exceeding chloride in content. Nonsugar components are not removed in process; they are at the same levels as in raw sugar, but the color is bleached.

Sulfitation sugar, the most common type of white sugar in the world, is therefore not suitable for industrial use or food and beverage manufacturers because it contains high ash, turbidity, and reducing sugars, and generally has a high sediment content. Higher grades of plantation white are made through addition of a carbonatation plant, where lime and CO_2 gas are reacted in the juice to form calcium carbonate, entrapping many nonsugar molecules during formation of the chalk crystals. Calcium carbonate is filtered off, entrapping more nonsugars, especially color, in the filter bed. By removing nonsugars from the stream and not recycling them, the carbonatation process improves the quality of the sugar. This process, with many variations, is common in India, Pakistan, China and Southeast Asia, and is discussed thoroughly in older literature (6,7). Powdered carbon treatment, before press filtration, is another additional process to improve quality and lower color. As demand for higher quality (refined quality) sugars increases in the cane-growing countries, there is increasing production of "improved" plantation white sugar. The best direct production sugars are made by the Blanco Directo process, where color precipitating reagents are used, again to remove nonsugars rather than bleach them. Blanco Directo entails syrup clarification and clarification of muds filtrate by phosphatation processes similar to those described.

Cane Sugar Refining

Refining cane sugar processes raw cane sugar into very high purity white and brown sugars and liquid products, including edible molasses. Content of water, ash, and reducing sugars is controlled. Products are of consistent quality and safe for home consumption and for the food and beverage industry. Refined products can be stored well for long periods; white refined sugars stored at ambient conditions for over 60 years show a slight increase in color as the only change. Traditionally, refineries have large packaging departments for their full range of products. The new white-end refineries, or a raw sugar factory or group of factories, tend to produce bulk white sugar only. These refineries have the cost benefit of returning their low grade material to a factory, rather than having to process it.

Refinery input (melt) is raw cane sugar at 96° to 98°Z polarization (% sucrose read by rotation of polarized light). The brown products have characteristic palatable cane and cane molasses flavors, not available from sugarbeet. A generalized refining scheme is shown in Figure 5. Details of unit processes are shown in Figure 4. Refineries are large processing plants operating around the clock typically for five (weekend shutdown) or 10 days (four-day shutdown). Fuel for freestanding refineries is fuel oil, natural gas, or coal, according to local availability; a few refineries have extended their power plants to generate extra electricity for the local grid; refineries attached to raw sugar factories use the factory's excess bagasse fuel.

The quality of incoming raw sugar is paramount for efficient operation. Polarization is a universal quality criterion. Color, ash (inorganic), invert sugar, moisture, dextran content, and grain size are other criteria that may be included in raw sugar purchase contracts.

Raw sugar is weighed into the refinery from rail car, ship, or raw sugar warehouse, and conveyed to the affination station, where it is mingled with a heavy syrup (80% solids content, or 80° Bx where Bx = Brix, wt %), then spun in basket centrifugals and washed with a spray of water to remove the added and

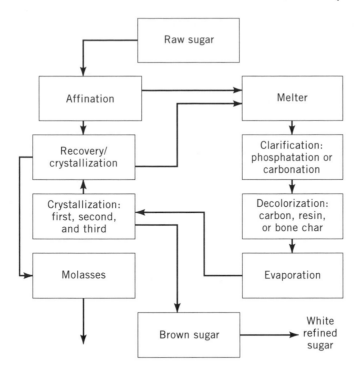

Fig. 5. Outline of a cane sugar refinery process.

the integral syrup coatings. The washed raw sugar is dissolved (melted) to give a washed sugar liquor of ca 70% solids content, which is pumped to clarification. Three types of clarification are in use.

Phosphatation. Phosphoric acid to give a concentration up to 400 mg/kg as P_2O_5 and calcium hydroxide as milk of lime or sugar solution of lime, up to pH 7.5–8.3, are combined with the sugar liquor in an aerated flotation clarifier. Calcium phosphate forms, occluding suspended solids and inorganics in its mass, and floats to the surface where it is scraped off by rotating blades. Clarified liquor (syrups are called liquors in refineries) is pumped out from the bottom of the clarifier. The process removes 25–40% color, ash, and turbidity from the sugar liquor (10).

Talo Phosphatation. Phosphatation is performed as described above with the addition of color-precipitating chemicals and a series of mud-desweetening steps, which remove a greater amount of color (30–50%), ash, and turbidity. Talo (a trademark of Tate & Lyle, plc, U.K.) phosphatation is the process mentioned above that is widely used in white end refineries. It has almost replaced traditional phosphatation (11).

Carbonatation. In this process, called carbonation in Europe, lime and carbon dioxide are mixed in liquor in a two-stage process similar to that for beet sugar processing but carried out on liquor of 65–70% solids (10).

Filtration. Any type of clarification is followed by filtration through leaf-type vertical or horizontal pressure filters. Carbonatated liquors, containing calcium carbonate, may require addition of diatomaceous earth as a filter precoat.

Phosphatated liquors are generally filtered with the addition of diatomaceous earth as precoat and body feed.

Decolorization. Filtration, often a refinery bottleneck, especially with poor-quality raw sugar, is followed by decolorization with bone char (traditional), granular activated carbon (now most common), ion-exchange resins, or any combination of these. Comparative merits and regeneration of these decolorizing systems are a frequent topic in the literature (6–8,11).

Crystallization. Decolorized liquor, or fine liquor of very pale yellow color, is evaporated further to 72–74% solids and sent to crystallization in a series of vacuum pans, as with raw cane sugar. Refinery strikes are designated 1, 2, 3, etc. Four to six white sugar strikes are common. The lowest grade runoff syrups are sent to a second series of pans and crystallized to improve sugar recovery in a process called remelt in the United States or recovery in the U.K. Brown low grade runoff syrups and refiners' final molasses are sold for food processing, brewing, and blending to make cane syrups and edible molasses.

Refined brown sugars, called soft sugars in the trade, are made by crystallizing sugar from a mixture of third and fourth runoff syrups and affination syrup (boiled brown sugars), or by coating white sugar crystals with a brown sugar liquor–caramel syrup (painted or coated brown sugar). Compositions of raw cane sugar, refined granulated, direct mill white, and Blanco Directo sugar are shown in Table 7. The white sugar from the centrifuges is dried in a rotary dryer using hot air. This dryer is universally misnamed the granulator because by drying in motion, it keeps the sugar crystals from sticking together, or keeps them granular. The hot sugar from the granulator is cooled in a similar rotary drum using cold air. Newest driers and coolers employ a fluidized-bed system (11).

Conditioning. After storage, sugar can become moist from water that has been trapped under the outside syrup coating of the crystal by the very high rate of crystallization and drying. After a few days, this moisture migrates outside the crystal and the sugar is wet again. This water can dissolve sugar in neighboring crystals and set up a hard cake of sugar. The moisture is removed by a process known as conditioning, in which the sugar is stored for about four days with a current of air passing through it to carry away the moisture. In one of many variants, a single silo is used with sugar being continuously added to the top and removed from the bottom, and a current of dry air blowing upward. In another system, the sugar is stored in a number of small bins. It is continuously transferred from bin to bin with dry air blowing around the conveyors that move the sugar.

Packing, Storing, and Shipping. Some refineries store bulk sugar and then package as needed, but more package the sugar and then warehouse the packages. The present trend is away from consumer-sized (<50-kg) packages and toward bulk shipments. There are various resale companies that buy bulk sugar and package it in small packages, or individual servings, for consumer distribution. Some refineries use their extensive packaging facilities to package other food products that require the same equipment.

Membrane Filtration Processes. Newest among cane sugar manufacturing systems are processes using membrane filtration to remove nonsucrose solids from juices and syrups. The low energy requirements, reduced effluent, and flexibility of throughput from these processes are the factors providing the impetus expected to result in viable membrane filtration factories by around the year 2005. There are two basic classes of membrane: plastic types with metal ions in the matrix, and ceramic types with a porous layer (stainless steel is a variation on ceramic), all with controlled porosity. The use of the plastic membranes, in combination with carbon-type adsorbent, to make white sugar directly from cane juice, without any sulfitation or other bleaching, has been reported (12). Ceramic membranes have been in use since 1993 in raw sugar manufacture in Hawaii (13) to make a very high quality raw sugar and a molasses that can be treated with ion-exclusion desugarization, described herein. Trials on all processes are being run throughout the sugarcane world.

Molasses Desugarization. The process of separating sucrose from final molasses by ion exclusion is common in beet sugar manufacture. Sugarbeet molasses contain about 50% sucrose on solids and only 1–2% invert; whereas sugarcane molasses contains 20–30% sucrose on solids and 15–25% invert. Separation of invert from the sucrose product fraction is expensive. It appears uneconomical to use this system to separate sucrose from cane molasses unless an invert syrup product fraction is also produced. This may be a salable product in cane-producing countries; it is not in the United States, where cheaper corn syrups have replaced liquid cane sugar and beet sugar products.

Economic Aspects

In sugarcane-growing countries, including the United States, a price or range is set by the government on raw sugar, and often also on cane and white sugar, to ensure a sufficient degree of domestic production. In many tropical countries, sugarcane cultivation and cane sugar production are principal sources of employment. Sugar produced for export is generally sold on long-term contracts, as with those for raw sugar to the United States. Sugar produced over domestic and contractual requirements is sold on the world market. Futures prices are listed on the U.S. and European futures exchanges for raw and white sugars classified by the various contractual specifications. Somewhat less than 10% of world production of cane and beet sugars ever reaches the world market; world market price is not relevant for national costs which are based on production. Production costs for cane sugar and beet sugar in the United States are published annually (1). Other producer countries' cost figures are not so readily available.

Increase in production of high quality white sugar in the tropics (white end refineries) has decreased the export of refined white sugars from Europe and sometimes from the United States. Total world production, and production in the United States and several other principal producer countries are given in Table 8. Consumption figures are listed in Table 9. All figures are given, as is traditional, in equivalent of 96 pol raw sugar value, where pol represents "polarization" or "degrees pol," a measure for sucrose (ie, 96% sucrose, or 96 g sucrose per 100 g product).

Noncentrifugal Sugar. In South and Central America, and in Asia, particularly India and Pakistan, there is considerable production of noncentrifugal cane sugar. Cane juice is simply boiled down in open vessels at atmospheric pressure until it begins to crystallize, and then scooped into molds, usually wooden,

Table 8. Cane Sugar[a] Production, World and Selected Countries[b], 10³ t

Area	Years	
	1994–1995[c]	1992–1993[c]
world	80,614 (69.4)	70,445 (61.5)
United States	4,130	3,980
Central America and Caribbean	11,491	12,072
South America	18,002	15,067
Australia	5,215	4,365
South Africa	1,780	1,600
People's Republic of China	4,710	6,827
India	15,850	11,525
the Philippines	1,650	2,130
Thailand	5,510	3,790

[a]Centrifugal raw value (96 pol).
[b]Ref. 14.
[c]Figures in parentheses represent percent of total cane and beet sugar production.

Table 9. Consumption of Sugar in Selected Countries[a,b]

Area	Total, 10³ t	Per capita, kg
Europe (>90% beet)	31,500	36.6
Great Britain (50% cane)	2,510	43.0
North America	9,680	33.4
Canada (90% cane)	1,230	42.2
United States (55% cane)	8,450	32.4
Central America	7,496	47.1
Cuba	850	77.6
Haiti	85	12.1
South America (95% cane)	13,217	42.1
Brazil	7,750	48.7
Bolivia	186	25.7
Asia	40,741	12.5
China (85% cane)	8,516	7.0
India	13,300	14.5
Singapore	152	53.9
Africa	9,616	13.5
Egypt	1,710	27.7
Ghana	92	5.4
South Africa	1,430	35.3
Australia	900	50.4
world (65% cane)	113,500	20.1

[a]Ref. 14.
[b]Data are for 1994. All figures are on a 96 pol basis. Unless otherwise noted, sugar is assumed to be cane in origin, although imported beet may be included in Africa and Asia.

where it hardens into cakes, cones, or whatever shape of mold has been chosen. This product is dark brown and contains all plant parts, soils, microorganisms, and solids that were in the cane juice. The product is a standard component of the daily diet for the low income populations where it is produced. The hard light to dark brown cakes are known as panela in South America, piloncillo in Mexico, panocha or pile in other Latin countries, gur or jaggery in the Indian subcontinent, and pingbian tong in China. Some $12-13 \times 10^6$ t of noncentrifugal sugar are produced each year, with India, at about 9×10^6 t, the primary producer. Colombia produces almost 10^6 t and Pakistan about 0.5×10^6 t. Because this sugar is produced from cane that could be sent to regular factories, there is some variability in production depending on the relative prices paid for cane by centrifugal and noncentrifugal sugar manufacturers.

Impact of HFCS. The U.S. sugar market changed dramatically in the late 1970s and early 1980s, with the introduction of high fructose corn syrup (HFCS). This liquid product is made from enzymatic hydrolysis of corn starch and processed, with chemical and enzymatic processes, to a range of low color liquid mixtures of fructose and glucose with sweetness equivalent to, or greater than, sucrose (see SUGAR, PROPERTIES OF SUCROSE). Because of other products from corn (by-product credits for major products of corn oil, gluten feed, and gluten meal), corn sweetener can be produced at a very cheap price (15). In the United States, some 3.5×10^6 t/yr of cane (and some beet) sugar were replaced by corn syrups; replacement was in beverages, canned goods, and other food products that could use a liquid sweetener. Almost all carbonated soft drinks in the United States have been made with corn syrups, not sugar, since the early 1980s. Because of this substitution, effects on sweetener markets by nonnutritive sweeteners have had little effect on the sugar market in the United States. However, nonnutritive sweeteners have seriously affected the sugar market in Europe.

Specifications and Standards

Specifications for raw cane sugar are set by purchase contracts. There are no international specifications, although the *Codex Alimentarius* is composing a draft specification. Because raw sugar is not sold as a food product in the United States (it is transported in bulk, like grain or coal), it is not subject to food regulations. Purchase contracts outside the United States are generally based only on pol; U.S. contracts are discussed in the literature (6).

For white sugars, there are no U.S. specifications or standards; specifications are decided among buyers and sellers. There are bottler's specifications for white sugars (generally outside the United States, because little sugar is used within the country by bottlers). Most white sugar buyer specifications emphasize color and turbidity or sediment. There are also limits on reducing sugars, ash, and moisture (2–6,8).

The European Union (EU) has a systematic classification of white sugars, shown in Table 10. *Codex Alimentarius* also has issued specifications for white sugars (17). The EU standards are widely used throughout Eastern Europe and Asia. Other countries, eg, Brazil and the People's Republic of China, have their own domestic specifications, which are also applied to imports.

Table 10. Quality Criteria for White Sugar, According to EC Sugar Market Regulations[a]

Quality criterion[b]	Grade[c] 1	Grade[c] 2	Grade[c] 3
sucrose content (polarization), °Z		99.7	99.7
moisture content, %	0.06	0.06	0.06
invert sugar content, %	0.04	0.04	0.04
color type, Brunswick unit	2	4.5	6
points[d]	4	9	12
ash content as conductivity, $	0.0108	0.0270	
points[e]	6	15	
color of solution, ICU	22.5	45	
points[f]	3	6	
points according to EC point system	8	22	

[a]Ref. 16.
[b]Quality criteria for all grades: sound, fair and marketable quality, dry, inhomogeneous granulated crystals, or free-flowing. All values are maximum except for sucrose content, which are minimum.
[c]Grade 2 is white sugar of standard quality.
[d]Where 0.5 unit = 1 point.
[e]Where 0.0018% = 1 point.
[f]Where 7.5 units = 1 point.

Health and Safety Factors

Sugar is one of the purest foods made, from natural sources, and has never been known to contain any toxic or harmful components. Intensive investigations by the U.S. Food and Drug Administration resulted in a book in 1986 on the health and safety factors of sugar (cane and beet) in the diet (18). The conclusion was that sugar has no deleterious effect on health in regard to heart disease, diabetes, or other metabolic disorder.

Sugar can, the report concluded, be a cause of dental cavities; rinsing the mouth with water after consuming a sugar product reduces this risk considerably. Dental cavities appear to be the only disease for which sucrose could be a cause.

Microbiological standards for sugars are as follows: (1) Canners' standards: for flat sour spores, an average of not more than 50 spores/10 g, with a maximum of 75 spores/10 g; for thermophilic anaerobic spores, present in not more than 60% in five samples; and for sulfate spoilage bacteria, present in not more than 40% in five samples and in any one sample to the extent of not more than 5 spores/10 g. (2) Carbonated beverage standards: not more than 200 mesophilic bacteria per 10 g, 10 yeast per 10 g, and 10 molds per 10 g. (3) "Bottler's" liquid sugar standards: not more than 100 mesophilic bacteria per 10 g (dry sugar), 10 yeast per 10 g (dry sugar), and 10 molds per 10 g (dry sugar). The reduction of water activity in highly concentrated sugar solutions retards microbiological growth on such products as jams, preserves, and canned fruit.

Cane Sugar Products

There are many variations on crystalline cane sugar from refineries, in addition to the direct production and noncentrifugal sugars described above.

Refined granulated sugar is the principal output of a cane sugar refinery. The particle size of the refined granulated sugar for table use varies from region to region. Different particle sizes have different names and are not standardized. Particle size is specified by the buyer, usually at a price premium. North American fine granulated averages 0.2–0.3-mm grain size, whereas standard European fine granulated averages 0.5–0.6 mm. Sugar of standard U.S. crystal size is known as caster sugar in the United Kingdom. Sugar crystals are separated into four to eight size groups by a series of vibrating screens, after the driers in the refinery.

Large-grain specialty sugars are used for candy and cookies. White large-grain sugar can be made only from the very purest of liquors; therefore, customers interested in the best sugar specify coarse grain. The highest quality best sugar is made by redissolving large-grain sugar and recrystallizing.

Fine-grain sugar, or fruit sugar, used because it is quick-dissolving, consists of small crystals obtained by screening.

Powdered sugar is made by grinding granulated sugar and adding 3% corn starch (in the United States) to help prevent caking. The fineness is designated by labels such as 4X, 6X, 10X. However, the label is misleading; 12X is not twice as fine as 6X. In other countries, calcium phosphate, or maltodextrins are used as hygroscopic additives.

Cubes are made by mixing a syrup with granulated sugar to the right consistency to form cubes. These are then dried. The process is expensive and the price of cubes is high relative to ordinary granulated sugar. Production of the cube is much greater in Europe and the Middle East than in North America. Many variations on the cubing process exist, from cutting up slabs of solidified sugar (the hardest cubes) to pressing and drying in various types of cube molds. Infrared drying is an effective modern addition.

Liquid sucrose and liquid invert, generally made by redissolving white sugar and inverting with invertase enzyme, are refinery products in Europe and outside the United States. In the United States they have been almost completely replaced by cheaper corn syrups made by enzymatic hydrolysis of starch and isomerization of glucose.

Brown sugar, including light and dark brown and occasional intermediate grades, comprises only a small part of the output of most refiners, ranging from only 3% in warm climates to perhaps 10% in cold regions. The area of highest brown sugar consumption in the world is British Columbia, Canada, where brown sugar accounts for 20% of total use. In this region, a favorite is a distinctly yellow sugar. Brown sugar is not raw sugar, but rather, as its manufacture is described herein (crystallization), it is refined. The difference between raw sugar and brown sugar is not so much the sucrose content, the color, or taste, but rather the absence of field soil, cane fiber, bacteria, yeasts, molds, and insect parts which may be present in raw sugars. Composition is outlined in Table 7.

Other Products.   Other products from sugarcane, in addition to cane sugar, are cane fiber (known as bagasse) and molasses, the final thick syrup from which no more sugar may be economically removed by crystallization. In some cane-growing countries, cane tops and leaves, separated during harvest, are used for cattle feed.

Bagasse. Cane fiber comes from a standard mill or diffuser at 50–55% moisture, and in most countries is used as fuel for the factory. In the People's Republic of China and some parts of India, sugarcane factories burn low grade coal, because wood is in short supply and bagasse fiber is used for paper or board manufacture. Excess bagasse is burned for cogeneration (8,19,20), or to run a refinery or distillery. Bagasse is also used in paper manufacture, for all grades from coarse brown to newspaper to fine papers, depending on other fibers and processing used. Some 7×10^6 t are used annually for pulp production for papers, particle boards, and fiber boards of various grades and durabilities. Bagasse has been used as a cellulose source for single cell protein production, and as animal feed. Feed quality is improved by steam hydrolysis/sodium hydroxide treatment of bagasse fiber (19,20). In the Dominican Republic, the United States, South Africa, and several countries in South America and Asia, bagasse, which contains 85–95% xylose, is treated by steam hydrolysis and subsequent dehydration to produce furfural; an estimated 90,000 t furfural is produced annually in this manner. Diacetyl (artificial butter flavor) is a by-product of this process in South Africa (20).

Molasses. The final molasses product from sugarcane factories is black-strap molasses, containing 25–35% sucrose and 8–15% each glucose and fructose. Because of the high mineral (primarily KCl) and browning polymer content, blackstrap is too bitter for human consumption; most is used for animal feed, alone or as an ingredient, and it is traded in international commerce for this purpose. Refinery molasses, and blends of both factory and refinery with various lighter syrups, are the sources of a wide range of food-grade molasses, known as treacle in Europe. Molasses is fermented to ethanol at sugarcane factories in almost all cane-growing areas outside the United States, for industrial alcohol. Molasses is the basis for almost all rum production (some rum is produced directly from sugarcane juice in the French-speaking Caribbean), and for other beverage alcohol, in Asian countries. Molasses has been used as a carbon source in a multitude of chemical and microbial reactions; it is usually the sugars in molasses that serve as the carbon source; hence, these products are included herein. Chemical and fermentation reactions can cause problems in storage, if molasses is put into storage too hot: it should always be at a temperature under 45°C.

High test molasses is not a residual material, but cane juice, sometimes partly clarified, concentrated by evaporation, with at least half its sucrose hydrolyzed to invert (glucose and fructose) by heating at the low juice pH (5.5).

Condensed molasses solubles (CMS) is a product made by drying molasses (spray or drum drying) on a neutral carrier; CMS is a more portable and storable form of molasses for animal feed.

Sucrochemistry. A wide range of fermentation and chemical products can be made from sucrose either per se or in juice or molasses. Products and substrates depend on the economics of each area. There is an extensive literature on the subject (19,20). Among the classes of products chemically derived from sucrose are the following. (*1*) Ethers (alkyl, benzyl, silyl, allyl, alkyl, and the internal ethers or anhydro derivatives, which last have generated a new sweetener. (*2*) Esters of fatty acids, including surfactants, emulsifiers, coatings, and a new fat substitute. (*3*) Other esters, eg, sulfuric acid esters that polymerize well; sulfate esters, including an antiulcerative; and other mixed esters. (*4*) Acetals, thioacetals, and ketals that act as intermediates and may have biocide activity. (*5*) Oxidation products, most of which are products of the hydrolyzed monosaccharide products and reduction products, including mannitol and sorbitol, and reductive aminolysis products, including methyl piperazine. (*6*) Halogen and sulfur derivatives and metal complexes, some with applications in water-soluble agricultural chemicals. (*7*) Polymers and resins: polycarbonates, phenolic resins, carbonate–, urea–, and melamine–formaldehyde resins, acrylics, and polyurethanes. Some of the many classes of compounds and products to be made from the sugarcane plant have been outlined (19); most of these are made from cane sugar. Because sugarcane is the most efficient plant at converting photosynthesis into chemical bonds, it can be the basis of a renewable resource economy.

BIBLIOGRAPHY

"Sugar Manufacture" in *ECT* 1st ed., Vol. 13, pp. 203–227, by L. A. Wills, Consultant, "Cane Sugar" under "Sugar," in *ECT* 2nd ed., Vol. 19, pp. 166–203, by H. G. Gerstner, Colonial Sugars Company; in *ECT* 3rd ed., Vol. 21, pp. 878–903, by F. G. Carpenter, U.S. Department of Agriculture.

1. D. J. Heinz, ed., *Sugarcane Improvement through Breeding*, Elsevier, Amsterdam, the Netherlands, 1987, p. 7.
2. M. Mathlouthi and P. Reiser, eds., *Sucrose: Properties and Applications*, Blackie and Son, Ltd., London, 1995.
3. M. A. Clarke, *Proc. Com. Int. Tech. Sucr.*, 424–443 (1995).
4. W. R. Junk and H. M. Pancoast, *Handbook of Sugars*, 2nd ed., AVI Publishing Co., Westport, Conn., 1980.
5. N. L. Pennington and C. W. Baker, eds., *Sugar, A User's Guide to Sucrose*, Van Nostrand Reinhold, New York, 1990.
6. G. P. Meade and J. P. C. Chen, eds., *Cane Sugar Handbook*, 11th ed., Wiley-Interscience, New York, 1985.
7. P. Honig, *Principles of Sugar Technology*, Vol. 1, Elsevier, Amsterdam, the Netherlands, 1953.
8. M. A. Clarke and M. A. Godshall, eds., *Chemistry and Processing of Sugarbeet and Sugarcane*, Elsevier, Amsterdam, the Netherlands, 1988.
9. M. A. Clarke and B. L. Legendre, *Proceedings of the South African Sugar Technologists Association*, 1996, in press.
10. M. C. Bennett, *Proc. Tech. Sess. Cane Sugar Refin. Res.*, 62–75 (1972).
11. M. A. Clarke, *Sugar y Azucar Yearbook*, Ruspam Publications, Fort Lee, N.J., 1983, pp. 71–93.
12. S. Galt, K. B. McReynolds and J. P. Monclin, *Proceedings of Workshop on Products of Sugarbeet and Sugarcane*, Sugar Processing Research Institute, Inc., New Orleans, La., 1994, p. 179; J. P. Monclin and C. Willett, *Proceedings of Workshop on Separation Processes in the Sugar Industry*, Sugar Processing Research Institute, Inc., New Orleans, La., 1996, pp. 16–28.
13. R. Kwok, *Proceedings of a Workshop on Separation Processes in the Sugar Industry*, Sugar Processing Research Institute, Inc., New Orleans, La., 1996, pp. 87–99.
14. *Sugar Economy, 1995/96*, Bartens, Berlin, Germany, 1996, pp. 15–25.
15. *Sugar and Sweetener Reports*, Economic Research Service, U.S. Dept. of Agriculture, Washington, D.C.

16. European Economic Community, *Council Directive of Dec. 11, 1973* (73/437/EC), Off. J.E.C. no. L356, 27.12.73, pp. 71–78; *First Conversion Directive of July 26, 1979* (79/786/EC), Off. J.E.C. no. L.239, 22.9.79, pp. 24–52.

17. *Codex Alimentarius*, Joint FAO/with Food Standards, White Sugar.

18. W. H. Glinsmann, H. Iransquin, and Y. K. Park, *J. Nutr.* **1116**(11-Suppl.), S1–S216 (1986).

19. J. M. Paturau, *By-Products of the Cane Sugar Industry*, 3rd ed., Elsevier, Amsterdam, the Netherlands, 1989.

20. M. A. Clarke, ed., *Proceedings of Workshop on Products of Sugarbeet and Sugarcane*, Sugar Processing Research Institute, Inc., New Orleans, La., 1996.

MARGARET A. CLARKE
Sugar Processing Research Institute, Inc.

BEET SUGAR

For the 1995 crop year, beet sugar accounted for about one-third of the world's sugar production, ie, 40×10^6 metric tons of the total of 119×10^6 t (1). The European Union (EU) supplied 40% of the subtotal, the United States 10%, and the Ukraine, Turkey, the Russian Federation, and Poland together accounted for a total of 25%. Beet sugar is also produced in China, Egypt, Morocco, Iraq, Iran, Finland, Argentina, Chile, Canada, and most of the former Soviet bloc countries. With the exceptions of the Ukraine and the EU, all of the beet sugar is consumed in the country of origin.

The growth of the beet sugar industry can be seen by comparison of the 1995 figures with this 1910 quote:

> "The past year began with hysterical clamor for a change in the protective tariff on sugar, and other troubles that threatened the life of the beet sugar industry and which created doubts as to its stability and future growth. During the first six months of last year there were no new beet sugar factories promoted and there was not much work done to develop new sugar territory" (2).

The technologies described in this article represent the best practices of U.S. agricultural and processing facilities. Compared with European production, U.S. output of sugar per hectare has lagged both in absolute terms and in rate of increase (Fig. 1); both trends are expected to continue, and both represent sustained rapid progress derived from technical advances (3). The main factor tending to diverge European and U.S. trends is the increasing percentage of U.S. sugar beet crops produced on land that is not irrigated, leading to lower yields more than balanced by lower production costs. Two other factors, high energy costs and high profit margins protected by a series of quotas and tariffs, have assured a constant flow of capital to EU factories. Thus, most of the advances in processing technology and often the manufactured equipment originates in Europe.

For the 1994 crops, 33 beet sugar factories in the United States processed 28.8×10^6 t of beets from 586,000 ha into 4.23×10^6 t (93×10^6 Cwt) of sugar;

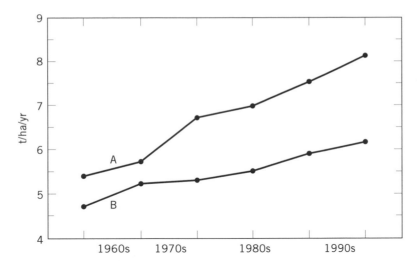

Fig. 1. Recovered beet sugar per hectare, 1960s–1990s, where A represents the European Union and B, the United States.

56% of the domestic production and 47% of the U.S. market. Cwt or hundred-weight of sugar is 100 lbs and is the common unit of commerce in the United States. This is a record level of production, an increase of 12% from the 1990 and 1991 crops. The average factory has a daily processing capacity of 4885 t compared to <3700 t in 1982. Since 1994, three factories in California have been closed and the construction of one new factory has been announced for central Washington State. European factories process 10,000–15,000 t/d.

Processing strategies vary considerably. The simplest scheme is to process the daily delivery of sugar beets directly to sugar, pulp, and molasses, storing beets only long enough to maintain a 24-hour operation. Such facilities produce sugar and by-products only when the crop is being harvested and must maintain an inventory of semifinished product(s) for year-round deliveries to customers. Elaborations of this simple scheme are aimed at extending the number of days of operation, increasing the annual production, and thus minimizing the potentially long periods of idle equipment. (*1*) Within a few hundred kilometers, locate microclimates and develop areas for growing beets which extend the number of harvest days. This is most practical in California, having a variety of microclimates. (*2*) Store beets during the relatively short harvest period to extend the number of processing days past the harvest period. Under favorable storage conditions of the cooler northern climates, exposed piles of beets can add another 120–150 days to the processing season. If several days of intensely cold weather occur early enough in the Fall, forced ventilation of beet piles can be used to freeze the entire pile and may add an additional five or six weeks to the schedule. Finally, the construction of covered sheds with internally circulating air systems affords preservation of frozen piles of beets well into the Spring months. (*3*) Increase the daily beet processing capacity and store the in-process purified, concentrated (thick) juice in large storage tanks. The final processing steps of crystallization take place during a "juice campaign" when beets are no longer available. (*4*) Install one of several similar systems to chromatographically separate sucrose from molasses, and use this thick juice to add production days to the calendar. Commonly, several factories ship molasses to such a facility which may operate 300+ days each year. This is also an effective tactic to increase the daily production rate of sugar even if the number of days during which beets are sliced remains the same.

Agricultural Practices

The sugar beet, *Beta vulgaris*, is a hearty biennial which produces crops of commercial impact in a wide range of climates, from the irrigated deserts of California's Imperial Valley, the high plains of Texas, the eastern slopes of the Rocky Mountains, the Great Lakes region, and Idaho's Snake River Valley, to the rich soils of the Red River Valley of North Dakota, Minnesota, and Manitoba, Canada, where a growing season as short as 100 days supports profitable crop yields. The crop is harvested at or near the time of the first hard frost (28°C) which terminates the photosynthetic production of sucrose and may threaten crop loss by freezing it in the ground. Whereas the Eastern and Idaho crops have to work around the cold winters, the nemesis of the Imperial Valley crop is the summer heat; the crops are planted in September and harvested from April through July. In northern California, the crop may be allowed to winter over for a Spring harvest, which must be completed before warm days trigger the seed production process that consumes much of the stored sucrose.

Besides traditional farmers' luck, a successful crop depends on seed quality and varietal characteristics, weed and pest control, timely irrigations or timely rains (not all beet crops are grown on irrigated land), disease control, crop rotations of at least three years, and a nitrogen management program designed to limit the amount of leaf growth to the minimum necessary to cover the rows and take full advantage of available sunlight. Most states that benefit from a healthy beet sugar growing–production system support the agricultural aspects through their University Extensions services, often in concert with USDA resources.

For a typical crop planted in the Spring and harvested in the Fall the following hold: for each hectare of land, 0.5 to 1 kg of seed (100,000 seeds/kg) are planted at 10 cm spacing in rows 55 cm apart that have been pre-fertilized. Nitrogen is by far the most important component of this fertilization treatment: too little, and the crop cannot thrive; too much, and the result is lush large crops with modest sugar content. The labor-intensive hoeing of beets has been displaced by planting to a stand of 65,000 to 85,000 plants per hectare. Mechanical thinning of beets is relatively rare.

The crop may be treated with pesticide 2 to 15 times, depending on insect and disease pressures in the area. Weed control is maintained by a combination of cultivation and pre-plant and post-emergence herbicides, and is not necessary after rows have been completely covered with a leaf canopy. All chemicals must be approved specifically for use on sugar beets by government regulatory agencies; the approved list gets shorter each year. The last irrigation usually occurs about six weeks prior to harvest, when the 1–2 kg roots are lifted with mechanical harvesters which defoliate the tops and leave them in the field.

Beet Receiving, Storage, and Handling Before Processing

Beets are loaded into side-dump or end-dump trucks in the field and taken to a receiving station, ideally located within 25 km but sometimes as remotely as several hundred km from the field. The receiving station may be the factory itself, an outside staging or piling ground which reloads the beets at a later date, or a rail car-loading facility which reloads the beets into open hopper cars for transport to the factory. At the receiving station the beets are unloaded from the truck and passed over a series of rotating grab-rolls arranged to allow trash, dirt, and small pieces of beets to fall out of the main stream. This first separation is especially important if the beets are destined for storage of more than a day or two.

Prolonged storage of sugar beets extends the factory processing campaign well past the harvest period sometimes by as much as seven months. During this storage period natural respiratory processes consume some of the sugar content of the root, reducing its commercial value. This loss of sugar can be further aggravated by yeast and mold infections beginning on bruised surfaces of beets and thriving when much dirt and trash are included in the storage piles. Because both processes are exothermic and accelerated by increased temperatures, poor storage conditions and practices can result in worthless piles of decaying material.

Commercial strategies for maintaining effective storage are (1) careful monitoring for unremoved leaves, trash, dirt, and early signs of rot or frost damage as the crop is received; (2) initially piling roots with temperatures between 0 and 5°C (never >10°C); (3) building large piles (Fig. 2) to stabilize temperatures and minimize surface area exposed to the elements (the largest piles are 70 m wide, 150 m long, and as high as 9 m; heights >7 m risk retarding natural ventilation which dissipates the heat of respiration); (4) carefully monitoring the condition of the piles to detect hot spots that can be processed immediately or discarded; (5) protecting the piles by covering with plastic sheets or straw; (6) mechanical ventilation through half- or full-round culverts placed under the beets as they are piled or in large sheds (60 m wide, 150 m long, and 10 m high at the sides) with an elaborate underground ventilation air recirculation–venting system; (7) using mechanical ventilation to deep-freeze the beets and stop respiration altogether (freezing usually requires four or five days of uninterrupted ambient temperatures <15°C). Once freezing begins, rewarming or thawing leads to spoilage.

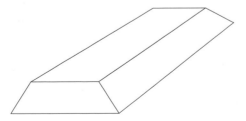

Fig. 2. Typical shape of a pile of stored sugar beets.

Processing of Beets to Sugar

A factory capable of processing 4500 t/d (capacities range from <3000 t/d to 12,000 t/d) requires about 200 truckloads per day, moving to the factory, to feed the around-the-clock operation. Planning and logistics of beet receiving and transportation is an important element of smooth factory operation.

Whether beets are processed on the day of harvest or several months afterward, they arrive at the factory and are dropped into a cement trough of moving water which flumes them past a series of weed rakes and trash collectors. The neutral buoyancy of beets facilitates removal of stones and dirt. Spent flume water passes through a mud-removing clarifier 8–16 m in diameter and is reused, sometimes through a series of recirculating ponds. This water always contains some sugar and is unsuitable for discharge because of its high BOD levels. Such material is usually placed in holding ponds to allow the BOD levels to drop to acceptable levels, often leading to objectionable odors, particularly near residential areas. Some factories use anaerobic digesters to treat these waters and to avoid censure (Fig. 3).

Beets are either elevated or pumped to an agitated beet washer, after which they are rinsed with clean water. This rinse water is fed back to the washer from which it overflows back to the flume system, providing constant blow down. The beets are fed from bins to slicing machines configured either as a horizontal rotating disk or a rotating drum where the beets and the blades are on the inside and the slices, called cossettes, fall to the outside. For both configurations, one knife makes a serrated cut across the face of the beet, followed by a second offset knife making a second serrated cut yielding elongated pieces with a diamond-shaped cross section 3–5 mm on a side and 4–10 cm in length. The knives can be adjusted in blocks to vary the thickness and length of the cossettes (Fig. 4). Long cossettes are always desirable and thickness is kept to the minimum which still allows the beet pieces to maintain integrity in the extraction step. Under the most unfavorable conditions where the beets have been frozen or decomposed and fall apart easily, the knives may be separated so much that slabs are produced.

Continuous Countercurrent Extraction of Sucrose

The extraction or diffusion process and the associated equipment usually define the overall beet processing capacity of a given factory. The diffuser, the cen-terpiece of this unit operation, may be capable of a throughput of more than 7000 t/d. The largest factories have multiple diffusers arranged to handle paral-lel flows of cossettes. For the many types of diffusers, the principle of operations is the same: cossettes are physically transported through the diffuser while wa-ter, which is introduced at the exit end and flows in the opposite direction as the cossettes, is constantly being mixed with the cossettes exit through screens from the end at which the cossettes are introduced. The amount of water introduced (draft) exceeds the weight of the cossettes by 5 to 20%; high draft leads to more efficient extraction of sucrose but at the costs of more energy spent on drying and evaporation and possibly higher extraction of impurities.

Extraction processes applied to sound, unfrozen beets are designed to allow the sucrose to diffuse from the beet cell wall mass and leave the intercellular

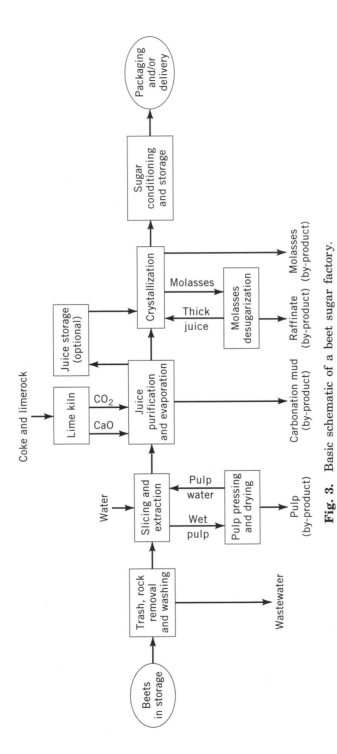

Fig. 3. Basic schematic of a beet sugar factory.

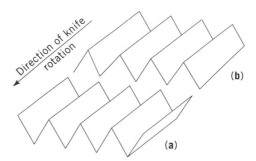

Fig. 4.   Offset arrangement of slicer knives where (**a**) shows the first cutting edge, and (**b**), the second.

material within the cells. Partial denaturing of the cell walls is accomplished by pre-scalding or heating cossettes after they leave the slicers, enough to allow free passage of the solution phase without rupturing the cell. Temperatures within the diffuser are maintained by using hot process water and indirect heating with low pressure steam. Maximum temperatures range from 50 to 70°C, depending on the integrity of the cossettes, the choice of diffuser design, and available heat-exchange capacity.

The three most common diffuser configurations are a vertical cylinder in which the semifluidized cossettes are scrolled upward (tower), a pair of upward-moving inclined twin-screw scrolls with cascading juices (slope), and a horizontal rotating drum equipped with offset compartments which allow the cossettes to fall forward as the drum turns (Raffinerie Tirlemontoise (RT) horizontal). Residence time within all of these diffusers is typically 45 to 60 minutes.

Spent cossettes (pulp) exit the diffuser with a moisture content of ca 92% and a sucrose content of ca 1%. To maximize the amount of sucrose returned to the process and minimize the amount of energy required to dry the pulp, this wet pulp is pressed in tapered twin-screw presses fitted with perforated side screens. On exiting the presses the moisture content of the pulp is ca 75% and the sucrose content ca 1%, which means that ca 2% of the sucrose that enters the factory with the cossettes leaves with the wet pulp (pulp loss). Water from the pressing step is passed through Dutch States Mines (DSM) screens and returned to the diffuser. The pressed pulp is normally mixed with molasses, dried to ca 10% moisture in rotating drum driers, stored, and sold as cattle feed, either as free shreds or pellets. Occasionally some of the pressed pulp is sold as-is to local feed operations. Processors in California use large cement strips, often disused airport runways, to solar dry wet pulp and save energy costs.

Improvements in pulp pressing technology have enabled manufacturers to reduce the moisture content of pressed pulp from 80% (moisture-to-solids ratio 4:1) to 75% (3:1 ratio), which has reduced the drying requirements by one-fourth. Beet pulp is much more difficult to press than most vegetable products, but there are indications that moisture in the mid-60% (ratios <2:1) range may be achievable with properly designed equipment.

The diffusion process has not been designed to ensure sterility, although temperatures above 65°C significantly retard microbial activity. Sulfur dioxide, thiocarbamates, glutaraldehyde, sodium bisulfite, and chlorine dioxide are all used, occasionally disregarding their redox incompatibilities, to knock down or control infections. The most common addition point is to the water from the pulp presses as it is returned to the diffuser. Surfactants are almost always used to control foaming in the diffusion process.

The raw juice exiting the diffuser is a murky dark gray solution occluded by colloidal materials from the ruptured beet tissue, small pieces of cossettes, and fine soil that escaped the fluming and washing processes. It is microbiologically and chemically unstable and unsuitable for concentration and crystallization. Common parlance assigns juice to more dilute process streams and syrup (or occasionally liquid) to steams with solids concentrations >70%.

Juice Purification

Raw juice is heated, treated sequentially with lime (CaO) and carbon dioxide, and filtered. This accomplishes three objectives: (1) microbial activity is terminated; (2) the thin juice produced is clear and only lightly colored; and (3) the juice is chemically stabilized so that subsequent processing steps of evaporation and crystallization do not result in uncontrolled hydrolysis of sucrose, scaling of heating surfaces, or coprecipitation of material other than sucrose.

Active lime and CO_2 are produced as needed by calcining lime rock in kilns fired with gas or metallurgical coke. Typically the uniformly sized lime rock and coke, in a 12:1 ratio, are dumped in the top of a vertical kiln and withdrawn from the bottom. They travel through fire and cooling zones so that the rock reaches a maximum of ca 1000°C without overburning which leads to unreactive CaO. Carbon dioxide, both from the lime rock and the combustion process, is withdrawn from the top of the kiln, washed, and sent to the process, while the CaO is mixed with dilute in-process streams (sweet water of ca 4–8% solids content) to form a CaO–water slurry (milk of lime) of ca 30% solids content. This is metered to the various points in the process where needed. Overall consumption of lime rock ranges from 2 to 5% rock on beets by weight. Alternatives to the coke-fired vertical kiln are firing the kiln with oil or natural gas and the gas-fired horizontal rotary kilns used in the cement industry.

The equipment and complexity of the juice purification scheme depend on the nature and variability of the raw material and on the performance of the preceding washing, slicing, and diffusion steps. Some of the more elaborate systems have been designed to be reconfigured on the run as the nature of the beets changes during the campaign.

Unit Operations. The chemistries elaborated by all of these systems are described by seven unit operations (Fig. 5). The first six, the use of lime and carbon dioxide as clarification agents, were laid out during the first half of the twentieth century and only the application technology has changed since, mainly from small batch processes designed to handle 1000 liters in a few hours to continuous systems capable of processing up to 10,000 L/min.

Pre-liming. Lime slurry, 0.25% lime on juice (0.250 g of CaO/100 g juice), is added to bring the pH of the mixture into the alkaline range. Insoluble calcium salts are precipitated as finely dispersed colloids. Calcium carbonate in the form of recycled first carbonation sludge is added to provide colloid absorption and stabilization. Temperature may be cool (50°C) or hot (80°C) depending on the

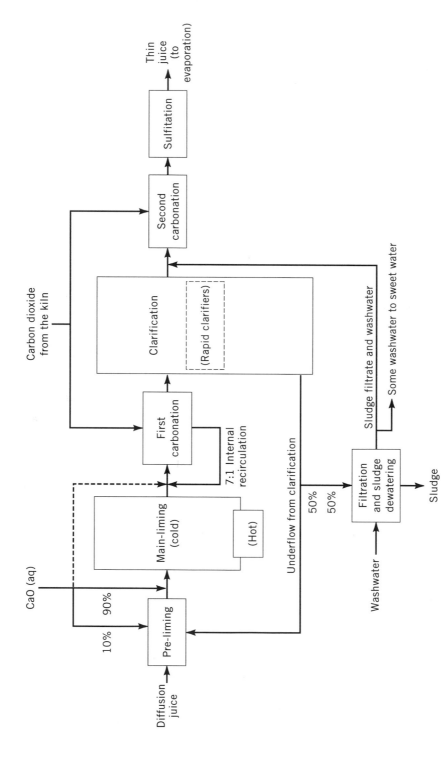

Fig. 5. Juice purification unit operation schematic where retention times are proportional to the size of the boxes and (– – –) are common optional processes.

temperature of the next step, or occasionally on the type of diffusion equipment. Retention time is 15 to 30 min.

Main Liming.   A further 1.50% CaO on juice is added and the juice is brought to its maximum alkalinity and pH. Conditions may be cool (50°C for ≤60 min, followed by 5–10 min at 80°C) or hot (80°C for 10–15 min). The invert sugars (glucose and fructose) are converted to organic acids, which do not form insoluble calcium salts. If this reaction does not take place in the early stages of the process, before pH stabilization, it almost always occurs later and is characterized by dropping pH and rising colors throughout the evaporation crystallization steps.

First Carbonation.   The process stream pOH is raised to 3.0 with carbon dioxide. Juice is recycled either internally or in a separate vessel to provide seed for calcium carbonate growth. Retention time is 15–20 min at 80–85°C. pOH of the juice purification process streams is more descriptive than pH for two reasons: first, all of the important solution chemistry depends on reactions of the hydroxyl ion rather than of the hydrogen ion; and second, the nature of the CO_3^{2-}–H_2O–Ca^{2+} equilibria results in a pOH which is independent of the temperature of the solution. All of the temperature effects on the dissociation constant of water are reflected by the pH.

Clarification.   Clarification is also referred to as sludge separation. First carbonation effluent is passed through a continuous clarifier where the precipitated calcium carbonate is allowed to settle while the clear juice overflows to second carbonation. Clarifiers may be rapid (10–15 min with increased flocculant, 10 ppm vs 2–3 ppm) or slow (45–60 min). The sludge that is not recycled to the pre-limer is dewatered by vacuum filtration. The sludge by-product is stored in large piles and may be sold as a soil amendment; the filtrate is returned to the process. The Spreckels factory (Mendota, California) reburns this sludge in a rotary kiln to reconvert the calcium carbonate back to lime.

Second Carbonation.   Calcium is reduced to the practical minimum by the addition of carbon dioxide at a pOH of 4.5 at a temperature of as near to 100°C as possible. This is the maximum temperature in the purification process and the retention time is only long enough to effect the pOH adjustment (5 min). The sludge from this unit operation is much less in amount than for first carbonation and is easily removed by in-line filters. The filtration is made even easier by the fact that the precipitate is almost pure calcium carbonate not fouled by the colloids found in first carbonation sludge.

Sulfitation.   Sulfur dioxide is added to a level of about 150 ppm on juice to discourage further color-forming Malliard reactions by tying up the small amount of invert sugars as bisulfite addition compounds.

Juice Purification Chemistry.   Lime in juice purification serves as a source of calcium, a source of alkalinity, and a source of calcium carbonate which serves as the clarification–filtration medium.

As a source of calcium, lime reacts with nonsucrose components to form a precipitate. Roughly one-third of these materials, by weight, form precipitates with calcium. Examples of such reactants are proteins, citrate, sulfate, saponins, phosphate, pectins, oxalate, and sulfite. The other two-thirds of the nonsucrose components remain in solution and salt-in sucrose reducing yields during subsequent crystallization steps. Diffusion juice components which do not precipitate with calcium include sodium, ammonia, nitrite, potassium, nitrate, betaine, magnesium, chloride, lactate, acetate, raffinose, glucose, glutamine, fructose, levans, and dextrans.

The stoichiometric ratio of lime to removable nonsucrose components is nearly five or six to one. This excess calcium is removed by the addition of carbon dioxide to form calcium carbonate, which is the primary clarification agent. Suspended or colloidal materials are adsorbed onto freshly precipitated calcium carbonate. These suspended or colloidal materials include not only those in the diffusion juice but also finely suspended precipitates of calcium salts. In sucrose solutions, $CaCO_3$ precipitate has a slight positive charge which is most effective at agglomerating negatively charged juice colloids. If the beets have been grown in clay-bearing soils, and washing does not adequately remove adhering soils, some of the clay may be present in juice as positively charged colloids. Unfortunately, these positively charged clay colloids have exactly the wrong chemistry for easy agglomeration and must be removed by the bulk filtration of the calcium carbonate.

High alkalinities of limed juice serve several functions. Foremost is to retard sucrose hydrolysis, one of the oldest reactions in the literature of chemical kinetics (6). Sucrose hydrolysis proceeds much more slowly at a moderately high pH than at an even slightly acidic pH.

Sucrose hydrolysis: reaction and kinetics

$$\text{sucrose} + \text{water} \xrightarrow[\text{salts}]{[H^+] \text{ or } [^-OH] \text{ heat}} \text{glucose} + \text{fructose}$$

$$\text{rate} = k(\text{sucrose})$$

$$k = k_{[H^+]}[H^+] + k_{[^-OH]}[^-OH]^{0.3} + k_{[salt]}[\text{salt}]$$

Not only is sucrose yield directly reduced as this reaction proceeds, but more nonsucrose components are formed. If 1% of the sucrose in a juice is hydrolyzed, it turns into ~1% nonsucrose components and the resultant loss to extraction is 2.5% (1% directly + 1.5% reduced crystallization yield). Each metric ton of the newly formed nonsucrose material salts 1.5 t of sucrose into molasses. The reaction has commercial significance at levels of only a few hundred parts per million.

High alkalinity not only protects sucrose from acid hydrolysis, but also helps produce a stable thin juice by acting on specific nonsucroses. The invert sugars glucose and fructose are oxidized to organic anions; lactate is the most common. If this reaction does not occur early in the sugar production process, glucose and fructose participate in Malliard reactions with free amino groups on other nonsucrose components, followed by Amadori rearrangements (7), and eventually produce medium to high molecular weight colorants. Excessive amounts of these colorants in crystallizing syrups frustrate the production of white sugar and lead to rework.

Glutamine, the most abundant amino acid in sugar beets and raw juice, is converted to 2-pyrollidinone-5-carboxylic acid and ammonia. The reaction is promoted by heat and the ammonia is driven off by the high pH of the process streams. The odor of ammonia is often easily detectable in beet sugar factories. Healthy, mature beets generally contain manageable amounts of glutamine, for which all purification systems have been designed. Immature beets contain appreciably higher levels of glutamine which leave residual amounts in thin juice. During evaporation ammonia is driven off and pH drops, sometimes leading to slightly acidic syrups. The most common way to counter this pH drop is by the addition of soda ash to second carbonation. Unfortunately, this addition increases the nonsucrose level of the juice.

Nonsucrose Components from Storage or Damage of Beets. Some non-sucrose components are associated with the conditions under which the beets have been stored prior to processing, as respiration products or products of microbial attack. In either case they directly and indirectly reduce sucrose yield and may cause other processing problems. Glucose and fructose have already been discussed and can derive from either source.

Raffinose, a nonreducing galactose–glucose–fructose trisaccharide, is observed when the whole beet is subjected to long periods of cool weather, either in the ground or in storage. Betaine, zwitterionic trimethyl glycine that the sugar beet uses to regulate osmotic pressure within its cells, seems to increase just prior to freezing. Levans and dextrans, the slimy products of microbial attack, are removed in carbonation with great difficulty. Dextrans have been shown to ruin the chemistry calcium carbonate precipitation in carbonation, retarding co-agulation with colloids (8). Usually, the amount of calcium carbonate (and hence the amount of lime) must be greatly increased to provide enough surface area to deal with these nonsucroses. Their presence is usually associated with degraded beets; although they can be a sign of factory process infections.

Nitrite is usually one indicator of the infection level in the diffuser. Exposure of nitrite to sulfur dioxide, either as a diffusion additive or later to thin juice, results in the production of potassium imidodisulfonate which precipitates when sugar is later crystallized, a cause of turbid or cloudy sugar.

Removal of Calcium Prior to Evaporation and Crystallization. The second carbonation step is designed to minimize the amount of calcium in thin juice by removing it as $CaCO_3$, the solubility of which is minimized by high temperature, high carbonate concentrations, and low nonsucrose concentrations. Counterbalancing these factors is the high solubility of calcium bicarbonate, which predominates as the pH drops below 8.4. The addition of CO_2 provides carbonate ion, but also lowers the pH. Upstream infection problems produce acids leading to lower juice pH. After prolonged heating the bicarbonate eventually expels CO_2 and produces carbonate, leading to $CaCO_3$ scaling on evaporation and heating surfaces, a common consequence and a significant cost increase in processing degraded beets. The minimum calcium content may vary from 20 ppm to more than 350 ppm depending on the raw material. Juice softening with resins, commonplace in European factories, has been installed in about one-third of U.S. factories, justified not only by process advantages but also by the need for molasses having low calcium/magnesium levels for molasses desugarization operations.

Crystallization and Recovery of Sucrose

The three-boiling scheme, typical of U.S. beet sugar production, is shown in Figure 6. Incoming thick juice is combined with recycled lower grade sugars, filtered, and crystallized under vacuum to yield about half the sucrose separated from the mother liquor and washed in batch centrifugals. The process is repeated

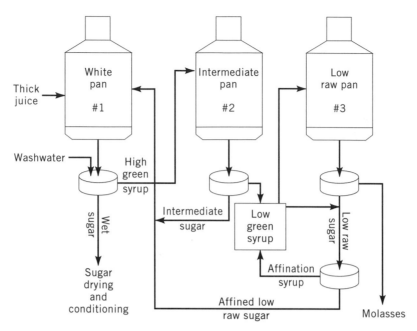

Fig. 6. The three-boiling beet sugar crystallization scheme.

two more times using continuous centrifugals for the separation of syrups and crystals; only the first crop of crystals is used for finished product. The third crop is washed with a slightly higher purity syrup (affined) to reduce the load of nonsugars and color. Both the second and third crops are returned to the first stage. After the liquor has been concentrated to about 10% past supersaturation, crystal growth is initiated by fully seeding each pan with enough crystals, 1–5 μm in diameter, to account for the finished pan.

All of these processes are carried out under kinetically controlled conditions and vacuum pan operations may be followed by a crystallizer in which the massecuite (crystals/mother liquor) is allowed more time for crystal growth, sometimes for as long as 48 hours for the third stage. Crystal growth is enhanced by higher purities, lower viscosities, higher concentrations, agitation of the crystallizing mass, higher temperatures, and time. The first boiling typically takes $1\frac{1}{2}$ to 2 hours; the second, 4 hours; and the final boiling, 8 hours or longer plus at least 12 hours of crystallizer time. Surfactants designed to reduce viscosities are commonly added to each stage. The batches are referred to as strikes, a holdover from the practice of initiating crystallization by hitting the pan with a large hammer or iron bar. Some older equipment may show signs of this abuse.

Crystallization batches range from 30,000 to 60,000 liters for each pan. Continuous centrifugals are typically used for second, third, and affination steps; continuous vacuum pans are less common but are used in the U.S. for intermediate strikes. Most horizontal batch crystallizers have been replaced by continuous units, and all are designed for controlled cooling of the massecuite to maintain supersaturation.

Balancing these crystallization steps to maximize the yield and maintain high product quality is a significant operational challenge, especially when confronted with changing thick juice quality, ie, color, purity, and pH stability (Table 1). These values reveal the critical dependence of performance on the quality of the thick juice and, by inference, on the nature of the beets. Calculations have assumed the same yield for each of the boilings. The higher nonsucrose loading has reduced the final yield by 5% for the low purity thick juice. For the same 100 kg of sugar entering crystallization, each stage of the low purity scenario has to handle 40–50% more material because of increased concentrations of returned sugar from the second and third stages. Because the process is controlled by kinetics, either pan yields must be sacrificed or overall throughput must be reduced; either choice has costly results. Low purity juices tend to be less stable with respect to color, pH, and calcium precipitation, and the choice is always between bad and worse.

Table 1. Material Balance and Purity Effects of Thick Juice Purity[a]

Material	Low purity juice		High purity juice	
	Sugar, kg	Purity, %	Sugar, kg	Purity, %
thick juice	100.0	88.0	100.0	92.0
white pan	208.2	93.5	143.8	93.9
intermediate pan	72.8	83.4	50.3	84.5
low pan	32.8	69.1	22.6	70.8
molasses	16.2	54.3	11.2	56.3
product	83.8		88.8	

[a]Purity = sugar content as percent of total dissolved solids content.

Conditioning and Storage of Sucrose

Washed, wet sugar contains about 1% moisture which must be reduced to about 0.03% without glazing over the crystals. This is accomplished in a rotating drum granulator–cooler in which warm air is passed over the crystals as they roll down the length of the drum. This is followed by cooler air to stabilize the crystals at <35°C before conditioning. Best practices allow sugar to cure for 24 hours before long-term storage or shipping as bulk sugar in rail cars or trucks. The most common customer complaint for every beet sugar producer is hard and lumpy sugar caused chiefly by failure of these processes.

Because beets are processed for only a portion of the year, a factory may store as much as half of the production sugar in order to provide continuous distribution to customers. This is accomplished either in sets of tall, vertical cement silos, each capable of holding 6800 t, or single large-diameter Weibul-type curing silos of 23,000 t capacity. The cement silos are theoretically first-in first-out configuration in which the sugar is dropped in the top and withdrawn from the bottom; in practice, sugar tends to funnel through the center of the silo, resulting in a last-in first-out scheme. A Weibul silo has a central distribution column with a rotating arm which scatters the sugar allowing it to fall through dehumidified air in even layers about the bin. Withdrawing sugar is accomplished by using the same arm to rake sugar from the top toward the center where it falls to the bottom into an annular opening around the central column. This last-in first-out system provides the best conditioning for the product which must be stored the longest.

Molasses Desugarization

Chromatographic separation of diluted molasses streams into a high purity fraction suitable for concentration and crystallization and a second low purity by-product, which can be concentrated and sold as an animal feed product, has been employed in Finland since the 1970s and in the United States since the mid-1980s. Since the early 1990s, production of sugar from beet molasses has almost tripled, and the trend is expected to continue for the next two years to consume most of the domestic beet molasses (Fig. 7) (3,9).

Strong cationic sulfonic acid resin beads of very uniform size (325 μm) are used as a stationary phase. Their active sites exist in a mixed sodium–potassium form depending on the makeup of the molasses being separated. Sugar has a slight affinity for these resins, betaine an even higher affinity, and most ions have little or no affinity, hence the term ion exclusion for this process. This places two fairly strict requirements on the feed molasses: the molasses must have low concentrations of calcium and magnesium (<500 ppm), which not only compete for the active sites but may precipitate within the resin bed, and the molasses constituents should be as uniform as possible so the nature of the resin is not constantly changing.

The separation uses either of two modes of operation: (1) a pulse method in which batches of feed are sequentially placed on the column and eluted with water, taking product and by-product fractions as desired, and (2) a simulated moving bed (SMB) configuration in which the feed is continually placed on the column at the point where its purity matches the separation. The pulse method provides more flexibility, less dilution, and better separations but suffers from a

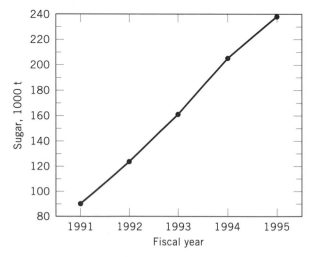

Fig. 7. U.S. production of beet sugar from molasses desugarization processes. Values for 1995 are estimated.

waiting time for the previous pulse to clear. The steady-state method has higher capacity and is easier to manage, but requires more valves on a series of sections in order to move the point of addition. The SMB separation is most effective when only two components are involved, ie, sucrose and salt or sucrose and betaine. Under normal operating conditions, 15 m of column length is required to economically separate molasses components on this 325-μm resin.

The product has purities typically in the 90–92% range and can be combined with thin juice, concentrated and crystallized, or concentrated and stored for later use. Crystallizing the desugarization thick juice apart from the normal beet campaign may be desired because the secondary molasses produced after the separation contains the nonsucrose components, which are the most difficult to separate from sucrose and perhaps should be set aside and sold instead of resubmitted to the columns.

The desugarization by-product is normally sold as a low value molasses. Pulse method systems also produce a relatively high value betaine-rich (at least 50% on solids) fraction. The concentrated betaine-rich by-product is used as a custom animal feed, whose European markets are well established and may provide a future opportunity in the U.S. feed industry. Beet sugar molasses contains from 3 to 6% betaine, by weight, about three-quarters of which may be recoverable as a potential by-product ($\sim$40–50% purity).

Marketing and Economic Aspects

U.S. beet sugar producers share the domestic sucrose market with sugar produced from domestic sugar cane and imported raw sugar. The grower–processor beet contract is highly participatory. The processor determines the amount of acreage that a factory can support, based on historical yields and daily factory processing capacity. The grower and processor enter into a contract, before seeds are planted, in which the grower agrees to plant, nurture, harvest, and deliver the beets whereas the processor agrees to purchase the entire crop. Payment is based on the sugar content of the beets and the net selling (after selling expenses) price (NSP) that the sugar actually fetches in the market. The fluctuations in the market are shared between the sugar beet growers and the processors ($\sim$60:40), and the variations in factory performance are borne by the processor.

The payment formula assumes that 85% of the sugar in the beet can be extracted and then \$0.60 out of every sales dollar is returned to the grower. The details vary somewhat from region to region with allowances for such things as early deliveries, losses during beet storage, actual factory performance, transportation and retransportation costs, and beet quality parameters. The quality parameter premiums may include extra high sugar content and low levels of nonsucrose components which translate to higher in-factory sucrose yields. The average return to the grower is closer to \$0.63 on the dollar.

Grower-owned cooperatives produce about 40% of the beet sugar in the United States. The revenues returned to these growers is also about \$0.60 on the sales dollar, although the contractual arrangements and payment formulas are much different, ie, basically all residual revenues after operating and selling expenses have been paid are distributed among the growers according to the amount of sugar delivered to the factories.

The U.S. government supports the domestic beet sugar industry through low interest loans to processors, secured by finished sugar as collateral. The rate, $0.18/lb ($0.40/kg) raw value, is established by legislative mandate. The U.S. Department of Agriculture restricts the amount of raw sugar to keep the domestic prices high enough to avoid forfeiture of sugar. If the amount of imported sugar drops below a predetermined minimum quota level, the department has the authority to enforce market allocations and force beet sugar processors to withhold sugar from the market at their own expense. The quota distributes the U.S. sugar import needs among raw sugar exporting trading partners. The loans and market allocations were mandated by the 1990 Farm Bill and have been reestablished in the 1995 version; the quota is part of the General Agreement on Tariffs and Trade and other trade agreements such as the Caribbean Basin Initiative and NAFTA.

Economics of Beet Sugar Production. Figure 8 presents the material and financial balances of production of 2500 consumer packages of sugar from 1 hectare. It starts with $99 worth of seed, which leads to $1771 worth of beet payments, from which $3169 worth of bulk sugar, pulp, and molasses is extracted, $3737 wholesale value after packaging, and finally $5187 on the grocers' shelves. About one-third of the increase in value from $99 to $5200 is realized by the grower, 40% by the processor, and, for this example, 28% by the wholesaling–retailing grocer.

With the current adequate supply of domestic sugar, the opportunity to increase net revenues by increasing prices is limited or diminishing. Growers and sugar producers must cooperate to maximize crop and factory yields while reducing their own costs. Margins are so slim that gains made by one group at the expense of the other surely weaken the relationship. Since 1970 this has led to the emergence of grower-owned cooperatives which combine the first three steps of the value chain (see Fig. 8), and which accounts for 40% of U.S. beet sugar production. This trend is expected to continue, not only by growers purchasing existing operating facilities but also through formal partnershipping, marketing

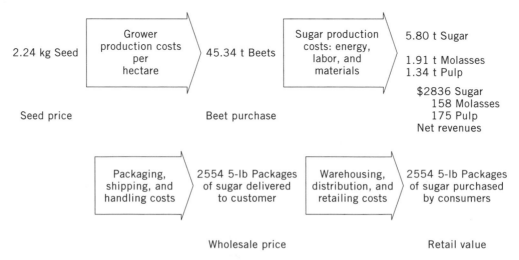

Fig. 8. Typical value chain for 1 ha U.S. beet sugar production. See text.

agreements, and complex risk and profit sharing agreements between growers, producers, and perhaps even customers.

   The Domestic Sugar Market.   Figure 9 (10) traces the market trends for sugar in the United States by food use customer group. Overall the market is growing at a rate of 3–4% per year, with confectionery, baking and cereal, and ice cream and dairy products accounting for most of the growth. The sharp decline in the beverage segment between 1980 and 1985 represents the replacement of sucrose by high fructose corn sweeteners as the sweetener of choice. The increasing use of nonnutritive sweeteners seems to have created new markets rather than displacing existing sucrose markets.

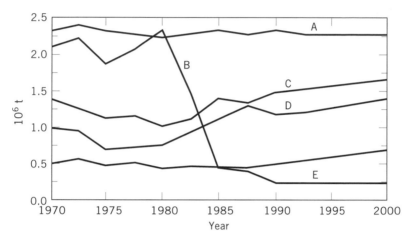

Fig. 9.   U.S. sugar deliveries by customer group where A represents the consumer; B, beverages; C, baking and cereal; D, confectionery; and E, ice cream and dairy.

Product Quality and Requirements

The most common parameters used to measure product quality are moisture, color, granulation, sediment, and ash.

   Moisture.   Moisture is usually determined by a vacuum oven-dry method at 80°C. Moisture levels of more than 0.05% are likely to lead to caking or lumping problems which can make storage and transfer of bulk sugar difficult. The usual standard is 0.03%, which manufacturers can easily meet. Care must be taken to avoid temperature differentials in storage which cause moisture to migrate and establish pockets of unacceptably high moisture levels.

   Color.   Color is usually specified as white and measured as a solution color using the specific absorbance at 420 nm. If the measurement is made on a filtered (0.45 μm) solution, it is reported as International Commission for Uniform Methods of Sugar Analysis (ICUMSA) units. More commonly, a second transmittance reading is taken at 720 nm, a turbidity correction made, and the result reported in Reference Basis Units (RBUs). Turbidity can also be used as a quality parameter. Using either scale, sugar begins to be noticeably off-white at about 50 units. The upper limit is usually 35 units, which corresponds to a straw-colored solution at 50% solids. High color in sugar is a harbinger of other problems such as foaming, off-odors, cloudiness or floc, or generally poor performance in production which is why many customers who use sugar in highly colored or opaque products insist on low color sugar.

Granulation.   Granulation is important to customers who do not want too much dust (fines), lumps, or grittiness, and/or who have a very specific need to mix the sugar with other dry ingredients or to turn it into a fondant. Customers in the latter categories are willing to pay a premium for specially screened sugar. Beet sugar is commonly produced with crystal sizes in the 0.400 mm range with a coefficient of variation of 25–30%. The size distribution is determined on a stack of three to eight sieves of decreasing sizes using U.S. Sieve values for reference.

Sediment.   Sediment is most commonly used as an operational check of filter efficiency and leakage, although some customers, especially those who intend to melt the sugar into clear solutions, write sediment restrictions. The measurement is normally done by passing the 50% solution used for the color determination through a half black–half white filter pad and visually counting the white and black specks.

Ash.   This can be determined by a gravimetric method using sulfuric acid to digest the sugar followed by burning in a muffle oven at 650°C. Measuring conductivity on the 50% solution and then referencing this value to the sulfated ash method is much more common. Typical values are 0.003–0.008%, with the upper limit on specifications usually written at 0.015%. Ash is a good indication of the general level of impurities in beet sugar and unacceptably high ash levels usually are accompanied by other problems. This is not an important parameter for cane sugar which can be quite acceptable with ash levels of more than 0.035% (see SUGAR, CANE SUGAR).

BIBLIOGRAPHY

"Beet Sugar" under "Sugar Manufacture" in *ECT* 1st ed., Vol. 13, pp. 217–227, by L. A. Willis, Consultant; "Beet Sugar" in *ECT* 2nd ed., Vol. 19, pp. 203–220, by R. A. McGinnis, Spreckels Sugar Co.; "Beet Sugar" under "Sugar" in *ECT* 3rd ed., Vol. 21, pp. 904–920, by R. A. McGinnis, Consultant.

1. *Sugar and Sweetener Situation and Outlook Report*, SSSV21N1, U.S. Dept. of Agriculture, Economic Research Service, Washington, D.C., Mar. 1996.
2. "The American Beet Sugar Industry in 1909," *Sugar Beet Gazette*, 1910.
3. Ref. 1, Feb. 1995.
4. R. A. McGinnis, ed., *Beet Sugar Technology*, 3rd ed., Beet Sugar Development Foundation, Denver, Colo., 1983, for an overview of the processes involved in all but molasses desugarization.
5. M. A. Clarke and M. A. Godshall, *Chemistry and Processing of Sugarbeet and Sugarcane*, Elsevier Science Publishers BV, Amsterdam, the Netherlands, 1988, p. 9.
6. L. Wilhelmy, *Pogg. Ann.* **81**, 413, 499 (1850).
7. K. Olsson, P. Pernemalm, and O. Theander, *Acta Chem. Scand. B*, **32**, 249–250 (1978).
8. J. F. T. Oldfield, J. Dutton, and H. Teague, *Proceedings of the 20th Technical Conference of the British Sugar Corp.*, British Sugar Corp., Ltd., Cambridge, U.K., 1967.
9. B. Pynnonen, *Proceedings of the Symposium on Industrial-Scale Process Chromatography Separations*, New Orleans, La., Mar. 27–30, 1996.

10. S. Vuilleumier, *U.S. Sugar Deliveries by Customer Group*, McKeany-Flavell Co., Inc., Oakland, Calif., June 1995.

MICHAEL CLEARY
Imperial Holly Corporation

SUGAR DERIVATIVES

Sucrose, commonly known as sugar, has been used as a natural sweetening agent for almost 4000 years. It is isolated from sugarbeet (*Beta vulgaris*) in Europe and from sugarcane (*Saccharum officinarum*) in the tropics. Its total world production in 1994–1995 was 116 million metric tons.

Sucrochemistry has seen rapid advances since the 1960s. Whereas well-characterized sucrose derivatives numbered only about 15 in 1965, over 300 well-identified sucrose compounds have been described more recently in the literature (1).

Structure of Sucrose

(1)

Sucrose is systematically named as α-D-glucopyranosyl β-D-fructofuranoside (**1**) [*57-50-1*]. It exists in two crystalline forms: a stable A-form crystallized from water, mp 184–185°C, and a metastable B-form recrystallized from methanol, mp 169–170°C (1). It is a nonreducing disaccharide with eight hydroxyl groups, of which three are primary (6, 1′, and 6′) and the remaining five are secondary (2, 3, 4, 3′, 4′). The carbon atoms in sucrose are numbered as primed and unprimed in (**1**). The correct ring structure of sucrose was first deduced by methylation studies (2,3). The configurations at the glycosidic centers were determined as a result of specific enzymic hydrolysis studies (4) and x-ray (5,6) and neutron diffraction data (7,8) (see CARBOHYDRATES).

Conformation. Neutron diffraction studies of sucrose revealed the presence of two strong intramolecular hydrogen bonds: O-2–HO-1′ and O-5–HO-6′ in the crystal form (7,8). These interactions hold the molecule in a well-ordered and rigid conformation. The two rings are disposed at an angle close to 90°, with the glucopyranosyl and fructofuranosyl residues adapting $^4C_1$ chair and $_3T^4$ twist conformations, respectively.

The conformation of sucrose in solution has been extensively investigated. Hard-sphere exoanomeric (HSEA) methods, supported by ^{1}H- and ^{13}C-nmr data, revealed that sucrose in dilute solution had a similar conformation to the crystal state but with only one intramolecular H bond between O-2 and HO-1′ (9). Further support of this conformation was provided by ^{13}C-nmr spin relaxation studies (10,11). An additional conformation containing a hydrogen bond between HO-3′ and O-2 was shown to be important (12). Using x-ray and Raman techniques, no evidence for intramolecular hydrogen bonds in dilute solution (<0.7 M) was found, but at saturated solution the conformation resembled that of the crystal structure (13). Further nmr studies on sucrose in water have indicated that the O-2–HO-1′ hydrogen bonding was not permanently maintained (14). The solid state ^{13}C-nmr spectra of partially deuterated crystalline sucrose has been reported (15). The spectrum showed some marked differences over the solution spectrum, not the least that C-1′ was observed at lower field with C-6 and C-6′ resonances at higher field. However, these differences are not unexpected considering the differences in electronic environments on going from solid to solution states.

Ethers

Trityl Ethers. Treatment of sucrose with four molar equivalents of chlorotriphenylmethyl chloride (trityl chloride) in pyridine gives, after acetylation and chromatography, 6,1′,6′-tri-O-tritylsucrose [35674-14-7] and 6,6′-di-O-tritylsucrose [35674-15-8] in 50 and 30% yield, respectively (16). Conventional acetylation of 6,1′,6′-tri-O-tritylsucrose, followed by detritylation and concomitant C-4 to C-6 acetyl migration using aqueous acetic acid, yields a pentaacetate, which on chlorination using thionyl chloride in pyridine and deacetylation produces 4,1′,6′-trichloro-4,1′,6′-trideoxygalactosucrose [56038-13-2] (sucralose), a low calorie sweetener (17).

Methyl Ethers. Methylation of sucrose is generally conducted under basic conditions. Etherification occurs initially at the most acidic hydroxyl groups, HO-2, HO-1′, and HO-3′, followed by the least hindered groups, HO-6 and HO-6′. Several reagents have found use in the methylation of sucrose, including dimethyl sulfate–sodium hydroxide (18,19), methyl iodide–silver oxide–acetone, methyl iodide–sodium hydride in N,N-dimethylformamide (DMF), and diazomethane–boron trifluoride etherate (20). The last reagent is particularly useful for compounds where mild conditions are necessary to prevent acyl migration (20).

Other Alkyl Ethers. Sucrose has been selectively etherified by electrochemical means to generate a sucrose anion followed by reaction with an alkyl halide (21,22). The benzylation of sucrose using this technique gives 2-O-benzyl- (49%), 1′-O-benzyl- (41%), and 3′-O-benzyl- (10%) sucrose (22). The benzylation of sucrose with benzyl bromide and silver oxide in DMF also produces the

2-O-benzyl ether as the principal product, but smaller proportions of 1′- and 3′-ethers (23). Octadienyl ether derivatives of sucrose, intermediates for polymers, have been prepared by a palladium-catalyzed telomerization reaction with butadiene in 2-propanol−water (24).

Silyl Ethers.   The preparation of per-O-trimethylsilyl ethers of sucrose is generally achieved by reaction with chlorotrimethylsilane and/or hexamethyldisilazane in pyridine (25,26). However, this reaction is not selective and in general per-trimethylsilyl ethers are only used as derivatives for gas chromatographic studies.

Sterically hindered silyl ethers such as *tert*-butyldimethylsilyl, *tert*-butyldiphenylsilyl, and tricyclohexylsilyl have been proposed as alternatives to trityl ethers. Reaction of sucrose with 3.5 molar equivalents of *tert*-butyldimethylsilyl chloride produces the 6,1′,6′-tri-O-silyl derivative in good yield (27). Silylation of sucrose with 0.65 equivalents of *tert*-butyldimethylsilyl chloride in pyridine gives the corresponding 6′-, 6,6′-di-, and 6,1′,6′-tri-O-*tert*-butyldimethylsilyl ethers in yields of 10.5, 36.4, and 33.5%, respectively. Monosubstitution at C-6 and C-1′ under the conditions employed was not observed. When sucrose was treated with one molar equivalent of the more sterically hindered *tert*-butyldiphenylsilyl chloride in pyridine, 6′-O-*tert*-butyldiphenylsilylsucrose was isolated in 49% yield (28). 6,6′-Di-O-*tert*-butyldiphenylsilylsucrose (78%) is obtained as the principal product when three molar equivalents of the silylating reagent are used. The 6,1′,6′-tri-O-*tert*-butyldiphenylsilylsucrose is the principal product on treatment of sucrose with 4.6 molar equivalents of the silylating reagent. These results clearly show that HO-6′ is the most reactive site toward silylation.

These hindered silyl ethers are generally more stable to acid hydrolysis than their trityl ether equivalents and can be removed using tetrabutylammonium fluoride. However, deprotection can be difficult and if there are ester groups they can hydrolyze and/or migrate. These difficulties and the relative expense of the reagents mean that trityl ethers have seen more use as selective protecting groups in industrial sucrochemistry.

Cyclic Acetals.   One of the most significant developments in the chemistry of sucrose was the synthesis of cyclic acetals which, despite many attempts, were not synthesized until 1974. The first synthesis of 4,6-O-benzylidenesucrose was achieved from the reaction of sucrose with α,α-dibromotoluene in pyridine (29). Since then, many new acetalating reagents have been used to give a variety of sucrose acetals, generally by transacetalation reactions.

Treatment of sucrose with 2,2-dimethoxypropane, DMF, and toluene-p-sulfonic acid gives 4,6-O-isopropylidenesucrose and 4,6:2,1′-di-O-isopropylidenesucrose (30,31). The 4,6-mono- and 4,6:1′2-diacetals are obtained in 60 and 70% yields, respectively, using the kinetic acetalating reagent, 2-methoxypropene (32). The unique eight-membered 2,1′-cyclic acetal bridges the two rings in sucrose, is more stable to acid than the 4,6-acetal linkage, and has been effective in providing access to selective reactions at C-2 and 1′ positions in sucrose. Interesting 8- and 12-membered ring cyclic acetals of sucrose have been synthesized by using 2,2-dimethoxydiphenylsilane, DMF, and toluene-p-sulfonic acid to give the 2,1′- and 2,1′:6,6′-diphenylsilylene derivatives in 45 and 10% yields, respectively (33).

Esters

Acetates. Because of the significant interest in selective acetylation reactions of sucrose, the need for a convenient and unambiguous method of identification has been recognized (34,35). The position of an acetyl group in a partially acetylated sucrose derivative can be ascertained by comparison of its ^{1}H-nmr acetyl methyl proton resonances after per-deuterioacetylation with those of the assigned octaacetate spectrum. The synthesis of partially acetylated sucroses has generally been achieved either by way of selectively protected derivatives such as trityl ethers and cyclic acetals or by direct selective acetylation and deacetylation reactions.

6-O-Acetylsucrose [63648-81-7] has been prepared in 40% yield by direct acetylation of sucrose using acetic anhydride in pyridine at −40°C (36). The 6-ester has subsequently been obtained in greater than 90% yield, by way of 4,6-cyclic orthoacetate. Other selective methods for the 6-acylated derivatives include the use of alkyl tin reagents such as dibutyl tin oxide (37) and of dibutyl stannolane derivatives (38). Selective acetylation of sucrose by an enzymic process has also been described. Treatment of sucrose with isopropenyl acetate in pyridine in the presence of Lipase P Amano gave, after chromatography, 6-O-acetylsucrose (33%) and 4′,6-di-O-acetylsucrose (8%). The latter compound has been obtained in 47% yield by the prolonged treatment (39).

The selective deacetylation of sucrose octaacetate [126-14-7] provides another route to partially acetylated sucrose derivatives. The selective chemical deacetylation of sucrose octaacetate has been the subject of much investigation (40–44). Reaction of the octaacetate absorbed on aluminium oxide produces heptaacetates that have the 6′-OH, 4-OH and 4′-OH groups free, but in low yield (40). This method has been modified (41–43); methanolic solutions of sucrose octaacetate and heptaacetates are adsorbed onto alumina impregnated with potassium carbonate. The various reactions produce heptacetates that have free hydroxyl groups at C-1′,3′,4′,6′ positions (41); hexaacetates with free hydroxyl groups at C-3′,4′ and C-1′,3′ (42); and pentaacetates with free hydroxyl groups at C-3′,4′,6′, C-1′,3′,4′, and C-2,3′,4′ positions (43), in low to moderate yields. More recently, sucrose octaacetate has been deacetylated using primary amines in the absence of solvents, to produce 2,3,4,6,1′,6′-hexa-O-acetylsucrose in 22% yield (44).

Sucrose octaacetate has been selectively deacetylated with a number of lipases and proteases in buffer solutions or biphasic media to produce 2,3,4,6,1′,3′,6′-hepta-O-acetylsucrose (45), 2,3,4,3′,4′-penta-O-acetylsucrose [34382-02-2], and 2,3,6,1′,6′-penta-O-acetylsucrose [35867-25-5] (46–51). Wheat germ lipase exclusively deacetylates in the fructose ring to produce 2,3,4,6,1′,3′-hexa-O-acetylsucrose and 2,3,4,6,1′-penta-O-acetylsucrose (52). Enzymic deacetylation of sucrose octaacetate with various lipases and proteases in organic solvents containing minimal amounts of water has been investigated (53). Prior to use, commercial enzymes are precipitated from a pH-adjusted buffer, dried, and then rehydrated. Identified were a 3:1 mixture of 3′- and 1′-hydroxy-heptaacetates using lipase AY 30 (*Candida cylindracea*), the 3′-hydroxy-heptaacetate from porcine pancreatic lipase, the 6′-hydroxyheptaacetate using lipase SP-435 (*Candida antartica*), and a 1:1 mixture of 4- and 6-hydroxy-heptaacetates with lipase AP 12 (*Asperilligus niger*). Of the protease enzymes, protease N (*Bacillus subtilis*) produces a 1:1:1 mixture of 3'-, 1'-, and 6'-hydroxy-heptaacetates. The 6'-hydroxy-heptaacetate was the predominant product with the serine protease proleather (*Bacillus subtilis*) and alcanase (*Bacillus licheniformis*). On prolonged treatment with the latter enzymes, a mixture containing unreacted starting material, 1'- and 6'-hydroxy-heptaacetates, 1',6'-dihydroxy-hexaacetate, and the 6,1',6'- and 4,1',6'-trihydroxy-pentacetates were identified. The 4,1',6'-trihydroxy-pentaacetate was reportedly isolated as a single compound but no yield was given (53).

Benzoates. The selective debenzoylation of sucrose octabenzoate [2425-84-5] using isopropylamine in the absence of solvents caused deacylation in the furanose ring to give 2,3,4,6,1',3',6'-hepta- and 2,3,4,6,1',6'-hexa-*O*-benzoyl-sucroses in 24.1 and 25.4% after 21 and 80 hours, respectively (54). The unambiguous assignment of partially benzoylated sucrose derivatives was accomplished by specific isotopic labeling techniques (54). Identification of any benzoylated sucrose derivative can thus be achieved by comparison of its ^{13}C-nmr carbonyl carbon resonances with those of the assigned octabenzoate derivative after benzoylation with 10 atom % benzoyl–carbonyl ^{13}C chloride in pyridine.

Reaction of 4,6:1',2-di-*O*-isopropylidenesucrose in pyridine–chloroform with 3.3 molar equivalents of benzoyl chloride at 0°C eventually produced 3',6'-di-*O*-benzoylsucrose (36%) as the major and 3',4',6'- and 3,3',6'-tribenzoates as the minor products. The relative reactivities of the hydroxyl groups toward benzoylation was HO-3' ≈ HO-6' > HO-4' > HO-3 (55).

Pivalates. The selective pivaloylation of sucrose with pivaloyl (2,2-dimethylpropionyl) chloride has been thoroughly investigated (56). The reactivity of sucrose toward pivaloylation was shown to be significantly different from other sulfonic or carboxylic acid chlorides. For example, reaction of sucrose with four molar equivalent of toluene-*p*-sulfonyl chloride in pyridine revealed, based on product isolation, the reactivity order of O-6 ≈ O-6' > O-1' > O-2 (57). In contrast, a reactivity order for the pivaloylation reaction, under similar reaction conditions, was observed to be O-6 ≈ O-6' > O-1' > O-4. Two divergent routes to sucrose octapivalate by way of this reaction have been suggested, each the result of different reactivities of the partially pivalated derivatives toward further acylation: (*1*) 6,6'-OH > 1'-OH > 4'-OH > 2-OH > 4-OH > 3'-OH > 3-OH, and (*2*) 6,6'-OH > 1'-OH > 3'-OH > 3-OH > 4'-OH > 2-OH and 4-OH.

Fatty Acid Esters. There has been much interest in the selective esterification of sucrose with fatty acids, primarily to produce surfactants, emulsifiers, detergents, and fat replacers (58–66), which comprise an ever-increasing market. These derivatives can be produced on an industrial scale by solventless or melt reactions. The transesterification reaction of sucrose with fatty acid methyl esters or triglycerides in the presence of a base, such as potassium carbonate, at 130–150°C, has been reported to give the corresponding fatty acid esters (58). Reactions using methyl esters are conducted under reduced pressure, thus removing methanol and driving the equilibrium to favor the formation of the sucrose fatty acid derivative.

The monofatty acid esters are surfactants and emulsifiers and have the advantage that they are biodegradable, nontoxic, edible, and can inhibit the growth of microorganisms in some cases (67). Produced by Dai-Ichi Kogyo Seiyaku

Company, Ltd. in Kyoto and Mitsubishi Corporation in Tokyo, Japan, they are approved by the U.S. FDA as food additives. Sucrose monofatty acid esters are used in food formulations and, because of their excellent skin compatibility, find application in shampoos and cosmetics (qv). They are also used on fruits and vegetables as edible semipermeable coatings to retard ripening and reduce wastage resulting from rotting.

A commercially interesting low calorie fat has been produced from sucrose. Proctor & Gamble has patented a mixture of penta- to octafatty acid ester derivatives of sucrose under the brand name Olestra. It was approved by the FDA in January 1996 for use as up to 100% replacement for the oil used in preparing savory snacks and biscuits. Olestra, a viscous, bland-tasting liquid insoluble in water, has an appearance and color similar to refined edible vegetable oils. It is basically inert from a toxicity point of view as it is not metabolized or absorbed. It absorbs cholesterol (low density lipoprotein) and removes certain fat-soluble vitamins (A, D, E, and K). Hence, Olestra has to be supplemented with these vitamins. No standard LD_{50} tests have been performed on Olestra; however, several chronic and subchronic studies were performed at levels of 15% in the diet, and no evidence of toxicity was found. No threshold limit value (TLV), expressed as a maximum exposure per m^3 of air, has been established, but it is estimated to be similar to that of an inert lipid material at 5 mg/m^3.

Olestra is prepared by a solventless transesterification process in which sucrose is treated with methyl ester of fatty acids in the presence of sodium methoxide between 100–180°C for 14 hours (68). The manufacturing process involves removal of the unreacted fatty acid esters by enzymic hydrolysis with lipases, refining, bleaching, and deodorizing to produce sucrose polyesters containing five or more fatty acid ester groups.

Other Carboxylic Esters. Selective 2-*O*-acylation of sucrose has been achieved by way of the 2-oxyanion compound. Treatment of sucrose in DMF with 3-lauryl-, 3-stearyl-, 3-hydrocinnamoyl-, and 3-(4-phenylbutyryl)-thiazolidine-2-thione derivatives and sodium hydride produced the corresponding 2-*O*-acyl derivatives in good yield (69). Syntheses of 6-*O*-acylsucroses were also achieved by acylation with 3-acylthiazolidine-2-thione and 3-acyl-5-methyl-1,3,4-thiadiazole-2(3*H*)-thione derivatives in the presence of sodium hydride in DMF, followed by acyl migration using 1,8-diazabicyclo[5.4.0]undec-7-ene (DBU) or aqueous triethylamine. 6-*O*-Acylsucroses were obtained directly when only DBU was used (70).

Enzymatic acylation reactions offer considerable promise in the synthesis of specific ester derivatives of sucrose. For example, reaction of sucrose with an activated alkyl ester in *N*,*N*-dimethylformamide in the presence of subtilisin gave 1′-*O*-butyrylsucrose, which on further treatment with an activated fatty acid ester in acetone in the presence of lipase *C. viscosum* produced the 1′,6-diester derivative (71,72).

Orthoesters. The value of cyclic orthoesters as intermediates for selective acylation of carbohydrates has been demonstrated (73). Treatment of sucrose with trimethylorthoacetate and DMF in the presence of toluene-*p*-sulfonic acid followed by acid hydrolysis gave the 6-*O*-acetylsucrose as the major and the 4-*O*-acetylsucrose [63648-80-6] as the minor component. The latter compound underwent acetyl migration from C-4 to C-6 when treated with an organic base, such as *tert*-butylamine, in DMF to give sucrose 6-acetate in >90% yield (74). When the kinetic reagent 2,2-dimethoxyethene was used, 4,6-*O*-(1-methoxyethylidene) sucrose [*116015-72-6*] (**2**), the intermediate for 6-*O*-acetylsucrose was obtained in near quantitative yield (75). The synthesis of 4,6-*O*-(1-ethoxy-2-propenylidene)sucrose has also been reported (76). Mild hydrolysis of this unsaturated cyclic orthoester derivative produced 4-*O*- and 6-*O*-acrolylsucrose for polymerization studies. The use of orthoester derivatives in the synthesis of 6-*O*-acetyl-2,3,4-tri-*O*-(3*S*-methylpentanoyl)sucrose, a precursor of tobacco flavor, has been described (77).

(2)

Phosphate Esters. The phosphorylation of sucrose using sodium metaphosphate has been reported (78). Lyophilization of a sodium metaphosphate solution of sucrose at pH 5 for 20 hours followed by storage at 80°C for five days produced a mixture of sucrose monophosphates. These products were isolated by preparative hplc, with a calculated yield of 27% based on all organic phosphate as sucrose monoesters. Small proportions of glucose and fructose were also formed.

Sulfonate Esters. Sucrose sulfonates are valuable intermediates for the synthesis of epoxides and derivatives containing halogens, nitrogen, and sulfur. In addition, the sulfonation reaction has been used to determine the relative reactivity of the hydroxyl groups in sucrose. The general order of reactivity in sucrose toward the esterification reaction is OH-6 ≈ OH-6′ > OH-1′ > HO-2.

The selective tosylation of sucrose using limited amounts of tosyl chloride has been studied. Treatment of sucrose with three molar equivalents of toluene-*p*-sulfonyl chloride (tosyl chloride) in pyridine at 0°C for six days produced crystalline 6,6′-di-*O*-tosylsucrose (18%) (57,79), a mixed syrupy tritosylate fraction (57,80), and a small amount of 2,6,1′,6′-tetra-*O*-tosylsucrose (80). The tritosylate fraction gave, after high pressure liquid chromatography, 6,1′,6′-tri-*O*-tosylate (26%) and 2,6,6′-tri-*O*-tosylate (7%). Tetramolar tosylation of sucrose in pyridine at 0°C gave 6,1′,6′-tri-*O*-tosylate (40%) and 2,6,1′,6′-tetra-*O*-tosylsucrose (32%) (81). The use of sterically hindered sulfonates such as 2,4,6-trimethyl- (82,83) or 2,4,6-triisopropyl- (84) benzenesulphonyl chloride (trimsyl and tripsyl chloride, respectively) offered greater selectivity, resulting into crystalline 6,1′,6′-tri-*O*-trimsylsucrose (50%) and 6,1′,6′-tri-*O*-tripsylsucrose (54%) (82–84).

Deoxyhalogeno Derivatives

The application of bimolecular, nucleophilic substitution (S_N2) reactions to sucrose sulfonates has led to a number of deoxhalogeno derivatives. Selective displacement reactions of tosyl (79,85), mesyl (86), and tripsyl (84,87) derivatives of sucrose with different nucleophiles have been reported. The order of reactivity of the sulfonate groups in sucrose toward S_N2 reaction has been found to be $6 > 6' > 4 > 1'$.

Direct halogenation of sucrose has also been achieved using a combination of DMF–methanesulfonyl chloride (88), sulfuryl chloride–pyridine (89), carbon tetrachloride–triphenylphosphine–pyridine (90), and thionyl chloride–pyridine–1,1,2-trichloroethane (91). Treatment of sucrose with carbon tetrachloride–triphenylphosphine–pyridine at 70°C for 2 h gave 6,6'-dichloro-6,6'-dideoxysucrose in 92% yield. The greater reactivity of the 6 and 6' primary hydroxyl groups has been associated with a bulky halogenating complex formed from triphenylphosphine dihalide ($(C_6H_5)_3P{=}CX_2$) and pyridine (90).

The first S_N2 displacement reaction at C-2 position in carbohydrates was achieved during the study of sulfuryl chloride reaction with sucrose (92). Treatment of 3,4,6,3',4',6'-hexa-O-acetylsucrose 2,1'-bis(chlorosulfate) with lithium chloride in hexamethylphosphoric triamide at 80°C for 20 h led to the corresponding 2,1'-manno derivative in 73% yield.

4,1',6'-Trichloro-4,1',6'-trideoxygalactosucrose (sucralose) (**3**) has 650 times the sweetness of sucrose. It was discovered by the carbohydrate chemistry research groups of Philip Lyle Memorial Research Laboratory in Reading and Queen Elizabeth College in London, England (17). It is poorly absorbed by the intestines and passes unchanged through the body. An average daily intake (ADI) of 3.5 mg per kg body weight has initially been recommended for sucralose. It is being developed and marketed by Tate & Lyle in England in collaboration with Johnson & Johnson in the United States. Sucralose has been approved for use in food in Canada, Australia, and the CIS, and is awaiting approval as a food and drink additive by the FDA in the United States and by European and other health authorities.

(**3**)

An economic synthesis of (**3**) has been patented (74,91). The process involves (*1*) synthesis of sucrose 6-acetate by way of sucrose 4,6-cyclic orthoacetate (**2**), and (**2**) selective chlorination using thionyl chloride–pyridine–1,1,2-trichloroethane, followed by removal of the acetate group.

Anhydrides and Epoxides

Anhydride derivatives of sucrose are generally synthesized by intramolecular nucleophilic displacement reactions of the respective sulfonate or deoxyhalogeno derivatives. Synthesis of 3,6- (88), 2,1'- (93), 2,3- (94), and 3',4'- (95,96) anhydrides have been described. The base-catalyzed reaction of 2-*O*-tosyl-6,1',6'-tri-*O*-tritylsucrose produced the expected 2,3-mannoepoxide as the principal product. A small proportion of the 3,4-altroepoxide was also isolated and occurred as a result of migration of the epoxide ring (94). A facile synthesis of sucrose 3',4'-epoxide, α-D-glucopyranosyl-3,4-epoxy-β-D-lyxohexulofuranoside (**4**) in 42% yield has been achieved, using triphenylphosphine and diethyl azodicarboxylate in DMF and incorporating acetic acid to prevent the formation of 3,6- and 1',4'-anhydro rings (96). The sugar epoxides are important intermediates for a variety of derivatives. For example, ring-opening reactions of sucrose hexaacetate 3',4'-epoxide (**4**) have led to a number of C-4'-substituted sucrose compounds, eg (**5**), where R = N$_3$ or Cl (97).

(4)   (5)

Nitrogen-Containing Compounds

The aminodeoxy derivatives of carbohydrates are of interest because they are components of such biologically active materials as antibiotics, glycoproteins, and bacterial polysaccharides. They are usually synthesized by catalytic reduction of the corresponding azido derivatives. Treatment of peracetylated sucrose-3',4'-lyxo- (**4**) and sucrose-3',4'-sorboepoxides with lithium azide in aqueous ethanol in the presence of ammonium chloride, at 80°C for 72 h, gives stereoselectively 4'-azido-4'-deoxysucrose (**5**, R = N$_3$, 63%) and 2,3,4,6-tetra-*O*-acetyl-α-D-glucopyranosyl-4'-azido-4'-deoxy-β-D-sorbofuranoside (82%), respectively (97). The 4'-azido derivatives have then been converted to the corresponding 4'-amines by catalytic hydrogenation. The ring-opening reaction of sucrose 2,3-manno-epoxide with azide as the nucleophile resulted in axial attack to produce 3-azido-3-deoxy-α-D-altropyranosyl β-D-fructofuranoside (94). 6,6'-Diazido-6,6'-dideoxysucrose has been used as an intermediate for the synthesis of 1-deoxymannonojirimycin (98). The 6,6'-diazido compound was hydrolyzed using ion-exchange resin (Amberlite IR 120 H^+) in water to produce a mixture of 6-azido-6-deoxy-D-glucose and 6-azido-6-deoxy-D-fructose. The latter compound was separated by chromatographic method in 62% yield. The corresponding crystalline glucose derivative was isolated in 64%. The 6-azido-6-deoxy-D-fructose was converted to 1,5-dideoxy-1,5-imino-D-mannitol (78%) by reductive cyclization in methanol–water in the presence of palladium-on-carbon. The 6-azido-6-deoxy-D-glucose was partially transformed to 6-azido-6-deoxy-D-fructose using glucose isomerase.

Morpholinoglucopyranosides have been synthesized from sucrose by selective lead tetraacetate oxidation of the fructofuranosyl ring to a dialdehyde (**6**). This product was subjected to reductive amination with sodium borohydride and a primary amine such as benzylamine to produce the N-benzylmorpholino derivative (**7**) (99).

The dialdehyde also underwent a Fischer cyclization with nitromethane and base to produce a mixture of four diasteriomeric nitropyranosides. The principal product was isolated by chromatography, hydrogenated using Raney nickel and then N-acetylated to give α-D-glucopyranosyl-4-acetamido-4-deoxy-β-D-glucoheptulopyranoside (**8**) (100). The oxidation of sucrose with sodium periodate cleaved both the rings to give the corresponding tetraaldehyde, which on treatment with nitromethane and sodium methoxide produced 3-nitro-3-deoxy-α-D-glucopyranosyl-4-nitro-4-deoxy-β-D-glucoheptulopyranoside as the principal product (100).

Sulfur-Containing Compounds

The reaction of sucrose 2,3-manno-epoxide with potassium thioacetate and ammonium chloride in aqueous ethanol gave the expected 3-S-acetyl-3-thio-altropyranoside (101). Treatment of 6,6'-dibromo-6,6'-dideoxysucrose hexaacetate with potassium thioacetate and N,N-dimethylthiocarbamate gave the corresponding derivatives of 6,6'-dithiosucrose. The air oxidation of 6,6'-dithiolsucrose gave the bridged 6,6'-episulfide. A detailed conformational study of sucrose 6,6'-dithiol and sucrose 6,6'-episulfide revealed that they are similar but distinguishable (102).

Oxidation Products

Sucrose can undergo two distinct types of oxidation reaction: (1) conventional oxidation of primary hydroxyl groups to aldehydes or carboxylic acid residues and a secondary hydroxyl to a ketone, and (2) oxidative cleavage of vicinal diols to produce dialdehyde species. The catalytic oxidation of carbohydrates with oxygen and noble-metal catalysts such as platinum and palladium is a well-known and selective reaction (103). The catalytic oxidation of sucrose with platinum and oxygen at pH 7 and at 100°C gave selectively the 6- and 6'-mono- and 6,6'-dicarboxylic acid derivatives. The 1'-OH group under these conditions was not oxidized (104). When the reaction was performed at pH 9 and 100°C, some oxidation occurred at the C-1' position to produce sucrose 6,6'-dicarboxylate as the major and sucrose 6,1',6'-tricarboxylate as the minor product. The reaction rate was dependent on temperature; oxidation was slow below 80°C (105). An industrial synthesis of sucrose 6,1',6'-tricarboxylate (9) (35%), using platinum on carbon in alkaline solution, has been claimed (106). The carboxylate derivatives of sucrose are of interest as chelators, detergent builders, and for application in food and drink formulations.

(9)

Sucrose on treatment with lead tetraacetate undergoes oxidative cleavage of the C-3' and C-4' bond to produce the corresponding 2',5'-dialdehydo derivative. Oxidation with sodium periodate cleaved both rings of the sucrose molecule to give the tetraaldehydo derivative and one molar equivalent of formic acid. The periodate oxidation of sucrose in water and aqueous DMF has been studied (107–109). Oxidation at 25°C and pH 7 in aqueous solution produces the dialdehyde derivatives; no initial selectivity in favor of either of the ring is observed

(107). The initial oxidation at C-2 and C-3 is rapidly followed by oxidation at C-4 with the formation of formic acid. The temperature and pH have a strong influence on the selectivity. Increasing the temperature to 75°C and decreasing the pH to 5 favor the formation of the dialdehyde resulting from oxidation in the glucose ring (108). The oxidation of the glucosyl moiety is also favored in 50% aqueous DMF, but the rate of reaction is much slower than in water alone (109).

Compounds from Enzymic Isomerization

The synthesis of some commercially important bulk sweeteners such as isomaltulose (Palatinose), isomaltitol (Palatinit), and Actilight (formerly Neosugar) has been achieved by enzymatic transformations of sucrose.

Palatinose, 6-O-(α-D-glucopyranosyl)-D-fructose, is produced from sucrose using an immobilized α-glucosyl transferase enzyme from *Protanimobacter rubrum* (110). It is produced by Mitsui Sugar Company in Tokyo, Japan. The annual production is roughly 10,000 metric tons per year. The main market for Palatinose is in Japan, where it is used as a noncariogenic sweetener promoting bifidogenus flora. Palatinose is a free-flowing, nonhygroscopic, crystalline material (mp 123–124°C). Its sweetness intensity is 42% of that of sucrose. It is not utilized by the microbial flora of the mouth and consequently no organic acid or polysaccharides are formed. It is hydrolyzed in the small intestine and has an energetic value of 16.7 kJ/g (4 kcal/g).

Palatinit is produced by catalytic hydrogenation of Palatinose, which is a mixture of 6-O-(α-D-glucopyranosyl)-D-mannitol (**10**) and 6-O-(α-D-glucopyranosyl)-D-sorbitol (**11**) (111). The process steps involve catalytic hydrogenation (Raney nickel catalyst), filtration and ion-exchange treatment to remove the catalyst, evaporation, and crystallization from hot water. The sweetness of Palatinit is neutral, whereas that of sucrose is round and balanced. Palatinit, which has a caloric value of 8.36 kJ/g (2 kcal/g), is claimed to be a noncariogenic and a suitable sweetener for diabetics. It is produced by the Sudzucker company of Germany. The annual production is 20,000 metric tons.

(10)   (11)

Actilight, a mixture of D-glucose, sucrose, and fructooligosaccharides with one to three fructofuranosyl residues linked by way of β-(1 → 2) bonds to the fructosyl moiety of sucrose, is commercially produced by microbial fermentation of sucrose using fructosyl transferase enzyme from *Aspergillus niger* (112). These products are claimed to be noncariogenic, reduced caloric sweeteners and promote bifidogenus flora. In Japan, Actilight is produced by Meiji Seika Company, in

France by Beghin Meiji Industries, and in the United States by Meiji Seika and Golden Technologies Company. However, it has not yet been approved as a food additive in the United States.

Leucrose, 6-O-(α-D-glucopyranosyl)-β-D-fructopyranose [7158-70-5], is synthesized from sucrose using a dextranase enzyme from *Leuconostoc mesenteriodes* and a small proportion of fructose (2%). Pfeifer & Langen of Germany have developed a production process for leucrose that involves extraction of the enzyme, treatment with 65% aqueous solution of sucrose and fructose (1:2 wt/wt) at 25°C, separation of the product from fructose by ion-exchange column chromatography, and crystallization. The product has not yet been launched on the market as of this writing (1996).

Polymeric Intermediates

A series of reactive sucrose derivatives as intermediates to a variety of different polymers has been reported. They are still a chemical curiosity and their commercial potential has yet to be established as of this writing. Interesting polymers or polymer intermediates such as sucrose methylacrylate gels, chelating resins, sucrose derivatives with carbonic acid amide groups or N-methylated groups as condensation components for formaldehyde, sucrose with photoactive groups, and sucrose derivatives with primary amino groups and their fatty acid amides have all been reported (113).

Monomethylacryloyl and vinylbenzyl derivatives of sucrose have been prepared as intermediates for polymers, and preparation of a range of copolymers of styrene and O-methylacryloylsucrose has been described (114). Synthesis of 4- and 6-O-acryloylsucrose has been achieved by acid-catalyzed hydrolysis of 4,6-O-(1-ethoxy-2-propenylidene)sucrose (76). These acryloyl derivatives have been polymerized and copolymerized with styrene (qv).

Sucrose Economics

The total world production of centrifugal sucrose in 1994–1995 exceeded 116 million metric tons; total production in each region is shown in Table 1.

The three single biggest producers were India, Brazil, and the United States, with 15.85, 12.6, and 7.24 $\times$ 10^9 kg, respectively. The average world

Table 1. World Production of White Sucrose for 1994–1995[a]

Area	Sucrose production, 10^6 t
North America	7.498
Central America	11.491
South America	18.542
Europe	27.502
Africa	7.508
Asia	37.782
Oceania	5.798
Total	*116.121*

[a]Sugar Economy (Berlin, Germany).

market price for raw sucrose in 1994–1995 was \$0.27/kg. A comparison of the price of sucrose and other sweeteners (qv) is given in Table 2.

Sucrose occupies a unique position in the sweetener market (Table 3). The total market share of sucrose as a sweetener is 85%, compared to other sweeteners such as high fructose corn syrup (HFCS) at 7%, alditols at 4%, and synthetic sweeteners (aspartame, acesulfame-K, saccharin, and cyclamate) at 4%. The world consumption of sugar has kept pace with the production. The rapid rise in the synthetic sweetener market during 1975–1995 appears to have reached a maximum.

Table 2. Price of Sucrose, Alditols, and Synthetic Sweeteners

Product	\$/kg			Sweetness
	Europe	United States	World	
sucrose	1	0.69	0.27	1
HFCS-55[a]	0.83	0.43		0.95
sorbitol			3	0.6
xylitol			6	0.1
isomalt			5	0.5
aspartame			65	200
acesulfame-K			80	200
saccharin			4.5	300
cyclamate			4.5	25

[a]HFCS = high fructose corn syrup.

Table 3. World Consumption of Sweeteners, 10^9 kg

Product	1975	1995
sucrose	76.4	116
high fructose corn syrup	0.7	10.1
synthetics	<4	10

BIBLIOGRAPHY

"Sugar Derivatives" in *ECT* 1st ed., Vol. 13, pp. 261–270, by J. L. Hickson, Sugar Research Foundation, Inc.; in *ECT* 2nd ed., Vol. 19, pp. 221–233, by J. L. Hickson, International Sugar Research Foundation, Inc.; in *ECT* 3rd ed., Vol. 21, pp. 921–939, by K. J. Parker, Tate & Lyle, Ltd.

1. R. Khan, *Adv. Carbohydr. Chem. Biochem.* **33**, 235 (1976); *Pure Appl. Chem.* **56**, 833 (1984); R. Khan, in M. Mathlouthi and P. Reiser, eds., *Sucrose Properties and Applications*, Blackie Academic & Professional, London, U.K., 1995, pp. 264–278; C. E. James, L. Hough, and R. Khan, in W. Herz and co-workers, eds., *Fortschritte der Chemie Organischer Naturstoffe*, Springer-Verlag, Vienna, Austria, 1989, pp. 117–184.
2. W. N. Haworth and E. L. Hirst, *J. Chem. Soc.* 185 (1926).
3. W. N. Haworth, E. L. Hirst, and A. Learner, *J. Chem. Soc.* 2432 (1927).

4. H. H. Schlubach and G. Rauchalles, *Ber.* **58**, 1842 (1925).
5. C. A. Beevers and W. Cochran, *Proc. Royal Soc. London Ser. A*, **190**, 257 (1947).
6. C. A. Beevers and co-workers, *Acta Crystallogr.* **5**, 689 (1952).
7. G. M. Brown and H. A. Levy, *Science* **141**, 921 (1963).
8. G. M. Brown and H. A. Levy, *Acta Crystallogr.* **B29**, 790 (1973).
9. K. Bock and R. U. Lemeiux, *Carbohydr. Res.* **100**, 63 (1982).
10. D. C. McCain and J. L. Markley, *Carbohydr. Res.* **152**, 73 (1986).
11. D. C. McCain and J. L. Markley, *J. Am. Chem. Soc.* **108**, 4259 (1986).
12. J. C. Christofides and D. B. Davies, *Carbohydr. Res.* **163**, 269 (1987).
13. M. Mathlouthi and co-workers, *Carbohydr. Res.* **81**, 213 (1980).
14. C. H. du Penhoat and co-workers, *J. Am. Chem. Soc.* **113**, 3720 (1991).
15. P. E. Pfeffer, L. Odier, and R. L. Dudley, *J. Carbohydr. Chem.* **9**, 619 (1990).
16. L. Hough, K. S. Mufti, and R. Khan, *Carbohydr. Res.* **21**, 144 (1972).
17. Brit. Pat. 1,543,167 (Mar. 28, 1979), L. Hough and co-workers (to Tate & Lyle Ltd.).
18. H. Bredereck, G. Hagelloch, and E. Hambsch, *Chem. Ber.* **87**, 35 (1954).
19. E. G. V. Percival, *J. Chem. Soc.* 648 (1935).
20. M. G. Lindley, G. G. Birch, and R. Khan, *Carbohydr. Res.* **4**, 360 (1975).
21. P. Wolf, H. Polligkeit, and C. H. Hamann, *DECHEMA Monogr.* **124**, 649 (1991).
22. C. H. Hamann and co-workers, *J. Carbohydr. Chem.* **12**, 173 (1993).
23. F. W. Lichtenthaler and co-workers, *Stärke*, **44**, 445 (1992).
24. K. Hill, B. Gruber, and K. J. Weese, *Tetrahedron Lett.* **35**, 4541 (1994).
25. F. A. Henglein and co-workers, *Makromol. Chem.* **24**, 1 (1957).
26. C. D. Chang and H. B. Hass, *J. Org. Chem.* **23**, 773 (1958).
27. F. Franke and R. D. Guthrie, *Aust J. Chem.* **31**, 1285 (1978).
28. H. Karl, C. K. Lee, and R. Khan, *Carbohydr. Res.* **101**, 31 (1982).
29. R. Khan, *Carbohydr. Res.* **32**, 375 (1974).
30. R. Khan, K. S. Mufti, and M. R. Jenner, *Carbohydr. Res.* **65**, 109 (1978).
31. R. Khan and K. S. Mufti, *Carbohydr. Res.* **43**, 247 (1975).
32. E. Fanton and co-workers, *J. Org. Chem.* **46**, 4057 (1981).
33. M. R. Jenner and R. Khan, *J. Chem. Soc. Chem. Commun.* 50 (1980).
34. T. Suami and T. Otake *Bull. Chem. Soc. Jpn.* **43**, 1219 (1970).
35. E. B. Rathbone, *Carbohydr. Res.* **205**, 402 (1990).
36. Brit. Pat. 2,079,749B (May 31, 1984), R. Khan and K. S. Mufti (to Tate & Lyle Ltd.).
37. Eur. Pat. Appl. EP 475,619 (Mar. 18, 1992), N. M. Vernon and R. E. Wingard Jr. (to McNeil-PPC Inc.).
38. Eur. Pat. Appl. EP 448,413 (Sept. 25, 1991), R. E. Walkup, N. M. Vernon, and R. E. Wingard Jr. (to Noramco Inc.).
39. U.S. Pat. 5,128,248 (July 7, 1992), J. S. Dordick, A. J. Hacking, and R. Khan (to Tate & Lyle Ltd.).
40. J. M. Ballard, L. Hough, and A. C. Richardson, *Carbohydr. Res.* **24**, 152 (1972).
41. K. Capek and co-workers, *Collect. Czech. Chem. Commun.* **51**, 1476 (1986).
42. K. Capek and co-workers, *Collect. Czech. Chem. Commun.* **50**, 2191 (1985).
43. K. Capek, T. Vydra, and P. Sedmera, *Collect. Czech. Chem. Commun.* **53**, 1317 (1988).
44. A. H. Haines, P. A. Konowicz, and H. F. Jones, *Carbohydr. Res.* **205**, 406 (1990).
45. Fr. Pat. 2,634,497 (Jan. 26, 1990), A. Guibert and J. Mentech (to Beghin-Say S.A.).
46. K.-Y. Chang, S.-H. Wu, and K.-T. Wang, *J. Carbohydr. Chem.* **10**, 251 (1991).
47. K.-Y. Chang, S.-H. Wu, and K.-T. Wang, *Carbohydr. Res.* **222**, 121 (1991).
48. G.-T. Ong, S.-H. Wu, and K.-T. Wang, *Bioorg. Med. Chem. Lett.* **2**, 161 (1992).
49. S.-H. Wu and co-workers, *J. Chin. Chem. Soc.* **39**, 675 (1992).
50. S. Bornemann and co-workers, *Biocatalysis*, **7**, 1 (1992).
51. M. A. Cruces and co-workers, *Ann. N.Y. Acad. Sci.* **672**, 436 (1992).
52. M. Kloosterman and co-workers, *J. Carbohydr. Chem.* **8**, 693 (1989).

53. D. C. Palmer and F. Terradas, *Tetrahedron Lett.* **35**, 1673 (1994).
54. A. H. Haines and co-workers, *Carbohydr. Res.* **205**, 53 (1990).
55. D. M. Clode and co-workers, *Carbohydr. Res.* **161**, 139 (1988).
56. M. S. Chowdhary, L. Hough, and A. C. Richardson, *J. Chem. Soc. Perkin I*, 419 (1984).
57. R. U. Lemeiux and J. P. Barrette, *Can. J. Chem.* **38**, 656 (1960).
58. U.S. Pat. 3,996,206 (Dec. 7, 1976), K. J. Parker, R. A. Khan, and K. S. Mufti (to Tate & Lyle Ltd.).
59. Eur. Pat. Appl. EP 254,376 (Jan. 27, 1988), P. Van der Plank and A. Rozendaal (to Unilever NV, Unilever Plc.).
60. Eur. Pat. Appl. EP 319,091 and 319,092 (June 7, 1989), G. J. van Lookeren (to Unilever NV, Unilever Plc.).
61. Eur. Pat. Appl. EP 322,971 (July 5, 1989), G. W. M. Willemse (to Unilever NV, Unilever Plc.).
62. Eur. Pat. Appl. EP 233,856 (Aug. 26, 1987), C. A. Bernhardt (to Proctor & Gamble Co.).
63. Eur. Pat. Appl. EP 236,288 (Sept. 9, 1987), C. A. Bernhardt (to Proctor & Gamble Co.).
64. Eur. Pat. Appl. EP 271,951 (June 22, 1988), S. A. McCoy, P. M. Self, and B. L. Madison (to Proctor & Gamble Co.).
65. Eur. Pat. Appl. EP 285,187 (Oct. 5, 1988), J. L. Y. Kong-Chen (to Proctor & Gamble Co.).
66. Eur. Pat. Appl. EP 346,845 (Dec. 20, 1989). K. Masaoka and Y. Kasori (to Mitsubishi Kasei Corp.).
67. Y. Ando and co-workers, *Report Hokkaido Inst. Hygiene*, **33**, 1 (1983).
68. U.S. Pat. 3,600,186 (Aug. 17, 1971), F. H. Mattson and R. A. Volpenheim (to Proctor & Gamble Co.).
69. C. Chauvin, K. Baczko, and D. Plusquellec, *J. Org. Chem.* **58**, 2291 (1993).
70. K. Baczko and co-workers, *Carbohydr. Res.* **269**, 79 (1995).
71. S. Riva and co-workers, *J. Am. Chem. Soc.* **110**, 589 (1988).
72. G. Carrera and co-workers, *J. Chem. Soc. Perkin 1*, 1057 (1989).
73. P. J. Garegg and S. Oscarson, *Carbohydr. Res.* **136**, 207 (1985).
74. Eur. Pat. Appl. EP 260,979 (Mar. 23, 1988), P. J. Simpson (to Tate & Lyle Ltd.).
75. Brit. Pat. Appl. 92,106,756 (May 19, 1992), R. Khan and co-workers (to Tate & Lyle Ltd.).
76. E. Fanton and co-workers, *Carbohydr. Res.* **240**, 143 (1993).
77. P. J. Garegg, S. Oscarson, and H. Ritzen, *Carbohydr. Res.* **181**, 89 (1988).
78. E. Tarelli and S. F. Wheeler, *Carbohydr. Res.* **269**, 359 (1995).
79. C. H. Bolton, L. Hough, and R. Khan, *Carbohydr. Res.* **21**, 133 (1972).
80. D. H. Ball, F. H. Bisset, and R. C. Chalk, *Carbohydr. Res.* **55**, 149 (1977).
81. J. M. Ballard and co-workers, *Carbohydr. Res.* **83**, 138 (1980).
82. S. E. Creasy and R. D. Guthrie, *J. Chem. Soc. Perkin 1*, 1373 (1974).
83. L. Hough, S. P. Phadnis, and E. Tarelli, *Carbohydr. Res.* **44**, C12 (1975).
84. R. G. Almquist and E. J. Reist, *Carbohydr. Res.* **46**, 33 (1976).
85. L. Hough and K. S. Mufti, *Carbohydr. Res.* **25**, 497 (1972).
86. L. Hough and K. S. Mufti, *Carbohydr. Res.* **27**, 47 (1973).
87. R. D. Guthrie and J. D. Watters, *Aust. J. Chem.* **33**, 2487 (1980).
88. R. Khan, M. R. Jenner, and K. S. Mufti, *Carbohydr. Res.* **39**, 253 (1975).
89. L. Hough, *Chem. Soc. Rev.* **14**, 357 (1985).
90. A. K. M. Anisuzziman and R. L. Whistler, *Carbohydr. Res.* **61**, 511 (1978).
91. Brit. Pat. Appl. 222,827A (Mar. 21, 1990), R. Khan and co-workers (to Tate & Lyle Ltd.).

92. R. Khan, M. R. Jenner, and H. Lindseth, *Carbohydr. Res.* **78**, 173 (1980).
93. M. K. Gurjar, L. Hough, and A. C. Richardson, *Carbohydr. Res.* **78**, C21 (1980).
94. M. K. Gurjar and co-workers, *Carbohydr. Res.* **150**, 53 (1986).
95. R. Khan, M. R. Jenner, and H. Lindseth, *Carbohydr. Res.* **78**, 99 (1978).
96. R. D. Guthrie and co-workers, *Carbohydr. Res.* **121**, 109 (1983).
97. R. Khan and co-workers, *Carbohydr. Res.* **162**, 199 (1987).
98. A. de Raadt and A. E. Stütz, *Tetrahedron Lett.* **33**, 189 (1993).
99. K. J. Hale, L. Hough, and A. C. Richardson, *Chem. Ind.* 268 (1988).
100. K. J. Hale, L. Hough, and A. C. Richardson, *Tetrahedron Lett.* **28**, 891 (1987).
101. L. Hough and co-workers, *Carbohydr. Res.* **174**, 145 (1988).
102. W. J. Lees and G. M. Whitesides, *J. Am. Chem. Soc.* **115**, 1860 (1993).
103. K. Heyns and H. Paulsen, *Adv. Carbohydr. Chem.* **17**, 169 (1962).
104. L. A. Edye, G. V. Meehan, and G. N. Richards, *J. Carbohydr. Chem.* **10**, 11 (1991).
105. L. A. Edye, G. V. Meehan, and G. N. Richards, *J. Carbohydr. Chem.* **13**, 273 (1994).
106. Ger. Offen. DE 3,900,677 (July 19, 1990), E. I. Leupold and co-workers (to Hoechst A.G.).
107. D. deWit and co-workers, *Recl. Trav. Chim. Pays-Bas*, **108**, 335 (1989).
108. D. deWit and co-workers, *Recl. Trav. Chim. Pays-Bas*, **109**, 518 (1990).
109. D. deWit and co-workers, *Carbohydr. Res.* **226**, 253 (1992).
110. Brit. Pat. Appl. 2,063,268 (Nov. 4, 1980), C. Bucke and P. S. J. Cheetham (to Tate & Lyle Ltd.).
111. H. Schiweck and co-workers, in W. Lichtenthaler, ed., *Carbohydrates as Organic Raw Materials*, VCH Verlagsgesellschaft, Wienheim, Germany, 1991, pp. 57–94.
112. S. Fuji and K. Komoto, *Zuckerind.* **116**, 197 (1991).
113. H. Gruber and G. Greber, in Ref. 111, pp. 95–116.
114. N. D. Sachinvala, W. P. Niemczura, and M. H. Litt, *Carbohydr. Res.* **218**, 237 (1991).

RIAZ KHAN
PAUL A. KONOWICZ
POLYtech

SUGAR ECONOMICS

Among the various types of sugars marketed, the mass consumption of sucrose far outweighs by volume the consumption of other products. Sucrose, also called table sugar, is a disaccharide which, upon hydrolysis, yields two monosaccharides: glucose and fructose. Various types of sugar vary in sweetness (1).

The creation of sugar in the leaves of natural plants is considered the most efficient way of capturing solar energy. Table 1 illustrates the area of land needed to capture 27.7×10^3 kJ (6.61×10^3 kcal) of solar energy (2) in various food products.

Sucrose (table sugar) is a white crystalline substance. It is marketed in the form of various shades of brown-colored crystals (brown sugar), as white powder (powdered sugar), and also as high density syrup (liquid sugar). There are many distinguishing properties of sugar: pleasant sweetness, nutritional value (easy digestive resorption), good texturing for bakery products, good preservative qualities in jelly products, etc. One important feature of sugar as a product manufactured for food that is greatly overlooked is its unlimited shelf life.

Table 1. Land Area Needed for Various Food Products[a]

Food products	Needed land, hectare[b]
sugar	0.4
potatoes	1.2
corn	2.4
wheat	2.4
hogs	5.3
whole milk	6.7
eggs	21
chickens	25
steers	45

[a]See text.
[b]To convert hectare to acres, multiply by 2.5.

When properly manufactured, cane and beet sugars are identical in physical, chemical, electric, and thermodynamic properties (see SUGAR, BEET SUGAR; SUGAR, CANE SUGAR).

In addition to providing essential nutritional value to support life, the sugar industry has made a profound impact in history and in the social evolution of humankind. Industrial production of sugar (sucrose) is based on sugarcane and sugar beet processing. Sugarcane is grown and processed in tropical and subtropical countries. These regions are inside a belt whose northern border crosses the North American continent at southern California and South Carolina, and the European continent at the southern border of Spain, the Asian continent at northern Arabia, Pakistan, and south China, and the Pacific Ocean at the 37th parallel. In the southern hemisphere, the borderline goes through south Brazil, crosses the Atlantic Ocean, the African continent at Natal, the Indian Ocean, below the northern coast of Australia, and the Pacific Ocean at the 34th parallel. All regions outside this belt, in the northern and southern hemispheres, depend on sugar beet processing for sugar manufacture.

Despite the fact that refined cane and beet sugar are physically and compositionally identical, the histories of their development are very different. The story of both sugars, cane and beet, is woven into historic tales of adventure and discovery. In trade and commerce, today as in the past both industries play major roles. In the literature there are many records of the fascinating role of sugar in the destinies of nations in war and peace (3).

Sugar Economics and Statistics

By the beginning of the twentieth century, the sugar industry had evolved to the point of playing a significant role in supplying the caloric needs of the world's population. By the end of 1914 the world cane sugar output stood at 11,523,158 t, while the beet sugar output had risen to 9,051,767 t, bringing the total world sugar production to 20,574,925 t/yr. During the next twenty years with increase of population and consumption, world sugar output continued to rise. Consequently, in the period of 1929–1930 cane sugar production reached 17,381,000 t, whereas beet sugar held its ground at 9,359,000 t, for a total world sugar output of 26,740,000 t/yr.

The effect of World War II was very drastically felt in the sugar industry. Raw materials, machine parts, and maintenance items needed to maintain sugar processing were diverted to the war industry. Many sugar industry experts were conscripted into the armed forces. In the European theater of war, factories and agriculture were destroyed. At the end of World War II sugar production barely totalled 19,162,000 t/yr.

The post-war period and the decades following it created very favorable conditions for the healthy recovery of both the cane and beet sugar industries. In the very early years of rebuilding, the war-torn European industry accepted the concept of creating very high capacity processing facilities with extensive automation. In realizing this concept, many small factories were closed and abandoned. Unfortunately for countries coming under Soviet dominance, this concept was not applied. The sugar industry in Eastern and Central Europe still suffers from the negative consequences of this failure. Nevertheless, during the past half-century the global recovery and ensuing progress of the sugar industry have been dramatic (4).

Although a great expansion of the world sugar industry occurred, U.S. sugar processing suffered three setbacks resulting from the appearance of corn sweeteners, noncaloric sweeteners, and the effects of health movements and fads. Despite caloric sweetener consumption increases in the United States from 56.3 to 65.1 kg per capita during the time period from 1980–1992, the U.S. sugar industry lost ground, moving from 38 to 29.3 kg in per capita consumption. Two new products expanded into the market corn-based and the artificial (noncaloric) sweeteners. These products' consumption rose from 20.9 to 35.4 and 4.4 to 11 kg per capita, respectively (5,6). However, sugar remains irreplaceable in many bakery, candy, confectionery, and other products. Further, factual claims of major nutrition science centers and the FDA that the sugar is a GRAS and healthy product when used in moderation have helped to consolidate the position of the U.S. sugar industry (Fig. 1).

In each country of the modern world, sugar production and trade play major social, political, and economical roles. In order to regulate and protect export, import, stocks, subsidies, tariffs, etc, an enormous number of laws have been enacted and agreements concluded on sugar, both domestically and internationally. Also, insofar as sugar is a basic staple for a large population, each country keeps a watchful eye on sugar production, consumption, and price, which vary widely from country to country, as shown in Table 2 (7,8).

During the years 1984–1994, world sugar consumption showed a constantly increasing trend, whereas production trailed slightly (9). The sugar disappearance (trade term) in the time period increased from 98,226,900 to 113,270,100 t/yr (+15,043,200 t/yr). Production did not keep pace and rose from

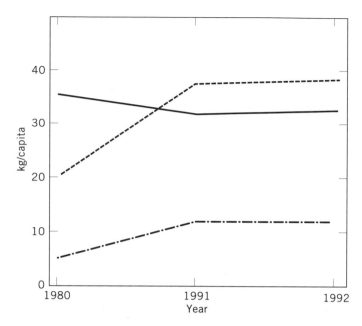

Fig. 1. U.S. sweetener consumption, 1980–1992, where (——) represents sucrose; (———), corn sweeteners; and (·—·—), noncaloric sweeteners.

Table 2. World Sugar Production, Consumption, and Retail Prices, 1993–1995

Country	Production, t × 10³	Consumption, t × 10³	Price, U.S. $/kg
Japan	857	2598	2.29
Norway		175	1.45
Taiwan	502	561	1.45
Austria	520	408	1.36
Ireland	192	153	1.23
Spain	1345	1299	1.14
Bulgaria[a]	9	200	1.10
the Netherlands	1230	673	1.10
Portugal	3	329	1.10
Brazil[a,b]	10112	7840	1.03
United Kingdom	1561	2493	0.97
United States	6214	8561	0.86
Argentina	1093	1236	0.81
Hungary	282	431	0.79
South Africa	1259	1330	0.73
Peru[a]	516	695	0.64
Poland	2170	1606	0.61
Canada	124	1163	0.59
Turkey	2192	1855	0.53

82

Table 2. (*Continued*)

Country	Production, t × 10³	Consumption, t × 10³	Price, U.S. $/kg
Egypt[a]	1092	1649	0.51
India[a]	10450	13184	0.48
France	4772	2559	1.50
Denmark	566	254	1.45
Russia[a]	2697	5404	1.43
Finland	154	244	1.34
Switzerland	152	317	1.21
Germany	4740	3103	1.12
Greece	315	359	1.10
Belgium	1133	558	1.10
Pakistan[a]	3176	2844	1.08
Italy	1541	1870	0.99
Sweden	414	391	0.95
China[a]	6541	7874	0.86
Australia	4483	855	0.79
Philippines	1860	1791	0.75
Mexico[a]	3859	4352	0.66
Indonesia[a]	2496	2916	0.61
Colombia[a]	1918	1198	0.59
Thailand[a]	4009	1457	0.55
Venezuela	505	741	0.53
Costa Rica	328	194	0.48
Guatemala[a]	1147	399	0.46

[a]Sugar of lesser quality: plantation white cane sugar, grayish beet sugar.
[b]A large volume of sugar is used for alcohol production.

100,436,500 to only 110,784,200 t/yr. This created a shortage in the ending stocks of 4,695,500 t of sugar per year (Fig. 2).

A 1995 report by Landell Mills Commodities Studies (Oxford, England) (7), pointed out that the United States sweetener industry (cane, beet, HFCS) provides 420,000 jobs a year and has a positive economic impact of $26.2 billion. By interpolating these data, it is likely that the world sweetener industry provides jobs to more than 15 million persons and affects the world economy by more than $100 billion (see SWEETENERS).

During the past two centuries the world sugar industry has evolved into a large and powerful business that has as a mission the responsibility for contributing to the supply of food that sustains life in humans. As of 1995, sugar was produced and refined in 126 countries in 1613 raw cane mills, 139 refineries, and 899 beet sugar factories (Table 3). The ever-increasing world population makes inevitable the future rationalization of sources and means for food supply. The world sugar industry will play an increasingly important role in realizing that goal.

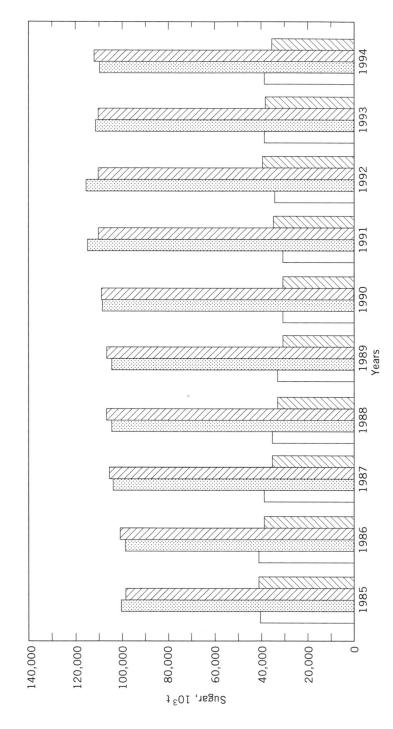

Fig. 2. World sugar balance, where ☐ represents opening stocks; ▨, production; ▨, consumption; and ▨, ending stocks (9,10).

Table 3. World Sugar Processing Facilities, 1995

Country	Cane	Refineries	Beet	Country	Cane	Refineries	Beet
Afganistan			1	Hungary			12
Albania			2	India	413		
Algeria	4			Indonesia	69		
Angola	3			Iran	2		37
Argentina	23			Iraq	1		1
Australia	28	4		Ireland			2
Austria			3	Italy			31
Bangladesh	16			Jamaica	9		
Barbados	3			Japan	22	31	8
Belarus			4	Kazakstan			9
Belgium			11	Kenya	8		
Belize	1			Republic of		4	
Bolivia	6			Korea			
Brazil	207	14		Kyrgyzstan			3
Bulgaria			7	Latvia			3
Burkina	1			Lebanon		4	
Burundi	1			Lithuania			5
Cameron	3			Madagascar	5		
Canada		5	2	Malawi	2		
Central African	1			Malaysia	3	2	
Republic				Mali	2		
Chad	1			Martinique	1		
Chile			5	Mauritius	18		
China	147		29	Mexico	65		
Colombia	13			Moldova			12
Congo	1			Morocco	3	2	11
Costa Rica	21			Mozambique	6		
Cote D'Ivoire	4			Myanmar	7		
Croatia			3	Nepal	6		
Cuba	56	11		the Netherlands			6
Czech Republic			40	New Zealand		1	
Denmark			4	Nicaragua	7		
Dominican	14			Nigeria	7		
Republic				Pakistan	64		4
Ecuador	6			Panama	4		
Egypt	9	2	1	Papua, New	1		
El Salvador	10	2		Guinea			
Ethiopia	2			Paraguay	7		
Germany		1	41	Peru	13	6	
Fiji	5			Philippines	41	8	
Finland		1	3	Poland		1	79
France		4	47	Portugal	1	2	1
Gabon	1			Puerto Rico	4	2	
Greece			5	Reunion	4		
Grenada	1			Romania			33
Guadeloupe	4			Russia		3	97
Guatemala	19			Rwanda	1		
Guyana	10			Saudi Arabia		1	
Haiti	4			Senegal	1		
Honduras	8			Sierra Leone	1		

Table 3. (*Continued*)

Country	Cane	Refineries	Beet	Country	Cane	Refineries	Beet
Singapore		1		Thailand	51		
Slovakia			10	Trinidad and	2		
Slovenia			1	Tobago			
Somalia	2			Tunisia		1	3
South Africa	15			Turkey			27
Spain	2		21	Uganda	2		
Sri Lanka	4			Ukraine		4	198
St. Chris.-Nevis	1			United Kingdom		2	10
Sudan	5			United States	38	13	34
Suriname	1			Uruguay	2	1	
Swaziland	3			Venezuela	21		
Sweden		1	4	Vietnam	8	2	
Switzerland			2	Yugoslavia			21
Syria			6	Zaire	1		
Taiwan	16			Zambia	1	1	
Tanzania	5			Zimbabwe	2	2	
				Totals	*1613*	*139*	*899*

BIBLIOGRAPHY

"Sugar Economics" in *ECT* 2nd ed., Vol. 19, pp. 233–236, by J. L. Hickson, International Sugar Research Foundation, Inc.; "Sugar Economics" under "Sugar" in *ECT* 3rd ed., Vol. 21, pp. 939–943, by K. Badenhop, B. W. Dyer & Co.

1. S. K. Susic and E. M. Guralj, *Osnovi Tehnologije Secera*, Belgrade, Yugoslavia, 1975, p. 2.
2. F. J. Stare, *Sugar Hearing*, New York, 1973.
3. L. Toth and A. B. Rizzuto, *Sugar—A User's Guide to Sucrose*, Van Nostrand Reinhold, New York, 1993, Chapt. 1.
4. F. O. Lichts, *World Sugar Statistics 1993/94, Stats 10*, Ratzeburg, Germany, 1994.
5. M. Blamberg, *Domino Sugar Report*, Domino Sugar Corp., May 1993.
6. *Food Rev. USDA*, 45 (May–Aug. 1994).
7. *Landell Mills Commodities Studies*, Oxford, U.K., 1995.
8. Ref. 4, Stats 11.
9. Ref. 4, Stats 19.
10. Ref. 4, Stats 14, 15, 17, and 18.

LASZLO TOTH
The Western Sugar Company

SPECIAL SUGARS

Although sucrose is commercially the most important sugar, there are also special sugars with special applications, among which fructose is the most important.

Fructose

D-Fructose [57-48-7] (levulose, fruit sugar) is a monosaccharide constituting one-half of the sucrose molecule. It was first isolated from hydrolyzed cane sugar (invert sugar) in the late nineteenth century (1,2). Fructose constitutes 4–8 wt % (dry sugar basis (dsb)) of many fruits, where it primarily occurs with glucose (dextrose) and sucrose (see CARBOHYDRATES; SWEETENERS). It also makes up 50 wt % (dsb) of honey (3,4).

Despite this ubiquity, fructose remained a noncommercial product until the 1980s because of the expense involved in its isolation and the care required for its handling. The development of technologies for preparing fructose from glucose in the isomerized mixture led to a greater availability of pure, crystalline fructose in the 1970s (5–7). However, the price for pure fructose was high enough in 1981 that the product was not competitive with sucrose and corn syrups as a commercial sweetener (see SYRUPS). With the entry of corn wet-milling companies into the crystalline fructose market in the late 1980s, raw material economies and enlarged manufacturing scale led to a nearly 10-fold production increase within a five-year period, making fructose prices competitive with other sweeteners for specific applications.

Pure D-fructose is a white, hygroscopic, crystalline substance and should not be confused with the high fructose corn syrups (HFCS) which may contain 42–90 wt % fructose and 23–29% water (8,9). The nonfructose part of these syrups is glucose (dextrose) plus small amounts of glucose oligomers and polymers. Fructose is highly soluble in water; at 20°C it is 79% soluble, compared with only 47% for glucose and 67% for sucrose.

(1)   (2)

The sweetness of fructose is 1.3–1.8 times that of sucrose (10). This property makes fructose attractive as an alternative for sucrose and other commercially available sweeteners. Fructose is probably sweetest in comparison with sucrose when cold and freshly made up in low concentrations at a slightly acidic pH (5). This relative sweetness difference is commonly attributed to changes in fructose structure when cold (β-D-fructopyranose (1), sweet) as compared to the structure when the sweetener is warm (β-D-fructofuranose (2), less sweet).

Based on nmr spectroscopy and sensory panel evaluation of sweetness, however, it has been observed that the absolute sweetness of fructose is the same at 5°C as at 50°C, and is not dependent on anomeric distribution (11). Rather, it may be the sweetness of sucrose, which changes with temperature, that gives fructose sweetness the appearance of becoming sweeter at low temperatures.

Also notable is the unique sweetness response profile of fructose compared to other sweeteners (3,4). In comparison with dextrose and sucrose, the sweetness of fructose is more quickly perceived on the tongue, reaches its intensity peak earlier, and dissipates more rapidly. Thus, the sweetness of fructose enhances many food flavor systems, eg, fruits, chocolate, and spices such as cinnamon, cloves, and salt. By virtue of its early perception and rapid diminution, fructose does not have the flavor-masking property of other common sugars.

The sweetness of fructose is enhanced by synergistic combinations with sucrose (12) and high intensity sweeteners (13), eg, aspartame, saccharin, acesulfame K, and sucralose. Information on food application is available (14,15). Fructose also reduces the starch gelatinization temperature relative to sucrose in baking applications (16–18).

Fructose possesses colligative properties that distinguish it from sucrose, glucose, and other nutritive sweeteners. It is one of the more effective monosaccharide humectants, binding moisture and lowering water activity, A_w, in food applications, thereby rendering the food products less susceptible to microbial growth and more stable to moisture loss (3,4). Ratios of fructose and higher molecular weight saccharides, oligosaccharides, and polysaccharides can be balanced to give increased control over freezing temperatures and storage stability in frozen products.

Fructose is a highly reactive molecule. When stored in solution at high temperatures, fructose not only browns rapidly but also polymerizes to dianhydrides [38837-99-9], [50692-21-2], [50692-22-3], [50692-23-4], [50692-24-5]. Fructose also reacts rapidly with amines and proteins in the nonenzymatic or Maillard browning reaction (5). This is a valued attribute in baked food products where crust color is important. An appreciation of these properties allows the judicious choice of conditions under which fructose can be used successfully in food applications.

Because of its relatively high degree of sweetness, fructose has been the object of commercial production for decades. Early attempts to isolate fructose from either hydrolyzed sucrose or hydrolyzed fructose polymers, eg, inulin (Jerusalem artichoke), did not prove economically competitive against the very low cost for sucrose processed from sugarcane or sugar beets.

Commercial quantities of crystalline fructose initially became available when the Finnish Sugar Company developed ion-exchange methods first for hydrolyzing sucrose and then for separating the hydrolysate into the constituents, ie, glucose and fructose. The latter step involves the calcium form of a sulfonated-polystyrene ion-exchange resin. Further economies in production were realized when the same company developed a method for crystallizing fructose from an aqueous rather than a water–alcohol solution (5).

More recent technologies also involve ion-exchange separation of fructose from glucose in a mixture obtained through the isomerization of glucose by means of immobilized glucose isomerase or microbial cells containing the enzyme (7), a technique pioneered by Hoffmann-La Roche. Another procedure for making crystalline fructose has been detailed (19), in which glucose (dextrose) is oxidized by glucose-2-oxidase to glucosone, which is then selectively hydrogenated to fructose. This procedure has the advantage of not requiring isomer separation in order to isolate the crystalline product.

Prior to 1970, commercially available crystalline fructose in Europe cost ca $17.6/kg and was produced at an annual rate of ca 7500 metric tons. By 1981, the amount of crystalline fructose available had risen to ca 20,000 metric tons, largely the result of the increased capacity of Hoffman-La Roche's plant in Thomson, Illinois (20). Prices in Europe and in the United States as of this writing (1996) were ca $8.80/kg and $2.20/kg, respectively. At these high prices relative to sugar, crystalline fructose use is relegated to high margin health foods and specialty dietary products, which account for a minute percentage of nutritively sweetened foods and beverages.

With the acquisition of the Thomson plant and technology, and the construction of their own plant in Lafayette, Indiana, in the late 1980s, the A. E. Stanley Manufacturing Company dramatically increased crystalline fructose production. On the strength of a growing appreciation for crystalline fructose's unique physical and functional properties, its competitive pricing, and its successful penetration of specific mainstream food applications, worldwide crystalline fructose production grew to more than 50,000 metric tons by 1992. In the same time period, crystalline fructose prices fell dramatically to ca $0.88/kg, a price equivalent to $0.75/kg sucrose on a sweetness basis (4).

Early applications of crystalline fructose focused on foods for special dietary applications, primarily calorie reduction and diabetes control. The latter application sought to capitalize on a significantly lower serum glucose level and insulin response in subjects with noninsulin-dependent diabetes mellitus (21,22) and insulin-dependent diabetes (23). However, because fructose is a nutritive sweetener and because dietary fructose conversion to glucose in the liver requires insulin in the same way as dietary glucose or sucrose, recommendations for its use are the same as for other nutritive sugars (24). Review of the health effects of dietary fructose is available (25).

Fructose has in the 1990s been successfully incorporated into formulas for the preparation of light and reduced calorie beverages and sports beverages; table syrup and table top sweeteners; baked goods; dairy products, including yogurt and chocolate milk; jams, jellies, and preserves; dry mix beverages, puddings, gelatins, and cake mixes; confectionery caramel fillings and starch-based jelly candies; and frozen dairy products and novelties (see FOOD PROCESSING).

Because of its hygroscopicity, fructose must be properly dried, packaged, and stored to prevent lumping and preserve free-flowing handling. Recommended storage and bulk handling conditions call for conditioned air at a relative humidity of less than 50% and a maximum temperature of 24°C (9).

Maltose

Maltose [69-79-4] (malt sugar) is a disaccharide, 4-O-α-D-glucopyranosyl-D-glucose (**3**), comprising two molecules of glucose (dextrose). Although occurring in some plants and fruits (26,27), it is more frequently recognized as a structural component of starch. Pure maltose is isolated with difficulty from a directed starch hydrolysate, ie, high maltose corn syrup, by precipitation with ethanol. Purification can be achieved by way of the β-maltose octaacetate. Removal of the acetate groups allows crystallization of the monohydrate of β-maltose. Commercial maltose typically contains 5–6 wt % of the trisaccharide maltotriose with traces of glucose (28). High maltose syrups from starch typically contain ca 8–9 wt % glucose, 40–80 wt % maltose, with higher saccharides as the remainder (29,30).

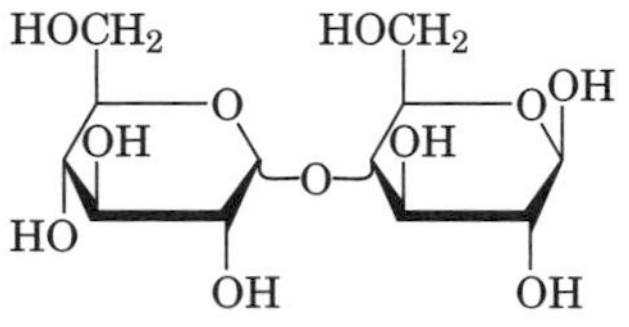

(3)

Such syrups are used in the preparation of confections, preserves, and other foodstuffs. The maltose in malt syrups is important in brewing (see BEER). Intravenous feeding (primarily in Europe and Japan) and sports beverage formulations take advantage of the fact that energy release from maltose becomes accessible to the body at a slower rate than energy supplied by monosaccharides (31).

Important physical and functional properties of maltose and maltose syrups include sweetness, viscosity, color stability, humectancy, freezing point depression, and promotion of beneficial human intestinal microflora growth. Maltose possesses ca 30–40% of the sweetness of sucrose in the pure state (32).

Hydrogenation of high maltose syrups gives a mixture of sugar alcohols, from which maltitol [585-88-6] (4) can be isolated in crystalline form. Maltitol is almost as sweet as sucrose (0.9 times) and has been promoted as a sweetener in various food applications (33).

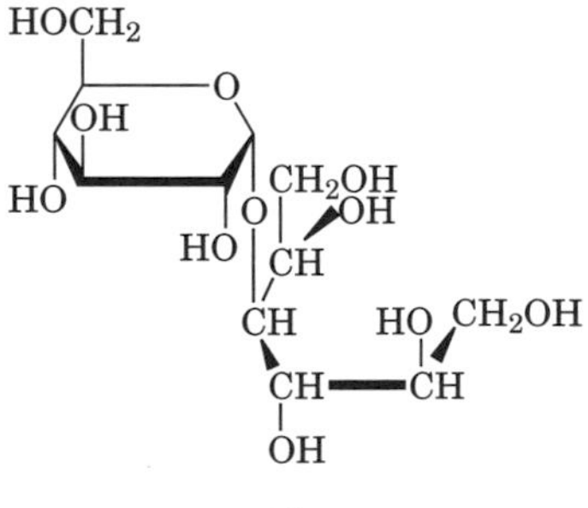

(4)

Lactose

Lactose [63-42-3] (milk sugar) is the only commercially available sugar that is derived from animal rather than plant sources. It is a disaccharide consisting of one galactose and one glucose moiety, 4-O-β-D-galactopyranosyl-D-glucose (**5**). The concentration of lactose in milk products ranges from 4.8 wt % in whole milk to 73.5 wt % in sweet dried whey (34). There have been reports of the presence of lactose in plant materials, eg, sapote and acacia, but this has not been confirmed (35,36).

(**5**)

Lactose is isolated commercially as the crystalline α-monohydrate from the whey by-products of cheese or caseinate production. It is available in varying degrees of purity. Fermentation grade is 98 wt % pure, whereas USP lactose is refined to 99.8 wt % purity (37). Although the α-monohydrate is the commercially available form of lactose, the sugar can be crystallized at high temperature to give the β-anhydride [56907-28-9]. The sugar is not very soluble in water (ca 22 g/100 g water at 25°C), nor is it very sweet (ca one-fifth the sweetening power of sucrose) (38). Lactose is a reducing sugar that reacts with amines and amino compounds with resultant browning.

Uses of lactose production by application include baby and infant formulations (30%), human food (30%), pharmaceuticals (25%), and fermentation and animal feed (15%) (39). It is used as a diluent in tablets and capsules to correct the balance between carbohydrate and proteins in cow-milk-based breast milk replacers, and to increase osmotic property or viscosity without adding excessive sweetness. It has also been used as a carrier for flavorings, volatile aromas, and synthetic sweeteners. Physiologically, lactose promotes the absorption of calcium, phosphorus, and essential trace minerals; has low cariogenicity; and is more slowly and gradually absorbed than sucrose, therefore of potential benefit to diabetes mellitus patients (38).

Lactose, and the lactose in substances such as milk and whey, has been hydrolyzed commercially by enzymes to yield products that can be tolerated physiologically much more easily by people who have a lactose intolerance (40–42).

BIBLIOGRAPHY

"Commercial Sugars" in *ECT* 1st ed., Vol. 13, pp. 244–251, by J. L. Hickson, Sugar Research Foundation; "Special Sugars" in *ECT* 2nd ed., Vol. 19, pp. 237–242, by J. L. Hickson, International Sugar Research Foundation, Inc.; in *ECT* 3rd ed., Vol. 21, pp. 944–948, by G. N. Bollenback, The Sugar Association, Inc.

1. C. P. Barry and J. Honeyman, *Adv. Carbohydr. Chem.* **7**, 53 (1952).

2. T. E. Doty and E. Vanninen, *Food Technol. (Chicago)*, **26**, 24 (1975).

3. J. S. White and D. W. Parke, *Cereal Foods World*, **34**(5), 392 (1989).

4. L. M. Hanover, in F. W. Schenck and R. E. Hebeda, eds., *Starch Hydrolysis Products: Worldwide Technology, Production and Applications*, VCH Publishers, Inc., New York, 1992, p. 201.

5. E. Vanninen and T. Doty, in G. G. Birch and K. J. Parker, eds., *Sugar: Science and Technology*, Applied Science Publications Ltd., London, U.K., 1979, p. 311.

6. T. Doty, in P. Koivistoinen and L. Hyvonen, eds., *Carbohydrate Sweeteners in Foods and Nutrition*, Academic Press, New York, 1980, p. 259.

7. W. P. Chen, *Process Biochem.* **15**, 36 (1980).

8. J. S. White, in Ref. 4, p. 177.

9. L. M. Hanover and J. S. White, *Am. J. Clin. Nutr.* **58**(5), 724S (1993).

10. R. S. Shallenberger, *J. Food Sci.* **28**, 584 (1963).

11. D. C. White, *Proceedings of the American Chemical Society Low Calorie Sweetener and Carbohydrate Symposium*, Los Angeles, Calif., 1988.

12. T. E. Doty and E. Vanninen, *Food Technol.* **11**, 34 (1975).

13. P. Van Tornout, J. Pelgroms, and J. Van Der Meeren, *J. Food Sci.* **50**, 469 (1985).

14. U.S. Pat. 4,676,991 (1987), C. K. Batterman, M. E. Augustine, and J. R. Dial (to A. E. Staley Manufacturing Co.).

15. World Pat. 88/08674 (1988), C. K. Batterman (to A. E. Staley Manufacturing Co.).

16. R. D. Spies and R. C. Hosney, *Cereal Chem.* **59**(2), 128 (1982).

17. D. C. White and G. N. Lauer, *Cereal Foods World* **35**(8), 728 (1990).

18. S. D. Horton, G. N. Lauer, and J. S. White, in Ref. 17, p. 734.

19. Eur. Pat. Appl. (Nov. 6, 1979), S. L. Neidelman and W. F. Amon (to Cetus Corp.).

20. *Chem. Week*, 49 (Oct. 7, 1981).

21. S. Akgun and N. H. Ertel, *Diabetes Care*, **8**, 279 (1985).

22. P. A. Crapo, O. G. Kolterman, and R. R. Henry, *Diabetes Care*, **9**, 111 (1986).

23. J. P. Bantle, D. C. Laine, and J. W. Thomas, *JAMA*, **256**, 3241 (1986).

24. American Dietetic Association, *J. Am. Diet. Assoc.* **93**(7), 816 (1993).

25. A. L. Forbes and B. A. Bowman, eds., *Am. J. Clin. Nutr.* **58**(5) (1993).

26. J. H. Pazur, in W. Pigman and D. Horton, eds, *The Carbohydrates*, 2nd ed., Vol. IIA, Academic Press, New York, 1970, p. 107.

27. R. S. Shallenberger, in H. L. Sipple and K. W. McNutt, eds., *Sugars in Nutrition*, Academic Press, New York, 1974, p. 74.

28. *Maltose Monohydrate*, NRC, Catalog No. M-105, Pfanstiehl Laboratories, Inc., Waukegan, Ill., 1981.

29. H. M. Pancoast and W. R. Junk, *Handbook of Sugars*, 2nd ed., Avi Publishing Co., Inc., Westport, Conn., 1980, p. 178.

30. S. M. Cantor, in Ref. 27, p. 116.

31. I. MacDonald, in Ref. 27, p. 310.

32. M. Okada and T. Nakakuki, in Ref. 4, p. 335.

33. C. A. M. Hough, in C. A. M. Hough, K. J. Parker, and A. J. Vlitos, eds., *Developments in Sweeteners*, Vol. 1, Applied Science Publishers, Ltd., London, 1979, p. 75.

34. T. A. Nickerson, *Food Technol. (Chicago)*, **32**, 40 (1978).

35. J. H. Pazur, in Ref. 26, p. 104.

36. R. S. Shallenberger, in Ref. 27, p. 71.

37. *Lactose*, Foremost Foods Co., Foremost-McKesson, Inc., San Francisco, Calif., 1970.

38. J. J. Dijksterhuis, in E. H. Reimerdes, ed., *Lactose as a Food Ingredient*, Expoconsult Publishers, the Netherlands, 1990, p. 5.

39. W. A. Roelfsema and co-workers, in B. Elvers, S. Hawkins, and G. Schulz, eds., *Ullman's Encyclopedia of Industrial Chemistry*, 5th ed., VCH Publishers, Inc., Germany, 1990, p. 107.

40. N. S. Shah and T. A. Nickerson, *J. Food Sci.* **439**, 1575 (1978).
41. V. H. Holsinger, *Food Technol.* (*Chicago*), **32**, 35 (1978).
42. S. M. Cantor, in Ref. 27, p. 368.

J. S. WHITE
White Technical Research Group

SUGAR ALCOHOLS

The sugar alcohols bear a close relationship to the simple sugars from which they are formed by reduction and from which their names are often derived (see CARBOHYDRATES). The polyols discussed herein contain straight carbon chains, each carbon atom usually bearing a hydroxyl group. Also included are polyols derived from disaccharides. Most of the sugar alcohols have the general formula $HOCH_2(CHOH)_n$—CH_2OH, where $n = 2$–5. They are classified according to the number of hydroxyl groups as tetritols, pentitols, hexitols, and heptitols. Polyols from aldoses are sometimes called alditols. Each class contains stereoisomers. Counting meso and optically active forms, there are three tetritols, four pentitols, ten hexitols, and sixteen heptitols, all of which are known either from natural occurrence or through synthesis (Fig. 1). Of the straight-chain

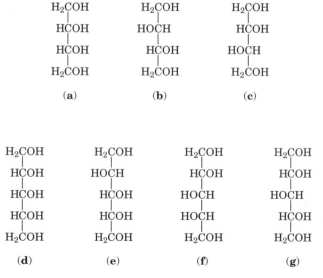

Fig. 1. Structures of the tetritols (**a**) erythritol, (**b**) D-threitol, and (**c**) L-threitol; and the pentitols (**d**) ribitol, (**e**) D-arabinitol, (**f**) L-arabinitol, and (**g**) xylitol.

polyols, sorbitol [50-70-4] and mannitol have the greatest industrial significance. However, maltitol, isomalt, and xylitol are finding increasing acceptance in applications in the United States and Canada.

Physical Properties

In general, these polyols are water-soluble, crystalline compounds with small optical rotations in water and a slightly sweet to very sweet taste. Selected physical properties of many of the sugar alcohols are listed in Table 1.

Polymorphism has been observed for both D-mannitol (6–9) and sorbitol (10). Three different forms exist for each hexitol. Bond lengths of crystalline pentitols and hexitols are all similar. The average C–C distance is 152 pm; the average C–O distance is 143 pm. Conformations in the crystal structures of sugar alcohols are rationalized by Jeffrey's rule that "the carbon chain adopts the extended, planar zigzag form when the configurations at alternate carbon centers are different, and is bent and nonplanar when they are the same" (9). Conformations are adopted which avoid parallel C–O bonds on alternate carbon atoms. Very little, if any, intramolecular hydrogen bonding exists in the crystalline sugar alcohols, but an extensive network of intermolecular hydrogen bonds has been found. Usually each hydroxyl group is involved in two hydrogen bonds, one as a donor and one as an acceptor (11).

The small optical rotations of the alditols arise from the low energy barrier for rotation about C–C bonds, permitting easy interconversion and the existence of mixtures of rotational isomers (rotamers) in solution (12).

The weakly acidic character of acyclic polyhydric alcohols increases with the number of hydroxyl groups, as indicated by the pK_a values in aqueous solution at 18°C (13).

Alcohol	pK_a	Ref.
glycerol	14.16	13
erythritol	13.90	13
xylitol	13.73	13
sorbitol	13.57	14
mannitol	13.50	14
dulcitol	13.46	13

At 60°C, the pK_a value of sorbitol is 13.00 (15).

In aqueous solution, sugar alcohols influence the structure of water, presumably by hydration of the polyol hydroxyl groups through hydrogen bonding, as indicated by effects on solution compressibility (16), vapor pressure (17), enthalpies of solution (18), dielectric constant (19), and Ag–AgCl electrode potential (20). Compressibility measurements indicate that mannitol in aqueous solution is hydrated with two molecules of water at 25°C (21). Osmotic coefficients are related to the number of hydrophilic groups per molecule, those of sorbitol being larger than those of erythritol (22).

Table 1. Physical Properties of the Sugar Alcohols

Sugar alcohol	CAS Registry Number	Melting point, °C	Optical activity in H$_2$O, $[\alpha]_D^{20\text{-}25}$	Solubility, g/100 g H$_2$O[a]	Heat of solution, J/g[b]	Heat of combustion, constant vol, kJ/mol[b]
Tetritols						
erythritol	[149-32-6]	120	meso	61.5	23.3[c]	−2091.6[d]
threitol	[7493-90-5]					
D-threitol	[2418-52-2]	88.5−90	+4.3	very sol		
L-threitol	[2319-57-5]	88.5−90	−4.3			
D,L-threitol	[6968-16-7]	69−70				
Pentitols						
ribitol	[488-81-3]	102	meso	very sol		
arabinitol	[2152-56-9]					
D-arabinitol	[488-82-4]	103	+131[e]	very sol		
L-arabinitol	[7643-75-6]	102−103	−130[e]			−2559.4[f]
D,L-arabinitol	[6018-27-5]	105				
xylitol	[87-99-0]	61−61.5 (metastable) 93−94.5 (stable)	meso	179	−153.1	−2584.5[g]
Hexitols						
allitol	[488-44-8]	155	meso	very sol		
dulcitol (galactitol)	[608-66-2]	189	meso	3.2[h]		−3013.7[d]
glucitol	[26566-34-7]					
sorbitol (D-glucitol)	[50-70-4]	93 (metastable) 97.7 (stable)	−1.985	235	−111.5	−3025.5[i]
L-glucitol	[6706-59-8]	89−91	+1.7			
D,L-glucitol	[60660-56-2]	135−137				
D-mannitol	[69-65-8]	166	−0.4	22	−120.9	−3017.1[i]
L-mannitol	[643-01-6]	162−163				
D,L-mannitol	[133-43-7]	168				
altritol	[5552-13-6]					
D-altritol	[17308-29-1]	88−89	+3.2	very sol		
L-altritol	[60660-58-4]	87−88	−2.9			
D,L-altritol	[60660-57-3]	95−96				
iditol	[24557-79-7]					
D-iditol	[23878-23-3]	73.5−75.0	+3.5			
L-iditol	[488-45-9]	75.7−76.7	−3.5	449		
Disaccharide alcohols						
maltitol	[585-88-6]	147−150	+90	175	−78.2	
lactitol	[585-86-4]	146	+14		−52.7	
isomalt	[64519-82-0]	145−150		24.5	−38.5	

[a] At 25°C unless otherwise indicated. [b] To convert J to cal, divide by 4.184. [c] Ref. 1. [d] Ref. 2. [e] In aqueous molybdic acid. [f] Ref. 3. [g] Ref. 4. [h] At 15°C. [i] Ref. 5.

Occurrence

D-Arabinitol (lyxitol [*488-82-4*]) (Fig. 1e) is found in lichens; in a variety of fungi; in the urediospores of wheat stem rust (23); in the dried herbiage of the Peruvian shrub, pichi, along with D-mannitol, dulcitol, and perseitol (24); and in the avocado (25). It is formed by fermentation of glucose (26–28) and in 40% yields using blackstrap molasses (29). Studies with ^{14}C-labeled glucose show that the yeast converts glucose C–1 to C–1 and C–5 in D-arabinitol and glucose C–2 to C–1, C–2, and C–4 of D-arabinitol (30). D-Arabinitol is formed by catalytic hydrogenation of D-arabinose in the presence of Raney nickel (31) and from the γ-lactones of D-arabinonic and D-lyxonic acids by reduction with sodium borohydride (32,33). L-Arabinitol (Fig. 1f) is synthesized by the reduction of L-arabinose which is abundant in nature (34,35).

D,L-Arabinitol can be prepared by the action of hydrogen peroxide in the presence of formic acid on divinyl carbinol (36) and, together with ribitol (Fig. 1d), from D,L-erythron-4-pentyne-1,2,3-triol, $HOCH_2CHOHCHOHC \equiv CH$ (37).

Xylitol (Fig. 1g) is found in the primrose (38) and in minor quantity in mushrooms (39). It can be obtained from glucose in 11.6% overall yield by a sequential fermentation process through D-arabinitol and D-xylulose (28).

Xylitol is synthesized by reduction of D-xylose catalytically (40), electrolytically (41), and by sodium amalgam (42). D-Xylose is obtained by hydrolysis of xylan and other hemicellulosic substances obtained from such sources as wood, corn cobs (43), almond shells, hazelnuts, or olive waste (44). Isolation of xylose is not necessary; xylitol results from hydrogenation of the solution obtained by acid hydrolysis of cottonseed hulls (45).

Xylitol also is obtained by sodium borohydride reduction of D-xylonic acid γ-lactone (32) and from glucose by a series of transformations through diacetone glucose (46).

Sorbitol (D-glycitol) (Fig. 2b) was discovered initially in the fresh juice of mountain ash berries in 1872. It is found in the fruits of apples, plums, pears, cherries, dates, peaches, and apricots; in the exudate of flowers of apples, pears, and cherries; and in the leaves and bark of apples, plums, prunes, the genus *Fraxinus*, and the genus *Euonymus*. Small amounts are found in the plane tree, the African snowdrop tree, and in various algae. Because sorbitol occurs to a very small extent in grapes, assay of the sorbitol content of wine (qv) has been used to detect its adulteration with other fruit wines or apple cider. An anhydride of sorbitol, polygallitol (1,5-sorbitan), is found in the *Polygala* shrub (47). Sorbitol occurs in the overwintering eggs of the European red mite, the latter affording it protection against freezing (48).

Sorbitol is synthesized commercially by high pressure hydrogenation of glucose, usually using a nickel catalyst. Catalyst promoters include magnesium salts (49), nickel phosphate, and iron (50). Other heterogenous catalysts used for glucose hydrogenation include cobalt, platinum, palladium, and ruthenium (51). Reduction of glucose to sorbitol also can be effected using ruthenium dichlorotriphenyl phosphine as a homogenous hydrogenation catalyst, preferably in the presence of a strong acid such as HCl (52). To form sorbitol, glucose is usually hydrogenated in the pH range of 4–8. Under alkaline conditions, glucose

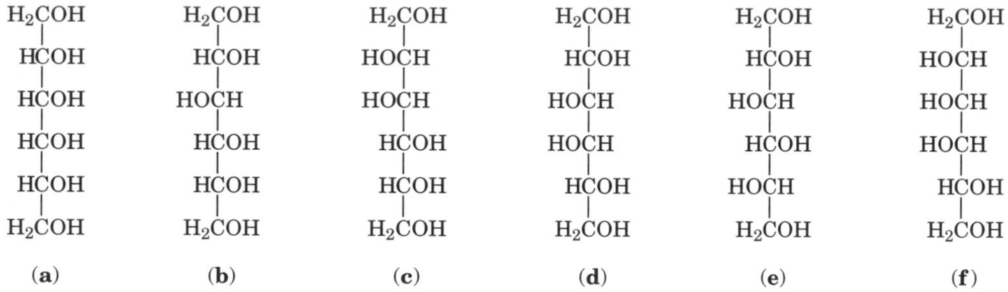

```
 H2COH        H2COH        H2COH        H2COH        H2COH        H2COH
  |            |            |            |            |            |
 HCOH         HCOH         HOCH         HCOH         HCOH         HOCH
  |            |            |            |            |            |
 HCOH         HOCH         HOCH         HOCH         HOCH         HOCH
  |            |            |            |            |            |
 HCOH         HCOH         HCOH         HOCH         HOCH         HOCH
  |            |            |            |            |            |
 HCOH         HCOH         HCOH         HCOH         HOCH         HCOH
  |            |            |            |            |            |
 H2COH        H2COH        H2COH        H2COH        H2COH        H2COH

  (a)          (b)          (c)          (d)          (e)          (f)
```

Fig. 2. Structures of hexitols (**a**) allitol, (**b**) sorbitol (D-glucitol), (**c**) D-mannitol, (**d**) dulcitol, (**e**) L-iditol, and (**f**) D-altritol.

isomerizes to fructose and mannose; hydrogenation of the fructose and mannose yields mannitol as well as sorbitol. In addition, under alkaline conditions the Cannizzaro reaction occurs and sorbitol and gluconic acid are formed. Gluconic acid formation during hydrogenation can be minimized if anion exchange resins in the basic form are the source of alkalinity (53). Although aqueous solutions

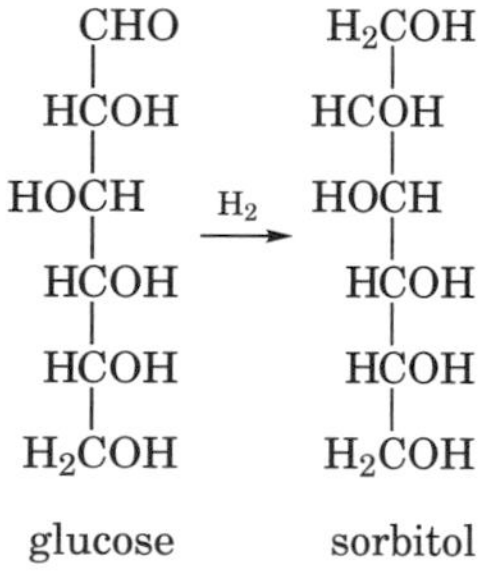

```
   CHO          H2COH
    |            |
   HCOH         HCOH
    |            |
   HOCH   H2    HOCH
    |    --->   |
   HCOH         HCOH
    |            |
   HCOH         HCOH
    |            |
   H2COH        H2COH

  glucose      sorbitol
```

are customarily used, the monomethyl ethers of ethylene glycol or diethylene glycol are satisfactory solvents (54). Electrolytic reduction of glucose was used formerly for the manufacture of sorbitol (55). Both the γ- and δ-lactones of D-gluconic acid may be reduced to sorbitol by sodium borohydride (32). Sorbitol results from simultaneous hydrolysis and hydrogenation of starch (56), cotton cellulose (57), or sucrose (58).

D-Mannitol is widespread in nature. It is found to a significant extent in the exudates of trees and shrubs such as the plane tree (80–90%), manna ash (30–50%), and olive tree. The manna ash, *Fraxinus rotundifolis*, formerly was cultivated in Sicily for the mannitol content of its sap. Mannitol occurs in the fruit, leaves, and other parts of various plants. This hexitol is present in pumpkin, hedge parsley, onions, celery, strawberries, the genus *Euonymus*, the genus *Hebe*, the cocoa bean, grasses, lilac, *Digitalis purpurea*, mistletoe, and lichens. Mannitol occurs in marine algae, especially brown seaweed, with a seasonal variation in mannitol content which can reach over 20% in the summer and autumn (59). It is found in the mycelia of many fungi and is present in the fresh mushroom to the extent of about 1.0% (60). There is a direct relation between mushroom yield and mannitol content (61). Microbial formation of D-mannitol occurs with fungi (62,63) or bacteria (64), starting with glucose, fructose, sucrose, or the tubers of Jerusalem artichokes (65). The precursor of mannitol in its biosynthesis in the mushroom *Agaricus bisporus* is fructose (66). Mannitol is produced from glucose in 44% yield after six days by aerobic fermentation with *Aspergillus candidus* (67) and in 30% yield after 10 days with *Torulopsis mannitofaciens* (68). It is formed by submerged culture fermentation of fructose with *Penicillium chrysogenum* in 7.3% conversion (69). Using sodium acetate as the sole carbon source, *Aspergillus niger* forms D-mannitol as well as D-arabinitol, erythritol, and glycerol (70). Small quantities of mannitol are found in wine (71).

Reduction of D-mannose with sodium borohydride or electrolysis leads to D-mannitol in good yield. Pure D-mannose is not yet commercially available but it can be obtained by acid hydrolysis of the mannan of ivory nutmeal in 35% yield (72) and from spent sulfite liquor or prehydrolysis extracts from conifers through the sodium bisulfite mannose adduct or methyl α-D-mannoside (73,74). Reduction of fructose leads to sorbitol and D-mannitol in equal parts. Sucrose, on reduction under hydrolyzing conditions, also yields the same products in the ratio of three parts of sorbitol to one of D-mannitol. Commercially, D-mannitol is obtained by the reduction of invert sugar. In alkaline media, glucose, fructose, and mannose are interconverted (75,76). All of the mannose formed can be reduced to mannitol. Mannitol can be prepared by hydrogenation of starch hydrolyzates in alkaline media in the presence of Raney nickel (77). Mannitol, because of its lower solubility, is usually separated from sorbitol by crystallization from aqueous solution. The two hexitols also can be separated chromatographically on a column of calcium poly(styrene-sulfonate), which preferentially retains sorbitol (78). Both the γ- and δ-lactones of D-mannonic acid are reduced catalytically to D-mannitol (79).

L-Mannitol does not occur naturally but is obtained by the reduction of L-mannose or L-mannonic acid lactone (80). It can be synthesized from the relatively abundant L-arabinose through the L-mannose and L-glucose cyanohydrins, conversion to the phenylhydrazines which are separated, liberation of L-mannose, and reduction with sodium borohydride (81). Another synthesis is from L-inositol (obtained from its monomethyl ether, quebrachitol) through the diacetonate, periodate oxidation to the blocked dialdehyde, reduction, and removal of the acetone blocking groups (82).

D,L-Mannitol has been obtained by sodium amalgam reduction of D,L-mannose. The identical hexitol is formed from the formaldehyde polymer, acrose, by conversion through its osazone and osone to D,L-fructose (α-acrose) followed by reduction (83).

Dulcitol (galactitol) is found in red seaweed, in shrubs of the *Euonymus* genus (84,85), the genus *Hippocratea*, the genus *Adenanthera*, the physic nut, the parasitic herb *Cuscuta reflexa*, and in the mannas from a wide variety of plants. It is produced by the action of yeasts (86) and is found in beer (87). D-Galactose is reduced to dulcitol catalytically (40), electrolytically (88), and chemically (89). Reduction of hydrolyzed lactose leads to the formation of both dulcitol and sorbitol (90,91). Dulcitol, which is relatively insoluble, is isolated by crystallization from the aqueous reduction mixture. Prehydrolysates of larch wood, which contain large quantities of galactose as well as glucose, arabinose, and xylose, can be simultaneously hydrolyzed and hydrogenated to a product containing 80% dulcitol, affording a pure product after repeated crystallization (92). Dulcitol is formed together with D-galactonic acid by treatment of D-galactose with alkali in the presence of Raney nickel (93). Reduction of D-galactonic acid γ-lactone with sodium borohydride leads to dulcitol (94). Photolysis of 1-deoxy-1-S-ethyl-1-thio-D-galactitol, and its sulfoxide in methanol forms dulcitol (95). Oxidation of a mixture and meso and racemic 1,5-hexadiene-3,4-diol with hydrogen peroxide and formic acid (trans addition of hydroxyl) leads predominantly to dulcitol, with some D,L-iditol (36,96).

Maltitol (4-O-α-D-glucopyranosyl-D-glucitol) formed by catalytic hydrogenation of maltose (97), has been obtained both as a noncrystalline powder and a viscous liquid (98). Structures of disaccharide alcohols are shown in Figure 3.

Lactitol (4-O-β-D-galactopyranosyl-D-glucitol) is obtained by sodium borohydride reduction (99,100) or catalytic hydrogenation (101) of lactose. Potentially large quantities of this sugar alcohol are available from lactose obtained from whey.

Isomalt, a mixture of α-D-glucopyranosyl-1,1-D-mannitol dihydrate and α-D-glucopyranosyl-1,6-D-sorbitol, is obtained by hydrogenating isomaltulose which is enzymatically derived from sucrose (102).

Hydrogenated starch hydrolysates (HSH) is a term used to describe a range of products which do not contain sorbitol or maltitol as a primary component, ie,

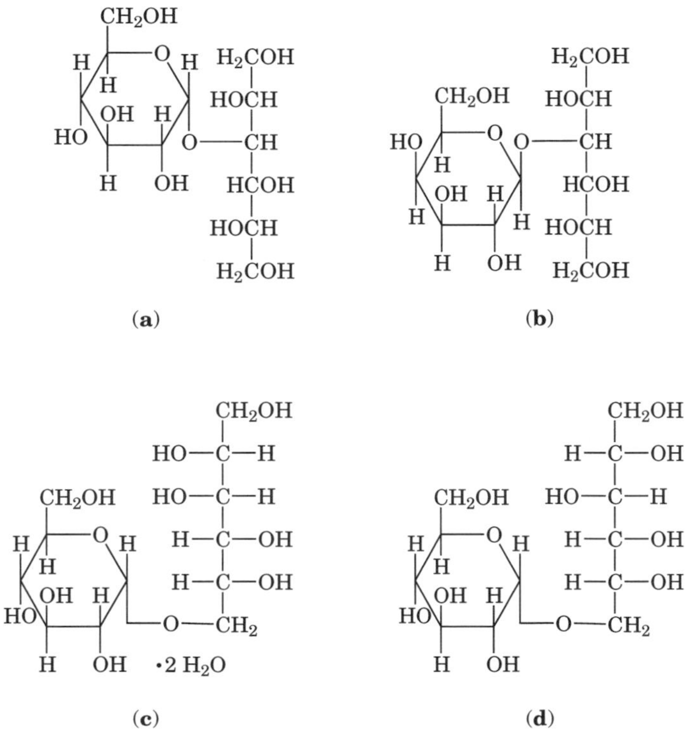

Fig. 3. Structures of disaccharide alcohols (**a**) maltitol, (**b**) lactitol, (**c**) α-D-glucopyranosyl-1,1-D-mannitol(dihydrate), and (**d**) α-D-glucopyranosyl-1,6-D-sorbitol.

at least 50%. HSH syrups contain a distribution of sorbitol, maltitol, and other hydrogenated oligo and polysaccharides. Syrups containing maltitol at a level of at least 50% are referred to as Maltitol syrups or Maltitol solutions.

Chemical Properties

Anhydrization. The sugar alcohols can lose one or more molecules of water internally, usually under the influence of acids, to form cyclic ethers. Nomenclature is illustrated by the hexitol derivatives. Monoanhydro internal ethers are called hexitans, and the dianhydro derivatives are called hexides. The main dehydration involves loss of water from the primary hydroxyl groups and the resulting formation of tetrahydrofuran derivatives having the configuration of the starting alditol, as shown for sorbitol (**1**) in Figure 4 (103–105). Small amounts of the 2,5-anhydrides form at the same time with inversion at the 2- or 5-positions, eg, 2,5-anhydro-L-iditol (**4**) from sorbitol. Hexitols anhydrize faster than pentitols or tetritols. Rates of dehydration depend on configuration and are slower where a noninvolved hydroxyl can interact with a leaving group.

Loss of a second molecule of water occurs on heating (**1**) (106,107), (**2**), or (**3**) (108) with concentrated sulfuric or hydrochloric acid forming 1,4:3,6-dianhydro-D-glucitol (isosorbide) (**5**). Mannitol and iditol anhydrize under similar conditions to isomannide (**6**) and isoidide (**7**), respectively. In (**6**), both hydroxyl groups are oriented toward each other (endo); in (**7**), both are oriented away from each other (exo); and in isosorbide (**5**), one hydroxyl is endo and the other exo. Xylitol loses two moles of water to form 1,3:2,5-dianhydroxylitol (109).

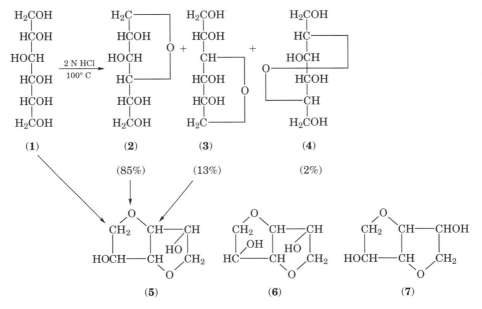

Fig. 4. Anhydrization of sorbitol (**1**) to 1,4-sorbitan (**2**), 3,6-sorbitan (**3**), 2,5-ahy-dro-L-iditol (**4**), and isosorbide (**5**). Also shown are isomannide (**6**) and isoidide (**7**), which arise from anhydrization of mannitol and iditol, respectively.

Esterification. Both partial and complete esters of sugar alcohols are known. The most important method for the preparation of partial fatty esters involves the interaction of polyols and fatty acids at 180–250°C (110). During direct esterification of the sugar alcohols, anhydrization occurs to varying degrees depending upon the conditions. Thus, esterification of sorbitol with stearic acid leads to a mixture of stearates of sorbitan and isosorbide as well as of sorbitol. Unanhydrized esters may be prepared by reaction with acid anhydrides or acid chlorides or by ester interchange reactions. In general, use of an excess of these reagents leads to esterification of all hydroxyl groups. Sorbitol hexanicotinate is prepared from the action of nicotinic acid chloride and sorbitol in the presence of pyridine (111). Completely substituted formate esters result from reaction of pentitols and hexitols and concentrated formic acid in the presence of phosphorus pentoxide (112). Primary hydroxyl groups react with esterifying reagents appreciably more rapidly than do secondary hydroxyls. As a consequence, it is possible to prepare ester derivatives involving only the primary hydroxyls, as in the formation of D-mannitol 1-monolaurate and 1,6-dilaurate by reaction of D-mannitol with lauroyl chloride at 100°C (113). The endo hydroxyl of isosorbide, which is involved in intramolecular hydrogen bond formation, is more easily esterified than the exo hydroxyl (114).

Cyclic carbonates result from polyols by transesterification using organic carbonates (115). Thus sorbitol and diphenylcarbonate in the presence of dibutyl tin oxide at 140–150°C form sorbitol tricarbonate in quantitative yield (116).

Mannitol hexanitrate is obtained by nitration of mannitol with mixed nitric and sulfuric acids. Similarly, nitration of sorbitol using mixed acid produces the hexanitrate when the reaction is conducted at 0–3°C and at −10 to −75°C, the main product is sorbitol pentanitrate (117). Xylitol, ribitol, and L-arabinitol are converted to the pentanitrates by fuming nitric acid and acetic anhydride (118). Phosphate esters of sugar alcohols are obtained by the action of phosphorus oxychloride (119) and by alcoholysis of organic phosphates (120). The 1,6-dibenzene sulfonate of D-mannitol is obtained by the action of benzene sulfonyl chloride in pyridine at 0°C (121). To obtain 1,6-dimethanesulfonyl-D-mannitol free from anhydrides and other by-products, after similar sulfonation with methane sulfonyl chloride and pyridine the remaining hydroxyl groups are acetylated with acetic anhydride and the insoluble acetyl derivative is separated, followed by deacetylation with hydrogen chloride in methanol (122). Alkyl sulfate esters of polyhydric alcohols result from the action of sulfur trioxide–trialkyl phosphates as in the reaction of sorbitol at 34–40°C with sulfur trioxide–triethyl phosphate to form sorbitol hexa(ethylsulfate) (123).

Etherification. The reaction of alkyl halides with sugar polyols in the presence of aqueous alkaline reagents generally results in partial etherification. Thus, a tetraallyl ether is formed on reaction of D-mannitol with allyl bromide in the presence of 20% sodium hydroxide at 75°C (124). Treatment of this partial ether with metallic sodium to form an alcoholate, followed by reaction with additional allyl bromide, leads to hexaallyl D-mannitol (125). Complete methylation of D-mannitol occurs, however, by the action of dimethyl sulfate and sodium hydroxide (126). A mixture of tetra- and pentabutyloxymethyl ethers of D-mannitol results from the action of butyl chloromethyl ether (127). Completely substituted trimethylsilyl derivatives of polyols, distillable *in vacuo*, are prepared by interaction with trimethylchlorosilane in the presence of pyridine (128). Hexa-vinylmannitol is obtained from D-mannitol and acetylene at 25.31 MPa (250 atm) and 160°C (129).

Reaction of olefin oxides (epoxides) to produce poly(oxyalkylene) ether derivatives is the etherification of polyols of greatest commercial importance. Epoxides used include ethylene oxide, propylene oxide, and epichlorohydrin. The products of oxyalkylation have the same number of hydroxyl groups per mole as the starting polyol. Examples include the poly(oxypropylene) ethers of sorbitol (130) and lactitol (131), usually formed in the presence of an alkaline catalyst such as potassium hydroxide. Reaction of epichlorohydrin and isosorbide leads to the bisglycidyl ether (132). A polysubstituted carboxyethyl ether of mannitol has been obtained by the interaction of mannitol with acrylonitrile followed by hydrolysis of the intermediate cyanoethyl ether (133).

Acetal Formation.   In common with other glycols, the sugar alcohols react with aldehydes and ketones to yield cyclic acetals and ketals. Five-membered rings are formed from adjacent hydroxyls and six-membered rings result from 1,3-hydroxyls. From the hexitols, mono-, di-, or triacetals or ketals may be obtained. Acetal formation is extensively used to protect hydroxyl groups during transformations in the polyol and carbohydrate series since conditions for the formation and removal of cyclic acetal linkages are relatively mild and proceed without inversion of configuration at asymmetric centers.

Oxidation.   Sorbitol is oxidized by fermentation with *Acetobacter suboxydans* to L-sorbose, an intermediate in the synthesis of ascorbic acid (134,135) (see VITAMINS). The same organism, *Acetobacter xylinium* and related bacteria convert erythritol to L-erythrulose (136), D-mannitol to D-fructose (137), and al-litol to L-ribohexulose (138). These results are generalized in the case of the more specific *Acetobacter suboxydans* (139) by the Hudson-Bertrand rules which state that in a *cis*-glycol having a D-configuration, a secondary hydroxyl adjacent to a primary hydroxyl is oxidized to a ketone (140). A similar stereospecific oxidation is that of an L-secondary hydroxyl adjacent to a primary hydroxyl in a polyol by diphosphopyridine nucleotide, which is catalyzed by L-iditol dehydrogenase; sor-bitol is oxidized to fructose, L-iditol to sorbose, xylitol to D-xylulose, and ribitol to D-ribulose (141) (see MICROBIAL TRANSFORMATIONS).

Careful oxidation using aqueous bromine produces mixtures of aldoses and ketoses. Thus, sorbitol is converted to a mixture of D-glucose, D-fructose, L-glu-cose, and L-sorbose (142). Aldoses and ketoses also result from ozone oxidation of sorbitol and mannitol (143). Reducing sugars are formed from the action of hydrogen peroxide in the presence of ferrous salts on erythritol, D-mannitol, dulcitol, or sorbitol. Ribitol, D-arabinitol, and xylitol are oxidized by mercury(II) acetate to mixtures of 2- and 3-pentuloses (144,145). Permanganate, man-ganese dioxide, chromic acid, and nitric acid completely oxidize polyols to carbon dioxide.

Reduction.   Sorbitol and mannitol are each converted by the action of concentrated hydriodic acid to secondary iodides (2- and 3-iodohexanes). The results of this reduction were used in early proofs of structure of glucose and fructose (146). Catalytic hydrogenolysis of the polyols results in breaking of both carbon-to-carbon and carbon-to-oxygen bonds. Thus, hydrogenolysis of sorbitol leads to the formation of ethylene glycol, propylene glycol, glycerol, erythritol, and xylitol (147,148). Ethylene glycol, propylene glycol, glycerol, erythritol, and monohydric alcohols are formed by the hydrogenolysis of xylitol (149).

Metal Complexes. The sugar alcohols form complexes in solution with most metal ions. In aqueous solution, calcium, strontium, and barium are more strongly complexed than are sodium, potassium, and magnesium (150). Solid magnesium complexes of sorbitol and other polyols are prepared with magnesium ethoxide (151); sorbitol alcoholate complexes of sodium and lithium precipitate from anhydrous ethanol (152). Polyol and carbohydrate complexes of alkali and alkaline earth metals have been reviewed (153). The stability of some polyol metal complexes is such that precipitation of the metal hydroxide from solution is inhibited. Thus, addition of sorbitol or D-mannitol to an aqueous solution of sodium ferric tartrate prevents precipitation of ferric oxide (154); D-mannitol prevents precipitation of titanium(IV) hydroxide (155), and at pH values above 12, precipitation of calcium, strontium, and cupric hydroxides is partially inhibited by sorbitol (156). Copper, iron, and cobalt all complex with cellobiitol (157). Tetritols, pentitols, and hexitols complex with copper in aqueous solutions of both cupric acetate and basic cupric acetate (158). Separation of mannitol from a mixture with dulcitol (resulting from reduction of wood sugar mixtures containing mannose and galactose) is achieved by selective complexation of dulcitol with ferric salts (159) (see also ALKOXIDES, METAL; CHELATING AGENTS).

Isomerization

Isomerization of sorbitol, D-mannitol, L-iditol, and dulcitol occurs in aqueous solution in the presence of hydrogen under pressure and a nickel–kieselguhr catalyst at 130–190°C (160). In the case of the first three, a quasiequilibrium composition is obtained regardless of starting material. Equilibrium concentrations are 41.4% sorbitol, 31.5% D-mannitol, 26.5% L-iditol, and 0.6% dulcitol. In the presence of the same catalyst, the isohexides establish an equilibrium at 220–240°C and 15.2 MPa (150 atm) of hydrogen pressure, having the composition 57% isoidide, 36% isosorbide, and 7% isomannide (161).

Analysis

Analytical separation of the sugar alcohols from each other and from similar materials, such as carbohydrates, is done chromatographically (see CHROMATOGRAPHY). Sorbitol, for example, is readily separated from glycerol, erythritol, and other polyols by chromatography on paper using butanol–water as developing solution (162), and on thin layers of silicic acid using butanol–acetic acid–ethyl ether–water as developing solution (163). Improved separations of sugar alcohols and carbohydrates have been obtained by including boric acid (164,165) or phenylboronic acid (166) in the solvent, thereby forming esters with the polyols. Paper electrophoresis of polyols and carbohydrates in molybdate (167), tellurate, germinate (168), or stannate (169) also enables the obtaining of some useful separations. Although polyols and carbohydrates can be separated by column chromatography on silicate absorbents (170), partition chromatography on ion exchange resins is of particular value for quantitative separations of these classes of materials (171). Use of cation exchange resins in the lithium form and anion exchange resins in the sulfate (172) or molybdate (173) form permit the separation of pentitols, hexitols, heptitols, and disaccharide alcohols as well as related carbohydrates (see ION EXCHANGE). Direct gas chromatographic determination of free sugar alcohols in biological media such as fermentation (qv) cultures has been reported (174). A mixture of glycerol, erythritol, D-arabinitol, and xylitol can be separated and each polyol determined by this procedure. Erythritol and other sugar alcohols are used as stationary phases for the gas chromatographic separation of volatiles in beer (qv) and wine (qv) (174).

Separated polyols are detected by a variety of reagents, including ammoniacal silver nitrate (175), concentrated sulfuric acid, potassium permanganate (163), lead tetraacetate, and potassium telluratocuprate (176). A mixture of sodium metaperiodate and potassium permanganate can be used to detect as little as 5–8 μg of mannitol or erythritol (177).

Conversion to acetates, trifluoroacetates (178), butyl boronates (179) trimethylsilyl derivatives, or cyclic acetals offers a means both for identifying individual compounds and for separating mixtures of polyols, chiefly by gas–liquid chromatography (glc). Thus, sorbitol in bakery products is converted to the hexaacetate, separated, and determined by glc using a flame ionization detector (180); aqueous solutions of sorbitol and mannitol are similarly separated and determined (181). Sorbitol may be identified by formation of its monobenzylidene derivative (182) and mannitol by conversion to its hexaacetate (183).

The sugar alcohols can be determined by periodate oxidation, followed by measurement of the formaldehyde or formic acid liberated (184) or titration of the excess periodate (185). Small quantities of sorbitol in biological fluids have been determined by this method (186). Measurement of the heat liberated in periodate oxidation of sorbitol in foodstuffs is the basis of a thermometric determination (187). Sorbitol is determined in wine and vinegar by precipitation with o-chlorobenzaldehyde forming the tris(o-chlorobenzylidene) derivative (188). Sugar alcohols may be analyzed colorimetrically after reaction with p-hydroxybenzaldehyde or p-dimethylaminobenzaldehyde, thiourea, and concentrated sulfuric acid (189). After complexation with copper, sorbitol and mannitol are determined by iodometric titration of excess cupric ion (190). Enzymatic assays have been described for several polyols, including sorbitol (191), ribitol (192), and erythritol (193). Although nonspecific, one of the most valuable procedures for the quantitative analysis of polyols is the determination of hydroxyl number. This method involves reaction with acetic anhydride, followed by measurement of the acetic acid liberated.

Polarimetric analysis of sorbitol and mannitol in the presence of each other and of sugars is possible because of their enhanced optical rotation when molybdate complexes are formed and the higher rotation of the mannitol molybdate complex under conditions of low acidity (194). The concentration of a pure solution of sorbitol may be determined by means of the refractometer (195). Mass spectra of trimethylsilyl ethers of sugar alcohols provide unambiguous identification of tetritols, pentitols, and hexitols and permit determination of molecular weight (196).

Manufacture of Sorbitol, Mannitol, and Xylitol

Sorbitol is manufactured by catalytic hydrogenation of glucose using either batch or continuous hydrogenation procedures. Corn sugar (qv) is the most important raw material, but other sources of glucose, such as hydrolyzed starch (qv), also may be used (197). Both supported nickel and Raney nickel are used as catalysts (198). In the continuous procedure, a 50% solution of dextrose in water is prepared and transferred to a mixing tank. The catalyst, nickel on diatomaceous earth, is added to the glucose solution and the resulting slurry is pumped to the reactor after being heated to about 140°C. Hydrogen is introduced into the reactor at a pressure of approximately 12.7 MPa (125 atm) concurrently with the sugar solution. Spent catalyst is collected on a filter and is separately regenerated for reuse. The sorbitol solution is purified in two steps: (1) by passage through an ion-exchange resin bed to remove gluconate as well as other ions, and (2) by treatment with activated carbon to remove trace organic impurities. The solution of pure sorbitol is concentrated in a continuous evaporator to a solution containing 70% solids, which is sold under the trademark SORBO (a registered trademark of SPI Polyols, Inc.), meeting USP 23 standards. Crystalline sorbitol is obtained by further concentration and crystallization (qv). It is sold in a variety of particle-size distributions, the one chosen depending on the application.

When invert sugar is used as a starting material, sorbitol and mannitol are produced simultaneously. Mannitol crystallizes from solution after the hydrogenation step owing to its lower solubility in water. It is sold as a white, crystalline powder or free-flowing granules (directly compressible grades) meeting USP 23 standards.

Extraction of mannitol from seaweed is a method of lesser importance commercially. In one method starting with this source, whole seaweed is steeped at 20°C in water which has been acidified to pH 2 with sulfuric acid. After filtration, the extract is neutralized and made alkaline with lime, precipitating calcium and magnesium sulfates together with some colloids. The filtrate or centrifugate from this operation is dialyzed; mannitol and mineral salts pass through the semipermeable membrane. The dialyzate is concentrated and mannitol crystallizes from solution on cooling. If the weight ratio of mannitol to alkali chlorides in the dialyzate is less than 1, this ratio is attained by addition of pure mannitol before crystallization is begun (199,200).

Xylose is obtained from sulfite liquors, particularly from hardwoods, such as birch, by methanol extraction of concentrates or dried sulfite lyes, ultrafiltration (qv) and reverse osmosis (qv), ion exchange, ion exclusion, or combinations of these treatments (201). Hydrogenation of xylose is carried out in aqueous solution, usually at basic pH. The Raney nickel catalyst has a loading of 2% at 125°C and 3.5 MPa (515 psi) (202,203).

Economic Aspects

The 1995 Canadian and United States sugar alcohol (polyol) production is shown in Table 2. The market share of each is also given. Liquids comprise 48%; crystalline product comprises 39%; and mannitol comprises 13% of the polyol market. An estimate of total U.S. sorbitol capacity for 1995 on a 70% solution

Table 2. 1995 United States and Canada Polyols Product Market Share

Products	Quantity, $t \times 10^3$	Sales, $\$ \times 10^6$	Market share, %
liquids	171	98.4	47.7
crystalline	51.0	80.9	39.2
mannitol	7.1	27.1	13.1
Total	229.1	206.4	100.0

basis was 498,000 t. ADM, Decatur, Ill., produced 68,200 t; Ethichem, Easton, Pa., 13,600 t; Lonza, Mapleton, Ill., 45,400 t; Roquette America, Gurnee, Ill., 68,200 t; and SPI Polyols, New Castle, Del., 75,000 t (204). Hoffman-LaRoche, which produces sorbitol for captive usage in the manufacture of Vitamin C (see VITAMINS), produced about 27,300 t in 1995.

The 1995 estimate of polyol market share by industry is shown in Table 3. The volume of polyols used in confectionery and oral care products is almost equivalent, at 24% and 26%, respectively. These two applications alone comprise 50% of the polyol usage (see DENTIFRICES).

Table 3. 1995 Polyols Market Share by Industry

Market segment	Volume, $t \times 10^3$	Sales	
		$\$ \times 10^6$	%
confectionery	55.6	84.6	41.0
oral care	58.5	30.3	14.7
pharmaceutical	21.7	25.4	12.3
food	15.4	15.7	7.6
vitamin C	30.0	15.0	7.3
industrial	20.8	13.3	6.4
surimi	9.0	11.6	5.6
surfactants	17.1	9.4	4.6
cosmetic	1.6	1.1	0.0
Total	229.9	206.4	100.0

Biological Properties

Toxicity.   Sugar alcohols are classified as relatively harmless. Acute oral toxicity values in mice for mannitol and sorbitol (5) are given in Table 4. The acute oral LD_{50} value for xylitol in mice is 25.7 g/kg (205). Ingestion of 10 g/d of either mannitol or sorbitol by a normal human subject for one month resulted in no untoward effects (206). Xylitol given to healthy humans for 21 d in increasing doses up to 75 g/d produced no adverse effects (207). The limiting dose of xylitol for production of diarrhea in humans is 20–30 g (4), but tolerance usually develops on continued administration (207).

Table 4. Acute Oral Toxicity of Mannitol and Sorbitol in Mice, LD_{50}, g/kg

Mice	Mannitol	Sorbitol
male	22.2	23.2
female	22.0	25.7

Laxation Thresholds. All sugar alcohols have the potential to cause diarrhea or flatulence owing to their slow absorption from the small intestine. Sorbitol and mannitol have laxation thresholds of 50 and 20 g/d, respectively, and where it is reasonably foreseeable that consumption will exceed these levels, foods must bear the statement "Excess consumption may have a laxation effect", per 21 CFR 184.1835 (Sorbitol) and 180.25 (mannitol). Maltitol, isomalt, lactitol, maltitol solution (syrup), and hydrogenated starch hydrolysates (HSH) are reported by manufacturers to have laxation thresholds of 100 g/d, 50 g/d, 20 g/d, 75 g/d, and 75 g/d, respectively. These sugar alcohols are not FDA-regulated as of 1996.

Blood Glucose and Insulin Response. In humans, ingestion of sugar alcohols has shown a significantly reduced rise in blood glucose and insulin response, owing to slow absorption by the body. As a result, many foods based on sugar alcohols have been used safely in the diets of diabetics (208).

Absorption of mannitol (209), sorbitol (210), and xylitol (4) from the intestinal tract is relatively slow, compared to that of glucose. In humans, approximately 65% of orally administered mannitol is absorbed in the dose range of 40–100 g. About one-third of the absorbed mannitol is excreted in the urine. The remainder is oxidized to carbon dioxide (211).

After an oral dose of 35 g of sorbitol, normal or mildly diabetic human subjects excreted 1.5–2.7% in the urine and oxidized 80–87% to carbon dioxide (212). Human subjects, after oral ingestion of up to 220 g of xylitol per day, excreted less than 1% of the dose in the urine (207), indicating efficient metabolism similar to that shown in the results of animal experiments (213). The first metabolic product from mannitol and sorbitol is fructose (214), and from xylitol it is D-xylulose (213). Fructose, sorbitol, and xylitol are principally metabolized in the liver, independent of insulin (215). Although fructose, sorbitol, and xylitol are used as glucose precursors by the liver (and the subsequent metabolism of glucose requires insulin), blood glucose concentration is increased only slightly following oral or intravenous administration of the substances (216).

Anticariogenicity. Sugar alcohols are not fermented to release acids that may cause tooth decay by the oral bacteria which metabolize sugars and starches (208). As a result, use of sugar alcohols in sugar-free chewing gum, pressed mints, confections, and toothpaste has been widely accepted.

Caloric Values. Absorption and metabolism data for each of the sugar alcohols has been reviewed by the Federation of American Societies for Experimental Biology (FASEB). Caloric values are lower than 4 cal/g (16.7 J/g), the caloric value of sugar. The values reported for each of the sugar alcohols are sorbitol, 2.6 cal/g (10.9 J/g); mannitol, 1.6 cal/g (6.7 J/g); maltitol, 3.0 cal/g (12.6 J/g); lactitol, 2.0 cal/g (8.4 J/g); isomalt, 2.0 cal/g (8.4 J/g); xylitol, 2.4 cal/g (10.0 J/g); and hydrogenated starch hydrolysate (HSH), 3.0 cal/g (12.6 J/g) (217). Manufacturers have received no letters from the FDA objecting to the use of these reduced caloric values.

Uses

The hexitols and their derivatives are used in many fields, including foods, pharmaceuticals (qv), cosmetics (qv), textiles (qv), and polymers.

Aqueous sorbitol solutions are hygroscopic and are used as humectants, softeners, and plasticizers in many different types of formulation. The hygroscopicity of sorbitol solutions is less than that of glycerol but greater than that of sugar solutions. In crystalline form, sorbitol does not absorb moisture greatly below the level of about 70% relative humidity. Above this level, sorbitol is deliquescent and dissolves in absorbed water. Mannitol is considerably less hygroscopic in its crystalline form. Many applications of mannitol take advantage of its low hygroscopicity and its resistance to occlusion of water.

Sweetness is often an important characteristic of sugar alcohols in food and pharmaceutical applications. The property of sweetness is measured in a variety of ways and has a corresponding variability in ratings (218). Based on one or more test methods, erythritol and xylitol are similar to or sweeter than sucrose (218,219). Sorbitol is about 60% as sweet as sucrose, and mannitol, D-arabinitol, ribitol, maltitol, isomalt, and lactitol are generally comparable to sorbitol (see SWEETENERS).

The partial fatty acid esters of the hexitols, usually anhydrized, find extensive use in surface-active applications, such as emulsification, wetting, detergency, and solubilization. Anhydrized sorbitol or mannitol moieties are versatile building blocks which confer hydrophilic properties on surfactants containing them. The less expensive sorbitol derivatives are used more extensively than the analogous mannitol compounds. Fatty acid esters of hexitans tend to be oil-soluble and to form water-in-oil (w/o) emulsions. Hydrophilic character of sorbitan fatty esters is enhanced by attachment of poly(oxyethylene) chains. As the poly(oxyethylene) chain length is increased, the tendency for water-solubility increases and oil-solubility decreases. Hexitan fatty esters having long poly(oxyethylene) chains tend to form o/w emulsions.

Foods. Sugar alcohols can replace sugar as a bulking agent in foods. As a result, sugar alcohols have increased the number and variety of sugar-free foods which utilize their functional advantages of anticariogenicity, insulin-independent metabolism, and lower caloric values. Approvals for the use of intense sweeteners in a variety of foods from beverages to confectionery products has enabled sweetness levels of sugar-free foods, utilizing sugar alcohols, to be increased for consumer acceptance.

Sorbitol is affirmed Generally Recognized As Safe (GRAS) by the FDA. Usage levels in foods, outlined by 21 CFR 184.1835, include 99% in hard candy and cough drops, 75% in chewing gum, 98% in soft candy, 30% in baked goods and baking mixes, 17% in frozen dairy desserts and mixes, and 12% in all other foods. The labels of foods where the reasonably foreseeable consumption may result in a daily ingestion of 50 g/d of sorbitol shall bear the statement, "Excess consumption may have a laxative effect". In these applications, sorbitol provides bulk, texture, and sweetness to the system.

Mannitol is a food additive permitted in food on an interim basis. Usage levels in foods, outlined by 21 CFR 180.25, include 98% in pressed mints and 5% in all other candy and cough drops, 31% in chewing gum, 4% in soft candy, 8% in confections and frostings, 15% in nonstandardized jams and jellies, and 2.5% in all other foods. The label of food where the reasonably foreseeable consumption may result in a daily ingestion of 20 grams of mannitol shall bear the statement "Excess consumption may have a laxative effect". Because mannitol is nonhygroscopic, it has been the bulking agent of choice for sugar-free chocolate-flavored coatings and compound coatings. Where the cooling effect of mannitol is undesirable, maltitol is used as a replacement. Mannitol is also used as a plasticizer in sugar-free chewing gum and as a dusting agent on the gum to prevent it from sticking to the wrapper.

Xylitol is approved, according to 21 CFR 172.395, for special dietary uses at levels not greater than that required to produce its intended effect. Xylitol is used in sugar-free chewing gum to provide sweetness, softness and a cooling effect.

Maltitol, lactitol, isomalt, maltitol solutions (syrups), and hydrogenated starch hydrolysates (HSH) have GRAS petitions filed with the FDA and are being sold commercially under self-determined GRAS status. Maltitol, owing to its lower negative heat of solution, is often preferred over mannitol as the bulking agent for sugar-free chocolate-flavored coatings and compound coatings. Isomalt, maltitol solutions (syrups), and HSH are used in sugar-free candies produced by conventional roping and stamping equipment.

Sugar alcohols have also found application in foods containing sugars. Sorbitol is an effective cryoprotectant in surimi, preventing denaturation of the muscle protein during frozen storage.

In artificially sweetened canned fruit, addition of sorbitol syrup provides body. Sorbitol has the property of reducing the undesirable aftertaste of saccharin (220). It sequesters metal ions in canned soft drinks as well as iron and copper ions in wines, thereby preventing the occurrence of cloudiness in compounds of these metals (221). Spray-drying a solution of mannitol and acetaldehyde gives a nonhygroscopic powder useful as a flavor-enhancer for fruit-flavored gelatins or beverages (222). D-Arabinitol has been formulated in jams and other foods as a sweetening agent of low caloric value (223). Maltitol and lactitol increase viscosity and confer sweetness in beverages and other foods (98,224). Carotenoids (225) and edible fats (226) are stabilized by sorbitol, which also prolongs the storage life of sterilized milk concentrates (227). Mixtures of sorbitol or mannitol and a fat, applied as an aqueous syrup to snacks and cereals, confer a rich mouth-feel and keep these foods crisp when they are immersed in milk (228). Both mannitol and sorbitol, at 1% concentration, inhibit the growth of A. niger in an intermediate-moisture (15–40%) food system containing raisins, peanuts, chicken, and nonfat dry milk (229). Freezer burn of rapidly frozen livers is prevented or greatly reduced by dipping in 25% sorbitol solution (230). Incorporation of sorbitol into frankfurter meat improves color, taste, shelf life, and facilitates removal of the casings (231). Sorbitol, sometimes in combination with propylene glycol (232), texturizes pet foods by its humectant properties (see FEEDS AND FEED ADDITIVES).

In candy manufacture, sorbitol is used in conjunction with sugars to increase shelf life. It is used in making fudge, candy cream centers, soft and grained marshmallows, and in other types of candy where softness depends upon the type of crystalline structure. The function of sorbitol in this application is in retarding the solidification of sugar often associated with staleness in such candy. In butter creams, an additional benefit is involved in flavor improvement by virtue of the sequestering action it has on trace metals. Sorbitol may be used in diabetic chocolates. Crystalline sorbitol or crystallized blends of sorbitol and mannitol constitute a sugarless confection (233), as does sorbitol containing up to 5% citric acid (234).

Sorbitol is used as a humectant and softener in shredded coconut, where it has a decided advantage over the invert sugar often used, because darkening of the product does not occur. A small quantity of sorbitol, as the 70% aqueous solution, added to peanut butter has been shown to reduce dryness and crumbliness and improve spreadability. Nuts coated with a blend of mannitol and sorbitol, from an aqueous solution or slurry of the hexitols, are roasted at 177–205°C. The hexitol coating on the nuts immobilizes salt applied during cooling and does not flake off (235,236). No oxidative degradation of the hexitols occurs on prolonged heating, and there is no significant decrease in hexitol content. Only traces of mannitans and sorbitans are formed (237).

Sorbitan fatty esters and their poly(oxyethylene) derivatives are used as shortening emulsifiers. In cakes and cake mixes, emulsification of the shortening is improved, resulting in better cake volume, texture, grain, and eating qualities. These emulsifiers are used in icings and icing bases as well as in pressure-packed synthetic cream-type toppings. In ice-cream, they confer improved body and texture and provide dryness and improved aeration. Poly-(oxyethylene(20)) sorbitan monooleate and the corresponding tristearate are used as emulsifiers either separately or as blends in ice cream and other frozen desserts. Crystallization of cane sugar is accelerated, the fluidity of the crystallizing mass improved, and the sugar yield increased by addition of poly(oxyethylene(20)) sorbitan monostearate.

Pharmaceuticals.   Mannitol finds its principal use in pharmaceutical applications (see PHARMACEUTICALS). It is used as a base in chewable, multilayer, and press-coated tablets of vitamins, antacids, aspirin, and other pharmaceuticals, sometimes in combination with sucrose or lactose. It provides a sweet taste, disintegrates smoothly, and masks the unpleasant taste of drugs such as aspirin. Tablets containing mannitol retain little moisture because of the low affinity of mannitol for water, making it an excellent excipient and thus suitable for use with moisture-sensitive actives. Mannitol is available as a powder for wet granulation tableting and in a granular form for direct compression tableting.

Sorbitol solution finds use as a bodying agent in pharmaceutical syrups and elixirs. The use of sorbitol in cough syrups reduces the tendency of bottle caps to stick because of crystallization of the sugar present (238). Sorbitol solution is also used as the base for sugar-free cough syrups, which have increased in popularity, particularly in the pediatric market. Enhanced stability is conferred by sorbitol in aqueous preparations of medications such as vitamin B_{12} (239), procaine penicillin (240), and aspirin (241). Stable suspensions of biologicals, such as smallpox vaccine, are obtained in a medium containing polydimethylsiloxane, mannitol, and sorbitol (242). Inclusion of sorbitol in aqueous suspensions of magnesium hydroxide prevents flocculation and coagulation, even when they are subjected to freeze–thaw cycles. A gel base in which other ingredients can be incorporated to make w/o creams is produced by combining 70% sorbitol solution with a lipophilic surfactant. Crystalline sorbitol is used as an excipient, where it gives a cool and pleasing taste from its endothermal heat of solution. Use of crystalline and powdered sorbitol enables preparation of troches with different degrees of hardness, by direct compression (243). Sorbitol is used in enema solutions (244).

A major pharmaceutical use of poly(oxyethylene) sorbitan fatty acid esters is in the solubilization of the oil-soluble vitamins A and D. In this way, multivitamin preparations can be made which combine both water- and oil-soluble vitamins in a palatable form.

Sorbitan sesquioleate emulsions of petrolatum and wax are used as ointment vehicles in skin treatment. In topical applications, the inclusion of both sorbitan fatty esters and their poly(oxyethylene) derivatives modifies the rate of release and promotes the absorption of antibiotics, antiseptics, local anesthetics, vasoconstrictors, and other medications from suppositories, ointments, and lotions. Poly(oxyethylene(20)) sorbitan monooleate, also known as Polysorbate 80 (USP 23), has been used to promote absorption of ingested fats from the intestine (245).

Manufacture of vitamin C starts with the conversion of sorbitol to L-sorbose. Sorbitol and xylitol have been used for parenteral nutrition following severe injury, burns, or surgery (246). An iron–sorbitol–citric acid complex is an intramuscular hematinic (247). Mannitol administered intravenously (248) and isosorbide administered orally (249) are osmotic diuretics. Mannitol hexanitrate and isosorbide dinitrate are antianginal drugs (see CARDIOVASCULAR AGENTS).

Cosmetics. Sorbitol is widely used in cosmetic applications, both as a humectant, in which case it retards the loss of water from o/w type creams, and as an emollient (see COSMETICS). Spreadability and lubricity of the emulsions are enhanced by aqueous sorbitol solution (250). Sorbitol is useful in both brushless and lather-type shaving creams as a humectant and plasticizer. It is incorporated in many toothpaste formulas of both regular and fluoride types, to the extent of 25–35%. This usage is based on its ability to act as an economic, viscous vehicle, and on its alkali stability, humectant, and plasticizing properties. Xylitol has been used as a humectant in toothpaste, both alone and in combination with sorbitol and glycerol (251) (see DENTIFRICES). Sorbitol solutions are used in mouthwash formulations. Because sorbitol resists fermentation by oral microorganisms, its use as a mouthwash does not increase the incidence of dental caries. Tests on patients having carious lesions have shown that sorbitol resists fermentation longer than glycerol.

Emulsions of fatty- and petroleum-based substances, both oils and waxes, of the o/w type are made by using blends of sorbitan fatty esters and their poly(oxyethylene) derivatives. Mixtures of poly(oxyethylene(20)) sorbitan monostearate (Polysorbate 60) and sorbitan monostearate are typical examples of blends used for lotions and creams. Both sorbitan fatty acid esters and their poly(oxyethylene) derivatives are particularly advantageous in cosmetic uses because of their very low skin irritant properties. Sorbitan fatty ester emulsifiers for w/o emulsions of mineral oil are used in hair preparations of both the lotion and cream type. Poly(oxyethylene(20)) sorbitan monolaurate is useful in shampoo formulations (see HAIR PREPARATIONS). Poly(oxyethylene) sorbitan surfactants are also used for solubilization of essential oils in the preparation of colognes and after-shave lotions.

Textiles. Sorbitol sequesters iron and copper ions in strongly alkaline textile bleaching or scouring solutions (see TEXTILES). In compositions for conferring permanent wash-and-wear properties on cotton fabrics, sorbitol is a scavenger for unreacted formaldehyde (252) and a plasticizer in soil-resistant and soil-release finishes (253).

Sorbitan fatty acid esters and their poly(oxyethylene) derivatives are used both to emulsify textile-treating chemicals and, by themselves, as finishes for textile processing. Poly(oxyethylene(20)) sorbitan monolaurate and its homologues are used as antistatic agents on textiles. Friction of yarns and fibers can be controlled by applying blends of sorbitan fatty acid esters with other surfactants. Sorbitan monopalmitate, monooleate, and monostearate, together with other surfactants, serve as fabric softeners and textile size plasticizers. The condensation product of sorbitol and epichlorohydrin, after reaction with the diethylene triamine diamide of stearic acid, is a component of a textile-softening composition (254). Poly(oxyethylene) derivatives of sorbitol and other polyols are used in viscose spinning baths to improve the properties of viscose rayon, extruded in the form of filaments and films. Sorbitan monooleate is used in dry-cleaning detergents, often in conjunction with an anionic surfactant.

Polymers. In combination with various metal salts, sorbitol is used as a stabilizer against heat and light in poly(vinyl chloride) (qv) resins and, with a phenolic antioxidant, as a stabilizer in uncured styrene–butadiene rubber (qv) compositions and in polyolefins (see HEAT STABILIZERS OLEFIN POLYMERS; RUBBER COMPOUNDING; (UV STABILIZERS)). Heat-sealable films are prepared from a dispersion of sorbitol and starch in water (255). Incorporation of sorbitol in collagen films greatly restricts their permeability to carbon dioxide (256).

Sorbitol, together with other polyhydric alcohols such as glycerol or pentaerythritol, can serve as the polyol component of alkyd resins and rosin esters for use in protective coatings and core binders (see ALKYD RESINS). The poly(oxypropylene) derivatives of sorbitol have found extensive use as polyol components of polyurethane resins, particularly for rigid urethane foams. In this application, sorbitol derivatives having short poly(oxypropylene) chain lengths, eg, poly(oxypropylene(10)) sorbitol, are combined with a diisocyanate such as toluene dissocyanate, together with blowing agents, catalysts, and other additives, to prepare foams. As the number of poly(oxypropylene) units attached to sorbitol increases above about 14, the products become suitable for use in semirigid and resilient foams, coatings, and elastomers (see URETHANE POLYMERS). Sorbitan fatty esters may be used as plasticizers, lubricants, and antifog agents for vinyl resins and other polymers. In suspension polymerization of vinyl chloride, use of sorbitan monolaurate, monopalmitate, and monostearate makes the primary particles of poly(vinyl chloride) more spherical, and controls their size and specific surface, as well as the porosity of the polymer.

Miscellaneous Uses. Sorbitol is used in flexible glues, cork binders, and printers' rollers, frequently in combination with glycerol, to confer strength and flexibility, as well as stability to humidity change. The high viscosity imparted by sorbitol in these applications is often desirable because it improves the mechanical strength and temperature resistance of the products (257). Glue-type products in which sorbitol is used include bookbinding and magazine and paper tape adhesives. Sorbitol, in conjunction with sugars, glycerol, or propylene glycol, has been used in tobacco as a component of casing solutions, to add moisture-retention properties. Sorbitol solution is used as a moistening agent in synthetic sponges. Clouds and fog disintegrate on introduction of finely divided sorbitol, usually together with inorganic substances such as silica or calcium phosphate (258). As a component in alkaline etching baths for aluminum, sorbitol helps eliminate scale formation on the surfaces of aluminum and aluminum alloys (see DISPERSANTS).

Mannitol and dulcitol are used as components of bacteriological media. Blood is protected during freezing, storing, and thawing by adding 15–20% of mannitol (259). Dulcitol, mannitol, and sorbitol protect freeze-dried bacterial cultures during storage (260), and animal semen is preserved by the addition of small quantities of sorbitol and mannitol together with other materials (261).

Poly(oxyethylene) derivatives of sorbitol and sorbitan fatty acid esters, usually in blends with anionic surfactants, are used as emulsifiers for insecticides, herbicides (qv), and other pesticides (qv). Blends of sorbitan surfactants are also used to suspend pigments in water-based paints and, in pigment pastes, to keep the dispersions uniform. Sorbitan monooleate functions as a corrosion inhibitor. Poly(oxyethylene(20)) sorbitan monopalmitate and monostearate sprayed on glass jars minimize breakage and surface damage (262). Glass fibers are protected by a coating containing poly(oxyethylene(20)) sorbitan monooleate (263). Oil slicks on sea water are dispersed when sprayed with mixtures of sorbitan monooleate and poly(oxyethylene(20)) sorbitan monooleate (102) (see SURFACTANTS).

Organogels can be formed from ethylene glycol, nitrobenzene, vegetable oils, and other organic liquids, using di- or tribenzylidene–sorbitol as gelling agents. In the explosives industry, mannitol hexanitrate is used as an initiator in blasting caps. Nitration of glycerin–ethylene glycol solutions of sorbitol yields low freezing, liquid, high explosive mixtures of value for dynamite formulas (see EXPLOSIVES AND PROPELLANTS).

BIBLIOGRAPHY

"Alcohols, Higher Polyhydric" in *ECT* 1st ed., Vol. 1, pp. 321–333, by R. M. Goepp, Jr., M. T. Sanders, and S. Soltzberg, Atlas Powder Co.; "Alcohols, Polyhydric–Sugar Alcohols" in *ECT* 2nd ed., Vol. 1, pp. 569–588, by F. R. Benson, Atlas Chemical Industries, Inc.; in *ECT* 3rd ed., Vol. 1, pp. 754–778, by F. R. Benson, ICI United States, Inc.

1. G. S. Parks and K. E. Manchester, *J. Am. Chem. Soc.* **74**, 3435 (1952).
2. G. S. Parks and co-workers, *J. Am. Chem. Soc.* **68**, 2524 (1946).
3. F. Stohmann and H. Langbein, *J. Prakt. Chem.* **45**(2), 305 (1892).
4. K. Lang, *Klin. Wochenschr.* **49**, 233 (1971).
5. ICI United States, Inc., unpublished research data.
6. F. T. Jones and K. S. Lee, *Microscope* **18**, 279 (1970).
7. H. M. Berman, G. A. Jeffrey, and R. D. Rosenstein, *Acta Crystallogr.* **24B**, 442 (1968).
8. H. S. Kim, G. A. Jeffrey, and R. D. Rosenstein, *Acta Crystallogr.* **24B**, 1449 (1968).
9. G. A. Jeffrey and H. S. Kim, *Carbohydr. Res.* **14**, 207 (1970).
10. Y. J. Park, G. A. Jeffrey, and W. C. Hamilton, *Acta Crystallogr.* **27B**, 2393 (1971).
11. G. A. Jeffrey, *Carbohydr. Res.* **28**, 233 (1973).
12. W. Kauzmann, F. B. Clough, and I. Tobias, *Tetrahedron* **13**, 57 (1961).
13. L. A. Mai, *J. Gen. Chem. USSR (Eng.)* **28**, 1304 (1958).

14. J. Thamsen, *Acta Chem. Scand.* **6**, 270 (1952).

15. P. Rys and H. Zollinger, *Helv. Chim. Acta* **49**, 1406 (1966).

16. F. Franks, J. R. Ravenhill, and D. S. Reid, *J. Solution Chem.* **1**, 3 (1972).

17. V. E. Bower and R. A. Robinson, *J. Phys. Chem.* **67**, 1540 (1963).

18. D. P. Wilson and W.-Y. Wen, *J. Phys. Chem.* **80**, 431 (1976).

19. F. Franks, D. S. Reid, and A. Suggett, *J. Solution Chem.* **2**, 99 (1973).

20. R. Gary and R. A. Robinson, *J. Chem. Eng. Data* **9**, 376 (1964).

21. S. Goto and T. Isemura, *Bull. Chem. Soc. Jpn.* **37**, 1697 (1964).

22. O. D. Bonner and W. H. Breazeale, *J. Chem. Eng. Data* **10**, 325 (1965).

23. N. Prentice and L. S. Cuendet, *Nature (London)* **174**, 1151 (1954).

24. N. K. Richtmyer, *Carbohydr. Res.* **12**, 233 (1970).

25. *Ibid.*, p. 135.

26. D. M. Holligan and D. H. Lewis, *J. Gen. Microbiol.* **75**, 155 (1973).

27. J. F. T. Spencer, J. M. Roxburgh, and H. R. Sallans, *J. Agric. Food Chem.* **5**, 64 (1957).

28. H. Onishi and T. Suzuki, *Appl. Microbiol.* **18**, 1031 (1969).

29. G. J. Hajny, *Appl. Microbiol.* **12**, 87 (1964).

30. J. F. T. Spencer and co-workers, *Can. J. Biochem. Physiol.* **34**, 495 (1956).

31. L. Hough and R. S. Theobald, in R. L. Whistler, ed., *Methods in Carbohydrate Chemistry*, Vol. I, Academic Press, Inc., New York, 1962, pp. 94–98.

32. H. L. Frush and H. S. Isbell, *J. Am. Chem. Soc.* **78**, 2844 (1956).

33. M. L. Wolfrom and K. Anno, *J. Am. Chem. Soc.* **74**, 5583 (1952).

34. M. Abdel-Akher, J. K. Hamilton, and F. Smith, *J. Am. Chem. Soc.* **73**, 4691 (1951).

35. R. Grewe and H. Pachaly, *Chem. Ber.* **87**, 46 (1953).

36. J. Wiemann and J. Gardan, *Bull. Soc. Chim. Fr.* 1546 (1955).

37. R. A. Raphael, *J. Chem. Soc.* 544 (1949).

38. R. Begbie and N. K. Richtmyer, *Carbohydr. Res.* **2**, 272 (1966).

39. K. Kratzl, H. Silbernagel, and K. H. Baessler, *Monatsh. Chem.* **94**, 106 (1963).

40. J. V. Karabinos and A. T. Balun, *J. Am. Chem. Soc.* **75**, 4501 (1953).

41. U.S. Pat. 1,612,361 (Dec. 28, 1926), H. J. Creighton (to Atlas Powder Co.).

42. E. Fischer, *Ber. Dtsch. Chem. Ges.* **27**, 2487 (1894).

43. Swiss Pat. 514,675 (Dec. 15, 1971), H. J. Schultze and R. Mondry (to Inventa AG fuer Forschung and Patentverwertung).

44. Fr. Pat. 2,047,193 (Mar. 12, 1971), L. Nobile.

45. U.S. Pat. 3,558,725 (Jan. 26, 1971), S. Kohno, I. Yamatsu, and S. Ueyama (to Eisai Kabushiki Kaisha).

46. J. Kiss, R. D'Souza, and P. Taschner, *Helv. Chim. Acta* **58**, 311 (1975).

47. K. Takiura and co-workers, *Yakugaku Zasshi* **94**, 998 (1974).

48. L. Soemme, *Can. J. Zool.* **43**, 881 (1965).

49. M. A. Phillips, *Br. Chem. Eng.* **8**, 767 (1963).

50. U.S. Pat. 3,538,019 (Nov. 3, 1970), R. J. Capik and L. W. Wright (to Atlas Chemical Industries).

51. U.S. Pat. 2,868,847 (Jan. 13, 1959), G. G. Boyers (to Engelhard Industries, Inc.).

52. U.S. Pat. 3,935,284 (Jan. 27, 1976), W. M. Kruse (to ICI United States Inc.).

53. A. Jacot-Guillarmod, A. von Bézard, and C. H. Haselbach, *Helv. Chim. Acta* **46**, 45 (1963).

54. U.S. Pat. 2,983,734 (May 9, 1961), D. E. Sargent (to General Electric Co.).

55. H. J. Creighton, *Trans. Electrochem. Soc.* **75**, 301 (1939).

56. U.S. Pat. 2,609,399 (Sept. 2, 1952), C. M. H. Kool, H. A. V. Westen, and L. Hartstra (to NVWA Scholten's Chemische Fabriken).

57. N. A. Vasyunina and co-workers, *Zh. Prikl. Khim.* 37, 2725 (1964).

58. R. Montgomery and L. F. Wiggins, *J. Chem. Soc.* 433 (1947).

59. W. A. P. Black, E. T. Dewar, and F. N. Woodward, *J. Appl. Chem. (London)* **1**, 414 (1951).
60. J. E. W. McConnell and W. B. Esselen, *Food Res.* **12**, 118 (1947).
61. G. K. Parrish, R. B. Beelman, and L. R. Kneebone, *Hortic. Sci.* **11**, 32 (1976).
62. R. R. Vega and D. Le Tourneau, *Phytopathology* **61**, 339 (1971).
63. D. Le Tourneau, *Trans. Br. Mycol. Soc.* **62**, 619 (1974).
64. J. W. Foster, *Chemical Activities of Fungi*, Academic Press, Inc., New York, 1949, pp. 470–472.
65. Ger. Pat. 871,736 (Mar. 26, 1953), H. H. Schlubach.
66. G. Duetsch and D. Rast, *Arch. Mikrobiol.* **65**, 195 (1969).
67. K. L. Smiley, M. C. Cadmus, and P. Liepins, *Biotechnol. Bioeng.* **9**, 365 (1967).
68. U.S. Pat. 3,622,456 (Nov. 23, 1971), H. Onishi (to Noda Institute for Scientific Research).
69. M. Abdel-Akher, I. O. Foda, and A. S. El-Nawawy, *J. Chem. U.A.R.* **10**, 355 (1967).
70. S. A. Barker, A. Gomez-Sanchez, and M. Stacey, *J. Chem. Soc.* 2583 (1958).
71. H. Thaler and G. Lippke, *Dtsch. Lebensm. Rundsch.* **66**, 96 (1970).
72. H. S. Isbell and H. L. Frush, in Ref. 31, pp. 145–147.
73. U.S. Pat. 3,677,818 (July 18, 1972), R. L. Casebier and co-workers (to International Telephone and Telegraph Corp.).
74. F. W. Herrick and co-workers, *Appl. Polym. Symp.* **28**, 93 (1975).
75. C. A. Lobry de Bruyn and W. Alberda van Ekenstein, *Rec. Trav. Chim. Pays Bas* **14**, 203 (1895).
76. M. L. Wolfrom and W. L. Lewis, *J. Am. Chem. Soc.* **50**, 837 (1928).
77. Neth. Pat. Appl. 65 15,786 (Oct. 3, 1966) (to Hefti Corp.).
78. U.S. Pat. 3,864,406 (Feb. 4, 1975), A. J. Melaja and L. Hämäläinen (to Suomen Sokeri Osakeyhtio).
79. J. W. E. Glattfeld and G. W. Schimpff, *J. Am. Chem. Soc.* **57**, 2204 (1935).
80. E. Baer and H. O. L. Fischer, *J. Am. Chem. Soc.* **61**, 761 (1939).
81. R. Kuhn and P. Klesse, *Chem. Ber.* **91**, 1989 (1958).
82. S. J. Angyal and R. M. Hoskinson, in Ref. 31, Vol. II, pp. 87–89.
83. E. Fischer, *Ber. Dtsch. Chem. Ges.* **23**, 2114 (1890).
84. W. Baker, *Nature (London)* **164**, 1093 (1949).
85. H. Rogerson, *J. Chem. Soc.* **101**, 1040 (1912).
86. H. Onishi and T. Suzuki, *J. Bacteriol.* **95**, 1745 (1968).
87. G. W. Hay and F. Smith, *Am. Soc. Brew. Chem. Proc.* 127 (1962).
88. K. Ashida, *Nippon Nogei Kagaku Kaishi* **23**, 167 (1949).
89. M. L. Wolfrom and A. Thompson, in Ref. 31, Vol. II, pp. 65–68.
90. L. Hough, J. K. N. Jones, and E. L. Richards, *Chem. Ind. (London)* 1064 (1953).
91. Brit. Pat. 941,797 (Nov. 13, 1963) (to Atlas Chemical Industries, Inc.).
92. L. A. Belozerova, *Tr. Leningrad. Lesotekh. Akad.* **105**, 89 (1966).
93. M. Delepine and A. Horeau, *Bull. Soc. Chim. Fr.* **4**(5), 1524 (1937).
94. M. L. Wolfrom and H. B. Wood, *J. Am. Chem. Soc.* **73**, 2933 (1951).
95. D. Horton and J. S. Jewell, *J. Org. Chem.* **31**, 509 (1966).
96. J. Wiemann and J. Gardan, *Bull. Soc. Chim. Fr.* 433 (1958).
97. P. Karrer and J. Büchi, *Helv. Chim. Acta* **20**, 86 (1937).
98. U.S. Pat. 3,741,776 (June 26, 1973), M. Mitsuhashi, M. Hirao, and K. Sugimoto (to Hayashibara Co.).
99. M. Abdel-Akher, J. K. Hamilton, and F. Smith, *J. Am. Chem. Soc.* **73**, 4691 (1951).
100. E. Dryselius, B. Lindberg, and O. Theander, *Acta Chem. Scand.* **11**, 663 (1957).
101. M. L. Wolfrom and co-workers, *J. Am. Chem. Soc.* **60**, 571 (1938).
102. Ger. Pat. 1,911,943 (Nov. 6, 1969), G. P. Conevari (to Esso Research and Engineering Co.).

103. S. Soltzberg, R. M. Goepp, and W. Freudenberg, *J. Am. Chem. Soc.* **68**, 919 (1946).
104. B. G. Hudson and R. Barker, *J. Org. Chem.* **32**, 3650 (1967).
105. R. Barker, *J. Org. Chem.* **35**, 461 (1970).
106. R. C. Hockett and co-workers, *J. Am. Chem. Soc.* **68**, 930 (1946).
107. T. Y. Shen, in Ref. 31, Vol. II, pp. 191–192.
108. R. C. Hockett and co-workers, *J. Am. Chem. Soc.* **68**, 927 (1946).
109. G. E. Ustyuzhanin and co-workers, *J. Gen. Chem.* (*Eng.*) **34**, 3966 (1964).
110. F. R. Benson in M. Schick, ed., *Non-ionic Surfactants*, Marcel Dekker, Inc., New York, 1967, pp. 261–266.
111. C. S. Tamo and G. F. DiPaco, *Boll. Chim. Farm.* **100**, 723 (1961).
112. L. Selleby and B. Wickberg, *Acta Chem. Scand.* **12**, 624 (1958).
113. E. Reinefeld and G. Klauenberg, *Tenside* **5**, 266 (1968).
114. K. W. Buck and co-workers, *J. Chem. Soc.* 4171 (1963).
115. L. Hough, J. E. Priddle, and R. S. Theobald, *J. Chem. Soc.* 1934 (1962).
116. U.S. Pat. 3,663, 569 (May 16, 1972), B. W. Lew (to Atlas Chemical Industries, Inc.).
117. T. Urbanski and S. Kwiatkowska, *Rocz. Chem.* **25**, 312 (1951).
118. I. G. Wright and L. D. Hayward, *Can. J. Chem.* **38**, 316 (1960).
119. Fr. Pat. 1,351,134 (Jan. 31, 1964) (to Colonial Sugar Refining Co. Ltd.).
120. T. Ukita and R. Takashita, *Chem. Pharm. Bull.* **9**, 606 (1961).
121. G. S. Skinner, L. A. Henderson, and C. G. Gustafson, *J. Am. Chem. Soc.* **80**, 3788 (1958).
122. Brit. Pat. 891,466 (Mar. 14, 1962) (to National Research Development Corp.).
123. U.S. Pat. 3,872,060 (Mar. 18, 1975), N. I. Burke (to Tee-Pak, Inc.).
124. A. N. Wrigley and E. Yanovsky, *J. Am. Chem. Soc.* **70**, 2194 (1948).
125. P. L. Nichols and E. Yanovsky, *J. Am. Chem. Soc.* **67**, 46 (1945).
126. W. Freudenberg and J. T. Sheehan, *J. Am. Chem. Soc.* **62**, 558 (1940).
127. G. R. Ames, H. M. Blackmore, and T. A. King, *J. Appl. Chem.* **14**, 503 (1964).
128. M. M. Sprung and L. S. Nelson, *J. Org. Chem.* **20**, 1750 (1955).
129. B. I. Mikhant'ev, V. L. Lapenko, and L. P. Pavlov, *Zh. Obshch. Khim.* **32**, 2505 (1962).
130. J. E. Wilson and R. H. Fowler, *Science* **128**, 3335 (1958).
131. U.S. Pat. 3,234,151 (Feb. 8, 1966), G. J. Stockburger, L. W. Wright, and J. D. Brandner (to Atlas Chemical Industries, Inc.).
132. U.S. Pat. 3,272,845 (Sept. 13, 1966), J. D. Zech and J. W. Le Maistre (to Atlas Chemical Industries, Inc.).
133. U.S. Pat. 2,598,174 (May 27, 1952), W. M. Hutchinson (to Phillips Petroleum Co.).
134. P. A. Wells and co-workers, *Ind. Eng. Chem.* **31**, 1518 (1939).
135. L. B. Lockwood, in Ref. 72, pp. 151–153.
136. H. J. Haas and B. Matz, *Ann.* 342 (1974).
137. L. T. Sniegoski, H. L. Frush, and H. S. Isbell, *J. Res. Nat. Bur. Stand. Sec. A* **65**, 441 (1961).
138. J. G. Carr and co-workers, *Phytochemistry* **7**, 1 (1968).
139. R. M. Hann, E. B. Tilden, and C. S. Hudson, *J. Am. Chem. Soc.* **60**, 1201 (1938).
140. G. Bertrand, *Ann. Chim.* (*Paris*) **3**(8), 181 (1904).
141. M. G. Smith, *Biochem. J.* **83**, 135 (1962).
142. E. Votocek and R. Lukes, *Rec. Trav. Chim. Pays Bas* **44**, 345 (1925).
143. C. D. Szymanski, *J. Appl. Polym. Sci.* **8**, 1597 (1964).
144. R. J. Stoodley, *Can. J. Chem.* **39**, 2593 (1961).
145. L. Stankovic, K. Linek, and M. Fedoronko, *Carbohydr. Res.* **10**, 579 (1969).
146. W. Pigman and D. Horton, *The Carbohydrates—Chemistry and Biochemistry*, 2nd ed., Vol. IA, Academic Press, Inc., New York, 1972, p. 8.
147. I. T. Clark, *Ind. Eng. Chem.* **50**, 1125 (1958).

148. G. van Ling and J. C. Vlugter, *J. Appl. Chem.* **19**, 43 (1969).
149. N. A. Vasyunina and co-workers, *Khim. Prom.* 82 (1962).
150. J. A. Mills, *Biochem. Biophys. Res. Commun.* **6**, 418 (1962).
151. U.S. Pat. 3,803,250 (Apr. 9, 1974), L. A. Hartmann (to ICI America, Inc.).
152. J. A. Rendleman, Jr., *J. Org. Chem.* **31**, 1845 (1966).
153. J. A. Rendleman, Jr., *Adv. Carbohydr. Chem.* **21**, 209 (1966).
154. G. F. Bayer, J. W. Green, and D. C. Johnson, *Tappi* **48**, 557 (1965).
155. C. Liegeois, *Bull. Soc. Chim. Fr.* 4081 (1972).
156. J. E. Hodge, E. C. Nelson, and B. F. Moy, *Agric. Food Chem.* **11**, 126 (1963).
157. O. Samuelson and L. Stolpe, *Sven. Papperstidn.* **77**, 513 (1974).
158. E. J. Bourne, F. Searle, and H. Weigel, *Carbohydr. Res.* **16**, 185 (1971).
159. U.S. Pat. 3,944,625 (Mar. 16, 1976), J. A. Neal (to Georgia Pacific Corp.).
160. L. W. Wright and L. A. Hartmann, *J. Org. Chem.* **26**, 1588 (1961).
161. L. W. Wright and J. D. Brandner, *J. Org. Chem.* **29**, 2979 (1964).
162. C. F. Smullin, L. Hartmann, and R. S. Stetzler, *J. Am. Oil Chem. Soc.* **35**, 179 (1958).
163. G. W. Hay, B. A. Lewis, and F. Smith, *J. Chromatogr.* **11**, 479 (1963).
164. B. P. Kremer, *J. Chromatogr.* **110**, 171 (1975).
165. J. F. Robyt, *Carbohydr. Res.* **40**, 373 (1975).
166. E. J. Bourne, E. M. Lees, and H. Weigel, *J. Chromatogr.* **11**, 253 (1963).
167. E. J. Bourne, D. H. Hutson, and H. Weigel, *J. Chem. Soc.* 4252 (1960).
168. W. J. Popiel, *Chem. Ind. (London)* 434 (1961).
169. E. M. Lees and H. Weigel, *J. Chromatogr.* **16**, 360 (1964).
170. B. W. Lew, M. L. Wolfrom, and R. M. Goepp, Jr., *J. Am. Chem. Soc.* **68**, 1449 (1946).
171. O. Samuelson in Ref. 31, Vol. VI, pp. 65–75.
172. J. Havlicek and O. Samuelson, *Chromatographia* **7**, 361 (1974).
173. S. A. Barker and co-workers, *Anal. Biochem.* **26**, 219 (1968).
174. L. Dooms, D. Decleck, and H. Verachtert, *J. Chromatogr.* **42**, 349 (1969).
175. L. Hough, *Nature (London)* **165**, 400 (1950).
176. J. Kocourek and co-workers, *J. Chromatogr.* **14**, 228 (1964).
177. R. V. Lemieux and H. F. Bauer, *Anal. Chem.* **26**, 920 (1954).
178. J. Shapira, *Nature (London)* **222**, 792 (1969).
179. M. P. Rabinowitz, P. Reisberg, and J. I. Bodin, *J. Pharm. Sci.* **63**, 1601 (1974).
180. H. K. Huntley, *J. Assoc. Off. Anal. Chem.* **51**, 1272 (1968); **56**, 66 (1973).
181. G. Manius and co-workers, *J. Pharm. Sci.* **61**, 1831 (1972).
182. *United States Pharmacopeia, 19th Revision*, United States Pharmacopeial Convention, Inc., Rockville, Md., 1975, p. 574.
183. Ref. 182, p. 295.
184. R. Belcher, G. Dryhurst, and A. M. G. MacDonald, *J. Chem. Soc.* 3964 (1965).
185. H. K. Hundley and D. D. Hughes, *J. Assoc. Off. Anal. Chem.* **49**, 1180 (1966).
186. J. M. Bailey, *J. Lab. Clin. Med.* **54**, 158 (1959).
187. L. S. Bark, D. Griffin, and P. Prachuabpaibul, *Analyst* **101**, 306 (1976).
188. F. C. Minsker, *J. Assoc. Off. Agric. Chem.* **42**, 316 (1959); **45**, 562 (1962).
189. H. D. Graham, *J. Food Sci.* **30**, 846 (1965).
190. G. Floret and M. V. Massa, *Ann. Pharm. Fr.* **24**, 201 (1966).
191. H. U. Bergmeyer, W. Gruber, and I. Gutman, in H. U. Bergmeyer, ed., *Methoden an Enzymologie Analyse 3*, Vol. 2, Verlag Chemie, Weinheim-Bergstrasse, Germany, 1974, pp. 1268–1271.
192. D. R. D. Shaw and D. Mirelman, *Anal. Biochem.* **38**, 299 (1970).
193. R. J. Sturgeon, *Carbohydr. Res.* **17**, 115 (1971).
194. J. Dokladalova and R. P. Upton, *J. Assoc. Off. Agric. Chem.* **56**, 1382 (1973).
195. P. Rovesti, *Chimica (Milan)* **34**, 309 (1958).
196. G. Peterson, *Tetrahedron* **25**, 4437 (1969).

197. Ger. Pat. 892,590 (Oct. 8, 1953) (to Deutsche Gold- and Silber-Scheideanstalt vomal Roessler).
198. M. A. Phillips, *Br. Chem. Eng.* **8**, 767 (1963).
199. Fr. Pat. 1,074,755 (Oct. 8, 1954) (to Association de recherches de l'industrie des algues marines).
200. Brit. Pat. 757,463 (Sept. 19, 1956) (to Association de recherches de l'industrie des algues marines).
201. J. W. Collins and co-workers, *Abstracts of Papers 170th ACS National Meeting, Chicago, Ill., August 1975, Carbohydrate Division, Paper No. 28.*
202. J. Wisniak, *Ind. Eng. Chem. Prod. Res. Dev.* **13**, 75 (1974).
203. Swiss Pat. 514,675 (Dec. 15, 1971) (to Inventa AG).
204. *Chem. Mktg. Reporter,*
205. W. Kieckebusch, W. Gziem, and K. Lang, *Klin. Wochenschr.* **39**, 447 (1961).
206. F. Ellis and J. C. Krantz, *J. Biol. Chem.* **141**, 147 (1941).
207. U. C. Dubach, E. Feiner, and I. Forgo, *Schweiz. Med. Wochenschr.* **99**, 190 (1969).
208. *Isomalt Infopac*, 3rd ed., Palatinit Sussungsmittel GmbH, 1990.
209. W. R. Todd, J. Myers, and E. S. West, *J. Biol. Chem.* **127**, 275 (1939).
210. A. N. Wick, M. C. Almen, and L. Joseph, *J. Am. Pharm. Assoc.* **40**, 542 (1951).
211. S. M. Nasrallah and F. L. Iber, *Am. J. Med. Sci.* **258**, 80 (1969).
212. L. H. Adcock and C. H. Gray, *Biochem. J.* **65**, 554 (1957).
213. O. Touster in H. L. Sipple and K. W. McNutt, eds., *Sugars in Nutrition*, Academic Press, Inc., New York, 1974, pp. 229–239.
214. O. Touster and D. R. D. Shaw, *Physiol. Rev.* **42**, 181 (1962).
215. A. Wretland, *Nutr. Metab.* **18**(Suppl. 1), 242 (1975).
216. H. Förster, in Ref. 212, pp. 259–280.
217. *Sorbitol Brochure*, Calorie Control Council, 1995.
218. H. R. Moskowitz, in Ref. 212, pp. 37–64.
219. M. G. Lindley, G. G. Birch, and R. Khan, *J. Sci. Food Agric.* **27**, 140 (1976).
220. U.S. Pat. 2,608,489 (Aug. 26, 1952), H. W. Walker (to Ditex Foods, Inc.).
221. H. W. Berg and C. S. Ough, *Wines Vines*, 27 (Jan. 1962).
222. U.S. Pat. 3,314,803 (Apr. 18, 1967), C. Dame, Jr. and R. E. Smiles (to General Foods Corp.).
223. Brit. Pat. 884,961 (Dec. 20, 1961) (to the Distillers Co., Ltd.).
224. Brit. Pat. 1,253,300 (Nov. 10, 1971), K. Hayashibara.
225. Brit. Pat. 887,883 (Jan. 24, 1962) (to Hoffmann-LaRoche & Co.).
226. J. C. Cowan, P. M. Cooney, and C. D. Evans, *J. Am. Oil Chem. Soc.* **39**, 6 (1962).
227. A. Leviton and M. J. Pallansch, *J. Dairy Sci.* **45**, 1045 (1962).
228. U.S. Pat. 3,769,438 (Oct. 30, 1973), D. T. Rusch and M. J. Lynch (to ICI America Inc.).
229. K. M. Alcott and T. P. Labuza, *J. Food Sci.* **40**, 137 (1975).
230. G. Kaess and J. F. Weidemann, *Food Technol.* **16**, 83 (1962).
231. U.S. Pat. 3,561,978 (Feb. 9, 1971), A. S. Geisler and G. C. Papalexis (to Technical Oil Products, Inc.).
232. U.S. Pat. 3,202,514 (Aug. 24, 1965), H. M. Burgess and R. W. Mellentin (to General Foods Corp.).
233. U.S. Pat. 3,438,787 (Apr. 15, 1969), J. W. Du Ross (to Atlas Chemical Industries, Inc.).
234. U.S. Pat. 3,738,845 (June 12, 1973), J. T. Liebrand (to Pfizer Inc.).
235. U.S. Pat. 3,671,266 (June 20, 1972), I. Cooper, W. A. Parker, and D. Melnick (to CPC International Inc.).
236. R. B. Alfin-Slater and co-workers, *J. Am. Oil Chem. Soc.* **50**, 348 (1973).
237. A. E. Waltking and co-workers, *J. Am. Oil Chem. Soc.* **50**, 353 (1973).

238. D. R. Ward, L. B. Lathrop, and M. J. Lynch, *Drug Cosmet. Ind.* **99**(1), 48, 157 (1966).
239. M. Barr, S. R. Kohn, and L. F. Tice, *J. Am. Pharm. Assoc. Sci. Ed.* **46**(11), 650 (1957).
240. G. R. Sabatini and J. J. Galesich, *J. Am. Pharm. Assoc. Pract. Pharm. Ed.* **17**, 806 (1956).
241. S. M. Blaug and J. W. Wesolowski, *J. Am. Pharm. Assoc. Sci. Ed.* **48**(12), 691 (1959).
242. U.S. Pat. 3,577,524 (May 4, 1971), W. D. Pratt (to American Cyanamid Co.).
243. U.S. Pat. 3,200,039 (Aug. 10, 1965), H. Thompson, Jr. (to Chas. Pfizer and Co.).
244. U.S. Pat. 2,992,970 (July 18, 1961), V. H. Baptist and A. Cherkin (to Don Baxter, Inc.).
245. C. M. Jones and co-workers, *Ann. Intern. Med.* **29**, 1 (1948).
246. K. H. Bäessler and K. Schultis, in H. Ghadimi, ed., *Total Parenteral Nutrition: Premises and Promises*, John Wiley & Sons, Inc., New York, 1975, pp. 65–83.
247. S. Lindvall and N. S. E. Andersson, *Br. J. Pharmacol.* **17**, 358 (1961).
248. H. E. Ginn, in Ref. 212, pp. 607–612.
249. F. J. Troncale and co-workers, *Am. J. Med. Sci.* **251**, 188 (1966).
250. S. J. Strianse, *Am. Perfum. Cosmet.* **77**(10), 31 (1962).
251. U.S. Pat. 3,932,604 (Jan. 13, 1976), J. B. Barth (to Colgate-Palmolive Co.).
252. U.S. Pat. 3,622,261 (Nov. 23, 1971), J. F. Cotton, J. W. Reed, and W. C. Monk (to West Point-Pepperell, Inc.).
253. U.S. Pat. 3,582,257 (June 1, 1971), J. J. Hirshfeld and B. J. Reuben (to Monsanto Co.).
254. U.S. Pat. 3,170,876 (Feb. 23, 1965), R. A. Olney (to Atlas Chemical Industries, Inc.).
255. U.S. Pat. 3,071,485 (Jan. 1, 1963), O. B. Wurzburg and W. Herbst (to National Starch and Chemical Corp.).
256. E. R. Lieberman and S. G. Gilbert, *J. Polym. Sci.* **41**, 33 (1973).
257. W. C. Griffin and E. G. Almy, *Ind. Eng. Chem.* **37**, 948 (1945).
258. Ger. Offen. 2,260,706 (June 28, 1973), R. Kuehne and co-workers (to Farbwerke Hoechst AG).
259. U.S. Pat. 3,177,117 (Apr. 6, 1965), J. F. Saunders (to the United States of America as represented by the Secretary of the Navy).
260. K. F. Redway and S. P. Lapage, *Cryobiology* **11**, 73 (1974).
261. Ger. Pat. 1,164,022 (Feb. 27, 1964), F. Smith and E. F. Graham (to the University of Minnesota).
262. Brit. Pat. 1,150,957 (May 7, 1969), A. B. Scholes (to Ball Brothers Co.).
263. U.S. Pat. 3,462,254 (Aug. 19, 1969), A. Marzocchi and G. E. Remmel (to Owens-Corning Fiberglass Corp.).

MARY E. LAWSON
SPI Polyols, Inc.

SULFA DRUGS.   See ANTIBACTERIAL AGENTS, SYNTHETIC–SULFONAMIDES.

SULFAMATES.   See SULFAMIC ACID AND SULFAMATES.

SULFAMIC ACID AND SULFAMATES

Sulfamic acid [5329-14-6] (amidosulfuric acid), HSO_3NH_2, molecular weight 97.09, is a monobasic, inorganic, dry acid and the monoamide of sulfuric acid. Sulfamic acid is produced and sold in the form of water-soluble crystals. This acid was known and prepared in laboratories for nearly a hundred years before it became a commercially available product. The first preparation of this acid occurred around 1836 (1). Later work resulted in identification and preparation of sulfamic acid in its pure form (2). In 1936, a practical process which became the basis for commercial preparation was developed (3,4). This process, involving the reaction of urea with sulfur trioxide and sulfuric acid, continues to be the main method for production of sulfamic acid.

Sulfamic acid has a unique combination of properties that makes it particularly well suited for scale removal and chemical cleaning operations, the main commercial applications. Sulfamic acid is also used in sulfation reactions, pH adjustment, preparation of synthetic sweeteners (qv), and a variety of chemical processing applications. Salts of sulfamic acid are used in electroplating (qv) and electroforming operations as well as for manufacturing flame retardants (qv) and weed and brush killers (see HERBICIDES).

Properties

Sulfamic Acid.   *Physical.*   Sulfamic acid is a dry acid having orthorhombic crystals. The pure crystals are nonvolatile, nonhygroscopic, colorless, and odorless. The acid is highly stable up to its melting point and may be kept for years without change in properties. Selected physical properties of sulfamic acid are listed in Table 1. Other properties are available in the literature (5–8).

Chemical Properties.   Selected chemical properties of sulfamic acid are listed in Table 2; other properties are listed in Reference 9.

Whereas sulfamic acid is a relatively strong acid, corrosion rates are low in comparison to other acids (Table 3). The low corrosion rate can be further reduced by addition of corrosion inhibitors (see CORROSION AND CORROSION CONTROL).

Inorganic Reactions.   Thermal decomposition of liquid sulfamic acid begins at 209°C. At 260°C, sulfur dioxide, sulfur trioxide, nitrogen, water, and traces of other products, chiefly nitrogen compounds, result.

Aqueous sulfamic acid solutions are quite stable at room temperature. At higher temperatures, however, acidic solutions and the ammonium salt hydrolyze to sulfates. Rates increase rapidly with temperature elevation, lower pH, and increased concentrations. These hydrolysis reactions are exothermic. Concentrated solutions heated in closed containers or in vessels having adequate venting can generate sufficient internal pressure to cause container rupture. An ammonium sulfamate, 60 wt % aqueous solution exhibits runaway hydrolysis when heated to 200°C at pH 5 or to 130°C at pH 2. The danger is minimized in a well-vented container, however, because the 60 wt % solution boils at 107°C (8,10). Hydrol-

Table 1. Physical Properties of Sulfamic Acid

Property	Value
mol wt	97.09
mp, °C	205
decomposition temperature, °C	209
density at 25°C, g/cm^3	2.126
refractive indexes, 25±3°C	
α	1.533
β	1.563
γ	1.568
solubility, wt %	
aqueous	
at 0°C	12.08
20°C	17.57
40°C	22.77
60°C	27.06
80°C	32.01
nonaqueous, at 25°C	
formamide	16.67
methanol	4.12
ethanol (2% benzene)	1.67
acetone	0.40
ether	0.01
71.8% sulfuric acid	0.00

Table 2. Chemical Properties of Sulfamic Acid

Property	Value
dissociation constant, at 25°C	0.101
heat of formation, kJ/mol[a]	−685.9
heat of solution, kJ/mol[a]	19.10
pH of aqueous solutions, at 25°C	
1.00 N	0.41
0.75 N	0.50
0.50 N	0.63
0.25 N	0.87
0.10 N	1.18
0.05 N	1.41
0.01 N	2.02

[a]To convert J to cal, divide by 4.184.

ysis reactions are:

$$HSO_3NH_2 + H_2O \longrightarrow NH_4HSO_4$$

$$NH_4SO_3NH_2 + H_2O \longrightarrow (NH_4)_2SO_4$$

Alkali metal sulfamates are stable in neutral or alkaline solutions even at boiling temperatures. Rates of hydrolysis for sulfamic acid in aqueous solutions have been measured at different conditions (Table 4) (8,10).

Table 3. 100-Day Metal Corrosion Rates in Aqueous Acid[a], mm

Metal	Sulfamic acid		Hydrochloric acid	
	20°C	40°C	20°C	40°C
iron	0.76	2.42	3.5	7.4
304 stainless steel	0.0001	0.0001	0.11	0.40
316 stainless steel	0.0000	0.0000	0.02	0.25
copper	0.013	0.036	0.53	1.63
aluminum	0.04	0.22	3.04	3.24
brass	0.014	0.032	0.098	0.037
gunmetal	0.002	0.022	0.29	1.40

[a]5-wt % solution.

Table 4. Hydrolysis of Aqueous Sulfamic Acid at 80°C

Time, h	Quantity of HSO_3NH_2 lost to hydrolysis, wt %		
	1% solution	10% solution	30% solution
1	4.5	7.8	7.9
2	9.1	15.1	15.1
3.1	13.3	22.7[a]	22.0
5	16.9	28.3	27.5
6	20.6	34.3	32.8
7	24.2	39.5	37.5
8	27.3	43.7	

[a]Value given is at pH = 4.2.

Sulfamic acid readily forms various metal sulfamates by reaction with the metal or the respective carbonates, oxides, or hydroxides. The ammonium salt is formed by neutralizing the acid with ammonium hydroxide:

$$Zn + 2\,HSO_3NH_2 \longrightarrow Zn(SO_3NH_2)_2 + H_2$$

$$CaCO_3 + 2\,HSO_3NH_2 \longrightarrow Ca(SO_3NH_2)_2 + H_2O + CO_2$$

$$FeO + 2\,HSO_3NH_2 \longrightarrow Fe(SO_3NH_2)_2 + H_2O$$

$$Ni(OH)_2 + 2\,HSO_3NH_2 \longrightarrow Ni(SO_3NH_2)_2 + 2\,H_2O$$

$$NH_4OH + HSO_3NH_2 \longrightarrow NH_4SO_3NH_2 + H_2O$$

Nitrous acid reacts very rapidly and quantitatively with sulfamic acid:

$$HSO_3NH_2 + HNO_2 \longrightarrow H_2SO_4 + H_2O + N_2$$

This reaction can be used for the quantitative analysis of nitrites (3,11). The reaction of sulfamic acid and concentrated nitric acid gives nitrous oxide (12,13):

$$HSO_3NH_2 + HNO_3 \longrightarrow H_2SO_4 + H_2O + N_2O$$

Chlorine, bromine, and chlorates oxidize sulfamic acid to sulfuric acid and nitrogen (1,14):

$$2 \, HSO_3NH_2 + KClO_3 \longrightarrow 2 \, H_2SO_4 + N_2 + KCl + H_2O$$

Chromic acid, permanganic acid, and ferric chloride do not oxidize sulfamic acid.

Hypochlorous acid reacts at low temperatures to form N-chlorosulfamic acid [17172-27-9] and N,N-dichlorosulfamic acid [17085-87-9]:

$$HSO_3NH_2 + HOCl \longrightarrow HSO_3NHCl + H_2O$$

$$HSO_3NH_2 + 2 \, HOCl \longrightarrow HSO_3NCl_2 + 2 \, H_2O$$

A similar reaction occurs when sodium is present with the formation of sodium N-chlorosulfamate [13637-90-6] and sodium N,N-dichlorosulfamate [13637-67-7]:

$$NaSO_3NH_2 + 2 \, NaOCl \longrightarrow NaSO_3NCl_2 + 2 \, NaOH$$

Sulfamoyl chloride [7778-42-9] forms from reaction with thionyl chloride (15,16):

$$HSO_3NH_2 + SOCl_2 \longrightarrow NH_2SO_2Cl + HCl + SO_2$$

Certain metal iodides, eg, sodium, potassium, cesium, or rubidium, react with sulfamic acid to form the corresponding alkali metal triodides (17):

$$2 \, HSO_3NH_2 + 3 \, CsI + H_2O \longrightarrow CsI_3 + Cs_2SO_4 + (NH_4)_2SO_3$$

An exception exists to the monobasic nature of sulfamic acid when it dissolves in liquid ammonia. Sodium, potassium, etc. add both to the amido and sulfonic portions of the molecule to give salts, such as $NaSO_3NHNa$.

Sodium sulfate and sulfamic acid form the complex $6 \, HSO_3NH_2 \cdot 5 \, Na_2SO_4 \cdot 15 \, H_2O$. Phosphorus pentachloride reacts with sulfamic acid to form the following complex (18):

$$HSO_3NH_2 + 2 \, PCl_5 \longrightarrow PCl_3 \cdot ClSO_2NH + Cl_2 + HCl + POCl_3$$

Sulfamic acid and its salts retard the precipitation of barium sulfate and prevent precipitation of silver and mercury salts by alkali. It has been suggested that salts of the type $AgNHSO_3K$ [15293-60-4] form with elemental metals or salts of mercury, gold, and silver (19). Upon heating such solutions, the metal deposits slowly in mirror form on the wall of a glass container. Studies of chemical and electrochemical behavior of various metals in sulfamic acid solutions are described in Reference 20.

Organic Reactions.   Primary alcohols react with sulfamic acid to form alkyl ammonium sulfate salts (21–23):

$$ROH + HSO_3NH_2 \longrightarrow ROSO_2O \cdot NH_4$$

Sulfation by sulfamic acid has been used in the preparation of detergents from dodecyl, oleyl, and other higher alcohols. It is also used in sulfating phenols and phenol–ethylene oxide condensation products. Secondary alcohols react in the presence of an amide catalyst, eg, acetamide or urea (24). Pyridine has also been used. Tertiary alcohols do not react. Reactions with phenols yield phenyl ammonium sulfates. These reactions include those of naphthols, cresol, anisole, anethole, pyrocatechol, and hydroquinone. Ammonium aryl sulfates are formed as intermediates and sulfonates are formed by subsequent rearrangement (25,26).

$$C_6H_5OH + NH_2SO_3H \longrightarrow NH_4SO_3-O-C_6H_5$$

$$NH_4SO_3-O-C_6H_5 \longrightarrow NH_4SO_3-C_6H_4OH$$

$$1,4\text{-}C_6H_4(OH)_2 + 2\ NH_2SO_3H \longrightarrow (NH_4SO_3O)_2C_6H_4$$

$$(NH_4SO_3O)_2C_6H_4 \longrightarrow (NH_4SO_3)_2C_6H_2(OH)_2$$

Studies of sulfoesterification of higher alcohols with sulfamic acid are described in Reference 26.

Amides react in certain cases to form ammonium salts of sulfonated amides (22). For example, treatment with benzamide yields ammonium *N*-benzoylsulfamate [*83930-12-5*], $C_6H_5CONHSO_3 \cdot NH_4$, and treatment with ammonium sulfamate yields diammonium imidodisulfonate [*13597-84-1*], $HN(SO_2ONH_4)_2$. Ammonium sulfamate or sulfamic acid and ammonium carbonate dehydrate liquid or solid amides to nitriles (27).

Primary, secondary, and tertiary amines react with sulfamic acid to form ammonium salts (28):

$$HSO_3NH_2 + RNH_2 \longrightarrow RNH_2 \cdot HSO_3NH_2$$

$$HSO_3NH_2 + R_3N \longrightarrow R_3N \cdot HSO_3NH_2$$

Guanidine salts can be prepared by reaction of thiocyanates and sulfamates (22).

Aldehydes form addition products with sulfamic acid salts. These are stable in neutral or slightly alkaline solutions but are hydrolyzed in acid and strongly alkaline solutions. With formaldehyde, the calcium salt of the methylol (hydroxymethyl) derivative [*82770-57-8*], $Ca(O_3SNHCH_2OH)_2$, is obtained as a crystalline solid.

Cadmium, cobalt, copper, and nickel sulfamates react with lower aliphatic aldehydes. These stable compositions are suitable for use in electroplating solutions for deposition of the respective metal (see ELECTROPLATING).

Fatty acid acyl halides react with sulfamic acid (29).

The *N*-alkyl and *N*-cyclohexyl derivatives of sulfamic acid are comparatively stable. The *N*-aryl derivatives are very unstable and can only be isolated in the salt form. A series of thiazolylsulfamic acids has been prepared.

Cellulose sulfated using sulfamic acid degrades less than if sulfated using sulfuric acid (23). Cellulose esters of sulfamic acids are formed by the reaction of sulfamyl halides in the presence of tertiary organic bases (see CELLULOSE ESTERS).

Other organic reactions of sulfamic acid are described in the literature (30,31).

Sulfamates.   Sulfamates are formed readily by the reaction of sulfamic acid and the appropriate metal or its oxide, hydroxide, or carbonate. Approximate heats of neutralization are -54.61 kJ/mol (-13.05 kcal/mol) for the NaOH reaction and -47.83 kJ/mol (-11.43 kcal/mol) for NH_4OH at 26–30°C. Sulfamates prepared from weak bases form acidic solutions, whereas those prepared from strong bases produce neutral solutions. The pH of 5 wt % solution of ammonium sulfamate is 5.2. Crystals of ammonium sulfamate deliquesce at relative humidity of 70% and higher. Both ammonium sulfamate [*7773-06-0*] and potassium sulfamate [*13823-50-2*] liberate ammonia at elevated temperatures and form the corresponding imidodisulfonate (12). Inorganic sulfamates are quite water-soluble, except for the basic mercury salt. Some relative solubilities of sulfamates at 25°C in 100 g of water are ammonium, 103 g; sodium, 106 g; magnesium, 119 g; calcium, 67 g; barium, 34.2 g; zinc, 115 g; and lead, 218 g. The properties of a number of sulfamates may be found in the literature (see Table 5).

Manufacture

Sulfamic Acid.   Sulfamic acid is manufactured by the reaction of urea and fuming sulfuric acid (3,4,44). This reaction is considered to take place in two steps involving an aminocarbonylsulfamic acid intermediate:

$$NH_2CONH_2 + SO_3 \longrightarrow [HSO_3NHCONH_2]$$

$$[HSO_3NHCONH_2] + H_2SO_4 \longrightarrow 2\ HSO_3NH_2 + CO_2$$

The overall reaction is as follows:

$$NH_2CONH_2 + SO_3 + H_2SO_4 \longrightarrow 2\ HSO_3NH_2 + CO_2$$

Urea reacts with fuming sulfuric acid in an exothermic reaction that needs agitation and cooling. After completion of the reaction, excess sulfur trioxide is removed by dilution or by other methods (45,46). A flow diagram of the process is shown in Figure 1.

In the sulfamic acid process, electrical energy is needed for removal of reaction heat, filtration, fluid transportation, etc. Consumption is about 300 kWh/t of sulfamic acid. Consumption of steam, used for the heat exchanger, crystallizer, and drier, is from 1000 to 1500 kg/t of sulfamic acid.

The reaction between urea and fuming sulfuric acid is rapid and exothermic. It may proceed with violent boiling unless the reaction temperature is controlled.

Table 5. Literature References for Properties of Sulfamates

Compound	CAS Registry Number	References
ammonium	[7773-06-0]	32
aluminum	[10101-13-0]	2,32
barium	[13770-86-0]	27,32,33
cadmium	[14017-36-8]	32
calcium	[13770-92-8]	32
cobalt	[14017-41-5]	32,34
copper	[14017-38-0]	27,32,35−37
iron	[14017-39-1]	12,32
lead	[32849-69-7]	27,32,35
lithium	[21856-68-8]	32
magnesium	[13770-91-7]	32,38
manganese	[83929-95-7]	32
nickel	[13770-89-3]	32,36
potassium	[13823-50-2]	32
silver	[14325-99-6]	12,15,19,27,32,33
sodium	[13845-18-6]	32
strontium	[83929-96-8]	32
thallium	[21856-70-2]	32
uranyl	[82783-83-3]	32
zinc	[13770-90-6]	32
1:1 anilinium	[10310-62-0]	22,35,39−41
1:1 amylaminium	[82323-98-6]	40
1:1 benzylaminium	[82323-99-7]	42
1:1 hydroxylaminium	[82324-00-3]	12,22
1:1 trimethylaminium	[6427-17-4]	38
1:1 hydrazinium	[39935-03-0]	2,12,35,40
α-naphthylaminium	[83929-97-9]	39,40
β-naphthylaminium	[83929-98-0]	39,40
1:1 piperidinium	[82324-01-4]	43
1:1 p-toluidinium	[68734-85-0]	39,40

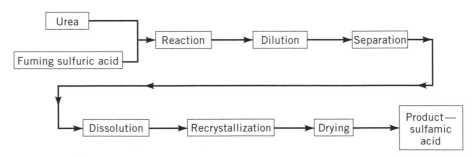

Fig. 1. Schematic flow diagram of the sulfamic acid process.

The reactants are strongly acidic. Therefore, operators should wear suitable protective gear to guard against chemical hazard. Special stainless steel, rubber lining, fiber-reinforced plastics, and polyvinyl chloride and carbon equipment are used.

The reaction takes place at atmospheric pressure. For stable control of the reaction rate, the reaction is first carried out at a temperature of 50°C and then at 60°C. Overall, this batch reaction takes about 9 hours. After completion of reaction, the slurry is diluted to about 70% sulfuric acid solution, and crude sulfamic acid crystals are separated by centrifuge. The crystals are dissolved in mother liquor to make a saturated solution at 60°C and the solution is concentrated under vacuum at 40°C. Purified sulfamic acid is obtained by recrystallization.

The yield of sulfamic acid varies with operating conditions. An overall yield of about 90% is obtained in the Nissan process. Losses are approximately 6% in the reaction stage, 3% in recrystallization, and 1% in other sections. The manufacture of ammonium sulfamate also is described in German and Japanese patents (47,48).

By-Products and Waste Disposal. A by-product of sulfamic acid manufacturing is fuming sulfuric acid or dilute sulfuric acid. The amount of sulfuric acid (as 100% H_2SO_4) is $1-1.5$ times as much by weight as the sulfamic acid product. This by-product also contains ammonium salts and is therefore normally used as raw material for fertilizer (see FERTILIZERS).

The off-gas from the reactor contains CO_2, SO_3, and H_2SO_4. The SO_3 is removed by absorption (qv) into concentrated sulfuric acid solution or by other means. The CO_2 and H_2SO_4 vapor is removed by absorption into water or alkaline solution.

Economic Aspects

Until the 1970s, the main production countries of sulfamic acid were the United States, several European countries, and Japan. The large amounts of dilute sulfuric acid by-product generated led to the difficult situation of by-product acid disposal. Concomitantly, the start of chemical production in developing Asian countries caused successional sulfamic acid production withdrawal in the 1980s. As of the mid-1990s production countries are Japan, Taiwan, Indonesia, India, and China. The 1995 world production capacity was ca 96,000 metric tons.

In Japan, sulfamic acid is produced and supplied in crystal form. It is packaged in 25-kg net weight paper bags and in 600-kg, 700-kg, and 750-kg resinous flexible containers. The truckload price (fob Japan) is $1–2/kg. Three principal uses of sulfamic acid are in chemical cleaning, as sulfonation reagent, and for use in synthetic sweetener.

Ammonium sulfamate is also produced and commercially available in Japan. It is packed in 25-kg net weight paper bags and 500-kg resinous flexible containers. The truckload price (fob Japan) is $1.5–$3/kg. Other sulfamates, eg, nickel sulfamate and aluminum sulfamate, are commercially available. The primary salts manufactured from sulfamic acid in the United States are the ammonium and nickel sulfamates. These salts of sulfamic acid are used mainly in electroplating.

Specifications and Analysis

Sulfamic Acid. Specifications and typical analyses of commercially available sulfamic acid are listed in Table 6.

The assay determination of sulfamic acid is made by titration of an accurately prepared sulfamic acid solution using sodium nitrite solution and an external potassium iodide starch-paste indicator. It is based on the reaction

$$HSO_3NH_2 + NaNO_2 \longrightarrow NaHSO_4 + H_2O + N_2$$

Standard $1/10\ N$ nitrite is used to titrate a solution prepared by dissolving 10–100 mg of sulfamic acid and about 6 mL of $(1 + 1)$ H_2SO_4 in 300 mL of distilled water at 40–50°C. At the end point, the colorless external potassium iodide starch-paste indicator changes to blue. A 1-mL solution of $1/10N$ $NaNO_2$ is equivalent to 9.709 mg of sulfamic acid. The $1/10\ N$ nitrite titrant solution is standardized using primary standard-grade sulfamic acid. For sulfamate assay determination, the same procedure is used as for sulfamic acid.

For crystal sulfamic acid assay, a simplified procedure of neutralization titration with sodium hydroxide solution may be used. At the end point, Bromothymol Blue (BTB) indicator changes color from yellow to yellowish green. A 1-mL solution of $1/2N$ NaOH is equivalent to 0.0485 g of sulfamic acid.

Table 6. Specifications and Analyses of Commercially Available Sulfamic Acid[a]

| | Crystalline material | |
Property	Specifications[b]	Typical analyses[c]
sulfamic acid, wt %	99.5[d]	99.9
ignition residue, wt %	0.05	0.0002
Fe, wt %	0.001	0.0001
heavy metal (as Pb), wt %	0.001	0.0001
chloride, wt %	0.002	0.0001
sulfate, wt %	0.05	0.01
nitrate, wt %	0.0001	<0.0001
solution appearance	transparent and colorless	transparent and colorless
moisture, wt %	0.1	0.01
particle size		
1680 μm stop, wt %	1.0	0.6
250 μm pass, wt %	30.0	24.6

[a]Values for product manufactured by Nissan Chemical Industries, Ltd.
[b]Values given are maximum unless otherwise noted.
[c]Values given are typical. Numbers for individual batches may vary.
[d]Value is minimum.

Health and Safety Factors

Contact with sulfamic acid and its solutions can cause eye burns and irritate the nose, throat, and skin. Workers should wear cup-type, rubber, or soft plastic-framed goggles, equipped with approved impact-resistant glass or plastic lenses. Goggles should be carefully fitted to ensure maximum protection and comfort.

Exposure to the skin can be minimized by wearing rubber gloves when handling sulfamic acid and its solutions, and hands should be washed thoroughly after handling. Breathing of the dust should be avoided and adequate ventilation should be provided.

In case of eye contact, the eyes should be flushed immediately using plenty of water for at least 15 min and a physician should be consulted. Exposed skin should also be flushed copiously with water. Anyone who has ingested the acid should immediately drink large amounts of water and consult a physician.

Toxicity data for the acid are as follows: oral LD_{50} (rats), 1600 mg/kg; and oral LD_{50} (mice), 3100 mg/kg (49). The physiological effects of sulfamic acid and ammonium sulfamate are described in Reference 50.

Uses

Sulfamic Acid. *Removal of Residues from Industrial Processing Equipment.* Properties of sulfamic acid that make it particularly well suited for scale-removal operations and chemical cleaning include the following: the acid is available in dry form, permitting convenient transportation, storage, handling, and packaging operations; it is a strong acid in aqueous solutions and is effective in solubilizing hard-water scales; sulfamic acid is nonvolatile and chloride-free; aqueous solutions do not emit objectionable or corrosive fumes and can be used on stainless steel generating no chloride stress corrosion; the acid is readily reactive with most deposits to form highly water-soluble compounds that are water rinsable; redeposition of dissolved solids is minimized. Sulfamic acid is less corrosive than many other strong acids on most materials of construction (see Table 3). Moreover, sulfamic acid is compatible with corrosion inhibitors, wetting agents, pH indicators, and other components of dry-cleaning formulations. The acid does not have a high toxicity rating, but all suppliers recommend caution in handling and recommend personal protection to avoid any injury.

Acid cleaners based on sulfamic acid are used in a large variety of applications, eg, air-conditioning systems; marine equipment, including salt water stills; wells (water, oil, and gas); household equipment, eg, copper-ware, steam irons, humidifiers, dishwashers, toilet bowls, and brick and other masonry; tartar removal of false teeth (50); dairy equipment, eg, pasteurizers, evaporators, preheaters, and storage tanks; industrial boilers, condensers, heat exchangers, and preheaters; food-processing equipment; brewery equipment (see BEER); sugar evaporators; and paper-mill equipment (see also EVAPORATION; METAL SURFACE TREATMENTS; PULP).

Manufacture of Dyes and Pigments and pH Adjustment. Use of sulfamic acid in the manufacture of dyes and pigments involves removal of excess nitrite from diazotization reactions (see AZO DYES) and is based on the following reaction:

$$NO_2^- + HSO_3NH_2 \longrightarrow HSO_4^- + N_2 + H_2O$$

Sulfamic acid is also used in some dyeing operations for pH adjustment; however, it is useful in lowering pH levels in a variety of other systems. The low pH persists at elevated temperatures and there are no objectionable fumes.

Paper-Pulp Bleaching. Sulfamic acid additions to chlorination bleaching stages are effective in reducing pulp-strength degradation associated with high temperatures (52) (see BLEACHING AGENTS). Other benefits are noted when sulfamic acid is added to the hypochlorite bleaching stage (53), including reduction of pulp-strength losses as a result of high temperature or low pH; increased production by means of higher temperatures and lower pHs at the same pulp-strength level; savings in chemical costs, eg, lower consumption of buffer, caustic soda, and higher priced bleaching agents; and improved efficiency through reducing effects of variation in temperature and pH.

Chlorine Vehicle and Stabilizer. Sulfamic acid reacts with hypochlorous acid to produce N-chlorosulfamic acids, compounds in which the chlorine is still active but more stable than in hypochlorite form. The commercial interest in this area is for chlorinated water systems in paper mills, ie, for slimicides, cooling towers, and similar applications (54) (see INDUSTRIAL ANTIMICROBIAL AGENTS).

Analytical and Laboratory Operations. Sulfamic acid has been recommended as a reference standard in acidimetry (55). It can be purified by recrystallization to give a stable product that is 99.95 wt % pure. The reaction with nitrite as used in the sulfamic acid analytical method has also been adapted for determination of nitrites with the acid as the reagent. This reaction is used commercially in other systems for removal of nitrous acid impurities, eg, in sulfuric and hydrochloric acid purification operations.

Sulfation and Sulfamation. Sulfamic acid can be regarded as an ammonia–SO_3 complex and has been used thus commercially, always in anhydrous systems. Sulfation of mono-, ie, primary and secondary, alcohols; polyhydric alcohols; unsaturated alcohols; phenols; and phenolethylene oxide condensation products has been performed with sulfamic acid (see SULFONATION AND SULFATION). The best-known application of sulfamic acid for sulfamation is the preparation of sodium cyclohexylsulfamate [*139-05-9*], which is a synthetic sweetener (see SWEETENERS).

Sulfamates. *Ammonium Sulfamate.* A number of flame retardants used for cellulosic materials, including fabrics and paper products, are based on ammonium sulfamate (56). These products are water-soluble and therefore nondurable if treated fabrics are washed or exposed to weathering conditions. For most fabric and paper constructions, efficient flame retardancy can be provided with no apparent effect on color or appearance and without stiffening or adverse effects on the feel of the fabrics. A wide variety of materials are treated, including hazardous work-area clothing, drapes, curtains, decorative materials, blankets, sheets, and specialty industrial papers (57).

Ammonium sulfamate is highly effective in nonselective herbicides to control weeds, brush, stumps, and trees (58) (see HERBICIDES).

Other. Nickel sulfamate is made by the combination of nickel carbonate and sulfamic acid. It is almost exclusively used in the plating industry, with its solutions used for both plating and electroforming. The principal value of this system is low internal stress in deposits and high plating rates. Other sulfamates used in plating solutions include the salts of cobalt, cadmium, ferrous iron, and lead (see ELECTROPLATING). Ferrous sulfamate is used in nuclear fuel processing solutions (59). Certain amine sulfamates impart softening properties to papers and textiles. Such materials exhibit long effective service, particularly at low humidity (see QUATERNARY AMMONIUM COMPOUNDS).

Calcium sulfamate and magnesium sulfamate are used effectively as a stiffening promoter of concrete and hydraulic cement (see CEMENT). Compared to calcium chloride [10043-52-4], which is commonly used as stiffening promoter, these sulfamates do not contain chloride and are not alkaline. Therefore highly durable concrete is made using these sulfamates (60,61). These sulfamates and sulfamic acid are also used with the ground injection material the main component of which is water glass, to help easy adjustment of gelation time and increasing firmness of ground (62–64). Crystallization of ammonium sulfate [7783-20-2] from its mother liquor with addition of guanidine sulfamate [51528-20-2] as crystal habit modifier produces bigger and spherically shaped granular crystals (65,66). Sulfamates of alkaline metals are used as additives in chromium tanning of hides (67).

BIBLIOGRAPHY

"Sulfamic Acid" in *ECT* 1st ed., Vol. 13, pp. 285–294, by G. G. Torrey, E. I. du Pont de Nemours & Co., Inc., Grasselli Chemicals Dept.; "Sulfamic Acid and Sulfamates" in *ECT* 2nd ed., Vol. 19, pp. 242–249, by D. Santmyers and R. Aarons, E. I. du Pont de Nemours & Co., Inc.; in *ECT* 3rd ed., Vol. 21, pp. 949–960, by E. B. Bell, E. I. du Pont de Nemours & Co., Inc.

1. H. Rose, *Pogg. Anal.* **33**, 235 (1834); **42**, 415 (1837); **61**, 397 (1844).
2. E. Berglund, *Ber.* **9**, 1896 (1876).
3. P. Baumgarten and I. Marggraff, *Ber.* **63**, 1019 (1930).
4. P. Baumgarten, *Ber.* **69**, 1929 (1936); U.S. Pat. 2,102,350 (Dec. 14, 1937), P. Baumgarten (to DuPont).
5. J. Donnay and H. Ondik, *Crystal Data Determinative Tables*, Vol. 2, 3rd ed., U.S. Dept. of Commerce, National Bureau of Standards, Joint Committee on Powder Standards, Washington, D.C., 1975, pp. 1–202.
6. A. Cameron and F. Duncanson, *Acta Crystallogr.* **1332**, 1563 (1976).
7. P. G. Sears and co-workers, *J. Inorg. Nucl. Chem.* **35**, 2087 (1973).
8. M. E. Cupery, *Ind. Eng. Chem.* **30**, 627 (June 1938).
9. J. Kurtz and J. Farrar, *J. Am. Chem. Soc.* **91**, 6057 (1969); G. Nash, *J. Chem. Eng. Data* **13**, 271 (1968).
10. J. K. Hunt, *Chem. Eng. News* **30**, 707 (1952); J. Notley, *J. Appl. Chem. Biotechnol.* **23**, 717 (Oct. 10, 1973); A. Tsypin and E. Fomenko, *Tr. Gos. Nauchno-Issled. Proektn. Inst. Azotn. Promsti. Prod. Org. Sint.* **27**, 21 (1974).
11. F. L. Hahn and P. Baumgarten, *Ber.* **63**, 3028 (1930).
12. E. Divers and T. Haga, *J. Chem. Soc.* **69**, 1634 (1896).
13. F. Ephraim and E. Lasocky, *Ber.* **44**, 395 (1911); P. Baumgarten, *Ber.* **71**, 80 (1938).
14. W. Traube and E. von Drathen, *Ber.* **51**, 111 (1918).
15. F. Ephraim and M. Gurewitsch, *Ber.* **43**, 138 (1910).
16. H. Denivelle, *Bull. Soc. Chim.* **3**, 2150 (1936).
17. P. Sakellaridis, *Bull. Soc. Chem. Biol.* **9–10**, 610 (Sept.–Oct. 1951).
18. P. Baumgarten, I. Marggstaff, *Ber.* **64**, 1582 (1931).
19. K. Hofmann and co-workers, *Ber.* **45**, 1731 (1912).
20. O. Tubertini and co-workers, *Ann. Chim.* (Rome) **57**, 555 (1967); R. Piontelli and co-workers, *Electrochim. Met.* **3**(1); **42**(4) (1968); **2**(2), 141 (1967); A. LaVecchia, *Elec-*

trochim. Met. **3**(1), 71 (1968); *Symposium on Sulfamic Acid and Its Electrometallurgic Applications* (*Proceedings*), Polytechnic School of Milan, Milan, Italy, May 25, 1966.

21. U.S. Pat. 1,931,962 (Oct. 24, 1933) and Ger. Pat. 558,296 (Aug. 22, 1930), K. Marx., K. Brodersen, and M. Quaedvlieg (to I. G. Farben).

22. Ger. Pat. 565,040 (Nov. 25, 1932), K. Brodersen and M. Quaedvlieg (to I. G. Farben).

23. W. N. Carton, *Anal. Chem.* **23**, 1016 (1951); H. Dietz and co-workers, *Agric. News Lett. E. I. du Pont de Nemours & Co., Inc.* **9**, 35 (1941).

24. U.S. Pat. 2,452,943 (Nov. 2, 1948), J. Malkemus and co-workers (to Colgate Palmolive Peet); Ger. Pat. 3,372,170 (Oct. 26, 1963), H. Remy (to Hoechst); U.S. Pat. 3,395,170 (June 28, 1966), J. Walts and L. Schenck (to General Aniline & Film).

25. K. Hofmann and E. Biesalski, *Ber.* **45**, 1394 (1912).

26. A. Quilico, *Gazz. Chim. Ital.* **37**, 793 (1927); *Atti. Accad. Lincei* **141**, 512 (1927).

27. J. Boivin, *Can. J. Res. Sect. B* **28**, 671 (1950).

28. L. Goodson, *J. Am. Chem. Soc.* **69**, 1230 (1947).

29. Brit. Pat. 372,389 (Apr. 28, 1932), J. Johnson (to I. G. Farben).

30. L. Audrieth, M. Sveda, H. Sisler, and M. Butler, *Chem. Rev.* **26**, 49 (1940).

31. K. Andersen, *Comprehensive Organic Chemistry*, Vol. 3, Pergamon Press, Oxford, U.K., 1979, p. 363.

32. E. Berglund, *Bull. Soc. Chim.* **29**, 422 (1878); *Lunds Univ. Acta* **13**, 4 (1875).

33. P. Eitner, *Ber.* **26**, 2833 (1893).

34. F. Ephraim and W. Flugel, *Helv. Chim. Acta* **7**, 724 (1924).

35. C. Paal and F. Kretschmer, *Ber.* **27**, 1241 (1894).

36. A. Callegari, *Gazzetta* **36**(2), 63 (1906).

37. M. Delepine and R. Demars, *Bull. Sci. Pharmacol.* **29**, 14 (1922).

38. G. Thies, Ph.d. dissertation, Frederick-Wilhelms University, Berlin, 1935.

39. A. Quilico, *Gazz. Chim. Ital.* **56**, 620 (1926).

40. C. Paal and H. Janicke, *Ber.* **28**, 3160 (1895); C. Paal and S. Daybeck, *Ber.* **30**, 880 (1897).

41. C. Paal, *Ber.* **34**, 2748 (1901).

42. C. Paal and L. Lowitsch, *Ber.* **30**, 869 (1897).

43. C. Paal and M. Hubaleck, *Ber.* **34**, 2757 (1901).

44. U.S. Pat. 2,880,064 (Mar. 31, 1959), M. Harbaugh and G. Pierce (to E. I. du Pont de Nemours & Co., Inc.); U.S. Pat. 2,191,754 (Mar. 27, 1940), M. Cupery (to E. I. du Pont de Nemours & Co., Inc.).

45. K. Toyokura and co-workers, *J. Chem. Eng. Jpn.* (Feb. 1979); Jpn. Pat. 70 24,649 (Nov. 27, 1967), S. Ito (to Bur. Ind. Tech.).

46. Ger. Pat. 2,637,948 (Aug. 24, 1976), R. Graeser and co-workers (to Hoechst); Ger. Pat. 2,106,019 (Feb. 9, 1971), R. Graeser and co-workers (to Hoechst); Brit. Pat. 1,068,942 (Aug. 9, 1962), W. Morris (to Seery, Defense); Brit. Pat. 1,062,329 (Dec. 22, 1964), A. Sowerby (to Marchon Products).

47. Ger. Pat. 1,915,723 (Mar. 27, 1969), H. Hofmeister (to Hoechst); Ger. Pat. 2,850,903 (Nov. 24, 1978), G. Muenster (to Hoechst).

48. Jpn. Pat. 7100,816 (Jan. 21, 1964), N. Sasaki and co-workers (to Mitsui Toatsu); Jpn. Pat. 6928,374 (Aug. 28, 1964), Yamaguchi and co-workers (to Nitto Chem.); Jpn. Pat. 3484,193 (Apr. 10, 1964), Azakmi and co-workers (to Toyo Koatsu).

49. *NIOSH 1979 Registry of Toxic Effects of Chemical Substances*, Vol. 2, U.S. Department of Health, and Human Services, Washington, D.C., 1979, p. 286.

50. A. M. Ambrose, *J. Ind. Hyg. Toxicol.* **25**(1), 26 (1943).

51. Jpn. Pat. 225,116 (Oct. 6, 1986), N. Brudney and A. A. Levy (to Richardson GmbH).

52. R. Tobar, *TAPPI* **47**, 688 (1964); U.S. Pat. 3,308,012 (Mar. 7, 1967), R. Tobar (to E. I. du Pont de Nemours & Co., Inc.).

53. U.S. Pat. 3,177,111 (Apr. 6, 1965), L. Larsen (to Weyerhaeuser).

54. U.S. Pat. 3,328,294 (June 27, 1967), R. Self and co-workers (to Mead); U.S. Pat. 3,749,672 (July 31, 1973), W. Golton and A. Rutkiewic (to E. I. du Pont de Nemours & Co., Inc.); U.S. Pat. 3,767,586 (Oct. 23, 1973), A. Rutkiewic (to E. I. du Pont de Nemours & Co., Inc.).

55. M. Caso and M. Cefola, *Anal. Chim. Acta* **21**, 205 (1959).

56. U.S. Pat. 2,723,212 (Nov. 8, 1955), R. Aarons and D. Wilson (to E. I. du Pont de Nemours & Co., Inc.).

57. U.S. Pat. 2,526,462 (Oct. 17, 1950), O. Edelstein (to Pond Lily).

58. U.S. Pat. 2,277,744 (Mar. 31, 1942), M. Cupery and A. Tanberg (to E. I. du Pont de Nemours & Co., Inc.); U.S. Pat. 2,368,274–276 (Jan. 30, 1945), R. Torley (to American Cyanamid); U.S. Pat. 2,709,648 (May 31, 1955), T. Ryker and D. Wolf (to E. I. du Pont de Nemours & Co., Inc.).

59. N. Bibler, *Nucl. Technol.*, 34 (Aug. 1977); L. Gray, *Nucl. Technol.*, 40 (Sept. 1978); R. Walser, U.S. Atomic Energy Commission Report, ARH-SA-69, 1970.

60. Jpn. Pat. 3,039 (Jan. 6, 1989), S. Kobayashi and A. Fukazawa (to Nissan Chemical Industries).

61. Jpn. Pat. 246,164 (Oct. 2, 1989), S. Kobayashi and J. Utida (to Nissan Chemical Industries).

62. Jpn. Pat. 168,485 (July 12, 1988), Y. Watanabe, E. Okumura, and M. Ando (to Nissan Chemical Industries).

63. Jpn. Pat. 312,289 (Dec. 20, 1988), S. Shimada and K. Kashiwabara (to Kyokado Engineering).

64. Jpn. Pat. 203,493 (Aug. 16, 1989), M. Ohta, E. Okumura, H. Kimura, and T. Kawahigashi (to Nissan Chemical Industries).

65. Jpn. Pat. 191,831 (July 12, 1994), H. Shibayama, K. Tonooka, M. Tsuzaki, T. Yamamoto, and K. Ura (to Kawasaki Seitetsu).

66. Jpn. Pat. 61,811 (Mar. 7, 1995), H. Shibayama, M. Tsuzaki, S. Fujiwara, and K. Ura (to Kawasaki Seitetsu).

67. Jpn. Pat. 89,600 (Apr. 20, 1988), H. Becker, W. Lotz, and K. Keller (to Hoechst AG).

KATSUMASA YOSHIKUBO
MICHIO SUZUKI
Nissan Chemical Industries, Ltd.

SULFANILAMIDE. See ANTIBACTERIAL AGENTS, SYNTHETIC—SULFONAMIDES.

SULFANLILIC ACID. See AMINES, AROMATIC—ANILINE AND ITS DERIVATIVES; SULFONATION AND SULFATION.

SULFATED ACIDS, ALCOHOLS, OILS, ETC. See SULFONATION AND SULFATION; SURFACTANTS.

SULFATION. See SULFONATION AND SULFATION.

SULFIDES.   See Sulfur compounds.

SULFITE PROCESS.   See Pulp.

SULFITES.   See Barium compounds; Sulfur compounds; Sulfuric and
Sulfurous esters.

SULFOALKYLATION.   See Sulfonation and sulfation.

SULFOCHLORINATION.   See Sulfonation and sulfation; Sulfurization and sulfurchlorination.

SULFOLANE AND
SULFONES

Sulfolane

Sulfolane was first described in the chemical literature in 1916. It has been
noted for its exceptional chemical and thermal stability and unusual solvent
properties. The search for a commercial process to produce sulfolane began about
1940. Market development quantities became available in 1959. Since then, both
the use of and the applications for sulfolane have increased dramatically.

    Properties.   Sulfolane [126-33-0], $C_4H_8SO_2$ (**1**), also known as tetrahydro-
thiophene-1,1-dioxide and tetramethylene sulfone, is a colorless, highly polar,
water-soluble compound. Physical properties are given in Table 1 (1).

$$\begin{array}{c}
H_2C - CH_2 \\
| \quad\quad | \\
H_2C \quad\ CH_2 \\
\diagdown\ /\ \\
S \\
/\!/\ \diagdown \\
O \quad\ O
\end{array}$$

(**1**)

Table 1. Physical Properties of Sulfolane

Property	Value
molecular weight	120.17
boiling point, °C	287.3
melting point, °C	28.5
specific gravity	
30/30	1.266
100/4	1.201
density, 15°C, g/cm^3	1.276
flash point, °C	165–178
viscosity, mPa·s(=cP)	
at 30°C	10.5
100°C	2.5
200°C	1.0
refractive index, n_D, at 30°C	1.48
heat of fusion, kJ/kg[a]	11.44
dielectric constant	43.3
surface tension, at 30°C, mN/m(=dyn/cm)	35.5

[a]To convert J to cal, divide by 4.184.

The thermal stability of sulfolane is summarized as follows (1):

Temperature, °C	Quantity of SO_2 liberated from 250 mL sulfolane, mg/h
180	0.6
200	2.8
220	3.3
240	24.1

Whereas sulfolane is relatively stable to about 220°C, above that temperature it starts to break down, presumably to sulfur dioxide and a polymeric material. Sulfolane, also stable in the presence of various chemical substances as shown in Table 2 (2), is relatively inert except toward sulfur and aluminum chloride. Despite this relative chemical inertness, sulfolane does undergo certain reactions, for example, halogenations, ring cleavage by alkali metals, ring additions catalyzed by alkali metals, reaction with Grignard reagents, and formation of weak chemical complexes.

Halogenation. Chlorine can be added to sulfolane by a uv-initiated process to give 3-chlorosulfolane [3844-04-0], 3,4-dichlorosulfolane [3001-57-8], and 3,3,4-trichlorosulfolane [42829-14-1] (3,4).

Bromination of sulfolane by BrCl under uv irradiation gives 2-bromosulfolane [29325-66-4], which reacts further to give *cis*-2,5-dibromosulfolane [30186-52-8] (5). Continued irradiation converts the *cis*-isomer to *trans*-2,5-dibromosulfolane [30186-54-0], which yields first the *trans*-2,4 isomer [30186-53-9] and then the *trans*-3,4 isomer [15091-30-2] upon further irradiation.

Ring Cleavage by Alkali Metals. The sulfolane ring can be cleaved by sodium or potassium metal in xylene at 66–140°C, by sodium amide in liquid ammonia

Table 2. Thermal Stability of Sulfolane in the Presence of Various Substances[a]

Additive	Additive quantity,[b] wt %	Reflux time, h	Sulfolane recovery, wt %	Remarks
none		5	>98	darkened after 30 min
aluminum chloride	40	5	56	blackened with evolution of heat and HCl
potassium carbonate	40	5	>98	darkened after 30 min
sodium acetate	40	5	>98	darkened after 30 min
sodium hydroxide[c]	100	6	>98	darkened after 2 h
sulfur	10	7	25	H_2S evolved, black sludge as residue
sulfuric acid[d]	25	6[e]	89	darkened after 30 min

[a]Ref. 2.
[b]Based on sulfolane at 100%.
[c]25 wt % solution
[d]93 wt % solution
[e]At 140–150°C

at −33°C, and by sodium ethoxide at 240–250°C (6). The reaction products for the alkali–metal ring cleavage are dimeric bis-1,8-octanesulfinate salts under static reaction conditions and butanesulfinate salts under stirred reaction conditions. Ring cleavage by sodium amide in liquid ammonia gives sodium 3-butanesulfinate, and ring cleavage by sodium ethoxide gives sodium butadienesulfinate. The C–S bond can also be cleaved by ultrasonically dispersed potassium in toluene, then methylated with iodomethane (7–9).

Ring Additions Catalyzed by Alkali Metals. The addition of tributyltin chloride and olefins such as styrene, isoprene, or butadiene to sulfolane is catalyzed by alkali metals, including sodium and lithium, and by sodium amide (10–13). The addition of tributyltin chloride to sulfolane in the presence of sodium amide results in the formation of 2,5-bis(tributyltin)sulfolane [41392-14-7]. The addition of styrene to sulfolane in the presence of sodium yields 67% monostyrenated and 17% distyrenated sulfolane. Under similar conditions, isoprene gives 63% mono- and 7% disubstituted product. The reaction of butyllithium and bromoalkyls gives 2-alkyl derivatives of sulfolane (14).

Grignard Reactions. Sulfolane and its alkyl homologues react with ethylmagnesium bromide in ether, benzene, or tetrahydrofuran (15,16). The reaction products are sulfolanyl 2-mono- [82770-58-9] and 2,5-dimagnesium bromides [82770-59-0].

Lewis Acid Complexes. Sulfolane complexes with Lewis acids, such as boron trifluoride or phosphorus pentafluoride (17). For example, at room temperature, sulfolane and boron trifluoride combine in a 1:1 mole ratio with the evolution of heat to give a white, hygroscopic solid which melts at 37°C. The reaction of sulfolane with methyl fluoride and antimony pentafluoride in liquid sulfur dioxide gives crystalline tetrahydro-1-methoxythiophenium-1-oxide hexafluoroantimonate, the first example of an alkoxysulfoxonium salt (18).

Production. Sulfolane is produced domestically by the Phillips Chemical Company (Borger, Texas). Industrially, sulfolane is synthesized by hydrogenating 3-sulfolene [77-79-2] (2,5-dihydrothiophene-1,1-dioxide) (**2**), the reaction product of butadiene and sulfur dioxide:

$$SO_2 + CH_2{=}CH{-}CH{=}CH_2 \rightleftharpoons \underset{(\mathbf{2})}{\underset{\underset{O\;\;O}{\overset{\diagdown\!\!/}{S}}}{\overset{\overset{HC{=}CH}{\overset{|\quad|}{H_2C\quad CH_2}}}{}}} \xrightarrow{H_2} \underset{(\mathbf{1})}{\underset{\underset{O\;\;O}{\overset{\diagdown\!\!/}{S}}}{\overset{\overset{H_2C{-}CH_2}{\overset{|\qquad|}{H_2C\quad CH_2}}}{}}}$$

Commercially, sulfolane is available as a crystalline anhydrous material, and containing 3 wt % deionized water as a freezing point depressant, as Sulfolane-W.

Toxicity. No mortality in laboratory animals was induced in percutaneous doses up to 3.8 g/kg body weight (19,20). Subcutaneous administration of sulfolane gives LD_{50} values for rats, mice, and rabbits of 1.606, 1.360, and 1.900–3.500 g/kg body weight, respectively (21). LD_{50} values for sulfolane by oral administration to laboratory animals were 1.9–5.0 g/kg body weight (19–23). In most cases, the cause of death was believed to result from convulsions, which lead to anoxia.

When administered intraperitoneally, sulfolane is excreted both unchanged and as 3-hydroxysulfolane [13031-76-0] (24). Sulfolane injected intraperitoneally in mice and rats at 200–800 mg/kg at ambient temperatures of 15 and 25°C caused a dose-related inhibition of the metabolic rate and hypoactivity, accompanied by hypothermia 60 min after injection. Despite their hypothermic condition these animals did not select a warm ambient temperature. Because sulfolane toxicity appears to be greater upon increased tissue temperature, the behavior of these animals seeking lower environmental temperature appears to enhance their chance of survival (25–28).

Subcutaneous injection of 100–750 mg/kg sulfolane at an ambient temperature of 10°C also caused a dose-dependent decrease in colonic temperature of rabbits. Metabolic rate remained unchanged during the initial phase of the hypothermia for all dose groups; but peripheral vasodilatation, as indicated by an increase in ear skin temperature, was seen at the higher dose levels (29).

Intracerebroventricular injection of sulfolane in dosages of 300, 1000, and 3000 μg caused the preoptic/anterior hypothalamic area temperature to rise 0.23, 0.47, and 0.56%, respectively. This hyperthermia was considered significant at the 3000-μg dosage (30).

Sulfolane causes minimal and transient eye and skin irritation (19,20). Inhalation of sulfolane vapors in a saturated atmosphere is not considered biologically significant. However, when aerosol dispersions have been used to elevate atmospheric concentration, blood changes and convulsions have been observed in laboratory animals (22,31). Convulsions caused by sulfolane injected intraperitoneally have also been studied (32).

Uses. *Extractive Solvent. Aromatic Hydrocarbons.* Sulfolane is used principally as a solvent for extraction of benzene, toluene, and xylene from mixtures containing aliphatic hydrocarbons (33–37). The sulfolane process was introduced in 1959 by Shell Development Company, and that process is licensed by Universal Oil Products. A sulfolane extraction process is also licensed by the Atlantic Richfield Company. In 1994, worldwide consumption was estimated at ca 6974 t/yr of sulfolane for 137 sulfolane extraction units (see BTX PROCESSES; EXTRACTION, LIQUID–LIQUID; XYLENES AND ETHYLBENZENE).

In general, the sulfolane extraction unit consists of four basic parts: extractor, extractive stripper, extract recovery column, and water–wash tower. The hydrocarbon feed is first contacted with sulfolane in the extractor, where the aromatics and some light nonaromatics dissolve in the sulfolane. The rich solvent then passes to the extractive stripper where the light nonaromatics are stripped. The bottom stream, which consists of sulfolane and aromatic components, and which at this point is essentially free of nonaromatics, enters the recovery column where the aromatics are removed. The sulfolane is returned to the extractor. The nonaromatic raffinate obtained initially from the extractor is contacted with water in the wash tower to remove dissolved sulfolane, which is subsequently recovered in the extract recovery column. Benzene and toluene recoveries in the process are routinely greater than 99%, and xylene recoveries exceed 95%.

Normal and Branched Aliphatic Hydrocarbons. The urea-adduction method for separating normal and branched aliphatic hydrocarbons can be carried out in sulfolane (38,39). The process obviates the necessity of handling and washing the solid urea–normal paraffin adduct formed when a solution of urea in sulfolane is contacted with the hydrocarbon mixture. Overall recovery by this process is typically 85%; normal paraffin purity is 98%.

Fatty Acids and Fatty Acid Esters. Sulfolane exhibits selective solvency for fatty acids and fatty acid esters which depends on the molecular weight and degree of fatty acid unsaturation (40–42). Applications for this process are enriching the unsaturation level in animal and vegetable fatty oils to provide products with better properties for use in paint, synthetic resins, food products, plastics, and soaps.

Wood Delignification. The production of wood pulp (qv) for the paper (qv) industry consists of removing lignin (qv) from wood chips, thus freeing the cellulose fibers. An aqueous solution containing 30–70 wt % sulfolane efficiently extracts the lignin from aspen, Western hemlock, and Southern pine wood chips. Pulp yields are from 50–75% (43,44).

Miscellaneous Extractions. Additional extractive separations using sulfolane involve (*1*) mercaptans and sulfides from sour petroleum (45); (*2*) *t*-butylstyrene from *t*-butylethylbenzene (46); (*3*) mixtures of close boiling chlorosilanes (47); and (*4*) aromatics from kerosene (48–50), naphtha (49,51–53), and aviation turbine fuel (54).

Extractive Distillation Solvent. Extractive distillation is a technique for separating components in narrow boiling range mixtures which are difficult to separate by ordinary fractionation. The process consists of allowing a higher boiling liquid that has a special affinity for one or more of the components in the mixture to flow downward in a distillation column countercurrent to the ascending vapors and thereby to enhance differences in volatility of components of the mixture. Sulfolane is a suitable extractive–distillation solvent for carrying out the separation of close boiling alcohols (55) and chlorosilanes (56,57); mono- and diolefins (58), such as isoprene and butadiene; electrochemical fluorination products; water from organic acids (59–62); ethers (63–68); ketones (69,70); esters (71); cycloalkanes from alkanes (72); and aromatic hydrocarbons (73–84) (see DISTILLATION, AZEOTROPIC AND EXTRACTIVE).

Gas Treating. Another large commercial use for sulfolane is the removal of acidic components, eg, H_2S, CO_2, COS, CS_2, and mercaptans, from sour gas streams. The process, known as the Sulfinol process, was introduced by Shell in 1963 and consists of contacting the gas stream with a mixture of sulfolane, an alkanolamine (usually diisopropanolamine), and water (85–107). The acid gases are absorbed chemically by the amine and absorbed physically by sulfolane, which results in the advantages of high acid gas loading and ease of solvent regeneration.

Other gas-treating processes involving sulfolane are (*1*) hydrogen selenide removal from gasification of coal, shale, or tar sands (qv) (108); (*2*) olefin removal from alkanes (109); (*3*) nitrogen, helium, and argon removal from natural gas (110); (*4*) atmospheric CO_2 removal in nuclear submarines; (*5*) ammonia and H_2S removal from waste streams; (*6*) H_2S, HCl, N_2O, and CO_2 removal from various streams (111–120); and (*7*) H_2S and SO_2 removal from gas mixtures (121–123). The last process differs from the Sulfinol process in that the H_2S and SO_2 are converted directly to high purity, elemental sulfur (see SULFUR REMOVAL AND RECOVERY).

Polymer Solvent. Sulfolane is a solvent for a variety of polymers, including polyacrylonitrile (PAN), poly(vinylidene cyanide), poly(vinyl chloride) (PVC), poly(vinyl fluoride), and polysulfones (124–129). Sulfolane solutions of PAN, poly(vinylidene cyanide), and PVC have been patented for fiber-spinning processes, in which the relatively low solution viscosity, good thermal stability, and comparatively low solvent toxicity of sulfolane are advantageous. Powdered perfluorocarbon copolymers bearing sulfo or carboxy groups have been prepared by precipitation from sulfolane solution with toluene at temperatures below 300°C. Particle sizes of 0.5–100 μm result.

Polymer Plasticizer. Nylon, cellulose, and cellulose esters can be plasticized using sulfolane to improve flexibility and to increase elongation of the polymer (130,131). More importantly, sulfolane is a preferred plasticizer for the synthesis of cellulose hollow fibers, which are used as permeability membranes in reverse osmosis (qv) cells (131–133) (see HOLLOW-FIBER MEMBRANES). In the preparation of the hollow fibers, a molten mixture of sulfolane and cellulose triacetate is extruded through a die to form the hollow fiber. The sulfolane is subsequently extracted from the fiber with water to give a permeable, plasticizer-free, hollow fiber.

Polymerization Solvent. Sulfolane can be used alone or in combination with a cosolvent as a polymerization solvent for polyureas, polysulfones, polysiloxanes, polyether polyols, polybenzimidazoles, polyphenylene ethers, poly(1,4-benzamide) (poly(imino-1,4-phenylenecarbonyl)), silylated poly(amides), poly(arylene ether ketones), polythioamides, and poly(vinylnaphthalene/fumaronitrile) initiated by laser (134–144). Advantages of using sulfolane as a polymerization solvent include increased polymerization rate, ease of polymer purification, better solubilizing characteristics, and improved thermal stability. The increased polymerization rate has been attributed not only to an increase in the reaction temperature because of the higher boiling point of sulfolane, but also to a decrease in the activation energy of polymerization as a result of the contribution from the sulfonic group of the solvent.

Electronic and Electrical Applications. Sulfolane has been tested quite extensively as the solvent in batteries (qv), particularly for lithium batteries. This is because of its high dielectric constant, low volatility, excellent solubilizing characteristics, and aprotic nature. These batteries usually consist of anode, cathode polymeric material, aprotic solvent (sulfolane), and ionizable salt (145–156). Sulfolane has also been patented for use in a wide variety of other electronic and electrical applications, eg, as a coil-insulating component, solvent in electronic display devices, as capacitor impregnants, and as a solvent in electroplating baths (157–161).

Miscellaneous Uses. Textile applications for sulfolane include preparation of concentrated, storage-stable basic dyes; fabric treating prior to dyeing to improve the dye adsorption; and fiber treating to improve the tensile strength, pilling resistance, and drawing properties (162–164) (see TEXTILES). The curing time of polysulfide-based sealants and fluoropolymer rubbers decreases significantly upon the incorporation of small amounts of sulfolane into the formulation (165,166). Sulfolane also exhibits catalytic activity for some reactions, increasing both the conversion and the selectivity. Examples of systems where sulfolane functions catalytically are the synthesis of 1,4-dicyanobutene and substituted ketones (167,168). Microemulsions containing sulfolane together with a cationic surfactant such as cetyltrimethylammonium bromide are useful for the detoxification of pesticides and chemical warfare agents by enabling improved removal or destruction through solubilization, oxidation, or hydrolysis (169). Sulfolane has also been used as a cosurfactant in systems for enhanced petroleum recovery (170).

Sulfones

3-Sulfolene (**2**) is the next most commercially important sulfone after sulfolane. Besides its precursor role in sulfolane manufacture, 3-sulfolene is an intermediate in the synthesis of sulfolanyl ethers, which are used as hydraulic fluid additives (see HYDRAULIC FLUIDS). 3-Sulfolene or its derivatives also have been used in cosmetics (qv) and slimicides. Selected physical properties of 3-sulfolene are listed in Table 3.

Table 3. Physical Properties of 3-Sulfolene

Property	Value
molecular weight	118.154
specific gravity, 70°C	1.314
melting point, °C	65
boiling point	dec
flash point, °C	113

Other sulfones with commercial potential include dimethyl sulfone [67-71-0] (sulfonylbismethane) (**4**), diiodomethyl p-tolylsulfone [20018-09-1] (**5**), 4,4'-dihydroxydiphenyl sulfone [80-09-1] (Bisphenol S) (**6**), bis(p-chlorophenyl) sulfone [80-07-9] (**7**), and 4,4'-bis(p-chlorophenylsulfonyl) biphenyl [22287-56-5] (**8**). Dimethyl sulfone (**4**) has been patented as a solvent for extractive distillation, electroplating baths, inks, and adhesives (76,171–173). Diiodomethyl p-tolylsulfone (**5**) has been patented for use as an antifungal preservative (174–176); and 4,4'-dihydroxydiphenyl sulfone (**6**) has been patented for a variety of applications, eg, as an electroplating solvent, a washfastening agent, and a component in a phenolic resin (177,178). The two chloro-containing sulfones have been patented in the preparation of high temperature engineering plastics (qv), (**7**) and (**8**), (179,180).

(**4**)   (**5**)   (**6**)

(**7**)   (**8**)

BIBLIOGRAPHY

"Sulfolane" in *ECT* 2nd ed., Vol. 19, pp. 250–254, by G. S. Morrow, Shell Chemical Co.; "Sulfolanes and Sulfones" in *ECT* 3rd ed., Vol. 21, pp. 961–968, by M. Lindstrom and R. Williams, Phillips Research Center.

1. *Technical Information on Sulfolane*, Bulletin 524, Phillips Chemical Co., Bartlesville, Okla.
2. T. E. Jordon and F. Kipnis, *Ind. Eng. Chem.* **41**, 2635 (1949).
3. V. I. Dornov and co-workers, *Khim. Seraorg. Soedin. Soderzh. Naftyakh Nefteprod.* **8**, 133 (1968).
4. V. I. Dornov and V. A. Snegotskaya, *Khim. Geterotsikl. Soedin. Sb.* **3**, 5 (1971).
5. V. I. Dornov and co-workers, *Zh. Org. Khim.* **6**, 2029 (1970).
6. V. I. Dornov and co-workers, *Khim. Seraorg. Soedin. Soderzh. Naftyakh Nefteprod.* **8**, 144 (1968).
7. T. S. Chou and co-workers, *Tetrahedron Lett.* **32**, 3551–3554 (1991).
8. T. S. Chou and co-workers, *J. Chin. Chem. Soc.* **38**, 283–287 (1991).
9. T. S. Chou and M. L. Sin, *Tetrahedron Lett.*, **26**, 4495–4498 (1985).
10. Ger. Offen. 2,246,939 (Apr. 5, 1973), D. J. Peterson, J. F. Ward, and R. A. D'Amico (to Procter & Gamble Co.).
11. U.S. Pat. 3,988,144 (Oct. 26, 1976), D. J. Peterson, J. F. Ward, and R. A. D'Amico (to Procter & Gamble Co.).
12. E. M. Asatryan and co-workers, *Arm. Khim. Zh.* **29**, 553 (1976).

13. E. M. Asatryan and co-workers, *Tezisy Dokl. Molodezhnaya Konf. Org. Sint. Fioorg. Khim.* **44** (1976).
14. L. Frazier and D. P. Claypool, *Synth. Commun.* **18**, 583–588 (1988).
15. T. E. Bezmenova and co-workers, *Khim. Seraorg. Soedin. Soderzh. Neftyakh Nefteprod.* **8**, 140 (1968).
16. U.S. Pat. 2,656,362 (Oct. 26, 1953), H. E. Faith (to Allied Laboratories, Inc.).
17. J. J. Jones, *Inorg. Chem.* **5**, 1229 (1966).
18. M. M. Abdel-Malik and co-workers, *Can. J. Chem.* **62**, 69–73 (1984).
19. V. K. H. Brown, L. W. Ferrigan, and D. E. Stevenson, *Br. J. Industr. Med.* **23**, 302 (1966).
20. H. F. Smyth and co-workers, *Am. Ind. Hyg. Assoc. J.* **30**, 470 (1969).
21. M. E. Anderson and co-workers, *Res. Commun. Chem. Pathol. Pharmacol.* **15**, 571 (1976).
22. M. E. Andersen and co-workers, *Toxicol. Appl. Pharmacol.* **40**, 463 (1977).
23. J. J. Roberts and G. P. Warwick, *Biochem. Pharmacol.* **6**, 217 (1961).
24. G. R. Middleton, R. W. Young, and L. J. Jenkins, Jr., *Annual Research Report*, Armed Forces Radiobiology Research Institute, National Technical Information Service, Springfield, Va., 1974, pp. 77–79.
25. C. J. Gordon and co-workers, *Arch. Toxicol.* **56**, 123–127 (1984).
26. P. H. Ruppert and R. S. Dyer, *Toxicol. Lett.* **28**, 111–116 (1985).
27. C. J. Gordon and co-workers, *J. Toxicol. Environ. Health* **16**, 461–468 (1985).
28. C. J. Gordon and co-workers, *Environ. Res.* **40**, 92–97 (1986).
29. F. S. Mohler and C. J. Gordon, *Arch. Toxicol.* **62**, 216–219 (1988).
30. F. S. Mohler and C. J. Gordon, *Neurotoxicology* **10**, 53–62 (1989).
31. Z. K. Filippova, *Khim. Seraorg. Soedin. Soderzh. Neftyakh. Nefteprod.* **8**, 701 (1968).
32. L. J. Burdette and R. S. Dyer, *Neurobehav. Toxicol. Teratol.* **8**, 621 (1986).
33. D. B. Broughton and G. F. Asselin, *7th World Petrol. Congr. Proc.* **4**, 65 (1968).
34. F. S. Beardmore and W. C. G. Kosters, *J. Inst. Pet. London* **49**(469), 1 (1963).
35. M. A. Plummer, *Hydrocarbon Process.* **52**(6), 91 (1973).
36. M. A. Plummer, *Hydrocarbon Process.* **59**(9), 203 (1980).
37. *Aromatics Extraction Process*, publication EM-9964, Atlantic Richfield Co.
38. U.S. Pat. 3,617,499 (Nov. 2, 1971), D. M. Little (to Phillips Petroleum Co.).
39. U.S. Pat. 3,645,889 (Feb. 29, 1972), D. O. Hanson (to Phillips Petroleum Co.).
40. U.S. Pat. 2,360,860 (Oct. 24, 1944), R. C. Morris and E. C. Shokal (to Shell Development Co.).
41. J. Wisniak, *Br. Chem. Eng.* **15**(1), 76 (1970).
42. Eur. Pat. 576,191 (Dec. 29, 1993), J. B. Cloughley and co-workers (to Scotia Holdings PLC, UK).
43. L. P. Clermont, *TAPPI* **53**, 2243 (1970).
44. L. D. Starr and co-workers, *TAPPI Alkaline Pulping Conference Preprints*, 1975, p. 195.
45. B. B. Agrawal and co-workers, *Erdoel Kohle, Erdgas, Petrochem.* **40**, 489–490 (1987).
46. U.S. Pat. 4,543,438 (Sept. 24, 1985), J. D. Henery and co-workers (to El Paso Products Co.).
47. U.S. Pat. 4,402,796 (Sept. 6, 1983), O. W. Marko and S. F. Rentsch (to Dow Corning Corp.).
48. Ind. Pat. 170,747 (May 9, 1992), S. K. Joshi and co-workers (to Council of Scientific and Industrial Research, India).
49. M. K. Khanna and B. S. Rawat, *Res. Ind.* **37**, 67–72 (1992).
50. A. K. Jain and S. J. Chopra in X. Chou, ed., *Proceedings of an International Conference on Petroleum Refining and Petrochemical Processing*, International Academic Publishers, Beijing, China, 1991, pp. 368–374.

51. F. A. Poposka and co-workers, *J. Serb. Chem. Soc.*, **56**, 489–498 (1991).

52. M. S. H. Fandary and co-workers, *Solvent Extr. Ion Exch.* **7**, 677–703 (1989).

53. R. Krishna and co-workers, *Indian J. Technol.* **25**, 602–606 (1987).

54. R. Mehrotra and co-workers, *Fluid Phase Equilib.* **32**, 17–25 (1986).

55. U.S. Pat. 5,160,414 (Nov. 3, 1992), F. M. Lee and co-workers (to Phillips Petroleum Co.).

56. U.S. Pat. 4,411,740 (Oct. 25, 1983), O. L. Flaningam and R. L. Halm (to Dow Corning Corp.).

57. U.S. Pat. 4,402,797 (Sept. 6, 1983), R. L. Halm and S. F. Rentsch (to Dow Corning Corp.).

58. U.S. Pat. 5,151,161 (Sept. 29, 1992), F. M. Lee and R. E. Brown (to Phillips Petroleum Co.).

59. U.S. Pat. 5,167,774 (Dec. 1, 1992), L. Berg.

60. U.S. Pat. 5,154,800 (Oct. 13, 1992), L. Berg.

61. U.S. Pat. 4,735,690 (Apr. 5, 1988), L. Berg and A. I. Yeh.

62. U.S. Pat. 4,642,166 (Feb. 10, 1987), L. Berg and A. I. Yeh.

63. U.S. Pat. 5,228,957 (July 20, 1993), L. Berg.

64. U.S. Pat. 5,160,414 (Nov. 3, 1992), F. M. Lee and co-workers (to Phillips Petroleum Co.).

65. L. Berg and A. I. Yeh, *AIChE J.*, **30**, 871–874 (1984).

66. U.S. Pat. 4,459,179 (July 10, 1984), L. Berg and A. I. Yeh.

67. L. Berg and A. I. Yeh, *Chem. Eng. Commun.*, **29**, 283–289 (1984).

68. U.S. Pat. 4,459,178 (July 10, 1984), L. Berg and A. I. Yeh.

69. U.S. Pat. 4,957,595 (Sept. 18, 1990), L. Berg and G. Bentu.

70. U.S. Pat. 4,948,471 (Aug. 14, 1990), L. Berg and G. Bentu.

71. U.S. Pat. 5,240,567 (Aug. 31, 1993), L. Berg and R. W. Wytcherley.

72. U.S. Pat. 4,944,849 (July 31, 1990), F. M. Lee (to Phillips Petroleum Co.).

73. U.S. Pat. 2,570,205 (Oct. 9, 1951), C. S. Carlson, E. Smith, and P. V. Smith, Jr. (to Standard Oil Development Co.).

74. B. S. Rawat, K. L. Mallik, and I. B. Gulati, *J. Appl. Chem. Biotechnol.* **22**, 1001 (1972).

75. U.S. Pat. 3,024,028 (May 17, 1977), D. M. Haskell (to Phillips Petroleum Co.).

76. U.S. Pat. 4,054,613 (Oct. 18, 1977), D. M. Haskell, E. E. Hopper, and B. L. Munro (to Phillips Petroleum Co.).

77. U.S. Pat. 4,090,923 (May 23, 1978), D. M. Haskell and C. O. Carter (to Phillips Petroleum Co.).

78. U.S. Pat. 3,689,373 (May 5, 1972), W. M. Hutchinson (to Phillips Petroleum Co.).

79. Brit. Pat. 1,333,039 (Oct. 10, 1973), M. F. Kelly and K. D. Uitta (to Universal Oil Products).

80. Can. Pat. 962,212 (Feb. 4, 1975), H. L. Thompson (to Universal Oil Products).

81. Brit. Pat. 1,392,735 (May 26, 1975), H. L. Thompson (to Universal Oil Products).

82. U.S. Pat. 5,310,480 (May 10, 1994), J. A. Vidueira (to UOP Inc.).

83. U.S. Pat. 5,069,757 (Dec. 3, 1991), R. E. Brown (to Phillips Petroleum).

84. F. M. Lee and D. M. Coombs, *Ind. Eng. Chem. Res.*, **26**, 564–573 (1987).

85. C. L. Dunn, *Hydrocarbon Process. Pet. Refiner* **43**, 150 (1964).

86. U.S. Pat. 3,630,666 (Dec. 28, 1971), L. V. Kunkel (to Amoco Production Co.).

87. U.S. Pat. 4,025,322 (May 24, 1977), E. J. Fisch (to Shell Oil Co.).

88. U.S. Pat. 4,100,257 (July 11, 1978), G. Sartori and D. W. Savage (to Exxon Research and Engineering Co.).

89. U.S. Pat. 3,965,244 (June 22, 1976), J. A. Sykes, Jr. (to Shell Oil Co.).

90. F. Murrieta-Guevara and co-workers, *Fluid Phase Equilib.* **95**, 163–174 (1994).

91. F. T. Okimoto, *Proc. Annu. Conv. Gas Process. Assoc.*, **72**, 149–151 (1993).

92. F. Murrieta-Guevara and co-workers, *Fluid Phase Equilib.* **86**, 225–231 (1993).

93. R. J. MacGregor and A. E. Mather, *Can. J. Chem. Eng.* **69**, 1357–1366 (1991).

94. F. J. Littel and co-workers, *J. Chem. Eng. Data* **37**, 49–55 (1992).

95. T. T. Teng and A. E. Mather, *Fas Sep. Purif.* **5**, 29–34 (1991).

96. A. E. Mather and co-workers, *Ind. Eng. Chem. Res.* **30**, 1213–1217 (1991).

97. Eur. Pat. 399,608 (Nov. 28, 1990), V. K. Chopra and S. P. Kaija (to Shell Internationale Research Maatschappij BV, the Netherlands).

98. A. E. Mather and co-workers, *Fluid Phase Equilib.* **56**, 313–324 (1990).

99. K. Yogish, *Can. J. Chem. Eng.* **68**, 511–512 (1990).

100. F. Murrieta-Guevara and co-workers, *Fluid Phase Equilib.* **53**, 1–6 (1989).

101. Brit. Pat. 2,202,522 (Sept. 28, 1988), J. G. Christy and co-workers (to Shell Internationale Research Maatschappij BV, the Netherlands).

102. A. E. Mather and B. E. Roberts, *Chem. Eng. Commun.* **72**, 201–211 (1988).

103. U.S. Pat. 4,749,555 (June 7, 1988), W. V. Bush (to Shell Oil Co.).

104. R. J. Demyanovich and S. Lynn, *Ind. Eng. Chem. Res.* **26**, 548–555 (1987).

105. Brit. Pat. 2,188,455 (Nov. 2, 1983), E. J. Fisch (to Shell Internationale Research Maatschappij BV, the Netherlands).

106. Brit. Pat. 2,072,525 (Oct. 7, 1981), R. Cornelisse (to Shell Internationale Research Maatschappij BV, the Netherlands).

107. Eur. Pat. 13049 (July 9, 1980), E. J. Van De Kraats and R. C. Darton (to Shell Internationale Research Maatschappij BV, the Netherlands).

108. U.S. Pat. 4,792,405 (Dec. 20, 1988), D. C. Baker (to Shell Oil Co.).

109. WO Pat. 8,702,031 (Apr. 9, 1987), Y. R. Mehra (to El Paso Hydrocarbons Co.).

110. WO Pat. 8,700,518 (Jan. 29, 1987), Y. R. Mehra (to El Paso Hydrocarbons Co.).

111. A. E. Mather and Yi-Gui Li, *Fluid Phase Equilib.* **96**, 119–142 (1994).

112. P. G. T. Fogg, *Solubility Data Ser.*, **50**, 358–383 (1992).

113. A. E. Mather and co-workers, *J. Chem. Technol. Biotechnol.* **51**, 197–208 (1991).

114. K. Yogish, *J. Chem. Eng. Jpn.*, **24**, 135–137 (1991).

115. A. E. Mather and co-workers, *Fluid Phase Equilib.* **56**, 313–324 (1990).

116. P. G. T. Foog, *Solubility Data Ser.* **42**, 330–341 (1989).

117. F. Murrieta-Guevara and co-workers, *Fluid Phase Equilib.* **44**, 105–115 (1988).

118. P. G. T. Foog, *Solubility Data Ser.* **32**, 137–326 (1988).

119. D. M. Kassim and co-workers, *Fluid Phase Equilib.* **41**, 287–294 (1988).

120. A. E. Mather and B. E. Roberts, *Can. J. Chem. Eng.* **66**, 519–520 (1988).

121. P. R. Gustafson and R. R. Miller, *Naval Research Laboratories Report 6926*, July 1969.

122. U.S. Pat. 3,551,102 (Dec. 29, 1970), G. R. Hettick and D. M. Little (to Phillips Petroleum Co.).

123. Ger. Pat. 2,108,284 (Aug. 31, 1972), G. Schulze (to Badische Anilin-und-Soda-Fabrik AG).

124. U.S. Pat. 2,706,674 (Apr. 19, 1955), G. M. Rothrock (to E. I. du Pont de Nemours & Co., Inc.).

125. U.S. Pat. 2,548,169 (Apr. 10, 1951), F. F. Miller (to B. F. Goodrich Co.).

126. U.S. Pat. 2,617,777 (Nov. 11, 1952), E. Heisenberg and J. Kleine (to Vereinigte Glanzstoff-Fabriken).

127. U.S. Pat. 2,953,818 (Sept. 27, 1960), L. R. Barton (to E. I. du Pont de Nemours & Co., Inc.).

128. U.S. Pat. 3,474,030 (Oct. 21, 1969), D. M. Little (to Phillips Petroleum Co.).

129. U.S. Pat. 4,540,716 (Sept. 10, 1985), M. J. Covitch and G. G. Sweetapple (to Diamond Shamrock Chemicals Co.).

130. U.S. Pat. 3,361,697 (Jan. 2, 1968), W. E. Garrison and T. J. Hyde (to E. I. du Pont de Nemours & Co., Inc.).

131. U.S. Pat. 2,471,272 (May 24, 1949), G. W. Hooker and N. R. Peterson (to The Dow Chemical Co.).

132. U.S. Pat. 3,619,459 (Nov. 9, 1971), P. G. Schrader (to The Dow Chemical Co.).

133. T. E. Davis and G. W. Skiens, *Polym. Prepr. Am. Chem. Soc. Div. Polym Chem.* **12**, 378 (1971).

134. U.S. Pat. 3,476,709 (Nov. 4, 1969), F. B. Jones (to Phillips Petroleum Co.).

135. U.S. Pat. 2,703,793 (Mar. 8, 1955), M. A. Naylor, Jr. (to E. I. du Pont de Nemours & Co., Inc.).

136. U.S. Pat. 3,175,994 (Mar. 30, 1965), A. Katchman and G. D. Cooper (to General Electric Co.).

137. Jpn. Pat. 70 06,533 (Mar. 5, 1970), M. Ikeda and co-workers (to Takeda Chemical).

138. U.S. Pat. 3,784,517 (Jan. 8, 1974), F. L. Hedberg and C. S. Marvel (to U.S. Dept. of the Air Force).

139. Jpn. Pat. 74 02,359 (Jan. 19, 1974), S. Izawa and K. Mizushiro (to Asahi Dow, Ltd.).

140. U.S. Pat. 3,951,914 (Apr. 20, 1976), S. L. Kwolek (to E. I. du Pont de Nemours & Co., Inc.).

141. M. Kakimoto, Y. Oishi, and Y. Imai, *Makromol. Chem., Rapid Commun.* **6**, 557–562 (1985).

142. WO Pat 8,403,891 (Oct. 11, 1984), V. Janson and H. C. Gors (to Raychem. Corp.).

143. Y. Saegusa, S. Iida, S. Nakamura, N. Chau, and Y. Iwakura, *J. Polym. Sci., Polym. Chem. Ed.* **22**, 1017–1023 (1984).

144. M. A. Williamson, J. D. Smith, P. M. Castle, and R. N. Kauffman, *J. Polym. Sci., Polym. Chem. Ed.* **20**, 1875–1884 (1982).

145. U.S. Pat. 5,252,413 (Oct. 12, 1993), M. Alamgir and K. M. Abraham (to EIC Laboratories, Inc.).

146. EP Pat. 529,802 (Mar. 3, 1993), A. Webber (to Eveready Battery Co.).

147. EP Pat. 346,675 (Jan. 10, 1990), N. Furukawa, S. Yoshimura, and M. Takahashi (to Sanyo Electric Co. Ltd., Japan).

148. S. J. Visco, M. Liu, and L. C. De Jonghe, *J. Electrochem. Soc.* **137**, 1191–1192 (1990).

149. EP Pat. 331,342 (Sept. 6, 1989), P. Cheshire and J. E. Przeworski (to Imperial Chemical Industries PLC, U.K.).

150. M. Maxfield, T. R. Jow, M. G. Sewchok, and L. W. Shacklette, *J. Power Sources*, **26**, 93–102 (1989).

151. V. R. Koch, L. A. Dominey, J. L. Goldman, and M. E. Langmuir, *J. Power Sources* **20**, 287–291 (1987).

152. U.S. Pat. 4,670,363 (June 2, 1987), T. A. Whitney and D. L. Foster (to Duracell, Inc.).

153. Y. Matsuda, M. Morita, K. Yamada, and K. Hirai, *J. Electrochem. Soc.* **132**, 2538–2543 (1985).

154. J. S. Foos, L. S. Rembetsy, and S. B. Brummer, Report, LBL-18414, Order No. DE85004345, National Technical Information Service (NTIS), 62 pp. 1984.

155. S. P. S. Yen, B. Carter, D. Shen, and R. Somoano, *Proc. Power Sources Symp.* **30**, 71–73 (1982).

156. G. E. Blomgren, *Proc. Symp. Power Sources Biomed. Implantable Appl. Ambient Temp. Lithium Batt.* **80-4**, 368–377 (1980).

157. Ger. Pat. 2,739,571 (Mar. 23, 1978), L. G. Versberg (to ASEA AB).

158. U.S. Pat. 3,891,458 (June 24, 1975), M. Wisenberg (to Electrochimica Corp.).

159. U.S. Pat. 4,110,015 (Aug. 29, 1978), T. B. Reddy (to American Cyanamid Co.).

160. R. Tobazeon and E. Gartner, *IEE Conf. Publ. London* **129**, 225 (1975).

161. Ger. Pat. 2,207,703 (Aug. 31, 1972), F. Huba and J. E. Bride (to Diamond Shamrock Corp.).

162. Brit. Pat. 1,349,511 (April 3, 1974), J. F. Dawson and J. Schofield (to Yorkshire Chemicals, Ltd.).

163. Jpn. Pat. 77 25,173 (Feb. 24, 1977), M. Ono and co-workers (to Mitsubishi Rayon Co., Ltd.).

164. Brit. Pat. 1,263,082 (Feb. 9, 1972), R. G. Roberts (to Courtaulds, Ltd.).
165. U.S. Pat. Off. T 974,003 (Sept. 1978), E. G. Miller, Def. Pub.
166. Ger. Pat. 2,449,095 (April 17, 1975), R. E. Kolb (to 3M Co.).
167. Ger. Pat. 2,256,039 (May 17, 1973), O. T. Onsager (to Halcon International Inc.).
168. Jpn. Pat. 74 25,925 (July 4, 1974), T. Kawaguchi and co-workers (to Kurary Co., Ltd.).
169. R. P. Seiders, *ACS Symp. Ser.* **272**, 265–273 (1985).
170. U.S. Pat. 4,485,872 (Dec. 4, 1984), P. R. Stapp (to Phillips Petroleum Co.).
171. U.S. Pat. 4,046,647 (Sept. 6, 1977), E. P. Harbulak (to M and T Chemicals, Inc.).
172. Jpn. Pat. 80 50,072 (Apr. 11, 1980), K. Hirano (to Pilot Ink Co., Ltd.).
173. Jpn. Pat. 80 66,980 (May 20, 1980), A. Yamada and K. Kimma (to Toagosei Chemical Industry Co., Ltd.).
174. U.S. Pat. 4,078,888 (Mar. 4, 1978), F. W. Arbir and F. C. Becker (to Abbott Laboratories).
175. *Fed. Reg.* **44**, 52189 (Sept. 1979).
176. Jpn. Pat. 78 73,434 (June 29, 1978), K. Kariyone and co-workers (to Fujisawa Pharmaceutical Co., Ltd.).
177. Brit. Pat. Appl. 2,001,679 (Feb. 7, 1979), S. A. Lipowski (to Diamond Shamrock Corp.).
178. Jpn. Pat. 79 28,357 (Mar. 2, 1979), J. Hiroshima and I. Kai (to Asahi Organic Chemicals Industry Co., Ltd.).
179. U.S. Pat. 4,016,145 (Apr. 5, 1977), R. W. Campbell (to Phillips Petroleum Co.).
180. U.S. Pat. 5,079,079 (Jan. 7, 1992), M. L. Stone, R. L. Bobsein, and H. F. Efner (to Phillips Petroleum Co.).

General Reference

J. A. Reddick and W. E. Bunger, *Organic Solvents*, 3rd ed., Vol. 2, Wiley-Interscience, New York, 1970, pp. 467–468.

EARL CLARK
Phillips Research Center

SULFONAMIDES.   See ANTIBACTERIAL AGENTS, SYNTHETIC–SULFON-
AMIDES.

SULFONATION AND SULFATION

Sulfonation and sulfation, chemical methods for introducing the SO_3 group into organic entities, are related and usually treated jointly. It is estimated that these methods were utilized in approximately 1000 operational plants throughout the world, including about 170 U.S. plants as of the mid-1990s (1,2).

In sulfonation, an SO_3 group is introduced into an organic molecule to give a product having a sulfonate, CSO_3, moiety. The compound may be a sulfonic acid, a salt, or a sulfonyl halide requiring subsequent alkaline hydrolysis. Aromatic hydrocarbons are generally directly sulfonated using sulfur trioxide, oleum, or sulfuric acid. Sulfonation of unsaturated hydrocarbons may utilize sulfur trioxide, metal sulfites, or bisulfites. The latter two reagents produce the corresponding hydrocarbon metal sulfonate salts in processes referred to as sulfitation and bisulfitation, respectively. Organic halides react with aqueous sodium sulfite to produce the corresponding organic sodium sulfonate. This is the Strecker reaction. In instances where the sulfur atom at a lower valance is attached to a carbon atom, the sulfonation process entails oxidation. Thus the reaction of a paraffin hydrocarbon with sulfur dioxide and oxygen is referred to as sulfoxidation; the reaction of sulfur dioxide and chlorine is called chlorosulfonation. The sulfonate group may also be introduced into an organic molecule by indirect methods through a primary reaction, eg, esterification, with another organic molecule already having an attached sulfonate group.

Sulfation is defined as any process of introducing an SO_3 group into an organic compound to produce the characteristic $C-OSO_3$ configuration. Typically, sulfation of alcohols utilizes chlorosulfuric acid or sulfur trioxide reagents. Unlike the sulfonates, which show remarkable stability even after prolonged heat, sulfated products are unstable toward acid hydrolysis. Hence, alcohol sulfuric esters are immediately neutralized after sulfation in order to preserve a high sulfation yield.

In sulfamation, also termed N-sulfonation, compounds of the general structure R_2NSO_3H are formed as well as their corresponding salts, acid halides, and esters. The reagents are sulfamic acid (amido–sulfuric acid), SO_3–pyridine complex, SO_3–tertiary amine complexes, aliphatic amine–SO_3 adducts, and chlorine isocyanate–SO_3 complexes (3).

Uses for Derived Products and Sulfonation Technology

Sulfonation and sulfation processes are utilized in the production of water-soluble anionic surfactants (qv) as principal ingredients in formulated light-duty and heavy-duty detergents, liquid hand cleansers, general household and personal care products (see COSMETICS), and dental care products (see DENTIFRICES). Generally household and personal care products must have light color, little unreacted material (free oil), low inorganic salt, and negligible odor. Bleach activators represent a more recent anionic development. Industrial sulfonated and sulfated product applications include emulsifiers for emulsion polymerizations, pesticide/herbicide emulsifiers, concrete additives, demulsifiers, textile wetting agents, dry-cleaning detergents, leather (qv) tanning agents, metal cleaners, corrosion inhibitors, and oil-production chemicals. Other commercially significant product sectors include lube additives, pesticide chemicals, medicinals, sweeteners (qv), cyclic intermediates, dye and pigment products, and ion-exchange (qv) resins.

Sulfonation and sulfation processes are important tools for organic synthesis of specific molecules and positional isomers. For example, the introduction of a sulfonic acid group into a specific position in an aromatic nucleus followed by substitution of a hydroxyl group by alkali fusion represents a basic industrial process. Other aromatic ring constituents, their isomeric positions, and steric factors can dramatically increase or decrease rates and feasibility of sulfonation. For example, a methyl group on an aromatic ring enhances electrophilicity thus increasing sulfonatability in the ortho–para positions. Conversely, the presence of a nitro group decreases aromatic electrophilicity, dramatically decreasing the sulfonating rate which occurs at the meta position.

Application chemists are most interested in physical and functional properties contributed by the sulfonate moiety, such as solubility, emulsification, wetting, foaming, and detersive properties. Products can be designed to meet various criteria including water solubility, water dispersibility, and oil solubility. The polar SO_3 moiety contributes detersive properties to lube oil sulfonates and dry-cleaning sulfonates.

Process Selection and Options.   Because of the diversity of feedstocks, no one process fits all needs. An acceptable sulfonation/sulfation process requires (1) the proper reagent for the chemistry involved and the ability to obtain high product yields; (2) consistency with environmental regulations such that minimal and disposable by-products are formed; (3) an adequate cooling system to control the reaction and to remove significant heat of reaction; (4) intimate mixing or agitation of often highly viscous reactants to provide adequate contact time; (5) products of satisfactory yields and marketable quality; and (6) acceptable economics.

Viscosity constraints may play a significant role not only dictating agitation/mixing requirements but also seriously affecting heat-exchange efficiency. For example, in the case of SO_3 sulfonation of various surfactant feedstocks, viscosity generally undergoes a 15- to 350-fold increase on reaction, significantly reducing heat-transfer coefficients. In some instances where the feedstock has a high melt point or either the feedstock or its sulfonic acid exhibits high viscosities under nominal sulfonation reaction temperatures, the use of higher temperature cooling media, such as hot water or steam cooling, has been utilized (4). The reactants and reaction products may be miscible or immiscible. In some instances, solvents may be required.

Industrial Changes Affecting Sulfonation/Sulfation Operations.   Changes since the 1980s impacting sulfonation/sulfation operations on a worldwide basis include maturation of the sulfonation industry and its technology, development of a highly competitive and reactive market environment, and environmental and safety regulations affecting the types of manufacturing operations, plant locations, by-product generation and disposal, plant emissions, and other discharge streams. SO_3 has become the preferred sulfonation reagent and a significant trend has developed toward utilizing *in situ* sulfur burning and gaseous SO_3-generating equipment integrated into continuous sulfonation plants. This provides lower cost SO_3, eliminating reliance on liquid SO_3 sources and the need for plant-site liquid SO_3 storage and handling facilities. Continuous falling film SO_3 sulfonation has become the method of choice for sulfonating liquid flowable feedstocks where sizeable product demand warrants continuous production. Raw material quality has improved, and there has been a greater emphasis placed on consistently producing high quality products. Computer-aided process control (qv) together with the use of automated analytical instruments has led to monitoring of raw materials and products, analysis of competitive products, and better understanding of the chemistry of sulfonation processing.

Clean air restrictions on air pollution (qv) from the spray drying of traditional solid detergent products and the move to provide consumer friendly (smaller) product packaging, have resulted in the development and marketing of ultradetergent concentrates in the form of concentrated liquids, flowable pastes, and agglomerated, compacted, and extruded solid products. Hence, the use of direct neutralization procedures without separation of spent acid from oleum sulfonation of detergent alkylate producing substantial inactive sodium sulfate filler, highly desirable for spray dried products, is being surplanted by high active SO_3-derived sulfonates having low levels of sodium sulfate. Consumer interest in detergent products derived from natural, renewable, oleochemical sources has increased the use of coconut and palm oil-derived fatty alcohol sulfates, and sulfo fatty methyl ester products. Moreover, the U.S. Department of Energy is promoting the development and use of more energy-efficient clothes washing machines which use colder ($\leq 10°C$, similar to European systems) water. This last change may increase the use of linear alkylbenzene sulfonate (LAS) and other lower foaming detergents (5). Books and reviews concerning sulfonation technology are available (1,6–13). Reference 1 is germane to the detergent industry.

Economic Aspects

It is estimated that in the United States during 1991, ca 1.92×10^6 metric tons of sulfonated, sulfated, and sulfamated products were manufactured. This represented a market value of approximately 3.4×10^9 and constituted ca 4.5% of the annual value of all synthetic organic chemical production (Table 1) (2). By considering the additional value of reagents and feedstocks, the total estimated value of raw materials and derived sulfonated products amounted to ca 5.0×10^9 or 6.6% of the value of all synthetic organic chemical production for 1991. Worldwide annual production of sulfonated and sulfated products was estimated to exceed 5.0×10^6 metric tons in the early 1990s (see Table 1) (1). Anionic surfactants comprised 79.2% of the total estimated volume of sulfonated products

Table 1. 1991 U.S. Production Volume and Value of Sulfonated, Sulfated, and Sulfamated Products[a,b]

Product class	Value, $/kg	Volume		Value		Change from 1982–1991, %	
		$t \times 10^3$	%	$\$ \times 10^3$	%	Volume	Value
surfactants	1.10	1516.8	79.2	1670.0	49.2	46.7[c]	6.1[c]
dyes	7.13	59.0	3.1	394.0	11.6	6.7	16.2
pigments	16.32	22.8	1.2	386.0	11.4	48.7	−5.1
medicinals	17.86	14.9	0.8	386.0	11.4	65.2	7.5
pesticides	9.03	34.4	1.7	354.1	10.4	−30.4	50.0
polymers/resins	1.14	3.1	0.2	31.7	0.9	53.3	4.7
cyclic intermediates	0.66	265.0	13.8	174.9	5.1	29.6	0.0
Total	*1.59[d]*	*1916.0*	*100.0*	*3396.9*	*100.0*	*219.8*	*79.4*

[a]Ref. 2.
[b]Products on 100% active basis.
[c]All classes of surfactants.
[d]Weighted average.

in the United States in 1991, representing 49.2% on a total products value basis. Whereas dyes and pigments together represent only 4.3% on a volume basis, because of their relatively high unit values, they jointly represent 23% on a total sulfonated products value basis. Similarly, although the production volumes of medicinals and pesticides (qv) are relatively low, these chemicals contribute significantly on a product value basis. Over 800 U.S. manufacturers are involved in production of the product classes shown, only a fraction of which produce sulfonated products.

Volume and product value for sulfonated product sectors between 1982 and 1991 increased substantially for most product groups. Pesticides (qv), however, experienced a volume decline, although the value increased 50%. Dyes had a nominal volume growth of only 6.7%.

Table 2 provides product volume, value, and growth data for U.S. sulfonated and sulfated surfactant products by classes and groups between 1982 and 1991. Whereas total anionic surfactant volume growth increased 38.9%, its composite product value grew a mere 7.8%, again reflecting strongly competitive markets. During this 10-year period sulfonated products grew at a 29.8% rate, whereas the sulfated products group grew at a 70.6% pace. In spite of significant product growth, sulfated products value decreased 36.4%. The volume contribution of sulfated products to total anionics over the 10-year period increased from 27.4 to 38.6%. In 1991, the three largest anionic products groups on a volume basis were alkylbenzene sulfonates (26.3%), lignosulfonates (23.1%), and ethoxylated alcohol sulfates (16.9%), representing a total of 66.3% by volume for all anionic surfactants. A relatively high proportion of anionic feedstocks are synthetically derived from petrochemicals, although there has been an increased production of fatty alcohols derived from natural replenishable oleochemical sources. Sulfonated and sulfated fatty acid, ester, amide, and animal or vegetable oil-derived products constitute only a few percent of total anionic products.

Reagents

Reagents for direct sulfonation and sulfation reactions are listed in Table 3, arranged according to perceived relative reactivity. The data includes 1994 U.S. costs, number of manufacturers, general usage, advantages, disadvantages, and applications. Since Fremy first batch-sulfonated olive oil with sulfuric acid in 1831 (14), sulfonation and sulfation reactions had mainly been conducted using sulfuric acid or oleum reagents. Each requires the use of several moles of reagent per mole of feedstock. Chlorosulfuric acid (qv) and SO_3 react stoichiometrically, however, and hence eventually gained commercial acceptance.

By 1987, sulfur trioxide reagent use in the United States exceeded that of oleum for sulfonation. Sulfur trioxide source is divided between liquid SO_3 and *in situ* sulfur burning. The latter is integrated into sulfonation production facilities.

Liquid SO_3 is commercially available as both unstabilized and stabilized liquids. Unstabilized liquid SO_3 can be utilized without problem as long as moisture is excluded, and it is maintained at ca 27–32°C. Stabilized liquid SO_3 has an advantage in that should the liquid freeze (16.8°C), in the absence of moisture pickup, the SO_3 remains in the gamma-isomer form and is readily

Table 2. Volume, Value, and Growth of U.S. Sulfonated and Sulfated Surfactant Products by Class[a]

Product class	1982, t	1991, t	1991 Product volume		Value, $/kg	1991 Product value		
			% Total	10-yr growth		$ × 10³	% Total	10-yr growth, %
sulfonated products, total	*848,499*	*1,101,132*	*72.6*	*29.8*	*0.93[b]*	*1,025,953*	*61.4*	*9.4*
alkylbenzene sulfonates, total	*260,069*	*398,532*	*26.3*	*53.2*	*1.30*	*516,896*	*30.9*	*26.2*
benzene, toluene, cumene, xylene sulfonates	49,120	70,243	4.6	43.0	0.81	56,897	3.4	-6.4
lignosulfonates	320,914	350,011	23.1	9.4	0.16	62,062	3.7	0
naphthalene sulfonates[c]	47,887	55,356	3.7	15.6	1.24	68,655	4.1	9.7
sulfosuccinic acid ester derivatives	8,440	14,997	1.0	77.7	2.79	41,850	2.5	44.6
sulfosuccinamic acid derivatives	1,072	1,653	0.1	53.9	1.24	2,046	0.1	-36.1
taurine derivatives	1,014	5,014	0.3	393.0	1.62	8,100	0.5	-49.2
oil-soluble sulfonates	89,585	78,087	5.1	-12.8	1.68	131,210	7.8	20.9
mixed linear olefin sulfonates	13,325	13,855	0.9	4.0	1.63	22,584	1.4	6.1
other sulfonic acids, salts	70,398	113,385	7.5	80.7	1.02	115,653	6.9	-15.7
sulfuric acid ester salts, total	*243,626*	*415,629*	*27.4*	*70.6*	*1.55[b]*	*644,275*	*38.6*	*-36.4*
sulfated alcohols, salts	118,936	142,279	9.4	19.6	1.93	274,598	16.4	-14.2
sulfated alcohol ethoxylates, salts	105,485	256,650	16.9	48.5	1.36	349,820	20.9	-11.7
sulfated natural fats, oils, salts	14,251	12,501	0.8	-12.3	1.01	12,625	0.7	-16.1
sulfated fatty acids, esters, amides, salts	4,954	4,199	0.3	-15.2	1.71	7,182	0.4	39.0
total sulfonated and sulfated products	*1,092,125*	*1,516,761*	*100.0*	*38.9*	*1.10[b]*	*1,670,178*	*100.0*	*7.8*
sulfonates, % total	77.7	72.6	72.6				61.4	9.4
sulfates, % total	22.3	27.4	27.4				38.6	-36.4

[a]Estimates based on U.S. International Trade Commission Reports between 1982 and 1991.
[b]Value is average.
[c]Includes naphthalene sulfonates and naphthalene sulfonate formaldehyde condensates.

Table 3. Reagents for Direct Sulfonation and Sulfation Reactions[a]

Reagent	Formula	Physical form	Cost, $/kg	Cost for SO_3, $/kg	U.S. mfg	U.S. plants
sulfur trioxide						
liquid	SO_3	liquid	0.20	0.20	3	4
gas	SO_3	gas, 3–8% SO_3	0.20	0.20	3	4
sulfur burning	SO_3	gas *in situ*, 3–8% SO_3	0.07	0.07	molten sulfur, 84	molten sulfur, 213
chlorosulfuric acid	$ClSO_3H$	liquid	0.39	0.57	2	2
oleum	$H_2SO_4 \cdot SO_3$	liquid	0.11	0.12	ca 70	ca 120
sulfuric acid	H_2SO_4[c]	liquid	0.08	0.10	ca 70	ca 120
sodium bisulfite	$NaHSO_3$	solid, 38% liquid	0.63	0.82	8	8
sodium sulfite	Na_2SO_3	solid, 38% liquid	0.64	1.00	6	6
sodium bisulfite, hydroperoxide catalyst	$NaHSO_3$, O_2	solid, 38% liquid	0.63	0.82	8	8
sulfamic acid	H_2NSO_3H	solid	0.95	1.15	3	3
sulfuryl chloride	SO_2Cl_2	liquid	0.90		1	1
sulfur dioxide and chlorine	SO_2, Cl_2	gases	0.26, 0.27		7, 24	11, 53
sulfur dioxide and oxygen	SO_2, O_2	gases	0.26, 0.03		7	11

[a] In order of descending reactivity.
[b] L = limited, Lo = low, M = moderate, S = substantial, VL = very limited, and WU = widely used.
[c] 93–100%.

remeltable (Table 4). Freezing of unstabilized liquid SO_3 results in higher melting beta- and gamma-forms and the somewhat dangerous pressure release on melting. Commercial stabilizers include B_2O_3 and B_2O_3 esters, borontrifluoride, CCl_4, methane sulfonic acid, methane sulfonyl chloride, and phosphorous oxychloride (1). Gaseous SO_3 can also be obtained by stripping 70% oleum (70%SO_3:30%H_2SO_4) or by utilizing SO_3 converter gas (6–8% SO_3) from H_2SO_4

Table 3. (Continued)

1994 usage[b]	Advantages	Disadvantages	Applications
VL	low cost, concentrated reagent	extremely reactive; charring	very few
WU	low cost, stoichiometric reactions; preferred reagent	requires significant dry diluent gas; mole ratio sensitive; liquid storage	most every sulfonation and sulfation reaction
WU	lowest cost SO_3 produced *in situ*; preferred reagent	catalyst requires startup time; higher investment cost	most every sulfonation and sulfation reaction
L	stoichiometric reactions	expensive; produces HCl gas, disposal problem	alcohol sulfation, dyes, etc
WU	low cost	reactions not stoichiometric; 3–4 mol generally required	dyes, alkylated aromatic sulfonation; continuous sulfation of alcohols
M	low cost, easily handable	reactions not stoichiometric; generally requires 3–4 mol	hydrotrope sulfonation of aromatics using azeotropic water removal, etc
S	simple processing	higher cost, except for sulfur burning	sulfosuccinates, lignin, olefins, Streker reaction
Lo	simple processing	higher cost	Streker reaction, etc
VL	sulfonation of olefins	requires hydroperoxide catalyst; costly	sulfonation of olefinic hydrocarbons producing primary paraffin sulfonation
VL	stoichiometric reaction, mild, simple	high cost; limited to NH_4 salt; heating to ca 150°C	small specialties, sulfations
VL	few	expensive, usually required catalyst	chlorosulfonation reactions; mostly research
VL	few, relatively inexpensive	not generally stoichiometric; need catalyst	chlorosulfonation of paraffins, produces HCl
VL	few, inexpensive	not stoichiometric; requires catalyst	sulfoxidation of paraffins

production. As of 1996, U.S. suppliers of liquid SO_3 were DuPont, Rhône-Poulenc, and Pressure Vessels Service.

Sulfur trioxide is an extremely strong electrophile that rapidly seeks to enter into transient or permanent relationships or reactions with organics containing electron donor elements, such as oxygen, nitrogen, halogen, and phosphorus. In some instances, SO_3 may first form a transient intermediate adduct at some moderate temperature, which at some higher temperature becomes unstable, liberating SO_3. This subsequently may react to produce a stable sulfonated product, often accompanied by a difficult to control strong or violent exotherm. The use

Table 4. Polymers of Sulfur Trioxide

Parameter	γ	β	α
probable structure[a]	$3\,SO_3 \rightleftharpoons O_2S\!\begin{smallmatrix}O-SO_2\\ \\O\\ \\O-SO_2\end{smallmatrix}$	$-OSOSOSOS-$ (chain with $\parallel O$ above and below each S)	similar to β-form, chains joined in layered structure
physical form	liquid or vitreous	silky fibers	fibrous needles
equilibrium melting point, °C	16.8[b]	32.5	62.3
vapor pressure, kPa[c]			
at −3.9°C	28.3	24.1	3.4
23.9°C	190.3	166.2	62.0
51.7°C	908.0	908.0	699.1
79.4°C	3280.6	3280.6	3280.6

[a]As suggested by electron diffraction patterns, ir, and Raman spectra.
[b]Bp = 44.7°C.
[c]To convert kPa to psi, multiply by 0.145.

of excess SO_3 sulfonation reagent leads to undesirable side reactions and darker colored products.

The extreme and violent reactivity of liquid SO_3 can be moderated by using (1) liquid solvents such as halogenated hydrocarbons, eg, methylene chloride or ethylene dichloride, or liquid SO_2, (2) H_2SO_4 acid heel, ie, *in situ* oleum, gaseous SO_3 sulfonation system, (3) vaporization of liquid SO_3, and (4) vaporization of liquid SO_3 and dilution with dry gases such as air, N_2, or SO_2. Stabilizers in liquid SO_3 can serve as catalysts which can accelerate SO_3 reactions, particularly with the halogenated solvents. A mixture of liquid-stabilized SO_3 and ethylene dichloride reportedly exploded violently after being left standing at room temperature (15). Such reactions can be substantially avoided by distilling SO_3 from its stabilizer before mixing with such solvents, and using such mixtures immediately thereafter. Liquid sulfur trioxide secured from chemical supply houses undoubtedly contains stabilizers.

Sulfur trioxide reactivity can also be moderated through the use of SO_3 adducts. The reactivity of such complexes is inversely proportional to their stability, and consequently they can be selected for a wide variety of conditions. Whereas moderating SO_3 reactivity by adducting agents is generally beneficial, the agents add cost and may contribute to odor and possible toxicity problems in derived products. Cellulosic material has been sulfated with SO_3–trimethylamine adduct in aqueous media at 0 to 5°C (16). Sulfur trioxide–triethyl phosphate has been used to sulfonate alkenes to the corresponding alkene sulfonate (17). Sulfur trioxide–pyridine adduct sulfates oleyl alcohol with no attack of the double bond (18).

Table 5 summarizes frequently used sulfuric acid, oleum, and liquid SO_3 sulfonating agents and their properties. Oleum, chlorosulfuric acid, and liquid SO_3 sulfonation reagents are all classified as hazardous and toxic chemicals, mandating special handling, storage, and usage procedures. These last are readily available from respective suppliers. These reagents react violently with water,

Table 5. Composition of Sulfuric Acid, Oleum, and Liquid SO$_3$ Sulfonating Reagents

No.	Designation	Free SO$_3$, %	Total SO$_3$, %	H$_2$SO$_4$, %	Equivalence to 100% H$_2$SO$_4$, %	Sp gr
1	66° Baumé[a]		76.08	93.2	93.2	1.8354
2	H$_2$SO$_4$, conc		78.37	96.0	96.0	1.8427
3	H$_2$SO$_4$, conc		80.0	98.0	98.0	1.8437
4	100% H$_2$SO$_4$		81.63	100.0	100.0	1.8391
5	100.6% H$_2$SO$_4$ oleum	3	82.18	97.0	100.6	1.855
6	20%	20	85.30	80.0	104.5	1.915
7	70%[b]	70	94.49	30.0	115.75	1.982
8	liquid SO$_3$[c]	99.8	99.80		122.5	1.9224

[a]Baumé concentration is based on the hydrometric method; °Bé = 145 − 145/sp gr.
[b]70% Oleum is used to provide gaseous SO$_3$ via stripping.
[c]Stabilized liquid SO$_3$ contains 0.2% stabilizer.

and in the event of spillage a toxic sulfuric acid fog or mist results from contact with atmospheric moisture.

A comparison of differences in the use of H$_2$SO$_4$–oleum vs gaseous SO$_3$ sulfonating agents is presented in Table 6. A listing of frequently used indirect means of introducing the sulfonate group through other reactions is assembled in Table 7. Comparative heats of sulfonation and sulfation reactions are listed in Table 8 for significant anionic species using SO$_3$ and oleum reagents (1).

Sulfonation

All sulfonation is concerned with generating a carbon sulfur(VI) bond (19) in the most controlled manner possible using some form of the sulfur trioxide moiety. The sulfonation of aromatic compounds such as benzene, dodecylbenzene, toluene (qv), xylenes, cumene (qv), or ethylbenzene are all examples of important commercial processes (see BTX PROCESSING; XYLENES AND ETHYLBENZENE). Sulfonation can be carried out in a number of ways using the reagents listed in Table 3. Sulfur trioxide is a much more reactive sulfonating reagent than any of its derivatives (see Tables 3 and 5) or adducts. Care should be taken with all sulfonating reagents owing to the general exothermic nature of the reaction. For example, $\Delta H = -170$ kJ/mol (40.6 kcal/mol) for dodecylbenzene sulfonic acid (see Table 8) (20).

The variety of reagents available makes possible the conversion of a wide range of aromatics into sulfonic acids (21–66). The reactivity of compounds that are activated toward electrophilic attack are so high that often alternative reagents are used in order to minimize undesirable by-products largely formed owing to excessive heating. By-products are so common to sulfonation that considerable time and expense is often utilized to develop methods for the removal of the excess heat from the crude product mixture during the reaction.

Aromatic Compounds. The accepted general mechanism (38–40,51) for the reaction of an aromatic compound with sulfur trioxide involves an activated

Table 6. Comparison of Sulfuric Acid and Gaseous SO₃ Sulfonation Reagents for Sulfonating Aromatic Hydrocarbons[a]

Comparative factors	Sulfuric acid/oleum	SO₃ (g)
aromatic sulfonation	water by-product produced	addition Rx; no water produced
mole ratio, Rx completion	~3–4 mol; excess reagent not critical	~1 mol; excess reagent quite critical
reagent miscibility	immiscible with organics; 2 immiscible liquids Rx	liquid–gas 2-phase Rx
solubility, organic solvents	immiscible	miscible
mechanical agitation	essential	not needed using high velocity gaseous falling film reaction
reaction temperature	varied (0–50°C), basis of product color quality; solvents often used to reduce viscosity	falling film process (short contact times) allows higher reaction tempera- ture profile, avoiding need for solvents
reaction mixture viscosity	relatively low	higher
reaction rate	slow	instantaneous
heat of sulfonation, alkylbenzene kJ/mol[b]	−112	−170
heat input	heat for completion	strongly exothermic, no heat required
heat exchange	low temperature Rx dictates refrigerated brine cooling system	film sulfonation uses ambient H_2O cooling; highly efficient
side reactions	minor	often extensive
derived product color	lighter	darker, except falling film system
final derived sulfonic acid	Rx mixture separates, H_2O often added to facilitate; color bodies partially removed into separated spent H_2SO_4	Rx mixture homogenous
generation of spent acid/1.0 kg sulfonic acid for disposal, kg	0.75–1.14	0
reagent boiling point, °C	290–317	44.5

[a]Rx = reaction.
[b]To convert J to cal, divide by 4.184.

intermediate as shown in equation 1.

$$R\!-\!C_6H_5 + SO_3 \longrightarrow [R\!-\!C_6H_5SO_3]^* \longrightarrow R\!-\!C_6H_4SO_3H \qquad (1)$$

The intermediate is believed to form through a π-complex, which collapses into a σ-complex, which then rearomatizes upon proton removal from the sp^3 carbon. This is followed by addition of a proton to the oxygen of the alkylbenzene sulfonate.

Table 7. Indirect Sulfonating Reagents

Reagent name	Chemical structure	Reaction
sodium isethionate	$HOCH_2CH_2SO_3Na$	esterification
N-methyl taurine	$CH_3\overset{\underset{\mid}{H}}{N}CH_2CH_2SO_3Na$	amidification
sodium chloromethyl sulfonate	$ClCH_2SO_3Na$	condensation
sulfoacetic acid	HSO_3CH_2COOH	esterification
phenol sulfonic acid	HO—⟨benzene⟩—SO_3H	esterification
carbyl sulfate	$H_2C\overset{CH_2-SO_2}{\underset{O-SO_2}{\big\backslash\!\!\diagup}}O$	condensation
formaldehyde sodium bisulfite	$HOCH_2SO_3Na$	sulfoalkylation
1,3-propane sultone	⌐—O—⌐ $CH_2CH_2CH_2SO_2$	sulfoalkylation
sodium ethene sulfonate	$CH_2{=}CHSO_3Na$	polymerization, sulfoalkylation
4-styrene sulfonic acid	$CH_2{=}CH$—⟨benzene⟩—SO_3H	polymerization
4-sulfophthalic anhydride	⟨benzene⟩$\overset{C=O}{\underset{C=O}{\big\rangle}O}$, HO_3S	condensation, esterification, etc
2-acrylamido-2-methyl propane sulfonic acid	$CH_2{=}\overset{\underset{\mid}{H}}{C}{-}\overset{\underset{}{O\atop\|}}{C}{-}\overset{\underset{\mid}{H}}{N}{-}CH_2{-}$ $CH(CH_3){-}CH_2{-}SO_3H$	polymerization
γ-sulfobutyrolactone	⌐—O—⌐ $CH_2CH_2CH(SO_3H)C{=}O$	condensation
acrylonitrile sodium sulfonate	$N\overset{\underset{}{O\atop\|}}{C}CCH_2SO_3Na$	hydrolysis
3-sulfopropionic anhydride	⌐—O—⌐ $SO_2CH_2CH_2C{=}O$	sulfoalkylation
1,4-butane sultone	⌐—O—⌐ $CH_2CH_2CH_2CH_2SO_2$	sulfoalkylation

Benzene.   The reaction of sulfur trioxide and benzene in an inert solvent is very fast at low temperatures. Yields of 90% benzenesulfonic acid can be expected. Increased yields of about 95% can be realized when the solvent is sulfur dioxide. In contrast, the use of concentrated sulfuric acid causes the sulfonation reaction to reach reflux equilibrium after almost 30 hours at only an 80% yield. The by-product is water, which dilutes the sulfuric acid establishing an equilibrium.

Table 8. Heats of Sulfonation and Sulfation Reactions Using Gaseous SO$_3$ and Oleum Reagents[a]

Organic feedstock	Reagent	Reaction	$\Delta H = $ kJ/mol[b]
alkylbenzenes	gaseous SO$_3$	sulfonation	-170
primary fatty alcohols	gaseous SO$_3$	sulfation	-150
ethoxylated alcohols	gaseous SO$_3$	sulfation	-150
α-olefins	gaseous SO$_3$	sulfonation	-210
alkylbenzenes	oleum	sulfonation	-112

[a]Ref. 1.
[b]To convert J to cal, divide by 4.184.

Dyes, Dye Intermediates, and Naphthalene.   Several thousand different synthetic dyes are known, having a total worldwide consumption of 298 million kg/yr (see DYES AND DYE INTERMEDIATES). Many dyes contain some form of sulfonate as $-SO_3H$, $-SO_3Na$, or $-SO_2NH_2$. Acid dyes, solvent dyes, basic dyes, disperse dyes, fiber-reactive dyes, and vat dyes can have one or more sulfonic acid groups incorporated into their molecular structure. The raw materials used for the manufacture of dyes are mainly aromatic hydrocarbons (67–74) and include benzene, toluene, naphthalene, anthracene, pyrene, phenol (qv), pyridine, and carbazole. Anthraquinone sulfonic acid is an important dye intermediate and is prepared by sulfonation of anthraquinone using sulfur trioxide and sulfuric acid.

There are three main uses for naphthalene sulfonic acid derivatives (75–79): as naphthalenic tanning material; alkyl naphthalene sulfonates for industrial applications as nondetergent wetting agents; and as dye intermediates. Consumption of naphthalene sulfonates as surfactants accounts for a large portion of usage. Naphthalene sulfonate–formaldehyde condensates are also used as concrete additives to enhance flow properties. Demand for naphthalene sulfonates in surfactants and dispersent applications, particularly in concrete, was expected to increase into the twenty-first century. Consumption as of 1995 was 16×10^6 kg/yr.

Alkylated Aromatics.   The world's largest volume synthetic surfactant is linear alkylbenzene sulfonate (LAS), which was developed as a biodegradable replacement for nonlinear alkylbenzene sulfonates (BAB). LAS is derived from the sulfonation of linear alkylbenzene (LAB). Detergent sulfonates use LAB in the 236 to 262 molecular weight range, having a $C_{11}-C_{13}$ alkyl group. The simplest sulfonation route uses 100% sulfuric acid. The by-product is water, which dilutes the sulfuric acid establishing an equilibrium so that less than a 100% yield of sulfonic acid is obtained. A large excess of sulfuric acid is therefore required. The favored routes use oleum (10–25% sulfur trioxide in sulfuric acid) or an SO$_3$–air mixture. The latter method is superior to either the sulfuric acid or oleum route, producing lighter colored sulfonic acid and reduced sulfate. In all these processes, sulfur trioxide is the sulfonating agent and the primary product is the *para*-alkylbenzene sulfonic acid, which may be neutralized using any of a variety of bases. Examples include sodium, calcium, magnesium, ammonia, isopropylamine, or triethanolamine. Sodium is the most common salt produced, because of its cost and performance. Neutralized slurries can show a pH drift when the sulfonic anhydrides and pyrosulfonic acids in the parent acid are not converted back to sulfonic acid. Treatment using a small amount of water eliminates this problem and also prevents darkening of the sulfonic acid.

U.S. production of LAS was estimated at more than 346×10^6 kg/yr in 1992. The largest captive producers include Colgate-Palmolive Company, The Dial Corporation, Lever Brothers Company, and The Procter & Gamble Company. Merchant producers include Stepan Company, Pilot Chemical Company, Vista Chemical Company, and Witco Corporation. The alkylbenzenesulfonic acids produced are of 97–98% purity, have low color, low free oil, and low free sulfuric acid. Whereas batch sulfonation (37) is still used at many companies and may be required for specific feedstocks, the continuous process using SO_3 generally is the preferred method of production as of the mid-1990s. The continuous process is utilized by Stepan Company, The Procter & Gamble Company, Colgate-Palmolive Company, The Dial Corporation, Unilever United States, Inc., Vista Chemical Company, and Witco Corporation. The use of LAS may decline slightly owing to greater hard water tolerance and better mildness of some competitive products.

Sulfonated toluene, xylene, and cumene, neutralized to the corresponding ammonium or sodium salts, are important industrially as hydrotropes or coupling agents in the manufacture of liquid cleaners and other surfactant compositions (19). They also serve as crisping agents in drum and spray drying operations. The sulfonation of toluene (27,39) or xylene with sulfuric acid is made practical by utilization of azeotropic water removal during the reaction, thus forcing the sulfonation to completion. Toluene sulfonation (80) yields 79% *para-*, 13% *ortho-*, and 8% *meta*-toluene sulfonic acid at 100°C. Isomer ratios change with temperature owing to the Jacobsen rearrangement.

Sulfitation and Bisulfitation of Unsaturated Hydrocarbons. Sulfites and bisulfites react with compounds such as olefins, epoxides, aldehydes, ketones, alkynes, aziridines, and episulfides to give aliphatic sulfonates or hydroxysulfonates. These compounds can be used as intermediates in the synthesis of a variety of organic compounds.

Sulfosuccinates and Sulfosuccinamates. The principal sulfonating reagent in these cases is the bisulfite molecule which readily attacks electron-deficient carbon centers. The starting materials are all electron-deficient double bonds, made so by their attachment to two vinylically situated electron-withdrawing groups; ie, carboxyl groups or their ester/amide derivatives, eg, maleic acid, fumaric acid, etc. Often the mono- or diester/diamide derivatives are made to react with the aqueous bisulfite giving the resulting sulfonated product. Variations in the choice of starting material can give a broad spectrum of products of widely varying chemical and physical properties. These compounds find application in household, toiletry, and cosmetic products.

Unsaturated Hydrocarbons. The reaction of long-chain, ie, C_{12}–C_{18}, α-olefins with strong sulfonating agents leads to surface-active materials (see SURFACTANTS). The overall product of sulfonation, termed α-olefin sulfonate (AOS), is really a mixture containing both alkenesulfonates (65–70%) and hydroxyalkanesulfonates (20–25%), along with small amounts of disulfonated products (7–10%) (81,82). The composition of the final product varies as a result of manufacturing conditions, and it is possible to exercise a limited amount of control over the final product mixture. Compared to batch sulfonation, the process design advantages offered by continuous thin-film SO_3 reactors have a maximum impact when applied to AOS production. Reactor designs include those of Allied, Ballestra, Chemithon, Lion Fat & Oil, Meccaniche Moderne, Mazzoni, and Stepan. Only with such equipment can AOS be economically produced, giving a balance of good color and low free oil.

The exothermic (-210 kJ/mol (-50.2 kcal/mol)) nature (20) of the SO_3-olefin reaction makes a neat process impractical. Heat removal can be accomplished using an evaporating–refluxing solvent, or by conduction using high surface area cooling equipment. To prevent degradation and minimize byproducts, SO_3 is applied in a diluted form, usually with dry air or nitrogen, at levels of 10 vol % or less. Generally, continuous thin-film sulfonators give the best product mixtures. Batch processes employ SO_3 complexed with a Lewis base and/or a solvent system, such as liquid SO_2. Sulfonation of α-olefins may also be carried out with chlorosulfonic acid.

The initial SO_3 sulfonation reaction (83–89) involves the formation of a carbon–sulfur bond at the terminal carbon of the olefin in accordance with Markovnikov's rule, to make the four-membered β-sultone ring (90–97), which is believed to occur through a concerted $\pi 2s + \pi 2s$ cyclo-addition mechanism (96). β-Sultone can further react with more SO_3 to form a cyclic pyrosulfonate ester. Other names (98–100) include pyrosultone, carbyl sulfate, or cyclic sulfonate–sulfate anhydride. The pyrosultone is metastable and can decompose upon aging to alkene sulfonic acid with the release of SO_3. This SO_3 is free to react with any remaining olefin or with the double bond of alkenesulfonic acids to form disulfonic acids. However, there is evidence that no alkenesulfonic acid is present before neutralization and hydrolysis (81). The decomposition of pyrosultones, as well as general equilibration of the crude product mixture, moves the process into a phase often referred to as digestion or conditioning. In the digestion phase, highly strained β-sultones (101) isomerize to a mixture of n-alkenesulfonic acids, as well as five- or six-membered ring structures referred to as γ-sultones or δ-sulfones, respectively. These rings are typically the products of carbocation rearrangements, from zwitterionic intermediates, via 1,2-hydride shifts. The carbocation rearrangements are driven by the formation of cationic sites which are more remote from the highly electron-withdrawing and powerfully destabilizing sulfonate group. If digestion is not carried out, the neutralized product contains a high proportion of 2-hydroxyalkanesulfonate, which is very insoluble. Approximate digestion conditions for β-sultone removal are from 30–50°C for 1–30 minutes. Continued digestion only increases the ratio of δ-sultone to γ-sultone, as the six-membered ring is thermally favored. This may not be desired because neutralized products contain the same ratio of δ-sultone to γ-sultone but in the form of 4-hydroxyalkanesulfonate and 3-hydroxyalkanesulfonate.

Conditions for hydrolysis (82) of the intermediate sultone mixture also help modify the ratio of alkenesulfonate to n-hydroxyalkanesulfonate, distribution of alkenesulfonate positional isomers, and completeness of conversion. Caustic hydrolysis using a slight stoichiometric excess of base is employed to ensure alkaline conditions throughout the hydrolysis phase of AOS production. The rate of hydrolysis depends a great deal on temperature. The δ-sultone requires the most time for conversion to 4-hydroxyalkanesulfonate. β-Sultones and γ-sultones hydrolyze so rapidly to 2-hydroxyalkanesulfonate and 3-hydroxyalkanesulfonate that temperatures below 100°C can be used. δ-Sultone completely hydrolyzes between 120 and 175°C in 1–30 minutes. The quality of the final product mixture is ultimately determined by the choice of conditions.

AOS prepared from α-olefins in the C_{12}–C_{18} range are most suitable for detergent applications. Evaluations of solubility, wetting, and foaming of C_{12}–C_{18} 3-hydroxyalkanesulfonate and 4-hydroxyalkanesulfonate show a decrease in solubility but an increase in foaming and wetting with increased molecular weight (82). Generally, alkenesulfonates show better detergency and foaming than n-hydroxyalkanesulfonates. Sulfonates of branched and internal olefins are poorer detergents but have good wetting properties.

Sulfonation of internal olefins has always been regarded as a difficult process, with only 80 or 90% yields expected. Quite often, large excesses of SO_3 are employed and the resulting material has an undesirable color (102–111). The crude sulfonated product mixture is processed in much the same way as are α-olefins, ie, digestion, neutralization, and hydrolysis. However, there are substantial differences in kinetics owing to the hindered nature of the internal olefin or subsequent β-sultone. Generally, β-sultone stability (112) is on the order of two hours at 35°C, and appears to rearrange almost exclusively to the γ-sultone. However, heating the acid to 150°C for several hours converts most of the γ-sultone to δ-sultone. Newer conditions have been established that improve on sulfonation yields so that these are comparable to those of α-olefins (112,113).

Sulfoxidation of Paraffins. The sulfoxidation of paraffins, known since the 1940s, is made possible by the use of free-radical chemistry with sulfur dioxide and oxygen (114–120). Free radicals can be generated in any number of ways. The most common method is uv-irradiation of the mixture throughout the reaction. The process produces random substitution and significant disulfonation, producing almost as much sulfuric acid (121) as it does sulfonic acid. One of the primary problems associated with generating the free radicals by uv-irradiation, is that colored materials are deposited on the light source impairing the illumination and retarding the process (122). Paraffin sulfonates, sometimes referred to as secondary alkanesulfonates (SAS), have application in liquid and heavy-duty solid detergents in Western Europe.

Sulfochlorination of Paraffins. The sulfonation of paraffins using a mixture of sulfur dioxide and chlorine in the presence of light has been around since the 1930s and is known as the Reed reaction (123). This process is made possible by the use of free-radical chemistry and has had limited use in the United States. Other countries have had active research into process optimization (124,125).

Buckminsterfullerene. Sulfonated fullerenes, a more recent development, are only a laboratory curiosity as of the mid-1990s. These are generally prepared in solvent from the cycloaddition of SO_3 to the C_{60} homologue, either by the use of liquid SO_3 or oleum (126–128). The reaction occurs at room temperature. The color changes from burgundy to pale orange or orange-red, depending on the concentration. The resulting product is very soluble in solvents such as acetone.

Fatty Acid Esters. Fatty acid ester sulfonates are manufactured by reaction of the corresponding hydrogenated (usually methyl) ester and a strong sulfonating agent, such as sulfur trioxide, in order to sulfonate on the alpha-position of the ester (129). The procedure for the reaction and equipment requirements are not much different from those for the production of LAS, with the exception of stoichiometry. In order to achieve a high degree of sulfonation, sulfur trioxide in excess of 1:1 stoichiometry may be used (130–132). Usually, the ratio of SO_3:ester is between 1.5 and 1, after which, a digestion period is necessary at a higher temperature, in order to achieve full conversion. One of the principal byproducts for this process is the α-sulfoalkane carboxylic diacid, which is easily re-esterified upon the addition of alcohol under acidic conditions to give the desired fatty acid ester sulfonic acid. Neutralization follows with bleaching to give the final desired product. Sulfonation increases the stability of the ester linkage and in a pH range of 5–10, α-sulfomethyl tallowate is stable even when boiled.

The mechanism for sulfonation of hydrogenated fatty esters is accepted as a two-stage process. A rapid sequence of reactions leads to the formation of intermediates having approximately 2:1 stoichiometry of sulfur trioxide to ester. In the subsequent slower and higher temperature aging step, the SO_3 is released for further reactions and the starting material conversion proceeds to completion (133).

Sodium fatty acid ester sulfonates are known to be highly attractive as surfactants. These have good wetting ability and excellent calcium ion stability as well as high detergency without phosphates, and are used in powders or liquids. They can also be used in the textile industry, emulsion polymerization, cosmetics, and metal surface fields. Moreover, they are attractive because they are produced from renewable natural resources and their biodegradability is almost as good as alkyl sulfates (134–137).

Petroleum and Related Feedstocks. *Mahogany Sulfonates.* By 1875 petroleum (qv) was being refined by treatment with sulfuric acid. Thus petroleum sulfonate by-products became the first petrochemical product. Since that time, by-product petroleum sulfonates have gradually found utilization in a great many applications, including as lubricant additives for high performance engines; as emulsifiers, flotation agents, and corrosion inhibitors; and for enhanced oil recovery. The importance of petroleum sulfonates has grown to the point where these compounds are produced as coproducts, or even as primary petrochemicals.

Petroleum sulfonates have traditionally been produced by both batch and continuous treatment of petroleum oils with oleum. These processes have been covered in several reviews (138,139). Natural petroleum sulfonates are coproducts in the manufacture of a variety of refined oils, most notably white (mineral) oils, lube oils, and process oils (plasticizer oils for rubber compounding). The feedstocks are selected primarily on the basis of the desired characteristics of the refined oils which generally contain 15–30% aromatics.

Typically, the oil is subjected to several successive oleum contacts ("treats") at ca 60–65°C. This charge ratio of feedstock:oleum is generally empirically determined, but depends in part on the degree of aromatic removal required for the product. In the case of white oil manufacture, 100 parts of feedstock are typically treated with ca 50 parts of 20% oleum (140). The resulting reaction mixture forms two layers that are separated by gravity or centrifugation. These layers have been referred to by color: the upper sulfonic acid oil layer is a rich mahogany; the lower acid sludge layer is green. Hence, the terms mahogany acids and green acids arose. The green acids or sludge consist predominantly of water-soluble polysulfonic acids and spent sulfuric acid. The sludge layer has found some industrial use as a demulsifier in ammonia production and for other applications. It is often burned for recovery of energy and SO_2. After sludge separation, the crude mahogany acid is usually neutralized using aqueous caustic, extracted using 2-propanol to facilitate partial oil reduction, desalted, and stripped of solvents.

Mahogany sulfonate composition typically is 62% sulfonate, 33.5% oil, 4% H_2O, and ca 0.5% Na_2SO_4. The equivalent weight of the mahogany sulfonate depends on the specific feedstock and the severity of the oleum treats. Typically this value is in the 420 to 550 range. Higher boiling lube stocks, such as bright stock, produce a sodium sulfonate equivalent weight of 600 to 700. The derived oil-soluble sulfonates are predominantly monosulfonates, but probably contain 2–5% disulfonates depending on feedstocks and procedures employed. The composition of the feedstocks and their resultant sulfonates are complex multiring structures, including aromatic and saturated ring constituents, typically consisting of about three to four cyclic aromatic and saturated ring components (141). Such multiring compounds provide multiple sites for polysulfonate formation. Suppression of polysulfonation is a challenge in sulfonating petroleum. The reaction of petroleum with oleum or SO_3 leads not only to sulfonation, but to a host of side reactions including oxidation (producing SO_2), polymerization, isomerization, cyclization, disproportionation, and dehydrogenation.

Sulfonates for Lube Additives. Most petroleum sulfonates used as lube additives are based on calcium or magnesium salts. These salts can be produced by direct neutralization of the sulfonic acid with $Ca(OH)_2$ or $Mg(OH)_2$, or by use of a metathesis process involving the sodium salt:

$$2\,RSO_3Na + CaCl_2 \longrightarrow (RSO_3)_2Ca + 2\,NaCl \tag{2}$$

This reaction is generally conducted in a hydrocarbon solvent such as heptane, and an aqueous solution of $CaCl_2$, with good mixing, resulting in a brine layer that is subsequently separated for disposal.

In some cases, a mixture of natural petroleum feedstock is preblended with synthetic alkylated aromatics, such as detergent aromatic alkylate bottoms or with first-intent synthetic mono- or dialkylated aromatics, selected to provide a suitable molecular weight for cosulfonation and subsequent processing. The use of blended feedstocks may eliminate the need for conducting an oil extraction–concentrating step, particularly for a typical 40% Ca or Mg petroleum sulfonated product.

The presence of polysulfonates in petroleum sulfonates used in lube formulations has a destabilizing effect on the formulation stability and function of the sulfonate in motor oils, etc. Special techniques are utilized to help reduce the carryover of residual sludge components, including the use of hydrocarbon solvents such as hexane or heptane to facilitate separation of sludge, often with centrifugation. Other desludging procedures include water wash, H_2SO_4 wash, clay percolation, and filtration.

Shell is the sole principal U.S. manufacturer of petroleum sulfonate having an estimated annual plant capacity of ca 27,000 metric tons; Witco and Pennrico-Morco are believed to supply a total of ca 7,000 metric tons annually.

Sulfonation With Oleum vs SO₃. The use of oleum has been compared to that of SO_3 for the batch sulfonation of eight feedstocks for the manufacture of white oils (142). The averaged results were as follows:

Sulfonating agent	Monosulfonic acid yield, %	Total sludge yield, %	Calculated polysulfonic acid, %	Ratio of mono:polysulfonate
20% oleum	2.9	75.9	23.8	1.0:8.2
SO₃	6.2	43.7	25.4	1.0:4.0

A significantly higher yield of monosulfonic acid was obtained when using SO_3. Less total sludge and calculated polysulfonic acid were also produced.

The equivalent weight distribution of natural petroleum sulfonates depends on the boiling range of the aromatic components in the feedstock, but generally consists of a broad continuum of molecular weight components (139). For many applications it is precisely this property of derived petroleum sulfonates that provides the unique properties, such as emulsification. Conversely, most oil-soluble synthetic sulfonates have much more limited components and molecular weight distribution.

Industry Changes. Factors impacting petroleum sulfonation operations since the late 1970s include the many significant changes and modernizations petroleum refineries have undergone leading to the closing of many refineries practicing oil sulfonation processes; white oil manufacturing technology has eliminated sulfonation and thus sludge disposal to utilize the more cost-efficient hydrogenation process (143); a principal shift has developed in the use of first-intent oil-soluble synthetic alkylated aromatic sulfonates in place of the traditional petroleum sulfonates for lube additives, and the synthetic sulfonates are made by continuous SO_3 sulfonation processes; and the large projected need for petroleum sulfonates for enhanced oil recovery processes has ceased owing to a significant and prolonged drop in crude oil market prices. Hence there has been a significant drop in the production of natural petroleum sulfonates.

Oil-Soluble Lube Sulfonates. Neutral or low base calcium and magnesium oil-soluble sulfonates are generally used as components in lubricants as dispersants and detergents to suspend oil-degradation products (sludge and carbon), and to help keep engine parts clean. Over-based or high alkalinity calcium and magnesium oil-soluble sulfonates are added to the lubricant formulation to neutralize the organic acids formed in the hot engine environment to prevent corrosion from acid attack, and to serve as a corrosion inhibitor. Low (**1**) and high (**2**) base lube sulfonates, where M is Ca or Mg, are typified as follows:

$$\left(R-\bigcirc-SO_3^- \quad M^{2+} - O_3S-\bigcirc-R\right) \cdot \left(M(OH)_2\right)_{0.15}$$

(**1**)

$$\left(R-\langle\bigcirc\rangle-SO_3^- \quad M^{2+} - O_3S-\langle\bigcirc\rangle-R \right) \cdot \left(MCO_3 \right)_{12-20}$$

(**2**)

Most alkaline-earth lube sulfonates are produced by direct neutralization of the sulfonic acid using the alkaline-earth oxide or hydroxide and a selected alcohol often aided by a relatively low boiling hydrocarbon solvent. High base lube sulfonates are usually produced from the low base sulfonate by direct carbonation of excess alkaline-earth oxide and CO_2 addition. A carbonation promoter is generally used to facilitate the process. The alkalinity of a lube sulfonate is expressed as titratable base number (TBN) equivalent to milligrams of KOH per gram of sample. High base sulfonates generally have alkalinities ranging from 100 to 400 TBN, representing 12.5–50 wt % as $CaCO_3$ typically suspended in sulfonate and oil. Such high TBN sulfonates are carefully filtered to remove particulates and give optically clear products. These products suspend a substantial amount of $CaCO_3$ (rock) in a perfectly clear oil-based stable microemulsion.

A significant proportion of lube sulfonates are based on high molecular weight synthetic alkylated aromatic feedstocks. Sulfonation of these feedstocks are generally conducted using high dilution, high velocity SO_3 continuous systems, such as the Chemithon jet-impact reactor or the falling film process. Exxon and Pilot Chemical Company utilize low temperature SO_3 sulfonation with SO_2 solvent.

If the sulfonation of specially designed high quality, first-intent, synthetic feedstock is properly SO_3-sulfonated and digested, it appears possible to eliminate a solvent desludging step and to directly produce a high active lube sulfonate, also eliminating the need to extract oil and concentrate the sulfonate as typically practiced for petroleum-derived sulfonates (144). To fulfill the foregoing, it is necessary to produce an alkylate where the lowest molecular weight component when monosulfonated is totally oil soluble. This equates to a minimal monoalkylate molecular weight of about 332 ($C_6H_5-C_{18}H_{37}$). Both polypropylene- and polyethylene-based alkylates are used for production of synthetic lube sulfonates. Most alkylated aromatic feedstocks are mixtures of mono- and dialkylates.

p-Dialkylbenzene isomer derived from branched-chain olefins is sterically hindered and resists sulfonation. Vigorous SO_3-sulfonation leads to dealkylation. Exxon is believed to utilize a dimerization or oligomerization process to convert polypropylene trimer, tetramer, hexamer, and octamer into two benzene monoalkylates having average molecular weights of ca 390 and 420, respectively (144). These are both branched-chain monoalkylates and are readily sulfonatable yielding sulfonic acids of ca 470 and 500 mol wt. Ethyl, Chevron, and Vista produce benzene alkylates based on long-chain alpha-olefin fractions. The formulated lubricant market is international in scope, dominated by Lubrizol, Ethyl, Exxon, Chevron, and Shell.

Petroleum Sulfonation Process Developments. Crude oil and/or topped crude oil were sulfonated using SO_3 by Marathon Oil Company at their Robinson, Illinois refinery and used directly in an adjacent oil field for enhanced oil recovery (EOR) (145–147). Stepan Company developed technology and a large plant for the manufacture of first-intent petroleum sulfonates using available refinery feedstocks by adapting their continuous falling film SO_3 sulfonation process (148,149). The improved technology utilized feedstock additives which provided for maintaining film integrity generally operating with a high temperature cooling system. The process provided a system for a highly selective monosulfonation reaction. The products were directly suitable for EOR applications (150–152). Another patented process covering the manufacture of petroleum sulfonates suitable for EOR is found in the literature (153). Several Russian patents have been issued for sulfonation of petroleum oils with SO_3 (154–156), the latter using three to four falling film sulfonators in series for prolonging diluted gaseous SO_3 contact with the petroleum film.

A Polish patent discloses a process for manufacture of petroleum sulfonates (157). Another Russian patent was reported for the continuous SO_3-sulfonation of alkylbenzene and petroleum feedstocks (158). A process patent has been issued for the sulfonation of viscous feedstocks utilizing an elongated tubular SO_3 film sulfonator with a pressurized high velocity SO_3 gas mixture and cooling with hot water. The process is suitable for biphenylalkylates, diphenylalkyl ethers, and petroleum (159). A process was disclosed wherein petroleum sulfonates were heated with NaOH to convert the sulfonate to a phenate and thereafter ethoxylating the phenolic group to produce a nonionic suitable for EOR (160). In the sulfonation of petroleum oils considerable oxidation occurs generating large amounts of SO_2. A patent was issued utilizing a sulfur burner catalytic SO_3-generator system for treatment of effluent from the sulfonation plant and for recovery and re-use of the generated SO_3 (161). A process for the production of petroleum sulfonates suitable for EOR based on separating and refining of lube oil extracts has been reported (162). A Russian publication has disclosed the SO_2 sulfonation of basic oils from western Siberian petroleum oils (163).

Sulfonates for Enhanced Oil Recovery. The use of hydrocarbon sulfonates for reducing the capillary forces in porous media containing crude oil and water phases was known as far back as 1927–1931 (164,165). Interfacial tensions between 10^{-9} and 10^{-11} N/m or less were established as necessary for the mobilization and recovery of crude oil (166–169).

Micellar/polymer (MP) chemical enhanced oil recovery systems have demonstrated the greatest potential of all of the recovery systems under study (170) and equivalent oil recovery for mahogany and first-intent petroleum sulfonates has been shown (171). Many somewhat different sulfonate, ie, slug, formulations, slug sizes (pore volumes), and recovery design systems were employed. Most of these field tests were deemed technically successful, but uneconomical based on prevailing oil market prices (172,173).

Amoco developed polybutene olefin sulfonate for EOR (174). Exxon utilized a synthetic alcohol alkoxysulfate surfactant in a 104,000 ppm high brine Loudon, Illinois micellar polymer small field pilot test which was technically quite successful (175). This surfactant was selected because oil reservoirs have brine salinities varying from 0 to 200,000 ppm at temperatures between 10 and 100°C. Petroleum sulfonate applicability is limited to about 70,000 ppm salinity reservoirs, even with the use of more soluble cosurfactants, unless an effective low salinity preflush is feasible.

Combining alkali, surfactant, and polymer (ASP) methods may result in decreased chemical costs. The ASP surfactant slug generally contains about 0.8 wt % alkali as Na_2CO_3, $NaHCO_3$, Na_2SiO_3, or $NaOH$; ca 0.1 wt % active surfactant and about 1000 mg/L polymer. A micellar polymer slug typically utilized 3–10% active surfactant. A 1987 field test of the ASP process in Wyoming indicated an incremental oil recovery of 20% of the original in-place oil, at an estimated chemical cost of less than $15.60/m^3 ($2.50/bbl) (176).

A second field evaluation of the ASP process has been initiated in Wyoming. Additionally, an ASP field project has been designed for the Peoples' Republic of China. The applicability of the ASP process to a variety of reservoirs has yet to be fully determined. Application of alkali and alkali polymer flooding has been limited to crude oils having discernible acid numbers, wherein the alkali produced crude oil soaps which in combination with alkali resulted in providing low interfacial tensions. The ASP process appears to be suitable for crude oils with nil acid numbers (177), and hence should have broad applicability.

Thermally stable foam additives, such as alkylaryl sulfonates and C_{16}–C_{18} alpha-olefin sulfonates, are being used in EOR steam flooding for heavy oil production. The foam is used to increase reservoir sweep efficiency (178,179). Foaming agents are under evaluation in chemical CO_2 EOR flooding to reduce CO_2 channeling and thus increase sweep efficiency (180).

An assessment of the toxicity potential of chemicals used in EOR has been reviewed (181). A series of first-intent petroleum sulfonates derived from petroleum fractions were synthesized along with a series of oil-soluble synthetic sulfonates. Their properties and preliminary oil recoveries have been reported (182).

Lignin. Lignosulfonates are complex polymeric materials obtained as by-products of wood (qv) pulping where lignin (qv) is treated with sulfite reagents under various conditions (see PULP). Only a fraction of the potentially available lignosulfonate is recovered owing to limited markets, low market prices, and the cost of refining and further upgrading of this by-product. Lignin by-products have historically been discharged into waterways. Pollution regulations require recovery and reprocessing of these materials. Lignin polymers contain substantial amounts of guaiacyl units, followed by p-hydroxyphenyl and syringyl units. Two principal wood pulping processes are utilized: the sulfite process and the kraft process. Sulfonation of lignin mainly occurs on the substituted phenyl–propene precursors at the alpha-carbon next to the aromatic ring.

Acid Sulfite Process. In the acid sulfite process woodchips and sawdust from hardwood sources are heated under pressure with a mixture of sulfurous acid and metal sulfite, eg, calcium, magnesium, sodium, or ammonium bisulfite. Hydrolysis and sulfonation convert lignin into lignosulfonate of molecular weight between 200 and 100,000. Structures are reported to be linear at molecular weights below about 4000, but coiled and solvated at higher molecular weights. The degree of sulfonation appears to increase with decreasing molecular weight. In processing, hydrogen, oxygen, and sulfur are added in the proportion of 4:4:1 (183). The wood material is generally digested at 125–145°C for 8–24 h during which the lignosulfonate is solubilized. The liquor is then filtered, removing the cellulosic fiber to 58% crude lignosulfonate which may be burned for fuel value, particularly for NH_4^+ and Ca^{2+} salts and recovery of SO_2 for recycling; or upgraded, spray dried, or further chemically treated to produce chemical derivatives. Although the sulfite process has largely been displaced by the kraft process, most lignosulfonates are derived from the sulfite process because higher yields are obtained. A flow diagram for this complex process has been published (184). Every metric ton of chemically produced pulp generates approximately 0.5 t of lignin liquors containing crude lignosulfonate.

Kraft Process.   Wood derived from softwood sources is generally heated under pressure with a 10–20% mixture of NaOH and Na_2S for only 4–6 h at ca 165–175°C. These conditions accelerate delignification in the pulping process. In this process, kraft lignin is filtered and separated from the black liquor, subsequently suspended in aqueous media, acidified, and subjected to H_2SO_3 sulfonation at relatively high temperatures and pressures. Westvaco Corporation is reported to be the only significant producer of lignosulfonate utilizing this process.

Research Trends.   Research since the late 1970s appears to have centered on further process modifications and optimization of conventional processes, reduction in pollution, evaluation of potentially more efficient processing technology, increased use of less expensive softwood sources to extend the available hardwood supplies, and accommodation of recycled newspaper pulp (see RECYCLING, PAPER).

Several studies have been published on the utilization of softwood sources for sulfite pulping (185–191). Sulfonation yields of 95–98% were obtained using $NaHSO_3$ and NaOH at between 120 and 150°C, and sulfonate contents up to almost 2% (as sulfur). Kinetic studies indicate that sulfonation of lignin at pH 7 is governed by a nucleophilic attack of sulfite on lignin. Sulfonation at neutral pH is activated primarily by phenolic groups para to the propane side chain (185). Chemimechanical pulping (CMP) and chemithermomechanical (CTM) pulping pretreatment before sulfite sulfonation of wood chips produced improved pulping properties of breaking strength, specific scattering coefficients, and web strength when sulfonation yields of 1.0–1.6% (as sulfur) were attained. Sulfitation at pH 4 proceeds by a quinomethide intermediate (188). Sulfonation of aspen wood chips using $NaOHSO_2$ at 120–170°C resulted in yields of 87–92%, with 0.5–0.6% sulfur content. Tensile and tear indexes were high, but brightness was somewhat lower (189). A study of the topochemistry of sulfonated black spruce established that cell wall lignin was found to contain 31% more sulfur per Ca unit than middle lamella lignin (190). In another study, softwood liquors were shown to contain more organic acids and were sulfonated to a greater degree (192). An investigation of the selective sulfonation of spruce wood chips showed excellent pulp properties for a narrow range of sulfur concentration. A scanning electron microscope was used to study morphological parts of the wood (193). A patent has been issued covering a sulfitation pulping process producing a tactile tissue softness based on hardwood chips (194). Another sulfitation pulping study showed that a preliminary treatment of pine wood chips with $NaHSO_3$ resulted in a more effective delignification during high yield sulfite pulping with better mechanical properties (195). The bleachability and brightness stability were improved for biomechanical pulp derived from aspen wood by using alkaline H_2O_2 or $Na_2S_2O_4$ (196).

A substantial number of publications and patents deal with improvement in the kraft pulping lignosulfonate process (197–207). A method for solubilizing kraft process lignin was reported wherein sulfonation was on the aromatic nucleus using a ferrous compound, $Na_2S_2P_8-Na_2SO_3$ combination (197). In refining pulp from wood, the extent of pulp fiber damage can be diminished by pretreatment (198). Oxidative pretreatment of pine wood facilitates delignification (199). A process has been patented in which lignins are selectively isolated having molecular weights of over 5000 followed by reaction with Na_2SO_3 and an aldehyde (201). The derived lignosulfonate is useful as a dye dispersant. A method has been patented for producing amine salts of lignosulfonate (202). A pretreatment of cold soda pulp using HNO_3 and Cl_2 has been reported (203). A patent has been issued for the sulfation of lignin using organic amine SO_3 complex reagent (204). A process patent for sulfonating lignin using oleum below 40°C has been issued (205). In another study, SO_2 emissions were decreased by improving the combustion operation (207).

Other Lignin Sulfonation/Sulfation Developments. The use of lignin treatment procedures involving steam explosion techniques has received substantial attention (208–214). A process patent claims the use of sulfamic acid as an accelerating agent to separate lignin from cellulosic components (215). Another patent claims a process of alkoxylating alkali lignin followed by sulfation with SO_3 or $ClSO_3H$ (216). Another discloses a process for producing oil-soluble sulfonated surfactants useful for oil recovery from lignin and alkylphenol (217). Most paper pulping processes generate SO_2 for use in sulfonation by burning molten sulfur.

Lignosulfonate Uses. Marketable lignosulfonates include ammonium, aluminum, calcium, chrome, ferrochrome, magnesium, potassium, sodium, and amine salts, and various combinations. Consumption of lignosulfonates may be divided into the following uses: animal feed pellets (15%), concrete additives (14%), road dust control (19%), oil-well drilling muds (4%), pesticide dispersant (3%), and other uses (45%) (218). Vanillin (3-methoxy-4-hydroxybenzaldehyde) is manufactured almost exclusively from lignosulfonate raw materials (219,220). The use of lignosulfonate was shown to be a sacrificial adsorbate for reducing petroleum sulfonate adsorption by 50% in enhanced oil recovery (221). However, it is effective only as a preflush, not as a cosurfactant in the sulfonate slug.

Styrene and Vinyl Monomer, Polymer, and Copolymer Sulfonates. The incorporation of sulfonates into polymeric material can occur either after polymerization or at the monomer stage (222). The sulfonic acid group is strongly acidic and can therefore be used to functionalize the polymer backbone to the desired degree. Depending on the molar fraction in the polymer, as well as on the macromolecular structure, the sulfonic acid group strongly interacts with water to bring about polymer swelling or gel formation, to a point where the complete dissolution of the polymer is possible (222). The ability of sulfonic acids to exchange counterions has made these polymers prominent in industrial water treatment (qv) applications, such as ion-exchange (qv) resins for demineralization, membranes for reverse osmosis or Donan dialysis, separators in electrochemical cells, and selective membranes of many types (222). Being strongly polar, the sulfonate functionality is used in such diverse areas as textile fiber dyeability, in thickeners and flocculants, rubber modifiers, and adhesive promoters (222). In addition, sulfonic acid derivatives, such as sulfonyl chlorides, amides, and anhydrides, make for an even wider range of uses.

The simplest monomer, ethylenesulfonic acid, is made by elimination from sodium hydroxyethyl sulfonate and polyphosphoric acid. Ethylenesulfonic acid is readily polymerized alone or can be incorporated as a copolymer using such monomers as acrylamide, allyl acrylamide, sodium acrylate, acrylonitrile, methylacrylic acid, and vinyl acetate (222). Styrene and isobutene fail to copolymerize with ethylene sulfonic acid.

The monomer 4-styrenesulfonic acid was prepared by dehydrohalogenation of p-bromoethylbenzene–sulfonyl chloride. The potassium salt can be polymerized in aqueous solution (222). The sulfonation of cross-linked polystyrene beads is being carried out in industry with concentrated sulfuric acid.

Sulfonated polyalkenes were prepared by using a triethyl phosphate–sulfur trioxide complex as the sulfonating reagent along with a solvent at low temperature. Sulfonation takes place at the α-position of the double bond with no cross-linking (222).

Sulfation

Sulfation (20) is the generation of an oxygen sulfur(IV) bond, where the oxygen is attached to the carbon backbone, in the most controlled manner possible, using some form of sulfur trioxide moiety. When sulfating alcohols, the reaction is strongly exothermic, $\Delta H = -150$ kJ/mol (-35.8 kcal/mol) (20). The mechanism for alcohol sulfation is believed to occur through a metastable pyrosulfate species which decomposes rapidly to alkyl hydrogen pyrosulfate, $ROSO_2OSO_3H$, which reacts with a second alcohol to produce an alcohol sulfate product mixture. Examples of feedstocks for such a process include alkenes, alcohols, or phenols. Unlike the sulfonates, which exhibit excellent stability to hydrolysis, the alcohol sulfates are readily susceptible to hydrolysis in acidic media. The thermal breakdown of alkyl sulfates under anhydrous conditions produces a mixture of products including the parent alcohol; dialkylsulfates, $ROSO_2OR$; dialkyl ether, ROR; and isomeric alcohols, olefins, and esters, $ROSO_3R$ (20). High processing temperatures must be avoided and the product neutralized soon after formation. Best product is obtained at sulfation temperatures of from 35 to 50°C and neutralization within one minute after sulfation (20). The sulfation of fatty alcohols and fatty polyalkoxylates has produced a substantial body of commercial detergents and emulsifiers.

Linear ethoxylates are the preferred raw materials for production of ether sulfates used in detergent formulations because of uniformity, high purity, and biodegradability. The alkyl chain is usually in the C_{12} to C_{13} range having a molar ethylene oxide:alcohol ratio of anywhere from 1:1 to 7:1. Propoxylates, ethoxylates, and mixed alkoxylates of aliphatic alcohols or alkyl phenols are sulfated for use in specialty applications.

Alcohols and Alkoxylates. Alkyl and alkoxyalkyl sulfates can be produced from the corresponding alcohols or nonionics by reaction with a wide variety of reagents including chlorosulfuric acid, sulfur trioxide, sulfuric acid, and sulfamic acid [5329-14-6] (223–233). The products are similar in wetting time, detergency, and foam generation. Chlorosulfuric acid gives slightly better colors but requires disposal of HCl. Sulfation using SO_3 requires controlled reaction temperatures and fast reaction times for best results. The preferred method of sulfation uses some form of continuous thin-film reactor. The reaction time is generally 0.1–0.5 seconds and the heat generated in the film is rapidly removed by the cooled reaction surface. In the United States, alkyl and alkoxyalkyl sulfates are mainly produced by the thin-film method. An important undesirable side reaction of ethoxylated alcohol sulfation is dioxane formation which can range from traces to hundreds or even thousands of ppm (mg/kg) depending on raw material quality and sulfation/neutralization conditions (20,234–236). Dioxane forms by the chemical cleavage of two molecules of ethylene oxide from the parent ethoxylated alcohol. Dioxane formation is favored by an excess of SO_3, high temperatures, long aging times, moisture in the feedstock, branching of alkyl chains, and longer ethylene oxide chains in the ethoxylated alcohol feedstock (20). Process conditions, which have been established to minimize dioxane formation, are a mole ratio of SO_3 to alcohol ethoxylate of 1.01–1.02:1; SO_3 in air, 3 vol % max; and the lowest temperatures possible for feedstock, reaction zone, and aging phase (20). Other examples of sulfated alcohols are mono- and difatty glyceryl ester sulfates, which are sulfated by either chlorosulfuric acid or sulfur trioxide (237–239).

Fats and Oils. Fats and oils (6) are traditionally sulfated using concentrated sulfuric acid. These are produced by the sulfation of hydroxyl groups and/or double bonds on the fatty acid portion of the triglyceride. Reactions across a double bond are very fast, whereas sulfation of the hydroxyl group is much slower. Yet 12-hydroxyoleic acid sulfates almost exclusively at the hydroxyl group. The product is generally a complex mixture of sulfated di- and monoglycerides, and even free fatty acids. Other feeds are castor oil, fish oil, tallow, and sperm oil.

These products have gradually been supplanted by sulfated fatty alcohols, sulfated monoglycerides, and sulfonated alkylbenzenes. Other sulfating agents include chlorosulfuric acid, SO_3, or sulfamic acid (19). Sulfated oils are used primarily in the emulsification of water-immiscible liquids such as cutting oil or mineral oil. The products are also used in textile wetting, cleaning, and finishing agents, or in oil sprays for insecticides and disinfectants.

Carbohydrates. Carbohydrates (240–244) of any form are easily sulfated in the presence of solvent, using sulfating reagents such as SO_3–pyridine, SO_3–triethylamine, SO_3–trimethylamine, or chlorosulfonic acid–pyridine. As an example, starch (qv) is sulfated using SO_3–trimethylamine at 0 to 5°C in aqueous media (16). Sulfated carbohydrate products find some use in industry as thickening agents.

Alkenes. The sulfation of low molecular weight alkenes using concentrated sulfuric acid is amenable to continuous operation. Good agitation is required and the reaction is performed at 70–80°C. Dialkyl sulfates are also formed. Longer (C_{12}–C_{18}) carbon chain alkenes yield detergent products. Order of addition, temperature, and stoichiometry are all important to this reaction. For example, the addition of 96% sulfuric acid to 1-dodecene at 0°C yields mainly dialkyl sulfate; further addition gives an 80% yield of 2-monoalkylsulfate. If the olefin is added to the acid, random isomers are produced.

Sulfamation

Sulfamation is the formation (245) of a nitrogen sulfur(VI) bond by the reaction of an amine and sulfur trioxide, or one of the many adduct forms of SO_3.

Heating an amine with sulfamic acid is an alternative method. A practical example of sulfamation is the artificial sweetener sodium cyclohexylsulfamate [*139-05-9*], produced from the reaction of cyclohexylamine and sulfur trioxide (246,247) (see SWEETENERS). Sulfamic acid is prepared from urea and oleum (248). Whereas sulfamation is not greatly used commercially, sulfamic acid has various applications (see SULFAMIC ACID AND SULFAMATES) (249–253).

Industrial Processes

A wide array of industrial processes is suitable for the manufacture of sulfated and sulfonated products. Process selection is dependent on the specific chemistry involved, choice and cost of reagents, physical properties of feedstocks and derived products, product volume requirements, operational mode (batch, continuous), quality of derived products, and possible generation and disposal of by-products as well as operating and equipment investment costs. Another important consideration is the location of the sulfonation plant relative to raw material suppliers, particularly for the more limited liquid SO_3 supplier's plants. On the other hand, molten sulfur used for *in situ* sulfur burning and gaseous SO_3 generation is readily available throughout the United States and worldwide. Another consideration for process selection is plant versatility in sulfonating a variety of feedstocks.

The handling of highly acidic sulfonation reagents and the actual sulfonation processing conditions for the production of acidic reaction products and by-products present a number of corrosion problems which must be carefully addressed. Special stainless steel alloys or glass-lined equipment are often used, although the latter generally has poorer heat-exchange properties. All environment regulations or restrictions must also be met. For example, in utilizing $ClSO_3H$ reagent, HCl gaseous by-product is generated requiring its recovery by adsorption or neutralization.

The viscosity of sulfonation and sulfation reaction mixtures increases with conversion, often producing extremely high viscosities. Figure 1 provides temperature–viscosity curves for oleum and SO_3-derived products. Sulfonation process design must accommodate such viscosities.

Photochemical Sulfonation Processes. Chlorosulfonation and sulfoxidation processes have been primarily used for the production of paraffin sulfonates. Both are photochemical reactions, each requiring specialized equipment. Undesirable polysulfonate generation greatly limits the extent of sulfonation and consequently such processing involves recovery and recycling of large quantities of unreacted feedstocks. Hoechst operates several substantial sized sulfoxidation plants in Europe for the production of C_{13}–C_{17} paraffin sulfonates.

Sulfamic Acid Batch Sulfation/Neutralization Process. The process for sulfating alcohols using sulfamic acid represents perhaps the simplest process. Minimal equipment is required. The sulfation and *in situ* neutralization of alcohols such as nonylphenol ethoxylate is conducted using a stainless steel or glass-lined stirred kettle by first charging with the alcohol, followed by addition of a molar quantity of solid crystalline sulfamic acid, mixing and heating to 130–160°C using a N_2 atmosphere (to preserve color), followed by partial cooling and subsequent addition of appropriate solvents. The reaction produces the

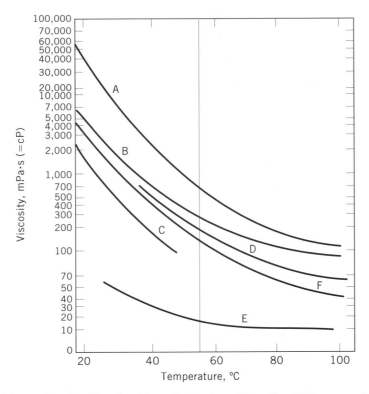

Fig. 1. Sulfonated and sulfated acid products viscosities after 98% conversions at varying temperatures where the vertical line indicates the maximum temperature for batch sulfonation using SO_3 to minimize color deterioration; lines A–C represent branched C_{12} alkyl benzene (BAB):sulfonic acid from SO_3, oleum (settled), and oleum (whole mixture), respectively; lines D and E, lauryl alcohol 3-ethoxylate sulfuric ester (SO_3) and lauryl alcohol sulfuric ester (SO_3), respectively; and line F, linear C_{12} alkyl benzene (LAB) sulfonic acid (SO_3).

ammonium salt of the sulfated alcohol. This process has been largely surplanted by SO_3 processes.

Sulfitation and Bisulfitation Sulfonation Processes. These reactions are generally conducted in a batch stirred tank system. The sulfite reagent is usually added as a concentrated aqueous solution and intimately mixed with the organic feedstock with heating, often under pressure. Because these are generally immiscible liquid or liquid–solid phase reactions, a cosolubilizer, hydrotrope, or surfactant may be added to facilitate reaction. At the larger paper pulping mills, molten sulfur is frequently burned on-site to produce SO_2 to economically make sulfurous acid and/or metal bisulfites for sulfonating lignin.

Processes for Sulfation of Fatty Alcohols with ClSO$_3$H. Lauryl alcohol is batch sulfated by gradual addition of $ClSO_3H$ to lauryl alcohol in a glass-lined stirred reactor over about a 2.5-h period at a temperature of 26–32°C. Gaseous HCl is expelled, aided by a slow continuous N_2 purge. The process, utilized by Stepan Company 1954–1962, requires a refrigerated cooling system. Lauryl alcohol has been sulfated using a cyclic loop batch $ClSO_3H$ sulfation process (254).

A continuous $ClSO_3H$ sulfation process has been patented and used by Henkel for the production of fatty alcohol sulfates (255). Fatty alcohol and $ClSO_3H$ are continuously injected into the bottom of a 1-cm annular space within a concentric cooled vertically tapered spiral reactor. The reaction is conducted at about 30°C with the reaction mixture propelled upwardly owing to the HCl gas generated by the reaction. Product residence time is estimated to be 1–2 min. This process produces excellent quality products, but appears to require refrigerated cooling. The substantially lower cost for liquid SO_3, *in situ* sulfur burner-generated gaseous SO_3 reagents, and problems and additional costs associated with handling and recovering the HCl by-product, has largely displaced the use of $ClSO_3H$ for sulfation of fatty alcohols.

Batch Stirred Tank H_2SO_4/Oleum Aromatic Sulfonation Processes. Low molecular weight aromatic hydrocarbons, such as benzene, toluene, xylene, and cumene, are sulfonated using molar quantities of 98–100% H_2SO_4 in stirred glass-lined reactors. A condenser and Dean-Stark-type separator trap are installed on the reactor to provide for the azeotropic distillation and condensation of aromatic and water from the reaction, for removal of water and for recycling aromatic. Sulfone by-product is removed from the neutralized sulfonate by extraction/washing with aromatic which is recycled.

Polypropylene-derived branched alkyl (C_{12}) benzene (BAB) has been batch sulfonated using 60–70% oleum in liquid SO_2 solvent at temperatures of −1 to −8°C, with SO_2 serving as a self-refrigerant and viscosity reducer in the process. After sulfonation and digestion, SO_2 is stripped, recovered, and recycled (256).

Details for the nonsolvent batch oleum sulfonation process for the production of BAB sulfonic acid have been described, including an excellent critique of processing variables (257). Relatively low reaction temperatures (ca 25–30°C) are necessary in order to obtain acceptable colored sulfonate, which requires refrigerated cooling (Table 9, example D).

Table 9 provides a detailed comparative summary for branched alkylbenzene (BAB) and linear alkyl benzene (LAB) detergent alkylate sulfonation in both batch and continuous modes. Essentially the same conversion of both branched chain and linear detergent alkylate to sulfonic acid is attainable using 98% H_2SO_4, 100.6% H_2SO_4, 104.5% H_2SO_4, and gaseous SO_3 in batch and continuous sulfonation processes utilizing the specified conditions. Composition of the neat sulfonation mixtures with and without water addition and spent acid separation show the corresponding compositions of the acids (lines 10–14 and lines 19–23, respectively) and for the NaOH-neutralized products (line 24). These comparisons clearly show the advantages inherent to SO_3 sulfonation processes (Table 9, example F).

Continuous Oleum Sulfonation Processes. Procter & Gamble reportedly operated a continuous oleum sulfation and sulfonation cyclic loop (dominant bath) process starting in 1937 (258). Detergent alkylate is contacted with oleum in a stream of freshly produced sulfonic acid (20:1 recycle at 55°C) in a gear pump, then through a heat exchanger in one loop, followed by a five minute pipeline digestion. One process option proceeded with addition of water to the sulfonation mixture in a second recycle loop (20:1) through a heat exchanger followed by continuous gravity separation of spent acid and neutralization of the sulfonic acid. A second process option included a tandem fatty alcohol oleum sulfation

Table 9. Comparative Summary of Various Batch and Continuous Detergent Alkylate Sulfonation Processes using H_2SO_4, Oleum, and Gaseous SO_3

Line	Sulfonation process examples	A	B	C	D	E	F
1	process source	Conoco	Stepan	Stepan	Kircher[a]	Chemithon	Stepan
2	sulfonation mode	batch	batch	batch	batch	continuous	continuous
3	sulfonation reagent	98% H_2SO_4	100.6% H_2SO_4	104.5% H_2SO_4	104.5% H_2SO_4	104.5% H_2SO_4	SO_3 film
4	alkylate type	BAB[b]	BAB[b]	BAB[b]	BAB[b]	LAB[c]	LAB[c]
	reagent:alkylate						
5	wt ratio	1.66:1.0	1.48:1.0	1.15:1.0	1.25:1.0	1.18:1.0	0.344:1.0
6	mol ratio	4.0:1.0	3.72:1.0	2.8:1.0	3.16:1.0	3.0:1.0	1.02:1.0
7	sulfonation temp, °C	66	54	28	25	54	54[d]
8	digestion temp, °C	66	54	28	25	54	54
9	H_2SO_4 conc after digestion, %	91.9	94.7	97.2	97.6	97.3	1.3
10	composition Rx[e] mix after digestion						
11	active sulfonic acid, %	49.4	53.4	61.3	58.3	61.3	96.5
12	H_2SO_4 (as 100%), %	45.9	43.8	36.8	40.0	37.0	1.3
13	water, %	4.1	2.6	1.1	0.8	1.0	1.0[f]
14	free oil, %	1.5	0.6	0.8	0.9	0.7	1.2
15	calcd % alkylate Rx[e] (SO_3), %	0.0	13.2	68.2	74.1	70.2	100.0
16	neutralized Na sulfonate, Na_2SO_4	44.3:55.7	47.3:52.7	50.7:49.3	51.8:48.2	62.3:37.7	98.2:1.8
17	kg H_2O added by kg Rx[e] mix[g]	0.13	0.052	0.096	0.144	0.133	
18	spent acid kg/kg sulfonic acid	1.14	0.77	0.75	0.79	0.70	
19	final sulfonic acid comp						
20	sulfonic acid, %	87.5	86.0	87.0	88.5	86.8	
21	H_2SO_4 (as 100%), %	7.5	8.6	7.9	7.1	8.9	
22	water, %	3.9	4.2	3.9	3.3	3.4	
23	free oil, %	1.1	1.2	1.2	1.1	1.0	
24	Na sulfonate:Na_2SO_4, wt ratio	89.6:10.4	88.0:12.0	89.0:11.0	90.2:9.8	88.0:12.0	
25	~5% active Klett color	30	35	40	40	30	20

[a] Ref. 257. [b] BAB = branch (polypropylene tetramer) alkylbenzene. [c] LAB = linear alkylbenzene. [d] Sulfonic acid temperature exiting reactor. [e] Rx = reaction. [f] 1.0 wt % water added after digestion to break anhydrides and pyroacids. [g] If phase separation used.

175

reaction conducted in the continuous stream of freshly produced and digested sulfonic acid within a third recycle cooling loop. Immediate neutralization of the sulfonation–sulfation mixture followed. The neutralized product was high in Na_2SO_4 which served as filler for the subsequently formulated and spray dried detergent. This process avoided the use of a refrigerated cooling system.

Chemithon secured patents on a somewhat similar oleum-type continuous sulfonation and sulfation process (259,260). So-called dominant bath sulfonation processes function well because they appear to operate in or near the homogeneous area of the phase diagram (1) and heat removal is readily controllable with relatively small incremental reaction.

Continuous single-pass oleum sulfonation of detergent alkylate has been conducted using a Votator (Girdler Corporation) scraping blade cooling heat exchanger operating at 500–900 rpm (261). Although it generally was believed that oleum sulfonation was a relatively slow reaction, particularly between two immiscible liquid reactants, the reaction was reportedly 90% complete within 30 s at ca 27°C using refrigerated cooling, followed by a 30 min digestion at ca 33°C (262). This underscores the importance of mixing. Meccaniche Moderne and Ballestra have continuous oleum sulfonation processes based on stirred tank sulfonation, digestion, spent acid separation, and neutralization.

Since the 1950s, oleum detergent sulfonation processes have been gradually replaced by continuous SO_3 sulfonation processes, particularly when making liquid detergents. SO_3 processes are preferred because high purity, high active products are produced directly with low inorganic by-products. The oleum process is encumbered with its generation of significant spent acid unless it is neutralized directly and utilized as filler in spray dried detergent products. The trend to eliminate detergent spray drying has catalyzed a shift to higher active, low salt, compacted, agglomerated, and extruded solid products derived from SO_3-based sulfonation processes.

Batch Stirred Tank SO_3 Sulfonation Processes. If the color of the derived sulfonate is not critical, such as in the production of oil-soluble ag-emulsifiers, a simple batch sulfonation procedure can be employed based on vaporizing liquid SO_3 (Ninol Labs, 1952) (13,263). Pilot Chemical Company adapted the original Morrisroe 60–70% oleum–SO_2 solvent sulfonation process (256) to utilize 92% liquid SO_3–8% liquid SO_2 mixtures, and more recently using 100% liquid SO_3. This cold sulfonation low viscosity sulfonation process produces excellent quality products, and reportedly has also been adapted for continuous processing as well. The derived sulfonic acid must be stripped of SO_2 solvent after completing sulfonation and digestion.

Conoco operated a stirred tank Pfaudler glass-lined reactor for the batch SO_3 sulfonation of detergent alkylate. The plant utilized over-the-fence SO_3 converter gas (8% SO_3 in dry air) having ~6–8 h batch cycles (264). Allied Chemical Company provided details for batch SO_3 sulfonation (265,266) and Conoco also published their procedure for SO_3 batch sulfonation (267). Andrew Jergens Company patented a cyclic batch sulfonation and sulfation process introducing nondiluted SO_3 vapors into a venturi contacter that emitted reaction product into a stirred reservoir tank where it was recycled from the reservoir vessel through a heat exchanger and back to the venturi in the cyclic loop. The unit operated in a vacuum (268). Derived color quality was unspecified.

Continuous SO₃ Single-Pass Sulfonation Processes. Conoco, Allied Chemical Company, and Gurdler Corporation conducted extensive studies using horizontally operating high speed scraping blade Votator heat exchangers, both for cyclic batch and single-pass SO₃ sulfonation. Cooling was only available for the outer wall of the annular space where reaction was conducted (269). This process has apparently only been used commercially for the production of petroleum sulfonates (270).

In 1960 Stepan Company became the first to develop and commercially operate a continuous falling film SO₃ sulfonation process using its proprietary design multitubular unit for the production of alkyl benzene sulfonates, fatty alcohol sulfates, and alkylphenolethoxy sulfates (4,13). The process has subsequently been adapted to produce alcohol ethoxysulfates, alpha-olefin sulfonates, and petroleum sulfonates. The company operates about 12 falling film SO₃ sulfonation units in the United States with a combined estimated annual capacity of about 500,000 t as active LAS. All falling film SO₃ sulfonation units operate cocurrently with both organic feedstocks and high velocity diluted SO₃ gas being introduced downwardly at the top of the reactors.

Table 10 provides a summary of estimated operating rates, sulfonator design characteristics, and dimensions based on published data (1) and manufacturers' technical bulletins. Continuous falling film processes are of two basic designs: concentric annular reactors (two vertical adjacent reaction surfaces) and vertical multitubular reactors. Sulfonation units can be supplied for utilizing liquid SO₃ reagent or designed for *in situ* sulfur burning and SO₃ generating equipment.

Chemithon's original falling film units utilized a moving rotor operating between the two concentric vertical reactor walls at the top of the unit to apply a uniform film of organic (271). Units available as of the mid-1990s have eliminated the rotor system and utilize calibrated and replaceable metering flanges to control the flow of organic. Chemithon's film reactors are of a somewhat unique design in utilizing a narrower reactor wall spacing (gap) providing for essentially complete reaction to occur within their short 2-m length of tubes. The relative hot sulfonic acid (ca 75°C) is immediately quench cooled with previously produced, cooled, and recirculated acid. Organic residence time in the reactor is estimated to be only 10–15 s, but acid remains in the quench cooling loop substantially longer, although it can be varied from a few to 15 minutes. Data in Table 10 for Chemithon's annular film reactor indicates that it utilizes a very high estimated gas velocity (~75 m/s (280 km/h)). Chemithon increases reactor design capacity by increasing the diameter of its concentric reactor tubes while maintaining the same gap. Derived products meet the highest industry quality standards.

Meccaniche Moderne (Busto Arsizo, Italy) acquired rights to the original Allied Chemical Company's concentric SO₃ sulfonation unit (272). Table 10 shows comparative data on Allied's U.S. design, and that for Meccaniche Modern's concentric design, indicating equivalency. This unit has a substantial space between reactor walls, hence it requires a 6 m length to complete reaction. The process has been scaled up to 4.0 t/h capacity.

Meccaniche Moderne multitube falling film continuous SO₃ sulfonation units are designed similarly to a heat-exchange tube bundle. They maintain essentially the same organic loading and gas velocity as for their concentric

Table 10. Estimated Typical Operating Conditions and Rates of Commercial Continuous SO$_3$ Falling Film Sulfonation Processes for Sulfonation of LAB and Lauryl Alcohol-3 Mol Ethoxylate Feedstocks[a]

Falling film SO$_3$ suppliers	Allied Chemical	Meccaniche Moderene	Chemithon Corp.	Ballestra SpA	Mazzoni Spa	Meccaniche Moderene
commercial designation		monotube reactor	annular film reactor	Sulphorex F	Sulpho film reactor	multitube
reactor type	concentric	concentric	concentric	multitube	multitube	multitube
LAB–sulfonic acid, kg/h	755	2,540	5,214	6,040	5,964	1,980
annual production rate LAB–sulfonic acid 8,000 h, metric tons[b]	1,510	20,000	41,712	48,318	47,200	15,840
reaction tubes	2	2	2	142	142	45
tube(s) diameter, cm	25.40, 21.90	102.8, 99.3	121.9, 120.5	2.54	2.54	2.50
concentric reactor wall spacing, cm	3.50	3.50	1.44			
gap (1/2 spacing), cm	1.75	1.65	0.72			
tube length, m	6.7	6.0	2.0	6.0	6.0	6.0
reactor tubes circumference, m	1.49	6.35	7.62	7.98	7.98	3.53
organic loading, kg/(h·mm)	0.38	0.4	0.4	0.4	0.4	0.3
calculated SO$_3$ conc, vol/vol, %	3.3	3.3	4.0	4.0	3.0[c]	5.0
nominal reaction gas velocity,[d] m/s	33.4	34.1	74.9	39.1	51.5	32.0
gas velocity, km/h (mph)	120 (74.7)	124.6 (77.5)	269 (167)	140.0 (87.5)	185.4 (115.2)	115 (71.5)
estimated gaseous residence time, s[e]	0.20	0.17	0.09	0.15	0.12	0.19
cooling jacket zones	3	3[f]	2	3	3	3
cooling system surface, m^2	10.0	38.1	15.2	67.9	67.9	21.2
surface per kg/min LAB sulfonic acid, m^2	0.80	0.67	0.225	0.675	0.82	0.643
post-reactor quench	no	no	yes	no	no	no
recommended alcohol-3M-ethoxylate loading, kg/(h·mm)	0.3	0.3	0.28	0.341	0.318	
calculated corresponding SO$_3$ conc, vol/vol, %[g]	1.96	1.9	2.6	2.5	1.8	

[a]Estimates based on published and industry information sources for production of linear alkyl (C$_{12}$) benzene alkylate sulfonic acid and lauryl-3 mol ethoxy sulfate. [b]Suppliers' process equipment and rates; higher capacity units may be available. [c]SO$_3$ concentration includes sum of initial reaction gas plus equalizer diluent air. [d]Nominal reaction gas velocity calculated in absence of organics. [e]Typical gaseous residence time calculated in absence of organics. [f]Multiple zones for outer reactor wall, one for inner wall. [g]At constant air supply.

reactor unit. The organic is fed to each tube through a mechanically calibrated slot between the internal organic distributor and the gas injector nozzle. Their unique reactor design provides for relatively easy removal of each tube, and reactor tubes can be blocked off by blank flanges at the head of the reactor, thus reducing unit capacity if desired. The lower ends of reactor tubes have double seals and packing glands designed to prevent leakage (273). Excellent quality products are reported.

Mazzoni SpA (Busto Arsizo, Italy) multitube falling film continuous SO_3 sulfonation units also utilize 6-m length tubes. A moderate venturi-like restriction at the top of each tube provides for equal gas distribution. The organic feed is distributed to each tube through stationary nozzles made of agate stone. The Mazzoni design features the introduction of an equalizer (dry) air stream at the head of each tube to press the organic flow against each reactor wall and help ensure the proper balancing of reaction gas to each reactor tube, should individual tube back-pressures differ (274). It has also been suggested that the equalizer air entering the tube first, parallel to the organic film, may slightly delay reaction and perhaps mute or dampen peak reaction temperature, but this has not been clearly established (1). The unit utilizes a double high efficiency liquid–gas separator system. It has been reported that Mazzoni and Ballestra have merged.

Ballestra SpA (Milan, Italy) multitube falling film continuous SO_3 sulfonation units utilize 6-m length reaction tubes, arranged similarly to a heat-exchanger tube bundle. The reactor tubes are enlarged at the top of the unit to accommodate the organic feed system and the gas inlet nozzle which has the same inlet diameter as the remaining portion of the reactor tube. Hence, the gas inlet contributes no pressure drop to the tube. This reactor design is claimed to provide self-compensation (self-equilibration) for balancing reaction gas–liquid mol ratio in each individual tube (275,276). Organic feed enters into each tube through a calibrated slot and flows around the protruding gas inlet. Excellent derived product quality is reported.

Lion Fat & Oil Company (Tokyo, Japan) concentric annular falling film continuous SO_3 sulfonation unit appears to be a modification of a Chemithon-type unit. It was reportedly designed primarily to sulfonate alpha-olefin which exhibits the highest heat of reaction and is the most heat sensitive of conventional feedstocks. The process introduced a substantial volume of a secondary dry air stream somewhat similar to an equalizer air stream (274) and driving gas stream (276). The air curtain introduced next to the downward flowing feedstock film is purported to set up a resistance, or slightly delay diffusion of the SO_3–air reaction gas to the organic film. The SO_3 reaction gas is introduced downwardly in the middle of the annular space between the two reaction walls. It is suggested that this system somewhat dampens the characteristic high peak temperatures typically generated in most other film sulfonation systems. The patent claims sulfonation under substantially isothermal conditions, although examples were run at only a 0.03 kg/(h·mm) organic loading which is 13.3 times lower (thinner) than typical 0.4 kg/(h·mm) for most other units (277,278). The quality of α-olefin sulfonate (AOS) is slightly better than from competitive units.

Summary of Characteristics of Falling Film Continuous SO_3 Sulfonation Processes. Both concentric and multitubular reactor systems supplied by competing manufacturers have surprisingly similar operating characteristics:

organic feedstock loading of ca 0.4 kg/(h·mm) (circumference) for LAB, and ca 0.3 kg/(h·mm) for alcohol ethoxylates; an SO_3 concentration of 3.3–5.0 vol % SO_3 for LAB sulfonation, and 2–3% SO_3 for alcohol ethoxylate sulfation; estimated gas reactor residence times of 0.09–0.2 s; reactor cooling surface of 0.7–0.8 m^2/kg LAB sulfonic acid/min (Chemithon unit = 0.23 m^2/(kg·min) cooling surface, however, it also utilizes the quench cooling heat exchanger at the base of the reactor to remove additional heat of reaction); and estimated gas velocities of 32–75 m/s (115–269 km/h (72–167 mph)) are employed. As noted, most falling film continuous SO_3 sulfonation units operate in the range of hurricane wind velocities (121–322 km/h) (1,13). Liquid residence times are estimated to be ca 10–30 s.

Organic Film and High Velocity Gas. Organic film loading of 0.4 kg/(h·mm) is equivalent to 0.32 mm average thickness at the point of entry, increasing to about 1 mm average film thickness at peak temperature where the reaction is generally ca 45% completed, and finally to about 2 mm average film thickness as the acid exits the reactor (1). Falling film continuous SO_3 sulfonation systems operating at or near hurricane wind velocities continually generate significant sized waves churning and dashing the organic film as it moves down the reactor somewhat like hurricane winds blowing over a body of water, although higher viscosity liquids are involved (Fig. 2).

Heat of Sulfonation. The use of high velocity SO_3 reaction gas in falling film sulfonation processes results in a rapid reaction producing a momentary relatively high peak temperature in the film within the first 10–20% of the reaction surface length. This is subsequently reduced as the film proceeds and is cooled traveling down the tube. The process works so effectively because the viscosity of the reaction mixtures is substantially reduced at these higher temperatures (see Fig. 1) aiding the mixing and enhancing rapid heat exchange primarily through film cooling at the reactor wall. Even though the peak temperature may be about 90°C for a short time (ca 0.02 s gas residence time), it generally does not appear to contribute to product color deterioration as experienced in batch

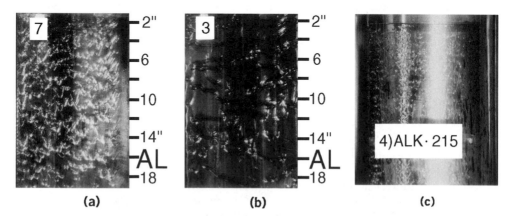

(a)   (b)   (c)

Fig. 2. High speed photos of organic film–high velocity air dynamics in falling film sulfonation process: (**a**) and (**b**) are vertical flat plate organic–air dynamics where (**a**) shows BAB–air at top, (**b**) BAB sulfonic acid–air at bottom of reactor; (**c**) simulated Allied-type concentric reactor inner cylindrical reaction surface showing BAB–high velocity air (see Table 10) (1 in. = 2.54 cm).

sulfonation processes for temperatures above 55°C (see Fig. 1). However, olefin sulfonation is an exception, exhibiting thermal sensitivity owing to the high heat of sulfonation (see Table 8) and low feedstock molecular weight.

Product Quality. Under ideal plant operating conditions, the quality of products derived from continuous SO_3 film sulfonator units is unlikely to be significantly different (1). Typical LAB sulfonic acid composition is ca 96.6% active sulfonic acid, 1.2% free oil, 1.2% H_2SO_4, and 1.0% water, the last added for stabilization purposes to break sulfonic acid anhydrides and pyro-acids (279). Klett color for a 5% active solution is typically 15–25.

Dioxane Problem. The sulfation of ethoxylated feedstocks with SO_3 results in the production of small quantities of dioxane, a toxic by-product. The level of dioxane produced in falling film SO_3 sulfation processes can be minimized by reducing the SO_3 mole ratio, utilizing more dilute SO_3 gas such as 2–3% SO_3, using higher velocity gas, reducing product throughput thereby reducing peak reaction temperature, and minimizing sulfated acid residence time in equipment prior to neutralization. Utilizing the above processing techniques for the production of 2–3 mol ethoxysulfates with free oils of 2.0% on a 100% active basis, dioxane levels of <50 ppm on a 100% active basis appear attainable (1). Lower dioxane levels require expensive steam and/or wipe film stripping procedures.

Processing Features. Process control of reactant's mol ratio is critical for optimal derived product quality because under-sulfonation usually produces unacceptable product, whereas over-sulfonation leads to undesirable side-reaction products and increased color. Comparison of product quality derived from the various equipment suppliers' units may be obscured by varying production design rates, production guarantee rates, quality of feedstocks utilized, actual operation of the plant, and by different analytical methods used in monitoring production, particularly for so-called product free oil. The azeotropic distillation method recovers only volatile components in the free oil by extraction methods which are not always quantitative, especially for ethoxylated feedstocks. The ion-exchange method is most reliable. Several publications report on mathematical modeling of falling film SO_3 sulfonation processing (1,280–283).

SO_3 Diluent Gas. In the continuous SO_3 sulfonation of LAB or BAB using 5% SO_3 in 95% dry diluent gas, the volume ratio of liquid feedstock to reaction gas is ca 1:1576, confirming the true nature of film reaction involved in the process. Continuous SO_3 processes based on dilute SO_3 gas systems utilize large volumes of air which must be compressed and dried preferably to a −60 to −70°C dew point. Such equipment and its operation are costly. Efforts have been made to recycle the reactor gas effluent which contains small quantities of SO_2, SO_3–H_2SO_4 mist, and organics (279,284). Various demister and air-filtration systems have been commercially evaluated but to date (ca 1997) none have been totally successful in producing high quantity, light-colored detergent products using recycled air (13). Studies have shown that as little as 0.002% (0.5–1.0 μm) of charged feedstock or sulfonated organics carried in the exiting and filtered reactor gas stream is sufficient to contribute product discoloration problems on gas recycling. In addition, SO_2 buildup soon becomes significant (13,263). Air recycling does not appear viable for conventional sulfur burning sulfonation systems.

Continuous Falling Film Sulfonation Process Flow.   Process flow diagrams, particularly for processes based on vaporizing liquid SO_3, reactor design details, and a critique of process operating details, are available (1,13). Figure 3 provides a process flow diagram for a typical falling film continuous SO_3 sulfonation plant with *in situ* sulfur burner SO_3 generating equipment. Air is, A, compressed; B, cooled; and C, dried to a dew point of -60 to $-70°C$, then supplied to the sulfur burner together with molten sulfur. SO_2 generated by the sulfur burner, D, is partially cooled (to $420°C$), E, and sent to the V_2O_5 fixed-bed catalytic three-stage converter, F, generating approximately 6–7% SO_3 gas stream with a typical 98.5–99% conversion. Most plants are equipped with an SO_3 adsorber system, G, capable of adsorbing SO_3 from the complete SO_3–air gas stream and is used on plant startups or during power failure periods. This generates H_2SO_4 during said times and the stripped air then passes through a packed (dilute NaOH) scrubbing tower, H, and is vented.

After the SO_3 converter has stabilized, the 6–7% SO_3 gas stream can be further diluted with dry air, I, to provide the SO_3 reaction gas at a prescribed concentration, ca 4 vol % for LAB sulfonation and ca 2.5% for alcohol ethoxylate sulfation. The molten sulfur is accurately measured and controlled by mass flow meters. The organic feedstock is also accurately controlled by mass flow meters and a variable speed-driven gear pump. The high velocity SO_3 reaction gas and organic feedstock are introduced into the top of the sulfonation reactor, J, in cocurrent downward flow where the reaction product and gas are separated in a cyclone separator, K, then pumped to a cooler, L, and circulated back into a quench cooling reservoir at the base of the reactor, unique to Chemithon concentric reactor systems. The gas stream from the cyclone separator, M, is sent to an electrostatic precipitator (ESP), N, which removes entrained acidic organics, and then sent to the packed tower, H, where SO_2 and any SO_3 traces are adsorbed in a dilute NaOH solution and finally vented, O. Even a 99% conversion of SO_2 to SO_3 contributes ca 500 ppm SO_2 to the effluent gas.

In some competitive sulfonation systems, a more efficient liquid–gas separator system is employed, thereby eliminating the need for the ESP unit, and utilizing a demister unit along with the dilute caustic scrubber. The ESP system must be bypassed for safety reasons when used with feedstocks having relatively low boiling volatiles. Cooled sulfonic acid is continuously removed from the quench cooling loop, generally at about 45–50°C, and pumped through digestors, P, which provide for a 10–30 min (adjustable) digestion period to help complete the reaction, after which about 1% water is injected through an in-line mixing system, Q, to destroy residual pyro-acids and sulfonic acid anhydrides and to help stabilize acid color (279). The water injection system is bypassed when alcohol sulfation is being conducted.

Optimization of alkylbenzene sulfonic acid quality requires a subtle balancing between controlling mol ratio and maximizing digestion in order to achieve highest conversion, lowest free oil and H_2SO_4, with lightest derived product color (279). The resultant acid is then sent to the loop neutralization system, R, comprising a positive displacement pump, in-line alkaline mixing system, and cooling heat-exchanger loop.

Sulfonation Plant Operations and Gas Effluent.   Standards governing U.S. sulfonation plant gas effluents differ depending on whether or not the plant

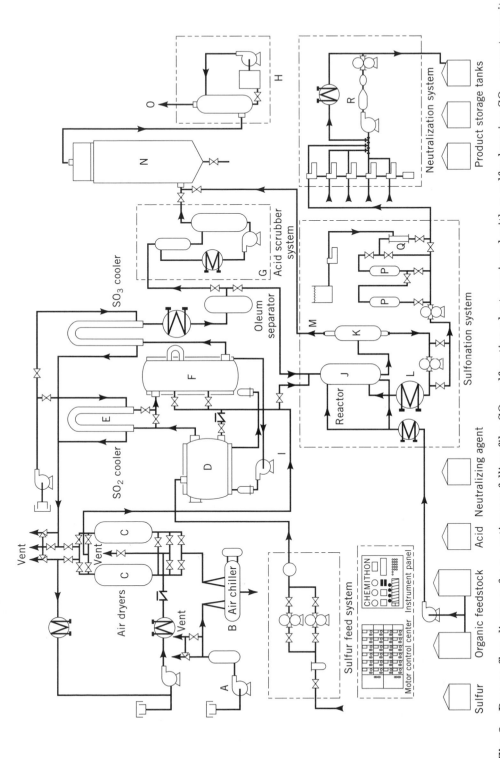

Fig. 3. Process flow diagram for a continuous falling film SO_3 sulfonation plant, equipped with a sulfur-burning SO_3 converter unit. See text. Courtesy of Chemithon Corp.

is equipped with a H_2SO_4 scrubbing system for adsorption of SO_3 gas (see Fig. 3). The installation of the SO_3 adsorber system qualifies the plant as a sulfuric production plant which has stringent regulations. Limitations and typical effluent from the sulfonation system are as follows:

Material	Requirements	Typical
SO_2	≤ 5 ppm	<1 ppm
$SO_3–H_2SO_4$	≤ 10 mg/m^3	<8 mg/m^3
organic mist	≤ 20 mg/m^3	<5 mg/m^3
opacity	$\leq 20\%$	$<5\%$

Most sulfonation plants monitor and control operations by computer. Sulfur-burning catalytic SO_3-generating equipment may require a 1–2 h stabilization period on startup. The unit can be kept in a standby position by maintaining heat to the unit when it is off-line. Liquid SO_3-based sulfonation plants do not require such a stabilization period and hence are more flexible to operate than sulfur-burning sulfonation plants.

Other Continuous SO$_3$ Processes.   The Ballestra multistirred tank continuous SO_3 cascade Sulphorex sulfonation system generally employs four stirred reactors so as to cascade from one to the next. SO_3 dry air mixture is specially proportioned to enter into each reactor, such as 50, 25, 20, and 5%, respectively, through a sparger system. Total product residence time is estimated at 90 minutes on average, hence this system is primarily recommended for sulfonation of detergent alkylates and fatty methyl esters (1). A continuous falling film SO_3 sulfonation process was patented based on a multitabular design, utilizing SO_3 gas without diluent gas and operating under subatmospheric conditions (285). However, patent examples failed to disclose critical derived product color properties.

The Albright & Wilson continuous SO_3 loop sulfonation system claims use of a static mixer or mechanical mixer system under conditions of turbulent flow at the point of contact with undiluted gaseous SO_3 (286,287). Patent examples include xylene, toluene, and fatty alcohol, reaction recycle is between 40:1 and 2000:1, but no product color data is provided.

The Procter & Gamble continuous dominant bath SO_3 sulfonation process utilizes 7.5% SO_3 dry gas stream introduced into a zone of turbulent flow. This flow is generated by either an in-line high speed mixer or venturi nozzle, which operates in a recycle cooling loop (5:1 to 30:1 ratio) (288). A continuous sulfonation process was disclosed wherein organic feedstock and a sulfonating agent including SO_3 was injected as a mist onto a rotating cylindrical cooling surface (289).

The Stratford Engineering Company's (Kansas City, Missouri) continuous SO_3 organic mist sulfonation uses a high speed atomizing rotor to horizontally disperse the organic feedstock stream impinging on the reactor walls in the presence of SO_3 gas to effect sulfonation of petroleum feedstocks (290).

Chemithon Corporation's continuous SO_3 jet-impact sulfonation process is a commercially available unit and appears to be a combination of the diluted gaseous SO_3–organic mist and SO_3–falling film sulfonation systems. Organic feedstock and a $4-10\%$ SO_3 dry air stream are rapidly mixed in a venturi nozzle and impinged onto a downwardly flowing stream of recycling reaction mixture operating in a cooling loop (291). The quality of derived products is reported to be only moderately less than those from falling film sulfonation processes. The unit has been utilized for various products including production of oil-soluble sulfonates.

A conically fabricated rotating continuous sulfonation system has been disclosed for the gaseous SO_3 sulfonation of organics flowing down the reactor walls with cooling under centrifugal action (292). More recently, two other rotating reaction systems have been claimed for the continuous SO_3 sulfonation of organics flowing under centrifugal action using undiluted gaseous SO_3 under vacuum (293,294). As yet, neither system has been used commercially.

SO_3-based sulfonation processes, particularly continuous falling film sulfonation processes, comprise ca 65% of total U.S. sulfonation plant capacity. This is expected to increase, particularly based on sulfur burner SO_3 generation sulfonation systems. Falling film continuous SO_3 sulfonation processes dominate because of greater versatility, lower reagent costs, elimination of by-product streams, and capability for direct production of high purity, high quality products (13). Details of a small laboratory apparatus for the experimental falling film continuous SO_3 sulfonation have been published (295).

BIBLIOGRAPHY

"Sulfonation and Sulfation" in *ECT* 1st ed., Vol. 13, pp. 317–337, by E. E. Gilbert, Allied Chemical Corp.; in *ECT* 2nd ed., Vol. 19, pp. 279–310, by E. E. Gilbert, Allied Chemical Corp.; in *ECT* 3rd ed., Vol. 22, pp. 1–45, by E. A. Knaggs, M. L. Nussbaum, and A. Shultz, Stepan Chemical Co.

1. W. H. deGroot, *Sulphonation Technology in the Detergent Industry*, Kluwer Academic Publishers, Dorrecht, the Netherlands, 1991.
2. U.S. International Trade Commission Report, *Synthetic Organic Chemicals, U.S. Production & Sales*, Government Printing Office, Washington, D.C. and estimated product segments, 1991.
3. R. Apel and W. Senkpiel, *Chem. Ber.* **91**, 1195 (1958).
4. U.S. Pat. 3,169,142 (Feb. 9, 1965), E. A. Knaggs and M. L. Nussbaum (to Stepan Co.).
5. *C & E News*, 30 (Jan. 25, 1995).
6. E. E. Gilbert, *Sulfonation and Related Reactions*, Interscience Publishers, New York, 1965; reprinted by R. E. Kreiger Publ. Co., Melbourne, FL.
7. E. E. Gilbert, *Chem. Rev.* **62**, 549 (1962).
8. H. Cerfontain, *Mechanistic Aspects in Aromatic Sulfonation and Desulfonation*, Interscience Publishers, New York, 1965.
9. W. M. Linfield, *Anionic Surfactants*, Parts I & II, Vol. 7, *Surfactant Science Series*, Marcel Dekker, Inc., New York, 1976.
10. J. Falbe, ed., *Surfactants in Commercial Products: Theory, Technology and Application*, Springer-Verlag, New York, 1986.
11. S. Patai and Z. Rappoport, eds., *The Chemistry of Sulphonic Acids, Esters and Their Derivatives*, John Wiley & Sons, Ltd., Chichester, U.K., 1991.
12. K. K. Andersen, in D. N. Jones, ed., *Sulphonic Acids and Their Derivatives*, Vol. 3, Pergamon Press, Oxford, U.K., 1991.
13. E. A. Knaggs, *Chemtech*, 436–445 (July 1992).
14. B. Levitt, *Oil Fat and Soap*, Chemical Publishing Co., New York, 1951, p. 33.

15. *Ullmann Encyclopedia*, 4th ed., Vol. 8, VCH Publishers, Weinheim, Germany, p. 414; E. E. Gilbert, *C&ENews*, 2 (Oct. 9, 1989).
16. U.S. Pat. 2,786,833 (Mar. 26, 1957), O. B. Wurzburg and co-workers (to National Starch Co.).
17. A. F. Turbak, *Ind. Eng. Chem., Prod. Res. Dev.* **1**, 275 (1962).
18. A. J. Stirton and co-workers, *J.A.O.C.S.* **29**, 198 (1952).
19. J. Hoyle, in S. Patai and Z. Rappoport, eds., *The Chemistry of Functional Groups*, John Wiley & Sons, Ltd., Chichester, U.K., 1991, Chapt. 10.
20. Ref. 1, Chapt. 4.
21. H. Cerfontain and A. Koeberg-Telder, *Recl. Trav. Chim. Pays-Bas*, **107**, 543 (1988).
22. *Ibid.*, p. 583.
23. P. de Wit and H. Cerfontain, *Recl. Trav. Chim. Pays-Bas*, **107**, 121 (1988).
24. U.S. Pat. 4,347,147 (Aug. 31, 1982), R. J. Allain and D. W. Fong (to Nalco Chemical Co.).
25. R. Varadaraj, P. Valint, Jr., J. Bock, S. Zushma, and N. Brons, *Langmuir*, **7**, 658 (1991).
26. B. Besergil and B. M. Baysal, *Ind. Eng. Chem. Res.* **29**, 667 (1990).
27. H. R. W. Ansink and H. Cerfontain, *Recl. Trav. Chim. Pays-Bas*, **111**, 183 (1992).
28. R. J. Cremlyn, F. J. Swinbourne, P. Fitzgerald, N. Godfrey, P. Hedges, J. Lapthorne, and C. Mizon, *Ind. J. Chem.* **23B**, 962 (1984).
29. H. R. W. Ansink and H. Cerfontain, *Phosphorus, Sulfur, and Silicon*, **63**, 335 (1991).
30. Ref. 23, p. 418.
31. U.S. Pat. 4,778,629 (Oct. 18, 1988), F.-F. Grabley, G. Reinhardt, G. Bader, and W. Rupp (to Hoechst Aktiengesellschaft).
32. U.S. Pat. 4,931,563 (June 5, 1990), S. A. Madison and L. M. Ilardi (to Lever Brothers Co.).
33. U.S. Pat. 5,072,034 (Dec. 10, 1991), P. Neumann, H. Eilingsfeld, and A. Aumueller (to BASF Aktiengesellschaft).
34. U.S. Pat. 5,124,475 (June 23, 1992), M. J. Nepras, P. F. Heid, M. I. Levinson, R. J. Bernhardt, and J. A. Hartlage (to Stepan Co.).
35. U.S. Pat. 4,721,805 (Jan. 26, 1988), M. Nussbaum (to Stepan Co.).
36. U.S. Pat. 4,692,279 (Sept. 8, 1987), M. Nussbaum (to Stepan Co.).
37. P. R. L. Grosjean and H. Sawistowski, *Int. Solv. Extr. Conf. Proc.* **1**, 80 (1980).
38. D. Zhao, J. Zhang, and C. Luo, *Huaxue Fanying Gongcheng Yu Gongyi*, **7**, 191 (1991).
39. M. Sohrabi, *Chimia*, **37**, 465 (1983).
40. A. El Homis, C. Guiglion, and R. Guiraud, *J. Chim. Phys.-Chim. Biol.* **80**, 383 (1983).
41. Eur. Pat. 0 583 960 (Aug. 13, 1993), E. V-P. Tao and W. D. Miller (to Eli Lilly and Co.).
42. H. Cerfontain, H. R. W. Ansink, N. J. Coenjaarts, E. J. de Graaf, and A. Koeberg-Telder, *Recl. Trav. Chim. Pays-Bas*, **108**, 445 (1989).
43. R. J. Cremlyn, J. M. Lynch, and F. J. Swinbourne, *Phosphorus, Sulfur, and Silicon*, **57**, 173 (1991).
44. H. C. A. van Lindert, A. Koeberg-Telder, and H. Cerfontain, *Recl. Trav. Chim. Pays-Bas*, **111**, 379 (1992).
45. P. de Wit and H. Cerfontain, *Recl. Trav. Chim. Pays-Bas*, **104**, 25 (1985).
46. A. Previero, J-C. Cavadore, J. Torreilles, and M-A. Coletti-Previero, *Biochim. Biophys. Acta*, **581**, 276 (1979).
47. U.S. Pat. 4,764,307 (Aug. 16, 1988), N. J. Stewart (to The British Petroleum Co.).
48. M. A. Munoz, M. Balon, C. Carmona, J. Hidalgo, and M. L. Poveda, *Heterocycles*, **27**, 2067 (1988).
49. H. R. W. Ansink, E. Zelvelder, and H. Cerfontain, *Recl. Trav. Chim. Pays-Bas*, **112**, 210 (1993).
50. H. Cerfontain, A. Koegerg-Telder, K. Laali, H. J. A. Lambrechts, and P. de Wit, *J. R. Neth. Chem. Soc.* **101**, 390 (1982).

51. F. A. Carey and R. J. Sundberg, *Advanced Organic Chemistry, Part A: Structure and Mechanism*, Plenum Press, New York, 1977, Chapt. 9.

52. U.S. Pat. 4,468,476 (Aug. 28, 1984), K. Yang and J. A. Reedy (to Conoco, Inc.).

53. U.S. Pat. 4,556,733 (Dec. 3, 1985), T. A. Sullivan and V. G. Witterholt (to E. I. du Pont de Nemours and Co., Inc.).

54. V. L. Krasnov, N. K. Tulegonova, and I. V. Bodrikov, *Zh. Org. Khimii*, **15**, 1997 (1979).

55. B. G. Ugarkar, B. V. Badami, and G. S. Puranik, *Arch. Pharm.* **312**, 977 (1979).

56. R. J. Cremlyn, J. P. Bassin, S. Farouk, M. Potterton, and T. Mattu, *Phosphorus, Sulfur, and Silicon*, **73**, 107 (1992).

57. S. Hertzberg and S. Liaaen-Jensen, *Acta. Chem. Scand. B*, **39**, 629 (1985).

58. A. Piasecki, *Synth. Comm.* **22**, 445 (1992).

59. R. J. Cremlyn, F. J. Swinbourne, P. A. Carter, and L. Ellis, *Phosphorus, Sulfur, and Silicon*, **47**, 267 (1990).

60. R. J. Cremlyn and D. Saunders, *Phosphorus, Sulfur, and Silicon*, **81**, 73 (1993).

61. Eur. Pat. 0 293 672 (May 18, 1988), S. R. Sandler (to Pennwalt Corp.).

62. U.S. Pat. 4,876,047 (Oct. 24, 1989), G. Benn, D. Farrar, and P. Flesher (to Allied Colloids Limited).

63. U.S. Pat. 5,091,548 (Feb. 25, 1992), P. Gibraltar, J. S. Ku, and T. M. Schmitt (to BASF Corp.).

64. U.S. Pat. 4,987,249 (Jan. 22, 1991), S. R. Sandler (to Atochem North America, Inc.).

65. U.S. Pat. 4,737,354 (Apr. 12, 1988), H. Fuchs, F-J. Weiss, E. Thomas, and J. Ritz (to BASF Aktiengesellschaft).

66. Eur. Pat. 0 289 952 (Apr. 29, 1988), S. R. Sandler (to Pennwalt Corp.).

67. U.S. Pat. 5,189,206 (Feb. 23, 1993), P. Herzig (to Ciba-Geigy Corp.).

68. U.S. Pat. 4,555,580 (Nov. 26, 1985), D-I. Schutze, R. Schmitz, and K. Wunderlich (to Bayer Aktiengesellschaft).

69. U.S. Pat. 4,386,966 (June 7, 1983), P. H. Fitzgerald (to E. I. du Pont de Nemours and Co., Inc.).

70. Eur. Pat. 0 580 194 (June 10, 1993), A. Marcotullio, C. Corno, and E. Donati (to Eniricerche SpA and Snaprogetti SpA).

71. A. Koeberg-Telder and H. Cerfontain, *J. Org. Chem.* **51**, 2563 (1986).

72. H. Cerfontain, A. Koeberg-Telder, K. Laali, and H. J. A. Lambrechts, *J. Org. Chem.* **47**, 4069 (1982).

73. J. P. Bassin, R. J. Cremlyn, and F. J. Swinbourne, Phosphorus, Sulfur, and Silicon, **72**, 157 (1992).

74. P. de Wit, H. Cerfontain, and A. Fischer, *Can. J. Chem.* **63**, 2294 (1985).

75. H. R. W. Ansink, E. J. de Graaf, E. Zelvelder, and H. Cerfontain, *Recl. Trav. Chim. Pays-Bas*, **111**, 499 (1992).

76. D. Becker and H. J. E. Loewenthal, *Tetrahedron*, **48**, 2515 (1992).

77. P. Beltrame, G. Bottaccio, P. Carniti, and G. Felicioli, *Ind. Eng. Chem. Res.* **31**, 787 (1992).

78. U.S. Pat. 5,110,981 (May 5, 1992), N. Milstein (to Henkel Corp.).

79. Ger. Pat. 4,218,725 (Dec. 9, 1993), F. Pieschel, E. Lange, and F. Reimers (to Chemie AG Bitterfeld).

80. S. R. Sandler and W. Karo, in A. T. Blomquist and H. H. Wasserman, eds., *Organic Functional Group Preparations*, Vol. 12, Academic Press, Inc., New York, 1968, Chapt. 21, pp. 506–509.

81. S. J. Herron, *Chim. Oggi*, 19 (July/Aug. 1993).

82. H. A. Green, in W. M. Linfield, ed., *Olefin Sulfonates*, Part II, Vol. 7, Mercel Dekker, New York, 1976, Chapt. 10.

83. N. S. Zefirov, N. V. Zyk, Y. A. Lapin, and A. G. Kutateladze, *Sulfur Lett.* **8**, 143 (1988).

84. D. W. Roberts, S. Sztanko, and D. L. Williams, *Tenside, Surfactants, Deterg.* **18**, 113 (1981).
85. U.S. Pat. 5,075,041 (Dec. 24, 1991), E. F. Lutz (to Shell Oil Co.).
86. U.S. Pat. 4,235,804 (Nov. 25, 1980), C. G. Krespan (to E. I. du Pont de Nemours & Co., Inc.).
87. N. S. Zefirov, N. V. Zyk, S. I. Kolbasenko, and A. G. Kutateladze, *Sulfur Lett.* **2**, 95 (1984).
88. R. Herke and K. Rasheed, *J. Am. Oil Chem. Soc.* **69**, 47 (1992).
89. D. W. Roberts, J. G. Lawrence, I. A. Fairweather, C. J. Clemett, and C. D. Saul, *Tenside, Surfactants, Deterg.* **27**, 82 (1990).
90. N. S. Zefirov, N. V. Zyk, S. I. Kolbasenko, and A. G. Kutateladze, *J. Org. Chem.* **50**, 4539 (1985).
91. R. M. Schonk, B. H. Bakker, and H. Cerfontain, *Recl. Trav. Chim. Pays-Bas*, **111**, 49 (1992).
92. D. W. Roberts, D. L. Williams, and D. Bethell, *J. Chem. Soc. Perkin Trans. II*, 389 (1985).
93. J. M. Canich, M. M. Ludvig, G. L. Gard, and J. M. Shreeve, *Inorg. Chem.* **23**, 4403 (1984).
94. R. M. Schonk, B. H. Bakker, and H. Cerfontain, *Phosphorus, Sulfur, and Silicon*, **59**, 173 (1991).
95. B. H. Bakker and H. Cerfontain, *Tetrahedron Lett.* **28**, 1699 (1987).
96. D. W. Roberts, P. S. Jackson, C. D. Saul, and C. J. Clemett, *Tetrahedron Lett.* **28**, 3383 (1987).
97. R. J. Terjeson, J. Mohtasham, and G. L. Gard, *Inorg. Chem.* **27**, 2916 (1988).
98. H. Cerfontain, A. Koeberg-Telder, H. J. A. Lambrechts, W. Lindner, S-Z. Zhang, and E. Vogel, *J. Org. Chem.* **52**, 3373 (1987).
99. B. H. Bakker and H. Cerfontain, *Tetrahedron Lett.* **28**, 1703 (1987).
100. Ref. 91, p. 389.
101. A. J. Buglass and J. G. Tillett, in Ref. 11, Chapt. 19.
102. D. W. Roberts and D. L. Williams, *Tenside, Surfactants, Deterg.* **22**, 4 (1985).
103. P. Radici, L. Cavalli, and C. Maraschin, *Tenside, Surfactants, Deterg.* **31**, 299 (1994).
104. Eur. Pat. 0 351 928 (July 19, 1989), J. Stapersma and R. Van Ginkel (to Shell Internationale Research Maatschappij BV).
105. CA Pat. 2,001,276 (Oct. 24, 1989), D. W. Roberts and P. S. Jackson (to Unilever PLC UK).
106. U.S. Pat. 4,248,793 (Feb. 3, 1981), S. Sekiguchi, K. Nagano, Y. Miyawaki, and K. Kitano (to The Lion Fat and Oil Co., Ltd.).
107. B. H. Bakker, H. Cerfontain, and P. M. Tomassen, *J. Org. Chem.* **54**, 1680 (1989).
108. B. H. Bakker, R. M. Schonk, and H. Cerfontain, *Recl. Trav. Chim. Pays-Bas*, **109**, 485 (1990).
109. R. M. Schonk, I. Lembeck, B. H. Bakker, and H. Cerfontain, *Recl. Trav. Chim. Pays-Bas*, **112**, 247 (1993).
110. Jpn. Pat. 01272564 (Oct. 31, 1989), Y. Endo (to Lion Corp.).
111. Eur. Pat. 0 355 675 (Aug. 16, 1989), A. Behler and H. Eierdanz (to Henkel Kommanditgesellschaft auf Aktien).
112. D. W. Roberts and D. L. Williams, *J. Am. Oil Chem. Soc.* **67**, 1020 (1990).
113. J. Stapersma, H. H. Deuling, and R. van Ginkel, *J. Am. Oil Chem. Soc.* **69**, 39 (1992).
114. U. Deister and P. Warneck, *J. Phys. Chem.* **94**, 2191 (1990).
115. PL Pat. 0157473 (June 30, 1992), W. Pasiuk-Bronikowsak, T. Bronikowski, and M. Ulejczyk (to Polska Akademia Nauk, Instytut Chemii Fizycnej).
116. Jpn. Pat. 61267545 (Nov. 17, 86), Y. Sato, K. Matsuda, and N. Ozaki (to Nippon Mining Co., Ltd.).
117. Jpn. Pat. 62033144 (Feb. 13, 1987), K. Matsuda, Y. Sato, and T. Suzuka (to Nippon Mining Co., Ltd.).

118. I. Stanciu, L. Papahagi, R. Avram, C. Cristescu, I. Duta, and D. Savu, *Rev. Chim.* **41**, 885 (1990).

119. Jpn. Pat. 01221359 (Sept. 4, 1989) (to Keishitsu Ryubun Sh).

120. N. Noriuchi, *Kitasato Arch. Exp. Med.* **59**, 79 (1986).

121. U.S. Pat. 4,454,075 (June 12, 1984), B. Mees and H. Ramloch (to Hoechst Aktiengesellschaft).

122. Eur. Pat. 0 185 851 (Sept. 11, 1985), Y. Sato, K. Matsuda, H. Ozald, T. Suzuka, and M. Yamana (to Nippon Mining Co. Limited).

123. U.S. Pat. 2,046,090 (June 30, 1936), C. F. Reed (to C. Horn).

124. DD Pat. 0160830 (Apr. 11, 1984), E. S. Balakirev, A. I. Gershenovi, and L. S. H. Genin (to Balakirev ES).

125. SU Pat. 2007392 (Feb. 15, 1994), M. B. Kats, V. A. Serov, V. G. Sidorov (to Volg Khimprom Production Association).

126. Eur. Pat. 0 575 129 (June 14, 1993), G. P. Miller, L. Y. Chiang, and J. M. Millar (to Exxon Research and Engineering Co.).

127. G. P. Miller, M. A. Buretea, M. M. Bernardo, C. S. Hsu, and H. L. Fang, *J. Chem. Soc., Chem. Commun.* **13**, 1549 (1994).

128. R. Malhotra, S. C. Narang, A. Nigam, S. Ganapathiappan, S. Ventura, A. Satyam, T. Bhardawaj, and D. C. Lorents, *Mat. Res. Soc. Symp. Proc.* **247**, 301 (1992).

129. UK Pat. 2 166 152 (Oct. 17, 1984), D. W. Roberts (to Unilever PLC).

130. Eur. Pat. 0 130 753 (June 22, 1984), S. Sekiguchi, K. Kitano, and K. Nagano (to Lion Corp.).

131. CA Pat. 2,000,278 (Oct. 6, 1989), B. Fabry (to Henkel kGaA).

132. U.S. Pat. 5,319,117 (June 7, 1994), B. Fabry, M. Schaefer, H. Anzinger (to Henkel Kommanditgesellschaftr auf Aktein).

133. H. Yoshimura, Y. Mandai, and S. Hashimoto, *J. Jpn. Oil Chem. Soc.* **41**, 21 (1992).

134. J. Steber and P. Wierich, *Tenside, Surfactants, Deterg.* **26**, 406 (1989).

135. T. Satsuki, K. Umehara, and Y. Yoneyama, *J. Am. Oil Chem. Soc.* **69**, 672 (1992).

136. T. Inagaki, *Proc. World Conf. Oleochem.*, 269–278 (1991).

137. E. A. Knaggs, J. A. Yeager, L. Varenyi, and E. Fischer, *J. Am. Oil Chem. Soc.* **42**, 805 (1965).

138. E. Meyer, *Petroleum Sulfonates—Chemical Amphibians*, Sonneborn Chemical & Refining Corp., New York, 1960.

139. C. Bluestein and B. Bluestein, in Ref. 82, Chapt. 9.

140. T. F. Brown, *Inst. Pet. Rev.* **9**, 314 (1955).

141. A. B. Brown and J. O. Knoblock, *Symposium on Composition of Petroleum Oils*, ASTM Technical Publication No. 224, American Society for Testing and Materials, Philadelphia, Pa., June 1958, pp. 213–229.

142. H. Kaye and co-workers, *Proceedings of 5th World Petroleum Congress*, Section III, Paper 25, New York, 1959, p. 321.

143. *Chem. Mark. Rep.*, 61 (Jan. 27, 1986).

144. U.S. Pat. 4,153,627 (May 8, 1979), P. Delbende and J. P. Herand (to Exxon Res. & Eng. Co.).

145. U.S. Pat. 3,956,372 (May 11, 1976), J. R. Coleman, Jr. and co-workers (to Marathon Oil Co.).

146. U.S. Pat. 4,144,266 (Mar. 13, 1978), M. A. Plummer and co-workers (to Marathon Oil Co.).

147. U.S. Pat. 4,147,638 (Apr. 3, 1979), M. A. Plummer and co-workers (to Marathon Oil Co.).

148. E. A. Knaggs, "The Role of the Independent Surfactant Manufacturer in Tertiary Oil Recovery", paper presented at *ACS Chemical Marketing and Economics Symposium*, Stepan Co., New York, Apr. 5–8, 1976; reprinted.

149. E. A. Knaggs, M. Nussbaum, J. Carlson, and co-workers, "Petroleum Sulfonate Utilization in Enhanced Oil Recovery," *Society of Petroleum Engineers 51st Annual Meeting*, SPE paper 6006, New Orleans, La., Oct. 3–6, 1976.

150. U.S. Pat. 4,148,821 (Apr. 10, 1979), M. L. Nussbaum and E. A. Knaggs (to Stepan Co.).

151. U.S. Pat. 4,177,207 (Dec. 4, 1979), M. L. Nussbaum and E. A. Knaggs (to Stepan Co.).

152. U.S. Pat. 4,252,192 (Feb. 24, 1981), M. L. Nussbaum and E. A. Knaggs (to Stepan Co.).

153. U.S. Pat. 4,847,018 (July 11, 1989), J. Koepke and W. C. Hsieh (to Union Oil Co.).

154. Russ. Pat. 1,072,417 (Oct. 23, 1992), A. M. Yatsenko and co-workers (to Yatsenko AM).

155. Russ. Pat. 1,086,731A-1 (Oct. 23, 1992), V. I. Rudoman, A. Melnik, and co-workers (to Ordzhonikidze Chemical/Pharmaceutical Institute).

156. Russ. Pat. 1,039,156A-1 (Oct. 15, 1992), A. Melnik and co-workers (to Forge Press Equipment des Bur).

157. P.L. Pat. 142,986B1 (Dec. 31, 1987), F. Steinmec and co-workers (Inst. Tech. Nafty, Pol.).

158. Russ. Pat. 808,496A-1 (Oct. 15, 1992), Y. A. Bochkareu, A. Melnik, and co-workers (to Bochkarev yu A).

159. U.S. Pat. 5,136,088 (Aug. 4, 1992), D. E. Farmer, N. Foster, and co-workers (to Chemithon Corp.).

160. U.S. Pat. 4,891,177 (Jan. 2, 1990), C. A. Stratton and N. R. Stratton (to M. R. Stratton).

161. U.S. Pat. 4,251,456 (Feb. 17, 1981), B. Brooks (to Chemithon Corp.).

162. U.S. Pat. 4,382,895 (May 10, 1983), E. L. Cole and R. Suggitt (to Texaco, Inc.).

163. L. A. Potolovskii and co-workers, *Khim. Tekhnol. Topl. Masel* (1), 40–43 (1980).

164. U.S. Pat. 1,651,311 (Nov. 29, 1927), H. Atkinson (to F. K. Holmestead).

165. U.S. Pat. 1,822,271 (Sept. 8, 1931), G. W. Coggeshall.

166. O. R. Wagner and R. O. Leach, *SPE J.*, 335–344 (Dec. 1966).

167. W. R. Foster, *J. Pet. Tech.* **25**, 205–210 (Feb. 1973).

168. P. Dunlop and co-workers, "Aqueous Surfactant Solutions Which Exhibit Ultra-Low Tensions at the Oil-Water Interface," Paper presented at the *165th National Meeting ACS*, Dallas, Tex., Apr. 8–13, 1973.

169. J. Taber and co-workers, *Am. Inst. of Chem. Eng. Symp.* **69**(127), 53 (1973).

170. *Enhanced Oil Recovery*, National Petroleum Council Report, Washington, D.C., 1984.

171. U.S. DOE Report, SAN 1395-2, U.S. Dept. of the Environment, Washington, D.C., Nov. 1978, p. 11.

172. V. A. Kuuskraa, "The Status and Potential of Enhanced Oil Recovery," SPE/DOE 14951, Paper presented at *5th EOR Symposium*, Tulsa, Okla., Apr. 20–23, 1986.

173. U.S. DOE Report, Niper-698, *Chemical EOR Workshop*, Houston, Tex., June 23–24, 1993.

174. H. R. Froning and L. E. Treiber, *SPE EOR Symposium*, Tulsa, Okla., Mar. 22–24, 1976, Proceedings, Paper SPE 5816.

175. E. D. Holstein, "Status and Outlook for EOR by Chemical Injection," No. 371-C, paper presented at the *American Petroleum Institute Meeting*, Dallas, Tex., Apr. 4, 1982.

176. M. Pitts, H. Surkalo, and W. Mundorf, *West Kiehl Alkaline–Surfactant–Polymer Field Project*, DOE/BC/14860-5 Report, U.S. Dept. of the Environment, Washington, D.C., Nov. 1994.

177. P. J. Schuler and co-workers, No. 14934, *SPE/DOE 5th EOR Symposium*, Tulsa, Okla., Apr. 20–23, 1986, pp. 135–152.

178. T. W. Patzek and co-workers, No. 17380, *SPE/DOE 6th EOR Symposium*, Tulsa, Okla., Apr. 17–20, 1988, pp. 663–676.

179. H. M. Muijs and co-workers, *SPE/DOE 6th EOR Symposium*, No. 17361, Tulsa, Okla., Apr. 17–20, 1988, pp. 905–914.

180. H. L. Hoefner and co-workers, *DOE/SPE 9th EOR Symposium*, Vol. II, No. 27787, Tulsa, Okla., Apr. 17–20, 1994, pp. 41–56.
181. *Toxicity of Chemical Compounds Used in EOR*, Report DOE/BC/10014-5, U.S. DOE, Washington, D.C., Sept. 1980.
182. *Large Scale Samples of Sulfonates for Lab Study in Tertiary Oil Recovery*, Report FE-2605-20, U.S. DOE, Washington, D.C., May 1979.
183. I. A. Pearl, *The Chemistry of Lignon*, Marcel Dekker, Inc., New York, 1967.
184. "Lignosulfonates," in *Chemical Economics Handbook*, SRI International, Menlo Park, Calif., Feb. 1993, p. 671.5000++.
185. D. Atack and co-workers, *Pulp Pap. Can.* (1981); Convention Issue, T103–T110 (1982).
186. C. Heitner and co-workers, *J. Wood Chem. Tech.* **2**(2), 169–185 (1982).
187. C. Heitner and co-workers, *Sven Paperstidning* **85**(12), R78–R86 (1982).
188. R. Beatson and co-workers, *J. Wood Chem. Tech.* **4**(4), 439–457 (1984).
189. R. Franzen and K. Li, *J. Pulp Paper Sci.*, J 44–J 50 (Mar. 1984).
190. R. Beatson and co-workers, *Tappi J.*, 82–85 (Mar. 1984).
191. M. Lucander, *Tappi J.* **71**(1), 118–124.
192. T. N. Bugaeva and co-workers, *Khim. Tekhnol. Tsellyul. Ee Proizvodnykh*, 106–110 (1985).
193. U. Westermark and co-workers, *Nordic Pulp Paper Res. J.* (4) (1987).
194. U.S. Pat. 4,634,499 (Jan. 6, 1987), R. S. Ampulski (to Procter & Gamble Co.).
195. R. S. Iozev and co-workers, *Tsellyul., Bum., Karton* (3), 24–25 (1993).
196. M. Sykes, *Tappi J.* **76**(11), 121–126 (1993).
197. G. Meshitsuka and co-workers, *Kami Pa Glkyoshi* **34**(11), 743–749 (1980).
198. J. Janson and co-workers, *Pulp Paper Can.* 82(4), 51–52, 55, 57, 60, 63 (1980).
199. J. Guerer and co-workers, *Int. J. Biol. Chem. Phys. Technol. Wood*, **36**(3), 123–130 (1982).
200. S. M. Saad, *Ind. Pulp Paper* **39**(2), 5–10, 25 (1984).
201. U.S. Pat. 4,521,336 (June 4, 1985), P. Dilling (to Westvaco Corp.).
202. U.S. Pat. 4,748,235 (May 31, 1988), P. Dilling (to Westvaco Corp.).
203. S. Singh and co-workers, *Nord. Pulp Pap. Res. J.* **6**(1), 27–29, 41 (1991).
204. U.S. Pat. 5,013,825 (May 7, 1991), P. Dilling (to Westvaco Corp.).
205. U.S. Pat. 5,043,433 (Aug. 27, 1991), P. Dilling (to Westvaco Corp.).
206. W. Poon and co-workers, *Tappi J.* **76**(7), 187–193 (1993).
207. H. Bjoerklund and co-workers, *Pulp Pap. Can.* **92**(8), 44–46 (1991).
208. T. A. Clark and co-workers, *J. Wood Chem. Tech.* **9**(2), 135–166 (1989).
209. H. Zhan and co-workers, *Zhonggud Zaozhi Xuebao* **6**, 38–46 (1991).
210. Can. Pat. Appl. Ca 2025522 AA 920318, D. Hornsey and G. H. Robert.
211. M. Barbe and co-workers, *Pulp. Pap. Can.* **91**(12), 142–144, 146–151 (1990).
212. Y. Ben and co-workers, *J. Wood Chem. Tech.* **13**(3), 349–369 (1993).
213. J. D. Taylor and E. Yu, *Chemtech.* (2), 38 (1995).
214. F. Carrasco, *Pulp Pap. Can.* **94**(3), 44–49 (1993).
215. U.K. Pat. Appl. GB 2,076,034 A (May 16, 1980), S. Moon and co-workers (to Natural Fibres Ltd.).
216. Swed. Pat. 452470B (Nov. 30, 1987), N. Schonfeldt and co-workers (to Korsnaes-Marma AB, Sweden).
217. U.S. Pat. 5,095,985A (Mar. 17, 1992), D. G. Naae and co-workers (to Texaco, Inc. USA).
218. *SRI International Chemical Economics Handbook*, SRI, Menlo Park, Calif., Feb. 1993, p. 671.5000D.
219. U.S. Pat. 3,054,659 (Sept. 18, 1962), D. Craig and co-workers (to Ontario Paper Co.).
220. U.S. Pat. 3,054,825 (Sept. 18, 1962), D. Craig and co-workers (to Ontario Paper Co.).
221. S. Hong and co-workers, No. 12699, *Proceedings of SPE/DOE*, Tulsa, Okla., Apr. 15, 1984, pp. 273–286.

222. D. M. Vofsi, in Ref. 11, Chapt. 20.

223. N. R. Pai, *Text. Dyer Printer*, **14**, 8, 27 (1981).

224. *Ibid.*, p. 7.

225. A. Brandstrom, G. Strandlund, and P-O. Lagerstrom, *Acta. Chem. Scand., Ser. B*, **34**, 467 (1980).

226. J. A. Mol and T. J. Visser, *Endocrinology*, **117**, 1 (1985).

227. W-H. Ho, *J. Chin. Chem. Soc.* **34**, 257 (1987).

228. C. Lee, E. A. O'Rear, J. H. Harwell, and J. A. Sheffield, *J. Colloid Interface Sci.* **137**, 296 (1990).

229. U.S. Pat. 4,650,888 (Mar. 17, 1987), K. Ochi, Y. Hinohara, and I. Matsunaga (to Chugai Seiyaku Kabushiki Kaisha).

230. U.S. Pat. 5,034,555 (July 23, 1991), A. J. O'Lenick, Jr. (to LCE Partnership).

231. U.S. Pat. 5,268,504 (Dec. 7, 1993), K. M. Webber (to Exxon Production Research Co.).

232. U.S. Pat. 5,037,992 (Aug. 6, 1991), J. F. Ward and R. S. Matthews (to The Procter & Gamble Co.).

233. U.S. Pat. 4,948,535 (Aug. 14, 1990), H. Stuhler and K. Dullinger (to Hoechst Aktiengesellschaft).

234. U.S. Pat. 4,954,646 (Sept. 4, 1990), R. Aigner, G. Muller, R. Muller, and H. Reuner (to Hoechst Aktiengesellschaft).

235. Eur. Pat. 0 377 916 (Dec. 12, 1989), J. K. Smid, R. H. Van der Veen, and J. G. Verschuur (to Stamicarbon BV).

236. Aus. Pat. 29586/89 (Feb. 3, 1989), R. Aigner, G. Muller, R. Muller, H. Reuner (to Hoechst Aktiengesellschaft).

237. U.S. Pat. 4,832,876 (May 23, 1989), F. U. Ahmed (to Colgate-Polmolive Co.).

238. U.S. Pat. 5,322,957 (June 21, 1994), B. Fabry, U. Ploog, A. Behler, and D. Feustel (to Henkel Kommanditgesellschaft auf Aktien).

239. World Pat. 91/18871 (June 6, 1990), B. Fabry, B. Gruber, F. Tucker, and B. Giesen (to Henkel Kommanditgesellschaft auf Aktien).

240. Jpn. Pat. 04005297 (Jan. 9, 1992) (to Kao Corp.).

241. Ger. Pat. 4006841 (May 3, 1990), B. Fabry, M. Weuthen, and A. Schumacher (to Henkel K-GaA).

242. U.S. Pat. 4,814,437 (Feb. 25, 1988), A. N. de Belder, L. G. Ahrgren, T. Malson (to Pharmacia AB).

243. L. Ahrgren, A. N. de Belder, and T. Malson, *Carbohydrate Polym.* **16**, 211 (1991).

244. T. Bocker, T. K. Lindhorst, J. Thiem, and V. Vill, *Carbohydrate Res.* **230**, 245 (1992).

245. L. F. Audrieth and M. Sveda, *J. Org. Chem.* **9**, 89 (1944).

246. Brit. Pat. 1,568,812 (1960), G. Smith and W. Laird (to Dawson International, Ltd.).

247. U.S. Pat. 909,011H (Apr. 3, 1973), J. Schwier.

248. CA Pat. 802,106 (Dec. 24, 1968), A. Sowerby (to Electric Reduction of Canada Ltd.).

249. U.S. Pat. 3,308,012 (Mar. 7, 1967), U. Ramon (to E. I. du Pont de Nemours & Co., Inc.).

250. Ger. Pat. 1,805,262 (Aug. 7, 1969), I. Michelson (to American Safety Equipment).

251. Brit. Pat. 1,395,497 (May 26, 1975), A. Calder and J. Whetstone (to Imperial Chem. Ind., Ltd.).

252. U.S. Pat. 3,226,430 (Dec. 28, 1965), M. Mhatre (to Abbott Laboratories).

253. R. Patterson, *Sugar J.* **38**(1), 28 (1975).

254. Brit. Pat. 789,199 (Aug. 6, 1958).

255. U.S. Pat. 2,931,822 (Apr. 5, 1960), G. Tischbirek (to Henkel).

256. U.S. Pat. 2,703,788 (Mar. 8, 1955), J. J. Morrisroe (to Purex Corp.).

257. J. E. Kircher and co-workers, *Ind. Eng. Chem.* **46**(9), 1925 (1954).

258. W. S. Fedor, B. Strain, and co-workers, *Ind. Eng. Chem.* **51**, 13–18 (1959).

259. U.S. Pat. 3,024,258 (Mar. 6, 1962), R. J. Brooks and B. Brooks (to Chemithon Corp.).

260. U.S. Pat. 3,058,920 (Oct. 16, 1962), R. J. Brooks and B. Brooks (to Chemithon Corp.).

261. H. Huber, *JAOCS* **33**, 33 (1956).
262. W. Faith and co-workers, *Industrial Chemicals*, John Wiley & Sons, Inc., New York, 1957, p. 6.
263. E. A. Knaggs and M. L. Nussbaum, *Soap Chem. Spec.* **38**(5), 237; (6), 165, 179; (7), 145, 149 (1962).
264. *Chem. Proc.*, 68–70 (Nov. 1961).
265. E. E. Gilbert and co-workers, *Ind. Eng. Chem.* **45**, 2065–2072 (1953).
266. *Ibid.* **50**, 276–284 (1958).
267. K. Gerhart, *JAOCS* **31**, 200 (1954).
268. U.S. Pat. 3,232,976 (Feb. 1, 1966), J. W. Lohr (to Andrew Jergens Co.).
269. U.S. Pat. 2,763,199 (Oct. 23, 1956), H. E. Luntz and co-workers.
270. *Can. Chem. Proc.*, 72 (Aug. 1957).
271. U.S. Pat. 3,427,342 (Feb. 11, 1969), R. J. Brooks and co-workers (to Chemithon Corp.).
272. U.S. Pat. 3,328,460 (June 27, 1967), J. E. Vander Mey (to Allied Chemical Co.).
273. Eur. Pat. Appl. 0570844 A1 (Nov. 24, 1993), C. Pisoni (to Meccaniche Moderne SrI).
274. U.S. Pat. 3,931,273 (Jan. 6, 1976), A. Lanteri (to Mazzoni, SpA).
275. Brit. Pat. GB 2043067B (May 1, 1983), M. Ballestra and G. Moretti (to Ballestra SpA).
276. M. Ballestra and G. Moretti, *High Quality Surfactants from SO₃ Sulphonation with Multi-tube Film Reactors*, Chimicaoggi, Luglio, Italy, 1984, p. 41.
277. U.S. Pat. 2,923,728 (Feb. 2, 1960), K. Falk and W. Taplin (to DuPont).
278. U.S. Pat. 3,925,441 (Dec. 9, 1975), S. Toyoda and T. Ogoshi (to Lion).
279. E. E. Gilbert and B. Veldhuis, *Ind. Eng. Chem.* **47**, 2300 (Nov. 1955).
280. G. R. Johnson and B. L. Crynes, *Ind. Eng. Chem. Proc. Dev.* **13**, 6 (1974).
281. E. J. Davis and co-workers, *Chem. Eng. Sci.* **34**, 539–550 (1979).
282. J. J. Fok and co-workers, Technical data, Gronigen University, Dept. of Chemical Engineering,
283. J. Gutierrez-Gonzalez and co-workers, *Ind. Eng. Chem. Res.* **27**(a), 1701–1707 (1988).
284. E. Gilbert and G. Flint, *Ind. Eng. Chem.* **50**, 276–284 (1958).
285. U.S. Pat. 3,535,339 (Oct. 20, 1970), H. Beyer and C. W. Motl (to Procter & Gamble Co.).
286. Brit. Pat. 1,563,994 (Apr. 2, 1980))to Albright & Wilson, Ltd.).
287. U.S. Pat. 4,226,796 (Oct. 7, 1980), B. J. Akrod (to Albright & Wilson, Ltd.).
288. Brit. Pat. 2,155,474A (Mar. 13, 1984), D. O. Robinette (to Procter & Gamble Co.).
289. U.S. Pat. 2,697,301 (Dec. 14, 1954), G. L. Hervert (to UOP Co.).
290. U.S. Pat. 3,607,087 (Sept. 21, 1971), W. A. Graham (to Stratford Engineering Co.).
291. U.S. Pat. 4,185,030 (Jan. 22, 1980), B. Brooks and R. Brooks (to Chemithon Corp.).
292. Aus. Pat. 24,395 (Apr. 16, 1959).
293. U.S. Pat. 4,163,751 (Aug. 7, 1979), J. Vander Mey (to Allied Chemical Co.).
294. U.S. Pat. 4,335,079 (June 15, 1982), J. Vander Mey.
295. R. C. Hurlbert and co-workers, *Soap Chem. Spec.* (May–June 1967).

EDWARD A. KNAGGS
Consultant

MARSHALL J. NEPRAS
Stepan Company

SULFONIC ACIDS

Sulfonic (or sulphonic) acids are classically defined as a group of organic acids which contain one or more sulfonic, $-SO_3H$, groups. The focus of this chapter is an overview of organic sulfonic acids which have the formula RSO_3H, where the R-group may be derived from many different sources. Typical R-groups are alkane, alkene, alkyne, and arene. The R-group may contain a wide variety of secondary functionalities such as amine, amide, carboxylic acid, ester, ether, ketone, nitrile, phenol, etc. Sulfonic acid derivatives are used industrially on a large scale in the manufacture of surfactants, dyes, inks, dispersing agents, and polymers. Sulfonic acids also find broad application as catalysts in both alkylation (qv) processes and general organic reactions. Sulfonic acid derivatives have found more esoteric applications in the areas of custom organic syntheses and biological research. Sulfonic acids derivatives, where the R-group is derived from an inorganic source such as a halide, oxygen (ie, sulfate), or amine (ie, sulfamic acid), are not discussed herein to a great extent. These last are often referred to as sulfuric acid derivatives (see CHLOROSULFURIC ACID).

Physical Properties

The physical properties of sulfonic acids vary greatly depending on the nature of the R-group. Sulfonic acids are found in both the solid and liquid forms at room temperature. No examples of gaseous sulfonic acids are known as of the mid-1990s. Sulfonic acids can be described as having similar acidity characteristics to sulfuric acid. Sulfonic acids are prone to thermal decomposition, ie, desulfonation, at elevated temperatures. However, several of the alkane-derived sulfonic acids show excellent thermal stability, as shown in Table 1. Arene-based sulfonic acids are thermally unstable. These must be distilled under extreme vacuum

Table 1. Physical Properties of Sulfonic Acids[a]

Acid	CAS Registry Number	Mp, °C	Bp,[b] °C	Density d_4^{25}, g/cm^3
methanesulfonic acid	[75-75-2]	20	122	1.48
ethanesulfonic acid	[594-45-6]	−17	123	1.33
propanesulfonic acid	[28553-80-2]	−37	159	1.19
butanesulfonic acid	[30734-86-2]	−15	149	1.19
pentanesulfonic acid	[35452-30-3]	−16	163	1.12
hexanesulfonic acid	[13595-73-8]	16	174	1.10
benzenesulfonic acid	[98-11-3]	44	172[c]	
p-toluenesulfonic acid	[104-15-4]	106	182[c]	
1-naphthalenesulfonic acid	[85-47-2]	78	dec	
2-naphthalenesulfonic acid	[120-18-3]	91	dec	1.44
trifluoromethanesulfonic acid	[1493-13-6]	none	162[d]	1.70

[a]Refs. 1 and 2.
[b]At 133 Pa (1 mm Hg) unless otherwise noted.
[c]At 13.3 Pa (0.1 mm Hg).
[d]At 101.3 kPa = 760 mm Hg.

conditions using a minimal amount of heating to avoid thermal decomposition. Polyaromatic compounds, such as 1- and 2-naphthalenesulfonic acids [83-47-2] and [120-18-3], respectively, readily decompose upon attempted distillation even at very high vacuum.

Sulfonic acids are such strong acids that in general they can be considered greater than 99% ionized. The pK_a value for sulfuric acid is -2.8 as compared to the pK_a values of -1.92, -1.68, and -2.8 for methanesulfonic acid, ethanesulfonic acid, and benzene sulfonic acid, respectively (3). Trifluoromethanesulfonic acid [1493-13-6] has a pK_a of less than -2.8, making it one of the strongest acids known (4,5). Trifluoromethanesulfonic acid is also one of the most robust sulfonic acids. Heating this material to 350°C causes no thermal breakdown (6).

The x-ray crystal structures of the many sulfonic acid derivatives, such as methanesulfonic acid, have been determined (7). A large amount of theoretical work has been compiled for simple sulfonic acid radical, anion, and cation species, including energies, dipole moments, optimized bond lengths and angles, Mulliken atomic charges, d-orbital occupancies, ionization potential, proton affinities, and homolytic hydrogen atom bond dissociation energies (8). Experimental data have also been compiled for sulfonic acids on physical properties such as heats of formation and heats of combustion (9).

Chemical Properties

Sulfonic acids are prepared on a commercial scale by the sulfonation of organic substrates using a variety of sulfonating agents, including sulfur trioxide [7446-11-9] (diluted in air), sulfur trioxide (in sulfur dioxide [7446-09-5]), sulfuric acid [7664-93-9], oleum [8014-95-7] (fuming sulfuric acid), chlorosulfuric acid [7790-94-5], sulfamic acid [5329-14-6], trialkylamine–sulfur trioxide complexes, and sulfite ions. Other methods of sulfonic acid production, which are practiced on an industrial scale, include the oxidation of thiols (qv), sulfide, disulfides, sulfoxides (qv), sulfones, and sulfinic acids (see SULFONATION AND SULFATION). A preparative review of sulfonic acids has been compiled and the variety of functionalities that may be present in a sulfonic acid derivative is wide. Sulfonic acids having amide, arene, alkane, alkene, alkyne, ester, hydroxyl, and nitrile functionality are known (10) (see SULFUR COMPOUNDS). This diversity leads to very rich reaction chemistry involving sulfonic acids and their corresponding sulfonates.

General Reaction Chemistry of Sulfonic Acids. Sulfonic acids may be used to produce sulfonic acid esters, which are derived from epoxides, olefins, alkynes, allenes, and ketenes, as shown in Figure 1 (10). Sulfonic acids may be converted to sulfonamides via reaction with an amine in the presence of phosphorus oxychloride [10025-87-3], POCl$_3$ (11). Because sulfonic acids are generally not converted directly to sulfonamides, the reaction most likely involves a sulfonyl chloride intermediate. Phosphorus pentachloride [10026-13-8] and phosphorus pentabromide [7789-69-7] can be used to convert sulfonic acids to the corresponding sulfonyl halides (12,13). The conversion may also be accomplished by continuous electrolysis of thiols or disulfides in the presence of aqueous HCl [7647-01-0] (14) or by direct sulfonation with chlorosulfuric acid. Sulfonyl fluorides are typically prepared by direct sulfonation with fluorosulfuric acid

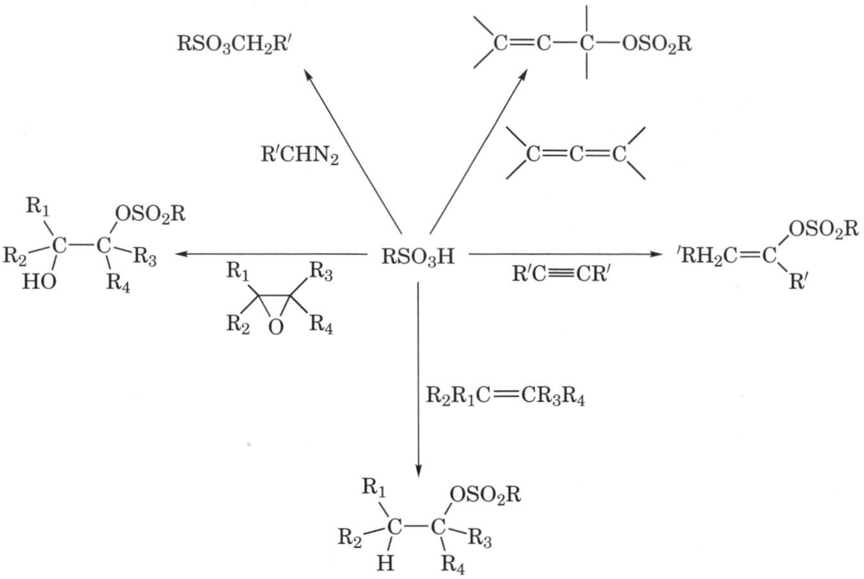

Fig. 1. Reaction chemistry of sulfonic acids.

[*7789-21-1*], or by reaction of the sulfonic acid or sulfonate with fluorosulfuric acid. Halogenation of sulfonic acids, which avoids production of a sulfonyl halide, can be achieved under oxidative halogenation conditions (15).

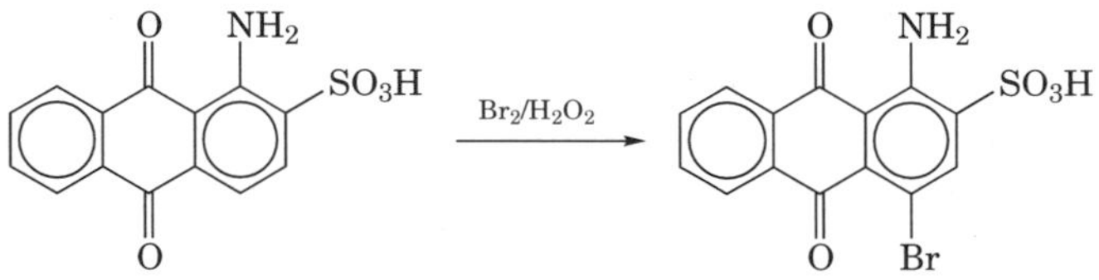

Sulfonic acids are prone to reduction with iodine [*7553-56-2*] in the presence of triphenylphosphine [*603-35-0*] to produce the corresponding iodides. This type of reduction is also facile with alkyl sulfonates (16). Aromatic sulfonic acids may also be reduced electrochemically to give the parent arene. However, sulfonic acids, when reduced with iodine and phosphorus [*7723-14-0*], produce thiols (qv). Amination of sulfonates has also been reported, in which the carbon–sulfur bond is cleaved (17). Ortho-lithiation of sulfonic acid lithium salts has proven to be a useful technique for organic syntheses, but has little commercial importance. Optically active sulfonates have been used in asymmetric syntheses to selectively *O*-alkylate alcohols and phenols, typically on a laboratory scale. Aromatic sulfonates are cleaved, ie, desulfonated, by uv radiation to give the parent aromatic compound and a coupling product of the aromatic compound, as shown, where Ar represents an aryl group (18).

$$\text{ArSO}_3\text{H} \xrightarrow[(-\text{SO}_3)]{h\nu} \text{ArH} + \text{Ar}\!-\!\text{Ar}$$

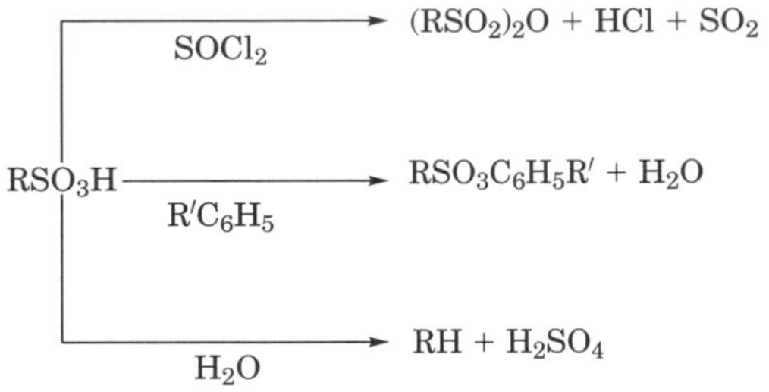

$$(RSO_2)_2O + HCl + SO_2$$

$$SOCl_2$$

$$RSO_3H \longrightarrow RSO_3C_6H_5R' + H_2O$$

$$R'C_6H_5$$

$$RH + H_2SO_4$$

$$H_2O$$

Fig. 2. Sulfonic acid transformations.

Sulfonic acids may be subjected to a variety of transformation conditions, as shown in Figure 2. Sulfonic acids can be used to produce sulfonic anhydrides by treatment with a dehydrating agent, such as thionyl chloride [7719-09-7]. This transformation is also accomplished using phosphorus pentoxide [1314-56-3]. Sulfonic anhydrides, particularly aromatic sulfonic anhydrides, are often produced *in situ* during sulfonation with sulfur trioxide. Under dehydrating conditions, sulfonic acids react with substituted aromatic compounds to give sulfone derivatives.

Sulfonic acids may be hydrolytically cleaved, using high temperatures and pressures, to drive the reaction to completion. As would be expected, each sulfonic acid has its own unique hydrolytic desulfonation temperature. Lower alkane sulfonic acids possess excellent hydrolytic stability, as compared to aromatic sulfonic acids which are readily hydrolyzed. Hydrolytic desulfonation finds use in the separation of isomers of xylene sulfonic acids and other substituted mono-, di-, and polysulfonic acids.

The cleavage products of several sulfonates are utilized on an industrial scale (Fig. 3). The fusion of aromatic sulfonates with sodium hydroxide [1310-73-2] and other caustic alkalies produces phenolic salts (see ALKYLPHENOLS; PHENOL). Chlorinated aromatics are produced by treatment of an aromatic sulfonate with hydrochloric acid and sodium chlorate [7775-09-9]. Nitriles (qv) (see SUPPLEMENT) can be produced by reaction of a sulfonate with a cyanide salt. Arenesulfonates can be converted to amines with the use of ammonia. This transformation is also rather facile using mono- and dialkylamines.

Miscellaneous Reactions. Aromatic sulfonic acid derivatives can be nitrated using nitric acid [52583-42-3], HNO_3, in H_2SO_4 (19). Sultones may be

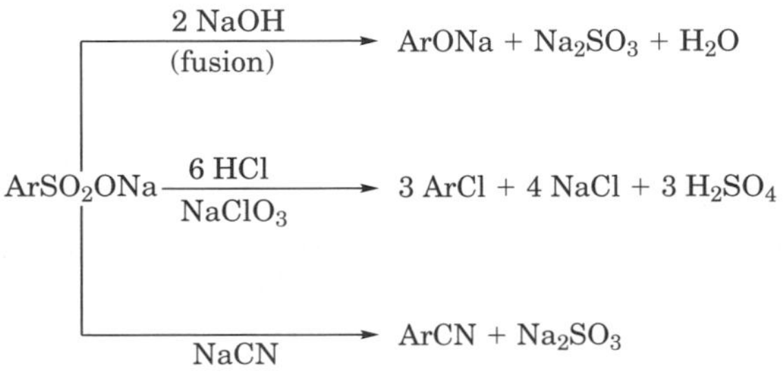

$$2\ NaOH$$
$$(fusion)$$
$$ArONa + Na_2SO_3 + H_2O$$

$$6\ HCl$$
$$ArSO_2ONa \xrightarrow{\quad\quad} 3\ ArCl + 4\ NaCl + 3\ H_2SO_4$$
$$NaClO_3$$

$$ArCN + Na_2SO_3$$
$$NaCN$$

Fig. 3. Cleavage reactions of sulfonates where Ar is an aryl group.

treated with hydrazine derivatives to give the corresponding ring-opened sulfonic acid (20).

Catalytic reduction of secondary functionalities in sulfonates, in which the sulfonate moiety is unchanged, is accomplished using standard hydrogenation techniques (21). Sulfonic acids may be converted to the corresponding silyl esters in very high yields (22).

Production

At the end of 1994, there were four primary methods of sulfonic acid production in the United States. The methods were falling film sulfonation (791,518 metric tons), representing 71% of production and using gaseous SO_3; oleum sulfonation (219,085 t), representing 19% of production; chlorosulfuric acid sulfonation (88,452 t), representing 8% of production; and SO_3 solvent-based sulfonation (26,308 t), representing 2% of production in SO_2 (22,23). These 1994 U.S. acid production numbers do not include minor sulfonic acid production via the use of low throughput continuous, batch SO_3 (100 wt %), H_2SO_4 (20–30 wt %), oleum (30 wt %), and sulfamic acid.

The vast majority of sulfonic acids were produced in the United States using continuous falling film sulfonation technology which utilizes vaporized SO_3 mixed with air. This technology dominates the sulfonation industry owing to the capability of high product throughput and low by-product waste streams. Of the other three methods of sulfonic acid production, the use of sulfur trioxide, where sulfur dioxide is the solvent, is perhaps the most tedious, owing to handling, recovery, and safety issues inherent in the use of the solvent.

Economic Aspects

In 1991, over 1×10^6 t sulfonic acids were produced in the United States (24). The materials, for the most part, were used as intermediates for the manufacture of sulfonates in the detergent market, dye manufacture, dispersing agents, catalysts, polymers, etc. Production of dodecylbenzenesulfonic acids derivatives dominated the sulfonic acid market (Table 2). These had a 38% overall share. The differences between the production tons and the tons sold is accounted for by in-plant use by various manufacturers verses merchant market production.

Table 2. U.S. Production, Sales, and Value of Sulfonic Acids[a]

					Production	
Sulfonic acid[b]	Production, t	% of Total	Sales, t	Av. price, $/t	t	% of Total
dodecylbenzenesulfonic acid	157,205	15.7	105,201	790	83,361,000	124,191,950
Ca salt	2,431	0.2	1,743	3,140	5,476,000	7,633,340
IPA salt	4,258	0.4	3,937	1,540	6,073,000	6,557,320
K salt	18					
Na salts	216,334	21.6	24,997	1,630	40,638,000	352,624,420
TEA salt	1,548	0.2	1,531	1,770	2,715,000	2,739,960
liginsulfonic acid						
Ca salt	292,780	29.3	283,561	110	31,607,000	32,205,800
Na salt	87,816	8.8	87,140	340	29,982,000	29,857,440
tridecylbenzenesulfonic acid, Na salt	11,738	1.2	709	1,650	1,167,000	19,367,700
xylenesulfonic acid, Na salt	34,076	3.4	29,973	730	21,888,000	24,875,480
all other sulfonic acids and salts	192,451	19.2	116,438	1,020	118,775,000	196,300,020
Totals	*1,000,655*	*100.0*	*655,230*		*341,682,000*	*796,353,430*

[a]Ref. 24.
[b]IPA = isopropylamine; TEA = triethanolamine.

199

As of the end of 1991, there were over 60 manufacturers of sulfonic acids and sulfonates in the United States (25). In 1995, Stepan Company was the largest merchant manufacturer of sulfonic acids in the United States, having a total sulfonation–sulfation capacity of over 522,000 t (26). In 1993, the largest volume sulfonic acid produced in the United States was linear alkylbenzenesulfonic acid, which translated to 346,000 t of linear alkylbenzenesulfonates, accounting for 25% of U.S. consumption of surfactants (qv) (23). Table 3 shows a summary of the major sulfonic acid and sulfonate producers in the United States as of the end of 1991. Of these, the largest nonmerchant market producers of sulfonic acids were The Procter & Gamble Company, Colgate-Palmolive Company, and Lever Brothers Company. Elf Atochem North America is the sole U.S. producer of methanesulfonic acid.

Quality and Storage

The quality of sulfonic acids produced as intermediates on an industrial scale is important to detergent manufacturers. Parameters such as color, water, free oil (unsulfonated material), and acid value (actual sulfonic acid) are all factors that determine the quality of a sulfonic acid. The quality of the feedstock prior to sulfonation, such as iodine value, water content, and sulfonatability, affects the quality of the sulfonic acid produced. Sulfonation conditions, such as temperature, molar ratio, rate, etc, also affect the quality of sulfonic acid.

Anhydrous sulfonic acids, particularly linear alkylbenzenesulfonic acids, are typically stored in stainless steel containers, preferably type 304 or 316 stainless steel. Use of other metals, such as mild steel, contaminates the acid with iron (qv), causing a darkening of the acid over time (27). The materials are usually viscous oils which may be stored and handled at 30–35°C for up to two months (27). All other detergent-grade sulfonic acids, eg, alcohol sulfates, alcohol ether sulfates, alpha-olefin sulfonates, and alpha-sulfomethyl esters, are not stored owing to instability. These are neutralized to the desired salt.

Analytical and Test Methods

Modern analytical techniques have been developed for complete characterization and evaluation of a wide variety of sulfonic acids and sulfonates. The analytical methods for free sulfonic acids and sulfonate salts have been compiled (28). Titration is the most straightforward method of evaluating sulfonic acids produced on either a laboratory or an industrial scale (29,30). Spectroscopic methods for sulfonic acid analysis include ultraviolet spectroscopy, infrared spectroscopy, and ^{1}H and ^{13}C nmr spectroscopy (31). Chromatographic separation techniques, such as gc and gc/ms, are not used for free sulfonic acids owing to characteristics such as acidity and lack of volatility. Typically the sulfonic acid must be neutralized, derivatized, and/or desulfonated (pyrolysis or chemical) prior to evaluation (32). Mixtures of sulfonic acids are typically separated on the laboratory scale using standard chromatographic methods, such as column chromatography, paper chromatography, tlc, and hplc (33). Separation of sulfonic acid mixtures on the industrial scale is not common. Neutralized sulfonic acid mixtures have been analyzed using infrared and ^{1}H and ^{13}C nmr (30). Modern separation techniques of

Table 3. U.S. Sulfonic Acid Producers as of 1991[a,b]

Sulfonic acid type	American Cyanamid Co.	Burlington Industries	CNC International Inc.	Colgate-Palmolive Co.	Chemithon Corp.	Dial Corp.	Dow Chemical Co.	Eastern Color & Chemical	E. I. du Pont de Nemours & Co.	Exxon Chemical Americas	Finetex, Inc.	Georgia-Pacific Corp.	Grant Industries	Griffex Chemical
alkylbenzenesulfonic acids and salts		+	+	+	+	+	+	+		+			+	
benzenesulfonic acid														
cumene sulfonic acid and salts														
toluene sulfonic acid and salts														
xylenesulfonic acid and salts									+					
ligninsulfonates												+		
naphthalenesulfonic acid and salts								+	+	+			+	
sulfosuccinamic acid and derivatives	+								+					+
taurine-based sulfonic acids and salts			+						+		+		+	
other sulfonic acids with amide linkage														
sulfosuccinic acid esters	+	+	+					+		+	+			+
other sulfonic acids with ester linkage											+			
sulfonic acids with ether linkage														
mixed alkane sulfonic acids and salts														
mixed linear olefin sulfonates														
water-soluble petroleumsulfonic acids														
all other sulfonic acids														

[a] Ref. 24.
[b] A list of other, smaller sulfonic manufacturers is also available (24).

Table 3. (Continued)

Company										
Harcros Chemicals, Inc.	+									
Henkel Corp.	+	+	+	+	+	+	+			
Hoechst Celanese Corp.	+					+	+			
ICI Americas, Inc.	+			+	+					
J. L. Prescott Co.	+									
Lever Brothers Co.	+							+		
Lignotech (U.S.), Inc.				+						
Lonza, Inc.	+									
Mona Industries					+		+			
National Starch & Chem.							+			
Pilot Chem. Co.	+			+				+		+
Procter & Gamble Co.	+		+					+		
Rhône-Poulenc, Inc.	+				+		+	+	+	
Ruetgers-Nease Chemical Co.	+	+	+	+						
Sandoz Chemical Corp.				+						
Stepan Co.	+	+		+				+	+ +	+
Vista Chemical, Inc.	+		+							
Witco Corp.	+ +		+ +		+		+		+ +	
Westvaco Corp.				+		+			+	

202

sulfonates include liquid chromatography and ion chromatography (34–36) (see
CHROMATOGRAPHY).

Health and Safety Factors

In general, unneutralized sulfonic acids are regarded as moderate to highly toxic
substances, as shown in Table 4. However, slight detoxification, via the introduction of a sulfonic acid moiety, is observed for nitrobenzene and aminobenzene.
This effect has been explained by the fact that sulfonated materials are more
hydrophilic than the parent compounds and thus are more readily excreted from
the body in the urine (40). Sulfonic acids emit toxic SO_x fumes upon heating to
decomposition (38). Halogenated sulfonic acids, such as trifluoromethane sulfonic
acid, also release toxic halogen-containing fumes when heated to decomposition

Table 4. Toxicity Profiles for Sulfonic Acids[a]

Compound	CAS Registry Number	Toxicity profile
benzenesulfonic acid	[98-11-3]	highly irritating to skin, eyes, and mucous membranes; oral LD_{50} (rat) = 890 mg/kg vs 3800 mg/kg for benzene
dodecylbenzene sulfonic acid	[27176-87-0]	moderately toxic by ingestion; oral LD_{50} (rat) = 650 mg/kg vs 34,000 mg/kg for dodecylbenzene
dodecylbenzene sulfonic acid (sodium salt)	[25155-30-0]	skin and eye irritant; oral LD_{50} (rat) = 1,260 mg/kg
m-nitrobenzenesulfonic acid (sodium salt)	[98-47-5]	oral LD_{50} (rat) = 11,000 mg/kg vs 640 mg/kg for nitrobenzene
m-aminobenzenesulfonic acid	[121-47-1]	oral LD_{50} (rat) = 12,000 mg/kg vs 440 mg/kg for aminobenzene
methanesulfonic acid	[75-75-2]	strong irritant to eyes, lungs, and mucous membranes
chlorosulfuric acid (chlorosulfonic acid)	[7790-94-5]	highly irritating and corrosive to eyes, skin, mucous membranes; poison; acute toxic effects
fluorosulfuric acid (fluorosulfonic acid)	[7789-21-1]	highly irritating to skin, mucous membranes; aquatic toxicity rating TL = 100–110 ppm after 96 h
trifluoromethanesulfonic acid	[1493-13-6]	oral LD_{50} (mouse) = 112 mg/kg
p-toluenesulfonic acid	[104-15-4]	highly irritating to skin, mucous membranes; oral LD_{50} (rat) = 2,480 mg/kg vs 5,000 mg/kg for toluene
1-naphthalenesulfonic acid	[85-47-2]	oral LD_{50} (rat) = 420 mg/kg vs 1,780 mg/kg for naphthalene
2-naphthalenesulfonic acid	[120-18-3]	oral LD_{50} (rat) = 400 mg/kg vs 1,780 mg/kg for naphthalene
3,5-dimethylbenzenesulfonic acid (m-xylene-m-sulfonic acid)	[18023-22-8]	ip LD_{50} (mouse) = 500 mg/kg

[a]Refs. 37–39.

(38). Sulfonic acid esters, especially those derived from low molecular weight alcohols, should not be heated above 120°C, as these materials have been reported to explosively decompose (41). Sulfonic acids have essentially the same corrosive characteristics as does concentrated sulfuric acid. Detergent-based sulfonic acids pose a contact hazard, as they are very corrosive to the skin.

When sulfonic acids are neutralized to sulfonic acid salts, the materials become relatively innocuous and low in toxicity, as compared to the parent sulfonic acid (see Table 4). The neutralized materials cause considerably less eye and skin irritation. The most toxic route of entry for sulfonic acid salts is ingestion (39). The toxicity of neutralized sulfonic acids, especially detergent sulfonates, has been directly related to the foaming capability of the material. In general, the higher the foaming power, the lower the toxicity (39). Sulfonates that are absorbed into a living system are readily distributed and excreted. A strong binding of sulfonates, such as the sodium salts of methanesulfonic and benzenesulfonic acids, to proteins has been observed (39). The mechanism of linear alkylbenzene sulfonate degradation has been shown to produce alkylbenzene and sulfate products (42). The alkyl chains are further degraded by standard C-2-β-oxidation. Mutagenic behavior has been observed for materials such as alkyl-substituted butanethiolsulfonates, owing to a strong alkylation capability (43,44). The mutagenicity of sulfonic acid-based azo dyes has also been extensively studied (45).

Environmental Issues

Linear alkylbenzenesulfonic acid is the largest intermediate used for surfactant production in the world. In the United States it has been determined that 2.6 g/d of material is used per inhabitant (46). Owing to the large volumes of production and consumption of linear alkylbenzenesulfonate, much attention has been paid to its biodegradation and a series of evaluations have been performed to thoroughly study its behavior in the environment (47–56). Much less attention has been paid to the environmental impact of other sulfonic acid-based materials.

Linear alkylbenzenesulfonate showed no deleterious effect on agricultural crops exposed to this material (54,55). Kinetics of biodegradation have been studied in both wastewater treatment systems and natural degradation systems (48,57,58). Studies have concluded that linear alkylbenzenesulfonate does not pose a risk to the environment (50). Linear alkylbenzenesulfonate has a half-life of approximately one day in sewage sludge and natural water sources and a half-life of one to three weeks in soils. Aquatic environmental safety assessment has also shown that the material does not pose a hazard to the aquatic environment (56).

Qualitative and quantitative methods for aromatic sulfonic acid determination in wastewater have been developed (59,60). Standard test methods have also been established for biodegradability of alkylbenzenesulfonates (61). These methods, which involve two tests, have been established to determine if linear alkylbenzenesulfonate is sufficiently removed via standard sewage treatment. The tests are a presumptive test, involving yeast cultures, and a confirming test, involving activated sludge. As a rule of thumb, if the presumptive test method produces a ≥90% reduction in linear alkylbenzenesulfonate, the material may be considered sufficiently biodegradable, and no further treatment is needed prior to release into the environment. A <80% reduction in linear alkylbenzene-sulfonate requires further testing to determine if a material has a unsatisfactory biodegradation profile. More recently, studies involving the monitoring of linear alkylbenzenesulfonate degradation in lagoon-type wastewater treatment facilities have shown >97% consumption, well above the required level of 90% consumption (62).

Branched alkylbenzenesulfonic acids and the corresponding sulfonates have been essentially eliminated from use in the commercial laundry detergent market. The biodegradation profile of branched alkylbenzenesulfonates was found to be unacceptable owing to several problems, the most important of which was slow and incomplete breakdown (49). Residual components caused foaming in sewage treatment plants and in natural waterways. Undegraded material was finding its way into freshwater sources and negatively effecting aquatic animal and vegetable life.

Uses

Surfactants and Detergents Uses.   Perhaps the largest use of sulfonic acids is the manufacture of surfactants (qv) and surfactant formulations. This is primarily owing to the dominance of linear alkylbenzenesulfonic acid production for detergent manufacture. In almost all cases, the parent sulfonic acid is an intermediate which is converted to a sulfonate prior to use. The largest volume uses for sulfonic acid intermediates are the manufacture of heavy-duty liquid and powder detergents, light-duty liquid detergents, hand soaps (see SOAP), and shampoos (see HAIR PREPARATIONS). The anionic components of these materials are based primarily on linear alkylbenzenesulfonates, alkenyl sulfonates, and sulfo fatty alkyl esters (63).

Specialty sulfonic acid-based surfactants make up a rather large portion of surfactant production in the United States. Approximately 136,000 metric tons of specialty sulfonic acid-based surfactants were produced in 1992, which included alpha-olefin sulfonates, sulfobetaines, sulfosuccinates, and alkyl diphenyl ether disulfonates (64). These materials found use in the areas of household cleaning products, cosmetics (qv), toiletries, emulsion polymerization, and agricultural chemical manufacture.

Lignosulfates, a complex mixture containing sulfonated lignin, are used as dispersing agents, wetting agents, binding agents, and sequestering agents (see LIGNIN) (65). Dry forms of the materials are used as road binders, concrete additives, animal feed additives, and in vanillin (qv) production (66). Lignin sulfonates are typically produced by the sulfonation/neutralization of lignin or as a by-product of sulfite wood pulping (see PULP).

Naphthalenic, lignin, and melamine-based sulfonic acids are used as dispersion and wetting agents in industry. The condensation product (**1**) of formaldehyde [50-00-0] and 2-naphthalene sulfonic acid sodium salt [532-02-5] has been widely used as a cement dispersant for increasing flowability and strength (67). The sulfonate (**1**) is also widely used as a dispersing agent in dyestuff manufacture and high temperature dyeing of polyester fibers with disperse dyes and vat dyeing of cotton fibers (68). In 1989, (**1**) was the major synthetic organic leather tanning agent produced in the United States (69). A derivative of (**1**) based on

4-aminobenzenesulfonic acid has also been produced on the commercial scale (70).

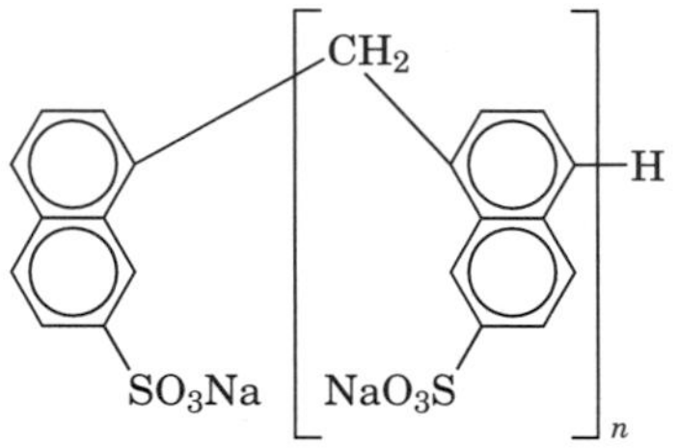

(**1**)

Other commercial naphthalene-based sulfonic acids, such as dinonylnaphthalene sulfonic acid, are used as phase-transfer catalysts and acid reaction catalysts in organic solvents (71). Dinonylnaphthalene sulfonic acid is an example of a water-insoluble synthetic sulfonic acid.

Benzenedisulfonic acid [*831-59-4*] (disodium salt), produced by the neutralization of the disulfonic acid with sodium sulfite [*7757-83-7*], is used in the manufacture of resorcinol [*108-46-3*] (1,3-benzenediol) (**2**), a chemical component found in rubber products and wood adhesives (72). The disodium salt is fused with sodium hydroxide, dissolved in water, and acidified to produce resorcinol, which is isolated via extraction (73).

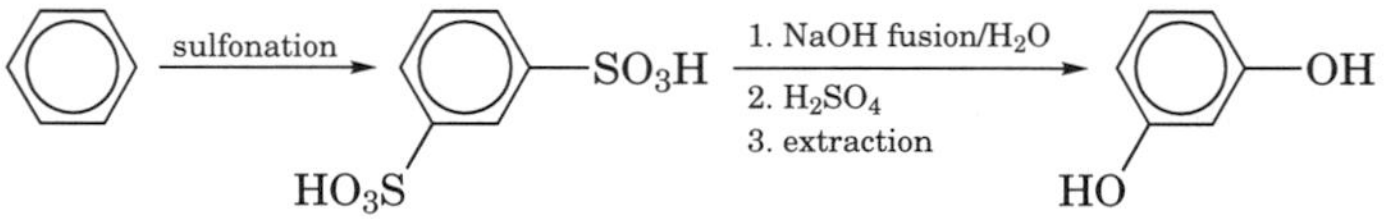

(**2**)

Sulfonic Acid-Based Dyestuffs. Sulfonic acid-derived dyes are utilized industrially in the areas of textiles (qv), paper, cosmetics (qv), foods, detergents, soaps, leather, and inks, both as reactive and disperse dyes. Of the principal classes of dyes, sulfonic acid derivatives find utility in the areas of acid, azoic, direct, disperse, and fiber-reactive dyes. In 1994, 120,930 t of synthetic dyes were manufactured in the United States, of which 5,600 t were acidic (74). The three largest manufacturers of sulfonic acid-based dyes for use in the United States are BASF, Bayer, and Ciba-Geigy.

Sulfonic acid-based azo dyes (qv) and intermediates are characterized by the presence of one or more azo, R–N═N–R, groups. An example of a water-soluble polysulfonic acid-based azo dye (**3**), where M is a metal, typically Na, for use in ink-jet printers, is shown (75).

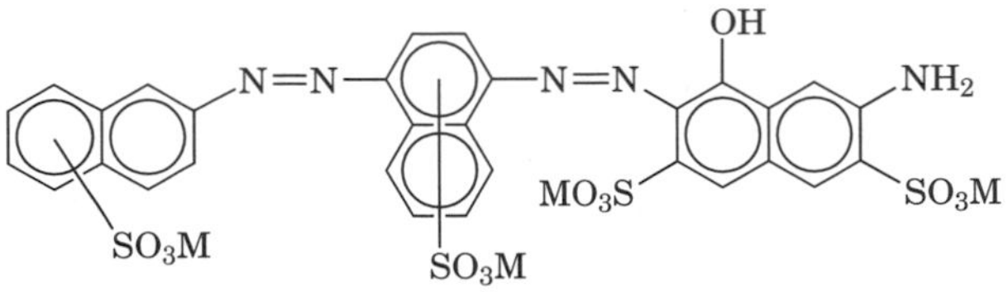

(3)

The material, made by a two-step diazotization of each naphthalenic sulfonic acid derivative, is typically used in the form of the neutralized sodium salt. A similar sulfonic acid-based azo dye (4) which falls into the class of reactive dyes is also shown (76). This compound, made similarly to (3), is used as a blue dyestuff for cotton and wool.

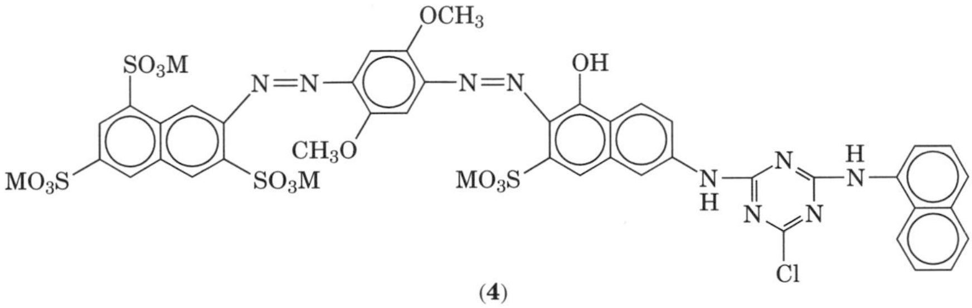

(4)

Changes in the backbone of the sulfonic acid azo dyes often produce drastic changes in properties of the materials. The disulfonic acid (5) is somewhat similar to (3), but is used to color leather red (77). More esoteric dyes have also been developed based on sulfonic acid metal complexes and chitosan-derived materials (78,79).

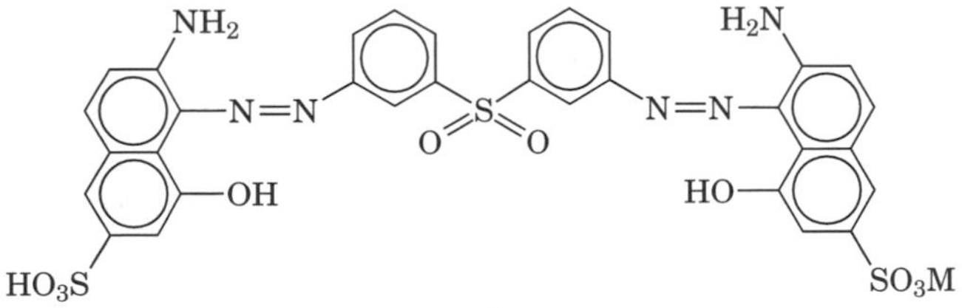

(5)

Amide-Based Sulfonic Acids.   The most important amide-based sulfonic acids are the alkenylamidoalkanesulfonic acids. These materials have been extensively described in the literature. A variety of examples are given in Table 5. Acrylamidoalkanesulfonic acids are typically prepared using technology originally disclosed by Lubrizol Corporation in 1970 (80). The chemistry involves an

Table 5. Alkenylamidoalkanesulfonic Acids,

$$R-\underset{\underset{H}{N}}{\overset{\overset{O}{\|}}{C}}-\underset{CH}{\overset{CH_2}{}}$$

Chemical name	R	CAS Registry Number
2-acrylamidopropanesulfonic acid	CH₃ CH₂ / CH SO₃H	[33028-26-1]
2-acrylamido-2-methylpropanesulfonic acid	CH₃, CH₃, CH₂, C, SO₃H	[15214-89-8]
3-acrylamido-2,4,4-trimethylpentane-sulfonic acid	CH₃ CH CH₂ C CH SO₃H CH₃ CH₃ CH₃	[79647-72-6]
2-acrylamido-2-(p-tolyl)ethanesulfonic acid	CH₃ (p-tolyl) C SO₃H H CH₂	[79647-73-7]
2-acrylamido-2-pyridylethanesulfonic acid	N (pyridyl) C SO₃H H CH₂	[79647-74-8]

initial reaction of an olefin, which contains at least one allylic proton, with an acyl hydrogen sulfate source, to produce a sulfonated intermediate. This intermediate subsequently reacts with water, acrylonitrile, and sulfuric acid.

Lubrizol Corporation manufactures and markets one of the most commercially successful acrylamidoalkanesulfonic acids, 2-acrylamido-2-methylpropanesulfonic acid, under the trademark of AMPS monomer (81). The material, a highly reactive, water-soluble sulfonic acid, is also soluble in many polar organic solvents. It shows excellent hydrolytic stability properties, along with thermal stability. AMPS monomer contributes lubricity and resistance to divalent cation precipitation in high performance polymer applications. Acrylamidoalkanesulfonic acids are used in a wide variety of applications. AMPS monomer in particular is used in surfactant applications, brass electroplating, printing inks, clear antifog coatings, permanent press and soil-release agents, textile sizes, and dye receptivity agents. Acrylamidoalkanesulfonic acids have also found application as leather finishing agents (82), hydraulic cement admixtures (83), curing accelerators for aminoplast resins (84), acrylic thermosetting coatings (85), and enhanced petroleum recovery (86).

Other amide-based sulfonic acid derivatives find a wide variety of commercial uses. Sulfonic acid amide derivatives, such as N-(2-chlorophenyl)-1-chloromethane sulfonamide [30064-44-9], are utilized in insecticide compositions (87). Acrylimidoaminoethanesulfonic acids and their derivatives are used as flocculating agents (qv) and are also used in the preparation of polymers. These materials are prepared by the sequential reaction of acrylonitrile [107-13-1] with SO_3, 2,2-dimethylpropane, and NH_3 [7664-41-7] at low temperatures (-30 to $-75°C$) (88). Amidoalkanesulfonic acids are used as dispersing agents for calcium soaps. They are typically prepared by the reaction of nitriles with olefins and concentrated sulfuric acid or oleum (89). Derivatives of these materials have also been prepared by treatment of nitriles with 2-hydroxyalkanesulfonic acids in concentrated sulfuric acid, in the presence of various vinyl compounds, to give polymeric dispersing agents (90).

Fluorinated and Chlorfluorinated Sulfonic Acids. The synthesis of chlorinated and fluorinated sulfonic acids has been extensively reviewed (91,92). The literature discusses the reaction of dialkyl sulfides and disulfides, sulfoxides and sulfones, alkanesulfonyl halides, alkanesulfonic acids and alkanethiols with oxygen, hydrogen chloride, hydrogen fluoride, and oxygen–chloride–hydrogen fluoride mixtures over metal halide catalysts, such as $CuCl_2$–KCl and $MnBr_2$–$LaCl_3$, to give the respective acids. The reaction of Cl_2CHSH with oxygen–hydrogen fluoride over $CuCl_2$–KCl gives CF_3SO_3H, $ClCF_2SO_3H$ [73043-98-8], Cl_2CFSO_3H [77801-23-1], CHF_2SO_3H [40856-11-9], and $ClCHFSO_3H$ [40856-08-4] (see FLUORINE COMPOUNDS).

Trifluoromethanesulfonic acid, also known as triflic acid [1493-13-6] is widely used in organic syntheses and has been thoroughly reviewed (93,94). It was first prepared in 1954 via the oxidation of bis(trifluoromethylthio)mercury with hydrogen peroxide [7722-84-1] (95). Several other routes of preparation have been disclosed (96–98). The acid exhibits excellent thermal and hydrolytic stability, it is not readily oxidized or reduced, nor is it prone to fluoride anion generation.

Trifluoromethanesulfonic acid is used for the polymerization of aromatic olefins (99). Highly branched paraffin hydrocarbons having high octane values can be prepared, using catalytic amounts of trifluoromethanesulfonic acid, via the alkylation of isoparaffin hydrocarbons with olefins for use as fuel additives (100). Catalytic amounts of trifluoromethanesulfonic acid are used to produce hydrocarbon-based oils from phenol [108-95-2] and coal (101). Trifluoromethanesulfonic acid has been widely utilized as a catalyst in Friedel-Crafts alkylation and acylation reactions, as well as a catalyst in organometallic chemistry. Cyclic and straight-chain polyethers are also prepared via the trifluoromethanesulfonic acid-catalyzed polymerization of tetrahydrofuran [109-99-9] (102). Trifluoromethanesulfonic acid is used to produce p-type semiconductors (qv) and various other conducting polymers via the doping of polyacetylene (103–105).

A compound closely related to trifluoromethane sulfonic acid, pentafluoroethanesulfonic acid (pentflic acid), $CF_3CF_2SO_2H$, has also been prepared and utilized as an organic catalyst (106). The material is generated via the lithiation of CF_3CF_2I, followed by sulfonylation with SO_2, oxidation with H_2O_2, and hydrolysis. Trifluoromethanesulfonic acid can be combined with antimony pentafluoride [7783-70-2] to form Magic Acid, a superacid catalyst. Magic Acid has been used as a catalyst to alkylate benzene and in the hydroisomerization of paraffin-based hydrocarbons for use in motor fuels (107,108). The superacid catalyst is also utilized in the polymerization of formaldehyde with carbon monoxide (109). One of the most widely used sulfonic acid-based catalysts is Nafion-H [*66796-30-3*] (**6**), manufactured by E. I. du Pont de Nemours & Co., Inc. It is a perfluorinated ion-exchange polymeric sulfonic acid, available in both the powdered and membrane forms, with a wide variety of commercial catalytic applications (110). This particular polymer is advantageous owing to its inertness to strong acids, strong bases, and reducing and oxidizing agents.

$$\text{--}(\text{CF}_2\text{CF}_2)_n \text{---} \text{CF}_2 \text{---} \text{CF} \text{--})_{\overline{x}}$$
$$|$$
$$\text{O}(\text{CF}_2\text{CF}(\text{CF}_3)\text{O})_m \text{CF}_2\text{CF}_2\text{SO}_3\text{H}$$

(**6**)

The sulfonic acid moiety has been incorporated into a variety of nonfluorinated polymeric materials (111). Chain-end sulfonated polymers are produced by the reaction of sultones with polymeric organolithiums (112). Polymeric sulfonic acids such as these are incorporated in positive-working photoresist compositions (113).

Biological Uses

Taurine [*107-35-7*] (2-aminoethanesulfonic acid), is the only known naturally occurring sulfonic acid. The material is an essential amino acid for cats and is used extensively by Ralston Purina Company as a food supplement in cat food manufacture. Approximately 5,000–6,000 t of taurine (synthetic and natural) were produced in 1993; 50% for pet food manufacture, 50% in pharmaceutical applications (114).

Sulfonic acids have found greatly expanded usage in biological applications. Whereas many of these sulfonic acids are not produced commercially on a large scale, these compounds are important. A review of the biological activity of sulfonic acids has been written (115). Whereas the toxicity of sulfonic acids is in general rather high, several sulfonic acids are beneficially utilized *in vivo*. Taurocholic acid (**7**) is an important bile component, aiding in the digestion of fat (116).

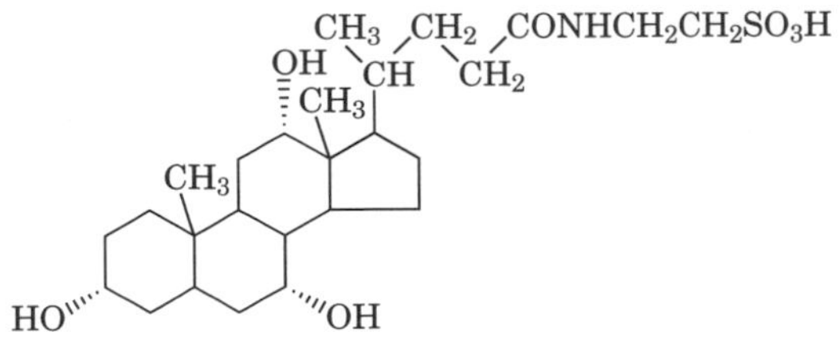

$$CH_3 \quad CH_2 \quad CONHCH_2CH_2SO_3H$$
$$OH \quad CH \quad CH_2$$
$$CH_3$$

CH_3

HO''''   ''''OH

(**7**)

Sulfonic acids are beginning to find widespread, specialized use both *in vitro* and *in vivo*. 8-Anilino-1-naphthalenesulfonic acid [*82-76-8*] (**8**) is used as a fluorescent probe for the study of proteins (117). Saclofen (3-amino-2-(4-chlorophenyl)propanesulfonic acid) (**9**) is a powerful antagonist of GABA at the GABAB receptor site (118). Potent inhibition of the herpes simplex virus has been observed using biphenyl disulfonic acid urea copolymers (119). Sulfonic acid derivatives have been shown to be potent antihuman immunodeficiency virus (anti-HIV) agents (120,121) (see ANTIVIRAL AGENTS). Sulfonated polystyrene (mol wt = 8000) (**10**) has shown potent HIV-inhibitory properties, without being overly toxic to host cells, during *in vitro* HIV-1 and HIV-2 reverse transcriptase inhibitory activity studies (122,123). Sulfonic acid-based azo dye derivatives have been patented for HIV treatment, although the use of these materials is limited (124).

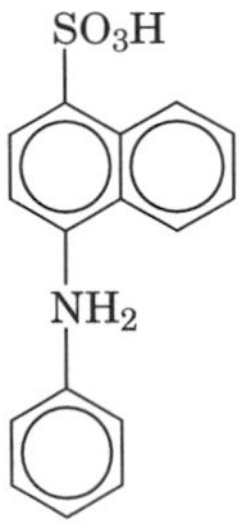

SO_3H

NH_2

(**8**)

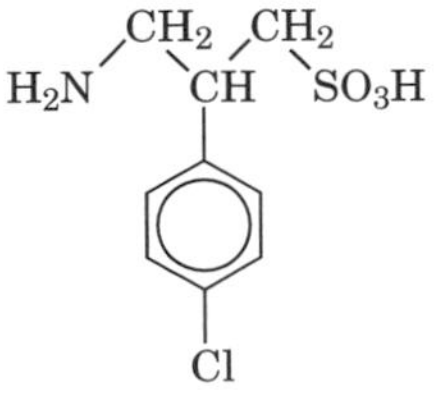

H_2N CH_2 CH_2
    CH SO_3H

Cl

(**9**)

CH_2 CH_2 CH_2
CH   CH   CH $\Big)_C$

SO_3H SO_3H SO_3H

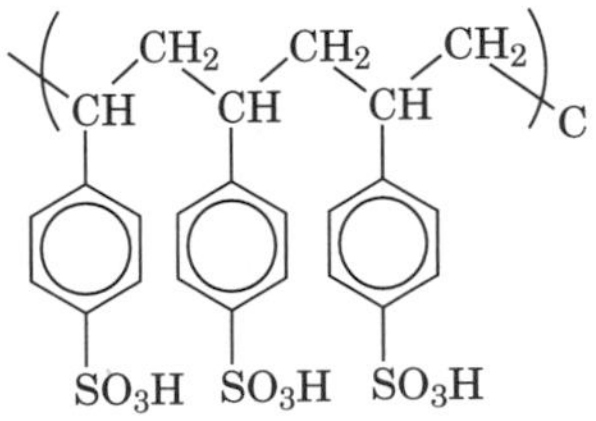

(**10**)

Other Applications. Hydroxylamine-*O*-sulfonic acid [*2950-43-8*] has many applications in the area of organic synthesis. The use of this material for organic transformations has been thoroughly reviewed (125,126). The preparation of the acid involves the reaction of hydroxylamine [*5470-11-1*] with oleum in the presence of ammonium sulfate [*7783-20-2*] (127). The acid has found application in the preparation of hydrazines from amines, aliphatic amines from activated methylene compounds, aromatic amines from activated aromatic compounds, amides from esters, and oximes. It is also an important reagent in reductive deamination and specialty nitrile production.

Two important widely used sulfonic acids are known as Twitchell's reagents, or as in Russia, the Petrov catalysts. These reagents are based on benzene or naphthalene (**11**) and (**12**), [*3055-92-3*] and [*82415-39-2*], respectively. The materials are typically made by the coupling of an unsaturated fatty acid with benzene or naphthalene in the presence of concentrated sulfuric acid (128). These sulfonic acids have been used extensively in the hydrolysis of fats and oils, such as beef tallow (129), coconut oil (130,131), fatty methyl esters (132), and various other fats and oils (133–135). Twitchell reagents have also found use as acidic esterification catalysts (136) and dispersing agents (137).

$$HO_3S-\langle\bigcirc\rangle-(CH_2)_nCOOH$$

(**11**)   (**12**)

Petroleum sulfonates have found wide usage in enhanced oil recovery technology (138). This technology involves the sulfonation, often in several steps using several different sulfonating agents, of a petroleum-based feedstock, which contains a wide mixture of aromatics, polycyclic aromatics, and paraffins, to produce a mixture of sulfonic acids which are neutralized to the corresponding sulfonates. The sulfonates are then used in the chemical flooding methods via injection into an existing oil well for enhancing the recovery of oil which is entrained in rock below the surface. These sulfonates can be inexpensively produced and are effective in the desired lowering of surface tension (oil in water) (see Petroleum, enhanced oil recovery).

The uses of methane sulfonic acid are broad (see Sulfur compounds). A variety of barium sulfonates have found use in antifriction lubricants for high speed bearing applications (see Bearing materials) (139). Calcium and sodium salts of sulfonated olefins, esters, or oils are used for the enhancement of extreme pressure properties of grease and gear lubricants (see Lubrication and lubricants) (140,141). These sulfonic acid salts are also used as combustion aids for gas oils and fuel oils, and also as detergent–dispersant additives for lubricants (142). Zirconium salts of alkyl and alkylaryl sulfonic acids are used in fuel oils to reduce particulate emissions when the fuel oils are burned (143). Magnesium salts of sulfonic acids are broadly used in lubricating oils to reduce the wear of moving parts (144). The calcium, magnesium, sodium, triethanolamine, and isopropylamine salts of alkylaryl sulfonic acids are useful in oil slick dispersion, especially at low seawater temperatures (145).

Sulfonic acid salts have found widespread use in the area of corrosion inhibition. Lubrizol Corporation produces a wide variety of sulfonic acids, particularly in the form of magnesium salts, for use in lubricant formulations, anticorrosion coatings, greases, and resins (146,147). Petroleum sulfonates are used in epoxy resin elastomers to improve anticorrosion properties of coatings and sealants (qv) (148,149).

BIBLIOGRAPHY

"Sulfonic Acids" in *ECT* 1st ed., Vol. 13, pp. 346–353, by A. A. Harban and C. E. Johnson, Standard Oil Co. (Indiana); in *ECT* 2nd ed., Vol. 19, pp. 311–319, by E. E. Gilbert, Allied Chemical Corp.; in *ECT*, 3rd ed., Vol. 22, pp. 45–63, by A. Schultz, Stepan Chemical Co.

1. R. Weast and J. Grasselli, *Handbook of Data on Organic Compounds*, 2nd ed., CRC Press, Boca Raton, Fla., 1985, pp. 930, 2955.
2. S. Budavari, ed., *The Merck Index*, 11th ed., Merck & Co., Inc., Rahway, N.J., 1989, pp. 167, 938, 1009, 1501.
3. S. Patai, Z. Rappoport, and C. J. M. Stirling, eds., *The Chemistry of Sulphones and Sulphoxides*, John Wiley & Sons, Inc., New York, 1988, Chapt. 7, pp. 262–266, for reviews of the acidities of various sulfonic acids.
4. G. Olah, G. Prakash, and J. Sommer, *Superacids*, John Wiley & Sons, Inc., New York, 1985.
5. N. Issacs, *Physical Organic Chemistry*, Longman Scientific and Technical, U.K., 1987.
6. R. E. Bank and R. N. Hazeldine, *The Chemistry of Organic Sulfur Compounds*, Vol. 2, Pergamon Press, Inc., New York, 1966, pp. 165–176.
7. Ref. 3, Chapt. 12.
8. S. Patai and Z. Rappoport, *The Chemistry of Sulphonic Acids, Esters and Their Derivatives*, John Wiley & Sons, Inc., New York, 1991, pp. 1–60, and references therein.
9. *Ibid.*, pp. 283–316.
10. S. R. Sandler and W. Karo, *Organic Functional Group Preparation*, Vol. I, Academic Press, Inc., New York, 1983, pp. 619–639; Ref. 8, Chapt. 10, pp. 366–373 for an overview of conversion of sulfonic acids to esters.
11. B. Helferich and R. Hoffmann, *Ann. Chem.* **657**, 86 (1962).
12. A. Barco, S. Benetti, G. P. Pollini, and R. Taddia, *Synthesis*, 877 (1974).
13. A. H. Kohlhase, *J. Am. Chem. Soc.* **54**, 2441 (1932).
14. Eur. Pat. 331,864 (Sept. 13, 1989), D. M. Gardner and G. A. Wheaton (to Pennwalt Corp., USA).
15. S. V. Gerasimov and V. D. Filimonov, *Zh. Prikl. Khim.* (*St. Petersburg*), **67**, 1043 (1994).
16. S. Oae and H. Togo, *Synthesis*, 371 (1981).
17. J. F. Bunnett, T. K. Brotherton, and S. M. Williamson, *Org. Synth.*, **V**, 816 (1973).
18. Ref. 8, Chapt. 14, pp. 578–581, for a review of photochemistry of sulfonic acids and sulfonates.
19. Czech. Pat. 256,226 (Dec. 15, 1988), J. Cerney and J. Zalouded (to the Government of Czechoslavakia).
20. I. Zeid, I. Ismail, H. Abd El-Bary, and F. Abdel-Azeim, *Liebigs Ann. Chem.* **5**, 481 (1987).
21. W. Kaminski and T. Paryjczak, *Zesz. Nauk.-Politech. Lodz. Chem.* **41**, 237 (1987).
22. A. Palomo and J. Cabre, *Afinidad*, **44**, 234 (1987).
23. *Chemical Economics Handbook*, SRI International, Menlo Park, Calif., Dec. 1994, pp. 583-8000Y.
24. United States International Trade Commission, *Synthetic Organic Chemicals, United States Production and Sales, 1991*, USITC Publication 2607, Washington, D.C., Feb. 1993, pp. 12-3, 12-11–12-14.
25. *Ibid.*, pp. 12-14.
26. Ref. 23, Feb. 1996, pp. 583.900 H.
27. W. Groot, *Sulfonation Technology in the Detergent Industry*, Kluwar Academic Publishers, Boston, Mass., 1991, pp. 181–194.

28. Ref. 8, Chapt. 9.
29. *Ibid.*, p. 325, and references cited therein.
30. M. Sak-Bosnar, L. Zelenka, and M. Budmir, *Tensire Surf. Det.* **29**, 289 (1992).
31. Ref. 8, p. 329, and references cited therein.
32. L. Ng and M. Hupe, *J. Chromatog.* **513**, 61 (1990).
33. Ref. 8, p. 330, and references cited therein.
34. D. Zhou and D. Pietrzyk, *Anal. Chem.* **64**, 1003 (1992).
35. W. Senden and R. Riemersma, *Tenside Surf. Det.* **27**, 46 (1990).
36. C. Hoeft and R. Zollars, *J. Liq. Chromatog.* **17**, 2691 (1994).
37. Ref. 2, pp. 334, 654.
38. R. Lewis, *Sax's Dangerous Properties of Industrial Materials*, 8th ed., Vols. II and III, Van Nostrand Reinhold, New York, 1992, pp. 363, 878, 495, 1496, 2488, 3314, 3315, 3381, 3522.
39. G. Clayton and F. Clayton, eds., *Patty's Industrial Hygiene and Toxicology*, 3rd ed., Vol. 2A, John Wiley & Sons, Inc., New York, 1981, pp. 2089–2091.
40. "Aryl Sulfonic Acids and Salts," *Information Profiles on Potential Occupational Hazards*, Vol. II, *Chemical Classes*, Center for Chemical Hazard Assessment, Syracuse Research Corp., U.S. Dept. of Commerce, Washington, D.C., 1979.
41. D. Collin and D. Wilson, *Chem. Ind. (London)*, **2**, 60 (1987).
42. R. B. Cain and D. R. Farr, *Biochem. J.* **106**, 859 (1968).
43. S. Sternberg, F. Philips, J. Scholler, *Ann. N.Y. Acad. Sci.* **68**, 811 (1958).
44. R. Montesan and co-workers, *Brit. J. Cancer*, **29**, 50 (1974).
45. K. T. Chung, *Mutation Res.* **3**, 201 (1992).
46. L. Huber, *Tenside Surf. Det.* **26**, 71 (1989).
47. J. L. Berna and co-workers, *Growth and Developments in Linear Alkylbenzene Technology: Thirty Years of Innovation and More to Come*, Petresa and UOP Pre-Publication Communication, 1994.
48. R. J. Larson and co-workers, *J. Am. Oil. Chem. Soc.* **70**, 645 (1993).
49. J. L. G. De Almeida, M. Dufaux, Y. Ben Taarit, and C. Naccache, *J. Am. Oil. Chem. Soc.* **71**, 675 (1994), and references cited therein.
50. *ECOSOL*, a sector group of CEFIC, European Chemical Industry Council Literature, 1992.
51. J. L. Berna and co-workers, *Tenside Suf. Det.* **26**, 101 (1989).
52. H. A. Painter and T. Zabel, *Tenside Surf. Det.* **26**, 108 (1989).
53. R. J. Larson, T. W. Federle, R. J. Shimp, and R. M. Ventullo, *Tenside Surf. Det.* **26**, 116 (1989).
54. K. Figge and P. Schoberl, *Tenside Surf. Det.* **26**, 122 (1989).
55. J. Waters and co-workers, *Tenside Surf. Det.* **26**, 129 (1989).
56. W. Gledhill, V. Saeger, and M. Trehy, *Environ. Toxicol. Chem.* **10**, 169 (1991).
57. J. M. Auiroga, D. Sales, and A. Gomer-Parra, *Toxicol. Environ. Chem.* **37**, 85 (1992).
58. T. Yakabe and co-workers, *Chemosphere*, **24**, 696 (1992).
59. A. Marcomini, M. Zanetta, and A. Sfriso, *Chem. Listy*, **87**, 15 (1993).
60. H. Fr. Schroeder, *J. Chromatog.* **647**, 219 (1993).
61. ASTM D2667-89, ASTM, Philadelphia, Pa., 1989, and references cited therein.
62. A. Moreno and co-workers, *Water Res.* **28**, 2183 (1994).
63. E. A. Knaggs, *CHEMTECH*, 436–445 (July 1992), for a review of major surfactant sulfonic acids.
64. *Specialty Surfactants: United States*, Kline & Co., Inc., 1993, pp. 5-1–5-50.
65. Ref. 23, Feb. 1993, pp. 671.5000 E.
66. *Ibid.*, pp. 651.5000 V–671.5001 F.
67. T. Mizunuma, M. Iizuda, and K. Izumi, *Industrial Applications of Surfactants II*, Royal Chemistry Society, London, 1990, pp. 101–113.

68. V. A. Shenai and N. R. Pai, *Text. Dyer Printer*, **12**, 30 (1979).
69. Ref. 23, Jan. 1992, pp. 593-5000 D–E.
70. U.S. Pat. 5,233,012 (Aug. 3, 1993), T. Date and co-workers (to Sanyo Kokusaku Pulp).
71. L. V. Gallacher, *Solution Behavior of Surfactants*: *Theoretical Applications and Aspects*, Vol. 2, Plenum Press, New York, 1982, pp. 791–801.
72. Ref. 23, May 1994, pp. 691.7000 A–I.
73. Rom. Pat. 103,354 (Nov. 9, 1993), T. Suteu and co-workers (to Combinatul Chimic, Fagaras, Romania).
74. A. Leder, K. Shariq, and Y. Ishikawa, in Ref. 23, Apr. 1996, pp. 520.5001 X.
75. Jpn. Pat. 62,252,465 (Nov. 4, 1987), Y. Suga, K. Shirota, M. Kobayashi, and T. Kimura (to Canon KK).
76. Jpn. Pat. 62,084,159 (Apr. 17, 1987), T. Hibara and Y. Sanada (to Mitsubishi Chemical Industries Co., Ltd.).
77. Ger. Pat. EP598,244 (May 25, 1994), J. P. Dix, G. Lamm, H. Reichelt, and G. Zeidler (to BASF, A-G).
78. Jpn. Pat. 04,085,367 (Mar. 18, 1992), S. Shimoyama and co-workers (to Ihara Chemical Kogyo KK, Japan; Den Material KK; Cosme Techno KK).
79. Ger. Pat. 301,408 (Dec. 24, 1992), K. R. Groth, F. Kleine, E. Schurig, and W. Dassler (to Chemie A-G Bitterfeld-Wolfen).
80. U.S. Pat. 3,506,707 (Apr. 14, 1970), L. E. Miller and D. L. Murfin (to Lubrizol Corp.).
81. "AMPS® Monomer," *Chem. Eng. News*, 31 (May 15, 1995).
82. Jpn. Pat. 04,089,900 (Mar. 24, 1992), M. Yoshida and M. Takahashi (to Lion KK).
83. Jpn. Pat. 01,176,255 (July 12, 1989), S. Kobayashi (to Nissan Chemical Industries, Ltd.).
84. Ger. Pat. 3,218,231 (Nov. 17, 1983), S. Piesch and P. Doerries (to Cassella A-G).
85. U.S. Pat. 5,093,425 (Mar. 3, 1992), G. Craun and B. Kunz (to Glidden Co.).
86. U.S. Pat. 4,404,111 (Sept. 13, 1983), L. Bi, M. Dillon, and C. Sharik (to Atlantic Richfield Co.).
87. Jpn. Pat. 49,001,849 (Jan. 17, 1974), T. Kitagaki, Y. Okauda, and H. Ito (to Kumiai Chemical Industry Co., Ltd.).
88. U.S. Pat. 93-25,478 (Mar. 3, 1993), M. J. Virnig (to Henkel Corp.).
89. Ger. Pat. 2,105,030 (Aug. 19, 1971), D. Hoke (to Lubrizol Corp.).
90. Jpn. Pat. 82-107,538 (June 24, 1982), H. Itoh, A. Nitta, and H. Kamio (to Mitsui Toatsu Chemicals, Inc.).
91. R. E. Bank and R. N. Hazeldine, *The Chemistry of Organic Sulfur Compounds*, Vol. 2, Pergamon Press, Inc., New York, 1966, pp. 165–176.
92. Ref. 3, Chapt. 21, p. 904, for a review of the preparation of perfluoroalkanesulfonic acids and derivatives.
93. P. Stang and M. White, *Aldrichimica Acta* **16**, 15 (1983).
94. R. Howells and J. Cown, *Chem. Rev.* **77**, 69 (1977).
95. R. Haszeldine and J. Kidd, *J. Chem. Soc.*, 4228 (1954).
96. M. Schmeisser, P. Sartori, and B. Lippsmeier, *Chem. Ber.* **103**, 868 (1970).
97. R. Haszeldine and J. Kidd, *J. Chem. Soc.*, 2901 (1955).
98. R. Haszeldine and co-workers, *Chem. Commun.*, 249 (1972).
99. Jpn. Pat. 8,057,524 (1981), (to Nippon Petrochemicals Co., Ltd. Jpn. Kokai Tokyo Koho).
100. U.S. Pat. 3,970,721 (July 20, 1976), J. Brockington and R. Bennett (to Texaco, Inc.).
101. U.S. Pat. 4,090,944 (1978), R. Moore and J. Cox.
102. Jpn. Pat. 7,722,098 (1977), S. Maeda, A. Kondo, K. Odabe, and Y. Nakanishi.
103. Jpn. Pat. Appl. 79/36,285 (Mar. 29, 1979) (to Showa Denko).
104. Ger. Pat. 3,018,389 (1981), V. Muench, H. Naarmann, and K. Penzein.
105. EP Pat. Appl. 44,935 (Feb. 3, 1982), V. Muench, M. Nourmann, and K. Penzein.

106. G. A. Olah, T. Weber, D. Bellew, and O. Farooq, *Synthesis*, **6**, 463 (1989).
107. R. Miethchen and co-workers, *J. Prakt. Chem.* **313**, 383 (1977).
108. U.S. Pat. 3,878,261 (Apr. 15, 1975), L. E. Gardner (to Phillips Petroleum Co.).
109. Jpn. Pat. 58,052,316 (Mar. 28, 1983), (to the Agency of Industrial Sciences and Technology).
110. S. Sondheimer and N. Bunce, *J. Macromol. Sci., Rev. Macromol. Chem. Phys.* **26**, 351 (1986), for a review of the structure, chemical reactions, and industrial uses of Nafion-H.
111. Ref. 3, Chapt. 20, p. 879, for a complete review of polymers containing sulfonic acid functionality.
112. R. Quirk and co-workers, *Makromol. Chem., Macromol. Symp.* **32**, 47 (1990).
113. Jpn. Pat. 02,105,156 (Apr. 17, 1990), T. Aoso, A. Umehara, and N. Aotani (to Fuji Photo Film Co., Ltd.).
114. N. Borgesou, S. Bizzari, M. Janshekar, and N. Takei, in Ref. 23, Sept. 1995, pp. 502.5005 S.
115. Ref. 3, Chapt. 18, pp. 767–787, for a review of the biological activity of sulfonic acids.
116. *Ibid.*, p. 776.
117. S. Ganguly and J. J. Ghosh, *Ind. J. Biochem. Biophys.* **17**, 213 (1980).
118. G. Abbenante and R. H. Prager, *Aust. J. Chem.* **45**, 1801 (1992).
119. S. P. Ahmed and co-workers, *Antivir. Chem. Chemother.* **6**, 34 (1995).
120. Jpn. Pat. 06,287,142 (Oct. 11, 1994), T. Kotani, K. Morita, and M. Tonomura (to Kanebo Ltd.).
121. Z. Osawa and co-workers, *Carbohydr. Polym.* **21**, 283 (1993).
122. G. Tan and co-workers, *Biochim. Biophys. Acta*, **1181**, 183 (1993).
123. P. P. Mohan, D. Schols, M. Baba, and E. De Clercq, *Antiviral Res.* **18**, 139 (1992).
124. U.S. Pat. 715,652 (Feb. 1, 1992), R. Haugwitz and L. Zalkow (to U.S. Dept. of Health and Human Services).
125. F. Wallace, *Org. Prep. Proced. Int.* **14**, 265 (1982).
126. R. Wallace, *Aldrichimica Acta*, **13**, 3 (1980).
127. Jpn. Pat. Kokai Tokyo Koho 79,149,400 (1979).
128. P. H. Groggins, *Unit Processes in Organic Syntheses*, 3rd ed., McGraw-Hill Book Co., Inc., New York and London, 1947, pp. 689–691.
129. Brit. Pat. 1,288,634 (Sept. 9, 1972), A. W. Routledge (to Albright and Wilson Ltd.).
130. S. A. Majid and M. A. Hossain, *Bangladesh J. Sci. Ind. Res.* **14**(1–2), 142 (1979).
131. S. A. Majid and M. A. Hossain, *Bangladesh J. Sci. Ind. Res.* **11**(1–4), 44 (1976).
132. S. D. Vaidya, V. V. R. Subrahmanyam, and J. G. Kane, *Ind. J. Technol.* **13**, 528 (1975).
133. Braz. Pat. 7,500,927 (Aug. 31, 1976), Z. C. Xavier and A. Di Donato (to the Government of Brazil).
134. S. A. Majid and M. A. Hossain, *Bangladesh J. Sci. Ind. Res.* **15**(1–4), 107 (1980).
135. G. Ramakrishna, C. Mallaiah, E. L. N. Sarma, and S. D. T. Rao, *Chem. Age India*, **21**, 605 (1970).
136. S. A. Majid, M. A. Hossain, and M. Shahabuddin, *Bangladesh J. Sci. Ind. Res.* **17**(1–2), 98 (1982).
137. Ger. Pat. DD 138,832 (Nov. 21, 1979), G. West and co-workers (to VEB Filmfabrik Wolfen).
138. N. C. Kothiyal and B. P. Pandey, *Himalayan Chem. Pharm. Bull.* **10**, 16–24 (1993).
139. U.S. Pat. 5,059,336 (Oct. 22, 1991), M. Naka and co-workers (to Nippon Oil Co. Ltd.).
140. Can. Pat. 1,290,741 (Oct. 15, 1991), J. N. Vinci (to Lubrizol Corp.).
141. Can. Pat. 1,221,677 (May 12, 1987), C. R. Sloan (to Pro-Long Tech.).
142. U.S. Pat. 4,737,298 (Apr. 12, 1988), M. Born and co-workers (to Institüt Fancais du Petrole).

143. Can. Pat. 1,187,285 (May 21, 1985), N. Feldman (to Exxon Research and Engineering Co.).
144. U.S. Pat. 4,647,387 (Mar. 3, 1987) (to Witco Corp.).
145. U.S. Pat. 4,597,893 (July 1, 1986), D. C. Byford and co-workers (to British Petroleum PLC).
146. U.S. Pat. 4,322,479 (Mar. 30, 1982), J. W. Forsberg (to Lubrizol Corp.).
147. U.S. Pat. 4,322,478 (Mar. 30, 1982), J. W. Forsberg (to Lubrizol Corp.).
148. U.S. Pat. 4,386,173 (May 31, 1983), Y. F. Chang (to Ford Motor Co.).
149. U.S. Pat. 4,799,553 (Jan. 24, 1989), Y. Wu (to Phillips Petroleum Co.).

PAUL S. TULLY
Stepan Company

SULFOXIDES

Sulfoxides are compounds that contain a sulfinyl group covalently bonded at the sulfur atom to two carbon atoms. They have the general formula $RS(O)R'$, $ArS(O)Ar'$, and $ArS(O)R$, where Ar and Ar' = aryl. Sulfoxides represent an intermediate oxidation level between sulfides and sulfones. The naturally occurring sulfoxides often are accompanied by the corresponding sulfides or sulfones. The only commercially important sulfoxide is the simplest member, dimethyl sulfoxide [67-68-5] (DMSO) or sulfinylbismethane.

Sulfoxides occur widely in small concentrations in plant and animal tissues, eg, allyl vinyl sulfoxide [81898-53-5] in garlic oil and 2,2'-sulfinyl-bisethanol [3085-45-8] as fatty esters in the adrenal cortex (1,2). Homologous methylsulfinylalkyl isothiocyanates, which are represented by the formula $CH_3SO(CH_2)_nNCS$, where $n = 3$ [37791-20-1], 4 [4478-93-7], 5 [646-23-1], 8 [75272-81-0], 9 [39036-83-4], or 10 [39036-84-5], have been isolated from a number of mustard oils in which they occur as glucosides (3). Two methylsulfinyl amino acids have also been reported: methionine sulfoxide [454-41-1] from cockroaches and the sulfoxide of S-methylcysteine, 3-(methylsulfinyl)alanine [4740-94-7]. The latter is the dominant sulfur-containing amino acid in turnips and may account in part for their characteristic odor (4).

Dimethyl sulfoxide occurs widely at levels of ≤3 ppm. It has been isolated from spearmint oil, corn, barley, malt, alfalfa, beets, cabbage, cucumbers, oats, onion, Swiss chard, tomatoes, raspberries, beer, coffee, milk, and tea (5). It is a common constituent of natural waters, and it occurs in seawater in the zone of light penetration where it may represent a product of algal metabolism (6). Its occurrence in rainwater may result from oxidation of atmospheric dimethyl sulfide, which occurs as part of the natural transfer of sulfur of biological origin (7,8).

Properties

For the most part, sulfoxides are crystalline, colorless substances, although the lower aliphatic sulfoxides melt at relatively low temperatures (Table 1). The lower aliphatic sulfoxides are water soluble, but as a class the sulfoxides are not soluble in water. They are soluble in dilute acids and a few are soluble in alkaline solution. They dissolve to a variable extent in organic solvents, depending on associated functional groups. Because of the very polar sulfoxide group, they generally are high boiling and when distillable require reduced pressure. DMSO is a colorless liquid and its properties are listed in Table 2. Dimethyl sulfoxide generally undergoes typical sulfoxide reactions. It is used herein as an illustrative example.

Thermal Stability. Dimethyl sulfoxide decomposes slowly at 189°C to a mixture of products that includes methanethiol, formaldehyde, water, bis(methylthio)methane, dimethyl disulfide, dimethyl sulfone, and dimethyl sulfide. The decomposition is accelerated by acids, glycols, or amides (30). This product mixture suggests a sequence in which DMSO initially undergoes a Pummerer reaction to give (methylthio)methanol, which is labile and reacts according to equations 1–3. Disproportionation (eq. 4) also occurs to a small extent:

$$CH_3SCH_2OH \rightleftharpoons CH_3SH + HCHO \qquad (1)$$

$$2\ CH_3SH + HCHO \rightleftharpoons CH_3SCH_2SCH_3 + H_2O \qquad (2)$$

$$2\ CH_3SH + (CH_3)_2SO \longrightarrow CH_3SSCH_3 + CH_3SCH_3 + H_2O \qquad (3)$$

$$2\ (CH_3)_2SO \longrightarrow CH_3SO_2CH_3 + CH_3SCH_3 \qquad (4)$$

Oxidation. Sulfoxides are readily converted to sulfones, usually in high yield, by a number of strong oxidizing agents, eg, potassium permanganate, hypochlorites, hydrogen peroxide, ozone, selenium dioxide, or hot nitric acid

Table 1. Melting and Boiling Points of Sulfoxides

Name	CAS Registry Number	Formula	Mp, °C	Bp, °C	Ref.
sulfinylbismethane	[67-68-5]	$(CH_3)_2SO$	18.55	189.0	9
1,1'-sulfinylbisethane	[70-29-1]	$(C_2H_5)_2SO$	15	88–90[a]	10
1,1'-sulfinylbispropane	[4253-91-2]	$(n\text{-}C_3H_7)_2SO$	18		10
1,1'-sulfinylbisbutane	[2168-93-6]	$(n\text{-}C_4H_9)_2SO$	32		11
1,1'-sulfinylbis(2-chloroethane)	[5819-08-9]	$(ClCH_2CH_2)_2SO$	110.2		12
1,1'-sulfinylbisbenzene	[945-51-7]	$(C_6H_5)_2SO$	70.5	340 dec[b]	13
methylsulfinylbenzene	[1193-82-4]	$C_6H_5S(O)CH_3$	30–30.5	139–140	14,15
phenylmethylsulfinylbenzene	[833-82-9]	$C_6H_5S(O)CH_2C_6H_5$	125.5		16
1,1'-sulfinylbis(methylenebenzene)	[621-08-9]	$(C_6H_5CH_2)_2SO$	135		17

[a]At 2.0 kPa (15 mm Hg).
[b]Slowly.

Table 2. Properties of Dimethyl Sulfoxide

Property	Value				Ref.
boiling point, °C	189.0				9
conductivity, at 20°C, S/cm	3×10^{-8}				9
dielectric constant, at 25°C, 10 MHz	46.7				18
dipole moment, C·m[a]	1.4×10^{-29}				9
entropy of fusion, J/(mol·K)[b]	45.12				19
free energy of formation gas, C_{graph}, S_2(g), at 25°C, kJ/mol[b]	115.7				20
freezing point, °C	18.55				9
heat capacity, J/(mol·K)[b]					
liquid at 25°C	153.2				19
ideal gas, (T, K), C_p	$6.94 + 5.6 \times 10^{-2}\,T$ $-0.227 \times 10^{-4}\,T^2$				20
heat of formation liquid, C_{graph}, S_{rhomb}, at 18°C, kJ/mol[b]	-199.6				21
heat of fusion, kJ/mol[b]	13.5				22
heat of vaporization, at 70°C, kJ/mol[b]	47.3				23
molal fp constant, °C/(mol·kg)	4.07				24
pK_a	35.1				25
pK_{BH^+} (aq sulfuric acid)	-2.7				26
refractive index, n_D^{25}	1.4768				27
flash point, open cup, °C	95				23
autoignition temperature in air, °C	300–302				23
flammability limits in air, vol %					
lower, 100°C	3–3.5				23
upper, 180°C	42–63				23
	25°C	*35°C*	*45°C*	*100°C*	
vapor pressure, kPa[c]	0.080	0.159	0.303	4.0	23,28
density, g/cm³	1.0955	1.0855	1.0757		29
viscosity, mPa·s(=cP)	1.996	1.654	1.396	0.68	29

[a]To convert C·m to debye, divide by 3.336×10^{-30}.
[b]To convert J to cal, divide by 4.184.
[c]To convert kPa to mm Hg, multiply by 7.5.

(31). Side reactions producing sulfonic acids sometimes occur during oxidation with nitric acid. Treatment of DMSO under strongly alkaline conditions with either sodium hypochlorite or hypobromite results in oxidation accompanied by halogenation; the hexahalodimethyl sulfone forms in high yield (eq. 5) (32):

$$CH_3SOCH_3 + 6\,NaOCl + 0.5\,O_2 \longrightarrow CCl_3SO_2CCl_3 + 6\,NaOH \qquad (5)$$

The moderate resistance of DMSO to oxidation permits it to be used as a solvent for oxidations with lead tetraacetate or the 2-nitropropane anion (33,34). Dichromate oxidation and permanganate oxidation have been used for quantitative determination of DMSO (35,36).

Reduction. Dimethyl sulfoxide is reduced to dimethyl sulfide by a number of strong reducing agents, including aluminum hydrides, hydriodic acid, diborane, thiols, phosphine derivatives, and zinc in sulfuric acid (31). Quantitative procedures for determining DMSO have been based on its reduction by stannous chloride in hydrochloric acid, titanium trichloride in dilute hydrochloric acid, or sodium borohydride reduction followed by gas–liquid chromatography for low levels in aqueous systems (6,37,38). However, DMSO is sufficiently resistant to reduction to function as a solvent for polarography from a +0.3 V anode potential to a −2.8 V cathode potential, both being relative to a calomel electrode, with ammonium perchlorate electrolyte (39).

Dimethyl sulfoxide is reduced to dimethyl sulfide in a variety of organic reactions which entail initial attachment of an electrophilic activating reagent to the oxygen atom of DMSO to give (**1**) as shown in equation 6. This is followed by displacement of the oxygen-containing groups from the sulfur by a nucleophile and leads to a sulfonium intermediate (**2**) (eq. 7), which usually reacts further to give the following products (40):

$$(CH_3)_2SO + \text{electrophile} \longrightarrow (CH_3)_2\overset{+}{S}O\text{-electrophile} \qquad (6)$$

$$(\mathbf{1})$$

$$\text{nucleophile} + (CH_3)_2\overset{+}{S}O\text{-electrophile} \longrightarrow \text{nucleophile-}\overset{+}{S}(CH_3)_2 + (O\text{-electrophile})^- \qquad (7)$$

$$(\mathbf{2})$$

Compounds, eg, phenacyl halides, benzyl halides, alkyl iodides, or alkyl esters of sulfonic acids, react with DMSO at 100–120°C to give aldehydes (qv) and ketones (qv) in 50–85% yields (eq. 8) (41):

$$C_6H_5CH_2Cl + (CH_3)_2SO \longrightarrow [C_6H_5CH_2OS(CH_3)_2]^+Cl^- \longrightarrow CH_3SCH_3 + HCl + C_6H_5CHO$$

$$(8)$$

By a suitable choice of activating reagents, primary and secondary alcohols can be selectively oxidized to carbonyl compounds in good yields at room temperatures. Typical activating reagents are acetic anhydride, sulfur trioxide–pyridine, dicyclohexyl carbodiimide, and phosphorus pentoxide (40).

In the alcohol oxidations, the sulfonium intermediate (**2**, nucleophile = $R_2C(OH)$) loses a proton and dimethyl sulfide to give the carbonyl compound (42). The most common mechanism for the decomposition of (**2**) is attack by a mild base to remove a proton from one of the methyl groups. Subsequent cyclic collapse leads to the carbonyl compound and dimethyl sulfide (eq. 9):

$$R_2C(OH)\overset{+}{\underset{\overset{|}{CH_3}}{S}}CH_3 \xrightarrow{\ B^-\ } R_2C(OH)\overset{+}{\underset{\overset{|}{CH_2^-}}{S}}CH_3 \longrightarrow R_2C{=}O + (CH_3)_2S \qquad (9)$$

$$(\mathbf{2})$$

Because of the mild reaction conditions and the reaction path, these oxidations are very selective and often are used to oxidize hydroxyl groups in molecules containing sensitive groups that would also react with the common oxidizing agents.

Carbon–Sulfur Cleavage. The carbon–sulfur bond of DMSO is broken in a number of reactions. Attempts to form the DMSO anion by the reaction of DMSO with sodium result in cleavage accompanied by methane evolution (eqs. 10 and 11) (43):

$$CH_3\overset{\displaystyle O}{\overset{\displaystyle \|}{S}}CH_3 + 2\ Na^0 \longrightarrow CH_3SONa + CH_3Na \tag{10}$$

$$CH_3Na + CH_3\overset{\displaystyle O}{\overset{\displaystyle \|}{S}}CH_3 \longrightarrow CH_4 + CH_3\overset{\displaystyle O}{\overset{\displaystyle \|}{S}}CH_2Na \tag{11}$$

Sulfoxides containing β-hydrogen atoms, eg, di-t-butylsulfoxide [2211-92-9], react with strongly basic systems, eg, potassium t-butoxide, in DMSO by sulfenic acid elimination to produce olefins (eq.12) (44):

$$[(CH_3)_3C]_2SO \xrightarrow[\text{DMSO}]{t\text{-}C_4H_9OK} (CH_3)_2C{=\!=}CH_2 + (CH_3)_3CSOH \tag{12}$$

In other cases, sulfenic acid elimination can involve γ-hydrogen atoms with the formation of cyclopropane derivatives. γ-Elimination is favored when DMSO is the reaction solvent. An example involving 1-methylsulfinyl-2-ethyl-3-phenyl propane [14198-15-3] is shown in equation 13 (45):

$$C_6H_5CH_2CH(C_2H_5)CH_2SOCH_3 \xrightarrow[70^\circ C]{NaCH_2SOCH_3,\ DMSO} C_6H_5CH{\Big\langle}\begin{matrix}CH_2\\ |\\ CHC_2H_5\end{matrix} + CH_3SO^-Na^+ \tag{13}$$

Pummerer Reactions. Acetic anhydride at 70°C converts ethyl phenyl-sulfinylacetate [54882-04-1] (**3**) to the α-acetoxy sulfide (**4**) in 70% yield (eq. 13) (46):

$$\underset{(\textbf{3})}{C_6H_5\overset{\displaystyle O}{\overset{\displaystyle \|}{S}}CH_2\overset{\displaystyle O}{\overset{\displaystyle \|}{C}}OC_2H_5} + (CH_3\overset{\displaystyle O}{\overset{\displaystyle \|}{C}})_2O \longrightarrow \underset{(\textbf{4})}{C_6H_5\overset{\displaystyle O}{\overset{\displaystyle \|}{S}}CH(O\overset{\displaystyle O}{\overset{\displaystyle \|}{C}}CH_3)\overset{\displaystyle O}{\overset{\displaystyle \|}{C}}OC_2H_5} \tag{14}$$

The reaction is quite general and usually results in 75–90% yields (47). All reactions in which a sulfoxide containing at least one α-hydrogen is reduced to the sulfide and also oxidized at the α-carbon are referred to as Pummerer reactions (48).

The initiating step in these reactions is the attachment of a group to the sulfoxide oxygen to produce an activated intermediate (**5**). Suitable groups are proton, acyl, alkyl, or almost any of the groups that also initiate the oxidations of alcohols with DMSO (40,48). In a reaction, eg, the one between DMSO and acetic anhydride, the second step is removal of a proton from an α-carbon to give an ylide (**6**). Release of an acetate ion generates the sulfur-stabilized carbonium ion (**7**), and the addition of acetate ion to the carbonium ion (**7**) results in the product (eq. 15):

$$
\underset{(\mathbf{5})}{\underset{+}{CH_3\overset{\displaystyle \overset{\textstyle O}{\|}}{\underset{|}{S}}CH_3}} \quad \xrightarrow{-H^+} \quad \underset{(\mathbf{6})}{CH_3\overset{\displaystyle \overset{\textstyle O}{\|}}{\underset{|}{S}}CH_2} \quad \longrightarrow \quad \underset{(\mathbf{7})}{CH_3\overset{+}{S}{=}CH_2} \quad \xrightarrow{CH_3COO^-} \quad CH_3SCH_2\overset{\displaystyle \overset{\textstyle O}{\|}}{O}CCH_3 \tag{15}
$$

Both inorganic and organic acid chlorides react vigorously with DMSO to give α-chloromethyl methyl sulfide (**9**) when conditions are sufficiently controlled. In these reactions, a chloride ion displacement of the oxygen-containing group from the initial activated intermediate gives the labile chlorodimethylsulfonium ion (**8**). This sulfonium salt follows the reaction pattern shown for the acetoxydimethylsulfonium ion (**5**) to give the product (**9**). The sequence involving thionyl chloride is shown in equation 16:

$$
\underset{+}{CH_3\overset{\displaystyle \overset{\textstyle O}{\|}}{\underset{|}{S}}CH_3} \quad \xrightarrow{Cl^-} \quad CH_3\overset{+}{\underset{|}{\underset{Cl}{S}}}CH_3 \quad \longrightarrow \quad CH_3SCH_2Cl \tag{16}
$$

$$
\qquad\qquad\qquad\qquad (\mathbf{8}) \qquad\qquad\qquad (\mathbf{9})
$$

When DMSO is mixed with concentrated hydrochloric acid, protonated DMSO is in equilibrium with the chlorodimethylsulfonium ion. Pummerer reactions and subsequent reaction of the initial products give a complex mixture of products including formaldehyde, bis(methylthio)methane, methanethiol, dimethyl disulfide, dimethyl sulfide, and others.

Methylsulfinyl Carbanion.   The activating influence of the sulfinyl group on α-hydrogens, considerably less than that of a carbonyl group, is nonetheless sufficient to result in a pK_a of 35.1 for DMSO. Consequently, strong bases, eg, sodium hydride or sodium amide, react with DMSO producing solutions of methylsulfinyl carbanion [*15590-23-5*], known as the dimsyl ion, which are synthetically useful (49). The solutions also provide a strongly basic reagent for generating other carbanions. The dimsyl ion shows the expected nucleophilicity of carbanions and serves as a source of methylsulfinylmethyl groups (49). Thus, with alkyl halides or sulfonate esters, sulfoxides are obtained (eq. 17), carbonyl compounds yield β-hydroxysulfoxides (eq. 18), and esters give β-ketosulfoxides (eq. 19) (49):

$$
n\text{-}C_4H_9Br \; + \; CH_3\overset{\displaystyle \overset{\textstyle O}{\|}}{S}CH_2^- \quad \longrightarrow \quad C_5H_{11}\overset{\displaystyle \overset{\textstyle O}{\|}}{S}CH_3 \; + \; Br^- \tag{17}
$$

$$
[\textit{1561-74-6}]
$$

$$(C_6H_5)_2CO + CH_3\overset{\overset{\displaystyle O}{\|}}{S}CH_2^- \xrightarrow{H^+} (C_6H_5)_2\overset{\overset{\displaystyle OH}{|}}{C}CH_2\overset{\overset{\displaystyle O}{\|}}{S}CH_3 \qquad (18)$$

$$[2863\text{-}39\text{-}0]$$

$$C_6H_5\overset{\overset{\displaystyle O}{\|}}{C}OC_2H_5 + CH_3\overset{\overset{\displaystyle O}{\|}}{S}CH_2^- \longrightarrow C_6H_5\overset{\overset{\displaystyle O}{\|}}{C}CH_2\overset{\overset{\displaystyle O}{\|}}{S}CH_3 + C_2H_5O^- \qquad (19)$$

$$[2813\text{-}22\text{-}1]$$

The dimsyl ion also adds to carbon–carbon double bonds, and if the mixture is heated for several hours, methanesulfenate is eliminated. The overall result is methylation, and for compounds such as quinoline or isoquinoline (eq. 20), yields are nearly quantitative (50). The reaction sequence for isoquinoline to 1-methylisoquinoline is as follows:

$$(20)$$

Care is required in running these reactions because the decomposition of the intermediate sulfoxide and of dimsyl sodium during the heating in the strongly alkaline system is exothermic and also produces a precipitate which can interfere with heat removal. Explosions have occurred (51).

Methoxydimethylsulfonium and Trimethylsulfoxonium Salts. Alkylating agents react with DMSO at the oxygen. For example, methyl iodide gives methoxydimethylsulfonium iodide (**10**) as the initial product. The alkoxysulfonium salts are quite reactive and, upon continued heating, either decompose to give carbonyl compounds or rearrange to the more stable trimethylsulfoxonium salts, eg, (**11**) (eq. 21) (52):

$$(CH_3)_2SO + CH_3I \longrightarrow [(CH_3)_2SOCH_3]^+I^- \longrightarrow [(CH_3)_3SO]^+I^- \qquad (21)$$

$$\qquad\qquad\qquad\qquad (\mathbf{10}) \qquad\qquad\qquad (\mathbf{11})$$

Trimethylsulfoxonium iodide (**11**) is of interest because treatment with sodium hydride or dimsyl sodium produces dimethylsulfoxonium methylide [5367-24-8] (**12**) (eq. 22), which is an excellent reagent for introducing a methylene group into a variety of structures (53):

$$(\mathbf{11}) \xrightarrow{NaH} (CH_3)_2\overset{\overset{\displaystyle O^+}{|}}{S}CH_2^- \qquad (22)$$

$$(\mathbf{12})$$

Many aldehydes and ketones react with (**12**) to give better than 75% yields of epoxides (eq. 23):

$$(\mathbf{12}) + (C_6H_5)_2C{=\!\!=}O \longrightarrow (C_6H_5)_2C{\overset{O}{\underset{CH_2}{\diagdown|}}} + (CH_3)_2SO \tag{23}$$

In similar reactions, (**12**) with carbon–carbon double bonds that are conjugated with carbonyl groups gives cyclopropane derivatives (eq. 24) (48):

$$(\mathbf{12}) + C_6H_5CH{=\!\!=}CH\overset{O}{\overset{\|}{C}}C_6H_5 \longrightarrow C_6H_5\underset{H}{\overset{CH_2}{\overset{/\,\backslash}{C}}}\!\!-\!\!\underset{H}{\overset{}{C}}\!\!-\!\!\overset{O}{\overset{\|}{C}}C_6H_5 \tag{24}$$

Complexes. The sulfoxides have a high (ca 4) dipole moment, which is characteristic of the sulfinyl group, and a basicity about the same as that of alcohols. They are strong hydrogen-bond acceptors. They would be expected, therefore, to solvate ions with electrophilic character, and a large number of DMSO complexes of metal ions have been reported (54). The bonding to the metal is through the oxygen except for platinum(II), palladium(II), and rhodium(II) complexes where metal–sulfur bonds occur. The strength of the solvates is commonly about the same as that of the corresponding hydrate, and exchange of ligands is readily accomplished.

The strong tendency of the DMSO oxygen to act as a hydrogen-bond acceptor leads also to a number of organic complexes. Chloroform forms both 1:1 and 1:2 complexes (55). Pyrrole and phenol give 1:1 complexes (56,57). In solutions of monosaccharides in DMSO, the anomer with cis hydroxyls on the first and second carbons is stabilized (58). Complexes with organic molecules that do not involve hydrogen bonding also occur, eg, the 1:1 complex of DMSO and 4-chlorobenzonitrile and other nitriles (59,60). Strong 1:1 complexes form with nitrogen tetroxide and sulfur trioxide (61,62). Charge-transfer complexes occur with cyanogen iodide, tetracyanoethylene, and oxygen (63–65).

Synthesis and Manufacture

The sulfoxides are most frequently synthesized by oxidation of the sulfides (66,67). A broad group of oxidizing agents can be used and, because the oxidation to the sulfoxide is considerably more rapid than further oxidation to the sulfone, a proper choice of reagent quantity and conditions leads to high sulfoxide yields; eg, hydrogen peroxide in stoichiometric amounts can give 75–90% sulfoxide yields (68). Nitrogen tetroxide in a solvent, eg, carbon tetrachloride at temperatures below ca 0°C, selectively gives the sulfoxide in yields of up to 95% (69). Oxidations using sodium metaperiodate in aqueous or water–methanol solutions at ice-bath temperatures give high sulfoxide yields without sulfone formation (70). Oxidation using tertiary amine–bromine complexes in aqueous acetic acid gives yields above 70% and has been suggested as a convenient procedure for preparing ^{18}O-labeled sulfoxides when $H_2{}^{18}O$–acetic acid is used as the reaction solvent (71). There are a number of sulfoxide synthesis procedures in which the appropriate fragments are joined to give the product.

$$\text{RO}\overset{\displaystyle O}{\overset{\|}{\text{S}}}\text{OR} + 2 \text{ R}'\text{MgX} \longrightarrow \text{R}'\overset{\displaystyle O}{\overset{\|}{\text{S}}}\text{R}' + 2 \text{ ROMgX} \tag{25}$$

$$\text{Ar}\overset{\displaystyle O}{\overset{\|}{\text{S}}}\text{OR} + \text{R}'\text{MgX} \longrightarrow \text{Ar}\overset{\displaystyle O}{\overset{\|}{\text{S}}}\text{R}' + \text{ROMgX} \tag{26}$$

$$2 \text{ CH}_3\text{OC}_6\text{H}_5 + \text{SO}_2 \xrightarrow{\text{AlCl}_3} \text{CH}_3\text{OC}_6\text{H}_4\overset{\displaystyle O}{\overset{\|}{\text{S}}}\text{C}_6\text{H}_4\text{OCH}_3 + \text{H}_2\text{O} \tag{27}$$

$$2 \text{ C}_6\text{H}_6 + \text{SOCl}_2 \xrightarrow{\text{AlCl}_3} \text{C}_6\text{H}_5\overset{\displaystyle O}{\overset{\|}{\text{S}}}\text{C}_6\text{H}_5 + 2 \text{ HCl} \tag{28}$$

The reaction of Grignard reagents with sulfite esters gives 40–70% yields (eq. 25), and with arylsulfinate esters the yields are ca 55% (eq. 26) (72). Optically active sulfoxides are synthesized in good yield from the reaction of optically active sulfinate esters using Grignard reagents (73). Sulfoxides are obtained in greater than 50% yield by alkylating sulfenates with alkyl bromides (74). Diaryl sulfoxides are also obtained by Friedel-Crafts syntheses, eg, the reaction of anisole with sulfur dioxide and aluminum chloride to obtain 40% bis(4-methoxyphenyl) sulfoxide [1774-36-3] (eq. 27) or the reaction of benzene and thionyl chloride and aluminum chloride to obtain 51% diphenyl sulfoxide [945-57-7] (eq. 28) (75). A comprehensive review of procedures for synthesizing sulfoxides is available (76).

Dimethyl Sulfoxide. Dimethyl sulfoxide is manufactured from dimethyl sulfide (DMS), which is obtained either by processing spent liquors from the kraft pulping process or by the reaction of methanol or dimethyl ether with hydrogen sulfide. In the kraft pulping process, the spent liquors are normally concentrated to ca 50% solids and burned to recover inorganic chemicals and heat values (see PULP). The lignin in the liquor contains aromatic methoxyl groups, which are cleaved by sulfide ions to produce dimethyl sulfide when the concentrated liquor is processed in a reactor at 200–250°C (77). The synthesis of dimethyl sulfide from methanol and hydrogen sulfide is accomplished by a vapor-phase reaction over a catalyst at above 300°C.

Dimethyl sulfide has been oxidized to DMSO by several procedures. In pilot-plant quantities, the oxidation was accomplished with nitric acid, but this route has been supplanted by oxidation with nitrogen dioxide or oxygen containing minor amounts of nitrogen dioxide. The oxidation using nitrogen dioxide is diagrammed in Figure 1 (78). Dimethyl sulfide is oxidized with a DMSO solution of nitrogen dioxide in a reactor at 40–50°C. The reactor contents pass into a zone at 100°C where excess dimethyl sulfide is sparged from the crude DMSO with nitrogen; the crude DMSO is then neutralized and distilled. The flow of nitrogen dioxide into the reactor is kept insufficient to oxidize all of the dimethyl sulfide so that all the nitrogen dioxide is converted to nitric oxide, which is quite insoluble in DMSO and escapes in the exit-gas stream. The gas stream passes through a heat exchanger for condensation of some of the dimethyl sulfide which is recycled to the reactor. The gases then are conducted to a second reactor where an excess of nitrogen dioxide converts all of the remaining sulfide to the

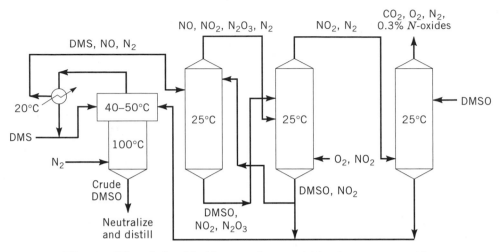

Fig. 1. Dimethyl sulfoxide manufacture with nitrogen tetroxide.

sulfoxide. The gases from this reactor contain substantially no organic matter and are oxidized with oxygen in a third reactor to regenerate the nitrogen dioxide. The gases finally pass through a DMSO scrubber for removal of nitrogen dioxide before venting to the atmosphere.

Processes involving oxygen and nitrogen oxides as catalysts have been operated commercially using either vapor- or liquid-phase reactors. The vapor-phase reactors require particularly close control because of the wide explosive limit of dimethyl sulfide in oxygen (1–83.5 vol %); plants in operation use liquid-phase reactions. Figure 2 is a schematic diagram for the liquid-phase process.

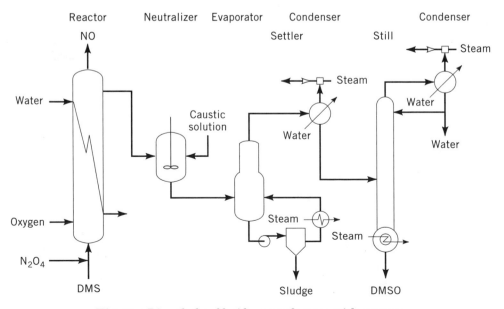

Fig. 2. Dimethyl sulfoxide manufacture with oxygen.

The product stream from the reactor is neutralized with aqueous caustic and is vacuum-evaporated, and the DMSO is dried in a distillation column to obtain the product.

Dimethyl sulfoxide is produced in commercial quantities in the United States by Crown Zellerbach Corporation. A typical analysis of industrial DMSO is given in Table 3 (23).

Table 3. Analysis of Industrial Dimethyl Sulfoxide

Properties	Analysis
DMSO assay, wt %	99.9
water content, wt %	0.1
color	water-white
other impurities	negligible
nonvolatiles	negligible

Health and Safety Factors

Dimethyl sulfoxide is a relatively stable solvent of low toxicity. The LD_{50} for single-does oral administration to rats is ca 17,400–28,300 mg/kg. Dimethyl sulfoxide by itself presents less hazard than many chemicals and solvents commonly used in industry. However, DMSO can penetrate the skin and may carry with it certain chemicals with which it is combined under certain conditions. Normal protective measures should be followed in the laboratory. If large quantities are handled where splashing and accidental contact may occur, protective clothing is recommended including suitable gloves and eye protectants. Butyl rubber gloves are more suitable than other types of material in terms of resisting penetration by DMSO solutions (79).

Dimethyl sulfoxide has received considerable attention as a useful agent in medicine (80,81). Compositions containing it are marketed in Germany and Austria as prescription items for analgesic uses. In the United States in 1981, the possible efficacy of DMSO as a drug was examined under jurisdiction of the FDA, but this agency has not released DMSO for drug use by humans other than for a painful bladder condition called interstitial cystitis. In veterinary medicine, DMSO is used for horses and dogs as a topical application to reduce swelling resulting from injury or trauma (see VETERINARY DRUGS).

Uses of Dimethyl Sulfoxide

Polymerization and Spinning Solvent. Dimethyl sulfoxide is used as a solvent for the polymerization of acrylonitrile and other vinyl monomers, eg, methyl methacrylate and styrene (82,83). The low incidence of transfer from the growing chain to DMSO leads to high molecular weights. Copolymerization reactions of acrylonitrile with other vinyl monomers are also run in DMSO. Monomer mixtures of acrylonitrile, styrene, vinylidene chloride, methallylsulfonic acid, styrenesulfonic acid, etc, are polymerized in DMSO–water (84). In some cases, the fibers are spun from the reaction solutions into DMSO–water baths.

Dimethyl sulfoxide can also be used as a reaction solvent for other polymerizations. Ethylene oxide is rapidly and completely polymerized in DMSO (85).

Diisocyanates and polyols or polyamines dissolve and react in DMSO to form solutions of polyurethanes (86) (see SOLVENTS, INDUSTRIAL).

Solvent for Displacement Reactions.   As the most polar of the common aprotic solvents, DMSO is a favored solvent for displacement reactions because of its high dielectric constant and because anions are less solvated in it (87). Rates for these reactions are sometimes a thousand times faster in DMSO than in alcohols. Suitable nucleophiles include acetylide ion, alkoxide ion, hydroxide ion, azide ion, carbanions, carboxylate ions, cyanide ion, halide ions, mercaptide ions, phenoxide ions, nitrite ions, and thiocyanate ions (31). Rates of displacement by amides or amines are also greater in DMSO than in alcohol or aqueous solutions. Dimethyl sulfoxide is used as the reaction solvent in the manufacture of high performance, polyaryl ether polymers by reaction of bis(4,4'-chlorophenyl) sulfone with the disodium salts of dihydroxyphenols, eg, bisphenol A or 4,4'-sulfonylbisphenol (88). These and related reactions are made more economical by efficient recycling of DMSO (89). Nucleophilic displacement of activated aromatic nitro groups with aryloxy anion in DMSO is a versatile and useful reaction for the synthesis of aromatic ethers and polyethers (90).

Solvent for Base-Catalyzed Reactions.   The ability of hydroxide or alkoxide ions to remove protons is enhanced by DMSO instead of water or alcohols (91). The equilibrium change is also accompanied by a rate increase of 10^5 or more (92). Thus, reactions in which proton removal is rate-determining are favorably accomplished in DMSO. These include olefin isomerizations, elimination reactions to produce olefins, racemizations, and H–D exchange reactions.

Extraction Solvent.   Dimethyl sulfoxide is immiscible with alkanes but is a good solvent for most unsaturated and polar compounds. Thus, it can be used to separate olefins from paraffins (93). It is used in the Institute Français du Pétrole (IFP) process for extracting aromatic hydrocarbons from refinery streams (94). It is also used in the analytical procedure for determining polynuclear hydrocarbons in food additives (qv) of petroleum origin (95).

Solvent for Electrolytic Reactions.   Dimethyl sulfoxide has been widely used as a solvent for polarographic studies and a more negative cathode potential can be used in it than in water. In DMSO, cations can be successfully reduced to metals that react with water. Thus, the following metals have been electrodeposited from their salts in DMSO: cerium, actinides, iron, nickel, cobalt, and manganese as amorphous deposits; zinc, cadmium, tin, and bismuth as crystalline deposits; and chromium, silver, lead, copper, and titanium (96–103). Generally, no metal less noble than zinc can be deposited from DMSO.

Cellulose Solvent.   Although DMSO by itself does not dissolve cellulose, the following binary and ternary systems are cellulose solvents: DMSO–methylamine, DMSO–sulfur trioxide, DMSO–carbon disulfide–amine, DMSO–ammonia–sodamide, DMSO–dinitrogen tetroxide, DMSO–paraformaldehyde, and DMSO–sulfur dioxide–ammonia (104,105). At least a ratio of three moles of active agent per mole of glucose unit is necessary for complete dissolution (104). Although only 80% of cellulose (qv) dissolves in DMSO–methylamine under cold anhydrous conditions, DMSO–dinitrogen tetroxide is a better solvent, particularly when a small quantity of water is added (106). Most of these systems can be used to produce cellulose fibers. The DMSO–paraformaldehyde system does not degrade cellulose, and it can form solutions containing up to 10 wt %

cellulose (107). It is believed that a methylolcellulose compound forms and is stable for extended periods of storage at ambient conditions (107). Regenerated cellulose articles, eg, films and fibers, can be prepared by contacting the DMSO–paraformaldehyde solutions with methanol and water (107,108).

Pesticide Solvent. The majority of organic fungicides, insecticides, and herbicides (qv) are soluble in DMSO, including such difficult-to-solvate materials as the substituted ureas and carbamates (see FUNGICIDES, AGRICULTURAL; INSECT CONTROL TECHNOLOGY; PESTICIDES). Dimethyl sulfoxide forms cosolvent systems of enhanced solubility properties with many solvents (109).

Clean-Up Solvent. Dimethyl sulfoxide is used to remove urethane polymers and other difficult-to-solvate materials from processing equipment.

BIBLIOGRAPHY

"Sulfoxides" in *ECT* 1st ed., Vol. 13, pp. 353–357, by E. G. Rietz, Chicago City Colleges, Wright Branch; in *ECT* 2nd ed., Vol. 19, pp. 320–337, by W. S. MacGregor, Crown Zellerbach Corp.; in *ECT* 3rd ed., Vol. 22, pp. 64–77, by W. S. MacGregor and J. V. Orle, Crown Zellerbach Corp.

1. A. Zwergal, *Pharmazie* **7**, 245 (1952).
2. T. Reichstein and A. Goldschneidt, *Helv. Chim. Acta* **19**, 401 (1936).
3. A. Kjaer and B. Christensen, *Acta Chem. Scand.* **12**, 833 (1958).
4. C. J. Morris and J. F. Thompson, *J. Am. Chem. Soc.* **78**, 1605 (1955).
5. T. W. Pearson, H. J. Dawson, and H. B. Lackey, *J. Agric. Food Chem.* **29**, 1089 (1981).
6. M. O. Andreae, *Anal. Chem.* **52**, 150 (1980).
7. M. D. Bentley, I. R. Douglass, J. A. Lacadie, and D. R. Whittier, *J. Air. Pollut. Control Assoc.* **22**, 359 (1972).
8. J. E. Lovelock, R. J. Maggs, and R. A. Rasmussen, *Nature (London)* **237**, 452 (1972).
9. H. L. Schlafer and W. Schaffernicht, *Angew. Chem.* **72**, 618 (1960).
10. W. Strecker and R. Spitaler, *Chem. Ber.* **59**, 1754 (1926).
11. N. Grabowsky, *Ann. Chem.* **175**, 348, 351 (1875).
12. H. Mohler, *Helv. Chim. Acta* **20**, 1188 (1937).
13. F. Krafft and R. E. Lyons, *Chem. Ber.* **29**, 435 (1896).
14. D. Barnard, J. M. Fabian, and H. P. Koch, *J. Chem. Soc.*, 2442 (1949).
15. H. Bohme, H. Fischer, and R. Frank, *Ann. Chem.* **563**, 54 (1949).
16. R. Pummerer, *Chem. Ber.* **43**, 1401 (1910).
17. H. Rheinboldt and E. Giescrecht, *J. Am. Chem. Soc.* **68**, 2671 (1946).
18. R. A. Hovermale and P. G. Sears, *J. Phys. Chem.* **60**, 1579 (1956).
19. H. L. Clever and E. F. Westrum, *J. Phys. Chem.* **74**, 1309 (1970).
20. H. Mackle and P. A. G. O'Hare, *Trans. Faraday Soc.* **58**, 1912 (1962).
21. T. B. Douglas, *J. Am. Chem. Soc.* **68**, 1072 (1946).
22. E. E. Weaver and W. Keim, *Proc. Indiana Acad. Sci. Monogr.* **70**, 123 (1961).
23. *Dimethyl Sulfoxide*, Technical bulletin, Crown Zellerbach Corp., Chemical Products Division, Orchards, Wash., 1966.
24. R. Garnsey and J. E. Prue, *Trans. Faraday Soc.* **64**, 1206 (1968).
25. W. N. Olmstead, Z. Margolin, and F. G. Bordwell, *J. Org. Chem.* **45**, 3295 (1980).
26. P. Haake, D. A. Tysse, S. R. Alpha, J. Kleckner, and R. D. Cook, *Q. Rep. Sulfur Chem.* **3**(2), 105 (1968).
27. R. G. LeBel and D. A. I. Goring, *J. Chem. Eng. Data* **7**, 100 (1962).
28. T. B. Douglas, *J. Am. Chem. Soc.* **70**, 2001 (1948).

29. P. G. Sears, W. D. Siegfried, and D. E. Sands, *J. Chem. Eng. Data* **9**, 261 (1964).
30. V. J. Traynelis and W. L. Hergenrother, *J. Org. Chem.* **29**, 221 (1964).
31. J. Drabonicz, T. Numata, and S. Oae, *Org. Prep. Proced. Int.* **9**(2), 63 (1977).
32. U.S. Pat. 3,304,331 (Feb. 14, 1967), C. DiSanto (to Stauffer Chemical Co.).
33. V. Zitko and C. T. Bishop, *Can. J. Chem.* **44**, 1749 (1966).
34. B. H. Klanderman, *J. Org. Chem.* **31**, 2618 (1966).
35. K. Stelmach, *Chem. Anal. (Warsaw)* **11**, 627 (1966).
36. L. H. Krull and M. Friedman, *J. Chromatogr.* **26**, 336 (1967).
37. E. Glynn, *Analyst* **72**, 248 (1947).
38. R. R. Legault and K. Groves, *Anal. Chem.* **29**, 1495 (1957).
39. J. L. Jones and H. A. Fritsche, Jr., *J. Electroanal. Chem.* **12**, 334 (1966).
40. A. J. Mancuso and D. Swern, *Synthesis*, 165 (1981).
41. N. Kornblum, W. J. Jones, and G. J. Anderson, *J. Am. Chem. Soc.* **81**, 4113 (1959).
42. M. G. Burdon and J. G. Moffatt, *J. Am. Chem. Soc.* **89**, 4725 (1967).
43. D. E. O'Connor and W. I. Lyness, *J. Org. Chem.* **30**, 1620 (1965).
44. J. E. Hofmann, T. J. Wallace, P. A. Argabright, and A. Shriesheim, *Chem. Ind. (London)*, 1243 (1963).
45. R. Baker and M. J. Spillett, *Chem. Commun.*, 757 (1966).
46. R. Pummerer, *Chem. Ber.* **43**, 1401 (1910).
47. L. Horner and P. Kaiser, *Ann. Chem.* **631**, 198 (1960).
48. T. Durst, in E. C. Taylor and H. Wynberg, eds., *Advances in Organic Chemistry: Methods and Results*, Vol. 6, Interscience Publishers, a division of John Wiley & Sons, Inc., New York, 1969, pp. 285–388.
49. E. J. Corey and M. Chaykovsky, *J. Am. Chem. Soc.* **87**, 1345 (1965).
50. G. A. Russell and S. A. Wiener, *J. Org. Chem.* **31**, 248 (1966).
51. F. A. French, *Chem. Eng. News* **44**, 48 (1966).
52. S. G. Smith and S. Winstein, *Tetrahedron* **3**, 317, 319 (1958).
53. E. J. Corey and M. Chaykovsky, *J. Am. Chem. Soc.* **87**, 1353 (1965).
54. W. L. Reynolds, in S. J. Lippard, ed., *Progress in Inorganic Chemistry*, Vol. 12, Interscience Publishers, a division of John Wiley & Sons, Inc., New York, 1970, pp. 1–99.
55. A. L. McCellan, S. W. Nicksic, and J. C. Guffy, *J. Mol. Spectrosc.* **11**, 340 (1963).
56. D. M. Porter and W. S. Brey, Jr., *J. Phys. Chem.* **72**, 650 (1968).
57. R. S. Drago, B. Wayland, and R. L. Carlson, *J. Am. Chem. Soc.* **85**, 3125 (1963).
58. V. S. R. Rao and J. F. Foster, *J. Phys. Chem.* **69**, 656 (1965).
59. C. D. Ritchie and A. Pratt, *J. Phys. Chem.* **67**, 2498 (1963).
60. C. D. Ritchie and A. Pratt, *J. Am. Chem. Soc.* **86**, 1571 (1964).
61. C. C. Addison and J. C. Sheldon, *J. Chem. Soc.*, 2705 (1956).
62. R. L. Whistler, A. H. King, G. Ruffini, and F. A. Lucas, *Arch. Biochem. Biophys.* **121**, 358 (1967).
63. E. Augdahl and P. Klaeboe, *Acta Chem. Scand.* **18**, 18 (1964).
64. F. E. Stewart, M. Eisner, and W. R. Carper, *J. Chem. Phys.* **44**, 2866 (1966).
65. T. Sato, H. Inoue, and K. Hata, *Bull. Chem. Soc. Jpn.* **40**, 1502 (1967).
66. T. Durst, in D. N. Jones, ed., *Comprehensive Organic Chemistry*, Vol. 3, Pergamon Press, Oxford, U.K., 1979, pp. 121–156.
67. E. E. Reid, *Organic Chemistry of Bivalent Sulfur*, Vol. 2, Chemical Publishing Co., New York, 1960, pp. 64–66.
68. D. Jerchel, L. Dippelhofer, and D. Renner, *Chem. Ber.* **87**, 947 (1954).
69. L. Horner and F. Hubenett, *Ann. Chem.* **579**, 193 (1953).
70. N. J. Leonard and C. R. Johnson, *J. Org. Chem.* **27**, 282 (1962).
71. S. Oae, Y. Ohmishi, S. Kozuka, and W. Tagaki, *Bull. Chem. Soc. Jpn.* **39**, 364 (1966).
72. H. Hepworth and H. W. Clapham, *J. Chem. Soc.* **119**, 1188 (1921).

73. K. K. Andersen, W. Gaffield, N. E. Papnikolaow, J. W. Foley, and R. I. Perkins, *J. Am. Chem. Soc.* **86**, 5637 (1964).

74. D. E. O'Connor and W. I. Lyness, *J. Org. Chem.* **30**, 1620 (1965).

75. R. L. Shriner, H. C. Struck, and W. J. Jorison, *J. Am. Chem. Soc.* **52**, 2060 (1930).

76. J. Drabowicz and M. Mikokajczyk, *Org. Prep. Proced. Int.* **14**(1–2), 45 (1982).

77. W. M. Hearon, W. S. MacGregor, and D. W. Goheen, *Tappi* **45**(1), 28A, 30A, 34A, 36A (1962).

78. H. Pruckner, *Erdoel Kohle Erdgas Petrochemie* **16**, 188 (1963).

79. "Dimethyl Sulfoxide," *A Summary of Toxicological and Safety Information*, Crown Zellerbach Corp., Chemical Products Division, Orchards, Wash., 1972.

80. *Ann. N.Y. Acad. Sci.* **141**, 1 (Mar. 15, 1967).

81. S. W. Jacob, E. E. Rosenbaum, and D. C. Wood, eds., *Dimethyl Sulfoxide*, Vol. 1, Marcel Dekker, Inc., New York, 1971.

82. T. Ouchi, A. Tatsumi, and M. Imoto, *J. Polym. Sci. Polym. Chem. Ed.* **16**, 707 (1978).

83. C. H. Bamford and A. N. Ferrar, *Proc. R. Soc. London Ser. A* **321**, 425 (1971).

84. U.S. Pat. 3,781,248 (Dec. 25, 1973), H. Sakai and co-workers (to Toray Industries, Inc.).

85. K. S. Kazanskii, A. S. Yanov, and S. A. Dubrovsky, *Makromol. Chem.* **179**, 969 (1978).

86. U.S. Pat. 3,658,731 (Apr. 25, 1972), T. Richardson and G. O. Hustad (to Wisconsin Alumni Research Foundation).

87. E. Buncel and H. Wilson, in V. Gold and D. Bethell, eds., *Adv. Phys. Org. Chem.*, Vol. 14, Academic Press, Inc., New York, 1977, pp. 133–202.

88. U.S. Pat. 4,175,175 (Nov. 20, 1979), R. N. Johnson and H. G. Farnham (to Union Carbide Corp.).

89. *Dimethyl Sulfoxide Recovery and Environmental Engineering*, Crown Zellerbach Corp., Chemical Products Division, Orchards, Wash., 1974.

90. T. Takekoshi, J. G. Wirth, D. R. Heath, J. E. Kochanowski, J. S. Manello, and M. J. Webber, *J. Polym. Sci. Polym. Chem. Ed.* **18**, 3069 (1980).

91. R. Steward, *Q. Rep. Sulfur Chem.* **3**(2), 99 (1968).

92. D. J. Cram and L. Gosser, *J. Am. Chem. Soc.* **86**, 5457 (1964).

93. U.S. Pat. 4,267,034 (May 12, 1981), C. O. Carter (to Phillips Petroleum Co.).

94. P. J. Bailes, *Chem. Ind. (London)*, 69 (1977).

95. E. O. Haenni, F. L. Joe, Jr., J. W. Howard, and R. L. Leibel, *J. Assoc. Off. Agric. Chem.* **45**(1), 59 (1962).

96. J. A. Porter, *AEC Research and Development Report DP-389*, July 1959.

97. T. H. Handley and J. H. Cooper, *Anal. Chem.* **41**, 381 (1969).

98. S. Morisaki, N. Baba, and S. Tajima, *Denki Kagaku* **38**, 746 (1970).

99. U.S. Pat. 3,772,170 (Nov. 13, 1973), N. R. Bharucha.

100. S. Nakagawa, Z. Takehara, and Y. Yoshizawa, *Denki Kagaku* **41**, 880 (1973).

101. V. V. Kuznetov and co-workers, *Zashch. Met.* **1**, 631 (1975).

102. V. V. Kuznetov and co-workers, *Izv. Sev. Kavk. Nauchno. Tsentra Vyssh. Shk. Ser. Estestv. Nauk.* **4**, 47 (1976).

103. E. Santos and F. Dyment, *Plating (East Orange, N.J.)* **60**, 821 (1973).

104. B. Philipp, H. Schleicher, and W. Wagenknecht, *Chemtech*, 702 (Nov. 1977).

105. A. F. Turbak, R. B. Hammer, R. E. Davies, and H. L. Hergert, *Chemtech*, 51 (Jan. 1980).

106. U.S. Pat. 4,076,933 (Feb. 28, 1978), A. F. Turbak, R. B. Hammer, N. A. Portnoy, and A. C. West (to International Telephone and Telegraph Corp.).

107. U.S. Pat. 4,097,666 (June 27, 1978), D. C. Johnson and M. D. Nicholson (to The Institute of Paper Chemistry).

108. R. B. Hammer, M. E. O'Shaughnessey, E. R. Strauch, and A. F. Turbak, *J. Appl. Polym. Sci.* **23**, 485 (1979).

109. *DMSO—The Agricultural Chemical Solvent*, Crown Zellerbach Corp., Chemical Products Division, Orchards, Wash., Oct. 1973.

General References

References 9, 31, 48, 54, 66, 81, and 87 contain reviews of DMSO chemistry.

W. W. Epstein and F. W. Sweat, *Chem. Rev.* **67**(3), 247 (1967).

D. Martin and H. G. Hauthal, *Dimethyl Sulfoxide*, Halsted Press, a division of John Wiley & Sons, Inc., New York, 1975.

B. S. Thyagarajan and N. Kharasch, *Intrascience Sulfur Reports*, Vol. 1, The Chemistry of DMSO, Intrascience Research Foundation, Santa Monica, Calif., 1966.

SULFUR

Sulfur [*7704-34-9*], S, a nonmetallic element, is the second element of Group 16 (VIA) of the Periodic Table, coming below oxygen and above selenium. In massive elemental form, sulfur is often referred to as brimstone. Sulfur is one of the most important raw materials of the chemical industry. It is of prime importance to the fertilizer industry (see FERTILIZERS) and its consumption is generally regarded as one of the best measures of a nation's industrial development and economic activity (see SULFUR COMPOUNDS; SULFUR REMOVAL AND RECOVERY; SULFURIC ACID AND SULFUR TRIOXIDE).

Sulfur has been known since antiquity. Early humans used sulfur to color cave drawings, employed sulfur fumes to kill insects and to fumigate, and knew about sulfur's color-removing or bleaching action. Mystical powers were attributed to the ethereal blue flame and pungent odor given off by burning the yellow rock. Medicinal use of sulfur was known to the Egyptians and Greeks. One contemporary use was developed as early as 500 BC, when the Chinese used sulfur as an ingredient of gunpowder. Although the modern history of sulfur may have begun with Lavoisier's proof in the late-eighteenth century that sulfur is an element, the first commercial sulfur was produced in Italy early in the fifteenth century. Sulfur production become Italy's main industry when, in 1735, development of a process to make sulfuric acid from sulfur was commercialized. When a French company gained an effective monopoly on the Sicilian deposits in 1839 and tripled the price, other countries, particularly England and the United States, developed internal sources of sulfur and sulfuric acid. Consumers in many countries quickly learned that sulfuric acid could be made from sulfur dioxide obtained from the roasting of iron pyrites, obviating the need for Sicilian sulfur.

The United States continued to depend on foreign sources of elemental sulfur even after the mineral was discovered in the United States in 1867 by oil prospectors investigating a salt dome in Calcasieu Parish, Louisiana. Various attempts were made to sink mine shafts. Realizing that conventional mining processes would be too uneconomical to compete with Sicilian sulfur, H. Frasch conceived of melting the sulfur underground by injecting superheated water into the formation and then lifting the melted sulfur to the surface using a sucker-rod pump. In 1894, the first flow of molten sulfur was pumped from the Calcasieu Parish deposit. In 1902, the Frasch process was successfully commercialized. This mining method later became important in the development and production of sulfur not only from the Texas–Louisiana salt dome area, but also from areas in western Texas, Mexico, Poland, and Iraq.

As of the 1990s, sulfur recovered as a by-product, involuntary sulfur, accounts for a larger portion of world supply than does mined or voluntary material. Sulfur is obtained from hydrogen sulfide, which evolves when natural gas (see GAS, NATURAL), crude petroleum (qv), tar sands (qv), oil shales (qv), coal (qv), and geothermal brines (see GEOTHERMAL ENERGY) are desulfurized (see SULFUR REMOVAL AND RECOVERY). Other sources of sulfur include metal sulfides such as pyrites; sulfate materials, including gypsum (see CALCIUM COMPOUNDS); and elemental sulfur in native and volcanic deposits mined in the traditional manner.

Sulfur constitutes about 0.052 wt % of the earth's crust. The forms in which it is ordinarily found include elemental or native sulfur in unconsolidated volcanic rocks, in anhydrite over salt-dome structures, and in bedded anhydrite or gypsum evaporate basin formations; combined sulfur in metal sulfide ores and mineral sulfates; hydrogen sulfide in natural gas; organic sulfur compounds in petroleum and tar sands; and a combination of both pyritic and organic sulfur compounds in coal (qv).

Properties

Allotropy. Sulfur occurs in a number of different allotropic modifications, that is, in various molecular aggregations which differ in solubility, specific gravity, crystalline form, etc. Like many other substances, sulfur also exhibits dynamic allotropy, ie, the various allotropes exist together in equilibrium in definite proportions, depending on the temperature and pressure. The molecular formulas for the various allotropes are $S-S_n$, where n is a large but unidentified number, such as $n \geq 10^6$. The particular allotropes that may be present in a given sample of sulfur depend to a large extent on its thermal history, the amount and type of foreign substance present, and the length of time that has passed for equilibrium to be attained.

In the solid and liquid states, the principal allotropes are designated traditionally as $S\lambda$, $S\mu$, and $S\pi$. Of these, only $S\lambda$ is stable in the solid state. Upon solidification of molten sulfur, $S\pi$ rapidly changes into $S\mu$, which is converted into $S\lambda$, although at a much slower rate. The molecular structure of $S\pi$ is that of an octatomic sulfur chain (1,2). The symbol $S\mu$ designates long, polymerized chains of elemental sulfur. $S\lambda$ is perhaps the most characteristic molecular form of sulfur, namely, that of a crown-shaped, octatomic sulfur ring designated in more recent literature as $S_{(r/8)}$ (3). The allotropes have different solubility in carbon disulfide. $S\pi$ and $S\lambda$ are soluble in carbon disulfide, whereas $S\mu$ does not dissolve in this solvent.

Sulfur crystallizes in at least two distinct systems: the rhombic and the monoclinic forms. Rhombic sulfur, $S\alpha$, is stable at atmospheric pressures up to 95.5°C, at which transition to monoclinic sulfur, $S\beta$, takes place. Monoclinic sulfur is then stable up to its natural melting point of 114.5°C. The basic molecular unit of both of these crystalline forms of sulfur is the octatomic sulfur ring $S_{(r/8)}$. Other forms of solid sulfur include hexatomic sulfur as well as numerous modifications of catenapolysulfur (2,4).

The molecular constitution of liquid sulfur undergoes significant and reversible changes with temperature variations. These changes are evidenced by the characteristic temperature dependence of the physical properties of sulfur. In most studies of liquid sulfur, some striking changes in its physical properties are observed at about 160°C. For example, the viscosity of purified sulfur, which at 120°C is about 11 mPa·s(=cP), drops to a minimum of 6.7 mPa·s at about 157°C, and then begins to rise. At 159–160°C, the viscosity of liquid sulfur rises sharply, increasing to 30 mPa·s at 160°C and reaching a maximum of about 93 Pa·s (930 P) at 187°C. Above this temperature, the viscosity gradually drops off again to about 2 Pa·s (20 P) at 306°C. A qualitative exploration of these viscosity changes in terms of the allotropy of sulfur implies that below 159°C, sulfur consists mainly of S_8 rings. A normal decrease of viscosity with rising temperature is observed. The sudden increase in the viscosity of sulfur above 159°C is attributed to the formation of polymeric sulfur chain molecules. Then, as the temperature rises further, the concentration of polymeric sulfur continues to increase, but the opposing effect of decreasing chain length resulting from thermal sulfur–sulfur bond scission causes a gradual decrease in viscosity in the temperature range between 187°C and the boiling point of sulfur. The chemical equilibria between the various forms in molten sulfur have been extensively investigated (2,3,5). A critical review of the literature concerning the molecular composition of molten sulfur is also available (6). Experiments that added much to the knowledge of the species present under different time–temperature parameters have been described (7,8). Previous theories concerning the polymerization of S_8 were shown to be in disagreement with well-established experimental facts and are considered unsatisfactory.

The molecular composition of sulfur vapor is a complex function of temperature and pressure. Vapor pressure measurements have been interpreted in terms of an equilibrium between several molecular species (9,10). Mass spectrometric data for sulfur vapor indicate the presence of all possible S_n molecules from S_2 to S_8 and negligible concentrations of S_9 and S_{10} (11). In general, octatomic sulfur is the predominant molecular constituent of sulfur vapor at low temperatures, but the equilibrium shifts toward smaller molecular species with increasing temperature and decreasing pressure.

Constants and Chemical Properties. The constants of sulfur are presented in Table 1. Two freezing points are given for each of the two crystalline modifications. When the liquid phase consists solely of octatomic sulfur rings, the temperature ranges at which the various modifications form are called the ideal freezing points. The temperatures at which the crystalline forms are in equilibrium with liquid sulfur containing equilibrium amounts of $S\pi$ and $S\mu$ are called natural freezing points.

There are four stable isotopes of sulfur: ^{32}S, ^{33}S, ^{34}S, and ^{36}S, which have relative abundances of 95.1, 0.74, 4.2, and 0.016%, respectively. The relative abundance of the various isotopes varies slightly, depending on the source of the sulfur; the ratio of ^{32}S to ^{34}S is 21.61–22.60. Three radioactive isotopes of

Table 1. Physical Constants of Sulfur

Property	Value Ideal	Value Natural	Reference
freezing point of solid phase, °C			
rhombic	112.8	110.2	
monoclinic	119.3	114.5	
boiling point, °C		444.6	12
density of solid phase, 20°C, g/cm^3			
rhombic		2.07	
monoclinic		1.96	
amorphous		1.92	
density of liquid, g/cm^3			
125°C		1.7988	
130°C		1.7947	
140°C		1.7865	
150°C		1.7784	
density of vapor, 444.6°C and 101.3 kPa(= 1 atm), g/L		3.64	
refractive index, n_D^{110}		1.929	
vapor pressure^a, P in Pa, T in K			
rhombic, 20–80°C	$\log P = 16.557 - 5166/T$		13
monoclinic, 96–116°C	$\log P = 16.257 - 5082/T$		14
liquid			
120–325°C	$\log P = 19.6 - 0.0062238T - 5405.1/T$		12
325–550°C	$\log P = 12.3256 - 3268.2/T$		
surface tension, mN/m(=dyn/cm)			
120°C		60.83	15
150°C		57.67	
critical temperature, °C		1040	
critical pressure, MPa^b		11.75	
critical volume, mL/g		2.48	
specific heat, C_p, J/(kg·K)^c			
rhombic, 24.9–95.5°C	$C_p = 468 + 0.814T$		
monoclinic, −4.5 to 118.9°C	$C_p = 465 - 0.908T$		
liquid, Sλ, 118.9–444.6°C	$C_p = 706 - 0.65T$		
gas, S, 25–1727°C	$C_p = 709 - 0.034T - 3.5 \times 10^6 T^{-2}$		
gas, S$_2$, 25–1727°C	$C_p = 558 + 0.018T - 5.2 \times 10^6 T^{-2}$		
heat of transformation (rhombic to monoclinic) at 95.5°C, J/g^c		11.25	
heat of fusion, J/g^c			
112.8°C S$_{rhombic}$ → Sλ (l)		49.8	
118.9°C S$_{monoclinic}$ → Sλ (l)		38.5	
linear thermal expansion of rhombic sulfur			
0–13°C		4.567×10^{-5}	
13–50°C		7.433×10^{-5}	
50–78°C		8.633×10^{-5}	
78–97°C		20.67×10^{-5}	
98–110°C		103.2×10^{-5}	

Table 1. (*Continued*)

Property	Value Ideal	Value Natural	Reference
latent heat of vaporization, L, J/g[c]	L^d	L^e	12
200°C	308.6		
300°C	289.3		
400°C	286.4	278.0	
420°C	287.6	276.3	
440°C	290.1	274.6	
460°C	293.1	273.0	
electrical resistivity, ohm·cm			16
20°C		1.9×10^{17}	
110°C		4.8×10^{12}	
400°C		8.3×10^{6}	
magnetic susceptibility, m^3/mol[f]			16
rhombic, 18°C		1.539	
monoclinic, 112°C		1.539	
liquid, 220°C		1.539	
standard reduction potential, S/S^{2-}, V		−0.508	16

[a]To convert log P_{Pa} to log P_{psi}, subtract 3.8384 from the constant.
[b]To convert MPa to psi, multiply by 145.
[c]To convert J to cal, divide by 4.184.
[d]Includes heat of dissociation to S_2 present in vapor.
[e]Minus heat of dissociation to S_2 present in vapor.
[f]To convert m^3/mol to emu/mol (cgs unit), divide by $4\pi \times 10^6$.

masses 31, 35, and 37 having half-lives of 2.6 s, 87 d, and 5 min, respectively, have been generated artificially.

Sulfur falls between oxygen and selenium in Group 16 and resembles oxygen in its chemical reactions with most of the elements. The normal orbital electron structure (17) is of the arrangement $1s^2\ 2s^2\ 2p^1\ 3s^2\ 3p^4$. Sulfur has valences of −2, +2, +3, +4, and +6. Selenium is a closely related element having a similar group of valences and analogous allotropy. Sulfur is between phosphorus and chlorine in the third Period of the Periodic Table. Although the properties of sulfur are generally those to be expected from its position in the Table, an exception is that its melting point is higher than expected, probably because of its complex molecular structure (17).

Sulfur is insoluble in water but soluble to varying degrees in many organic solvents, such as carbon disulfide, benzene, warm aniline, warm carbon tetrachloride, and liquid ammonia (18). Carbon disulfide is the most commonly used solvent for sulfur.

Sulfur combines directly and usually energetically with almost all of the elements. Exceptions include gold, platinum, iridium, and the helium-group gases (19). In the presence of oxygen or dry air, sulfur is very slowly oxidized to sulfur dioxide. When burned in air, it forms predominantly sulfur dioxide with small amounts of sulfur trioxide. When burned in the presence of moist air, sulfurous acid and sulfuric acids are slowly generated.

Hydrochloric acid reactions with sulfur only in the presence of iron to form hydrogen sulfide. Sulfur dioxide forms when sulfur is heated with concentrated sulfuric acid at 200°C. Dilute nitric acid up to 40% concentration has little effect, but sulfur is oxidized by concentrated nitric acid in the presence of bromine with a strongly exothermic reaction (19).

Sulfur combines directly with hydrogen at 150–200°C to form hydrogen sulfide. Molten sulfur reacts with hydrogen to form hydrogen polysulfides. At red heat, sulfur and carbon unite to form carbon disulfide. This is a commercially important reaction in Europe, although natural gas is used to produce carbon disulfide in the United States. In aqueous solutions of alkali carbonates and alkali and alkaline-earth hydroxides, sulfur reacts to form sulfides, polysulfides, thiosulfates, and sulfites.

At room temperature, sulfur unites readily with copper, silver, and mercury and vigorously with sodium, potassium, calcium, strontium, and barium to form sulfides. Iron, chromium, tungsten, nickel, and cobalt react much less readily. In a finely divided state, zinc, tin, iron, and aluminum react with sulfur on heating (19).

Various sulfides of the halogens are formed by direct combination of sulfur with fluorine, bromine, and chlorine. No evident reaction occurs with iodine; instead, the elements remain as components of a mixture. Mixtures of sulfur and potassium chlorate, or sulfur and powdered zinc, are highly explosive.

Sulfur is involved in numerous organic reactions (20). When dissolved in amines, chemical interaction between sulfur and the solvent results in the formation of colored species ranging from deep yellow to orange and green (see SULFUR DYES). Many organic reactions involving sulfur are commercially significant. Sulfur is important in the manufacture of lubricants, plastics, pharmaceuticals (qv), dyes, and rubber goods (see DYES AND DYE INTERMEDIATES; LUBRICATION AND LUBRICANTS; PLASTICS PROCESSING; RUBBER CHEMICALS).

Sulfur is not considered corrosive to the usual construction materials. Dry, molten sulfur is handled satisfactorily in mild steel or cast-iron equipment. However, acid-generating impurities, which may be introduced in handling and storage, create corrosive conditions. The exposure of sulfur to moisture and air causes the formation of acids which attack many metals. To combat such corrosion difficulties, protective coatings of organic compounds, cement, or sprayed resistant metals are often applied to exposed steel surfaces, including pipe and equipment used in handling liquid sulfur, and to structural members that come in contact with solid sulfur. Also practical in some applications is the use of resistant metal alloys, particularly those of aluminum and stainless steel. Naturalization of the generated acids by the addition of basic chemicals is sometimes employed.

Elemental Sulfur

Sources of sulfur are called voluntary if sulfur is considered to be the principal, and often the only, product. Sulfur has also been recovered as a by-product from various process operations. Such sulfur is termed involuntary sulfur and accounts for the largest portion of world sulfur production (see SULFUR REMOVAL AND RECOVERY).

Occurrence. *Salt-Dome Sulfur Deposits.* The sulfur deposits associated with salt domes in the Gulf Coast regions of the southern United States and Mexico have historically been the primary sources of U.S. sulfur. These remain an important segment of both U.S. and world sulfur supply. Although the reserves are finite, many are large and voluntary productive capacity ensures the importance of these sources for some time to come. In 1994, the output from the salt domes in the U.S. was about 2.09 million metric tons (21).

Salt domes of the U.S. Gulf Coast are vertical structures, usually circular in outline, with steeply dipping flanks, and composed of coarsely crystalline halite, ie, NaCl, interspersed with anhydrite, $CaSO_4$. The cap rock that surmounts the salt dome consists of anhydrite in contact with the salt and gypsum, $CaSO_4 \cdot 2H_2O$, derived from the anhydrite. Limestone in the form of fine gray carbonate interspersed with vugs, seams, fissures, and cavities is frequently associated with the gypsum and anhydrite formations. It may be present as a stratum overlying these formations, as lenticular beds covering part of them, or as disseminated lenses and nodules included in the upper part of the cap rock.

The sulfur occurs as well-developed crystal aggregates in veins and vugs or as disseminated particles in the porous limestone and gypsum section of the cap rock. Several theories have been proposed for the occurrence of sulfur in salt domes. One theory suggests the formation of limestone and hydrogen sulfide from anhydrite in the presence of reducing agents. This reaction, however, requires temperatures of about 650°C and, although oil or other hydrocarbons may be present to act as reducing agents, the temperature actually attained is not sufficient to support this theory. In 1946, the presence of anaerobic, sulfate-reducing bacteria was discovered in cap rock. The ability of these bacteria to promote reaction at normal temperatures is recognized as the more likely origin of sulfur. Anaerobic bacteria consume hydrocarbons as a source of energy, but combine sulfur instead of oxygen with the hydrogen. The hydrocarbon-fueled bacteria reduce anhydrite to hydrogen sulfide, calcium carbonate, and water. The hydrogen sulfide remains dissolved in the formation waters until it precipitates as crystalline sulfur through various oxidation reactions, possibly initiated by oxygen and carbon dioxide dissolved in water percolating from the upper sediments. In 1966, 329 salt-dome structures were identified by the U.S. Bureau of Mines in the U.S. Gulf Coast area and offshore tidelands. Of these, 27 have been commercial sulfur producers. As of this writing (ca 1996), one of the salt-dome sulfur deposits, Main Pass operated by Freeport Sulphur Company in Louisiana, was being commercially mined by the Frasch Process. The other 26 are no longer mined. Some could be reactivated if justified by economic circumstances.

Evaporite Basin Sulfur Deposits. Elemental sulfur occurs in another type of subsurface deposit similar to the salt-dome structures in that the sulfur is associated with anhydrite or gypsum. The deposits are sedimentary, however, and occur in huge evaporite basins. It is believed that the sulfur in these deposits, like that in the Gulf Coast salt domes, was derived by hydrocarbon reduction of the sulfate material and assisted by anaerobic bacteria. The sulfur deposits in Italy (Sicily), Poland, Iraq, the CIS, and the United States (western Texas) are included in this category.

Mining techniques similar to the Frasch salt-dome mining systems have been applied successfully. These developments and particularly those in western Texas, Poland, and Iraq have significantly contributed to world sulfur production and reserves. Hot-water mining of the Polish deposits began in 1966 at Gryzbow. In 1979, production at Gryzbow and at another deposit, Jeziorko, was nearly five million metric tons. By 1995, production declined to ~2 million metric tons. The new Osiek Mine officially opened near Grzybow in September, 1993 after several months of test production. The cost of production there was reported to have been cut about 50% by using a system to recycle hot water from nearby power stations, cutting labor costs, and closing nonproductive facilities. Production ceased at the Machow Mine in 1992. However, final transfer of the operation to local authorities has been delayed. Environmental problems continue at Machow, and the cost of remediation, including the elimination of hydrogen sulfide emissions from the pit has been estimated at more than $200 million. Production of sulfur from the Culberson mine (Freeport Sulphur Company) in western Texas began in 1968. The Culberson mine was the only one of this type in operation in the United States as of the mid-1990s. During the late 1980s, this mine was the world's largest sulfur producer, exceeding two million metric tons per year during peak production. In Iraq, the Mishraq deposit has the potential for contributing over one million metric tons per year. Salt domes and similar sulfur-bearing structures occur in regions other than the United States, Mexico, Poland, Iraq, and the CIS, but sulfur deposits that could be economically productive have not been discovered in any of them.

Sulfur is recovered by Frasch and open-pit mining in several locations in the CIS, mainly in the Ukraine, Turkmenistan, and Russia. The largest deposits lie within the Ukraine Republic on the Polish–Ukraine border, where there are three mines, Yazov, Nemorov, and Tazdahl. Mining of sulfur in the Ukraine is by open-pit and the Frasch process. However, Frasch production has not been very successful because of the low porosity of the sulfur-bearing ore. There are problems also with product quality owing to bitumen contamination. In addition, significant environmental problems have also constrained production. In Turkmenistan, Guardak is believed to be the only Frasch mine outside the Ukraine. In Russia, there are some small operations mining sulfur from underground volcanic deposits.

Production of Frasch and native sulfur probably peaked in the CIS at about 3.2 million metric tons. It has declined since and was reported as 400,000 metric tons in 1995. About 200,000 metric tons was produced in the Ukraine, with the remainder from Turkmenistan and Russia.

Volcanic and Native Sulfur Deposits. Elemental sulfur occurs in other types of surface or underground deposits throughout the world, but seldom in sufficient concentration to be commercially important. Sulfur of volcanic origin occurs in many parts of the world. These deposits originated from gases emitted from active craters, solfataras, or hot springs, which contain deposited sulfur in fractures of rock, by replacement in the rock itself or in the sediments of lake beds. Volcanic deposits usually occur in tufas, lava flows, and similar volcanic rocks but also in sedimentary and intrusive formations. Scattered deposits of this type have been discovered throughout the mountain ranges bordering the Pacific Ocean, particularly in Japan, the Philippine Islands, and Central and South America. Some volcanic deposits are worked profitably and are important in the countries in which this type of deposit is found. The Japanese deposits are among this group and have had a long and productive history of considerable tonnages. Most volcanic deposits, however, are in isolated regions and at high elevations where production and transportation costs are prohibitive.

Extraction.   *Frasch Process.*   In the Frasch process, large quantities of hot water are introduced through wells drilled into buried deposits of native sulfur. The heat from the water melts the sulfur in the vicinity of the wells; the melted sulfur is then removed to the surface as molten elemental sulfur of high purity. Economical operation of a salt dome or a subsurface sulfur deposit by the Frasch process requires a porous sulfur-bearing limestone, a large and dependable supply of water, and a source of inexpensive fuel. A power plant is required, in which the necessary volume of hot water is produced, as is compressed air for pumping molten sulfur from the wells and electric power for drilling, lighting, operating maintenance equipment, loading sulfur for shipment, and similar operations.

A typical setting of equipment for a sulfur well and the principles of mining are illustrated schematically in Figure 1. First, a hole is drilled to the bottom layer of the salt-dome cap rock with equipment of the same type as that used in oil fields. Three concentric pipes within a protective casing are placed in the hole. A 20-cm pipe inside an outer casing is sunk through the cap rock to the bottom of the sulfur deposit. Its lower end is perforated with small holes. Then, a 10-cm pipe is lowered to within a short distance of the bottom. Last and innermost is a 2.5-cm pipe, which is lowered more than halfway to the bottom of the well.

Water-heated under pressure to 160°C is pumped down the space between the 20-cm and 10-cm pipes and, during the initial heating period, also down the 10-cm pipe. The superheated water flows out the holes at the bottom into the porous sulfur-bearing formation (Fig. 1a). When the temperature of the sulfur-bearing formation exceeds the melting point of sulfur, the liquid sulfur, being approximately twice as heavy as water, percolates downward through the porous limestone to form a pool at the bottom of the well. A heating period of 24 h or longer is required to accumulate a liquid sulfur pool of sufficient size, and then pumping of hot water down the 10-cm pipeline is stopped. Static pressure of the hot water pumped into the formation then forces the liquid more than 100-m up into the 10-cm pipe (Fig. 1b). Compressed air forced down the 2.5-cm pipe aerates and lightens the liquid sulfur so that it rises to the surface (Fig. 1c). Injection of hot water is continued down the 20-cm pipe to maintain the sulfur melting process, and the compressed-air volume is adjusted to equalize the sulfur pumping rate with the sulfur melting rate. If the pumping rate exceeds the melting rate, the sulfur pool is depleted and the well produces water. At this point, the compressed-air flow is stopped, and hot water is again injected until the liquid sulfur pool is reestablished.

The sulfur-bearing cap rock, being an enclosed formation, is essentially the equivalent of a pressure vessel. Hot water, pumped into the formation to melt sulfur, must be withdrawn after cooling at approximately the same rate as it is put in, otherwise the pressure in the formation would increase to the point where further water injection would be impossible. Bleedwater wells, used to extract water from the formations, usually are located on the flanks of the dome away from the mining area where the water temperature is lowest. The water is treated to remove soluble sulfides and other impurities before being discharged to disposal ditches or canals.

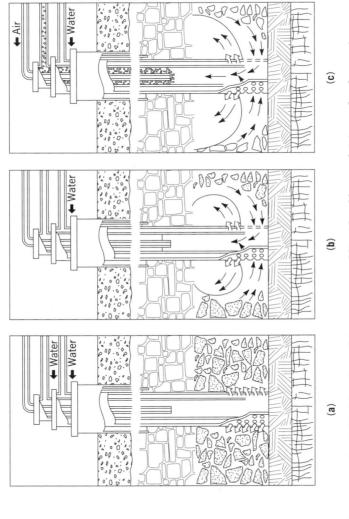

Fig. 1. The Frasch process: (**a**) initial heating; (**b**) movement of liquid sulfur; and (**c**) result of pumping compressed air. The thinner arrows indicate the flow of molten sulfur. See text. Courtesy of Freeport Minerals Company.

241

On the surface, the liquid sulfur moves through steam-heated lines to a separator where the air is removed. Depending on the mine location, the liquid sulfur may be pumped to storage vats to be solidified, to tanks for storage as a liquid, to pipelines, or to thermally insulated barges for transport to a central shipping terminal.

Sulfur wells that are favorably located produce continuously over long periods. Some may last a year or more; other may be abandoned within a few weeks, because denseness of the rock formation may retard circulation of hot water and molten sulfur. The extraction of sulfur weakens the rock formation and subsidence may follow. This may break the pipes in the well, ending productivity. Although subsidence is desirable in mining, wells may be lost as a result. The advantage of subsidence is that the volume of exhausted formation through which hot water can circulate is reduced. After caving, the crushed exhausted formation is relatively impervious and therefore confines the circulation of hot water to the more porous sulfur-bearing parts of the deposit.

Directional drilling techniques were an important advance in sulfur mining methods. Casings are placed in the hole in such a manner as to extend into the sulfur formation somewhat horizontally. Thus, substantial amounts of sulfur in the deposit overlie the hole. Also, the casings in the subterranean volume likely to be affected by subsidence are parallel to the expected earth movement and therefore are less affected by shear. The result is better utilization of heating water and longer well life. Directional drilling has also permitted efficient reworking of areas exhausted to vertical mining techniques.

Another advance in sulfur mining technology has been the development of a method involving seawater in the Frasch process, making it feasible to mine deposits distant from freshwater supplies. In such a plant, seawater is first deoxygenated by bringing it in direct contact with combustion gases in a packed tower. The seawater is preheated by these gases and its temperature raised to 106°C in indirect heat exchangers by means of steam (qv) furnished by high pressure boilers. Condensate from the heat exchangers is recycled to the boilers; this limits freshwater requirements to leaks and other small losses in the system. Production from several sulfur mines involves seawater from both stationary systems and portable, barge-mounted power plants (see also MINERALS RECOVERY AND PROCESSING).

Hydrodynamic Process.   The hydrodynamic process is similar to the Frasch process in that superheated water is used to melt the underground sulfur. However, the techniques involved are different. The process was developed in Poland to exploit sulfur deposits, which, because of thin bedding, wide dispersion, and frequent impermeability, did not appear to be amenable to production by the Frasch method, which requires some degree of deposit isolation. The techniques employed include the use of explosives to control permeability or to create sealed-off gases and the calculated manipulation of underground water pressure, temperature, and flow conditions by control at injection and breakwater points. The system requires a constant rate of fuel consumption throughout the life of the mine and improves the rate of sulfur recovery. Hydrodynamic mining is used in Poland and in Iraq where the mining area was developed with Polish assistance.

Volcanic and Other Surface Deposits.   Sulfur is recovered from volcanic and other surface deposits by a number of different processes, including distillation, flotation, autoclaving, filtration, solvent extraction, or a combination of several of these processes. The Japanese sulfur deposits are reached by tunnel, and mining is done by the room-and-pillar, chamber-and-pillar with filling, and cut-and-fill systems. Sulfur was historically extracted from the ore by a distillation process performed in rows of cast-iron pots, each containing about 180 kg of ore. Each row of pots is connected to a condensation chamber outside the furnace. A short length of pipe connects each pot with a condenser. Brick flues connect combustion gases under the pots. Sulfur vapor flows from the pots to the condensation chamber where the liquid sulfur is collected. The Japanese ore contains 25–35 wt % sulfur. This method has been superseded by other sources of sulfur production.

The sulfur deposits in Italy have been worked since ancient times. Originally, sulfur was removed by piling the ore in central heaps, covering it with earth, and then igniting the pile. By this method, 30–50 wt % of the sulfur was burned to provide heat for melting the remainder of the sulfur in the ore. Less than 50 wt % of the sulfur originally contained in the ore was recovered. In about 1880, the first Gill gas furnace was installed. The original furnace had two chambers arranged so that the heat from burning ore in one chamber passed through ore in the other chamber to melt a considerable portion of the sulfur. When the sulfur in the first chamber burned, the chamber was refilled. When the partially extracted ore in the first chamber burned, it was refilled and the partially extracted ore in the other chamber was ignited. This method involved better utilization of the heat of combustion. Later, furnaces contained as many as six chambers and permitted up to 80% sulfur recovery (22).

Extensive experimentation has led to numerous patents for various thermal processes for extracting sulfur from ores, either as elemental sulfur or as SO_2, but very few of these processes have been operated commercially. The proposed processes involve shaft furnaces, multiple-hearth furnaces, rotary kilns, and fluidized-bed roasters. In all of these, ground ore is heated with oxygen-free, hot combustion gases to distill elemental sulfur; or the sulfur in the ore is burned with air to yield SO_2 for sulfuric acid production. In 1953, a commercial plant was brought on-stream at the Yerrington, Nevada, copper mine for recovering the sulfur as SO_2 from the Leviathan deposit of low grade sulfur ore in Alpine Country, California. The process consisted of four fluid-bed reactors, in which the ore was roasted in air to produce SO_2 for a contact sulfuric acid plant (23). For the production of elemental sulfur, the use of oxygen-free, hot combustion gases in a fluidized bed has been proposed to distill sulfur from the ore as a vapor, which is then condensed to liquid sulfur.

Various processes have been proposed and tested for the recovery of sulfur from native ores by solvent extraction, and many patents have been issued. Carbon disulfide, the best solvent for sulfur, has often been suggested for extraction of sulfur from ore. Some plants in Italy, Germany, South America, and the United States have used carbon disulfide for this purpose, but the cost of the solvent, the high losses, and its flammability detract from low operating costs. Many other solvents have been tried, including hot caustic solution, chlorinated hydrocarbons, ammonium sulfide, xylene, kerosene, and various high boiling oils. The sulfur is recovered either by volatilizing the solvent or by crystallizing the sulfur.

Various combinations of autoclaving, filtration, and centrifuging are used in some processes to recover sulfur from ore. One such process, involving continuous autoclaving, flotation (qv), and filtration (qv), was first used commercially at a plant in Columbia (24). The ore is finely ground to less than 625-μm (28-mesh) and suspended in water to form a slurry of about 30% solids, which is pumped continuously to a three-compartment, agitated autoclave. In the autoclave, the slurry is heated and the sulfur melts by steam injection into the bottom of each compartment. Agitation causes the sulfur to coalesce into globules that separate from the gangue. Hot slurry from the autoclave flows into a quench pot to be cooled by water injection, and the sudden cooling solidifies the separated sulfur particles. The cooled slurry is throttled to atmospheric pressure and flows into a 625-μm (28-mesh) vibrating screen. The oversize material passes directly to a sulfur melter, whereas the underflow from the screen passes to a flotation circuit for separation of the smaller sulfur particles. This gives a concentrate of 90–95 wt % sulfur which then passes to the sulfur melter. Melted sulfur is pumped through a filter for removal of gangue.

One more variation to the many methods proposed for sulfur extraction is the fire-flood method. It is a modern version of the Sicilian method, by which a portion of the sulfur is burned to melt the remainder. It would be done *in situ* and is said to offer cost advantages, to work in almost any type of zone formation, and to produce better sweep efficiency than other systems. The recovery stream would be about 20 wt % sulfur as SO_2 and 80 wt % elemental sulfur. The method was laboratory-tested in the late 1960s and patents were issued. However, it was not commercially exploited because sulfur prices dropped.

Sulfide Ores

Occurrence.   The metal sulfides, which are scattered throughout most of the world, have been an important source of elemental sulfur. The potential for recovery from metal sulfides exists, although these sources are less attractive economically and technologically than other sources of sulfur. Nevertheless sulfide ores are an important source of sulfur in other forms, such as sulfur dioxide and sulfuric acid.

Some of the most important metal sulfides are pyrite [1309-36-0], FeS_2; chalcopyrite [1308-56-1], $CuFeS_2$; pyrrhotite [1310-50-5], $Fe_{n-1}S_n$; sphalerite [12169-28-7], ZnS; galena [12179-39-4], PbS; arsenopyrite [1303-18-0], $FeS_2 \cdot FeAs_2$; and pentlandite [53809-86-2], $(Fe,Ni)_9S_8$. Sulfide deposits often occur in massive lenses, but may occur in tabular shape, in veins, or in a disseminated state. The deposits may be of igneous, metamorphic, or sedimentary origin.

Pyrite is the most abundant of the metal sulfides. For many years, until the Frasch process was developed, pyrite was the main source of sulfur and, for much of the first half of the twentieth century, comprised over 50% of world sulfur production. Pyrite reserves are distributed throughout the world and known deposits have been mined in about 30 countries. Possibly the largest pyrite reserves in the world are located in southern Spain, Portugal, and the CIS. Large deposits are also in Canada, Cyprus, Finland, Italy, Japan, Norway,

South Africa, Sweden, Turkey, the United States, and Yugoslavia. However, the three main regional producers of pyrites continue to be Western Europe; Eastern Europe, including the CIS; and China.

Pyrites production is the main source of sulfuric acid for both fertilizer and nonfertilizer uses in China and has been increasing steadily. However, production has been declining steadily in all other regions. This trend seems likely to continue.

In the 1980s, pyrites, as a percentage of total sulfur produced, comprised nearly 18%; in the 1990s, however, pyrites are only 13% and are expected to fall to around 10% by the year 2005. In 1995, 68% of world pyrites production was in China. By 2005, it is forecast that China will produce about 81% of the world total.

Pyrometallurgical Processes. *Orkla Process.* A process for recovering sulfur from cuprous pyrite was developed by the Orkla Mining Company in Norway (25). The sulfur output of 80,000–100,000 t/yr furnished an important part of European requirements until 1962, when the smelter was shut down. The process was once used on a much smaller scale in Portugal and Spain and probably in the CIS (26). The Orkla process involves recovery of about 80% of the sulfur contents of a pyritic copper ore by direct smelting in the presence of a carbonaceous reducing agent.

Noranda Process. When pyrites are heated to about 540°C in the absence of oxygen, about half of the sulfur content in the pyrites evolves in the elemental form. Noranda Mines Ltd. and Battelle Memorial Institute developed a process based on this property to recover elemental sulfur from pyrite (27). The first commercial plant was built at Welland, Ontario, in 1954 but operated on an experimental basis for only a few years before being closed for economic reasons.

Outokumpu Process. Outokumpu Base Metals Oy, Finland's largest mining and metallurgical company, discovered a complex ore body at Pyhasalmi, Finland, containing pyrite, sphalerite, chalcopyrite, barite, and small amounts of pyrrhotite, arsenopyrite, and molybdenite. The ore can be beneficiated by flotation to obtain pyrite concentrate as well as copper and zinc concentrates. A process was developed to treat the pyrite concentrate in a flash smelter for recovery of elemental sulfur and iron cinder. The commercial smelter located at Kokkolla began operating in 1962 (28). In 1977, production of elemental sulfur was stopped, although sulfur dioxide is still produced and sold for sulfuric acid production. Similarly, the Outokumpu process was used to recover elemental sulfur at a plant in Botswana, but as of the 1990s, the sulfur is recovered as $SO_2-H_2SO_4$.

Hydrometallurgical Processes. Recovery of sulfur in the processing of nonferrous metal sulfides has been in the form of SO_2 and/or H_2SO_4 when smelter (pyrometallurgical) operations are employed. However, there have been accounts of processes, mainly hydrometallurgical, in which sulfur is recovered in the elemental form (see METALLURGY, EXTRACTIVE).

One, the CLEAR process, was investigated by Duval Corporation near Tucson, Arizona (29). It involves leaching copper concentrated with a metal chloride solution, separation of the copper by electrolysis, and regeneration of the leach solution in a continuous process carried out in a closed system. Elemental sulfur is recovered. Not far from the Duval plant, Cyprus Mines Corporation operated a process known as Cymet. Sulfide concentrates undergo a two-step chloride solution leaching and are crystallized to obtain cuprous chloride crystals. Elemental sulfur is removed during this stage of the process.

Another process, which also generates elemental sulfur as a by-product, has been patented by Envirotech Research Center in Salt Lake City (29). In the Electroslurry process, a ball mill finely grinds a chalcopyrite concentrate, which reacts with an acidic copper sulfate solution for iron removal. The liquor is electrolyzed and the iron is oxidized to the ferric form. This latter step leaches copper from the copper sulfide for deposition on the cathode. Elemental sulfur is recovered at the same time.

A pressure leaching system to handle copper sulfide called the Sherritt-Cominco (SC) copper process was developed by these two Canadian firms. Pilot-plant testing was completed in 1976 (29), but commercial application of this technology has not been achieved.

Sulfates

Occurrence.   The largest untapped source of sulfur occurs in the ocean as dissolved sulfates of calcium, magnesium, and potassium (see OCEAN RAW MATERIALS). The average sulfur concentration in seawater is 880 ppm. Thus, 1 km^3 of seawater contains about 0.86×10^6 t of elemental sulfur in the form of sulfates. Natural and by-product gypsum, $CaSO_4 \cdot 2H_2O$, and anhydrite, $CaSO_4$, rank second only to the oceans as potential sources of sulfur. Mineral deposits of gypsum and anhydrite are widely distributed in extremely large quantities. Gypsum is a by-product waste material from several manufacturing processes; most notable is the waste gypsum produced in manufacturing phosphoric acid from phosphate rock and sulfuric acid (see PHOSPHORIC ACID AND THE PHOSPHATES).

Extraction.   Although many processes have been developed to recover elemental sulfur from gypsum or anhydrite, high capital and operating costs have precluded widespread use of these processes and are expected to continue to do so while less expensive sources remain available. Obtaining sulfur from gypsum processes has been attractive during periods when sulfur has been in short supply and energy costs remained relatively low. Because these processes require large amounts of energy when energy costs are high, sulfur extraction is unlikely to be competitive. However, gypsum and anhydrite remain economical sources of sulfur in other forms, including sulfuric acid, cement, and ammonium sulfate, in areas such as India where sulfur must otherwise be imported.

Thermal Reduction of Gypsum.   The initial work involving the thermochemical technique was carried out in Germany and more recently by the U.S. Bureau of Mines (USBM), which did research work on two processes for the recovery of elemental sulfur from gypsum at the Salt Lake City Metallurgy Research Center in the late 1960s (30). Both processes involved reduction roasting of gypsum using coal or reducing gases at 900–950°C to produce calcium sulfide. Process one involved carbonation of a water slurry of calcium sulfide with CO_2-bearing flue gases from the reduction kiln to precipitate calcium carbonate and to evolve hydrogen sulfide. The latter could be converted to sulfur in a standard Claus unit. Process two made use of a countercurrent ion-exchange system and sodium chloride to produce by-product sodium carbonate and calcium chloride as well as elemental sulfur. Three metric tons each of the by-product were produced per metric ton of sulfur.

Neither of these processes has been commercialized, although some aspects of the methodology were incorporated into a plant operated for a short time by the Elcor Company (31). This company, which operated briefly in western Texas in 1968 using natural gypsum, is the only one known to have commercially attempted to recover elemental sulfur from this material by a two-step thermal process. The Elcor plant was shut down shortly after it began operation. Although most technical problems were said to have been solved, production costs were prohibitive.

Phosphogypsum.   Phosphogypsum is produced in tremendous quantities in the manufacture of phosphate fertilizers (qv). A process used by Fertilizer India, Ltd. (Planning and Development) involved a shaft kiln (31). Following bench-scale tests, tests on a larger scale were conducted at the Regional Research Laboratory at Jorhat (Assam Province). The feed to the top of the kiln consisted of a nodulized mixture of phosphogypsum, pulverized coke, and clay additives. Air was introduced to the bottom of the kiln such that the temperature in the hottest zone was maintained at 1100–1200°C. Under these conditions, calcium sulfate is reduced to sulfur dioxide, which then reacts to yield elemental sulfur. Although the process was technically feasible, it was found to be uneconomical. As of 1994, no commercial process existed for economical sulfur recovery from phosphogypsum.

Bacteriological Sulfur.   Anaerobic, sulfate-reducing bacteria burn hydrocarbons as a source of energy, but combine sulfur instead of oxygen with the hydrogen to form hydrogen sulfide. Several experimenters have tried to utilize this knowledge in a controlled process for producing sulfur from gypsum or anhydrite (32). This process requires a strain of sulfate-reducing bacteria, an organic substrate whose hydrocarbons provide food for the bacteria, and close control of environmental conditions in order to obtain maximum sulfur yields.

Finely ground gypsum is fed into a stirred reaction tank containing the organic substrate and the bacteria. The substrate can be a petroleum fraction, although sewage, spent sulfite liquor, molasses, or brewery waste can also be used. The advantage of a petroleum-based substrate is that its composition can be more closely controlled. Air must be excluded from the system because the bacteria are anaerobic. A hydrogen-purging system keeps air out and at the same time promotes fermentation.

Carbon dioxide generated by the fermentation process must be removed to help maintain the pH of the solution at pH 7.6–8.0. Carbon dioxide also inhibits the activity of the bacteria. The oxidation reduction potential is kept at 100–200 mV. The ideal temperature in the reactor varies with different strains in the bacteria but generally is 25–35°C.

As the reaction proceeds, a part of the mix is continuously withdrawn from the tank and is centrifuged, and the solids removed by centrifuging are resuspended in the reactor. Filtrate from the centrifuge goes to a stripping tower for removal of dissolved carbon dioxide and hydrogen sulfide, which is combined with the carbon dioxide and hydrogen sulfide gases passing from the top of the reactor. The combined gases are passed through a scrubbing tower for removal of the carbon dioxide and recovery of the hydrogen sulfide, which is fed to a conventional recovery unit for conversion to elemental sulfur. There is also the possibility of recovering other organic co-products, such as vitamins (qv) and steroids (qv). The rate at which bacteria reduce gypsum to hydrogen sulfide is quite slow, necessitating many large reaction tanks. A 300-t/d plant is estimated to need 10 3785-m^3 (10^6-gal) reactor tanks. An organism of the *Desulfovibrio* genus has been used to make hydrogen sulfide in early experiments. Subculturing and selectively reisolating active organisms could lead to a strain with higher activity. By such techniques, a 1000-fold increase in activity after 40 generations was achieved in the 1960s (32) (see GENETIC ENGINEERING, MICROBES).

Production

Sulfur is produced from a variety of sources using many different techniques in many countries around the world. Worldwide changes have affected not only the sources of sulfur, but also the amounts consumed. Sulfur sources in the United States underwent significant changes during the 1980s. Voluntary sulfur from the Frasch process (mines) supplied only 25% of the sulfur in the United States in 1995, compared to about 53% in 1980, whereas recovered or involuntary sulfur supplied 63% of the sulfur in the United States in 1995, compared to 34% in 1980. About 12% is supplied from other forms, primarily by metallurgy (21,33).

Recovered elemental sulfur, a nondiscretionary by-product from petroleum (qv) refining, natural gas processing (see GAS, NATURAL), and coking plants, was produced primarily to comply with environmental regulations that were applied directly to emissions from the processing facility or indirectly by restricting the sulfur content of the fuels sold or used by the facility. Recovered elemental sulfur was produced by 59 companies at 150 plants in 26 states, one plant in Puerto Rico, and one plant in the U.S. Virgin Islands. Most of these plants were relatively small, with only 22 reporting an annual production exceeding 100,000 metric tons. By source, 52% was produced at three coking plants and 86 refineries or satellite plants treating refinery gases. The remainder was produced by 27 companies at 61 natural gas treatment plants.

Table 2 shows the estimated annual world sulfur production capacity in all forms. Figure 2 shows actual annual world sulfur production by type. The actual 1993 sulfur production was considerably less than the reported production capacity partly owing to geographic regions such as the CIS undergoing political changes. Similarly, sulfur production in other regions of the world is less than reported capacity for a variety of other factors such as overestimation of recovered sulfur production from clean air requirements and lower production from voluntary sources to balance market demand. Another factor usually ignored in estimated capacity is the subtle changes in sulfur contents of energy sources.

Economic Aspects

At the beginning of the twentieth century, the world's sulfur demand of about two million metric tons was met by sulfur produced from elemental deposits in Sicily, Italy, and from pyrite mined on the Iberian Peninsula. By 1995, sulfur was recovered in more than 78 countries.

Table 2. 1993 World Sulfur[a] Production Capacity[b,c], t × 10³

Country	Capacity
North America	
Canada	9,900
United States	11,170
Total	*21,070*
Latin America	
Brazil	500
Chile	440
Mexico	2,775
other	675
Total	*4,390*
Africa	
South Africa	1,000
other	270
Total	*1,270*
Europe	
Belgium	400
CIS	11,000
Finland	620
France	1,850
Germany	2,530
Italy	830
Netherlands	350
Poland	4,900
Spain	1,550
Sweden	544
Turkey	613
United Kingdom	500
Yugoslavia	750
other	1,724
Total	*28,161*
Asia	
China[d]	5,450
Iran	500
Iraq	1,600
Japan	4,100
Kuwait	330
Oceania	330
Saudi Arabia	1,780
other	1,400
Total	*15,490*
Total world	*70,381*

[a]Sulfur in all forms.
[b]Includes capacity at operating plants as well as plants on standby basis.
[c]Ref. 21.
[d]Estimated.

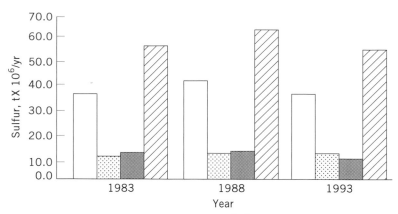

Fig. 2. World sulfur production by type, where (▢) represents brimstone; (▨), SOF; (▉), pyrites; and (▨), total sulfur (33).

Sulfur consumption reached peak levels by the beginning of the 1990s. The apparent annual consumption of sulfur in all forms in the United States nearly reached 13.2 million metric tons by 1995. World sulfur production increased steadily from 53.6 million metric tons in 1984 to an all-time high of 60.1 million metric tons in 1989, declining to 54.6 million metric tons in 1995.

Among the most important changes affecting the sulfur industry during the early 1990s were those that occurred in Eastern Europe and Asia, especially in China and India. As demand for food and fertilizers increased in Asia, so did the demand for sulfur.

The recovery of elemental sulfur from natural-gas processing and petroleum refining began to increase so that by 1980 output was four million metric tons in the United States, six million metric tons in Canada, and two million metric tons in France. As a result of environmental regulations in the mid-1990s, production from nondiscretionary sources of sulfur accounts for a greater percentage of the total world production than those of mined or discretionary sources of sulfur. This is expected to continue into the twenty-first century. Increases in the short-term are expected because of greater emphasis on reducing the amount of sulfur in hydrocarbon fuels and because of the focus on recycling (qv) and other process changes to reduce the amounts of process waste for treatment and disposal. Sulfur and sulfuric acid sold or used in the United States in the early 1990s are summarized in Table 3.

Operating Factors. Increasing environmental concerns and subsequent governmental regulations have had a large impact on the sulfur industry. Even before the U.S. Clean Air Act was enacted in 1977, oil refineries and natural-gas processors realized the necessity of removing sulfur from both upstream products and off-gases. The USBM began collecting data on recovered sulfur in 1938. Although references were made to this type of sulfur prior to that time, no official data are available to quantify recovery. At first, recovered sulfur was considered a waste material, not a commercial by-product. As time went on, the importance of recovered sulfur increased as sulfur demand increased faster

Table 3. Sulfur and Sulfuric Acid Sold or Used in the United States, t × 10³

Use	Elemental sulfur[a]		Sulfuric acid[b]		Total	
	1992	1993	1992	1993	1992	1993
ores						
copper			648	696	648	696
uranium and vanadium			8	1	8	1
other			46	49	46	49
pulp mills and paper products	27		296	304	323	304
inorganic pigments, paints and allied products, industrial organic chemicals, and other chemical products[c]	140	74	425	317	565	391
other inorganic chemicals	124	122	192	232	316	354
synthetic rubber and other plastic materials and synthetics	60	64	278	259	338[r]	323
cellulosic fibers, including rayon			43	51	43	51
drugs			15	15	15	15
soaps and detergents			50	45	50	45
industrial organic chemicals			196	82	196	82
fertilizers						
nitrogenous			227	123	227	123
phosphatic			8,300	7,906	8,300	7,906

Table 3. (Continued)

Use	Elemental sulfur[a]		Sulfur acid[b]		Total	
	1992	1993	1992	1993	1992	1993
pesticides			3	7	3	7
other agricultural chemicals	756	914	38	30	794	944
explosives			12	9	12	9
water-treating compounds			131	94	131	94
other chemical products			146	147	146	147
petroleum refining and other	308	571	385	388	693	959
petroleum and coal products						
nonferrous metals			30	26	30	26
other primary metals			1	2	1	2
storage batteries (acid)			31	28	31	28
exported sulfuric acid			14	10	14	10
total identified[d]	*1,416*	*1,744*	*11,547*	*11,062*	*12,963*	*12,806*
unidentified	*669*	*1,008*	*793*	*824*	*1,462*	*1,832*
Total[d]	*2,084*	*2,753*	*12,340*	*11,886*	*14,424*	*14,639*

[a]Does not include elemental sulfur used for production of sulfuric acid.
[b]Sulfur equivalent.
[c]No elemental sulfur was used in inorganic pigments and paints and allied products.
[d]Data may not add to totals shown because of independent rounding.

252

than the supply of Frasch and other native sulfur. Recovered sulfur became the primary domestic source of elemental sulfur in 1982.

The U.S. Clean Air Act set limits on the quantity of pollutants that could be released into the atmosphere. Sulfur dioxide was identified as one of the most common pollutants and also one of the principal contributors of acid rain, known to damage both natural and artificial environments. Gas-cleaning apparatus removed as much sulfur dioxide as was technologically possible from off-gases. Petroleum refiners and gas processors have recovered increasingly greater percentages of sulfur from the gas stream as elemental sulfur; metal smelters (especially nonferrous metals) have recovered by-product sulfuric acid; and coal and oil burning electric power plants have recovered by-product gypsum (calcium sulfate). Each industry has cut emissions dramatically. The type of by-product recovered is determined by the sulfur content of the gas stream; elemental sulfur is recovered when the sulfur dioxide content was relatively high.

The 1990 Amendments to the U.S. Clean Air Act require a 50% reduction of sulfur dioxide emissions by the year 2000. Electric power stations are believed to be the source of 70% of all sulfur dioxide emissions (see POWER GENERATION). As of the mid-1990s, no utilities were recovering commercial quantities of elemental sulfur in the United States. Two projects had been announced: Tampa Electric Company's plan to recover 75,000–90,000 metric tons of sulfuric acid (25,000–30,000 metric tons sulfur equivalent) annually at its power plant in Polk County, Florida, and a full-scale sulfur recovery system to be installed at PSI Energy's Wabash River generating station in Terre Haute, Indiana. Completed in 1995, the Terre Haute plant should recover about 14,000 t/yr of elemental sulfur.

Existing technology to recover elemental sulfur from power plant off-gases has a cost estimated to be 50% higher than the cost of recovering by-product gypsum, much of which is disposed as waste. As landfill costs become higher, elemental sulfur recovery is expected to become a more attractive alternative to by-product gypsum production.

Over the years, larger quantities of sulfur have been recovered for a number of reasons. These include increased petroleum refining and natural-gas processing, more stringent limitations on sulfur dioxide emissions, and higher sulfur contents of the crude oil refined. Another contributing factor is the lower sulfur content limits set on petroleum-based fuels.

Because sulfur supplies, either as elemental sulfur or by-product sulfuric acid, have grown owing to increased environmental awareness, demand for sulfur has decreased in some consuming industries for the same reason. Industries such as titanium dioxide productions, which traditionally utilized sulfuric acid, have concerted to more environmentally friendly processes. In addition, many consumers who continue to use sulfuric acid are putting an emphasis on regenerating or recycling spent acid.

Another area where improved air quality has impacted on sulfur use is in agriculture. As sulfur dioxide emissions have decreased, sulfur content of soils has also decreased. Sulfur, recognized as the fourth most important plant nutrient, is necessary for the most efficient use of other nutrients and optimum plant growth. Because many soils are becoming sulfur-deficient, a demand for sulfur-containing fertilizers has been created. Farmers must therefore apply a nutrient that previously was freely available through atmospheric deposition and low grade fertilizers.

Increased environmental awareness continues to create new challenges as well as a variety of new market opportunities for sulfur producers. Further pollution reduction requirements continue to increase gradually nondiscretionary supplies of sulfur and sulfur products. At the same time, recycling and reengineering have caused slight decreases in demand. These trends are likely to continue in the future.

Similar to other commercial commodities, sulfur is stockpiled or vatted when production exceeds demand. Inventories of elemental sulfur held by U.S. producers peaked at 5.6 million metric tons in 1977; inventories held by Canadian producers in Alberta, Canada, peaked at 20.6 million metric tons in 1978 and 1979. By 1995, annual U.S. production of sulfur in all forms had grown to 11.5 million metric tons and apparent consumption of sulfur in all forms was about 13.2 million metric tons; the annual growth rate was about 3% during the 1970s and 1980s. In North America, discretionary sulfur output decreased to about 2.9 million metric tons in 1995 as overall world demand for sulfur declined. During this same period, nondiscretionary sulfur output constantly increased (21,33).

In the latter part of the 1970s, world demand for sulfur began to rise, primarily to meet increased use of sulfur as sulfuric acid in processing phosphatic fertilizers. At the same time, difficulties acquiring equipment for shipping sulfur from Canada and Poland, both principal suppliers, to world markets as well as the war between Iran and Iraq brought about a worldwide shortage of sulfur. Much of the demand was met with sulfur from Frasch producers' inventories in the U.S. Gulf Coast. As shipping difficulties were overcome in Canada, stockpiles were depleted by increased shipments from producers's inventories in Alberta. Throughout the 1980s, world sulfur inventories continued to decline.

The decrease in world sulfur inventories ended in the period 1990–1992. From 1991 to 1992, sulfur inventories remained relatively stable. However, world sulfur inventories in 1993 increased sharply, to an estimated 11.8 million metric tons. This increase was caused by a sharp fall in world demand for phosphate fertilizers, which, because of market conditions, led to a large increase in vatting sulfur, especially in Canada. Figure 3 shows world sulfur inventory levels in the main producing countries or regions from 1980 through 1994.

The world's largest sulfur inventories are still in Canada. By the end of 1994, after significant vatting, stocks increased by approximately 2.2×10^6 to 7.8×10^6 t. The United States, which had 4.2 million metric tons of sulfur inventories in 1982, reduced sulfur inventories to the lowest levels in a decade during 1992, a record year for phosphate fertilizer exports. This changed during 1993–1994, when phosphate fertilizer production eased and sulfur stocks increased to 1.1 million metric tons. Sulfur inventories in Poland and West Asia have also declined slightly (33).

In addition to domestic production of Frasch and recovered elemental sulfur, U.S. requirements for sulfur are met with by-product sulfuric acid from copper, lead, molybdenum, and zinc smelting operations as well as imports from Canada and Mexico. By-product sulfur is also recovered as sulfur dioxide and hydrogen sulfide (see SULFUR REMOVAL AND RECOVERY).

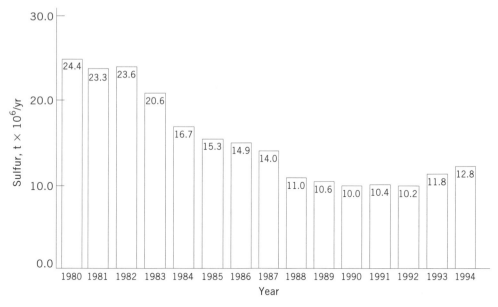

Fig. 3. World sulfur inventories (33).

Sulfur Terminology

Many terms are used to describe the commercial forms of sulfur. The most common of these terms, along with brief descriptions and typical uses or references, are as follows.

Term	Description and use
amorphous sulfur	see *insoluble sulfur*
bright sulfur	crude sulfur free of discoloring impurities; bright yellow
brimstone	see *crude sulfur*
broken rock sulfur	sulfur broken and sold as a mixture of lumps and fines; see *refined sulfur*
broken sulfur	solid crude sulfur crushed to <2.38 mm (−8 mesh)
colloidal sulfur	finely divided sulfur typically suspended in water; can be prepared by physical or chemical means; uses are mainly pharmaceutical
crude sulfur	commercial nomenclature for elemental sulfur; may be bright or dark but is free of arsenic, selenium, and tellurium
dark sulfur	crude sulfur discolored by minor quantities of hydrocarbons having up to 0.3 wt % carbon content
dusting sulfur	finely divided crude or refined sulfur prepared for pesticidal uses

Term	Description and use
elemental sulfur	processed sulfur in the elemental form produced from native sulfur or combined sulfur sources, generally having a minimum sulfur content of 99.5 wt %
flour sulfur	crude sulfur ground through 50–74 μm sieves (300–200 mesh), depending on the brand; used in rubber vulcanization, dyes, gun powder, agricultural dusts, dusting and wettable sulfur
flowable sulfur	synonymous with colloidal sulfur but used more frequently to describe agricultural sulfur; uses are mainly pesticidal
flowers of sulfur (sublimed sulfur)	powdered form of sulfur produced by sublimation; may contain up to 30% of the amorphous allotrope; used in rubber vulcanization, agricultural dusts, pharmaceutical products, stock feeds
formed sulfur	sulfur formed to a specific shape, such as prills, granules, pellets, pastilles, or flakes; see *prilled sulfur*
Frasch sulfur	elemental sulfur produced from native sulfur sources by the Frasch mining process
ground sulfur	solid sulfur ground into different physical size particles to serve various applications; may be combined with additives for special properties such as reduced dusting or improved dispersion
insoluble sulfur (Crystex)	produced by extracting flowers of sulfur with CS_2 or by fast quenching liquid or vaporized sulfur and extracting with CS_2 to remove soluble sulfur allotropes; used in rubber vulcanization, rubber cements, cutting oils, high pressure lubricants
lac sulfur	precipitated from polysulfide solutions by sulfuric acid; contains up to 45 wt % calcium sulfated liquid sulfur; uses are almost entirely pharmaceutical
liquid sulfur	see *molten sulfur*
molten sulfur	crude sulfur in the liquid phase
native sulfur	sulfur that occurs in nature in the elemental form
precipitated sulfur	precipitated from polysulfide solution by hydrochloric acid and washed to remove all calcium chloride; uses are almost entirely pharmaceutical
prilled sulfur	solid crude sulfur in pellets; produced by cooling molten sulfur with air or water
recovered sulfur	elemental sulfur produced from combined sulfur sources by any method
refined sulfur	elemental sulfur produced by distilling crude sulfur; purity not less than 99.8%; when burned in small quantities, refined sulfur provides sulfur dioxide for fumigation, sugar and starch refining, preserving and bleaching

Term	Description and use
roll sulfur	refined sulfur cast into convenient sizes; uses include chemical manufacturing, burned for curing, fumigating, and preserving or bleaching effects; see *refined sulfur*
rubbermaker's sulfur	ground sulfur of various fineness having special specifications for low acid ash, and moisture contents
run-of-mine sulfur	mined by the hot-water process; solid shipments may contain 50 wt % fines; lump diameters up to 20 cm or more
screened com- mercial sulfur	run-of-mine sulfur having particle size determined by screening, generally 9.5 and 1.68 mm (2 and 12 mesh), plus associated fines
slated sulfur	solid crude sulfur in the form of slate-like lumps; produced by allowing molten sulfur to solidify on a moving belt
specialty sulfur	prepared or refined grades of elemental sulfur that include amorphous, colloidal, flower, precipitated, wettable, or paste sulfur
spray sulfur	finely divided sulfur combined with various wetting agents and water; prepared for pesticidal uses; see *wettable sulfur*
wettable sulfur	powdered sulfur that has been treated for easy dispersion in water

Analysis

Elemental sulfur in either its ore or its refined state can generally be recognized by its characteristic yellow color or by the generation of sulfur dioxide when it is burned in air. Its presence in an elemental state or in a compound can be detected by heating the material with sodium carbonate and rubbing the fused product on a wet piece of silver metal. A black discoloration of the silver indicates the presence of sulfur. The test is quite sensitive. Several other methods for detecting small amounts of elemental sulfur have also been developed (34).

Quantitatively, sulfur in a free or combined state is generally determined by oxidizing it to a soluble sulfate, by fusion with an alkali carbonate if necessary, and precipitating it as insoluble barium sulfate. Oxidation can be effected with such agents as concentrated or fuming nitric acid, bromine, sodium peroxide, potassium nitrate, or potassium chlorate. Free sulfur is normally determined by solution in carbon disulfide, the latter being distilled from the extract. This method is not useful if the sample contains polymeric sulfur.

Generally, crude sulfur contains small percentages of carbonaceous matter. The amount of this impurity is usually determined by combustion, which requires an exacting technique. The carbonaceous matter is oxidized to carbon dioxide and water; the carbon dioxide is subsequently absorbed (18). Automated, on-stream determination of impurities in molten sulfur has been accomplished by infrared spectrophotometry (35).

The moisture content of crude sulfur is determined by the differential weight of a known sample before and after drying at about 110°C. Acid content is determined by volumetric titration with a standard base. Nonvolatile impurities or ash are determined by burning the sulfur from a known sample and igniting the residue to remove the residual carbon and other volatiles.

The National Safety Council, National Fire Protection Association, and other similar organizations publish technical information that describes general safety practices for use during the testing, handling, storage, and transport of sulfur (21,36–40). Each of these publications include a list of references for additional health and safety information.

Uses

Sulfur is unusual compared to most large mineral commodities in that the largest portion of sulfur is used as a chemical reagent rather than as a component of a finished product. Its predominant use as a process chemical generally requires that it first be converted to an intermediate chemical product prior to use in industry. In most of the ensuing chemical reactions between these sulfur-containing intermediate products and other minerals and chemicals, the sulfur values are not retained. Rather, the sulfur values are most often discarded as a component of the waste product.

Sulfuric acid is the most important sulfur-containing intermediate product. More than 85% of the sulfur consumed in the world is either converted to sulfuric acid or produced directly as such (see SULFURIC ACID AND SULFUR TRIOXIDE). Worldwide, well over half of the sulfuric acid is used in the manufacture of phosphatic fertilizers and ammonium sulfate for fertilizers. The sulfur source may be voluntary elemental, such as from the Frasch process; recovered elemental from natural gas or petroleum; or sulfur dioxide from smelter operations.

As of the mid-1990s, the largest demand for sulfur has been for agricultural purposes. The principal use is for phosphatic fertilizer processing (see Table 3). Other uses have been in petroleum refining; leaching of copper and uranium ores; and production of organic and inorganic chemicals, paints and pigments, pulp and paper, and synthetic materials.

Agriculture. Sulfur is one of the elements essential for plant growth. Its functions within the plant are related closely to those of nitrogen and the two nutrients are synergistic. Sulfur is required for plant growth in quantities equal to, and sometimes exceeding, those of phosphorus. Sulfur has a variety of vital functions within the plant's biochemistry. It is a principal constituent of amino acids (qv) such as cysteine and methionine. It is also essential in the formation of enzymes, vitamins (qv) such as biotin and thiamine, and a variety of other important compounds in the plant, including chlorophyll. When sulfur is deficient, both plant yield and quality suffer. Plants that are sulfur-deficient are characteristically small and spindly. The younger leaves are light-green to yellowish, and in the case of legumes, nodulation of the roots is reduced. The oil content of seeds is diminished and the maturity of fruits is delayed in the absence of adequate sulfur.

Plant nutrient sulfur has been growing in importance worldwide as food production trends increase while overall incidental sulfur inputs diminish. Increasing crop production, reduced sulfur dioxide emissions, and shifts in fertilizer sources have led to a global increase of crop nutritional sulfur deficiencies. Despite the vital role of sulfur in crop nutrition, most of the growth in world fertilizer consumption has been in sulfur-free nitrogen and phosphorus fertilizers (see FERTILIZERS).

Agriculture is the largest industry for sulfur consumption. Historically, the production of phosphate fertilizers has driven the sulfur market. Phosphate fertilizers account for approximately 60% of the sulfur consumed globally. Thus, although sulfur is an important plant nutrient in itself, its greatest use in the fertilizer industry is as sulfuric acid, which is needed to break down the chemical and physical structure of phosphate rock to make the phosphate content more available to plant life. Other mineral acids, as well as high temperatures, also have the ability to achieve this result. Because of market price and availability, sulfuric acid is the most economic method. About 90% of sulfur used in the fertilizer industry is for the production of phosphate fertilizers. Based on this technology, the phosphate fertilizer industry is expected to continue to depend on sulfur and sulfuric acid as a raw material.

Another fertilizer use is in the production of ammonium sulfate by reaction of sulfuric acid and ammonia. Some of this production is deliberate, but the greater portion is by-product material resulting from coke-oven operations, synthetic fiber manufacture, and, to a lesser extent, utility stack-gas scrubbing (see AMMONIUM COMPOUNDS). Furthermore, a small amount of sulfuric acid is used, mainly in Europe, to produce potassium sulfate using the Mannheim process. Finally, sulfur has been safely used for centuries in agriculture as a natural fungicide and pesticide, and as an amendment to ameliorate soils and irrigation waters (see FUNGICIDES, AGRICULTURAL; PESTICIDES; SOIL CHEMISTRY OF PESTICIDES).

Plastics and Other Synthetic Products. Sulfur is used in the production of a wide range of synthetics, including cellulose acetate, cellophane, rayon, viscose products, fibers, and textiles. These uses may account for 2% of sulfur demand in developed countries. Sulfur intermediates for these manufacturing processes are equally divided between carbon disulfide and sulfuric acid.

Paper Products. Paper (qv) products account for about 2% of sulfur demand. The largest single segment of demand is in the manufacture of wood pulp by the sulfite process (see PULP). In this process, the main sulfur intermediate is sulfur dioxide, which is generally produced at the plant site by burning elemental sulfur. Some sulfur dioxide, however, is produced as a by-product at smelter operations, purified and liquefied, and shipped to the pulp mills. The sulfur dioxide is converted to sulfurous acid, and the salt of this acid is a principal component of the cooking liquor for the sulfite process.

Paints. Paints account for perhaps 3% of sulfur consumption (see PAINT). The main sulfur use is for the production of titanium dioxide pigment by the sulfate process. Sulfuric acid reacts with ilmenite or titanium slag and the sulfur remains as a ferrous sulfate waste product. Difficulties with this process have led to the development of the chloride process (see PIGMENTS, INORGANIC; TITANIUM COMPOUNDS).

Nonferrous Metal Production. Nonferrous metal production, which includes the leaching of copper and uranium ores with sulfuric acid, accounts for about 6% of U.S. sulfur consumption and probably about the same in other developed countries. In the case of copper, sulfuric acid is used for the extraction of the metal from deposits, mine dumps, and wastes, in which the copper contents are too low to justify concentration by conventional flotation techniques or the recovery of copper from ores containing copper carbonate and silicate minerals that cannot be readily treated by flotation (qv) processes. The sulfuric acid required for copper leaching is usually the by-product acid produced by copper smelters (see METALLURGY, EXTRACTIVE; MINERALS RECOVERY AND PROCESSING).

Sulfuric acid is the most commonly used reagent for the recovery of uranium from ores, and vanadium is often recovered as a coproduct. The sulfuric acid used is either the by-product sulfuric acid produced at smelters or sulfuric acid produced from elemental sulfur.

Petroleum Refining. Petroleum refining includes not only the refining of petroleum but associated chemical processes where process streams may serve both the refinery and the chemical complex. Sulfuric acid requirements for these processes account for approximately 6% of sulfur demand. About 60% of the sulfuric acid used in petroleum refining is returned as spent acid for reclaiming; therefore, the demand for new sulfuric acid is about 3% of the total sulfur demand. The principal use for sulfuric acid is as a catalyst for alkylation (qv), a process by which liquid high octane gasoline components having very good stability are produced from a combination of gaseous streams. Sulfuric acid and hydrofluoric acid are competing catalysts in this process. Sulfuric acid for refinery processes is manufactured from recovered sulfur produced at the refinery and from contaminated acid (acid sludge) returned to the acid plants for reconstitution (see PETROLEUM).

Iron and Steel Production. Consumption of sulfur in the iron (qv) and steel (qv) industry in the form of sulfuric acid accounts for about 2% of sulfur demand. The sulfuric acid is used as a pickling agent to removal mill scale, rust, dirt, and grease from the surface of steel products prior to further processing (see METAL SURFACE TREATMENTS). The sulfuric acid pickling process faces increasing competition from hydrochloric acid pickling, largely because of the problem of disposing of the ferrous sulfate waste product. Hydrochloric acid is expected to replace sulfuric acid for pickling over the long term. There are no well-defined sources of sulfuric acid for steel pickling. Rather, it is generally obtained from merchant sulfuric acid plants that use the cheapest form of sulfur available in the area where it is produced.

Soil and Water Treatment. Agricultural soils and waters in the irrigated, arid regions of the world frequently benefit by the application of sulfur and its compounds. These benefits result from improvements in water penetration and movement and from increased availability of phosphorus and certain micronutrients which are otherwise unavailable owing to high soil pH. There are extensive areas of such soils in the western provinces of Canada and in many areas west of the Mississippi River in the United States. Throughout the other parts of the world, large tracts of arid, irrigated soils exist, notably in the Middle East, northern Africa, Australia, and many places in both Eastern and Western Europe (41,42).

Animal Nutrition. Sulfur in the diets of ruminant animals is beneficial to the animals' growth (see FEEDS AND FEED ADDITIVES). Sulfur increases feed intake, cellulose and dry matter digestion, and the synthesis of microbial protein. This results in increased meat, milk, and wool production (43). The special uses for sulfur in agriculture demonstrate a significant and continuing need for increased use of sulfur (44).

Highway Construction. The preparation and use of sulfur–asphalt (SA) paving materials have been reviewed (45,46). In the 1930s, asphalt (qv) was easily available and priced lower than sulfur. As of the 1990s, this is no longer the case. There are four different types of sulfur paving materials.

Sulfur as an Additive for Asphalt. Sulfur-extended asphalt (SEA) binders are formulated by replacing some of the asphalt cement (AC) in conventional binders with sulfur. Binders that have sulfur asphalt weight ratios as high as 50:50 have been used, but most binders contain about 30 wt % sulfur. Greater latitude in design is possible for SEA paving materials, which are three-component systems, whereas conventional asphalt paving materials are two-component systems. Introduction of sulfur can provide some substantial benefits. At temperatures above 130°C, SEA binders have lower viscosities than conventional asphalt. The lower viscosity enables the plant to produce and compact the mix at lower temperatures than with conventional mix. The SEA materials have flatter viscosity curves than asphalt cements and are more viscous than asphalt at summer road temperatures. This causes the pavement to be more resistant to rutting and deformation during hot weather. The SEA binders have increased stability (47), and SEA is very resistant to gasoline, diesel fuel, and other solvents (48).

During 1974–1985, about 200 sulfur–asphalt roads were constructed worldwide, half of which were in the United States. All U.S. SEA experimental sections designed and constructed according to standard practices using standard materials are performing as well as the control sections of conventional asphalt in these experimental projects (49).

The physical properties of elemental sulfur can be modified by its reaction with various organic and inorganic compounds. Many of the resulting sulfur products tend to have properties similar to paving asphalt (49,50).

Sand–Asphalt–Sulfur. Sulfur can be utilized as a means of upgrading poorly graded aggregates. Sand–asphalt–sulfur (SAS) does not provide an appreciable reduction of asphalt content of the paving material, but the utilization of sand rather than crushed stone is economically important. This technology was developed by Shell Canada Resources, Ltd., under the trade name Thermopave, which is a high quality paving material (51–54). After being tested in various provinces of Canada, SAS roads were constructed in the U.S.

Recycling of Commercial Asphalt Materials. For many years, repair and maintenance of roads and highways has consisted of putting a layer of asphalt concrete over the existing pavement. In many areas, the road level is at the maximum height for proper drainage and to satisfy safety considerations. As a result, these roads must have exposed surfaces removed before additional paving can take place. The increasing cost of both asphalt and aggregate has presented a greater incentive to reuse rather than to discard the removed material.

Many existing roads fail because the asphalt becomes stiff and brittle. If the materials are too stiff, additives that lower the viscosity must be used. The feasibility of using sulfur to soften or reduce the viscosity of the oxidized binder in recycled pavements has been successfully demonstrated by the U.S. Bureau of Mines and others (55–57).

Sulfur Concrete. Sulfur concrete (SC) is a mixture of sulfur and fine and coarse aggregates. These materials are heated to about 140°C, placed, and then allowed to cool and solidify into a rigid, concrete-like material. Concretes prepared with sulfur as the binder have mechanical properties comparable to portland cement concretes. Methods of preparing sulfur concretes that are not subject to rapid deterioration have been developed and evaluated in Austria, Canada, Denmark, Poland, and the United States. A modified sulfur binder called sulfur polymer cement (SPC), which contains 3–5 wt % organic material, has been developed for mixing with aggregates and the resulting SC products maintain substantially all of the initial strength gain, even after 300 freeze–thaw cycles (49).

Sulfur concretes are used in many specialty areas where Portland cement concretes are not completely satisfactory. Because SC can be formulated to resist deterioration and failure from mineral acid and salt solutions, it is used for construction of tanks, electrolytic cells, thickeners, industrial flooring, pipe, and others. In addition, SC is under investigation for many other prospective uses (58,59) (see CEMENT).

Sulfur Polymer Cement. SPC has been proven effective in reducing leach rates of reactive heavy metals to the extent that some wastes can be managed solely as low level waste (LLW). When SPC is combined with mercury and lead oxides (both toxic metals), it interacts chemically to form mercury sulfide, HgS, and lead sulfide, PbS, both of which are insoluble in water. A dried sulfur residue from petroleum refining that contained 600-ppm vanadium (a carcinogen) was chemically modified using dicyclopentadiene and oligomer of cyclopentadiene and used to make SC (58). This material was examined by the California Department of Health Services (Cal EPA) and the leachable level of vanadium had been reduced to 8.3 ppm, well below the soluble threshold limit concentration of 24 ppm (59).

Sulfur polymer cement shows promise as an encapsulation and stabilization agent for use with low level radioactive and mixed wastes. Use of SPC allows accommodation of larger percentages of waste than PCC. As of this writing (1997), SPC-treated waste forms have met requirements of both the Nuclear Regulatory Commission (NRC) and the Environmental Protection Agency (EPA).

Coatings. Sulfur coating formulations have been used on a concrete warehouse in which potash is stored (58). Adhesion to this nonporous concrete wall has been related to the amount and type of plasticizer. Glass and other fibers, fillers, and fine aggregates affect strength and abrasion resistance but not adhesion. The use of a sulfur spray coating to stabilize a hilltop which was eroding and sliding onto a road and work area below has been described (60). Approximately 8100 m^2 on top of the hill was spray-coated to stabilize the hilltop. The coating has prevented further erosion.

Mortar Substitute. In the early 1960s, the idea of using sulfur coatings to replace mortar in construction was advanced (61). Bricks, blocks, and similar materials are stacked one on the other and, because the blocks are dry, can be moved and adjusted until the desired wall configuration is achieved. The blocks are thus bonded by applying the coating to the wall surfaces only. The coating mixture is applied at 120–150°C and contains 90–95 wt % sulfur and small percentages of fiber and additives. The coating solidifies almost immediately to form a hard, impervious surface. In 1963, the first building of this type was constructed at the Southwest Research Institute. The building has been in continuous use and the coating has not presented any problems.

In 1972, the Agency for International Development (AID) cooperated with the Sulphur Institute to prepare a program to evaluate the feasibility of using the sulfur surface-bond technique in developing countries. Afterwards AID awarded a contract to Southwest Research Institute to construct houses in South America and Africa using this technique. Under the contract, about 60 houses have been completed (62).

Foams. Sulfur can be foamed into a lightweight insulation that compares favorably with many organic foams and other insulating materials used in construction. It has been evaluated as thermal insulation for highways and other applications to prevent frost damage (63) (see FOAMED PLASTICS; INSULATION, THERMAL).

Sulfur Impregnation. Impregnation of bonded, abrasive grinding wheels using sulfur improves strength and provides both lubricating and cooling qualities during grinding operations. Sulfur-impregnated wheels are well suited for grinding tough materials, including stainless steel, brass, bronze, and nickel. The impregnated wheels cut faster and prevent the welding of metal chips. In more difficult jobs, including gear grinding and surface grinding, the sulfur-impregnated wheels have four to eight times the life span of nonimpregnated wheels (64) (see ABRASIVES).

Sulfur can be used to impregnate ceramic tile. Sulfur impregnation reduces water absorption and makes the tile frost-resistant when used on exterior surfaces, including floors, roofs, or entrances to buildings (65). Investigators at the Institute of Paper Chemistry have described the production of sulfur-impregnated paperboard for use in temporary housing and techniques for manufacturing sulfurized, corrugated container board suitable for self-sustaining shipping containers (66,67). Sulfurized container board is shock-resistant and maintains its strength under hot, humid conditions which can cause normal boxboard to lose over half of its strength (65) (see PAPER).

Other Uses. Other uses include intermediate chemical products. Overall, these uses account for 15–20% of sulfur consumption, largely in the form of sulfuric acid but also some elemental sulfur that is used directly, as in rubber vulcanization. Sulfur is also converted to sulfur trioxide and thiosulfate for use in improving the efficiency of electrostatic precipitators and limestone/lime wet flue-gas desulfurization systems at power stations (68). These miscellaneous uses, especially those involving sulfuric acid, are intimately associated with practically all elements of the industrial and chemical complexes worldwide.

BIBLIOGRAPHY

"Sulfur" in *ECT* 1st ed., Vol. 13, pp. 358–373, by F. L. Jackson, E. C. Thaete, Jr., L. B. Gittinger, Jr., L. A. Nelson, Jr., and H. Blanchet, Freeport Sulphur Co.; in *ECT* 2nd ed., Vol. 19, pp. 337–366, by P. T. Comiskey, L. B. Citlinger, Jr., L. F. Good, F. L. Jackson, L. A. Nelson, Jr., and T. K. Wiewiorowski, Freeport Sulphur Co., and M. D. Barnes, Covenant College; in *ECT* 3rd ed., Vol. 22, pp. 78–106, by D. W. Bixby and H. L. Fike, The Sulphur Institute, J. E. Shelton, U.S. Bureau of Mines, and T. K. Wiewiorowski, Freeport Minerals Co.

1. P. W. Shenk and V. Thummler, *Z. Elektrochem.* **63**, 1002 (1959).
2. B. Meyer, *Chem. Rev.* **76**, 367 (1976).
3. T. K. Wiewiorowski and F. J. Touro, *J. Phys. Chem.* **70**, 3528 (1966).
4. M. Schmidt, *Angew. Chem.* **12**, 445 (1973).
5. B. Meyer, *Sulfur, Energy, and Environment*, Elsevier Science Publishing Co., Inc., New York, 1977.
6. V. R. Steudel, *Z. Anorg. Alig. Chem.* **478**, 139 (1981).
7. V. R. Steudel and H. J. Mausle, *Z. Anorg. Alig. Chem.* **478**, 156 (1981).
8. H. J. Mausle and V. R. Steudel, *Z. Anorg. Alig. Chem.* **478**, 177 (1981).
9. C. Preuner and W. Schupp, *Z. Phys. Chem. (Leipzig)*, **68**, 129 (1909).
10. H. Braune, S. Peter, and V. Neveling, *Z. Naturforsch Teil A*, **6**, 32 (1951).
11. J. Berkowitz and J. R. Marquart, *J. Chem. Phys.* **39**, 275 (1963).
12. W. A. West and A. W. C. Menzies, *J. Phys. Chem.* **33**, 1880 (1929).
13. G. Fouritier, *Compt. Rend.* **218**, 194 (1944).
14. K. Newmann, *Z. Phys. Chem.* **A171**, 416 (1934).
15. R. Farelli, *J. Am. Chem. Soc.* **72**, 4016 (1950).
16. E. W. Washburn, ed., *International Tables of the Numerical Data of Physics, Chemistry and Technology* (*ICT*), Vols. 1–7, McGraw-Hill Book Co., Inc., New York, 1926–1930.
17. P. C. L. Thorne and A. M. Ward, *Inorganic Chemistry*, 3rd ed., Nordeman Publishing Co., Inc., New York, 1939, p. 6.
18. W. N. Tuller, ed., *The Sulphur Data Book*, McGraw-Hill Book Co., Inc., New York, 1954.
19. J. W. Mellor, *A Comprehensive Treatise on Inorganic and Theoretical Chemistry*, Vol. 10, John Wiley & Sons, Inc., New York, 1930, p. 27.
20. A. Senning, *Die Schwefelurg Organischer Verbindungen*, Chemisches Institut der Universitat Aarhus, Denmark (1971).
21. *U.S. Bureau of Mines Annual Report*, U.S. Bureau of Mines, Washington, D.C., 1994.
22. D. B. Mason, *Ind. Eng. Chem.* **30**, 740 (1938).
23. R. B. Thompson and D. MacAskill, *Chem. Eng. Prog.* **51**, 3691 (1955).
24. T. P. Forbath, *Trans. AIME*, **196**, 811 (1953).
25. T. Kaier, *Eng. Min. J.* **155**(7), 88 (1954).
26. H. R. Potts and E. G. Lawford, *Trans. AIME*, **58**, 1 (1949).
27. *Eng. Min. J.* **155**(9), 142 (1954).
28. G. O. Argall, Jr., *World Min.* 18 (Mar. 1967).
29. *Chem. Eng.*, 31 (Dec. 5, 1980).
30. *Sulphur*, **80** (1969).
31. *Sulphur*, **147** (1980).

32. *Chem. Eng. News*, 21 (Mar. 20, 1967).

33. *Sulphur Outlook*, The Sulphur Institute, Washington, D.C., 1995 and 1996.

34. F. Feigl, *Spot Tests in Inorganic Analysis*, Elsevier Science Publishing Co., Inc., New York, 1958, pp. 372–375.

35. R. F. Matson, T. K. Wiewiorowski, and D. E. Schof, Jr., *Chem. Eng. Prog.* **61**, 9, 67 (1965).

36. *Handling and Storage of Solid Sulfur*, Data Sheet 612, National Safety Council, 1991.

37. *Properties and Essential Information for Safe Handling and Use of Sulfur*, Chemical Safety Data Sheet SD-74, Manufacturing Chemists Association (CMA), 1959.

38. J. D. Beaton, S. L. Tisdale, and J. Platou, *Crop Response to Sulphur in North America*, Technical Bulletin 18, The Sulphur Institute, Washington, D.C., 1971.

39. *Handling Liquid Sulfur*, Data Sheet 592, National Safety Council, 1993.

40. *Prevention of Sulfur Fires and Explosions*, NFPA No. 655, National Fire Protection Association, 1993.

41. S. L. Tisdale, *Sulphur Inst. J.* **6**(1), 2 (1970).

42. L. K. Stromberg and S. L. Tisdale, *Treating Irrigated Arid-Land Soils with Acid-Forming Sulphur Compounds*, Technical Bulletin 24, The Sulphur Institute, Washington, D.C., 1979.

43. S. L. Tisdale, *Sulphur in Forage Quality and Ruminant Nutrition*, Technical Bulletin 22, The Sulphur Institute, Washington, D.C., 1977.

44. J. D. Beaton and S. L. Tisdale, *Potential Plant Nutrient Consumption in North America*, Technical Bulletin 16, The Sulphur Institute, Washington, D.C., 1969.

45. H. L. Fike, *Some Potential Applications of Sulphur, Sulphur Research Trends*, American Chemical Society, Washington, D.C., 1972.

46. D. Y. Lee, *Modification of Asphalt and Asphalt Paving Mixtures by Sulphur Additives*, Interim Report, Engineering Research Institute, Iowa State University, Ames, Iowa, 1971.

47. *Mix Design Methods for Asphalt Concrete (MS-2)*, The Asphalt Institute, 1974.

48. W. C. McBee and T. A. Sullivan, *I&EC Prod. Res. Dev.* Vol **16**, 93 (1977).

49. W. C. McBee, T. A. Sullivan, and H. L. Fike, *Sulfur Construction Materials*, Bulletin 678, U.S. Bureau of Mines, Washington, D.C., 1985.

50. H. J. Lentz and E. A. Harrigan, *Laboratory Evaluation of Sulphlex-233: Binder Properties and Mix Design*, FHWA/RD-80/146, U.S. GPO 1981-0-725-620/1163, 1980.

51. W. C. McBee, T. A. Sullivan, and B. W. Jong, *Modified-Sulfur Cements for Use in Concretes, Flexible Pavings, Coatings, and Grouts*, RI 8545, U.S. Bureau of Mines, Washington, D.C., 1981.

52. R. A. Burgess and I. Deme, in J. R. West, ed., *Sulphur in Asphalt Paving Mixes*, American Chemical Society, Washington, D.C., 1975.

53. I. Deme, *Proceedings of International Road Federation World Meeting*, Munich, Germany, 1973.

54. T. A. Sullivan, W. C. McBee, and K. L. Rasmussen, *Studies of Sand–Sulphur–Asphalt Paving Materials*, RI 8087, U.S. Bureau of Mines, Washington, D.C., 1975.

55. D. Saylak and co-workers, *Beneficial Uses of Sulfur in Sulfur–Asphalt Pavements*, American Chemical Society, Washington, D.C., 1975.

56. W. C. McBee, T. A. Sullivan, and D. Saylak, *Recycling Old Asphalt Pavements with Sulfur*, ASTM STP622, American Society for Testing Materials, Philadelphia, Pa., 1978.

57. D. L. Strand, *Rural Urban Roads*, **18**(7), 42 (1980).

58. J. S. Platou, ed., *Sulphur Research and Development*, The Sulphur Institute, Washington, D.C., 1981.

59. Treatment Standards of Liquid Redox Waste in California, State of California Department of Health Services, Toxic Substances Control Program, Alternative Technology

Division, June 1990; *Sulphur Polymer Cement Concrete, Design and Construction Manual*, The Sulphur Institute, Washington, D.C., 1994.

60. J. E. Paulson and co-workers, *Sulfur Composites as Protective Coatings and Construction Materials*, American Chemical Society, Washington, D.C., 1978.
61. U.S. Pat. 3,306,000 (1967), M. D. Barnes (to Research Corp.).
62. H. L. Fike and M. Conitz, *Surface Bond Construction Materials*, Envo Publishing Co., Lehigh Valley, Pa., 1976.
63. G. L. Woo and co-workers, *Sulfur Foam and Commercial-Scale Field Application Equipment*, American Chemical Society, Washington, D.C., 1978, pp. 227–240.
64. U.S. Pat. 3,341,355 (Sept. 12, 1967), T. P. Gallager (to Fuller-Merriam Co.).
65. U.S. Pat. 3,208,190 (Sept. 25, 1965), M. J. Kakos and J. V. Fitzgerald (to Tile Council of America).
66. J. A. Van der Akker and W. A. Wink, *Pap. Ind. Pap. World*, **30**(2), 231 (1948).
67. U.S. Pat. 2,568,349 (Sept. 18, 1951), R. C. McKee (not assigned).
68. D. R. Owens and co-workers, *Power*, (May 19, 1988).

General References

E. P. Johnson, *Chem. Eng.* 34 (Nov. 2, 1981).
Chem. Eng. News, 20 (May 24, 1982).
P. A. Gallagher, *Proc. R. Dublin Soc. Ser. B*, **2**(20), (1969).
M. Braud, *Sulphur Inst. J.* **5**(4), 3 (1969–1970).
A. C. Ludwig, B. B. Cerhardt, and J. M. Dale, *Material and Techniques for Improving the Engineering Properties of Sulphur*, FHWA/RD-80-023, U.S. GPO 1980-625-892/2032, 1980.
H. L. Fike, *Proceedings of Symposium on New Uses for Sulphur and Pyrites*, Madrid, Spain, The Sulphur Institute, Washington, D.C., 1976, pp. 215–230.

The Sulphur Institute

SULFUR COMPOUNDS

Carbon Sulfides

The only commercial carbon sulfide is carbon disulfide (qv) [75-15-0], CS_2. There are several unstable carbon sulfides. Carbon subsulfide [12976-57-2], C_3S_2, is a red liquid (mp $-0.5°C$, bp $60-70°C$ at 1.6 kPa (12 mm Hg)) produced by the action of an electric arc on carbon disulfide (1–4). The structure has been shown to be $S=C=C=C=S$ on the basis of its reactions to form malonic acid derivatives and on the basis of physical measurements. It is unstable and decomposes in a few weeks at room temperature; it decomposes explosively when heated rapidly at $100-120°C$ with formation of a black polymeric substance $(C_3S_2)_x$ (5,6). Dilute solutions in CS_2 are fairly stable, but photochemical polymerization to $(C_3S_2)_x$ occurs.

The reaction of C_3S_2 with 2-aminopyridine and N-phenylbenzamidine yields the higher condensed derivatives; C_3S_2 also reacts with β-aminocrotonate to yield $[H_2NC(CH_3)=C(COOC_2H_5)]_2S$ (7).

Carbon monosulfide [2944-05-0], CS, is an unstable gas produced by the decomposition of carbon disulfide at low pressure in a silent electrical discharge or photolytically (1–3) in the presence or absence of sulfur (3). It decomposes with a half-life of seconds or minutes to a black solid of uncertain composition (1–3). The monosulfide can be stabilized in a CS_2 matrix at $-196°C$, and many stable coordination complexes of CS with metals have been prepared by indirect means (8).

Carbon monosulfide, CS, which can be produced by passing CS_2 through a high voltage discharge, can react with C_6H_5SCl at low temperatures in toluene to give $C_6H_5SC(S)Cl$. In a similar manner CS reacts with S_2Cl_2 by double insertion to yield $ClC(S)SSC(S)Cl$ (9). Carbon monosulfide can also react with amines or thiols in toluene or DMF (10). A review of the synthetic utility of CS has been published (11).

Incompletely Characterized Carbon Sulfides. A poorly characterized black solid, known as carsul, occurs as a residue in sulfur distillation or as a precipitate in molten Frasch sulfur (12,13). Although this material may approach the composition of a carbon sulfide, it is more likely also to contain some chemically bound hydrogen and possibly other elements. Carbon–sulfur surface compounds of the formula C_xS, where x is greater than 4, are prepared by reaction of carbonaceous materials with a sulfur-containing gas, eg, sulfur dioxide at $400-700°C$. They are useful as cathodes in electrical cells (14). Other poorly characterized, thermally stable black solids with sulfur contents ranging as high as 46 wt % have been reported as products of high temperature reactions of sulfur with charcoal, cellulose, and various carbon compounds. These, considered as carbon sulfides by some investigators, may contain other elements as well (15,16).

Carbonyl Sulfide. *Physical Properties.* Carbonyl sulfide [463-58-1] (carbon oxysulfide), COS, is a colorless gas that is odorless when pure; however, it has been described as having a foul odor. Physical constants and thermodynamic properties are listed in Table 1 (17,18). The vapor pressure has been fitted to an equation, and a detailed study has been made of the phase equilibria of

Table 1. Physical and Thermodynamic Properties of Carbonyl Sulfide

Property	Value	Ref.
mol wt	60.074	17
mp, °C	−138.8	18
bp, °C	−50.2	18
ΔH fusion, at 134.3 K, kJ/mol[a]	4.727	18
ΔH vaporization, at 222.87 K, kJ/mol[a]	18.57	18
density at 220 K, 101.3 kPa(=1 atm), g/cm^3	1.19	18
sp gr, gas at 298 K (air = 1)	2.10	18
critical temperature, °C	105	18
critical pressure, kPa[b]	5946	18
critical volume, cm^3/mol	138	18
triple point, K	134.3	18
ΔG formation, at 25°C, kJ/mol[a]	−169.3	2
ΔH formation, at 25°C, kJ/mol[a]	−142.1	2
S formation, at 25°C, J/(mol·K)[a]	231.5	2
C_p, at 25°C, J/(mol·K)[a]	41.5	2
autoignition temperature in air, °C	ca 250	18
flammability limits in H_2O-saturated air at 17.9°C, vol %		
upper limit	9.6	
lower limit	33.2	18
solubility in water at 101.3 kPa(=1 atm), vol %		
0°C	0.356	18
10°C	0.224	
20°C	0.149	

[a]To convert J to cal, divide by 4.184.
[b]To convert kPa to psi, multiply by 0.145.

the carbonyl sulfide–propane system, which is important in the purification of propane fuel (19,20). Carbonyl sulfide can be adsorbed on molecular sieves (qv) as a means for removal from propane (21). This approach has been compared to the use of various solvents and reagents (22).

Chemical Properties. Reviews of carbonyl sulfide chemistry are available (18,23,24). Carbonyl sulfide is a stable compound and can be stored under pressure in steel cylinders as compressed gas in equilibrium with liquid. At ca 600°C carbonyl sulfide disproportionates to carbon dioxide and carbon disulfide; at ca 900°C it dissociates to carbon monoxide and sulfur. It burns with a blue flame to carbon dioxide and sulfur dioxide. Carbonyl sulfide reacts only slowly with water to form carbon dioxide and hydrogen sulfide. Much technology has been developed for hydrolysis of carbonyl sulfide in gas streams to permit the more ready removal of the sulfur content as hydrogen sulfide. The commercial processes have been reviewed in detail (25). Typical catalysts for the gas-phase hydrolysis are based on copper sulfide, chromium oxide, chromia–alumina, and platinum.

Alternative means for removal of carbonyl sulfide for gas streams involve hydrogenation. For example, the Beavon process for removal of sulfur compounds remaining in Claus unit tail gases involves hydrolysis and hydrogenation over cobalt molybdate catalyst resulting in the conversion of carbonyl sulfide, carbon disulfide, and other sulfur compounds to hydrogen sulfide (25).

Carbonyl sulfide reacts slowly with aqueous alkali metal hydroxides, which can therefore be used to free carbonyl sulfide from acidic gases. The half-life at pH 12 (0.01 N KOH) is 3 min at 22°C. The product initially formed from carbonyl sulfide and alkali is the thiocarbonate [*534-18-9*] which then breaks down to carbonate and sulfide:

$$COS + 2\,NaOH \longrightarrow NaS\overset{\displaystyle O}{\overset{\displaystyle \|}{C}}ONa + H_2O$$

$$NaS\overset{\displaystyle O}{\overset{\displaystyle \|}{C}}ONa + 2\,NaOH \longrightarrow Na_2CO_3 + Na_2S + H_2O$$

To effectively remove carbonyl sulfide from a gas stream, special alkaline scrubbing liquors are used. These contain sodium aluminate or sodium plumbite, or they are made of alkalies with a hydrolysis catalyst based on Zn, Fe, Ni, or Cu. Diethanolamine, diglycolamine, or other alkanolamines (qv) mixed with water remove carbonyl sulfide from sour, ie, acid-gas-containing, gas streams (25,26) (see CARBON DIOXIDE).

Carbonyl sulfide reacts with chlorine forming phosgene (qv) and sulfur dichloride [*10545-99-0*], and with ammonia forming urea and ammonium sulfide [*12135-76-1*]. Carbonyl sulfide attacks metals, eg, copper, in the presence of moisture and is thought to be involved in atmospheric sulfur corrosion (27,28). Its presence in propane gas at levels above a few ppm may cause the gas to fail the copper-corrosion test.

It reacts with alcohols in the presence of base to form monothiocarbonate esters:

$$ROH + COS + NaOH \longrightarrow RO\overset{\displaystyle O}{\overset{\displaystyle \|}{C}}SNa + H_2O$$

Carbonyl sulfide reacts with amines to form thiocarbamates:

$$2\,R_2NH + COS \longrightarrow [R_2N\overset{\displaystyle O}{\overset{\displaystyle \|}{C}}S]^-[NH_2R_2]^+$$

With amines such as those used in gas sweetening, it reacts forming mainly ureas:

$$COS + 2\,HOCH_2CH_2OCH_2CH_2NH_2 \longrightarrow (HOCH_2CH_2OCH_2CH_2NH)_2CO + H_2S$$

The photolysis of carbonyl sulfide is a laboratory method for the production of monoatomic sulfur, a short-lived species (29).

Occurrence and Preparation. Carbonyl sulfide is formed by many high temperature reactions of carbon compounds with donors of oxygen and sulfur. A principal route is the following reaction (30):

$$CO + S^0 \rightleftharpoons COS$$

This equilibrium favors COS up to ca 500°C. At higher temperatures, COS dissociates increasingly, eg, to 64% at 900°C. The reaction may be run at 65–200°C to produce carbonyl sulfide if an alkaline catalyst is used (31). A Rhône-Poulenc patent describes the manufacture of carbonyl sulfide by the reaction of methanol with sulfur at 500–800°C (32).

Other important reactions yielding carbonyl sulfide follow (33,34):

Reaction	Remarks
$CO_2 + H_2S \longrightarrow COS + H_2O$	in Claus furnace and over various catalysts
$CO_2 + CS_2 \rightleftharpoons 2\,COS$	at ca 500°C, catalyzed by silica
$CS_2 + SO_3 \longrightarrow COS + SO_2 + S^0$	above 400°C as gases, at low temperatures as liquids
$HSCN + H_2O \longrightarrow COS + NH_3$	in H_2SO_4 at 20°C
$FeS_2 + CO \longrightarrow FeS + COS$	fast above 800°C

Carbonyl sulfide occurs as a by-product in the manufacture of carbon disulfide and is an impurity in some natural gases, in many manufactured fuel gases and refinery gases, and in combustion products of sulfur-containing fuels (25). It tends to be concentrated in the propane fraction in gas fractionation; an amine sweetening process is needed to remove it.

Carbonyl sulfide is overall the most abundant sulfur-bearing compound in the earth's atmosphere: 430–570 parts per trillion (10^{12}), although it is exceeded by H_2S and SO_2 in some industrial urban atmospheres (27). Carbonyl sulfide is believed to originate from microbes, volcanoes, and the burning of vegetation, as well as from industrial processes. It may be the main cause of atmospheric sulfur corrosion (28).

Production, Shipment, and Specifications. Carbonyl sulfide is available in 97% min purity in cylinders up to 31.8 kg contained weight. It is shipped as a flammable gas. There appears to be no full-scale commercial production of carbonyl sulfide in the United States.

Analytical Methods. Detection of carbonyl sulfide in air can be done by gas chromatography or by combustion to sulfur dioxide and determination of the latter. Where hydrogen sulfide and carbonyl sulfide occur together, the carbonyl sulfide can be determined by combustion after hydrogen sulfide is absorbed by lead acetate, which does not absorb carbonyl sulfide (35).

Health and Safety Factors. Carbonyl sulfide is dangerously poisonous, more so because it is practically odorless when pure. It is lethal to rats at 2900 ppm. Studies show an LD_{50} (rat, ip) of 22.5 mg/kg. The mechanism of toxic action appears to involve breakdown to hydrogen sulfide (36). It acts principally on the central nervous system with death resulting mainly from respiratory paralysis. Little is known regarding the health effects of subacute or chronic exposure to carbonyl sulfide; a 400-μg/m^3 max level has been suggested until more data are available (37). Carbon oxysulfide has a reported inhalation toxicity in mice LD_{50} (mouse) = 2900 ppm (37).

Uses. There may be some captive use of carbonyl sulfide for production of certain thiocarbamate herbicides (qv). One patent (38) describes the reaction of diethylamine with carbonyl sulfide to form a thiocarbamate salt which is then alkylated with 4-chlorobenzyl halide to produce S-(4-chlorobenzyl) N,N-diethylthiocarbamate [28249-77-6], ie, benthiocarb. Carbonyl sulfide is also reported to be useful for the preparation of aliphatic polyureas. In these preparations, potassium thiocyanate and sulfuric acid are used to first generate carbonyl sulfide, COS, which then reacts with a diamine:

$$2 \text{ KCNS} + 2 \text{ H}_2\text{SO}_4 + 2 \text{ H}_2\text{O} \longrightarrow 2 \text{ COS} + (\text{NH}_4)_2\text{SO}_4 + \text{K}_2\text{SO}_4$$

$$\text{H}_2\text{NRNH}_2 + \text{COS} \longrightarrow \text{H}_3\overset{+}{\text{N}}\text{RNH}\overset{\overset{\text{O}}{\|}}{\text{C}}\text{S}^- \longrightarrow \text{(\!-\!RNH}\overset{\overset{\text{O}}{\|}}{\text{C}}\text{NH\!-\!)}_{\overline{n}} + \text{H}_2\text{S}$$

Monoureas can also be prepared using COS and an alkylamine, or an arylamine in the absence or presence of a catalyst, where R = H, alkyl, or aryl (39):

$$2 \text{ RNH}_2 + \text{CO} + \text{S}^0 \longrightarrow \text{RNH}\overset{\overset{\text{O}}{\|}}{\text{C}}\text{NHR}$$

The above reaction produces COS as an intermediate which can be isolated when a catalytic amount of selenium is present (40).

Thiophosgene

Physical Properties. Thiophosgene [463-71-8] (thiocarbonyl chloride), $CSCl_2$, is a malodorous, red-yellow liquid (bp 73.5°C, d_{20}^{15} 1.509, n_D^{20} 1.5442). It is only slightly soluble with decomposition in water, but it is soluble in ether and various organic solvents.

Chemical Properties. Thiophosgene is more resistant to hydrolysis than its oxygen analogue, phosgene, but it is slowly hydrolyzed to carbon dioxide, hydrogen sulfide, and hydrochloric acid. It can be oxidized to a lacrimatory thiophosgene S-oxide (41). Its utility in organic synthesis has been reviewed (42,43).

Thiophosgene reacts with alcohols and phenols to form chlorothionoformates or thiocarbonates. The most studied reactions of thiophosgene are with primary amines to give isothiocyanates and with secondary amines to give thiocarbamyl chlorides:

$$\text{RNH}_2 + \text{CSCl}_2 \longrightarrow \text{RN}{=}\text{C}{=}\text{S} + 2 \text{ HCl}$$

$$2 \text{ R}_2\text{NH} + \text{CSCl}_2 \longrightarrow \text{R}_2\text{N}\overset{\overset{\text{S}}{\|}}{\text{C}}\text{Cl} + \text{R}_2\text{NH} \cdot \text{HCl}$$

The reaction of thiophosgene with various bisphenols using phase-transfer catalysis gives polythiocarbonates (44):

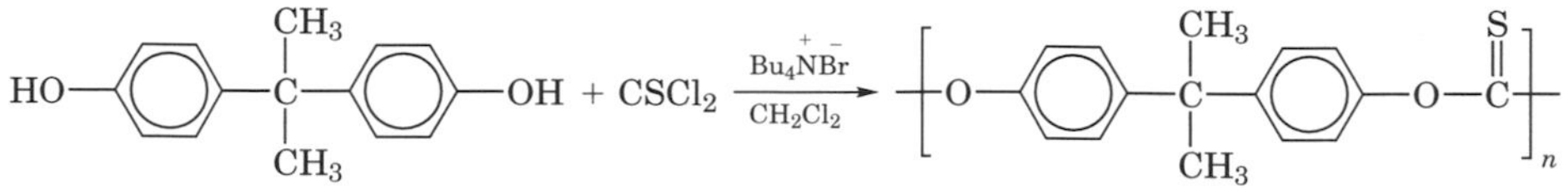

Most of the reactions of thiophosgene involve the expected chemistry of an acid chloride, in which the chlorine atoms are replaceable by various nucleophiles. A reaction involving the $C{=}S$ bond is the Diels-Alder addition:

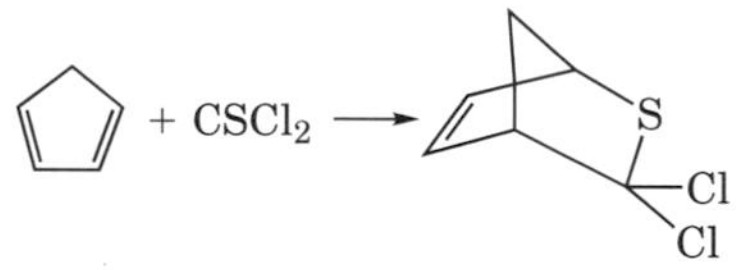

Preparation.   Thiophosgene forms from the reaction of carbon tetrachloride with hydrogen sulfide, sulfur, or various sulfides at elevated temperatures. Of more preparative value is the reduction of trichloromethanesulfenyl chloride [594-42-3] by various reducing agents, eg, tin and hydrochloric acid, stannous chloride, iron and acetic acid, phosphorus, copper, sulfur dioxide with iodine catalyst, or hydrogen sulfide over charcoal or silica gel catalyst (42,43).

There is no full-scale U.S. commercial production of thiophosgene, but it is available in glass ampuls from laboratory reagent suppliers. Thiophosgene may be produced in Israel as an intermediate for tolnaftate (2-naphthyl N-methyl-N-m-tolylthiocarbamate) [2398-96-1], an antifungal drug.

Health and Safety Factors.   Thiophosgene has an LD_{50} (rat, oral) of 929 mg/kg and an LC_{50} (inhalation, rat) of 370 mg/m^3 (45). It has both irritant and systemic toxic properties.

The MSDS (46) for thiophosgene describes it as highly toxic, corrosive lachrymator and moisture sensitive compound. It may be fatal if inhaled, swallowed, or absorbed through the skin. When using this material one should wear the appropriate NIOSH/OSHA-approved respirator, chemical-resistant gloves, safety goggles, and other protective clothing. It should be used only in a chemical fume hood.

Trichloromethanesulfenyl Chloride

Physical Properties.   Trichloromethanesulfenyl chloride [594-42-3] (perchloromethyl mercaptan, a misnomer but used as the common commercial name), CCl_3SCl, is a strongly acrid, pale yellow liquid, boiling at 149°C with some decomposition at atmospheric pressure, 68°C at 6.93 kPa (52 mm Hg), and 25°C at 0.8 kPa (6 mm Hg); sp gr (20°C/4°C) 1.6996; n_D^{23} 1.541 (47,48). It slowly hydrolyzes and is soluble in most organic solvents. It has a vapor density of 6.414

(air $= 1$) (49) and a vapor pressure of 3.0 mm Hg at 20°C (50). It is insoluble in water, and is nonflammable but supports combustion (49).

Chemical Properties. A detailed review of the chemistry of trichloromethanesulfenyl chloride has been published (51). It is stable for prolonged periods at ambient temperature but decomposes slowly at its atmospheric boiling point forming sulfur monochloride, carbon tetrachloride, carbon disulfide, and polymeric oils. Storage in contact with iron causes it to decompose to carbon tetrachloride and sulfur monochloride. It is hydrolyzed only slowly by water at room temperature in a complex series of steps, proceeding by way of the unstable sulfenic acid and an isolable thiophosgene S-oxide [24768-49-8], $Cl_2C{=}S{=}O$, yielding a complex product mixture including hydrochloric acid, phosgene, thiophosgene, trichloromethanesulfonyl chloride, chlorocarbonyl-sulfenyl chloride [2757-23-5], and bis(trichloromethyl) disulfide [15110-08-4] (52). At 160°C, hydrolysis is rapid and leads ultimately to the formation of carbon dioxide, hydrochloric acid, and sulfur. Trichloromethanesulfenyl chloride reacts rapidly with sodium hydroxide forming sodium dichloromethanesulfinate [36829-83-1], $CHCl_2SO_2Na$ (53).

The oxidation of trichloromethanesulfenyl chloride by nitric acid or oxidative chlorination in the presence of water yields trichloromethanesulfonyl chloride [2547-61-7], CCl_3SO_2Cl, which is a lacrimatory solid (mp 140–142.5°C), which is surprisingly stable to hydrolysis and can be steam-distilled. Trichloromethanesulfenyl chloride can be reduced to thiophosgene by metals in the presence of acid and by various other reducing agents. The sulfur-bonded chlorine of trichloromethanesulfenyl chloride is most easily displaced by nucleophilic reagents, but under some conditions, the carbon-bound chlorines are also reactive (54).

$$CCl_3SCl + 2\ RNH_2 \longrightarrow CCl_3SNHR + RNH_2{\cdot}HCl$$

$$CCl_3SCl + RONa \longrightarrow CCl_3SOR + NaCl$$

$$CCl_3SCl + RSH \longrightarrow CCl_3SSR + HCl$$

$$CCl_3SCl + C_6H_5N(CH_3)_2 \longrightarrow p\text{-}CCl_3SC_6H_4N(CH_3)_2{\cdot}HCl$$

$$CCl_3SCl + 4\ RONa \longrightarrow (RO)_4C + 4\ NaCl + S^0$$

$$CCl_3SCl + KF \longrightarrow CFCl_2SCl + KCl$$

Trichloromethanesulfenyl chloride can be added across double bonds under either free-radical or cationic-initiated conditions (55,56):

$$CCl_3SCl + R_2C{=}CR_2 \longrightarrow CCl_3SCR_2CClR_2$$

Aryl chlorothioformates are prepared, for example, by reaction of a mixture of 3-$(CH_3)_3CC_6H_4OH$ and $NaHSO_3$, CCl_3SCl, KI, and concentrated H_2SO_4 in CCl_4 plus H_2O for 10 hours at room temperature to give a 63% yield of $(CH_3)_3CC_6H_4OCSCl$ (57). Phenylthiocarbamoyl chlorides are useful as intermediates for pharmaceuticals and agrochemicals were prepared by the reaction of CCl_3SCl with aromatic amines (58).

Manufacture.   Trichloromethanesulfenyl chloride is made commercially by chlorination of carbon disulfide with the careful exclusion of iron or other metals, which catalyze the chlorinolysis of the C—S bond to produce carbon tetrachloride. Various catalysts, notably iodine and activated carbon, are effective. The product is purified by fractional distillation to a minimum purity of 95%. Continuous processes have been described wherein carbon disulfide chlorination takes place on a granular charcoal column (59,60). A series of patents describes means for yield improvement by chlorination in the presence of difunctional carbonyl compounds, phosphonates, phosphonites, phosphites, phosphates, or lead acetate (61).

Shipment and Storage.   Much of the trichloromethanesulfenyl chloride produced is used captively. However, it can be shipped in drums. Standard materials for storage and handling are nickel, Monel, and glass. Trichloromethanesulfenyl chloride is not corrosive to nickel and Monel when dry, but care must be taken to prevent its contact with moisture.

Economic Aspects.   The U.S. manufacture of trichloromethanesulfenyl chloride is carried out by Zeneca, Inc. (Perry, Ohio). Most of the usage is captive. The 1995 price was ca $2.50/kg.

Analytical Methods.   A method has been described for gas chromatographic analysis of trichloromethanesulfenyl chloride as well as of other volatile sulfur compounds (62). A method has been recommended for determining small amounts of trichloromethanesulfenyl chloride in air or water on the basis of a color-forming reaction with resorcinol (63).

Health and Safety Factors.   Trichloromethanesulfenyl chloride is extremely toxic and mutagenic (64). It has an LD_{50} in rabbits (percutaneous) of 1410 mg/kg (65,66) and an LC_{50} in rats (male) of 11 ppm/h (66). The LD_{50} oral in rats is 83 mg/kg (64). There are hazards involved from inhalation and possible skin contact. Severe local irritation can result from contact of the liquid or vapor with the skin, eyes, mucous membranes, and upper respiratory tract. The irritating nature of the vapor may also serve as a warning of its presence. The MSDS should be consulted and all work carried out with good ventilation in a laboratory fume hood with the operator wearing proper gloves, suitable protective clothing, and respiratory equipment. The TLV of trichloromethanesulfenyl chloride is an 8-h time-weighted average exposure of 0.1 ppm (0.8 mg/m^3).

Uses.   The principal commercial application for trichloromethanesulfenyl chloride is as an intermediate for the manufacture of fungicides, the most important being captan (*N*-(trichloromethylthio)-4-cyclohexene-1,2-dicarboximide) [*133-06-2*] (**1**) and folpet (*N*-(trichloromethylthio)phthalimide) [*133-07-3*] (**2**) (see FUNGICIDES, AGRICULTURAL):

(**1**)   (**2**)

A commercial soil or seed treatment fungicide, 5-ethoxy-3-trichloromethyl-1,2,4-thiadiazole [2593-15-9] (etridiazole) is prepared from 2-chloro-3-trichloro-methyl-1,2,4-thiadiazole [5848-93-1], a product of trichloromethanesulfenyl chloride and trichloroacetamidine (68).

A fluorinated analogue, CFCl$_2$SCl, is made in Europe from trichloro-methanesulfenyl chloride and is used for production of the fungicides dichlorofluanid [1085-98-9] (**3**) and tolylfluanid [731-27-1] (**4**) (69):

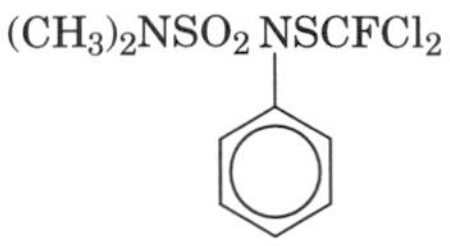

(CH$_3$)$_2$NSO$_2$ NSCFCl$_2$ (CH$_3$)$_2$NSO$_2$ NSCFCl$_2$

CH$_3$

(**3**) 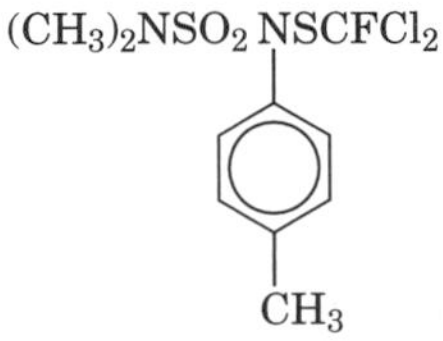   (**4**)

Trichloromethanesulfenyl chloride has also been used for the preparation of lubricant additives (see LUBRICATION AND LUBRICANTS). Some higher homologues and analogues of trichloromethanesulfenyl chloride have been reported, ie, trichlorovinylsulfenyl chloride [19411-15-5], 1,1,2,2-tetrachloroethanesulfenyl chloride [1185-09-7], 1,2,2,2-tetrachloroethanesulfenyl chloride [920-62-7], and pentachloroethanesulfenyl chloride [5940-94-3] (70–74). A commercial fungicide, Captafol [2425-06-1], is the reaction product of 1,1,2,2-tetrachloroethanesulfenyl chloride with tetrahydrophthalimide (75).

Trichloromethanesulfenyl chloride is useful in the preparation of various series of sulfur-containing heterocycles. A commercially significant example is the preparation of 1,2,4-thiadiazoles.

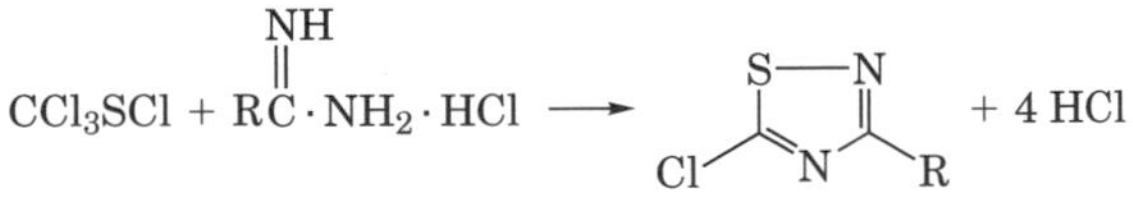

NH
‖
CCl$_3$SCl + RC·NH$_2$·HCl ⟶   (structure)   + 4 HCl

Hydrogen Sulfide

Hydrogen sulfide [7783-06-4] is present in the gases from many volcanoes, sulfur springs, undersea vents, swamps, and stagnant bodies of water. Bacterial reduction of sulfates and bacterial decomposition of proteins forms hydrogen sulfide. The role of hydrogen sulfide in the natural sulfur cycles is available (76). Of greater importance as a source of sulfur are sour gases, which occur in large amounts in several locations. The concentrations of hydrogen sulfide in these sour gases are as follows (77):

Location	H_2S, wt %
France (Elf Aquitaine, Lacq)	15.0–16.0
Germany (Varnhorn)	22.4
Canada (Harmattan, Alberta)	53.5
Canada (Bearberry, Alberta)	90.0
Canada (Panther River, Alberta)	70–80
United States (Smackover, Miss.)	25–45
Russia (Astrakhan)	22.5
People's Republic of China (Zhaolanzhuang)	60–90

Hydrogen sulfide is a by-product of many industrial operations, eg, coking and the hydrodesulfurization of crude oil and of coal. Hydrodesulfurization is increasing in importance as the use of high sulfur crude oil becomes increasingly necessary (see PETROLEUM, REFINERY PROCESSES). A large future source of hydrogen sulfide may result if coal liquefaction attains commercial importance (see COAL CONVERSION PROCESSES).

Physical Properties. Hydrogen sulfide, H_2S, is a colorless gas having a characteristic rotten-egg odor. The physical properties of hydrogen sulfide are given in Table 2.

Detailed studies of the hydrogen sulfide–nitrogen system (85–87) and those of the hydrogen sulfide–water system have been reviewed (84,88). At low temperatures and high pressure, hydrogen sulfide forms a crystalline hexahydrate [66230-40-8]. Hydrogen sulfide is soluble in certain polar organic solvents, notably methanol, acetone, propylene carbonate, sulfolane, tributyl phosphate, various glycols, and glycol ethers (25,89). N-Methylpyrrolidinone is an exceptionally good solvent, dissolving 49 mL/g at 20°C at atmospheric pressure (90). Hydrogen sulfide is less soluble in nonpolar solvents; for example, the solubility of hydrogen sulfide gas at 20°C is 8.9 mL/g in hexane and 16.6 mL/g in benzene (76). Hydrogen sulfide is very soluble in alkanolamines, which are used as scrubbing solvents for removal of hydrogen sulfide from gas streams (25,91–93). The dissolution of hydrogen sulfide in amines results in the formation of a salt, which generally dissociates upon heating. Detailed studies have been published on the systems hydrogen sulfide–ethanolamine, –diethanolamine, –Sulfinol (a mixture of alkanolamines, sulfolane, and water) and –diglycolamine (94–97) (see SULFUR REMOVAL AND RECOVERY).

Because of its low dielectric constant, liquid hydrogen sulfide is a poor solvent for ionic salts, eg, NaCl, but it does dissolve appreciable quantities of anhydrous $AlCl_3$, $ZnCl_2$, $FeCl_3$, PCl_3, $SiCl_4$, and SO_2. Liquid hydrogen sulfide or hydrogen sulfide-containing gases under pressure dissolve sulfur. At equilibrium H_2S pressure, the solubility of sulfur in liquid H_2S at −45, 0, and 40°C is 0.261, 0.566, and 0.920 wt %, respectively (98). The equilibria among H_2S_x, H_2S, and sulfur have been studied (99,100).

Chemical Properties. Although hydrogen sulfide is thermodynamically stable, it can dissociate at very high temperatures. The decomposition thermodynamics and kinetics have been reviewed and the equilibrium constant for the

Table 2. Physical and Thermodynamic Properties of Hydrogen Sulfide

Property	Value	Ref.
mol wt	34.08	78
mp, °C	−85.53	79
bp, °C	−60.31	79
ΔH fusion, kJ/mol[a]	2.375	78
ΔH vaporization, kJ/mol[a]	18.67	78
density at −60.31°C, kg/m^3	949.6	80
sp gr, gas (air = 1)	1.182[b]	81
critical temperature, °C	100.38	82
critical pressure, kPa[c]	9006	82
critical density, kg/m^3	346.0	82
ΔG formation, kJ/mol[a]	−33.6	17
ΔH formation, at 25°C, kJ/mol[a]	−20.6	17
S formation, at 25°C, J/(mol·K)[a]	205.7	17
C_p, at 27°C, J/(mol·K)[a]	34.2	17
autoignition temperature in air, °C	ca 260	83
explosive range in air at 20°C, vol %		
upper limit	46	83
lower limit	4.3	
vapor pressure, kPa[c]		
−60°C	102.9	79,84
−40°C	257.9	79
−20°C	562.0	79
0°C	1049	79
20°C	1814	79
40°C	2937	79
60°C	4480	79
solubility in water[d], g/100 g soln		81
0°C	0.710	
10°C	0.530	
20°C	0.398	

[a]To convert J to cal, divide by 4.184.
[b]Based on air = 79 mol % nitrogen plus 21 mol % oxygen.
[c]To convert kPa to psi, multiply by 0.145; to convert kPa to bars, divide by 100.
[d]At 101.3 kPa(=1 atm) total pressure.

reaction has been determined (101,102):

$$2 \text{ H}_2\text{S} \rightleftharpoons 2 \text{ H}_2 + \text{S}_2$$

Above ca 850°C, the experimental hydrogen yield agrees well with that calculated by thermodynamic studies and a decomposition enthalpy of 16 kJ/mol (3.8 kcal/mol) is obtained and is in close agreement with theory. Below 850°C, equilibrium occurs quite slowly without a catalyst. With silica as catalyst and, to a greater degree, with cobalt molybdate or sulfided platinum as catalyst, greater hydrogen production occurs at 450−850°C than is theoretically predicted on the basis of the above equation. For example, at 477°C, 3.6 wt % H$_2$ results at equilibrium over cobalt molybdate. These enhanced hydrogen yields are explained on the basis of the formation of sulfur species other than S$_2$ (102).

The dissociation of hydrogen sulfide has been proposed as a potential hydrogen source for refineries, which normally need hydrogen and have ample hydrogen sulfide. Progress has been made in processes involving transition-metal catalysts, eg, Mo, W, or Ru sulfides, and processes involving the shifting of the equilibrium by removal of the sulfur (100,103). In the presence of a specially designed photoelectrolytic catalyst, visible light cleaves hydrogen sulfide to hydrogen and sulfur (104).

Hydrogen sulfide is oxidized by a number of oxidizing agents, as shown in Table 3. The actual products formed, where alternatives are indicated, are functions of the quantity of oxidant as well as of reaction conditions.

Certain of the above reactions are of practical importance. The oxidation of hydrogen sulfide in a flame is one means for producing the sulfur dioxide required for a sulfuric acid plant. Oxidation of hydrogen sulfide by sulfur dioxide is the basis of the Claus process for sulfur recovery. The Claus reaction can also take place under milder conditions in the presence of water, which catalyzes the reaction. However, the oxidation of hydrogen sulfide by sulfur dioxide in water is a complex process leading to the formation of sulfur and polythionic acids, the mixture known as Wackenroeder's liquid (105).

Oxidation of hydrosulfide salts to sulfur by water-soluble quinones is the basis of the Stretford and the Takahax processes used in sulfur recovery. Oxidation of hydrosulfide to sulfur by sodium vanadate, which can be regenerated by air, is a key reaction of the Beavon process, also used commercially in sulfur recovery. Oxidation of hydrogen sulfide until no odor is detectable proceeds with air over a cobalt molybdate catalyst (106). The reaction of hydrogen sulfide with I_2 is quantitative and is used for analytical determination of H_2S. The oxidation of hydrogen sulfide by hydrogen peroxide in aqueous sodium hydroxide is a means for abatement of hydrogen sulfide in steam from geothermal operations or

Table 3. Oxidation Reactions of Hydrogen Sulfide

Oxidizing agent	Conditions	Sulfur-containing products
O_2	flame	SO_2, some SO_3
	flame or furnace, catalyst (Claus process)	sulfur
	aqueous soln of H_2S, catalyst	sulfur
H_2O_2	neutral; alkaline solution	sulfur; $S_2O_3^{2-}$, SO_4^{2-}
Na_2O_2	dry, elevated temperature	Na_2S, Na_2S_x
O_3	aqueous	sulfur, H_2SO_4
SO_2	elevated temperature, catalyst	sulfur
	aqueous soln	sulfur, polythionic acids
H_2SO_4	concentrated acid	sulfur, SO_2
HNO_3	aqueous	H_2SO_4
NO	silica gel catalyst	sulfur
NO_2^-	aqueous at pH 5–7	sulfur, NO
	aqueous at pH 8–9	sulfur, NH_3
Cl_2	gaseous, excess Cl_2	SCl_2
	gaseous, excess H_2S	sulfur
	aqueous soln, excess Cl_2	H_2SO_4
I_2	aqueous soln	sulfur (quantitative)
Fe^{3+}	aqueous soln	sulfur, FeS

steam condensate from power plants, sewage, and industrial waste systems (107) (see GEOTHERMAL ENERGY). Iron salts are effective catalysts for this reaction.

Anhydrous gaseous or liquid hydrogen sulfide is practically nonacidic, but aqueous solutions are weakly acid. The K_a for the first hydrogen is 9.1×10^{-8} at 18°C; for the second, is 1.2×10^{-15}. Reaction of hydrogen sulfide with one molar equivalent of sodium hydroxide gives sodium hydrosulfide; with two molar equivalents of sodium hydroxide, sodium sulfide forms. Hydrogen sulfide reacts with sodium carbonate to produce sodium hydrosulfide and sodium bicarbonate, a reaction which is reversible upon heating. Carbon dioxide can liberate hydrogen sulfide from sodium hydrosulfide. Various metal oxides and hydroxides react with hydrogen sulfide forming sulfides. A useful application is the removal of hydrogen sulfide by reaction with iron oxide (25):

$$Fe_2O_3 \cdot H_2O + 3\,H_2S \longrightarrow Fe_2S_3 + 4\,H_2O$$

Anhydrous hydrogen sulfide does not react at ordinary temperatures with metals, eg, mercury, silver, or copper. However, in the presence of air and moisture, the reaction is rapid, leading to tarnishing in the case of silver and copper.

$$4\,Cu\ (or\ Ag) + 2\,H_2S \longrightarrow 2\,Cu_2S\ (or\ Ag_2S) + 2\,H_2O$$

Hydrogen sulfide causes the precipitation of sulfides from many heavy-metal salts. The classical qualitative analysis scheme depends on precipitation of the sulfides of Hg, Pb, Bi, Cu, Cd, As, Sb, and Sn under acid conditions and the sulfides of Co, Ni, Mn, Zn, and Fe under ammoniacal conditions.

Hydrogen sulfide reacts with molten sulfur and depresses the viscosity of the latter, particularly at 130–180°C; infrared spectral studies show that polysulfanes, H_2S_x, form. The average chain length of these polysulfanes is shorter than the equilibrium chain length of molten sulfur alone at the same temperature; consequently, the viscosity of the molten sulfur is markedly reduced (99,100).

Hydrogen sulfide reacts with olefins under various conditions forming mercaptans and sulfides (108,109). With ethylene it can react to ultimately give diethyl sulfide (110). With unsymmetrical olefins, the direction of addition can be controlled by the choice of either a free-radical initiator, including ultraviolet light, or an acidic catalyst (110):

$$CH_2{=}CH_2 + H_2S \longrightarrow CH_3CH_2SH \xrightarrow{CH_2=CH_2} (C_2H_5)_2S$$

In the presence of sulfur, a disulfide can be produced (111):

$$2\,CH_2{=}CH_2 + H_2S \longrightarrow 2\,CH_3CH_2SH \xrightarrow{S} C_2H_5SSC_2H_5$$

When free-radical initiation is used, cocatalysts, eg, phosphites (112), and uv photoinitiators such as acetophenone derivatives (113) can be used to increase the rate and conversion of the olefins to the desired mercaptans.

If olefins with electron-withdrawing substituents are involved, the addition can be conducted with a basic catalyst.

$$H_2S + CH_2\!\!=\!\!CHCOOR \xrightarrow{\text{base}} HSCH_2CH_2COOR$$

$$H_2S + 2\, CH_2\!\!=\!\!CHCOOR \xrightarrow{\text{base}} S(CH_2CH_2COOR)_2$$

Hydrogen sulfide reacts with alcohols under acid-catalyzed conditions, usually with solid acidic catalysts at elevated temperatures:

$$CH_3OH + H_2S \longrightarrow H_2O + CH_3SH \xrightarrow{CH_3OH} CH_3SCH_3 + H_2O$$

The catalyzed reaction of methanol and H_2S followed by reaction with sulfur yields dimethyl disulfide in good yield (114).

$$2\, CH_3OH + 2\, H_2S \xrightarrow{-2\, H_2O} 2\, CH_3SH \xrightarrow{S} CH_3SSCH_3 + H_2S$$

At 550–600°C, hydrogen sulfide reacts with chloroaromatic compounds forming thiophenols and diaryl sulfides (115):

$$H_2S + C_6H_5Cl \longrightarrow HCl + C_6H_5SH \text{ (mainly)} + (C_6H_5)_2S$$

It also reacts with epoxides forming 2-hydroxyalkyl thiols, eg, 2-mercaptoethanol [60-24-2] and bis(2-hydroxyalkyl) sulfides such as bis(2-hydroxyethyl) sulfide [111-48-8]; both products are made commercially:

$$H_2S + CH_2\overset{O}{\overset{\diagup\diagdown}{-\!\!\!-}}CH_2 \longrightarrow HSCH_2CH_2OH$$

$$H_2S + 2\, CH_2\overset{O}{\overset{\diagup\diagdown}{-\!\!\!-}}CH_2 \longrightarrow S(CH_2CH_2OH)_2$$

Hydrogen sulfide reacts with nitriles in the presence of a basic catalyst forming thioamides. A commercial example is its addition to cyanamide with the formation of thiourea [62-56-6]:

$$H_2S + NH_2C\!\!\equiv\!\!N \longrightarrow NH_2\overset{\overset{\textstyle S}{\|}}{C}NH_2$$

Manufacture. Small cylinders of hydrogen sulfide are readily available for laboratory purposes, but the gas can also be easily synthesized by action of dilute sulfuric or hydrochloric acid on iron sulfide, calcium sulfide [20548-54-3], zinc sulfide [1314-98-3], or sodium hydrosulfide [16721-80-5]. The reaction usually is run in a Kipp generator, which regulates the addition of the acid to maintain a steady hydrogen sulfide pressure. Small laboratory quantities of hydrogen sulfide can be easily formed by heating at 280–320°C a mixture of sulfur and a hydrogen-rich, nonvolatile aliphatic substance, eg, paraffin. Gas evolution proceeds more smoothly if asbestos or diatomaceous earth is also present.

Commercial-scale processes have been developed for the production of hydrogen sulfide from heavy fuel oils and sulfur as well as from methane, water vapor, and sulfur. The latter process can be carried out in two steps: reaction of methane with sulfur to form carbon disulfide and hydrogen sulfide followed by hydrolysis of carbon disulfide (116).

Hydrogen sulfide has been produced in commercial quantities by the direct combination of the elements. The reaction of hydrogen and sulfur vapor proceeds at ca 500°C in the presence of a catalyst, eg, bauxite, an aluminosilicate, or cobalt molybdate. This process yields hydrogen sulfide that is of good purity and is suitable for preparation of sodium sulfide and sodium hydrosulfide (see SODIUM COMPOUNDS). Most hydrogen sulfide used commercially is either a by-product or is obtained from sour natural gas.

Recovery from Gas Streams. The crude oil refined in the United States contains varying amounts of sulfur, eg, 0.04 wt % in Pennsylvania crude to ca 5 wt % in heavy Mississippi crude. Hydrodesulfurization is becoming increasingly important as a refinery operation. More than 90% of the sulfur in crude oils is accounted for in the gas–oil and coke-distillate fractions. The sulfur compounds are mostly acyclic and cyclic sulfides; many of the latter are highly stable aromatic types, eg, benzothiophenes. Their removal is accomplished by passing the sulfur-rich fractions through a fixed-bed catalyst with hydrogen, which is generally by-product hydrogen from catalytic reforming. Besides conversion of 80–90% of the sulfur compounds to hydrogen sulfide, the hydrocarbon saturation is increased. The hydrogen sulfide is usually converted by Claus process oxidation to sulfur (see SULFUR REMOVAL AND RECOVERY).

Corrosivity. Anhydrous hydrogen sulfide has a low general corrosivity toward carbon steel, aluminum, Inconel, Stellite, and 300-series stainless steels at moderate temperatures. Temperatures greater than ca 260°C can produce severe sulfidation of carbon steel. Alternative candidates for hydrogen sulfide service at higher temperatures and concentrations include 5Cr–0.5Mo and 9Cr–1Mo alloy steels, 400-series stainless steels, 300-series stainless steels, and Inconel. Stainless steels of low carbon content in the annealed condition are preferable.

Wet hydrogen sulfide can be quite corrosive to carbon steel; corrosion rates can exceed 2.5 mm/yr. The actual corrosion rate depends on temperature and hydrogen sulfide concentration. All common metallic materials of construction, ie, carbon steels, Cr–Mo alloy steels, stainless steels, nickel-based alloys, and aluminum, can be used in wet hydrogen sulfide service depending on the process environment and the desired service life. In addition to general corrosion, wet hydrogen sulfide service can cause sulfide stress cracking. The atomic hydrogen created in the corrosion reaction diffuses into the metal which causes cracking. An important factor in reducing the likelihood of sulfide stress cracking is to limit the hardness of the base metal, weld metal, and base metal's heat-affected zone. An extensive compendium of hydrogen sulfide corrosion information has been published, and standards for sulfide-resistant metals for oil field use have been proposed (117,118). High nickel stainless steels and nickel-based alloys are the most resistant to sulfide attack under simulated deep oil- and gas-well environments (119). In the case of hydrogen sulfide copper or copper-bearing alloys are not to be used on any process equipment and associated instruments, controls, or electrical–mechanical equipment.

Economic Aspects. Most hydrogen sulfide is made and used captively or sold by pipeline at prices which are highly variable, depending on locality. Production in the United States exceeds 1.1×10^6 t/yr. It has been estimated that 2.4×10^6 t/yr of sulfur are recovered from H_2S-containing refinery streams and 1.8×10^6 t/yr of sulfur are recovered from H_2S-containing natural gas (120).

Sulfuric acid production in 1995 was estimated to consume 110,000 metric tons of H_2S, (121).

Analytical Methods. A method recommended by NIOSH for measurement of hydrogen sulfide in air involves drawing a known volume of air through a desiccant tube containing sodium sulfate to remove water vapor. A tube containing a molecular sieve is then used to trap hydrogen sulfide, which is subsequently desorbed thermally and analyzed by gas chromatography. This method is useful for analysis to $15-60$ mg/m^3 (122). An alternative method involves aspirating a measured volume of air through an alkaline suspension of cadmium hydroxide. Cadmium sulfide precipitates, and the collected sulfide is determined by spectrophotometric measurement of the methylene blue produced by reaction of the sulfide with an acid solution of N,N-dimethyl-p-phenylenediamine and ferric chloride. This method is useful to $8.5-63$ mg/m^3 (123).

A simple laboratory method for detecting hydrogen sulfide is the use of lead acetate paper, which darkens upon formation of lead sulfide. Instruments containing a moving tape treated with lead acetate and having an optical device for measuring the reflectance of the tape have been devised for monitoring the presence of hydrogen sulfide. Detector tubes, eg, Draeger tubes, are available for fast, semiquantitative measurement of hydrogen sulfide in air. Electronic devices based on solid-state sensors (qv) have also become available for continuous monitoring of hydrogen sulfide (124).

Health and Safety Factors. Hydrogen sulfide has an extremely high acute toxicity and has caused many deaths both in the workplace and in areas of natural accumulation, eg, cisterns and sewers (125,126). Hydrogen sulfide can be hazardous to workers in the gas, oil chemical, geothermal energy, and viscose rayon industries and workers in sewer systems, tanneries, mining, drilling, smelting, animal-waste disposal, and on fishing boats. Brief exposure to hydrogen sulfide at a concentration of 140 mg/m^3 causes conjunctivitis and keratitis (eye damage), and exposures at above ca 280 mg/m^3 cause unconsciousness, respiratory paralysis, and death. There is no conclusive evidence of adverse health effects from repeated long-term exposure to hydrogen sulfide at low concentrations, although there is some evidence pointing to nervous system, cardiovascular, gastrointestinal, and ocular disorders. Symptoms of low level exposure can include headache, nausea, insomnia, fatigue, and inflammation of the eyes and mucous membranes. Irritation of the eyes and respiratory system has been reported at concentrations below 1 ppm.

Hydrogen sulfide is especially dangerous when it occurs in low lying areas or confined workspaces or when it exists in high concentrations under pressure.

Because of corrosion and embrittlement, hydrogen sulfide lines and fittings are prone to develop leaks. Hydrogen sulfide is fast acting and the exposed person may become unconscious quickly, with no opportunity to escape the contaminated space. Moreover, hydrogen sulfide has a deceptively sweet smell at 30–100 ppm and deadens the sense of smell above this range; therefore, its odor is an unreliable indicator of dangerous concentrations. Self-contained breathing apparatus or a supplied-air respirator is necessary for workers entering areas known or suspected to contain toxic levels of hydrogen sulfide. A summary of the effects of hydrogen sulfide exposure on humans is given in Table 4.

Because many lethal accidents have occurred when using or generating hydrogen sulfide, it is imperative that before using it the Material Safety Data Sheet (MSDS) for this product be thoroughly read and understood (127). It is also advisable to read a hydrogen sulfide users manual (128). This manual discusses physical properties, health hazards, material of construction, storage, shipping container, recommended unloading, emergencies, and it has an MSDS, as well as several appendices for use when rail/tank car use is required. OSHA exposure limits for H_2S are 10 ppm (14 mg/m^3) OSHA/TWA; 10 ppm is the NIOSH recommended 10-minute ceiling.

Protective measures involve prompt detection and adequate ventilation. Continuous monitoring is recommended to signal an evacuation alarm if the

Table 4. Effects of Hydrogen Sulfide Inhalation on Humans[a]

Number of people exposed	Concentration, mg/m^3	Duration of exposure	Effect
1	17,000		death
1	2,800–5,600	<20 min	death
10	1,400	<1 min	unconsciousness, abnormal ECG; death of one person
342	1,400–2,800	<20 min	hospitalization of 320, death of 22 including 13 in hospital, residual nervous system damage in 4
5	1,400	instant	unconsciousness, death
1	1,400	<25 min	unconsciousness, low blood pressure, pulmonary edema, convulsions, hematuria
4	400–760		unconsciousness
1	320	20 min	unconsciousness, arm cramps, low blood pressure
78	20–35		burning eyes in 25, headache in 32, loss of appetite in 31, weight loss in 20, dizziness in >19
6500	15–20	4–7 h	conjunctivitis
population of Terre Haute, Indiana	0.003–11	intermittent air pollution episodes over a 2-mo period	nausea in 13, headache, shortness of breath in 4, sleep disturbance in 5, throat and eye irritation in 5

[a]Ref. 125.

workplace concentration exceeds 70 mg/m^3 (50 ppm) and a warning alert if it is present at 15–70 mg/m^3 (10–50 ppm).

Alberta, which has a major sulfur industry, permits the long-term presence of 0.004 mg/m^3. Various states have passed emission standards for effluent air or gas from stationary sources. A good summary of these regulations as of 1979 is available (126). In addition to processes to control emissions by scrubbing hydrogen sulfide from gas streams, considerable progress has been made in controlling smaller hydrogen sulfide emissions from sewage systems and industrial waste systems by use of hydrogen peroxide as an oxidant.

Uses.   Most of the hydrogen sulfide recovered as a by-product is converted to elemental sulfur by the Claus process or to sulfuric acid where there is a market for the acid near the source of the hydrogen sulfide (see SULFURIC ACID AND SULFUR TRIOXIDE). Hydrogen sulfide is also used to prepare various inorganic sulfides, notably sodium sulfide and sodium hydrosulfide, which are used in the manufacture of dyes (qv), rubber chemicals (qv), pesticides, polymers, plastics additives, leather (qv), and pharmaceuticals (qv). A large amount of sodium hydrosulfide or sodium sulfide is used and largely recycled in kraft pulping; hydrogen sulfide can be used for replenishing the sulfide content (see PULP). An important industrial application is the reaction of hydrogen sulfide with alcohol or olefins (alkenes) to produce thiols (qv) or mercaptans. Hydrogen sulfide is also used for presulfiding petroleum cracking catalysts.

In metallurgy, hydrogen sulfide is used to precipitate copper sulfide from nickel–copper-containing ore leach solutions in Alberta, Canada, or to precipitate nickel and cobalt sulfides from sulfuric acid leaching of laterite ores in Moa Bay, Cuba (120) (see METALLURGY, EXTRACTIVE METALLURGY). Hydrogen sulfide is also used in the production of heavy water for the nuclear industry (84).

Hydrogen Polysulfides

Individual hydrogen polysulfides (sulfanes) have been characterized from H_2S_2 [*13465-07-1*] up to at least H_2S_8 [*12026-49-2*] (129). These are of no commercial utility by themselves, although sodium and calcium polysulfides, which are made by addition of sulfur to the corresponding monosulfides, are used commercially. The atmospheric boiling point of H_2S_2 is 70.7°C and the boiling point of H_2S_3 [*13845-23-3*] is 69°C at 0.3 kPa (2 mm Hg). The higher hydrogen polysulfides have been separated by fractional distillation at 13 mPa (10^{-4} mm Hg). All are unstable. The reaction of Na_2S_n (from $Na_2S + S_8$) with SCl_2 in anhydrous and air-free C_1–C_5 alcohols is reported to yield sulfanes (H_2S_n where $n = 2–16$) and these were analyzed by hplc, using a C_{18}-bonded phase and methanol as the eluent (130). The sulfanes decompose slowly under argon yielding H_2S and S_n homocycles. With alkali they decompose rapidly to H_2S and S_8. The high resolution ftir spectrum of H_2S_2 has been reported (131). Hydrogen polysulfides are extremely unstable even to traces of alkalies, and must be kept in HCl-treated glass or silica vessels because ordinary glass is too alkaline. A mixture of hydrogen polysulfides can be prepared by addition of an alkali polysulfide to acid (132).

The H_2S_x sulfanes are the subject of several reviews (129,133). Except for hydrogen sulfide these have no practical utility. Sodium tetrasulfide

[12034-39-8] is available commercially as a 40 wt % aqueous solution and is used to dehair hides in tanneries, as an ore flotation agent, in the preparation of sulfur dyes (qv), and for metal sulfide finishes (see LEATHER; MINERAL RECOVERY AND PROCESSING).

The practical importance of the higher sulfanes relates to their formation in sour-gas wells from sulfur and hydrogen sulfide under pressure and their subsequent decomposition which causes well plugging (134). The formation of high sulfanes in the recovery of sulfur by the Claus process also may lead to persistance of traces of hydrogen sulfide in the sulfur thus produced (100). Quantitative determination of H_2S_x and H_2S in Claus process sulfur requires the use of a catalyst, eg, PbS, to accelerate the breakdown of H_2S_x (135).

Sulfur Halides and Oxyhalides

Sulfur forms several series of halides with all of the halogens except iodine. The fluorides include the commercially important sulfur hexafluoride [2551-62-4] (see FLUORINE COMPOUNDS, INORGANIC).

Sulfur Monochloride. *Physical Properties.* Sulfur monochloride [10025-67-9], S_2Cl_2, is a yellow-orange liquid with a characteristic pungent odor. It was first discovered as a chlorination product of sulfur in 1810. Table 5 provides a list of the physical properties.

Chemical Properties. The chemistry of the sulfur chlorides has been reviewed (141,142). Sulfur monochloride is stable at ambient temperature but

Table 5. Physical and Thermodynamic Properties of Sulfur Monochloride, S_2Cl_2

Property	Value	Ref.
mol wt	135.03	136–137
mp, °C		
commercial	−76	136,137
pure	−76.1	136,137
bp, °C	137.8	136,137
specific gravity, 20°C	1.68	136,137
refractive index, n_D^{20}	1.670	136,137
flash point, Cleveland open cup, °C	130	136,137
vapor pressure, kPa^a		136,137
0°C	0.479	136,137
21.1°C	1.513	136,137
37.7°C	3.5	136,137
solubility in water	insoluble	136,137
ΔH formation, at 25°C, kJ/mol^b		
liquid	−32	138,139
gas	−58	138,139
molar heat capacity, J/(mol·K)^b, °C	115	140
specific heat at 25°C, J/g^b	0.92	140
heat of hydrolysis, J/mol^b	146.44	140
heat of vaporization, J/mol^b	36,401	140

^{a}To convert kPa to mm Hg, multiply by 7.5.
^{b}To convert J to cal, divide by 4.184.

undergoes exchange with dissolved sulfur at 100°C, indicating reversible dissociation. When distilled at its atmospheric boiling point, it undergoes some decomposition to the dichloride, but decomposition is avoided with distillation at ca 6.7 kPa (50 mm Hg). At above 300°C, substantial dissociation to S_2 and Cl_2 occurs. Sulfur monochloride is noncombustible at ambient temperature, but at elevated temperatures it decomposes to chlorine and sulfur (137). The sulfur then is capable of burning to sulfur dioxide and a small proportion of sulfur trioxide.

Sulfur monochloride is hydrolyzed at a moderate rate by water at room temperature but rapidly at higher temperatures. In the vapor state, the hydrolysis rate is slow and involves disproportionation of the primary hydrolysis products:

$$S_2Cl_2 + 2\,H_2O \longrightarrow 2\,HCl + SO_2 + H_2S$$

In solution, the hydrogen sulfide and sulfur dioxide [7446-09-5] thus formed react to producing polythionic acids and elemental sulfur.

Various reducing agents, eg, hydrogen iodide, can abstract chlorine from sulfur monochloride leaving elemental sulfur:

$$S_2Cl_2 + 2\,HI \longrightarrow I_2 + 2\,HCl + 2\,S^0$$

This reaction is useful analytically.

Some metallic oxides and sulfides are chlorinated by sulfur chlorides:

$$ZnS + S_2Cl_2 \longrightarrow ZnCl_2 + 3\,S^0$$

Sulfur trioxide [7446-11-9] reacts with sulfur monochloride to produce pyrosulfuryl chloride [7791-27-7], $ClSO_2OSO_2Cl$.

Hydrogen sulfide can react with S_2Cl_2 to produce, depending on conditions, a mixture of sulfanes (H_2S_x where $x > 1$) or dichlorosulfanes (S_xCl_2 where $x > 2$). These compounds tend to be unstable at ambient temperatures.

Numerous organic reactions of sulfur monochloride are of practical and commercial importance. Of particular importance is the reaction of sulfur monochloride with olefins to yield various types of addition products (142). With ethylene, the severe vesicant bis(2-chloroethyl) sulfide [505-60-2] (mustard gas) forms with elemental sulfur and polysulfides (see CHEMICALS IN WAR). Propylene reacts similarly:

$$2\,CH_2{=}CH_2 + S_2Cl_2 \longrightarrow (ClCH_2CH_2)_2S + S$$

Higher olefins appear to yield mainly bis(2-chloralkyl) disulfides.

The reaction of S_2Cl_2 with 1,5-hexadiene gives cyclic sulfide containing chlorines I and II which at 20°C give upon isomerization the stable mixture of 35% (I) and 65% (II) (143).

$$\text{I} \qquad\qquad\qquad \text{II}$$

Sulfurized olefins (S_2Cl_2 plus isobutene) are further reacted with S and Na_2S to give products useful as extreme pressure lubricant additives (144,145). The reaction of unsaturated natural oils with sulfur monochloride gives resinous products known as Factice, which are useful as art-gum erasers and rubber additives (146,147). The addition reaction of sulfur monochloride with unsaturated polymers, eg, natural rubber, produces cross-links and thus serves as a means for vulcanizing rubber at moderate temperatures. The photochemical cross-linking of polyethylene has also been reported (148).

Using sulfur trioxide plus chlorine, or sulfur dioxide plus chlorine, sulfur monochloride yields thionyl chloride [7719-09-7], $SOCl_2$. Various nucleophilic reactions can displace the chlorine atoms of sulfur monochloride:

$$S_2Cl_2 + 2\ RONa \longrightarrow ROSSOR + 2\ NaCl$$

$$S_2Cl_2 + 2\ RSH \longrightarrow RS_4R + 2\ HCl$$

$$S_2Cl_2 + 4\ R_2NH \longrightarrow R_2NSSNR_2 + 2\ R_2NH{\cdot}HCl$$

The reaction with morpholine is used to make a commercial vulcanizing agent.

Reaction with anilines yields the Herz compounds, which are intermediates for certain sulfur dyes (qv) as well as vulcanizing agents for rubber (149).

Manufacture. Sulfur monochloride is made commercially by direct chlorination of sulfur, usually in a heel of sulfur chloride from a previous batch. The chlorination appears to proceed stepwise through higher sulfur chlorides (S_xCl_2, where $x > 2$). If conducted too quickly, the chlorination may yield products containing SCl_2 and S_xCl_2 as well as S_2Cl_2. A catalyst, eg, iron, iodine, or a trace of ferric chloride, facilitates the reaction. The manufacture in the absence of Fe and Fe salts at 32–100°C has also been reported (149–151).

The commercial manufacture of carbon tetrachloride by chlorination of carbon disulfide yields sulfur monochloride.

$$CS_2 + 3\ Cl_2 \longrightarrow CCl_4 + S_3Cl_2$$

The sulfur monochloride formed in this reaction can be treated with additional carbon disulfide in the presence of a catalyst to yield carbon tetrachloride and sulfur. Alternatively, the sulfur monochloride and carbon tetrachloride can be

$$CS_2 + 2 S_2Cl_2 \longrightarrow CCl_4 + 6 S^0$$

separated by careful distillation (qv) in an inert gas stream (152). Modern processes utilize continuous introduction of chlorine into liquid sulfur at ca 240°C and 101 kPa (1 atm). The commercial product is assayed for purity by specific gravity determination. A typical commercial specification is 98.0 wt % sulfur monochloride minimum, 51.4 wt % total chlorine.

Shipment and Storage.   Sulfur monochloride is minimally corrosive to carbon steel and iron when dry. If it is necessary to avoid discoloration caused by iron sulfide formation or chloride stress cracking, 310 stainless steel should be used. Sulfur monochloride is shipped in tank cars, tank trucks, and steel drums. When wet, it behaves like hydrochloric acid and attacks steel, cast iron, aluminum, stainless steels, copper and copper alloys, and many nickel-based materials. Alloys of 62 Ni–28 Mo and 54 Ni–15 Cr–16 Mo are useful under these conditions. Under DOT HM-181 sulfur monochloride is classified as a Poison Inhalation Hazard (PIH) Zone B, as well as a Corrosive Material (DOT Hazard Class B). Shipment information is available (140).

Uses.   The reaction of S_2Cl_2 with aromatic compounds can yield disulfides or mixtures of mono-, di-, and polysulfides.

$$S_2Cl_2 + 2 C_6H_6 \longrightarrow (C_6H_5)_2S_x$$

A similar reaction with phenols is employed to make commercial vulcanizing agents and antioxidants (see ANTIOXIDANTS; ANTIOZONANTS).

Sulfur monochloride also reacts with substituted phenols to give condensation products useful for rubber compounding (qv) (150). Elf Atochem NA manufactures products known as Vultacs by the reaction of amylphenols and sulfur monochloride (153).

Sulfur monochloride can act as a chlorinating agent in some reactions. For example, *O,O*-dialkyl phosphorodithioic acids [756-80-9] are converted to *O,O*-dialkyl phosphorochloridothioate:

$$\underset{(CH_3O)_2\overset{\displaystyle\overset{S}{\|}}{P}SH}{} + S_2Cl_2 \longrightarrow \underset{(CH_3O)_2\overset{\displaystyle\overset{S}{\|}}{P}Cl}{} + HCl + 3 S^0$$

Carboxylic acids can be converted to acid chlorides or anhydrides:

$$RCOOH + S_2Cl_2 \longrightarrow R\overset{\displaystyle\overset{O}{\|}}{C}Cl$$

$$2 RCOOH + S_2Cl_2 \longrightarrow (R\overset{\displaystyle\overset{O}{\|}}{C})_2O$$

The principal commercial uses of sulfur monochloride are in the manufacture of lubricant additives and vulcanizing agents for rubber (147,154,155) (see

LUBRICATION AND LUBRICANTS; RUBBER CHEMICALS). The preparation of additives for wear and load-bearing improvement of lubricating oils is generally carried out in two steps and the technology is described in numerous patents (155) (see SULFURIZATION AND SULFCHLORINATION).

Economic Aspects. The price of sulfur monochloride in 1995 was \$0.25/kg bulk.

Health and Safety Factors. Sulfur monochloride is highly toxic and irritating by inhalation, and is corrosive to skin and eyes (156). The OSHA permissible exposure limit is 1 ppm (6 mg/m^3). Pulmonary edema may result from inhalation. Because its vapor cannot be tolerated even at low concentrations, its presence serves as a warning factor. Sulfur monochloride is not highly flammable, having flash points of 118°C (closed-cup) and 130°C (open-cup) and an auto-ignition temperature of 234°C.

Sulfur Dichloride. *Physical Properties.* Sulfur dichloride [*10545-99-1*], SCl_2, is a reddish or yellow fuming liquid which decomposes in moist air with the evolution of hydrogen chloride (157). Pure sulfur dichloride is unstable and is supplied commercially as a 72–82 wt % SCl_2 mixture, with sulfur monochloride comprising the remaining percentage. The melting point is reported in the range −121.5 to −61°C, and the boiling point with decomposition is 59°C. The sp gr is 1.621 at 15°C. Distillation of sulfur dichloride with minimal decomposition and improved storability requires the addition of a stabilizer, eg, PCl_3 (ca 0.1%).

The solubility of sulfur dichloride in water is not meaningful because it reacts rapidly. It is slightly soluble in aliphatic hydrocarbons and very soluble in benzene and carbon tetrachloride. The heat of formation is −22 kJ/mol (−5.3 kcal/mol) for the gas at 25°C (138).

Chemical Properties. Sulfur dichloride in the liquid state at ambient temperature is in equilibrium with sulfur monochloride and dissolved chlorine:

$$2\,SCl_2 \rightleftharpoons S_2Cl_2 + Cl_2$$

The equilibrium constant is 0.013 at 18°C. Sulfur dichloride reacts violently with water, forming hydrogen chloride, sulfur dioxide, hydrogen sulfide, sulfur, and a mixture of thionic acids.

Sulfur dichloride is oxidized by sulfur trioxide or chlorosulfuric acid [*7790-94-5*] (qv) to form thionyl chloride:

$$SCl_2 + SO_3 \longrightarrow SOCl_2 + SO_2$$

$$SCl_2 + ClSO_3H \longrightarrow SOCl_2 + SO_2 + HCl$$

Economic Aspects. The price of sulfur dichloride was \$0.26/kg bulk in 1995. Akzo Nobel is the only U.S. producer with merchant sales. Whereas some companies still produce sulfur dichloride for captive use (158), Occidental Chemical Company ended sulfur dichloride production in late 1993.

Chemical Reactions. Sulfur dichloride reacts with an excess of sulfur trioxide forming pyrosulfuryl chloride:

$$SCl_2 + 3\,SO_3 \longrightarrow S_2O_5Cl_2 + 2\,SO_2$$

Sulfur dichloride undergoes many of the same reactions with organic compounds as described for sulfur monochloride. Addition to olefins affords a route to bis(2-chloroalkyl) sulfides and, in certain cases, heterocyclic sulfides (159,160).

$$RCH{=}CH_2 + SCl_2 \longrightarrow (RCHClCH_2)_2S$$

With a few olefins, the addition can yield sulfenyl chlorides, eg, 2-dichloroethane sulfenyl chloride [2441-27-2]:

$$CHCl{=}CHCl + SCl_2 \longrightarrow Cl_2CHCHClSCl$$

Sulfur dichloride reacts with hexafluoropropene in the presence of fluorosulfonic acid at 30–60°C to give the following (161):

$$CF_3CF{=}CF_2 + SCl_2 + HSO_3F \longrightarrow CF_3CFCF_2OSO_2F + CF_3CFClCF_2OSO_2F \ (8.6\%)$$
$$\underset{\text{SCl (75\%)}}{|}$$

Manufacture.   The manufacture of sulfur dichloride is similar to that of sulfur monochloride, except that the last stage of chlorination proceeds slowly and must be conducted at temperatures below 40°C. The preparation of a high assay sulfur dichloride requires special techniques, eg, continuous chlorination during distillation or distillation with traces of phosphorus trichloride or phosphorus pentasulfide [1314-80-3] (162–164). Crude product containing 80 wt % sulfur dichloride to which is added 0.1 wt % phosphorus trichloride can be distilled to yield a 98–99 wt % pure sulfur dichloride, which can be stored for weeks at room temperature without appreciable change.

Shipment and Storage.   Sulfur dichloride, if kept dry, is noncorrosive at ambient temperatures, thus carbon steel and iron can be used in the construction of tanks, piping, and drums. However, when water or humidity is present, materials resistant to hydrochloric acid must be used, eg, glass-lined pipe, Teflon, titanium, Hastelloy C, or possibly a chemically resistant, glass-reinforced polyester. Threaded pipe joints should be assembled with Teflon tape. Hoses should be constructed with a Teflon inner lining with the outer tube constructed of Neoprene or braided 316 stainless steel protected by an adequate thickness of Teflon. Sulfur dichloride should be stored away from heat and away from direct rays of the sum. Toluene and sulfur dichloride react exothermically when catalyzed by iron or ferric chloride. Safety precautions should be followed when such a mixture is present (165).

Uses. Sulfur dichloride is used as a chlorinating agent in the manufacture of parathion [56-38-2] insecticide intermediates (see INSECT CONTROL TECHNOLOGY):

$$(CH_3O)_2 \overset{\overset{S}{\|}}{P}SH + SCl_2 \longrightarrow (CH_3O)_2 \overset{\overset{S}{\|}}{P}Cl + 2\ S^0 + HCl$$

It is also useful in the rapid vulcanization of rubber, eg, in the preparation of thin rubber goods by coating molds or fabrics with rubber latex. Rubber substitutes, such as white Factice, are made from unsaturated oils and sulfur dichloride. The cross-linking ability of sulfur dichloride is also utilized to modify drying oils for varnishes and inks.

An insecticide intermediate, 4,4'-thiobisphenol [2664-63-3] is made from sulfur dichloride and phenol:

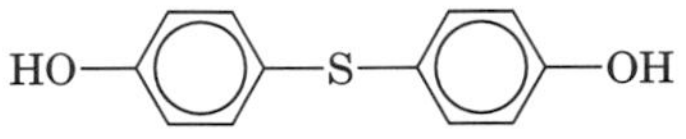

Antioxidants used in lubricants are also made by reaction of sulfur dichloride with phenols:

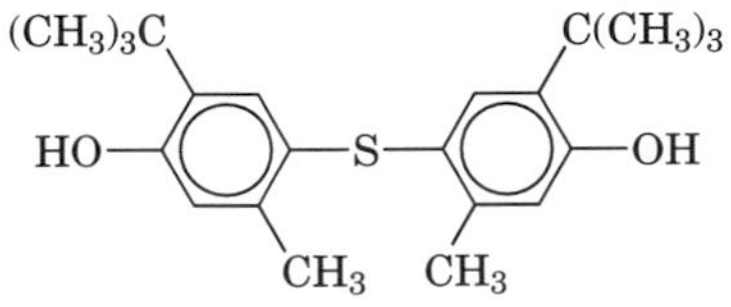

The addition of sulfur dichloride to 1,2-dichloroethylene yields 1,2,2-trichloroethanesulfenyl chloride which, after further chlorination to 1,1,2,2-tetrachloroethanesulfenyl chloride [1185-09-7], is used to produce the fungicide captafol [2425-06-1] (Difolatan). Lubricating oil additives of types similar to those produced using sulfur monochloride are a significant application for sulfur dichloride (166). The chemistry of sulfur dichloride in the synthesis of sulfur-containing heteroatomic rings has been reviewed (167).

Oxyhalides

Thionyl Chloride. *Physical Properties.* Thionyl chloride [7719-09-7], $SOCl_2$, is a colorless fuming liquid with a choking odor. Selected physical and thermodynamic properties are listed in Table 6. Thionyl chloride is miscible with many organic solvents including chlorinated hydrocarbons and aromatic hydrocarbons. It reacts quickly with water to form HCl and SO_2. Thionyl chloride is stable at room temperature; however, slight decomposition occurs just slightly above its boiling point, so prolonged refluxing should be avoided. It decomposes

Table 6. Physical and Thermodynamic Properties of Thionyl Chloride

Property	Value	Ref.
mol wt	118.98	168
mp, °C	−104.5	168
bp, °C	76	168
specific gravity, 25°C	1.63	168
molar heat capacity, J/(mol·K)^a	120.5	168
latent heat of vaporization, kJ/mol^a	31.	168
viscosity, mPa·s(=cP)		
0°C	0.81	168
25°C	0.60	168
refractive index, n_{D}^{20}	1.517	168
electrical conductivity at 25°C, S/cm(=mho/cm)	2×10^{-6}	168
vapor pressure, kPa^b		
−20°C	1.5	169
0°C	4.5	
5°C	5.9	
20°C	11.6	
26°C	14.7	
ΔH formation at 25°C, kJ/mol^a		
liquid	−246	17
gas	−213	17
ΔG formation gas, at 25°C, kJ/mol^a	−198	17
S formation, J/(mol·K)^a	309	17
C_p, at 25°C J/(mol·K)^a		
liquid	121	17
gas	66.5	17

^{a}To convert J to cal, divide by 4.184.
^{b}To convert kPa to mm Hg, multiply by 7.5.

fairly rapidly above 150°C and completely decomposes at 500°C to chlorine, sulfur dioxide, and sulfur monochloride. Thionyl chloride is nonflammable.

Chemical Properties. Thionyl chloride chemistry has been reviewed (169–173). Significant inorganic reactions of thionyl chloride include its reactions with sulfur trioxide to form pyrosulfuryl chloride and with hydrogen bromide to form thionyl bromide [507-16-4]. With many metal oxides it forms the corresponding metal chloride plus sulfur dioxide and therefore affords a convenient means for preparing anhydrous metal chlorides.

The reactions of thionyl chloride with organic compounds having hydroxyl groups are important. Alkyl chlorides, alkyl sulfites, or alkyl chlorosulfites form from its reaction with aliphatic alcohols, depending on reaction conditions, stoichiometry, and the alcohol structure:

$$\mathrm{ROH} + \mathrm{SOCl_2} \longrightarrow \mathrm{RCl} + \mathrm{SO_2} + \mathrm{HCl}$$

$$\mathrm{ROH} + \mathrm{SOCl_2} \longrightarrow \mathrm{RO\overset{\overset{\displaystyle O}{\|}}{S}Cl} + \mathrm{HCl}$$

$$2\,\mathrm{ROH} + \mathrm{SOCl_2} \longrightarrow \mathrm{RO\overset{\overset{\displaystyle O}{\|}}{S}OR} + 2\,\mathrm{HCl}$$

These reactions can be catalyzed by bases, eg, pyridine, or by Lewis acids, eg, zinc chloride. In the case of asymmetric alcohols, steric control, ie, inversion, racemization, or retention of configuration at the reaction site, can be achieved by the choice of reaction conditions (173,174). Some alcohols dehydrate to olefins when treated with thionyl chloride and pyridine.

With phenols, thionyl chloride forms the aryl chlorides only in exceptional cases, eg, with trinitrophenol (picric acid). The reaction of thionyl chloride with primary amines produces thionylamines.

$$RNH_2 + SOCl_2 \longrightarrow RN{=}S{=}O + 2\,HCl$$

Sulfamic acids yield sulfamyl chlorides, which are useful intermediates for herbicide synthesis (175):

$$RNHSO_3H + SOCl_2 \longrightarrow RNHSO_2Cl + HCl + SO_2$$

The reaction of thionyl chloride with carboxylic acids yields either acid chlorides or, in some cases, anhydrides depending on the stoichiometry:

$$RCOOH + SOCl_2 \longrightarrow R\overset{\overset{\displaystyle O}{\|}}{C}Cl + HCl + SO_2$$

$$2\,RCOOH + SOCl_2 \longrightarrow (R\overset{\overset{\displaystyle O}{\|}}{C})_2O + 2\,HCl + SO_2$$

This route to acid chlorides is often preferred over the alternative use of phosphorus trichloride because the by-products, SO_2 and HCl, are gaseous and easily removed. On the other hand, the use of phosphorus trichloride yields phosphorous acid as a by-product. This can decompose exothermically with evolution of toxic and flammable phosphine if overheated, however, phosphorous acid is saleable as a valuable by-product on a commercial scale.

Manufacture. Thionyl chloride may be made by any of the following reactions:

$$SCl_2 + SO_3 \longrightarrow SOCl_2 + SO_2$$

$$SCl_2 + SO_2 + Cl_2 \longrightarrow 2\,SOCl_2$$

$$SCl_2 + SO_2Cl_2 \longrightarrow 2\,SOCl_2$$

The sulfur dichloride can be fed as such or produced directly in the reactor by reaction of chlorine with sulfur monochloride.

In a batch process (176), a glass-lined jacketed iron vessel is charged with either sulfur monochloride or sulfur dichloride and about 1% of antimony trichloride as a catalyst. Chlorine is introduced into the reactor near the bottom. Liquid oleum is added to the reactor at such a rate that the temperature of the reaction mass is held at ca 25°C by the use of cooling water in the jacket.

When the batch is completed, a slight excess of oleum and chlorine is added to reduce to a minimum the residual SCl_2. Because thionyl chloride combines readily with sulfur trioxide to form the relatively stable pyrosulfuryl chloride, it is necessary to maintain the concentration of sulfur trioxide in the reaction mass at a low level; hence, the addition of oleum to sulfur chloride rather than the reverse. When all of the reactants are added, heat is applied to the jacket of the reactor and the batch is refluxed until most of the sulfur dioxide, hydrogen chloride, and chlorine are eliminated. The thionyl chloride is then distilled from the reactor.

In another process sulfur monochloride, sulfur dioxide, and chlorine are allowed to react at 200°C in the presence of an activated carbon catalyst (177):

$$S_2Cl_2 + 2\,SO_2 + 3\,Cl_2 \longrightarrow 4\,SOCl_2$$

Using a 0–10% excess of chlorine and 100% excess of sulfur dioxide, conversions of around 50% are obtained. The liquids in the reaction product are condensed and separated, the sulfur mono- and dichloride are returned for further reaction, and the excess gases are also recycled, producing an ultimate yield near 100% on all reactants.

At present, thionyl chloride is produced commercially by the continuous reaction of sulfur dioxide (or sulfur trioxide) with sulfur monochloride (or sulfur dichloride) mixed with excess chlorine. The reaction is conducted in the gaseous phase at elevated temperature over activated carbon (178). Unreacted sulfur dioxide is mixed with the stoichiometric amount of chlorine and allowed to react at low temperature over activated carbon to form sulfuryl chloride, which is fed back to the main thionyl chloride reactor.

A number of processes have been devised for purifying thionyl chloride. A recommended laboratory method involves distillation from quinoline and boiled linseed oil. Commercial processes involve adding various high boiling olefins such as styrene (qv) to react with the sulfur chlorides to form adducts that remain in the distillation residue when the thionyl chloride is redistilled (179). Alternatively, sulfur can be fed into the top of the distillation column to react with the sulfur dichloride (180). Commercial thionyl chloride has a purity of 98–99.6% minimum, having sulfur dioxide, sulfur chlorides, and sulfuryl chloride as possible impurities. These can be determined by gas chromatography (181).

Storage and Shipping.   Thionyl chloride is classified as a corrosive material; it is shipped and stored in glass-lined steel tanks, zinc-coated (galvanized) steel drums, or special plastic drums having a DOT permit. Stainless steel (304L or 316L) may be used if all traces of moisture are excluded. Thionyl chloride stored in tanks or drums should be kept cool and away from direct sunlight and water.

Economic Aspects.   The price of thionyl chloride in mid-1995 was $1.21/kg. As of late 1996 there was only one U.S. producer, Bayer Chemical (Baytown, Texas). Oxychem discontinued production in 1992.

Health and Safety Factors.   Thionyl chloride is a reactive acid chloride which can cause severe burns to the skin and eyes and acute respiratory tract injury upon vapor inhalation. The hydrolysis products, ie, hydrogen chloride and sulfur dioxide, are believed to be the primary irritants. Depending on the extent of inhalation exposure, symptoms can range from coughing to pulmonary edema (182). The LC_{50} (rat, inhalation) is 500 ppm (1 h), the DOT label is Corrosive, Poison, and the OSHA PEL is 1 ppm (183). The safety aspects of lithium batteries (qv) containing thionyl chloride have been reviewed (184,185).

Uses. A principal use of thionyl chloride is in the conversion of acids to acid chlorides, which are employed in many syntheses of herbicides (qv), surfactants (qv), drugs, vitamins (qv), and dyestuffs. Possible larger-scale applications are in the preparation of engineering thermoplastics of the polyarylate type made from iso- and terephthaloyl chlorides, which can be made from the corresponding acids plus thionyl chloride (186) (see ENGINEERING PLASTICS).

The reactive intermediate, $(C_2H_5)_2NCH_2CH_2Cl \cdot HCl$, which is used to produce cationic starch, is made by the reaction of $(C_2H_5)_2NCH_2CH_2OH$ with thionyl chloride. A synthetic sweetener (qv), sucralose [56038-13-2], is made by the reaction of sucrose or an acetate thereof with thionyl chloride to replace three hydroxy groups by chlorines (187,188).

An intermediate for the herbicide bentazone [25057-89-0] (5) can be prepared by the following reactions (174):

$$(CH_3)_2CHNH_2 + SO_3 \longrightarrow (CH_3)_2CHNHSO_3H + SOCl_2 \longrightarrow (CH_3)_2CHNHSO_2Cl$$

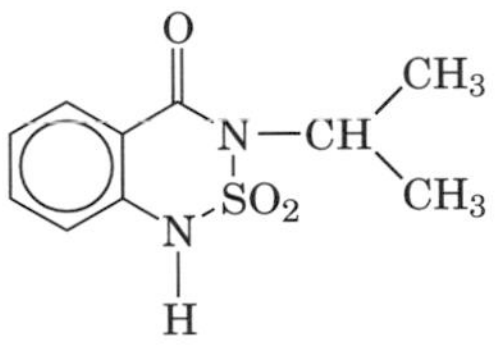

(5)

An example of a sulfite ester made from thionyl chloride is the commercial insecticide endosulfan [115-29-7]. A stepwise reaction of thionyl chloride with two different alcohols yields the commercial miticide, propargite [2312-35-8] (189). Thionyl chloride also has applications as a co-reactant in sulfonations and chlorosulfonations. A patent describes the use of thionyl chloride in the preparation of a key intermediate, bis(4-chlorophenyl) sulfone [80-07-9], which is used to make a commercial polysulfone engineering thermoplastic (see POLYMERS CONTAINING SULFUR, POLYSULFONE) (190). The sulfone group is derived from chlorosulfonic acid; the thionyl chloride may be considered a co-reactant which removes water (see SULFOLANES AND SULFONES).

High energy density batteries with long shelf life, developed originally for military use, are based on lithium and thionyl chloride. These batteries are used in backup or standby power sources for computer, missile, and telephone systems (191,192).

Sulfuryl Chloride. *Physical Properties.* Sulfuryl chloride [7791-25-5], SO_2Cl_2, is a colorless to light yellow liquid with a pungent odor. Physical and thermodynamic properties are listed in Table 7. Sulfuryl chloride dissolves sulfur dioxide, bromine, iodine, and ferric chloride. Various quaternary alkyl-

Table 7. Physical and Thermodynamic Properties of Sulfuryl Chloride

Property	Value	Ref.
mol wt	134.968	17
mp, last crystal point, °C	−54	193
bp, °C	69.1	193
density, at 25°C, g/cm^3	1.6570	194
latent heat of vaporization, kJ/mol^a	27.95	193
surface tension at 23.5°C, mN/m(=dyn/cm)	35.26	194
viscosity, mPa·s(=cP)		
0°C	0.918	193
37.8°C	0.595	193
refractive index, n_{D}^{20}	1.443	193
vapor pressure, kPa^b		
0°C	5.45	193
18°C	12.69	
68.7°C	99.3	
coefficient of expansion, 0–38°C, °C^{-1}	0.0012	193
electrical conductivity, S/cm	3×10^{-8}	193
ΔH formation, at 25°C, kJ/mol^a		
liquid	−394	17
gas	−364	17
ΔG formation gas, at 25°C, kJ/mol^a	−320	17
S gas, at 25°C, J/(mol·K)^a	311	17
C_p, at 25°C, J/(mol·K)^a		
liquid	134	17
gas	77	17

^{a}To convert J to cal, divide by 4.184.
^{b}To convert kPa to mm Hg, multiply by 7.5.

ammonium salts dissolve in sulfuryl chloride to produce highly conductive solutions. Sulfuryl chloride is miscible with acetic acid and ether but not with hexane (193,194).

Chemical Properties. The chemistry of sulfuryl chloride has been reviewed (170,172,195). It is stable at room temperature but readily dissociates to sulfur dioxide and chlorine when heated. The equilibrium constant has the following values (194):

$$K = \frac{[p\mathrm{SO_2}][p\mathrm{Cl_2}]}{[p\mathrm{SO_2Cl_2}]}$$

T, °C	30	40	50	102	159	191
	2.92	5.13	8.48	240	902	1330

where p = pressure in kPa (to convert to atm, divide by 101.3).

The decomposition of sulfuryl chloride is accelerated by light and catalyzed by aluminum chloride and charcoal. Many of the reactions of sulfuryl chloride are explainable on the basis of its dissociation products. Sulfuryl chloride reacts with sulfur at 200°C or at ambient temperature in the presence of aluminum chloride producing sulfur monochloride. It liberates bromine or iodine from bromides or iodides. Sulfuryl chloride does not mix readily with water and hydrolyzes rather slowly.

$$SO_2Cl_2 + 2\,H_2O \longrightarrow H_2SO_4 + 2\,HCl$$

The reaction of sulfuryl chloride with a stoichiometric amount of sulfuric acid produces chlorosulfuric acid [7790-94-5] (chlorosulfonic acid):

$$SO_2Cl_2 + H_2SO_4 \longrightarrow ClSO_3H$$

This latter reaction is reversible. Sulfuryl chloride can be fractionally distilled from boiling chlorosulfonic acid in the presence of a catalyst, eg, a mercuric salt.

Iodine reacts with sulfuryl chloride in the presence of aluminum chloride as catalyst-forming iodine chlorides. Sulfuryl chloride reacts with anhydrous ammonia yielding a series of sulfamides of the general formula $NH_2SO_2(NHSO_2)_nNH_2$, where $n \geq 0$. A cyclic compound of the formula $(SO_2NH)_3$ [13954-94-4] is also produced.

The organic chemistry of sulfuryl chloride involves its use in chlorination and sulfonation (172,175,196,197). As a chlorinating agent, sulfuryl chloride is often more selective than elemental chlorine. The use of sulfuryl chloride as a chlorinating agent often allows more convenient handling and measurement as well as better temperature control because of the lower heat of reaction as compared with chlorine. Sulfuryl chloride sometimes affords better selectivity than chlorine in chlorination of active methylene compounds (198–200):

$$SO_2Cl_2 + CH_3COCH_2COOC_2H_5 \longrightarrow CH_3COCHClCOOC_2H_5 + HCl + SO_2$$

Alkanes can be simultaneously chlorinated and chlorosulfonated. This commercially useful reaction has been applied to polyethylene (201–203). Aromatics can be chlorinated on the ring, and in the presence of a free-radical initiator alkylaromatic compounds can be chlorinated selectively in the side chain. Ring chlorination can be selective. A patent shows chlorination of 2,5-di- to 2,4,5-trichlorophenoxyacetic acid free of the toxic tetrachlorodibenzodioxins (204). With alkenes, depending on conditions, the chlorination can be additive or substitutive. The addition of sulfuryl chloride can occur to form a 2-chloroalkanesulfonyl chloride (205).

$$RCH{=}CH_2 + SO_2Cl_2 \longrightarrow RCHClCH_2SO_2Cl$$

However, when sulfur monochloride is used as the catalyst, chlorosulfites form.

$$RCH{=}CH_2 + SO_2Cl_2 \xrightarrow{S_2Cl_2} RCHClCH_2O\overset{\displaystyle O}{\overset{\|}{S}}Cl$$

With alkylamines, sulfuryl chloride can produce alkylsulfamoyl chlorides, which are useful intermediates in herbicide syntheses (175).

$$RNH_2 \cdot HCl + SO_2Cl_2 \longrightarrow RNHSO_2Cl + 2\ HCl$$

Manufacture. The preparation of sulfuryl chloride is carried out by feeding dry sulfur dioxide and chlorine into a water-cooled glass-lined steel vessel containing a catalyst, eg, activated charcoal. Alternatively, chlorine is passed into liquefied sulfur dioxide at ca 0°C in the presence of a dissolved catalyst, eg, camphor, a terpene hydrocarbon, an ether, or an ester. The sulfuryl chloride is purified by distillation; the commercial product is typically 99 wt % pure, as measured by ASTM distillation method D850.

Shipment and Storage. To harmonize the DOT regulations with the UN regulations, sulfuryl chloride must be shipped in pressure-rated vessels. A double-wall disposable drum is required for drum shipments. With caution, dilute alkali can be used to hydrolyze and dispose of residual sulfuryl chloride. Dry steel tanks can be used to store sulfuryl chloride but some corrosion may occur; lead, nickel, glass, Teflon, 310 stainless steel, or baked phenolics (tested for this service) are preferred. It is important to prevent exposure to atmospheric moisture and vent any hydrogen chloride pressure that builds up. Sulfuryl chloride should also be protected from prolonged exposure to strong light.

Economic Aspects. The truckload price of sulfuryl chloride in mid-1995 was $1/kg. Occidental Chemical Company (Niagara Falls, New York) is the only merchant producer. A large amount is made and used captively by DuPont for manufacture of chlorosulfonated elastomer.

Health and Safety Factors. Sulfuryl chloride is both corrosive to the skin and toxic upon inhalation. The TLV suggested by the manufacturer is 1 ppm. The vapors irritate the eyes and upper respiratory tract, causing prompt symptoms ranging from coughing to extreme bronchial irritation and pulmonary edema. The DOT label is Corrosive, Poison.

Uses. Uses of sulfuryl chloride include the manufacture of chlorophenols, eg, chlorothymol for use as disinfectants. It is also used in the manufacture of alpha-chlorinated acetoacetic derivatives, eg, $CH_3COCHClCOOC_2H_5$, which are precursors for important substituted imidazole drugs, phosphate insecticides, and fungicides (198–199). Other herbicide synthesis uses of sulfuryl chloride in which compounds with SO_2N linkages are produced have been reported (175). Sulfuryl chloride is used captively by DuPont on a large scale in the manufacture of chlorosulfonated polyethylene [9008-08-6] (see ELASTOMERS, SYNTHETIC–CHLOROSULFONATED POLYETHYLENE). Process improvements have been the subject of DuPont and Toyo Soda patents (201–203).

Sulfur Nitrides

The sulfur nitrides have been the subject of several reviews (206–208). Although no commercial applications have as yet been developed for these compounds, some interest was stimulated by the discovery that polythiazyl, a polymeric sulfur nitride, $(SN)_x$, with metallic luster, is electroconductive (see INORGANIC

HIGH POLYMERS) (208,209). Other sulfur nitrides are unstable. Tetrasulfur nitride is explosive and shock-sensitive.

Sulfur Oxides

Numerous oxides of sulfur have been reported and those that have been characterized are SO [13827-32-2], S_2O [20901-21-7], S_nO ($n = 6-10$), SO_2, SO_3, and SO_4 [12772-98-4]. Among these, SO_2 and SO_3 are of principal importance. Sulfur oxide chemistry has been reviewed (210–212). Sulfur trioxide, SO_3, is discussed elsewhere (see SULFURIC ACID AND SULFUR TRIOXIDE).

Sulfur Dioxide. *Physical Properties.* Sulfur dioxide [7446-09-5], SO_2, is a colorless gas with a characteristic pungent, choking odor. Its physical and thermodynamic properties are listed in Table 8. Heat capacity, vapor pressure, heat of vaporization, density, surface tension, viscosity, thermal conductivity, heat of formation, and free energy of formation as functions of temperature are available (213), as is a detailed discussion of the sulfur dioxide–water system (215).

Liquid sulfur dioxide expands by ca 10% when warmed from 20 to 60°C under pressure. Pure liquid sulfur dioxide is a poor conductor of electricity, but high conductivity solutions of some salts in sulfur dioxide can be made (216). Liquid sulfur dioxide is only slightly miscible with water. The gas is soluble to the extent of 36 volumes per volume of water at 20°C, but it is very soluble (several hundred volumes per volume of solvent) in a number of organic solvents, eg, acetone, other ketones, and formic acid. Sulfur dioxide is less soluble in nonpolar solvents (215,217,218). The use of sulfur dioxide as a solvent and reaction medium has been reviewed (216,219).

Inorganic Chemistry. Sulfur dioxide is extremely stable to heat, even up to 2000°C. It is not explosive or flammable in admixture with air. The oxidation of sulfur dioxide by air or pure oxygen is a reaction of great commercial importance and is commonly conducted at 400–700°C in the presence of a catalyst, eg, vanadium oxide. The oxidation of sulfur dioxide to sulfur trioxide for manufacture of sulfuric acid has been discussed (see SULFURIC ACID AND SULFUR TRIOXIDE). For oxidation of sulfur dioxide at low levels, for example, in smelter off-gases, various catalysts as well as chemical oxidants such as Caro's acid, H_2SO_5, are used (220). In certain sulfur dioxide recovery processes, oxidation of sulfur dioxide is carried out on activated carbon (qv), which also serves as an adsorbent for the sulfur trioxide and sulfuric acid produced (221). The oxidation of sulfur dioxide to sulfuric acid and sulfates in the atmosphere is important with regard to air pollution studies (222–224). Radicals, eg, HO, HO_2, and CH_3O_2, appear to be the principal species responsible for the homogeneous oxidation of sulfur dioxide in the atmosphere, which can occur at rates as high as 4%/h (224).

Oxidation of sulfur dioxide in aqueous solution, as in clouds, can be catalyzed synergistically by iron and manganese (225). Ammonia can be used to scrub sulfur dioxide from gas streams in the presence of air. The product is largely ammonium sulfate formed by oxidation in the absence of any catalyst (226). The oxidation of SO_2 catalyzed by nitrogen oxides was important in the early processes for manufacture of sulfuric acid (qv). Sulfur dioxide reacts with

Table 8. Physical and Thermodynamic Properties of Sulfur Dioxide

Property	Value	Ref.
mol wt	64.06	17
mp, °C	−72.7	213
bp, °C	−10.02	210
ΔH fusion, kJ/mol[a]	7.40	210
ΔH vaporization, at −10.0°C, kJ/mol[a]	24.92	210
vapor density, at 0°C, 101.3 kPa(=1 atm), air = 1	2.263	210
liquid density, at −20°C, g/cm³	1.50	213
critical temperature, °C	157.6	213
critical pressure, kPa[b]	7911	213
critical volume, cm³/g	122.0	213
ΔG formation gas, at 25°C, kJ/mol[a]	−300.19	17
ΔH formation gas, at 25°C, kJ/mol[a]	−296.82	17
S formation gas, at 25°C, J/(mol·K)[a]	248.1	17
C_p gas, at 25°C, J/(mol·K)[a]	39.9	17
dielectric constant at −16.5°C	17.27	210
dipole moment at 25°C, C·m[c]	3.87×10^{-30}	
vapor pressure, kPa[b]		
10°C	230	214
20°C	330	
30°C	462	
40°C	630	
solubility in water, at 101.3 kPa, g/100 g H₂O		215
0°C	22.971	
10°C	16.413	
20°C	11.577	
30°C	8.247	
40°C	5.881	

[a]To convert J to cal, divide by 4.184.
[b]To convert kPa to psi, multiply by 0.145.
[c]To convert C·m to D, divide by 3.336×10^{-30}.

chlorine or bromine forming sulfuryl chloride or bromide [507-16-4].

$$SO_2 + Cl_2 \longrightarrow SO_2Cl_2$$

Reduction of sulfur dioxide to sulfur includes an industrially important group of reactions (227). Hydrogen sulfide reduces sulfur dioxide even at ambient temperature in the presence of water, but in the dry state and in the absence of a catalyst, a temperature of ca 300°C is required.

$$SO_2 + 2\,H_2S \longrightarrow 3/x\,S_x + 2\,H_2O$$

This reaction is catalyzed by silica, bauxite, and various metal sulfides. The usual catalyst is activated alumina, which also catalyzes the reduction by methane (228). Molybdenum compounds on alumina are especially effective catalysts for the hydrogen sulfide reaction (229).

The Claus process, which involves the reaction of sulfur dioxide with hydrogen sulfide to produce sulfur in a furnace, is important in the production of sulfur from sour natural gas or by-product sulfur-containing gases (see SULFUR REMOVAL AND RECOVERY).

When the Claus reaction is carried out in aqueous solution, the chemistry is complex and involves polythionic acid intermediates (105,211). A modification of the Claus process (by Shell) uses hydrogen or a mixture of hydrogen and carbon monoxide to reduce sulfur dioxide, carbonyl sulfide, carbon disulfide, and sulfur mixtures that occur in Claus process off-gases to hydrogen sulfide over a cobalt molybdate catalyst at ca 300°C (230). Reformed natural gas reduces sulfur dioxide in a process developed by Asarco (231).

Reduction of sulfur dioxide by methane is the basis of an Allied process for converting by-product sulfur dioxide to sulfur (232). The reaction is carried out in the gas phase over a catalyst. Reduction of sulfur dioxide to sulfur by carbon in the form of coal has been developed as the Resox process (233). The reduction, which is conducted at 550–800°C, appears to be promoted by the simultaneous reaction of the coal with steam. The reduction of sulfur dioxide by carbon monoxide tends to give carbonyl sulfide [463-58-1] rather than sulfur over cobalt molybdate, but special catalysts, eg, lanthanum titanate, have the ability to direct the reaction toward producing sulfur (234).

With hot metals, sulfur dioxide usually forms both metal sulfides as well as metal oxides. In aqueous solution, sulfur dioxide is reduced by certain metals or by borohydrides to dithionites.

Sulfur dioxide dissolves in water, forming a weak acid solution of sulfurous acid, H_2SO_3; however, the pure substance H_2SO_3 has never been isolated. At 101.3 kPa (1 atm) sulfur dioxide pressure, the solubility of sulfur dioxide is 18.5% at 0°C and 5.1% at 40°C. Lower solubilities are observed when other diluent gases, eg, air, are present. At 25°C, the hypothetical H_2SO_3 has a first ionization constant of 1.72×10^{-2} and a second ionization constant of 1.1×10^{-9}. At low temperatures, concentrated sulfurous acid affords a crystalline hydrate, $SO_2 \cdot 6H_2O$ [24402-69-5].

The absorption of sulfur dioxide in alkaline (even weakly alkaline) aqueous solutions affords sulfites, bisulfites, and metabisulfites. The chemistry of the interaction of sulfur dioxide with alkaline substances, either in solution, slurry, or solid form, is also of great technological importance in connection with air pollution control and sulfur recovery (25,227,235–241). Even weak bases such as zinc oxide absorb sulfur dioxide. A slurry of zinc oxide in a smelter can be used to remove sulfur dioxide and the resultant product can be recycled to the roaster (242).

The anaerobic reaction of sulfur dioxide with aqueous ammonia produces a solution of ammonium sulfite [10192-30-0]. This reaction proceeds efficiently, even with a gas stream containing as little as 1 wt % sulfur dioxide. The sulfur dioxide can be regenerated at a high concentration by acidulation or by stream stripping of the ammonium sulfite solution, or the sulfite can be made to precipitate and the ammonia recovered by addition of lime (243). The process can also be modified to produce ammonium sulfate for use as fertilizer (244) (see FERTILIZERS). In a variant of this process, the use of electron-beam radiation catalyzes the oxidation of sulfur dioxide in the presence of ammonia to form ammonium sulfate (245).

Organic Chemistry. The organic chemistry of sulfur dioxide, particularly as it relates to food applications, has been discussed (246). Although no reaction takes place with saturated hydrocarbons at moderate temperatures, the simultaneous passage of sulfur dioxide and oxygen into an alkane in the presence of a free-radical initiator or ultraviolet light affords a sulfonic acid such as hexanesulfonic acid [13595-73-8]. This is the so-called sulfoxidation reaction (247):

$$2\,C_6H_{14} + 2\,SO_2 + O_2 \longrightarrow 2\,C_6H_{13}SO_3H + H_2O$$

Simultaneous treatment of an alkane with sulfur dioxide and chlorine affords a sulfonyl chloride, eg, hexylsulfonyl chloride [14532-24-2], and is referred to as chlorosulfonation or the Reed reaction (247,248).

$$C_6H_{14} + SO_2 + Cl_2 \longrightarrow C_6H_{13}SO_2Cl + HCl$$

Isomer mixtures are generally obtained. Chlorosulfonation is used to produce chlorosulfonated polyethylene, a curable thermoplastic. Preformed sulfuryl chloride may also be used.

The reaction of sulfur dioxide with olefins under free-radical-catalyzed conditions produces copolymers which, in most cases, are of an alternating 1:1 type (249,250):

$$SO_2 + C_4H_9CH{=}CH_2 \longrightarrow \underset{\substack{|\\ C_4H_9}}{\left(\!CHCH_2SO_2\!\right)_x}$$

Although these polymers have inadequate stability and processibility for most plastics applications, the ability to undergo scission back to the gaseous monomers has afforded some utility in fabrication of electron-beam resists for photolithography. Polybutene sulfone (251) and polyhexene sulfone (252) have been developed for this small-volume but high value application.

Sulfur dioxide acts as a dienophile in the Diels-Alder reaction with many dienes (253,254) and this reaction is conducted on a commercial scale with butadiene. The initial adduct, sulfolene [77-79-2] is hydrogenated to a solvent, sulfolane [126-33-0], which is useful for selective extraction of aromatic hydrocarbons from

$$CH_2{=}CH{-}CH{=}CH_2 + SO_2 \longrightarrow \underset{\substack{CH_2\\ \diagdown\!SO_2\!\diagup\\ CH_2}}{\overset{CH{=}CH}{|\quad|}} \xrightarrow{H_2} \underset{\substack{CH_2\\ \diagdown\!SO_2\!\diagup\\ CH_2}}{\overset{CH_2{-}CH_2}{|\quad|}}$$

a refinery stream and as a reaction solvent. Sulfur dioxide reacts with Grignard reagents forming sulfinic acid salts.

$$2\,RMgCl + 2\,SO_2 \longrightarrow (RSO_2)_2Mg + MgCl_2$$

Other reactions of sulfur dioxide forming sulfinic acids or sulfones have been reviewed (254).

Manufacture. *Combustion of Sulfur.* For most chemical process applications requiring sulfur dioxide gas or sulfurous acid, sulfur dioxide is prepared by the burning of sulfur or pyrite [1309-36-0], FeS_2. A variety of sulfur and pyrite burners have been developed for sulfuric acid and for the pulp (qv) and paper (qv) industries, which produce and immediately consume about 90% of the captive sulfur dioxide produced in the United States. Information on the European sulfur-to-sulfuric acid technology (with emphasis on Lurgi) is available (255).

The production of sulfur dioxide gas by combustion of Frasch-process sulfur and sulfur recovered from natural gas or oil-refinery gases is relatively simple and is the preferred method, except in cases in which economic considerations favor the use of pyrite, such as in some European countries. For most applications, with the exception of sulfuric acid manufacture, as high a sulfur dioxide content as possible is desired. For sulfur burners utilizing air, the theoretical maximum is 21 vol % SO_2, and under satisfactory conditions 14–20 vol % can be achieved. The various sulfur burners commercially available are rated by their manufacturers as capable of producing gas in the range of 5–18 wt % sulfur dioxide. In the upper range of sulfur dioxide concentration, theoretical flame temperatures for the combustion of sulfur in air are 1200–1600°C. One kilogram of sulfur requires one kilogram of oxygen for its complete combustion. This is the amount contained in 3.35 m^3 air. The reaction produces 4500 kJ/kg (1942 Btu/lb) of sulfur dioxide from sulfur. The types of burners used are also dependent on scale, with typical usages as follows: high pressure spray (multiple units), 35,000 kg/h; multifluid atomization (multiple units), 35,000 kg/h; spinning cup (single unit), 35,000 kg/h; rotary kiln (single unit), ≤900 kg/h; and pan-type burner (single unit), ≤610 kg/h.

Large sulfuric acid plants are based on spray burners, where the sulfur is pumped at 1030–1240 kPa (150–180 psig) through several nozzles into a refractory-lined combustion chamber. An improved nozzle, resistant to plugging or fouling, has been introduced (256). The combustion chambers are typically horizontal baffle-fitted refractory-lined vessels. The largest plants in fertilizer complexes burn up to 50 t/h of sulfur.

Spinning-cup atomizers are used in some plants to provide finer atomization, allowing smaller burner chambers and easier turndown, but with the burden of added rotating equipment. Rotary kiln burners were once popular to burn lower quality sulfur, but few are still in operation. Spray burners can be operated intermittently and used at higher rates than rotary burners.

Pan and cascade burners are generally more limited in flexibility and are useful only where low sulfur dioxide concentrations are desired. Gases from sulfur burners also contain small amounts of sulfur trioxide, hence the moisture content of the air used can be important in achieving a corrosion-free operation. Continuous operation at temperatures above the condensation point of the product gases is advisable where exposure to steel (qv) surfaces is involved. Pressure atomizing-spray burners, which are particularly suitable when high capacities are needed, are offered by the designers of sulfuric acid plants.

Air-atomizing sulfur burners can use smaller combustion chambers at the expense of requiring more power to compress the air. Air is preheated in a jacket, permitting faster combustion. The combustion volume can be as low as 0.085 m^3 (3 ft^3) to burn 900 kg/d of sulfur (257). These burners can deliver high strength gases of 17–19 vol % SO$_2$ and are used in plants that produce liquid SO$_2$ and in pulp mills. Enriched air–oxygen can be used as the atomizing fluid in sprays that use external mixing at the point of atomization. A particularly efficient and clog-resistant spray nozzle uses a combination of air atomization and effervescent atomization (257).

The spinning-cup sulfur burner is suitable for very large capacities, ie, up to 39 t/h of sulfur in a single spinning cup. It is similar to spinning-cup oil burners, except that it is modified to prevent solidification of sulfur. The spinning cup provides centrifugal motion to the sulfur, which is then atomized by a flow of air in an annular space surrounding the cup. This burner also sprays the sulfur into a brick-lined combustion chamber.

Lurgi, Monsanto, and others have developed substoichiometric sulfur burners. The use of pure or enriched oxygen for burning sulfur is advocated by various oxygen suppliers, who provide burner and chamber designs for this purpose. With the improvements in hollow-fiber separation of air, the economic advantage of enriched oxygen for burning sulfur is increasing. Submerged combustion of molten sulfur with oxygen has been in use since 1989, and fits well into refinery Claus unit operation (258).

Other burners are used for low capacity operations. A cascade or checker burner, in which molten sulfur flows down through brick checkerwork countercurrent to a flow of air, is used in small units with a sulfur trioxide converter to condition gases entering electrostatic precipitators at boiler plants operating on low sulfur coal. A small pan burner, which is fed with solid, low carbon sulfur, is used to produce sulfur dioxide for solution in irrigation water to control the pH and maintain porosity in the soil. The same type of burner is used to disinfect wastewater; in this case sulfur dioxide is used instead of chlorine.

Burning Pyrites.   The burning of pyrite is considerably more difficult to control than the burning of sulfur, although many of the difficulties have been overcome in mechanical pyrite burners. The pyrite is burned on multiple trays which are subject to mechanical raking. The theoretical maximum SO$_2$ content is 16.2 wt %, and levels of 10–14 wt % are generally attained. As much as 13 wt % of the sulfur content of the pyrite can be converted to sulfur trioxide in these burners. In most applications, the separation of dust is necessary when sulfur dioxide is made from pyrite. Several methods can be employed for this, but for many purposes the use of water-spray towers is the most satisfactory. The latter method also removes some of the sulfur trioxide and cools the gas. For most applications, burner gases need no further treatment other than cooling to permit their absorption either by water or by alkaline solutions or slurries.

A number of plants burn pyrite or zinc, lead, copper, and nickel sulfides in fluid-bed roasters. The pyrite is fed onto a grate where it and its calcines are kept as a fluid by an upward flow of air through openings in the grate (see FLUIDIZATION). Steam-generating coils are installed above the grate to remove excess heat and produce steam while maintaining the fluidized bed at the desired temperature. Use of oxygen instead of air in a specially designed fluidized-bed apparatus makes possible the manufacture of high strength sulfur dioxide from sulfide ore (259).

A technique known as flash roasting was developed in the 1920s. Finely ground pyrites or other sulfide ores fall through a chamber in which air is introduced countercurrently from the bottom. The particles burn at 1000–1100°C yielding sulfur dioxide free of sulfur trioxide at a concentration suitable for making sulfuric acid or for sulfite pulping (260).

A basic research study on combustion of sulfur led to the postulation that sulfur trioxide may actually be the primary combustion product and that sulfur dioxide may then be produced by the further reaction of sulfur trioxide with sulfur vapor in the oxygen-deficient region of the flame (261).

Recovery from Flue Gases. Recovery of sulfur dioxide from flue gases has been described (25,93,227). The stack gas from smelting often contains sufficient sulfur dioxide (ca 6 wt %) for economic conversion to sulfuric acid; the lower concentration in power plant stack gases generally requires some method for concentrating the sulfur dioxide.

Production from Oleum. Production of SO_2 from oleum was developed at Stauffer Chemical Company and is used commercially by Rhône-Poulenc to produce liquid sulfur dioxide. It can be integrated with an existing oleum operation or with a concentrated sulfuric acid-consuming operation.

$$2\,SO_3 + S^0 \longrightarrow 3\,SO_2$$

A flow diagram of the plant is shown in Figure 1. Approximately 30 wt % oleum is pumped into a reboiler; 2, and SO_3 is evolved. The resultant 20 wt % oleum is returned to the oleum towers of an acid plant. For heat conservation, the oleum streams are interchanged in a heat exchanger, 1. The SO_3 stream passes through molten sulfur in reactor 3 where the reaction takes place, and molten sulfur is continuously fed to reactor 3. Periodically, sludge is withdrawn from the reactor; this occurs every six to eight months if bright sulfur is used. The SO_2 vapors pass through a sulfuric acid tower for removal of trace SO_3. Cooling, 9; gas cleaning, 10; compressing, 11; and water condensation, 13, complete the process of producing liquid SO_2 for storage. The reactor, 3, is cooled and/or heated for startup by a closed-water circulation loop, 4–6. The sulfur dioxide produced is 99.99% pure. Rhône-Poulenc operates three commercial plants of this type. Development of gas-cleaning techniques and proper economical materials of construction has resulted in over 96% on-stream time.

Spent Sulfuric Acid. Spent sulfuric acid recovered from petrochemical and refinery processes can be fed to a high temperature furnace at 870–1260°C, where it is transformed into sulfur dioxide, water, and other gaseous products. After suitable scrubbing and drying, the gases are passed to a conventional contact sulfuric acid plant (263).

Corrosivity. Almost all common materials of construction are resistant to commercial dry liquid sulfur dioxide, dry sulfur dioxide gas, and hot sulfur dioxide gas containing water at above the dew point (264,265). These include cast iron, carbon steel, copper, brass, and aluminum. Where hot gases or hot solutions are involved, the temperature resistance, particularly for plastics and resins, and the resistance to thermal shock, particularly for ceramic, glass, and stone, should be taken into account. The latter materials are inert to wet gas, sulfurous acid,

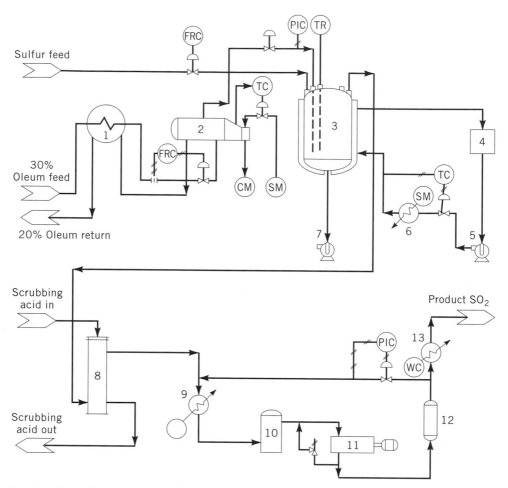

Fig. 1. Flow diagram of production of sulfur dioxide from oleum: 1, 30% oleum exchanger; 2, SO₃ vaporizer; 3, reactor; 4, coolant surge tank; 5, coolant circulating pump; 6, coolant exchangers; 7, sludge and acid pump; 8, scrubber; 9, SO₂ cooler; 10, gas cleaner; 11, SO₂ compressor; 12, pulsation damper; and 13, SO₂ condenser. CM is the condensate; FRC, flow recording controller; PIC, pressure indicating controller; SM, steam; TC, temperature recorder; and WC, cooling water (262).

and sulfite solutions. Carbon, graphite, and impregnated carbon are suitable for practically all types of sulfur dioxide service. Lead is also resistant to sulfur dioxide and sulfites under most conditions. Aluminum is resistant under a variety of conditions and is favored in some food-industry applications involving wet sulfur dioxide. Organic coatings are generally resistant but may fail if gas diffusion through the film is appreciable. Among the organic materials, hard rubber has been satisfactory in sulfurous acid at moderate temperatures, and butyl rubber may perform similarly.

Iron, steel, nickel, copper–nickel alloys, and Inconel Ni–Cr–Fe are satisfactory for dry or hot sulfur dioxide, but are readily corroded below the dew point or by wet sulfur dioxide gas, sulfurous acid, and sulfites. Inconel is especially resistant to very hot sulfur dioxide gas. Metals best suited to a wide variety of wet, dry, and hot sulfur dioxide, sulfurous acid, and sulfite service are nickel–chromium alloys, eg, Worthite and Durimet 20, and several of the austenitic stainless steels. Type 304 may be satisfactory for mild conditions but types 316 and 317 are usually required for more severe applications at high temperatures. When sulfuric acid is also present, the 20-grade stainless steels may be needed. Crucible SC-1 stainless steel is said to be useful for sulfur dioxide scrubbers (266).

Liquid sulfur dioxide discolors iron, copper, and brass at ca 300 ppm moisture and produces light scale at ca 0.1 wt % moisture and serious corrosion at ca 0.2 wt % or higher moisture content. Copper and brass can be used to handle wet sulfur dioxide where some corrosion can be tolerated, or where the moisture level is low. Wooden tanks are widely used for sulfurous acid preparation, handling, and storage. Sulfite pulp digestors are made of steel lined with acid-resistant brick.

Corrosion by atmospheric sulfur dioxide should be considered in the development and evaluation of protective coatings (267,268). Sulfur dioxide and sulfuric acid therefrom are highly damaging to carbonate building stones (269).

Shipment and Storage. Liquid sulfur dioxide is commonly shipped in North America using 55- and 90-t tank cars, 20-ton tank trucks, 1-ton cylinders, and 150-lb cylinders. Cylinders made of specified steel are affixed with the green label for nonflammable gases. The DOT classification is Poison Gas, Inhalation Hazard. Purchasers of tank-car quantities are required to have adequate storage facilities for prompt transfer.

All shipping containers for liquid sulfur dioxide are arranged so that withdrawal or transfer of the contents can be effected either as a gas or as a liquid. In general, the pressure of the sulfur dioxide in its container is used for the transfer. The most convenient method for attaining the required pressure differential in the case of cylinders and drums is heating. Because these containers are equipped with fusible plugs, it is recommended that the temperature never be allowed to exceed 52°C either in storage or during heating for transfer. In order to minimize the danger of a container of sulfur dioxide becoming full of liquid and failing as a result of the development of hydrostatic pressure with rising temperature, the maximum allowable sulfur dioxide capacity by weight of any container has been set by DOT at 1.25 times the water capacity by weight. It is common practice to fill containers to only 1.15 times the water capacity.

Economic Aspects. Merchant sulfur dioxide is produced by eight North American manufacturers; the total was about 410,000 metric tons in 1994 (310,000 in the United States, 90,000 in Canada). The largest producers in the United States are Rhône-Poulenc (from sulfur trioxide reduction by sulfur) and Hoechst Celanese. There is also a larger captive production. Growth of merchant sulfur dioxide is projected at 2–3%/yr. The mid-1995 price was $0.25/kg.

Uses are estimated to be manufacture of hydrosulfites and other chemicals, 40%; pulp and paper, 23%; food and agriculture (mainly corn processing), 14%; water and waste treatment, 9%; metal and ore refining, 6%; oil recovery and refining, 4%; and miscellaneous, including sulfonation of oils and as a reducing agent or antioxidant, 4% (270,271).

Grades and Specifications. The main grade of liquid sulfur dioxide is known as the technical, industrial, or commercial grade. This grade contains a minimum of 99.98 wt % sulfur dioxide and is a water-white liquid free of sulfur trioxide and sulfuric acid. It contains only a trace at most of nonvolatile residue. Its most important specification is the moisture content, which is generally set at 100 ppm maximum. The only other grade sold is the refrigeration grade of liquid sulfur dioxide, a premium grade having the same purity and specifications as the industrial grade, except for the moisture content, which is specified as 50 ppm maximum. At least one manufacturer sells a single grade for which specifications have been established as follows: color, APHA 25 max; nonvolatile residue, 25 ppm max; and moisture, 50 ppm max.

Analytical Methods. The official NIOSH recommended method for determining sulfur dioxide in air consists of drawing a known prefiltered volume of air through a bubbler containing hydrogen peroxide, thus oxidizing the sulfur dioxide to sulfuric acid. Isopropyl alcohol is then added to the contents in the bubbler and the pH of the sample is adjusted with dilute perchloric acid. The resultant solution is then titrated for sulfate with 0.005 M barium perchlorate, and Thorin is used as the indicator.

There is an end-point color change from yellow or yellow-orange to pink. This method is useful over the range 6.6–26.8 mg/m^3 and is useful for compliance with the OSHA standard (272).

A method suitable for analysis of sulfur dioxide in ambient air and sensitive to 0.003–5 ppm involves aspirating a measured air sample through a solution of potassium or sodium tetrachloromercurate, with the resultant formation of a dichlorosulfitomercurate. Ethylenediaminetetraacetic acid (EDTA) disodium salt is added to this solution to complex heavy metals which can interfere by oxidation of the sulfur dioxide. The sample is also treated with 0.6 wt % sulfamic acid to destroy any nitrite anions. Then the sample is treated with formaldehyde and specially purified acid-bleached rosaniline containing phosphoric acid to control pH. This reacts with the dichlorosulfitomercurate to form an intensely colored rosaniline–methanesulfonic acid. The pH of the solution is adjusted to 1.6 ± 0.1 with phosphoric acid, and the absorbance is read spectrophotometrically at 548 nm (273).

The Reich test is used to estimate sulfur dioxide content of a gas by measuring the volume of gas required to decolorize a standard iodine solution (274). Equipment has been developed commercially for continuous monitoring of stack gas by measuring the near-ultraviolet absorption bands of sulfur dioxide (275–277). The determination of sulfur dioxide in food is conducted by distilling the sulfur dioxide from the acidulated sample into a solution of hydrogen peroxide, followed by acidimetric titration of the sulfuric acid thus produced (278). Analytical methods for sulfur dioxide have been reviewed (279).

Interest in trace determination of sulfur dioxide in the atmosphere has led to a number of instrumental methods such as chemiluminescence in a hydrogen-rich flame (280), matrix isolation and Fourier-transform infrared spectroscopy (281), colorimetric determination (282), conductimetric sensing after passage through a special membrane (283), and high pressure chemical ionization mass spectrometry (284). Continuous monitoring methods have been compared using ultraviolet, infrared, and electrochemical methods (285,286). Other continuous monitoring means have been described using chemiluminescence (287) and solid electrolyte sensors, said to be rugged enough for stack gas monitoring (288).

In continuous monitoring as much as 30% of the sulfur dioxide may be lost in condensed moisture (289).

Health, Safety, and Environmental Factors. Sulfur dioxide has only a moderate acute toxicity (183). The lowest published human lethal concentration is 1000 ppm for 10 months. The lowest published human toxic concentration by inhalation is 3 ppm for 5 days or 12 ppm for 1 hour. The lowest published human lethal concentration is 3000 ppm for 5 months. In solution (as sulfurous acid), the lowest published toxic dose is 500 μg/kg causing gastrointestinal disturbances. Considerable data is available by other modes of exposure and to other species; NIOSH standards are a time-weighted average of 2 ppm and a short-term exposure limit of 5 ppm (183).

Sulfur dioxide shows some mutagenic effects in microorganisms and fruit flies. Human lymphocyte DNA damage has been observed. It is an equivocal tumorigenic agent by RTECS criteria (183).

Sulfur dioxide is a strong irritant; concentrations even as low as 2 ppm can have a respiratory irritant, choking, and sneeze/cough inducing effect, but some persons can acclimatize to 20–30 ppm. The symptoms at 50 ppm are sufficiently disagreeable that most persons would not tolerate them for more than a few minutes. Acutely toxic levels can cause suppurative bronchitis and asthma-like or influenza-like symptoms. At very high levels, asphyxia leading to death, or chemical bronchopneumonia may develop which can be fatal after several days. Overexposure to sulfur dioxide has formerly been widespread in the smelting and paper industries. However, animal and human studies suggest that sulfur dioxide has only a low degree of chronic toxicity at subacute levels.

Sulfur dioxide occurs in industrial and urban atmospheres at 1 ppb–1 ppm and in remote areas of the earth at 50–120 ppt (27). Plants and animals have a natural tolerance to low levels of sulfur dioxide. Natural sources include volcanoes and volcanic vents, decaying organic matter, and solar action on seawater (28,290,291). Sulfur dioxide is believed to be the main sulfur species produced by oxidation of dimethyl sulfide that is emitted from the ocean.

Reviews and critical assessments are available regarding the role of sulfur dioxide and sulfuric acid in the ecosystem (292–304). The status of knowledge regarding source–receptor relationships, climatological factors, and atmospheric chemistry as a background for regulatory policy has been reviewed (292). The 1990 Clean Air Act Amendments are intended to cut annual sulfur dioxide emissions 40% from 1980 levels, and thus minimize the potential effects of acid rain, which can occur even at great distances from the source. The program formulated under this legislation allows market-based trading of emission allowances and flexibility in choice of technologies to reduce emissions, so as to ameliorate lake and stream acidification and damage to plants, structures, and humans. Estimated emissions of sulfur dioxide in the United States declined from 25.7×10^6 t in 1980 to 23.3×10^6 t in 1990. Downward trends in atmospheric sulfate were noted. However, acidification continues to effect sensitive forest, soil, and aquatic ecosystems. Human health data appear inconclusive at other than extreme levels. Regulatory action to limit sulfur emissions has already had a significant impact on the capital and operating costs of power generation (qv), the selection of fuel, and the coal mining, gas, and petroleum industries. The benefits of proposed regulation must be weighed against high costs and economic disruptions.

Beneficial and Harmful Effects. At low levels, sulfur dioxide in the atmosphere is not harmful to crops, but damage can occur at excessive levels (305–309). Crops differ greatly in their sensitivity. Forest damage attributed to acid rain is often cited but the observed symptoms seem to have multiple causes and the contribution of sulfur acids is unspecified. The sulfur in precipitation is, up to a point, beneficial to plant growth because sulfur is an essential nutrient. Lessening the sulfur content of the atmosphere requires that supplementary sulfur be provided in fertilizer to some crops; some crops already require supplementary sulfur. Sulfur dioxide itself has been found useful in drip irrigation systems (310,311) and in calcareous soils (308). Small field generators have been developed for this purpose.

Sulfur Dioxide Emissions and Control. A substantial part of the sulfur dioxide in the atmosphere is the result of burning sulfur-containing fuel, notably coal, and smelting sulfide ores. Methods for controlling sulfur dioxide emissions have been reviewed (312–314) (see also AIR POLLUTION CONTROL METHODS; COAL CONVERSION PROCESSES, CLEANING AND DESULFURIZATION; EXHAUST CONTROL, INDUSTRIAL; SULFUR REMOVAL AND RECOVERY).

Uses. The dominant use of sulfur dioxide is as a captive intermediate for production of sulfuric acid. There is also substantial captive production in the pulp and paper industry for sulfite pulping, and it is used as an intermediate for on-site production of bleaches, eg, chlorine dioxide or sodium hydrosulfite (see BLEACHING AGENTS). There is a substantial merchant market for sulfur dioxide in the paper and pulp industry. Sulfur dioxide is used for the production of chlorine dioxide at the paper (qv) mill site by reduction of sodium chlorate in sulfuric acid solution and also for production of sodium dithionite by the reaction of sodium borohydride with sulfur dioxide (315). This last application was growing rapidly in North America as of the late 1990s.

A smaller but important use for sulfur dioxide is for stabilization of pulp (qv) brightness after hydrogen peroxide bleaching of mechanical pulps. Sulfur dioxide neutralizes the alkalinity and destroys any excess hydrogen peroxide, which if left in the pulp would cause it to lose brightness.

In food processing (qv), sulfur dioxide has a wide range of applications (316,317) as a fumigant, preservative, bleach, and steeping agent for grain and dried fruit. Because of the sensitivity of some persons to sulfur dioxide, it has been banned for use on fresh produce by the U.S. FDA. In the manufacture of wine (qv), a small amount of sulfur dioxide is added to destroy bacteria, molds, and wild yeasts without harming yeasts needed for the desired fermentation. Wine casks are also sterilized by sulfur dioxide. The formation of nitrosamines in beer is prevented by sulfur dioxide treatment in the malting process (318). In molasses manufacture, sugar refining, and soy protein processing, sulfur dioxide is used for bleaching as well as for the prevention of microbial growth (319). In making high fructose corn syrups, sodium bisulfite from sulfur dioxide is added to the enzymatic isomerization step to prevent undesired microbial action. In the earlier stages, where corn is steeped in preparation to wet milling, a 0.1–0.2 wt % sulfur dioxide solution is used to swell and soften kernels and prevent growth of microorganisms. Sulfur dioxide is used throughout corn syrup processing to prevent microbial growth and retard nonenzymatic development of color by the Maillard reaction (see FOOD PROCESSING; SYRUPS).

In water treatment, sulfur dioxide is often used to reduce residual chlorine from disinfection and oxidation (320). This technology is used in potable water treatment, in sewage treatment, and especially in industrial wastewater treatment (see WATER, INDUSTRIAL WATER TREATMENT; WATER, MUNICIPAL WATER TREATMENT; WATER, SEWAGE). Although sodium metabisulfite [7681-57-4] and sodium sulfite [7757-83-7] may be used, the use of liquid sulfur dioxide has convenience and cost advantages and lends itself to automation. Therefore, the larger plants tend to use sulfur dioxide rather than sulfite or bisulfite salts. Sulfur dioxide is also employed as a reducing agent for conversion of chromates to less toxic chromic compounds in plating-shop effluents.

In petroleum technology, sulfur dioxide, or sodium sulfite, is used as an oxygen scavenger (321). This use is particularly important in secondary and tertiary oil recovery processes involving flooding of underground oil formations using water or aqueous solutions (see PETROLEUM, ENHANCED OIL RECOVERY). To prevent corrosion in piping and storage systems it is important to lower the oxygen concentration of the water to 50–100 ppb. The rate of oxygen reduction is increased by adding a cobalt compound as catalyst and maintaining a pH of 8.5–10 (322). Sulfur dioxide and sodium sulfites are also used in oil refining and other industrial processes, notably in boiler waters, as oxygen scavengers.

Another sulfur dioxide application in oil refining is as a selective extraction solvent in the Edeleanu process (323), wherein aromatic components are extracted from a kerosene stream by sulfur dioxide, leaving a purified stream of saturated aliphatic hydrocarbons which are relatively insoluble in sulfur dioxide. Sulfur dioxide acts as a cocatalyst or catalyst modifier in certain processes for oxidation of o-xylene or naphthalene to phthalic anhydride (324,325).

In mineral technology, sulfur dioxide and sulfites are used as flotation depressants for sulfide ores. In electrowinning of copper from leach solutions from ores containing iron, sulfur dioxide prereduces ferric to ferrous ions to improve current efficiency and copper cathode quality. Sulfur dioxide also initiates precipitation of metallic selenium from selenous acid, a by-product of copper metallurgy (326).

In kaolin (clay) processing, sulfur dioxide reduces colored impurities, eg, iron compounds. In the bromine industry, sulfur dioxide is used as an antioxidant in spent brine to be reinjected underground. In agriculture, especially in California, sulfur dioxide is used to increase water penetration and the availability of soil nutrients by virtue of its ability to acidulate saline–alkali soils (327). It is also useful for cleaning ferric and manganese oxide deposits from tile drains (328).

In magnesium casting, sulfur dioxide is employed as an inert blanketing gas. Another foundry application is as a rapid curing catalyst for furfuryl resins in cores. Surprisingly, in view of the many efforts to remove sulfur dioxide from flue gases, there are situations where sulfur dioxide is deliberately introduced. In power plants burning low sulfur coal and where particulate stack emissions are a problem, a controlled amount of sulfur dioxide injection improves particulate removal.

Sulfur dioxide is useful as a solvent for sulfur trioxide in sulfonation reactions; for example, in the large-scale production of alkylbenzenesulfonate surfactant (329). A newer use for sulfur dioxide is in cyanide detoxification in connection with cyanide leaching of precious metals from mine dumps.

Sulfur Oxygen Acids and Their Salts

Sulfuric acid, H_2SO_4, the most important commercial sulfur compound (see SULFURIC ACID AND SULFUR TRIOXIDE), and peroxymonosulfuric acid [7722-86-3] (Caro's acid), H_2SO_5, are discussed elsewhere (see PEROXIDES AND PEROXIDE COMPOUNDS, INORGANIC). The lower valent sulfur acids are not stable species at ordinary temperatures. Dithionous acid [15959-26-9], $H_2S_2O_4$, sulfoxylic acid [20196-46-7], H_2SO_2, and thiosulfuric acid [13686-28-7], $H_2S_2O_3$ are unstable species. A discussion of efforts to isolate and characterize the unstable sulfur acids is given (330).

Sodium Sulfite.   *Physical Properties.*   Anhydrous sodium sulfite [7757-83-7], Na_2SO_3, is an odorless, crystalline solid and most commercial grades other than by-product materials are colorless or off-white (331–334). It melts only with decomposition. The specific gravity of the pure solid is 2.633 (15.4°C). Sodium sulfite is quite soluble in water. It has a maximum solubility of 28 g/100 g sol at 33.4°C; at higher and lower temperatures, it is less soluble in water. Below this temperature, the heptahydrate crystallizes; above this temperature, the anhydrous salt crystallizes. Sodium sulfite is soluble in glycerol but insoluble in alcohol, acetone, and most other organic solvents.

Chemical Properties.   Anhydrous sodium sulfite is stable in dry air at ambient temperatures or at 100°C, but in moist air it undergoes rapid oxidation to sodium sulfate [7757-82-6]. On heating to 600°C, sodium sulfite disproportionates to sodium sulfate and sodium sulfide [1313-82-2]. Above 900°C, the decomposition products are sodium oxide and sulfur dioxide. At 600°C, it forms sodium sulfide upon reduction with carbon (332).

Aqueous solutions of sodium sulfite are alkaline and have a pH of ca 9.8 at 1 wt %. The solutions are oxidized readily by air. The redox potential is a function of pH, as would be expected from the following equation:

$$SO_3^{2-} + 2\,OH^- \longrightarrow SO_4^{2-} + H_2O + 2\,e^- \qquad E^0 = +0.90 \text{ V}$$

Although the usual product of oxidation of sulfite by air is sulfate, at high pH substantial amounts of the strong oxidant Na_2SO_5 can form (335).

Sodium sulfite undergoes addition of sulfur to form sodium thiosulfate. At acidic pH, the chemistry of sodium sulfite is that of bisulfite, metabisulfite, and sulfur dioxide.

Manufacture.   In a typical process, a solution of sodium carbonate is allowed to percolate downward through a series of absorption towers through which sulfur dioxide is passed countercurrently. The solution leaving the towers is chiefly sodium bisulfite of typically 27 wt % combined sulfur dioxide content. The solution is then run into a stirred vessel where aqueous sodium carbonate or sodium hydroxide is added to the point where the bisulfite is fully converted to sulfite. The solution may be filtered if necessary to attain the required product grade. A pure grade of anhydrous sodium sulfite can then be crystallized above 40°C because the solubility decreases with increasing temperature.

In a patented process, a stirred suspension of sodium sulfite is continuously treated with aqueous sodium hydroxide and a sulfur dioxide-containing gas at 60–85°C, and 96% pure anhydrous sodium sulfite is removed by filtration (336). In another continuous one-step process, substantially anhydrous sodium carbonate and sulfur dioxide are concurrently introduced into a saturated solution of sodium sulfite at pH 6.5–7.6 and above 35°C with continuous removal of sodium sulfite (337).

Various processes have been disclosed wherein moist solid sodium pyrosulfite [7681-57-4] is stirred in a steam-heated vessel with sodium carbonate. The exothermic reaction at 80–110°C results in the drying of the product. A lower grade of sodium sulfite is produced commercially in the United States as a by-product of the sulfonation–caustic cleavage route to resorcinol (333).

Shipment and Storage. Anhydrous sodium sulfite is supplied in 22.7- and 45.4-kg moistureproof paper bags or 45.4- and 159-kg fiber drums. Most sodium sulfite is shipped by rail in hopper cars. Sodium sulfite should be protected from moisture during storage. When dry it is quite stable, but when wet it is oxidized by air.

Economic Aspects. In the United States, sodium sulfite is produced for merchant sales by General Chemical at ~36,000 t capacity (Claymont, Delaware), Solvay at ~55,000 t capacity (Green River, Wyoming), and Olympic at ~8,000 t capacity (Tacoma, Washington). Indspec (Petrolia, Pennsylvania) makes sodium sulfite at ~50,000 t capacity as a coproduct of resorcinol manufacture by the benzenedisulfonic acid route. There are other large (mostly captive) producers. Pulp uses are 60%; water treatment, 15%; photography, 10%; and miscellaneous, including textile bleaching, food, chemical intermediates, ore flotation, and mineral recovery, 15%. Overall growth estimated in 1993 was 3–4%, stimulated by water treatment, such as boiler water deoxygenation. Further growth was expected as a reductant for hydrogen peroxide in nonchlorine bleaching of pulp (qv). The price for the technical-grade solid sodium sulfite in mid-1995 was $0.66/kg (338).

Grades and Specifications. The commercial grades of sodium sulfite available in the United States are (*1*) photographic grade: white crystalline material, typically 97.0 wt % minimum Na_2SO_3 assay, 0.5 wt % maximum insoluble matter, 0.15 wt % maximum alkalinity as Na_2CO_3, 20 ppm maximum iron, 20 ppm maximum heavy metals (as Pb), 0.01 wt % maximum thiosulfate (as $S_2O_3^{2-}$); (*2*) anhydrous technical grade: white granular material, typically 95.0–97.5 wt % Na_2SO_3 minimum assay; (*3*) anhydrous *Food Chemicals Codex* grade: 95 wt % minimum (typically 98.6 wt %) Na_2SO_3 assay, 3 ppm mixture arsenic, 10 ppm maximum heavy metals (as Pb), 30 ppm maximum selenium; (*4*) anhydrous technical (by-product) grade: pink to red-brown powder, 91 wt % Na_2SO_3 minimum, 5 wt % sodium sulfate, and ca 3 wt % sodium carbonate (used in sulfite pulping and oxygen scavenging from boiler and process water); and (*5*) aqueous solution: 22 wt % minimum Na_2SO_3, pH 8.8, specific gravity (25°C/25°C) 1.27. The aqueous solution is a convenient form for use under conditions where there are no facilities for handling and dissolving solid salt (331–334).

Analytical Methods. A classical and still widely employed analytical method is iodimetric titration. This is suitable for determination of sodium sulfite, for example, in boiler water. Standard potassium iodate–potassium iodide solution is commonly used as the titrant with a starch or starch-substitute indicator. Sodium bisulfite occurring as an impurity in sodium sulfite can be determined by addition of hydrogen peroxide to oxidize the bisulfite to bisulfate, followed by titration with standard sodium hydroxide (279).

Health and Safety Factors. Although sodium sulfite has no detectible odor, its dust and solutions are irritating to the skin, eyes, and mucous membranes. The ingestion of sodium sulfite causes gastric irritation resulting from the liberation of sulfurous acid. Large doses can cause violent colic and diarrhea as well as disturbance of the circulatory and nervous system. Sodium sulfite has an acute oral LD_{50} (mouse) of 820 mg/kg. It is weakly mutagenic in some test microorganisms and mammalian cells. There is inadequate evidence of any carcinogenicity (183).

Uses. Pulp Manufacture. Sodium sulfite is utilized in neutral semi-chemical pulping, acid sulfite pulping, high yield sulfite cooling, and some kraft pulping processes (339). Many pulp mills prepare their own sulfite and recycle as much as possible, but use of merchant sodium sulfite by pulp mills is substantial. Much of the by-product sodium sulfite from resorcinol manufacture goes into pulp applications as well as a substantial fraction of the lower assay manufactured sodium sulfite.

Water Treatment. Sodium sulfite is an agent in the reduction of chlorine or oxygen in water. Dissolved oxygen in boiler water tends to enhance pitting and other types of corrosion. In boilers operated at below 4.82 MPa (700 psi), a residual concentration of 30 ppm of sodium sulfite is generally effective. Catalytic amounts of cobalt are often added to accelerate the reaction of oxygen with sulfite (321,322) (see WATER, INDUSTRIAL WATER TREATMENT).

In waterflooding of oil fields, oxygen dissolved in the water can contribute to pitting of pipes and clogging of injection wells and the oil-bearing rock with iron corrosion products. A residual concentration of ca 10 ppm sodium sulfite is usually effective in preventing these problems. Sodium sulfite is also used to remove oxygen from drilling muds (see PETROLEUM, DRILLING FLUIDS).

In removing excess free chlorine from municipal or industrial water and from wastewater, sodium sulfite competes with bisulfite or sulfur dioxide. Other commercial applications of sodium sulfite in wastewater treatment include the reduction of hexavalent chromium to the less toxic Cr^{3+} salts as well as the precipitation of silver and mercury.

Photography. Sodium sulfite is useful as a reducing agent in certain photographic fixing baths, developers, hardeners, and intensifiers (334). However, the principal use is as a film preservative and discoloration preventative (see PHOTOGRAPHY).

Miscellaneous. In ore flotation, sodium sulfite functions as a selective depressant. In textile processing, sodium sulfite is used as a bleach for wood (qv) and polyamide fibers and as an antichlor after the use of chlorine bleach. Synthetic applications of sodium sulfite include production of sodium thiosulfite by addition of sulfur and the introduction of sulfonate groups into dyestuffs and other organic products. Sodium sulfite is useful as a scavenger for formaldehyde in aminoplast−wood compositions, and as a buffer in chrome tanning of leather.

Sodium Bisulfite. Sodium bisulfite, $NaHSO_3$, exists in solution but is not a stable compound in the solid state. The anhydrous sodium bisulfite of commerce consists of sodium metabisulfite, $Na_2S_2O_5$. Aqueous sodium bisulfite solution, having specific gravity 1.36 and containing the equivalent of 26–27 wt % SO_2, is a commercial product.

Sodium Metabisulfite. *Physical Properties.* Sodium metabisulfite (sodium pyrosulfite, sodium bisulfite (a misnomer)), $Na_2S_2O_5$, is a white granular or powdered salt (specific gravity 1.48) and is storable when kept dry and protected from air. In the presence of traces of water it develops an odor of sulfur dioxide and in moist air it decomposes with loss of part of its SO_2 content and by oxidation to sodium sulfate. Dry sodium metabisulfite is more stable to oxidation than dry sodium sulfite. At low temperatures, sodium metabisulfite forms hydrates with 6 and 7 moles of water. The solubility of sodium metabisulfite in water is 39.5 wt % at 20°C, 41.6 wt % at 40°C, and 44.6 wt % at 60°C (340). Sodium metabisulfite is fairly soluble in glycerol and slightly soluble in alcohol.

Chemical Properties. The chemistry of sodium metabisulfite is essentially that of the sulfite–bisulfite–metabisulfite–sulfurous acid system. The relative proportions of each species depend on the pH. The pH of a sodium bisulfite solution obtained by dissolving 10 wt % sodium metabisulfite in water at 20°C is 4.9; at 30 wt %, the pH is 4.4.

Manufacture. Aqueous sodium hydroxide, sodium bicarbonate, sodium carbonate, or sodium sulfite solution are treated with sulfur dioxide to produce sodium metabisulfite solution. In one operation, the mother liquor from the previous batch is reinforced with additional sodium carbonate, which need not be totally in solution, and then is treated with sulfur dioxide (341,342). In some plants, the reaction is conducted in a series of two or more stainless steel vessels or columns in which the sulfur dioxide is passed countercurrent to the alkali. The solution is cooled and the sodium metabisulfite is removed by centrifuging or filtration. Rapid drying, eg, in a stream-heated shelf dryer or a flash dryer, avoids excessive decomposition or oxidation to which moist sodium metabisulfite is susceptible.

Shipment and Storage. Sodium metabisulfite can be stored under air at ambient temperatures, but under humid conditions the product cakes and the available SO_2 content decreases as a result of oxidation. Therefore, storage should be under cool, dry conditions. The product is shipped in 22.7- and 45.4-kg polyethylene-lined (moistureproof) bags and 45.4- and 181-kg fiber drums. Steel drums are used for export. The solution is shipped in tank cars and trucks. Dry sodium metabisulfite can be handled in iron or steel equipment. Sodium bisulfite solutions can be handled in 316 and 347 stainless steels, lead, rubber, wood, Haveg, glass-reinforced polyester, or cross-linked polyethylene.

Economic Aspects. U.S. production of sodium metabisulfite is estimated to be well in excess of 45,000 t, but statistics are confused by some commingling with sodium sulfite. The principal U.S. producers are Rhône-Poulenc and General Chemical. The price in mid-1995 was $0.63/kg for anhydrous sodium bisulfite.

Grades and Specifications. Sodium metabisulfite is available in photographic, food and NF, and technical grades (340). Typical analyses for the food and NF grade are 99 wt % $Na_2S_2O_5$, 0.6 wt % Na_2SO_4; 0.4 wt % Na_2SO_3, 3 ppm iron, 0.2 ppm arsenic, 0.1 ppm lead, and 2 ppm selenium. Typical analyses for the technical grade are 98 wt % $Na_2S_2O_5$, 1.1 wt % Na_2SO_4, 0.8 wt % Na_2SO_3, and 4 ppm iron. An aqueous solution typically assaying 20–42% $NaHSO_3$ is sold in tank trucks.

Health and Safety Factors. Sodium metabisulfite is nonflammable, but when strongly heated it releases sulfur dioxide. The oral acute toxicity is slight and the LD_{50} (rat, oral) is 2 g/kg. Sodium bisulfite appears to be weakly mutagenic to some bacteria, in rodent embryos, and in a human lymphocyte test. There is inadequate evidence for carcinogenicity (183,343).

The solid product and its aqueous solutions are mildly acidic and irritate the skin, eyes, and mucous membranes. The solid material when moist generates the pungent, irritating odor of sulfur dioxide. Food-grade sodium metabisulfite is permitted in those foods that are not recognized as sources of vitamin B_1, with which sulfur dioxide reacts (316) (see VITAMINS, THIAMINE).

Uses. Sodium metabisulfite is extensively used as a food preservative and bleach in the same applications as sulfur dioxide. Because sodium metabisulfite is most effective at low pH, the active agent is probably sulfur dioxide or sulfurous acid (340). Sodium bisulfite (sodium metabisulfite) is used in photography (qv), as a reductant, and as a preservative for thiosulfate fixing baths. Other reducing agent applications include reduction of chromate in plating effluents to less toxic chromium salts, which can then precipitate upon addition of lime (344,345). Sodium metabisulfite is also used to reduce chlorine in industrial process water and wastewater. In the textile industry, sodium bisulfite containing metabisulfite is used as a bleach, especially for wool (qv); as an antichlor after bleaching of nylon, for reducing vat dyes; and in rendering certain other dyes soluble. It is a less powerful reductant than sodium dithionite [7775-14-6].

In tanneries, sodium bisulfite is used to accelerate the unhairing action of lime. It is also used as a chemical reagent in the synthesis of surfactants (qv). Addition to alpha-olefins under radical catalyzed conditions yields sodium alkylsulfonates (wetting agents). The addition of sodium bisulfite under base-catalyzed conditions to dialkyl maleates yields the sulfosuccinates.

The reversible addition of sodium bisulfite to carbonyl groups is used in the purification of aldehydes. Sodium bisulfite also is employed in polymer and synthetic fiber manufacture in several ways. In free-radical polymerization of vinyl and diene monomers, sodium bisulfite or metabisulfite is frequently used as the reducing component of a so-called redox initiator (see INITIATORS). Sodium bisulfite is also used as a color preventative and is added as such during the coagulation of crepe rubber.

Sodium Dithionite. *Physical Properties.* Sodium dithionite (sodium hydrosulfite, sodium sulfoxylate), $Na_2S_2O_4$, is a colorless solid and is soluble in water to the extent of 22 g/100 g of water at 20°C.

Chemical Properties. Anhydrous sodium dithionite is combustible and can decompose exothermically if subjected to moisture. Sulfur dioxide is given off violently if the dry salt is heated above 190°C. At room temperature, in the absence of oxygen, alkaline (pH 9–12) aqueous solutions of dithionite decompose slowly over a matter of days. Increased temperature dramatically increases the decomposition rate. A representation of the decomposition chemistry is as follows:

$$2\ S_2O_4^{2-} + H_2O \longrightarrow 2\ HSO_3^- + S_2O_3^{2-}$$

The decomposition of dithionite in aqueous solution is accelerated by thiosulfate, polysulfide, and acids. The addition of mineral acid to a dithionite solution produces first a red color which turns yellow on standing; subsequently, sulfur precipitates and evolution of sulfur dioxide takes place (346). Sodium dithionite is stabilized by sodium polyphosphate, sodium carbonate, and sodium salts of organic acids (347).

Sodium dithionite is most stable and effective as a reducing agent in alkaline solutions, although with excess strong alkali the following reaction occurs:

$$3\,Na_2S_2O_4 + 6\,NaOH \longrightarrow 5\,Na_2SO_3 + Na_2S + 3\,H_2O$$

Dithionite is a stronger reducing agent than sulfite. Many metal ions, eg, Cu^+, Ag^+, Pb^{2+}, Sb^{3+}, and Bi^{3+}, are reduced to the metal, whereas TiO^{2+} is reduced to Ti^{3+} (346). Dithionite readily reduces iodine, peroxides, ferric salts, and oxygen. Some of the decolorizing applications of dithionite, eg, in clay bleaching, are based on the reduction of ferric iron.

Addition of sodium dithionite to formaldehyde yields the sodium salt of hydroxymethanesulfinic acid [79-25-4], $HOCH_2SO_2Na$, which retains the useful reducing character of the sodium dithionite although somewhat attenuated in reactivity. The most important organic chemistry of sodium dithionite involves its use in reducing dyes, eg, anthraquinone vat dyes, sulfur dyes, and indigo, to their soluble leuco forms (see DYES, ANTHRAQUINONE). Dithionite can reduce various chromophores that are not reduced by sulfite. Dithionite can be used for the reduction of aldehydes and ketones to alcohols (348). Quantitative studies have been made of the reduction potential of dithionite as a function of pH and the concentration of other salts (349,350).

Manufacture. A review of older manufacturing processes is available (351). Commercial processes for production of sodium dithionite are based on reduction of sulfite or bisulfite. In order of approximate worldwide production, commercial processes are formate reduction, zinc reduction, sodium amalgam reduction (352–354), and electrochemical reduction. The formate process (355–357) was commercialized during the 1970s and has become the predominant manufacturing technology. The zinc process (358) continues to decrease in popularity because of environmental concerns regarding zinc. The amalgam process was commercialized in the late 1960s but has limited growth because it must be tied to a sodium amalgam source. The electrochemical process, commercialized in the late 1980s, is the newest available technology and utilizes only caustic and sulfur dioxide as raw materials (359). Anhydrous or solution product can be manufactured by all processes; however, the formate and zinc processes typically produce dry product, the amalgam and electrochemical processes typically produce solution product.

Sodium dithionite solution can be produced on-site utilizing a mixed sodium borohydride–sodium hydroxide solution to reduce sodium bisulfite. This process has developed, in part, because of the availability of low cost sulfur dioxide or bisulfite at some paper mills. Improved yields, above 90% dithionite based on borohydride, can be obtained by the use of a specific mixing sequence and an optimized pH profile (360,361). Electrochemical technology is also being offered for on-site production of sodium hydrosulfite solution (362).

Shipment and Storage.   Anhydrous sodium dithionite is shipped in sealed, water-tight containers and must be stored in dry and cool locations. Standard containers are metal drums and semibulk containers. Properly stored dry product can have an effective shelf-life of at least 6–12 months. Dissolving dry product should be accomplished by adding the dry product to the solution to avoid potential ignition of the dry product. Several designs for automated mixing have been developed to minimize problems with dry dissolving. The solution product has grown significantly because of health, flammability, storage, and handling issues of the dry product. Dithionite solution can be stored in dedicated, insulated tanks that may be refrigerated for longer-term storage. Commercial solution product is stored with an inert atmosphere or vapor barrier.

Economic Aspects.   U.S. capacity for production of merchant sodium dithionite (solids basis) was estimated at 93,000 metric tons in 1994. There are three North American producers of sodium dithionite. Hoechst Celanese is the largest producer (68,000 tons capacity) with two formate production locations and one zinc process location. Olin (25,000 t capacity) produces solution product only at two locations using both the amalgam and electrochemical processes. In 1994, Vulcan started a small solution plant in Wisconsin using the Olin electrochemical process. In addition, it is estimated that 13,000 t/yr is produced at U.S. pulp mills using the Borol process from sulfur dioxide and sodium borohydride. Growth is estimated at 2–3%/yr. The price in mid-1995 was ~$1.40/kg (363,364).

Grades.   There are three primary commercial sodium dithionite products: 88 min wt % anhydrous product, 70 wt % dry product (often blended with other stabilizers or additives), and 125 g/L stabilized solution.

Analytical Methods.   Various analytical methods involve titration with oxidants, eg, hexacyanoferrate (ferricyanide), which oxidize dithionites to sulfite. Iodimetric titration to sulfate in the presence of formaldehyde enables dithionite to be distinguished from sulfite because aldehyde adducts of sulfite are not oxidized by iodine. Reductive bleaching of dyes can be used to determine dithionite, the extent of reduction being determined photometrically. Methods for determining mixtures of dithionite, sulfite, and thiosulfates have been reviewed (365). Analysis of dithionite particularly for thiosulfate, a frequent and undesirable impurity, can be done easily by liquid chromatography (366).

Health and Safety Factors.   Dry sodium dithionite, when exposed to moist air, heats and can ignite spontaneously. In case of fire, the burning material must be deluged with water, as too little water may be worse than none at all. Carbon dioxide and dry extinguishers are ineffective because dithionite provides its own oxygen for combustion. Large amounts of sulfur dioxide are liberated, further complicating fire-fighting efforts. Self-contained breathing units should be worn by all fire fighters.

Sodium dithionite is considered only moderately toxic. The solution is reported to have an LD_{50} (rat, oral) of about 5 g/kg. As with sulfites, fairly large doses of sodium dithionite can probably be tolerated because oxidation to sulfate occurs. However, irritation of the stomach by the liberated sulfurous acid is expected. As a food additive, sodium dithionite is generally recognized as safe (GRAS) (367).

Uses. Textile applications have historically been primary uses for dithionite, including dye reduction, dye stripping from fabric, bleaching, and equipment cleaning. These applications have not grown as rapidly and consume ca 35% of the North American consumption. Indigo dyeing of cotton denims has been an important application. Conditions have been discussed for efficient use of sodium hydrosulfite in commercial vat dyeing processes (368).

Pulp and paper bleaching applications have grown and in 1995 represented about half of the North American dithionite usage, mainly to brighten mechanical pulps. There has also been significant growth in color stripping of secondary fibers and some use in polish bleaching of chemical pulps. The factors controlling bleaching with hydrosulfite have been described (369–371). Clay bleaching applications represent about 10% of North American consumption.

Miscellaneous uses include reductive bleaching of glue, gelatin, soap, oils, food products, and oxygen scavenging in water (used for high pressure boilers or for synthetic rubber polymerization). Because the toxicity of sodium dithionite is only slight, it is permitted in various food-contact applications. Users of dithionites have largely converted from zinc to sodium dithionite, to avoid water pollution problems.

Zinc Dithionite. Zinc dithionite [7779-86-4], ZnS_2O_4, is a white, water-soluble powder. Although it exhibits somewhat greater stability in aqueous solution compared to sodium dithionite at a given temperature and pH, it is no longer used in the United States because of regulatory constraints on pollution of water by zinc.

Sodium and Zinc Formaldehyde Sulfoxylates. Although free sulfoxylic acid [20196-46-7], H_2SO_2, has not been isolated and its salts are in doubt, organic derivatives, which may be viewed as adducts of sulfoxylic acid, are commercially made. The latter are mainly sodium formaldehyde sulfoxylate [149-44-0], $HOCH_2SO_2Na$ (commercially sold as the dihydrate) and zinc formaldehyde sulfoxylate [24887-06-7] (351). These compounds are water-soluble reducing agents with uses similar to the dithionites but are more stable. They can be used in reducing and bleaching applications at lower pH values and at somewhat higher temperatures than the dithionites. For example, the formaldehyde sulfoxylates are useful at pH 3.4 as compared to an optimum pH of ca 6 for zinc dithionite and ca 9.5 for sodium dithionite. The most stable compound of this series is basic zinc formaldehyde sulfoxylate, which can be used as a reductant at ca 100°C, compared to ca 50°C for sodium dithionite.

In addition to applications in dyeing, sodium formaldehyde sulfoxylate is used as a component of the redox system in emulsion polymerization of styrene–butadiene rubber recipes.

Thiocyanic Acid and Its Salts

Free thiocyanic acid [463-56-9], HSCN, can be isolated from its salts, but is not an article of commerce because of its instability, although dilute solutions can be stored briefly. Commercial derivatives of thiocyanic acid are principally ammonium, sodium, and potassium thiocyanates, as well as several organic thiocyanates. The chemistry and biochemistry of thiocyanic acid and its derivatives have been reviewed extensively (372–374).

Ammonium Thiocyanate. *Physical Properties.* Ammonium thiocyanate [1762-95-4], NH_4SCN, is a hygroscopic crystalline solid which deliquesces at high humidities (375,376). It melts at 149°C with partial isomerization to thiourea. It is soluble in water to the extent of 65 wt % at 25°C and 77 wt % at 60°C. It is also soluble to 35 wt % in methanol and 20 wt % in ethanol at 25°C. It is highly soluble in liquid ammonia and liquid sulfur dioxide, and moderately soluble in acetonitrile.

Chemical Properties. Ammonium thiocyanate rearranges upon heating to an equilibrium mixture with thiourea: 30.3 wt % thiourea at 150°C, 25.3 wt % thiourea at 180°C (373,375). At 190–200°C, dry ammonium thiocyanate decomposes to hydrogen sulfide, ammonia, and carbon disulfide, leaving guanidine thiocyanate [56960-89-5] as a residue. Aqueous solutions of ammonium thiocyanate are weakly acidic; a 5 wt % solution has a pH of 4–6.

Thiocyanates are rather stable to air, oxidation, and dilute nitric acid. Of considerable practical importance are the reactions of thiocyanate with metal cations. Silver, mercury, lead, and cuprous thiocyanates precipitate. Many metals form complexes. The deep red complex of ferric iron with thiocyanate, $[Fe(SCN)_6]^{3-}$, is an effective indicator for either ion. Various metal thiocyanate complexes with transition metals can be extracted into organic solvents.

The organic chemistry of thiocyanates is notably that of nucleophilic displacement of alkyl halides by thiocyanate anion to form alkyl thiocyanates:

$$RCH_2Cl + SCN^- \longrightarrow RCH_2SCN + Cl^-$$

In certain instances, the thiocyanates, eg, propyl thiocyanate [764-49-8] can rearrange to isothiocyanates, eg, propyl isothiocyanate [57-06-7]:

Ammonium thiocyanate reacts with amines yielding thioureas, such as naphthyl thiourea [86-88-4]:

Manufacture. An extensive technology was developed initially in the 1930s for isolation of ammonium thiocyanate from coke-oven gases, but this technology is no longer practiced in the United States (372). However, such thiocyanate recovery processes are used industrially in Europe. Likewise, the direct sulfurization of cyanides to thiocyanates is not practiced commercially in the United States. The principal route used in the United States is the reaction of carbon disulfide with aqueous ammonia, which proceeds by way of ammonium dithiocarbamate [513-74-6]. Upon heating, the ammonium dithiocarbamate decomposes to ammonium thiocyanate and hydrogen sulfide.

$$CS_2 + 2\,NH_3\,(aq) \longrightarrow NH_2\overset{\overset{\displaystyle S}{\|}}{C}SNH_4 \longrightarrow NH_4SCN + H_2S$$

In a typical batch operation, carbon disulfide is added to four molar equivalents of 25–30 wt % aqueous ammonia in a stirred vessel, which is kept closed for the first one to two hours. The reaction is moderately exothermic and requires cooling. After two to three hours, when substantially all of the disulfide has reacted, the reaction mixture is heated to decompose dithiocarbamate and trithiocarbonate and vented to an absorption system to collect ammonia, hydrogen sulfide, and any unreacted carbon disulfide.

This reaction can also be run in a continuous fashion. In the initial reactor, agitation is needed until the carbon disulfide liquid phase reacts fully. The solution can then be vented to a tower where ammonia and hydrogen sulfide are stripped countercurrently by a flow of steam from boiling ammonium thiocyanate solution. Ammonium sulfide solution is made as a by-product. The stripped ammonium thiocyanate solution is normally boiled to a strength of 55–60 wt %, and much of it is sold at this concentration. The balance is concentrated and cooled to produce crystals, which are removed by centrifugation.

Shipment and Storage. The crystalline material is shipped as a nonhazardous material, in polyethylene-lined fiber drums. The solution can be shipped in drums or bulk. Suitable materials of construction for handling ammonium thiocyanate are aluminum, 316 stainless steel, rubber, poly(vinyl chloride), and glass-reinforced epoxy. Steel, 304 stainless steel, and copper alloys should be avoided (375,376).

Economic Aspects. Ammonium thiocyanate capacity in the United States is well in excess of 9000 t/yr on a 100% solids basis, but production in the 1990s is substantially less than this capacity. In the United States only Witco produces ammonium thiocyanate. Production growth is small. The price of ammonium thiocyanate in mid-1995 was $2.02/kg.

Grades and Specifications. A technical crystal of 98% minimum assay with ca 2 wt % maximum water content is commercially available, as well as a 50–55 wt % aqueous solution (375,376). The latter is the predominant product.

Analytical Methods. Thiocyanate is quantitatively precipitated as silver thiocyanate, and thus can be conveniently titrated with silver nitrate. In the presence of a ferric salt, a red-brown color, produced by the ferric thiocyanate compex, indicates the end point.

Health and Safety Factors. The lowest published human oral toxic dose is 430 mg/kg, causing nervous system disturbances and gastrointestinal symptoms. The LD_{50} (rat, oral) is 750 mg/kg (183). Thiocyanates are destroyed readily by soil bacteria and by biological treatment systems in which the organisms become acclimatized to thiocyanate. Pyrolysis products and combustion products can include toxic hydrogen cyanide, hydrogen sulfide, sulfur oxides, and nitrogen oxides.

Uses. Ammonium thiocyanate is a chemical intermediate for the synthesis of several proprietary agricultural chemicals, mainly herbicides. Its use as a nonselective herbicide is obsolete, although it is still employed as an adjuvant for aminotriazole in nonselective herbicides, which are used principally for brush and perennial weed control. It also is used in photography as a stabilizing agent. It makes undeveloped silver halide substantially insensitive to light. An old use that is practiced is the incorporation of ca 0.1 wt % ammonium thiocyanate in ammoniacal nitrogen fertilizer solutions to inhibit the corrosion of steel. It is also used in various rustproofing compositions. Ammonium thiocyanate was used in the separation of hafnium from zirconium; hafnium is extracted as a thiocyanate complex into an organic phase. This use has declined with the nuclear industry in the United States.

There are many smaller specialized uses for ammonium thiocyanate, including stabilization of glue formulations, as an ingredient in antibiotic fermentations, and as an adjuvant in textile dyeing and printing. A rapidly developing newer use is as a tracer in oil fields. The flow pattern of injected water in enhanced oil recovery operations can be followed by taking water samples from producing wells and analyzing them for thiocyanate by colorimetric measurement of the red ferric complex.

Sodium and Potassium Thiocyanates. *Physical and Chemical Properties.* Sodium thiocyanate [540-72-7], NaSCN, is a colorless deliquescent crystalline solid (mp 323°C). It is soluble in water to the extent of 58 wt % NaSCN at 25°C and 69 wt % at 100°C. It is also highly soluble in methanol and ethanol, and moderately soluble in acetone. Potassium thiocyanate [333-20-0], KSCN, is also a colorless crystalline solid (mp 172°C) and is soluble in water to the extent of 217 g/100 g of water at 20°C and in acetone and alcohols. Much of the chemistry of sodium and potassium thiocyanates is that of the thiocyanate anion (372–375).

Manufacture, Shipment, and Analysis. In the United States, sodium and potassium thiocyanates are made by adding caustic soda or potash to ammonium thiocyanate, followed by evaporation of the ammonia and water. The products are sold either as 50–55 wt % aqueous solutions, in the case of sodium thiocyanate, or as the crystalline solids with one grade containing 5 wt % water and a higher assay grade containing a maximum of 2 wt % water. In Europe, the thiocyanates may be made by direct sulfurization of the corresponding cyanide. The acute LD_{50} (rat, oral) of sodium thiocyanate is 764 mg/kg, accompanied by convulsions and respiratory failure; LD_{50} (mouse, oral) is 362 mg/kg. The lowest published toxic dose for potassium thiocyanate is 80–428 mg/kg, with hallucinations, convulsions, or muscular weakness. The acute LD_{50} (rat, oral) for potassium thiocyanate is 854 mg/kg, with convulsions and respiratory failure.

Shipping, analysis, and safety factors are similar to those of ammonium thiocyanate, except that the alkali thiocyanates are more thermally stable. Sodium thiocyanate is best handled in 316 stainless steel. At room temperature where some iron contamination can be tolerated, type 304 can be used. Aluminum corrodes more rapidly than stainless steel, but some alloys can be used below 60°C. Some but not all rubber equipment is satisfactory.

Economic Aspects. Capacity for sodium thiocyanate in the United States is substantially the same as that for ammonium thiocyanate because both products can be made in the same plants, but production is estimated at only slightly over 1000 t. The rate of growth is slight. The price on a 100 wt % basis in 1995 was $2.10/kg. Most sodium thiocyanate is sold as the solution. Potassium thiocyanate is a much lower volume product.

Uses. The largest use for sodium thiocyanate is as the 50–60 wt % aqueous solution, as a component of the spinning solvent for acrylic fibers (see FIBERS, ACRYLIC; ACRYLONITRILE POLYMERS). Other textile applications are as a fiber swelling agent and as a dyeing and printing assist. A newer commercial use for sodium thiocyanate is as an additive to cement in order to impart early strength to concrete (376).

Sodium thiocyanate and other thiocyanate salts are used to prepare organic thiocyanates, eg, methylene dithiocyanate [6317-18-6]; Verichem's N-948, a broad-spectrum industrial biocide used as a slimicide in paper production, also for controlling microbial growth in cooling water, cutting oils and fluids, hides, oil-field injection waters, drilling muds, adhesives, paints, and other polymer emulsions; thiocyanomethylthiobenzothiazole [21564-17-0] (**6**) (Busan 30A, Buckman Laboratories, Inc.), a fungicide used as a seed and bulb treatment as a contact fungicide for several crops; and as a preservative in paint.

(**6**)

Lesser amounts of sodium thiocyanate are used in color toning photographic paper, as a stabilizer in rapid film development, and as a sensitizing agent in color negative-film emulsions. It is also used as a brightener in copper electroplating.

Methanesulfonyl Chloride

Physical Properties. Methanesulfonyl chloride [124-63-0] (MSC), CH_3SO_2Cl, is a clear liquid, and is soluble in a wide variety of organic solvents, eg, methanol and acetone (Table 9).

Chemical Properties. Methanesulfonyl chloride (MSC) is a reactive chemical which allows introduction of the mesyl group, $CH_3SO_2^-$, into a wide range of substrates. MSC undergoes free-radical-initiated addition to olefins to produce chloro-substituted sulfones (377–379) (eq. 1). With strong bases, MSC can undergo dehydrochlorination to a transitory reactive intermediate, $CH_2=SO_2$, which can dimerize or undergo various addition reactions (380–382). MSC undergoes reactions with alcohols, amines, active methylene compounds (in the presence of bases), and aromatic hydrocarbons (in the presence of Friedel-Crafts catalysts) to replace, generally, a hydrogen atom by a methanesulfonyl group (382–401).

Table 9. Physical Properties of MSC

Property	Value
mol wt	114.6
bp, at 97.3 kPa (730 mm Hg), °C	161
refractive index, at 25°C, n_D^{20}	1.45
freezing point, °C	−33
viscosity, at 25°C, mm^2/s (=cSt)	1.33
specific gravity	
20°C	1.475
30°C	1.467
flash point, COC^a, °C	>230
coefficient of cubical expansion, °C^{-1}	0.00082
solubility in water	insoluble (hydrolyzes)
latent heat of vaporization^b, kJ/kg^c	349

^{a}Cleveland open cup.
^{b}Calculated value at the boiling point.
^{c}To convert kJ/kg to Btu/lb, multiply by 0.4303.

The CH_3SO_2-O- group in methanesulfonate esters is a good leaving group, facilitating nucleophilic displacements. The utility of MSC for methanesulfonylation is well known in the agricultural, pharmaceutical, and specialty chemical industries. Methanesulfonyl chloride is a product of Elf Atochem North America (386–388).

$$CH_3SO_2Cl + \quad \underset{/}{\overset{\backslash}{}}C{=}C\overset{/}{\underset{\backslash}{}} \quad \xrightarrow[\text{Cu}^+, \text{R}_3\text{N}]{\text{peroxide or}} \quad CH_3SO_2\overset{|}{C}{-}\overset{|}{C}Cl \tag{1}$$

Manufacture. Methanesulfonyl chloride is made commercially either by the chlorination of methyl mercaptan or by the sulfochlorination of methane. The product is available in 99.5% assay purity by Elf Atochem NA in the United States or by Elf Atochem SA in Europe.

Shipment and Storage. The weight per gallon is 5.6 kg. MSG is shipped and stored in 55-gallon (0.208 m^3) steel drums or 5-gallon (0.02 m^3) steel pails with a polyethylene liner. Anhydrous MSC is also corrosive toward titanium, titanium–palladium, and zirconium as measured in metal strip tests at 50°C. It is classified as a corrosive liquid.

Health and Safety Factors. MSC has a vapor toxicity on mice of LD$_{50}$ 4.7 mg/L. It is a lachrymator and in order to prevent contact with eyes, goggles should be worn. It is also corrosive to skin and therefore chemically resistant gloves and protective clothing should be worn to prevent contact with skin. Containers should only be opened where there is adequate ventilation.

Uses. Most applications of MSC are for intermediates in the pharmaceutical, photographic, fiber, dye, and agricultural industries. There also are miscellaneous uses as a stabilizer, catalyst, curing agent, and chlorination agent.

Some important applications of MSC are shown in equations 2–4 (388,400). A pharmaceutical intermediate, 2-azabicyclo[2.2.1]hept-5-en-3-one is produced by equations 3 and 4.

MSC for asymmetric synthesis

$$
CH_3SO_2Cl + HO-\underset{\underset{OC_2H_5}{|}}{\overset{\overset{CH_3}{|}}{\underset{|}{C}}}-H \xrightarrow[\text{acetone}]{\text{pyridine}} CH_3SO_3-\underset{\underset{OC_2H_5}{|}}{\overset{\overset{CH_3}{|}}{\underset{|}{C}}}-H \xrightarrow[-CH_3SO_3^-]{Nu^-} H-\underset{\underset{OC_2H_5}{|}}{\overset{\overset{CH_3}{|}}{\underset{|}{C}}}-Nu \qquad (2)
$$

$$(S) \qquad\qquad\qquad (S) \qquad\qquad\qquad (R)$$

Methanesulfonyl cyanide as a dienophile

$$
CH_3SO_2Cl \xrightarrow{Na_2SO_3-NaHCO_3} CH_3SO_2Na \xrightarrow{ClCN} CH_3SO_2CN \qquad (3)
$$

$$(4)$$

Methanesulfonic Acid

Physical Properties. Methanesulfonic acid [75-75-2] (MSA), CH_3SO_3H, is a clear, colorless, strong organic acid available in bulk quantities from Elf Atochem North America as a 70% solution and on an anhydrous basis (100%). MSA is soluble in water and in many organic solvents. Its physical properties are described in Table 10.

Chemical Properties. MSA combines high acid strength with low molecular weight. Its pK_a (laser Raman spectroscopy) is −1.9, about twice the acid

Table 10. Physical Properties of MSA[a]

Property	Value	
	70%	99.5%[b]
formula weight		96.1
MSA content, %	69.5–70.5	99.5[b]
appearance	clear	clear
color, APHA	200	150
freezing point, °C	−60	+19
boiling point[c], °C		122
refractive index, at 25°C		1.4308
density, at 25°C, g/cm³		1.47–1.483

[a]Refs. 402 and 403.
[b]Value is minimum.
[c]At 0.133 kPa (1.0 mm Hg).

strength of HCl and half the strength of sulfuric acid. MSA finds use as catalyst for esterification, alkylation, and in the polymerization and curing of coatings (402,404,405). The anhydrous acid is also useful as a solvent.

The metal salts of MSA are highly soluble in water as well as in some organic solvents, making MSA useful in electroplating operations. For example, lead sulfate is insoluble in water, whereas lead methanesulfonate (lead mesylate) is water soluble.

MSA also finds use in preparing biological and agricultural chemicals, textile treatment chemicals, and for plastics and polymers. Various high molecular weight polymers are soluble in MSA, it is biodegradable and can be recycled, it has an advantage in many applications in that it is a nonoxidizing acid which results in no charring and minimal byproducts, and because of this it is a preferred acid for catalyzing esterifications to give low color ester products.

Specifically MSA has been found to be more effective than p-toluenesulfonic acid and sulfuric acid in preparing dioctyl phthalate (405). A U.S. patent also discloses its use to prepare light-colored fatty esters (406). It is also important as a catalyst to prepare acrylates, methacrylates, adipates, phthalates, trimellitates, thioglycolates, and other esters.

The use of MSA as catalyst to prepare 2-alkylphenols and 2,6-dialkylphenol has been described (407). MSA has also been used as an aromatic alkylation agent (408).

MSA and other lower alkanesulfonic acids are useful for plating of lead, nickel, cadmium, silver, and zinc (409). MSA also finds use in plating of tin, copper, lead, and other metals. It is also used in printed circuit board manufacture. In metal finishing the metal coating can be stripped chemically or electrolytically with MSA. MSA also finds use in polymers and as a polymer solvent and as a catalyst for polymerization of monomers such as acrylonitrile. MSA also finds use in ion-exchange resin regeneration because of the high solubility of many metal salts in aqueous solutions.

The popularity of MSA as an electrolyte in electrochemical applications has developed as a result of the following unique physical and chemical properties: (1) exhibits low corrosivity and is easy to handle, (2) nonoxidizing, (3) manufacturing process yields a high purity acid, (4) exceptional electrical conductivity, (5) high solubility of metal salts permits broad applications, (6) MSA-based formulations are simpler, (7) biodegradable, and (8) highly stable to heat and electrical current.

Health and Safety Factors.   MSA is a strong toxic acid and is corrosive to skin. The acute oral toxicity of the sodium salt in mice LD_{50} is 6.2 g/kg. The 1976 edition of the NIOSH Registry of Toxic Effects of Chemical Substances lists certain reaction products of MSA as having suspected mutagenic, teratogenic, and carcinogenic activity (410).

Manufacture.   Methanesulfonic acid is made commercially by oxidation of methyl mercaptan by chlorine in aqueous hydrochloric acid to give methanesulfonyl chloride which is then hydrolyzed to MSA.

Shipping and Storage.   MSA is shipped in tank trucks and in plastic 55-gallon drums or smaller containers with polyethylene inserts. The freight classification is Alkyl Sulfonic Acid, Liquid; 8 Corrosive Material, UN 2586, Chemical N01BN.

BIBLIOGRAPHY

"Sulfur Compounds—Structures" and "Sulfur Compounds, Inorganic" in *ECT* 1st ed., Vol. 13, pp. 374–430, by P. Macaluso, Stauffer Chemical Co.; "Sulfur Compounds" in *ECT* 2nd ed., Vol. 19, pp. 367–424, by P. Macaluso, Stauffer Chemical Co.; in *ECT* 3rd ed., Vol. 22, pp. 107–167, by E. D. Weil, Stauffer Chemical Co.

1. G. Gattow and W. Behrendt, in A. Senning, ed., *Topics in Sulfur Chemistry*, Vol. 2, Georg Thieme Verlag, Stuttgart, Germany, 1977, pp. 197–206.
2. *Gmelins Handbuch der Anorganischen Chemie*, 8th ed., System No. 14, Part D4, Springer-Verlag, Berlin, Heidelberg, 1977, pp. 1–28.
3. E. M. Strauss and R. Steudel, *Z. Naturforsch, B: Chem. Sci.*, **42**(6), 682 (1987).
4. W. Stadlbauer and T. Kappe, *Chem. Ztg.* **101**(3), 137 (1977).
5. A. O. Diallo and J. H. Dixmier, *C. R. Acad. Sci. Ser. C* **263**, 375 (1966).
6. A. P. Ginsberg, J. L. Lundberg, and W. E. Silverthorn, *Inorg. Chem.* **10**, 2079 (1971).
7. W. Studlbauer, F. Kappe, and E. Ziegler, *Z. Naturforsch., B: Anorg. Chem., Org. Chem.* **33B**(1), 89 (1978).
8. M. Herberhold, *Nachr. Chem. Tech. Lab.* **29**, 365 (1981).
9. J. K. Klabunde, M. P. Kramer, A. Senning, and E. K. Moltzen, *J. Am. Chem. Soc.* **106**, 263 (1984).
10. E. K. Moltzen, M. P. Kramer, A. Senning, and K. J. Klabunde, *J. Org. Chem.* **52**, 1156 (1987).
11. M. P. Kramer, *Diss. Abstr. Int. B* **47**(2), 623 (1986); E. K. Moltzen, A. Senning, M. P. Kramer, and K. J. Klabunde, *J. Org. Chem.* **49**, 3854 (1984).
12. J. B. Hyne, *Understanding Sulphur*, paper presented at Sulphur 81, International Conference, Calgary, Alberta, Canada, May 25–28, 1981.
13. A. R. Shirley, Jr., and R. S. Meline, *Adv. Chem. Ser.* **140**, 38 (1975).
14. U.S. Pat. 4,143,214 (Mar. 6, 1979), C. H. Chang and J. M. Longo (to Exxon Research and Engineering Co.).
15. J. P. Wibaud, *Recl. Trav. Chim. Pays-Bas* **41**, 153 (1922).
16. R. Ciusa, *Gazz. Chim. Ital.* **55**, 385 (1925), and references therein.
17. D. D. Wagman, W. H. Evans, V. B. Parker, I. Halow, S. M. Bailey, and R. H. Schumm, *Selected Values of Chemical Thermodynamic Properties*, NBS Technical Note 270-3, National Bureau of Standards, U.S. Dept. of Commerce, Washington, D.C., 1968.
18. Ref. 2, Part D5, pp. 2–140.
19. D. B. Robinson and N. H. Senturk, *J. Chem. Thermodyn.* **11**, 461 (1979).
20. R. D. Miranda, D. B. Robinson, and H. Kalra, *J. Chem. Eng. Data* **21**(1), 62 (1976).
21. M. B. Mick, *Hydrocarbon Process.* **55**(7), 137 (1976).
22. S. Weber and G. McClure, *The Cosden/Malaprop Process for Light Hydrocarbon Desulfurization*, Annual Meeting Paper AM-81-49, National Petroleum Refiners' Association, Washington, D.C., 1981.
23. R. J. Ferm, *Chem. Rev.* **57**, 621 (1957).
24. *Carbon Disulfide, Carbonyl Sulfide: Literature Review and Environmental Assessment*, NTIS Report No. PB-257947, report for Environmental Protection Agency, U.S. Dept. of Commerce by Stanford Research Institute, Menlo Park, Calif., 1976.
25. A. Kohl and F. Riesenfeld, *Gas Purification*, 2nd ed., Gulf Publishing Co., Houston, Tex., 1974, pp. 607–633.
26. K. Nelson and L. Wolfe, *Oil Gas J.*, 183 (July 27, 1981).
27. P. J. Crutzen, L. E. Heidt, J. P. Krasnec, W. H. Pollock, and W. Seller, *Nature (London)* **282**, 253 (1979).
28. T. E. Graedel, G. W. Kammlot, and J. P. Franey, *Science* **212**, 663 (1981).

29. O. P. Strausz, in A. Senning, ed., *Sulfur in Organic and Inorganic Chemistry*, Vol. 2, Marcel Dekker, New York, 1972, pp. 1–12.

30. S. M. Kluczewski, K. A. Brown, J. N. B. Bell, F. J. Sandalls, B. M. R. Jones, and M. J. Minski, *J. Labelled Compounds Radiopharm.* **21**(5), 485 (1984).

31. U.S. Pat. 3,235,333 (Feb. 15, 1966), E. A. Swakon and E. Field (to Standard Oil Co. of Indiana).

32. U.S. Pat. 4,007,254 (Feb. 8, 1977), R. Buathier, A. Combes, F. Pierrot, and H. Guerpillon (to Rhône-Poulenc Industries).

33. A. Attar, *Fuel* **57**, 201 (1978).

34. R. Steudel, *Z. Anorg. Allg. Chem.* **346**, 255 (1966).

35. R. G. Confer and R. S. Brief, *Am. Ind. Hyg. Assoc. J.*, 843 (Nov. 1975).

36. C. P. Changelis and R. A. Neal, *Toxicol. Appl. Pharmacol.* **55**, 198 (1980).

37. N. I. Sax, *Dangerous Properties of Industrial Materials*, 5th ed., Van Nostrand Rheinhold Co., New York, 1979, p. 470; *Dangerous Prop. Ind. Mater. Rep.* **12**(3), 349–354 (1992).

38. Jpn. Kokai Tokkyo Koho JP 57026657 (Feb. 12, 1982) (to Ihara Chemical Industry, Co., Ltd.).

39. S. R. Sandler and W. Karo, *Organic Functional Group Preparations*, 2nd ed., Vol. II, Academic Press, Inc., New York, 1986, pp. 152, 167–170.

40. T. Mizuno, I. Nishiguchi, T. Hirashima, and N. Sonoda, *Heteroat. Chem.* **1**(2), 157 (1990).

41. B. Zwanenburg, L. Thijs, and S. Strating, *Tetrahedron Lett.* **51**, 4461 (1969).

42. H. Tilles, in N. Kharasch and C. Y. Meyers, eds., *The Chemistry of Organic Sulfur Compounds*, Vol. 2, Pergamon Press, Oxford, U.K., 1966, pp. 311–336; H. Maegerlein, G. Meyer, and H. D. Rupp, *Synthesis* (1), 26 (1974); Jpn. Pat. 03060419A2 (Mar. 15, 1991), T. Kagawa, T. Morooka, T. Uotani, and K. Tsuzuki.

43. S. Sharma, *Synthesis*, (11), 803 (Nov. 1978); S. Sharma, *Sulfur Rep.* **5**(1), 1 (1986); A. Jackson, *Spec. Chem.* **14**(3), 210,214 (1994).

44. L. H. Tagle, F. R. Diaz, J. C. Vega, and P. F. Alquinta, *Makromol. Chem.* **186**(5), 915 (1985); L. H. Tagle, F. R. Diaz, and R. Ruenzalida, *J. Macromol. Sci., Pure Appl. Chem.* **A31**(3), 283 (1994).

45. *Registry of Toxic Effects of Chemical Substances*, U.S. Dept. of Health and Human Services, Public Health Service, Center for Disease Control, National Institute for Occupational Safety and Health, Cincinnati, Ohio, Sept. 1980.

46. *Thiophosgene*, MSDS, No. T9136, Aldrich Chemical Co., Milwaukee, Wis., May 11, 1995.

47. *Perchloromethyl Mercaptan*, Product Report, Stauffer Chemical Co., New York, 1966.

48. *Perchloromethyl Mercaptan*, Stauffer Chemical Co., Agricultural Chemical Division, Chemical Intermediates Data Sheet, Westport, Conn., Nov. 1978.

49. R. J. Lewis, Sr., *Hanley's, Condensed Chemical Dictionary*, 12th ed., Van Nostrand-Reinhold Publishers, New York, 1993, p. 1171.

50. R. C. Weber, P. A. Parker, and M. Bowser, *Vapor Pressure Distribution of Selected Organic Chemicals*, USEPA No. 600/2-81-021, U.S. EPA, Washington, D.C., 1981, p. 39.

51. F. A. Drahowzal, in N. Kharasch, ed., *Organic Sulfur Compounds*, Vol. 1. Pergamon Press, Oxford, U.K., 1961, pp. 361–374.

52. J. Silhánek and M. Zbirovský, *Collect. Czech. Chem. Commun.* **42**, 2518 (1977).

53. J. Silhánek and M. Zbirovský, *Int. J. Sulfur Chem.* **8**, 423 (1973).

54. E. Kuhle, *Synthesis*, 561 (Nov. 1970).

55. H. Kloosterziel, *Q. Rep. Sulfur Chem.* **2**, 353 (1967).

56. H. Senning, *Chem. Rev.* **65**, 385 (1965).

57. Jpn. Pat. 61229861A2 (Oct. 14, 1986), K. Tsuzuki and T. Uotani (to Toyo Soda Mfg. Co.).

58. Jpn. Pat. 61233662A2 (Oct. 17, 1986), K. Tsuzuki and T. Uotani (to Toyo Soda Mfg. Co.).
59. U.S. Pat. 3,673,246 (June 27, 1972), G. Meyer and H. Mägerlein (to Glanzstoff A.-G.); H. Mägerlein, G. Meyer, and H. Rupp, *Synthesis*, 478 (Sept. 1971).
60. U.S. Pat. 3,808,270 (Apr. 30, 1974), H. Rupp, G. Meyer, H. Zengel, and H. Mägerlein (to Akzo NV).
61. U.S. Pat. 4,092,357 (May 30, 1978), C. C. Greco and E. N. Walsh (to Stauffer Chemical Co.); U.S. Pat. 4,093,651 (June 6, 1978), M. L. Honig, C. C. Greco, and E. N. Walsh (to Stauffer Chemical Co.); U.S. Pat. 4,100,190 (July 11, 1978), V. C. Martines and R. G. Campbell (to Stauffer Chemical Co.); U.S. Pat. 4,101,446 (July 18, 1978), M. L. Honig (to Stauffer Chemical Co.).
62. L. Kremer and L. D. Spicer, *Anal. Chem.* **45**, 1963 (1973).
63. A. Hellwig and D. Hempel, *Z. Chem.* **7**, 315 (1967).
64. *Documentation of TLVS*, 5th ed., ACGIH, Cincinnati, Ohio, 1986, p. 466.1.
65. E. H. Vernot, J. D. MacEwen, C. C. Haun, and E. R. Kinkaid, *Toxicol. Appl. Pharmacol.* **42**, 417 (1977).
66. E. H. Vernot, J. D. Macewen, C. C. Haun, and E. R. Kirkead, *Toxicol. Appl. Pharmacol.* **42**, 417 (1977).
67. *Hazard Information Review No. IR-216*, prepared by Enviro Control, Inc., for TSCA Interagency Testing Committee, Washington, D.C., Nov. 25, 1980; *Fed. Reg.* **54**, 2920 (Nov. 19, 1989).
68. M. Sittig, *Pesticide Manufacturing and Toxic Materials Control Encyclopedia*, Noyes Data Corp., Park Ridge, N.J., 1980, pp. 399–400.
69. F. Kuhle, E. Klauke, and F. Grewe, *Angew. Chem.* **76**, 807 (1964).
70. U.S. Pat. 3,296,302 (Jan. 3, 1967), E. D. Weil, K. J. Smith, and E. J. Geering (to Hooker Chemical Corp.).
71. U.S. Pat. 3,200,146 (Aug. 10, 1965), E. D. Weil, E. J. Geering, and K. J. Smith (to Hooker Chemical Corp.).
72. Belg. Pat. 613,887 (Feb. 28, 1962), (to California Research Corp.).
73. U.S. Pat. 3,259,653 (July 5, 1966), E. D. Weil, E. J. Geering, and K. J. Smith (to Hooker Chemical Corp.).
74. U. S. Pats. 3,144,482; 3,144,483 (Aug. 11, 1964), R. B. Flay (to California Research Corp.).
75. U.S. Pat. 3,178,447 (Apr. 13, 1965), G. K. Kohn (to California Research Corp.).
76. B. Meyer, *Sulfur, Energy, and Environment*, Elsevier Scientific Publishing Co., Amsterdam, the Netherlands, 1977.
77. J. B. Hyne, *Recent Developments in Sulfur Production from Hydrogen Sulfide Containing Gases*, paper presented at 181st National Meeting, ACS, Atlanta, Ga., Mar. 29–Apr. 3, 1981.
78. M. Schmidt and W. Siebert, in J. C. Bailar, Jr., H. J. Emeleus, R. Nyholm, and A. F. Trotman-Dickenson, eds., *Comprehensive Inorganic Chemistry*, Pergamon Press, Oxford, U.K., 1973, pp. 826–837.
79. *Selective Values of Properties of Chemical Compounds*, Thermodynamics Research Center Project, Texas A&M University, College Station, Tex., Dec. 31, 1970.
80. J. J. Lagowski, "The Chemistry of Nonaqueous Solvents," *Inert Aprotic and Acidic Solvents*, Vol. 3, Academic Press, Inc., New York, 1970; see also Chapt. 4.
81. *Gmelins Handbuch der Anorganischen Chemie*, 8th ed., System No. 9, Part B1, Verlag Chemie, Weinheim/Bergstrasse, Germany, 1953, pp. 1–125.
82. J. C. Bailar, H. J. Emeleus, R. Nyholm, and A. F. Trotman-Dickenson, *Comprehensive Inorganic Chemistry*, Vol. 2, 1973, Pergamon Press, Oxford, U.K., pp. 826–883; D. Ambrose, *NPL Rep. Chem.* **107** (Feb. 1980).
83. *Hydrogen Sulfide, Product Safety Information Sheet*, Stauffer Chemical Co., Industrial Chemical Division, Westport, Conn., 1973.

84. H. J. Neuberg, J. F. Athcriey, and L. G. Walker, *Girdler-Sulfide Process Physical Properties*, Atomic Energy of Canada Report AECL-5702, Chalk River Nuclear Laboratories, Chalk River, Ontario, Canada, 1977.
85. G. J. Besserer and D. B. Robinson, *J. Chem. Eng. Data* **20**, 157 (1975).
86. H. Kalra, T. R. Krishnan, and D. B. Robinson, *J. Chem. Eng. Data* **21**, 222 (1976).
87. D. B. Robinson, G. P. Hamaliuk, T. R. Krishnan, and P. R. Bishnoi, *J. Chem. Eng. Data* **20**, 153 (1975).
88. M. P. Burgess and R. P. Germann, *AIChE J.* **15**, 272 (1969).
89. D. K. Judd, *Hydrocarbon Proc.* **57**, 122 (Apr. 1978).
90. N. L. Yarym-Agaev, V. G. Matvienko, and N. V. Povalyaeva, *J. Appl. Chem. USSR* **53**, 1810 (1980).
91. C. Ouwerkerk, *Hydrocarbon Proc.* **57**, 89 (Apr. 1978).
92. F. C. Vidaurri and L. C. Kahre, *Hydrocarbon Proc.* **56**, 333 (Nov. 1977).
93. B. L. Crynes, *Chemical Reactions as a Means of Separation*, Marcel Dekker, Inc., New York, 1977, pp. 155–197.
94. J. I. Lee, F. D. Otto, and A. E. Mather, *J. Chem. Eng. Data* **20**, 161 (1975).
95. J. D. Lawson and A. W. Garst, *J. Chem. Eng. Data* **21**, 20 (1976).
96. E. E. Isaacs, F. D. Otto, and A. E. Mather, *J. Chem. Eng. Data* **22**, 317 (1977).
97. J. L. Martin, F. D. Otto, and A. E. Mather, *J. Chem. Eng. Data* **23**, 163 (1978).
98. J. J. Smith, D. Swenson, and B. Meyer, *J. Chem. Eng. Data* **15**(1), 144 (1970).
99. T. K. Wiewiorowski and F. J. Touro, in A. V. Tobolsky, ed., *The Chemistry of Sulfides*, Interscience Publishers, a division of John Wiley & Sons, Inc., New York, 1968.
100. M. E. D. Raymont and R. K. Kerr, *Alberta Sulphur Res. Quar. Bull.* **12**(4), 38 (1976).
101. M. E. D. Raymont, *Alberta Sulphur Res. Quar. Bull.* **8**(1), 24 (1971).
102. M. E. D. Raymont, *Alberta Sulphur Res. Quar. Bull.* **10**(4), 1 (1974); M. E. D. Raymont and J. B. Hyne, *Alberta Sulphur Res. Quar. Bull.* **12**(2), 3 (1975); **12**(3), 1 (1975).
103. U.S. Pat. 3,962,409 (June 8, 1976), Y. Kutera, N. Todo, and N. Fukuda (to Agency of Industrial Science and Technology).
104. M. Grätzel, *Chem. Eng. News* **59**(30), 40 (July 27, 1981); N. G. Vilesov, *J. Appl. Chem. USSR* (Engl. trans. ed.) **53**, 1763 (1981).
105. D. Lyons and G. Nickless, in G. Nickless, ed., *Inorganic Sulphur Chemistry*, Elsevier Publishing Co., Amsterdam, the Netherlands, 1968, pp. 509–533.
106. R. A. Ross and M. R. Jeanes, *Ind. Eng. Chem. Prod. Res. Dev.* **13**(2), 102 (1974).
107. H. M. Castrantas, *J. Pet. Technol.*, 914 (May 1981).
108. E. E. Reid, *Organic Chemistry of Bivalent Sulfur*, Vol. 1, Chemical Publishing Co., Inc., New York, 1958, pp. 18–21.
109. A. Schöberl and A. Wagner, *Houben-Weyl, Methoden der Organischen Chemie*, 4th ed., Vol. 5, Georg Thieme Verlag, Stuttgart, Germany, 1962, pp. 20–21; S. R. Sandler and W. Karo, *Org. Funct. Group Prep.* **1**, 586, 592 (1983); Y. Labat, *Phos., Sulfur Sil.* **74**(1–4), 173 (1993); R. V. Srinivas and G. T. Carroll, in G. R. Lappin and J. D. Sauer, eds., *Alpha Olefins Applications Handbook*, Marcel Dekker, Inc., New York, 1989, Chapt. 14, pp. 375–398.
110. U.S. Pat. 4,568,767 (Feb. 4, 1986), E. J. Dzierza and B. Buchholz (to Elf Atochem NA).
111. U.S. Pat. 4,937,385 (June 26, 1990), B. Buchholz, E. Dzierza, and R. B. Hager (to Elf Atochem NA).
112. Can. Pat. 902,113 (June 6, 1972), D. J. Martin and E. D. Weil (to Stauffer Chemical Co.).
113. U.S. Pat. 4,140,604 (1979), D. A. Dimmig (to Elf Atochem NA).
114. U.S. Pat. 5,026,915 (June 25, 1991), J. R. Baltrus, B. Buchholz, and E. J. Dzierza (to Elf Atochem North America).

115. M. G. Voronkov and E. N. Deryagina, *Phosphorus Sulfur* **7**, 123 (1979).
116. R. F. Bacon and E. S. Boe, *Ind. Eng. Chem.* **37**, 469 (1945); U.S. Pat. 5,089,246 (Feb. 18, 1992), G. R. Schatz (to Elf Atochem NA).
117. R. N. Tuttle and R. D. Kane, eds., *H₂S Corrosion in Oil and Gas Production*, National Association of Corrosion Engineers, Houston, Tex., 1981.
118. *Sulfide Stress Cracking Resistant Metallic Material for Oil Field Equipment*, NACE Standard MR-01-75, 1980 rev., Technical Practices Committee, National Association of Corrosion Engineers, Houston, Tex., 1980.
119. A. I. Asphahani, *Corrosion (Houston)* **37**, 327 (1981).
120. M. C. Manderson and C. D. Cooper, *Sulfur Supply and Demand and Its Relationship to New Energy Sources*, in Ref. 59, Apr. 2, 1981.
121. B. Heydorn, M. Anderson, Y. Yoshida, and F. P. Kalt, *Chemical Economics Handbook*, No. 780.0000, SRI International, Menlo Park, Calif., Mar. 1992, p. 780,0002 K.
122. *NIOSH Manual of Analytical Methods*, Method No. P & CAM 296, Publication No. 80-125, Vol. 6, U.S. Dept. of Health and Human Services, Washington, D.C., 1980.
123. *NIOSH Manual of Analytical Methods*, Method No. 54, Vol. 2, U.S. Dept. of Health and Human Services, Washington, D.C., 1977.
124. *Sulphur* **124**, 47 (May–June 1976); **128**, 43 (Jan.–Feb. 1977).
125. *Occupational Exposure to Hydrogen Sulfide*, Criteria Document, Dept. of Health, Education, and Welfare (NIOSH) Publication No. 77-158; National Technical Information Service Publication No. PB274196, National Institute of Occupational Safety and Health, Washington, D.C., May 1977.
126. *Hydrogen Sulfide*, Report of the Subcommittee on Hydrogen Sulfide, Committee of Medical and Biologic Effects of Environmental Pollution, Division of Medical Science, National Research Council, University Park Press, Baltimore, Md., 1979.
127. *Hydrogen Sulfide*, MSDS, Matheson Gas Products, Montgomeryville, Pa., 1994.
128. *Hydrogen Sulfide User's Manual*, Issue 3, Sheritt Gordon Ltd/Thio-Pet Chemical Ltd., Fort Saskatchewan, Alberta, Canada, Jan. 1993.
129. K. W. C. Burton and P. Machmer, in Ref. 84, pp. 336–366.
130. H. J. Moeckel, *Fresenius' Z. Anal. Chem.* **318**, 116 (1984).
131. P. Mittler, K. M. T. Yamada, G. Winnewisser, and M. Birk, *J. Mol. Spectrosc.* **164**, 390 (1994).
132. H. F. Meissner, E. R. Conway, and H. S. Mickley, *Ind. Eng. Chem.* **48**, 1347 (1956).
133. E. Muller and J. B. Hyne, *Can. J. Chem.* **46**, 2341 (1968).
134. J. B. Hyne, *Chem. Can.*, 26 (May 1978).
135. T. H. Ledford, *Anal. Chem.* **53**, 908 (1981).
136. *Sulfur Chlorides*, Bulletin B-11317-75, Stauffer Chemical Co., Industrial Chemical Division, Westport, Conn., 1975.
137. *Gmelins Handbuch der Anorganischen Chemie*, Vol. 9, Part B(3), Verlag Chemie, Weinheim/Bergstrasse, 1963, pp. 1748–1790.
138. G. A. Takacs, *J. Chem. Eng. Data* **23**, 174 (1978).
139. K. C. Mills, *Thermodynamic Data for Inorganic Sulphides, Selenides, and Tellurides*, Butterworths, Ltd., London.
140. *Sulfur Products Handbook on Sulfur Monochloride and Sulfur Chloride*, Bulletin SPE-SUL-HB 10/9, Oxychem Basic Chemicals Group, Occidental Chemical Corp., Dallas, Tex., 1993, p. 3.
141. L. A. Wiles and Z. S. Ariyan, *Chem. Ind. (London)*, 2102 (Dec. 22, 1962).
142. M. Muhlstadt and D. Martinetz, *Z. Chem.* **14**, 297 (1974).
143. G. A. Tolstikov, R. G. Kantyukova, L. V. Spirkhin, L. M. Khalilov, and A. A. Panasenko, *Zh. Org. Khim.* **19**, 48 (1983).
144. U.S. Pat. 4,594,274 (1990), E. F. Zaweski and J. G. Jolly.
145. U.S. Pat. 4,966,720 (Oct. 30, 1990), D. J. DeGonia and P. G. Griffin (to Ethyl Petroleum Additives, Inc.).

146. W. Hofmann, *Vulcanization and Vulcanizing Agents*, Palmerton Publishing Co., New York, 1965, p. 79.

147. S. M. Erhan and R. Kleiman, *J. Am. Oil Chem. Soc.* **67**, 670 (1990).

148. P. V. Zamotaev, O. P. Mityukhin, and A. A. Usenko, *Vysokomol. Soedin., Ser. B,* **34**(6), 18 (1992).

149. Jpn. Kokai Tokkyo Koho JP 59 18740 A2 (Jan. 31, 1984) (to Sumitomo Chem. Co. Ltd.).

150. Sumitomo Chem. Co. Ltd., Jpn. Kokai Tokkyo Koho JP 59 18729 A2 (Jan. 31, 1984); *Chem. Abstr.*, **100**, 193369.

151. JP 04317401 A2 (Nov. 9, 1992), Y. Ida, K. Hamada, H. Kodu, and K. Kato.

152. U.S. Pat. 3,884,985 (May 20, 1975), K. P. Hoffman (to Stauffer Chemical Co.).

153. U.S. Pats. 3,919,171 (Nov. 11, 1975) and 3,992,362 (Nov. 10, 1976), L. M. Martin (to Pennwalt Corp., now Elf Atochem NA).

154. E. R. Braithwaite, ed., *Lubrication and Lubricants*, Elsevier Publishing Co., New York, 1967.

155. U.S. Pat. 2,708,199 (May 10, 1955), H. H. Eby (to Continental Oil Co.); U.S. Pat. 3,346,549 (Oct. 10, 1967), J. F. Ford and E. S. Forbes (to British Petroleum Co., Ltd.); U.S. Pat. 3,471,404 (Oct. 7, 1969), H. Myers (to Mobil Oil Corp.); U.S. Pat. 4,204,969 (May 27, 1980), A. G. Papay and J. P. O'Brien (to Edwin Cooper, Inc.); U.S. Pat. 4,225,488 (Sept. 30, 1980), A. G. Horodysky and P. S. Landis (to Mobil Oil Corp.); U.S. Pat. 4,240,958 (Dec. 23, 1980), M. Braid (to Mobil Oil Corp.) are representative patents.

156. N. I. Sax and R. J. Lewis, Sr., eds., *Danger. Prop. Ind. Mater. Rep.* **5**(6), 90–92 (1985).

157. *Sulfur Dichloride*, Product Safety Information Data Sheet, Stauffer Chemical Co., Industrial Chemical Division, Westport, Conn., 1973.

158. *Chem. Mark. Rep.* **245**(16), 19 (Apr. 18, 1994).

159. E. J. Corey and E. Block, *J. Org. Chem.* **31**, 1663 (1966); E. D. Weil, K. S. Smith, and R. Gruber, *J. Org. Chem.* **31**, 1669 (1966); F. Lautenschlaeger, *J. Org. Chem.* **31**, 1679 (1966).

160. F. Lautenschlaeger, in Ref. 99, pp. 73–81.

161. A. V. Fokin, A. I. Rapkin, and V. I. Matveenko, *Izv. Akad. Nauk. SSSR, Ser. Khim.* (6), 1455 (1985).

162. U.S. Pat. 3,219,413 (Nov. 23, 1965), K. E. Kunkel and D. S. Rosenberg (to Hooker Chemical Corp.).

163. R. J. Rosser and F. R. Whitt, *J. Appl. Chem.* **10**, 229 (1960).

164. U.S. Pats. 3,071,441 and 3,071,442 (Jan. 1, 1963), J. H. Schmadebeck (to Hooker Chemical Corp.).

165. J. C. Caporossi, *Chem. Eng. News* **66**(32), 2 (1988).

166. Eur. Pat. Appl. EP 311450A2 (Apr. 12, 1989), D. J. Martella and J. J. Jaruzelski.

167. Y. Ohshiro and M. Komatsu, *Kagaka Zokan (Kyoto)*, (115), 33 (1988).

168. *Thionyl Chloride*, Product Information Bulletin, Bayer Corp., Inorganic Chemicals Division, Pittsburgh, Pa., 1995.

169. Ref. 137, pp. 1791–1802.

170. L. F. Fieser and M. Fieser, *Reagents for Organic Synthesis*, John Wiley & Sons, Inc., New York, 1967, pp. 1158–1163.

171. M. Davis, H. Skuta, and A. J. Krusback, in N. Kharasch, ed., *Mechanisms of Reactions of Sulfur Compounds*, Vol. 5, Gordon & Breach, Ltd., London, 1970, p. 1.

172. F. J. Dinan and J. F. Bieron, *A Survey of Reactions of Thionyl Chloride, Sulfuryl Chloride and Sulfur Chlorides*, Occidental Chemical Corp., Niagara Falls, N.Y., 1990.

173. S. S. Pizey, *Synthetic Reagents*, Vol. 1, John Wiley & Sons, Inc., New York, 1974, pp. 321–357.

174. T. G. Squires, W. W. Schmidt, and C. S. McCandlish, Jr., *J. Org. Chem.* **40**, 134 (1975).

175. G. Hambrecht, K. Konig, and G. Stubenrauch, *Angew. Chem. Int. Ed. Engl.* **20**, 151 (1981).
176. U.S. Pat. 2,362,057 (Nov. 7, 1944), J. P. Edwards (to Hooker Electrochemical Co.).
177. U.S. Pat. 2,431,823 (Dec. 2, 1947), A. Pechukas (to Pittsburgh Plate Glass Co.).
178. Ger. Pat. 939,571 (Feb. 23, 1956), H. Jonas and P. Lueg (to Farbenfabriken Bayer A-G).
179. U.S. Pat. 3,156,529 (Nov. 10, 1964), D. S. Rosenberg and H. Flaxman (to Hooker Chemical Corp.).
180. U.S. Pat. 3,155,457 (Nov. 3, 1964), K. E. Kunkel (to Hooker Chemical Corp.).
181. W. Czerwinski and W. Gromotowicz, *J. Chromatogr.* (520), 163–168 (1990).
182. S. Konichezkv. A. Schnattner, T. Ezri, P. Bokenboim, and D. Geva, *Chest* **104**(3), 971–973 (1993).
183. *Registry of Toxic Effects of Chemical Substances*, National Institute for Occupational Safety and Health, Washington, D.C., Issue 95-1, Feb. 1995; CD-ROM U.S. database provided by Canadian Centre for Occupational Health and Safety, Ottawa.
184. G. Eichinger, *J. Power Sources* **43**(1–3), 259–266 (1993).
185. A. M. Ducatman, B. S. Ducatman, and J. A. Barnes, *J. Occup. Med.* **30**(4), 309–311 (1988).
186. U.S. Pat. 4,229,565 (Oct. 21, 1980), H. C. Gardner and M. Matzner (to Union Carbide Corp.).
187. Brit. Pat. Appl. 2,222,827 (Mar. 21, 1990), R. A. Khan, G. H. Sankey, P. J. Simpson, and N. M. Vernon (to Tate and Lyle PLC).
188. Eur. Pat. Appl. 354,050 (Feb. 7, 1990), N. J. Homer, G. Jackson, G. H. Sankey, and P. J. Simpson (to Tate and Lyle PLC).
189. U.S. Pat. 3,272,854 (Sept. 13, 1966), R. A. Covey, A. E. Smith, and W. L. Hubbard (to Uniroyal Corp.).
190. U.S. Pat. 4,172,852 (Oct. 30, 1979), W. F. Ark, J. H. Kawakami, and U. A. Steiner (to Union Carbide Corp.).
191. C. R. Schlaikjer, *J. Power Sources* **26**(1–2), 161 (1989).
192. G. E. Blomgren, *J. Power Sources* **26**(1–2), 51 (1989).
193. *Sulfuryl Chloride*, data bulletin, Occidental Chemical Corp., Basic Chemicals Group, Dallas, Tex., 1994.
194. Ref. 137, pp. 1802–1821.
195. I. Tabushi and H. Kitaguchi, in J. S. Pizey, ed., *Synthetic Reagents*, Vol. 4, Ellis Horwood, distributed by John Wiley & Sons, Inc., Chichester, U.K., 1981, pp. 336–396.
196. Ref. 170, pp. 1128–1131.
197. R. Stroh and W. Hahn, in Ref. 109, Vol. 5/3, pp. 873–890.
198. U.S. Pat. 3,399,214 (Aug. 27, 1968), M. Kulka, D. S. Thiara, and W. A. Harrison (to Uniroyal, Inc.).
199. T. F. Kozlova, G. B. Shaklova, V. F. Belugin, V. G. Zhelonkin, and N. V. Sedov, *Sov. Chem. Ind.* **6**, 394 (1971).
200. D. Masilamani and M. M. Rogić, *J. Org. Chem.* **46**, 4484 (1981).
201. U.S. Pat. 4,871,815 (Oct. 3, 1989), T. Nakagawa, M. Narui, and Y. Sakanaka (to Toyo Soda Manufacturing Co., Ltd.).
202. U.S. Pat. 4,663,396 (May 5, 1987), T. Nakagawa, M. Narui, and Y. Sakanaka (to Toyo Soda Manufacturing Co., Ltd.).
203. U.S. Pat. 4,145,491 (Mar. 9, 1979), D. J. Ryan (to E. I. du Pont de Nemours & Co., Inc.).
204. U.S. Pat. 4,345,097 (Aug. 17, 1982), K. J. Howard and A. E. Sidwell (to Vertac Chemical Corp.).
205. U.S. Pat. 4,902,393 (Feb. 20, 1990), D. J. Muller (to Hüls AG).
206. H. W. Roesky, *Angew. Chem. Int. Ed. Engl.* **18**, 91 (1979).

207. M. Becke-Goehring, in *Gmelins Handbuch der Anorganischen Chemie*, 8th ed., Vol. 32, Part 1, Springer-Verlag, Berlin, 1977.

208. M. M. Labes, P. Love, and L. F. Nichols, *Chem. Rev.* **79**, 1 (1979).

209. M. J. Cohen, A. F. Garito, A. J. Heager, A. G. MacDiarmid, C. M. Mikulski, M. S. Saran, and J. Kleppinger, *J. Am. Chem. Soc.* **98**, 3844 (1976).

210. P. W. Schenk and R. Steudel, in Ref. 105, pp. 369–418.

211. R. Steudel, *Phosphorus and Sulfur* **23**(1–3), 33–64 (1985).

212. P. W. Schenk and R. Steudel, *Angew. Chem. Int. Ed. Engl.* **4**, 402 (1965).

213. C. L. Yaws, K. Y. Li, and C. H. Kuo, *Chem. Eng.*, 85 (July 8, 1974).

214. N. B. Vargaftik, *Tables on the Thermophysical Properties of Liquids and Gases*, Halsted Press, a division of John Wiley & Sons, Inc., New York, 1975, p. 503.

215. Ref. 137, pp. 1131–1293.

216. D. F. Burow, in J. J. Lagowski, ed., *Chemistry of Nonaqueous Solvents*, Vol. 3, Academic Press, New York, 1970, pp. 137–185.

217. Y. Xu, R. P. Schutte and L. G. Hepler, *Can. J. Chem. Eng.* **70**, 569–573 (1992).

218. G. H. Härtel, *J. Chem. Eng. Data* **30**, 57–61 (1985).

219. L. F. Audrieth, *Liquid Sulfur Dioxide—A Novel Reaction Medium*, republished by Rhône-Poulenc Basic Chemicals Co., Shelton, Conn.

220. *Sulphur*, 29–38 (Mar.–Apr. 1995).

221. G. N. Brown, S. L. Torrence, A. J. Repik, J. L. Stryker, and F. J. Ball, *Sulfur Dioxide Processing*, reprint manual published by American Institute of Chemical Engineering, New York, 1975.

222. E. Altwicker, *Adv. Environ. Sci. Eng.* **3**, 80 (1980).

223. J. O. Nriagu, ed., *Sulfur in the Environment*, John Wiley & Sons, Inc., New York, 1978.

224. J. G. Calvert, F. Su, J. W. Bottenheim, and O. P. Strausz, *Atmos. Environ.* **12**, 197 (1978).

225. L. R. Martin and T. W. Good, *Atmos. Environ.* **25A**(10), 2395–2399 (1991).

226. A. Saleem, K. E. Janssen, and P. A. Treland, *Sulphur* (230), 39–44 (Jan.–Feb. 1994).

227. J. B. Pfeiffer, ed., *Sulfur Removal and Recovery*, Advances in Chemistry Series 139, American Chemical Society, Washington, D.C., 1975.

228. J. Sarlis and D. Berk, *Ind. Eng. Chem. Res.* **27**, 1951–1954 (1988).

229. S. C. Paik and J. S. Chung, *Appl. Catal. B* **5**(3), 233–243 (1995).

230. C. D. Swaim, Jr., in Ref. 227, pp. 111–119.

231. J. M. Henderson and J. B. Pfeiffer, in Ref. 227, pp. 35–47.

232. W. D. Hunter, Jr., J. C. Fedoruk, A. W. Michener, and J. E. Harris, in Ref. 227, pp. 23–34.

233. P. Steiner, H. Juntgen, and K. Knoblauch, in Ref. 227, pp. 180–191.

234. J. Happel, M. Hnatow, L. Bajars, and M. Kundrath, *Ind. Eng. Chem. Prod. Res. Dev.* **14**, 155 (1975).

235. *Sulphur* **153**, 41 (Mar.–Apr. 1981).

236. R. B. Engdahl and H. S. Rosenberg, *Chemtech*, 118 (Feb. 1978).

237. C. G. Cornell and D. A. Dahlstrom, *Chem. Eng. Prog.* **69**(12), 47 (1973).

238. W. J. Osborne and C. B. Earl, in Ref. 230, pp. 158–163.

239. L. Korosy, H. L. Gewanter, F. S. Chalmers, and S. Vasan, in Ref. 227, pp. 192–211.

240. U.S. Pat. 3,911,093 (Oct. 7, 1975), F. G. Sherif, J. S. Hayford, and J. E. Blanch (to Stauffer Chemical Co.).

241. J. P. Hayford, L. P. VanBrocklin, and M. A. Kuck, *Chem. Eng. Prog.* **69**(12), 54 (1973).

242. *Sulphur* (232), 59–60 (May–June 1994).

243. Ref. 227, pp. 14, 16, 108–110.

244. Ref. 25, pp. 267–278.

245. N. W. Frank, *Rad. Phys. Chem.* **40**(4), 267–272 (1992).

246. L. C. Schroeter, *Sulfur Dioxide*, Pergamon Press, New York, 1966.

247. F. Asinger, *Paraffins—Chemistry and Technology*, Pergamon Press, Oxford, U.K., 1968, pp. 645–668, 483–571.

248. D. Estel, K. Mateew, W. Pritzkow, W. Schmidt-Renner, *J. Prakt. Chem.* **323**, 262 (1981).

249. E. M. Fettes and F. O. Davis, in N. G. Gaylord, ed., *Polyethers, III, Polyalkylene Sulfides and Other Polythioethers*, Interscience Publishers, New York, 1962, pp. 225–270.

250. C. S. Marvel and E. D. Weil, *J. Am. Chem. Soc.* **76**, 61 (1954), and references therein.

251. M. J. Bowden and L. F. Thompson, in *Polymeric Materials—Science and Engineering, Proceedings of the ACS Division of Polymeric Materials Science and Engineering* Vol. 68, ACS, Washington, D.C., 1993, p. 46.

252. H. Ito, L. A. Pederson, S. A. MacDonald, Y. Y. Chang, J. R. Lyerla, and C. G. Willson, *J. Electrochem. Soc.* **135**(6), 1504–1508 (1988).

253. S. R. Sandler and W. Karo, *Organic Functional Group Preparations*, Vol. I, Academic Press, Inc., New York, 1983, pp. 632–633.

254. B. Deguin and P. Vogel, *J. Am. Chem. Soc.* **114**, 9210–9211 (1992).

255. U. H. Sander, H. Fischer, U. Rothe, and R. Kola, in A. I. More, ed., *Sulfur, Sulfur Dioxide and Sulfuric Acid*, English ed., British Sulphur Corp. Ltd., London, 1984.

256. *Sulphur* (226), 19–21 (May 1992).

257. A. H. Lefebvre, X. F. Wang, and C. A. Martin, *J. Propulsion* **4**(4), 293–298 (1988).

258. R. L. Schendel, *Oil Gas J.* **91**(39), 63–66 (1993).

259. U.S. Pat. 3,632,312 (Jan. 4, 1972), W. W. Jukkola (to Dorr-Oliver, Inc.).

260. C. D. Jentz, *Pulp Paper Can.*, 101 (1941).

261. A. Urbanek and M. Burgiell, *Proceedings of the 5th International Sulphur Conference, Nov. 1981*, British Sulphur Corp., Ltd., London, 1982.

262. *Process for Production of Liquid Sulfur Dioxide*, Bulletin of Stauffer Chemical Co., Licensing Administrator, Westport, Conn., 1970 (successor: Rhône-Poulenc Licensing Dept., Shelton, Conn.).

263. U.S. Pat. 4,177,248 (Dec. 4, 1979), W. T. Richard (to Stauffer Chemical Co.).

264. N. E. Hamner, ed., *Corrosion Data Survey*, 5th ed., National Association of Corrosion Engineers, Houston, Tex., 1975.

265. *Mater. Eng.*, 14 (May 1979).

266. S. L. Sakol and R. A. Schwartz, *Sulfur Dioxide Processing*, reprint manual, American Institute of Chemical Engineers, New York, 1975.

267. L. Igetoft, *Ind. Eng. Chem. Prod. Res. Dev.* **24**, 375–378 (1985).

268. M. J. Justo and M. G. S. Ferreira, *Corrosion Science* **29**(11–12), 1353–1369 (1989); **34**(4), 533–545 (1993).

269. F. Schuster, M. M. Reddy, and S. I. Sherwood, *Mater. Perf.* **33**(1), 76–80 (1994).

270. "Chemical Profile on Sulfur Dioxide," *Chem. Mark. Rep.* 33 (Aug. 22, 1994).

271. R. A. Belbutowski, paper at *Sulfur Markets Symposium*, Sulfur Institute, Washington, D.C., Mar. 24, 1994.

272. *NIOSH Manual of Analytical Methods*, Method No. S308, Vol. 2, U.S. Dept. of Health and Human Services, Public Health Service, Washington, D.C., 1977.

273. *NIOSH Manual of Analytical Methods*, Method No. P & CAM 160, Vol. 1, U.S. Dept. of Health, Education, and Welfare, Public Health Service, Washington, D.C., 1977.

274. *Sulphur* **131**, 52 (July–Aug. 1977); J. R. Donovan, E. D. Kennedy, D. R. McAlister, and R. M. Smith, *Chem. Eng. Prog.*, 89 (June 1977).

275. R. S. Saltzman, *Challenges in Stack Sulfur Emissions Measurements*, paper presented at 30th Annual Petroleum Mechanical Engineering Conference, Tulsa, Okla., Sept. 21–25, 1975.

276. C. J. Lang, R. S. Saltzman, and G. G. DeHaas, *TAPPI Environmental Conference*, 1975 preprints, Technical Association of the Pulp and Paper Industry, Atlanta, Ga., May 1975, pp. 129–145.

277. D. G. Barrett and J. R. Small, *Chem. Eng. Prog.* **69**(12), 35–38 (1973).

278. *Food Chemicals Codex*, 3rd ed., National Academy of Sciences, Washington, D.C., 1993.

279. F. D. Snell and L. S. Ettre, *Encyclopedia of Industrial Chemical Analysis*, Vol. 18, John Wiley & Sons, Inc., New York, 1973, pp. 408ff.

280. J. M. Roberts, *Chemtracts: Anal. Phys. Chem.* **1**(5), 328–330 (1989).

281. D. W. T. Griffith and G. Schuster, *J. Atmos. Chem.* **5**(1), 59–81 (1987).

282. B. S. M. Kumar and N. Balasubramanian, *J. AOAC Int.* **75**(6), 1006–1010 (1992).

283. J. S. Symanski and S. Bruckenstein, *Anal. Chem.* **58**(8), 1771–1777 (1986).

284. F. L. Eisele and H. Berresheim, *Anal. Chem.* **64**, 283–288 (1992).

285. J. R. Cochrane, A. W. Ferguson, and D. K. Harris, *Chem. Eng.*, 4–9 (June 1993).

286. *Sulphur* (228), 37–45 (Sept.–Oct. 1993).

287. D. E. Schorran, C. Fought, D. F. Miller, W. G. Coulombe, R. E. Keislar, R. Benner, and D. Stedman, *Environ. Sci. Technol.* **28**(7), 1307–1311 (1994).

288. J. E. Jones, *Pollution Eng.*, 38–40 (June 1985).

289. W. Freitag, in J. Jahnke, ed., *An Operators Guide to Eliminating Bias in CEM Systems*, U.S. EPA Contract Report 68-D2-0168, Washington, D.C., Nov. 1994, pp. 3–12.

290. P. J. Wallace and T. M. Gerlach, *Science* **265**, 497–499 (1994).

291. D. Möller, *Atmos. Environ.* **18**(1), 29–39 (1984).

292. P. M. Irving, ed., U.S. National Acid Precipitation Assessment Program (NAPAP), *Acidic Deposition: State of Science and Technology*, Vols. I–IV, Integrated Assessment Report, U.S. National Acid Precipitation Assessment Program, National Academy of Sciences, Supt. Documents, Government Printing Office, Washington, D.C., 1991.

293. *U.S. National Acid Precipitation Assessment Program (NAPAP) 1992 Report to Congress*, Government Printing Office, Washington, D.C., June 1993.

294. G. M. Hidy, *Atmospheric Sulfur and Nitrogen Oxides: Eastern North American Source Receptor Relationships*, Academic Press, Orlando, Fla., 1994.

295. J. Greyson, *Carbon, Nitrogen and Sulfur Pollutants and Their Determination in Air and Water*, Marcel Dekker, Inc., New York, 1990.

296. J. N. Galloway, R. J. Charlson, M. O. Andreae, and H. Rodhe, eds., *The Biogeochemical Cycling of Sulfur and Nitrogen in the Remote Atmosphere*, Kluwer Academic Publishers, Inc., New York, 1985.

297. R. Markuszewski and T. D. Wheelock, eds., *Processing and Utilization of High-Sulfur Coals*, Vols. I–III, Elsevier Science, Inc., New York, 1990.

298. E. S. Saltzman and W. J. Cooper, eds., *Biogenic Sulfur in the Environment*, ACS Symposium Series No. 393, American Chemical Society, Washington, D.C., 1989.

299. P. A. Clark, B. E. A. Fisher, and R. A. Scriven, *Atmos. Environ.* **21**(5), 1125–1131 (1987).

300. B. E. A. Fisher and P. A. Clark, *Atmos. Environ.* **20**(11), 2219–2229 (1986).

301. H. Ellis and M. L. Bowman, *J. Environ. Eng.* **120**, 273–290 (1994).

302. *Subgroup on Metals of the Tri-Academy Committee on Acid Deposition, Acid Deposition: Effects on Geochemical Cycling and Biological Availability of Trace Elements*, National Academy Press, Washington, D.C., 1985.

303. Committee on Haze in National Parks and Wilderness Areas, National Research Council, *Protecting Visibility in National Parks and Wilderness Areas*, National Academy Press, Washington, D.C., 1993.

304. R. W. Howarth, J. W. Stewart, and M. V. Ivanov, *Sulfur Cycling on the Continents: Wetlands, Terrestrial Ecosystems and Associated Water Bodies*, Scientific Committee on Problems of the Environment Series, John Wiley & Sons, Inc., New York, 1992.

305. L. F. Pitelka, *EPRI J.*, 50–52 (Apr.–May 1987).

306. B. Rosenbaum, T. C. Strickland, and M. K. McDowell, *Water, Air Soil Poll.* **74**(3–4), 307–319 (1994).

307. J. J. Mackenzie and M. T. El-Ashry, *Technol. Rev.* **92**(3), 64–71 (1989).

308. M. A. Tabatatai, *Sulfur in Agriculture*, American Society of Agronomy, Madison, Wisc., 1986.

309. W. E. Winner, H. A. Mooney, and R. A. Goldstein, eds., *Sulfur Dioxide and Vegatation: Physiology, Ecology and Policy Issues*, Stanford University Press, Palo Alto, Calif., 1985.

310. C. M. Burt and C. Hash, *Proceedings of the 3rd International Drip/Trickle Irrigation Congress*, Nov. 1985, ASAE, St. Joseph, Mich., 1985.

311. J. M. Mee, M. Jahangir, M. Al-Mishal, I. Faruq, and S. Al-Salem, *Papers of 189th National Meeting American Chemical Society*, Div. Envir. Chem. **25**(1), 28–31 (1985).

312. J. Makansi, *Power*, 23–63 (Mar. 1993).

313. T. J. Manetsch, *IEEE Trans. Power Sys.* **9**, 1921–1926 (1994).

314. L. R. Roberts, K. N. T. Horne, A. Mann, and S. Keil, *Chem. Eng. Prog.*, 22–24 (Oct. 1985).

315. J. W. Gerrie, in *Proceedings of Canada Pulp and Paper Association 61st Annual Meeting*, Canada Pulp and Paper Association, Montreal, Quebec, 1975, T251–T254.

316. B. L. Wedzicha, ed., *Chemistry of Sulfur Dioxide in Foods*, Elsevier Science, Inc., New York, 1984.

317. B. L. Wedzicha, *Chem. Brit.* **27**(11), 1030–1032 (1991).

318. R. A. Scanlon, *CRC Crit. Rev. Food Technol.* **5**, 357 (1975).

319. W. R. Junk and H. M. Pancoast, *Handbook of Sugars*, AVI Publishing Co., Inc., Westport, Conn., 1973, p. 110ff.

320. *Dechlorination Using Sulfur Products*, Technical Information Bulletin TIR-17, Rhône-Poulenc Basic Chemicals Co., Shelton, Conn., 1992.

321. *Sodium Sulfite Solution 25-9 in Oxygen Scavenging*, Technical Information Bulletin TIR-13, Rhône-Poulenc Basic Chemicals Co., Shelton, Conn., 1991.

322. R. L. Miron, *Mater. Perform.*, 45 (June 1981).

323. P. J. Bailes, C. Hanson, and M. A. Hughes, *Chem. Eng.* **83**(10), 115–120 (1976).

324. Brit. Pat. 1,186,126 (Apr. 2, 1970), J. M. Maselli (to W. R. Grace & Co.).

325. C. N. Satterfield, *Heterogeneous Catalysis in Practice*, McGraw-Hill Book Co., New York, 1980, p. 203.

326. F. Habashi, *Sulphur Inst. J.* **12**(3–4), 15 (1976).

327. *Treating Irrigated Arid-Land Soils with Acid-Forming Sulphur Compounds*, Technical Bulletin No. 24, The Sulphur Institute, Washington, D.C., 1979.

328. L. R. Grass and A. J. MacKenzie, *Sulphur Inst. J.* **6**(1), 8 (1970).

329. T. Niichiro, *Synthesis* (12), 639–645 (1971).

330. *Gmelins Handbuch der Anorganischen Chemie*, 8th ed., System No. 9, Part B2, Verlag Chemie, Weinheim/Bergstrasse, Germany, 1960, pp. 373–613.

331. *Sodium Sulfite*, product brochure, General Chemical Co., Claymont, Del., 1993.

332. *Sodium Sulfite Anhydrous, Technical Grade and High Purity Grade*, data sheets, Solvay Minerals, Inc., Houston, Tex., 1995.

333. *Sodium Sulfite, Commercial Anhydrous*, data sheet, Indspec Chemical Corp., Pittsburgh, Pa., 1992.

334. *Sodium Sulfite*, product data sheet and product safety information sheets, Rhône-Poulenc Basic Chemicals Co., Shelton, Conn., 1992.

335. E. A. P. Devuyst, V. A. Ettel, and M. A. Mosoiu, *Chemtech*, 426 (July 1979).

336. Ger. Pat. 2,940,697 (Feb. 5, 1981), A. Metzger and G. Muenster (to Hoechst A-G).

337. U.S. Pat. 4,003,985 (Jan. 18, 1977), R. J. Hoffman, S. L. Bean, P. Seeling, and J. W. Swaine, Jr. (to Allied Chemical Corp.).

338. *Chem. Mark. Rep.* 9 (Apr. 12, 1993).

339. M. J. Kocurek, O. V. Ingruber, and A. Wong, *Sulfite Science and Technology*, Vols. 1–4, 3rd ed., Technical Association of the Pulp and Paper Industry, Atlanta, Ga., 1985.

340. *Sodium Metabisulfite, Solid SO₂*, Technical Information sheet, Rhône-Poulenc Basic Chemicals Co., Shelton, Conn., 1991.

341. Can. Pat. 3,057,418 (July 23, 1992), W. H. Bortle, S. L. Bean, and M. D. Dulik (to General Chemical Co.).

342. U.S. Pat. 4,844,880 (July 4, 1989), S. L. Bean, M. D. Dulik, and R. J. Wilson (to HMC Patents Holding Co., Inc.).

343. *Sodium Bisulfite*, CHRIS Hazardous Chemical Data, Coast Guard, U.S. Dept. of Transportation, Washington, D.C., data sheet revised 1978.

344. *Treatment of Chromium Waste Liquors*, General Chemical Corp., Claymont, Del., 1989.

345. J. Litt, *Metal Finish.* **52**, 61 (Nov. 1971).

346. D. Lyons and G. Nickless, in Ref. 84, pp. 199–239.

347. L. W. Codd and co-workers, eds., *Chemical Technology, An Encyclopedic Treatment*, Vol. 1, Barnes & Noble Books, New York, 1968, p. 461.

348. J. G. deVries and R. M. Kellogg, *J. Org. Chem.* **45**, 4126 (1980).

349. G. Valensi, *Cent. Belge. Etude Corros. Rapp. Tech.* **121**, 207/1–207/22 (1973).

350. A. Teder, *Acta Chem. Scand.* **27**, 705 (1973).

351. A. Janson and F. Scholtz, in W. Forst, ed., *Ullmanns Encyklopädie Technischen Chemie*, 3rd ed., Vol. 15, Urban & Schwartzenberg, München-Berlin, Germany, 1964, pp. 480–488.

352. U.S. Pat. 3,523,069 (Jan. 29, 1969), C. W. Oloman (to British Columbia Research Council).

353. M. S. Spencer, P. J. Carnell, and W. J. Skinner, *Ind. Eng. Chem. Proc. Res. Dev.* **8**, 184 (1969).

354. C. Oloman, *J. Electrochem. Soc.* **117**, 1604 (1970).

355. Ger. Pat. 2,748,935 (May 3, 1979), W. Ostertag, G. Ertl, G. Wunsch, V. Kiener, E. Voelkl, and S. Schreiner (to BASF A-G).

356. G. Ertl, V. Kiener, W. Ostertag, G. Wunsch, and K. Bittler, *Angew. Chem.* **91**, 333 (1979).

357. U.S. Pat. 3,927,190 (Dec. 16, 1975), T. Kato, H. Okazaki, A. Sugio, and Y. Yushukawa (to Mitsubishi Gas Chemical Co., Ltd.).

358. Brit. Pat. 1,145,824 (Apr. 21, 1965) (to Electric Reduction Co. of Canada, Ltd.).

359. U.S. Pat. 4,992,147 (Feb. 12, 1991), R. E. Bolick, II, D. W. Cawlfield, and J. M. French (to Olin Corp.).

360. U.S. Pat. 5,336,479 (Aug. 9, 1994), D. C. Munroe (to Morton International, Inc.).

361. D. Munroe, O. Bhatia, J. Westling, M. McCrary, A. Casassa, and S. Masse, *Proceedings of the TAPPI 1994 Pulping Conference*, Book 1, TAPPI Press, Atlanta, Ga., 1994, pp. 83–86.

362. U.S. Pat. 4,892,636 (Jan. 9, 1990), R. E. Bolick, II, D. W. Cawlfield, and K. E. Woodard, Jr. (to Olin Corp.).

363. *Chemical Economics Handbook*, Stanford Research Institute, Menlo Park, Calif., 1992, pp. 780.4000 B–C.

364. *Chem. Mark. Rep.* 49 (Oct. 10, 1994).

365. L. V. Hoff, in J. H. Karchmer, ed., *The Analytical Chemistry of Sulfur and its Compounds*, Part 1, Interscience, a division of John Wiley & Sons, Inc., New York, 1970, pp. 245–247.

366. K. J. Stutts, *Anal. Chem.* **59**, 543–544 (1987).

367. *Fed. Reg.*, 6117 (Jan. 25, 1980).

368. J. N. Etters, *Am. Dyestuffs Rep.* **78**, 18 (1989).
369. A. R. Beaudry and N. R. Ducharme, *Preprints of Annual Meeting, Technical Section, Canadian Pulp and Paper Association*, Pt. B, Canadian Pulp and Paper Association, Montreal, 1993, pp. B13–B26.
370. R. Barton, C. Tredway, M. Ellis, and E. Sullivan, in R. A. Leask, ed., *Pulp and Paper Manufacture*, 3rd ed., Vol. 2, *Mechanical Pulping*, TAPPI, Atlanta, Ga., and CPPA, Montreal, Quebec, 1986, Chapt. XIX, pp. 227–237.
371. *Two-Stage Bleaching with V-BRITE Bleach, and V-BRITE Bleaching Compounds*, Product Application bulletins 10172 and 10174, Hoechst Celanese Corp., Charlotte, N.C., 1992.
372. A. A. Newman, ed., *Chemistry and Biochemistry of Thiocyanic Acid and Its Derivatives*, Academic Press, Inc., New York, 1975.
373. Ref. 2, Part D5, pp. 148–209.
374. Ref. 253, pp. 369–376.
375. *Ammonium Thiocyanate, Potassium Thiocyanate, and Sodium Thiocyanate*, Product bulletins, Argus Chemical Division, Witco Chemical Co., Marshall, Tex.
376. C. K. Nmai, M. A. Bury, and H. Farzam, *Concrete Int.* **16**(4), 22–25 (1994).
377. H. Goldwhite, M. S. Gibson, and C. Harris, *Tetrahedron* **21**(10), 2743 (1965).
378. Fr. Pat. 1,409,516 (1965), H. W. Moore, F. F. Rust, and H. S. Klein.
379. M. Asscher and D. Vofsi, *J. Chem. Soc.*, 4962 (1964).
380. I. J. Borowitz, *J. Am. Chem. Soc.* **86**, 1146 (1963).
381. G. Opitz, M. Kleeman, D. Bucher, G. Walz, and R. Rieth, *Angew. Chem.* **78**(11), 604 (1966).
382. E. Block, *Reactions of Organosulfur Compounds*, Academic Press, Inc., New York, 1978, p. 278.
383. W. E. Truce and C. W. Vriesen, *J. Am. Chem. Soc.* **75**, 5032 (1953).
384. R. H. Wiley and H. Kraus, *J. Org. Chem.* **22**, 994–995 (1957).
385. J. P. Horwitz, J. Chua, and M. Noel, *J. Org. Chem.* **29**, 2076 (1969).
386. *Methane Sulfonyl Chloride (MSC)—Properties, Reactions and Applications*, Technical Bulletin S-104, Elf Atochem North America, Philadelphia, Pa.
387. *Concerning Methane Sulfonyl Chloride*, Technical Bulletin, Elf Atochem North America, Philadelphia, Pa.
388. *Use of Methanesulfonyl Chloride (MSC) in the Synthesis of Chiral Compounds*, Technical Bulletin A-70-11, Elf Atochem North America, Philadelphia, Pa., 1992.
389. U.S. Pat. 2,825,736 (Mar. 4, 1958), A. C. Cope and N. Burg (to Merck and Co.).
390. K. Weinges, *Ann.* **615**, 203–209 (1958).
391. L. Toldy and I. Fabricius, *Chem. Ind. (London)*, 665 (1957),
392. U.S. Pat. 2,703,808 (Mar. 8, 1955), E. R. Buchman (to Research Corp.).
393. R. F. Schwenker, Jr., and E. Pacsu, *Ind. Eng. Chem.* **50**, 91–96 (1958).
394. Belg. Pat. 613,335 (Feb. 15, 1962), C. Gyogyszer and V. T. Gyara.
395. J. H. Looker and C. H. Hayes, *J. Am. Chem. Soc.* **79**, 745–747 (1957).
396. J. H. Looker, C. H. Hayes, and D. N. Thatcher, *J. Am. Chem. Soc.* **79**, 741–744 (1957).
397. U.S. Pat. 3,072,688 (Jan. 8, 1963), H.-J. E. Hess (to Chas. Pfizer & Co.).
398. N. L. Wendler and D. Taub, *J. Org. Chem.* **25**, 1828–1829 (1960).
399. A. G. Kastava, *Zhur. Obshchei Khim.* **18**, 729 (1948).
400. G. J. Griffiths and F. E. Previdoli, *J. Org. Chem.* **58**, 6129 (1993).
401. R. P. Glinski, M. S. Khan, and R. L. Kalman, *J. Org. Chem.* **38**, 4299 (1973).
402. *Methane Sulfonic Acid (MSA), Properties, Reactions and Applications*, Technical Bulletin PB-70-1A, Elf Atochem North America, Philadelphia, Pa., 1993.
403. D. B. Roitman, J. McAlister, and F. L. Oaks, *J. Chem. Eng. Data* **39**, 56 (1994).
404. *Methane Sulfonic Acid for Catalyzed Coatings*, Technical Bulletin, Elf Atochem North America, Philadelphia, Pa., 1993.

405. *Preparation of DOP*, Technical Bulletin, Elf Atochem North America, Philadelphia, Pa.
406. U.S. Pat. 3,071,604 (Jan. 1, 1963), A. G. Mohan and W. Christian (to Nopco Chemical Co.).
407. U.S. Pat. 3,116,336 (Dec. 31, 1963), J. L. Van Winkle (to Shell Development Co.).
408. U.S. Pat. 2,525,942 (Oct. 17, 1950), W. A. Proell (to Standard Oil Co. of Indiana).
409. *MSA Zinc Plating Process*, Technical Bulletin, Elf Atochem North America, Philadelphia, Pa.
410. P. D. Lawley, in C. E. Searle, ed., *Chemical Carcinogens*, ACS Monograph 182, American Chemical Society, Washington, D.C., 1984.

EDWARD D. WEIL
Polytechnic University

STANLEY R. SANDLER
Elf Atochem North America, Inc.

SULFUR DIOXIDE. See SULFUR COMPOUNDS.

SULFUR DYES

Sulfur dyes are used mainly for dyeing textile cellulosic materials or blends of cellulosic fibers (qv) with synthetic fibers such as acrylic fibers, polyamides (nylons), and polyesters. They are also used for silk (qv) and paper (qv) in limited quantities for specific applications. Solubilized sulfur dyes are used on certain types of leathers (qv).

From an applications point of view, the sulfur dyes are between vat, direct, and fiber-reactive dyes. They give good to moderate lightfastness and good wetfastness at low cost and rapid processing (see DYES, APPLICATION AND EVALUATION).

Traditionally, these dyes are applied from a dyebath containing sodium sulfide. However, development in dyeing techniques and manufacture has led to the use of sodium sulfhydrate, sodium polysulfide, sodium dithionite, thiourea dioxide, and glucose as reducing agents. In the reduced state, the dyes have affinity for cellulose (qv) and are subsequently exhausted on the substrate with common salt or sodium sulfate and fixed by oxidation.

The range of colors covers all hue classification groups except a true red. As a rule, the hues are dull compared with other dye classes. Black is the most important, followed by blues, olives, and browns (see DYES AND DYE INTERMEDIATES).

The first sulfur dyes were produced in 1873 by heating organic cellulose-containing material, such as wood sawdust, humus, bran, cotton waste, and waste paper, with alkali sulfides and polysulfide (1). These dyes were dark and hygroscopic and had a bad odor. Composition varied and they were easily soluble in water giving greenish dyeings, both from alkali and alkali sulfide baths. The dyes were fixed on cotton (qv) fiber and eventually turned brown on exposure to air or chemical oxidation with bichromate solution. The valuable tinctorial properties of these dyes, in addition to their low price, led to their use in the cotton dyeing industry. They were sold under the name of *Cachou de Laval*. The main constituent was lignin (qv), which is used to produce CI Sulfur Brown 1.

The quality of sulfur dyes has greatly improved, mainly because well-defined organic compounds are used as starting materials. Owing to better techniques, sulfur dyes have good storage stability and fastness properties, and are generally easily adaptable to modern dyeing methods. An overview of the development of sulfur dyes is given in Table 1. In the 1980s and 1990s, reduction of sulfide levels has been achieved.

Chemical Properties

Classification and Structure.   Little is known about the structure of sulfur dyes, and therefore, they are classified according to the chemical structure of the starting materials.

The process of sulfurization is usually carried out by a sulfur bake, in which the dry organic starting material is heated with sulfur between 160 and 320°C; a polysulfide bake, which includes sodium sulfide; a polysulfide melt, in which aqueous sodium polysulfide and the organic starting material are heated under reflux or under pressure in a closed vessel; or a solvent melt, in which butanol, Cellosolve, or dioxitol are used alone or together with water. In the last two methods, hydrotropes may be added to enhance the solubility of the starting material. The hydrotropes improve yield and quality of the final dyestuff.

The temperature and duration of heating have a marked effect on both the shade and properties of the dye and conditions have to be carefully controlled in order to achieve uniformity. The precipitation by air oxidation must also be controlled to prevent variations in shade.

Sulfur Bake.   The yellow, orange, and brown sulfur dyes belong to this group. The dyes are usually made from aromatic amines, diamines, and their acyl and nuclear alkyl derivatives. These may be used in admixture with nitroanilines and nitrophenols or aminophenols to give the desired shade. The color formed is said to be the result of the formation of the thiazole chromophore, evident in dye structure (**1**).

Investigation into the structure of the sulfur-bake dye Immedial Yellow GG, CI Sulfur Yellow 4 [*1326-75-6*] (CI 53160), by chemical degradation of the dye and confirmatory synthesis of the postulated structures showed that a mixture of four dyes was obtained when benzidine and 4-(6-methyl-2-benzothiazolyl)aniline were baked with sulfur. The original British patent for Sulfur Yellow 4 dates back to 1906 (19). The principal structure obtained is (**1**).

Table 1. Historical Events in the Development of Sulfur Dyes

Year	Development	Product	CI designation Name	CI designation Number	CAS Registry Number	Ref.
1873	heating organic wastes with alkali sulfides/alkali polysulfides	*Cachou de Laval*	Sulfur Brown 1	53000	[1326-37-0]	1
1893	heating benzene and naphthalene derivatives with alkali polysulfides	Vidal Black	Sulfur Black 3	53180	[1326-81-4]	2
1897	sulfurization of diphenylamine derivatives	Immedial Black V	Sulfur Black 9	53230	[1326-97-2]	3
1899	inexpensive black from 2,4-dinitro-phenol by use of reflux method of sulfurization	Sulfur Black T	Sulfur Black 1	53185	[1326-82-5]	4
1900	use of leucoindophenols	Immedial Pure Blue	Sulfur Blue 9	53430	[1327-56-6]	5
	first reddish (bordeaux) dyes from 8-amino-2-hydroxyphenazine	Immedial Maroon B	Sulfur Red 3	53710	[1327-84-0]	6
1902	commercial orange and yellow dyes from toluene-2,4-diamine	Immedial Orange C	Sulfur Orange 1	53050	[1326-49-4]	7
1904	clear green dye from phenyl-peri acid indophenol	Thionol Brilliant Green GG	Sulfur Green 3	53570	[1327-73-7]	8
1908–1909	chlorine fast blue dyes from carbazole and *N*-ethylcarbazole indophenols	Hydron Blue R / Hydron Blue G	Vat Blue 43 / Vat Blue 42	53630 / 53640	[1327-79-4] / [1327-81-7]	9
1926	chlorine fast black dye from *p*-(2-naphthylamino)phenol	Indocarbon CL	Sulfur Black 11	53290	[1327-14-6]	10
1932–1934	clear violet-to-blue dyes from substituted trichlorophenoxazones	Immedial New Blue 5RCF	Sulfur Blue 12	53800	[1327-96-4]	11
		Immedial Bordeaux 3BL	Sulfur Red 7	53810	[1327-97-5]	
1935–1938	fast browns from decacyclene (**8**) and its polynitro derivatives, using S_2Cl_2 as sulfurization agent or baking with sulfur	Immedial Katechu 4RL	Sulfur Brown 52	53320	[1327-18-0]	12
		Immedial Yellow Brown GL	Sulfur Brown 60	53325	[1327-20-4]	
		Immedial Brown GGL	Sulfur Brown 51	53327	[1327-22-6]	

Table 1. (*Continued*)

Year	Development	Product	CI designation Name	Number	CAS Registry Number	Ref.
	ready-to-dye liquid dyes from commercial powders/press cake by solution in sodium sulfide–sodium sulfhydrate mixture	Sodyesul Black 4GCF	Leuco Sulfur Black 1	53185	[66241-11-0]	13
1940–1941	synthetic methods introduced for conversion of organic pigment-type dyes to sulfide soluble dyes	Thionol Ultra Green B	Sulfur Green 14 Leuco Sulfur Red 14		[12227-06-4]	14,15
1945	sulfurization with aluminum chloride–sulfur monochloride complex of organic dyes and pigments					16
1948	water-soluble dyes by conversion of conventional sulfur dyes with alkali sulfites/bisulfites to thiosulfonic acid derivatives	Thionol Black BM	Solubilized S Black 1	53186	[1326-83-6]	
1959	condensation of intermediates containing NH$_2$ groups with cyanuric chloride; further conversion to thiol groups					17
1960	Inthion and Dykolite synthetic dyes; Bunte salts (alkyl- and arylthiosulfato dyes) of nonsulfurized type	Inthion Brilliant Blue I3G	Condense S Blue 2	18790	[12224-49-6]	
1966		Dykolite Brilliant Orange 3G	Condense S Orange 2			
1975	fast brown dyes from Indanthrene-type vat dyes by sulfurization with sulfur in chlorosulfonic acid					18

343

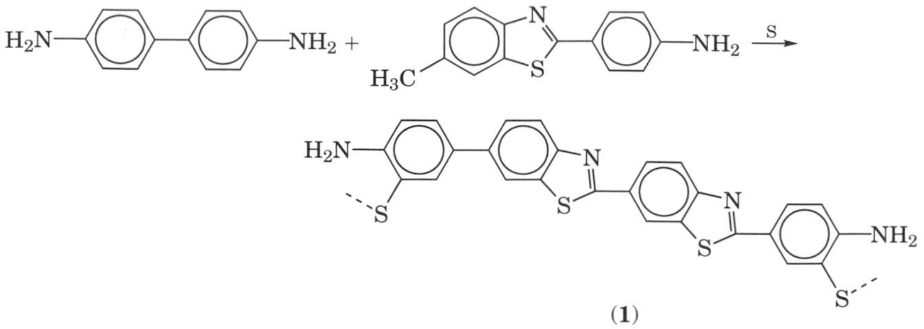

(1)

The dye has been degraded by a fusion with caustic potash and the degradation products identified as various o-anilinyl mercaptans. They were identified and characterized by condensation with monochloroacetic acid to give the thioglycolic acids which, on acidification, were converted to well-defined crystalline lactams (**2–4**) together with a small amount of p-aminobenzoic acid.

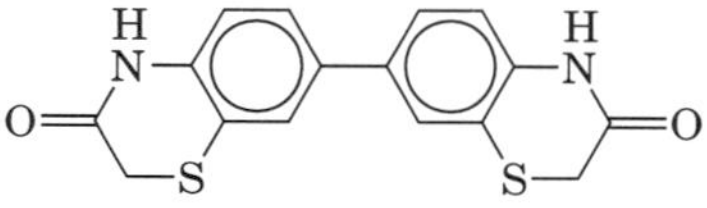

(2) 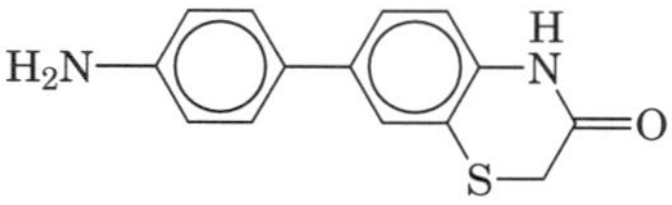   (3) 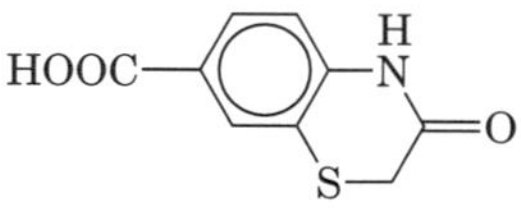   (4)

Investigation of Immedial Orange C, CI Sulfur Orange 1 [1326-49-4] (CI 53050), produced from 2,4-diaminotoluene (MTD) revealed ca 6–8 mols of MTD linked by thiazole rings. Whether the linkage is linear (**5**) or branched (**6**) was not ascertained with certainty but again showed that the thiazole chromophore is present.

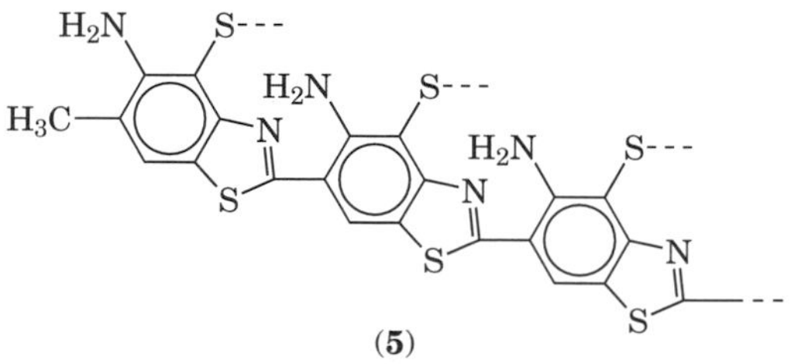

(5)

The solubility in alkali sulfide solutions stems from the presence of disulfide groups –S–S– in ortho position to the terminal amino groups. These disulfide groups are reduced to the mercapto groups –SH, which are soluble in alkali.

(**6**)

Polysulfide Bake. Although most dyes made formerly by this method are made by a polysulfide melt, some dyes where certain nitro and phenolic bodies prevent color formation in the sulfur bake are still made by this method. Pre-reduction of nitro groups substantially reduces the explosion hazard associated with heating these compounds with sulfur alone. Included in this class are also the exceptionally lightfast sulfur dyes derived from the polynitrodecacyclenes. Decacyclene (**8**) itself is made by a sulfur bake of acenaphthene (**7**) (20). Deca-cyclene is sulfur-baked to give CI Sulfur Brown 52. However, this color cannot be obtained from acenaphthene in a one-step sulfur bake because a significant amount of diacenaphtho[1,2-*b*;1′,2″-*d*]thiophene (**9**) is formed which, on sulfur baking, gives a dark brown dye of inferior lightfastness. The structure of the sulfur dyes derived from these polynuclear compounds is unknown.

(**7**)   (**8**)   (**9**)

Polysulfide Melt. CI Sulfur Black 1 [*1326-82-5*] (CI 53185), derived from 2,4-dinitrophenol, is the most important dye in this group which also includes the indophenol-type intermediates. The latter are applied in the stable leuco form. The derived dyes are usually confined to violet, blue, and green shades. Other members of this group are intermediates capable of forming quinoneimine (**10**) or phenazone structures (**11**) that produce red-brown or Bordeaux shades:

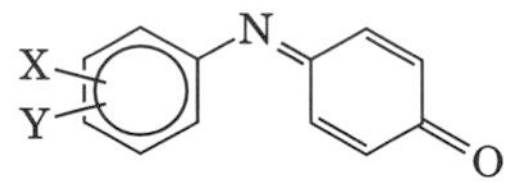

(10) 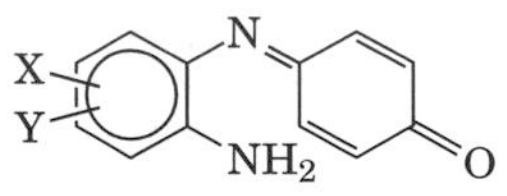 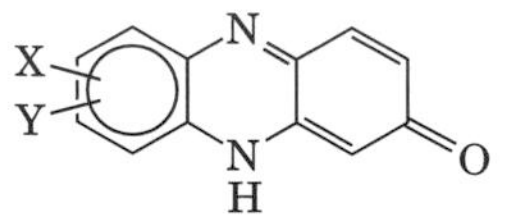   (11)

With the exception of the phenazones, these compounds lead to formation of the thiazinone ring structure (**12**) which closely resembles the phenazone ring system.

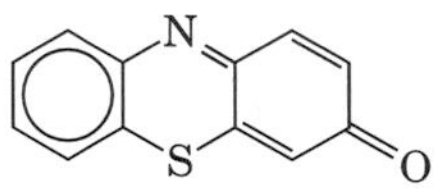

(**12**)

The structure of indophenol-type sulfur dyes was studied (21) on CI Sulfur Blue 9 made from the indophenol (**13**). This compound was purified and brominated to a tetrabromo derivative (**14**) identical with that obtained on bromination of Methylene Violet [2516-05-4] (**15**).

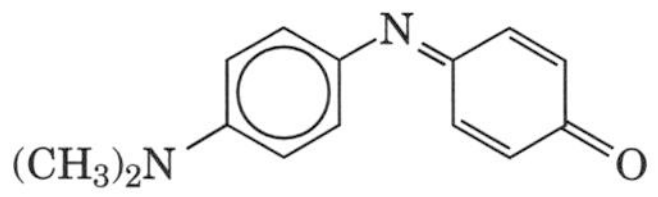

(**13**)

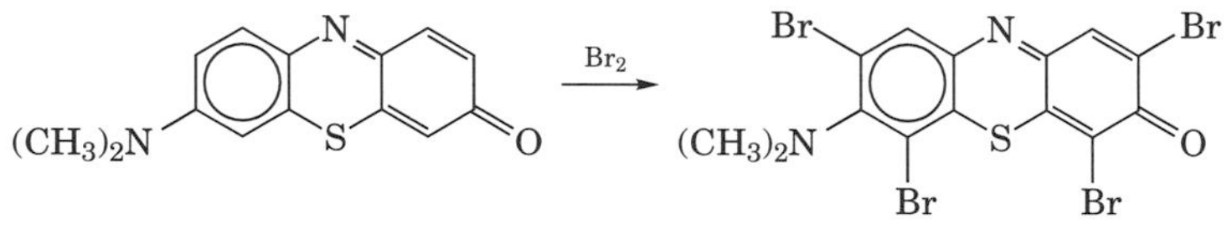

(**15**)   (**14**)

Thus, the presence of a thiazine ring in CI Sulfur Blue 9 was conclusively proved. The thiazine ring is the fundamental chromophore that accounts for the high color value of both the sulfur dye and Methylene Blue [61-73-4], including their ability to form pale yellow leuco forms on reduction. Methylene Violet (**15**) is obtained from Methylene Blue (**16**) by hydrolysis in boiling alkali.

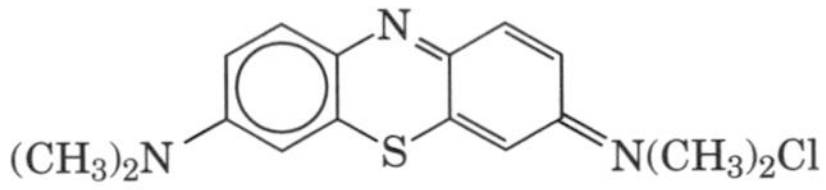

(16)

This work supports the suggestion that sulfur dyes are complex thiazines derived generally from diphenylamine units (22). Depending on the conditions of the thionation, two or more thiazine units could be linked by sulfur to produce a typical sulfur blue (**17**, R = OH or NH$_2$).

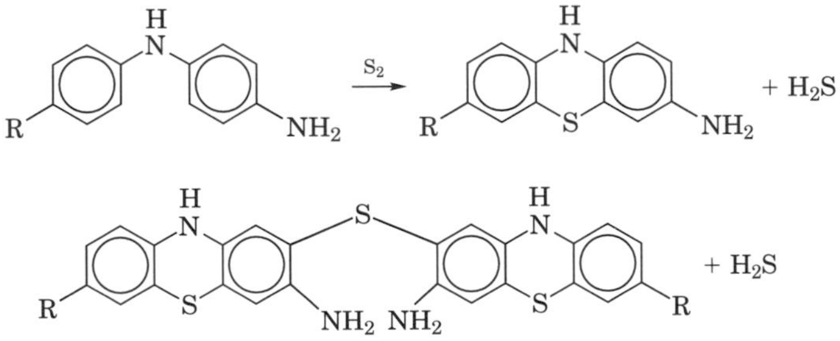

(17)

On more severe thionation, a third thiazine ring is formed to give a sulfur black. However, if hydroxyl groups instead of amino groups are attached at positions 2 and 2′, no ring closure would take place and the blue dye would be stable to heat. These formulas are general expressions for the nuclear structures of the blue-to-black sulfur dyes; they do not take into consideration the quinonoid formation of each dye and other aspects.

Further work elucidated the structure of the dyes in this group by synthesizing trichloroarylthiazinones (**18**) from an *o*-aminoaryl mercaptan (**19**) and chloranil (**20**); treatment with sodium disulfide replaced the halogens by mercapto groups.

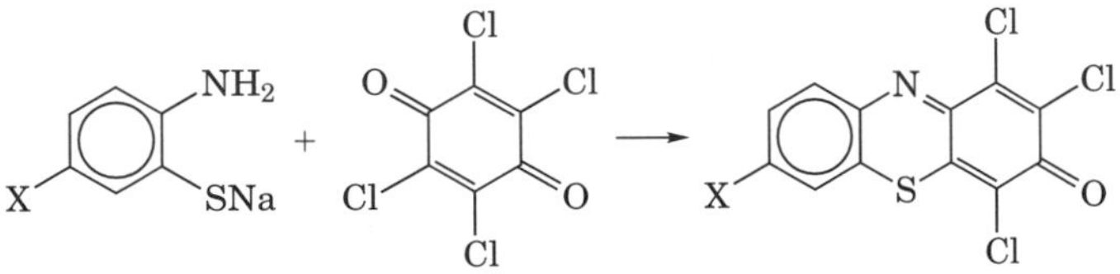

    (19)   **(20)**   **(18)**

By this method, sulfur dyes derived from 4-hydroxydiphenylamine are seen to be essentially identical. Similarly, the sulfur dye from 7-anilino-9-methyl-1,2,4-

trichlorophenothiazin-3-one (**21**) and the sulfur dye from 4-anilino-4'-hydroxy-2-methyldiphenylamine (**22**) were compared and found to be almost identical, as was the related trichlorophenothiazinone derivative (**23**) from carbazole when compared with Hydron Blue indophenol, CI Vat Blue 43 (CI 53630) (**24**).

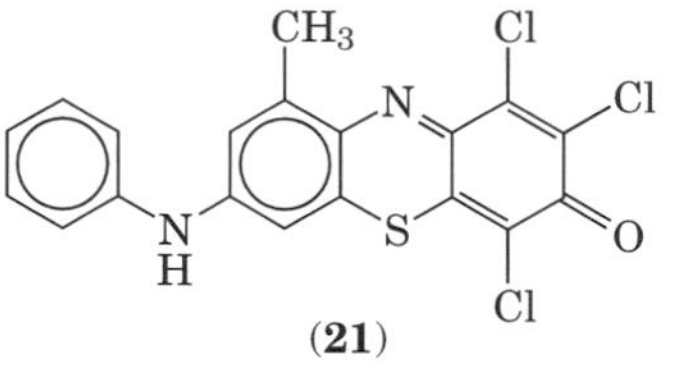

(**21**) 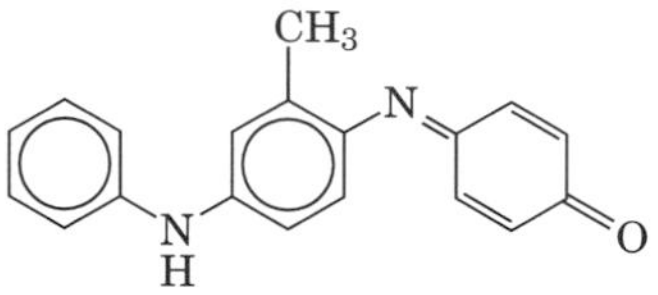   (**22**)

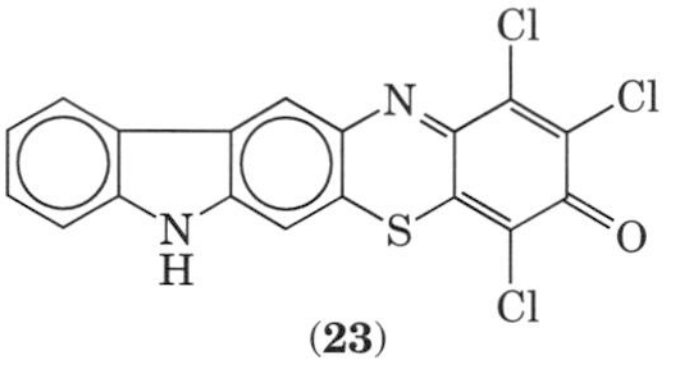

(**23**) 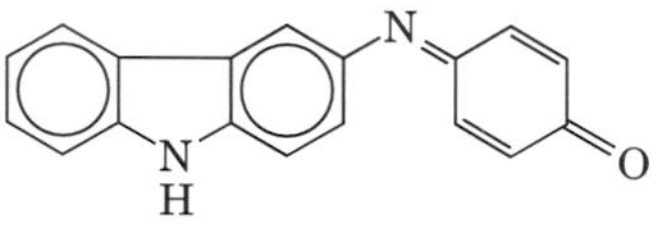   (**24**)

These formulas do not account for all the sulfur or oxygen found in the elemental analysis of the above dyes and formulas have been suggested whereby loosely bound sulfur is depicted as thiozonide or polysulfide sulfur. Oxygen is contained in sulfoxide groups (Fig. 1). For example, formula (**25**) has been suggested for CI Vat Blue 43 (23), formula (**26**) for a green dye from 4-amino-4'-hydroxydiphenylamine (24) and formula (**27**) for CI Sulfur Black 11 from 4-hydroxyphenyl-2-naphthylamine (25) (see Fig. 1). Although the structure of CI Sulfur Black 1, the most important of the sulfur dyes, has been investigated, it has not been established. Starting with 2,4-dinitrophenol, it was found that in color formation one mol ammonia is split off from 3 mol dinitrophenol, indicating that the structure consists of three benzene rings, an important factor that appears to be overlooked in the structural formulas published (26–29).

Sulfurized Vat Dyes. These dyes occupy an intermediate position between the true vat colors and sulfur dyes because, like vat dyes, they are dyed preferentially from a sodium dithionite–caustic soda bath. However, some dyes of this class can also be dyed from an alkali sulfide bath or a combination of the two, depending on the dyeing method used and the nature of the substrate to be dyed. This has led to some confusion because CI Vat Blue 42 and 43 are listed in the constitution section of the *Colour Index* under sulfur dyes. Although inferior to true vat dyes in fastness properties, they offer the advantage of better fastness, especially to chlorine, than conventional sulfur dyes.

In dyehouses where sulfide effluent is a problem, sulfur dyes of good chlorine fastness that dye satisfactorily from a dithionite–caustic alkali bath offer an advantage. Included in this group are CI Sulfur Black 11, CI Sulfur Red 10 [*1326-96-1*] (CI 53228), CI Sulfur Brown 96 [*1326-96-1*] (CI 53228), CI Vat Blue 42, and CI Vat Blue 43. The shades of these dyes can be brightened by the addition of vat dyes thus increasing the fastness of the resulting dye.

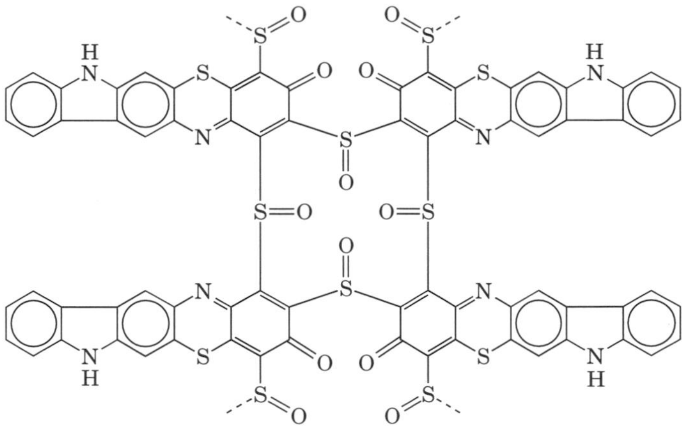

(25)

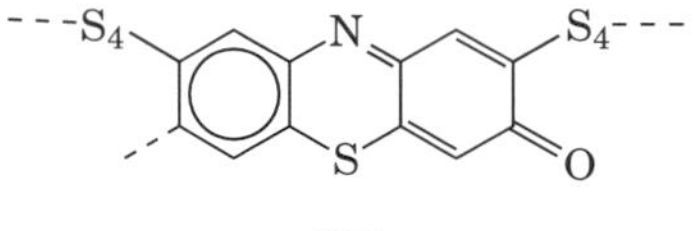

(26)

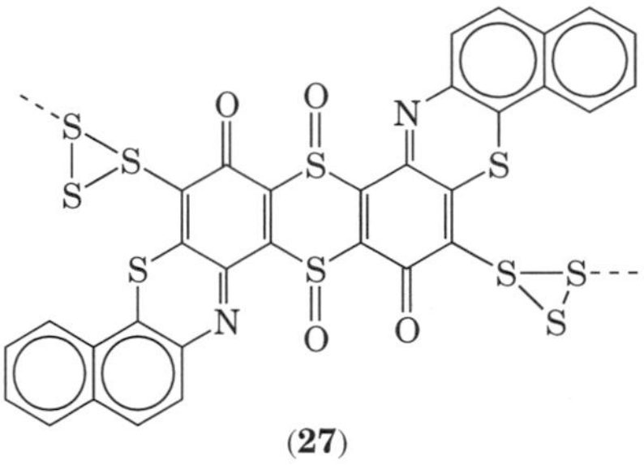

(27)

Fig. 1. Sulfur dyes incorporating sulfoxide, $>$S$=$O (25), polysulfur (26), and thio-zonoide, —S—S— (27) structures.

The structure of the sulfurized vat dyes is uncertain but in the presence of amino or methyl groups, thiazole-ring formation is possible. Examples have been confined to the dyes whose intermediates have been subjected to the conventional sulfur or polysulfide bake (Tables 2–5).

Manufacture

Sulfur Bake. Thionation of the various aromatic amino or nitro compounds with sulfur was formerly carried out in iron pans fitted with agitators and heated by gas fire. However, some bakes begin to stiffen up as the reaction progresses preventing further agitation; consequently, the baking process is not uniform throughout. Baking pans have been replaced by more efficient iron cylinder rotary bakers. The bakers, usually of 500–1000 kg capacity, are heated directly by gas jets or hot flue gases from a fire source and rotated at 2–10 rpm on hollow-end trunnions supported by self-aligning bearing rollers. The off-gases from the reaction are led through a catch pot to a scrubber unit containing caustic soda solution. The spent H_2S is converted to Na_2S and NaHS for recycling. The

Table 2. Sulfur-Bake Dyes

Intermediates	Shade	CI designation		CAS Registry Number
		Name	Number	
(CH₃, NH₂, NH₂-substituted benzene)	orange	Sulfur Orange 1	53050	[1326-49-4]
(CH₃, NH₂, NH₂-substituted benzene and NHCH(=O) diamine)	yellowish brown	Sulfur Brown 26	53090	[1326-60-9]
(CH₃, NH₂, NH₂-substituted benzene and NH₂/NO₂ benzene)	yellowish brown	Sulfur Brown 12	53065	[1327-86-2]
(NH₂/NH₂ benzene and NH₂/CH₃ toluene)	olive	Sulfur Green 12	53045	[1236-48-3]
(CH₃, NHCH(=O), NH₂ benzenes)	reddish yellow	Sulfur Yellow 1	53040	[1326-47-2]
	brown	Sulfur Brown 52	53320	[1327-18-0]

bakers are rotated until the raw dye is ground to a powder. It is discharged and standardized to strength and shade purified by solution in either caustic soda or sodium sulfide. Insoluble matter is removed by filtration. The liquors are evaporated to dryness on a steam-heated rotating single- or double-drum dryer or the dye is precipitated by the addition of acid or sodium bisulfite or by blowing air into the alkaline brew. The H₂S generated is absorbed in caustic soda solution. The precipitated product is filtered, washed, and air blown dry before discharge. The final drying is usually carried out in fan-assisted steam-heated air ovens.

Fluid Polysulfide Melt.   Thionation of the various indophenols is carried out in covered, jacketed, stainless-steel kettles fitted with agitator, thermometer,

Table 3. Polysulfide-Bake Dyes

Intermediate(s)	Shade	CI designation		CAS Registry Number
		Name	Number	
	yellow	Sulfur Yellow 9	53010	[1326-40-5]
	dull reddish brown	Sulfur Brown 56	53722	[1327-87-3]
	olive dull green	Sulfur Green 11 Sulfur Green 1	53165 53166	[12262-52-1] [1326-77-8]
	brown	Sulfur Brown 31	53280	[1327-11-3]
	dull reddish brown	Sulfur Brown 7	53275	[1327-10-2]
	brown	Sulfur Brown 6	53335	[1327-25-9]
	dull green	Sulfur Green 9	53005	[1326-39-2]
	brownish olive	Sulfur Brown 4	53210	[1326-90-5]

reflux condenser, and facilities for removal and absorption of the hydrogen sulfide gas liberated during thionation. When flammable solvents are employed in the melt, flameproof equipment is used. For bright shades, pure intermediates are used and iron is excluded. The sodium polysulfide may be screened to remove its main impurity, ferrous sulfide, which interferes when copper salts are added to the melt and may also have a dulling effect on the final shade. Shades are brightened by water-miscible solvents such as ethanol, butanol, glycerol, ethylene glycol, and Cellosolve. In some cases, the thionation is carried out entirely in

Table 4. Polysulfide-Melt Dyes

Intermediates	Shade	CI designation Name	Number	CAS Registry Number
	bluish black	Sulfur Black 9	53230	[1327-56-6]
	greenish black	Sulfur Black 1	53185	[1326-82-5]
	bright blue	Sulfur Blue 9	53430	[1326-97-2]
	reddish blue–bluish violet	Sulfur Blue 7	53440	[1327-57-7]
	green bluish green	Sulfur Green 3 Sulfur Green 2	53570 53571	[1327-73-7] [1327-74-0]
	dull Bordeaux	Sulfur Red 3	53710	[1327-84-0]

352

Table 4. (Continued)

Intermediates	Shade	CI designation		CAS Registry Number
		Name	Number	
	Bordeaux	Sulfur Red 6	53720	[1327-85-1]
	blue–reddish navy	Vat Blue 42	53640	[1327-81-7]
	bluish black	Sulfur Black 11	53290	[1327-14-6]
	Bordeaux brown	Sulfur Red 10 Leuco Sulfur Brown 96	53228	[1326-96-1]
	reddish blue	Sulfur Blue 12	53800	[1327-96-4]
	dull bluish red	Sulfur Red 7	53810	[1327-97-5]

Table 5. Sulfurized Vat Dyes

Intermediate	Shade	CI designation		CAS Registry Number
		Name	Number	
![structure]—SO₃Hᵃ	yellowish brown	Vat dye	58820	[1328-11-6]
	olive	Vat Green 7	58825	[1328-12-7]
CH₃	yellowish orange	Vat Orange 21	69700	[1328-39-8]
CH₂Cl	yellow	Vat Yellow 21	69705	[1328-40-1]
CH₃	dull greenish blue green	Vat Blue 7 Vat dye	70305 70310	[6505-58-4]

ᵃMono or di.

solvent after first distilling off any water contained in the raw materials. In order to keep the melt fluid, hydrotropic substances such as urea, thiourea, or xylene sulfonates are added. Hydrotropic substances shorten thionation time and increase yield and brightness.

Metallic additions to the melt, usually in the form of copper sulfate, brighten the shade of certain dyes, such as the Bordeaux range made from phenazones and the greens made from the indophenols; the metal forms a complex with the dye. However, copper-containing dyes cannot be applied to material that requires vulcanization.

The composition of the polysulfide varies. It is generally applied in excess with the result that the final dye is wholly or partly in solution in leuco form. At this stage, it can be converted to a liquid form by suitable dilution and the addition of sodium hydrosulfide or diluted, if necessary filtered, and then precipitated by blowing air through the liquor or adding dilute acid. When stronger dyes are required, precipitation may be carried out by adding sodium nitrite directly to the final melt. The excess polysulfide is oxidized and ammonia liberated. Air blowing is usually continued in order to obtain the required brightness of shade. In some cases, caustic soda is added to partly dissolve the dye and air blowing is continued until a satisfactory shade is obtained. The dyestuff is then isolated by filtration, washed, air blown dry, discharged, and dried in a heated air oven.

Application

Sulfur dyes are applied to leuco form. In this form, the dye has affinity for the fiber. After the dye is completely absorbed by the fiber, it is reoxidized *in situ*. In dyes, such as the bright blues which contain quinonimine groups, further reduction takes place in a manner similar to the reduction of the keto group in vat dyes.

In the 1995 edition of the *Colour Index*, the sulfur dyes are classified according to application method and the structure of the intermediates, into ordinary or conventional dyes, leuco or prereduced dyes, and thiosulfonic derivatives of conventional dyes (solubilized sulfur dyes) (30).

The reducing agent traditionally employed with sulfur dyes is sodium sulfide, but sodium sulfhydrate, NaHS, together with a small quantity of alkali such as sodium carbonate or sodium hydroxide is also widely used. The dyebaths prepared in this way are less alkaline than those with sodium sulfide alone which facilitates the rinsing of the dyed goods. Effluent control has resulted in a search for alternative reducing agents (see DYES, ENVIRONMENTAL CHEMISTRY). Alkaline sodium dithionite (sodium hydrosulfite) can be used with some sulfur dyes, particularly the blues, but over-reduces red-browns such as CI Sulfur Brown 12 [*1327-86-2*], CI Sulfur Red 6 [*1327-85-1*], and similar types. Alkaline sodium formaldehyde sulfoxylate has also been employed with blues, but has the same drawbacks. Glucose and sodium hydroxide in almost boiling solution have been proposed for solubilized or dispersed sulfur dyes. Small amounts of sodium sulfide or sodium dithionite together with glucose and sodium hydroxide assist the reduction, but the pH should be maintained above 10.5 throughout the dyeing.

Sodium polysulfide solutions (2–5 g/L), as antioxidants, prevent bronzing. Glucose has been used for a similar purpose and sodium borohydride effectively prevents bronzing when dyeing sulfur black. Other dyebath additives, such as wetting and antifoam agents, may not be compatible with the reducing agents, so must be carefully selected.

The conventional sulfur dye powder is made into a paste with a small amount of soft water and an alkali-stable wetting agent. Boiling for a few minutes in a strong solution of sodium sulfide reduces the dye. The dissolved dye is diluted to the required dyebath volume. When dyeing pale shades, the final bath should contain at least 5 g/L sodium sulfide (60%), irrespective of the amount used to dissolve the dye.

The leuco sulfur dyes are in a prereduced form. They are produced in liquid form by dissolving paste from the parent sulfur dye in a reducing agent containing sodium sulfhydrate and alkali and, occasionally, other hydrotropic agents. These products were first introduced into the United States in 1936 and are available mainly from five manufacturers. They are true solutions requiring only to be diluted with water. Small amounts of reducing agents should be added depending on the dyeing method and the depth of shade required. Solubilized sulfur dyes have no real affinity for the fiber until they are converted into their leuco compounds by a reducing agent. After exhaustion on the fiber, rinsing and oxidation is effected in the same way as for the other types. Solubilized sulfur dyes can be used in aqueous solution to prepad the fabric which can then be treated with reducing agent, with intermediate drying in order to give more level dyeings. Cross-wound packages of yarn may be dyed by this two-stage method. The dyes may be applied in the conventional way by using a leuco dyebath from the start, but it is recommended that they be first dissolved in water only and that the reducing agent is added to the dyebath separately before dyeing.

In the pad–steam method, the padding liquor is prepared by adding the required quantity of prereduced liquid sulfur dye to water at 25–40°C together with additional chemicals and auxiliaries, depending on the dye and depth of shade desired. The padding liquor is fed into the trough of the dye padder which is fitted with a constant leveling device and temperature control. A padding mangle squeezes the well-prepared fabric evenly to provide a pickup of 50–80% of its own weight. The steamer, which should be free from air when operating, is generally set at 102–104°C. A dwell time of 60 s is usually adequate. The remainder of the equipment consists of washing and aftertreatment boxes through which the fabric passes successively before being dried. The continuous operation used in this process is depicted in Figure 2. A typical padding liquor recipe for a sulfur navy blue shade on 100% cotton drill is given in Table 6.

Blends of polyester with cotton (qv) or viscose are first dyed with disperse dyes, then with sulfur dyes (see FIBERS, POLYESTER; FIBERS, REGENERATED CELLULOSICS). Disperse and sulfur dyes can also be applied simultaneously in a pad–dry–thermofix/chemical reduction pad–steam sequence. In this case, the sulfur dyes cannot be used in their reduced form because of the effect of the sodium

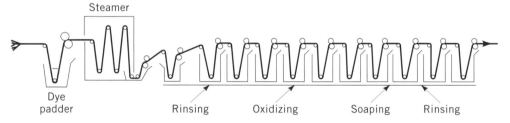

Fig. 2. Continuous operation of pad–steam dyeing.

Table 6. Padding Liquor Recipe at 30°C

Ingredient	g/L[a]
Sulphol Liquid Dark Blue QL (JR)	150
Leucad 71 sodium sulfhydrate, 35%	5
caustic soda 100%	0.5
anionic wetting agent	2.0[b]
EDTA[c]-based sequestrant	0.5

[a]Unless otherwise stated.
[b]mL/L.
[c]EDTA = ethylenediaminetetraacetic acid.

sulfide on the disperse dye. Therefore, this method is confined to the solubilized sulfur dyes or sulfur dyes in the dispersed form.

Solubilized sulfur dyes can also be applied without reducing agents (31). The dye, together with urea and thiourea or similar compounds, is padded on 100% cotton, then dried and thermofixed at 150–175°C. In the case of polyester–cotton, suitable disperse dyes can be added to the padding liquor and thermofixed at higher temperatures together with the solubilized sulfur dye (32). Other uses of finely dispersed sulfur dyes without reducing agents have been described (33).

Before the pressurized package dyeing machine became established, sulfur dyes were often employed for dyeing cotton warps in rope form (ball warps). The yarn in ropes or cables, which had been doubled into a W-form, was passed several times through a series of baths of reduced dye liquor and baths of rinsing water with intermediate squeezing. The depth of shade was gradually built up; the method resulted in a dyeing of excellent uniformity of shade across a woven fabric. This method is rarely used in the 1990s.

A variation is applied where warps for denims are first dyed with sulfur black or blue and then with indigo. The sulfur dye in its reduced form is added to the prewetting or scouring bath. The yarn is passed through and rinsed before passing to the part of the machine where the indigo is applied by the traditional method of successive stages of impregnation and air oxidation. The sulfur dye improves the appearance and reduces the quantity of indigo required.

Aftertreatment. Because the dye is applied by reduction and oxidation, many methods are available to obtain the correct hue. Air oxidation takes place gradually after the residual reducing agent has been rinsed away, but in general chemical oxidation is faster. The traditional oxidizing agents include sodium or potassium bichromate mixed with acetic acid; addition of copper sulfate slightly improves lightfastness. However, because of ecological restrictions on bichromates, other oxidizing agents are coming into use, such as hydrogen peroxide, sodium perborate, and products based on potassium iodate or sodium bromate mixed with acetic acid. Sodium chlorite is used in alkaline solution together with detergent. These reagents give different effects and are more suitable under some conditions than others. Generally speaking, hydrogen peroxide and sodium perborate give dyeings that are slightly less fast to wet processing than other products.

Aftertreatments include resin finishes, which improve fastness properties, and dye-fixing agents of the epichlorhydrin–organic amine type. These agents react with the dye to give condensation products that are not water soluble and hence more difficult to remove.

Tendering Effects. Cellulosic materials dyed with sulfur black have been known to suffer degradation by acid tendering when stored under moist warm conditions. This effect may result from the liberation of small quantities of sulfuric acid which occurs when some of the polysulfide links of the sulfur dye are ruptured. A buffer, such as sodium acetate, or a dilute alkali in the final rinse, especially after oxidation in acidic conditions, may prevent this occurrence. Copper salts should never be used with sulfur black dyes because they catalyze sulfuric acid generation. Few instances of tendering with sulfur dyes other than black occur and the problem is largely confined to cotton.

Economic Aspects

The low cost of sulfur dyes, coupled with good fastness properties and the ease of application, continues to ensure a high consumption. The number of manufacturers has fallen since the 1960s and production is mainly confined to the United States, U.K., Germany, and Spain. There is one principal producer in each of these countries. There are other, less well-known manufacturers in Russia, the People's Republic of China, South Korea, Japan, and Brazil (Table 7).

Few production figures have been available since 1966, when sulfur dyes represented 9.1% of total U.S. dye production and 15.8% of the dyes made for use on cellulosic fibers. World production is estimated at 110,000–120,000 t/yr. This is the highest percentage of any group of dyes. In terms of value, however, the picture is very different because the dyes are relatively inexpensive.

Table 7. Principal Sulfur Dye Manufacturers and Products

Range name	CI classification	Manufacturer	Country
Asathio	sulfur	Asahi Chemical Co.	Japan
Cassulfon	leuco sulfur, liquid	Dystar	Germany
Diresul	leuco sulfur, liquid	Clariant Corp.	Spain
Endurol	vat, sulfurized vat dyes	James Robinson, Ltd.	United Kingdom
Episol	solubilized sulfur[a]	James Robinson, Ltd.	United Kingdom
Hydron	vat, sulfurized vat dyes	Dystar	Germany
Hydrosol	solubilized sulfur	Dystar	Germany
Immedial	sulfur	Dystar	Germany
Indocarbon	sulfur, dispersed form	Dystar	Germany
Kayasol	solubilized sulfur	Nippon Kayaku Co., Ltd.	Japan
Kayaku Homodye	sulfur, dispersed form	Nippon Kayaku Co., Ltd.	Japan
Pirocard	sulfur	Clariant Corp.	Spain
Pirosol	solubilized sulfur	Clariant Corp.	Spain
Sodyeco Liquid	sulfur, sulfurized vat dyes, liquid	Clariant Corp.	United States
Sodyesul	leuco sulfur	Clariant Corp.	United States
Sodyevat	vat, sulfurized vat dyes, dispersed form	Clariant Corp.	United States
Sulphol	sulfur	James Robinson, Ltd.	United Kingdom
Sulphol Liquid	leuco sulfur, liquid	James Robinson, Ltd.	United Kingdom
Sulphosol	solubilized sulfur	James Robinson, Ltd.	United Kingdom
Youhaodron	vat, sulfurized vat	China National Chemicals Import and Export Corp.	People's Republic of China
Youhao Sulphur	sulfur		

[a] For leather.

Commercial Forms of Sulfur Dyes

Powders. Powders have, in the past, been the principal form in which sulfur dyes were sold. In general, they are made from the dried press cake, finely ground, and standardized with common salt, sodium sulfate, or soda ash. They are prepared for dyeing by making a paste with water, which is dissolved by boiling with the necessary amount of reducing agent and further addition of water.

Prereduced Powders. These are usually made from press cake paste to which a reducing agent has been added, such as sodium sulfide, sodium hydrosulfide, or sodium dithionite, which solubilize the dye in water. Before drying, the dye paste may be mixed with dispersing and stabilizing agents to aid application.

Grains. Grains are usually prereduced powders. The quantities of sodium sulfide, hydrosulfide, and mineral salts are adjusted to give a grainy product when the paste slurry is dried on a steam-heated drum dryer. Grains offer the advantage of not being dusty.

Dispersed Powders. The principal use of dispersed powders is in pad–dry–chemical pad–steam dyeing techniques. They are normally made from press cake by ball or bead milling to microparticle size in the presence of dispersing agents. The drying is strictly controlled and is carried out in the presence of anticoagulants to prevent aggregation of the dispersed dye particles.

Dispersed Pastes. The milled pastes vary in strength, but for ease of handling the consistency of the dispersed pastes generally permits pouring from the container. Because there is no drying step, dispersed pastes are usually cheaper than dispersed powders.

Liquids. Some liquid dyes are made directly from the thionation melt by additions of caustic soda and sodium hydrosulfide. Hydrotropic substances are sometimes added, either at the initial thionation stage or after the polysulfide melt is finished in order to keep the reduced dye in solution. Partly reduced liquids are also available. They are usually more concentrated than fully reduced liquids, thus saving packaging and transportation costs. However, they require a further addition of reducing agent to the dyebath in order to obtain full color value. On the other hand, fully reduced liquids are ready to use, because the amount of reducing agent for each dye has been carefully controlled in order to obtain maximum stability on storage and maximum color yield in use. They are not affected by low temperatures as are the dispersed pastes.

Water-Soluble Brands. Water-soluble sulfur dyes occur in both powder and liquid form. The liquid product is cheaper because no drying step is required. These dyes are the Bunte salts, or thiosulfuric acid derivatives of the sulfur dyes. They are made by warming the polysulfide-free pastes with sodium sulfite or bisulfite until they are water soluble. They are salted out from solution or isolated by drum drying the liquor. Aqueous solutions of these dyes show little or no affinity for cellulosic fiber until a reducing agent has been added and, therefore, they penetrate tightly woven materials and give level dyeings with less tendency to bronzing.

Health and Safety

During the 1980s and 1990s, much more emphasis has been placed on health and safety aspects within the chemical and dyestuff industries. As a consequence, benzidine and β-naphthylamine, which are known carcinogens, have been banned in many countries. In some cases, alternatives to these intermediates have been found in order to retain dyestuffs with similar shades and properties. For example, bright yellow to orange sulfur dyes are obtained from a sulfur bake of 4-(6-methyl-2-benzothiazolyl)aniline, N,N'-diformyl-p-phenylenediamine, and 2,2'- or 4,4'-diformyldiaminodiphenyl disulfide (34). These shades were previously obtainable only from mixtures containing benzidine. The handling of hazardous chemicals such as nitroanilines, dinitro- and diaminotoluenes, nitro- and dinitrophenols, and chlorodinitrobenzenes is kept to a minimum. In all cases, suitable protective gear is employed. Hydrogen sulfide is one of the chief hazards of sulfur dye manufacture and is usually removed by caustic soda scrubber units. Accidental escape of H_2S can be detected by a system that incorporates detector heads specifically for hydrogen sulfide. Process sheets should list all the hazards associated with each operation.

The effluent from the dye manufacture and textile dyeing industry is usually treated by aeration in large tank farms. The aeration equipment may be a floating rotary blade beater or vertical helix columns through which compressed air is passed or alternatively dissipated through venturi pipes. By these means, ca 10–12% of the oxygen in the air can be effectively utilized in the degradation of the sulfur dye waste, rendering it acceptable to municipal sewage treatment farms. Small dye houses utilize spent flue gases or hydrogen peroxide as well as ferrous salts to treat plant effluent. However, such an operation is more costly and mixing with spent bleach liquors may reduce costs. Anthraquinone sulfonic acids can be used as catalysts to assist aeration.

Uses

The sulfur dyes are widely used in piece dyeing of traditionally woven cotton goods such as drill and corduroy fabrics (see TEXTILES). The cellulosic portion of polyester–cotton and polyester–viscose blends is dyed with sulfur dyes. Their fastness matches that of the disperse dyes on the polyester portion, especially when it is taken into account that these fabrics are generally given a resin finish.

Yarn is dyed with sulfur dyes, although raw stock dyeing has declined in the 1990s. The dyeing of knitted fabrics, both 100% cotton or blends of cotton with synthetic fibers, is increasing. Although the problems of premature oxidation of the dyebath should not be underrated, sulfur dyes are successfully applied to this type of fabric on open and closed winches and on modern jet-dyeing machines (35). Piece dyeing on the jig uses considerable quantities of sulfur dyes. Conventional methods are used, but special care is necessary to ensure that the dyes are completely dissolved, preferably in a separate vessel, permitting the boiling of dye and reducing agent together.

Cotton yarn is dyed in package machines and the dye exhausted by increasing the temperature and adding salt. The dye must be completely dissolved when preparing the dyebaths to avoid contamination with undissolved dye in the yarn package. The increased availability of the prereduced liquid dyes and the improved quality of sodium sulfide have reduced this problem. Incorrectly dissolved dye was previously the cause of most faulty dyeings.

Continuous dyeing of piece goods by pad–steam methods is one of the principal outlets for sulfur dyes, mostly in prereduced liquid form. Nontextile uses for sulfur dyes are limited. Sulfur Black is used for dyeing paper, particularly for lamination applications. Leather dyeing is done with the solubilized sulfur Episol dyes. These dyes give well-penetrated dyeings with far better wetfastness than is usually obtained with acid or direct dyes. The dyes are applied to the neutralized leather at pH 7.5–8.0 without reducing agents, and are fixed in a manner similar to the fixation of acid dyes with formic acid at pH 3.8.

BIBLIOGRAPHY

"Sulfur Dyes" in *ECT* 1st ed., Vol. 13, pp. 445–458, by J. J. Ayo, Jr., and E. Kuhn, General Aniline & Film Corp; in *ECT* 2nd ed., Vol. 19, pp. 424–441, by D. G. Orton, Southern Dyestuff Co., a division of Martin Marietta Corp.; in *ECT* 3rd ed., Vol. 22, pp. 168–189, by R. A. Guest and W. E. Wood, James Robinson & Co., Inc.

1. E. Croissant and L. M. F. Bretonnière, *Bull. Soc. Ind. Mulhouse* **44**, 465 (1874); Brit. Pat. 1489 (Apr. 24, 1873).
2. Ger. Pat. 84,632 (Mar. 22, 1893); Ger. Pat. 85,330 (Dec. 10, 1893); Brit. Pat. 19,980 (1893); U.S. Pat. 53,248 (1893), R. Vidal (to Société Anonym Matières Colorantes).
3. Ger. Pat. 103,861 (Oct. 24, 1897); U.S. Pat. 610,541 (1897); Brit. Pat. 25,234 (1897), G. Kalischer (Cassella & Co.).
4. Ger. Pat. 127,835 (Dec. 7, 1899), Priebs and O. Kaltwasser (to Aktien-Gesellschaft, Berlin).
5. Ger. Pat. 134,947 (Aug. 19, 1900); U.S. Pat. 693,633 (Aug. 24, 1900), A. Weinberg and R. Herz (to Cassella & Co.).
6. Ger. Pat. 126,175 (Aug. 11, 1901); U.S. Pat. 701,435 (Jan. 5, 1901), A. Weinberg (to Cassella & Co.).
7. Ger. Pat. 139,430 (Jan. 12, 1903); Ger. Pat. 141,576 (May 7, 1902); Ger. Pat. 152,595 (May 7, 1902), A. Weinberg (to Cassella & Co.).
8. Ger. Pat. 162,156 (May 25, 1904), M. Boninger (to Sandoz).
9. Ger. Pat. 218,371 (Dec. 29, 1908); Brit. Pat. 2198 (Feb. 6, 1909), L. Haas and R. Herz.
10. Ger. Pat. 261,651 (Oct. 24, 1911) (to Aktien-Gesellschaft Für Aniline Fabrikation, Berlin).
11. Brit. Pat. 411,431 (Nov. 29, 1932), W. Zerweck (to I. G. Farbenindustrie).
12. Ger. Pat. 653,675 (Nov. 30, 1937); Ger. Pat. 655,487 (Jan. 17, 1938), W. Hagge and K. Haagen.
13. U.S. Pat. 2,130,415 (Sept. 20, 1938), A. J. Buchanan (to Southern Dyestuffs Corp.).
14. Brit. Pat. 547,853 (Mar. 12, 1941), N. H. Haddock (to ICI).
15. Brit. Pat. 541,146 (May 13, 1940), N. H. Haddock (to ICI).
16. U.S. Pat. 2,369,666 (Feb. 20, 1945), A. L. Fox (to E. I. du Pont de Nemours & Co., Inc.); Brit. Pat. 573,831 (Dec. 7, 1945) (to ICI).
17. H. Hiyama, *Synthetic Sulfur Dyes*, Osaka Municipal Technical Research Institute, Japan, 1959.
18. Ger. Offen. 2,347,537 (1975), E. Krusche and co-workers (to Cassella & Co.).
19. Brit. Pat. 4097/06 (1906), Schmidt (to Cassella).
20. Ger. Pat. 693,862 (July 19, 1936), R. Rieche and B. Schiedk (to I.G. Farbenindustrie); P. Bachmann, *Ber.* **36**, 965 (1903).
21. R. Gnehm and F. Kauffler, *Ber.* **37**, 2619, 3032 (1904).
22. *Mon. Scient.* **11**, 655 (1897); *Mon. Scient.* **17**, 427 (1903).
23. K. H. Shah, B. D. Tilak, and K. Venkataraman, *Proc. Indian Acad. Sci. Sect. A* **28**, 111 (1948).
24. W. N. Jones and E. E. Reid, *J. Am. Chem. Soc.* **54**, 4393 (1932).

25. H. E. Fierz-David and E. Merian, *Abr. der Chem. Tech. der Textilfasern* **146** (1948).
26. I. Chmelnitzkaja and V. Werchowskaja, *Anilinokrasochnaya Prom.* **5**, 67 (1935).
27. T. Kubota, *J. Chem. Soc. Jpn.* **55**, 565 (1934).
28. H. Hiyama, *J. Chem. Soc. Jpn. Ind. Chem. Sect.* **51**, 97 (1948).
29. A. Weinberg, *Ber.* **63**(A), 117 (1930).
30. *J. Soc. Dyers Colour.* **89**, 416, 420, 421.
31. C. Heid, *Z. ges Textilind.* **70**, 626 (1968).
32. Brit. Pat. 1,321,453 (Feb. 24, 1970), H. Smithson (to J. Robinson & Co., Ltd.).
33. EP-A 501,197 (1992), W. Bauer and co-workers (to Cassella & Co.).
34. Jpn. Appl. 71 9309 (Feb. 26, 1971), S. Yoshioka (to Nippon Kayaku & Co., Ltd.).
35. H. M. Tobin, *Am. Dyest. Rep.* **68**(9), 26 (Sept. 1979).

General References

H. E. Fierz-David, *Künstliche Organic Farbstoffe*, Julius Springer, Berlin, 1926.
O. Lange, *Die Schwelfelfarbstoffe, Ihre Herstellung und Verwendung*, 2nd ed., Spamer, Leipzig, Germany, 1925.
H. A. Lubs, *The Chemistry of Synthetic Dyes and Pigments*, Reinhold Publishing Corp., New York, 1955, Chapt. 6.
J. F. Thorpe and M. A. Whiteley, eds., *Thorpe's Dictionary of Applied Chemistry*, 4th ed., Vol. 11, Longmans, Green & Co., London, 1954.
K. Venkataraman, *The Chemistry of Synthetic Dyes and Pigments*, Vol. 2, 1952, Chapt. 35; Vol. 3, 1970, Chapt. 1; Vol. 7, 1974, Chapt. 1, Academic Press, London and New York.
M. L. Crossley and co-workers, *BIOS (British Intelligence Objectives Committee) Misc.*, Vol. 55, H. M. Stationery Office, London, 1946.
J. Avery and co-workers, *BIOS 983, 1946*; D. A. W. Adams and co-workers, *BIOS 1155*, H. M. Stationery Office, London, 1947.
FIAT (Field Information Agency Technical) Microfilm 764, H. M. Stationery Office, London, 1946, reels 82CC, 92AA, and 186C.
FIAT 1313, Vols. 2 and 3, H. M. Stationery Office, London, 1948.
C. Heid, K. Holoubek, and R. Klein, *Int. Textile Rep.* **54**, 1314 (1973).
W. Prenzel, *Chemiefasern + Text. Anwendungstech./Text. Ind.* **24/76**, 293 (1974).
Colour Index, 3rd ed., The Society of Dyers and Colourists, Bradford, U.K., and The American Association of Textile Chemists and Colorists, Research Triangle Park, N.C., 1975, rev. ed., 1982.
W. E. Wood, *Rev. Prog. Color. Relat. Tap.* **7**, 80 (1976).
Klein, *J.S.D.C.* **98**, 110 (1982).
Aeberhard, *Textilveredlung* **16**, 442 (1981).
Von der Eltz, *J.S.D.C.* **101**, 168 (1985).
J. Water Poll. Cont. Fed. **46**, 2778 (1974).

J. SENIOR
R. A. GUEST
W. E. WOOD
James Robinson, Ltd.

SULFURIC ACID AND
SULFUR TRIOXIDE

Sulfuric acid [7664-93-9], H_2SO_4, is a colorless, viscous liquid having a specific gravity of 1.8357 and a normal boiling point of approximately 274°C. Its anhydride, sulfur trioxide [7446-11-9], SO_3, is also a liquid, having a specific gravity of 1.857 and a normal boiling point of 44.8°C. Sulfuric acid is by far the largest-volume chemical commodity produced. It is sold or used commercially in a number of different concentrations, including 78 wt % (60° Bé), 93 wt % (66° Bé), 96 wt %, 98–99 wt %, 100%, and as various oleums, ie, fuming sulfuric acid [8014-95-7], $H_2SO_4 + SO_3$. Stabilized and unstabilized liquid SO_3 are items of commerce.

Sulfuric acid has many desirable properties that lead to its use in a wide variety of applications, including production of basic chemicals, steel (qv), copper (qv), fertilizers (qv), fibers (qv), plastics, gasoline (see GASOLINE AND OTHER MOTOR FUELS), explosives (see EXPLOSIVES AND PROPELLANTS), electronic chips, batteries (qv), and pharmaceuticals (qv). It typically is less costly than other acids; it can be readily handled in steel (qv) or common alloys at normal commercial concentrations. It is available and readily handled at concentrations >100 wt % (oleum). Sulfuric acid is a strong acid; it reacts readily with many organic compounds to produce useful products. Sulfuric acid forms a slightly soluble salt or precipitate with calcium oxide or hydroxide, the least expensive and most readily available base. This is a useful property when it comes to disposing of sulfuric acid. Concentrated sulfuric acid is also a good dehydrating agent and under some circumstances it functions as an oxidizing agent.

History

Sulfuric acid has been an important item of commerce since the early to middle 1700s. It has been known and used since the Middle Ages. In the eighteenth and nineteenth centuries, it was produced almost entirely by the chamber process, in which oxides of nitrogen (as nitrosyl compounds) are used as homogeneous catalysts for the oxidation of sulfur dioxide. The product made by this process is of rather low concentration (typically 60° Baumé, or 77–78 wt % H_2SO_4). This is not high enough for many of the commercial uses of the 1990s. The chamber process is therefore considered obsolete for primary sulfuric acid production. However, more recently, modifications to the chamber process have been used to produce sulfuric acid from metallurgical off-gases in several European plants (1,2).

During the first part of the twentieth century, the chamber process was gradually replaced by the contact process. The primary impetus for development of the contact process came from a need for high strength acid and oleum to make synthetic dyes and organic chemicals. The contact process employing platinum catalysts began to be used on a large scale late in the nineteenth century. The pace of its development was accelerated during World War I in order to provide concentrated mixtures of sulfuric and nitric acid for explosives production.

In 1875, a paper by Winkler awakened interest in the contact process, first patented in 1831. Winkler claimed that successful conversion of SO_2 to SO_3 could only be achieved with stoichiometric, undiluted ratios of SO_2 and O_2.

Although erroneous, this belief was widely accepted for more than 20 years and was employed by a number of firms. Meanwhile, other German firms expended a tremendous amount of time and money on research. This culminated in 1901 with Knietsch's lecture before the German Chemical Society (3) revealing some of the investigations carried out by the Badische Anilin-und-Soda-Fabrik. This revealed the abandonment of Winkler's theory and further described principles necessary for successful application of the contact process.

In 1915, an effective vanadium catalyst for the contact process was developed and used by Badische in Germany. This type of catalyst was employed in the United States starting in 1926 and gradually replaced platinum catalysts over the next few decades. Vanadium catalysts have the advantages of exhibiting superior resistance to poisoning and being relatively abundant and inexpensive, compared to platinum. After World War II, the typical size of individual contact plants increased dramatically in the United States and around the world in order to supply the rapidly increasing demands of the phosphate fertilizer industry. The largest sulfur burning plants as of the mid-1990s produce approximately 3300 metric tons of acid per day. Plants using sulfur in other forms, especially SO_2 from smelting operations (metallurgical plants), have also increased in size. One metallurgical plant has been built to produce 3500 metric tons of acid per day.

Another significant change in the contact process occurred in 1963, when Bayer AG announced the first large-scale use of the double-contact (double-absorption) process (4–7). In this process, SO_2 gas that has been partially converted to SO_3 by catalysis is cooled, passed through sulfuric acid to remove SO_3, reheated, and then passed through another one or two catalyst beds. Through these means, overall conversions can be increased from about 98% to >99.7%, thereby reducing emissions of unconverted SO_2 to the atmosphere. Because of worldwide pressures to reduce SO_2 emissions, most plants as of the mid-1990s utilize double-absorption. An early U.S. patent (8) disclosed the general concept of this process but apparently was not reduced to practice at that time.

Physical Properties

Sulfur Trioxide.   Pure sulfur trioxide [7446-11-9] at room temperature and atmospheric pressure is a colorless liquid that fumes in air. This material can exist in both monomeric and polymeric forms. In the gaseous and liquid state pure SO_3 is an equilibrium mixture of monomeric SO_3 and trimeric S_3O_9 (9), also called γ-SO_3.

$$3\,SO_3 \rightleftharpoons S_3O_9$$

In the gaseous state the equilibrium lies far to the left (10). In the liquid state the amount of S_3O_9 is reported by some investigators to be ~25% at 25°C (11). Others report that the liquid is primarily S_3O_9 (12). For both gas (10,13) and liquid (14–16), the degree of association increases with decreasing temperature.

If the SO_3 is pure, it freezes to γ-SO_3, also called ice-like SO_3, at 16.86°C (17). It is possible that some monomeric SO_3 may also be present in the crystal structure of the ice-like form (15,18–20).

Traces of moisture, ie, of H_2SO_4, as low as 10^{-3} mol % (15,21) cause liquid SO_3 to polymerize first to a low melting, asbestos-like form, β-SO_3. β-SO_3 forms crystals with a silky luster. Additional reaction involves cross-linking of β-SO_3 to form a high melting asbestos-like form, α-SO_3. Alpha-SO_3 crystals resemble ice needles. An additional form has been mentioned in the literature as vaselinartiges (22), ie, vaseline-like, or gelatinous (23–25). This form it has been speculated, is partially cross-linked β-SO_3. Both the alpha and beta forms melt to give γ-SO_3. The early literature used reversed nomenclature for α-, β-, and γ-SO_3 when α- was the ice-like form; β- the low melting, asbestos-like form; and γ- the high melting, asbestos-like form.

β-SO_3 consists of helical chain molecules (26) of unknown length (21). α-SO_3 is also a polymer similar to beta-SO_3, but probably in a layered cross-linked structure (20,21). Melting points, or more precisely triple points, of 32.5°C and 62.2°C have been given to the β- and α-polymers respectively (20,27). The presence of these values persists in product literature, but they are rarely observed in industrial practice. The β-polymer melts only slowly a few degrees above its reported melting point (28). The α-polymer can best be melted by heating under pressure to 80°C (23,28). Without pressure, the polymer sublimes. Initial melting of β- often leaves behind a residue which is much more difficult to melt. This residue is probably cross-linked β-polymer that may be a precursor to pure α-polymer (24). Because of the slowness of melting and the lack of a distinct melting point, the melting process is believed to be a slow depolymerization rather than a general disintegration of the entire polymer molecule (29).

Even in the presence of considerable moisture, solid polymer never forms above 30°C (23). Below 30°C, liquid stability decreases with increasing moisture and decreasing temperature (23). The actual formation of solid polymer has been hypothesized to involve the formation in the liquid of high molecular weight polysulfuric acids, followed by precipitation.

In perfectly dry SO_3 no sulfuric acid molecules would exist. But in the presence of even a trace of water, it seems likely that high molecular weight polysulfuric acids exist. As the higher polyacids form, the increasing molecular weight would be expected to decrease their solubility, leading to precipitation. Owing to kinetic considerations, lower temperatures also favor the formation and hence precipitation of higher polyacids (24). The formation of polyacids in solution may be relatively rapid and the rate of formation of solid beta-polymer controlled by the rate of nucleation (24). Temperatures below 0°C do not appear to accelerate polymer formation as is sometimes believed. Instead, samples quickly cooled to −30°C and −78°C, held, and then quickly rewarmed showed less polymer than SO_3 slowly cooled to 0°C, and then slowly rewarmed. The cycle times for all samples were the same (24).

A study on the thermodynamic properties of the three SO_3 phases is given in Reference 30. Table 1 presents a summary of the thermodynamic properties of pure sulfur trioxide. A significantly lower value has been reported for the heat of fusion of γ-SO_3, 24.05 kJ/kg (5.75 kcal/kg) (41) than that in Table 1, as have slightly different critical temperature, pressure, and density values (32).

Table 1. Properties of Sulfur Trioxide

Property	Value	Reference
critical temperature, °C	217.8	31
critical pressure, kPa^a	8208	31
critical density, g/cm^3	0.630	31
triple point temperature, γ-phase, °C	16.8	30
triple point pressure, γ-phase, kPa^a	21.13	27
normal boiling point, °C	44.8	27
melting point, γ-phase, °C	16.8	27
transition temperature, °C	−183.0	30
density, γ-phase, g/cm^3		
liquid at 20°C	1.9224	32
solid at −10°C	2.29	33
coefficient of thermal expansion at 18°C, °C^{-1}	0.002005	34
liquid heat capacity, at 30°C, kJ/(kg·°K)^b	3.222	32,35
heat of formation of gas at 25°C, kJ/mol^b	−395.76	36
free energy of formation of gas at 25°C, kJ/mol^b	−371.07	36
entropy of gas, at 25°C, kJ/(mol·K)^b	256.8	36
heat of dilution, kJ/kg^b	2.110	32
heat of fusion, kJ/kg^b		
α	322.4	32
β	150.7	32
γ	92.11	32
heat of sublimation, MJ/kg^b		
α	0.8518	30
β	0.7269	30
γ	0.7029	30
heat of vaporization, γ-liquid, MJ/kg	0.5843	30
diffusion in air, at 80°C, m/s	0.000013	37
liquid dielectric constant, at 18°C	3.11	38
electric conductivity	negligible	39,40

^{a}To convert kPa to psi, multiply by 0.145.
^{b}To convert J to cal, divide by 4.184.

Figure 1 shows the density of sulfur trioxide as a function of temperature. This curve (42) is a composite of data taken from the literature (43–46). The vapor pressures of sulfur trioxide's α-, β-, and γ-phases are presented in Figure 2 (47). Different values of SO$_3$ vapor pressure for α-, β-, and γ-phases have been reported in References 30 and 32 (Table 2).

The thermodynamic properties of sulfur trioxide, and of the oxidation reaction of sulfur dioxide are summarized in Tables 3 and 4, respectively. Thermodynamic data from Reference 49 are believed to be more accurate than those of Reference 48 at temperatures below about 435°C.

Sulfuric Acid.   Sulfuric acid is a dense, colorless liquid at room temperature, having specific gravity as shown in Figure 3 (50). Historically, the concentration of sulfuric acid has been reported as specific gravity (sp gr) in degrees

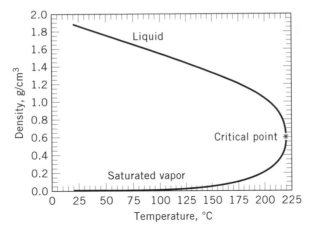

Fig. 1. Density of sulfur trioxide, where * represents the critical point (42–46).

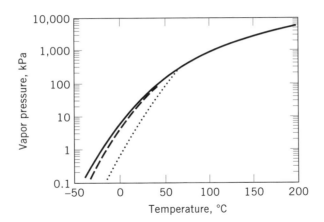

Fig. 2. Vapor pressure of sulfur trioxide: (——), γ-SO$_3$; (– – –), β-SO$_3$; and (••••), α-SO$_3$ (47). To convert kPa to psi, multiply by 0.145.

Table 2. Vapor Pressure of Sulfur Trioxide[a]

	Vapor pressure, Pa[b]		
Temperature, °C	α-SO$_3$	β-SO$_3$	γ-SO$_3$
0	773	4,266	5,999
25	9,732	45,860	57,730
50	86,660	126,700	126,700
75	400,000	400,000	400,000

[a]Ref. 32.
[b]To convert Pa to psi, divide by 6895.

Table 3. Thermodynamic Properties of Sulfur Trioxide[a]

Temperature, K	ΔH_f, kJ/mol[b]	ΔF_f, kJ/mol[b]
600	−460.2	−359.9
700	−459.6	−343.3
800	−458.9	−326.7
900	−457.9	−310.2
1000	−456.7	−293.9
1100	−455.5	−277.7
1200	−454.1	−261.5

[a]Ref. 48.
[b]To convert J to cal, divide by 4.148.

Table 4. Thermodynamic Properties of $SO_2 + 1/2\ O_2 \rightarrow SO_3$[a]

Temperature, K	ΔH_T, kJ/mol[b]	ΔF_T, kJ/mol[b]	K_p, Pa$^{-1/2}$[c]
600	−97.99	−41.59	13.13
700	−97.36	−32.30	0.7445
800	−96.57	−23.05	0.1006
900	−95.69	−13.97	0.02033
1000	−94.60	−4.94	0.005686
1100	−93.51	4.02	0.002025
1200	−92.30	12.84	0.000867

[a]Ref. 48.
[b]To convert J to cal, divide by 4.148.
[c]To convert Pa$^{-1/2}$ to atm$^{-1/2}$, multiply by 318.32.

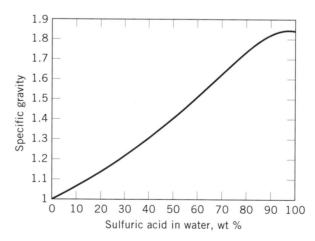

Fig. 3.   Specific gravity of sulfuric acid, 15°C/4°C (50).

Baumé. In the United States, the Baumé scale is calculated by the following formula:

$$°\text{Bé} = 145 - \left(\frac{145}{\text{sp gr}} \right)$$

In Germany and France the Baumé scale is calculated using 144.3 as the constant. The Baumé scale only includes the sulfuric acid concentration range of 0–93.19% H_2SO_4. Higher concentrations are not included in the Baumé scale because density is not a unique function of concentration between 93% and 100% acid. The density of sulfuric acid versus temperature and concentration is shown in Figure 4 (50).

Figures 5 and 6 present the electrical conductivity of sulfuric acid solutions (51,52). For sulfuric acid solutions in the 90–100% H_2SO_4 concentration range, the electrical conductivity measurements reported by Reference 52 are believed to be the best values; other conductivity data are also available (53,54).

The viscosity of sulfuric acid solutions is plotted in Figure 7 (55); other viscosity data may be found in References 54–60. Surface tension of sulfuric acid solutions is presented in Figure 8 (61). Surface tension of selected concentrations of sulfuric acid as a function of temperature up to the boiling point is given in Reference 62; other data are also available (58,59,63–65).

The index of refraction of sulfuric acid solutions (62) and additional related data (66), along with solubility data for oxygen in sulfuric acid solutions (67), are available in the literature. The solubility of sulfur dioxide in concentrated sulfuric acid is shown in Figure 9 (68); additional data are also available (69).

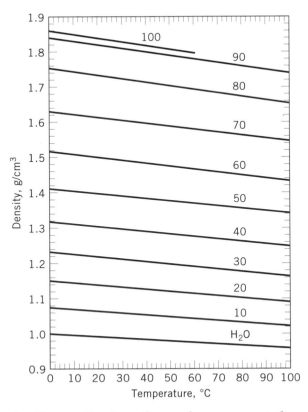

Fig. 4. Density of sulfuric acid, where the numbers represent the wt % of H_2SO_4 in H_2O (50).

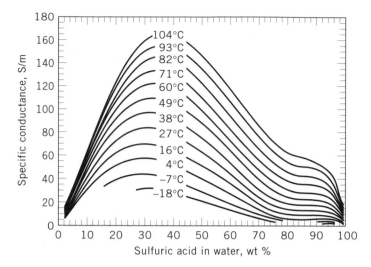

Fig. 5. Specific conductance of sulfuric acid (51).

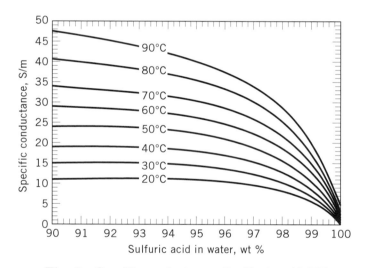

Fig. 6. Specific conductance of sulfuric acid (52).

Data on chemical properties such as self-dissociation constants for sulfuric and dideuterosulfuric acid (60,65,70,71), as well as an excellent graphical representation of physical property data of 100% H_2SO_4 (72), are available in the literature. Critical temperatures of sulfuric acid solutions are presented in Figure 10 (73).

Boiling points of sulfuric acid are given in Figure 11. There is some uncertainty in the data close to 100% H_2SO_4 (74). Freezing points also are not well established, in part because acid purity and cooling rates significantly affect the observed freezing points. Acid impurities lower the freezing point, and cooling rates may be such that subcooled liquid sulfuric acid is produced. Figure 12

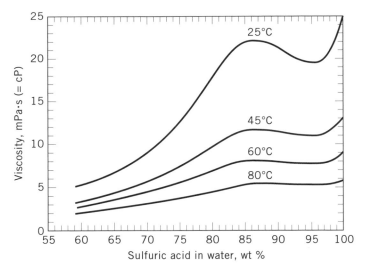

Fig. 7.   Viscosity of sulfuric acid (55).

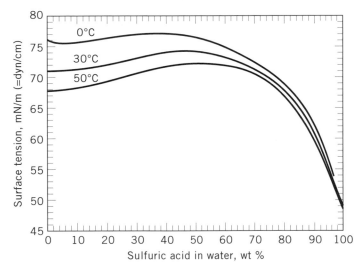

Fig. 8.   Surface tension of sulfuric acid (61).

shows freezing points (75). Additional freezing point data (76) and discussions (59,77) are available.

At atmospheric pressure, sulfuric acid has a maximum boiling azeotrope at approximately 98.48% (78,79). At 25°C, the minimum vapor pressure occurs at 99.4% (78). Data and a discussion on the azeotropic composition of sulfuric acid as a function of pressure can also be found in these two references. The vapor pressure exerted by sulfuric acid solutions below the azeotrope is primarily from water vapor; above the azeotropic concentration SO_3 is the primary component of the vapor phase. The vapor of sulfuric acid solutions between 85% H_2SO_4 and 35% free SO_3 is a mixture of sulfuric acid, water, and sulfur trioxide vapors.

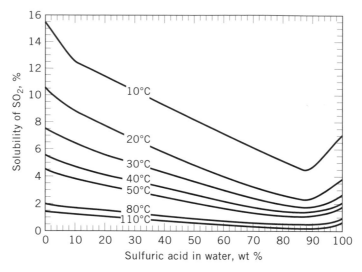

Fig. 9.   Solubility of SO_2 in sulfuric acid at SO_2 pressure of 101.3 kPa (1 atm) (68).

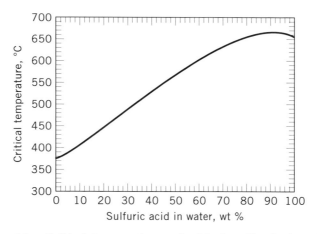

Fig. 10.   Critical temperatures of sulfuric acid solutions (73).

At the boiling point, sulfuric acid solutions containing <85% H_2SO_4 evaporate water exclusively; those containing >35% free SO_3 (oleum) evaporate exclusively sulfur trioxide.

   A tabulation of the partial pressures of sulfuric acid, water, and sulfur trioxide for sulfuric acid solutions can be found in Reference 80 from data reported in Reference 81. Figure 13 is a plot of total vapor pressure for 0–100% H_2SO_4 vs temperature. References 81 and 82 present thermodynamic modeling studies for vapor-phase chemical equilibrium and liquid-phase enthalpy concentration behavior for the sulfuric acid–water system. Vapor pressure, enthalpy, and dew point data are included. An excellent study of vapor–liquid equilibrium data are available (79).

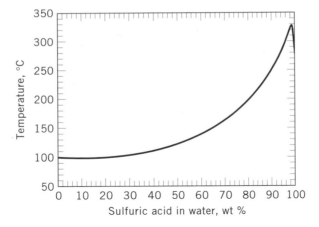

Fig. 11.   Normal boiling point of sulfuric acid (74).

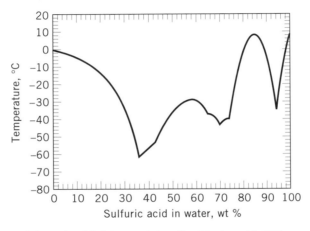

Fig. 12.   Melting points of sulfuric acid (75).

Figure 14 shows the heat of mixing of sulfuric acid and water (83). Additional data are in Reference 84.

Oleum.   Oleum strengths are usually reported as wt % free SO_3 or percent equivalent sulfuric acid. The formula for converting percent oleum to equivalent sulfuric acid is as follows:

$$\% \ H_2SO_4 = 100 + \% \ oleum/4.444$$

Thus, 20% oleum is equivalent to 104.5% H_2SO_4.

Oleum is generally thought of as a mixture of sulfuric acid and free sulfur trioxide. In various strength oleums the free SO_3 actually forms disulfuric acid, $H_2S_2O_7$, and trisulfuric acid, $H_2S_3O_{10}$ (85). The existence of higher molecular weight polyacids and their presence in significant amounts has been the subject of considerable controversy. An excellent review and discussion of the compositions of both the liquid and vapor phases of oleum, including a thermodynamic description of oleum with regard to the formation of $H_2S_2O_7$, is available (85).

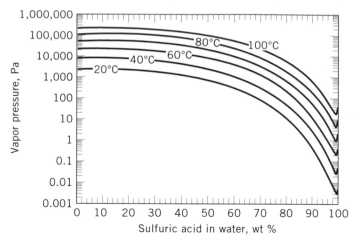

Fig. 13.   Vapor pressure of sulfuric acid (81). To convert Pa to mm Hg, multiply by 0.0075.

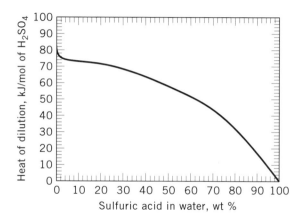

Fig. 14.   Heat of mixing of sulfuric acid from H₂O and H₂SO₄ at 25°C (83). To convert J to cal, divide by 4.184.

The density of oleum at 20°C (76) and at 25°C (39) has been reported. The boiling points of oleum are presented in Figure 15 (86). Freezing points are shown in Figure 16 (75,87). An excellent discussion on the crystallization points of oleum is available (69). The solubility of sulfur dioxide in oleum has been reported (68,69). Viscosity of oleum is summarized in Figure 17 (55); additional viscosity data are available (76).

A composite curve of heat of infinite dilution of oleum from reported data (3,88–90) is presented in a compiled form in the literature (91), where heats of formation of oleums from liquid or gaseous SO₃ are also reported (Tables 5 and 6). Heat of vaporization data are also available (92). Oleum heat capacity data are presented in Figure 18 (76); solubility data for SO₂ in oleum can be found in Reference 69.

There are significant differences in various data sets published for oleum vapor pressure. A review of existing vapor pressure data plus additional data

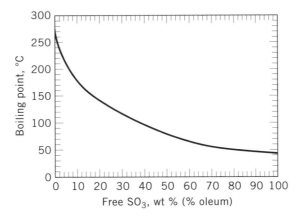

Fig. 15. Normal boiling points of oleum (86).

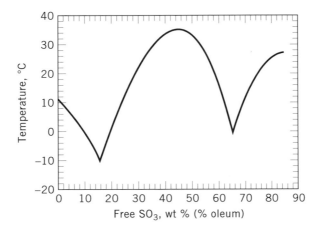

Fig. 16. Melting points of oleum. Refs. 75 (<65%) and 87 (>65%).

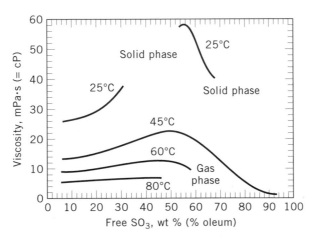

Fig. 17. Viscosity of oleum (55).

Table 5. Heat of Formation of Oleums from Liquid SO₃[a,b]

Free SO₃, %	Liquid phase	
	SO₃, MJ/kg[c]	H₂O, MJ/kg[c]
0	1.107	4.919
10	1.009	5.086
20	0.9067	5.260
30	0.8022	5.434
40	0.6978	5.632
50	0.5905	5.841
60	0.4808	6.061
70	0.3684	6.317
80	0.2535	6.654
90	0.1333	7.106
100	0	

[a]Ref. 91.
[b]Reaction $H_2O(l) + x\ SO_3(l) \rightarrow H_2SO_4 \cdot (1 - x)\ SO_3(l)$ where x = mol total SO_3/mol H_2O.
[c]To convert J to cal, divide by 4.184.

Table 6. Heat of Formation of Oleums from Gaseous SO₃[a,b]

Free SO₃, %	Liquid phase	
	SO₃, MJ/kg[c]	H₂O, MJ/kg[c]
0	1.645	7.311
10	1.547	7.803
20	1.445	8.384
30	1.340	9.080
40	1.236	9.986
50	1.129	11.17
60	1.019	12.84
70	0.9067	15.56
80	0.7918	20.79
90	0.6716	35.83
100	0.5383	

[a]Ref. 91.
[b]Reaction $H_2O(l) + x\ SO_3(g) \rightarrow H_2SO_4 \cdot (1 - x)\ SO_3(l)$ where x = mol total SO_3/mol H_2O.
[c]To convert J to cal, divide by 4.184.

from 10 to 8600 kPa (1.45 to 1247 psi) over the entire concentration range of oleum is available (93), including equations for vapor pressure versus temperature. Vapor pressure curves for oleum calculated from these equations are shown in Figure 19. Additional vapor pressure data from 0.06 to 14 kPa (0.5–110 torr) is given in the literature (92).

Manufacture

Sulfuric acid may be produced by the contact process from a wide range of sulfur-bearing raw materials by several different process variants, depending largely on the raw material used. In some cases sulfuric acid is made as a by-product of

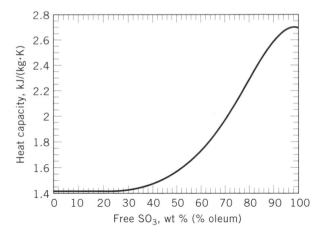

Fig. 18.   Heat capacity of oleum at 20°C (76). To convert J to cal, divide by 4.184.

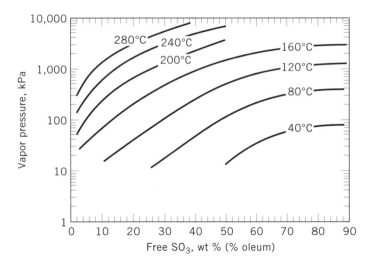

Fig. 19.   Vapor pressure of oleum (93). To convert kPa to psi, multiply by 0.145.

other operations, primarily as an economical or convenient means of minimizing air pollution (qv) or disposing of unwanted by-products.

The contact process remained virtually unchanged from its introduction in the 1800s until the 1960s, when the double absorption process was introduced to reduce atmospheric SO_2 emissions. Double absorption did not, however, substantially change the nature of the process or the process equipment. In the 1970s and 1980s the increased value of energy, and production of sulfuric acid from a variety of waste products, including off-gases and spent sulfuric acid, led to a number of process and equipment modifications (94,95).

The principal direct raw materials used to make sulfuric acid are elemental sulfur, spent (contaminated and diluted) sulfuric acid, and hydrogen sulfide. Elemental sulfur is by far the most widely used. In the past, iron pyrites or related compounds were often used; but as of the mid-1990s this type of raw material is not common except in southern Africa, China, Kazakhstan, Spain, Russia, and Ukraine (96). A large amount of sulfuric acid is also produced as a by-product of nonferrous metal smelting, ie, roasting sulfide ores of copper, lead, molybdenum, nickel, zinc, or others.

In all types of contact plants, the first steps in the process have the objective of producing a reasonably continuous, contaminant-free gas stream containing appreciable sulfur dioxide and some oxygen. This gas stream is preferably dry, but plants can be designed to handle wet gas directly from H_2S combustion. This requires careful design of equipment to minimize mist formation in the condensation–absorption portion of the plant. If the initial oxygen concentration of the process gas is low, additional air or oxygen must be added prior to or during catalytic oxidation to ensure that there is an excess over stoichiometric needs for conversion of SO_2 to SO_3.

The gas stream containing sulfur dioxide is either dried before passing to the catalytic oxidation step, or is oxidized in the presence of water vapor with subsequent acid condensation and removal. When acid is produced from elemental sulfur, the air used for sulfur burning is predried. In all cases, typical plant designs use sulfuric acid from the process as a drying agent. Wet catalytic oxidation is relatively uncommon. Some applications of Haldor Topsøe's WSA-2 wet gas catalysis process are described in the literature (97).

The sulfur trioxide produced by catalytic oxidation is absorbed in a circulating stream of 98–99% H_2SO_4 that is cooled to approximately 70–80°C. Water or weaker acid is added as needed to maintain acid concentration. Generally, sulfuric acid of approximately 98.5% concentration is used, because it is near the concentration of minimum total vapor pressure, ie, the sum of SO_3, H_2O, and H_2SO_4 partial pressures. At acid concentrations much below 98.5% H_2SO_4, relatively intractable aerosols of sulfuric acid mist particles are formed by vapor-phase reaction of SO_3 and H_2O. At much higher acid concentrations, the partial pressure of SO_3 becomes significant.

The catalytic oxidation of SO_2 to SO_3 is highly exothermic and, as expected, equilibrium becomes increasingly unfavorable for SO_3 formation as temperature increases above 410–430°C. Unfortunately, this is about the minimum temperature level required for typical commercial catalysts to function. Consequently, plant catalytic reactors (converters) are typically designed as multistage adiabatic units and have gas cooling between each stage. SO_2 concentrations in the gas stream range from 4–14 vol %; lower or higher concentrations are occasionally handled by special process or plant modifications.

In early years the contact process frequently employed only two or three catalyst stages (passes) to obtain overall SO_2 conversions of approximately 95–96%. Later, four pass converters were used to obtain conversions of from 97% to slightly better than 98%. For sulfur-burning plants, this typically resulted in sulfur dioxide stack emissions of 1500–2000 ppm.

In the early 1970s, air pollution requirements led to the adoption of the double contact or double absorption process, which provides overall conversions of better than 99.7%. The double absorption process employs the principle of intermediate removal of the reaction product, ie, SO_3, to obtain favorable equilibria and kinetics in later stages of the reaction. A few single absorption plants are still being built in some areas of the world, or where special circumstances exist, but most industrialized nations have emission standards that cannot be achieved without utilizing double absorption or tail-gas scrubbers. A discussion of sulfuric acid plant air emissions, control measures, and emissions calculations can be found in Reference 98.

Plants producing oleum or liquid SO_3 typically have one or two additional packed towers irrigated with oleum ahead of the normal SO_3 absorption towers. Partial absorption of SO_3 occurs in these towers, and sulfuric acid is added to maintain desired oleum concentrations. Normally, oleum up to about 35 wt % free SO_3 content can be made in a single tower; two towers are used for 40 wt % SO_3. Liquid SO_3 is produced by heating oleum in a boiler to generate SO_3 gas, which is then condensed. Oleums containing SO_3 >40 wt % are usually produced by mixing SO_3 with low concentration oleum.

Where elemental sulfur or hydrogen sulfide is used as raw material, considerable heat is evolved during the initial combustion. Additional heat is generated by catalytic oxidation to SO_3 and by the reaction of SO_3 and H_2O to form H_2SO_4. In such plants, much of the heat is typically used to produce steam, which can be utilized either for heating requirements in other processes or to generate power via turbines. In many cases, large plants of this type are essentially co-producers of steam or power and sulfuric acid; both products have significant economic value. Where spent acid is used as raw material, it usually is decomposed in furnaces fired by gas, oil, or other fuels (sometimes H_2S or sulfur), and the high temperature gas from such furnaces can also generate steam or power (see POWER GENERATION).

In general, plants using SO_2 gas derived from metallic sulfides, spent acids, or gypsum anhydrite purify the gas stream before drying it by cold, ie, wet, gas purification. Various equipment combinations including humidification towers, reverse jet scrubbers, packed gas cooling towers, impingement tray columns and electrostatic precipitators are used to clean the gas.

Plants that burn good quality elemental sulfur or H_2S gas generally have no facilities for purifying SO_2. Before the advent of relatively pure Frasch or recovered sulfur, however, hot gas purification was frequently used in which the SO_2 gas stream was passed through beds of granular solids to filter out fine dust particles just prior to its entering the converter.

Sulfur shipped as a solid frequently becomes contaminated with dirt and scale during shipping and handling. In areas of the world where solid sulfur is still handled, molten sulfur is frequently filtered prior to use as an alternative to, or in combination with, hot gas purification. Since the early 1970s, most sulfur used in the United States and Europe has been shipped and handled as a liquid containing very low ash concentrations, typically <0.005%. Using this type of raw material, neither sulfur filtration nor hot gas purification are essential, and are rarely used.

Tail gas scrubbers are sometimes used on single absorption plants to meet SO_2 emission requirements, most frequently as an add-on to an existing plant, rather than on a new plant. Ammonia (qv) scrubbing is most popular, but to achieve good economics the ammonia value must be recovered as a usable product, typically ammonium sulfate for fertilizer use. A number of other tail gas scrubbing processes are available, including use of hydrogen peroxide, sodium hydroxide, lime and soda ash. Other tail gas processes include active carbon for wet oxidation of SO_2, molecular sieve adsorbents (see MOLECULAR SIEVES), and the absorption and subsequent release of SO_2 from a sodium bisulfite solution.

Small amounts of sulfuric acid mist or aerosol are always formed in sulfuric acid plants whenever gas streams are cooled, or SO_3 and H_2O react, below the sulfuric acid dew point. The dew point varies with gas composition and pressure but typically is 80–170°C. Higher and lower dew point temperatures are possible depending on the SO_3 concentration and moisture content of the gas. Such mists are objectionable because of both corrosion in the process and stack emissions.

More recently, sulfuric acid mists have been satisfactorily controlled by passing gas streams through equipment containing beds or mats of small-diameter glass or Teflon fibers. Such units are called mist eliminators (see AIR POLLUTION CONTROL METHODS). Use of this type of equipment has been a significant factor in making the double absorption process economical and in reducing stack emissions of acid mist to tolerably low levels.

Coalescing demister pads have been used in some single absorption plants instead of packed fiber beds to remove mist from the stack gas. For submicrometer particle collection, these devices are not as efficient as packed fiber beds. Nevertheless, they have been used in some plants to obtain nearly invisible emissions. Successful use of a coalescing demister pad requires careful control of plant operating conditions to minimize mist formation.

Generation of Sulfur Dioxide Gas.   *Sulfur Burning.*   There is a trend toward very large single-train plants. Because of this the usual practice is to use horizontal, brick-lined combustion chambers with dried air and atomized molten sulfur introduced at one end. Atomization typically is accomplished either by pressure spray nozzles or by mechanically driven spinning cups. Because the degree of atomization is a key factor in producing efficient combustion, sulfur nozzle pressures are typically 2.76 MPa (150 psi) or higher. Sulfur burners are typically designed as proprietary items by companies specializing in acid plant design and construction. Some designs contain baffles or secondary air inlets to promote mixing and effective combustion.

In any procedure involving the handling of molten sulfur, the lines and spray nozzles must be steam-jacketed, and steam pressure must hold the molten sulfur within the range of 135–155°C, where its viscosity is at a minimum. Above 160°C, the viscosity rises sharply, and at 190°C its viscosity is 13,000 times that at 150°C.

The self-sustaining ignition temperature of pure sulfur is approximately 260°C but may be slightly higher for dark sulfur, ie, material containing organic impurities. Consequently, a source of ignition is not required if the combustion chamber is preheated to approximately 400–425°C before sulfur is admitted. When burning sulfur in air, SO_2 concentrations in the range of about 3–14 vol % can be produced by the burners described. Special burners capable of producing higher SO_2 concentrations are also available (see SULFUR COMPOUNDS).

The temperature of the gas leaving the sulfur burner is a good indication of SO_2 concentration, even though the thermocouples employed for temperature measurement (qv) frequently read somewhat lower than the true temperatures, because of radiation and convection errors. A temperature of 970°C corresponds to about 10.0 vol % SO_2, 1050°C to 11.0 vol % SO_2, and 1130°C to 12.0 vol % SO_2. Other temperatures and concentrations are in similar proportion.

At high flame temperatures, small amounts of nitrogen react with oxygen to form nitrogen oxides, NO_x, primarily nitric oxide, NO. The chemistry of these nitrogen oxides is complex. Ultimately, however, some form nitrosylsulfuric acid, which ends up either as trace amounts in product acids or, in considerably higher concentrations, as condensed acid collected at mist eliminators.

Sulfur burners are normally operated at moderate pressures, in the range of 135.8–170.3 kPa (5–10 psig), using air supplied by the main blower for the plant.

Spent Acid or H_2S Burning.   Burners for spent acid or hydrogen sulfide are generally similar to those used for elemental sulfur. There are, however, a few critical differences. Special types of nozzles are required both for H_2S, a gaseous fuel, and for the corrosive and viscous spent acids. In a few cases, spent acids may be so viscous that only a spinning cup can satisfactorily atomize them. Because combustion of H_2S is highly exothermic, careful design is necessary to avoid excessive temperatures.

Spent acid burning is actually a misnomer, for such acids are decomposed to SO_2 and H_2O at high temperatures in an endothermic reaction. Excess water in the acid is also vaporized. Acid decomposition and water vaporization require considerable heat. Any organic compounds present in the spent acid oxidize to produce some of the required heat. To supply the additional heat required, auxiliary fuels, eg, oil or gas, must be burned. When available, sulfur and H_2S are excellent auxiliary fuels.

Relatively high (typically 980–1200°C) temperatures are required to decompose spent acids at reasonable burner retention times. Temperatures depend on the type of spent acid. A wide variety of spent acids can be processed in this way, but costs escalate rapidly when the sulfuric acid concentration in spent acid (impurity-free basis) falls below about 75%. A few relatively uncontaminated spent acids can be reused without decomposition by evaporating the excess water in concentrators, or by mixing in fresh sulfuric acid of high concentration. Weak spent acids are frequently concentrated by evaporation prior to decomposition.

Because large amounts of water vapor are produced by combustion of H_2S or spent acids, ambient, not dried, air is supplied to the burners. In some cases, burners are operated at pressures slightly below atmospheric to pull in outside air; in other cases, preheated combustion air at low pressure may be supplied by ducts.

Ore Roasting, Sintering, or Smelting.   Generation of SO_2 at nonferrous metal smelters is determined primarily by the needs of the various metallurgical processes and only incidentally by requirements of the sulfuric acid process. Traditionally, sulfur recovery from copper (qv), nickel, lead (qv), and zinc (qv), smelters has been limited to treatment of gases from roasters, sintering machines, and converters. Roasters and sintering machines operate continuously and produce a fairly uniform but low concentration off-gas. Traditional converters, which operate as batch reactors, produce gases having varying concentrations of SO_2. Moreover, there is considerable time when the furnaces are off-line for charging and transferring metals. Treatment of these cyclical and low grade gases is costly and involves technical problems that make efficient acid production difficult.

More recently, intensive smelting processes have been developed which use highly oxygen-enriched air or even technically pure oxygen to minimize fuel consumption. The resulting gases are high (25–75%) in SO_2 but generally low in oxygen. These smelting processes are often combined with traditional batch converting and the combined gas flow to the acid plant is more uniform and has higher strength than when treating converting gases alone.

The most modern smelters (ca 1997) use continuous smelting and converting processes which utilize high levels of oxygen-enriched air to produce a uniform flow of high strength process gas. This allows efficient acid plant design including high levels of energy recovery that was formerly only possible in sulfur-burning acid plants (see METALLURGY, EXTRACTIVE METALLURGY).

Process Details and Flow Sheets. The stoichiometric relation between reactants and products for the contact process may be represented as follows:

$$SO_2 + 1/2\ O_2 \rightleftharpoons SO_3 \tag{1}$$

$$SO_3 + H_2O \longrightarrow H_2SO_4 \tag{2}$$

There are three important characteristics of the first of these equations. It is exothermic, reversible, and shows a decrease in molar volume on the right-hand side, ie, in the direction of the desired product. To improve equilibrium or driving force for the reaction, the sulfuric acid industry has attempted one or a combination of the following process design modifications: increasing concentration of SO_2 in the process gas stream; increasing concentration of O_2 in the process gas stream by air dilution or oxygen enrichment; increasing the number of catalyst beds; removing the SO_3 product by interpass absorption, known as the double absorption process; lowering catalytic converter inlet operating temperatures, ie, using better catalysts; and increasing the catalytic converter operating pressure (pressure plants).

Single Absorption Sulfur-Burning Plants. Single absorption sulfuric acid plants were standard in the industry for many years. These used either relatively low strength (approximately 8 vol %) SO_2 gas without air dilution, or air dilution designs and higher (approximately 10 vol %) inlet gas strength. Air dilution was a common design option using additional dry air, instead of heat exchangers, to cool the process gas entering the last one or two converter passes. The additional air improved conversion at the final converter pass by increasing oxygen concentration and reducing equivalent sulfur dioxide concentration of the process gas. Its chief advantage was reduced investment over designs with heat exchangers for interpass cooling.

Figure 20 shows a typical flow sheet for a four-pass, single absorption sulfur-burning plant without air dilution. Plants of this design burn sulfur to generate a process gas stream of about 7.5–9.0% SO_2. Typical converter operating conditions are shown in Table 7. The scheme shown produces a full range of products including 66° Bé and 98.5% acid and oleum. If 66° Bé acid product is not needed, a common pump tank is usually used for both the drying and absorbing towers. If oleum production is not needed, the equipment within the dashed box is omitted and process gas enters the absorbing tower directly from the economizer.

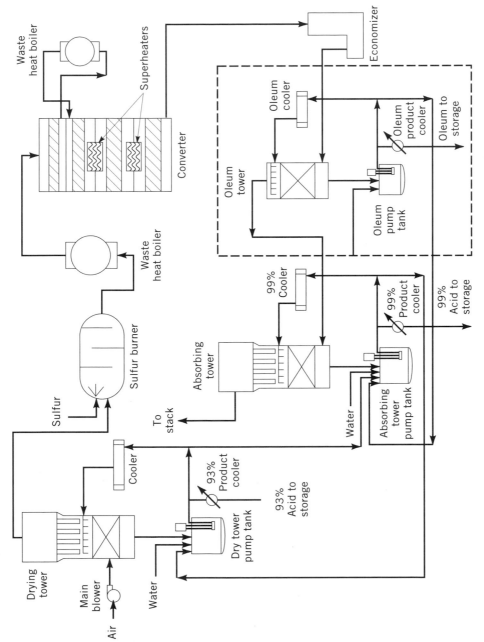

Fig. 20. Flow sheet for a single absorption sulfur-burning plant.

383

Table 7. Converter Conditions For a Single Absorption Plant[a]

Converter pass temperature, °C	1	2	3	4
inlet	410–445	430–450	430–435	425–430
outlet	595	500	450	430–435
ΔT	150–185	50–70	15–20	5

[a]Using 8 % SO_2 inlet gas.

Process air in sulfur-burning plants is dried by contacting it with 93–98 wt % sulfuric acid in a countercurrent packed tower. Dry process air is used to minimize sulfuric acid mist formation in downstream equipment, thus reducing corrosion problems and stack mist emissions.

Most of the heat of combustion from the sulfur burner is removed in a boiler, which reduces the process gas temperature to the desired converter inlet temperature. Typically, the inlet temperature (Table 7) to the first converter pass is dictated by catalyst performance, catalyst bed depth, and process gas strength. Standard, ie, sodium- or potassium-promoted, vanadium pentoxide catalysts do not have sustained catalytic activity at temperatures <400–410°C, although fresh catalyst may have an initial reaction ignition temperature as low as 385°C. Such low ignition temperatures cannot be sustained by conventional catalysts. (The catalyst ignition temperature is the temperature below which substantial catalytic conversion (approaching equilibrium) cannot be sustained in any given bed or pass.) Newer catalysts, promoted with cesium, have a considerably lower sustainable ignition temperature (approximately 375°C) and have proved useful in special situations (99).

SO_2 gas is catalytically oxidized to SO_3 in a fixed bed reactor (converter) which operates adiabatically in each catalyst pass. The heat of reaction raises the process gas temperature in the first pass to approximately 600°C (see Table 7). The temperature of hot gas exiting the first pass is then lowered to the desired second pass inlet temperature (430–450°C) by removing the heat of reaction in a steam superheater or second boiler.

In converter passes downstream of the first pass, exit temperatures are limited by thermodynamic equilibrium to around 500°C or less. To obtain optimum conversion, the heats of reaction from succeeding converter passes are removed by superheaters or air dilution. The temperature rise of the process gas is almost directly proportional to the SO_2 converted in each pass, even though SO_2 and O_2 concentrations can vary widely.

Gas leaving the converter is normally cooled to 180–250°C using boiler feedwater in an "economizer." This increases overall plant energy recovery and improves SO_3 absorption by lowering the process gas temperature entering the absorption tower. The process gas is not cooled to a lower temperature to avoid the possibility of corrosion from condensing sulfuric acid originating from trace water in the gas stream. In some cases, a gas cooler is used instead of an economizer.

Gas leaving the economizer flows to a packed tower where SO_3 is absorbed. Most plants do not produce oleum and need only one tower. Concentrated sulfuric acid circulates in the tower and cools the gas to about the acid inlet temperature. The typical acid inlet temperature for 98.5% sulfuric acid absorption towers is 70–80°C. The 98.5% sulfuric acid exits the absorption tower at 100–125°C, depending on acid circulation rate. Acid temperature rise within the tower comes from the heat of hydration of sulfur trioxide and sensible heat of the process gas. The hot product acid leaving the tower is cooled in heat exchangers before being recirculated or pumped into storage tanks.

Acid circulated over SO_3 absorbing towers is maintained at about 98.5% to minimize its vapor pressure. Where lower concentration product acid is desired, it is made either in separate dilution facilities, or in drying towers operated at 93–96% H_2SO_4.

The conversion efficiency of any sulfuric acid plant can be presented in terms of an equilibrium-stage process. Figure 21 presents the equilibrium-stage diagram of a single absorption sulfur-burning plant using 8% SO_2 burner gas. The slopes of the adiabatic temperature rise lines are directly proportional to the specific heat capacity of the process gas, which is reasonably constant for any degree of conversion.

The curve in Figure 21 represents SO_2 equilibrium conversions vs temperature for the initial SO_2 and O_2 gas concentrations. Each initial SO_2 gas concentration has its own characteristic equilibrium curve. For a given gas composition, the adiabatic temperature rise lines can approach the equilibrium curve but never cross it. The equilibrium curve limits conversion in a single absorption plant to slightly over 98% using a conventional catalyst. The double absorption process removes this limitation by removing the SO_3 from the gas stream, thereby altering the equilibrium curve.

Double-Absorption Plants. In the United States, newer sulfuric acid plants are required to limit SO_2 stack emissions to 2 kg of SO_2 per metric ton of 100% acid produced (4 lb$_m$/short ton; lb$_m$ = pounds mass). This is equivalent to a sulfur dioxide conversion efficiency of 99.7%. Acid plants used as pollution control

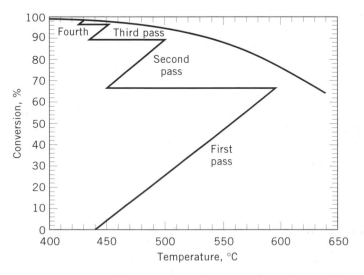

Fig. 21. Single absorption equilibrium-stage diagram where the equilibrium curve is for 8% SO_2, 12.9% O_2; the diagonal lines represent the adiabatic temperature rise of the process gas within each converter pass; the horizontal lines represent gas cooling between passes, where no appreciable conversion occurs.

devices, for example those associated with smelters, have different regulations. This high conversion efficiency is not economically achievable by single absorption plants using available catalysts, but it can be attained in double absorption plants when the catalyst is not seriously degraded.

A typical double absorption plant design uses intermediate SO_3 absorption after the second or, more commonly, the third converter pass. This is called the 3 + 1 configuration. As of the mid-1990s newer double absorption plants usually contained a total of four catalyst passes in a 3 + 1 configuration. Plants having five passes in a 3 + 2 configuration have also been built. Figure 22 presents a typical flow diagram for a sulfur-burning double absorption plant. Typical converter temperatures in a double absorption converter are given in Table 8.

Approximately 90–95% of total sulfur trioxide produced by the double absorption process is absorbed in the interpass absorption tower. The sulfur trioxide produced in subsequent converter passes is absorbed in the final absorbing tower. Interpass absorbing tower operation is similar to an absorbing tower in a single absorption plant. In both cases, acid irrigation rates are designed so that acid temperature exiting the tower is 100–125°C. The smaller amount of sulfur trioxide absorbed in the final absorption tower of double absorption plants typically raises its acid temperature to only ≤105°C. Interpass and final absorbing towers of double absorption plants are very similar in size because tower diameter is dependent on total gas throughput, not sulfur trioxide concentration.

Another large difference between single and double absorption processes is that, after interpass absorption, the process gas has to be reheated from approximately 80°C to approximately 425°C before reentering the converter. Reheating the process gas is accomplished in gas-to-gas heat exchangers (see Fig. 22), using some of the heat from the initial converter passes. The gas-to-gas heat exchangers are a primary cost item. All other plant operations are very similar to the corresponding single absorption processes.

Nonsulfur Burning Plants. Acid plants having a SO_2 source other than sulfur burning usually receive cold process gas that must be heated to the reaction temperature of approximately 425°C before it enters the converter. Reheating can be done by burning additional sulfur and adding the hot gas from the sulfur furnace to the main gas stream, or using gas-to-gas heat exchangers, using heat from the converter as the heat source. In a dual absorption plant without additional sulfur burning, nearly all of the heat of reaction in the converter is needed for heating and reheating the process gas.

Oleum Manufacture. To produce fuming sulfuric acid (oleum), SO_3 is absorbed in one or more special absorption towers irrigated by recirculated oleum. Because of oleum vapor pressure limitations the amount of SO_3 absorbed from

Table 8. Converter Conditions for a 3 + 1 Double Absorption Plant[a]

Converter pass temperature, °C	1	2	3	4
inlet	415–420	430–445	430–445	425–430
outlet	600–610	530	470	450
ΔT	185	90	30	20

[a]Using 11.5 % SO_2 inlet gas.

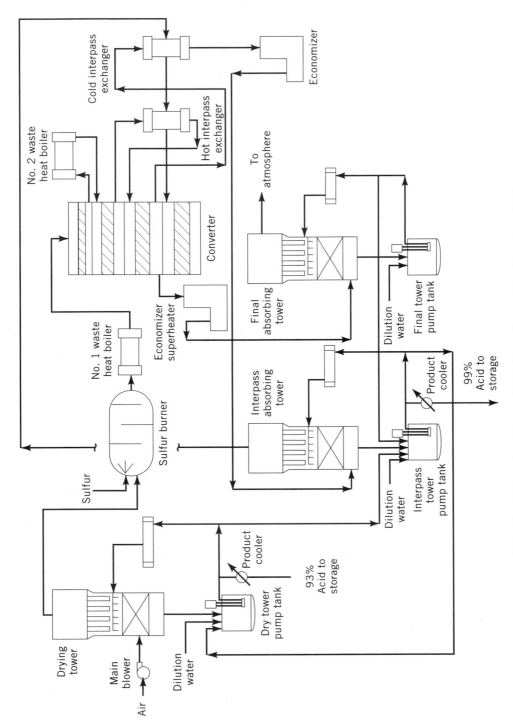

Fig. 22. Flow sheet for a dual absorption sulfur-burning plant.

387

the process gas is typically limited to less than 70%. Because absorption of SO_3 is incomplete, gas leaving the oleum tower must be processed in a nonfuming absorption tower.

The absorption of SO_3 for oleum production is carried out over a relatively narrow temperature range. The upper temperature is set to provide a reasonable partial pressure driving force for the oleum concentration used. The lower practical temperature limit is the freezing point of oleums, which is high enough to be a problem in shipping and handling as well. For some oleum uses it is practical to add small amounts of HNO_3 as an antifreeze (100).

Sulfur Trioxide. The anhydride of sulfuric acid, SO_3, is a strong organic sulfonating and dehydrating agent which has some specialized uses (see SULFONATION AND SULFATION). Its principal applications are in production of detergents and as a raw material for chlorosulfuric acid and 65% oleum. More recently, SO_3 gas has been added to cooled combustion gases at many coal-burning power plants to improve dust removal in electrostatic precipitators (see AIR POLLUTION CONTROL METHODS).

Liquid SO_3 is a difficult material to handle because of its relatively low (44.8°C) boiling point, its tendency to form solid polymers below 30°C, plus high reactivity with almost all organic substances and water. It reacts explosively with water because of a very high heat of reaction. In addition, trace amounts of water (sulfuric acid) act as polymerization catalysts for liquid SO_3 and produce a series of high molecular weight polymers having elevated melting points. When polymerization occurs, attempts to melt the solids can cause piping or equipment failures because high (exceeding atmospheric pressure) vapor pressures develop before melting occurs. Polymerization is promoted by cooling the liquid below 30°C. It can be inhibited by adding small amounts of various patented stabilizers to the liquid, such as 0.3% dimethyl sulfate with 0.005% boric oxide (28). Stabilized liquid SO_3 is an item of commerce, and instructions for its storage and use are available from suppliers.

Liquid SO_3 is usually produced by distilling SO_3 vapor from oleum and condensing it. This operation is normally carried out at a sulfuric acid plant where the stripped oleum can be readily refortified or reused. Eliminating all traces of sulfuric acid from the SO_3 vapor stream is important to minimize polymerization of the liquid condensate. When this is done, it is frequently possible to utilize unstabilized liquid SO_3 if precautions are taken to prevent it from freezing before use. At some plants, gaseous 100% SO_3 is utilized directly instead of producing liquid.

Equipment. *Absorption and Drying Towers.* Towers are typically carbon steel vessels lined with acid proof brick and mortar and packed with ceramic saddles (see ABSORPTION). More recently, all metal towers having no brick lining have been built from high silicon stainless steel alloys such as Sandvik SX or Saramet.

Various distributors have been used, including perforated pipes and trough and downcomer designs. Some designs are prone to plugging from packing chips and sulfates, and thus strainers are frequently used in acid lines. Distributors were formerly made out of cast iron and were limited in the number of distribution points per unit area that could be achieved. Designs using stainless steel alloys can provide up to four distribution points per square foot (43/m²).

Well-designed absorption or drying towers operate at absorption efficiencies >99.5%, typically 99.8 to >99.9%. Drying towers are typically designed using special glass fiber mesh pads at the gas exit to remove acid spray and large-mist particles. Sometimes, and especially when the main blower is downstream from the drying tower, more efficient packed-bed fiber mist eliminators are installed. Glass or Teflon packed-fiber bed mist eliminators are standard in interpass and final absorption towers where high efficiency collection of acid mist is more critical.

Acid Coolers. Cast iron trombone coolers, once the industry standard (101), are considered obsolete. In 1970, anodically passivated stainless steel shell and tube acid coolers became commercially available. Because these proved to have significant maintenance savings and other advantages, this type of cooler became widely used. Anodic passivation uses an impressed voltage from an external electrical power source to reduce metal corrosion. Anodic passivation and its application to sulfuric acid equipment such as stainless steel acid coolers and carbon steel storage tanks has been well studied (102–104). More recently, shell and tube coolers made from Sandvik SX or Saramet have been installed in several acid plants. These materials do not require anodic protection.

Plate and frame coolers using Hastelloy C-276 plates have been used successfully. Anodically protected plate coolers are available as well as plate coolers with plates welded together to minimize gasketing. Another promising development is the introduction of plate coolers made of Hastelloy D205 (105). This alloy has considerably better corrosion resistance to concentrated sulfuric acid at higher temperatures than does C-276. Because of the close clearance between plates, cooling water for plate coolers must be relatively clean.

Impervious (impregnated) graphite coolers are utilized in acid service in which the sulfuric acid concentration is 90–93% or less. When operated properly, graphite coolers provide excellent service in weak acids. They are not recommended for acid services over 93% concentration. Although graphite is relatively inert, the graphite impregnating agent is attacked by concentrated sulfuric acid. The main disadvantage of graphite coolers is brittleness. They are subject to damage by mechanical or hydraulic shocks. State-of-the-art acid plant coolers and the evolution of acid coolers are discussed in Reference 106.

Catalysts. Commercial sulfuric acid catalysts typically consist of vanadium and potassium salts supported on silica, usually diatomaceous earth (see DIATOMITE). Catalyst pellets are available in various formulations, shapes, and sizes depending on the manufacturer and the particular converter pass in which they are to be used. A detailed discussion of oxidation catalysts for sulfuric acid production is available (107).

Formerly, most catalysts were supplied as solid cylindrical extrudates or pellets ranging from 4–10 mm diameter. Ring-shaped catalysts or variations, including rings having longitudinal ribs, are used almost exclusively as of the 1990s, primarily as a means of saving energy via reduced gas pressure drop. The various ring-shaped pellets also have greater resistance to dust fouling. Ring catalysts also have somewhat higher activity per unit volume than pellet catalysts.

Increased catalyst-bed pressure drop caused by dust fouling reduces production of acid and significantly increases energy consumption by the plant's blower.

To avoid these problems, first converter-pass catalyst pellets are screened at every significant turnaround, typically every 12–24 months. Second-pass catalyst pellets need screening less frequently because the first converter-pass catalyst bed acts as a filter for the rest of the converter. Typical screening losses range from 10–15% of the catalyst bed per screening. Screening losses depend on screen mesh size and catalyst hardness, as well as on screening rate.

Under normal operating conditions, vanadium catalysts in first and second pass service slowly lose catalytic activity over time. Catalyst aging is a combination of a loss of catalytically active material from catalyst pellets and irreversible changes within the pellets. It is well-documented that at operating temperatures the active catalytic ingredient is a molten salt that migrates from catalyst pellets into adjacent dust. Catalyst aging is accelerated by increasing temperatures and temperature cycling. Exposure to moisture above the dew point is not detrimental to sulfuric acid catalyst. However exposure to moisture at temperatures below the dew point produces irreversible damage. Prolonged exposure to moisture reduces the vanadium to the $+3$ oxidation state, which is very difficult to reoxidize under converter conditions. Moisture also damages the binders that hold together the silica support. This reduces catalyst hardness resulting in higher than normal screening losses.

Catalytic Converters.   Converters of different engineering firms vary in design. Stainless steel converters are frequently preferred, although carbon steel converters are also used. Stainless steel offers better corrosion resistance and significantly higher strength at operating temperatures than carbon steel. In plants having high temperatures and high NO_x content, significant scaling has been observed in stainless steel converters, although significant metal loss has not been found. This problem appears to be more closely linked to high NO_x content than high temperature and is not sufficiently serious to result in abandonment of stainless steel designs.

Stainless steel designs generally use all-welded interior construction, including stainless screens for catalyst supports. Traditional carbon steel converter designs use steel shells, sometimes partly or wholly brick-lined, with cast iron and alloy internals. In the high temperature converter passes, carbon steel is protected from hot gases by brick lining or a sprayed aluminum coating. Both carbon and stainless steel converters are insulated to reduce heat losses. Essentially all designs use horizontal catalyst beds arranged one over another with gas flowing down through the catalyst.

A relatively new innovation in stainless steel converter design uses structurally shaped support grids and division plates (108). Two advantages are claimed for the structural shape of the support grids and division plates: improved resistance to temperature differentials during startup and a higher strength design requiring less metal.

Gas–Gas Heat Exchangers.   Gas-to-gas heat exchangers in double absorption plants are built of carbon steel or stainless steel. Typical tube diameters range from 37.5–100 mm, and the tubes run vertically in a typical shell-and-tube arrangement. To reduce corrosion and scaling, carbon steel tubes in high temperature service are Alonized, which is a proprietary aluminum alloy coating vapor diffused into the metal surface. Alonizing significantly increases the life of carbon steel with the added benefit of no loss of heat transfer from accumulation of corrosion scale.

Mist Eliminators. Most modern mist eliminators use fiber beds enclosed or supported by stainless steel or alloy-20 wire mesh or the like. Fibers are generally glass, or in some cases Teflon. Mist eliminator design, ie, fiber diameter, packing density, bed depth, etc, is determined by particle size and loading, which is in turn determined largely by the application.

Acid mist eliminators use three aerosol collection mechanisms: inertial impaction, interception, and Brownian motion. Inertial impaction works well for aerosols having particle diameters larger than 3 μm; Brownian motion and interception work well with aerosols having smaller particle diameters.

In drying towers of sulfur-burning plants, mesh pads or inertial impaction-type mist eliminators are usually adequate. High efficiency mist eliminators are usually used in drying towers of spent acid or metallurgical plants.

Packed fiber bed mist eliminators can be designed to operate at almost any desired particle collection efficiencies, depending on the allowable pressure drop and cost. A good discussion of sulfuric acid mist generation, control, and mist eliminator design is available (109,110).

Oleum Equipment and Piping. The traditional material of construction for oleum is carbon steel. Relatively low oleum velocities must be used in steel piping to prevent excessive corrosion. The corrosiveness of oleum decreases with increasing SO_3 concentration. For oleum concentrations <5% SO_3, carbon steel is not recommended because of excessive corrosion. Steel is borderline from 5% to about 15% SO_3, depending on temperature.

Cast iron is not a recommended construction material for oleum. It is well-documented that cast iron fails catastrophically by cracking in oleum service. The mechanism or the cause of cracking is not well understood, but failures result from a buildup of internal stresses within the cast iron. One notable exception is process iron, a proprietary cast iron of Chas. S. Lewis Co. (St. Louis, Missouri). Process iron has been used successfully in oleum service.

Standard stainless steels have significantly greater corrosion resistance to oleum than carbon steel, but higher price may make these materials less economical, except for special services such as valves, liquid distributors, oleum boilers, etc.

Spent Acid Regeneration Gas-Cleaning Equipment. Process gas leaving the spent acid regeneration furnace and waste heat boiler must be cleaned prior to entering the drying tower. Except where spent gases make up only a small portion of the total SO_2 stream, it is also necessary to remove as much water as possible from the spent gas to avoid water balance problems in the absorption portion of the plant. For many years spent gas purification equipment consisted of brick and lead-lined spray humidification towers, electrostatic precipitators, Karbate cooler/condensers, packed cooling towers, and impingement tray towers. Venturi scrubbers and cyclone separators were sometimes used to remove acid mist and spray instead of electrostatic precipitators.

In the late 1980s a new type of gas-cleaning technology, based on patented DuPont froth scrubbing technology, was successfully introduced by Monsanto Enviro-Chem (St. Louis, Missouri) (111). Trademarked DynaWave, the main application of this technology for sulfuric acid plants is the reverse jet scrubber in which a jet of scrubbing liquid is sprayed into the gas stream so as to balance the liquid and gas momentums. This produces a highly turbulent froth zone that efficiently cools and scrubs gas leaving the spent boiler at temperatures typically as high as 350°C.

DynaWave equipment is typically smaller than conventional equipment and is usually made of fiber glass reinforced plastic (FRP) at considerable capital savings over conventional materials. DynaWave is now in use at a number of plants worldwide.

Sulfuric Acid Piping.   The traditional material of construction for handling concentrated sulfuric acid (>70 wt %) for most of the twentieth century has been gray cast iron. Gray cast iron was the preferred material because of low price and availability, plus tolerable corrosion. The expected service life of gray cast-iron pipe is typically 7–15 yr, depending on the temperature, concentration, and velocity of the acid. Gray cast iron pipe and fittings should not be utilized immediately adjacent to anodically protected coolers. Gray iron pipe and fittings within 3–4 m of anodically protected acid coolers can be expected to fail catastrophically in three months to a year by spontaneous cracking similar to that observed in oleum service. Owing to its brittle nature, the ultimate failure of gray iron pipe may be catastrophic if thinning is not properly monitored.

In 1977 the U.S. water supply industry and the centrifugally cast iron pipe foundries made a change from gray cast iron to ductile cast iron pipe as the preferred material. In contrast to gray cast iron, ductile iron does not suffer from spontaneous catastrophic cracking in acid service near anodically protected equipment, or from catastrophic, brittle failure at the end of its useful life. Tests have shown that the expected life of ductile iron pipe in concentrated sulfuric acid service is about the same as gray cast iron pipe, with the possible exception of high velocity (>2.5 m/s) or other special services. Ductile iron pipe is available only in diameters of ≥75 mm. Small (<75 mm) diameter pipe is typically stainless steel or alloy 20, depending on concentration and temperature.

A few plants have been built using Sandvik SX, Saramet, or anodically protected 304 or 316 stainless steel piping. These options offer potentially higher reliability and lower iron content in product acid than ductile iron. The economics of these options vs ductile iron or other alloys are highly dependent on piping diameter and expected piping life. Impurities and acid velocities also are important in choosing the optimum piping material.

In the gas cleaning sections of spent acid or metallurgical sulfuric acid plants, the weak acid scrubbing circuit is typically handled by plastic or glass fiber reinforced plastic (FRP) pipe. The contaminants in weak acid usually vary too greatly to allow use of an economical alloy.

Carbon steel is not normally a suitable piping material for concentrated sulfuric acid because of high corrosion rates in flowing acid. However, where temperatures and flow rates are low, heavy wall steel pipe is sometimes used for transferring product acid.

Sulfuric Acid and Oleum Storage Tanks.   Carbon steel is used in concentrated sulfuric acid storage tanks because in quiescent and low temperature conditions its corrosion rate is acceptable. Carbon steel is not suitable for handling sulfuric acid in concentrations between 80–90% or less than 68% even under quiescent conditions, unless passivating agents are present. Failures of sulfuric acid storage tanks have occurred owing to the corrosive nature of the acid and the peculiarities of the corrosion phenomenon (112). Rigorous design, fabrication, and inspection is required for safe operation. Excellent guidelines exist (113) (see TANKS AND PRESSURE VESSELS).

Materials of Construction. Resistance of alloys to concentrated sulfuric acid corrosion increases with increasing chromium, molybdenum, copper, and silicon content. The corrosiveness of sulfuric acid solutions is highly dependent on concentration, temperature, acid velocity, and acid impurities. An excellent summary is available (114). Good general discussions of materials of construction used in modern sulfuric acid plants may be found in References 115 and 116. More detailed discussions are also available (117–121). For nickel-containing alloys Reference 122 is appropriate. An excellent compilation of the relatively scarce literature data on corrosion of alloys in liquid sulfur trioxide and oleum may be found in Reference 122.

One of the more resistant, versatile, and economical alloys available in wrought form is alloy 20, which is designated as CN-7M or Durimet 20 in cast form. Alloy 20 and similar alloys, eg, alloy 20-Cb3, can handle sulfuric acid at ambient and slightly elevated temperatures throughout the concentration range, including oleum. High nickel/chrome–moly alloys such as alloy C-276, alloy C-22 and alloy 59 also handle sulfuric acid at all concentrations. Resistance of these alloys is extended to higher temperatures for certain concentrations, especially <93% (123). At the high temperatures and concentrations found in contact plant absorbing towers, the Lewmet 55 (cast) and Lewmet 66 (wrought) alloys (Chas. S. Lewis Co., St. Louis, Missouri) are used for velocity-sensitive components such as orifice plates, pump parts, nozzles, and screens (122,124).

Since the mid-1980s the high (5–6% Si) silicon stainless steels have made inroads as materials of construction for pipe and equipment handling contact plant acid in the 93–99% concentration ranges at operating temperatures. These include Sandvik SX and Saramet. These materials are, however, unacceptable in oleum service or in acid service below about 90%. In the mid-1980s, Monsanto Enviro-Chem discovered an operating window at very high (approximately 170–200°C) temperatures and 99% concentration where high chromium stainless steels, such as type 310, can be used (125,126).

Haynes has more recently developed a wrought, easily formable high (5–6%) silicon nickel base alloy designated alloy D-205 which is being used for plate and frame heat-exchanger plates to cool 93% and 99% acid in contact plants. Reference 118 is an excellent compilation of isocorrosion diagrams for metals in sulfuric acid service. Special precautions should be observed when utilizing isocorrosion charts as guidelines for alloy selection. Metal skin temperature, not bulk acid temperature, should be used as the criterion of selection because metal skin temperatures may be significantly higher or lower than bulk acid temperatures. This factor is particularly important when designing heat-transfer equipment; heating coils have failed catastrophically because alloy selection was based on acid bulk temperatures instead of the higher metal skin temperature of the heating coil. In the case of acid cooling, an unnecessarily more expensive higher alloy may be chosen on the basis of bulk acid temperature.

Oxidizing contaminants, eg, nitric acid and ferric ions, may significantly alter the performance of alloys in sulfuric acid. For example, Hastelloy B-2 performs extremely well in concentrated sulfuric acid but corrodes rapidly in the presence of ferric ions, nitric acid, or free SO_3 in sulfuric acid. For this reason, Hastelloy B-2 is not recommended for oleum service.

Tantalum has excellent corrosion resistance to concentrated sulfuric acid at high temperatures, but very poor corrosion resistance to oleum. Tantalum's high price prevents it from becoming a common material of construction in concentrated sulfuric acid service. Zirconium has excellent corrosion resistance to sulfuric acid solutions up to the boiling point in the concentration range of 0–65 wt % H_2SO_4. Duriron, a cast high silicon iron, has excellent corrosion resistance to sulfuric acid at all concentrations up to the boiling point. It is, however, very brittle and use is limited because of its propensity to fracture when subjected to thermal or mechanical shock.

In the past, lead was widely used as a material of construction for sulfuric acid concentrations less than about 80 wt % and occasionally for concentrations up to 93 wt %, at temperatures <45°C. Because of lead's low physical strength and loss of strength at high temperatures, extensive external supports are normally required. Consequently, modern practice uses various acid-resisting plastic materials instead of lead wherever practical. Many plastics do not resist acid above 50–60 wt % H_2SO_4. The resistance and properties of plastics can vary widely depending on exact composition, degree of polymerization, etc. Hence, specific data or information should be obtained. Tetrafluoroethylene (TFE), fluorinated ethylene propylene (FEP), and perfluoroalkoxy polymer (PFA) materials, such as Teflon, are the only common plastics that resist all acid concentrations (with temperature limitations). Poly(vinylidene fluoride) (PVDF), poly(ethylene-cochlorotrifluoroethylene) (ECTFE) and various proprietary baked phenolic resins are reasonably effective up to 94–95 wt % H_2SO_4 at temperatures near ambient.

In the gas-cleaning sections of metallurgical and spent acid regeneration plants, extensive use is made of polyproplyene for tower packings and lined pipe, and of FRP for vessels, gas duct, and piping. Butyl rubber, Hypalon, and ethylene–propylene diene monomer (EPDM) have useful resistance in these systems as well as where elastomers are needed, eg, gaskets, pump linings, etc. Specific impurities or operating conditions may have a large impact on resistance.

Special Plant Designs. *Energy-Efficient Plants.* During the 1970s and 1980s a dramatic increase in energy cost, plus governmental regulation regarding cogeneration of electric power, led to significant changes in plant design. Many of these changes have become standard as of the 1990s. More recently, the need for additional electric generating capacity in many regions of the United States has become a force in the development of energy recovery (see ENERGY MANAGEMENT; PROCESS ENERGY CONSERVATION).

Design changes have included increased (up to 12% SO_2) gas strength; increased (up to 6.2 MPa (900 psi) and 480°C) steam pressure and superheat; low temperature economizers; suction side dry towers, ie, dry towers placed on the suction side of the main compressor; reduced plant pressure drop via new catalyst shapes and low pressure equipment design; and installation of turbogenerators to convert steam to electricity. These changes allowed sulfur-burning plants to recover about 70% of the available energy.

A significant development occurred in the mid-1980s with the introduction of the heat recovery system (HRS) developed by Monsanto Enviro-Chem (127,128). In the HRS process, absorbing towers operate at temperatures as high as 220°C, recovering the sensible heat of the gas stream and the heat of reaction as steam with pressures up to 1140 kPa (150 psig). This achieves thermal efficiencies of 90–95% compared to about 55% for plants in the early 1970s. A number of HRS plants have been built or retrofitted, including at least two in excess of 3000 t/d. The HRS process uses conventional stainless alloys and was made possible by the discovery that above 99% acid, a number of alloys have very low corrosion rates at temperatures as high as 200°C. The HRS process requires extremely careful control of acid concentration. Catastrophic corrosion rates can result if tower concentrations are significantly outside of the prescribed range. This requires rapid and proper response to boiler or other leaks that can introduce water to the process.

A further enhancement to the HRS process whereby the exhaust from a gas fired turbine is used to superheat steam from the HRS process is also possible (129). The superheated steam is then fed through a turbogenerator to produce additional electricity. This increases the efficiency of heat recovery of the turbine exhaust gas. With this arrangement, electric power generation of over 13.6 kW for 1 t/d (15 kW/STPD) is possible. Good general discussions on the sources of heat and the energy balance within a sulfuric acid plant are available (130,131).

Cement Plants. Calcium sulfate in the form of anhydrite; gypsum, $CaSO_4 \cdot 2H_2O$; or by-product gypsum from phosphate fertilizers is occasionally used to produce sulfuric acid and cement (qv). Approximately one kilogram of portland cement is produced for each kilogram of sulfuric acid. Because the capital requirement for such installations is approximately six to eight times the cost of an elemental sulfur burning plant, these are uneconomical except under special circumstances.

Oxygen-Enriched Processes. Typically, the cost of obtaining oxygen (qv) at high concentrations makes its use uneconomical at sulfuric acid plants unless special circumstances exist. Although proposed processes for oxygen enrichment have been published, perhaps only one plant of this type operated commercially for a sustained period. That plant, remodeled by Consolidated Mining and Smelting Company (Cominco) to use 25 vol % SO_2, 30 vol % O_2, balance nitrogen (132,133), is no longer in existence.

Oxygen-enriched air is sometimes used in spent acid decomposition furnaces to increase furnace capacity. Use of oxygen-enriched air reduces the amount of inerts in the gas stream in the furnace and gas purification equipment. This permits higher SO_2 throughput and helps both the heat and water balance in the conversion and absorption sections of the plant (134).

Sulfuric Acid Concentrators. Concentrators for increasing the strength of dilute sulfuric acid by removing water have been used since the early days of the industry. A two-volume text on this subject was published in 1924 (135); more recent discussions of the subject are also available (32,136–141).

The need for acid concentrators exists because many uses of sulfuric acid do not lead to its consumption. Instead, the acid is diluted and partially degraded and contaminated. In the past, large amounts of acid were disposed of either by using it in the phosphate fertilizer industry to dissolve phosphate rock or by neutralization and subsequent discharge to waterways.

Concern over contaminants entering the food chain through fertilizer removed the first option. Increased cost and regulation has all but removed the second. Thus concentration, or recycling, has become more attractive and in many cases even a necessity.

Concentrators frequently not only concentrate the acid, but may purify it as well, removing both organic and inorganic materials. Examples of some options available are discussed in Reference 140. Vacuum evaporation is widely used. Its main advantage is the ability to produce relatively high product acid concentrations at low operating temperatures, thus reducing corrosion. Flash evaporation is sometimes employed as an initial purification step (140,142). In addition, small amounts of organic contaminants are frequently partially oxidized and sometimes can be largely removed by treating with small amounts of hydrogen peroxide or nitric acid to accelerate oxidation (143) (see EVAPORATION).

Economic Aspects

Historically, consumption of sulfuric acid has been a good measure of a country's degree of industrialization and also a good indicator of general business conditions. This is far less valid in the 1990s, because of the heavy sulfuric acid usage by the phosphate fertilizer industry. Of total U.S. sulfuric acid consumption in 1994 of 42.5×10^6 metric tons, over 70% went into phosphate fertilizers as compared to 45% in 1970 and 64% in 1980 (144). Uses other than fertilizer have grown only slowly or declined. This trend is expected to continue. Production and consumption trends in the United States are shown in Tables 9 and 10.

Other uses of sulfuric acid are in production of textile fibers, explosives, alkylate for gasoline production, pulp and paper, detergents, inorganic pigments, other chemicals, and as a leaching agent for ores, a pickling agent for iron and

Table 9. U.S. Sulfuric Acid[a] Production by Sulfur Source[b], t $\times$ 10³

Sulfur source	1985	1990	1993
elemental sulfur	29,929	32,938	28,439
hydrogen sulfide	315	350	290
smelter gas			
copper	2,231	3,381	3,643
zinc	431	413	398
lead[c]	266	165	202
pyrites	610	0	0
other new sulfur sources	410	400	320
decomposition of spent acid	1,844	2,575	2,403
fortified acid	151		
Total	*36,187*	*40,222*	*35,695*

[a]On a 100% H_2SO_4 basis.
[b]Ref. 144.
[c]Includes molybdenum smelter gas.

Table 10. U.S. Consumption of Sulfuric Acid[a,b], t $\times$ 10³

Use	1985	1990	1993
phosphoric acid	24,865	28,419	27,253[c]
other fertilizers	612	517	408
caprolactam	640	800	800
petroleum alkylation	1,470	1,590	1,550
copper leaching	1,020	1,955	2,129
uranium and vanadium ore processing	62	92	3
other ore processing	54	272	150
titanium dioxide	320	350	350
hydrofluoric acid	505	441	436
aluminum sulfate	582	608	595
methyl methacrylate	636	876	806
cellulosic fibers and plastics	256	225	213
pulp and paper	794	887	930
all other	4,857	4,798	2,620
Total	*36,673*	*41,830*	*38,243*

[a]On a 100% H_2SO_4 basis.
[b]Ref. 144.
[c]Includes normal superphosphate.

steel, and as a component of lead storage batteries. Sulfur trioxide, either as liquid or from oleum, finds significant use as a sulfonating agent for surfactants.

The reduction in tetraethyl lead for gasoline production is expected to increase the demand for petroleum alkylate both in the U.S. and abroad. Alkylate producers have a choice of either a hydrofluoric acid or sulfuric acid process. Both processes are widely used. However, concerns over the safety or potential regulation of hydrofluoric acid seem likely to convince more refiners to use the sulfuric acid process for future alkylate capacity.

Additional areas for growth are expected to be in copper leaching, caprolactam, pulp and paper, methyl methacrylate, and batteries (144).

As of 1993–1994, over 70% of sulfuric acid production was not sold as such, but used captively to make other materials. At almost all large fertilizer plants, sulfuric acid is made on site, and by-product steam from these sulfur-burning plants is generally used for concentrating phosphoric acid in evaporators. Most of the fertilizer plants are located in Florida, Georgia, Idaho, Louisiana, and North Carolina. In the production of phosphate fertilizers, the primary role of sulfuric acid is to convert phosphate rock to phosphoric acid and solid calcium sulfates, which are removed by filtration.

Two principal factors affected the U.S. sulfuric acid industry in the 1980s. The first was the increased availability of recovered sulfur vs Frasch sulfur (see SULFUR REMOVAL AND RECOVERY). This occurred because of environmental concerns and regulations forcing more sulfur to be recovered at refineries, power plants, etc. The effect of this change was that the cost of sulfur in the marketplace became driven largely by the cost of nonsulfur industries, rather than by the traditional discretionary sulfur producers, and tended to stabilize U.S. sulfur prices.

The second principal event was the increased recovery of sulfur dioxide as sulfuric acid from lead, zinc, nickel, and copper smelters in the U.S., Canada, and Mexico, also resulting from environmental concerns. Much of the Canadian and Mexican production found its way into the U.S. market. Net U.S. acid imports increased 3.7 times from 1985 to 1990, and increased 4.9 times from 1985 to 1993. Together with increased U.S. smelter acid production, this added 3.4×10^6 t more of acid to the U.S. market in 1993 than in 1985. With relatively little growth in the sulfuric acid market, this influx caused many U.S. merchant market sulfur burning plants, particularly in the northeast and midwest to close permanently. More recently, the increased use of smelter acid for copper leaching has kept this acid out of the market, removing some of the pressure from the remaining merchant market sulfur-burning plants.

Fertilizer manufacturers generally benefited from the stabilized sulfur price and are somewhat insulated from the effect of imports because of the high value of steam (and electricity) produced by their sulfuric acid plants.

Sulfuric acid prices are usually quoted per ton of equivalent 100% H_2SO_4, even though actual assays may be less or more than 100%. In the early 1980s prices ranged from $82–104/t ($75–95/short ton). Beginning in about the mid-1980s, prices fell to $44–66/t ($40–60/short ton). As of mid-1995, prices increased to $55–83/t ($55–75/short ton), depending upon location within the United States. Projected total commercial value of sulfuric acid was 2.8×10^9 for 1995. Oleum typically sells at a premium of about $5 per metric ton (or a higher premium for concentrations >30% free SO_3). Small acid shipments or spot orders typically command a premium vs quoted contract prices in bulk.

The cost and price of sulfuric acid depend in large part on raw material cost and on freight costs. In many areas, the delivered cost of sulfur is the most important factor affecting sulfuric acid pricing. By-product raw material, ie, SO_2, costs at smelters are essentially zero, but the remote locations of many smelters make freight costs significant. Nevertheless, the nondiscretionary nature of smelter acid means that it must be sold if the smelter is to operate.

Worldwide sulfuric acid production figures are shown in Table 11. Future growth is expected to be relatively slow in Western Europe and higher than average in the Middle East and in Africa, particularly in North Africa, which has extensive phosphate rock deposits. Although significant new capacity was built in North Africa in the 1980s, significant additional growth is expected. Russia and The People's Republic of China continue to need more fertilizer than these countries produce, but expansion of sulfuric acid capacity has been limited by available capital. Worldwide, as third world economies improve, especially those in Eastern Europe and Asia, production of nondiscretionary acid is expected to increase, owing to increased emphasis on protecting the environment.

Because sulfuric acid has its greatest use in fertilizers, trends in that industry have a significant effect on the sulfuric acid business. Owing to a weak U.S. dollar in the early 1990s and high demand for fertilizer abroad, a considerable portion of U.S. phosphate fertilizer production was exported. High fertilizer exports are expected to continue until Third World countries can meet their own demands.

Table 11. World Production of Sulfuric Acid[a,b], t $\times$ 10[6]

Location	1985	1990	1993
North America	43.4	48.0	41.3
Central and South America	5.4	5.5	6.2
Western Europe	26.6	22.1	17.0
Eastern Europe	35.7	33.3	19.6
Africa	11.8	14.8	16.4
Asia	24.0	31.4	33.5
Oceania	2.4	1.7	1.3
Total world	*149.3*	*156.8*	*135.3*

[a]Ref. 144.
[b]On a 100% H_2SO_4 basis.

Analysis and Specifications

Specifications for sulfuric acid vary rather widely. Exceptions include the federal specifications for "Sulfuric Acid, Technical" and "Sulfuric Acid, Electrolyte (for storage batteries)" and the *Food Chemicals Codex* specification for sulfuric acid, frequently called food-grade acid (although industrywide, "food-grade" is nonspecific). Very little has been done to establish industry-wide analytical standards in the United States, except for development of the ASTM analytical methods, designated as E223-88 and summarized in Table 12.

Typical specifications for several common types or grades of acid are shown in Table 13. Similar limits are generally used for other sulfuric acid concentrations, with the exception of turbidity values for high strength acids (and oleum) and SO_2 and nitrate values in oleums. Because iron sulfate is relatively insoluble in concentrated acids, the turbidities of 98–99% H_2SO_4 and oleum may be higher than shown, even at acceptable total iron concentrations.

Sulfur dioxide concentrations in oleum are rarely specified or measured, but typical values are considerably higher than in acids of ≤99 wt % concentrations. This occurs because oleum is produced at relatively low temperatures in the presence of appreciable SO_2 in the gas phase, thus leading to high solubility.

Table 12. ASTM Analytical Methods for Sulfuric Acid[a]

Section number	Properties	Principle
1–7	general, reagents	
8–16	total acidity	NaOH titration using phenolphthalein
17–26	Baumé gravity	hydrometer measurement
27–33	nonvolatile matter	evaporation and weighing
34–43	iron	reduce, measure colorimetrically as orthophenanthroline complex
44–51	sulfur dioxide	remove by air sweep, absorb in alkali, add excess iodine, titrate
52–61	arsenic	evolve as arsine, absorb in pyridine and diethylthiocarbamate, measure color

[a]Ref. 145.

Table 13. Typical Sulfuric Acid Specifications[a]

| | Acid type | | | |
Property	Electrolyte[b,c] Class 1[f]	Technical[d] Class 1[g]	Food Chemicals Codex[h]	Technical[e] 66°Bé (93%)
H_2SO_4, wt %	93.2	93.0		93.2
sp gr, 15.5/15.5	1.8354	1.8347		1.835–1.837
nonvolatiles[i], %	0.03	0.025		0.02–0.03
As[i], ppm	1.0	2.5	3	
SO_2[i], ppm	40		40[j]	40–80
iron[i], ppm	50		200	50–100
heavy metals[i], ppm			20	
nitrate[i], ppm	5.0		10	5–20
color[i]	per test			100–200 APHA

[a]Ref. 146.
[b]Limits are also specified for platinum, organics, copper, zinc, antimony, selenium, nickel, manganese, ammonium, and chloride.
[c]Fed. Spec. O-S-801E.
[d]Fed. Spec. O-S-809E.
[e]Typical industry sulfuric acid.
[f]Three other classes of lower strength acids are included.
[g]One other class of lower strength acid is included.
[h]Ref. 147.
[i]Value given is maximum.
[j]Value is for reducing substances as SO_2.

It is not possible to strip SO_2 from oleums by air blowing, a technique that is frequently applied to product acids of ≤99% concentration.

Measurement and specification of nitrates or other nitrogen oxide compounds in sulfuric acid is a complex subject. The difficulty occurs because nitrogen oxides are usually present both as nitrous and nitric compounds, predominantly in the nitrous form. Hence, analytical procedures specific for nitrates only do not give a complete analysis.

A procedure to measure both types of nitrogen oxide compounds at the same time involves development of a pink color by mixing $FeSO_4$ with sulfuric acid, followed by measurement or comparison of color intensity. This general type of procedure and the possible alternatives are discussed in References 148–150.

Although color and turbidity of acid products are important properties, there is little standardization in such measurements. A frequently used procedure is to determine color and turbidity by comparison with standards originally developed by the American Public Health Association (APHA) for examination of water (151).

A number of different grades of acid are produced for specialized uses, such as reagent grade, food grade, and electrolyte grade. In addition, some producers offer special premium priced grades that contain little or no turbidity and color, or in some cases a maximum iron concentration of 10 ppm. Certain objectionable elements such as arsenic, lead, mercury, and selenium are not commonly specified for technical grade acid, but some producers attempt to hold each of these at <1 ppm, or <2–5 ppm in the case of lead, to minimize possible problems. Selenium is not usually present except at a few metallurgical-type plants or at plants using volcanic sulfur as raw material.

Metallurgical (smelter) plants and spent acid decomposition plants usually produce acid of good (low) color because the SO_2 feed gases are extensively purified prior to use. In some cases, however, and particularly at lead smelters, sufficient amounts of organic flotation agents are volatilized from sulfide ores to form brown or black acid. Such acid can be used in many applications, particularly for fertilizer production, without significant problems arising.

Descriptions of sulfuric acid analytical procedures not specified by ASTM are available (32,152). Federal specifications also describe the required method of analysis. Concentrations of 78 wt % and 93 wt % H_2SO_4 are commonly measured indirectly by determining specific gravity. Higher acid concentrations are normally determined by titration with a base, or by sonic velocity or other physical property for plant control. Sonic velocity has been found to be quite accurate for strength analysis of both fuming and nonfuming acid.

Health and Safety

Shipping and Handling. Sulfuric acid is injurious to the skin, mucosa, and eyes. Moreover, dangerous amounts of hydrogen may be evolved in reactions between weakened acid and metals. Sulfuric acid at high concentrations reacts vigorously with water, organic compounds, and reducing agents. Oleums and liquid SO_3 frequently react with explosive violence, particularly with water.

Those engaged in handling sulfuric acid should obtain detailed information on safe handling practices. Material Safety Data Sheets (MSDS) are available from U.S. and European manufacturers.

The *Code of Federal Regulations* (CFR) includes detailed rules for packaging and shipping H_2SO_4. Pertinent sections are 49 CFR 171.15–171.17, hazardous material incidents and discharges; 49 CFR 172.101–172.102, hazardous material tables; and the following references for specific materials:

Material	49 CFR Section	Label
sulfuric acid	173.202 and 173.242	Corrosive
oleum (<30%)	173.201 and 173.243	Corrosive
oleum (≥30%)	173.227 and 173.244 Poison Inhalation Hazard-Zone B	Corrosive and Poison
sulfur trioxide	173.227 and 173.244 Poison Inhalation Hazard-Zone B	Corrosive and Poison

Requirements for transport in ships or barges are outlined in Section 46 of the CFR.

Steel tank cars, often lined to minimize iron contamination, are usually employed for high concentrations of sulfuric acid. Bottom outlets or valves are not allowed, nor are internal steam coils. Tank contents must be unloaded via standpipe. Using air pressure to unload is not recommended for safety reasons, but if air pressure is used, gauge pressures should be held at <0.21 MPa (30 psi).

General handling precautions should be observed. Sulfuric acid must not come in contact with eyes, skin, or clothing. When handling containers or operating equipment containing sulfuric acid, equipment appropriate for exposure conditions should be worn. These include chemical splash goggles, face shield and chemical splash goggle combination (not face shield alone), rubber acid proof gloves, and a full acid-proof suit, hood, and boots. Personnel must avoid breathing mist or vapors. The acid should be handled only in areas having sufficient ventilation to prevent irritation. Alternatively, an appropriate NIOSH/MSHA approved respirator should be worn.

Acid containers must be kept closed. Water must not enter containers. When diluting the acid, always add the acid slowly with agitation to the surface of the aqueous solution to avoid violent spattering, boiling, and eruption. Never add the solution to the concentrated acid.

Handling containers and pipelines also requires that special precautions be observed. An emptied container retains vapor and product residue. Thus all labeled safeguards must be observed until the container is cleaned, reconditioned, or destroyed. Drums, if not self-venting, should be periodically vented to prevent accumulations of hydrogen. To avoid hydrogen explosions when welding, any vessel that has contained sulfuric acid must be thoroughly purged and tested for explosive conditions before welding commences. In dismantling lines and equipment, it should always be assumed that a spray of acid may occur, and suitable precautions should be taken. Iron or other solid sulfates may plug lines or retain pockets of acid. Tightening flange bolts on pipes filled with acid is dangerous because of the possibility of mechanical failures.

In case of physical contact with sulfuric acid, immediately flush eyes or skin with plenty of water for at least 15 min while removing contaminated clothing and shoes. Call a physician. Wash clothing before reuse and destroy contaminated shoes.

In the case of inhalation, remove the individual to fresh air. If necessary, give artificial respiration, preferably mouth to mouth. If breathing is labored, give oxygen. Call a physician.

In case of a spill or leak, keep people away and upwind of the spill. If it is necessary to enter the spill area, self-contained breathing apparatus and full protective clothing, including boots, must be worn. The area should be diked using sand or earth to contain the spill, the acid removed by vacuum truck, and the spill area flushed with water. Washings should be neutralized with lime or soda ash, and pollution control authorities notified of any runoff into streams or sewers and of any air pollution incidents. Safety showers with deluge heads, protected against freezing, should be readily available at appropriate locations in any plant producing or using sulfuric acid.

Sulfuric Acid Toxicity.   Sulfur trioxide does not exist in the atmosphere except in trace amounts because of its affinity for water. It rapidly combines with moisture in air to form sulfuric acid mist. Sulfuric acid aerosol or mist is a significantly more powerful pulmonary irritant than sulfur dioxide on a sulfur-equivalent basis (153–155). Physiological responses to sulfuric acid mist inhalation are highly dependent on the particle size of the aerosol (153,155–157). For a constant sulfuric acid aerosol concentration, the irritant action of the aerosol increases with aerosol particle size. Other factors affecting the physiological response of inhaling sulfuric acid mist are humidity, temperature, and previous exposure. Studies of prolonged exposure to sulfuric acid fumes have been performed on workers in plants manufacturing lead acid batteries. Prolonged exposure to mineral acid fumes causes the teeth of the exposed subject to deteriorate (158–160). Overexposure to sulfuric acid aerosols results in pulmonary edema, chronic pulmonary fibrosis, residual bronchiectasis, and pulmonary emphysema. Additional toxicological information may be found in the literature (161,162).

The International Agency for Research on Cancer (IARC) has classified "occupational exposure to strong inorganic acid mists, containing sulfuric acid," as a Category 1 carcinogen, ie, a substance that is carcinogenic to humans (163). The American Conference of Governmental Industrial Hygienists (ACGIH) has classified "sulfuric acid contained in strong inorganic acid mists" as a Suspected Human Carcinogen (category A2) (164). These classifications are for mists only and do not apply to sulfuric acid or sulfuric acid solutions. These classifications are highly controversial. The overall weight of evidence from all human studies has been evaluated (165). This study concluded, "Thus, the current epidemioligic data alone do not warrant classifying MSA (mists containing sulfuric acid) as a definite human carcinogen." The overall weight of evidence from animal studies has been evaluated (166). This study concluded, "No evidence of carcinogenic potential was found in these studies, although the investigations were compromised due to inadequate design and reporting."

The Threshold Limit Value (TLV) of sulfuric acid mist for humans agreed upon by the ACGIH, OSHA, and NIOSH is 1 mg/m^3 of air. Sulfuric acid aerosols below the TLV are commonly not detected by odor, taste, or irritation. A TLV of 1 mg/m^3 is recommended by the ACGIH to prevent pulmonary irritation and injury to the teeth at particle sizes likely to occur in industrial situations (167). As of February 1997, ACGIH was reviewing its TLV recommendation in light of its classification of sulfuric acid mists as an A2 carcinogen. The following data summarizes human response to various levels of concentration of sulfuric acid aerosols:

Concentration, mg H$_2$SO$_4$/m^3 of air	Response
0.5–2.0	barely noticeable irritation
3.0–4.0	coughing, easily noticeable
6.0–8.0	decidedly unpleasant, marked alterations in respiration

The American Industrial Hygiene Association lists the following Emergency Response Planning Guidelines (ERPGs) for sulfuric acid, oleum, and sulfur trioxide (168): ERPG-1, 2 mg/m^3; ERPG-2, 10 mg/m^3; and ERPG-3, 30 mg/m^3. The ERPG-1 is the maximum airborne concentration below which it is believed that nearly all individuals could be exposed for up to 1 h without experiencing other than mild transient adverse health effects or perceiving a clearly defined objectionable odor. The ERPG-2 is the maximum airborne concentration below which it is believed that nearly all individuals could be exposed for up to 1 h without experiencing or developing irreversible or other serious health effects or symptoms that could impair their abilities to take protective action. The ERPG-3 is the maximum airborne concentration below which it is believed that nearly all individuals could be exposed for up to 1 h without experiencing or developing life-threatening health effects.

BIBLIOGRAPHY

"Sulfuric Acid and Sulfur Trioxide" in *ECT* 1st ed., Vol. 13, pp. 458–506, by B. M. Carter and G. Flint (Sulfur Trioxide), General Chemical Division, Allied Chemical & Dye Corporation; in *ECT* 2nd ed., Vol. 19, pp. 441–482, by T. S. Harrer, Allied Chemical Corporation; in *ECT* 3rd ed., Vol. 22, pp. 190–232, by J. R. Donovan and J. M. Salamone, Monsanto Enviro-Chem Systems, Inc.

1. *Sulphur* (157), 34 (Nov.–Dec. 1981).
2. *Sulphur* (220), 20 (May–June 1992).
3. R. Knietsch, *Ber.* **34**, 4069 (1901).
4. *Chem. Eng. News* **42**(51), 42 (Dec. 21, 1964).
5. *Sulphur* (54), 30 (Oct. 1964).
6. U.S. Pat. 3,142,536 (July 28, 1964), H. Guth and co-workers (to Bayer AG).
7. U.S. Pat. 3,259,459 (July 5, 1966), W. Möller (to Bayer AG).
8. U.S. Pat. 1,789,460 (June 20, 1931), C. B. Clark (to General Chemical Corp.).
9. *Gmelins Handbuch der Anorganischen Chemie, Schwefel*, Ergänzunsband 3, Springer-Verlag, New York, 1980, p. 247.
10. R. J. Lovejoy and co-workers, *J. Chem. Phys.* **36**, 612 (1962).
11. G. E. Walrafan and T. F. Young, *Trans. Faraday Soc.* **56**, 1419 (1960).
12. R. J. Gillespie and E. A. Oubridge, *Proc. Chem. Soc.* 308 (1960).
13. K. Stopperka, *Z. Chem.* **6**, 153 (1966).
14. H. Gerding, W. J. Nijveld, and G. J. Muller, *Nature*, **137**, 1033 (1936).
15. H. Gerding and R. Gerding-Kroon, *Rec. Trav. Chim.* **56**, 794 (1937).
16. Ref. 9, p. 265.
17. P. W. Schenk and R. Steudel, in G. Nickless, ed., *Inorganic Sulphur Chemistry*, Elsevier, New York, 1968, p. 392.
18. Ref. 17, p. 391.
19. M. Schmidt and W. Siebert in A. F. Trotman-Dickenson, exec. ed., *Comprehensive Inorganic Chemistry*, Pergamon Press, Elmsford, New York, 1973.
20. E. S. Scott and L. F. Audrieth, *J. Chem. Educ.* **31**, 174 (1954).
21. Ref. 17, p. 388.
22. *Gmelins Handbuch der Anorganischen Chemie*, Teil B, Lief. 1, Verlag Chemie GmbH, Weinheim, Germany, 1953, p. 333.
23. C. F. P. Bevington and J. L. Pegler, *Chem. Soc. Special Publ.* (12) 283 (1958).
24. D. C. Abercromby, R. A. Hyne, and P. F. Tiley, *J. Chem. Soc.*, 5832 (1963).
25. J. H. Colewell and G. D. Halsey, Jr., *J. Phys. Chem.* **66**, 2179 (1954).
26. R. Westrik and C. H. MacGillavry, *Acta. Cryst.* **7**, 764 (1954).
27. A. Smits and P. Schoenmaker, *J. Chem. Soc.*, 1108 (1926).
28. E. E. Gilbert, *The Stabilization of Sulfur Trioxide*, Annales du Génie Chemique, Congrès International du Soufre, Toulouse, France, 1967.
29. Ref. 19, p. 866.
30. J. H. Colwell, *The Physical Properties of SO₃*, Ph.D. dissertation, Department of Chemistry, University of Washington, Seattle, Washington, 1961.
31. J. F. Mathews, *Chem. Rev.* **72**(1), 71 (1972).
32. W. W. Duecker and J. R. West, eds., *The Manufacture of Sulfuric Acid*, Reinhold Publishing Co., New York, 1959; reprinted by Robert E. Krieger Publishing Co., Huntington, N.Y., 1971, pp. 447–448.
33. R. Westrick and C. H. MacGillavry, *Rec. Trav. Chim.* **60**, 794 (1941).
34. D. M. Lichty, *J. Am. Chem. Soc.* **34**, 1440 (1912).

35. Ref. 22, pp. 337–353.
36. D. R. Stull, Dir., *JANAF Thermochemical Tables, 1965–1968*, National Bureau of Standards, U.S. Department of Commerce, Washington, D.C.
37. C. D. Spangenberg, *Chem. Eng.* **58**(9), 170 (1951).
38. A. A. Maryott and E. R. Smith, *Table of Dielectric Constants of Pure Liquids*, Circular 514, National Bureau of Standards, Washington, D.C., Aug. 10, 1951, p. 44.
39. R. Popiel, *J. Chem. Eng. Data*, **9**(2), 269 (1964).
40. P. Walden, *Z. Anorg. Chem.* **25**, 209 (1900).
41. R. C. Brasted, *Comprehensive Inorganic Chemistry*, Vol. 8, D. Van Nostrand Co., Inc., Princeton, N.J., 1961, p. 135.
42. A. L. Horvath, *Physical Properties of Inorganic Compounds*, Crane, Russak & Co., Inc., New York, 1975, p. 316.
43. A. Berthoud, *Helv. Chim. Acta* **5**, 513 (1922).
44. A. Berthoud, *J. Chim. Phys.* **20**, 77 (1923).
45. E. Lax and C. Synowietz, eds., *Taschenbuch fur Chemiker und Physiker*, 3rd ed., Springer-Verlag, Berlin, Germany, 1967.
46. R. Schenck, *Lieb. Ann.* **316**, 1 (1901).
47. D. R. Stull, *Ind. Eng. Chem.* **39**, 517 (1947).
48. W. H. Evans and D. D. Wagman, *J. Res. Natl. Bur. Stand.* **49**, 141 (1952).
49. G. K. Boreskov, *Catalysis in Sulfuric Acid Production* (Russian), Goskhimizdat, Moscow, USSR, 1954.
50. *International Critical Tables*, Vol. III, McGraw-Hill Book Co., Inc., New York, 1928, pp. 56–57.
51. H. E. Darling, *J. Chem. Eng. Data*, **9**(3), 421 (July 1964).
52. R. G. H. Record, *Instrum. Eng.* **4**(7), 131 (1967).
53. R. Haase, P. F. Sauerman, and K. H. Duecker, *Z. Phys. Chem.* **48**, 206 (1966).
54. N. N. Greenwood and A. Thomson, *J. Chem. Soc.* 3485 (1959).
55. N. F. Bright, H. Hutchinson, and D. Smith, *J. Soc. Chem. Ind.* **65**, 385 (1946).
56. Engineering Sciences Data Unit, *Approximate Data on the Viscosity of Some Common Liquids*, Item No. 66024, Royal Aeronautical Society, London, July 1966.
57. W. Tartakowskaja, J. Bondarenko, and L. Jameljanowa, *Acta Phys. Khim. USSR* **6**, 609 (1937).
58. M. Usanovic, T. Sumarokova, and V. Udovenko, *Zh. Obsh. Khim.* **9**, 1967 (1939).
59. M. Usanovic, T. Sumarokova, and V. Udovenko, *Acta Phys. Khim.* **11**, 505 (1939).
60. R. J. Gillespie and E. A. Robinson, in T. C. Waddington, eds., *Non-Aqueous Solvent Systems*, Academic Press, Inc., New York, 1965, pp. 123–128.
61. J. L. R. Morgan and C. E. Davis, *J. Am. Chem. Soc.* **38**, 555 (1916).
62. J. E. L. Maddock, *A Light Scattering Study of Growth of Sulfuric Acid Aerosol Particles*, Ph.D. dissertation, Department of Chemical Engineering and Chemical Technology, Imperial College of Science and Technology, London, June 1974.
63. L. Sabinina and L. Terpugow, *Z. Phys. Chem.* **173A**, 237 (1935).
64. P. Walden, *Z. Phys. Chem.* **65**, 129 (1909).
65. M. Liler, *Reaction Mechanisms in Sulfuric Acid and Other Strong Acid Solutions*, Academic Press, Inc., New York, 1971, p. 5.
66. E. E. Remsberg, D. Lavery, and B. Crawford, *J. Chem. Eng. Data*, **19**(3), 264 (1974).
67. W. F. Linke, *Solubilities-Inorganic and Metal-Organic Compounds*, 3rd ed., Vol. II, American Chemical Society, Washington, D.C., 1965, pp. 1229–1230.
68. A. G. Amelin, *Sulfuric Acid Technology* (Russian), Khimiya, Moscow, USSR, 1971, p. 458.
69. F. D. Miles and T. Carson, *J. Chem. Soc.* 786 (1946).
70. J. W. Mellor, *A Comprehensive Treatise on Inorganic and Theoretical Chemistry*, Vol. X, Longmans, Green & Co., New York, 1930, pp. 351–360, 384–425.
71. W. H. Lee in J. J. Lagowski, ed., *The Chemistry of Non-Aqueous Solvents*, Vol. II, Academic Press, Inc., New York, 1967.

72. Ref. 42, pp. 313–335.

73. J. E. Stuckey and C. H. Secoy, *J. Chem. Eng. Data*, **8**(3), 386 (July 1963).

74. Ref. 32, 1971, p. 411.

75. C. M. Gable, H. F. Betz, and S. H. Maron, *J. Am. Chem. Soc.* **72**, 1445 (1950).

76. Ref. 68, pp. 448–450.

77. A. Seidell, *Solubilities Inorganic and Metal–Organic Compounds*, Vol. I, 4th ed., D. Van Nostrand Co., Princeton, N.J., 1958, pp. 1161–1162.

78. J. E. Kunzler, *Anal. Chem.* **25**, 93 (1953).

79. H. Lennartz and co-workers, *Vaporization Equilibrium of the Water–Sulfuric Acid System*, Rep. Eur. 6783, Commission of European Communities, Hydrogen Energy Vector, Europe, 1980, pp. 60–70.

80. *Perry's Chemical Engineer's Handbook*, 6th ed., McGraw-Hill Book Co., Inc., New York, 1984, p. 3–69.

81. T. Vermeulen and co-workers, *Vapor–Liquid Equilibrium of the Sulfuric Acid / Water System*, AIChE meeting, Anaheim, Calif., (June 10, 1982).

82. A. Bosen and H. Engels, *Fluid Phase Equilibria* **43**, 213–230 (1988).

83. D. D. Wagman and co-workers, *J. Phys. Chem. Ref. Data*, **11**(Suppl. 2), (1982), American Chemical Society and the American Institute of Physics for the National Bureau of Standards, U.S. Department of Commerce, Washington, D.C.

84. J. E. Kunzler and W. F. Giauque, *J. Am. Chem. Soc.* **74**, 3472 (1952).

85. J. Nilges and J. Schrage, *Fluid Phase Equilibria* **68**, 247–261 (1991).

86. F. D. Miles, H. Niblock, and G. L. Wilson, *Trans. Faraday Soc.* **36**, 345 (1940).

87. J. H. Colwell and G. D. Halsey, *J. Phys. Chem.* **66**, 2179–2182 (1962).

88. F. R. Bichowski and F. D. Rossini, *Thermochemistry of Chemical Substances*, 1st ed., Reinhold Publishing Co., New York, 1936.

89. H. Howard, *J. Soc. Chem. Ind.* **29**(1), 3 (1910).

90. H. P. J. J. Thomsen, *Thermochemische Untersuchungen*, Vol. III, Barth, Leipzig, Germany, 1883.

91. C. V. Herrmann, *Ind. Eng. Chem.* **33**, 898 (June 1941).

92. J. C. D. Brand and A. J. Rutherford, *J. Chem. Soc.* **10**, 3916 (1952).

93. J. Schrage, *Fluid Phase Equilibria* **68**, 229–245 (1991).

94. *Sulphur*, (191), 30 (July–Aug. 1987).

95. *Sulphur*, (210), 24 (Sept.–Oct. 1990).

96. *World Sulphur and Sulphuric Acid Plant List & Atlas*, 6th ed., British Sulphur Publishing, London, 1993.

97. Ref. 95, p. 28.

98. T. L. Muller in A. J. Buonicore and W. T. Davis eds., *Air Pollution Engineering Manual*, Van Nostrand Reinhold, New York, 1992, p. 469.

99. A. Vavere and J. R. Horne, *SO_2 Emissions Reductions in Sulfuric Acid Plants*, AIChE meeting, Minneapolis, Minnesota, Aug. 1992.

100. A. M. Fairlie, *Sulfuric Acid Manufacture*, Reinhold Publishing Co., New York, 1936.

101. T. J. Browder in A. I. More, ed., "Making the Most of Sulfuric Acid," *Proceedings of The British Sulphur Corporation's Fifth International Conference*, British Sulphur Publishing, London, Nov. 1981, pp. 183–205.

102. J. D. Palmer, *Can. Chem. Process.* **60**(8), 35 (Aug. 1976).

103. P. D. Nolan, *Can. Chem. Process.* **61**(5), 40 (May 1977).

104. D. Fyfe and co-workers, *International Corrosion Forum: Corrosion '75*, Paper No. 63, Toronto, National Association of Corrosion Engineers (NACE), Houston, Texas, 1975.

105. B. Ornberg and P. Crook, *A New Nickel Allow and a New Semi-Welded Plate Heat Exchanger: The State of the Art for Tomorrow's Acid Coolers*, Sulphur '94 (British Sulphur Conference), Tampa, Fla., Nov. 1994.

106. *Sulphur*, (230), 23 (Jan.–Feb. 1994).

107. J. R. Donovan, R. D. Stolk, and M. L. Unland, in B. E. Leach, ed., *Applied Industrial Catalysis*, Vol. 2, Academic Press, Inc., New York, 1983, pp. 245–286.

108. G. M. Cameron, R. M. Fries, and S. M. Puricelli, *Demonstrated New Technologies for Sulphuric Acid Plants of the Nineties*, Sulphur '94 (British Sulphur Conference), Tampa, Fla., Nov. 1994.
109. R. Duros and E. D. Kennedy, *Chem. Eng. Prog.* **74**(9), 70 (Sept. 1978).
110. *Sulphur*, (206), 22 (Jan.–Feb. 1990).
111. J. E. McLean, J. R. Myers, and D. A. Schleiffarth, *Sulphur* No. 216, 41–43 (Sept.–Oct. 1991).
112. M. Tiivel and co-workers, *Carbon Steel Sulfuric Acid Storage Tank—Inspection Guidelines*, Marsulex, Inc., North York, Ontario, Canada, 1986.
113. NACE Standard PR0294 (latest revision), *Design, Fabrication and Inspection of Tanks for the Storage of Concentrated Sulfuric Acid and Oleum at Ambient Temperatures*, NACE International, Houston, Tex., 1994.
114. M. G. Fontana and N. D. Green, *Corrosion Engineering*, 2nd ed., McGraw-Hill Book Co., Inc., New York, 1978, pp. 223–241.
115. *Sulphur*, (201), 23 (Mar.–Apr. 1989).
116. *Sulphur*, (216), 27 (Sept.–Oct. 1991).
117. S. K. Brubaker, *Materials of Construction for Sulfuric Acid*, Process Industries Corrosion, National Association of Corrosion Engineers meeting, Houston, Tex., 1986.
118. B. D. Craig, *Handbook of Corrosion Data*, ASM International, Metals Park, Ohio, 1989.
119. S. K. Brubaker, in *Metals Handbook—9th Edition*, Vol. 13 (*Corrosion*), ASM International, Metals Park, Ohio, 1987, p. 1148.
120. NACE Standard RP0391, *Materials for the Handling and Storage of Concentrated (90–100%) Sulfuric Acid*, NACE International, Houston, Tex., 1991.
121. NACE Publication 5A151 (1985 revision), *Materials of Construction for Handling Sulfuric Acid*, NACE International, Houston, Tex., 1951.
122. *The Corrosion of Nickel-Containing Alloys in Sulfuric Acid and Related Compounds*, Corrosion Engineering Bulletin CEB-1, Inco Alloys International, Huntington, W.Va., 1983.
123. *Corrosion Resistance of Hastelloy Alloys*, Haynes International, Kokomo, Ind., 1980.
124. G. E. McClain, *Chem. Eng. Prog.* **78**(2) 48–50 (Feb. 1982).
125. U.S. Pat. 4,576,813, (Mar. 18, 1986), D. R. McAlister and S. A. Ziebold (to Monsanto Co.).
126. D. R. McAlister and co-workers, *A Major Breakthrough in Sulfuric Acid*, AIChE National Meeting, New Orleans, La., Apr. 1986.
127. *Sulphur*, (207), 51 (Mar.–Apr. 1990).
128. D. R. McAlister and co-workers, *Chem. Eng. Prog.* **82**(7), 34–38 (July 1986).
129. S. M. Puricelli, *Electrical Power—More than a Sulfuric Acid By-Product*, AIChE Meeting, New Orleans, La., Mar. 1992.
130. *Sulphur*, (147), 32 (Mar.–Apr. 1980).
131. U. H. F. Sander, U. Rothe, and R. Kola, in *Sulphur, Sulphur Dioxide, and Sulphuric Acid*, The British Sulphur Corp., London, 1984, pp. 309–315, 320–323.
132. A. F. Snowball, *Can. Chem. Process Ind.* **31**(12), 1110 (Dec. 1947).
133. S. D. Kirkpatrick, *Chem. Eng.* **55**(4), 96 (Apr. 1948).
134. M. G. Ding, *The Use of Oxygen or Oxygen Enriched Air for Sulphuric Acid Recovery*, Sulphur 1990 (British Sulphur Conference), Cancun, Mexico, Apr. 1990.
135. P. Parish and F. C. Snelling, *Sulfuric Acid Concentration*, 2 vols., Ernest Benn Ltd., London, 1924.
136. J. M. Connor, *Sulphur* (131), 39 (July–Aug. 1977).
137. G. M. Smith and E. Mantius, *Chem. Eng. Prog.* **74**(9), 78 (Sept. 1978).
138. U. Sander and G. Daradimos, *Chem. Eng. Prog.* **74**(9), 57 (Sept. 1978).
139. R. Al Samadi, C. M. Evans, I. M. Smith, *Sulphur* (207), 43–50 (Mar.–Apr. 1990).

140. *Chem. Eng.* **100**(4), 47 (Apr. 1993).

141. Ref. 131, pp. 350–370.

142. E. O. Jones and K. L. Kensington, *Spent acid recovery using WADR process system*, ACS meeting, Chicago, Ill., Aug. 19, 1993.

143. H. R. Kueng and P. Reimann, *Chem. Eng.* **89**(8), 72 (Apr. 1982).

144. Sulfuric Acid in *Chemical Economics Handbook*, SRI International, Menlo Park, Calif., Aug. 1995.

145. *ASTM E 223-88*, Vol. 15.05 of 1992 Standards, American Society for Testing and Materials, Philadelphia, Pa., 1992, pp. 279–287.

146. *Sulfuric Acid, Electrolyte for Storage Batteries*, Specification O-S-801E, Nov. 21, 1990; *Sulfuric Acid, Technical*, Specification O-S-809E, Nov. 16, 1990; as amended Oct. 21, 1992, Federal Supply Service, Ceneral Services Administration, Washington, D.C.

147. Sulfuric Acid, *Food Chemicals Codex*, National Academy Press, Washington, D.C., 1981, pp. 317–318.

148. F. Snell and C. Snell, *Colorimetric Methods of Analysis*, 3rd ed., Vol. 2, D. Van Nostrand, Inc., New York, 1949, p. 798.

149. G. Norwitz, *Analyst*, **87** (1039), 829 (1962).

150. A. Irudayasamy and A. R. Natarajan, in Ref. 149, pp. 831–832.

151. *Standard Methods for the Examination of Water and Wastewater*, 17th ed., American Public Health Association (APHA), Inc., Washington, D.C., 1989, pp. 2-11–2-16.

152. O. T. Fasullo, *Sulfuric Acid Use and Handling*, McGraw-Hill Book Co., Inc., New York, 1965.

153. *Air Quality Criteria for Sulfur Oxides*, National Air Pollution Control Administration, U.S. Dept. of Health, Education, and Welfare, Washington, D.C., Jan. 1969, pp. 89–102.

154. M. O. Amdur, L. Silverman, and P. Drinker, *Am. Med. Assoc. Arch. Ind. Hyg. Occup. Med.* **6**, 306 (Oct. 1952).

155. V. M. Sim and R. E. Pattle, *J. Am. Med. Assoc.* **165**, 1908 (Dec. 14, 1957).

156. R. E. Pattle, F. Burgess, and H. Cullumbine, *J. Pathol. Bacteriol.* **72**, 219 (1956).

157. M. O. Amdur, *Arch. Ind. Health* **18**, 407 (1958).

158. D. Malcolm and E. Paul, *Br. J. Ind. Med.* **18**, 63 (1961).

159. J. B. Lynch and J. Bell, *Br. J. Ind. Med.* **4**, 84 (1947).

160. H. J. Bruggen Cate, *Br. J. Ind. Med.* **25**, 249 (1968).

161. S. Chaney, W. Bloomquist, K. Muller, and G. Goldstein, *Arch. Environ. Health* **35**(4), 211 (July–Aug. 1980).

162. S. M. Horvath and co-workers, *Effects of Sulfuric Acid Mist Exposure on Pulmonary Function*, EPA-600/S1-81-044, issued as PB81-208977 by NTIS, Washington, D.C., June 1981.

163. *IARC Monograph Eval. Carcinogen. Risk*, IARC, **54**, 41–119 (1992).

164. "1996 TLVs and BELs," *American Conference of Governmental Industrial Hygienists*, Cincinnati, Ohio, 1996, p. 34.

165. N. Sathiakumar and co-workers, *Crit. Rev. Toxicol.* (July 1997).

166. J. A. Swenberg and R. O. Beauchamp, Jr., *Crit. Rev. Toxicol.* (July 1997).

167. Documentation of the Threshold Limit Values, 6th ed., *American Conference of Governmental Industrial Hygienists, Inc.*, Cincinnati, Ohio, 1986, p. 544.

168. *Emergency Response Planning Guidelines*, American Industrial Hygiene Association, Fairfax, Va., 1989.

THOMAS L. MULLER
E. I. du Pont de Nemours & Co., Inc.

SULFURIC AND SULFUROUS ESTERS

Sulfuric and sulfurous acids form a series of esters analogous to those from other acidic materials. The hydrogen of the acid is replaced by a carbon-containing group. Because two hydrogens are present in the sulfur-based acids, there are two series of esters. Replacement of one hydrogen results in an acid ester. If both hydrogens are replaced, whether with the same, with different, or with bifunctional substituents, symmetrical and unsymmetrical diesters form. The two series are represented by the following general formulas, where R is a carbon

$$
\begin{array}{cc}
\overset{\displaystyle O}{\underset{\displaystyle O}{\overset{\|}{\underset{\|}{ROSOR'}}}} & \overset{\displaystyle O}{\overset{\|}{ROSOR'}}
\end{array}
$$

group and R′ is a carbon group, hydrogen, or metal cation. In the acid ester series a chlorine or amine group may be present in place of the hydroxy group. These compounds can be used as intermediates in making diesters. The carbon groups most commonly present are those from short- and long-chain alcohols and from hydroxyaromatic and heterocyclic compounds. Table 1 illustrates the variety of known compounds and some of their properties.

Dimethyl sulfate, diethyl sulfate, and long-chain monoalkyl alkali metal sulfates are the compounds of practical interest. Dialkyl sulfates have been used to introduce a methyl or ethyl group into a wide variety of organic compounds. The alkylation chemistry of the lower alkyl hydrogen sulfates or their salts is of far lesser importance, but is closely related to that of dialkyl sulfates in that the monoalkyl esters are intermediate alkylating agents in those instances where both alkyl groups of the dialkyl sulfate are usable for alkylation. The long-chain monoalkyl hydrogen sulfates, in the form of their alkali metal or alkylammonium salts, are valuable as surface-active agents and are used commercially. The short-chain monoalkyl hydrogen sulfates are intermediates in commercial conversion routes from olefins to alcohols (see ETHANOL; PROPYL ALCOHOLS, ISO-PROPYL ALCOHOL; SULFONATION AND SULFATION; SURFACTANTS).

Physical Properties

The physical properties of sulfuric and sulfurous esters are best understood as resulting from the blending of the polar contribution of an acid or neutral sulfate group with the contribution of the usually nonpolar carbon group. The neutral sulfate group of esters is much less polar than the acid sulfate group, $-OSO_3H$. With lower alkyl groups, the polar effects dominate, whereas with higher alkyl groups, the nonpolar interactions of the alkyl dominate.

The lower alkyl hydrogen sulfates are moderately viscous liquids. Higher alkyl hydrogen sulfates are hygroscopic, low melting solids; their hygroscopic nature results from the acid group, and their solid state results from the alkyl group orientation. They do not have boiling points, but decompose on heating.

Table 1. Some Sulfuric and Sulfurous Acid Esters

Ester	CAS Registry Number	$Bp_{kPa}{}^{a}$, °C	Mp, °C
Sulfates, $(RO)_2SO_2$			
R			
open-chain			
methyl	[77-78-1, 75-93-4]	$188.8_{101.3}$, $69.70_{1.33}$	
ethyl	[64-67-5]	208, $89_{1.20}$	
n-propyl	[598-05-0]	$95_{0.670}$	
isopropyl	[2973-10-6]	$80_{0.530}$	
n-butyl	[625-22-9]	$103_{0.200}$	
ethyl *n*-butyl	[5867-95-8]	$117_{2.40}$	
chloromethyl	[73455-05-7]	$97_{1.87}$	
2-chloroethyl	[5411-48-3]	$150_{0.940}$	
n-decyl	[66186-16-1]		37.7
n-tetradecyl	[66186-19-4]		58.0
phenyl	[4074-56-0]	$194.6_{2.66}$	
cyclic			
ethylene	[1072-53-3]		99
1,3-propylene	[1073-05-8]		63
methylene (dimer)	[20757-83-9]		155
Sulfites, $(RO)_2SO$			
R			
open-chain			
methyl	[616-42-2]	$126-127_{101.3}$	
ethyl	[623-81-4]	$159-160_{101.3}$	
n-propyl	[623-98-3]	$82_{2.00}$	
isopropyl	[4773-13-1]	$78_{2.67}$	
n-butyl	[626-85-7]	$110_{1.73}$	
3-chloropropyl	[83929-99-1]	$162_{1.73}$	
phenyl	[4773-12-0]	$185_{2.00}$	13-16
cyclic			
ethylene	[3741-38-6]	$80_{3.72}$	
1,2-propylene	[1469-73-4]	$84_{3.92}$	
pentaerythritol,di	[3670-93-7]		154
Organo hydrogen sulfates, $(RO)SO_3H$			
R			
methyl	[75-93-4]	$130-140_{101.3}$ dec	
n-decyl	[142-98-3]	liquid	
n-dodecyl	[151-41-7]		25-27
n-tetradecyl	[4754-44-3]		37-39
n-hexadecyl	[143-02-2]		40-42
n-octadecyl	[143-03-3]		51-52
phenyl	[937-34-8]		
Organo halosulfates, $(RO)SO_2X$			
R, X			
methyl,chloro	[812-01-1]	$134-135_{101.3}$, $42_{2.13}$	
methyl,fluoro	[421-20-5]	$92_{101.3}$, $45_{21.4}$	
ethyl,chloro	[625-01-4]	$72_{15.6}$, $58_{2.66}$	
ethyl,fluoro	[371-69-7]	113_{100}, $21_{1.60}$	

Table 1. (*Continued*)

Ester	CAS Registry Number	Bp$_{kPa}$^{a}, °C	Mp, °C
n-propyl,chloro	[*819-52-3*]	53$_{1.33}$	
isopropyl,chloro	[*36610-67-0*]	50$_{0.67}$ dec	

Organo halosulfites, (RO)SOX

R, X			
methyl,chloro	[*13165-72-5*]	35$_{8.00}$	
ethyl,chloro	[*6378-11-6*]	50–53$_{8.00}$	
n-propyl,chloro	[*22598-38-5*]	78$_{10.0}$, 42$_{1.69}$	
isopropyl,chloro	[*22598-56-7*]	71–73$_{10.0}$, 55$_{5.23}$	

^{a}To convert kPa to mm Hg, multiply by 7.5; 101.3 kPa = 1 atm.

Stability depends mostly on purity, with purer materials having longer shelf lives (1). For the higher alkyl groups, the anhydrous compounds are soluble and the monohydrates are insoluble in ether. Solutions in water are strongly ionized and acidic. The lower dialkyl sulfates are liquids with faint but pleasant odors; *n*-nonyl and higher normal aliphatic and cyclic sulfates are solids.

The diesters are moderately polar and therefore are miscible with most common organic solvents. Solubility in water is low-to-insoluble, with dimethyl sulfate having a water solubility of 2.8 g/100 mL at 18.0°C (2).

The sulfites have some laboratory use, but are not commercially important and are less known. Monoesters of sulfurous acid are quite unstable, although salts have been identified. The diesters of sulfurous acid are mostly liquids with boiling points somewhat less than those of the corresponding sulfates.

Many esters of halosulfurous acid and of halosulfuric acid are known, ie, halosulfites and halosulfates, respectively. With the smaller alkyl groups, the compounds are high boiling liquids with a penetrating odor and a lacrimatory action. Esters of halosulfuric acid are more stable to moisture when stored than those of halosulfurous acid.

Selected properties of dimethyl and diethyl esters of sulfuric acid are listed in Table 2.

Table 2. Selected Physical Properties of Dimethyl and Diethyl Sulfates

Property	(CH$_3$O)$_2$SO$_2$	(C$_2$H$_5$O)$_2$SO$_2$
bp, at 101.3 kPa (= 1 atm), °C	188.8	208 dec
13.3 kPa		143
1.33 kPa		88
mp, °C	−31.8	−24.4
sp gr$_{20}^{20}$	1.328	1.1803
n_D^{20}	1.3874	1.3396
solubility in water, wt %	2.8 at 18°C	0.7 at 20°C
flash point (open cup), °C	116	113
autoignition temp, °C	188	
heat of formation (liquid), kJ/mol^{a,b}	687	757

^{a}To convert J to cal, divide by 4.184.
^{b}Ref. 3.

Chemical Properties

Sulfates. The chemistry of alkyl sulfates is dominated by two fundamental process types: reaction with nucleophiles and reaction as acids. Reaction with nucleophiles results in alkylation.

Alkylation. In alkylation, the dialkyl sulfates react much faster than do the alkyl halides, because the monoalkyl sulfate anion ($ROSO_3^-$) is more effective as a leaving group than a halide ion. The high rate is most apparent with small primary alkyl groups, eg, methyl and ethyl. Some leaving groups, such as the fluorinated sulfonate anion, eg, the triflate anion, $CF_3SO_3^-$, react even faster in ester form (4). Against phenoxide anion, the reaction rate is methyl triflate [333-27-7] $\gg$ dimethyl sulfate $\gg$ methyl p-toluenesulfonate [23373-38-8] (5). Dialkyl sulfates, as compared to alkyl chlorides, lack chloride ions in their products; chloride corrodes and requires the use of a gas instead of a liquid. The lower sulfates are much less expensive than lower bromides or iodides, and they also alkylate quickly.

Although the first alkyl group is the most reactive, the second alkyl group on the intermediate anion can also alkylate. The temperature must be higher than needed for the first alkyl, and conditions favoring hydrolysis of sulfate ester, eg, the presence of water and alkali, should be minimized. As a general rule, alkylations by alkyl chloride are intermediate in rate between those of the corresponding dialkyl sulfates and alkyl hydrogen sulfates. In most reported reactions, only one alkyl group reacts. If the alkylated molecule acquires a positive charge, the monoalkyl sulfate anion becomes the negative counterion. This is illustrated in the common reaction of quaternization of a tertiary amine, eg, pyridine.

$$C_5H_5N + CH_3OSO_2OCH_3 \longrightarrow C_5H_5NCH_3^+ + CH_3OSO_3^-$$

The reactions of these nucleophilic processes are usually S_N2 rather than S_N1. The reaction rate is methyl > ethyl > isopropyl, as with the alkyl halides. As the species to be alkylated becomes more nucleophilic, alkylation becomes faster, eg, a sulfur-containing anion alkylates more quickly than a phenolic anion.

Dimethyl sulfate and diethyl sulfate react with almost all types of compounds with unshared electron pairs, whether the reaction center is at oxygen, nitrogen, carbon, sulfur, phosphorus, some metals, or most other heteroatoms from Groups 15(VA) and 16(VIA). Alkylation at oxygen can occur with alcohols, phenols, or acids. Methylation of phenol to give the methyl ether or anisole is carried out commercially. With phenol alone, the reaction is slow with dimethyl sulfate at 100–120°C, giving anisole and some by-product dimethyl ether and sulfonation products (6). Reaction is much faster in basic medium, with phenolate ion as the reactant. The first methyl group reacts at 45–60°C upon addition of dimethyl sulfate to a mixture of molten phenol and sodium hydroxide containing very little water. Continuing the reaction by heating to 100°C results in the reaction of the second methyl group and a 95% yield of anisole (7–9). Ethylation of phenolate ion with diethyl sulfate requires higher temperatures, ie, ca 50–55°C, for the first ethyl group and 145°C for the second ethyl group. Numerous substituted phenols have been alkylated in either aqueous or alcoholic alkaline solutions, or dimethylformamide solution with K_2CO_3 (9–11). Thiophenols react faster than phenols in alkaline solution (12).

Alcohols react readily in alkaline solution. Conditions for pentaerythritol are typical (13). Use of alcoholic KOH in dimethyl sulfoxide gives fair to good ethyl ether yields with diethyl sulfate at 50–55°C (14).

Carboxylic acids react with the first alkyl group of dialkyl sulftes at 120°C to give esters in high yield, whereas the second alkyl group requires higher temperatures, ie, 200°C (15). The use of carboxylate salts of inorganic or organic bases gives fast esterification under neutral conditions (16). A highly branched acid, eg, trimethylacetic acid, converts readily to the methyl ester; dimethylformamide is a good medium (11,17). Good conditions for diethyl sulfate with lower acids have been determined (18). Dimethyl or diethyl carbonate can be made by heating the corresponding sulfate and alkali–metal carbonate at 150–210°C, which illustrates the high activity of the sulfates (19). Sulfonic acids and their salts alkylate at 100°C (20,21).

The nitrogen of aliphatic and aromatic amines is alkylated rapidly by alkyl sulfates yielding the usual mixtures. Most tertiary amines and nitrogen heterocycles are converted to quaternary ammonium salts, unless the nitrogen is of very low basicity, eg, in triphenylamine. The position of dimethyl sulfate-produced methylation of several heterocycles with more than one heteroatom has been examined (22). Acyl cyanamides can be methylated (23). Metal cyanates are converted to methyl isocyanate or ethyl isocyanate in high yields by heating the mixtures (24,25).

$$CH_3O\overset{\displaystyle O}{\overset{\displaystyle \|}{C}}NHCN + (CH_3O)_2SO_2 \xrightarrow[\text{(70\% overall)}]{\text{NaOH}} CH_3O\overset{\displaystyle O}{\overset{\displaystyle \|}{C}}N\underset{\displaystyle \underset{\displaystyle CH_3}{|}}{C}N$$

Carbon is alkylated in the form of enolates or as carbanions. The enolates are ambident in activity and can react at an oxygen or a carbon. For example, refluxing equimolar amounts of dimethyl sulfate and ethyl acetoacetate with potassium carbonate gives a 36% yield of the O-methylation product, ie, ethyl 3-methoxy-2-butenoate, and 30% of the C-methylation product, ie, ethyl 2-methyl-3-oxobutanoate (26). Generally, only one alkyl group of the sulfate reacts with beta-diketones, beta-ketoesters, or malonates (27). Factors affecting the O:C alkylation ratio have been extensively studied (28). Reaction in the presence of solid Al_2O_3 results mostly in C-alkylation of ethyl acetoacetate (29).

Carbanions in the form of phenyllithium, sodium naphthalene complex, sodium acetylide, or aromatic Grignard reagents react with alkyl sulfates to give a C-alkyl product (30–33). Grignard reagents require two moles of dimethyl sulfate for complete reaction.

Sulfur is reactive in many forms. Mercaptides are alkylated to thioethers, and thioethers react further to give sulfonium salts, $R_3S^+CH_3OSO_3^-$.

Suitable thiones also alkylate. Thioacridone (34) and thiourea (35) are examples: the first gives the alkylmetcaptoacridine and the second gives the isothiourea.

$$(H_2N)_2C{=}S \xrightarrow{(CH_3O)_2SO_2} \begin{array}{c} H_2N \\ \diagdown \\ HN \end{array}\!\!C{-}SCH_3$$

Several inorganic atoms have also been methylated by dimethyl sulfate. Sodium iodide gives methyl iodide, organoxytitanium dichloride gives several highly colored organoxytitanium methyl sulfates, titanium tetrachloride yields titanium chlorosulfates or methyl sulfates, and mossy tin gives dimethyltin sulfate (36–39).

These reactions involve mostly dimethyl and diethyl sulfate. Cyclic sulfates are also reactive, and several have been compared by determining reaction rates with a substituted pyridine or with water (40). In both cases, 1,2-ethylene sulfate is more reactive than 1,3-propylene sulfate or dimethyl or diethyl sulfates.

Hydrolysis. The hydrolysis of dialkyl and monoalkyl sulfates is a process of considerable interest commercially. Successful alkylation in water requires that the fast reaction of the first alkyl group with water and base be minimized. The very slow reaction of the second alkyl group results in poor utilization of the alkyl group and gives an increased organic load to a waste-disposal system. Data have accumulated since 1907 on hydrolysis in water under acid, neutral, and alkaline conditions, and best conditions and good values for rates have been reported and the subject reviewed (41–50).

Sulfates having alkyl groups from methyl to pentyl have been examined. With methyl as an example, the hydrolysis rate of dimethyl sulfate increases with the concentration of the sulfate. Typical rates in neutral water are first order and are 1.66×10^{-4} s^{-1} at 25°C and 6.14×10^{-4} s^{-1} at 35°C (46,47). Rates with alkali or acid depend on conditions (42,48). Rates for the monomethyl sulfate [512-42-5] are much slower, and are nearly second order in base. Values of the rate constant in dilute solution are 6.5×10^{-5} L/(mol·s) at 100°C and 4.64×10^{-4} L/(mol·s) at 138°C (44). At 138°C, first-order solvolysis is ca 2% of the total. Hydrolysis of the monoester is markedly promoted by increasing acid strength and it is first order. The rate at 80°C is 3.65×10^{-4} s^{-1} (45). Alkaline solvolysis has been studied by a calorimetric method (49). Heat of hydrolysis of dimethyl sulfate to the monoester under alkaline conditions is 106 kJ/mol (25 kcal/mol) (51).

Cyclic esters show accelerated hydrolysis rates. Ethylene sulfate compared to dimethyl sulfate is twice as fast in weak acid (first order) and 20 times as fast in weak alkali (second order) (50). Catechol sulfate [4074-55-9] is 2×10^7 times faster than diphenyl sulfate in alkaline solution (52). Alcoholysis rates of several dialkyl sulfates at 35–85°C are also known (53).

Studies of reaction mechanisms in ^{18}O-enriched water show the following: cleavage of dialkyl sulfates is primarily at the C—O bond under alkaline and acid conditions, and monoalkyl sulfates cleave at the C—O bond under alkaline conditions and at the S—O bond under acid conditions (45,54). An optically active half ester (*sec*-butyl sulfate [3004-76-0]) hydrolyzes at 100°C with inversion under alkaline conditions and with retention plus some racemization under acid conditions (55). Effects of solvent and substituted structure have been studied, with moist dioxane giving marked rate enhancement (44,56,57). Hydrolysis of monophenyl sulfate [4074-56-0] has been similarly examined (58).

Reactions other than those of the nucleophilic reactivity of alkyl sulfates involve reactions with hydrocarbons, thermal degradation, sulfonation, halogenation of the alkyl groups, and reduction of the sulfate groups. Aromatic hydrocarbons, eg, benzene and naphthalene, react with alkyl sulfates when catalyzed by aluminum chloride to give Friedel-Crafts-type alkylation product mixtures (59). Isobutane is readily alkylated by a dipropyl sulfate mixture from the reaction of propylene in propane with sulfuric acid (60).

Pyrolysis. Thermal stability of the dialkyl sulfate is of interest when using solutions in inert media, eg, hydrocarbons, or on storage of the neat compound. Heating dimethyl sulfate at ca 200°C produces dimethyl ether, SO_3, and other products (61). Higher alkyl sulfates, eg, the diethyl sulfates, decompose to give olefin and oxidation products rapidly at 220°C and slowly at 140°C (62). Heating higher monoalkyl hydrogen sulfates generally gives olefins also, but the monomethyl ester at 130–140°C gives dimethyl sulfate and sulfuric acid.

Sulfonation. Sulfonation is a common reaction with dialkyl sulfates, either by slow decomposition on heating with the release of SO_3 or by attack at the sulfur end of the O—S bond (63). Reaction products are usually the dimethyl ether, methanol, sulfonic acid, and methyl sulfonates, corresponding to both routes. Reactive aromatics are commonly those with higher reactivity to electrophilic substitution at temperatures >100°C. Triphenylamine, diphenylmethylamine, anisole, and diphenyl ether exhibit ring sulfonation at 150–160°C, 140°C, 155–160°C, and 180–190°C, respectively, but diphenyl ketone and benzyl methyl ether do not react up to 190°C. Diphenylamine methylates and then sulfonates. Catalysis of sulfonation of anthraquinone by dimethyl sulfate occurs with thallium(III) oxide or mercury(II) oxide at 170°C. Alkyl interchange also gives sulfation.

Miscellaneous. Halogenation of the methyl group of dimethyl sulfate by chlorine and by fluorine has been described (64,65). Reduction of dimethyl sulfate by hydriodic acid occurs in a way that is comparable to that shown by sulfuric acid and results in reduction to sulfur (66).

Sulfites. The literature concerning dialkyl sulfites is extensive, although less than for sulfates. Reactions involving alkylation are similar to those of sulfates. Sulfites also undergo elimination, transesterification, and isomerization. The last two parallel reactions of phosphites.

In alkylation, phenols and amines are alkylated by sulfites in high yield and quaternary salts readily form (67). Ethylene sulfite reacts yielding hydroxyethyl derivatives and SO_2 elimination, corresponding to its activity as an ethylene oxide precursor (68).

$$\text{ArOH} + \underset{\text{O}}{\overset{\text{O}}{\diagdown}}\text{S}{=}\text{O} \xrightarrow{\text{90–100°C}} \text{ArOCH}_2\text{CH}_2\text{OH} + \text{SO}_2$$

Reaction of carboxylate ion with *p*-nitrophenyl sulfites gives the carboxylate *p*-nitrophenyl esters. If the *p*-nitrophenyl sulfite is unsymmetrical

($O_2NC_6H_4OS(O)OR$, where R is ethyl or phenyl), carboxylate attacks the *p*-nitrophenyl side (69). Some amino acids react with methyl and benzyl sulfites in the presence of *p*-toluenesulfonic acid to give methyl and benzyl esters of the amino acids as *p*-toluenesulfonate salts (70). With alcohols, the conversion of benzil to a monoacetal upon addition of sulfuric acid to the benzil in methanol and dimethyl sulfite proceeds in high yield (71).

Hydrolysis of dialkyl sulfites under acidic and alkaline conditions, which is followed by the use of $^{18}OH_2$, proceeds by attack at sulfur to give S–O cleavage (72). The rate of hydrolysis is generally faster for cyclic and aryl sulfites than for dialkyl sulfites (73). Activation parameters of hydrolysis are known for some sulfites, and the increased rate for ethylene sulfite results from a reduced entropy of activation which results from a rigid ring structure (74).

Manufacture

Monoester Hydrogen Sulfates. The hydrogen sulfates are prepared by the action of a sulfating agent on the corresponding alcohol or phenol. The following reagents are used for sulfation: sulfur trioxide, sulfuric acid, chlorosulfuric acid, sulfur trioxide–amine complexes, and sulfamic acid (see SULFONATION AND SULFATION). The latter two produce the amine salt of the sulfate ester directly. The reaction of sulfuric acid with olefins also gives hydrogen sulfates. These reactions are commercially important for the preparation of ethyl, 2-propyl, 2-butyl, and longer-chain sodium alkyl sulfates. The last are useful as detergents (see ALCOHOLS, HIGHER ALIPHATIC–SURVEY AND NATURAL ALCOHOLS MANUFACTURE). With the exception of ethylene, the olefins give esters of secondary alcohols.

Methyl hydrogen sulfate is obtained in yields of 98–99% upon treating sulfur trioxide with methanol below 0°C (75). Ethyl hydrogen sulfate [540-82-9] is obtained by the sulfation of ethylene with sulfuric acid to the mixed sulfate, followed by the addition of water to the reaction mixture and hydrolysis of any diethyl sulfate formed. It is usually carried out with 96–98 wt % sulfuric acid at 70 and 80°C and 500–1500 kPa (5–15 atm). The kinetics of this reaction have been thoroughly investigated (76). At 150°C and higher pressures with only 70 wt % acid followed by addition of just enough water for hydrolysis, the otherwise necessary reconcentration of the spent sulfuric acid is eliminated (77). Because of the severe corrosion under these conditions, tantalum equipment is necessary. Ethyl hydrogen sulfate can also be prepared in 87% yield by heating sodium hydrogen sulfate with ethanol; a 74–86% yield is obtained by the reaction of ethanol with sulfur trioxide in liquid sulfur dioxide solution or by treating diethyl sulfate with ethanol and distilling the ether *in vacuo* (78,79).

$$(C_2H_5O)_2SO_2 + C_2H_5OH \longrightarrow (C_2H_5)_2O + C_2H_5OSO_3H$$

Propylene and butylene require much milder conditions for their sulfation with sulfuric acid. Butylene is sulfated at 30–50°C and 300–600 kPa (ca 3–6 atm) with 30–60 wt % sulfuric acid, and propylene is sulfated at 10–30°C and 500 kPa (ca 5 atm) with 65–85 wt % sulfuric acid. The rate of sulfation of propylene increases sharply with increasing pressure (80). It can also be increased by the addition of kerosene, which raises the concentration of olefin in the liquid phase (81).

Detergents have been manufactured from long-chain alkenes and sulfuric acid, especially those obtained from shale oil or cracking of petrolum wax. These are sulfated with 90–98 wt % acid at 10–15°C for a 5-min contact time and at an acid–alkene molar ratio of 2:1 (82). Dialkyl sulfate initially forms when 96 wt % acid is added to 1-dodecene at 0°C, but it is subsequently converted to the hydrogen sulfate in 80% yield upon the further addition of sulfuric acid. The yield can be increased to 90% by using 98 wt % sulfuric acid and pentane as the solvent at -15°C (83).

Long-chain alcohols, such as are obtained by the hydrogenation of coconut oil, polymerization of ethylene, or the oxo process (qv), are sulfated on a large scale with sulfur trioxide or chlorosulfuric acid to acid sulfates; the alkali salts are commercially important as surface-active agents (see SURFACTANTS). Poly(vinyl alcohol) can be sulfated in pyridine with chlorosulfuric acid to the hydrogen sulfate (84).

Some substituted alkyl hydrogen sulfates are readily prepared. For example, 2-chloroethyl hydrogen sulfate [36168-93-1] is obtained by treating ethylene chlorohydrin with sulfuric acid or amidosulfuric acid. Heating hydroxy sulfates of amino alcohols produces the corresponding sulfuric monoester (85).

Because phenolic compounds are easily sulfonated, their sulfation must be accomplished with milder sulfating agents, eg, complexes of sulfur trioxide or chlorosulfonic acid with trimethylamine, dimethylformamide, pyridine, or dimethylaniline, in anhydrous or aqueous medium below 100°C (86–89).

Salts of alkyl hydrogen sulfates that are free of sulfuric acid can be prepared from crude hydrogen sulfates containing $ROSO_3H$ and sulfuric acid in a two-step reaction. First, the crude product reacts in aqueous solution with the hydroxide or carbonate of a metal whose sulfate is insoluble in water. After filtration of the insoluble metal sulfate, the necessary amount of a water-soluble metal sulfate is added to the filtrate, which precipitates the insoluble metal sulfate, leaving the desired salt of the alkyl hydrogen sulfate in solution.

Diorgano Sulfates. Dialkyl sulfates up to octadecyl can be made from the alcohols by a general method involving the following reactions (90):

$$ROH + SO_2Cl_2 \longrightarrow ROSO_2Cl + HCl$$

$$2\,ROH + SOCl_2 \longrightarrow (RO)_2SO + 2\,HCl$$

$$ROSO_2Cl + (RO)_2SO \longrightarrow (RO)_2SO_2 + RCl + SO_2$$

Another method involves the reaction of dialkyl sulfites with sulfuryl chloride (91):

$$2\,(RO)_2SO + SO_2Cl_2 \longrightarrow (RO)_2SO_2 + 2\,ROSOCl$$

Mixed esters are synthesized by the reaction of one alkyl chlorosulfate with a different sodium alkoxide or a different dialkyl sulfite as follows (92):

$$ROSO_2Cl + NaOR' \longrightarrow ROSO_2OR' + NaCl$$

$$ROSO_2Cl + (R'O)_2SO \longrightarrow ROSO_2OR' + R'Cl + SO_2$$

The only commercially important dialkyl sulfates are dimethyl sulfate and diethyl sulfate. Estimated worldwide production in 1996 for dimethyl sulfate was 90,000 metric tons per year. Dimethyl sulfate was initially made by vacuum pyrolysis of methyl hydrogen sulfate:

$$2\ CH_3OSO_3H \longrightarrow (CH_3O)_2SO_2 + H_2SO_4$$

Later it was synthesized in a batch process from dimethyl ether and sulfur trioxide (93) and this combination was adapted for continuous operation. Gaseous dimethyl ether was bubbled at 15.4 kg/h into the bottom of a tower 20 cm in diameter and 365 cm high and filled with the reaction product dimethyl sulfate. Liquid sulfur trioxide was introduced at 26.5 kg/h at the top of the tower. The mildly exothermic reaction was controlled at 45–47°C, and the reaction product (96–97 wt % dimethyl sulfate, sulfuric acid, and methyl hydrogen sulfate) was continuously withdrawn and purified by vacuum distillation over sodium sulfate. The yield was almost quantitative, and the product was a clear, colorless, mobile liquid. A modified process is described in Reference 94. Properties are listed in Table 3.

Diethyl sulfate can be prepared by a variety of methods. When ethyl hydrogen sulfate is heated with sodium chloride to 80°C, hydrogen chloride is liberated. The resulting reaction mixture is then distilled at 1.33–2.00 kPa (10–15 mm Hg) at a maximum kettle temperature of 190°C to give diethyl sulfate in 90% yield (95).

$$2\ C_2H_5OSO_3H + NaCl \longrightarrow C_2H_5OSO_3H + C_2H_5OSO_3Na + HCl$$

$$C_2H_5OSO_3H + C_2H_5OSO_3Na \longrightarrow (C_2H_5O)_2SO_2 + NaHSO_4$$

Passing a stream of nitrogen at 95–100°C through a reaction mixture of ethyl ether and 30 wt % oleum prepared at 15°C results in the entrainment of diethyl sulfate. Continuous operation provides a >50% yield (96). The most economical process for the manufacture of diethyl sulfate starts with ethylene and 96 wt % sulfuric acid heated at 60°C. The resulting mixture of 43 wt %

Table 3. Properties of a Commercial Dimethyl Sulfate

Property	Typical value
acidity as H_2SO_4, %[a]	0.11
bp, at 101.3 kPa (= 1 atm), °C	
at 95%	190.4
97%	192.0
1–97%	5.0 range
color[b]	passes 0.0001 N iodine
sp gr$_{15.6}^{15.6}$	1.334

[a]Determined by nonaqueous titration.
[b]This test is a visual comparison of the color of dimethyl sulfate with that of a 0.0001 N iodine solution. Commercial dimethyl sulfate should be lighter in color.

diethyl sulfate, 45 wt % ethyl hydrogen sulfate, and 12 wt % sulfuric acid is heated with anhydrous sodium sulfate under vacuum, and diethyl sulfate is obtained in 86% yield; the commercial product is >99% pure (97).

Lower dialkyl sulfates were made from the alcohols in earlier work; the reaction mass at ca 100°C was stripped by a recirculated inert-gas stream, and the product was recovered by passage through a partial condenser. Yields of 90% for diethyl sulfate and 85% for dimethyl sulfate were reported (98).

In the reaction of ethylene with sulfuric acid, several side reactions can lead to yield losses. These involve oxidation, hydrolysis–dehydration, and polymerization, especially at sulfuric acid concentrations >98 wt %; the sulfur trioxide can oxidize by cyclic addition processes (99).

Dimethyl and diethyl sulfate are available in a variety of containers from 0.5-kg glass bottles to tank cars. Diethyl sulfate is somewhat less toxic than dimethyl sulfate and is considered noncorrosive, but dimethyl sulfate is classified as a corrosive liquid and ICC regulations must be observed. Mild steel to 306 stainless steel is used for large-volume storage. Dimethyl sulfate is manufactured by E. I. du Pont de Nemours & Co., Inc., and diethyl sulfate by Union Carbide Corporation. Other producers are Rhône-Poulenc and Hoechst-Celanese.

Cyclic sulfates can be prepared by a variety of methods. Ethylene sulfate is obtained in low yield from ethylene oxide and sulfur trioxide (100). Methylene sulfate is produced from formaldehyde and sulfur trioxide (101).

Oxidation of cyclic sulfites with permanganate in acetic acid solution gives cyclic sulfates (102). Heating monohydroxyalkyl hydrogen sulfates with thionyl chloride causes ring closure (103).

Acidolysis of cyclic sulfites with sulfuric acid and ester interchange with dimethyl sulfate produce cyclic sulfates (104).

$$\begin{matrix} \overset{O}{\overset{\|}{CH_2OCCH_3}} \\ | \\ CH_2OCCH_3 \\ \overset{\|}{O} \end{matrix} + (CH_3O)_2SO_2 \longrightarrow \begin{matrix} O \\ | \\ SO_2 \\ | \\ O \end{matrix} + 2\; CH_3\overset{O}{\overset{\|}{C}}OCH_3$$

Diorgano Sulfites. Symmetrical or mixed dialkyl sulfites are prepared by the stepwise reaction of thionyl chloride either with two molecules of an alcohol or with stoichiometric quantities of two alcohols in pyridine (105).

$$ROH + SOCl_2 + C_5H_5N \longrightarrow ROSOCl + C_5H_5N \cdot HCl$$

$$R'OH + ROSOCl + C_6H_5N \longrightarrow R'OSOOR + C_5H_5N \cdot HCl$$

An alternative route is the reaction of iodine or bromine in pyridine with liquid sulfur dioxide at 20°C, which gives good to high yields of sulfites (106).

Heating the adduct of ethylene oxide and sulfur dioxide with primary alcohols in the presence of alkali hydrides or a transition-metal halide yields dialkyl sulfites (107). Another method for the preparation of methyl alkyl sulfites consists of the reaction of diazomethane with alcoholic solutions of sulfur dioxide (108).

$$ROH + CH_2N_2 + SO_2 \longrightarrow ROSOOCH_3 + N_2$$

Cyclic sulfites can be prepared from the glycols and thionyl chloride in the presence of pyridine; this route is analogous to the preparation of dialkyl sulfites (109). Cyclic sulfites are also obtained when the polymerization product of alkylene oxides with sulfur dioxide is decomposed by heating (110). Ester exchange proceeds smoothly with dimethyl sulfite and glycerol or its chlorohydrins to yield the cyclic sulfites (111). Glycerol gives poor results with thionyl chloride.

Aromatic polysulfites can be produced if bisphenols, eg, bisphenol A, are heated with diphenyl sulfite in the presence of lithium hydride (112).

Halosulfates and Halosulfites. A general method for the preparation of alkyl halosulfates and halosulfites is the treatment of the alcohol with sulfuryl or thionyl chloride at low temperatures while passing an inert gas through the mixture to remove hydrogen chloride (113).

$$ROH + SO_2Cl_2 \longrightarrow ROSO_2Cl + HCl$$

$$ROH + SOCl_2 \longrightarrow ROSOCl + HCl$$

This method is also used with alcohols of the structure $Cl(CH_2)_nOH$ (114). Haloalkyl chlorosulfates are likewise obtained from the reaction of halogenated alkanes with sulfur trioxide or from the chlorination of cyclic sulfites (115,116). Chlorosilanes form chlorosulfate esters when treated with sulfur trioxide or chlorosulfuric acid (117). Another approach to halosulfates is based on the addition of chlorosulfuric or fluorosulfuric acid to alkenes in nonpolar solvents (118).

Health and Safety Factors

The most commonly used dialkyl sulfate is dimethyl sulfate. This is also the most hazardous in liquid and vapor forms. The hazard arises from its toxicity, high reactivity, and to some extent, combustibility. Dimethyl sulfate is corrosive and poisonous, and its effects may be either acute or chronic. Because it has an analgesic effect on many body tissues, even severe exposures may not be immediately painful. Dimethyl sulfate is particularly dangerous to the eyes and respiratory system. It causes severe burns, but symptoms may be delayed. Exposed workers must be immediately and properly treated to avoid either permanent injury to eyes and lungs or even death. Skin burns can also be severe. Ingestion causes convulsions and paralysis, with later damage to the kidneys, liver, and heart (119–121). Genetic effects have been extensively reviewed, and its mutagenicity is correlated with carcinogenicity (122,123).

According to the U.S. Department of Labor (OSHA), exposure to dimethyl sulfate shall not exceed an eight-hour time-weighted average of 1 ppm in air (119). Because both liquid and vapor can penetrate the skin and mucous membranes, control of vapor inhalation alone may not be sufficient to prevent absorption of an excessive dose. Dimethyl sulfate is listed as an industrial substance with suspected carcinogenic potential in humans (119). Thus, the ACGIH recommends a time-weighted average threshold limit value of 0.1 ppm, based on tests with laboratory animals.

Dimethyl sulfate lacks warning properties. It looks like water when spilled and has no distinctive odor; these characteristics mandate careful handling, but its high boiling point permits safe usage with careful attention to procedure. No work should be undertaken with dimethyl sulfate or diethyl sulfate until the worker has studied and understood the precautions, procedures, and exposure effects given in the material safety data sheets, in several government publications, and in the product bulletins for the materials (120,121,124–127). Protective equipment is essential. Clothing and gloves should be made of butyl rubber. All operations should take place in a hood or in an area with good ventilation. Concentrations in air of 10 ppm or higher are immediately dangerous to life and health, and self-contained breathing apparatus must be used (127). Exposure to 97 ppm for 10 min can be fatal. In the laboratory, only mechanical pipetting is safe. Protective sleeves, aprons, and face shields are used. If the liquid should contact the skin, the exposed areas should be flushed immediately with water and then soaped and rinsed thoroughly. The victim should receive medical attention. Treatment for other exposures is described in References 120 and 121.

Because dimethyl sulfate looks like water, operations are preferably not performed when water is present, eg, wet floors or rain. Any spills or leaks should not be left unattended; they should be contained, and runoff to sewers should be avoided. Minor spills should be flooded with water to dilute and hydrolyze the dimethyl sulfate. The area should then be covered with a dilute (2–5 wt %) caustic solution or a dilute (2–5 wt %) ammonia solution, or soda ash may be sprinkled over the neat liquid and the mix wetted with a gentle spray of water. The neutralizing agent should remain on the affected area for 24 h and then should be washed away. Only personnel wearing protective equipment should perform these operations. The product bulletins should be consulted for procedures to be followed for more severe spills. Concentrated ammonia should not be used with neat dimethyl sulfate because explosions have resulted after their contact (128).

In the laboratory, excess reagent in a product should be destroyed before workup. Addition of diluted aqueous ammonia is the most effective practice, if ammonia is otherwise acceptable. Combustibility is a minor problem. The open-cup flash point of 116°C for dimethyl sulfate is well above normal handling temperatures. Flammable, toxic vapors are given off at elevated temperatures.

The acute toxicities of various sulfates have been reported (126). Generally, dimethyl sulfate (LD_{50} 440 mg/kg in rats) is more toxic than diethyl sulfate (LD_{50} 880–1412 mg/kg in rats), which is more toxic than dibutyl sulfate (lowest observed toxic dose, 9500 mg/kg in rats). Ethylene sulfate is more toxic than dimethyl sulfate (40).

The monoalkyl derivatives in salt form appear to have low toxicity. The monomethyl sulfate sodium salt has an approximate oral lethal dose greater than 5000 mg/kg of body weight for rats (129). Monododecyl sulfate sodium salt is widely marketed as a detergent and shampoo ingredient (oral LD_{50} 1268 mg/kg for rats) (126). Both dimethyl sulfate and monomethyl sulfate occur in the environment in coal fly-ash and in airborne particulate matter (130).

Uses

The sulfuric acid esters as compared to the sulfurous esters are the most widely used. In nature they appear as solubilizing groups in detoxification–excretory mechanisms and as sulfated carbohydrate groups in modified proteins. The significant uses are alkylation, formation of long-chain alcohol monosulfates as surfactants, and formation of intermediates in preparation of some lower alcohols. Alkylation (qv) involves primarily dimethyl and diethyl sulfates in the preparation of a wide variety of intermediates and products, especially in the fields of dyes, agricultural chemicals, drugs, and other specialties. In particular, dimethyl sulfate is a powerful reagent yielding quaternary salts in the form of the methosulfates. The use of dimethyl sulfate instead of methyl iodide can be useful when alkylations are carried out in dimethylformide (131). Some quaternary ammonium salts are also surfactants and fabric softeners (132) (see QUATERNARY AMMONIUM COMPOUNDS).

Minor uses include those as dyes, intermediates, stabilizers, and some specialty polymers. It is not always clear which uses are commercial and which are only alleged. One type of fiber-reactive dye has a pendent alcohol sulfate group. This group reacts with cotton in an alkaline pad to produce both an ether–cellulose bond and thereby a washfast dye, eg, Remazol dyes (Hoechst). Methyl hydrogen sulfate was once used as a solvent in the bromination of indigo (133). Dimethyl sulfate is also used with boron compounds to stabilize liquid sulfur trioxide (134). Modified polymers, mostly in the form of polyquaternary salts, have some use as coagulants for slimes, suspensions, and emulsions and as antistatic agents (qv) (135) (see FLOCCULATING AGENTS).

Proposed uses of the esters are numerous, but their applications in the modification of cotton, in catalysis, and as solvents are most prominent. The introduction of the aminoethyl group into cotton by the reaction of aminoethyl sulfate on alkali-swollen cotton is well known (136). The basic group provides a dye site for acid dyes. Catalyst activity has been reported for ionic, organometallic, and free-radical routes. Acidic activity of monoalkyl sulfates is used to promote hydrolysis of poly(vinyl acetate) and to quench the base-catalyzed polymerization of epoxides (137,138). Some pyrolysis reactions are catalyzed by the presence of dimethyl sulfate, eg, in the formation of ketene from the thermal cracking of acetic acid or alkyl acetate and in the formation of 1,3-dihydroxypropene ether and acrolein acetals from 1,1,3-trihydroxypropane ethers at 200–450°C (139,140). The alkylation of phenol with isobutylenes is promoted by catalytic amounts of dimethyl sulfate (141). Some care is needed in interpreting the catalytic effect of dialkyl sulfates. Many reactions in which activity exists are probably susceptible to acid catalysis. The sulfate can cause the formation of acids either by hydrolysis or by pyrolysis. For example, dimethyl sulfate is a curing agent for furfuryl alcohol condensates, because the neutral ester is a source of sulfuric acid when heated to 180°C (142). Cationic activity in catalysis is shown in the promotion of tetrahydrofuran polymerization by ethyl chlorosulfate and in the initiation of polymerization of several heterocycles (143,144) (see POLYETHERS, TETRAHYDRO-FURAN AND OXETANE POLYMERS).

Organometallic usage is shown in the preparation of titanium- or vanadium-containing catalysts for the polymerization of styrene or butadiene by the reaction of dimethyl sulfate with the metal chloride (145). Free-radical activity is proposed for the quaternary product from dimethylaniline and dimethyl sulfate and for the product from 1,1,4,4-tetramethyl-2-tetrazene and dimethyl sulfate (146,147).

Several solvent uses have been proposed. Dimethyl sulfate has been used as a solvent for the study of Lewis acid–aromatic hydrocarbon complexes (148). It also is effective as an extraction solvent to separate phosphorus halide–hydrocarbon mixtures and aromatic hydrocarbons from aliphatics, and it acts as an electrolyte in electroplating iron (149–152). The toxicity of dimethyl sulfate precludes its use as a general-purpose solvent.

A mixture of dimethyl sulfate with SO_3 is probably dimethyl pyrosulfate [*10506-59-9*], $CH_3OSO_2OSO_2OCH_3$, and, with chlorobenzene, it yields the 4,4'-dichlorodiphenylsulfone (153). Trivalent rare earths can be separated by a slow release of acid into a solution of rare earth chelated with an ethylenediaminetetraacetic acid agent and iodate anion. As dimethyl sulfate slowly hydrolyzes and pH decreases, each metal is released from the chelate in turn and precipitates as the iodate, resulting in improved separations (154).

A number of mixed alkyl and aryl sulfites have been patented as insecticides and biocides (155) (see INSECT CONTROL TECHNOLOGY). Other proposed uses include a polysulfite as a sensitizer in photographic emulsion, diethyl sulfite for the removal of catalyst fragments in polypropylene, and ethylene sulfite as an accelerator for aminoplastic molding compositions (156–158) (see AMINO RESINS AND PLASTICS; PHOTOGRAPHY). A mixed, cyclic sulfite–carbonate ester of pentaerythritol has been polymerized to high molecular weight solids, which can be used in preparing laminates (159). Use of dimethyl sulfate as an electrolyte in high energy-density batteries has also been described (160) (see BATTERIES).

BIBLIOGRAPHY

"Sulfuric and Sulfurous Esters" in *ECT* 1st ed., Vol. 13, pp. 506–513, by J. M. Straley, Eastman Kodak Co.; in *ECT* 2nd ed., Vol. 19, pp. 483–498, by J. Fuchs, E. I. du Pont de Nemours & Co., Inc.; in *ECT* 3rd ed., Vol. 22, pp. 233–254, by W. B. McCormack and B. C. Lawes, E. I. du Pont de Nemours & Co., Inc.

1. E. W. Maurer, A. J. Stirtan, and J. K. Weil, *J. Am. Oil Chem. Soc.* **37**, 34 (1960); U.S. Pat. 3,133,946 (May 19, 1964), E. W. Maurer, A. J. Stirtan, and J. K. Weil (to U.S. Dept. of Agriculture).
2. *Dimethyl Sulfate, Properties, Uses, Storage and Handling*, bulletin, E. I. du Pont de Nemours & Co., Inc., Wilmington, Del., Jan. 1981.
3. H. Mackle and W. V. Steele, *Trans. Faraday Soc.* **65**, 2053 (1969).
4. E. S. Lewis, S. Kukes, and C. D. Slater, *J. Am. Chem. Soc.* **102**, 303, 1619 (1980).
5. E. S. Lewis and S. Vanderpool, *J. Am. Chem. Soc.* **99**, 1946 (1977).
6. L. J. Simon and M. Frejaqueous, *Compt. Rend.* **176**, 900 (1923).
7. H. F. Lewis and co-workers, *Ind. Eng. Chem.* **22**, 34 (1930).
8. E. Y. Wolford, *Ind. Eng. Chem.* **22**, 397 (1930).
9. G. H. Green and J. Kenyon, *J. Chem. Soc.*, 1589 (1950).
10. R. S. Mathur and R. H. Common, *Steroids* **10**, 547 (1967); J. Lille and co-workers, *Tr. Nauch Isslied. Inst. Slantsev (USSR)* (18), 113 (1969); O. P. Vig, *J. Indian Chem. Soc.* **52**, 442 (1975).

11. M. Pailer and P. Bergthaller, *Monatsh. Chem.* **99**, 103 (1968).

12. C. M. Suter and H. L. Hanson, *J. Am. Chem. Soc.* **54**, 4400 (1932).

13. L. Orthner and G. Freyss, *Lieb. Ann.* **484**, 146 (1930).

14. D. R. Benedict, T. A. Bianch, and L. A. Cate, *Synthesis*, 428 (1979).

15. L. J. Simon, *Compt. Rend.* **176**, 583 (1923).

16. F. H. Stodda, *J. Org. Chem.* **29**, 2490 (1964); A. Werner and W. Seybold, *Chem. Ber.* **37**, 3658 (1904); U. S. Pat. 1,924,615 (Aug. 29, 1933), S. A. Merley (to Doherty Research Co.); N. C. Jameison and D. F. Loncrini, *Chem. Ind.*, 522 (1979).

17. M. H. Richard, *Ann. Chim. Phys.* **21**, 336 (1910).

18. B. K. Zeindov and I. A. Panteeva, *Vses. Soveshch. Sin. Zhirozamen, Poverkhnostnoaktiv. Veschestuam Moyushch. Sredstvam, 3rd, Shebekino*, 225 (1965); *Maslozhir. Prom.* **33**, 40 (1967).

19. Hung. Teljes 11221 (Mar. 28, 1977), I. Weisz and co-workers.

20. A. Etienne, G. Lonchambon, and R. Garreau, *Bull. Soc. Chim. France*, 483 (1977).

21. A. Werner, *Lieb. Ann.* **321**, 269 (1902).

22. V. M. Reddy and K. K. Reddy, *Indian J. Chem.* **17B**, 353 (1979); I. Y. Shirobokov, *Zh. Org. Khim.* **16**, 788 (1980); A. McKillop and R. J. Kobylecki, *J. Org. Chem.* **39**, 2710 (1974).

23. U.S. Pat. 3,941,824 (Mar. 2, 1976), J. J. Fuchs (to E. I. du Pont de Nemours & Co., Inc.).

24. Ger. Pat. 2,828,259 (Jan. 10, 1980), G. Giesselmann, K. Guenther, and W. Fuenten (to Deutsche Gold und Silber-Scheideanstalt); Jpn. Pat. 72-37,615 (Sept. 22, 1972), H. Kadowaki, M. Kametani, and T. Nakamura (to Nitto Chemical Industries Co.).

25. M. H. Slotta and H. L. Gerhart, *Chem. Res.* **58B**, 1320 (1925).

26. P. S. Clezy, *Tetrahedron Lett.*, 741 (1966).

27. J. U. Nef, *Lieb. Ann.* **309**, 187 (1899).

28. E. M. Arnett and V. M. De Palma, *J. Am. Chem. Soc.* **99**, 5828 (1977); G. Bram, F. Guibe, and P. Sarthov, *Tetrahedron Lett.*, 4903 (1972); Y. Hara and M. Matsuda, *Bull. Chem. Soc. Jpn.* **49**, 1126 (1976); A. Brandstrom, *Acta Chem. Scand.* **30B**, 203 (1976).

29. G. Bram, T. Eillebeen-Kahn, and N. Geraghty, *Synth. Commun.* **10**, 279 (1980).

30. K. K. Anderson and S. W. Fenton, *J. Org. Chem.* **29**, 3270 (1960).

31. D. Lipkin, E. F. Jones, and F. Galiano, *Am. Chem. Soc. Div. Pet. Chem. Repr.* **4**(4), B14 (1959).

32. A. K. Kranzfelder and F. J. Sowa, *J. Am. Chem. Soc.* **59**, 1490 (1937).

33. C. M. Suter and H. L. Gerhart, *J. Am. Chem. Soc.* **57**, 107 (1935).

34. M. Vlassa, M. Kezdi, and I. Goia, *Synthesis*, 850 (1980).

35. U.S. Pat. 3,896,230 (July 22, 1975), H. L. Klopping (to E. I. du Pont de Nemours & Co., Inc.).

36. R. F. Weinland and K. Schmid, *Chem. Ber.* **38**, 2327 (1905).

37. C. Gopinathan and J. Gupta, *Indian J. Chem.* **3**, 231 (1965).

38. *Ibid.*, **11**, 948 (1973).

39. U.S. Pat. 3,711,524 (Jan. 16, 1973), J. R. Leebrick (Lucille Coon, part interest).

40. G. W. Fischer, R. Jentzsch, and V. Kasanzewa, *J. Prakt. Chem.* **317**, 943 (1975).

41. R. Kreman, *Monatsh. Chem.* **28**, 13 (1907).

42. H. F. Lewis, O. Mason, and R. Morgan, *Ind. Eng. Chem.* **16**, 811 (1924).

43. G. H. Green and J. Kenyon, *J. Chem. Soc.*, 1389 (1950).

44. G. M. Calhoun and A. L. Burnell, Jr., *J. Am. Chem. Soc.* **77**, 6441 (1955).

45. B. D. Batts, *J. Chem. Soc. B*, 551 (1966).

46. R. E. Robertson and S. E. Sugamon, *Can. J. Chem.* **44**, 1728 (1966).

47. V. A. Kolesnikov and co-workers, *Kinet. Katal.* **18**, 875 (1977).

48. J. Kaniewski and co-workers, *Pol. J. Chem.* **52**, 587 (1978).

49. Tn. N. Motorova and co-workers, *Khim. Sredstva Zashchity Rast., M.*, 146 (1980).
50. C. H. Bamford and C. F. H. Tipper, *Chemical Kinetics*, Vol. 10, Elsevier North-Holland, Inc., New York, 1972, p. 39.
51. E. T. Kaiser, M. Panar, and F. H. Westheimer, *J. Am. Chem. Soc.* **85**, 602 (1963).
52. E. T. Kaiser, I. R. Katz, and T. F. Wulfers, *J. Am. Chem. Soc.* **87**, 3781 (1965).
53. V. A. Kolesnikov, R. V. Efremov, and S. M. Danov, *Kinet. Catal.* **20**, 671 (1979).
54. I. Lauder, I. R. Wilson, and B. Zerner, *Aust. J. Chem.* **14**, 41 (1961).
55. R. L. Burwell, Jr., *J. Am. Chem. Soc.* **74**, 1462 (1952).
56. S. Burstein and S. Liebeman, *J. Am. Chem. Soc.* **80**, 5235 (1958).
57. B. D. Batts, *J. Chem. Soc. B*, 547 (1966).
58. J. L. Kice and J. M. Anderson, *J. Am. Chem. Soc.* **88**, 5242 (1966).
59. H. L. Kane and A. Lowry, *J. Am. Chem. Soc.* **58**, 2605 (1936); J. Epelberg and A. Lowy, *J. Am. Chem. Soc.* **63**, 101 (1941).
60. U.S. Pat. 3,665,050 (May 23, 1972), L. J. McGovern, C. L. West, and O. Webb (to Stratford Energy Corp.).
61. B. C. Lawes, DuPont, private communication, 1981; C. D. Hurd, *The Pyrolysis of Carbon Compounds*, Chemical Catalog Co., New York, 1929, p. 144.
62. J. U. Nef, *Lieb. Ann.* **318**, 43 (1901).
63. E. E. Gilbert, *Sulfonation and Related Reactions*, Wiley-Interscience, New York, 1965, pp. 23–24.
64. M. Volmer, *Bull. Soc. Chim. France* **27**, 681 (1920).
65. L. A. Harmon and R. J. Legon, *J. Chem. Soc. Perkin I*, 2675 (1979).
66. A. R. Vasudeva Murthy, *Proc. Indian Acad. Sci.* **47**, 11 (1953).
67. U.S. Pat. 3,168,546 (Feb. 2, 1969), A. Ballauf and co-workers (to Farbenfabriken Bayer).
68. W. W. Carlson and L. H. Cretcher, *J. Am. Chem. Soc.* **69**, 1952 (1947).
69. U.S. Pat. 2,917,502 (Dec. 15, 1959), S. R. Schwyzer (to Ciba Pharmaceutical Company).
70. J. M. Theabald, M. W. Williams, and G. T. Young, *J. Chem. Soc.*, 1927 (1963).
71. Ger. Pat. 2,365,497 (Apr. 24, 1975), J. Bruenisholz and A. Kirchmayr (to Ciba Geigy).
72. D. Kerr and I. Lauder, *Aust. J. Chem.* **15**, 561 (1962).
73. R. E. Davis, *J. Am. Chem. Soc.* **84**, 599 (1962); P. B. D. De La Mare, J. G. Tillett, and H. F. Van Woerden, *J. Chem. Soc.*, 4888 (1962).
74. P. A. Bristow and T. G. Tillett, *Chem. Commun.*, 1010 (1967).
75. K. A. J. Chamberlain and co-workers, *BIOS Final Report No. 1482*, 1946, p. 6.
76. H. G. Harris and D. M. Himmelblau, *J. Phys. Chem.* **67**, 802 (1963); H. G. Harris, Jr., and D. M. Himmelblau, *J. Chem. Eng. Data* **9**(1), 61 (1964).
77. Ger. Pat. 1,035,632 (Aug. 7, 1958), H. G. Van Raay (to Farbwerke Hoechst AG).
78. U.S. Pat. 3,024,263 (Mar. 6, 1962), M. Letherman (to U.S. Dept. of the Navy).
79. D. S. Breslow, R. R. Hough, and J. T. Fairclough, *J. Am. Chem. Soc.* **76**, 5361 (1954).
80. M. S. Nemtsov, *Khim. Promst. Kiev*, 633 (1960).
81. G. R. Schultze, J. Moos, and K. D. Ledwoch, *Erdoel Kohle* **11**, 12 (1958).
82. G. D. Inskeep and A. Mussard, *Ind. Eng. Chem.* **47**, 2 (1955); D. Steward and E. McNeill, *Chem. Age (London)* **63**, 48 (1950); S. F. Birch, *J. Inst. Pet.* **38**, 69 (1952); K. L. Butcher and G. M. Nickson, *Trans. Faraday Soc.* **54**, 1195 (1958).
83. E. Clippinger, *Ind. Eng. Chem. Prod. Res. Dev.* **3**, 3 (1964).
84. Ger. Pat. 1,086,434 (Aug. 4, 1960), J. Szita (to Farbenfabriken Bayer AG).
85. E. Cherbuliez and co-workers, *Helv. Chim. Acta* **48**, 830 (1965); U.S. Pat. 3,194,826 (July 13, 1965), A. Goldstein and F. A. Nowak, Jr. (to Chemirad Corp.).
86. J. Parrod and L. Robert, *Compt. Rend.* **230**, 450 (1950).
87. A. Butenandt and co-workers, *Z. Physiol. Chem.* **321**, 258 (1960).
88. G. N. Burkhart and A. Lapworth, *J. Chem. Soc.*, 684 (1926).

89. E. J. Fendler and J. H. Fendler, *J. Org. Chem.* **33**, 3852 (1968).

90. C. Barkenbus and J. J. Owen, *J. Am. Chem. Soc.* **56**, 1204 (1934); R. Levaillant, *Compt. Rend.* **200**, 940 (1935).

91. Ger. (East) Pat. 32,780 (Jan. 5, 1965), W. Lugenheim, E. Carstens, and H. Fuerst.

92. G. A. Sokol'skii, *Zh. Org. Khim.* **2**(6), 951 (1966).

93. J. Avery and co-workers, *BIOS Final Report No. 986*, pp. 175, 227, 1946–1948.

94. Czech. Pat. 157,946 (1976), M. Sadilek and M. Soulak.

95. Brit. Pat. 774,384 (May 8, 1957), E. Roberts (to Ministry of Supply).

96. U.S. Pat. 2,816,126 (Dec. 10, 1957), R. Evans and L. T. Hogarth (to United States of America).

97. Fr. Pat. 1,006,211 (Apr. 21, 1952) (to Société Anon. des Manufactures des Glaces et Produits Chimiques de Saint-Gobain, Chauny et Cirey).

98. Brit. Pat. 581,115 (Oct. 1, 1946), (to Nicholas Proprietary Ltd.).

99. D. S. Breslow, R. R. Hough, and J. T. Fairclough, *J. Am. Chem. Soc.* **76**, 5361 (1954); D. S. Breslow and R. R. Hough, *J. Am. Chem. Soc.* **79**, 5000 (1957); A. A. Goldberg, *J. Chem. Soc.*, 716 (1942).

100. U.S. Pat. 3,100,780 (Aug. 13, 1963), D. L. Klass (to Pure Oil Co.).

101. U.S. Pat. 2,805,228 (Sept. 3, 1963), J. L. Smith (to Eastman Kodak Co.).

102. J. Lichtenberger and J. Hincky, *Bull. Soc. Chim. France*, 1495 (1961); W. Baker and B. F. Burrows, *J. Chem. Soc.*, 2257 (1961); Brit. Pat. 944,406 (Dec. 11, 1963), L. F. Wiggins, C. C. Beard, and J. W. James (to Aspro-Nicholas Ltd.).

103. Ger. Pat. 1,049,870 (Feb. 5, 1959), J. Brunken and E. J. Poppe (to VEB Filmfabrik Agfa Wolfen).

104. Ger. (East) Pat. 18,485 (Apr. 4, 1960), J. Brunken and E. J. Poppe (to VEB Filmfabrik Agfa Wolfen).

105. W. Gerrard, *J. Chem. Soc.*, 99 (1939); L. Denivelle, *Compt. Rend.* **208**, 1024 (1939); Ger. Pat. 1,133,367 (July 19, 1962), H. F. Wilson (to Rohm & Haas Co.).

106. S. Hasegawa, M. Nojima, and N. Takura, *J. Chem. Soc. Perkin I*, 108 (1976).

107. Ger. Pat. 1,200,807 (Sept. 16, 1966), K. Stuerzer (to Th. Goldschmidt AG); Ger. Pat. 1,212,072 (Mar. 10, 1966), K. Stuerzer (to Th. Goldschmidt AG).

108. H. Hesse and S. Majumdar, *Chem. Ber.* **93**, 1129 (1960).

109. S. Hauptmann and K. Dietrich, *J. Prakt. Chem.* **19**, 174 (1963).

110. U.S. Pat. 3,022,315 (Feb. 20, 1962), W. A. Rogers, Jr., J. E. Woehst, and R. M. Smith (to Dow Chemical Co.); Ger. Pat. 1,217,970 (June 2, 1966), H. Distler and G. Dittus (to Badische Anilin- und Soda-Fabrik AG); Ger. Pat. 1,188,610 (Mar. 11, 1965), H. Hoefermann and H. Springmann (to Chemische Werke Huels AG).

111. H. F. van Woerden, C. F. van Valkenburg, and G. M. van Woerkam, *Rec. Trav. Chim. Pays-Bas* **86**, 601 (1967).

112. Ger. Pat. 1,213,612 (Mar. 31, 1966), K. Stuerzer (to Th. Goldschmidt AG).

113. R. Lavaillant, *Ann. Chim.* **6**, 459 (1936); W. Voss and E. Blanke, *Lieb. Ann.* **485**, 258 (1930).

114. Fr. Pat. 965,161 (Sept. 5, 1950), (to Société Anon. d'Innovations Chimiques dite: Sinnova ou Sadic).

115. F. G. Bordwell and G. W. Crosby, *J. Am. Chem. Soc.* **78**, 5367 (1956); U.S. Pat. 2,860,123 (Nov. 11, 1958), R. V. Jones (to Phillips Petroleum Co.).

116. U.S. Pat. 2,684,977 (July 27, 1954), M. J. Viard (to Société Anon. des Manufactures de Glaces et Produits Chimiques de Saint-Gobain, Chauny et Cirey).

117. M. Schmidt and H. Schmidbauer, *Chem. Ber.* **95**, 47 (1962).

118. U.S. Pat. 1,510,425 (1925), W. Traube; W. Traube and R. Justh, *Brennst. Chem.* **4**, 150 (1923).

119. G. D. Clayton and F. E. Clayton, eds., *Patty's Industrial Hygiene and Toxicology*, 3rd ed., Vols. 2A and 2B, Wiley-Interscience, New York, 1981, pp. 2094, 2892, 2928, and 2930.

120. *Dimethyl Sulfate–Properties, Uses, Storage and Handling*, Technical bulletin, DuPont, Wilmington, Del., Jan. 1981.

121. *Dimethyl Sulfate–Product Safety Bulletin*, Technical bulletin, DuPont, Wilmington, Del., Mar. 1980.

122. G. R. Hoffmann, *Mutat. Res.* **75**, 63 (1980).

123. R. F. Newbold and co-workers, *Nature (London)* **283**, 596 (1980).

124. *Dimethyl Sulfate, Material Safety Data Sheet*, DuPont, Wilmington, Del., 1980.

125. *Diethyl Sulfate, Material Safety Data Sheet*, Union Carbide, New York, 1976.

126. *Registry of Toxic Effects of Chemical Substances*, U.S. Department of Health and Human Services, Washington, D.C., 1979.

127. *NIOSH/OSHA–Pocket Guide to Chemical Hazards*, U.S. Dept. of Health and Human Services, U.S. Dept. of Labor, Washington, D.C., Sept. 1980.

128. H. Lindler, *Angew. Chem. Int. Ed.* **2**, 262 (1963).

129. Haskell Laboratory, E. I. du Pont de Nemours & Co., Inc., private communication, 1975.

130. D. J. Eatough and co-workers, *Environ. Sci. Technol.* **15**, 1502 (1981).

131. R. Kuhn and H. T. Trischmann, *Chem. Ber.* **96**, 284 (1963).

132. *McCutcheon's Detergents and Emulsifiers*, North American ed., MC Publishing Co., New Jersey, 1979, annual.

133. K. A. J. Chamberlain, A. M. North, and D. C. Wilson, *BIOS Final Report No. 1482.6*, 1946.

134. U.S. Pat. 3,160,474 (Dec. 8, 1964), W. G. Schnoor and A. W. Yodis (to Allied Chemical Corp.).

135. U.S. Pat. 4,137,164 (Jan. 30, 1979), A. T. Coscia and M. N. D. O'Connor (to American Cyanamid Co.); U.S. Pat. 3,993,615 (Nov. 23, 1976), S. B. Markofsky (to W. R. Grace & Co.); U.S. Pat. 2,723,256 (Nov. 8, 1955), M. Hayek (to E. I. du Pont de Nemours & Co., Inc.).

136. U. A. Reeves and J. D. Guthrie, *Text. Res. J.* **23**, 522 (1953).

137. Ger. Pat. 1,915,222 (Oct. 2, 1969), H. Nakamura and A. Saito (to Japan Synthetic Chemical Industries Co., Ltd.).

138. Ger. Pat. 2,656,727 (June 29, 1978), R. Gehm, K. H. Baumann, and W. Harder (to BASF AG).

139. K. K. Georgieff, *Can. J. Chem.* **30**, 332 (1952).

140. Brit. Pat. 695,789 (Aug. 19, 1953), R. H. Hall and E. S. Stern (to Distillers Co., Ltd.).

141. U.S. Pat. 3,116,336 (Dec. 31, 1963), J. L. Van Winkle (to Shell Oil Co.).

142. Ger. Pat. 911,659 (May 17, 1954), A. Schmidt (to Chemische Werke Huels AG).

143. S. Kobayuski and co-workers, *Bull. Chem. Soc. Jpn.* **46**, 3214 (1973).

144. T. Fujisawa, Y. Yokota, and T. Mukaiyama, *Bull. Chem. Soc. Jpn.* **40**, 147 (1967); Belg. Pat. 666,828 (Nov. 16, 1965), M. H. Litt, A. J. Levy, and T. G. Bassiri (to Allied Chemical Corp.); Ger. Pat. 1,206,585 (Dec. 9, 1965), W. Seeliger (to Chemische Werke Huels AG).

145. U.S. Pat. 3,389,128 (June 18, 1968), J. W. Bayer and W. C. Grinonneau (to Owens-Illinois, Inc.).

146. T. Otso and Y. Takemura, *Bull. Chem. Soc. Jpn.* **43**, 567 (1970).

147. K. Sugiyama, *Kinhi Daigako Kogahuba Kenkyu Hokoku* **13**, 27 (1979).

148. W. I. Aalbersberg and co-workers, *J. Chem. Soc.*, 3055 (1959).

149. U.S. Pat. 2,801,957 (Aug. 6, 1957), G. C. Ray (to Phillips Petroleum Co.).

150. U.S. Pat. 2,776,327 (Jan. 1, 1957), A. Schneider (to Sun Oil Co.).

151. P. Pascal and M. L. Quinet, *Ann. Chim. Anal. Chim. Appl.* **23**, 5 (1941).

152. V. Y. Rybkovskii and co-workers, *Tr. Kishinev S-kh. Inst.* **123**, 95 (1974).

153. U.S. Pat. 2,971,985 (Feb. 14, 1961), R. Joly, R. Bucourt, and C. Fabignon (to UCLAF, Paris); U.S. Pat. 3,355,497 (Nov. 28, 1967), E. Budnick (to Plains Chemical Co.).

154. F. H. Firsching and co-workers, *J. Inorg. Nucl. Chem.* **36**, 1655 (1974).

155. Jpn. Pat. 63-8245 (June 7, 1963), M. Nagasawa and F. Yamamato (to Ihara Agricultural Chemical Co.); U.S. Pat. 3,179,682 (Apr. 20, 1965), A. Covey, A. E. Smith, and W. L. Hubbard (to United States Rubber Co.).

156. Belg. Pat. 615,408 (Apr. 13, 1962), P. P. Chiesa, J. R. Dann, and J. W. Gates, Jr. (to Kodak Soc. Anon.).

157. Brit. Pat. 903,077 (Aug. 9, 1962), (to "Montecatini" Societa Generale per l'Industria Mineraria e Chimica).

158. Brit. Pat. 866,440 (Apr. 26, 1961), C. P. Vale, S. Gutter, and W. Wilson (to British Industrial Plastics Ltd.).

159. U.S. Pat. 3,251,857 (May 1, 1966), F. Hostettler and E. F. Cox (to Union Carbide Corporation).

160. N. P. Yao, E. D'Orsay, and D. N. Bennion, *J. Electrochem. Soc.* **115**, 999 (1968); S. C. S. Wang and D. N. Bennion, *J. Electrochem. Soc.* **128**, 1827 (1981).

General References

E. E. Gilbert, *Sulfonation and Related Reaction*, Wiley-Interscience, New York, 1965, Chapt. 6.

C. M. Suter, *Organic Chemistry of Sulfur*, John Wiley & Sons, Inc., New York, 1944, Chapt. 1.

W. B. McCormack
B. C. Lawes
E. I. du Pont de Nemours & Co., Inc.

SULFURIZATION AND SULFURCHLORINATION

Sulfur reacts with alkanes to either dehydrate (eq. 1), oxidize, forming carbon disulfide and hydrogen sulfide (eq. 2), or cyclize, forming thiophenes (eq. 3). The products of alkane sulfurization depend on the temperature, the time at the temperature, and the structure of the hydrocarbon (1).

$$\bigcirc + 3\,S^0 \xrightarrow[200°C]{} \bigcirc\!\!\!\!\bigcirc + 3\,H_2S \tag{1}$$

$$C_xH_y \xrightarrow[>500°C]{\text{sulfur}} x\,CS_2 + y/2\,H_2S \tag{2}$$

$$\begin{array}{c} X \quad\quad Y \\ \diagdown CH{-}CH\diagup \\ | \quad\quad | \\ R{-}CH_2 \;\; CH_2{-}R' \end{array} + 4\,S^0 \longrightarrow \begin{array}{c} X \quad\quad Y \\ \diagdown C{-}C\diagup \\ \diagup\!\!\diagup \quad\quad \\ R{-}C \quad C{-}R' \\ \diagdown_S\diagup \end{array} + 3\,H_2S \tag{3}$$

Generally, unsaturated compounds, eg, alkenes and natural fats and their derivatives, are much more reactive toward sulfur than alkanes. Sulfur reacts with unsaturated compounds at temperatures of 120–215°C, forming products that are usually dark and often viscous cross-linked mixtures of dithiole-3-thiones (eq. 4) (2) and sulfides (Table 1) (3).

$$\text{(structure)} \xrightarrow[\text{200°C}]{\text{sulfur}} \text{(structure)} \quad (4)$$

The mechanisms for the reaction of sulfur with alkanes and unsaturated compounds are highly speculative, being strongly influenced by the specific structure of the substrate and by the conditions (particularly temperature) of reaction. Alkane (4), olefin (5), animal fat (6), and vegetable oil (7) sulfurization have been extensively studied because these reactions are models for vulcanization. Moreover, the products are used as lubricant additives.

Sulfur reacts with mercaptans in the presences of basic catalysts at temperatures of 75–105°C, forming sulfides. These sulfides are usually light in color and are formed without cross-linking. The sulfurization of mercaptans leads to di-, tri-, or higher polysulfides, depending on the mole ratio used (eqs. 5 and 6). An extensive list of references to the sulfurization of mercaptans is available (8).

$$2\,R\!-\!S\!-\!H + 2\,S \xrightarrow[-H_2S]{\text{catalyst}} R\!-\!S_3\!-\!R \quad (5)$$

$$2\,R\!-\!S\!-\!H + 4\,S \xrightarrow[-H_2S]{\text{catalyst}} R\!-\!S_5\!-\!R \quad (6)$$

Table 1. Products of the Reaction of 1-Octene with Sulfur at 140°C

Product		Quantity, %
$\underset{\displaystyle C_6H_{13}\!-\!\overset{\displaystyle CH_3}{\overset{\displaystyle |}{CH}}\!-\!S_{\overline{x}}\!-\!CH_2\!-\!CH\!=\!CH\!-\!C_5H_{11}}{}$	av $x = 6.7$	25
$C_6H_{13}\!-\!CH\!-\!CH_2$ with S_a, S_b, $CH_2\!-\!CH\!-\!C_6H_{13}$	av $(a + b) = 6.7$	30
$C_8H_{17}\!-\!S\!-\!C_8H_{17}$ $+\ C_8H_{17}\!-\!S\!-\!C_8H_{15}$		15
$C_8H_{17}\!-\!S_a\!-\!C_8H_{16}\!-\!S_b\!-\!C_8H_{17}$	av $(a + b) = 4.7$	15
$C_6H_{13}\!-\!CH\!-\!CH_2$ with S, S, $CH_2\!-\!CH\!-\!C_6H_{13}$ $+\ C_8H_{17}\!-\!S\!-\!C_8H_{16}\!-\!S\!-\!C_8H_{17}$		15

Sulfurization of unsaturated compounds in the presence of hydrogen sulfide also affords polysulfides (9). It is postulated that this reaction forms the mercaptan *in situ*, which then further reacts to form the polysulfide (see SULFUR COMPOUNDS).

Sulfur monochloride [10025-67-9], S_2Cl_2, and sulfur dichloride, SCl_2, react with unsaturated materials, forming products that are cross-linked by sulfur but which also contain chlorine (eq. 7) (10).

$$2\ R_2C{=}CR_2 + S_xCl_2 \longrightarrow \quad \begin{matrix} R_2C - CR_2 \\ |\quad\ \ | \\ Cl\ \ \ S_x \\ |\ \ \\ R_2C - CR_2 \\ |\ \ \\ Cl \end{matrix} \qquad (7)$$

Sulfur monochloride and sulfur dichloride also react with mercaptans, yielding tetra- and trisulfides (eqs. 8 and 9) (11).

$$2\ R-S-H + S_2Cl_2 \longrightarrow R-S_4-R \qquad (8)$$

$$2\ R-S-H + SCl_2 \longrightarrow R-S_3-R \qquad (9)$$

Properties

Properties of typical commercial sulfurized unsaturated compounds and mercaptans are listed in Table 2.

Uses

Sulfurized and sulfurchlorinated unsaturated compounds and mercaptans are used as lubricant additives (antiwear, friction modification, load-carrying, extreme pressure and temperature, corrosion inhibition, and antioxidants), refinery catalyst regeneration compounds, steel processing (annealing) aids, and vulcanization catalysts (see LUBRICATION AND LUBRICANTS).

Manufacture

Sulfurization of unsaturated compounds and mercaptans is normally carried out at atmospheric pressure, in a mild or stainless steel, batch-reaction vessel equipped with an overhead condenser, nitrogen atmosphere, an agitator, heating media capable of 120–215°C temperatures and a scrubber (typically caustic bleach or diethanolamine) capable of handling hydrogen sulfide. If the reaction involves the use of H_2S as a reactant or the olefin or mercaptan is a low boiling material, a stainless steel pressurized vessel is recommended.

Table 2. Chemical and Physical Properties of Sulfurized and Sulfurchlorinated Unsaturated Compounds and Mercaptans

Product[a] designation	Sulfur, wt %	Chlorine, wt %	Active sulfur, wt %	Copper strip tarnish test, 10 wt % in oil	Viscosity, mm²/s (=cSt) 40°C	Viscosity, mm²/s (=cSt) 100°C	Density, at 25°C, g/cm³	Total acid number, mg KOH/g	Pour point, °C
Base 10-L[b]	10.0	0	0	1B	1100	80	0.98	25	18
Base 10-SE[c]	10.0	0	0	1B	20	4	0.94	5	10
Base 14-L[b]	14.0	0	4	4a	1700	100	0.99	25	27
Base 401[d]	40.0	0	25	4c	80	10	1.07	7	−50
Base L-66[e]	5.5	5.5	0	1B	1800	100	0.99	5	10
Sulperm 18[f]	18.0	0	8	4a	600	55	1.02	12	16
Sulperm 110f	10.0	0	0	1B	500	40	0.98	8	10
DTNPS[g]	37.0	0	25	4c	35		1.04	5	−50
DTBTS[h]	44.0	0	0	1B	3		1.00	0	<−60

[a]Products and properties submitted by Keil Chemical Division of Ferro Corp.

[b]Based on animal fat.

[c]Based on methyl ester of vegetable oil.

[d]Based on olefin.

[e]Sulfurchlorinated animal fat.

[f]Based on a mixture of animal and vegetable oils plus synthetic esters.

[g]Di-t-nonyl pentasulfide (DTNPS) is based on mercaptan.

[h]Di-t-butyl trisulfide (DTBTS) is based on mercaptan.

Sulfurchlorination of unsaturated compounds or mercaptans is normally carried out at atmospheric pressure in a glass-lined reaction vessel because of the potential to liberate HCl during the reaction. The sulfurchlorination vessel is equipped with a cooling jacket or coils (very exothermic reaction), a nitrogen or dry air sparging system, an overhead condenser, and a caustic or bleach scrubber. If one of the reactants (olefin or mercaptan) is a low boiling material, ie, isobutylene, a glass-lined pressure vessel is recommended.

Health and Safety Factors

Sulfurized and sulfurchlorinated unsaturated compounds or mercaptans are normally considered nonhazardous. These materials may, however, liberate H_2S and HCl at elevated (>200°C) temperatures and during combustion.

BIBLIOGRAPHY

"Sulfurization and Sulfurchlorination" in *ECT* 2nd ed., Vol. 19, pp. 498–506, by R. S. Dalter and I. Hechenbleikner, Carlisle Chemical Works, Inc.; in *ECT* 3rd ed., Vol. 22, pp. 255–266, by K. Kammann and G. E. Verdino, Keil Chemical Division, Ferro Corp.

1. S. Oae, *Organic Chemistry of Sulfur*, Plenum Press, New York, 1977.
2. N. Lozach and L. Legrande, *Compt. Rend.* **232**, 2330 (1951).
3. L. Bateman and co-workers, *J. Chem. Soc. London*, 2836–2846 (1958).
4. W. A. Pryor, *Mechanisms of Sulfur Reactions*, McGraw-Hill Book Co., Inc., New York, 1962.
5. L. Bateman, C. G. Moore, and M. Porter, *J. Chem. Soc. London*, 2866–2879 (1958).
6. A. Dorinson, *Lubric. Eng.* **39**, 519 (1983).
7. K. P. Kammann and A. I. Phillips, *J. Am. Oil Chem. Soc.* **62**, 917–923 (1985).
8. U.S. Pat. 5,146,000 (Sept. 8, 1992), N. Ozbalik (to Ethyl Corp.).
9. U.S. Pat. 2,061,019 (Nov. 17, 1936), A. S. Carter and F. B. Downing (to Du Pont).
10. U.S. Pat. 3,697,499 (Oct. 10, 1972), H. Meyers (to Mobil Oil Corp.).
11. Y. Minoura, *J. Soc. Rubber. Ind. Japan* **23**, 213–217 (1950).

MICHAEL P. DUNCAN
Keil Chemical Division
Ferro Corporation

SULFUR REMOVAL AND RECOVERY

Sulfur (qv) is among the most widely used chemicals and often considered to be one of the four basic raw materials of the chemical industry. In 1993, worldwide production of sulfur reached 55 million metric tons (1). Production of sulfuric acid consumes the vast majority (~90%) of sulfur (2) (see SULFURIC ACID AND SULFUR TRIOXIDE). This acid is a stepping stone in the production of other sulfur-containing compounds, most notably ammonium sulfate fertilizer which accounts for 60% of the total worldwide sulfur consumption (2) (see AMMONIUM COMPOUNDS; FERTILIZERS).

Sulfur can be produced directly via Frasch mining or conventional mining methods, or it can be recovered as a by-product from sulfur removal and recovery processes. Production of recovered sulfur has become more significant as increasingly sour feedstocks are utilized and environmental regulations concerning emissions and waste streams have continued to tighten worldwide. Whereas recovered sulfur represented only 5% of the total sulfur production in 1950, as of 1996 recovered sulfur represented approximately two-thirds of total sulfur production (1). Recovered sulfur could completely replace native sulfur production in the twenty-first century (2).

Sulfur recovery processes have historically been focused primarily on the removal and conversion to elemental sulfur of two sulfur species: hydrogen sulfide and sulfur dioxide. These species have garnered the most attention because they represent the largest source of potential sulfur emissions. Additionally, adverse effects of both have been well documented. Hydrogen sulfide is a known toxin and sulfur dioxide is a dangerous component of atmospheric air pollution (qv). A primary factor in the selection of a removal and recovery process is which of these two species, hydrogen sulfide or sulfur dioxide, is to be removed. Where both species are to be removed, hydrogen sulfide removal processes are generally preferred; thus sulfur dioxide and other sulfur compounds in untreated waste streams must first be converted to hydrogen sulfide.

Other factors which have a significant influence on process selection include absolute quantity of sulfur present, concentration of various sulfur species, the quantity and concentration of other components in the stream to be treated, quantity and conditions (temperature and pressure) of the stream to be treated, and, the location-specific environmental regulations governing overall sulfur recovery and allowable sulfur dioxide emissions (3).

Hydrogen Sulfide Removal and Recovery

Hydrogen sulfide [7783-06-4], H_2S, represents the largest source of recovered sulfur. Hydrogen sulfide is often present in natural gas or refinery streams, occurring naturally or as the by-product of processing operations such as hydrotreating. A number of processes have been developed to remove and recover hydrogen sulfide. These processes are generally categorized according to the primary mechanism, as adsorption (qv), absorption (qv), or conversion. It is also common for these processes to be classified on the basis of functionality, for example, as scavenger, Redox, Claus, etc.

Adsorption Processes. The processes based on adsorption of hydrogen sulfide onto a fixed bed of solid material are among the oldest types of gas treating applications (4). Two common sorbent materials for low concentration gas streams are iron oxide and zinc oxide.

Hydrogen sulfide reacts with iron oxide [1317-61-9] to form iron sulfide, according to the following chemical reaction:

$$2\,Fe_2O_3 + 6\,H_2S \longrightarrow 2\,Fe_2S_3 + 6\,H_2O \tag{1}$$

The sulfur is thus removed from the gas stream and trapped in the sorbent as iron sulfide [1317-37-9]. Over time all of the iron oxide becomes sulfided and the adsorptive capacity of the sorbent becomes exhausted. The bed can be partially regenerated by oxidation, as follows:

$$2\,Fe_2S_3 + 3\,O_2 \longrightarrow 2\,Fe_2O_3 + 6\,S \tag{2}$$

Typically, the iron oxide sorbent can only be regenerated a few times before it becomes plugged with sulfur [7704-34-9] and must be replaced. Also, owing to the pyrophoric nature of the iron sulfide, many operators choose not to regenerate the sorbent at all. Two or more adsorption vessels are generally provided, allowing continuous treatment during the time one vessel is being regenerated or replaced. The iron oxide process is shown schematically in Figure 1.

The zinc oxide [1314-13-2] process is similar to the iron oxide process. There is, however, one key distinction. The zinc sulfide [1314-98-3], ZnS, formed cannot be oxidized back to zinc oxide, and therefore the sorbent bed must be replaced once the capacity is fully utilized.

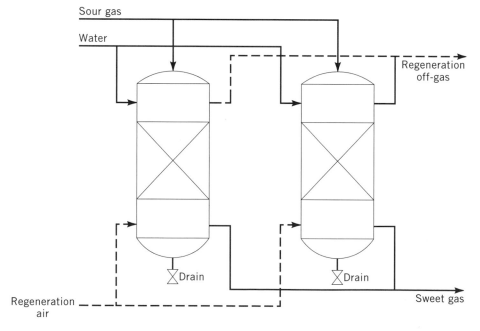

Fig. 1. Iron oxide process where ⊠ represents the iron oxide sorbent bed; (——), the adsorption system; and (– – –), the regeneration stream and off-gas. Vessels can be operated concurrently or independently.

The sulfur removed via these fixed-bed metal oxide processes is generally not recovered. Rather the sulfur and sorbent material both undergo disposal. Because the sorbent bed has a limited capacity and the sulfur is not recovered, the application of these processes is limited to gas streams of limited volumetric rate having low concentrations of hydrogen sulfide.

Processes featuring molecular sieves (qv) have been developed to extend the operating range of adsorption processes. Molecular sieves are crystalline structures manufactured having specific pore sizes and highly polarized local charges. Rather than relying on chemical reactions, as is the case with the metal oxide processes, the molecular sieve processes take advantage of the structure of the molecular sieve to attract and retain specific chemical species, eg, hydrogen sulfide. The manufacturing of the molecular sieves can be controlled to target the removal of certain components selectively.

Molecular sieves are typically regenerated using a slip stream of the treated gas at elevated temperature and reduced pressure. This regeneration step creates an enriched hydrogen sulfide stream which must then be further treated if the sulfur is to be recovered. A typical molecular sieve adsorption unit is shown schematically in Figure 2.

The advancement of integrated gasification combined cycle (IGCC) power plants has led to the development of moving or fluidized-bed adsorption bed processes featuring metal oxide sorbents able to withstand severe operating conditions. Zinc titanate has proven particularly well suited for the demands of hot gas cleanup. These systems rely on chemical reaction of the sorbent with hydrogen sulfide to form metal sulfides. The sorbent is regenerated with air, yielding a concentrated sulfur dioxide stream suitable for downstream sulfur recovery.

Absorption Processes.   Absorption-based processes are the most widely practiced hydrogen sulfide removal techniques. These processes use a liquid

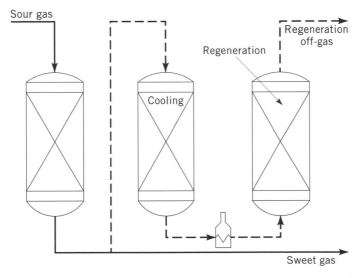

Fig. 2.   Molecular sieve process where ⊠ is a molecular sieve sorbent bed; (——), the adsorption system; and (– – –), the regeneration system.

solvent to absorb hydrogen sulfide and render a sweet gas stream. The hydrogen sulfide-rich solvent can be regenerated for reuse. This regeneration step produces an enriched hydrogen sulfide stream suitable for processing in downstream sulfur recovery unit. Many absorption (qv) processes remove carbon dioxide (qv), and to a lesser degree carbonyl sulfide (see SULFUR COMPOUNDS), carbon disulfide (qv), and mercaptans, along with the hydrogen sulfide. Hydrocarbons (qv) may also be absorbed in some processes.

Absorption processes are categorized based on the mechanism of absorption, as either chemical or physical. In addition, a number of hybrid absorption processes featuring both chemical and physical solvents have been developed.

Chemical Absorption.   In chemical absorption processes, the solvent reacts with the hydrogen sulfide and other species to form a new complex compounds which are held in the solvent. These reactions occur in an absorber tower where the solvent and the sour streams are contacted countercurrently across trays or packing. A regenerator, operating at higher temperature and lower pressure, is employed to release the hydrogen sulfide and other absorbed compounds from the solvent. The solvent is then cooled and returned to the absorber. A simplified schematic for a typical chemical absorption process is given in Figure 3. Two common chemical solvents are aqueous solutions of alkanolamines or alkali carbonate salts.

Alkanolamines.   Gas sweetening, ie, removal of hydrogen sulfide and carbon dioxide, using alkanolamines was patented in 1930. Several amine solvents are available as of the mid-1990s. The most widely used are mono-

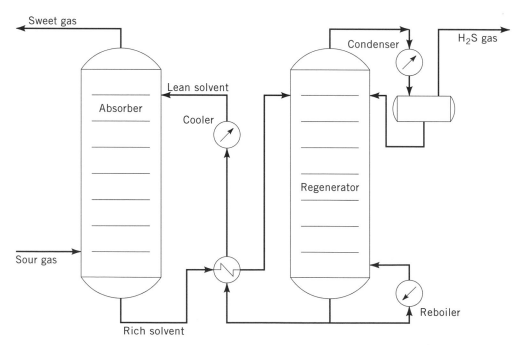

Fig. 3.   Flow diagram for a chemical absorption process where the horizontal lines within the towers represent trays or packing.

ethanolamine [141-43-5], diethanolamine [111-42-2], diglycolamine [929-06-6], and methyldiethanolamine [105-59-9]. Amine processes are generally applicable when hydrogen sulfide concentration in the feed gas is relatively low (eg, $\leq$ 5–25 mol %) and high purity (<4 ppmv hydrogen sulfide) must be achieved in the treated gas.

The primary reactions of amines and hydrogen sulfide and carbon dioxide are those of a base with acids:

$$2\ RNH_2 + H_2S \rightleftharpoons (RNH_3)_2S \tag{3}$$

$$(RNH_3)_2S + H_2S \rightleftharpoons 2\ RNH_3HS \tag{4}$$

$$2\ RNH_2 + CO_2 + H_2O \rightleftharpoons (RNH_3)_2CO_3 \tag{5}$$

$$(RNH_3)_2CO_3 + CO_2 + H_2O \rightleftharpoons 2\ RNH_3HCO_3 \tag{6}$$

Primary and secondary amines are generally nonselective in terms of hydrogen sulfide and carbon dioxide, exhibiting no preference for the absorption of one component over the other. Tertiary amines such as MDEA are selective. Although hydrogen sulfide and carbon dioxide have similar acidities, the reactions with a tertiary amine involve different chemical mechanisms. Hydrogen sulfide, a Brønstead acid, has a proton to donate and reacts extremely rapidly with the amine. Carbon dioxide, a Lewis acid which has electron pairs, but no proton, must first react with water to form the bicarbonate ion, which has a proton to donate, before reaction with the amine. This difference in reaction kinetics allows the use of tertiary amines to remove hydrogen sulfide preferentially over carbon dioxide.

Other components in the feed gas may react with and degrade the amine solution. Many of these latter reactions can be reversed by application of heat, as in a reclaimer. Some reaction products cannot be reclaimed, however. Thus to keep the concentration of these materials at an acceptable level, the solution must be purged and fresh amine added periodically. The principal sources of degradation products are the reactions with carbon dioxide, carbonyl sulfide, and carbon disulfide. In refineries, sour gas streams from vacuum distillation or from fluidized catalytic cracking (FCC) units can contain oxygen or sulfur dioxide which form heat-stable salts with the amine solution (see FLUIDIZATION; PETROLEUM).

The basic amine flow sheet has been optimized to provide energy savings (see ENERGY CONSERVATION). A process improvement to the traditional amine treating process was developed at the University of Manchester Institute of Science and Technology (UMIST) (6). This process uses a two-section absorber column. The main section carries out the bulk absorption of hydrogen sulfide accompanying carbon dioxide; the second section reduces the acid gas components to ultralow concentrations. Regeneration has also been separated into multiple sections, allowing solvent to be regenerated near thermodynamic optimum. The resultant process (Fig. 4) is more efficient than the traditional process configuration. Energy savings have been estimated to be near 50% and solvent circulation rates are significantly reduced. The UMIST technology can be applied to most proprietary chemical solvent processes.

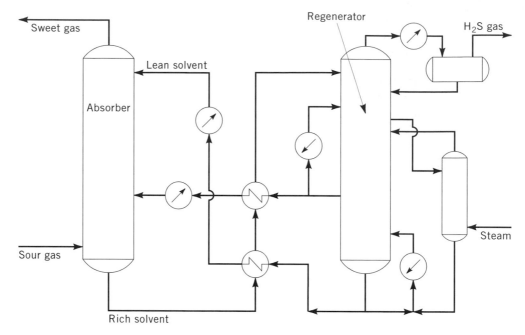

Fig. 4. UMIST absorption process.

The first alkanolamine to gain commercial acceptance was mono-ethanolamine (MEA). Consequently, MEA is the best known amine process. The MEA process is no longer protected under patent and therefore both the MEA solution and all design parameters may be readily obtained. MEA removes both hydrogen sulfide and carbon dioxide and does so nonselectively. Relative to other amine processes, MEA offers several advantages, primarily the lowest solvent cost and lowest hydrocarbon co-absorption. MEA does, however, typically require higher solvent circulation rates and higher regeneration energy than competing amine processes. In addition, the MEA process has shown a higher tendency toward corrosion and foaming. Thus it is no longer the most widely used.

Diethanolamine (DEA) has replaced MEA as the most widely used amine solvent. High load DEA technologies, such as that developed by Elf Aquitaine, permit the use of high (up to 40 wt % DEA) concentration solutions. The Elf Aquitaine–DEA process allows lower circulation rates, and has consequent reductions in capital and utility expenses. DEA tends to be more resistant to degradation by carbonyl sulfide and carbon disulfide than MEA. DEA is, however, susceptible to degradation by carbon dioxide.

The diglycolamine (DGA) process also allows high (up to 60 wt % DGA) solvent concentrations for reduced circulation rate and energy requirements. High solvent costs and a higher tendency to absorb heavy hydrocarbons have limited the use of this solvent.

The process based on methyldiethanolamine (MDEA), a tertiary amine, offers increased hydrogen sulfide selectivity. In addition, MDEA exhibits improved resistance to degradation and lower tendency for corrosion, relative to other amine systems. High solvent concentrations and lower heats of reaction, relative to MEA and DEA, can yield energy savings or capacity increases in existing facilities. Greater problems exist for heat-stable salts, particularly in tail gas cleanup units, and higher solvent costs have limited MDEA application.

Exxon's Flexsorb SE solvents achieve high hydrogen sulfide selectivity by virtue of their molecular structure. These solvents are sterically hindered secondary amines. A bulky molecule is used to shield the available hydrogen radical on the nitrogen atom and prevent the insertion of carbon dioxide. The reaction with hydrogen sulfide is not sensitive to the amine's structure, so the steric hindrance affords higher hydrogen sulfide selectivity.

Alkali Carbonates. The hot carbonate process was originally developed by the U.S. Bureau of Mines, using aqueous 25–30 wt % solutions of potassium carbonate for the removal of hydrogen sulfide and carbon dioxide.

$$K_2CO_3 \text{ (aq)} + CO_2 + H_2O \rightleftharpoons 2\,KHCO_3 \tag{7}$$

$$K_2CO_3 \text{ (aq)} + H_2S \rightleftharpoons KHCO_3 + KHS \tag{8}$$

Hot potassium carbonate is generally considered for bulk acid gas removal applications. Often a secondary removal step using an alkanolamine is also used to meet product specifications.

Hot potassium carbonate processes are intended for the removal of carbon dioxide, or the co-removal of hydrogen sulfide and carbon dioxide. As a result of the regeneration chemistry, these hot-pot processes are not suitable for the removal of hydrogen sulfide without significant carbon dioxide also in the untreated gas stream.

A number of processes have been developed using hot potassium carbonate plus an activator. The activator, which may be DEA, boric acid, or a hindered amine, serves to accelerate the rate of absorption, thus reducing absorber and regenerator sizes. Catacarb, Benefield, and Flexsorb HP are examples of proprietary processes of this type.

Physical Absorption. Whereas chemical absorption relies on solvent reactions to hold acid gas components in solution, physical absorption exploits gas–liquid solubilities. The amount of absorption for these solvents is directly proportional to the partial pressure of the acid gas components. Thus these processes are most applicable in situations involving high pressure feed streams containing significant concentrations of acid gas components. To favor absorption, lower temperatures are often employed. Some processes require refrigeration.

The process flow sheet for a physical absorption unit is similar to that of the chemical absorption processes (see Fig. 3), featuring an absorber and regenerator. One key difference is that the solvent is regenerated primarily through reduction in pressure, although heating or stripping may also be required. Generally, the regeneration energy requirements for physical absorption solvents are lower than those for chemical solvents.

High acid gas loadings are typically used, resulting in lower solvent circulation rates and reduced equipment sizes. This is one of the key advantages of physical absorption processes. Depending on the solvent, other chemical species may be absorbed, including carbon dioxide, organic sulfur species such as carbonyl sulfide, carbon disulfide, and mercaptans, and heavy hydrocarbons, if present in the feed stream in significant quantity. These hydrocarbons represent product loss and potential operating problems in downstream sulfur recovery processes.

The solvent can be tailored to provide selective acid gas removal based on the liquid–gas solubilities. For example, the Selexol process, licensed by Union Carbide Corporation, uses the dimethyl ether of polyethylene glycol (DMPEG) to provide high hydrogen sulfide selectivity. The solubility of hydrogen sulfide in DMPEG is 8–10 times that of carbon dioxide.

Cold methanol has proven to be an effective solvent for acid gas removal. Cold methanol is nonselective in terms of hydrogen sulfide and carbon dioxide. The carbon dioxide is released from solution easily by reduction in pressure. Steam heating is required to release the hydrogen sulfide. A cold methanol process is licensed by Lurgi as Rectisol and by the Institúte Francaise du Petrole (IFP) as IFPEXOL.

Hybrid Processes. A number of processes have been developed which use both chemical and physical absorption solvents to offer high purity treat gas and low energy solvent regeneration. The operation of these processes is usually similar to that of the individual chemical or physical absorption processes. The solvent composition is typically customized to meet the requirements of individual applications.

Shell's Sulfinol process is one example of the hybrid-type absorption process. The original Sulfinol process was based on a solution of diisopropanol-amine (DIPA), a chemical solvent, and sulfolane (tetrahydrothiophene dioxide), a physical solvent, and water. This composition is known as Sulfinol-D. Typical compositions might have been 40% DIPA, 40% sulfolane, and 20% water. The presence of the physical solvent greatly increased removal of any organic sulfur compounds. Additionally, regeneration was accomplished using lower energy input and tendency toward foaming was decreased. Shell also offers a Sulfinol formulation which replaces the DIPA with methyldiethanolamine (MDEA), known as Sulfinol-M. The use of MDEA affords a greater selectivity for hydrogen sulfide absorption. Exxon's Flexsorb PS and Union Carbide's Ucarsol LE processes are additional examples of this type of process.

Conversion Processes. Most of the adsorption and absorption processes remove hydrogen sulfide from sour gas streams thus producing both a sweetened product stream and an enriched hydrogen sulfide stream. In addition to the hydrogen sulfide, this latter stream can contain other co-absorbed species, potentially including carbon dioxide, hydrocarbons, and other sulfur compounds. Conversion processes treat the hydrogen sulfide stream to recover the sulfur as a salable product.

The Claus process is the most widely used to convert hydrogen sulfide to sulfur. The process, developed by C. F. Claus in 1883, was significantly modified in the late 1930s by I. G. Farbenindustrie AG, but did not become widely used until the 1950s. Figure 5 illustrates the basic process scheme. A Claus sulfur recovery unit consists of a combustion furnace, waste heat boiler, sulfur condenser, and a series of catalytic stages each of which employs reheat, catalyst bed, and sulfur condenser. Typically, two or three catalytic stages are employed.

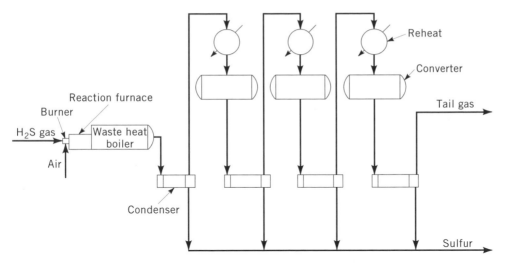

Fig. 5. Schematic of a three-stage Claus plant.

The Claus process converts hydrogen sulfide to elemental sulfur via a two-step reaction. The first step involves controlled combustion of the feed gas to convert approximately one-third of the hydrogen sulfide to sulfur dioxide (eq. 9) and noncatalytic reaction of unburned hydrogen sulfide with sulfur dioxide (eq. 10). In the second step, the Claus reaction, the hydrogen sulfide and sulfur dioxide react over a catalyst to produce sulfur and water (eq. 10). The principal reactions are as follow:

$$2\,H_2S + 3\,O_2 \longrightarrow 2\,SO_2 + 2\,H_2O \tag{9}$$

$$2\,H_2S + SO_2 \longrightarrow 3\,S^0 + 2\,H_2O \tag{10}$$

The amount of combustion air is tightly controlled to maximize sulfur recovery, ie, maintaining the appropriate reaction stoichiometry of 2:1 hydrogen sulfide to sulfur dioxide throughout downstream reactors. Typically, sulfur recoveries of up to 97% can be achieved (7). The recovery is heavily dependent on the concentration of hydrogen sulfide and contaminants, especially ammonia and heavy hydrocarbons, in the feed to the Claus unit.

The operating range of the Claus process covers feed streams having hydrogen sulfide concentrations of ca ≥25 mol %. The straight-thru configuration, in which all of the acid gas is fed directly to the burner, can be applied to streams containing a minimum of 55–60 mol % hydrogen sulfide. Processing streams having lesser amounts of hydrogen sulfide may result in combustion flame instability when this furnace configuration is employed. As a result, lean hydrogen sulfide streams often require preheating of acid gas feed and/or combustion air. Alternatively, a split-flow configuration is used to ensure stable operations. In this arrangement a portion of the feed, ~55–60% of the total, bypasses the furnace. These two combustion schemes are shown in Figure 6.

The use of oxygen enrichment of the combustion air gained attention in the late 1980s to mid-1990s. Oxygen enrichment can be used either to increase the

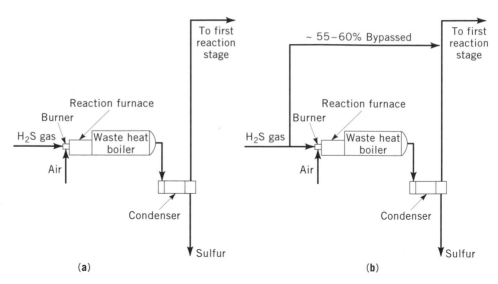

Fig. 6. Claus plant combustion schemes: (**a**) straight-thru and (**b**) split-flow processes.

processing capacity of an existing unit or to extend operation to low concentration hydrogen sulfide feeds (8). A number of additional benefits claimed for oxygen enrichment include more reliable operation, increased contaminant destruction, and improved sulfur recovery.

As the sulfur recovery offered by a Claus plant is limited in practice to about 97%, a number of processes have been commercialized to recover the residual sulfur present in the Claus plant tail gas. Comprimo's Superclaus process and Parson's Hi-Activity process are two examples of direct oxidation tail gas treating processes. Both consist of the replacement of the final Claus stage by, or the addition of, a reaction stage featuring proprietary catalyst to promote the direct oxidation of hydrogen sulfide in the Claus tail gas to sulfur selectively. Air is injected upstream of the reactor. The hydrogen sulfide and oxygen react over the catalyst via the following reaction:

$$2\,H_2S + O_2 \longrightarrow 2\,S^0 + 2\,H_2O \tag{11}$$

A sulfur condenser follows the reactor. These processes, ie, Superclaus or Parson's Hi-Activity process, can boost the overall sulfur recovery to up to 99.2%.

Another variety of tail gas treatment process is an extension of the Claus reaction by operating the final reactor at temperatures below the sulfur dew point. Sulfur production from the oxidation reaction of hydrogen sulfide with sulfur dioxide in the catalytic region is favored by decreasing temperature. Typically, in Claus plants the converters operate marginally above the dew point of sulfur to maximize conversion without depositing liquid sulfur on the bed. The cold bed sub-dew point processes deliberately operate at temperatures which result in sulfur condensing out of the vapor phase to be adsorbed on the catalyst. Multiple reactor vessels are employed to allow cycling of the reactor between adsorption/reaction service and desorption of the sulfur. These sub-dew point processes can increase the overall Claus plant sulfur recovery to up to 99%, as limited by equilibrium conversion and sulfur vapor pressure losses. Elf Aquitaine's Sulfreen process, Amoco's cold bed adsorption (CBA) process, and the Mineral and Chemical Resource Company (MCRC) process licensed by Delta Hudson are all variations on the cold bed sub-dew point process.

IFP developed a process suitable for application to Claus unit tail gas treatment which also extends the Claus reaction in the liquid phase. A process flow diagram is shown in Figure 7. The Claus plant tail gas is contacted countercurrently in a packed tower using a polyethylene glycol solution. A proprietary metal–salt catalyst is dissolved into the glycol solution. Hydrogen sulfide and sulfur dioxide are absorbed into the liquid phase and react, via the Claus reaction (eq. 10) to form elemental sulfur. The sulfur settles to the bottom of the tower and is drained off the boot. The sulfur-free catalyst and glycol solution are recycled to the top of the tower. The tower is maintained by steam condensate injection at a temperature marginally above the solidification temperature of the sulfur to maximize conversion. The IFP process can boost the overall sulfur recovery to 99%. Recovery of the IFP process is limited by sulfur vapor pressure losses, and Claus reaction equilibrium. Additionally, carbonyl sulfide and carbon disulfide in the Claus tail gas are unaffected by the process, and contribute directly to recovery losses.

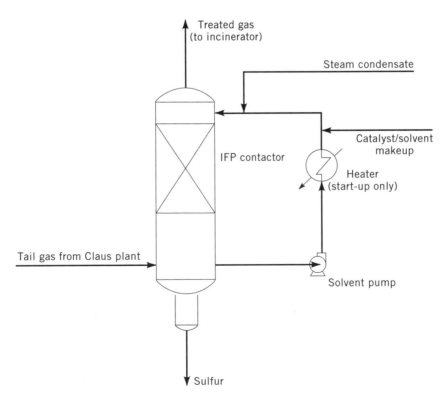

Fig. 7. The IFP process.

The Shell Claus off-gas treatment (SCOT) process and the Beavon process are similar, involving a hydrogenation step. In both processes the Claus tail gas reacts with a reducing gas to convert all sulfur species to hydrogen sulfide. The reaction step is followed by water quench in the SCOT and Beavon+MDEA (or formulated solvent) processes. Absorption of the hydrogen sulfide into an amine, usually MDEA, solution follows. The hydrogen sulfide is recycled to the Claus plant. Hydrogenation-based processes typically boost the overall recovery of the Claus unit to greater than 99.8%.

One variation of the Beavon process replaces the MDEA absorber with a Stretford unit. The Stretford process, developed by the North Western Gas Board (Manchester, England), is a hydrogen sulfide conversion process based on liquid-phase redox, ie, reduction–oxidation, reactions to form elemental sulfur. Hydrogen sulfide is absorbed and oxidized to sulfur by an aqueous solution of sodium carbonate, sodium vanadate, and an oxidation catalyst such as anthraquinone disulfonic acid or sodium anthraquinone disulfonate. The hydrogen sulfide reacts with the sodium carbonate to form sodium hydrosulfide. The vanadate oxidizes the hydrosulfide to sulfur. The sulfur is recovered off the oxidizing tank, as a slurry, and separated by filter or centrifuge to produce a sulfur cake. The vanadium solution is regenerated with air and returned to the absorber. The process flow diagram is given in Figure 8. The Stretford process was originally applied to treating brewery gases, sewage gases, and manufactured gases containing small quantities of hydrogen sulfide. The process was applied with limited success in sour natural gas sweetening. Stretford units ultimately found acceptance as a Claus unit tail gas cleanup process, but, more recently, concerns over the vanadium used in the process have limited its application.

The Lo-Cat process, licensed by US Filter Company, and Dow/Shell's SulFerox process are additional liquid redox processes. These processes have

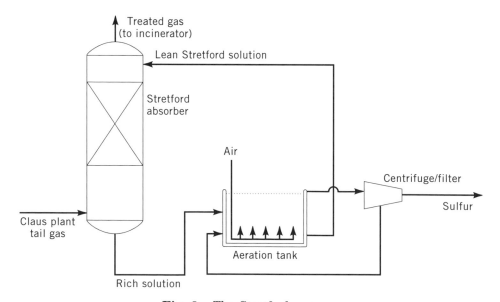

Fig. 8. The Stretford process.

replaced the vanadium oxidizing agents used in the Stretford process with iron. Organic chelating compounds are used to provide water-soluble organometallic complexes in the solution. As in the case of Stretford units, the solution is regenerated by contact with air.

NKK's Bio-SR process is another iron-based redox process which instead of chelates, uses *Thiobacillus ferroidans* bacteria to regenerate the solution (9). This process absorbs hydrogen sulfide from a gas stream into a ferric sulfate solution. The solution reacts with the hydrogen sulfide to produce elemental sulfur and ferrous sulfate. The sulfur is separated via mechanical means, such as filtering. The solution is regenerated to the active ferric form by the bacteria.

These redox processes are usually applicable for small sulfur capacities. The sulfur is typically produced as a slurry, and can be upgraded to cake or molten sulfur. At low pressures, the redox processes can replace the amine Claus and tail gas cleanup processes with a single step, yet obtain sulfur recoveries of 99%. At higher pressures, the redox processes experience sulfur plugging and foaming problems.

Other wet oxidation processes under development as of the mid-1990s include Marathon Oil's Hysulf process which uses an organic solvent to remove the hydrogen sulfide. One significant distinction of the Hysulf process is that in addition to sulfur, hydrogen is produced.

Idemitsu Kosan Company is also developing technology to recover both hydrogen and sulfur. An aqueous solution of ferric chloride is used to chemically absorb hydrogen sulfide. Sulfur is released by the absorption reaction and recovered from the solution. A low voltage electrolyzer produces hydrogen as the solution is regenerated.

Alberta Sulfur Research Ltd. is developing thermal cracking technology to produce sulfur and hydrogen from hydrogen sulfide. The process could be integrated into a conventional Claus unit by replacing the waste heat boiler with the thermal cracker. A portion of the hydrogen sulfide feed to the combustor is split off to feed the cracker. Sulfur and hydrogen would be separated and recovered from the uncracked hydrogen sulfide, which is recycled to the combustor. Another hydrogen sulfide cracking technology, using microwaves, is being studied by Argonne National Laboratories and the Kurchatov Institute (Russia).

A derivative of the Claus process is the Recycle Selectox process, developed by Parsons and Unocal and licensed through UOP. Once-Thru Selectox is suitable for very lean acid gas streams (1–5 mol % hydrogen sulfide), which cannot be effectively processed in a Claus unit. As shown in Figure 9, the process is similar to a standard Claus plant, except that the thermal combustor and waste heat boiler have been replaced with a catalytic reactor. The Selectox catalyst promotes the selective oxidation of hydrogen sulfide to sulfur dioxide, ie, hydrocarbons in the feed are not oxidized. These plants typically employ two Claus catalytic stages downstream of the Selectox reactor, to achieve an overall sulfur recovery of 90–95%.

For feeds greater than 5 mol % hydrogen sulfide, but less than 40 mol %, the Recycle Selectox process can be used. In this variation of the Selectox process, a portion of the process gas leaving the condenser downstream of the Selectox reactor is recycled so as to limit the outlet gas temperature from the reactor to <205°C.

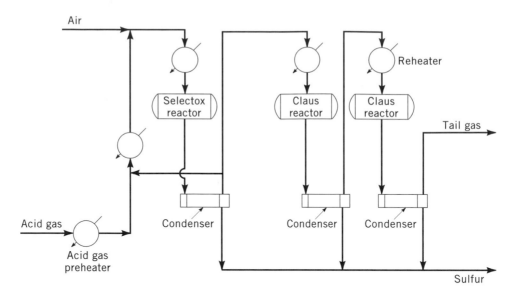

Fig. 9. The Selectox process.

Another variation of the Selectox process can be used with the Beavon process in tail gas treating. The hydrogenated Claus tail gas stream is sent to a Selectox reactor. Overall recoveries of up to 98.5% are possible. Use of Beavon/Selectox, however, typically costs more than use of Superclaus.

Sulfur Dioxide Removal and Recovery

As worldwide attention has been focused on the dangers of acid rain, the demand to reduce sulfur dioxide [7446-09-5] emissions has risen. Several processes have been developed to remove and recover sulfur dioxide. Sulfur can be recovered from sulfur dioxide as liquid sulfur dioxide, sulfuric acid, or elemental sulfur. As for the case of hydrogen sulfide, sulfur dioxide removal processes are categorized as adsorption, absorption, or conversion processes.

Adsorption Processes. Sulfur dioxide can be adsorbed on a solids bed, as can hydrogen sulfide. Fixed-, fluidized-, and moving-bed configurations have all been demonstrated. Copper oxide has been proven as an effective sorbent. Sulfur dioxide is adsorbed onto the sorbent via chemical reaction to form copper sulfate. The sulfated sorbent material can be regenerated by reduction using hydrogen, carbon monoxide, or methane to produce sulfur dioxide and hydrogen sulfide, which must be further processed to recover the sulfur. Removal efficiencies of greater than 95% are possible.

An additional benefit of adsorption-based sulfur dioxide removal processes is that nitrogen oxides, NO_x, are also removed by the sorbent. Nitrogen oxides desorb when the sorbent is heated using hot air.

A fluidized-bed process being developed by NOXSO for treating coal-fired furnace flue gases is illustrated schematically in Figure 10. Sulfur dioxide and nitrogen oxides from the ash-freed flue gas are adsorbed onto alumina beads impregnated with copper oxide in a fluidized bed. The nitrogen oxides are

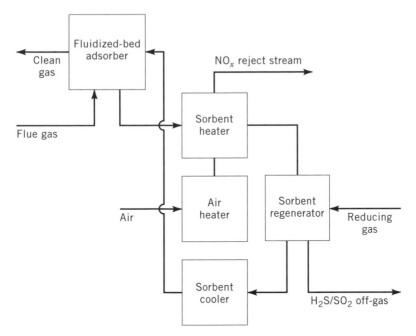

Fig. 10. Fluidized-bed SO_x/NO_x process.

desorbed at high temperature and recycled back to the coal-fired furnace. The sulfur is liberated as sulfur dioxide and hydrogen sulfide during the regeneration of the sorbent with reducing gases.

Absorption Processes. Most flue gas desulfurization (FGD) systems are based on absorption of the sulfur dioxide into a nonregenerable alkali-salt solvent. Sulfur absorbed using nonregenerable solvents is not recovered and the alkali sulfite–sulfate produced presents a disposal problem.

Wet lime–limestone processes are the most common FGD systems. Sulfur dioxide is removed from the flue gas by direct contact with an aqueous slurry of lime–limestone. Absorbed sulfur dioxide reacts with the lime–limestone to form calcium sulfite. The desulfurized flue gas can be directly vented to the atmosphere. Excess oxygen in the flue gas reacts with the calcium sulfite formed by the absorption reaction to form calcium sulfate. The sludge produced is typically dewatered and used as landfill. A secondary oxidation stage where oxygen is supplied via an external air stream may be employed to drive the calcium sulfite oxidation reaction to near completion. The sludge from the forced oxidation stage is suitable for sale as gypsum after dewatering. The overall reaction leading to sulfate formation may be represented as follows:

$$2\,CaCO_3 + 2\,SO_2 + 4\,H_2O + O_2 \longrightarrow 2\,CaSO_4 \cdot 4\,H_2O + 2\,CO_2 \qquad (12)$$

Other reagents can be employed in a wet scrubbing process, but application is typically limited by the higher costs.

The double-alkali process uses a two-stage process. Sulfur dioxide is absorbed by a solution of sodium sulfite or aluminum sulfate in a scrubbing tower.

The reagent is regenerated by reaction with lime or limestone, which forms the calcium sulfite sludge. The advantage of this double-alkali process is the separation of the absorption and sludge formation steps. Other sodium reagents have been employed, including sodium carbonate and sodium hydroxide. These processes produce aqueous solutions of sodium sulfite and sodium sulfate.

A variation of the nonregenerable absorption is the spray dry process. Lime slurry is sprayed through an atomizing nozzle into a tower where it countercurrently contacts the flue gas. The sulfur dioxide is absorbed and water in the slurry evaporated as calcium sulfite–sulfate collects as a powder at the bottom of the tower. The process requires less capital investment, but is less efficient than regular scrubbing operations.

Regenerable absorption processes have also been developed. In these processes, the solvent releases the sulfur dioxide in a regenerator and then is reused in the absorber. The Wellman-Lord process is typical of a regenerable process. Figure 11 illustrates the process flow scheme. Sulfur dioxide removal efficiency is from 95–98%. The gas is prescrubbed with water, then contacts a sodium sulfite solution in an absorber. The sulfur dioxide is absorbed into solution by the following reaction:

$$Na_2SO_3 + H_2O + SO_2 \rightleftharpoons 2\,NaHSO_3 \tag{13}$$

An evaporator–crystallizer is used to reverse the sodium bisulfite formation reaction and release the sulfur dioxide as a vapor. The regenerated sodium

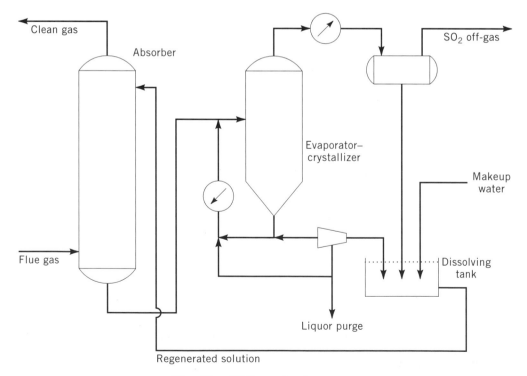

Fig. 11.   Wellman-Lord process.

sulfite, which crystallizes out of solution, is redissolved and returned to the absorber. The absorber overhead gas can be vented to the atmosphere. A concentrated sulfur dioxide stream is produced as a by-product of this process.

Elkem Technology's Elsorb process is similar to the Wellman-Lord process; however, it employs a different solvent. An aqueous solution of sodium phosphate is used.

The Cansolv process, licensed by Union Carbide Corporation, features an aqueous solution of a proprietary amine to absorb the sulfur dioxide. Dow Chemical was developing a similar process based on an amine solution as of this writing (1997).

Linde AG offers the Clintox process for sulfur dioxide removal. This process uses a physical solvent to absorb the sulfur dioxide. A concentrated sulfur dioxide stream is produced by regeneration. The Clintox process can be integrated with the Claus process by recovering sulfur dioxide from the incinerated tail gases and recycling the sulfur dioxide to the front of the Claus unit.

Conversion Processes. A number of options exist for handling concentrated sulfur dioxide streams. One option is the sale of a liquid sulfur dioxide product. Alternatively, the sulfur dioxide can be converted to elemental sulfur or to sulfuric acid.

In two processes under development as of 1997, the sulfur dioxide stream reacts with reducing gas over a proprietary catalyst to form elemental sulfur. Both processes have achieved a sulfur recovery of 96% in a single reactor. Multiple reactor systems are expected to achieve 99+% recovery of the feed sulfur. The direct sulfur recovery process (DSRP), under development at Research Triangle Institute, operates at high temperature and pressure. A similar process being developed at Lawrence Berkeley Laboratory is expected to operate near atmospheric pressure.

The Bio-FGD process converts sulfur dioxide to sulfur via wet reduction (10). The sulfur dioxide gas and an aqueous solution of sodium hydroxide are contacted in an absorber. The sodium hydroxide reacts with the sulfur dioxide to form sodium sulfite. A sulfate-reducing bacteria converts the sodium sulfite to hydrogen sulfide in an anaerobic biological reactor. In a second bioreactor, the hydrogen sulfide is converted to elemental sulfur by *Thiobacilli*. The sulfur from the aerobic second reactor is separated from the solution and processed as a sulfur cake or liquid. The process, developed by Paques BV and Hoogovens Technical Services Energy and Environment BV, can achieve 98% sulfur recovery. This process is similar to the Thiopaq Bioscrubber process for hydrogen sulfide removal offered by Paques.

The wet-gas sulfuric acid (WSA) process, developed by Haldor Topsoe, converts sulfur dioxide to sulfuric acid. In this process, sulfur dioxide is oxidized to sulfur trioxide over catalyst. The sulfur trioxide hydrolyzes to sulfuric acid in a glass-tube condenser. A more recent development for this process has been the coupling of the WSA process and nitrogen oxide removal technology. The resultant process, SNOX, treats streams containing nitrogen oxides and sulfur dioxide simultaneously. Upstream of the sulfur dioxide oxidation reactor, the gas reacts over a catalyst with injected ammonia to convert the nitrogen oxides to nitrogen and water. Removal efficiencies of 95% have been demonstrated using this process.

Economics of Sulfur Recovery

The primary driver in sulfur recovery applications is not economic potential, but rather environmental regulation. The capital investment required for sulfur recovery facilities is significant. Increasing pressure to maximize recovery and throughput at minimum investment is constantly being brought to bear on the chemical process industry.

BIBLIOGRAPHY

"Sulfur Removal" in *ECT* 3rd ed., Vol. 22, pp. 267–292, by J. R. West, Texagulf Inc.

1. V. Kwong and R. E. Meissner III, *Chem. Eng.* **102**(2), 74 (Feb. 1995).
2. G. Parkinson, G. Ondrey, and S. Moore, *Chem. Eng.* **101**(6), 31 (1994).
3. B. G. Goar and E. Nasato, *Oil & Gas J.* **92**(21), 62 (May 23, 1994).
4. Speight, *The Chemistry and Technology of Petroleum*, 2nd ed., Marcel Dekker, New York, 1991, p. 665.
5. Ref. 1, p. 76.
6. C. M. Caruna, *Chem. Eng. Prog.* **92**(2), 13–14 (Feb. 1996).
7. Ref. 6, p. 63.
8. R. W. Watson, R. Hull, and A. Sarssam, *HTI Quarterly*, 95 (Winter 1995/1996).
9. Ref. 2, p. 33.
10. Ref. 1, p. 82.

General References

Speight, *The Chemistry and Technology of Petroleum*, 2nd ed., Marcel Dekker, New York, 1991.
H. G. Paskall and J. A. Sames, *Sulphur Recovery*, Western Research, Calgary, Canada, 1988.
S. A. Newman, ed., *Acid and Sour Gas Treating Processes*, Gulf Publishing, Houston, Tex., 1985.
A. L. Kohl and F. C. Riesenfeld, *Gas Purification*, 3rd ed., Gulf Publishing, Houston, Tex., 1979.
B. L. Crynes, ed., *Chemical Reactions as a Means of Separation*, Marcel Dekker, New York, 1977.
R. N. Maddox, *Gas and Liquid Sweetening*, Campbell Petroleum Series, Norman, Okla., 1977.
P. A. Ferguson, *Hydrogen Sulfide Removal from Gases, Air, and Liquids*, Noyes Data Corp., Park Ridge, N.J., 1975.
J. B. Pfeiffer, ed., *Sulfur Removal and Recovery from Industrial Processes*, American Chemical Society, Washington, D.C., 1975.
A. V. Slack, *Sulfur Dioxide Removal from Waste Gases*, Noyes Data Corp., Park Ridge, N.J., 1971.
W. Haynes, *Brimstone: The Stone That Burns*, Van Nostrand Co., Princeton, N.J., 1959.
The Sulphur Industry, Texas Gulf Sulphur Co., Houston, Tex., 1959.
W. N. Tuller, ed., *The Sulfur Data Book*, McGraw-Hill Book Co., New York, 1954.
D. Leppin and D. A. Dalrynple, *Int. Gas Res. Conf. Proc.*, 298–312 (1996).
S. Savin and J. Nougayrede, *Int. Gas Res. Conf. Proc.*, 352–360 (1996).
A. Rehmat and co-workers, *1995 International Gas Research Conference*, Cannes, France, Nov. 6–9, 1995.
S. B. Jagtap and T. D. Wheelock, *ACS 211th National Meeting*, New Orleans, Mar. 24–28, 1996.

D. Leppin and D. Dalrymple, *GPA 74th Annual Convention*, San Antonio, Tex., Mar. 13–15, 1995; J. N. Iyengar, J. E. Johnson, and M. V. O'Neill, *ibid.*

P. D. Clark, E. Fitzpatrick, and K. L. Lesage, *AIChE 1995 Spring National Meeting*, Houston, Tex., Mar. 19–23, 1995.

V. Kwong, R. E. Meissner III, *45th Annual Okla. Univ. Laurance Reid Gas Conditioning Conference*, Norman, Ok, Feb. 26–Mar. 1, 1995.

W. W. Kensell and co-workers, *7th GRI Sulfur Recovery Conference*, 1995.

J. E. Stauffer, *Sulphur 94 International Conference*, Tampa, Fla., Nov. 6–9, 1994.

R. van Yperen and E. Boellaard, *6th International Preparation Of Catalysts Symposium*, Louvin-La-Neuve, Belgium, Sept. 5–8, 1994.

K. S. Fisher and B. J. Petrinec, *6th GRI Sulfur Recovery Conference*, Austin, Tex., May 15–17, 1994; T. J. Kenney and A. R. Khan, *ibid.*

L. H. Stern, D. K. Stevens, and W. Nehb, *GPA 73rd Annual Convention*, New Orleans, La., Mar. 7–9, 1994.

N. I. Dowling and J. B. Hyne, *44th Annual Oklahoma University Laurence Reid Gas Conditioning Conference*, Norman, Okla., Feb. 27–Mar. 2, 1994.

W. I. Echt and C. J. Wendt, *AIChE 1993 Spring National Meeting*, Houston, Tex., Mar. 28–Apr. 1, 1993.

S. K. Blevins, *43rd Annual Oklahoma University Laurence Reid Gas Conditioning Conference*, Norman, Okla., Mar. 1–2, 1993.

G. E. Gryka, *1992 International Gas Research Conference*, Orlando, Fla., Nov. 16–19, 1992.

M. P. Quinlan, *GPA 71st Annual Convention*, Anaheim, Calif., Mar. 16–18, 1992.

J. B. Hyne, *1991 Laurence Reid Gas Conditioning Conference*, Norman, Okla., Mar. 4–6, 1991.

M. Kitto, *Sulfur 90 Internation Conference*, Cancun, Mexico, Apr. 1–4, 1990.

H. Satoh, S. Kametani, and R. Yanagawa, *GRI Liquid Redox Sulfur Recovery Conference*, Austin, Tex., May 7–9, 1989.

D. R. Simbeck and B. L. Schulman, *AIChE 1986 Summer National Meeting*, Boston, Mass., Aug. 24–27, 1986.

M. C. Manderson and C. D. Cooper, *Sulfur 84 International Conference*, Calgary, Alberta, Canada, June 3–6, 1984.

J. R. Braithwaite, *Oil & Gas J.* 83–86 (Sept. 2, 1996).

Hydrocarbon Process. **75**(4), 105–150 (Apr. 1996).

C. M. Caruna, *Chem. Eng. Prog.* **92**(2), 11–15, 17, (Feb. 1996).

K. Petrov and S. Srinivasan, *Int. J. Hydrogen Energy*, **21**(3), 163–169 (1996).

R. W. Watson, R. Hull, and A. Sarssam, *HTI Quarterly*, 95–101 (Winter 1995/1996).

G. E. Halkos, *Energy Sources*, **17**(4), 391–412 (July–Aug. 1995).

V. Kwong and R. E. Meissner III, *Chem. Eng.* **102**(2), 74–83 (Feb. 1995).

C. Cilleruelo and E. Garcia, *Coal Sci. Tech.* **24**, 1883–1886 (1995).

Sulphur, **235**, 21–27, 59–68 (Nov.–Dec. 1994).

R. Siriwardane, J. A. Poston, and G. Evans, *Ind. & Eng. Chem. Res.* **33**(11), 2810–2818 (Nov. 1994).

B. G. Goar and E. Nasato, *Oil & Gas J.* **92**(21), 61–67 (May 23, 1994).

G. Parkinson, G. Ondrey, and S. Moore, *Chem. Eng.* **101**(6), 30–31, 33, 35 (1994).

W. J. Cook and M. Neyman, *US DOE Report DOE/MC/29470-3668*, U.S. Dept. of the Environment, Washington, D.C., Aug. 1993.

D. K. Stevens and W. H. Buckhannan, *Sulphur*, (225), 37–38, 41–42, 44–46, 48 (Mar.–Apr. 1993).

S. M. Campbell and A. S. Holmes, *Env. Protection*, 12–20 (Mar. 1993).

E. Sasaoka and M. Sakamoto, *Energy & Fuels*, **7**(5), 632–633 (Sept.–Oct. 1993).

L. C. Harrison and D. E. Ramshaw, *Hydrocarbon Proc.* **71**, 89–90 (Jan. 1992).

G. Samdani and E. Daniels, *Chem. Eng.* **98**(9), 41, 47 (Sept. 1991).

R. Smock, *Power Eng.*, 17–22 (Aug. 1991).

J. Borsboom and J. A. Lagas, *Hydrocarbon Tech. Int.*, 31–32, 34–36 (1991).

S. R. Gupta and D. W. Stanbridge, *Energy Prog.* **6**(4), 239–247 (Dec. 1986).

A. E. Chute, *Chem. Eng. Prog.*, 61–65 (Oct. 1982).

H. G. Paskall, *Sulphur*, 38–43 (May–June 1982).

M. Huval and H. Van De Venne, *Oil & Gas J.*, 91–99 (Aug. 17, 1981).

A. E. Chute, *Hydrocarbon Proc.*, 119–124 (Apr. 1977).

R. W. Tennyson and R. P. Schaaf, *Oil & Gas J.* **75**(2), 78 (1977).

B. G. Goar, *Hydrocarbon Proc.*, 248–252 (Sept. 1968).

D. Leppin and D. A. Dalrynple, *Int. Gas Res. Conf. Proc.*, 298–312 (1996).

S. Savin and J. Nougayrede, *Int. Gas Res. Conf. Proc.*, 352–360 (1996).

A. Rehmat and co-workers, *1995 International Gas Research Conference*, Cannes, France, Nov. 6–9, 1995.

S. B. Jagtap and T. D. Wheelock, *ACS 211th National Meeting*, New Orleans, La., Mar. 24–28, 1996.

C. M. Caruna, *Chem. Eng. Prog.* **92**(2), 11–15, 17 (Feb. 1996).

C. Cilleruelo, E. Garcia, and co-workers, *Coal Sci. Tech.* **24**, 1883–1886 (1995).

MICHAEL CAPONE
Exxon Engineering

SUNSCREEN AGENTS.   See COSMETICS.

SUPERACIDS.   See ANTIMONY COMPOUNDS; FLUORINE COMPOUNDS, INORGANIC

SUPERALLOYS.   See HIGH TEMPERATURE ALLOYS.

SUPERCOMPUTERS.   See COMPUTER TECHNOLOGY.

SUPERCRITICAL FLUIDS

Supercritical fluids (SCFs) have gained considerable attention for a variety of processes and technologies since the early 1970s. The supercritical region of a pure fluid, which may be defined as the area above both the critical pressure and critical temperature, is shown in Figure 1. A unique feature of supercritical fluids may be demonstrated by beginning with a subcritical liquid at point A on Figure 1. If the liquid is depressurized isothermally in a view cell from point A to point E, the presence of a meniscus is observed as the vapor pressure line is

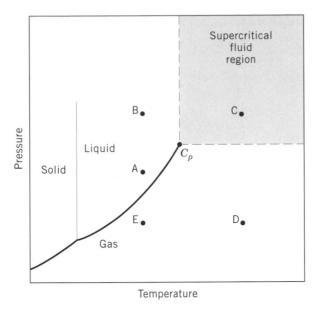

Fig. 1. Schematic pressure–temperature diagram for a pure material showing the supercritical fluid region, where C_p is the pure component critical point and dots A to E are points on the diagram. See text.

crossed. However, if the liquid takes the path of A–B–C–D–E, the fluid then passes from a liquid phase to a gas and no meniscus is seen. On this path, if one were looking only inside the view cell, one could not tell whether the component was in the gas, liquid, or fluid state. This A–B–C–D–E pathway is used in supercritical drying to avoid collapse of delicate microstructures by the strong surface tension forces that arise at liquid–vapor interfaces.

Frequently, the term compressed fluid, a more general expression than supercritical fluid, is used. A compressed fluid can be either a supercritical fluid, a near-critical fluid, an expanded liquid, or a highly compressed gas, depending on temperature, pressure, and composition.

From a historical point of view, supercritical fluids have been the subject of research since the early 1800s, perhaps as early as 1822, when Baron Cagniard de la Tour discovered the critical point of a compound (1). The phenomenon of solubility enhancement in dense gases was discovered in the late 1870s, when the effects of pressure on the solubility of the potassium iodide–ethanol system were observed (2). The technology progressed slowly until the late 1970s and early 1980s, when a number of processes were commercialized and research intensified. Many excellent books and review articles on supercritical fluid technology are available (1,3–7).

The supercritical fluid carbon dioxide, CO_2, is of particular interest. This compound has a mild (31°C) critical temperature (Table 1); it is nonflammable, nontoxic, and, especially when used to replace freons and certain organic solvents, environmentally friendly. Moreover, it can be obtained from existing industrial processes without further contribution to the greenhouse effect (see AIR POLLUTION). Carbon dioxide is fairly miscible with a variety of organic solvents, and is readily recovered after processing owing to its high volatility. It is a small

Table 1. Critical Properties for Common Supercritical Fluids[a]

Solvent	CAS Registry Number	T_c, °C	P_c, MPa[b]	ρ_c, g/cm^3
ethylene	[74-85-1]	9.3	5.04	0.22
xenon	[7440-63-3]	16.6	5.84	0.12
carbon dioxide	[124-38-9]	31.1	7.38	0.47
ethane	[64-17-5]	32.2	4.88	0.20
nitrous oxide	[10024-97-2]	36.5	7.17	0.45
propane	[74-98-6]	96.7	4.25	0.22
ammonia	[7664-41-7]	132.5	11.28	0.24
n-nutane	[106-97-8]	152.1	3.80	0.23
n-pentane	[109-66-0]	196.5	3.37	0.24
isopropanol	[67-63-0]	235.2	4.76	0.27
methanol	[67-56-1]	239.5	8.10	0.27
toluene	[108-88-3]	318.6	4.11	0.29
water	[7732-18-5]	374.2	22.05	0.32

[a]T_c = critical temperature; P_c = critical pressure; ρ_c = critical density.
[b]To convert MPa to psi, multiply by 145.

linear molecule and thus diffuses more quickly than bulkier conventional liquid solvents, especially in condensed phases such as polymers. Finally, CO_2 is the second least expensive solvent after water.

Water has an unusually high (374°C) critical temperature owing to its polarity. At supercritical conditions water can dissolve gases such as O_2 and nonpolar organic compounds as well as salts. This phenomenon is of interest for oxidation of toxic wastewater (see WASTE TREATMENTS, HAZARDOUS WASTE). Many of the other more commonly used supercritical fluids are listed in Table 1, which is useful as an initial screening for a potential supercritical solvent. The ultimate choice for a specific application, however, is likely to depend on additional factors such as safety, flammability, phase behavior, solubility, and expense.

Properties of Supercritical Fluids and Their Mixtures

Solvent Strength of Pure Fluids. The density of a pure fluid is extremely sensitive to pressure and temperature near the critical point, where the reduced pressure, P_r, equals the reduced temperature, T_r, = 1. This is shown for pure carbon dioxide in Figure 2. Consider the simple case of the solubility of a solid in this fluid. At ambient conditions, the density of the fluid is 0.002 g/cm^3. Thus the solubility of a solid in the gas is low and is given by the vapor pressure over the total pressure. The solubilities of liquids are similar. At the critical point, the density of CO_2 is 0.47 g/cm^3. This value is nearly comparable to that of organic liquids. The solubility of a solid can be 3–10 orders of magnitude higher in this more liquid-like CO_2.

The solvation strength of a given supercritical compressed fluid is related directly to the fluid density (8). Thus solvent strength may be manipulated over a wide range by making small changes in temperature and pressure. In general, the greater the density, the greater the ability of a given compressed fluid to solvate a component. One means of expressing solvent strength is by the solubility parameter, δ, which is the square root of the cohesive energy density

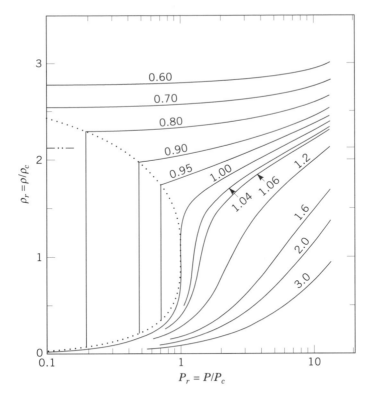

Fig. 2. Reduced density, ρ_r, versus reduced pressure, P_r, isotherms for pure carbon dioxide, where the numbers on the curves represent $T_r = T/T_c$ values. Critical properties may be found in Table 1.

and can be defined rigorously (9). A plot of the solubility parameter for CO_2 vs pressure resembles that of the density vs pressure (Fig. 2). Although the solubility parameter of CO_2 is larger than that of propane, a substantial portion of the value results from the large quadrupole moment exhibited by CO_2. The dispersion component of the solubility parameter is actually less than that of ethane. Another measure of the strength of van der Waals forces, the polarizability per volume, is also very small for CO_2: below that of ethane at comparable conditions. This makes CO_2 more like a fluorocarbon than a hydrocarbon with respect to solvent strength.

A particularly attractive and useful feature of supercritical fluids is that these materials can have properties somewhere between those of a gas and a liquid (Table 2). A supercritical fluid has more liquid-like densities, and subsequent solvation strengths, while possessing transport properties, ie, viscosities

Table 2. Comparison of Properties of Gases, Supercritical Fluids, and Liquids

Physical property	Gases	Supercritical fluids	Liquids
density, g/cm^3	0.001	0.2–1.0	0.6–1.6
diffusivity, cm^2/s	0.1	0.001	0.00001
viscosity, g/(cm·s)	0.0001	0.001	0.01

and diffusivities, that are more like gases. Thus, an SCF may diffuse into a matrix more quickly than a liquid solvent, yet still possess a liquid-like solvent strength for extracting a component from the matrix.

Physical properties of pure supercritical fluids may be found in many of the standard reference textbooks and journals (10). There are also computerized databases (qv) available for physical properties, eg, DIPPERS, DECHEMA, NIST 14, and SUPERTRAP (11).

Phase Behavior.   One of the pioneering works detailing the phase behavior of ternary systems of carbon dioxide was presented in the early 1950s (12) and consists of a compendium of the solubilities of over 260 compounds in liquid (21–26°C) carbon dioxide. This work contains 268 phase diagrams for ternary systems. Although the data reported are for liquid CO_2 at its vapor pressure, they yield a first approximation to solubilities that may be encountered in the supercritical region. Various additional sources of data are also available (1,4,7,13).

An understanding of the phase behavior of a particular system of interest is important because complex results can sometimes occur. A dramatic example, which occurs frequently for solubilities in supercritical systems, is the retrograde behavior. Figure 3 clearly shows the presence of a retrograde region. For an isobaric system at some pressure, such as 12.7 MPa (1841.5 psi), an increase in temperature of a solution of ethylene and naphthalene from 300 to 320 K results in an increase in the equilibrium solubility of naphthalene. This behavior is

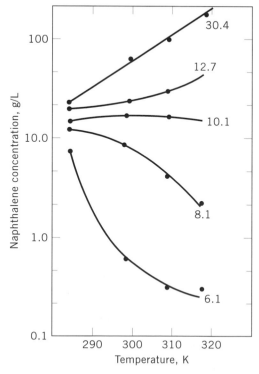

Fig. 3.   Solubility isobars for solid naphthalene [91-20-3] in ethylene [74-85-1], where the numbers represent pressures in MPa (14,15). To convert MPa to psi, multiply by 145.

typical of liquid solvent systems. For the same increase in temperature (300 to 320 K) but at a pressure of 8.1 MPa (1174.5 psi), the solubility of naphthalene decreases by nearly an order of magnitude. Because this latter behavior is the opposite of typical liquid solvents, it is termed retrograde solubility.

Either pressure or temperature may be used to control the solubility of solids in the vicinity of the mixture critical point. The appearance of the retrograde region occurs because of two competing effects of temperature. The first is on the vapor pressure of the solute; the second on the density of the supercritical solvent. At the higher pressures the dominant temperature effect is on the vapor pressure, giving positively sloped isobars. Near the critical pressure the density is more sensitive to temperature than at the higher pressures (see Fig. 2). At these near-critical pressures, the relatively rapid density decrease with increasing temperatures dominates the effect on solute vapor pressure, leading to negatively sloped isobars and retrograde behavior.

In terms of the solubilities of solutes in a supercritical phase, the following generalizations can be made. Solute solubilities in supercritical fluids approach and sometimes exceed those of liquid solvents as the SCF density increases. Solubilities typically increase as the pressure is increased. Increasing the temperature can cause increases, decreases, or no change in solute solubilities, depending on the temperature effect on solvent density and/or the solute vapor pressure. Also, at constant SCF density, a temperature increase increases the solute solubility (16).

To increase the solvation ability of carbon dioxide further, especially for high molecular weight or polar compounds, both nonpolar and polar cosolvents or modifiers may be added from 0 to 20 mol % (17). The cosolvent interacts more strongly with the solute than does CO_2. Small amounts of alkane cosolvents have been shown to increase significantly the solubility of hydrocarbons in supercritical CO_2 (18). Figure 4 shows that addition of as little as 3.5 mol % methanol to CO_2 increases the solubility of cholesterol [57-88-5] by an order of magnitude. When using cosolvents in supercritical fluid chromatography (sfc), the phase

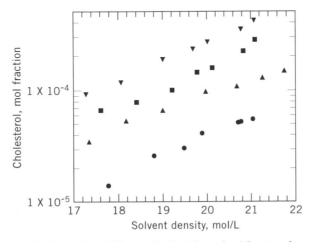

Fig. 4. Solubility of cholesterol in CO_2 at 35°C with and without polar cosolvents, where (•) represents pure CO_2; (■), CO_2 + 3.5 mol % acetone; (▲), CO_2 + 3.5 mol % ethanol; and (▼), CO_2 + 3.5 mol % methanol (19).

behavior of the cosolvent–supercritical fluid must be well characterized. Significant degradation in chromatographic performance can occur if the cosolvent is not miscible with CO_2 (20).

Compressed gases and fluids have the ability to dissolve in and expand organic liquid solvents at pressures typically between 5–10 MPa (50–100 bar). This expansion nearly always decreases the liquid's solvent strength. If enough compressed fluid is added, eventually the mixture solvent strength is comparable to that of the pure compressed fluid. In Figure 5, an organic solid is dissolved in toluene. As more compressed CO_2 is added to the solution, indicated by higher pressure, the mole fraction of the solute decreases (21). The decrease is partly a result of dilution, but it is clear that as the pressure exceeds about 5 MPa (725 psi), a dramatic change in the solute solubility occurs. For phenanthrene, there is a decrease of nearly two orders of magnitude. At about 6 MPa (870 psi), the mixture solvent strength approaches that of pure liquid CO_2. Knowledge of when a solute begins to precipitate can be important. Such information, for instance, helps to determine whether heavy hydrocarbons precipitate when using CO_2 injection in an oil reservoir.

Classification of Phase Boundaries for Binary Systems.   Six classes of binary diagrams have been identified. These are shown schematically in Figure 6. Classifications are typically based on pressure–temperature (P–T) projections of mixture critical curves and three-phase equilibria lines (1,5,22,23). Experimental data are usually obtained by a simple synthetic method in which the pressure and temperature of a homogeneous solution of known concentration are manipulated to precipitate a visually observed phase.

The Class I binary diagram is the simplest case (see Fig. 6a). The P–T diagram consists of a vapor–pressure curve (solid line) for each pure component, ending at the pure component critical point. The loci of critical points for the binary mixtures (shown by the dashed curve) are continuous from the critical point of component one, C_α, to the critical point of component two, C_β.

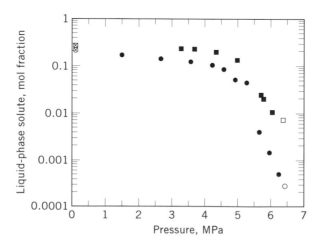

Fig. 5.   Solubility of naphthalene (squares) and phenanthrene (circles) in mixtures of toluene expanded with carbon dioxide at 25°C (■, ●), in pure toluene (⊠, ⊗), and in pure CO_2 at 6.4 MPa (□, ○) (21).

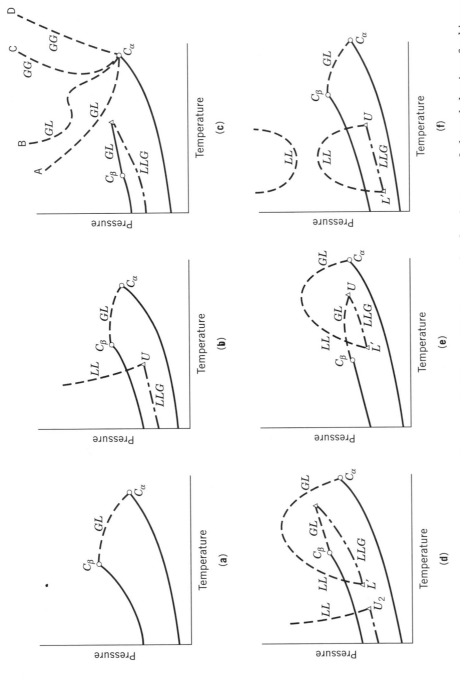

Fig. 6. Qualitative pressure–temperature diagrams depicting critical curves for the six types of phase behaviors for binary systems, where C_α or C_β corresponds to pure component critical point; G, vapor; L, liquid; U, upper critical end point; and L', lower critical end point. Dashed curves are critical lines or phase boundaries (5). **(a)** Class I, the Ar–Kr system; **(b)** Class II, the CO_2–C_8H_{18} system; **(c)** Class III, where the dashed lines A, B, C, and D correspond to the H_2–CO, CH_4–H_2S, He–H_2, and He–CH_4 system, respectively; **(d)** Class IV, the CH_4–C_6H_{16} system; **(e)** Class V, the C_2H_6–C_2H_5OH system; and **(f)** Class VI, the D_2O–2-methylpyridine system.

459

Additional binary mixtures that exhibit Class I behavior are CO_2–n-hexane and CO_2–benzene. More complicated behavior exists for other classes, including the appearance of upper critical solution temperature (UCST) lines, two-phase (liquid–liquid) immiscibility lines, and even three-phase (liquid–liquid–gas) immiscibility lines. More complete discussions are available (1,4,22). Additional simple binary system examples for Class III include CO_2–hexadecane and CO_2–H_2O; Class IV, CO_2–nitrobenzene; Class V, ethane–n-propanol; and Class VI, H_2O–n-butanol.

Polymers and Supercritical Fluids. Prior to the mid-1980s, little information was published regarding polymer processing with supercritical and near-critical fluids (1). In 1985, the solubilities of many polymers in near- and supercritical CO_2 were reported. These polymers were examined for their ability to increase viscosity in CO_2-enhanced oil recovery (24). Since then, a number of studies have examined solubilities of polymers in supercritical fluids (25–28). With the exception of a few polymers, such as polydimethylsiloxane and fluoropolymers, most high molecular weight polymers do not dissolve in neat CO_2 (25,27). However, although not soluble in a particular supercritical fluid, polymers can uptake a significant amount of the fluid (29). There has been a tremendous amount of effort in this area, likely spawned by increased interest in gas separations (qv) by polymeric membranes (see MEMBRANE TECHNOLOGY) (30). Typical equilibrium sorption isotherms for CO_2 in various polymers are available (31–33). Many of these data are for temperatures up to 35°C and for pressures up to the vapor pressure of CO_2 at 25°C. Data for the diffusivity variation of CO_2 in four different polymers as a function of pressure are available (34). The ability of supercritical CO_2 to be used as a polymer processing aid has been examined (35,36). One of these studies demonstrated that CO_2 could reduce the viscosity of a high molecular weight polydimethylsiloxane polymer over two orders of magnitude, primarily through the addition of free volume to the solution. In addition, CO_2 has been used commercially for blowing foams in extruder processes.

A supercritical or compressed gas can be sorbed into a polymer to act as a plasticizing agent. As the concentration of the compressed fluid is increased in the polymer phase, the sorption and subsequent swelling of an amorphous polymer can cause a glass-to-liquid phase transition. The glass-transition temperature (T_g) of a polymer can be depressed to below the normal T_g by 100°C or more (32,37). Certain polymers can exhibit an isobaric liquid-to-glass transition with a temperature increase, defined as retrograde vitrification (38) (Fig. 7). Continuing to increase the temperature results in a normal transition from a glass to a liquid state. The retrograde vitrification is caused by an increase in the solubility of CO_2 in the polymer at the lower temperatures. The T_g behavior may be exploited in polymer processing to produce extremely small voids only a few micrometers in diameter. As an example, Figure 8 shows how polymethylmethacrylate (PMMA) can be foamed at 40°C using sorbed CO_2 as the foaming agent (39). As the pressure decreases, the growing voids are quenched at the T_g. Foaming at room temperature or below is possible because CO_2 is capable of inducing a retrograde vitrification region in PMMA.

A crystalline or semicrystalline state in polymers can be induced by thermal changes from a melt or from a glass, by strain, by organic vapors, or by liquid solvents (40). Polymer crystallization can also be induced by compressed (or

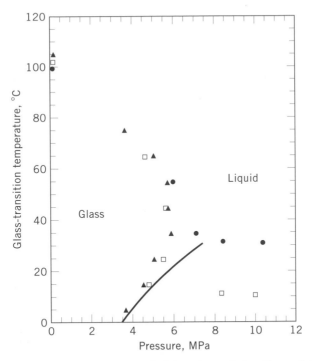

Fig. 7. Glass-transition temperatures of (▲) polymethylmethacrylate (PMMA); (□) polymethylmethacrylate-*co*-styrene (SMMA60); and (●) polystyrene (PS) as a function of carbon dioxide pressure, where the solid line represents CO_2 vapor pressure (37). To convert MPa to psi, multiply by 145.

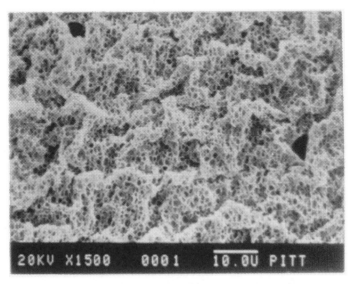

Fig. 8. Electron micrograph of a polymethylmethacrylate foamed by rapid pressure release with carbon dioxide at 40°C and 34.47 MPa (4998 psi) (39).

supercritical) gases, such as CO_2 (41). The plasticization of a polymer by CO_2 can increase the polymer segmental motions so that crystallization is kinetically possible. Because the amount of gas (or fluid) sorbed into the polymer is a direct function of the pressure, the rate and extent of crystallization may be controlled by controlling the supercritical fluid pressure. As a result of this ability to induce crystallization, a history effect may be introduced into polymers. This can be an important consideration for polymer processing and gas permeation membranes.

Dispersions in Supercritical Fluids. The ability to design surfactants for the interface between water (or organics) and supercritical fluids offers new opportunities in protein and polymer chemistry, separation science, reaction engineering, environmental science for waste minimization and treatment, and materials science. The design of surfactants for conventional reverse micelles and water-in-oil microemulsions is reasonably well understood for supercritical fluid alkane solvents (42–46). Microemulsions are thermodynamically stable and transparent dispersions of one phase within another. For CO_2, surfactant design is more difficult because the properties of CO_2 are much different from those of water or nonpolar organic solvents. Unlike water, carbon dioxide has no dipole moment. Even when highly compressed, CO_2 has far weaker van der Waals forces than hydrocarbon solvents, making CO_2 more like a fluorocarbon or fluoroether. Because of carbon dioxide's weak solvent strength, either lipophilic (high molecular weight) or hydrophilic materials are often insoluble in CO_2. It is possible, however, to form dispersions of either hydrophilic or lipophilic phases in a CO_2-continuous phase. Organic-in-CO_2 dispersions may be stabilized using surfactants containing CO_2-philic tails, such as fluorinated compounds (47). More recently, a stable microemulsion containing a water-like core was formed within a CO_2-continuous phase (48). The surfactant was an ammonium carboxylate perfluoropolyether (PFPE), $CF_3O(CF_2CF(CF_3)O)_3CF_2COO-NH_4^+$, commercially available in the COOH form, which has an average molecular weight of 740. The number of moles of water per mole of surfactant reached 20 in the one-phase region. The existence of bulk-like water was confirmed using uv–visible, Fourier transform infrared, fluorescence, and electron paramagnetic spectroscopy. According to fluorescence measurements, acrylodan-labeled bovine serum albumin (BSA), mol wt 67,000, is soluble in these water-in-CO_2 microemulsions. This protein appears to experience an environment similar to that of native BSA in buffer.

Molecular Modeling of Phase Behavior

Although modeling of supercritical phase behavior can sometimes be done using relatively simple thermodynamics, this is not the norm. Especially in the region of the critical point, extreme nonidealities occur and high compressibilities must be addressed. Several review papers and books discuss modeling of systems comprised of supercritical fluids and solid or liquid solutes (1,4–7,9,49,50).

One of the simplest cases of phase behavior modeling is that of solid–fluid equilibria for crystalline solids, in which the solubility of the fluid in the solid phase is negligible. Thermodynamic models are based on the principle that the fugacities (escaping tendencies) of component i, f_i, are equal for all phases at equilibrium under constant temperature and pressure (51). The solid-phase fugacity, f_2^s, can be represented by the following expression at temperature T:

$$f_2^s = P_2^s \phi_2^s \, \exp\left[\int_{P_2^s}^{P} \frac{v_2^s}{RT} \, dP \right] \tag{14}$$

where P_2^s is the vapor pressure and the superscript s denotes saturation; ϕ_2^s is the fugacity coefficient at the saturation pressure (usually unity); and v_2^s is the solid molar volume, which is frequently assumed to be pressure-independent. Equating the fugacity of the pure solid to the fugacity of the solid in the fluid phase, the following equation is obtained, after rearranging for the mole fraction of the solid component in the fluid phase.

$$y_2 = \frac{P_2^s}{P} \left[\frac{\phi_2^s \, \exp\left(\frac{v_2^s(P - P_2^s)}{RT}\right)}{\phi_2^F} \right] \tag{15}$$

The fugacity coefficient of the solid solute dissolved in the fluid phase (ϕ_2^F) has been obtained using cubic equations of state (52) and statistical mechanical perturbation theory (53). The enhancement factor, E, shown as the quantity in brackets in equation 2, is defined as the real solubility divided by the solubility in an ideal gas. The solubility in an ideal gas is simply the vapor pressure of the solid over the pressure. Enhancement factors of 10^4 are common for supercritical systems. Notable exceptions such as the squalane–carbon dioxide system may have enhancement factors greater than 10^{10}. Solubility data can be reduced to a simple form by plotting the logarithm of the enhancement factor vs density, resulting in a fairly linear relationship (52).

Specific chemical interactions, eg, associations resulting from hydrogen bonding or donor–acceptor interactions, can have a pronounced effect on SCF solution phase behavior. Hydrogen bonding among mixtures containing super-critical fluids is important to understand because of the increased interest in near- and supercritical water solutions, and in polar cosolvents and surfactants in other fluids such as CO_2 (54–58). The fluid density has been shown to have a significant effect on hydrogen bonding because of the compressible nature of supercritical fluids (59).

Various equations of state have been developed to treat association in su-percritical fluids. Two of the most often used are the statistical association fluid theory (SAFT) (60,61) and the lattice fluid hydrogen bonding model (LFHB) (62). These models include parameters that describe the enthalpy and entropy of as-sociation. The most detailed description of association in supercritical water has been obtained using molecular dynamics and Monte Carlo computer simulations (63), but this requires much larger amounts of computer time (64–66).

A number of theoretical models have been proposed to describe the phase behavior of polymer–supercritical fluid systems, eg, the SAFT and LFHB equa-tions of state, and mean-field lattice gas models (67–69). Many examples of polymer–supercritical fluid systems are discussed in the literature (1,3).

Experimental Techniques

The discussion on fluid properties and phase diagrams stresses the importance of physical confirmation of the actual phase behavior, lest unexpected behavior is encountered. A number of different experimental techniques are available for determining phase behavior. These include dynamic flow-through cells, static systems using visual observations in a variable-volume view cell, static systems with sampling for analysis, and the use of static or dynamic optical transmission cells for uv–visible, Fourier transform infrared spectroscopy (ftir), and Raman spectroscopy for analysis (1,5). Perhaps the most useful tool for examining phase behavior is the variable-volume view cell (Fig. 9) containing a piston to separate the pressurizing fluid from the sample. Using this single apparatus allows both manipulation of temperature, pressure, or composition and visual inspection for the appearance of unexpected phases. Samples can be removed for analysis, phase volumes can be measured to determine mixture composition and molar volumes (70), and phase boundaries can be measured. Many different configurations of view cells have been proposed. Some are capable of pressures in excess of 100 MPa (14,500 psi). The cell contents may be viewed safely through the sapphire window by use of a mirror, video camera, or borescope.

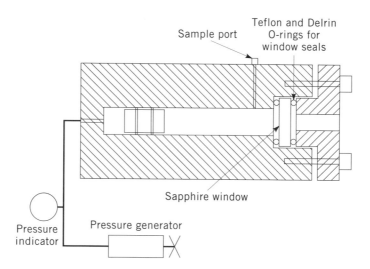

Fig. 9. Schematic drawing of a typical stainless steel variable-volume view cell having a movable internal piston. The outside diameter of the cell is 5.08 cm; the inside, 1.75 cm. Working pressure is 34 MPa (4,930 psi) at room temperature.

Processes and Applications

In a supercritical fluid process, advantages in process performance must exceed the penalties from the requirement for elevated pressures. It is prudent to undertake a formal hazard analysis of the SCF process to identify unknown and potentially dangerous design conditions (see HAZARD ANALYSIS AND RISK ASSESSMENT).

A variety of SCF processes have been commercialized. These include the Kerr-McGee Corporation residuum oil supercritical extraction (ROSE) process, which uses supercritical alkanes as the extracting solvent to separate heavy components in crude oil, ie, to carry out petroleum deasphalting; the high pressure polyethylene production process; processes for coffee decaffeination using CO_2, which are used by Kraft General Foods in Houston, Texas, and HAG GF AG in Bremen, Germany; the Takeda Chemical Industries process in Japan, which uses CO_2 as an SCF to remove acetone from antibodies; the supercritical water oxidation process, which is used for purifying aqueous wastes containing organics at Huntsman Corporation in Austin, Texas; the Union Carbide supercritical CO_2-assisted spray painting process (UNICARB), which can lower the emissions of volatile organic carbon (VOC); and supercritical fluid chromatography, which is used by Lee Scientific, ISCO Inc., Hewlett-Packard, and others.

Additional commercial processes are available for extraction of tea, hops, oriental herbs, tobacco leaves, and pharmaceuticals; CO_2-enhanced oil recovery; environmental applications such as extraction/flocculation of aqueous wastes; reactions with integrated separations such as aminations (ethylene and triethylaluminum), production of sec-butyl alcohol from isobutene, and conversion of chlorobenzene to phenol and diphenylene oxide in supercritical water; cleaning and drying precision parts and aerogel formation; as well as analytical supercritical extraction.

Many processes are being developed and planned in the environmental field, in materials processing, for chemical reactions, in food and pharmaceutical processing, and in other areas (72,73). These include precipitation processes such as the rapid expansion of supercritical solution (RESS), swelling of liquid solutions by adding compressed fluid (GAS), and spraying of liquid solutions into compressed fluids (PCA); reverse micelle separations; regeneration of adsorbents; cleaning of semiconductors, fabrics, and soil; extraction and fractionation utilizing chelating agents (qv) for heavy-metal extraction; removal of oils, cholesterol, etc, from food products, ethanol and other organics from water, and fractionation of oils and fats; polymerizations in SCFs, fractionation, crystallization, devolatilization, extraction of impurities, and impregnation of polymers; as well as separation of recycled polymers.

Supercritical Fluid Chromatography and Analytical Extraction. As an analytical tool, the use of supercritical fluids has gained broad acceptance. As in other SCF processes, density is used as the controlling feature. Many SCF chromatographic separations use a programmed density profile that is similar to conventional temperature ramping in gas chromatography. The mobile phase is typically the supercritical fluid and the stationary phase can be in a packed or capillary column. CO_2 is commonly used as the mobile phase, but cosolvents may be required for enhanced separations of polar compounds. Supercritical fluid chromatography can also be used effectively to fractionate oligomers and, to a lesser degree, high molecular weight polymers (1,3), which are too nonvolatile for conventional gas chromatography. Because the supercritical fluid density can be controlled accurately by manipulation of pressure and temperature conditions, fractionation of an oligomeric mixture or other mixtures can be tuned to give a desired separation (1,74).

Extraction of samples using supercritical fluids is becoming an accepted method for gas chromatography (gc) sample preparation, especially in the environmental field (75). For example, the U.S. Environmental Protection Agency has a method (SW 846 3561) that uses supercritical fluids as the extracting solvent for the removal of polyaromatic hydrocarbons (PAHs) from soil. The efficacy of extraction is comparable to organic solvent extraction but much less time is required. Extraction using an SCF can save 5.5 hours if used instead of a 14-step liquid separatory funnel extraction; it can save 48 hours if used instead of a Soxhlet extraction employing rotary evaporation (76).

Supercritical CO_2 has also been tested as a solvent for the removal of organic contaminants from soil. At 60°C and 41.4 MPa (6,000 psi), more than 95% of contaminants, such as diesel fuel and polychlorinated biphenyls (PCBs), may be removed from soil samples (77). Supercritical CO_2 can also extract from soil the following: hydrocarbons, polyaromatic hydrocarbons, chlorinated hydrocarbons, phenols, chlorinated phenols, and many pesticides (qv) and herbicides (qv). Sometimes a cosolvent is required for extracting the more polar contaminants (78).

Extractions. SCFs are used as extraction solvents in commercial food, pharmaceutical, environmental, and petroleum applications (1,5,6,16,23). An excellent overview of the patent literature up to 1991 is available (1).

Food Applications. Carbon dioxide, a nontoxic material, can be used to extract thermally labile food components at near-ambient temperatures. The food product is thus not contaminated with residual solvent, as is potentially the case when using conventional liquid solvents such as methylene chloride or hexane. In the food industry, CO_2 is not recorded as a foreign substance or additive. Supercritical solvents not only can remove oils, caffeine, or cholesterol from food substrates, but can also be used to fractionate mixtures such as glycerides and vegetable oils into numerous components.

Carbon dioxide is used to extract α-acids from hops. These acids impart a characteristic bitter taste to beer (72). Although the yields are similar to those using methylene chloride, the extract's color, composition, odor, and texture are more controllable, and extraction using CO_2 retains the aroma-producing essential oils. Other flavors, spices (see FLAVORS AND SPICES), and fragrances that have been extracted include lilac, essential oils, black pepper, nutmeg, vanilla, basil, ginger, paprika, rosemary, chamomile, and ground chilies. The extraction of citrus oils has also shown potential.

A process for the removal of caffeine from coffee beans using supercritical CO_2 was patented in the United States in 1974, and commercial plants went on-line in Germany in 1978 and in the United States in 1988. By using water-saturated CO_2 as a solvent, the caffeine content of coffee beans can be reduced from 3 to nearly 0.2% (79). Figure 10 shows a part of the Kraft General Foods patented process for green coffee bean decaffeination. Saturation of both the green (unroasted) coffee beans and the CO_2 with water has been found to improve the caffeine extraction rates. Also, increasing the temperature and pressure improves the partitioning of caffeine into the supercritical phase (80). Although the economics of coffee or tea decaffeination by SCFs have not been reported, the efficacy of the process is evident from Kraft General Foods' nearly 45.4×10^3 t/yr commercial coffee decaffeination plant production. Moreover,

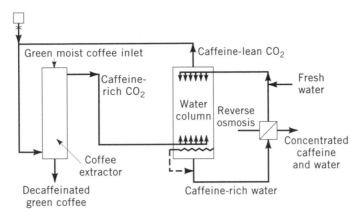

Fig. 10. Semicontinuous coffee decaffeination process using supercritical CO_2 (1).

only a third of the tea in Germany is decaffeinated using ethyl acetate. The rest is extracted using supercritical CO_2 (72).

The USDA and others have been studying the use of supercritical CO_2 for extraction and fractionation of fats and oils from food products. One of the areas studied has been the extraction of oils, particularly triglycerides, from soybeans, sunflower seeds, cottonseed, rapeseed, corn, peanuts, coconut, and rice (1,5,79). Some distinct advantages are obtained over extraction of these oils using an organic solvent such as hexane. Supercritical fluid extraction (SFE) provides an oil lower in iron and phosphorus compounds, lower in free fatty acid, and lighter in color. The reduction of fats in food products by SFE has been studied and includes a broad range of potential applications, such as oil removal from potato chips (1) and almonds by extracting up to 87% of the oil (81), lipid removal from fish and meat (82), and extraction of lipids, such as cholesterol, from eggs, milk, butter, and meat products (72,83). It is possible to remove 80–95% of the cholesterol from egg yolks and up to 90% of the fat (72,73). Similar results can be achieved for dried meats (82).

Pharmaceutical Applications. The pharmaceutical field is another area where SFE using CO_2 is particularly attractive, especially where solvent toxicity and the presence of a toxic residual solvent are concerns. Examples include extraction of vitamin E from soybean oil and a purification method for vitamin E (73). The latter process is attractive because vacuum distillation results in some thermal degradation of the product and a reduced yield. In some instances the effect of supercritical fluid solvents on the generally insoluble proteins, peptides, and amino acids is of concern (84). Humid supercritical CO_2 has been found to denature proteins. However, this has been attributed mainly to the presence of water and the processing temperature. In food products this denaturing may improve the nutritional value. Some enzymes, such as penicillin amidase, can be deactivated by dry supercritical carbon dioxide, but most others seem to survive without substantial damage. The solubilities of certain drugs in supercritical CO_2 and CO_2–cosolvent have been tested (85–87). Recrystallization of drugs via SCFs has also been demonstrated.

Environmental Applications. Waste minimization can be viewed as a key to a healthy U.S. chemical industry. The two most environmentally benign solvents, as well as the two most naturally abundant and least expensive, are water and compressed carbon dioxide. Carbon dioxide at pressures above 4 MPa (580 psia) is a leading replacement for organic solvents to minimize organic waste and reduce volatile organic carbon emissions. The replacement of organic solvents with CO_2 in both old and new technologies could reduce global warming by reducing emissions of volatile hydrocarbons and freons. Moreover, a small amount of the vast quantity of CO_2 produced industrially could be put to good use before it is vented to the atmosphere. Water at near-critical and supercritical conditions becomes a good solvent for most organic substances.

SCFs can be used to extract organic contaminants from aqueous mixtures and to regenerate adsorbents, such as activated carbon or porous polymeric resins (88). A related application is the separation of ethanol and water using CO_2. CF Systems Corporation has developed a process using supercritical propane to separate oils from refinery sludge and contaminated soil (89). The process can extract up to 99% of the liquid hydrocarbons from a wastestream, and a modification to the process uses CO_2 to treat wastewater. Because CO_2 extraction is limited to predominantly nonpolar compounds, cosolvents (or modifiers) are added to improve its solvation ability. Tri-n-butyl phosphate [*126-73-6*] (TBP) added to supercritical CO_2 was found to complex with uranium, potentially providing an organic solvent-free process for uranium extraction from spent fuel rods (90). Heavy metals also have the potential of being extracted from soil and other waste streams by CO_2 that contains a chelating agent (see CHELATING AGENTS). Chelating moieties having CO_2-philic oligomers that dissolve into CO_2 and aid in the extraction of metals from solid matrices have been developed (91).

Fractionation. Kerr-McGee developed the ROSE process for separating the heavy components of crude oil, eg, asphaltenes, resins, and oils, in the 1950s. This process was commercialized in the late 1970s, when crude oil and utility costs were no longer inexpensive. In the ROSE process (Fig. 11), residuum and pentane are mixed and the soluble resins and oils recovered in the supercriti-

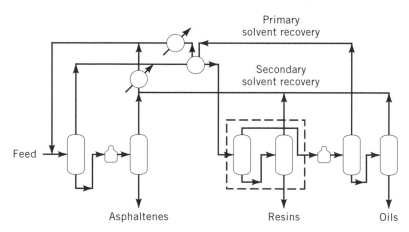

Fig. 11. Schematic of a residuum oil supercritical extraction (ROSE) process using compressed pentane to separate vacuum resids into asphaltenes (high molecular weight hydrocarbons), resins, and oils.

cal phase. By stepwise isobaric temperature increases, which decrease solvent density, the resin and oil fractions are precipitated sequentially.

Supercritical fluids can be used to fractionate low vapor pressure oils and polymers, which would be difficult using either high vacuum or molecular distillation, because the impurities have about the same volatility as the primary homologues. Furthermore, the solubility of these species is sufficiently high in liquid solvents so that the desired selectivities are difficult to achieve. In contrast, the solvent power of SCFs can be carefully controlled using pressure and temperature to fractionate synthetic or natural oils such as silicones, fluoroethers, α-olefins, copolymers, and petroleum pitch and resins (3,92,93). Fractionation with respect to chemical composition is possible by supercritical fluid solvent choice. Typically, without process optimization, narrow molecular weight distribution fractions are obtained from polydisperse parent materials. In an extreme example, poly(n-butylacrylate) was fractionated into 10 fractions using supercritical N_2O. These fractions had a polydispersity index (PDI), defined as $\overline{M}_w/\overline{M}_n$, ranging from 1.27 to 2.45; the parent material had a PDI of 20.3 (93). Preparation of narrow molecular weight distribution fractions (PDI approximately 1) should be of interest for low volatility applications, biomaterials, low PDI requirements, electronics, product tagging, and research studies.

Supercritical fluids can be used to induce phase separation. Addition of a light SCF to a polymer solvent solution was found to decrease the lower critical solution temperature for phase separation, in some cases by more than 100°C (1,94). The potential to fractionate polyethylene (95) or accomplish a fractional crystallization (21), both induced by the addition of a supercritical antisolvent, has been proposed. In the latter technique, existence of a pressure eutectic ridge was described, similar to a temperature eutectic trough in a temperature-cooled crystallization.

Reactions. Supercritical fluids are attractive as media for chemical reactions. Solvent properties such as solvent strength, viscosity, diffusivity, and dielectric constant may be adjusted over the continuum of gas-like to liquid-like densities by varying pressure and temperature. Subsequently, these changes can be used to affect reaction conditions. A review encompassing the majority of studies and applications of reactions in supercritical fluids is available (96).

Supercritical solvents can be used to adjust reaction rate constants (k) by as much as two orders of magnitude by small changes in the system pressure. Activation volumes (slopes of ln k vs P) as low as -6000 cm^3/mol were observed for a homogeneous reaction (97). Pressure effects can also be pronounced on reversible reactions (17). In one example the equilibrium constant was increased from two- to sixfold by increasing the solvent pressure. The choice of supercritical solvent can also dramatically affect an equilibrium constant. An obvious advantage of using supercritical fluid solvents as a media for chemical reactions is the adjustability of the reaction kinetics and equilibria owing to solvent effects.

Bioreactions. The use of supercritical fluids, and in particular CO_2, as a reaction media for enzymatic catalysis is growing. High diffusivities, low surface tensions, solubility control, low toxicity, and minimal problems with solvent residues all make SCFs attractive. In addition, other advantages for using enzymes in SCFs instead of water include reactions where water is a product, which can be driven to completion; increased solubilities of hydrophobic materials; increased biomolecular thermostability; and the potential to integrate both the reaction and separation bioprocesses into one step (98). There have been a number of biocatalysis reactions in SCFs reported (99–101). The use of lipases shows perhaps the most commercial promise, but there are a number of issues remaining unresolved, such as solvent–enzyme interactions and the influence of the reaction environment. A potential area for increased research is the synthesis of monodisperse biopolymers in supercritical fluids (102).

Polymerizations. Carbon dioxide can be used for homogeneous polymerization of a few soluble polymers such as fluoroacrylates and for heterogeneous polymerizations of many other polymers (25,27). The polymerization of polyethylene in supercritical ethylene monomer operates at over 270 MPa (39,150 psi) (1). The feasibility of free-radical polymerization of styrene in supercritical alkanes has been shown (103). As in the polyethylene process, polymer molecular weight may be controlled by the solvation power of the supercritical solvent. More recent work has exploited CO_2-philic groups, units that dissolve readily into the CO_2 solvent phase, such as siloxane, fluoroether, fluoroalkyl, fluoroacrylate, and phosphazene (25,91,104). The utility of these CO_2-philic groups has been demonstrated using the synthesis of a number of polymers in liquid and supercritical carbon dioxide, including fluoropolymers and tetrafluoroethylene; the cationic polymerization of vinyl and cyclic ethers; and precipitation polymerization of acrylic acid (25,26). An amide-endcapped poly(hexafluoropropylene oxide) surfactant has been synthesized (26) that can be successfully used for the inverse emulsion polymerization of acrylamide. In cases where polymerizations of polymers that are not significantly soluble in CO_2 are desired, it may be possible to utilize CO_2 as a solvent for the monomer and a swelling agent for a polymer substrate. As an example, styrene has been polymerized in supercritical CO_2-swollen poly(chlorotrifluoroethylene) (28). This is potentially a novel technique for the formation of polymer blends.

Supercritical Water. When water is raised to near and above its critical state, its fluid properties change dramatically. Supercritical water behaves like a "nonaqueous" fluid and has the ability to dissolve many nonpolar organic compounds such as alkanes and chlorinated biphenyls (PCBs). At high enough pressures, supercritical water is miscible with organic compounds and oxygen, making this an attractive medium for oxidation reactions (55). These properties have spawned new commercial processes collectively termed hydrothermal oxidation (HO), in which organic wastes in water are oxidized using air or oxygen at temperatures from 350 to 650°C and pressures from 13.8 to 68.9 MPa (2,000–10,000 psia) (Fig. 12). Much of the literature refers to this process as supercritical water oxidation (SCWO). There have been many studies of reactions and phase behavior of organic compounds in supercritical water. Literature examples include molecular simulations (66), substituent and solvent effects (106), kinetics and reaction mechanisms for organic oxidations in supercritical water (107), and hydrolysis and oxidation of glucose in supercritical water (108) (see HYDROTHERMAL PROCESSING).

Hydrothermal oxidation can be used to treat (detoxify) a wide variety of waste streams, including organic solvents; sewage sludge; fuels; contaminated oils, soils, and groundwater; process wastewater from chemical, petroleum, pulp and paper, textile, and other plants; and medical wastes, including fugitive

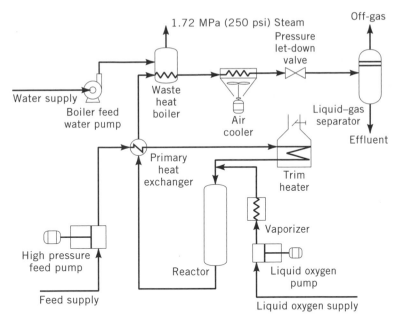

Fig. 12. Typical flow diagram of a hydrothermal oxidation process (HO), also known as supercritical water oxidation (SCWO) (73,105).

recombinant deoxyribonucleic acid (DNA). In HO the oxidation reaction occurs in a single phase, resulting in rapid oxidation rates. Typically 99.99% of the toxic organic compounds is converted to water and carbon dioxide. Nitrogen compounds produce mainly nitrogen gas, together with small amounts of nitrous oxides or ammonia. Halogens, phosphorus, and sulfur produce simple acids (55). Because HO reaction temperatures are typically 1000–2000°C lower than incinerators (qv), production of nitrogen oxides is at a minimum.

Using supercritical water is not without its drawbacks, two of which are the high pressures and temperatures involved. Another difficulty is the extreme corrosive nature of water at supercritical conditions. If halogenated organics are treated, special alloy reactors are required.

Huntsman Corporation's Austin, Texas research laboratory has incorporated an Eco Waste Technologies (EWT) hydrothermal oxidation unit (Fig. 12) to treat up to 19 L/min (lpm) of wastewater containing alcohols and amines (73,105). As of this writing (ca 1996), EWT was working on larger commercial units capable of 380 lpm or larger, and having a projected treatment cost of $0.03–0.05/L. This is similar to treatment costs in a cement kiln, without the potential for atmospheric pollution that a kiln has, and is nearly an order of magnitude less expensive than on-site incineration (105).

Materials. Supercritical fluids offer many opportunities in materials processing, such as crystallization, recrystallization, comminution, fiber formation, blend formation, and microcellular (foam) formation.

Rapid Expansion from Supercritical Solutions. Rapid expansion from supercritical solutions (RESS) across an orifice or nozzle is used commercially to precipitate solids. In this technique a solute, eg, drug, polymer, or crystalline compound, is dissolved in a near-critical or supercritical fluid. The homogeneous solution is depressurized rapidly by spraying the solution through a nozzle or orifice, which may be heated to prevent freezing owing to Joule-Thompson cooling. Modeling indicates that high ($>10^6$) supersaturations are attainable. Nucleation rates are on the order of $10^4/(cm^3 \cdot s)$ and density reduction time scales are as small as 10^{-6} seconds (109–111). By controlling upstream and downstream temperatures and pressures the desired precipitated morphology may be attained. Thus, particles, spheres, films, or fibers can be formed, depending on the material. RESS has been applied to inorganics and ceramics (qv) such as silica (SiO_2) and polycarbosilane (1), organics and pharmaceuticals (qv) (85,86), polymers (112), and two-solute systems (87). Polymer blend formation using RESS has been examined (113). Poly(ethyl methacrylate) and poly(methyl methacrylate) were precipitated into a homogeneous mixture from chlorodifluoromethane. Phase behavior measurements coupled with fluid dynamic modeling of the RESS process reveal that the location for the occurrence of phase separation with respect to the nozzle is a primary factor in determining whether particles or fibers are formed. An influential variable that can govern the transition between fibers and particles is the length-to-nozzle-diameter ratio, L/D, where a small ratio typically results in particles (111,112).

Gas Antisolvent Recrystallizations. A limitation to the RESS process can be the low solubility in the supercritical fluid. This is especially evident in polymer–supercritical fluid systems. In a novel process, sometimes termed gas antisolvent (GAS), a compressed fluid such as CO_2 can be rapidly added to a solution of a crystalline solid dissolved in an organic solvent (114). Carbon dioxide and most organic solvents exhibit full miscibility, whereas in this case the solid solutes had limited solubility in CO_2. Thus, CO_2 acts as an antisolvent to precipitate solid crystals. Using CO_2's adjustable solvent strength, the particle size and size distribution of final crystals may be finely controlled. Examples of GAS studies include the formation of monodisperse particles (<1 μm) of a difficult-to-comminute explosive (114); recrystallization of β-carotene and acetaminophen (86); salt nucleation and growth in supercritical water (115); and a study of the molecular thermodynamics of the GAS crystallization process (21).

Precipitation Using a Compressed Fluid Antisolvent. Another antisolvent process is precipitation using a compressed fluid antisolvent (PCA). An organic solution is sprayed through a nozzle into a compressed fluid (116), the solvent rapidly diffuses into CO_2 while the CO_2 swells the solution to precipitate the solute. Unlike the case for the GAS process, for PCA the solvent must dissolve in the compressed fluid. The PCA process has been used to form extremely small ($\sim$100 nm), monodisperse microspheres of polystyrene by spraying a dilute solution of polystyrene and toluene into compressed CO_2. Breakup of the liquid jet followed by rapid drying and vitrification of the polymer in the CO_2 phase results in the small uniform microspheres. Submicrometer and larger particles of a biodegradable polymer, poly(L-lactic acid), have also been formed by this process, leading to the potential formation of microspheres containing chemicals for controlled drug-release applications (see CONTROLLED RELEASE TECHNOLOGY) (117). Increasing polymer concentrations and changes to the spray configurations can vary the final polymer morphology from extremely small microspheres to microporous and hollow fibers, to oriented fibrils, to microballoons, and to microcellular microspheres (118–121). Figure 13 shows an interesting polymer formation resulting from precipitation from a compressed fluid antisolvent. Polyacrylonitrile (PAN) microfibrils were formed by spraying a dimethylformamide (DMF) solution into compressed CO_2. The latter acted as both the antisolvent and the drying agent. Spray recrystallization of protein powders to form uniform monodisperse microspheres has also been demonstrated (122).

The PCA process uses supercritical fluid drying to help preserve fine microstructures in the material. Supercritical fluid drying is a technique that has been used for many years to dry biological materials and, more recently, aerogels (qv). The original solvent is replaced by exchange with a supercritical fluid, such as CO_2, and the system is depressurized above the critical temperature of the SCF. SCFs have no vapor–liquid interface. Thus fine microstructures are preserved because of the absence of the strong vapor–liquid interfacial forces during drying.

Foaming and Extrusion of Polymers. Increasing government regulations concerning the use of chlorofluorocarbons (CFCs) have prompted the use of replacement solvents and processes for the formation of conventional foams (qv) and microcellular materials. Supercritical CO_2 or nitrogen have been used directly in the formation of microcellular thermoplastic parts (123). CO_2 dissolved in the polymer may be nucleated by two different methods: by increasing the system temperature to above the normal polymer glass-transition temperature (T_g), as is done in conventional foaming, or by rapid pressure release (124–126). The latter technique incorporates the ability of a supercritical fluid to depress the glass-transition temperature of the polymer, and can allow for the formation of polymer foams at near room temperatures. In a process piloted by Axiomatics Corporation, supercritical CO_2 is added directly into a polymer extruder and

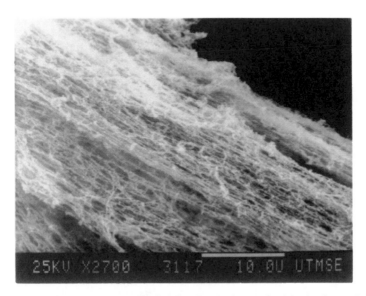

Fig. 13. Scanning electron micrograph of polyacrylonitrile fibrils formed by spraying a 0.05 wt % polyacrylonitrile in dimethylformamide solution into CO_2 through a 50-μm inner diameter, 18-cm-long nozzle at a temperature of 40°C, density of 0.66 g/mL, and solution flow rate of 0.36 mL/min (118).

used to produce foamed parts (73). Microcellular polymeric materials have also been formed by the copolymerization of methyl methacrylate and ethylene glycol dimethacrylate in supercritical fluids (127).

Other. Numerous other supercritical fluid processes and techniques have been and are continuing to be developed. The novel spray application UNI-CARB uses supercritical CO_2 to replace volatile diluents in coating formulations, thereby reducing the volatile organic compound emissions by up to 80% (128). Rapid expansion of CO_2 in the spray can also improve the quality of coatings, thus performance benefits are obtained in a process developed for environmental advantages.

Carbon dioxide can successfully regenerate pluronic block copolymer micelles loaded with model organic contaminants, which is part of a potential process to remove trace organic compounds from aqueous waste streams (129). Supercritical fluids have a demonstrated ability to plasticize polymers. Coupled with enhanced diffusion rates, it is possible to impregnate polymers with low molecular weight additives and monomers using supercritical CO_2 as both a swelling agent and a carrier solvent (130). A number of companies have commercialized, or are in the process of commercializing, cleaning processes that use supercritical CO_2 as the solvent (131). Flotation separation of mixtures of polyolefins and nonolefin thermoplastics by density difference using supercritical carbon dioxide, sulfur hexafluoride, and mixtures of the two shows great potential (132).

BIBLIOGRAPHY

"Supercritical Fluids" in *ECT* 3rd ed., Suppl. Vol., pp. 872–893, by K. Johnston, University of Texas at Austin.

1. M. A. McHugh and V. J. Krukonis, *Supercritical Fluid Extraction Principles and Practice*, 2nd ed., Butterworth-Heinemann, Boston, Mass., 1994.
2. J. B. Hannay and J. Hogarth, *Proc. R. Soc. London*, **29**, 324 (1879).
3. M. A. McHugh and V. J. Krukonis, in J. I. Kroschwitz, ed., *Encyclopedia of Polymer Science and Engineering*, 2nd ed., Vol. 16, John Wiley & Sons, Inc., 1989, p. 368.
4. T. J. Bruno and J. F. Ely, *Supercritical Fluid Technology: Reviews in Modern Theory and Applications*, CRC Press, Inc., Boca Raton, Fla., 1991.
5. M. E. Paulaitis and co-workers, eds., *Chemical Engineering at Supercritical-Fluid Conditions*, Ann Arbor Science Publishers, Mich., 1983.
6. E. Stahl, K.-W. Quirin, and D. Gerard, *Dense Gases for Extraction and Refining*, Springer-Verlag, New York, 1988.
7. R. J. Sadus, *High Pressure Phase Behaviour of Multicomponent Fluid Mixtures*, Elsevier, Amsterdam, the Netherlands, 1992.
8. S. K. Kumar and K. P. Johnston, *J. Supercritical Fluids*, **1**, 15 (1988).
9. K. P. Johnston, D. G. Peck, and S. Kim, *Ind. Eng. Chem. Res.* **28**, 1115 (1989).
10. R. C. Reid, J. M. Prausnitz, and B. E. Poling, *The Properties of Gases and Liquids*, 4th ed., McGraw-Hill Book Co., Inc., New York, 1987; *J. Phys. Chem. Ref. Data.*
11. D. G. Friend and M. L. Huber, *Int. J. Thermophysics*, **15**, 1279 (1994).
12. A. W. Francis, Jr., *J. Phys. Chem.* **58**, 1099 (1954).
13. S. Ohe, *Vapor–Liquid Equilibrium Data at High Pressure*, Elsevier, Amsterdam, the Netherlands, 1990.

14. Y. V. Tsekhanskaya, M. B. Iomtev, and E. V. Mushkina, *Russ. J. Phys. Chem.* **38**, 1173 (1964).

15. G. A. M. Diepen and F. E. C. Scheffer, *J. Am. Chem. Soc.* **70**, 4081 (1948); *J. Phys. Chem.* **57**, 575 (1953).

16. G. G. Hoyer, *Chemtech*, **15**(7), 440 (July 1985).

17. D. G. Peck, A. J. Mehta, and K. P. Johnston, *J. Phys. Chem.* **93**, 4297 (1989).

18. J. M. Dobbs, J. M. Wong, and K. P. Johnston, *J. Chem. Eng. Data*, **31**, 303 (1986).

19. J. M. Wong and K. P. Johnston, *Biotechnology Prog.* **2**, 29 (1986).

20. S. H. Page and co-workers, *J. Supercritical Fluids*, **6**, 95 (1993).

21. D. J. Dixon and K. P. Johnston, *AIChE J.* **37**, 1441 (1991).

22. J. S. Rowlinson, *Liquids and Liquid Mixtures*, 3rd ed., Butterworth, London, U.K., 1982.

23. G. M. Schneider, *Angew. Chem. Int. Ed. Engl.* **17**, 716 (1978).

24. J. P. Heller and co-workers, *Soc. Petroleum Engr. J.* **25**(5), 679 (Oct. 1985).

25. J. M. DeSimone, Z. Guan, and C. S. Elsbernd, *Science*, **257**, 945 (1992).

26. J. R. Combes, Z. Guan, and J. M. DeSimone, *Macromolecules*, **27**, 865 (1994).

27. F. A. Adamsky and E. J. Beckman, *Macromolecules*, **27**, 312 (1994).

28. J. J. Watkins and T. J. McCarthy, *Macromolecules*, **28**, 4067 (1995).

29. D. C. Bonner, *Polym. Engr. Sci.* **17**, 65 (1977).

30. W. J. Koros and R. T. Chern, in R. W. Rousseau, ed., *Handbook of Separation Process Technology*, John Wiley & Sons, Inc., New York, 1987.

31. W. J. Koros and M. W. Hellums, in Ref. 3, Suppl. Vol., pp. 724–802.

32. R. G. Wissinger and M. E. Paulaitis, *J. Polym. Sci. Part B Polym. Phys.* **25**, 2497 (1987).

33. A. C. Puleo, N. Muruganandam, and D. R. Paul, *J. Polym. Sci. Part B Polym. Phys.* **27**, 2385 (1989).

34. A. R. Berens and G. S. Huvard, in K. P. Johnston and J. M. L. Penninger, eds., *Supercritical Science and Technology*, American Chemical Society, Washington, D.C., 1989, p. 207.

35. A. Garg, L. Gerhardt, and E. Gulari, *Proceedings of the Second International Symposium on Supercritical Fluids*, Boston, Mass., 1991.

36. Y. Xiong and E. Kiran, *Polymer*, **36**, 4817 (1995).

37. P. D. Condo, D. R. Paul, and K. P. Johnston, *Macromolecules*, **27**, 365 (1994).

38. P. D. Condo and K. P. Johnston, *Macromolecules*, **25**, 6730 (1992).

39. S. K. Goel and E. J. Beckman, *Polym. Engr. Sci.* **34**, 1137 (1994).

40. A. B. Desai and G. L. Wilkes, *J. Polym. Sci.* **46**, 291 (1974).

41. J. S. Chiou, J. W. Barlow, and D. R. Paul, *J. Appl. Polym. Sci.* **30**, 3911 (1985).

42. G. J. McFann and K. P. Johnston, *J. Phys. Chem.* **95**, 4889 (1991).

43. R. M. Lemert, R. A. Fuller, and K. P. Johnston, *J. Phys. Chem.* **94**, 6021 (1990).

44. D. G. Peck, R. S. Schechter, and K. P. Johnston, *J. Phys. Chem.* **95**, 9541 (1991); D. G. Peck and K. P. Johnston, *Macromolecules*, **26**, 1537 (1993).

45. D. G. Peck and K. P. Johnston, *J. Phys. Chem.* **95**, 9549 (1991).

46. K. A. Bartscherer, H. Renon, and M. Minier, *Fluid-Phase Equilibria*, **107**, 93 (1995).

47. Z. Guan and J. M. DeSimone, *Macromolecules*, **27**, 5527 (1994).

48. K. P. Johnston and co-workers, *Science*, **271**, 624 (1996).

49. J. F. Brennecke and C. A. Eckert, *AICHE J.* **35**, 1409 (1989).

50. J. M. Wong, R. S. Pearlman, and K. P. Johnston, *J. Phys. Chem.* **89**, 2671 (1985).

51. J. M. Prausnitz, R. N. Lichtenthaler, and E. G. de Azevedo, *Molecular Thermodynamics of Fluid-Phase Equilibria*, 2nd ed., Prentice-Hall, Inc., Engelwood Cliffs, N.J., 1986.

52. K. P. Johnston and C. A. Eckert, *AICHE J.* **27**, 773 (1981).

53. K. P. Johnston, D. H. Ziger, and C. A. Eckert, *Ind. Eng. Chem. Fundam.* **21**, 191 (1982).

54. R. B. Gupta and K. P. Johnston, *Ind. Eng. Chem. Res.* **33**, 2819 (1994).

55. R. W. Shaw and co-workers, *Chem. Eng. News*, **69**(5), 26 (Dec. 23, 1991).
56. A. A. Chialvo and P. T. Cummings, *J. Chem. Phys.* **101**, 4466 (1994).
57. J. M. Walsh and co-workers, *Chem. Eng. Comm.* **86**, 124 (1989).
58. J. L. Fulton, G. G. Yee, and R. D. Smith, *J. Supercrit. Fluids*, **3**, 169 (1990).
59. S. G. Kazarian and co-workers, *J. Am. Chem. Soc.* **115**, 11099 (1993).
60. S. H. Huang and M. Radosz, *Ind. Eng. Chem. Res.* **30**, 1994 (1991).
61. B. Folie and M. Radosz, *Ind. Eng. Chem. Res.* **34**, 1501 (1995).
62. R. B. Gupta and co-workers, *AICHE J.* **38**, 1243 (1992).
63. L. W. Flanagin and co-workers, *J. Phys. Chem.* **99**, 5196 (1995).
64. A. G. Kalinichev and J. D. Bass, *Chem. Phys. Letters*, **231**, 301 (1994).
65. T. I. Mizan, P. E. Savage, and R. M. Ziff, *J. Phys. Chem.* **98**, 13067 (1994).
66. P. B. Balbuena, K. P. Johnston, and P. J. Rossky, *J. Am. Chem. Soc.* **116**, 2689 (1994).
67. C. Panayiotou and J. H. Vera, *Polym. J.* **14**, 681 (1982).
68. E. J. Beckman, R. Koningsveld, and R. S. Porter, *Macromolecules*, **23**, 2321 (1990).
69. I. C. Sanchez and R. H. Lacombe, *J. Phys. Chem.* **80**, 2352 (1976); *Macromolecules*, **11**, 1145 (1978).
70. J. R. DiAndreth, J. M. Ritter, and M. E. Paulaitis, *Ind. Eng. Chem. Res.* **26**, 337 (1987).
71. R. Randhava and S. Calderone, *Chem. Engr. Prog.* **81**, 59 (1985).
72. G. Parkinson and E. Johnson, *Chemical Eng.* **96**(7), 35–39 (July 1989).
73. S. Moore and co-workers, *Chemical Eng.* **101**(3), 32–35 (Mar. 1994).
74. F. P. Schmitz and E. Klesper, *J. Supercritical Fluids*, **3**, 29 (1990).
75. L. T. Taylor, *Analy. Chem.* **67**(11), 364A–370A (June 1, 1995).
76. ISCO Brochure SFE90-1, ISCO, Inc., Lincoln, Nebr., 1990.
77. *Chem. Eng.* **100**(5), 21–23 (May 1993).
78. A. Laitinen, A. Michaux, and O. Aaltonen, *Environ. Tech.* **15**, 715 (1994).
79. S. Vijayan, D. Singh, and B. E. Hickson, *Proc. 2nd Inter. Conf. Separ. Sci. Tech.* **2**, 442 (1989).
80. H. Peker and co-workers, *AICHE J.* **38**, 761 (1992).
81. C. A. Passey and M. Gros-Louis, *J. Supercritical Fluids*, **6**, 255 (1993).
82. A. D. Clarke, *Reciprocal Meat Conf. Proceed.* **44**, 101 (1991).
83. M. T. G. Hierro and G. Santa-Maria, *Food Chem.* **45**, 189 (1992).
84. J. K. P. Weder, *Café Cacao Thé*, **34**, 87 (1990).
85. K. A. Larson and M. L. King, *Biotech. Prog.* **2**, 73 (1986).
86. C. J. Chang and A. D. Randolph, *AICHE J.* **35**, 1876 (1989); *AICHE J.* **36**, 939 (1990).
87. J. W. Tom and P. G. Debenedetti, *Polym. Prep.* **33**(2), 104 (1992).
88. A. Akgerman and co-workers, in Ref. 4, pp. 479–509.
89. D. W. Hall, J. A. Sandrin, and R. E. McBride, *Environ. Prog.* **9**, 98 (1990).
90. *Chem. Eng.* **102**(6), 17 (June 1995).
91. A. V. Yazdi and E. J. Beckman, *J. Mater. Res.* **10**, 530 (1995).
92. I. Yilgör, J. E. McGrath, and V. J. Krukonis, *Polym. Bull.* **12**, 499 (1984).
93. K. M. Scholsky and co-workers, *J. Appl. Polym. Sci.* **33**, 2925 (1987).
94. C. A. Irani and C. Cozewith, *J. Appl. Polym. Sci.* **31**, 1879 (1986).
95. C. Chen, M. A. Duran, and M. Radosz, *Ind. Eng. Chem. Res.* **33**, 306 (1994).
96. P. E. Savage and co-workers, *AICHE J.* **41**, 1723 (1995).
97. K. P. Johnston and C. Haynes, *AICHE J.* **33**, 2017 (1987).
98. S. V. Kamat, E. J. Beckman, and A. J. Russell, *Crit. Rev. Biotech.* **15**, 41 (1995).
99. C. T. Lira and P. J. McCrackin, *Ind. Eng. Chem. Res.* **32**, 2608 (1993).
100. T. W. Randolph and co-workers, *Science*, **238**, 387 (1988).
101. D. C. Steytler, P. S. Moulson, and J. Reynolds, *Enzyme Microb. Technol.* **13**, 221 (1991).
102. A. J. Russell, E. J. Beckman, and A. K. Chaudhary, *Chemtech*, **24**(3), 33 (Mar. 1994).
103. V. P. Saraf and E. Kiran, *Polym. Prep.* **31**(1), 687 (1990).
104. W. H. Tuminello, G. T. Dee, and M. A. McHugh, *Macromolecules*, **28**, 1506 (1995).
105. *Oil Gas J.* **92**(44), 44–45 (Oct. 31, 1994).
106. M. T. Klein, Y. G. Mentha, and L. A. Torry, *Ind. Eng. Chem. Res.* **31**, 182 (1992).

107. P. E. Savage and M. A. Smith, *Environ. Sci. Tech.* **29**, 216 (1995).

108. H. R. Holgate, J. C. Meyer, and J. W. Tester, *AICHE J.* **41**, 637 (1995).

109. P. G. Debenedetti, *Chem. Eng. Sci.* **42**, 2203 (1987).

110. J. W. Tom and P. G. Debenedetti, *J. Aerosol. Sci.* **22**, 555 (1991).

111. A. K. Lele and A. D. Shine, *Ind. Eng. Chem. Res.* **33**, 1476 (1994).

112. S. Mawson and co-workers, *Macromolecules*, **28**, 3182 (1995).

113. S. N. Boen and co-workers, in I. Noda and D. N. Rubingh, eds., *Polymer Solutions, Blends, and Interfaces*, Elsevier Science Publishing Co., Inc., New York, 1992, p. 151.

114. P. M. Gallagher and co-workers, in Ref. 34, p. 334.

115. F. J. Armellini and J. W. Tester, in Ref. 35.

116. D. J. Dixon, K. P. Johnston, and R. A. Bodmeier, *AICHE J.* **39**, 127 (1993).

117. T. W. Randolph, A. D. Randolph, and M. Mebes, *Biotechnol. Prog.* **9**, 429 (1993).

118. G. Luna-Bárcenas and K. P. Johnston, *Polymer*, **36**, 3173 (1995).

119. S. D. Yeo and co-workers, *Macromolecules*, **28**, 1316 (1995).

120. D. J. Dixon and K. P. Johnston, *J. Appl. Polym. Sci.* **50**, 1929 (1993).

121. D. J. Dixon, G. Luna-Bárcenas, and K. P. Johnston, *Polymer*, **35**, 3998 (1994).

122. S.-D. Yeo and co-workers, *Biotechnol. Bioeng.* **41**, 341 (1993).

123. V. Kumar and N. P. Suh, *Polym. Eng. Sci.* **30**, 1323 (1990).

124. D. J. Dixon, "Formation of Polymeric Materials by Precipitation with a Compressed Fluid Antisolvent," Ph.D. Dissertation, University of Texas at Austin, Austin, Tex., 1992.

125. S. K. Goel and E. J. Beckman, *AICHE J.* **41**, 357 (1995).

126. M. Wessling and co-workers, *J. Appl. Poly. Sci.* **53**, 1497 (1994).

127. G. Srinivasan and J. R. Elliott, Jr., *Ind. Eng. Chem. Res.* **31**, 1414 (1992).

128. K. A. Nielsen and co-workers, SAE Tech. Paper No. 910091, 1991, Society of Automative Engineers, New York.

129. G. J. McFann and co-workers, *Ind. Eng. Chem. Res.* **32**, 2336 (1993).

130. A. R. Berens and co-workers, *J. Appl. Poly. Sci.* **46**, 231 (1992).

131. D. Hairston, *Chem. Eng.* **102**(1), 65–67 (Jan. 1995).

132. M. S. Super, R. M. Enick, and E. J. Beckman, in *Emerging Technology in Plastics Recycling*, American Chemical Society, Washington, D.C., 1992, p. 172.

DAVID J. DIXON
South Dakota School
of Mines & Technology

KEITH P. JOHNSTON
University of Texas at Austin

SUPEROXIDES.   See PEROXIDES AND PEROXIDE COMPOUNDS.

SUPERPHOSPHATE.   See FERTILIZERS; PHOSPHORIC ACID AND THE PHOSPHATES.

SURFACE AND INTERFACE ANALYSIS.   See SUPPLEMENT.

SURFACTANTS/DETERSIVE SYSTEMS.   See SURFACTANTS; DETERGENCY.

SURFACTANTS

The term surfactant is a contraction of surface-active agent. Coined in 1950, surfactant has become universally accepted to describe organic substances having certain characteristics in structure and properties. The term detergent is often used interchangeably with surfactant. As a designation for a substance capable of cleaning, detergent can also encompass inorganic substances when these do in fact perform a cleaning function. More often, however, detergent refers to a combination of surfactants and other substances, organic or inorganic, formulated to enhance functional performance, specifically cleaning, over that of the surfactant alone. It is so used herein.

Surfactants are characterized by the following features. Amphipathic structure: surfactant molecules are composed of groups of opposing solubility tendencies, typically an oil-soluble hydrocarbon chain and a water-soluble ionic group; solubility: a surfactant is soluble in at least one phase of a liquid system; adsorption at interfaces: at equilibrium, the concentration of a surfactant solute at a phase interface is greater than its concentration in the bulk of the solution; orientation at interfaces: surfactant molecules and ions form oriented monolayers at phase interfaces; micelle formation: surfactants form aggregates of molecules or ions called micelles when the concentration of the surfactant solute in the bulk of the solution exceeds a limiting value, the so-called critical micelle concentration (CMC), which is a fundamental characteristic of each solute–solvent system; and functional properties: surfactant solutions exhibit combinations of cleaning (detergency), foaming, wetting, emulsifying, solubilizing, and dispersing properties.

The presence of two structurally dissimilar groups within a single molecule is the most fundamental characteristic of surfactants. The surface behavior (surface activity) of the surfactant molecule is determined by the makeup of the individual groups, solubility properties, relative size, and location within the surfactant molecule. The term amphipathy has been proposed as "the occurrence in a single molecule or ion, with a suitable degree of separation, of one or more groups which have affinity (sympathy) for the phase in which the molecule or ion is dissolved, together with one or more groups which are antipathic to the medium (ie, which tend to be expelled by it)" (1).

Different designations describe the opposing groups within the surfactant molecules, eg, hydrophobic (water hating) and hydrophilic (water liking), lipophobic (fat hating) and lipophilic (fat liking), oleophobic (fat (oil) hating) and oleophilic (fat (oil) liking), and lyophobic (solvent hating) and lyophilic (solvent liking). The terms polar and nonpolar are also used to designate water-soluble and water-insoluble groups, respectively.

Surface activity is not limited to aqueous systems, however. All of the combinations of aqueous and nonaqueous phases are known to occur, but because water is present as the solvent phase in the overwhelming proportion of commercially important surfactant systems, its presence is assumed in much of the common terminology of industry. Thus, the water-soluble amphipathic groups are often referred to as solubilizing groups.

Until the end of the nineteenth century, soap (qv), the alkali metal salt of long-chain carboxylic acids, was the only synthetic surfactant. With the spread of chemical technology, particularly in the early twentieth century, technical shortcomings of soap, ie, its insolubility in hard water and its acidity, were acutely felt in the textile industry (see TEXTILES). Sulfonated oils were developed as hardness- and acid-stable dyeing and wetting assistants. Activity in Germany in the years following World War I led to many new surfactant structures, which can be viewed as the beginning of the modern surfactant industry. The growth of the petrochemical industry after World War II further aided the growth of the surfactant industry in providing high quality, relatively inexpensive starting materials (see PETROLEUM). However, periodic instances of crude oil shortage and attendant price increases coupled with simultaneous improvements in the production of oil-bearing seeds (see SOYBEANS AND OTHER OILSEEDS) have led to a reexamination of oleochemical feedstocks as potential sources for surfactant manufacture.

The shift to oleochemicals has been supported by increasing environmental concerns and a preference by some consumers, especially in Europe, for materials based on natural or renewable resources. Although linear alkylbenzenesulfonates (LASs) are petrochemically based, alcohol ethoxylates, alcohol ethoxysulfates, and primary alcohol sulfates are derived from long-chain alcohols that can be either petrochemically or oleochemically sourced. There has been debate over the relative advantages of natural (oleochemical) vs synthetic (petrochemical) based surfactants. However, detailed analyses have shown there is little objective benefit for one over the other.

Surfactants are classified depending on the charge of the surface-active moiety. In anionic surfactants, this moiety carries a negative charge, as in soap, $C_{17}H_{35}CO_2^-Na^+$. In cationic surfactants, the charge is positive, $(C_{18}H_{37})_2N^+ \cdot (CH_3)_2Cl^-$. In nonionic surfactants, as the name implies, there is no charge on the molecule. The solubilizing contribution can be supplied, eg, by a chain of ethylene oxide groups, $C_{15}H_{31}O(CH_2CH_2O)_7H$. Finally, in amphoteric surfactants, solubilization is provided by the presence of positive and negative charges in the molecule, as in $C_{12}H_{25}N^+(CH_3)_2CH_2CO_2^-$. In general, the hydrophobic group consists of a hydrocarbon chain containing ca 10–20 carbon atoms. The chain may be interrupted by oxygen atoms, a benzene ring, amides, esters, other functional groups, and/or double bonds. A propylene oxide hydrophobe can be considered a hydrocarbon chain in which every third methylene group is replaced by an oxygen atom. In some cases, the chain may carry substituents, most often halogens. Siloxane chains have also served as the hydrophobe in some surfactants.

Hydrophilic, solubilizing groups for anionic surfactants include carboxylates, sulfonates, sulfates, and phosphates. Cationics are solubilized by amine and ammonium groups. Ethylene oxide chains and hydroxyl groups are the solubilizing groups in nonionic surfactants. Amphoteric surfactants are solubilized by combinations of anionic and cationic solubilizing groups.

The molecular weight of surfactants may be as low as ca 200 up to the thousands for polymeric structures. A surfactant with a straight-chain C_{12}-hydrophobe and a solubilizing group is generally an effective structure. The optimum can be higher by several carbon atoms or even slightly lower than 12 depending on the nature of the polar group and the desired function of the surfactant.

In the application of surfactants, physical and use properties, precisely specified, are of primary concern. Chemical homogeneity is of little significance in practice. In fact, surfactants are generally polydisperse mixtures defined as "a preparation containing molecules which are all of the same type, but which vary only in chain length or in some other structural detail ... (as) distinguished from a mixture of two different molecular species" (2). Examples of such polydispersed mixtures include the natural fats as precursors of fatty acid-derived surfactant structures, eg, coconut oil contains glycerol esters of C_6-C_{18} fatty acids. The ratio of these fatty acids can vary, depending on the extent of distillative treatment to which the oil has been subjected. Nonionic surfactants of the alcohol ethoxylate type are polydisperse not only with respect to the hydrophobe but also in the number of ethylene oxide units attached. In the case of alkylbenzene, another surfactant raw material of great commercial significance, the alkyl substituent not only contains different chain lengths but also carries the benzene ring at all except the terminal positions along the hydrocarbon chain.

Commercial surfactants are complicated mixtures exceedingly difficult to separate into pure molecular species (3,4). Under these circumstances, the IU-PAC and *Chemical Abstracts* nomenclatures designed to describe pure molecular species are of little utility. The American Oil Chemists' Society (5) recognizes the value of names such as lauryl and myristyl as designating a mixture of chain lengths with a distribution around C_{12} and C_{14}, respectively. The terms coco or tallow describing a mixture of acids derived from coconut oil and tallow, respectively, are even less systematic but still useful. In analogy with the chemically precise acyl groups, the acid radicals from coconut and tallow fatty acids are sometimes referred to as cocoyl and tallowyl, respectively. Chemical names devised for ingredients of cosmetic formulations have attained general acceptance in the United States and are used on labels (6) (see COSMETICS).

Abbreviations of prominent use properties of the various classes of commercial surfactants are shown in Table 1. Antimicrobial activity includes germicidal, bactericidal, and bacteriostatic effects; emolliency describes lubrication or a soft feel imparted to skin by surfactants; a hair conditioner is a substantive surfactant applied from aqueous solution to impart a lubricating or antistatic effect; and opacifiers are used to thicken hand-dishwashing products and cosmetic preparations to convey an appearance of high concentration and to retard solvent drainage from foam.

A representative list of U.S. companies that market surfactants is given in Table 2.

Physical Chemistry of Interfaces

The usefulness of surfactants stems from the effects that they exert on the surface, interfacial, and bulk properties of their solutions and the materials their solutions come in contact with.

Phenomena at Liquid Interfaces.   The area of contact between two phases is called the interface; three phases can have only a line of contact, and only a point of mutual contact is possible between four or more phases. Combinations of phases encountered in surfactant systems are L–G, L–L–G, L–S–G, L–S–S–G, L–L, L–L–L, L–S–S, L–L–S–S–G, L–S, L–L–S, and L–L–S–G, where G = gas, L = liquid, and S = solid. An example of an L–L–S–G system is an aqueous surfactant solution containing an emulsified oil, suspended

Table 1. Abbreviations of Properties or Uses of Commercial Surfactants

Property	Abbreviation	Property	Abbreviation
antistatic reduction or elimination	A	intermediate for further synthesis	I
antimicrobial activity	AM	lime-soap dispersion	L
corrosion inhibition	C	opacification	O
demulsification	dE	oil recovery[a]	OR
defoaming	dF	penetrating power	p
detergency[b]	D	rewetting effectiveness[d]	rW
dispersing efficacy[b]	Di	solubilizing power[b]	S
emulsion polymerization[c]	eP	softening efficacy[e]	Sp
emulsification[b]	E	stabilizing power	St
emolliency	Em	suspending power	Su
foaming power[b]	F	textile assistance[f]	Ta
foam stabilization	Fs	viscosity building	V
hair conditioning	H	wetting power[b]	W

[a]Includes enhanced, tertiary.
[b]Key functional property.
[c]For polymers, paints, and coatings.
[d]For paper, textiles.
[e]For fabric, fiber, and yarn.
[f]For processing, fiber lubrication, and dyeing.

solid, and entrained air (see EMULSIONS; FOAMS). This embodies several conditions common to practical surfactant systems. First, because the surface area of a phase increases as particle size decreases, the emulsion, suspension, and entrained gas each have large areas of contact with the surfactant solution. Next, because interfaces can only exist between two phases, analysis of phenomena in the L–L–S–G system breaks down into a series of analyses, ie, surfactant solution to the emulsion, solid, and gas. It is also apparent that the surfactant must be stabilizing the system by preventing contact between the emulsified oil and dispersed solid. Finally, the dispersed phases are in equilibrium with each other through their common equilibrium with the surfactant solution.

Figures 1a and 1b represent typical gas–liquid and liquid–liquid interfaces at equilibrium. Assuming that gas, G, consists of air and vapor of the liquid, L, at equilibrium, there is continuous movement of liquid molecules through the gaseous interfacial region R_G because rates of evaporation and condensation at the interface I_G are equal (Fig. 1). Liquid molecules are also moving continuousy into and out of I_G through the liquid interfacial region, R_L. R_G and R_L represent nonhomogeneous transitional regions between the homogeneous phases, G and L. Systems are known in which R_G and R_L have thicknesses equivalent to two or more layers of molecules, but for most analyses the interface I_G can be considered as consisting of a single layer of molecules.

For thermodynamic treatment of surface phenomena, the thickness of the boundary regions can often be ignored or their effect eliminated by selection of a convenient location for the interface I_{GL}. The liquid–liquid interface, I_{LL} (Fig. 1b) is similarly associated with interfacial regions, R_A and R_B, which can be treated like the gas–liquid interface in most analyses. Because few liquids are completely immiscible, mutual saturation is taken as the equilibrium condition.

Table 2. Manufacturers of Surfactant Chemicals

Code	Manufacturer or source	Code	Manufacturer or source
ACT	Aceto Corp.	HUN	Hunstman Corp.
ACY	American Cyanamid Co.	HXL	Hexcel Corp., Hexcel Chemical
AKZ	Akzo Chemical Co.		Products
AMH	American Hoechst Celanese	ICI	ICI Americas Inc.
	Corp., Sp.	LNZ	Lonza, Inc.
ARI	Arizona Chemical Co.	MIL	Milliken and Co., Milliken
BIT	Boliden Intertrade Inc.		Chemicals Div.
CAL	Calgene Chemical Inc.	MOA	Mona Industries, Inc.
CAS	Cas Chemical, Inc.	MON	Monsanto Industrial Chemicals Co.
CGY	CIBA Specialty Chemical Corp.,	OC	Olin Corp., Olin Chemicals
	Consumer Care Div.	PPG	PPG Industries, Inc.
CHP	C. H. Patrick & Co., Inc.	PG	The Procter & Gamble Co.
CLA	Clariant Corp.	PIL	Pilot Chemical Co.
CRO	Croda, Inc.	QCC	Quad Chemical Corp.
CTL	Continental Chemical Corp.	REN	Ruetgers Nease Corp.
DEX	Dexter Chemical Corp.	RH	Rohm & Haas Co.
DUP	E. I. du Pont de Nemours	RHP	Rhône Poulenc, Inc.
	& Co., Inc.	SEC	Scher Chemicals, Inc.
EFH	E. F. Houghton & Co.	SHC	Shell Chemical Co.
EMK	Emkay Chemical Co.	SLC	Soluol Chemical Co., Inc.
ESS	Essential Industries, Inc.	STP	Stepan Co.
ETH	Ethox Chemicals Inc.	SYB	Sybron Chemicals Inc.
EXX	Exxon Chemical Co./Tomah	TCC	Texaco, Inc., Texaco Chemical Co.
	Products	TCM	Intex Products, Inc.
FET	Finetex, Inc.	UCC	Union Carbide Corp., Specialty
GWR	W. R. Grace & Co., Organic		Chemicals Div.
	Chemicals Div.	VIS	Condea Vista Corp.
HAL	C. P. Hall Co.	VNC	R. T. Vanderbilt Co.
HAR	Harcros Chemicals, Inc.	VND	Van Dyk & Co., Inc.
HCL	Hart Chemical, Ltd.	WTC	Witco Corp.
HNC	Henkel/Emery Corp.	WVA	Westvaco Corp., Polychemicals Div.
HPC	Hercules, Inc.	WYN	BASF Corp.
HRT	Hart Products Corp.		

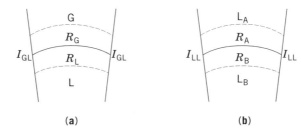

Fig. 1. (**a**) Gas–liquid (GL) interface; (**b**) liquid–liquid (LL) interface.

Surface Energy of Gas–Liquid Interfaces. Each unit of a liquid surface is associated with a quantity of free energy called the free-surface energy $(mJ/m^2 \ (=erg/cm^2))$. Work equivalent to the increase in free-surface energy must be added to a liquid to increase its surface area. A pure liquid tends toward the condition of minimum energy, which is the shape of minimum area. Thus, the shape of a free-falling droplet would be spherical under idealized conditions. The tendency of a pure liquid to minimize its surface can be explained on the basis of intermolecular attractive forces. Even though a liquid is a freely flowing fluid mass, the individual molecules are so close together that van der Waals-type forces are the dominant attraction between molecules. Thus, a molecule in the interior of a liquid is not subject to a net attractive force in any direction because the number of molecules in the sphere surrounding it is, on the average, the same in all directions. However, a molecule at the surface of the liquid is subject to a strong inward attraction because the number of molecules in the hemisphere of the vapor close enough to exert an attractive force pulling out of the liquid is small as compared to the number in the hemisphere of liquid pulling inward from the surface. A reasonable ratio of vapor-to-liquid molecules is 1:1000. This unbalanced attraction causes as many molecules as possible to move out of the surface into the bulk liquid, reducing its area to a minimum.

Surface Energy of Liquid–Liquid Interfaces. Each unit area of the interface between two mutually insoluble liquids is associated with a quantity of energy called the interfacial free energy, $(mJ/m^2 \ (=erg/cm^2))$. Expansion of the interfacial area requires addition of energy. Conversely, the interface tends to decrease the surface to minimum area. Breaking an emulsion is an example of a decrease in the area of a liquid–liquid interface. Molecules at a liquid–liquid interface are subject to a more balanced sphere of attraction than molecules at a gas–liquid interface because the number of molecules on both sides of a liquid–liquid interface is of the same order of magnitude. Thus, there is a strong force of attraction between two liquids in contact and the interfacial free energy is always less than the surface free energy of the more energetic liquid.

Surface and Interfacial Tension. Some properties of liquid surfaces are suggestive of a skin that exercises a contracting force or tension parallel to the surface. Mathematical models based on this effect have been used in explanation of surface phenomena, such as capillary rise. The terms surface tension (gas–liquid or gas–solid interface) and interfacial tension (liquid–liquid or liquid–solid) relate to these models which do not reflect the actual behavior of molecules and ions at interfaces. Surface tension is the force per unit length required to create a new unit area of gas–liquid surface $(mN/m \ (=dyn/cm))$. It is numerically equal to the free-surface energy. Similarly, interfacial tension is the force per unit length required to create a new unit area of liquid–liquid interface and is numerically equal to the interfacial free energy.

Energy of Adhesion. The interfacial energy between two mutually insoluble saturated liquids, A and B, is equal to the difference in the separately measured surface energies of each phase:

$$\gamma_{AB} = \gamma_A - \gamma_B \tag{1}$$

where γ is free-surface or interfacial energy. The term γ_{AB} represents the energy that must be added to the system to separate the liquids. The work of adhesion, W_A, between two liquids, ie, the energy necessary to increase the interfacial area by one square unit, is defined by the Dupré equation (eq. 2):

$$W_A = \gamma_A + \gamma_B - \gamma_{AB} \tag{2}$$

The quantity of energy required to separate the two liquids increases as the interfacial tension between them decreases; the lower the interfacial energy, the stronger the adhesion.

An attraction also exists at the interface between a liquid and an insoluble solid. The interfacial tension is lower than the sum of the surface tensions of the two phases. Dupré's equation is applicable to the solid–liquid interface and has the following form:

$$W_{SL} = \gamma_{SG} + \gamma_{LG} - \gamma_{SL} \tag{3}$$

However, the surface tension of the solid, γ_{SG}, and the solid–liquid interfacial tension, γ_{SL}, cannot be measured directly by simple means. The work of adhesion of the solid to the liquid W_{SL}, is usually determined by other techniques.

Contact Angle.   The line of contact between the three phases of a G–L–S system is the locus of all points from which the angle of contact between the liquid and the solid can be measured (Fig. 2). The drop of liquid, L, is resting on the solid, S, and both phases are exposed to the gas, G, at equilibrium saturation of the liquid in air (gas). The drop is assumed to be small enough for the flattening pressure of gravity to be negligible. The vector X_G is tangent to the liquid at its contact with the solid. The angle between the tangent and the surface of the solid is called the contact angle, θ. The equilibrium value of θ is an indicator of the energy relationships between liquid–liquid and liquid–solid interfaces. The contact angle is related to the free-surface energies through the Young-Dupré equation (eq. 4):

$$\gamma_{SG} = \gamma_{SL} + \gamma_{LG} \cos \theta \tag{4}$$

Combination with equation 2 for energy of adhesion of a solid–liquid interface (eq. 3) gives equation 5:

$$W_{SL} = \gamma_{LG} (1 + \cos \theta) \tag{5}$$

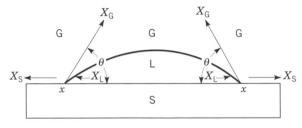

Fig. 2.   Contact angle at the oil–solid–air boundary line. See text.

When the contact angle is zero, ie, $\cos \theta = 1$, then $W_{SL} = 2\gamma_{LG}$. Under this condition, the attraction of the liquid for the solid is exactly equal to the attraction of the liquid for itself and complete wetting of the solid by the liquid is possible at equilibrium. If the liquid attracts the solid more strongly than it attracts itself, then $W_{SL} > 2\gamma_{SL}$ is a possible condition even though the negative contact angle that it implies is impossible. No contact angle (nil) is used to distinguish $W_{SL} > 2\gamma_{LG}$ from zero for $W_{SL} = 2\gamma_{LG}$. The droplet of Figure 2 has a circular line of contact with the solid which is shown in the cross-sectional diagram only at the intercepts, points x. The tangent to the droplet, X_G, is a vector that represents the tendency of the air (gas)–liquid interface toward the spherical shape of minimum energy content. There are two interfaces in the plane of the solid: liquid–solid, and gas (air)–solid. The liquid–solid interface tends to a minimum as represented by the vectors X_L parallel to the solid surface and directed toward the center of the circle of contact. The surface force represented by the vectors X_S is too weak to effect contraction of the solid surface. Nevertheless, such a force does exist and works toward minimizing the area of the gas–solid interface. The area of the gas–solid interface may be reduced by substituting a liquid–solid interface, which is the tendency represented by the vectors X_S. For a contact angle of zero, $\cos \theta = 1$, and at equilibrium, $X_S = X_G + X_L$. A contact angle of nil represents the condition $X_S > X_G + X_L$ in which equilibrium cannot be attained and the liquid continues to spread over the solid surface. Contact angles of 90° or greater are associated with low solid–liquid adhesion. In practical surfactant technology, this condition is apparent because of the lack of wetting of the solid by the liquid. A contact angle of 180°, which would indicate the complete absence of solid–liquid adhesion, is not known to exist. A hysteresis effect is nearly always observed in the measurement of contact angles, ie, the advancing angle is greater than the receding angle.

Spreading. The decrease in the free-surface energy of a system caused by the spreading of a liquid over the surface of a solid or another liquid is called the spreading coefficient. The total surface energy must be decreased by the spreading process which implies a greater attraction between unlike molecules than between like molecules. For a liquid–solid system, the decrease in the free energy, ϕ (spreading coefficient), as a result of spreading liquid, L, over solid, S, is equation 6:

$$\phi = \gamma_{SG} - \gamma_{LG} - \gamma_{SL} \tag{6}$$

Combining with equation 2 for solids, equation 3 gives equation 7:

$$\phi = W_{SL} - 2\gamma_{LG} \tag{7}$$

Thus, the decrease in free-surface energy of the total system when the liquid spreads over the solid equals the energy of adhesion W_{SL}, minus the energy of cohesion, $2\gamma_{LG}$. The energy of cohesion is the energy that must be added to a liquid system to separate a unit column of liquid by cross section, thus creating two units of new surface area. If ϕ is positive, the liquid spreads over the solid; the greater the positive value, the more readily spreading occurs. The spreading of a liquid over a solid proceeds slowly to equilibrium but in many practical systems capillary action or mechanical agitation speeds the wetting process. The spreading coefficient of a liquid–liquid system is calculated directly by substituting the surface and interfacial tensions in equation 8:

$$\phi = \gamma_o - \gamma_w - \gamma_{wo} \tag{8}$$

where γ_o represents the surface tension of the oil phase, γ_w the surface tension of the surfactant solution, and γ_{wo} the interfacial tension. The greater the positive value of ϕ, the greater the ease of spreading.

Effects of Surfactants on Solutions.   A surfactant changes the properties of a solvent in which it is dissolved to a much greater extent than is expected from its concentration effects. This marked effect is the result of adsorption at the solution's interfaces, orientation of the adsorbed surfactant ions or molecules, micelle formation in the bulk of the solution, and orientation of the surfactant ions or molecules in the micelles, which are caused by the amphipathic structure of a surfactant molecule. The magnitude of these effects depends to a large extent on the solubility balance of the molecule. An efficient surfactant is usually relatively insoluble as individual ions or molecules in the bulk of a solution, eg, 10^{-2} to 10^{-4} mol/L.

Figure 3, a schematic representation of adsorption and orientation in a solution of an anionic surfactant, shows a relatively greater number of adsorbed molecules in the G–L interface than in the bulk of the solution, a relatively greater number of adsorbed molecules on the S–L interface (beaker walls) than in the bulk of the solution, the adsorbed molecules are oriented with their lyophilic moieties in the solvent and their lyophobic moieties, respectively, at the G–L interface and in contact with the solid surface, the surfactant molecules aggregated in the micelles oriented with the lyophilic moieties exposed to the solvent and shielding the lyophobic moieties in the center of the micelles, and the adsorbed molecules in equilibrium with the micelles through the relatively

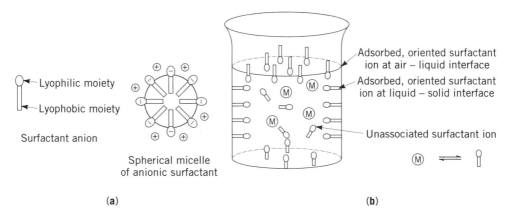

(a)   (b)

Fig. 3. Schematic diagram of anionic surfactant solution at equilibrium above its critical micelle concentration, where M = micelle and $\oplus$ are counterions: (**a**) components; (**b**) location of components at equilibrium.

small concentration of individual solute molecules in the bulk of the solution. If the G–L interface were replaced by an L–L interface, eg, an oil layer on an aqueous solution, the orientation of the ions in Figure 3 would not be changed.

Adsorption at Gas–Liquid and Solid–Liquid Interfaces. Positive adsorption, the concentration of one component of a solution at a phase boundary, results in a lowering of the free-surface energy of the solution. Accumulation of a surfactant at a solution interface means that the attractive forces between surfactant and solvent are less than the attraction between solvent molecules. As thermal diffusion brings surfactant molecules into the surface, accumulation occurs because the solute molecules cannot re-enter the solution against the stronger mutual attraction of the solvent molecules. Negative adsorption occurs when the attraction between solute and solvent molecules is greater than that between solvent molecules, and exists in concentrated aqueous solutions of inorganic compounds such as NaOH. It is associated with a surface tension slightly higher than for pure water.

The quantitative relationship between the degree of adsorption at a solution interface (7), G–L or L–L, and the lowering of the free-surface energy can be deduced by using an approximate form of the Gibbs adsorption isotherm (eq. 9), which is applicable to dilute binary solutions where the activity coefficient is unity and the radius of curvature of the surface is not too great:

$$\Gamma_s = -(C_s/\mathrm{RT})(d\gamma/dC_s) \qquad (9)$$

Equation 9 states that the surface excess of solute, Γ_s, is proportional to the concentration of solute, C_s, multiplied by the rate of change of surface tension, with respect to solute concentration, $d\gamma/dC_s$. The concentration of a surfactant in a G–L interface can be calculated from the linear segment of a plot of surface tension versus concentration and similarly for the concentration in an L–L interface from a plot of interfacial tension. In typical applications, the approximate form of the Gibbs equation was employed to calculate the area occupied by a series of sulfosuccinic ester molecules at the air–water interface (8) and the energies of adsorption at the air-water interface for a series of commercial nonionic surfactants (9).

The adsorbed layer at G–L or S–L surfaces in practical surfactant systems may have a complex composition. The adsorbed molecules or ions may be close-packed forming almost a condensed film with solvent molecules virtually excluded from the surface, or widely spaced and behave somewhat like a two-dimensional gas. The adsorbed film may be multilayer rather than monolayer. Counterions are sometimes present with the surfactant in the adsorbed layer. Mixed monolayers are known that involve molecular complexes, eg, one-to-one complexes of fatty alcohol sulfates with fatty alcohols (10), as well as complexes between fatty acids and fatty acid soaps (11). Competitive or preferential adsorption between multiple solutes at G–L and L–L interfaces is an important effect in foaming, foam stabilization, and defoaming (see DEFOAMERS).

Adsorption at Solid–Liquid Interface. Practical applications of surfactants usually involve some manner of surfactant adsorption on a solid surface. This adsorption is always associated with a decrease in free-surface energy, the magnitude of which must be determined indirectly. The force with which the adsorbate is held on the adsorbent may be roughly classified as physical, ionic, or chemical. Physical adsorption is a weak attraction caused primarily by van der Waals forces. Ionic adsorption occurs between charged sites on the substrate and oppositely charged surfactant ions, and is usually a strong attractive force. The term chemisorption is applied when the adsorbate is joined to the adsorbent by covalent bonds or forces of comparable strength. The surface condition of a solid adsorbent markedly affects its adsorption characteristics. Important considerations include smoothness, cleanliness, particle size, packing of powders, and presence of capillary systems. Typical substrates on which adsorption effects are important include metals, glass, plastics, textile fibers, sand, crushed minerals, plant foliage, paper, and the mixed solid dirt that collects on clothing, linen, walls, and floors. Analysis of the shape of adsorption isotherms has been the technique most widely used in investigations of adsorption mechanisms, but their interpretation has not been established for the numerous combinations of adsorbent–adsorbate and interactions in multicomponent surfactant systems (see ADSORPTION, LIQUID SEPARATION).

Physical and ionic adsorption may be either monolayer or multilayer (12). Capillary structures in which the diameters of the capillaries are small, ie, one to two molecular diameters, exhibit a marked hysteresis effect on desorption. Sorbed surfactant solutes do not necessarily cover all of a solid interface and their presence does not preclude adsorption of solvent molecules. The strength of surfactant sorption generally follows the order cationic > anionic > nonionic. Surfaces to which this rule applies include metals, glass, plastics, textiles (13), paper, and many minerals. The pH is an important modifying factor in the adsorption of all ionic surfactants but especially for amphoteric surfactants which are least soluble at their isoelectric point. The speed and degree of adsorption are increased by the presence of dissolved inorganic salts in surfactant solutions (14).

The state of the adsorbed layer has been studied by electron diffraction, x-ray diffraction, interferometer, and electron microscope. The contact angle of pure liquids on solid surfaces, ie, wetting behavior, has been used effectively in investigations of adsorbed monolayer properties (15–17). Adsorption can be demonstrated visually by dipping a clean glass plate, that is readily water wettable, into a solution of a cationic surfactant. On withdrawal the plate is water repellent because of the orientation of the adsorbed layer. When a complete and closely packed film of surfactant molecules or ions is adsorbed on a solid, the substrate no longer exerts any effect on the contact angle. Preferential adsorption or competition between surfactant solutes for the surface of solid substrates is an important effect and the basis of preferential wetting.

Micelles.   Surfactant molecules or ions at concentrations above a minimum value characteristic of each solvent-solute system associate into aggregates called micelles. The formation, structure, and behavior of micelles have been extensively investigated. The term critical micelle concentration (CMC) denotes the concentration at which micelles start to form in a system comprising solvent, surfactant, possibly other solutes, and a defined physical environment. If the properties of a surfactant solution are plotted as a function of the concentration of the surfactant, the properties usually vary linearly with increasing concentration up to the CMC at which point there is a break in the curve (18,19). This effect is illustrated schematically for aqueous solutions of sodium lauryl sulfate in Figure 4. Experimental data on this compound have been reported in a simi-

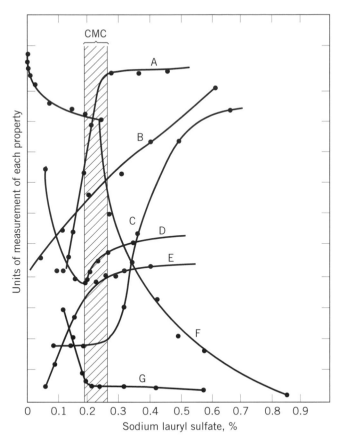

Fig. 4. Physical property curves for aqueous sodium lauryl sulfate where A is deter-gency; B, density change; C, conductivity (high frequency); D, surface tension; E, osmotic pressure; F, equivalent conductivity; and G, interfacial tension.

lar composite chart. Other properties have also been used for the determination of CMC, eg, refractive index, light scattering, dialysis, and dye solubilization. Although most studies of micellar size and properties have involved aqueous systems, nonaqueous systems have received some attention (20–24). The ori-entation of surfactant ions in an aqueous solution with the hydrophilic moieties exposed to water is shown in Figure 3. In a nonaqueous system, the orientation is reversed and the lipophilic moieties are exposed to the organic solvent. Although the CMC of surfactants in aqueous solutions depends on the structure of the compounds and the environment, for many anionic surfactants at low electrolyte concentrations and room temperature it is 10^{-3} to 10^{-2} mol/L (25); for nonionic surfactants under comparable conditions it is less (10^{-5} to 10^{-4} mol/L) (26).

Micelle size is expressed as the micellar molecular weight or, more gen-erally, the aggregation number, ie, the number of monomers making up the micelles. Micellar aggregation numbers generally lie between 20 and 100, for single-chain anionic and cationic surfactants (27). Large aggregation numbers (>1000) have been reported for nonionic micelles (28), especially as the cloud point is approached. Assuming that the radius of a micelle is just short of the extended monomer chain length, and that the interior of the micelle is filled with the monomer chain, the following relationships were calculated between aggregation number, m, and carbon chain length, n, for surfactants composed of paraffin (hydrocarbon) chains with simple anionic head groups (29):

n	12	14	16	18
m	33	46	60	78

In a surfactant series where the hydrophilic group is constant, but the size, ie, length, of the hydrophobic group is increased, CMC values decrease for both ionic and nonionic surfactants. At constant hydrophobe size, CMC values decrease with decreasing ethylene oxide content of nonionic surfactants. Increasing the electrolyte concentration lowers the CMC for both anionic and nonionic surfactants, especially for the former. The CMC of anionic surfactants increases with temperature, whereas the CMC of the nonionic surfactants decreases with temperature. Since the mid-1950s, micellar properties of homogeneous nonionic surfactants have been studied and in some cases compared with polydisperse nonionic surfactants (30). A quantative structure–property relationship has been successfully used to predict CMC's of a series of nonionic surfactants (31). A theoretical treatment to describe micellization of mixed nonionic systems has been proposed (32). Under the assumption of ideal mixing and a phase separation model, the theory gives excellent prediction for surface tensions and the CMC of the mixtures. The CMC and other micellar properties of homogeneous and polydisperse nonionic surfactants of comparable composition are similar. A detailed thermodynamic treatment of solutions of ideal multicomponent micelles has been described (33) (see THERMODYNAMICS). In principle, this treatment can be used to predict monomer concentration, CMC, and other micellar properties. However, its practical application presents substantial difficulties. Extensive CMC values for many surfactants under a variety of conditions have been compiled (34,35).

Small micelles in dilute solution close to the CMC are generally believed to be spherical. Under other conditions, micellar materials can assume structures such as oblate and prolate spheroids, vesicles (double layers), rods, and lamellae (36,37). All of these structures have been demonstrated under certain conditions, and a single surfactant can assume a number of structures, depending on surfactant, salt concentration, and temperature. In mixed surfactant solutions, micelles of each species may coexist, but usually mixed micelles are formed. Anionic-nonionic mixtures are of technical importance and their properties have been studied (38,39).

Micellar properties are affected by changes in the environment, eg, temperature, solvents, electrolytes, and solubilized components. These changes include complicated phase changes, viscosity effects, gel formation, and liquefication of liquid crystals. Of the simpler changes, high concentrations of water-soluble alcohols in aqueous solution often dissolve micelles and in nonaqueous solvents addition of water frequently causes a sharp increase in micellar size.

Measurement of Surface Activity. Each surface-active property can be measured in a variety of ways and the method of choice depends on the characteristics of the substance to be tested. The most frequently determined properties are surface tension (γ_{SG}, γ_{LG}), interfacial tension (γ_{LL}, γ_{LG}), contact angle (θ), and CMC.

Surface and Interfacial Tension. Theories and experimental methods for measuring these properties have been reviewed (40). The du Nouy tensiometer, ie, the ring method, is probably the most widely used instrument for the determination of static surface and interfacial tension (41). Other frequently used methods include drop weight (42), pendent drop (43), maximum bubble pressure (44), capillary rise, and sessile drops. Dynamic surface tension can be measured by a speedup of the bubbling rate in the maximum bubble-pressure method (45) or the change in wavelength and form when a jet of liquid issues from an elliptical orifice (46). Dynamic methods provide information on the time required for surfactant molecules to migrate to the interface. Interfacial tensions can also be measured by the spinning-drop method (47). This method is especially effective for measuring low interfacial tensions.

Contact Angle. A number of methods for direct determination of the contact angle have been proposed, but these measurements are difficult and the results not always reproducible. Roughness of the solid surface and the difference between advancing and receding angle lowers measurement precision. The tilted plate is the simplest and most direct method for this determination (48). Contact angles can also be measured by observing single oil drops on solid surfaces (17) or on single fibers (49).

Critical Micelle Concentration. The rate at which the properties of surfactant solutions vary with concentration changes at the concentration where micelle formation starts. Surface and interfacial tension, equivalent conductance (50), dye solubilization (51), iodine solubilization (52), and refractive index (53) are properties commonly used as the basis for methods of CMC determination.

Anionic Surfactants

Carboxylate, sulfonate, sulfate, and phosphate are the polar, solubilizing groups found in most anionic surfactants. In dilute solutions of soft water, these groups are combined with a 12–15 carbon chain hydrophobe for best surfactant properties. In neutral or acidic media, or in the presence of heavy-metal salts, eg, Ca, the carboxylate group loses most of its solubilizing power.

Of the cations (counterions) associated with polar groups, sodium and potassium impart water solubility, whereas calcium, barium, and magnesium promote oil solubility. Ammonium and substituted ammonium ions provide both water and oil solubility. Triethanolammonium is a commercially important example. Salts (anionic surfactants) of these ions are often used in emulsification. Higher ionic strength of the medium depresses surfactant solubility. To compensate for the loss of solubility, shorter hydrophobes are used for application in high ionic-strength media. The U.S. shipment of anionic surfactants in 1993 amounted to 49% of total surfactant production.

Carboxylates. Soaps represent most of the commercial carboxylates. The general structure of soap is $RCOO^- M^+$, where R is a straight hydrocarbon chain in the C_9–C_{21} range and M^+ is a metal or ammonium ion. Interruption of the chain by amino or amido linkages leads to other structures which account for the small volumes of the remaining commercial carboxylates.

Large volumes of soap are used in industrial applications as gelling agents for kerosene, paint driers, and as surfactants in emulsion polymerization (see DRIERS AND METALLIC SOAPS; EMULSIONS; SOAP). The first laundry powders developed were tallow soap-based, which showed excellent detergency in soft water. In the presence of polyvalent metal ions, eg, Ca^{2+}, Mg^{2+}, Ba^{2+}, or Fe^{3+}, in wash water, soap reacts to form lime soap or curds which are unsightly, reduce soap concentration, and adversely affect performance. Incorporation of lime soap dispersants disperses lime soap to near invisibility was one approach toward solvent this problem and a method for rating lime soap dispersion efficacy has been published (54). Another approach was to build soap powders with sodium carbonate or silicate. In the 1940s, neither approach had proven sufficiently effective or economical to prevent replacement of soap powders by syndets, a combination of synthetic surfactants, ie, alkylbenzenesulfonate, and builders, ie, pentasodium tripolyphosphate. When concern over water eutrophication resulted in a ban of phosphorus (qv) in laundry detergents, the possibility of returning to laundry powders based on a combination of soap and lime soap dispersants received much attention, particularly by the USDA (55). However, phosphate builders have been effectively replaced by combinations of zeolite, citrate, and polymers, coupled with rebalanced synthetic active systems, therefore soaps have not regained their former prominence. Nonetheless soap itself is generally present as a minor component of surfactants to control suds profile, reduce dye transfer, control powder properties, and act as a cosurfactant or cobuilder.

Carboxylates with a fluorinated alkyl chain are marketed by the 3M/Industrial Chemical Products Company under the trade name Fluorad surfactants. They also include other functional derivatives of fluorinated and perfluorinated alkyl chains. Replacement of hydrogens on the hydrophobe by fluorine atoms leads to surfactant molecules of unusually low surface tension. This property imparts excellent leveling effectiveness.

Polyalkoxycarboxylates. These surfactants are produced either by the reaction of sodium chloroacetate with an alcohol ethoxylate:

$$ROH + n \ CH_2\!-\!CH_2 \longrightarrow R(OC_2H_4)_nOH$$

$$R(OC_2H_4)_nOH + ClCH_2COONa \longrightarrow R(OC_2H_4)_nOCH_2COOH + NaCl$$

or from an acrylic ester and an alcohol alkoxylate.

$$ROH + n \ CH_2\!-\!CHR' \longrightarrow R(OC_2H_3R')_nOH$$

$$R(OC_2H_3R')OH + CH_2\!\!=\!\!CHCOOR'' \xrightarrow{\text{NaOH}} R(OC_2H_3R')_nO(CH_2)_2COONa + R''OH$$

These can also be successfully produced by direct oxidation of alcohol ethoxylates under carefully controlled conditions (56). Because of the presence of the ethylene oxide linkages, these products possess a higher aqueous solubility which manifests itself in greater compatibility with cationic surfactants and polyvalent cations. These materials are carboxylates analogous of alcohol ethoxy sulfates and structure-activity relationships for these alcohol ethoxy carboxylates parallel the analogous sulfates to a degree, finding applications as lime soap dispersants, wetting agents, and detergents in cosmetic formulations. Table 3 gives composition and properties of some representative commercial products.

N-Acylsarcosinates. Sodium N-lauroylsarcosinate [7631-98-3] is a good soap-like surfactant. Table 4 gives trade names and properties. The amido group in the hydrophobe chain lessens the interaction with hardness ions. N-Acylosarcosinates have been used in dentifrices (qv) where they are claimed to inactivate enzymes that convert glucose to lactic acid in the mouth (57). They are prepared from a fatty acid chloride and sarcosine:

$$RCOCl + CH_3NHCH_2COONa + NaOH \longrightarrow RCON(CH_3)CH_2COONa + NaCl + H_2O \quad (10)$$

Acylated Protein Hydrolysates. These surfactants are prepared by acylation of protein hydrolysates with fatty acids or acid chlorides. The hydrolysates are variable in composition, depending on the degree of hydrolysis. Collagen from leather (qv) processing is a common protein source. Acylated protein hydrolysates (Maypon, by Inotex Chemical Company) are mild surfactants recommended for personal-care products (see COSMETICS).

Sulfonates. The sulfonate group, $-SO_3M$, attached to an alkyl, aryl, or alkylaryl hydrophobe, is a highly effective solubilizing group. Sulfonic acid surfactants are strong, their salts are relatively unaffected by pH, they are stable to oxidation, and because of the strength of the C–S bond also stable to hydrolysis.

Table 3. Polyethoxycarboxylates, $R(OCH_2CH_2)_nCH_2COOM$

Trade name	R	n	M	Functions[a]
Neodox	$C_{12}-C_{13}$	7	Na	D,W
Sandopan DTC	tridecyl[b]	7	Na	L,W
Sandopan KST	cetyl	12	Na	D,L,W

[a]See Table 1.
[b]Branched chain.

Table 4. N-Acylsarcosinates, $RCON(CH_3)CH_2COOM$

Trade name	RCO–	M	Conc, wt %	Functions[a]	Manufacturer[b]
Hamposyl C	cocoyl	H	100	C,D,E,eP,Fs	GWR
Hamposyl C-30	cocoyl	Na	30	C,D,E,F,Fs	GWR
Hamposyl L	lauroyl	H	94	C,D,E,eP,Fs	GWR
Hamposyl L-30	lauroyl	Na	30	C,D,E,F,Fs	GWR
Hamposyl 0	oleoyl	H	94	C,D,E,eI,Fs	GWR
Maprosyl 30	lauroyl	Na	30	C,D,E,F	STP
Sarkosyl LC	cocoyl	H	94	C,D,E,F	CGY
Sarkosyl 0	oleoyl	H	94	C,D,E	CGY

[a]See Table 1.
[b]See Table 2.

Sulfonates interact moderately with the hardness, Ca^{2+} and Mg^{2+}, but significantly less so than carboxylates. Sulfates can be tailored for specific applications by introduction of double bonds or ester or amide groups, either into the hydrocarbon chain or as substituents. Because the introduction of the SO_3H function is inherently inexpensive, eg, by oleum, SO_3, SO_2Cl_2, or $NaHSO_3$, sulfonates are heavily represented among high volume surfactants (see SULFONATION AND SULFATION).

The first sulfonate was prepared from paraffin and SO_2Cl_2 by the Reed reaction (eq. 11) (58,59):

$$RH + SO_2 + Cl_2 \xrightarrow{uv} RSO_2Cl + HCl \xrightarrow[H_2O]{3\ NaOH} RSO_3Na + 2\ NaCl + 2\ H_2O \qquad (11)$$

Workup of the complex reaction mixture was difficult and expensive. As a result, these surfactants have been withdrawn from the U.S. market. In Europe, Hoechst AG produces secondary alkanesulfonates by a similar reaction (eq. 12):

$$RCH_2CH_3 + SO_2 + 1/2\ O_2 \xrightarrow{uv} RCH(CH_3)SO_2OH \xrightarrow{NaOH} RCH(CH_3)SO_3Na + H_2O \qquad (12)$$

These products are marketed under the trade name Hostapur. Hostapur SAS 60, a 60% active paste, is used at low levels in some European liquid hand-dishwashing products.

Alkylbenzenesulfonates. The alkylbenzenesulfonates (ABS) have the largest production volume and usage of the non-soap surfactants. In the 1940s ABS, a branched-chain alkylbenzenesulfonate, began to replace soap as the principal surfactant in U.S. household laundry products. The C_{12} ABS was derived from benzene and a propylene tetramer. The methyl branching in the alkylchain causes the ABS to be only slowly biodegradable. By the 1960s ABS volume had grown to a point that significant environmental buildup occurred and caused foam blankets on bodies of water. At the same time, commercialization of the Ziegler process for ethylene oligomerization provided a relatively economic route to straight chain hydrophobes that could be converted to detergents, alcohols, nonionic surfactants, and linear alkylbenzenesulfonates (LAS). By 1965, the U.S. detergent industry converted from hard (branched) to soft (linear) surfactants at an estimated cost of 150×10^6. Although LAS has replaced ABS in U.S. consumer products, the latter continues to be used in emerging and developing countries because of more attractive economics. However, here too it is being gradually phased out and replaced by LAS.

Alkylbenzenesulfonic acids are light-colored, viscous liquids whose alkali and ammonium salts are soluble in water. The salts are neutral, relatively unaffected by water hardness, and stable toward alkaline and acid hydrolysis. Calcium and magnesium alkylbenzenesulfonates are less water-soluble than sodium or ammonium salts. They are, however, soluble in hydrocarbons and are used extensively in these media. LAS salts are stable to oxidation and can be incorporated in formulations containing oxidizing agents. Pure alkylbenzenesulfonates have been synthesized for property studies (60,61).

Alkylbenzenesulfonates are effective surfactants, which respond well to builders and foam boosters in detergent formulations. These properties, together with low cost, availability, and consistent quality, account for their dominant position in household laundry products (62,63).

Alkylbenzenesulfonates are manufactured in a three-step process:

$$RR'CHCl + C_6H_6 \xrightarrow[\text{or AlCl}_3]{\text{HF}} RR'CHC_6H_5 + HCl \tag{13}$$

or

$$R'CH{=}CHR + C_6H_6 \xrightarrow[\text{or AlCl}_3]{\text{HF}} R'CH(CH_2R)C_6H_5 \tag{14}$$

$$RR'CHC_6H_5 \xrightarrow[\text{or oleum}]{\text{SO}_3/\text{air}} RR'CHC_6H_4SO_3H \tag{15}$$

$$RR'CHC_6H_4SO_3H \xrightarrow{\text{NaOH}} RR'CHC_6H_4SO_3Na \tag{16}$$

The product from reactions 13 or 14 is a secondary alkylbenzene in which the point of attachment of the benzene ring to the alkyl chain is located variably among the internal chain positions (see also ALKYLATION). The relation of performance properties to composition, ie, distribution of the benzene ring along the alkyl chain, has been investigated (64). U.S. production of alkylbenzenesulfonates in 1993 (344,000 metric tons) was slightly higher than in 1980 (294,000 t). Representative commercial alkylbenzenesulfonates are given in Table 5.

Short-Chain Alkylarenesulfonates. This group of sulfonates includes the sodium and ammonium salts of toluene-, xylene-, and isopropylbenzenesulfonic acids. The alkyl chain in these compounds is too short to confer surface activity. Industrially, however, these materials are important in their function as coupling agents, solubilizers, or hydrotropes. As such, they are widely used in liquid detergent compositions, for hand-dishwashing, laundering, and in hard surface cleaners where high concentrations of organic surfactants or inorganic compounds need to be kept in solution in the finished product.

Lignosulfonates. Lignosulfonates are by-products of the paper industry where they are manufactured by reaction of SO_2 and $Ca(HSO_3)_2$ with wood pulp (see LIGNIN; PAPER; WOOD). These are produced in large volume and find application as inexpensive dispersing agents, ore flotation agents, in oil-well mud drilling formulations, and in pelletizing animal feed products (see FEED AND FEED ADDITIVES). Lignosulfonates do not reduce surface tension to the extent that surfactants do, nor do they form oriented monolayers or micelles (65).

Naphthalenesulfonates. These surfactants are manufactured by sulfonation and neutralization of naphthalene, tetrahydronaphthalene, alkylnaphthalenes, and formaldehyde–naphthalene condensation products. The alkali metal salts are stable to hydrolysis and oxidation. They are highly water-soluble and hence are not strongly surface-active in soft water. Naphthalenesulfonates find widespread application as dispersants, wetting agents, stabilizers, and suspending agents for latex, carbon black, pigments, and other disperse systems.

Table 5. Alkylbenzenesulfonates, $RC_6H_4SO_3M^a$

Trade name	M^b	Conc, wt %	Functions[c]	Manufacturer[d]
Bio-Soft D-40	Na	40	D,E,F,W	STP
Bio-Soft D-62	Na	60	D,E,F,W	STP
Bio-Soft N-300	TEA	60	D,E,F,W	STP
Calsoft LAS-99[e]	H	98	D,E,I	PIL
DDBSA 99-B[f]	H	97	D,I	MON
Emkal NOBS[g]	Na	40	D	EMK
Naxel AAS-40S	Na	40	D	RNC
Naxel AAS-45S	Na	40–43	D	RNC
Naxel AAS-60S	TEA	60	D	RNC
Rhodacal 70/B	Ca	70	E	RHP
Rhodacal DS 10	Na	98	D	RHP
SA 597	H	97	D,I	VIS
S 697[h]	H	97	D,I	VIS
Sulfonate AAS-35S	Na	31.5	W	CLA
Sulfonate AAS-40FS	Na	40.0	W	CLA
Sulfonate AAS-45S	Na	41–43	F	CLA
Sulfonate AAS-60S	TEA	60	D,W	CLA
Sulfonate AAS-50MS	AM	40	W	CLA
Sulfonate AAS-75S	Ca	60	W	CLA
Sulfonate AAS-90	Na	85	D,W	CLA
Sulfonate AAS-70S	Naaa	60	W	CLA
Stepantan DS-40	Na	40	W,E,F	STP
Stepantan H-100	H	97	I,D,E,F,OR	STP
Sulfonate AA9	Na	90	W	BIT
Sulfonic Acid LS	H	96	I	HCL
Vista C-550	Na	50	D	VIS
Vista C-560	Na	60	D	VIS
Witconate 30DS	Na	30	D,I	WTC
Witconate 45BX	Na	42	D,I	WTC
Witconate 45DS	Na	43	D,F,W	WTC
Witconate 60B	Na	60	F,S,D	WTC
Witconate 60L	Ca	60	E	WTC
Witconate 60T	TEA	58	E,D	WTC
Witconate 79S	TEA	52	D,Di,E	WTC
Witconate 90F	Na	91	D,T,A	WTC
Witconate 90FH[i]	Na	91	D	WTC
Witconate 93S	amine	91	Di,E	WTC
Witconate 1298	H	97	I	WTC
Witconate 1298 H[i]	H	97	I	WTC
Witconate P1059[i]	AM	91	D	WTC
Witconate 1238	Na	39	D,F,W	WTC
Witconate 1250	Na	51	D,F,W	WTC
Witconate SK	Na	40	D,F	WTC

[a] R = dodecyl unless otherwise noted.
[b] TEA = triethanolammonium; AM = ammonium.
[c] See Table 1.
[d] See Table 2.
[e] R = alkyl.
[f] R = C_{10}–C_{13}.
[g] R = nonyl.   [h] R = tridecyl.   [i] R is branched.

Condensation products find extensive use in tanning leather (qv). Commercially available naphthalenesulfonates are given in Table 6.

α-Olefinsulfonates. α-Olefinsulfonates (AOS) are produced by reaction of SO_3–air with an α-olefin followed by intermediate neutralization of sultone intermediates. In its properties, AOS is generally similar to LAS. It is reported to be somewhat milder to skin and to biodegrade slightly more readily. Because of the simplicity and low cost of starting materials, α-olefinsulfonates have been of great interest to surfactant producers for many years. The principal applications include liquid soap formulations and cosmetic products. Commercial AOS is a multicomponent mixture in which the double bond is located at a number of positions along the chain. It contains significant proportions of hydroxyalkanesulfonate isomers in which the hydroxyl group is similarly distributed along the carbon chain. Representative commercial AOS products are shown in Table 7.

Petroleum Sulfonates. Petroleum sulfonates originally were produced as by-products of petroleum refining. Expanding usage and increasing complexity of specifications have caused petroleum sulfonates to be considered coproducts from the refining of certain petroleum fractions. Reactions of high boiling feedstock with oleum yields a crude acid mixture that separates into an upper brown, mahogany layer and a greenish, lower layer. Historically, the names mahogany acids and green acids originate with the color of these layers.

Green acids and products derived from them are high in water-soluble polysulfonic acids of low economic value. They are often burned to recover SO_3. Mahogany acids are predominantly monosulfonated and represent products of considerable value including the salts. When expanding usage led to a shortage of traditional petroleum sulfonates, synthetic alkylates such as detergent alkylate bottoms, were sulfonated to provide synthetic petroleum sulfonates. These account for a significant fraction of petroleum sulfonate production in the 1990s.

Petroleum sulfonates are widely used as solubilizers, dispersants (qv), emulsifiers, and corrosion inhibitors (see CORROSION AND CORROSION INHIBITORS).

Table 6. Naphthalenesulfonates, $RC_{10}H_6SO_3Na$

Trade name	R or description	Conc, wt %	Functions[a]	Manufacturer[b]
Aerosol OS	isopropyl	75	Di,W	ACY
Alkanol XC	solid	90	Di,Ta,W	DUP
Daxad II	CH_2O condensed	87	Di,E	GWR
Emkal NNS	nonyl	35	D,Di,E,W	EMK
Petro BA	alkyl	50	D,S,W	WTC

[a]See Table 1.
[b]See Table 2.

Table 7. α-Olefinsulfonates, $RCH=CHSO_3Na$

Trade name	R	Conc, wt %	Functions[a]	Manufacturer[b]
Bio-Terge AS-40	$C_{14}-C_{16}$	40	D,W,F	STP
Carsonol AOS-40	$C_{14}-C_{16}$	40	D,W	LNZ
Witconate AOS	$C_{10}-C_{12}$	40	D,W	WTC

[a]See Table 1.
[b]See Table 2.

More recently, they have emerged as the principal surfactant associated with expanding operations in enhanced oil recovery (66). Alkaline-earth salts of petroleum sulfonates are used in large volumes as additives in lubricating fluids for sludge dispersion, detergency, corrosion inhibition, and micellar solubilization of water. The chemistry and properties of petroleum sulfonates have been described (67,68). Principal U.S. manufacturers include Exxon and Shell, which produce natural petroleum sulfonates, and Pilot, which produces synthetics.

Sulfonates with Ester, Amide, or Ether Linkages. Dialkyl Sulfosuccinates. Introduced in 1939 by the American Cyanamid Company under the Aerosol trademark, dialkyl sulfosuccinates have become widely used specialty surfactants (Table 8) (9). Within the limitations in hydrolytic stability imposed by the presence of the ester groups, sulfosuccinates are mild, versatile surfactants used when strong wetting, detergency, rewetting, penetration, and solubilization effectiveness is needed.

The symmetrical diesters are produced by esterification of maleic anhydride, followed by addition of sodium bisulfite across the double bond (see SUCCINIC ACID AND SUCCINIC ANHYDRIDE):

$$
\begin{array}{c}
\overset{\displaystyle O}{\overset{\displaystyle \|}{\text{ROCCH}}} \\
\underset{\displaystyle \|}{\underset{\displaystyle O}{\text{ROCCH}}}
\end{array}
+ \text{NaHSO}_3 \longrightarrow
\begin{array}{c}
\overset{\displaystyle O}{\overset{\displaystyle \|}{\text{ROCCHSO}_3\text{Na}}} \\
\underset{\displaystyle \|}{\underset{\displaystyle O}{\text{ROCCH}_2}}
\end{array}
$$

Amidosulfonates.   Amidosulfonates or *N*-acyl-*N*-alkyltaurates, are derived from taurine, $H_2NCH_2CH_2SO_3Na$, and are effective surfactants and lime soap dispersants (Table 9). Because of high raw material cost, usage is relatively small. Technically, amidosulfonates are of interest because they are stable to hydrolysis, unaffected by hard water, and compatible with soap. They have been used in soap–surfactant toilet-bar formulations. With shorter, acyl groups, they make excellent wetting agents.

Table 8. Dialkyl Sulfosuccinates, ROOCCH$_2$CH(SO$_3$Na)COOR

Trade name	R	Conc, wt %	Functions[a]	Manufacturer[b]
Aerosol AY-100	amyl	100	Di,E,W	ACY
Aerosol OT-100	octyl	100	E,W	ACY
Alrowet D-65	octyl	65	Ta,W	CGY
Emcol 4500	octyl	70	D,S,W	WTC
Monawet MB-45	isobutyl	45	W	MOA
Monawet MT-70	tridecyl[c]	70	Di,eP,W	MOA
Geropon WT 27	octyl	70	E,rW,W	RHP
Schereowet DOS-70	octyl	70	E,W	SBC
Tex-Wet 1001	octyl	60	rW,Ta,W	TCM

[a]See Table 1.
[b]See Table 2.
[c]Branched.

Table 9. Sodium *N*-Acyl-*N*-Akyltaurates, RCONCH$_3$CH$_2$CH$_2$SO$_3$Na

Trade name	RCO–	Conc, wt %	Functions[a]	Manufacturer[b]
Geropon TC42	cocoyl	24	D,Di,F	RHP
Geropon T77	oleoyl	67	D,Di,F,W	RHP
Geropon TK32	tall oil	20	D,Di,Su	RHP
Tergenol S. Liq.	oleoyl	32	D,Ta	HRT

[a]See Table 1.
[b]See Table 2.

Sodium *N*-oleoyl-*N*-methyltaurate [*137-20-2*] is produced in three high yield steps:

$$H_2C\!-\!CH_2 + NaHSO_3 \longrightarrow HOCH_2CH_2SO_3Na$$
$$\overset{\diagdown \diagup}{O}$$

$$CH_3NH_2 + HOCH_2CH_2SO_3Na \longrightarrow CH_3NHCH_2CH_2SO_3Na + H_2O$$

$$C_{17}H_{33}COCl + CH_3NHCH_2CH_2SO_3Na \xrightarrow{NaOH} C_{17}H_{33}CON(CH_3)CH_2CH_2SO_3Na + HCl$$

2-Sulfoethyl Esters of Fatty Acids. Like *N*-acyl-*N*-alkyltaurates, these compounds, also known as acyl isethionates, are excellent surfactants, mild to skin, and unaffected by hardness ions. The presence of an ester linkage limits hydrolytic stability. This limitation, combined with relatively high raw material costs, has prevented widespread usage of the 2-sulfoethyl esters in consumer detergent products. The cocoyl derivative, however, has been the principal ingredient in detergent bars for personal use, and is finding increasing use in other personal wash applications.

2-Sulfoethyl esters are produced commercially in two steps:

$$\overset{O}{\overset{\diagup \diagdown}{CH_2\!-\!CH_2}} + NaHSO_3 \longrightarrow HOCH_2CH_2SO_3Na$$

$$RCOOCH_2CH_2SO_3Na + HCl \longrightarrow RCO_2CH_2CH_2SO_3Na$$

where R = linear long chain.

The viscous fatty acid chloride is mixed with solid sodium isothionate and is heated under vacuum and agitation without water or solvent; HCl is evolved, leaving the product as a finely divided powder. The direct esterification of fatty acid and sodium isethionate has been patented (69).

Sulfonates with ether linkages include ring-sulfonated alkylphenol ethoxylates and a disulfonated alkyldiphenyl oxide, Dowfax 2A1, and 3B2 (Dow Chemical Company). This surfactant is characterized by high solubility in salt solutions, strong acids, or bases. It is used in industrial and institutional cleaners.

Fatty Acid Ester Sulfonates. Fatty acid ester sulfonates (FAES) are generally produced from methyl esters, ie, methyl ester sulfonate (MES) and prepared via sulfonation, followed by bleaching and neutralization, in a relatively difficult and complex process:

$$
\underset{\substack{\text{O} \\ \parallel}}{\text{RCH}_2\text{COCH}_3} \xrightarrow{\text{SO}_3} \underset{\substack{\text{SO}_3\text{H} \\ | \;\; \text{O} \\ \;\;\; \parallel}}{\text{RCHCOCH}_3} + \underset{\substack{\text{SO}_3\text{H} \\ | \;\; \text{O} \\ \;\;\; \parallel}}{\text{RCHCOSO}_3\text{CH}_3} \xrightarrow[\text{H}_2\text{O}_2]{\text{CH}_3\text{OH}} \xrightarrow{\text{NaOH}} \underset{\substack{\text{SO}_3^-\text{Na}^+ \\ | \;\; \text{O} \\ \;\;\; \parallel}}{\text{RCHCOCH}_3} + \underset{\substack{\text{SO}_3^- \\ | \;\; \text{O} \\ \;\;\; \parallel}}{\text{RCHCO}^- 2\text{Na}^+}
$$

Excess methanol is added during the bleaching step to minimize formation of sodium fatty acid sulfonate (SFAS) (70).

Historically, these sulfonates have been difficult to produce in good quality and to formulate into laundry powders because of thermal and hydrolytic stability problems. Used basically as an anionic, oleochemically sourced replacement for LAS, FAE sulfonate benefits include good detergency at low concentration, low environmental load, and good supply of high quality material (71). To avoid possible thermal stability problems, these sulfonates have found application in liquid products. They are used in at least one U.S. dishwashing liquid, and the cocoyl derivative, Alpha-Step MC48, is marketed by Stepan Chemical Company.

Sulfates and Sulfated Products. The sulfate group, $-\text{OSO}_3\text{M}$, where M is a cation, represents the sulfuric acid half-ester of an alcohol and is more hydrophilic than the sulfonate group because of the presence of an additional oxygen atom. Attachment of the sulfate group to a carbon atom of the hydrophobe through the $\text{C}-\text{O}-\text{S}$ linkage limits hydrolytic stability, particularly under acidic conditions. Usage of sulfated alcohols and sulfated alcohol ethoxylates has expanded dramatically since the 1970s as the detergent industry reformulates consumer products to improve biodegradability, lower phosphate content, and move from powder to liquid.

Alcohol Sulfates. Commercial production of alcohol sulfates started in Europe in the 1930s and products have become well established in specialty markets. The hydrophobes are predominantly of petrochemical origin although in the 1990s some shift is evident to oleochemical hydrophobes obtained by reduction of fatty acids and esters. Secondary alcohol sulfates have never been used extensively in the United States. Alcohol sulfates from straight-chain primary alcohols are similar in performance properties, feel, and emolliency to soaps of corresponding chain lengths (see also ALCOHOLS, HIGHER ALIPHATIC). Increasing chain length requires higher temperatures to attain maximum detergency and wetting effects. Physical and interfacial properties of a series of purified alcohol sulfates have been described (72). Commercial alcohol sulfates are sold as nearly colorless solutions of concentrations 25–40% or as light-colored pastes. Purified sodium lauryl sulfate of activity 90–99% is available, primarily for use in dentifrices (qv).

Alcohol sulfates are stable to hard water. Magnesium lauryl sulfate forms voluminous foam with low water content. This is particularly useful in rug shampoos where soil is removed by vacuum pickup of foam generated by vigorous brushing with a minimum of detergent solution. Sensitivity to hydrolysis in hot alkaline and acidic media is one of the main disadvantages of alcohol sulfates. Nonetheless, versatility under use conditions is attested to by the variety of different salts offered commercially (Table 10). Principal applications include sham-

Table 10. Alkyl Sulfates, ROSO$_3$M^a

Trade name	M^b	Conc, wt %	Functions^c	Manufacturer^d
Avirol SL 2010	Na	30	D,Di,E,W	HNC
Avirol SA 4106^e	Na	45	E,Di,W	HNC
Avirol SA 4110^f	Na	30	E	HNC
Conco Sulfate A	NH$_4$	30	D	CTL
Conco Sulfate P	K	30	D	CTL
Conco Sulfate EP	DEA	35	D,W	CTL
Conco Sulfate WR Dry	Na	90	D,W	CTL
Duponol C	Na		D	WTC
Equex S	N	29	D,Di,W	PG
Carsonol TLS 65	TEA	65	D,F	LNZ
Carsonol MLS	Mg	30	D,W	LNZ
Witcolate WAC-LA	Na	29	D,W	WTC
Witcolate AM	NH$_4$	28.5	D,W	WTC
Witcolate TLS500	TEA	40	D,W	WTC
Rhodapon OLS^g	Na	33	eP	RHP
Standapol A	NH$_4$	28	D,F	HNC
Stepanol WA 100	Na	97	D,Di,E,F,W	STP
Stepanol MG	Mg	29	D,F,W	STP
Stepanol AM	NH$_4$	28	D,Di,F,W	STP
Stepanol WAT	TEA	40	D,F	STP
Witcolate D-510^e	Na	40	P,W	WTC

^{a}R = lauryl unless otherwise noted.
^{b}DEA = diethanolammonium; TEA = triethanolammonium.
^{c}See Table 1.
^{d}See Table 2.
^{e}R = 2-ethylhexyl.
^{f}R = 2-decyl.
^{g}R = octyl.

poos, dentifrices, textile processing, emulsion polymerization, and increasing use in laundry products, especially in Europe.

Ethoxylated and Sulfated Alcohols. Since the 1960s these products have experienced dramatic growth. Most of the increase has come from expanded usage in consumer detergent products, resulting from expanding availability of relatively low cost primary straight-chain alcohols from petrochemical sources; superior biodegradability of sulfated alcohol ethoxylates compared to corresponding sulfated alkylphenol ethoxylates; necessity of hardness-insensitive surfactants in heavy-duty laundry detergents brought about by phosphate content reduction; expansion of reliable, inexpensive sulfation technology; and addition of less expensive ethylene oxide which lowers overall costs.

Ethoxylated alcohol sulfates have several advantages over alcohol sulfates including lower sensitivity to hardness with respect to foaming and detersive effectiveness, less irritation to skin and eyes, and higher water solubility.

Sulfation of alcohol ethoxylates is carried out in high yields with oleum, chlorosulfonic acid, sulfamic (amidosulfuric) acid, and gaseous sulfur trioxide. The last is the preferred reagent. It is usually generated *in situ* by burning elemental sulfur (see SULFONATION AND SULFATION). In particular, SO$_3$ technology has displaced the use of sulfamic acid, an agent uniquely suited for the preparation of ammonium salts which are widely used in light-duty dishwashing formulations (see SULFAMIC ACID AND SULFAMATES).

$$RO{-}(CH_2CH_2O)_nH + H_2NSO_3H \longrightarrow RO{-}(CH_2CH_2O)_nSO_3NH_4$$

Sulfated alcohol ethoxylates generally contain from 10 to 40% ethylene oxide calculated on the weight of the starting alcohol. These are offered commercially as light-colored, odorless liquids containing 30–70% of the active ingredients. At high concentrations, sulfated alcohol ethoxylates become viscous and are diluted with ethanol for easier handling. The products are stable and compatible with builders in liquid and solid formulations. At very high alkalinity and in acid media, the sulfate linkage is susceptible to hydrolysis, particularly at elevated temperatures. In addition to light-duty liquid formulation, shampoos are widely formulated with sulfated alcohol ethoxylates. Their usage in laundry detergents has dramatically increased since the early 1980s, driven primarily by growth of heavy-duty liquids. Sulfated alcohol ethoxylates have replaced LAS to a significant degree in these applications, where hardness-tolerance, relative ease of formulation, and enzyme stability are more favorable. They also find application in textile processing and emulsion polymerization (Table 11) (72–74).

Table 11. Alcohols, Ethoxylated and Sulfated, R$-$(OCH$_2$CH$_2$$-$)$_nOSO_3$M^a

Trade name	R	M	Conc, wt %	Functions[b]	Manufacturer[c]
Alfonic 1412-A	$C_{12}-C_{14}{}^d$	NH$_4$	58	D,E,W	VIS
Alfonic 1412-55[e]	$C_{12}-C_{14}{}^d$	Na	38	D,E,W	VIS
Avirol SC-3002	lauryl	Na	29	eP	HNC
Calfoam NEl-60	lauryl	NH$_4$	57.5	D,E,F	PIL
Calfoam SEL-60	lauryl	Na	60	D,E,F	PIL
Calfoam SLS-30	lauryl	Na	30	D,E,F	PIL
Conco Sulfate 219	lauryl	Na	60	D,E,W	CTL
Neodol 25-3A	$C_{12}-C_{14}{}^f$	NH$_4$	59	D,E,F	SHC
Neodol 25-3S	$C_{12}-C_{14}{}^f$	Na	59	D,E,F	SHC
Rhodapex CD-128	lauryl	NH$_4$	56	D,Di,E,W	RHP
Richonol S-1285C	lauryl	Na	55	D,F	WTC
Richonol S-1300C	lauryl	NH$_4$	60	D,F	WTC
Richonol S-5260	lauryl	Na	27	D,F	WTC
Standapol ES-1	lauryl	Na	25	D	HNC
Standapol EA-3	lauryl	NH$_4$	27	D	HNC
Standapol EA-40	myristyl	NH$_4$	58	D	HNC
Steol CS-460	$C_{12}-C_{15}{}^d$	Na	60	D,E,F,W	STP
Steol CA-460	$C_{12}-C_{15}{}^d$	NH$_4$	60	D,E,F,W	STP
Sulfotex PAI	caprylyl/capryl	NH$_4$	46–50	D,E,F	HNC
Sulfotex PAI-S	caprylyl/capryl	Na	45–49	D,E,F	HNC
Witcolate AE-3	$C_{12}-C_{15}{}^d$	NH$_4$	59	D	WTC

$^a n = 3$ unless otherwise noted.
^{b}See Table 1.
^{c}See Table 2.
^{d}Linear.
$^e n = 5$.
^{f}Linear, primary.

Ethoxylated and Sulfated Alkylphenols. Because these alkylphenols degrade less readily than the sulfated alcohol ethoxylates, their anticipated expansion failed to materialize, although by 1965 they were widely used in retail detergent products. Sulfated alkylphenol ethoxylates are used in hospital cleaning products, textile processing, and emulsion polymerization. Sulfated alkyphenol ethoxylates are sold as colorless, odorless aqueous solutions at concentrations of ≥30%. The presence of ethylene oxide in the molecule increases resistance to hardness ions and reduces skin irritation. Representative commercial sulfated alkylphenol ethoxylates are given in Table 12.

Sulfated Acids, Amides, and Esters. Reaction with sulfuric acid may be carried out on fatty acids, alkanolamides, and short-chain esters of fatty acids. The disodium salt of sulfated oleic acid is a textile additive and an effective lime soap dispersant. A typical sulfated alkanolamide structure is $C_{11}H_{23}CONHCH_2CH_2OSO_3Na$. Others include the sulfates of mono and diethanolamides of fatty acids in the $C_{12}-C_{18}$ detergent range. The presence of the amide group confers added stability toward acid and base hydrolysis at elevated temperatures. Sulfated alkanolamides are effective cleaners with good lathering properties and find application as textile additives and in the formulation of industrial cleaners.

The manufacture of sulfated esters is a simple process requiring no complicated equipment. A triglyceride is first transesterified with a low molecular weight alcohol and then sulfated. A typical sulfated ester is $CH_3(CH_2)_7CH_2$-$CH(OSO_3Na)CH_2(CH_2)_5CH_2COOR$, where R is methyl, ethyl, propyl, butyl, or amyl. The sulfated esters are marketed as clear, yellow-brown viscous solutions. They are good wetting agents and are used as textile additives. They also possess good detersive and foaming properties. U.S. producers of this group of surfactants include Emkay, Hoechst AG, and Dexter Chemical Corporation.

Sulfated Natural Oils and Fats. Sulfated natural triglycerides were the first nonsoap commercial surfactants introduced in the middle of the nineteenth century. Since then sulfates of many vegetable, animal, and fish oils have been investigated (see also FATS AND FATTY OILS). With its hydroxyl group and a double bond, ricinoleic acid (12-hydroxy-9,10-octadecenoic acid) is an oil constituent particularly suited for sulfation. Its sulfate is known as turkey-red oil. Oleic acid is also suited for sulfation. Esters of these acids can be sulfated with a minimum of hydrolysis of the glyceride group. Polyunsaturated acids, with several double

Table 12. Alkylphenols, Ethoxylated and Sulfated, $RC_6H_4(OC_2H_4)_nOSO_3M$

Trade name	R	n	M	Conc, wt %	Function[a]	Manufacturer[b]
Rhodapex CO-433	nonyl	4	Na	28	D,Di,E,F,W	RHP
Rhodapex CO-436	nonyl	4	NH$_4$	58	D,Di,E,F,W	RHP
Abex EP-120	nonyl	4	NH$_4$	30	D,Di,E,F,W	RHP
Surfonic N-10	nonyl			100	E,W,S,Ta	TCC
Triton X-301	octyl[c]		Na	20	D,Di,F	UCC
Witcolate D51-51	alkyl[c]		Na	30	eP	WTC

[a]See Table 1.
[b]See Table 2.
[c]Branched.

bonds, lead to dark-colored sulfation products. The reaction with sulfuric acid proceeds through either the hydroxyl or the double bond. The sulfuric acid half ester thus formed is neutralized with caustic soda:

$$R'CH{=}CHR + H_2SO_4 \longrightarrow R'CH_2\overset{\overset{\displaystyle R}{|}}{C}HOSO_3H \qquad (17)$$

$$R'CH_2\overset{\overset{\displaystyle R}{|}}{C}HOH + H_2SO_4 \longrightarrow R'CH_2\overset{\overset{\displaystyle R}{|}}{C}HOSO_3H \qquad (18)$$

$$R'CH_2\overset{\overset{\displaystyle R}{|}}{C}HOSO_3H + NaOH \longrightarrow R'CH_2\overset{\overset{\displaystyle R}{|}}{C}HOSO_3Na + H_2O \qquad (19)$$

The first reaction is more important because double bonds are encountered more frequently than hydroxyl groups. Solubility in water increases with increasing sulfation, as does the danger of hydrolysis of the glyceride linkage. Sulfated oils are widely used as wetting and penetrating agents, emulsifiers, textile softeners, and lubricants. The products are usually sold under descriptions referring to the starting oils, eg, sulfated peanut oil, 50%.

Phosphate Esters.   Mono and diesters of orthophosphoric acid:

$$RO{-}\overset{\overset{\displaystyle HO}{|}}{\underset{\underset{\displaystyle HO}{|}}{P}}{=}O \text{ and } HO{-}\overset{\overset{\displaystyle RO}{|}}{\underset{\underset{\displaystyle RO}{|}}{P}}{=}O$$

and their salts have been useful surfactants since the 1940s (75). In contrast to sulfonates and sulfates, the resistance of alkyl phosphate esters to acids and hard water is poor. Calcium and magnesium salts are insoluble. In the acid form, the esters show limited water solubility, although their alkali metal salts are more soluble. The surface activity of phosphate esters is good, although in general it is somewhat lower than that of the corresponding phosphate-free precursors. Thus, a phosphated nonylphenol ethoxylated with 9 mol of ethylene oxide is less effective as a detergent in hard water than its nonionic precursor. At higher temperatures, however, the phosphate surfactant is significantly more effective (76).

Because of high costs and the limitations noted above, phosphate surfactants find application in specialty situations where such limitations are of no concern. As specialty surfactants, phosphate esters and their salts are remarkably versatile. Applications include emulsion polymerization of vinyl acetate and acrylates; dry-cleaning compositions where solubility in hydrocarbon solvents is a particular advantage; textile mill processing where stability and emulsifying power for oil and wax under highly alkaline conditions is necessary; and industrial cleaning compositions where tolerance for high concentrations of electrolyte and alkalinity is required. In addition, phosphate surfactants are used as corrosion inhibitors, in pesticide formulations, in papermaking, and as wetting and dispersing agents in drilling mud fluids.

Polyphosphoric acid, P_2O_5, $POCl_3$, and PCl_3 are suitable phosphorylating agents. Reaction of an alkyl sulfate with sodium pyrophosphate has also been reported for preparation of alkyl pyrophosphates (77). In general, phosphorylation leads to a mixture of reaction products that are sold without further separation. Thus, when lauryltri(ethyleneoxy)ethanol reacts with 0.3 mol of P_2O_5 at 50°C and is neutralized with 50% aqueous NaOH, the reaction mixture contains the following products:

Monoester, large fraction	$C_{12}H_{25}(-OCH_2CH_2-)_4OPO(OH)ONa$
Diester, medium fraction	$(C_{12}H_{25}(-OCH_2CH_2-)_4O)_2PO(ONa)$
Triester, small fraction	$(C_{12}H_{25}(-OCH_2CH_2-)_4O)_3PO$

Smaller amounts of the phosphate esters of the polyethylene glycol which is present in the ethoxylated alcohol and a trace of NaH_2PO_3 are also present. Table 13 gives a list of commercial phosphate surfactants.

Table 13. Phosphate Ester Surfactants

Component	Cation[a]	Manufacturer[b]
butyl phosphate	K,H	CHP,SYB,DUP
hexyl phosphate	H,K	ICI
2-ethylhexyl phosphate	H,Na	CHP,ETH,WTC,SYB,OC,RHP
octyl phosphate	H,K,A	DUP,HNK,WTC
decyl phosphate	H	DUP
octyldecyl phosphate	H	DUP
mixed alkyl phosphate	H,DEA	AMH,CTL,DUP
hexyl polyphosphate	K	DEX
octyl polyphosphate	H,K	DEX
glycerol monoester of mixed fatty acids[c]		WTC
2-ethylhexanol[d]	H	ETH,HNK,LNZ,PPG,WTC
deodecyl alcohol[d]	H	EXX,PHP
tridecyl alcohol[e]	H	MIL,WTC,DEX,EHT,LNZ,MOA,RHP
9-octadecenyl alcohol[d]	H	ETH,MOA,PHP
polyhydric alcohols[d]	H	ETH,RHP,WTC
phenol[d]	H	EMR,MOA,RH,WTC
octylphenol[d,e]	H,Mg	RHP,WTC
nonylphenol[d,e]	H	CTL,DEX,ESS,ETH,RHP,HRT,LNZ,OC, PPG,REN,WTC
dodecylphenol[d,e]	H	DEX,RHP,MOA
dinonylphenol[d,e]	H	ETH,RHP,WTC

[a]A = alkylammonium; DEA = diethanolammonium.
[b]See Table 2.
[c]Phosphated.
[d]Ethoxylated and phosphated.
[e]Branched.

Nonionic Surfactants

Unlike anionic or cationic surfactants, nonionic surfactants carry no discrete charge when dissolved in aqueous media. Hydrophilicity in nonionic surfactants is provided by hydrogen bonding with water molecules (78). Oxygen atoms and hydroxyl groups readily form strong hydrogen bonds, whereas ester and amide groups form hydrogen bonds less readily. Hydrogen bonding provides solubilization in neutral and alkaline media. In a strongly acid environment, oxygen atoms are protonated, providing a quasi-cationic character. Each oxygen atom makes a small contribution to water solubility; more than a single oxygen atom is therefore needed to solubilize a nonionic surfactant in water. Nonionic surfactants are compatible with ionic and amphoteric surfactants. Because a polyoxyethylene group can easily be introduced by reaction of ethylene oxide with any organic molecule containing an active hydrogen atom, a wide variety of hydrophobic structures can be solubilized by ethoxylation.

Polyoxyethylene Surfactants.   Polyoxyethylene-solubilized nonionics (ethoxylates) were introduced in the United States as textile chemicals shortly before 1940. The solubility of these compounds derives from recurring ether linkages in a polyoxyethylene chain $-O-CH_2CH_2-O-CH_2CH_2-O-CH_2CH_2-O-$. A single oxyethylene group, $-OCH_2CH_2-$ contributes slightly more to hydrophilicity than a single methylene, CH_2, contributes to hydrophobicity. To effect complete miscibility with water at room temperature, ca 60–75% by weight of polyoxyethylene content is needed on most hydrophobes. With increasing temperature, ethoxylates become less soluble as a result of decreased hydration and increasing micellar size (79,80). The temperature at which the appearance of a second phase is observable is the cloud point, a useful characteristic of nonionics independent of concentration in the range of 0.5–10 wt % (81). Small amounts of anionic surfactants may raise the cloud point by several degrees. Surface activity and performance efficiency of nonionics are usually greatest at temperatures just below the cloud point. The surface activity of ethoxylates is not adversely affected by water hardness. High electrolyte concentrations, particularly the presence of sodium ions, decrease the solubility of ethoxylates by a salting-out effect. Hydrochloric acid and calcium ions, however, increase water solubility. Ethoxylates are capable of solubilizing iodine in aqueous solution through formation of addition complexes. Ethoxylate–iodine complexes maintain the biocidal activity of iodine while reducing its toxicity to humans (82). Ethoxylates are moderate foamers and do not respond to conventional foam boosters (see FOAMS). Foaming shows a maximum as a function of ethylene oxide (qv) content (83). Low foaming nonionic surfactants are prepared by terminating the poyoxyethylene chain with less soluble groups such as polyoxypropylene and methyl groups. Ethoxylates can be prepared to attain almost any hydrophilic–hydrophobic balance. For incorporation into powdered products, they suffer from the disadvantage of being liquids or low melting waxes which complicates the manufacture of free-flowing, crisp powders. Solid products are manufactured with ethoxylates of high ethylene oxide content. The latter, however, are too water soluble to provide good surface activity.

Nonionic surfactants are often characterized in terms of their hydrophile–lipophile balance (HLB) number (see EMULSIONS). For simple alcohol ethoxylates, the HLB number may be calculated from the following:

$$HLB = E/5$$

where E is the weight percentage of ethylene oxide in the molecule. The functionality of nonionic surfactants depends on HLB as follows:

HLB range	Application
3–6	water-in-oil (w/o) emulsifier
7–9	wetting agent
8–15	oil-in-water (o/w) emulsifier
13–15	detergent
15–18	solubilizer

Ethoxylation. Base-catalyzed ethoxylation of aliphatic alcohols, alkylphenols, and fatty acids can be broken down into two stages: formation of a monoethoxy adduct and addition of ethylene oxide to the monoadduct to form the polyoxyethylene chain. The sequence of reactions is shown in equations 20–22:

$$RO^- + H_2C\overset{O}{\overbrace{\quad}}CH_2 \xrightarrow{\text{slow}} ROCH_2CH_2O^- \qquad (20)$$

$$ROH + ROCH_2CH_2O^- \xrightarrow{\text{fast}} RO^- + ROCH_2CH_2OH \qquad (21)$$

$$ROCH_2CH_2O^- + H_2C\overset{O}{\overbrace{\quad}}CH_2 \xrightarrow{\text{fast}} ROCH_2CH_2OCH_2CH_2O^- \qquad (22)$$

Equation 20 is the rate-controlling step. The reaction rate of the hydrophobes decreases in the order primary alcohols > phenols > carboxylic acids (84). With alkylphenols and carboxylates, buildup of polyadducts begins after the starting material has been completely converted to the monoadduct, reflecting the increased acid strengths of these hydrophobes over the alcohols. Polymerization continues until all ethylene oxide has reacted. Beyond formation of the monoadduct, reactivity is essentially independent of chain length. The effectiveness of ethoxylation catalysts increases with base strength. In practice, ratios of 0.005–0.05:1 mol of NaOH, KOH, or $NaOCH_3$ to alcohol are frequently used.

Alcohol Ethoxylates. These products have emerged as the principal nonionic surfactants in consumer detergent products (Table 14). Consistent quality, expanding production capacity of relatively inexpensive detergent range straight-chain, highly biodegradable alcohols, and increasing usage in laundry products, particularly heavy-duty liquids, are the principal factors underlying this growth.

Alcohol ethoxylates vary in physical form from liquids to waxes. With increasing ethylene oxide content, viscosity increases; a slight hydrophobe odor, present in the lower members of the series, decreases; specific gravity increases

Table 14. Alcohol Ethoxylates, $R\text{-}(OCH_2CH_2)_n OH$

Trade name	R	n	Concn, wt %	Functions[a]	Manufacturer[b]
Alfonic 610-50R	C_6,C_8,C_{10}[c]	3	100	D,E,W	VIS
Alfonic 1012-40	$C_{10}-C_{12}$[c]	2.5	100	E	VIS
Alfonic 1012-60	$C_{10}-C_{12}$[c]	5.7	100	D,E,W	VIS
Alfonic 1412-40	$C_{12}-C_{14}$[c]	3.0	100	D,E,W	VIS
Alfonic 1412-60	$C_{12}-C_{14}$[c]	7.0	100	D,E,W	VIS
Antarox BL-214			100	D,rW,Ta	RHP
Antarox BL-225			100	D,W	RHP
Antarox BL-240			100	D,Ta,W	RHP
Antarox BL-330			95	D,W	RHP
Antarox BL-344			90	D,W	RHP
Antarox LF-330			95	D,W	RHP
Antarox LF-344			90	D,W	RHP
Arosurf 66-E2	isostearyl	2	100	E,Di,S,St	WTC
Arosurf E 20	isostearyl	20	100	E,Di,S,St	WTC
Bio Soft EA-10	C_{10},C_{12}[c]		100	D,E	STP
Brij 30	lauryl	4	100	E	ICI
Brij 35	lauryl	23	100	E	ICI
Brij 52	cetyl	2	100	E	ICI
Brij 58	cetyl	20	100	E,S	ICI
Brij 72	stearyl	2	100	E	ICI
Brij 78	stearyl	20	100	E	ICI
Brij 92	oleyl	2	100	E	ICI
Brij 98	oleyl	20	100	E	ICI
Brij 700	stearyl	100			ICI
Brij 721	stearyl	21			ICI
Cerfak 1400			85	D,W	EFH
Rhodasurf BC-420	tridecyl[d]		100	E	RHP
Rhodasurf BC-720	tridecyl[d]		100	E,Fs,W	RHP
Rhodasurf ON-877	tridecyl[d]		70	Di,E,S	RHP
Examide DA			50	D,E,Ta	SLC
Neodol 25-3	$C_{12}-C_{15}$[c]	3	100	Di,I	SHC
Neodol 23-6.5	$C_{12}-C_{13}$[c]	6.5	100	D	SHC
Neodol 25-7	$C_{12}-C_{15}$[c]	7	100	D,W	SHC
Neodol 25-12	$C_{12}-C_{15}$[c]	12	100	D,E,W	SHC
Neodol 45-7	$C_{14}-C_{15}$[c]	7	100	D,W	SHC
Neodol 45-13	$C_{14}-C_{15}$[c]	13	100	D,E,W	SHC
Neodol 91-2.5	C_9-C_{11}[c]	2.5	100	Di,W	SHC
Neodol 91-6	C_9-C_{11}[c]	6	100	D,W	SHC
Neodol 91-8	C_9-C_{11}[c]	8	100	D,W	SHC
Plurafac A-38	alkyl		100	D,E,W	WYN
Poly Tergent J-200	tridecyl[d]		99	D,Di,E,S	OC
Poly Tergent J-500	tridecyl[d]		99	D,D,E,S	OC
Renex 30	tridecyl		100	D,Di,W	ICI
Rhodasurf L-4	lauryl	4	100	E	RHP
Rhodasurf L-25	lauryl	25	100	E	RHP
Stanoamul B-3	cetyl	30	100	E	HNC
Surfonic JL-80X	alkyl[c]		100	D,Di,E,W	TCC
Surfonic LF-17	alkyl[c]		100	D,W	TCC
Tergitol 25-L-3	$C_{12}-C_{15}$[c]	3	100	Di,I	UCC
Tergitol 25-L-7	$C_{12}-C_{15}$[c]	7	100	D,W	UCC

Table 14. (*Continued*)

Trade name	R	n	Concn, wt %	Functions[a]	Manufacturer[b]
Tergitol 25-L-12	$C_{12}-C_{15}$[c]	12	100	D,E,W	UCC
Tergitol 15-S-3	$C_{11}-C_{15}$[e]	3	100	I,Ta	UCC
Tergitol 15-S-5	$C_{11}-C_{15}$[e]	5	100	D,E,W	UCC
Tergitol 15-S-7	$C_{11}-C_{15}$[e]	7	100	D,E,W	UCC
Tergitol 15-S-9	$C_{11}-C_{15}$[e]	9	100	D,E,W	UCC
Tergitol 15-S-12	$C_{11}-C_{15}$[e]	12	100	D,E,W	UCC
Tergitol 15-S-15	$C_{11}-C_{15}$[e]	15	100	W,E	UCC
Tergitol 15-S-20	$C_{11}-C_{15}$[e]	20	100	E,Ta	UCC
Tergitol 15-S-30	$C_{11}-C_{15}$[e]	30	100	E,Ta	UCC
Tergitol TMN-3	trimethylnonyl[d]	3	100	Ta,W	UCC
Tergitol TMN-10	trimethylnonyl[d]	10	90	Ta,W	UCC
Trycol DA-4	isodecyl[d]	4	100	E,W	HNK
Trycol DA-6	isodecyl[d]	6	100	I,P,W	HNK
Trycol LAL-4	lauryl	4	100	E	HNK
Trycol LAL-8	lauryl	8	100	D,rW,W	HNK
Trycol LAL-12	lauryl	12	100	D,W	HNK
Trycol LAL-23	lauryl	23	100	S,Ta	HNK
Trycol OAL-23	oleyl	23	100	S,St,Ta	HNK
Trycol SAL-20	stearyl	20	100	E,St,Ta	HNK
Trycol TDA-3	tridecyl[d]	3	100	E	HNK
Trycol TDA-9	tridecyl[d]	9	100	E,I,W	HNK
Trycol TDA-12	tridecyl[d]	12	100	D,Fs	HNK
Trycol TDA-18	tridecyl[d]	18	100	Di,S,Ta	HNK
Volpo 3	oleyl	3	97	E,S	CRO
Volpo 10	oleyl	10	97	E,S	CRO
Volpo 20	oleyl	20	97	E,S	CRO

[a]See Table 1.
[b]See Table 2.
[c]Linear, primary.
[d]Branched.
[e]Linear, secondary.

from <1 to slightly <1.2; and water solubility increases, ie, ca 65–70 wt % of ethylene oxide content is needed for complete miscibility with water at room temperature. Alcohol ethoxylates are effective cleaners in domestic and industrial detergent products. Depending on ethylene content, alcohol ethoxylates provide excellent emulsifying, wetting, and dispersing performance.

In the ethoxylation reaction, primary alcohols react most readily, followed by secondary and tertiary alcohols. The rate of reaction of the primary hydroxyl with ethylene oxide is close to that of the ethylene oxide chain buildup reaction. As a result, the ethylene oxide chain is built up before all alcohol has been converted into the monoadduct and unethoxylated alcohol is present in the final ethoxylation product. Alcohol ethoxylates are polydisperse with respect to ethylene oxide chain length. An ethoxylate containing nominally eight ethylene oxide units actually contains significant amounts of other ethoxylates containing from 0–20 ethylene oxide units. The composition of the mixture has been reported as following a Weibull and Nycander distribution more closely than a Poisson distribution. Because the ethoxylates at the extremes of the 0–20 range are less useful than the nominal product, much effort has been expended to peak the distribution more narrowly around the target value. Additionally, the unethoxylated alcohol and lower end ethoxomers are relatively volatile and can cause undesirable tower pluming during production of a spray dried powder. At least one U.S. detergent powder has been spray dried using peaked ethoxylates (85). A U.S. patent has been issued claiming benefits in oily soil cleaning via use of peaked ethoxylates. Acid catalysts provide a sharper distribution but can lead to undesired side products such as dioxanes which must be removed. A 1980 U.S. patent claims a barium hydroxide catalyst promoted with phenolic compounds is capable of yielding a substantially narrow distribution of ethoxylates (86). This was followed by a number of other patents claiming a variety of catalysts to yield peaked ethoxylates (87).

Alkylphenol Ethoxylates.   In physical and performance properties, these products are similar to the alcohol ethoxylates. Table 15 indicates that most alkylphenol ethoxylates (APEs) are derived from alkylphenols carrying branched alkyl side chains, predominantly nonyl and octyl. Chain branching accounts for the lower rate of biodegradation of alkylphenol ethoxylates compared to that of claiming benefits in oily soil cleaning via use of peaked ethoxylates. On the other hand, chain branching increases water solubility and contributes to marginally superior detersive properties. Because the stringent biodegradability criteria adopted by the U.S. Soap and Detergent Association apply to consumer products only, it is not surprising that alkylphenol ethoxylates continue to be used extensively in industrial products. Not withstanding removal of APEs from consumer products by the principal U.S. detergent manufacturers, they have continued to receive criticism on environmental and toxicological grounds (88). Subsequent studies have shown that current APEs may not be accumulating in the environment as significantly (89). Nonetheless, work on environmentally occurring estrogen mimics has raised further concerns about APEs and biodegradation intermediates. They will probably be gradually phased out of the industrial and institution market (62).

Alkylphenol ethoxylates are chemically stable and highly versatile surfactants that find application in a large variety of industrial products including acid and alkaline metal cleaning formulations, hospital cleaners, herbicides (qv) and insecticides, oil-well drilling fluids, synthetic latices, and many others (see DISINFECTANTS AND ANTISEPTICS; ELASTOMERS, SYNTHETIC; INSECT CONTROL TECHNOLOGY; METAL SURFACE TREATMENTS; PESTICIDES; PETROLEUM, DRILLING FLUIDS).

Commercial alkylphenol ethoxylates are almost always produced by base-catalyzed ethoxylation of alkylphenols. Because phenols are more strongly acidic than alcohols, reaction with ethylene oxide to form the monoadduct is faster. The product, therefore, does not contain unreacted phenol. Thus, the distribution of individual ethoxylates in the commercial mixture is narrower, and alkylphenol ethoxylates are more soluble in water.

Carboxylic Acid Esters.   In the carboxylic acid ester series of surfactants, the hydrophobe, a naturally occurring fatty acid, is solubilized with the hydroxyl groups of polyols or the ether and terminal hydroxyl groups of ethylene oxide chains.

Table 15. Alkylphenol Ethoxylates, $RC_6H_4(OC_2H_4)_nOH$

Trade name	R^a	n	Concn, wt %	Functions[b]	Manufacturer[c]
Conco NI-40	nonyl		100	D,E,W	CTL
Hyonic PE-100	nonyl[d]	10	100	D	HNK
Igepal CA-630	octyl	9	100	E,D,Di	RHP
Igepal CO-210	nonyl[d]	1.5	100	D,Di,Fs,I	RHP
Igepal CO-530	nonyl	6	100	E,D,Di	RHP
Igepal CO-710	nonyl	10.5	100	D,W	RHP
Igepal CO-850	nonyl	20	100	D,E,S,W	RHP
Igepal CO-880	nonyl	30	100	D,Di,E,S,W	RHP
Igepal DM-730	dialkyl	24	100	D,E,S	RHP
Makon 4	nonyl	4	100	D,E,S	STP
Makon 6	nonyl	6	100	D,E,S	STP
Makon 8	nonyl	8	100	D,E,S	STP
Makon 10	nonyl	9	100	A,D,E,S,Ta	STP
Makon 12	nonyl	12	100	A,D,E,S,Ta	STP
Makon 14	nonyl	14	100	D,E	STP
Makon 30	nonyl	30	100	D,E	STP
Nonionic E-4	alkyl[d]		100	D,E,W	CAL
Nonionic E-10	alkyl[d]		100	D,E,W	CAL
Polytergent B-150	nonyl	4.5	100	D,Di,E,W	OC
Polytergent B-200	nonyl	6	100	D,Di,E,W	OC
Polytergent B-300	nonyl	9	100	D,Di,E,W	OC
Polytergent B-350	nonyl	10.5	100	D,Di,E,W	OC
Polytergent B-500	nonyl	15	100	D,Di,E,W	OC
Rexol 25/1	nonyl	1	100	D,Di,I,St	HCL
Rexol 25/507	nonyl	50	70	dE,S,St	HCL
Sterox DF	dodecyl	6.2	100	D,E	MON
Sterox DJ	dodecyl	10.3	100	D,F,W	MON
Sterox ND	nonyl	4	100	E,D	MON
Sterox NJ	nonyl	9.2	100	D,Di,E,W	MON
Sterox NK	nonyl	10.3	100	D,Di,E,W	MON
Surfonic N-10	nonyl	1	100	D,Di,E,dF	HUN
Surfonic N-31.5	nonyl	3.15	100	D,Di,E,dF	HUN
Surfonic N-40	nonyl	4	100	D,Di,E,W	HUN
Surfonic N-60	nonyl	6	100	D,Di,E,W	HUN
Surfonic N-95	nonyl	9.5	100	D,Di,E,W	HUN
Surfonic N-100	nonyl	10	100	D,Di,E,W	HUN
Surfonic N-120	nonyl	12	100	D,Di,E,W	HUN
Surfonic N-300	nonyl	30	100	D,W	HUN
T-DET N-4	nonyl	4	100	D,E	HAR
T-DET N-6	nonyl	6	100	D,Di,E,W	HAR
T-DET N-9.5	nonyl	9.5	100	D,Di,E,W	HAR
T-DET N-14	nonyl	14	100	D,Di,E,W	HAR
Tergitol NP-4	nonyl	4	100	D,E	UCC
Tergitol NP-7	nonyl	7	100	D,Di,E,W	UCC
Tergitol NP-9	nonyl	9	100	D,Di,E,W	UCC
Tergitol NP-40	nonyl	40	100	D,E,W	UCC
Triton N-57	nonyl	7–8	100	D,E	UCC
Triton N-111	nonyl	12–13	100	Di,E	UCC
Triton X-45	octyl	5	100	D,E	UCC
Triton X-102	octyl	12–13	100	D	UCC
Triton X-305	octyl	30	70	E,St	UCC
Triton X-705	octyl	70	70	E,St	UCC

[a] R is branched unless otherwise noted. [b] See Table 1. [c] See Table 2. [d] Nonbranched.

Glycerol Esters. Commercial glycerol esters, though named after the most abundant species, almost always are mixtures of isomeric mono- and diglycerides (see GLYCEROL). Trade names and compositions of typical commercial products are given in Table 16.

Saturated fatty acid mono- and diglycerides are light-colored solids melting between 25 and 85°C. The 1-monoglycerides are higher melting than the corresponding 2-monoglycerides. Glycerides of unsaturated fatty acids are liquid at room temperature. In general, the odor of the glycerides is characteristic of the fats from which they are derived. The two hydroxyl groups in monoglycerides are not sufficiently hydrophilic to carry even an easily solubilized acid like oleic acid into solution. Despite the lack of water solubility, the fatty acid mono- and diglycerides are commercially important and technically interesting surfactants.

In bread, cakes, and other bakery products, mono- and diglycerides are almost universally present as emulsifiers, dispersing agents, and lubricants (see BAKERY PROCESSES AND LEAVENING AGENTS). They are also used in candies, ice cream, yeast, butter, whipped toppings, and icings. Flavor oils for carbonated beverages (qv) and for bakery products are emulsified or solubilized by surfactant mixtures that include blends of mono- and diglycerides (see FLAVORS AND SPICES). Glycerol monostearate acts as an emulsifier and opacifier in cosmetic formulations. In cutting, drawing, and finishing of metal products, mono- and diglycerides find application as emulsifiers, lubricants, and corrosion inhibitors. They are also used as emulsifiers, dispersants, suspending agents, and grinding aids in the manufacture of paint (qv) and polymers (qv). The reaction of fats

Table 16. Glycerol Esters[a] of Fatty Acids[b]

Trade name	Fatty acid[c]	Manufacturer[d]
Aldo MC	coco	LNZ
Aldo ML	lauric	LNZ
Aldo MR	ricinoleic	LNZ
Aldo MS	stearic	LNZ
Cerasynt 945	stearic	VND
Emerest 2400	stearic	HNK
Witconol 2421	oleic	WTC
Flexricin 13	ricinoleic	CAS
Hallco CPH-34-N	lauric	HAL
Hodag GML	lauric	CAL
Hodag GMO	lauric	CAL
Hodag GMR	ricinoleic	CAL
Hodag E GMS	stearic	CAL
Kessco GDL	lauric[e]	STP
Kessco GMO	oleic	STP
Kessco GMS	stearic	AKZ
Witconol MST	stearic	WTC

[a]Ester type is mono- unless otherwise noted.
[b]Concentration = 100 wt %; function type is E[c].
[c]See Table 1.
[d]See Table 2.
[e]Ester type is di-.

with glycerol is the most important industrial method of preparing mono- and diglycerides. In this reaction, the fatty acid groups are redistributed randomly by heating at 180–250°C in the presence of an alkaline catalyst.

Polyoxyethylene Esters. This series of surfactants consists of polyoxyethylene (polyethylene glycol) esters of fatty acids and aliphatic carboxylic acids related to abietic acid (see RESINS, NATURAL). They differ markedly from mono- and diglycerides in properties and uses.

Commercial polyoxyethylene fatty acid esters contain varying proportions of monoesters, $RCO(OC_2H_4)_nOH$; diesters, $RCO(OC_2H_4)_nOOCR$; and polyglycol, $H(OC_2H_4)_nOH$. The composition of the mixture can be forced toward the mono- or diester by the ratio of reactants and process of manufacture. Commercial products are usually described by acid esterified, predominant ester species (mono- or di-), and either molecular weight of the polyglycol or number of ethylene oxide groups in the polyglycol chain. Composition and functions of typical polyoxyethylene esters are listed in Table 17.

The physical properties of the fatty acid ethoxylates depend on the nature of the fatty acid and even more on ethylene oxide content. As the latter increases, consistencies of the products change from free-flowing liquids to slurries to firm waxes (qv). At the same time, odor, which is characteristic of the fatty acid, decreases in intensity. Odor and color stability are important commercial properties, particularly in textile applications. Oleic acid esters, though possessing good functional properties, cannot be used because they tend to yellow on exposure to heat and air.

At room temperature, ca 60 wt % ethylene oxide is needed to solubilize the fatty acids. Surface activity of the ethoxylates is moderate and less than that of alcohol or alkylphenol ethoxylates (84). The ethoxylates are low foamers, a useful property in certain applications. Emulsification is the most important function. Its importance is reflected in the wide range of lipophilic solubilities available in the commercial products. Like all organic esters, fatty acid ethoxylates are susceptible to acid and alkaline hydrolysis.

Fatty acid ethoxylates are used extensively in the textile industry as emulsifiers for processing oils, antistatic agents (qv), softeners, and fiber lubricants, and as detergents in scouring operations. They also find application as emulsifiers in cosmetic preparations and pesticide formulations. Fatty acid ethoxylates are manufactured either by alkali-catalyzed reaction of fatty acids with ethylene oxide or by acid-catalyzed esterification of fatty acids with preformed poly(ethylene glycol). Deodorization steps are commonly incorporated into the manufacturing process.

The principal constituents of rosin (qv) are abietic and related acids. Tall oil (qv) is a mixture of unsaturated fatty and alicyclic acids of the abietic family. Refined tall oil may be high in rosin acids or unsaturated acids, depending on the refining process. Ethoxylates of rosin acids, eg, dehydroabietic acid, are similar to fatty acid ethoxylates in surfactant properties and manufacture, except for their stability to hydrolysis. No noticeable decomposition is observed when a rosin ester of this type is boiled for 15 min in 10% sulfuric acid or 25% sodium hydroxide (90). Steric hindrance of the carboxylate group associated with the alicyclic moiety has been suggested as the cause of this unexpectedly great hydrolytic stability.

Table 17. Polyoxyethylene Esters of Rosin, Tall Oil, and Fatty Acids[a]

Trade name	Ester type	Acid	Composition PEG[b]	POE[c]	Functions[d]	Manufacturer[e]
Emerest 2620	mono	lauric	200		E,dF,V	HNK
Emerest 2622	di	lauric	200		E	HNK
Emerest 2625	mono	isostearic	200		E,Ta	HNK
Emerest 2534	mono	pelargonic	300		Ta	HNK
Emerest 2647	sesqui	oleic	400		E,Ta	HNK
Emerest 2652	di	lauric	400		E,Ta	HNK
Emerest 2660	mono	oleic	600		E,Ta	HNK
Ethofat C/15	mono	coco		5	D,Di,E	AKZ
Ethofat C/25	mono	coco		15	D,Di,E	AKZ
Ethofat 142/20	mono	tall oil		10	D,Di,E	AKZ
Ethofat 242/25	mono	tall oil		15	D,Di,E	AKZ
Ethofat 0/15	mono	oleic		5	D,Di,E	AKZ
Ethofat 0/20	mono	oleic		10	D,Di,E	AKZ
Ethofat 60/15	mono	stearic		5	D,Di,E	AKZ
Ethofat 60/20	mono	stearic		10	D,Di,E	AKZ
Ethofat 60/25	mono	stearic		15	D,Di,E	AKZ
Hodag 40-L	mono	lauric	400		E,Sp,W	CAL
Hodag 40-R	mono	ricinoleic	400		E,Sp,W	CAL
Hodag 42-S	di	stearic	400		E,Sp,W	CAL
Hodag 60-L	mono	lauric	600		E,Sp,W	CAL
Hodag 60-S	mono	stearic	600		E,Sp,W	CAL
Hodag 62-0	di	oleic	600		E,Sp,W	CAL
Hodag 150-S	mono	stearic	1500		E,Sp,W	CAL
Myrj 45	mono	stearic		8	E	ICI
Myrj 52	mono	stearic		40	E	ICI
Nonisol 210	di	oleic	400		E	CGY
Nonisol 300	mono	stearic	400		E,St,W	CGY
Nopalcol 4-L	mono	lauric	400		E	HNK
Nopalcol 6-R	mono	ricinoleic	600		E	HNK
Nopalcol 10-CO[f]		castor	1000		E	HNK
Nopalcol 30-S	mono	stearic	400		E	HNK
Pegosperse 400MS	mono	stearic	400		A,dF,E	LNZ
Pegosperse 400 MO	mono	oleic	400		A,dF,E	LNZ
Pegosperse 400 DOT	di	tall oil	400		A,dF,E	LNZ
Pegosperse 600 DOT	di	tall oil	600		A,dF,E	LNZ
Pegosperse 400 ML	mono	lauric	400		A,dF,E	LNZ
Pegosperse 600 ML	mono	lauric	600		A,dF,E	LNZ
Pegosperse 1750 MS	mono	stearic	1750		A,dF,E	LNZ
Renex 20	mono	fatty, rosin		16	D,W	ICI
Surfactant AR 150	mono	rosin			D,E	HPC
Witconol H31 A	mono	oleic	400		E	WTC

[a]Concentration = 100 wt %, unless otherwise noted.
[b]Mol wt of esterified poly(ethylene glycol) (PEG).
[c]Number of ethylene oxide groups in esterified polyoxyethylene (POE).
[d]See Table 1.
[e]See Table 2.
[f]Concentration = 99 wt %.

Anhydrosorbitol Esters. Fatty acid esters of anhydrosorbitol (see ALCOHOLS, POLYHYDRIC) are the second largest class of carboxylic ester surfactants. The important commercial products are the mono-, di-, and triesters of sorbitan and fatty acids (Table 18). Sorbitan is a mixture of anhydrosorbitols, principally 1,4-sorbitan (**1**) and isosorbide (**2**):

(**1**)   (**2**)

Sorbitan oleate and the monolaurate are pale yellow liquids. Palmitates and stearates are light tan solids. Sorbitan esters are not soluble in water but dissolve in a wide range of mineral and vegetable oils. They are lipophilic emulsifiers, solubilizers, softeners, and fiber lubricants that find application in synthetic fiber manufacture, textile processing, and cosmetic products. Sorbitan esters

Table 18. Sorbitan Esters of Fatty and Tall Oil Acids[a]

Trade name	Ester type	Acid	Functions[b]	Manufacturer[c]
Armotan ML	mono	lauric	E,S	AKZ
Armotan MO	mono	oleic	E,S	AKZ
Armotan MS	mono	stearic	E,S	AKZ
Emsorb 2500	mono	oleic	E,C	HNK
Emsorb 2502	sesqui	oleic	E	HNK
Emsorb 2503	tri	oleic	E,Sp	HNK
Emsorb 2505	mono	stearic	E,Ta	HNK
Emsorb 2510	mono	palmitic	E	HNK
Emsorb 2515	mono	lauric	E	HNK
Glycomul SOC	mono/sesqui	oleic	E	LNZ
Glycomul TO	tri	oleic	E	LNZ
Glycomul TS	tri	stearic	E	LNZ
Hodag SML	mono	lauric	E	CAL
Hodag SMP	mono	palmitic	E	CAL
Hodag STS	tri	stearic	E	CAL
Hodag SMO	mono	oleic	E	CAL
Hodag STO	tri	oleic	E	CAL
Span 20	mono	lauric	E,V	ICI
Span 40	mono	palmitic	E,V	ICI
Span 60	mono	stearic	E,V	ICI
Span 65	tri	stearic	E,V	ICI
Span 80	mono	oleic	E,V	ICI
Span 85	tri	oleic	E,V	ICI

[a]Concentration = 100 wt %.
[b]See Table 1.
[c]See Table 2.

have been approved for human ingestion and are widely used as emulsifiers and solubilizers in foods, beverages, and pharmaceuticals.

Anhydrosorbitol esters are prepared commercially by direct esterification of sorbitol with a fatty acid at 225–250°C in the presence of an acidic catalyst. Internal ether formation and esterification take place under these conditions.

Ethoxylated Anhydrosorbitol Esters. Ethoxylation of sorbitan fatty acid esters leads to a series of more hydrophilic surfactants (Table 19). All hydroxyl groups of sorbitan can react with ethylene oxide. The structure of the principal component of a nominal polyoxyethylene (20) sorbitan monostearate illustrates the composition of these products, where $w + x + y + z = 20$.

$$
\begin{array}{c}
(\text{OCH}_2\text{CH}_2)_{\overline{x}}\text{OH} \\
|
\end{array}
$$

$$\text{HO}-(\text{CH}_2\text{CH}_2\text{O})_{w} \quad \text{CH}_2-(\text{OCH}_2\text{CH}_2)_{y}-\text{OCC}_{17}\text{H}_{35} \quad (\text{OCH}_2\text{CH}_2)_{z}-\text{OH}$$

Table 19. Polyoxyethylene Derivatives of Sorbitan Fatty Acid Esters[a]

Trade name	Ester type	Fatty acid	POE[b]	Functions[c]	Manufacturer[d]
Armotan PML-20	mono	lauric	20	A,E,S,Sp,Ta	AKZ
Armotan PMO-20	mono	oleic	20	A,E,S,Sp,Ta	AKZ
Armotan PMS-20	mono	stearic	20	A,E,S,Sp,Ta	AKZ
Emsorb 6900	mono	oleic	20	Di,E,S	HNK
Emsorb 6901	mono	oleic	5	E	HNK
Emsorb 6903	tri	oleic	20	E	HNK
Emsorb 6905	mono	stearic	20	E	HNK
Emsorb 6906	mono	stearic	4	E	HNK
Emsorb 6910	mono	palmitic	20	E	HNK
Emsorb 6915	mono	lauric	20	A,E,Ta	HNK
Glycosperse L-20	mono	lauric	20	E	LNZ
Glycosperse 0-5	mono	oleic	5	E	LNZ
Glycosperse 0-20	mono	oleic	20	E	LNZ
Glycosperse P-20	mono	palmitic	20	E	LNZ
Glycosperse S-20	mono	stearic	20	E	LNZ
Glycosperse TO-20	tri	oleic	20	E	LNZ
Glycosperse TS-20	tri	stearic	20	E	LNZ
Hodag PSML-20	mono	lauric	20	E	CAL
Hodag PSMP-20	mono	palmitic	20	E	CAL
Tween 20	mono	lauric	20	E,S,Ta	ICI
Tween 21	mono	lauric	4	E,W	ICI
Tween 40	mono	palmitic	20	TE,S,Ta	ICI
Tween 61	mono	stearic	4	E	ICI
Tween 81	mono	oleic	5	E	ICI
Tween 85	tri	oleic	20	E,Ta	ICI

[a]Concentration = 100 wt %.
[b]Number of ethylene oxide groups in esterified polyoxyethylene (POE).
[c]See Table 1.
[d]See Table 2.

Typical commercial ethoxylated sorbitan fatty acid esters are yellow liquids, except tristearates and the 4- and 5-mol ethylene oxide adducts which are light tan solids. These adducts, as well as the 20-mol adducts of the triesters, are insoluble but dispersible in water. The monoester 20-mol adducts are water soluble. Ethoxylated sorbitan esters are widely used as emulsifiers, antistatic agents, softeners, fiber lubricants, and solubilizers. In combination with the unethoxylated sorbitan esters or with mono- or diglycerides, these are often used as co-emulsifiers. The ethoxylated sorbitan esters are produced by heating sorbitan esters with ethylene oxide at 130–170°C in the presence of alkaline catalysts.

Natural Ethoxylated Fats, Oils, and Waxes. Castor oil (qv) is a triglyceride high in ricinoleic esters. Ethoxylation in the presence of an alkaline catalyst to a polyoxyethylene content of 60–70 wt % yields water-soluble surfactants (Table 20). Because alkaline catalysts also effect transesterification, ethoxylated castor oil surfactants are complex mixtures with components resulting from transesterification and subsequent ethoxylation at the available hydroxyl groups. The ethoxylates are pale amber liquids of specific gravity just above 1.0 at room temperature. They are hydrophilic emulsifiers, dispersants, lubricants, and solubilizers used as textile additives and finishing agents, as well as in paper (qv) and leather (qv) manufacture.

Glycol Esters of Fatty Acids. Ethylene glycol, diethylene glycol, and 1,2-propanediol esters of fatty acids constitute a relatively small group of ester surfactants. The mono- and dilaurates of the three glycol types are liquids; the stearates, solids. Glycol esters are strongly lipophilic emulsifiers, opacifiers, and plasticizers (qv) that are formulated in combination with hydrophilic emulsifiers. Glycol monoesters can be prepared by the alkali-catalyzed reaction of fatty acids with ethylene or propylene glycol or by esterification of fatty acids with a glycol. As expected, the products are generally mixtures. Principal components of representative commercial glycol esters are given in Table 21.

Table 20. Ethoxylated Castor Oils[a]

Trade name	POE[b]	Functions[c]	Manufacturer[d]
Alkamuls EL 980	200	Di,E,S,Ta	RHP
Alkamuls EL 719[e]	40	Di,E,S,Ta	RHP
Alkamuls EL 620	30	Di,E,Em,S,Ta	RHP
Surfactol-318		E,dF,Ta	CAS
Surfactol-380		E,Sp,Ta	CAS
Trylox CO-5	5	E,dF	HNK
Trylox CO-30	30	E,Sp,Ta	HNK
Trylox CO-40	40	E,S,Ta	HNK
Trylox CO-200	200	A,Sp,Ta	HNK
Trylox HCO-25[f]	25	E,Sp	HNK

[a]Concentration = 100 wt % unless otherwise noted.
[b]Number of ethylene oxide groups in esterified polyoxyethylene (POE).
[c]See Table 1.
[d]See Table 2.
[e]Concentration = 96 wt %.
[f]Hydrogenated castor oil.

Table 21. Glycol Esters of Fatty Acids[a]

Trade name	Glycol ester[b]	Functions[c]	Manufacturer[d]
Aldo PMS	PG monostearate	E	LNZ
Emerest 2350	EG monostearate	O	HNK
Emerest 2355	EG distearate	O	HNK
Emerest 2381	PG monostearate	O	HNK
CPH-53-N	PG monostearate	E	HAL
Hodag DGL	DEG monolaurate	E	CAL
Hodag DGO	DEG monooleate	E,Ta	CAL
Hodag DGS	DEG monostearate	E,Ta	CAL
Hodag PMS	PG monostearate	E	CAL
Kessco EGDS	EG distearate	E,O	STP
Kessco EGMS	EG monostearate	E,O	STP
Pegosperse 50 MS	EG monostearate	E	LNZ
Pegosperse 50 DS	EG distearate	E	LNZ
Pegosperse 100 O	DEG monooleate	E	LNZ
Pegosperse 100 S	DEG monostearate	E	LNZ
Witconol CAD	DEG stearate	E	WTC

[a]Concentration = 100 wt %.
[b]EG = ethylene glycol, DEG = diethylene glycol, PG = propylene glycol.
[c]See Table 1.
[d]See Table 2.

Lanolin alcohols are obtained by saponification of purified wool grease, a mixture of high molecular esters that is recovered in wool (qv) scouring. Ethoxylation of purified lanolin alcohols yields a full series of lipophilic and hydrophilic nonionic emulsifiers whose largest use is in cosmetic preparations. Manufacturers include Amerchol, Croda, ICI, Henkel Corporation, Westbrook Lanolin, Witco, and Pulcra, SA.

Alkyl Polyglycosides. The alkyl polyglycosides (APGs) are unusual in offering a hydrophile based on natural, ie, sugar (qv), chemistry:

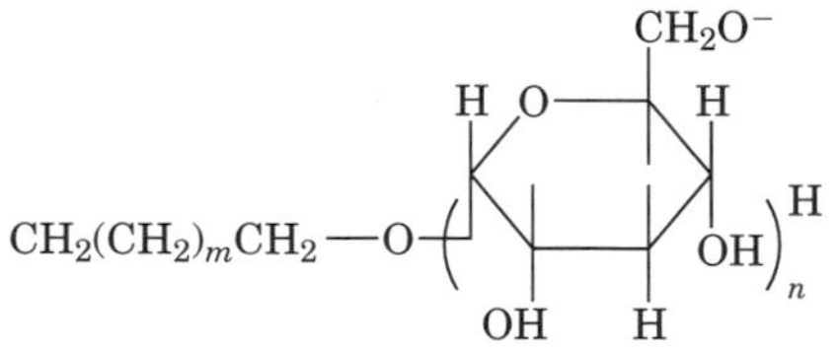

Generally, $m = 6-14$ and n averages between 1 and 2. These were introduced for industrial application in the early 1970s (91). However, they were not offered on a large scale or in good quality until the 1980s. Manufactured from alcohols and carbohydrates using processes based on Fisher glycosylation (92), the process of APG manufacture must be carefully controlled to avoid undesirable by-products. Despite many advantages (93), APGs have not been embraced by significant detergent manufacturers, although they have been used in dishwashing liquids (Henkel Corporation). An unattractive cost-performance profile has been a primary factor in the limited success of APG.

Carboxylic Amides. Carboxylic amide nonionic surfactants are condensation products of fatty acids and hydroxyalkyl amines.

Diethanolamine Condensates. The first fatty acid diethanolamides, sometimes referred to as Kritchevsky amides, offered commercially were of the 2:1 or regular type (94). Regular amides are obtained by heating one mole of fatty acid with two moles of diethanolamine. Nominally, the composition of regular diethanolamides is represented by $RCON(CH_2CH_2OH)_2 \cdot HN(CH_2CH_2OH)_2$. The actual composition has been reported as diethanolamide, $RCON(CH_2CH_2OH)_2$, 50%; diethanolamine, $HN(CH_2CH_2OH)_2$, 25%; amide ester, $RCON(CH_2CH_2OH)$-CH_2CH_2OOCR, 10%; amine ester, $RCOOCH_2CH_2NHCH_2CH_2OH$, 10%; and amine soap, $RCOO^-H_2N(CH_2CH_2OH)_2^+$, 5% (95).

The second type of diethanolamide is the 1:1 or superamide which contains components of the reaction of one mole fatty acid and one mole diethanolamine. A typical superamide composition is $\geq 90\%$ diethanolamide, 7% unreacted diethanolamine, and 2.5% amine and amide ester.

The presence of the large excess of diethanolamine and its derivatives in regular diethanolamides increases water solubility compared to superamides. The physical form and water solubility of the fatty acid diethanolamides are shown in Table 22. Although the superamides of coco and lauric acids and the regular amides of the higher fatty acids are insoluble in water, they are readily dispersed and solubilized in combinations with more hydrophilic surfactants.

Fatty acid diethanolamides are versatile and widely used surfactants (Table 23). Foam stabilization and detergency are the most important functional properties. In addition, they are effective in increasing viscosity of detergent solutions, and provide emolliency, corrosion inhibition, emulsification, wetting, and lime soap dispersion. Diethanolamides find application in liquid dishwashing detergents, shampoos, janitorial scrub soaps, textile processing, lubricating oils, and dry-cleaning detergents.

Regular fatty acid diethanolamides are prepared by heating fatty acid with diethanolamine at 160–180°C for 2–4 h. Superamides are prepared by heating a fatty acid methyl ester with an equimolar amount of diethanolamine at 100–110°C for 2–4 h; the methanol formed is distilled off (Table 23).

Monoalkanolamine Condensates. Coco, lauric, oleic, and stearic monoethanolamides and monoisopropanolamides are the principal surfactants in the monoalkanolamide group (Table 24). Monoalkanolamides are generally water-insoluble solids that are easily solubilized by hydrophilic surfactants. Except for solubility and viscosity, properties and uses are similar to the diethanolamides. Manufacturing processes and yields have been described (96).

Table 22. Physical Form and Solubility of Fatty Acid Diethanolamides

Fatty acid	Amine:acid	Physical form	Water solubility
coco	2:1	liquid	soluble
coco	1:1	paste	limited solubility
lauric	2:1	soft paste	soluble
oleic	1:1	paste	limited solubility
oleic	2:1	liquid	dispersible
stearic	2:1	paste	dispersible

Table 23. Fatty Acid Diethanolamides

Trade name	Fatty acid	Type[a]	Conc, wt %	Functions[b]	Manufacturer[c]
Aminol CA-2	ricinoleic	R	100	D,Sp	FET
Calamide C	coco	S	100	D,Fs	PIL
Calamide O	oleic	S	100	D,Fs	PIL
Calamide CW-100	coco	R	100	D,Fs	PIL
Emid 6511	lauric	S	100	Fs,V	HNK
Emid 6514	coco	S	100	Fs,V	HNK
Emid 6543	tallow	S	100	E,dF,Ta	HNK
Emid 6544	capric	R	100	Fs,W	HNK
Emid 6545	oleic	S	100	E,dF,Ta	HNK
Hartamide 9137	oleic	R	80	A,Fs,V	HCL
Hartamide LDA 70	lauric	S	100	Fs,V	HCL
Hartamide OD	coco	S	100	Fs,S	HCL
Monamine ACO-100	lauric	R	100	D,E,V	MOA
Monamine LM-100	lauric/myristic	R	100	D,Fs,V,W	MOA
Monamine T-100	tall oil	R	100	D,Fs,V,W	MOA
Monamid 150-AD	coco	S	100	Fs,V	MOA
Monamid 150-IS	isostearic	S	100	E,C	MOA
Ninol 201	oleic	R	100	Fs,V	STP
Ninol 55-LL	lauric	S	100	D,Fs,V	STP
Alkamide Rodea	ricinoleic	R	100	E,S	RHP
Schercomid CDA	coco	R	100	D,Di,Fs	SBC
Schercomid SL-ML	lauric/myristic	S	100	F,Fs,V	SBC
Standamid CD	capric	R	98	D,E,Fs,St,V	HNK
Standamid KD	coco	S	100	F,V	HNK
Varamide MA-1	coco	S	90	E,Fs,V	WTC
Varamide A-2	coco	R	100	D	WTC
Varamide ML-1	lauric	S	100	D,Fs	WTC
Witcamide 82	coco	S	100	V	WTC
Witcamide 5195	lauric	R		Fs,V	WTC

[a]R = regular; S = super.
[b]See Table 1.
[c]See Table 2.

Table 24. Monoalkanolamides of Fatty Acids[a]

Trade name	Fatty acid	Amine	Functions[b]	Manufacturer[c]
Emid 6500	coco	ethanol	Fs,V	HNK
Intermediate 300	tallow/coco	ethanol	Fs,St,V	WTC
Intermediate 325	coco	ethanol	Fs,St,V	WTC
Monamid LIPA	lauric	isopropanol	Fs,V	MOA
Monamid LMA	lauric	ethanol	Fs,V	MOA
Monamid S	stearic	ethanol	O	MOA
P&G Amide No. 27	coco	ethanol	Fs,St,V	PG
Schercomid CME	coco	ethanol	D,Fs	SBC
Witcamide CPA	coco	isopropanol	C,E,V	WTC
Witcamide 70[d]	stearic	ethanol	C,O,V	WTC

[a]Concentration = 100 wt % unless otherwise noted.
[b]See Table 1.
[c]See Table 2.
[d]Concentration = >90 wt %.

520

Polyoxyethylene Fatty Acid Amides. Ethoxylation of fatty acid amides with one or two mole ethylene oxide yields nominally the same compositions as are obtained by condensing mono- or diethanolamines with fatty acids. Ethoxylation of fatty acid amides with more than 1 mol ethylene oxide yields predominantly secondary amides because the second amide hydrogen is less reactive than the first and also less reactive than the hydroxyl hydrogen. Ethoxylation can be carried out by conventional technology. Although the products of condensation are generally preferred, the longer-chain ethylene oxide adducts of fatty amides sometimes exhibit special properties. Some representative products are given in Table 25.

Fatty Acid Glucamides. Fatty acyl glucamides (FAGA) or polyhydroxy-amides (PHA) have been adopted by detergent manufacturers in the United States and Europe (63). Whereas exact details of FAGA preparation are proprietary, FAGA is produced via reaction of fatty acid methyl ester with N-methyl glucamine and attendant elimination of methanol (97). The methyl ester would be produced via the standard route of transesterification with fatty triglycerides; the glucamine, via reaction between glucose and methylamine with attendant hydrogenation and elimination of water.

Fatty acid glucamides have appeared at significant levels in dishwashing liquids (Europe and U.S.) and heavy-duty liquids (U.S.). In U.S. hand dishwashing liquids their inclusion has been accompanied by reduction in amine oxide levels. In the case of U.S. heavy-duty liquids, inclusion of FAGAs has been accompanied by reduction or complete elimination of LAS (63). Potential benefits include improved mildness for dishwashing liquids and improved enzyme stability in fabric washing detergents. In some situations, nonionic FAGAs may appear to be simple replacements for LAS, a nonionic surfactant. However, in comparison to nonionic alcohol ethoxylates, FAGA has a more compact head group. This can give stronger interaction and more efficient packing with anionics such as alcohol sulfates and ethoxy sulfates. This could lead to improved formulation flexibility and stability. It could be argued that different packing parameters at soil interfaces could lead to improved detergency, although such benefits are often difficult to demonstrate under practical conditions. FAGAs have relatively high Krafft points which could lead to solubility problems. Accordingly, the FAGAs

Table 25. Polyoxyethylene Fatty Acid Amides[a]

Trade name	Fatty acid	POE[b]	Functions[c]	Manufacturer[d]
Amidox C-2	coco	2	D,E,Fs	STP
Amidox C-5	coco	5	D,E,Fs	STP
Amidox L-2	lauric	2	D,E,Fs	STP
Amidox L-5	lauric	5	D,E,Fs	STP
Ethomid HT/15	hydrogenated tallow	5	D,Di,E	AKZ
Ethomid HT/60	hydrogenated tallow	50	D,Di,E	AKZ
Ethomid O/17	oleic	5	D,Di,E	AKZ

[a]Concentration = 100 wt %.
[b]Number of ethylene oxide groups in polyoxyethylene (POE).
[c]See Table 1.
[d]See Table 2.

are best formulated with anionic surfactants. It has been estimated that the cost of FAGA is less than that of APG at comparable production levels (62).

Polyalkylene Oxide Block Copolymers. The higher alkylene oxides derived from propylene, butylene, styrene (qv), and cyclohexene react with active oxygens in a manner analogous to the reaction of ethylene oxide. Because the hydrophilic oxygen constitutes a smaller proportion of these molecules, the net effect is that the oxides, unlike ethylene oxide, are hydrophobic. The higher oxides are not used commercially as surfactant raw materials except for minor quantities that are employed as chain terminators in polyoxyethylene surfactants to lower the foaming tendency. The hydrophobic nature of propylene oxide units, $-CH(CH_3)CH_2O-$, has been utilized in several ways in the manufacture of surfactants. Manufacture, properties, and uses of poly(oxyethylene-co-oxypropylene) have been reviewed (98).

Poly(Oxyethylene-co-Oxypropylene) Nonionic Surfactants. A great variety of these surfactants is marketed by BASF Corporation under the Pluronic polyol trademark. The synthesis follows:

$$\underset{\underset{\text{HOCHCH}_2\text{OH}}{\overset{\text{CH}_3}{|}}}{} + (b-1)\ \underset{\text{CH}_3\text{CHCH}_2}{\overset{\overset{\text{O}}{\diagup\!\backslash}}{}} \xrightarrow{\text{OH}} \underset{\text{HO}\!\!-\!\!(\text{CHCH}_2\text{O})_b\!\!-\!\!\text{H}}{\overset{\text{CH}_3}{|}}$$

$$\underset{\text{HO}\!-\!(\text{CHCH}_2\text{O})_b\!-\!\text{H}}{\overset{\text{CH}_3}{|}} + 2a\ \underset{\text{CH}_2\!\!-\!\!\text{CH}_2}{\overset{\overset{\text{O}}{\diagup\!\backslash}}{}} \longrightarrow \underset{\text{HO}(\text{CH}_2\text{CH}_2\text{O})_a\!\!-\!\!(\text{CH}\!-\!\text{CH}_2\text{O})_b\!\!-\!\!(\text{CH}_2\text{CH}_2\text{O})_a\text{H}}{\overset{\text{CH}_3}{|}}$$

The structure of individual block polymers is determined by the nature of the initiator (1,2-propanediol above), the sequence of addition of propylene and ethylene oxides, and the percentage of propylene and ethylene oxides in the surfactant. Thus, when the order of addition is reversed, a different structure is obtained in which the hydrophobic moieties are on the outside of the molecule. With ethylene glycol as the initiator, the reactions are as follows:

$$\text{HOCH}_2\text{CH}_2\text{OH} + (n-1)\ \underset{\text{CH}_2\!\!-\!\!\text{CH}_2}{\overset{\overset{\text{O}}{\diagup\!\backslash}}{}} \longrightarrow \text{HO}\!\!-\!\!(\text{CH}_2\text{CH}_2\text{O})_n\!\!-\!\!\text{H}$$

$$\text{HO}\!\!-\!\!(\text{CH}_2\text{CH}_2\text{O})_n\!\!-\!\!\text{H} + 2b\ \underset{\text{CH}_3\text{CH}\!\!-\!\!\text{CH}_2}{\overset{\overset{\text{O}}{\diagup\!\backslash}}{}} \longrightarrow \underset{\text{H}\!\!-\!\!(\text{OCHCH}_2)_b\!\!-\!\!(\text{CCH}_2\text{CH}_2)_n\!\!-\!\!(\text{CH}_2\text{CHO})_b\!\!-\!\!\text{H}}{\overset{\text{CH}_3 \qquad\qquad\qquad\qquad \text{CH}_3}{|\qquad\qquad\qquad\qquad\quad |}}$$

A third series of surfactants can be prepared by simultaneous addition of ethylene oxide and propylene oxide. Initially, propylene oxide predominates in the mixed oxides. In the second reaction step, ethylene oxide predominates. Using a trifunctional initiator, eg, 1,1,1-trimethylolpropane, the final surfactant structure consists of three polyoxyalkylene chains each containing blocks of polyoxyethylene and polyoxypropylene, attached to a single carbon (99). Pluronic polyol surfactants vary from mobile liquids to flakeable solids. Solubility in water ranges from slightly soluble to completely miscible at the boil. Pluronic polyols with high ethylene oxide content exhibit no solution cloud point even at 100°C. In general, they are soluble in aromatic solvents but insoluble in kerosene or mineral oil. They exhibit low toxicity, are nonirritating to the skin, and almost tasteless. Chemical stability is comparable to that of the alcohol ethoxylates.

Block polymer nonionic surfactants are not strongly surface-active but exhibit commercially useful surfactant properties. Aqueous solutions characteristically foam less than those of other surfactant types. They act as detergents, wetting and rinsing agents, demulsifiers and emulsifiers, dispersants, and solubilizers. They are used in automatic dishwashing detergent compositions, cosmetic preparations, spin finishing compositions for textile processing, metal-cleaning formulations, papermaking, and other technologies. Block polymer nonionic surfactants are manufactured by Dow Chemical Company and BASF Corporation.

Cationic Surfactants

The hydrophobic moiety of a cationic surfactant carries a positive charge when dissolved in aqueous media, which resides on an amino or quaternary nitrogen. A single amino nitrogen is sufficiently hydrophilic to solubilize a detergent-range hydrophobe when protonated in dilute acidic solution; eg, laurylamine is soluble in dilute hydrochloric acid. For increased water solubility, additional primary, secondary, or tertiary amino groups can be introduced or the amino nitrogen can be quaternized with low molecular weight alkyl groups such as methyl or hydroxyethyl. Quaternary nitrogen compounds are strong bases that form essentially neutral salts with hydrochloric and sulfuric acids. Most quaternary nitrogen surfactants are soluble even in alkaline aqueous solutions. Polyoxyethylated amino surfactants behave like nonionic surfactants in alkaline solutions and like cationic surfactants in acid solutions.

Cationic surfactants are widely used in acidic aqueous and nonaqueous systems as textile softeners, conditioning agents, dispersants, emulsifiers, wetting agents, sanitizers, dye-fixing agents, foam stabilizers, and corrosion inhibitors (100). To some extent, the usage pattern mirrors that of the anionic surfactants in neutral and alkaline solutions. The positively charged cationic surfactants are more strongly adsorbed than anionic or nonionic surfactants on a variety of substrates including textiles, metal, glass (qv), plastics, minerals, and animal and human tissue, which can often carry a negative surface charge. Substantivity of cationic surfactants is the key property in many applications. In general, they are incompatible with anionic surfactants. Reaction of the two large, oppositely charged ions gives a salt insoluble in water. Ethoxylation moderates the tendency to form insoluble products with anionic surfactants.

Many benzenoid quaternary cationic surfactants possess germicidal, fungicidal, or algicidal activity. Solutions of such compounds, alone or in combination with nonionic surfactants, are used as detergent sanitizers in hospital maintenance. Classified as biocidal products, their labeling is regulated by the U.S. EPA. The 1993 U.S. shipments of cationic surfactants represented 16% of the total sales value of surfactant production. Some of this production is used for the preparation of more highly substituted derivatives (101).

Amines. *Oxygen-Free Amines.* Aliphatic mono-, di-, and polyamines derived from fatty and rosin acids make up this class of surfactants. Primary, secondary, and tertiary monoamines with C_{18} alkyl or alkenyl chains constitute the bulk of this class. The products are sold as acetates, naphthenates, or oleates. Principal uses are as ore-flotation agents, corrosion inhibitors, dispersing agents, wetting agents for asphalt (qv), and as intermediates for the production of more highly substituted derivatives. For ore flotation, corrosion inhibition, and incorporation into asphalt, the amines are generally marketed in proprietary formulations.

In addition to the mono- and dialkylamines, representative structures of this class of surfactants include N-alkyltrimethylene diamine, $RNH(CH_2)_3NH_2$, where the alkyl group is derived from coconut, tallow, and soybean oils; or is 9-octadecenyl, 2-alkyl-2-imidazoline (**3**), where R is heptadecyl, heptadecenyl, or mixed alkyl, and 1-(2-aminoethyl)-2-alkyl-2-imidazoline (**4**), where R is heptadecyl, 8-heptadecenyl, or mixed alkyl.

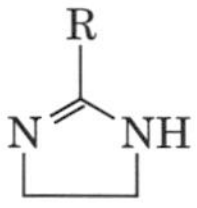

$$
\begin{array}{c}
R \\
\diagup \diagdown \\
N \qquad NH \\
|\_\_\_\_|
\end{array}
$$

(3)

$$
\begin{array}{c}
R \\
\diagup \diagdown \\
N \qquad NCH_2CH_2NH_2 \\
|\_\_\_\_|
\end{array}
$$

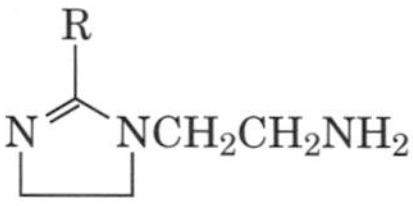

(4)

Oxygen-Containing Amines. This group includes amine oxides, ethoxylated alkylamines, 1-(2-hydroxyethyl)-2-imidazolines, and alkoxylates of ethylenediamine. Oxygen-containing amines are steadily increasing in economic importance.

Amine oxides were first patented in Germany in 1939. They were introduced into the United States in 1956, but did not attract widespread attention until 1961, when the first patents describing their use as surfactants were issued (100,102). The amine oxide group is polar, with the highest electron density around the oxygen atom. Amine oxides exhibit strong hydrogen bonding tendencies and are hygroscopic, cationic in acid solutions, and nonionic in neutral or alkaline solution (103):

$$
\begin{array}{ccc}
O & & OH \\
\uparrow & & | \\
RN(CH_3)_2 + H^+ & \rightleftharpoons & RN^+(CH_3)_2 \\
& & \text{hydroxyammonium ion}
\end{array}
$$

In acidic media, amine oxides and anionic surfactants form precipitates; the CMC is much greater than in neutral or alkaline media. Change in CMC parallels change from ionic to nonionic form. Amine oxides are stable in formulated detergent products and do not act as oxidizing agents. Composition and function of representative commercial amine oxides are given in Table 26.

Table 26. Amine Oxides, RN(CH$_3$)$_2$ or RN(CH$_2$CH$_2$OH)$_2$

$$\overset{O}{\underset{\uparrow}{}} \qquad \overset{O}{\underset{\uparrow}{}}$$

Trade name	R	(CH$_3$)$_2$ or (EO)$_2$[a]	Conc, wt %	Functions[b]	Manufacturer[c]
Ammonyx CO	cetyl	(CH$_3$)$_2$	30	D,E,F,Fs,W	STP
Ammonyx DMCD-40	lauryl	(CH$_3$)$_2$	40	D,F,Fs,W	STP
Ammonyx MO	myristyl	(CH$_3$)$_2$	30	F,Fs,W	STP
Ammonyx SO	stearyl	(CH$_3$)$_2$	25	E,H	STP
Aromox C/12	coco	(EO)$_2$	50	E,Fs	AKZ
Aromox DMCD	coco	(CH$_3$)$_2$	40	D,Fs	AKZ
Aromox DMHTD	hydrogenated tallow	(CH$_3$)$_2$	40	Fs	AKZ
Aromox DM16	hexadecyl	(EO)$_2$	40	Fs	AKZ
Aromox T/12	tallow	(EO)$_2$	50	E,Fs	AKZ
Barlox 10S	decyl	(CH$_3$)$_2$	30	D,Em,Fs,V	LNZ
Barlox 12	coco	(CH$_3$)$_2$	30	D,Em,Fs,V	LNZ
Barlox 14	myristyl	(CH$_3$)$_2$	30	D,Em,Fs,V	LNZ
Ammonix LO	lauryl	(CH$_3$)$_2$	30	D,Em,Fs,V	STP
Ammonix MO	myristyl	(CH$_3$)$_2$	30	D,Em,Fs,V	STP
Schercamox DML	lauryl	(CH$_3$)$_2$	30	D,Em,Fs,V	SBC
Schercamox DMM	myristyl	(CH$_3$)$_2$	30	D,Em,Fs,V	SBC
Standamox O1	oleyl	(CH$_3$)$_2$	55	AM,Em,Ta,V	HNK
Varox 185E	C$_{12-15}$ (oxypropyl)	(EO)$_2$	42	D,E,F,Fs	WTC
Varox 365	lauryl	(CH$_3$)$_2$	30	D,E,F,Fs	WTC

[a](EO)$_2$ = bis(2-hydroxyethyl).
[b]See Table 1.
[c]See Table 2.

Amine oxides have attracted widespread interest as replacements for alkanolamides as foam builders in liquid hand-dishwashing compositions. Although considerably more expensive than alkanolamides, amine oxides have maintained a position in hand-dishwashing formulations. In addition to foam boosting, amine oxides are effective shampoo surfactants, cotton (qv) detergents, wetting agents in concentrated electrolytes, and emulsifiers. They impart softness and body to fabrics and build viscosity in aqueous solutions, and find application in household and industrial detergents, detergent sanitizers, plating-bath additives, and textile mill processing formulations. Amine oxides are marketed as pale yellow aqueous solutions or light yellow pastes having concentrations of 30–55 wt %. They are prepared by oxidation of tertiary amines with hydrogen peroxide (104). Yields are above 85%.

Ethoxylation of alkyl amine ethoxylates is an economical route to obtain the variety of properties required by numerous and sometimes small-volume industrial uses of cationic surfactants. Commercial amine ethoxylates shown in Tables 27 and 28 are derived from linear alkyl amines, aliphatic t-alkyl amines, and rosin (dehydroabietyl) amines. Despite the variety of chemical structures, the amine ethoxylates tend to have similar properties. In general, they are yellow or amber liquids or yellowish low melting solids. Specific gravity at room

Table 27. Aliphatic and Rosin Amine Ethoxylates[a],

Trade name	Amine	$m + n$	Functions[b]	Manufacturer[c]
Ethomeen C/12	coco	2	A,Di,E,Ta	AKZ
Ethomeen C/15	coco	5	A,Di,E,Ta	AKZ
Ethomeen C/20	coco	10	A,Di,E,Ta	AKZ
Ethomeen C/25	coco	15	A,Di,E,Ta	AKZ
Ethomeen S/12	soybean	2	A,Di,E,Ta	AKZ
Ethomeen S/15	soybean	5	A,Di,E,Ta	AKZ
Ethomeen S/20	soybean	10	A,Di,E,Ta	AKZ
Ethomeen S/25	soybean	15	A,Di,E,Ta	AKZ
Ethomeen T/12	tallow	2	A,Di,E,Ta	AKZ
Ethomeen T/15	tallow	5	A,Di,E,Ta	AKZ
Ethomeen T/25	tallow	15	A,Di,E,Ta	AKZ
Ethomeen 18/12	stearyl	2	A,Di,E,Ta	AKZ
Ethomeen 18/15	stearyl	5	A,Di,E,Ta	AKZ
Ethomeen 18/20	stearyl	10	A,Di,E,Ta	AKZ
Triton RW-20	$t\text{-}C_{12}\text{-}C_{14}$[d]	2	D,E,OR	UCC
Triton RW-75	$t\text{-}C_{12}\text{-}C_{14}$[d]	7.5	D,E,OR	UCC
Triton RW-150	$t\text{-}C_{12}\text{-}C_{14}$[d]	15	D,E,OR	UCC
Trymeen CAM-10	coco	10	A,E	HNK
Trymeen SAM-50	stearyl	50	A,Ta	HNK
Trymeen TAM-15	tallow	15	Ta,Ta	HNK
Trymeen TAM-20	tallow	20	A,Sp,Ta	HNK
Trymeen TAM-25	tallow	25	A,Ta	HNK
Trymeen TAM-40	tallow	40	A,E,Ta	HNK

[a]Concentration = 100 wt %.
[b]See Table 1.
[c]See Table 2.
[d]Branched-chain.

Table 28. Fatty Alkyl 1,3-Propanediamine Ethoxylates,

$$RNCH_2CH_2CH_2N \begin{cases} (CH_2,CH_2O)_a\text{—}H \\ (CH_2,CH_2O)_b\text{—}H \end{cases}$$
$$\vert$$
$$(CH_2CH_2O)_c\text{—}H$$

Trade name	R	$a + b + c$	Conc, wt %	Functions[a]	Manufacturer[b]
Ethoduomeen T/13	tallow	3	100	Di,E	AKZ
Ethoduomeen T/20	tallow	10	95	Di,E,W	AKZ
Ethoduomeen T/25	tallow	15	95	Di,E,W	AKZ

[a]See Table 1.
[b]See Table 2.

temperature ranges from 0.9 to 1.15, and they are soluble in acidic media. Higher ethoxylation promotes solubility in neutral and alkaline media. The lower ethoxylates form insoluble salts with fatty acids and other anionic surfactants. Salts of higher ethoxylates are soluble, however. Oil solubility decreases with increasing ethylene oxide content but many ethoxylates with a fairly even hydrophilic–hydrophobic balance show appreciable oil solubility and are used as solutes in the oil phase.

Amine ethoxylates are used as acid thickeners, emulsifiers, dispersants, antistatic agents, textile softeners, and lubricants, and as wetting agents in both acidic and alkaline solutions. As corrosion inhibitors, they are used in aqueous acid solutions, refined petroleum products, and formulations for use in the production and refining of petroleum (qv). In addition, amine ethoxylates are applied as components in mill-processing formulations for fabrication of metal and textile products, as wetting agents for asphalt (qv), and as frothing agents for ore flotation (qv).

Manufacturing processes and equipment are similar to those employed for alcohol ethoxylate preparation. In the absence of steric hindrance, ethylene oxide reacts with both hydrogens of primary amines at relatively low temperatures (90–120°C) without added catalysts (105). When the nitrogen atom is hindered, as it is in the Triton RW products, only one of the amino hydrogens reacts with ethylene oxide. Once this reaction is complete, a basic catalyst is added and ethoxylation proceeds in the manner of the alcohol-based nonionics. In N-alkyl-1,3-propanediamine, all three amino hydrogens are available for reaction with ethylene oxide. N-Alkyl-1,3-propanediamines are prepared from fatty monoamines and acrylonitrile, followed by reduction of the resulting 3-cyanoethylalkyl amine.

Compositions and functions of typical commercial products in the 2-alkyl-1-(2-hydroxyethyl)-2-imidazolines series are given in Table 29. 2-Alkyl-1-(2-hydroxyethyl)-2-imidazolines are used in hydrocarbon and aqueous systems as antistatic agents, corrosion inhibitors, detergents, emulsifiers, softeners, and viscosity builders. They are prepared by heating the salt of a carboxylic acid with (2-hydroxyethyl)ethylenediamine at 150–160°C to form a substituted amide; 1 mol water is eliminated to form the substituted imidazoline with further heating at 180–200°C. Substituted imidazolines yield three series of cationic surfactants: by ethoxylation to form more hydrophilic products; quaternization with benzyl chloride, dimethyl sulfate, and other alkyl halides; and oxidation with hydrogen peroxide to amine oxides.

Table 29. 2-Alkyl-1-(2-Hydroxyethyl)-2-Imidazolines,

Trade name	Fatty acid	Conc, wt %	Functions[a]	Manufacturer[b]
Finazoline CA	coco	100	A,C,E,Sp,Ta	FET
Hodag Amine C-100-L	lauric	100	C,D,E	CAL
Hodag Amine C-100-O	oleic	100	C,D,E	CAL
Hodag Amine C-100-S	stearic	100	C,D,E	CAL
Monazoline CY	caprylic	90	A,AM,C,D,E,Sp,V	MOA
Monazoline C	coco	90	A,AM,C,D,E,Sp,V	MOA
Monazoline O	oleic	90	A,AM,C,D,E,Sp,V	MOA
Monazoline T	tall oil	90	A,AM,C,D,E,Sp,V	MOA
Varine C	coco	100	A,C,E	WTC
Varine O	oleic	100	A,C,E	WTC
Witcamine AL42-12	tall oil	87	A,C,E,I	WTC

[a]See Table 1.
[b]See Table 2.

Ethylenediamine Alkoxylates. The reaction 1,2-alkylene oxides with ethylenediamine forms the basis of a series of surfactants of the following general structure:

$$
\begin{array}{cc}
\text{H}\!-\!(\text{OCH}_2\text{CH}_2)_{\overline{y}}\!(\text{OC}_3\text{H}_6)_x & (\text{C}_3\text{H}_6\text{O})_{\overline{x}}\!(\text{CH}_2\text{CH}_2\text{O})_{\overline{y}}\!-\!\text{H} \\
& \!\diagdown\text{NCH}_2\text{CH}_2\text{N}\diagdown \\
\text{H}\!-\!(\text{OCH}_2\text{CH}_2)_{\overline{y}}\!(\text{OC}_3\text{H}_6)_{\overline{x}} & (\text{C}_3\text{H}_6\text{O})_{\overline{x}}\!(\text{CH}_2\text{CH}_2\text{O})_{\overline{y}}\!-\!\text{H}
\end{array}
$$

where x and y each varies from ca 4–100. As in the oxygen block polymer surfactants, the polyoxypropylene chain contributes hydrophobicity and the polyoxyethylene moiety, hydrophilicity. The two tertiary nitrogens contribute cationic character in the lower molecular weight polymers but are so highly shielded in the higher molecular weight range that the products behave essentially as nonionic compounds.

A number of these structures are offered commercially by BASF Corporation under the trade name Tetronic polyols. The products are similar to oxygen block polymers. Although not strongly surface active per se, they are useful as detergents, emulsifiers, demulsifiers, defoamers, corrosion inhibitors, and lime-soap dispersants. They are reported to confer antistatic properties to textiles and synthetic fibers.

Amines with Amide Linkages. Representatives of this group are prepared from carboxylic acids and di- and polyamines. The amide linkage connects the amine to relatively inexpensive hydrophobes. Formulas for typical amide amines are as follow:

$$\text{RCONHCH}_2\text{CH}_2\text{NHCH}_2\text{CH}_2\text{NH}_2\text{ORRCON(CH}_2\text{CH}_2\text{NH}_2)_2$$

where R is derived from coconut, oleic, stearic, or tall oil acids, and

$$\text{RCONHCH}_2\text{CH}_2\text{CH}_2\text{N(CH}_3)_2$$

$$\text{RCONHCH}_2\text{CH}_2\text{CH}_2\text{NHCH}_2\text{CH}_2\text{CH}_2\text{NH}_2\text{ORRCON(CH}_2\text{CH}_2\text{CH}_2\text{NH}_2)_2$$

$$\text{RCONHCH}_2\text{CH}_2\text{NHCH}_2\text{CH}_2\text{NHCH}_2\text{CH}_2\text{NHCH}_2\text{CH}_2\text{NH}_2$$

where R is derived from stearic acid. Ethoxylation of these structures leads to surfactants more soluble in water. Amide amines are dark-colored liquids that find application in corrosion inhibitors, petroleum demulsifiers, and metal-processing formulations.

Quaternary Ammonium Salts. The quaternary ammonium ion is a much stronger hydrophile than primary, secondary, or tertiary amino groups, strong enough to carry a hydrophobe into solution in the surfactant molecular weight range, even in alkaline media. The discrete positive charge on the quaternary ammonium ion promotes strong adsorption on negatively charged substrates, such as fabrics, and is the basis for the widespread use of these surfactants in domestic fabric-softening compositions (see QUATERNARY AMMONIUM COMPOUNDS). Representative commercial quaternary surfactants are given in Tables 30–33.

Table 30. Dialkyldimethylammonium Salts[a], RR'N⁺(CH₃)₂A

Trade name	R and R'	Functions[b]	Manufacturer[c]
Adogen 462	coco	C,Di,E	WTC
Adogen 470	tallow	Sp	WTC
Arquad 2C-75	coco	C,E	AKZ
Arquad 2HT-75	stearyl	Sp	AKZ
Kemamine Q9702C	hydrogenated tallow	E,H,Sp	WTC
Kemamine Q6502C	coco	C,Di,E	WTC
Kemamine Q2802C	behenyl	H,Sp	WTC
Variquat K300	coco	C,E	WTC
Varisoft 137[d]	hydrogenated tallow	Sp	WTC

[a]Concentration = 75 wt % and A = Cl unless otherwise noted.
[b]See Table 1.
[c]See Table 2.
[d]Concentration = 90 wt %; A = CH₃SO₄.

Table 31. Alkylbenzyldimethylammonium Chlorides, R(C₆H₅CH₂)N(CH₃)₂Cl

Trade name	R	Conc, wt %	Functions[a]	Manufacturer[b]
Barquat MB-50	myristyl	50	AM	LNZ
Barquat OJ-50	oleyl	50	AM	LNZ
BTC 50 USP	C₁₂,₁₄,₁₆,₁₈	50	AM	STP
BTC 824	C₁₄,₁₆,₁₈	50	AM	STP
Dehyquart	lauryl	34	AM	HNK
Kemamine BQ 2802C	behenyl	75	A,E,H	WTC
Kemamine BQ 9742C	tallow	75	Di,dE	WTC
Variquat 50ME	C₁₂,₁₄,₁₆	57.5	AM	WTC

[a]See Table 1.
[b]See Table 2.

Table 32. Alkyltrimethylammonium Salts, RN⁺(CH₃)₃A

Trade name	R	A	Conc, wt %	Functions[a]	Manufacturer[b]
Acetoquat CTAB	cetyl	Br	95	AM	ACT
Bromat	cetyl	Br	100	AM	HXL
Kemamine Q9743C	tallow	Cl	75	AM,dE,Sp	WTC

[a]See Table 1.
[b]See Table 2.

Table 33. Alkylpyridinium Halides, R—N⁺(C₆H₅)—X⁻

Trade name	R	X	Conc, wt %	Functions[a]	Manufacturer[b]
Acetoquat CPC	cetyl	Cl	100	AM	ACI
Acetoquat CPB	cetyl	Br	95	AM	ACI
Dehyquart crystals	lauryl	Cl	90–94	A,E,H	HNK

[a]See Table 1.
[b]See Table 2.

In addition to the structures in Tables 30–33, commercial quaternary surfactants include a variety of modifications, prepared, eg, by quaternization of simple tertiary amines:

$$RN(CH_2CH_2OH)_2 + R'X \longrightarrow R'R\overset{+}{N}(CH_2CH_2OH)_2X^-$$

$$R-\underset{\underset{CH_2CH_2OH}{|}}{\overset{N}{\diagdown}} + R'X \longrightarrow R-\underset{\underset{R'}{\diagup}\underset{CH_2CH_2OH}{\diagdown}}{\overset{N}{\diagdown}} X^-$$

and of amide–amines:

$$RCONHCH_2CH_2CH_2N(CH_3)_2 + R'X \longrightarrow RCONHCH_2CH_2CH_2NR'(CH_3)_2X$$

Quaternary Ammonium Esters. The most rapidly growing category of quaternary ammonium salts are the ester quaternaries which utilize the presence of one or more ester linkages in the molecule to improve biodegradability (106). The ester quaternary most commonly used in household softeners in the United States is *N,N*-di(tallowoyl-*oxy*-ethyl)-*N,N*-dimethyl ammonium chloride, or diethyl ester dimethyl ammonium chloride (DEEDMAC) (107). Its structure results from the incorporation of ethyl ester groups between tallow chains and nitrogen in the ditallow dimethyl ammonium chloride molecule. Hydrolysis of the ester linkages under either acidic or alkaline conditions increases the overall rate of biodegradation of the molecule. A related structure in which one of the methyl groups in DEEDMAC is replaced with hydroxyethyl is also being used in some commercial applications (108).

Amphoteric Surfactants

Amphoteric surfactants contain both an acidic and basic hydrophilic group. Ether or hydroxyl groups may also be present to enhance the hydrophilicity of the surfactant molecule. Examples of amphoteric surfactants include amino acids and their derivatives in which the nitrogen atom tends to become protonated with decreasing pH of the solution. Amino acid salts, under these conditions, contain both a positive and a negative charge on the same molecule. In alkylbetaines, $RN^+(CH_3)_2CH_2CO_2$, discrete opposing charges are present in the molecule at all pH values.

Amphoteric surfactants are generally considered specialty surfactants, however, usage has expanded significantly. In 1993 shipments represented 2.8% of the total value for U.S. surfactants. Amphoteric surfactants do not irritate skin and eyes, exhibit good surfactant properties over a wide pH range, and are compatible with anionic and cationic surfactants. A basic nitrogen and an acidic carboxylate group are the predominant functional groups. In some structures sulfonate or sulfate provide the negative charge. Among the simplest amphoteric structures are the alkylbetaines, prepared from alkyldimethylamines and sodium chloroacetate. Relatively inexpensive alkyldimethylamines containing an amide group can be prepared from fatty acids and 3-dimethylaminopropylamine. Reaction with sodium chloroacetate yields acylamidopropylbetaine. Commercial alkyl betaines and their amidopropyl analogues are described in Tables 34 and 35.

Imidazolinium Derivatives. Amphoteric imidazolinium derivatives are prepared from the 2-alkyl-1-(2-hydroxyethyl)-2-imidazolines and from sodium chloroacetate. The most likely structure of the reaction product is as follows (109):

$$\text{HOCH}_2\text{CH}_2\text{—N}\overset{\overset{\displaystyle R}{|}}{\underset{+}{\text{—N}}}\text{—CH}_2\text{CH}_2\text{CO}_2^-$$

Table 34. Alkylbetaines, $RN^+(CH_3)_2CH_2CO_2$

Trade name	R	Conc, wt %	Functions[a]	Manufacturer[b]
Emcol CC 37-18	coco	45	D,E,F,Fs,V	WTC
Lonzaine 12C	coco	35	E,F,H	LNZ
Lonzaine 16S	cetyl	35	E,F,H	LNZ
Varion CDG	lauryl	31	D,E,F,V	WTC
Velvetex AB-45	coco	43–45	H,V	HNK
Velvetex OLB-50	oleyl	48–52	H	HNK

[a]See Table 1.
[b]See Table 2.

Table 35. Amidopropylbetaines, $RCONHCH_2CH_2CH_2N^+(CH_3)_2CH_2CO_2$

Trade name	RCO	Conc, wt %	Functions[a]	Manufacturer[b]
Amphosol CA	cocoyl	30	D,F,Fs	STP
Lonzaine CO	cocoyl	35	D,E,F	LNZ
Monateric CAB-LC	cocoyl	30	D,E,H,W	MOA
Monateric LMAB	lauroyl	30	D,E,H,W	MOA
Schercotaine CAB	cocoyl	45	D,F,W	SBC
Schercotaine IAB	isostearyl	30	C,Sp,V	SBC
Schercotaine MAB	myristoyl	35	A,D,V,W	SBC
Schercotaine PAB	palmitoyl	35	C,V	SBC
Varion (CADG)-W	cocoyl	35–37	D,E,F,V	WTC
Velvetex BC 35	cocoyl	34–37	A,D,F,Fs,H	HNK

[a]See Table 1.
[b]See Table 2.

Unlike the 2-alkyl-2-imidazolines, this structure is stable and resistant to hydrolysis. After ring cleavage, reaction with sodium chloroacetate yields linear products:

$$RCONHCH_2CH_2N^+CH_2CH_2OHCl^+ + HCl + ClCH_2COONa$$

$$\longrightarrow RCONHCH_2CH_2N(CH_2CH_2OH)CH_2COONa$$

$$RCONHCH_2CH_2NHCH_2CH_2OH + 2\ ClCH_2COONa \longrightarrow RCONHCH_2CH_2\overset{\overset{\displaystyle CH_2COONa}{|}}{\underset{\underset{\displaystyle CH_2COONa}{|}}{N}}CH_2CH_2OH$$

Commercial imidazolinium derivatives are for the most part not clearly characterized as to their structure and are frequently marketed as proprietary products. As result, it is difficult to relate properties and uses to structural parameters. Imidazolinium derivatives are recommended as detergents, emulsifiers, wetting and hair conditioning agents, foaming agents, fabric softeners, and antistatic agents. There is some evidence that in cosmetic formulations certain imidazolinium derivatives reduce eye irritation caused by sulfate and sulfonate surfactants present in these products.

Production and Economic Aspects

Total U.S. surfactant shipments in 1993 were almost ~5×10^9 kg with a sales value of $\$3.5 \times 10^9$, a 3% increase over 1992 (110). Adjusted for inflation, the actual growth was about 2%, indicative of a mature market. Nonetheless, the market has changed and will continue to do so, as evidenced by the introduction of alkylpolyglucosides, alkylglucosamides, and ester quaternaries in the 1990s, and by intense patent activity. Over the three-year period from 1992 to 1994, 1500 world patents were filed for cleaning products, 33% of which dealt with surfactants (63).

Environmental profile has been and will continue to be an important determinant of a market acceptability of surfactants, but will not guarantee its success (111–113). Price and availability have been primary determinants of hydrophobe choice for alcohol-based feedstocks (114).

A reexamination of so-called renewability has shown that advantages for oleochemicals are not sufficiently clear (115), especially because manufacture of surfactants in the United States accounts for only 0.03% of annual crude oil consumption (62). On these bases, the primary determinants of surfactant choice will continue to be cost effectiveness and availability. The 1993 U.S. market has been estimated to be worth $\$3.7 \times 10^9$ (110). Approximately one-half was anionic surfactant ($\$1.8 \times 10^9$) and one-third nonionic surfactant ($\$1.2 \times 10^9$). The balance was made up by cationics ($\$1.2 \times 10^6$) and amphoterics ($\$600 \times 10^6$). The U.S. International Trade Commission (116) provides a minutely detailed breakdown of surfactant production.

Uses

Household and personal products dominate surfactant technology literature. Sales volume per product in these categories is large enough to justify research and investment in large manufacturing facilities. Industrial applications represent smaller sales volumes per product. Research and especially technical service is often conducted by the user or by a service company that supplies a variety of items to the consuming industry.

Household and Personal Products. Detergency, ie, cleaning, is the primary function of household and personal products. More recently, a secondary function, such as softening in combination with detergency in laundry detergents or conditioning in combination with detergency in shampoos, has been offered as an additional product benefit. In general, products have tended toward functional specialization (Table 36). In 1993, 48.5% of surfactants were used in household products and 8.5% in personal care products. An analysis for 1995 U.S. consumption of principal surfactants showed LAS represented 30% total consumption; AES, 19%; PAS, 16%; alcohol ethoxylates, 18%; and APE, 17% (62).

Industrial Uses. Surfactants are widely used outside the household for a variety of cleaning and other purposes. Often the volume or cost of the surfactant consumed in industrial applications is small compared to benefit. The industrial and institutional market accounted for 43% of 1993 usage (see Table 36).

Table 36. Surfactant Uses

Use	Function

household products[a]
    detergents, scouring powders,
      hard-surface cleaners, fabric
      softeners
	cleansing, abrasion, softening, antistatic

personal care products[a] soaps, shampoos, creams	cleansing, moisturizing, softening

agriculture
    phosphate fertilizers
	manufacturing cycle shortening and
	    prevention of caking during storage

spray applications of herbicides,
      insecticides, fungicides
	wet, disperse, suspend powdered pesticides;
	    emulsification of pesticide solutions;
	    promotion of wet spreading; penetration
	    of toxicant

building and construction
    paving
	improve asphalt-to-gravel and sand bond
	    (prevent stripping)

concrete	promote air entrainment for density, plasticity, insulating control

elastomers and plastics
    emulsion polymerization
	solubilization: monomer and catalyst
	    dissolved, react in surfactant micelles;
	    emulsification of monomers; stabilization
	    of latexes

foamed polymers	introduction of air and cell size control
latex adhesive	promote wetting and improving bond strength

Table 36. (*Continued*)

Use	Function
plastic articles	antistatic agents
plastic coating and laminating	wetting agents
food and beverages	
food-processing plants	clean and sanitize walls, floors, processing equipment
fruits and vegetables	remove pesticide residues and aid in wax coating
bakery products and ice cream	solubilize flavor oils, control consistency, retard staling
beverages	solubilize flavor oils
crystallization of sugar	improve washing and reduce processing time
cooking fats and oils	prevent spattering in frying because of superheating and sudden volatization of water
industrial cleaning	
janitorial supplies	detergents and sanitizers for walls, floors, windows
miscellaneous cleaning	detergents for railroad cars, trucks, airplanes, tunnel walls, engines, etc
descaling	wetting agents and corrosion inhibitors in acid cleaning of boiler tubes and heat exchangers
soft goods	detergents for laundry and dry cleaning
wax strippers	improve wetting and penetration of old finish
leather	
skins	detergent and emulsifier in degreasing
tanning	promote wetting and penetration
hides	emulsifiers in fat liquoring
dyeing	promote wetting, penetration, level dyeing
metals	
concentration of ores	wetting and foaming, ie, collectors and frothers in ore flotation
cutting and forming	wetting, emulsification, lubrication, corrosion inhibition in rolling oils, cutting oils, drawing lubricants, buffing, grinding compounds
casting	mold-release additives
rust and scale removal	wetting, foaming, corrosion inhibition in pickling (acid cleaning) and electrolytic cleaning
grease removal	emulsification and wetting in emulsion cleaning preparations
plating	wetting and foaming in electrolytic plating baths
mining	
mining and transporting coal and minerals	wetting and dedusting

Table 36. (*Continued*)

Use	Function
paper	
pulp treatment	deresinification, pitch dispersion, washing
paper machine	defoaming, felt washing, color leveling, dispensing
calender	wetting and leveling in coating and coloring operations
towels and pads	rewetting (improve moisture absorption)
paints and protective coating	
pigment preparation	flushing (promote preferential wetting by paint vehicle); dispersing and wetting of pigment during grinding
latex paints	emulsify oil or polymer, disperse pigment, stabilize latex, retard sedimentation and pigment separation, modify wetting and rheological properties
waxes and polishes	emulsify waxes, stabilize emulsions, and wet substrates in finishes for floor and automobiles; antistat
petroleum production and products	
drilling fluids	emulsify oils, disperse solids, modify rheological properties of drilling-and-completion fluids for oil and gas wells
mist drilling	convert intrusion water to foam in air drilling; emulsify and disperse sludge and sediment in clean-out wells; modify wetting of formation at producing zone
secondary and tertiary (enhanced) recovery	in flooding operations, release crude oil from formation surface (preferential wetting)
refined petroleum products	as detergent, sludge dispersant, corrosion inhibitor in fuel oils and turbine oils
textiles	
preparation of fibers and filaments	detergent and emulsifier in raw-wool scouring; dispersant in viscose rayon spin baths; lubricant and antistat in spinning hydrophobic filaments
gray goods preparations	wetting and detergency in slashing and sizing formulations; wetting and detergency in kier boiling and bleaching of cotton and carbonizing of wool; detergency in scouring piece goods; emulsification of processing oils
dyeing and printing	wetting, penetration, solubilization, emulsification, dye leveling, detergency, dispersion
finishing textiles	wetting and emulsification in finishing formulations; softening, lubricating, antistatic additives

[a]Includes extensive consumer products list.

BIBLIOGRAPHY

"Detergency" in *ECT* 1st ed., Vol. 4, pp. 938–960, by A. M. Schwartz, Harris Research Laboratories; "Detergents" in *ECT* 1st ed., Suppl. 1, pp. 190–223, by A. M. Schwartz, Harris Research Laboratories; "Detergency" in *ECT* 2nd ed., Vol. 6, pp. 853–895, by A. M. Schwartz, Harris Research Laboratories; "Surface Active Agents" in *ECT* 1st ed., Vol. 13, pp. 513–536, by D. M. Price, Oakite Products, Inc.; "Nonionic Surfactants" in *ECT* 1st ed., Suppl. 2, pp. 490–522, by R. L. Mayhew and F. E. Woodward, General Aniline & Film Corp.; "Surfactants" in *ECT* 2nd ed., Vol. 19, pp. 507–593, by C. E.. Stevens, Management Research Consultants; "Surfactants and Detersive Systems" in *ECT* 3rd ed., Vol. 332–432, by A. Cahn, Consultant, and J. L. Lynn, Jr., Lever Brothers Co.

1. G. S. Hartley, *Aqueous Solutions of Paraffin-Chain Salts*, Hermann & Cie, Paris, 1936, p. 45.
2. R. A. Gibbons, *Nature* **200**, 665 (Nov. 1963).
3. B. A. Mulley, in M. J. Schick, ed., *Surfactant Science*, Vol. 1, Marcel Dekker, Inc., New York, 1967, Chapt. 13, pp. 422–431.
4. J. Rubinfield, E. M. Emery, and H. D. Cross, *J. Am. Oil Chem. Soc.* **41**, 822 (1964).
5. H. P. Dupuy, *J. Am. Oil Chem. Soc.* **45**, 390A (1968).
6. *International Cosmetic Ingredient Dictionary*, 6th ed., CTFA Information Resources Dept., The Cosmetic, Toiletry, and Fragrance Association, Inc., Washington, D.C., 1995.
7. N. K. Adam, *The Physics and Chemistry of Surfaces*, 3rd ed., Oxford University Press, London, 1941, pp. 107–117.
8. C. R. Caryl, *Ind. Eng. Chem.* **33**, 731 (1941).
9. A. M. Mankowich, *J. Am. Oil Chem. Soc.* **43**, 615 (1966).
10. R. Matalon, *J. Colloid Sci.* **8**, 53 (1953).
11. E. D. Goddard and A. Ackilli, *J. Colloid Sci.* **18**, 585 (1963).
12. J. W. McBain and J. C. Henniker, *Colloid Chem.* **7**, 47 (1950).
13. A. S. Weatherburn and C. H. Bayley, *Textile Res. J.* **22**, 797 (1952).
14. L. H. Flett, L. F. Hoyt, and J. Walter, *Am. Dyestuff Rep.* **41**, 139 (1952).
15. H. W. Fox and W. A. Zisman, *J. Colloid Sci.* **7**, 428 (1952).
16. G. J. Kahan, *J. Colloid Sci.* **6**, 571 (1954).
17. E. D. Goddard, M. P. Aronson, and M. L. Gum, in T. F. Tadrus, ed., *The Effect of Polymers on Dispersion Properties*, Academic Press, Inc., New York, 1981.
18. W. C. Preston, *J. Phys. Colloid Chem.* **52**, 84 (1948).
19. K. Hess, W. Philipoff, and H. Kiessig, *Kolloid Z.* **88**, 49 (1939).
20. C. R. Singleterry, *J. Am. Oil Chem. Soc.* **32**, 446 (1955).
21. S. Kaufman and C. R. Singleterry, *J. Colloid Sci.* **10**, 139 (1955).
22. E. J. Fendler, J. H. Fendler, R. T. Medary, and O. A. El Seoud, *J. Phys. Chem.* **77**, 1432 (1973).
23. O. A. El Seoud, E. J. Fendler, J. H. Fendler, and R. T. Medary, *J. Phys. Chem.* **77**, 1976 (1973).
24. J. H. Fendler, E. J. Fendler, R. T. Medary, and O. A. El Seoud, *J. Chem. Soc. Faraday Trans. 1*, **69**, 280 (1973).
25. R. Goto and co-workers, *J. Chem. Soc. Jpn.* **75**, 73 (1950).
26. M. J. Schick and A. H. Gilbert, *J. Colloid Sci.* **20**, 464 (1965).
27. M. J. Rosen, *Surfactants and Interfacial Phenomena*, John Wiley & Sons, Inc., New York, 1978, pp. 90–92, and references therein.
28. R. R. Balmbra, J. S. Clunie, J. M. Corkill, and J. F. Goodman, *Trans. Faraday Soc.* **58**, 1661 (1962).
29. H. E. Garrett, in K. Durham, ed., *Surface Activity and Detergency*, MacMillan and Co., Ltd., London, 1961, pp. 37–38.

30. E. H. Crook, D. B. Fordyce, and G. F. Trebbi, *J. Phys. Chem.* **67**, 1987 (1963).

31. P. D. T. Huibers and co-workers, *Langmuir*, **12**, 1462–1470 (1996).

32. J. H. Clint, *J. Chem. Soc. Faraday Trans. 1*, **71**, 1327 (1975).

33. D. G. Hall, *Trans. Faraday Soc.* **66**, 1351, 1359 (1970).

34. P. Mukerjee and M. J. Mysels, *Critical Micelle Concentration of Aqueous Surfactant Systems*, NSROS-NBS 36, U.S. Dept. of Commerce, Washington, D.C., 1971.

35. N. M. van Os, J. R. Haak, and L. A. M. Rupert, eds., *Physico-Chemical Properties of Selected Anionic, Cationic, and Nonionic Surfactants*, Elsevier, Amsterdam, the Netherlands, 1993.

36. G. J. T. Tiddy, *Phys. Rep.* **57**, 1 (1980).

37. T. A. Bostock, M. P. McDonald, G. J. T. Tiddy, and L. Waring, *Surface Active Agents*, SCI, London, 1979, p. 181.

38. J. M. Corkill, J. F. Goodman, and J. R. Tate, *Trans. Faraday Soc.* **60**, 986 (1964).

39. M. J. Schick and D. J. Manning, *J. Am. Oil Chem. Soc.* **43**, 133 (1966).

40. W. D. Harkins, in A. Weissberger, ed., *Physical Methods of Organic Chemistry*, Vol. 1, Interscience Publishers, New York, 1945, Chapt. 6.

41. H. W. Fox and C. H. Chrisman, Jr., *J. Phys. Chem.* **56**, 284 (1952).

42. R. C. Brown and H. McCormack, *Philos. Mag.* **39**, 420 (1948).

43. S. Fordham, *Proc. R. Soc. London Ser. A.* **194**, 1 (1948).

44. A. S. Brown, R. U. Robinson, E. H. Sirois, H. G. Thiboult, W. McNeil, and A. Tofias, *J. Phys. Chem.* **56**, 701 (1952).

45. M. Picon, *Ann. Pharm. Fr.* **6**, 84 (1948).

46. K. L. Sutherland, *Aust. J. Chem.* **7**, 319 (1948).

47. J. L. Cayais, R. S. Schecter, and W. H. Wade, *The Measurement of Low Interfacial Tension via the Spinning Drop Technique*, Dept. of Chemistry and Chemical Engineering, The University of Texas at Austin.

48. Ref. 7, pp. 180–185, 413.

49. B. J. Carroll, *J. Colloid Interface Sci.* **57**, 488 (1976).

50. A. B. Scott and H. V. Tartar, *J. Am. Chem. Soc.* **65**, 692 (1943).

51. M. L. Corrin and W. D. Harkins, *J. Am. Chem. Soc.* **69**, 679 (1947).

52. S. Ross and J. P. Oliver, *J. Phys. Chem.* **63**, 1671 (1959).

53. H. B. Klevens, *J. Phys. Colloid Chem.* **52**, 130 (1948).

54. H. C. Borghetty and C. A. Bergman, *J. Am. Oil Chem. Soc.* **27**, 88 (1950).

55. W. R. Noble, R. G. Bistline, Jr., and W. M. Linfield, *Soap Cosmet. Chem. Spec.* **38**, 62 (July 1972).

56. U.S. Pat. 5,250,727 (Oct. 5, 1993), H. E. Fried (to Shell Oil Co.).

57. U.S. Pat. 2,689,170 (Sept. 14, 1954), W. J. King (to Colgate-Palmolive Co.).

58. U.S. Pats. 2,174,110 and 2,263,312 (Sept. 26, 1959), C. F. Reed (to E. I. du Pont de Nemours & Co., Inc.).

59. W. H. Lockwood, *Chem. Ind.* (*London*) **62**, 760 (1948).

60. J. M. Davis and E. F. Degering, *Proc. Indiana Acad. Sci.* **56**, 116 (1946).

61. W. E. Truce and J. F. Lyons, *J. Am. Chem. Soc.* **73**, 126 (1951).

62. J. S. Birtwistle, "The Global Outlook for Surfactants from a U.S. Perspective," paper presented at *Cesio '96 World Surfactant Congress*, Barcelona, Spain, June 1996.

63. D. J. Kitko, in R. Coffey, ed., *New Horizons, AOCS Press*, Champaign, Ill., 1996, p. 22.

64. Technical Information Bulletin, *Impact of LAB Composition on LAS Performance*, Condea Vista Co., Houston, Tex., 1996.

65. W. C. Browning, in F. Asinger, ed., *4th International Congress on Surface-Active Substances, Brussels*, Sect. A, Vol. 1, Gordon and Breach Science Publishers, New York, 1964, pp. 141–154.

66. E. A. Knaggs and J. W. Hodge, *Petroleum Sulfonates—Key Process Chemicals in Micellar Polymer Oil Recovery Systems*, American Chemical Society, Houston Chemical, Marketing & Economics Division, Houston, Tex., Mar. 27, 1980.

67. *Petroleum Sulfonates*, Technical Service Bulletin, No. 5734, Witco Chemical Corp., Division Sonneborn Chemical & Refining Corp., New York, 1973.

68. C. Bluestein and B. R. Bluestein, in W. M. Linfield, ed., *Surfactant Science Series*, Vol. 7, Part 2, Marcel Dekker, Inc., New York, 1976, pp. 315–343.

69. U.S. Pat. 3,394,155 (July 23, 1968), A. Cahn and H. Lemaire (to Lever Brothers Co.).

70. T. Satsuki, *Inform*, **3**, 1099 (1992).

71. T. Satsuki, in A. Cahn, ed., *Proceedings of the 3rd World Conference on Detergents: Global Perspectives*, AOCS Press, Champaign, Ill., 1994, pp. 135–140.

72. J. K. Weil, R. G. Bistline, and A. J. Stirton, *J. Phys. Chem.* **62**, 1083 (1958).

73. J. K. Weil, A. J. Stirton, R. G. Bistline, and E. W. Maurer, *J. Am. Oil Chem. Soc.* **36**, 241 (1959).

74. R. G. Bistline, A. J. Stirton, J. K. Weil, and E. W. Maurer, *J. Am. Oil Chem. Soc.* **34**, 516 (1957).

75. G. A. Kosalapoff, *Organophosphorus Compounds*, John Wiley & Sons, Inc., New York, 1950.

76. R. L. Mayhew and F. Krupin, *Soap Chem. Spec.* **38**(4), 55, 93, 95; (5) 80, 81, 167, 169 (1962).

77. A. M. Schwartz and J. W. Perry, *Surface Active Agents*, Vol. 1, Interscience Publishers, Inc., New York, 1949, p. 145.

78. L. N. Ferguson, *J. Am. Chem. Soc.* **77**, 5288 (1955).

79. H. Arai, *J. Colloid Interface Sci.* **23**, 348 (1957).

80. P. H. Elworthy and C. B. Macfarlane, *J. Chem. Soc.*, 311 (1964).

81. J. M. Cross, *Proc. Chem. Specialties Mfgs. Assoc.* **135**, 143 (June 1950).

82. U.S. Pat. 2,931,777 (Apr. 5, 1960), H. A. Shelanski (to GAF Corp.).

83. M. N. Fineman, G. L. Brown, and R. J. Meyers, *J. Phys. Chem.* **56**, 963 (1952).

84. A. N. Wrigley, F. D. Smith, and A. J. Stirton, *J. Am. Oil Chem. Soc.* **34**, 39 (1957).

85. U.S. Pat. 4,441,881 (Apr. 10, 1984), T. Padron and R. M. Ruppert (to Lever Brothers Co.).

86. U.S. Pat. 4,210,764 (July 1, 1980), K. Yang, G. L. Neild, and P. H. Washecheck (to Conoco, Inc.).

87. M. T. Cox, in Ref. 63, pp. 143–144.

88. A. Marcomini and co-workers, *Chemosphere*, **17**(5), 853 (1988).

89. C. G. Naylor, *Soap/Cosmet./Chem. Spec.* (Aug. 27, 1992).

90. W. B. Satkowski and W. B. Bennet, *Soap Chem. Spec.* **33**(7), 37 (1957).

91. M. F. Cox, in K. R. Lange, ed., *Detergents and Cleaners*, Hanser, N.Y., 1994, p. 81.

92. P. Schulz, *Chem Oggi*, **10**, 33 (1992).

93. A. D. Urfer and co-workers, in A. R. Baldwin, ed., *Proceedings of the 3rd World Conference on Detergents: Looking Forward to the 90's*, American Oil Chemists Society, Chicago, Ill., 1987, pp. 268–271.

94. U.S. Pat. 2,089,212 (Aug. 10, 1937), W. Krichevsky (to Ninol Laboratories).

95. W. S. Lennon and I. M. Rosenbaum, *Am. Perfum.* **72**(4), 76 (1958).

96. L. J. Garrison and J. H. Pasleau, *Deter. Age*, **98**, 27–29 (Jan. 1968).

97. H. Andree and B. Middelhauve, in Ref. 63, p. 98.

98. I. R. Schmolka, in Ref. 3, Chapt. 10.

99. I. R. Schmolka, *J. Am. Oil Chem. Soc.* **54**, 11 (1977).

100. U.S. Pat. 3,001,945 (Sept. 26, 1961), H. F. Drew and R. E. Zimmer (to Procter & Gamble Co.).

101. R. R. Egan, *J. Am. Oil Chem. Soc.* **45**, 481 (1968).

102. U.S. Pat. 2,999,068 (Sept. 5, 1961), W. Pilcher and S. L. Eton (to Procter & Gamble Co.).

103. K. W. Herrmann, *J. Phys. Chem.* **66**, 295 (1962).

104. D. B. Lake and G. L. K Hoh, *J. Am. Oil Chem. Soc.* **40**, 628 (1963).

105. H. L. Saunders, J. B. Braunwarth, R. B. McConnell, and R. A. Swenson, *J. Am. Oil Chem. Soc.* **46**, 167 (1969).

106. E. S. Baker, paper presented at annual *AOCS Meeting*, Atlanta, Ga., Apr. 1994.

107. U.S. Pat. 5,545,340 (Aug. 13, 1996), E. H. Wahl and co-workers (to Procter & Gamble Co.).

108. R. L. McConnell, *Inform*, **5**(1) (Jan. 1994).

109. B. C. Trivedi, A. J. Digioia, and P. J. Menardi, *J. Am. Oil Chem. Soc.* **58**, 754 (1981).

110. A. Teng, *Happi*, **31**(6), 45 (1994).

111. W. E. Bishop, C. C. Kuta, and C. A. Pittinger, "Life Cycle Analysis and Its Relevance to the Detergent Industry," paper presented *New Horizon's 92 CSMA/AOCS Detergent Industry Conference*, Bolton Landing, New York, Sept. 14, 1992.

112. C. Fusler, in Ref. 63, pp. 58–63.

113. *Tenside*, **32**(2) (1995), devoted to life cycle analysis.

114. W. J. B. Vogel, in Ref. 63, p. 123.

115. J. E. Heinze, "Sustainability of Petro Chemical and Oleo Chemical Based Surfactants," a paper presented at *Cesio '96 World Surfactants Conference*, Barcelona, Spain, June 1996.

116. *Synthetic Organic Chemicals, United States Production and Sales, 1993*, USITC Publ. 2810, United States International Trade Commission, U.S. Government Printing Office, Washington, D.C., 1994.

General References

A Handbook of Industry Terms, Soap and Detergent Association, New York, 1981.

A. W. Adamson, *Physical Chemistry of Surfaces*, 4th ed., John Wiley & Sons, Inc., New York, 1982.

M. Ash and I. Ash, *Handbook of Industrial Surfactants*, Gower, Aldershot, U.K., 1993.

A. R. Baldwin, ed., *Proceedings of the Second World Conference on Detergents: Looking Toward the 90's*, American Oil Chemists Society, Chicago, Ill., 1987.

M. Balsam and E. Sagarin, eds., *Cosmetics, Science and Technology*, 2nd ed., Vols. 1–3, Wiley-Interscience, New York, 1972 and 1974.

P. Becher, *Emulsions: Theory and Practice*, 2nd ed., Reinhold Publishing Corp., New York, 1977.

B. R. Bluestein and C. L. Hilton, eds., *Amphoteric Surfactants, Surfactant Science Series*, Vol. 12, Marcel Dekker, Inc., New York, 1982.

A. Cahn, ed., *Proceedings of the 3rd World Conference on Detergents: Global Perspectives*, AOCS Press, Champaign, Ill., 1994.

B. J. Carroll, in E. Matijevic, ed., *Surface and Colloid Science*, Vol. 9, John Wiley & Sons, Inc., New York, 1976.

R. Coffey, ed., *New Horizons*, AOCS Press, Champaign, Ill., 1996.

J. Cross, ed., *Anionic Surfactants, Chemical Analysis, Surfactant Science Series*, Vol. 8, Marcel Dekker, Inc., New York, 1977.

J. Cross, ed., *Nonionic Surfactants: Chemical Analysis, Surfactant Science Series*, Vol. 19, Marcel Dekker, Inc., New York, 1987.

J. Cross and E. J. Singer, *Cationic Surfactants: Analytical and Biological Evaluation*, Surfactant Science Series, Vol. 53, Marcel Dekker, Inc., New York, 1994.

CSMA Detergents Division, *Test Methods Compendium*, 2nd ed., CSMA, Inc., Washington, D.C., 1985.

W. Cutler and R. C. Davis, ed., *Detergency: Theory and Test Methods, Surfactant Science Series*, Vol. 5, Marcel Dekker, Inc., New York, 1972.

W. G. Cutler and E. Kissa, eds., *Detergency: Theory and Technology, Surfactant Science Series*, Vol. 20, Marcel Dekker, Inc., New York, 1987.

Detergents: In Depth, '89, Soap and Detergent Association, New York, 1989.

J. Falbe, ed., *Surfactants in Consumer Products: Theory, Technology, and Application*, Springer-Verlag, New York, 1987.

S. E. Friberg and B. Lindman, *Organized Solutions: Surfactants in Science and Technology, Surfactant Series*, Vol. 44, Marcel Dekker, Inc., New York, 1992.

P. R. Garrett, *Defoaming: Theory and Industrial Applications, Surfactant Science Series*, Vol. 45, Marcel Dekker, Inc., New York, 1993.

C. Gloxhuber and K. Kunslter, eds., *Anionic Surfactants: Biochemistry, Toxicology, Dermatology, Surfactant Science Series*, Vol. 43, Marcel Dekker, Inc., New York, 1992.

P. Hiemenz, *Principles of Colloid and Surface Chemistry*, Marcel Dekker, Inc., New York, 1977.

A. Huthig, *Waschmittelchemie*, Henkel & Cie, Heidelberg, Germany, 1978.

E. Jungerman, ed., *Cationic Surfactants, Surfactant Science Series*, Vol. 4, Marcel Dekker, Inc., New York, 1970.

A. Kitahara and W. Watanabe, eds., *Electrical Phenomena at Interfaces: Fundamentals, Measurements, and Applications, Surfactants Science Series*, Vol. 15, Marcel Dekker, Inc., New York, 1984.

K. R. Lange, ed., *Detergents and Cleaners*, Hanser Publications, New York, 1994.

W. M. Linfield, ed., *Anionic Surfactants: Surfactant Science Series*, Vol. 7, Marcel Dekker, Inc., New York, 1976.

E. H. Lucassen-Reynders, ed., *Anionic Surfactants: Physical Chemistry of Surfactant Actions, Surfactant Science Series*, Vol. 11, Marcel Dekker, Inc., Basel, Switzerland, 1981.

McCutcheon's Emulsifiers & Detergents, North American & International edition, Glen Rock, N.J., 1996.

McCutcheon's Emulsifiers & Functional Materials, Detergents, North American & International dition, Glen Rock, N.J., 1996.

C. A. Miller and P. Neogi, eds., *Interfacial Phenomena: Equilibrium and Dynamic Effects, Surfactant Science Series*, Vol. 17, Marcel Dekker, Inc., New York, 1985.

B. M. Milwidsky and D. M. Gabriel, *Detergent Analysis: A Handbook for Cost-Effective Quality Control*, Micelle Press, Cranford, N.J., 1989.

K. L. Mittal, ed., *Colloidal Dispersions and Micellar Behavior*, ACS Symposium Series 9, American Chemical Society, Washington, D.C., 1975.

K. L. Mittal, ed., *Micellization, Solubilization, and Microemulsions*, Vols. 1 and 2, Plenum Press, New York, 1977.

K. L. Mittal, ed., *Solution Chemistry of Surfactants*, Vols. 1 and 2, Plenum Press, New York, 1979.

K. L. Mittal and B. Lindman, eds., *Surfactants in Solution*, Vols. 1–3, Plenum Press, New York, 1984; Vols. 4–6, 1987; Vols. 7–10, 1990.

K. L. Mittal and D. O. Shah, *Surfactants in Solution*, Vol. 11, Plenum Press, New York, 1992.

D. Myers, *Surfactant Science & Technology*, VCH, New York, 1988.

I. Piirma, ed., *Polymeric Surfactants, Surfactant Science Series*, Vol. 42, Marcel Dekker, Inc., New York, 1992.

J. M. Richmond, ed., *Cationic Surfactants: Organic Chemistry, Surfactant Science Series*, Vol. 34, Marcel Dekker, Inc., New York, 1990.

M. J. Rosen, ed., *Structure/Performance Relationships in Surfactants*, American Chemical Society, Washington, D.C., 1984.

M. J. Rosen, *Surfactants and Interfacial Phenomena*, 2nd ed., Wiley-Interscience, New York, 1989.

D. N. Rubingh and P. M. Holland, eds., *Cationic Surfactants: Physical Chemistry, Surfactant Science Series*, Vol. 37, Marcel Dekker, Inc., New York, 1991.

M. J. Schick, ed., *Nonionic Surfactants, Surfactant Science Series*, Vol. 1, Marcel Dekker, Inc., New York, 1966.

M. J. Schick, ed., *Nonionic Surfactants, Physical Chemistry, Surfactant Science Series*, Vol. 23, Marcel Dekker, Inc., New York, 1987.

T. M. Schmitt, ed., *Analysis of Surfactants, Surfactant Science Series*, Vol. 40, Marcel Dekker, Inc., New York, 1992.

A. M. Schwartz, in E. Matijevic, ed., *Surface and Colloid Science*, Vol. 5, John Wiley & Sons, Inc., New York, 1972.

A. M. Schwartz and J. W. Perry, *Surface Active Agents: Their Chemistry and Technology*, Krieger, Huntington, N.Y., 1978.

A. M. Schwartz, J. W. Perry, and J. Berch, *Surface Active Agents and Detergents*, Vol. 2, Interscience Publishers, Inc., New York, 1958.

D. J. Shaw, *Introduction to Colloid and Surface Chemistry*, 3rd ed., Butterworth, London, 1981.

K. Shinoda, ed., *Solvent Properties of Surfactant Solutions, Surfactant Science Series*, Vol. 2, Marcel Dekker, Inc., New York, 1967.

L. Spitz, ed., *Soaps and Detergents*, AOCS Press, Champaign, Ill., 1996.

R. D. Swisher, ed., *Surfactant Biodegradation, Surfactant Science Series*, Vol. 3, Marcel Dekker, Inc., New York, 1970.

R. D. Swisher, ed., *Surfactant Biodegradation, Surfactant Science Series*, 2nd ed., Vol. 18, Marcel Dekker, Inc., New York, 1987.

T. F. Tadros, *Surfactants*, Academic Press, London, 1984.

R. Zana, ed., *Surfactant Solutions: New Methods of Investigation, Surfactant Science Series*, Vol. 22, Marcel Dekker, Inc., New York, 1986.

JESSE L. LYNN, JR.
BARBARA H. BORY
Lever Company

SUTURES

Surgical sutures are sterile, flexible strands used to close wounds or to tie off tubular structures such as blood vessels. Made of natural or synthetic fiber and usually attached to a needle, they are available in monofilament or multifilament forms. Sutures are classified by the *United States Pharmacopeia* (USP) (1) as either absorbable or nonabsorbable. The USP also categorizes sutures according to size (diameter) and lists certain performance requirements. Sutures are regulated by the Food and Drug Administration (FDA) as medical devices under the Food, Drug, and Cosmetics (FDC) Act of 1938, the Medical Device Act of 1976, and the Medical Device Reporting regulation of 1995.

Natural fibers (qv) have been used as sutures throughout recorded history. Whenever a fiber was discovered, surgeons would quickly evaluate its potential as a suture. As synthetic fibers such as nylon-6,6 and polyethylene terephthalate became available in the 1950s, surgeons began to use these to replace the commonly used natural fibers such as linen, cotton (qv), surgical gut (collagen), and silk (qv). However, only cotton and linen have been significantly displaced in the suture marketplace. Silk and surgical gut are still widely used by surgeons in the 1990s because of their desirable performance characteristics. Steel (qv) wire is also widely used as a suture.

Beginning in the 1960s, suture manufacturers started developing new fiber-forming polymers specifically for suture use. Synthetic fibers that were biocompatible, strong, and possessed good knot-tying characteristics were sought. The first of the new polymers to achieve commercial success as a suture was polyglycolic acid, which was introduced as Dexon in 1970. Polyglycolic acid-based sutures have achieved unparalleled success in the marketplace primarily because of the ability to be hydrolyzed and subsequently absorbed by the body. Sutures based on other polymers, such as nylon-6, polyethylene, polypropylene, polydioxanone, polybutylene terephthalate, polyglactin, polyglyconate, polybutester, polytetrafluoroethylene, poliglecaprone, lactomer, and glycomer, have achieved varying degrees of commercial success.

Sutures are sold in the United States under generic and/or proprietary names. The most widely sold sutures are listed in Table 1. The generic names for the synthetic polymeric sutures are those that have been approved by the United States Adopted Names Council (USAN).

The principal manufacturers of sutures in the United States are Deknatel in Fall River, Massachusetts; W. L. Gore & Associates in Flagstaff, Arizona; Ethicon in Sommerville, New Jersey; Sherwood-Davis & Geck in Danbury, Connecticut; and United States Surgical in Norwalk, Connecticut.

Sutures are required to hold tissues together until the tissues can heal adequately to support the tensions exerted on the wound during normal activity. Sutures can be used in skin, muscle, fat, organs, and vessels. Nonabsorbable sutures are designed to remain in the body for the life of the patient, and are indicated where permanent wound support is required. Absorbable sutures are designed to lose strength gradually over time by chemical reactions such as hydrolysis. These sutures are ultimately converted to soluble components that are then metabolized and excreted in urine or feces, or as carbon dioxide in expired air. Absorbable sutures are indicated only where temporary wound support is needed.

Because all sutures have finite breaking strengths, proper care must be exercised in use and size selection. Suture failure may entail serious consequences such as herniation of the incision or complete wound disruption (dehiscence). Failure usually occurs in one of three ways: breaking of sutures, tearing of tissues, or untying of knots. The risk of failure can be reduced by selecting the proper suture material and size, by using careful sewing technique, and by tying a knot of the proper configuration to keep it secure.

Monofilament and multifilament sutures behave very differently in surgery. Monofilaments, which pass easily through living tissue and generate little frictional resistance, contain no pores or interstices that might harbor infectious organisms. Multifilament sutures, on the other hand, are frequently coated to reduce frictional drag and damage to tissue, to fill the interstices be-

Table 1. Surgical Sutures Manufactured in the United States

Generic name	Trade name(s)[a]
cotton	
glycomer-631	Biosyn
lactomer	Polysorb
nylon-6	Ethilon
nylon-6,6	Nurolon, Dermalon, Surgilon, Bralon, Monosof
poliglecaprone-25	Monocryl
polybutester	Novafil
polydioxanone	PDS II
polyethylene terephthalate (polyester)	Silky II Polydek, Tevdek II, Ethibond Excel, Mersilene, TI.Cron, Surgidac
polyglactin-910	Vicryl, coated Vicryl
polyglycolic acid	Dexon II, Dexon "S"
polyglyconate	Maxon
polypropylene	Deklene, Prolene, Surgilene, Surgipro
expanded polytetrafluoroethylene (ePTFE)	Gore-Tex
silk	Sofsilk
steel	Flexon
surgical gut	Surgigut

[a]All trade names except polyester are registered trademarks.

tween fibers, and to ease the repositioning of already-tied knots. Coatings must be carefully chosen, because these can affect knot security and handling properties. The chemical and physical properties of sutures, suturing techniques, and tissue reactions to sutures have been the subject of many reviews (2–8).

The high strength fibers that are used to make sutures tend to be stiff, to take a set when packaged, and to exhibit memory, ie, the property of returning to the convoluted shape taken on in the package. Special processing steps have been developed for monofilaments to reduce the stiffness and memory without adversely affecting the strength (9,10). Braided or twisted multifilament sutures are inherently less stiff than monofilaments of comparable strength.

Sutures must be knotted in use, but the knot raises the local stress, so that the suture is most likely to fail at the knot. For this reason, the USP specifies minimum knot-pull tensile strength requirements for sutures. The USP tensile strength is determined by applying a load to the ends of a suture specimen that has been tied around a rubber tube using a surgeon's knot (Fig. 1). The strength is defined as the maximum load recorded before failure, in kilogram force. Depending on the characteristics of the suture material, the knot-pull strength may be as little as one-half the straight-pull tensile strength, which is the strength when tested without a knot.

Although the USP defines the knot-pull tensile strength test in terms of a surgeon's knot, most surgeons use a variety of other knots depending on the type of wound and its location in the body. For example, the sliding knot and the three-throw square knot (Fig. 2) are commonly used. The sliding knot is used more often because it has practical benefits in surgical technique. Additional throws on the knot are often recommended (11–13) to increase the knot-pull tensile strength.

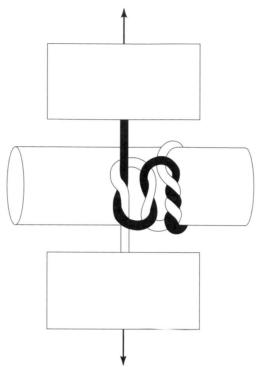

Fig. 1. Illustrations of the USP test for tensile strength. The suture is tied in a surgeon's knot around a short length of rubber tubing.

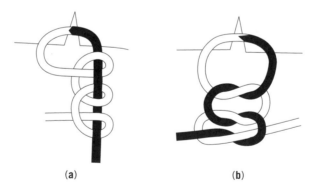

(a)   (b)

Fig. 2. Illustrations of (**a**) a three-throw sliding knot, and (**b**) a three-throw flat square knot.

Similar to any implanted foreign body, sutures produce an initial inflammatory response at the site of the injury. Granulation tissue consisting of fibrous connective tissue and new blood vessels gradually encapsulates the suture as the wound begins to heal. Using absorbable sutures, this tissue is ultimately resorbed as the suture disappears. Natural products such as silk and gut tend to provoke a more severe response than do the synthetic materials. Before marketing a new suture, manufacturers conduct preclinical studies to evaluate the safety and effectiveness of a suture. Most instances of biological incompatibility appear to be the result of the presence of low molecular weight components such as additives, residues of starting materials or processing chemicals, degradation products, or accidental contaminants. Sterilization by ethylene oxide leaves residues of ethylene oxide and its reaction products ethylene chlorohydrin and ethylene glycol. Sterilization by ionizing radiation results in varying degrees of degradation, depending on the chemical structure of the polymer. Some degradation products may be toxic. The suture polymer itself, however, is rarely responsible for toxic effects.

Absorbable Sutures

Absorbable sutures are classified by the USP into collagen and synthetic sutures. Synthetic absorbable sutures are available as braids or monofilaments. Absorbable sutures are only intended for indications where temporary wound support is needed.

Natural Absorbable Sutures. Natural absorbable sutures (collagen) are composed primarily of collagen and are sold as plain, chromic, and mild chromic surgical gut sutures. Contrary to popular belief, catgut sutures are not derived from cats. The term gut probably originated from the fact that collagen sutures are derived from the serosa layer of beef intestine (the gut) or the submucosa layer of sheep intestine, which are mechanically processed and bleached to remove fat, muscle, mucoproteins, and contaminants. The resulting tissue, called goldbeater, is sliced into thin ribbons and may be treated with chromium salts to cross-link the collagen protein chains. The ribbons are either left undyed or dyed black using pyrogallol and ferric ammonium citrate. They are then sorted by size and twisted together under tension to make multiple-plied strands in a range of sizes. After drying, the strands are polished (ground) to a diameter that meets USP standards, as shown in Table 2.

Table 2. USP Specifications for Collagen Sutures[a]

Size		Average diameter, mm		Average knot-pull strength[c]	
USP	Metric[b]	Minimum	Maximum	Kilogram-force	Newton
8-0	0.5	0.050	0.069	0.045	0.441
7-0	0.7	0.070	0.099	0.07	0.686
6-0	1	0.10	0.149	0.18	1.77
5-0	1.5	0.15	0.199	0.38	3.73
4-0	2	0.20	0.249	0.77	7.55
3-0	3	0.30	0.339	1.25	12.26
2-0	3.5	0.35	0.399	2.00	19.61
0	4	0.40	0.499	2.77	27.16
1	5	0.50	0.599	3.80	37.27
2	6	0.60	0.699	4.51	44.23
3	7	0.70	0.799	5.90	57.86
4	8	0.80	0.899	7.00	68.65

[a]Ref. 1.
[b]The metric size is 10 times the minimum diameter.
[c]Values are minimum average of 10 tests (see Fig. 1 for illustration of test method).

All surgical gut sutures lose strength gradually over a period of days, but retain some residual strength for several weeks. Absorption involves enzymatic digestion and is usually complete after several months (14), depending on the type of gut. Plain gut is absorbed more rapidly than chromicized gut. Strength loss and absorption are accelerated in the presence of infection and certain proteolytic enzymes. Animal collagen is a foreign protein. Surgical gut frequently provokes an intense inflammatory response.

Braided Synthetic Absorbable Sutures. Suture manufacturers have searched for many years to find a synthetic alternative to surgical gut. The first successful attempt to make a synthetic absorbable suture was the invention of polylactic acid [26023-30-3] suture (15). The polymer was made by the ring-opening polymerization of L-lactide [95-96-5] (1), the cyclic dimer of L-lactic acid.

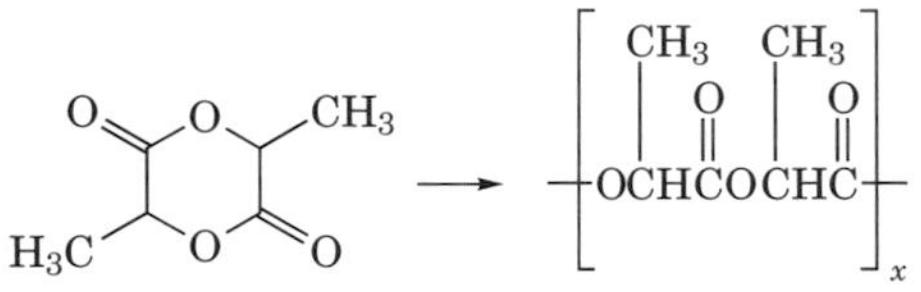

(1)

Polylactic acid sutures are slowly degraded by the following hydrolysis reaction shown and can take years to be completely absorbed (16). These sutures were never commercialized.

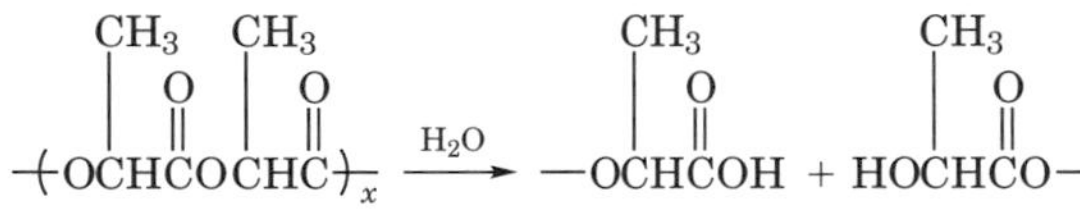

In 1967, a polyglycolic acid [26124-68-5] suture (17) made by the ring-opening polymerization of glycolide (2), the cyclic dimer of glycolic acid, was invented. Polyglycolic acid sutures are degraded by hydrolysis more rapidly than polylactic acid.

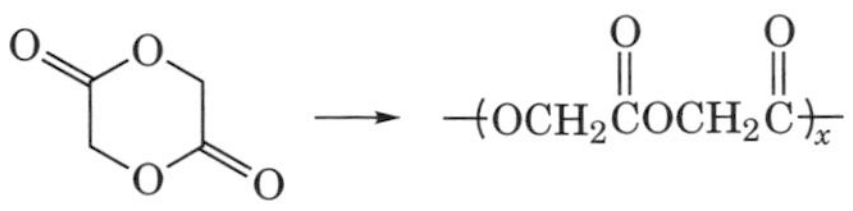

(2)

Polyglycolic acid sutures were commercialized by the Davis & Geck division of American Cyanamid in 1970 under the trade name Dexon and represented a milestone in suture development. This material was not only the first synthetic alternative to surgical gut, but it was also the first synthetic fiber specifically designed for suture use. A few years later Ethicon introduced Vicryl, a copolymer of glycolide and L-lactide having the generic name polyglactin-910 [26780-50-7]. Vicryl is nominally composed of 90% glycolide and 10% L-lactide. Lactomer, another copolymer of glycolide with lactide, was introduced as Polysorb by the United States Surgical Corporation in 1990. The composition of Lactomer is also within the 10% limit on lactide allowed by the FDA.

Polyglycolic acid and poly(glycolide–lactide) copolymers are made by bulk polymerization in the presence of a Lewis acid catalyst. The polymer is melt-spun into fine filaments (yarns) which are then braided to form sutures in a range of different diameters specified by the USP (Table 3). The larger sizes may contain a core of twisted or untwisted yarns in addition to the braided sleeve yarns. Most polyglycolic acid and poly(glycolide–lactide) sutures are coated with proprietary formulations to reduce drag as the suture is passed through tissue.

Polyglycolic acid sutures are sold in the United States under the trade names Dexon II and Dexon "S". Dexon II is coated with polycaprolate (**3**), a copolymer of caprolactone [502-44-3] and glycolide. The sutures are claimed to retain approximately 35% of their out-of-package tensile strength three weeks after implantation and to be absorbed between 60 and 90 days.

$$-\left(\left(OCH_2CH_2CH_2CH_2CH_2\overset{\overset{\displaystyle O}{\|}}{C}\right)_{\!k}\!\!\left(OCH_2\overset{\overset{\displaystyle O}{\|}}{C}OCH_2\overset{\overset{\displaystyle O}{\|}}{C}\right)_{\!l}\right)_{\!x}-$$

(**3**)

Polyglactin-910 suture is available uncoated or coated with a mixture of calcium stearate and polyglactin-370, which is a 30/70 copolymer of glycolide/lactide. The suture is sold under the trade names Vicryl and coated Vicryl.

Table 3. USP Specifications for Synthetic Absorbable Sutures

Size		Average diameter, mm		Average knot-pull strength[a]	
USP	Metric	Minimum	Maximum	Kilogram-force	Newton
10-0	0.2	0.020	0.029	0.025[b]	0.245[b]
9-0	0.3	0.030	0.039	0.050[b]	0.490[b]
8-0	0.4	0.040	0.049	0.07	0.686
7-0	0.5	0.050	0.069	0.14	1.37
6-0	0.7	0.070	0.099	0.25	2.45
5-0	1	0.10	0.149	0.68	6.67
4-0	1.5	0.15	0.199	0.95	9.32
3-0	2	0.20	0.249	1.77	17.36
2-0	3	0.30	0.339	2.68	26.28
0	3.5	0.35	0.399	3.90	38.25
1	4	0.40	0.499	5.08	49.82
2	5	0.50	0.599	6.35	62.27
3 and 4	6	0.60	0.699	7.29	71.49

[a]Values are minimum average of 10 tests.
[b]Measured by a straight-pull test.

It is claimed to retain approximately 50% of its out-of-package tensile strength three weeks after implantation and to be absorbed between 56 and 70 days. A more rapidly degrading version of polyglactin-910 is available under the trade name Vicryl Rapide. This suture is also coated with polyglactin-370 and calcium stearate. It is claimed to lose 50% of its out-of-package strength in five days and to be absorbed in 42 days.

Lactomer suture is coated with a copolymer of caprolactone and glycolide (**3**) and sold under the trade name Polysorb. It is claimed to retain approximately 30% of its out-of-package tensile strength three weeks after implantation and to be absorbed in 56 to 70 days.

Monofilament Synthetic Absorbable Sutures. Ethicon introduced the first monofilament synthetic absorbable suture in 1984 when it marketed PDS polydioxanone (**4**) sutures. The polymer is produced by the bulk polymerization of 2,5-*p*-dioxanone. The suture is distributed under the trade name PDS II. It is claimed to retain approximately 50% of its strength four weeks after implantation, 25% at six weeks, and to be absorbed within six months.

$$-\!\!\left(\text{OCH}_2\text{CH}_2\text{OCH}_2\overset{\displaystyle\overset{\text{O}}{\|}}{\text{C}}\right)_{\!\!x}-$$

(**4**)

Polyglyconate (**5**) is made by the bulk copolymerization of a mixture of 67% glycolide and 33% trimethylene carbonate. The suture is distributed under the trade names Maxon and Maxon CV. It is claimed to retain approximately 50% of its strength four weeks after implantation, 25% at six weeks, and to be essentially completely absorbed in six months.

$$-\!\!\left(\!\left(\text{OCH}_2\overset{\displaystyle\overset{\text{O}}{\|}}{\text{C}}\text{OCH}_2\overset{\displaystyle\overset{\text{O}}{\|}}{\text{C}}\text{O}\right)_{\!\!k}\!\!\left(\text{OCH}_2\text{CH}_2\text{CH}_2\text{O}\overset{\displaystyle\overset{\text{O}}{\|}}{\text{C}}\right)_{\!\!l}\right)_{\!\!x}-$$

(**5**)

Poliglecaprone-25 is a copolymer of 75% glycolide and 25% caprolactone. The molecular weight and the percentages of glycolide and caprolactone are not the same as the coating polymer on Dexon II, but the structure (**3**) is as shown for that polymer. The suture is distributed under the trade name Monocryl. It is claimed to retain approximately 20 to 30% of its strength two weeks after implantation and to be absorbed in 91 to 119 days.

Glycomer-631 (**6**) is a terpolymer produced by the bulk polymerization of a mixture of 60% glycolide, 14% 2,5-*p*-dioxanone, and 26% trimethylene carbonate. The suture is distributed under the trade name Biosyn. It is claimed to retain approximately 40% of its strength three weeks after implantation and to be absorbed completely in 90 to 110 days.

$$\small \text{+}\!\left(\!\text{+OCH}_2\overset{\overset{\displaystyle O}{\|}}{\text{C}}\text{OCH}_2\overset{\overset{\displaystyle O}{\|}}{\text{C}}\!\right)_{\!k}\!\!\left(\!\text{+OCH}_2\text{CH}_2\text{OCH}_2\overset{\overset{\displaystyle O}{\|}}{\text{C}}\!\right)_{\!l}\!\!\left(\!\text{+OCH}_2\text{CH}_2\text{CH}_2\text{O}\overset{\overset{\displaystyle O}{\|}}{\text{C}}\!\right)_{\!m}\!\!\right)_{\!x}$$

(6)

Nonabsorbable Sutures

Nonabsorbable Natural Sutures. Cotton and silk are the only nonabsorbable sutures made from natural fibers that are still available in the United States. Cotton suture is made from fibers harvested from various species of plants belonging to the genus *Gossipium*. The fiber is composed principally of cellulose. The seeds are separated from the cotton bolls, which are carded, combed, and spun into yarns that are then braided or twisted to form sutures in a range of sizes (Table 4). The suture is bleached with hydrogen peroxide and subsequently coated (finished or glaced) with starch and wax. The suture may be white or dyed blue with D&C Blue No. 9.

Silk (qv) suture is made from the threads spun by the silkworm *Bombyx mori*. The fiber is composed principally of the protein fibroin and has a natural coating composed of sericin gum. The gum is usually removed before braiding the silk yarns to make sutures in a range of sizes. Fine silk sutures may be made by simply twisting the gum-coated silk yarns to produce the desired diameter. White silk is undyed. Silk is either dyed black with logwood extract or blue

Table 4. USP Specifications for USP Class I Nonabsorbable Surgical Sutures

Size		Average diameter, mm		Average knot-pull strength[a]	
USP	Metric	Minimum	Maximum	Kilogram-force	Newton
12-0	0.01	0.001	0.009	0.001[b]	0.0098[b]
11-0	0.1	0.010	0.019	0.006[b]	0.059[b]
10-0	0.2	0.020	0.029	0.019[b]	0.186[b]
9-0	0.3	0.030	0.039	0.043[b]	0.422[b]
8-0	0.4	0.040	0.049	0.06	0.588
7-0	0.5	0.050	0.069	0.11	1.07
6-0	0.7	0.070	0.099	0.20	1.96
5-0	1	0.10	0.149	0.40	3.92
4-0	1.5	0.15	0.199	0.60	5.88
3-0	2	0.20	0.249	0.96	9.41
2-0	3	0.30	0.339	1.44	14.12
0	3.5	0.35	0.399	2.16	21.18
1	4	0.40	0.499	2.72	26.67
2	5	0.50	0.599	3.52	34.52
3 and 4	6	0.60	0.699	4.88	47.86
5	7	0.70	0.799	6.16	60.41
6	8	0.80	0.899	7.28	71.39
7	9	0.90	0.999	9.04	88.65

[a]Values are minimum average of 10 tests.
[b]Measured by a straight-pull test.

with D&C Blue No. 9. The suture may be uncoated or coated either with high molecular weight polydimethylsiloxane or with wax.

Although silk and cotton are classified as nonabsorbable sutures, these do lose strength gradually in living tissue and slowly break up after long periods of implantation (18). The USP specifications for Class I nonabsorbable sutures (silk or synthetic fibers) are shown in Table 4.

Braided Synthetic Nonabsorbable Sutures. Braided synthetic nonabsorbable sutures are made by melt-spinning thermoplastic polymers into fine filaments (yarns), and braiding them, with or without a core, to form multifilament sutures in a range of sizes. Nylon-6,6 [32131-17-2] (**7**) is a polyamide produced by the condensation polymerization of adipic acid and 1,6-hexanediamine.

$$\left[\!-\text{NHCH}_2\text{CH}_2\text{CH}_2\text{CH}_2\text{CH}_2\text{CH}_2\text{NH}\overset{\overset{\text{O}}{\|}}{\text{C}}\text{CH}_2\text{CH}_2\text{CH}_2\text{CH}_2\overset{\overset{\text{O}}{\|}}{\text{C}}\!-\right]_x$$

(**7**)

Braided synthetic nonabsorbable sutures may be coated using either high molecular weight polydimethylsiloxane or wax. The sutures are available undyed (clear monofilaments and white braids), or post-dyed black (with logwood extract), blue (FD&C Blue No. 2), or green (D&C Green No. 5). Braided nylon-6,6 sutures are sold under the trade names Bralon, Nurolon, and Surgilon. Although Nylon-6,6 is classified as a nonabsorbable suture, it does slowly hydrolyze in tissue and is not indicated for attachment of synthetic prostheses such as heart valves.

Polyethylene terephthalate [25038-59-9] (**8**) is a polyester produced by the condensation polymerization of dimethyl terephthalate and ethylene glycol. Polyethylene terephthalate sutures are available white (undyed), or dyed green with D&C Green No. 6, or blue with D&C Blue No. 6. These may be coated with polybutylene adipate (polybutilate), polyydimethylsiloxane, or polytetrafluoroethylene [9002-84-0]. The sutures are distributed under the trade names Ethibond Exel, Mersilene, Polydek, Silky II Polydek, Surgidac, Tevdek II, Polyester, and TI.Cron.

$$\left[\!-\text{OCH}_2\text{CH}_2\text{O}\overset{\overset{\text{O}}{\|}}{\text{C}}\!-\!\bigcirc\!-\!\overset{\overset{\text{O}}{\|}}{\text{C}}\!-\right]_x$$

(**8**)

Monofilament Synthetic Nonabsorbable Sutures. Monofilament synthetic nonabsorbable sutures are made from thermoplastic resins melt-spun to form monofilaments. Spinnarets of different capillary diameter are used to make a range of suture sizes.

Nylon-6 [25038-54-4] (**9**) is made by the bulk addition polymerization of caprolactam. Monofilament Nylon-6 sutures are available undyed (clear), or in post-dyed black (with logwood extract), blue (FD&C Blue No. 2), or green (D&C Green No. 5). Monofilament nylon-6 sutures are sold under the trade names Ethilon and Monosof; monofilament nylon-6,6 sutures, under the trade names Dermalon and Ophthalon; and monofilament polyethylene terephthalate sutures, under the trade name Surgidac.

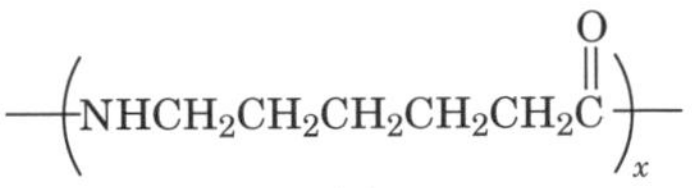

$$-\left(NHCH_2CH_2CH_2CH_2CH_2\overset{\overset{\displaystyle O}{\|}}{C}\right)_x-$$

(9)

Polybutester (**10**) is a polyether–ester produced by the condensation polymerization of dimethyl terephthalate, polytetramethylene ether glycol [25190-06-1], and 1,4-butanediol [110-63-4]. Polybutester sutures are available in clear, ie, undyed, or blue, ie, melt-pigmented with (phthalocyaninato(2-)) copper. Monofilament polybutester is sold under the trade name Novafil.

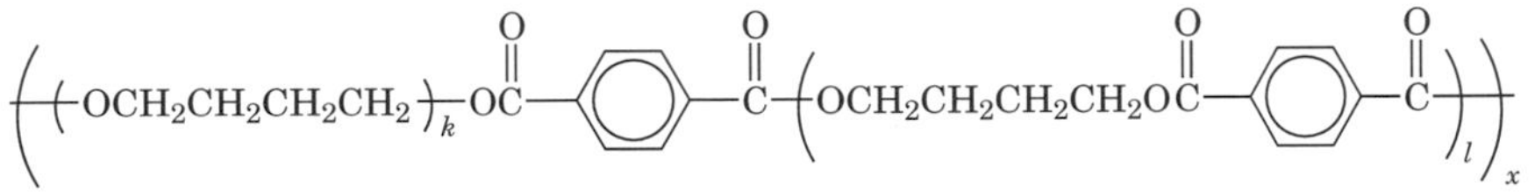

$$\left(\left(OCH_2CH_2CH_2CH_2\right)_{\!k}O\overset{\overset{\displaystyle O}{\|}}{C}-\!\!\bigcirc\!\!-\overset{\overset{\displaystyle O}{\|}}{C}\!\left(OCH_2CH_2CH_2CH_2O\,\overset{\overset{\displaystyle O}{\|}}{C}-\!\!\bigcirc\!\!-\overset{\overset{\displaystyle O}{\|}}{C}\right)_{\!l}\right)_{\!x}$$

(10)

Polypropylene [9003-07-0] is made by the polymerization of propylene gas. Polypropylene (qv) sutures are available in clear (undyed) or blue (melt-pigmented with [phthalocyaninato(2-)] copper). Monofilament polypropylene sutures are sold under the trade names Deklene, Prolene, Surgilene, and Surgipro.

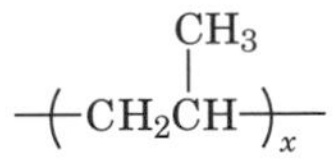

$$-\left(CH_2\overset{\overset{\displaystyle CH_3}{|}}{CH}\right)_x-$$

Polytetrafluoroethylene suture is composed of expanded polytetrafluoroethylene (ePTFE), resulting in a porous microstructure having longitudinally oriented nodes and fibrils. The suture is sold by W. L. Gore & Associates, Inc. under the trade name Gore-Tex Suture.

Steel [52013-36-2] suture is made from 316-L stainless steel wire. The suture may be monofilament, known as fixation wire, or multifilament twisted wires. The steel is heat-treated to improve ductility. The multifilament strands are either uncoated, or coated with Teflon (polytetrafluoroethylene) or Teflon-fluorinated ethylene–propylene copolymer.

Needles

Most sutures are equipped with needles. Surgical needles are made from heat-treatable stainless steel. They can be straight or bent to a 3/8, 1/2, or 5/8 curve. The points of the needles may be ground to a taper (round shaft) for use in tissues that are easy to penetrate, or ground to a triangular or polyhedral geometry to form cutting edges for use in tough tissue such as skin, sclera, or even bone.

The most widely used types of needle points are illustrated in Figure 3. Not illustrated is the blunt point, a taper needle having a rounded point sharp enough to penetrate skin, but not sharp enough to penetrate a rubber glove.

Needles are commonly attached to sutures by mechanical crimping. A suture strand is inserted into the hole or flange or swage at one end of the needle and the metal is crimped onto the strand. Needles may also be attached with adhesives (qv), or by means of a short length of shrink-fit tubing slipped over the ends of the needle and strand. The USP classifies sutures according to the strength of needle attachment. Sutures having standard needle attachment are strongly attached. Sutures having removable needle attachment can be deliberately separated from the needle, when necessary, by means of a quick tug. Needles are often treated with a curable silicone compound to reduce friction, making penetration easier.

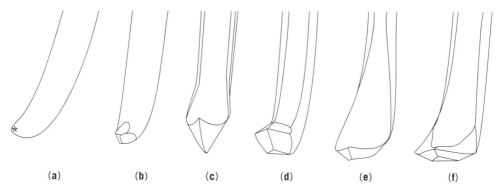

(a)	(b)	(c)	(d)	(e)	(f)

Fig. 3.   Illustrations of needle points: (**a**) taper point, which has a round body tapering to a point; (**b**) cutting taper point, which has a Y-shaped point on a tapered shaft; (**c**) cutting edge point, which is triangular, having one edge on the inside of the curve; or, as shown in this diagram, reverse cutting, which is triangular and has a flat side on the inside of the curve; (**d**) diamond cutting point, a diamond-shaped point having cutting edges in both the lateral and vertical planes on a flattened shaft; (**e**) lancet point, a reverse-cutting-point needle that has the outside edge ground flat; and (**f**) spatula point, which is a flattened hexagon having cutting edges in the lateral plane.

Packaging

A suture package must maintain sterility as well as the physical and chemical integrity of the suture–needle combination. The suture must be sealed in packaging that presents a barrier to infectious agents. In the case of absorbable sutures, the package must also present a barrier to water vapor. The package must be easy to open by the operating room personnel and easily delivered to the sterile field. This is usually accomplished by a dual-package system, although many of Ethicon's packages are single-barrier. The suture is placed inside a paper or plastic holder, which can then be aseptically transferred (dropped into) the sterile field. The inner holder is sealed inside an outer envelope that can be either a peelable aluminum foil or a Tyvek–Mylar (or Tyvek–paper) two-web peelable package.

Sterilization

Commercially available sutures are subjected to a validated sterilization process, and carefully sealed. Ethylene oxide gas as a sterilant for sutures has the advantage that it produces no significant adverse effects on physical properties (see STERILIZATION TECHNIQUES). Using impermeable packages such as aluminum foil, the gas must be introduced before the package is sealed. To reduce the level of residual ethylene oxide in the product, the package is allowed to breathe for a specified time, or evacuated to a specified pressure and time, before sealing. Sterilization by ionizing radiation (γ-radiation from cobalt-60 or β-radiation from an electron beam) offers the advantage that sutures may be sterilized after sealing the package, and that production lots may be released based on radiation dosimetry (dose release) after the sterilization cycle has been validated. However, ionizing radiation reduces the molecular weight of a polymer by the process of chain scission, or alters the properties by cross-linking the polymer. Consequently, synthetic absorbable sutures and polypropylene sutures, which are especially sensitive to radiation effects, are preferably sterilized by ethylene oxide.

Regulatory Requirements

Sutures are regulated by the FDA as medical devices intended for human use. The Federal Food, Drug, and Cosmetic Act of 1938 has been amended many times, in particular by the Medical Device Amendments of 1976. Devices have been under the jurisdiction of the FDA since 1938. Section 513 of the Act requires the FDA to classify medical devices (sutures) in one of three categories as follows.

Class I: General Controls. This category regulates devices for which performance standards or premarket approvals are not required. Class I controls include: A, adulteration (Section 501); B, misbranding (Section 502); C, registration and listing (Section 510); D, premarket notification (Section 510(k)); E, banned devices (Section 516); F, notification, replacement, or refund (Section 518); and G, good manufacturing practices (Section 520(f)). Only stainless steel sutures are classified as Class I devices.

Class II: Performance Standards. This category regulates devices for which General Controls are not sufficient to ensure safety and effectiveness. Class II controls include all the controls in Class I. In addition, manufacturers of sutures in Class II must give the FDA 90-days notice of their intent to market the suture, by way of a 510(k) filing providing data to show that the suture is "substantially equivalent" to an already-approved device. Most sutures were reclassified into Class II in 1989. Other than USP standards, performance standards for sutures have not actually been promulgated, but the FDA has taken the stand that sufficient information exists in the published literature to ensure the safety and effectiveness of sutures in this class.

Class III: Premarket Approval. Similar to a new drug approval, a premarket approval grants the applicant a license to market a specific well-characterized device. These devices are subject to the requirements of Section 515 of the Food, Drug, and Cosmetic Act. A post-amendment device is a device put in commercial distribution after May 28, 1976. If it is not substantially equivalent to a preamendment device it is automatically in Class III, and a premarket approval application (PMA) is required. The application must include reports of preclinical and clinical studies done in support of claims of safety and efficacy as well as any labeling claims made for the device. Once the PMA is submitted, the FDA determines whether the application includes the required information. If the PMA is suitable for scientific review, the FDA has 180 days from the filing date to approve or deny the application. Polybutester, polydioxanone, polyglyconate, and ePTFE sutures are all regulated as Class III devices.

The Medical Device Reporting Regulation (21 CFR 803) applied to sutures approved for commercial distribution. A report must be submitted to the FDA within 30 days after the manufacturer becomes aware of information that reasonably suggests that the suture may have either caused or contributed to death or serious injury, malfunctioned, or the potential of causing death or serious injury. However, if the event would cause the manufacturer to take immediate action to prevent a significant risk to health caused by the suture, then a report must be submitted within five working days.

USP Standards

Voluntary standards for sutures are published by the USP (1). Standard test methods and specifications for diameter, knot-pull tensile strength, straight-pull tensile strength, needle-holding strength, length (as a percentage of labeled length), extractable color, soluble chromium (gut sutures), sterility, and package labeling procedures are given. If a suture complies with all the requirements of the *Pharmacopeia* it may be labeled "USP". Sutures that meet all the requirements except maximum diameter, ie, oversize sutures, may be labeled "USP except for diameter." The package insert then shows a table of maximum overage in millimeters for each USP size suture supplied. Compliance with USP standards is a condition of approval for Class II (performance standards) and Class III (premarket approval) sutures.

The USP lists different specifications for minimum average knot-pull tensile strength of sutures of different chemical composition, eg, silk or synthetic fibers (USP Class I); cotton and linen or coated natural or synthetic fibers where the coating forms a casing of significant thickness but does not contribute appreciably to strength (USP Class II); stainless steel (USP Class III); surgical gut (collagen suture); and synthetic absorbable (synthetic suture). The USP specifications for USP Class I nonabsorbable surgical sutures (silk or synthetic fibers) are reproduced in Table 4.

Economic Aspects

In 1995 the U.S. market for sutures was approximately $695,000,000; the market worldwide was $1,450,000,000. The U.S. market can be broken down by suture type as follows:

Suture type	Approximate sales, $
coated braided synthetic absorbable	164,900,000
polypropylene monofilament	121,000,000
nylon monofilament	61,900,000
silk	53,600,000
surgical gut	52,100,000
coated braided polyethylene terephthalate	34,700,000
monofilament synthetic absorbable	23,800,000
braided nylon	9,600,000
steel	8,400,000
uncoated braided synthetic absorbable	8,200,000
uncoated braided polyethylene terephthalate	4,500,000
polybutester	2,500,000
cotton	80,000

The three-year annual growth rate for sales of all sutures, from 1992 through 1995, was −2.6%. The annual sales growth rate declined by 5.3% from 1994 to 1995. The decline in demand for sutures is predicted to continue owing to cost-containment programs in the hospitals, fewer surgical procedures, competition from other methods of wound closure such as staples, clips, and surgical tapes, and improved surgical procedures (minimally invasive surgery) using less suture. The market for surgical sutures is expected to remain flat in the United States and Europe, with moderate growth predicted for the Asia–Pacific region.

BIBLIOGRAPHY

"Sutures" in *ECT* 3rd ed., Vol. 22, pp. 433–447, by J. B. McPherson, American Cyanamid Co.

1. *The United States Pharmacopeia*, **23**, 1475–1477, 1835–1836, The United States Pharmacopoeial Convention, Inc., Rockville, Md., 1995.
2. D. Ratner, B. R. Nelson, and T. M. Johnson, *Seminars Derm.* **13**, 20–26 (1994).
3. B. C. Benicewicz and P. K. Hopper, *J. Bioactive Comp. Polym.* **6**, 64–94 (1991).
4. B. C. Benicewicz and P. K. Hopper, *J. Bioactive Comp. Polym.* **5**, 453–472 (1990).
5. D. J. Casey and O. G. Lewis, in A. F. von Recum, ed., *Handbook of Biomaterials Evaluation*, Macmillan Publishing Co., New York, 1986, pp. 86–94.
6. T. H. Barrows, *Clinical Materials*, **1**, 233–257 (1986).
7. C. C. Chu, *CRC Crit. Rev. Biocomp.* **1**, 261–322 (1985).
8. C. C. Chu, in M. Szycher, ed., *Biocompatible Polymers, Metals, and Composites*, Technomic Publishing Co., Lancaster, Pa., 1983, pp. 477–523.
9. U.S. Pat. 5,217,485 (June 8, 1993), C. K. Liu and J. C. Brewer (to United States Surgical Corp.).
10. U.S. Pat. 4,911,165 (Mar. 27, 1990), D. J. Lennard, E. V. Menezes, and R. Lilenfeld (to Ethicon, Inc.).
11. S. S. Kadirkamanathan and co-workers, *J. Am. Coll. Surg.* **182**, 46–54 (1996).
12. R. C. Dinsmore, *J. Am. Coll. Surg.* **180**, 689–699 (1995).
13. E. J. C. van Rijssel, J. B. Trimbos, and M. H. Booster, *Am. J. Obstet. Gynecol.* **162**, 93–97 (1990).
14. R. W. Postlethwait, *Ann. Surg.* **181**, 144 (1975).

15. U.S. Pat. 3,636,956 (Jan. 25, 1972), A. K. Schneider (to Ethicon, Inc.).

16. M. Vert and co-workers, in G. W. Hastings and P. Ducheyne, eds., *Macromolecular Biomaterials*, CRC Press, Boca Raton, Fla., 1984, pp. 119–142.

17. U.S. Pat. 3,297,033 (Jan. 10, 1967), E. E. Schmitt and R. A. Polistina (to American Cyanamid Co.).

18. R. W. Postlethwait, *Ann. Surg.* **171**, 892 (1970).

O. GRIFFIN LEWIS
Consultant

WALTER FABISIAK
Sherwood-Davis & Geck

SWEETENERS

Sugar [57-50-1] (sucrose) imparts a sweet taste that is quick, clean, and short-lived. These desirable qualities render sugar (qv) the gold standard for sweet taste. Sugar is also an important functional ingredient for preparing attractive foods. It provides the support for bulkiness, texture, preservation, flavor, and color. However, sugar is a nutritive sweetener. It is easily metabolized, yielding an energy of ca 4 kcal/g (16.7 kJ/g), a fact welcome by some people, but disliked by others. Furthermore, metabolism of sugar and other fermentable carbohydrates (qv) by the microorganisms in the oral cavity contributes to tooth decay. Thus, for obvious reasons, there is a strong demand for alternative sweeteners that possess all the advantages of sugar but do not demonstrate the disadvantages. As society's attitude continues to shift toward slimness and increased health, this demand for good alternative sweeteners is expected only to intensify.

As of this writing (ca 1996), an ideal alternative sweetener does not exist. There are, however, many sweet compounds in use, which generate less calories than sugar, albeit without all the advantages of sugar. These alternative sweeteners can be classified into two groups: nutritive and nonnutritive. Alternative nutritive sweeteners are less caloric than sugar, but retain many of sugar's desirable chemical and physical properties. Hence these are useful as bulking agents in sugar-free products. Principal examples of alternative nutritive sweeteners are the sugar alcohols (qv), eg, sorbitol [50-70-4]; mannitol [69-65-8]; xylitol [87-99-0]; maltitol [585-88-6]; lactitol [585-86-4]; erythritol [149-32-6]; hydrogenated starch hydrolysate; and isomalt, a mixture of glucosyl sorbitol [534-73-6] and glucosylmannitol [20942-99-8]. These alcohols are reduced saccharides resulting from catalytic hydrogenation and, for the most part, are less sweet and less caloric than sugar (ca 2.4 kcal/g (10.0 kJ/g)) (1) and mostly noncariogenic. Erythritol reportedly yields only 0.4 kcal/g (1.67 kJ/g). Sorbitol, mannitol, and xylitol are approved food additives in the United States.

Another example of an alternative nutritive sweetener is fructose [57-48-7]. Although naturally occurring and yielding ca 4 kcal/g (16.7 kJ/g), the same as sugar, fructose does not cause a fluctuation in blood sugar, ie, glucose levels after ingestion, making fructose a better choice for diabetics (2). Fructose is also more potent than sugar (ca 1.5 times) and therefore can be cost-effective for the food industry. The most popular form of fructose used in beverage products is a 55:45 mixture with glucose, usually referred to as high fructose corn syrup (HFCS) (see SYRUPS). Fructooligosaccharide, about half as sweet as sucrose, is a low calorie (ca 2 kcal/g (8.37 kJ/g)) sweetener approved for food use only in Japan (3).

Nonnutritive sweeteners are potently sweet in general and only minute quantities are required for sweetening foods. As such, foods containing nonnutritive sweeteners generate no or negligible calories from the sweeteners themselves, regardless of whether or not these sweeteners are caloric.

Sweetness potency denotes how many times a given compound is more potent than sugar on the same weight basis. For example, because it takes 0.75 g/L of aspartame, a nonnutritive sweetener, to match the sweetness of a 10% (100 g/L) sucrose solution, aspartame is assigned a 133(=100/0.75)-times (commonly written as 133X) sweetness potency. When compared to a lower concentration of sugar solution, the sweetness potency is usually much higher, eg, 180X for aspartame when compared with the sweetness of a 2% sucrose solution. Therefore, reported sweetness potencies must be interpreted carefully, preferably with the % concentration of the matching sucrose solution also indicated. Because sweet beverages commonly employ about 10% sucrose, sweetness intensity matching this sucrose solution is commonly used. The determination of sweetness potency is rather subjective and can be greatly affected by the sensitivity and experience of tasters; other ingredients in the solution, eg, pure water vs flavored beverage; texture of the food; pH; temperature of the samples; etc. Therefore, the published potency should be used only as a guideline and a food technologist should optimize the sweetener level in each product.

In addition to being both sweet and safe, a good alternative sweetener should have other qualities similar to those of sucrose. These include stability as a function of temperature and pH, clean sweet taste, quick onset, no lingering aftertaste, compatibility with other food ingredients, high water solubility, high dissolution rate, and ease of handling. It should also be nonhygroscopic, synergistic with other sweeteners, economical (same or cheaper than sugar based on sweetness equivalence), and have a high degree of consumer acceptance, eg, no perceived toxicity. Even an extremely potent sweetener must possess these qualities in order to be developed as a commercial product.

Another functionality that needs to be addressed from the processed food industry's point of view is maintenance of texture. For diet beverages, water is the bulking agent. Its weight in the formula does not change significantly regardless of what high potency sweetener is used. However, for foods that do not contain much water such as bread and cake, sugar or starch is generally employed as the bulking agent for texture support. If sugar or starch is to be replaced by a high potency sweetener, the formula needs to have a low calorie bulking agent replacement. The commonly used bulking agents are sugar alcohols and polydextrose [68424-04-4] (ca 1 kcal/g (4.18 kJ/g)). Both have shortcomings, eg, laxative side effect, and have to be carefully evaluated before incorporation in foods.

Nonnutritive Sweeteners

As of 1994, there were six principal nonnutritive sweeteners being used throughout the world. In descending order of usage by weight (estimated % of the total worldwide high potency sweetener usage and key countries wherein the sweetener is approved are indicated in parentheses), these were aspartame (41%; United States, Europe, Japan), saccharin (39%; United States, Europe, Japan), cyclamate (16%; Europe), acesulfame-K (2%; United States, Europe), stevioside (1%; Japan), and glycyrrhizin (1%; Japan) (4). Two sweeteners of commercial potential, sucralose and alitame, were awaiting approvals by the FDA as of 1996. Thaumatin, neohesperidin dihydrochalcone, and glycyrrhizin are flavor modifiers of GRAS (generally recognized as safe) status affirmed by Flavor and Extract Manufacturers Association (FEMA) of the United States.

L-Sugars, the enantiomers of naturally occurring D-sugars, have similar chemical, physical, and organoleptic properties, but generate no calories in animals. The stereochemistry does not permit L-sugars to bind the enzymes required for D-sugar metabolism. L-Sugars, which are nonnutritive sweeteners of low potency, have been proposed as bulking agents (5), but as of this writing a food additive petition on behalf of L-sugars had not been presented to the FDA.

Aspartame. Aspartame [22839-47-0; 53906-69-1] (APM, L-aspartyl-L-phenylalanine methyl ester) (1), also known under the trade names of Nutra-Sweet and EQUAL, is the most widely used nonnutritive sweetener worldwide. This dipeptide ester was synthesized as an intermediate for an antiulcer peptide at G. D. Searle in 1965. Although this compound was known in the literature, its sweet taste was serendipitously discovered when a chemist licked his finger which was contaminated with it. Many analogues, especially the more stable esters, were made (6) and their taste qualities and potencies determined. It was the first compound to be chosen for commercial development. Following the purchase of G. D. Searle by Monsanto, the aspartame business was split off to become a separate Monsanto subsidiary called the NutraSweet Company.

$$\text{HOOCCH}_2\overset{\displaystyle |}{\underset{\displaystyle \text{NH}_2}{\text{CH}}}\overset{\displaystyle \text{O}}{\overset{\displaystyle ||}{\text{C}}}\text{NH}\overset{\displaystyle |}{\underset{\displaystyle \text{CH}_2}{\text{CH}}}\overset{\displaystyle \text{O}}{\overset{\displaystyle ||}{\text{C}}}\text{OCH}_3$$

(1)

Aspartame was approved by the FDA in 1981 for use in dry goods. Two years later it was approved for use in carbonated beverages (qv). Additional approvals came in 1993 for baked goods, candies, and still beverages. Aspartame can be used legally in just about all food categories. Economically the most important use of nonnutritive sweeteners is in carbonated beverages. Aspartame has enjoyed great success in this category. In 1992, about 8040 t of aspartame (ca $110/kg ($50/lb)) was used in the United States (ca 83% of world consumption). Of this amount, approximately 85% was for beverages. Through price incentives and rebates, most diet foods that use aspartame contain NutraSweet's brand of aspartame as the sole sweetener, and the package carries a NutraSweet swirl logo. The U.S. patent for aspartame expired in December, 1992. Subsequently, the price has declined. The other principal producer of aspartame is the Holland Sweetener Company (trade name: SANECTA), a joint venture of Dutch and Japanese companies. Aspartame is also made by Ajinomoto (trade name: Pal Sweet) and Miwon, a Korean company.

In soft drinks, aspartame is 180 times more potent than sugar (10% sucrose sweetness equivalence). In water, it is ca 133X. Aspartame has a good sweet taste but its time–intensity profile is different from that of sugar. Aspartame has a slower onset and a lingering sweet taste. Many people also detect an off-taste, which may be remedied by blending aspartame with acesulfame-K. The sweetener mixture exhibits synergistic effect and has a more rounded taste that is much more appealing than either sweetener alone. Aspartame has also been reported to enhance some flavors of foods. Therefore, when converting a sugar-based food to aspartame, the amount of aspartame needed is not necessarily 1/180 of the weight of sugar. Product development using various trial amounts of aspartame is prudent.

Aspartame is caloric. As a peptide, it yields ca 4 kcal/g (16.7 kJ/g). However, because of its high sweetness potency, only a minute quantity is consumed, resulting in a negligible calorie contribution. Aspartame is also soluble in water at room temperature (ca 3 g/100 mL at pH 3) (7). The solubility is a function of pH and temperature. Higher temperature and lower pH increase the solubility. These conditions also increase the rate of decomposition. Aspartame, in its common fine powder form, does not have a high dissolution rate in water. This can be a problem for manufacturing diet beverages. Methods for improving the dissolution rate of aspartame by cogrinding it with food acids (8–11) or polysaccharides (9,12–14), cospray drying with edible bulking agents (15–17), agglomeration (18), and complexation with metal ions (19) have been described in patents.

The methyl ester group of aspartame is very susceptible to bond cleavage. When this cleavage takes place, the sweetness is lost. Avoidance of excessive heat exposure to aspartame is therefore desirable. The impact of heat degradation can be reduced by the encapsulation of aspartame in maltodextrin; fatty acids, eg, hydrogenated cotton seed oil; or other coatings for baking purposes (20). In tropical climates, addition of excess aspartame (a costly option) and rapid supermarket shelf turnover have been employed to maintain product quality. For the same reason, most diet drinks dispensed from a beverage dispenser, ie, soda fountain, contain a blend of aspartame and saccharin, instead of pure aspartame.

In water solution, the relationship between pH and stability of aspartame is a bell curve having maximum stability at pH 4.3 (Fig. 1). At higher or lower

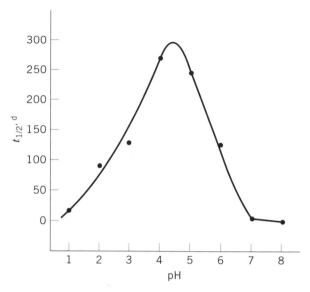

Fig. 1. Stability of aspartame in water at 25°, where $t_{1/2}$ is the half-life (21).

pH, the half-life of aspartame in water diminishes quickly. Most soda and fruit-flavored ready-to-drink beverages are formulated at a pH of approximately 3.0 and 3.0 to 4.5, respectively. To convert these beverages to aspartame-containing diet products, it would be advantageous to adjust the pH as close to 4.3 as possible without changing the original flavor and taste.

The principal pathway for the decomposition of aspartame begins with the cleavage of the ester bond, which may or may not be accompanied by cycliza-tion (Fig. 2). The resultant diketopiperazine and/or dipeptide can be further hy-drolyzed into individual amino acids (qv).

The rate of aspartame degradation in dry mixes is more dependent on the water activity than on the temperature (23). In dry mixes, aspartame may also engage in Maillard reactions with the aldehyde moieties of flavoring agents, resulting in the loss of sweetness and flavor. Use of the corresponding acetals of the flavor compounds to avoid this reaction has been reported (24).

In principle, aspartame is produced through the coupling of two amino acid moieties. One moiety consists of L-phenylalanine methyl ester hydrochloride (**2**) made by treating the amino acid in methanol and hydrochloric acid; the other is aspartic acid anhydride hydrochloride or formic acid salt. The coupling reaction generates two positional isomers, α and β.

Methods (25,26) to increase the ratio of the desired α-isomer (**1**) versus the unsweet β-isomer [22839-61-8] (**3**) exist and are proprietary. The isomers can be separated by subjecting the solution of the final step to hydrochloric acid. The desired α-isomer hydrochloride salt crystallizes out of the solution; the β-isomer remains. There are many patented synthetic processes. The large-scale synthesis of aspartame has been discussed (27–47).

More than a thousand analogues of aspartame have been synthesized in the hopes of producing a second-generation dipeptide sweetener. Compounds

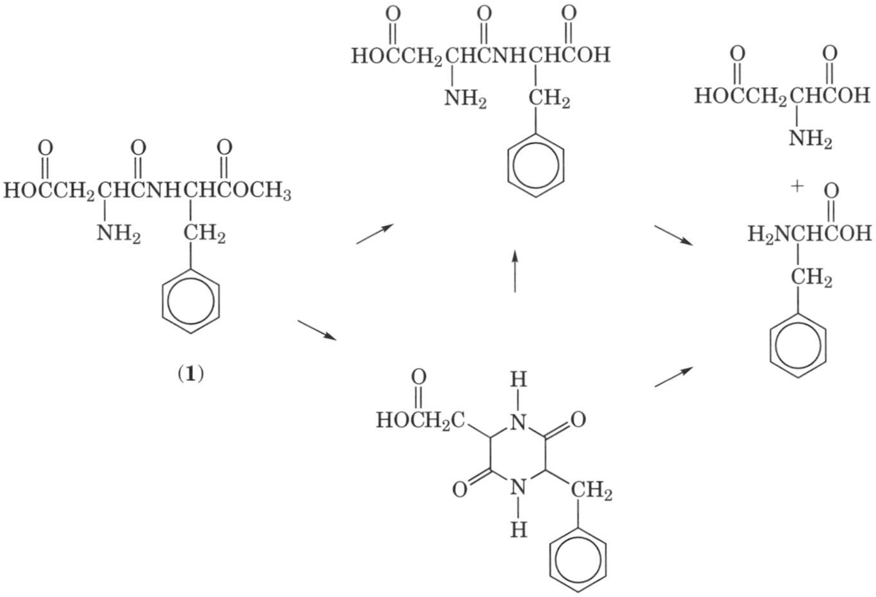

Fig. 2. Decomposition of aspartame to diketopiperazine and/or aspartyl-phenylalanine and then to the amino acids aspartic acid and phenylalanine (22). Courtesy of Marcel Dekker, Inc.

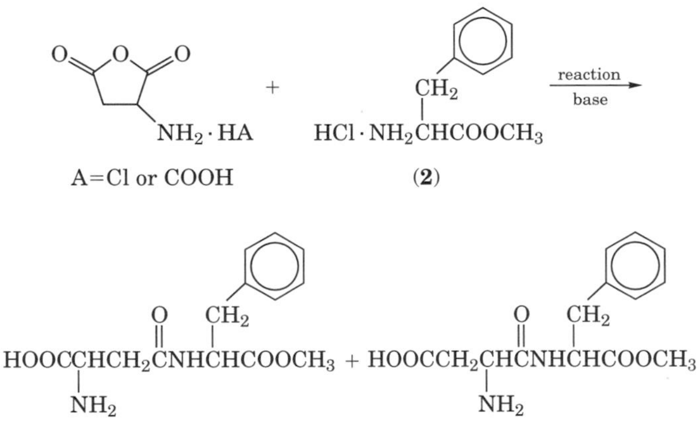

representative of a variety of structural subclasses of aspartyl dipeptides are shown in Table 1. Based on structure–activity relationship analyses, a pattern of structural requirements for binding with the sweetness receptor has emerged (Fig. 3). Most dipeptide analogues have been eliminated from development as commercial sweeteners owing to toxicity problems, unacceptable taste quality or functionality, or other reasons. Alitame [*80863-62-3*] is the only other dipeptide sweetener being developed for commercial use.

Table 1. Aspartame Derivatives

Compound	CAS Registry Number	Potency	Reference
Asp-D-Ser-O-CH$_2$CH$_2$CH$_3$	[52946-03-9]	320X	48
Asp-L-Ser-(O—[cyclopentyl: CH$_3$, CH$_3$, H$_3$C, CH$_3$]—)OCH$_3$		800X	49
Asp-D-Ala-O-β(+)-fenchyl	[107797-74-0]	2,500X	50
Asp-NHC(CH$_3$)$_2$COO-β(+)-fenchyl	[117306-08-8]	2,000X	51
Asp-NHCH(COOCH$_3$)COO-β(+)-fenchyl	[61091-21-2]	33,000X	52
Asp-NHC(C$_6$H$_5$)HCOO-β(+)-fenchyl	[103365-67-9]	3,700X	53
Asp-NHCHCOOCH$_3$ (CH$_2$-2(R)-$endo$-bornyl)	[130790-14-6]	1,930X	54
Asp-NH[cyclopropyl] COOCH$_2$CH$_2$CH$_3$		275X	55
Asp-NHCHNHC(=O)—[cyclopentyl: CH$_3$, CH$_3$, CH$_3$, CH$_3$, CH$_3$]		900X	56
Asp-D-Ala-C(=O)-NH—[thietane ring: H$_3$C, CH$_3$, H$_3$C, CH$_3$, S] (alitame)	[80863-62-3]	2,000X	
Asp-NH—[C(H$_5$C$_2$)(H)—C(=O)—NH—C(C$_6$H$_5$)(H)—C(H$_5$C$_2$)]		2,500X	57
HOOCCH$_2$CHC(=O)-L-Phe-OCH$_3$ (NHC(=O)NHCH)	[82778-17-4]	200X	58
HOOCCH$_2$CHC(=O)-L-Phe-OCH$_3$ (NHC(=O)NH—C$_6$H$_4$—CN)		10,000X	59
HOOCCH$_2$CHC(=O)-L-Phe-OCH$_3$ (NHC(=S)NH—C$_6$H$_4$—NO$_2$)		50,000X	59
HOOCCH$_2$CHCNH—C$_6$H$_4$—CN (NHC(=O)CF$_3$)	[39219-30-2]	3,000X	60
D-Ala-L-Asp-L-Phe-OCH$_3$		170X	61

562

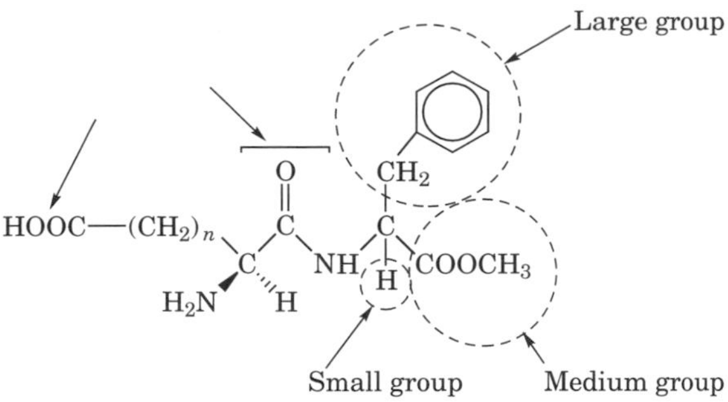

Fig. 3. Structure–activity summary of dipeptide sweeteners, where n may be 0 or 1 (62). There are no known replacements for the acid or amide groups denoted by arrows, although thioamide has some sweetness. If the NH_2 is replaced by NHC(O)R, the potency is increased when R = NH—⟨○⟩—X, where X = CN, NO_2, etc. Courtesy of Chapman and Hall.

The safety of aspartame for human consumption has been studied extensively. The results of these studies have satisfied the FDA. However, because phenylalanine is a metabolite of aspartame, people who lack the ability to metabolize this amino acid should refrain from using aspartame. Any aspartame-containing diet food must indicate that the product contains phenylalanine.

Acesulfame-K. Acesulfame-K [*55589-62-3*] (**4**), the potassium salt of acesulfame [*33665-90-6*] (6-methyl-1,2,3-oxathiazin-4(3*H*)-one 2,2-dioxide), is a sweetener that resembles saccharin in structure and taste profile. 5,6-Dimethyl-1,2,3-oxathiazine-4(3*H*)-one 2,2-dioxide, the first of many sweet compounds belonging to the dihydrooxathiazinone dioxide class, was discovered accidentally in 1967 (63). From these many sweet compounds, acesulfame was chosen for commercialization. To improve water solubility, the potassium salt was made. Acesulfame-K (trade name: Sunette) was approved for dry product use in the United States in 1988 and in Canada in October, 1994. Later, it was approved by the FDA for additional food categories such as yogurts, frozen and refrigerated desserts, and baked goods.

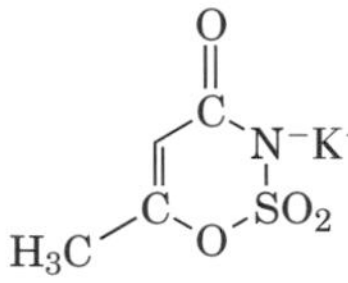

(**4**)

Acesulfame-K is a white crystalline powder having a long (six years or more) shelf life. It readily dissolves in water (270 g/L at 20°C). Like saccharin, acesulfame-K is stable to heat over a wide range of pH. At higher concentrations, there is a detectable bitter and metallic off-taste similar to saccharin. Use of the sodium salt of ferulic acid [*437-98-4*] (FEMA no. 3812) to reduce the bitter aftertaste of acesulfame-K has been described (64). The sweetness potency of acesulfame-K (100 to 200X, depending on the matching sucrose concentration) (63) is considered to be about half that of saccharin, which is about the same as that of aspartame.

Acesulfame-K–aspartame blends exhibit a significant synergistic effect (Fig. 4) (65,66). This synergy provides large cost savings for the diet foods industry. The blend also has a more rounded taste. Each sweetener apparently masks the off-taste associated with the other. Increased blend usage is expected.

Many synthetic processes have been described for acesulfame. One involves the condensation of a halosulfonyl isocyanate and an acetylene or a ketone (67,68). The fluorosulfonyl isocyanate can be prepared by reaction of sulfuryl diisocyanate with fluorosulfonic acid (69).

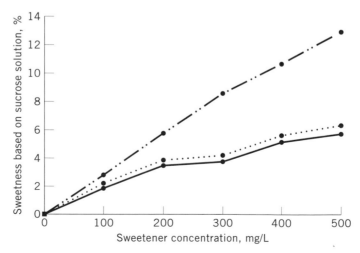

Saccharin. Saccharin [81-07-2], 3-oxo-2,3-dihydro-1,2-benzisothiazole 1,1-dioxide (o-sulfobenzimide or o-benzosulfimide), (**5**) was accidentally discovered to

Fig. 4. Isosweet blends where the solid line represents acesulfame-K, the dotted line aspartame, and (–••–) a 1:1 blend (65). Courtesy of American Association of Cereal Chemists.

be a sweet compound in 1878. A pilot plant was set up in New York to manufacture saccharin, which was displayed in a London exposition in 1885 (70). Since that time, saccharin has been used in many parts of the world.

(5)

M = H	saccharin
M = Na$^+$	sodium saccharin
M = (Ca)$^+_{1/2}$	calcium saccharin

In 1969, a chronic toxicity study on a cyclamate:saccharin (10:1) blend indicated bladder cancer problems in rats. Cyclamate was soon banned by the FDA, but saccharin remained an approved sweetener. In 1977, the FDA proposed a ban on saccharin because of the discovery of bladder tumors in some male rats fed with high doses of saccharin. Because no other nonnutritive sweetener was available at that time, the proposed ban faced strong opposition. Legislation to stay the ban has been passed in the U.S. Congress periodically. In December, 1991, the FDA withdrew its proposed ban. All saccharin-containing packaged products are required to carry a warning label indicating that saccharin has been determined to cause cancer in laboratory animals.

The main utility of saccharin had been in beverages and as a table-top sweetener. Upon the approval of aspartame for carbonated beverages in 1983, aspartame displaced saccharin in most canned and bottled soft drinks. However, saccharin is still used, usually blended with aspartame, in carbonated soft drinks dispensed from soda fountains.

Saccharin is acidic and not very soluble in water. For improved solubility, the food industry prefers the sodium or calcium [6485-34-3] salt. Sodium saccharin [128-44-9] is so widely used that it is often referred to simply as saccharin. The aqueous solubilities of both salts are about the same, ie, 0.67 g/mL. Saccharin, stable to heat over a wide pH range, can withstand most food processing (qv) conditions. Interactions between saccharin and other food ingredients have not been reported.

Saccharin imparts a sweetness that is pleasant at the onset but is followed by a lingering, bitter aftertaste. Sensitivity to this bitterness varies from person to person. At high concentration, however, most people can detect the rather unpleasant aftertaste. Saccharin is synergistic with other sweeteners of different chemical classes. For example, saccharin–cyclamate, saccharin–aspartame, saccharin–sucralose, and saccharin–alitame combinations all exert synergy to various degrees. The blends, as a rule, exhibit less aftertaste than each of the component sweeteners by themselves.

Saccharin is the most economical sweetener available. It is 300 times (8% sucrose solution sweetness equivalence) more potent than sugar and its price in 1996 was about \$6.05/kg, ca \$0.02/(kg·sweet unit). Sugar, on the other hand, was ca \$0.77/kg, which is 39 times more expensive than saccharin on equal sweetness basis. Consequently, the low cost and high stability of saccharin render it the sweetener of choice for dentifrices (qv), other toiletry products, and pharmaceuticals (qv).

The original process (71) for saccharin synthesis required the separation of ortho- and para-isomeric intermediates. In early 1950, a newer process for saccharin production was developed and the Maumee Chemical Company was subsequently formed. The Maumee process uses anthranilic acid [118-92-3] as the starting material. After the merger of PMC Specialties Group and the Maumee Chemical Company, the process was improved to a one-pot continuous process using methyl anthranilate [134-20-3] as the starting material (70). As of this writing, PMC is the sole producer of saccharin in the United States. Saccharin is also produced in Japan, China, Korea, and Taiwan.

Many analogues of saccharin have been synthesized since its discovery. With the exception of one compound, thieno[3,4-d]isothiazolone dioxide [59337-79-0], 1000X, this effort has not generated more potent compounds. Acesulfame-K could be considered a ring-modification derivative of saccharin, however.

Cyclamate. Sodium cyclamate [139-05-9] (**6**), the sodium salt of cyclamic acid [100-88-9], was so widely used that it was often just called cyclamate. The other common salt, calcium cyclamate [139-06-0], is useful in low sodium diets.

Cyclamate was first synthesized in 1937. Like the other sweeteners, its sweet taste was accidentally discovered (71,72). The FDA in 1958 classified sodium cyclamate as a GRAS sweetener. In 1966, however, it was reported that cyclamate could be metabolized by bacteria in the intestine to cyclohexylamine [108-91-8] (**7**), which had been shown to cause chromosome damage in animals. In 1969, a two-year chronic toxicity study with a sodium cyclamate–sodium saccharin (10:1) mixture found bladder tumors in rats. The FDA took cyclamate off the GRAS list, banning it from foods and beverages but permitting its sale in pharmacies. In 1970, after a congressional investigation, the FDA banned the use of cyclamate entirely. The culprit for bladder tumor was mainly assigned to cyclohexylamine and therefore saccharin was not affected.

Abbott Laboratories, which has conducted additional toxicity and carcinogenicity studies with cyclamate, a 10:1 mixture of cyclamate–saccharin, and cyclohexylamine, claimed to be unable to confirm the 1969 findings. Abbott then filed a food additive petition for cyclamate in 1973, which was denied by the

FDA in 1980. In 1982, the Calorie Control Council and Abbott Laboratories filed a second food additive petition containing the results of additional safety studies (73). That petition was still pending as of 1996. Cyclamate is, however, allowed for use in any or all three categories, ie, food, beverage, and tabletop, in about 50 countries. Sweet 'n Low, known in the United States as a saccharin-based table-top sweetener, contains exclusively cyclamate in Canada.

Cyclamate is about 30 times (8% sucrose solution sweetness equivalence) more potent than sugar. Its bitter aftertaste is minor compared to saccharin and acesulfame-K. The mixture of cyclamate and saccharin, especially in a 10:1 ratio, imparts both a more rounded taste and a 10–20% synergy. Cyclamate (**6**) is manufactured by sulfonation of cyclohexylamine (**7**). Many reagents can be used, including sulfamic acid, salts of sulfamic acid, and sulfur trioxide (74–77).

Stevioside. Stevioside [57817-89-7] (**8**) is a naturally occurring sweetener (ca 300X, 4% sucrose solution sweetness equivalence) extracted from a South American plant, *Stevia rebaudiana* Bertoni. The dried leaves, the water extract of leaves, and the refined chemical ingredients, eg, (**8**) and rebaudioside A [58543-16-1] (**9**), can all be used as sweetening agents. These are collectively referred to as stevia. Discovered in Paraguay and Brazil, the plant was identified in the early 1970s as a plant of high economical value and transported to Japan for cultivation, where the commercialization of stevia leaves as a natural sweetener became a success. In 1987, a total of 1700 metric tons of stevia leaves were utilized in Japan, representing an estimated 190 metric tons of pure stevioside (78). Stevia is cultivated primarily in southern China, Taiwan, Thailand, and Malaysia.

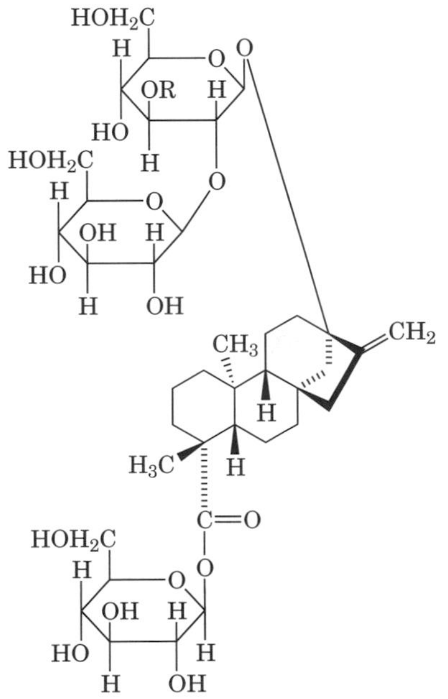

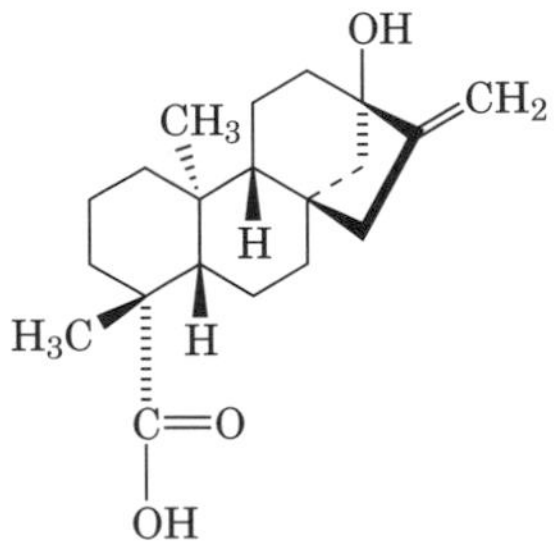

(**10**)

(**8**) R = H

(**9**) R =

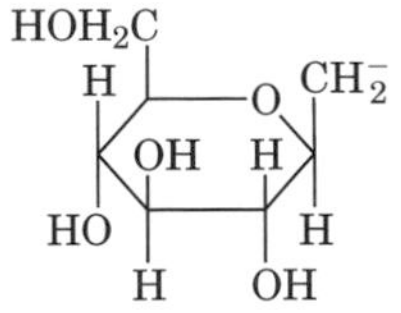

Stevioside and rebaudioside A are diterpene glycosides. The sweetness is tainted with a bitter and undesirable aftertaste. The time–intensity profile is characteristic of naturally occurring sweeteners: slow onset but lingering. The aglycone moiety, steviol [471-80-7] (**10**), which is the principal metabolite, has been reported to be mutagenic (79). Wide use of stevia in Japan for over 20 years did not produce any known deleterious side effects. However, because no food additive petition has been presented to the FDA, stevioside and related materials cannot be used in the United States. An import alert against stevia was issued by the FDA in 1991. In 1995, however, the FDA revised this import alert to allow the importation and use of stevia as a diet supplement (80), but not as a sweetener or an ingredient for foods. Several comprehensive reviews of stevia are available (81,82).

Glycyrrhizin. Glycyrrhizin (**11**), also known as glycyrrhizic acid [1405-86-3], is a glycoside isolated from the roots of licorice, *Glycyrrhiza glabra* L. For improved water solubility, an ammoniated salt is commonly used. This can be in the form of either ammonium glycyrrhizinate or monoammonium glycyrrhizinate.

(**11**)

The sweetness potency of glycyrrhizin is about 33X (10% sucrose solution sweetness equivalence) (3). Its taste, however, is accompanied by a characteristic licorice flavor, making it incompatible with many other food ingredients. The time–intensity profile is similar to that of other naturally occurring high potency sweeteners: slow onset followed by lingering aftertaste. It is claimed to be heat-stable. Ammonium glycyrrhizinate, which tends to precipitate below pH 4.5, is affirmed in the United States as a GRAS flavoring agent (FEMA no. 2528). It is not approved for use as a sweetener.

Glycyrrhiza root extracts containing at least 90 wt % pure glycyrrhizin are widely used in Japan, second only to stevia sweeteners. Glycyrrhizin exerts pharmacological effects, eg, edema and hypertension, on account of which the Japanese government has urged people to curtail consumption to less than 200 mg/d of glycyrrhizin in drug formulations (83). Several reviews of glycyrrhizins are available (83–86).

Sucralose. Sucralose [56038-13-2] is a trichlorodisaccharide sweetener developed by the British sugar company Tate & Lyle during the 1970s (87–89). It was licensed to McNeil Specialty Products Company (a Johnson & Johnson subsidiary) in the United States. A food additive petition was filed with the FDA in 1987 (90). As of December, 1996, the petition was still pending. Sucralose was approved for use as a sweetener by Canada in 1991, by Australia, Mexico, and Russia in 1993, by Romania in 1994, and by New Zealand in 1996.

The disaccharide structure of (**12**) (trade name: SPLENDA) is emphasized by the manufacturer as responsible for a taste quality and time–intensity profile closer to that of sucrose than any other high potency sweetener. The sweetness potency at the 10% sucrose solution sweetness equivalence is between 450 and 500X, or about two and one-half times that of aspartame. When compared to a 2% sugar solution, the potency of sucralose can be as high as 750X. A moderate degree of synergy between sucralose and other nonnutritive (91) or nutritive (92) sweeteners has been reported.

Sucralose is quite stable to heat over a wide range of pH. However, the pure white dry powder, when stored at high temperature, can discolor owing to release of small quantities of HCl. This can be remedied by blending it with maltodextrin (93) and other diluents. The commercial product can be a powder or a 25% concentrate in water, buffered at pH 4.4. The latter solution may be stored for up to one year at 40°C. At lower pH, there is minimal decomposition. For example, in a pH 3.0 cola carbonated soft drink stored at 40°C, there is less than 10% decomposition after six months. The degradation products are reported to be the respective chlorinated monosaccharides, 4-chloro-4-deoxy-galactose (**13**) and 1,6-dichloro-1,6-dideoxy-fructose (**14**) (94).

(**12**) (**13**) (**14**)

The synthesis of sucralose (Fig. 5) (88,93,94) employs sucrose (glucosyl-fructose) (**15**) as the starting material. Through discriminating blocking, ie, tritylating the primary alcohols followed by acetylating the secondary alcohols, and deblocking steps, all three primary alcohols become free while all secondary hydroxyls are blocked. The acetyl group on the C-4 position of glucose moiety migrates to the primary alcohol of the glucose moiety by refluxing at 125°C. The three hydroxyl groups are then converted to the chloro functional groups. There is a stereochemical inversion at the C-4 position of the glucose moiety. Thus the product is a trichloro-galactosyl-fructose. Many other halogenated sucrose analogues are also sweet (88).

Alitame. A new group of aspartyl-dipeptide sweeteners became known to the public in 1983 (95). Alitame [*80863-62-3*], L-aspartyl-D-alanine *N*-(2,2,4,4-tetramethylthietan-3-yl)amide (**16**), was selected for commercial development. In 1986 Pfizer filed a food additive petition with the FDA. As of December, 1996, it was still pending. Alitame was approved for use as a sweetener by Australia

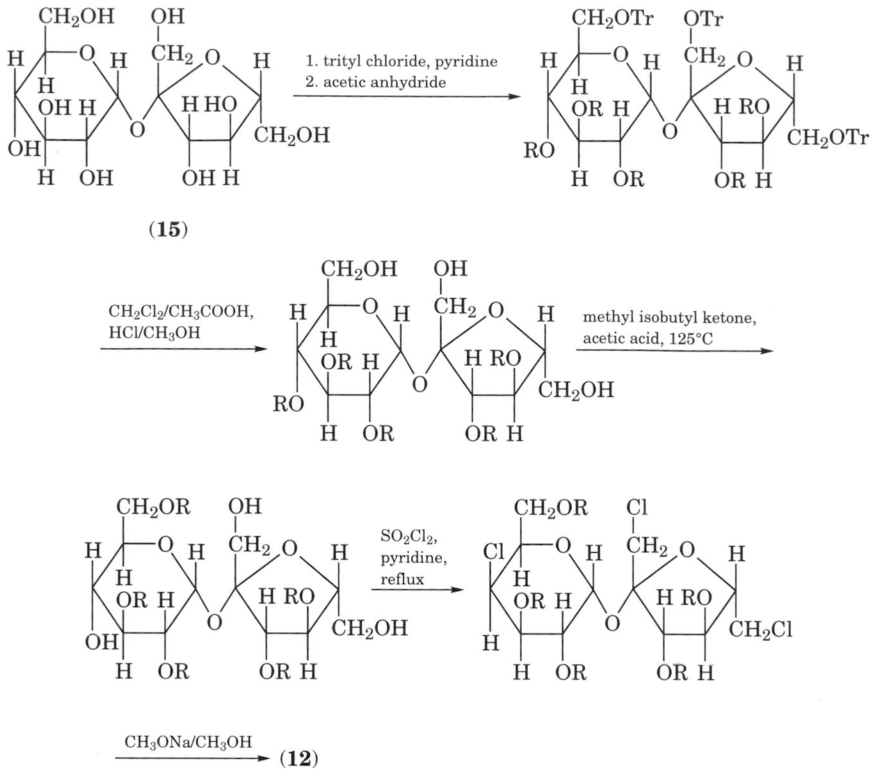

Fig. 5. Synthesis of sucralose, where Tr is the trityl group and R is $-\overset{\overset{\displaystyle O}{\|}}{C}CH_3$ (94). Courtesy of Marcel Dekker, Inc.

in 1993, by China, Mexico, and New Zealand in 1994, by Indonesia in 1995, and by Colombia in 1996.

Alitame (trade name: Aclame) is a water-soluble, crystalline powder of high sweetness potency (2000X, 10% sucrose solution sweetness equivalence). The sweet taste is clean, and the time–intensity profile is similar to that of aspartame. Because it is a sterically hindered amide rather than an ester, alitame is expected to be more stable than aspartame. At pH 2 to 4, the half-life of alitame in solution is reported to be twice that of aspartame. The main decomposition pathways (Fig. 6) include conversion to the unsweet β-aspartic isomer (**17**) and hydrolysis to aspartic acid and alanine amide (96). No cyclization to diketopiperazine or hydrolysis of the alanine amide bond has been reported. Alitame-sweetened beverages, particularly colas, that have a pH below 4.0 can develop an off-flavor which can be avoided or minimized by the addition of edetic acid (EDTA) [60-00-4] (97).

Although the exact pathway for manufacturing alitame is proprietary, one of the routes for small-scale synthesis has been given (95). This 1983 Pfizer patent lists many active analogues and serves as a good reference for the structure–activity relationship. An alitame (dipeptide) and saccharin (heterocycle) blend imparts synergistic sweetness, whereas the alitame–aspartame blend

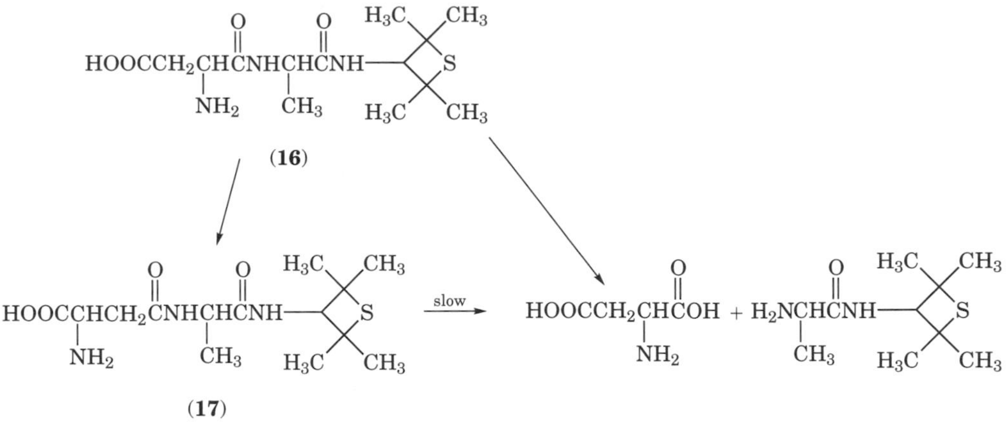

H₃C CH₃

$$\text{HOOCCH}_2\overset{\underset{|}{\text{NH}_2}}{\text{CH}}\overset{\overset{\text{O}}{\|}}{\text{C}}\text{NH}\overset{\underset{|}{\text{CH}_3}}{\text{CH}}\overset{\overset{\text{O}}{\|}}{\text{C}}\text{NH}\!\!-\!\!\square\!\!-\!\!\text{S}$$

(16)

slow →

(17)

Fig. 6. Main degradation pathways of alitame (96). Courtesy of Marcel Dekker, Inc.

(both dipeptides) does not, which suggests that synergy takes place only between structurally unrelated sweeteners.

Neohesperidin Dihydrochalcone. In the 1960s, there was a strong effort by the U.S. Department of Agriculture (USDA) to study the structure–activity relationship of citrus-derived chemicals. The goal was to reduce the bitter taste of citrus juices derived from bitter principles such as naringin [10236-47-2] (18), neohesperidin [13241-33-3] (19), and limonin [1180-71-8]. Neohesperidin is a glycoside composed of a flavanone and a disaccharide (glucose and L-rhamnose [3615-41-6]). Upon treatment with potassium hydroxide, the flavanone ring opens up to yield a chalcone. Catalytic hydrogenation of this chalcone produces neohesperidin dihydrochalcone [20702-77-6] (NHDC) (20), which tastes sweet.

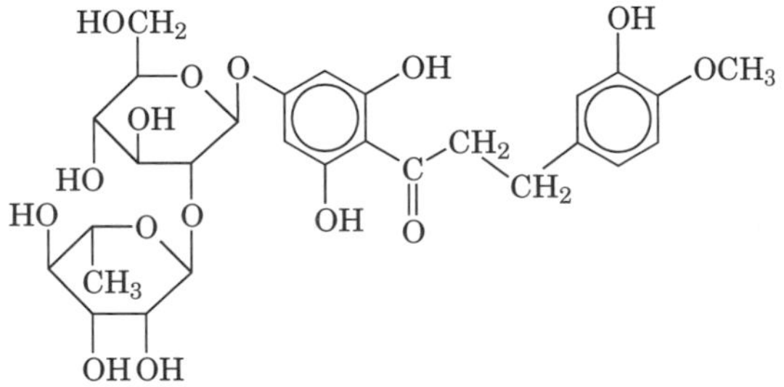

(20)

Many other dihydrochalcones have been made, but most of the toxicological studies have been conducted using NHDC and thus (20) has been petitioned and allowed for use. Neohesperidin is best isolated from the bitter orange (Seville orange), but it can also be synthesized from (18) and isovanillin [621-59-0] (21) (Fig. 7) (98).

NHDC imparts a sweetness that has a much slower onset and much greater lingering than sucrose. There is a slight aftertaste. The sweet potency at the

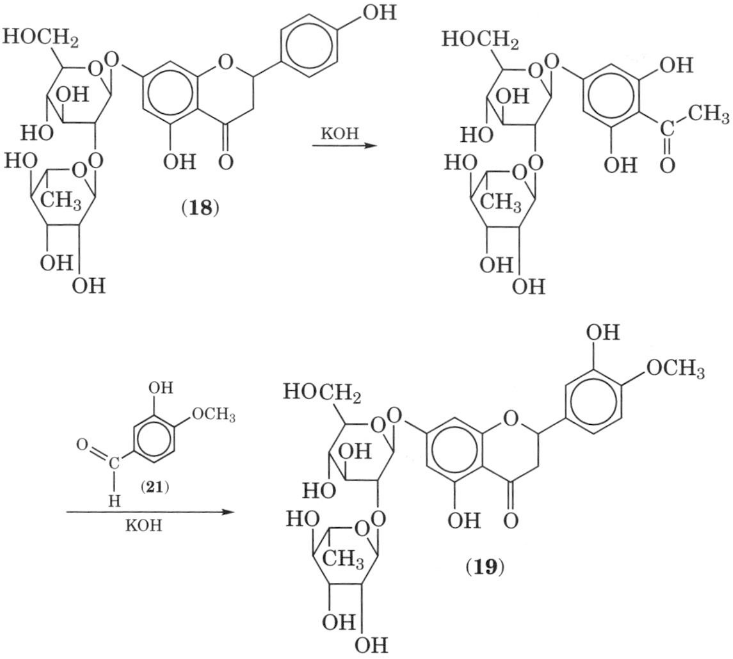

Fig. 7. Synthesis of neohesperidin (**19**) from naringin (**18**) (98). Courtesy of Marcel Dekker, Inc.

10% sucrose solution sweetness equivalence is about 300X. The most significant advantage of (**20**) is its ability to reduce the bitterness of the citrus bitter principles: naringin (**18**) and limonin. For example, at 5% sucrose equivalence, NHDC increased the threshold for (**18**) from 20 mg/kg to 49 mg/kg (98).

NHDC is an off-white powder having low solubility in water (0.5 g/L at room temperature). It is quite stable over a broad range of pH and temperature. Under extreme conditions, hydrolysis of the ether linkage between the saccharides and the aglycone can take place. However, the aglycone itself is reported to be sweet. NHDC is allowed for use as a sweetener by the European Union Sweeteners Directive in 1994. It has not been approved as a sweetener in the United States. In 1993, however, it was affirmed by FEMA as a GRAS flavor modifier (FEMA no. 3811) for many food categories.

Thaumatin. Thaumatin [53850-34-3] is a mixture of proteins extracted from the fruit of a West African plant, *Thaumatococcus daniellii* (Bennett) Benth. Work at Unilever showed that the aqueous extract contains two principal proteins: thaumatin I and thaumatin II. Thaumatin I, mol wt 22,209, contains 207 amino acids in a single chain that is cross-linked with eight disulfide bridges. Thaumatin II has the same number of amino acids, but there are five sequence differences. Production of thaumatins via genetic engineering technology has been reported (99).

Thaumatin (trade name: Talin) is a very potent sweetener (ca 2000X, 10% sucrose solution sweetness equivalence). However, its potency is overshadowed by inferior taste qualities. The onset of sweetness is very slow, and after reaching the maximum sweetness, a very long-lingering sweetness combined with an unpleasant aftertaste follows. Primarily owing to this poor taste quality, thaumatin is not considered a practically useful sweetener. It is, however, used as a flavor enhancer, especially in products such as chewing gum. Thaumatin and thaumatin B-recombinant were affirmed GRAS flavors (FEMA no. 3732 and 3814, respectively). They are not approved as sweeteners in the United States.

As a protein, thaumatin is remarkably water-soluble (up to 60%) and is stable to heat at low pH. It has been reported that a thaumatin solution at pH less than 5.5 can be heated at 100°C for several hours without loss of sweetness. Comprehensive reviews on thaumatin as sweetener are available (100,101).

Discovery of New Sweeteners

Discovering new sweeteners is similar to the discovery of new drugs and includes candidates from serendipity or from screenings of plants the activities of which are described anecdotally. Structure–activity relationship (SAR) studies are quickly implemented to design and synthesize compounds related to lead candidates. Biological activities, ie, sweetness, are measured, and common stereochemical features contributing to the activity are deduced. New analogues based on these features are synthesized and the activities compared. After several rounds of SAR studies, a few compounds of desirable potencies are selected for additional biological studies, including toxicity tests. The compound best fitting the criteria, eg, sweetness potency, quality, stability, safety, and economics, is then set forth and developed for commercial uses. Compounds reported to be sweet are listed in Table 2; structures are given in Figure 8.

Table 2. Other Compounds Reported to Be Sweet

Compounds	CAS Registry Number	Structure number	Potency	Reference
abrusoside D	[125003-00-1]	(22)	75X	102
baiyunoside	[86450-75-1]	(23)	500X	103
brazzein[a]	[160047-05-2]		200X	104
chloroform	[67-66-3]		40X	105
6-chloro-D-tryptophan	[17808-35-4]	(24)	1300X	106
dulcin	[150-69-6]	(25)	250X	107
glycergic acid	[84215-86-1]	(26)	500X	108
hernandulcin	[108944-70-3]	(27)	1000X	109
hydrofluorene sweeteners	[34069-54-0]	(28)	1400X	110
mogroside V	[88901-36-4]	(29)	256X	111
monatin	[146142-94-1]	(30)	800X	112
monellin[a,b]	[9062-83-3]		2500X	113
osladin	[33650-66-7]	(31)	3000X	103
P-4000	[553-79-7]	(32)	4000X	114
pentadin[a]	[61391-05-7]		500X	113
perillartine	[30950-27-7]	(33)	2000X	115
phyllodulcin	[55555-33-4]	(34)	400X	116
SRI oxime V	[59691-20-2]	(35)	450X	115
suosan	[140-46-5]	(36)	700X	107

[a]Materials is a protein.
[b]The mabinlins, another group of proteins, are also sweet.

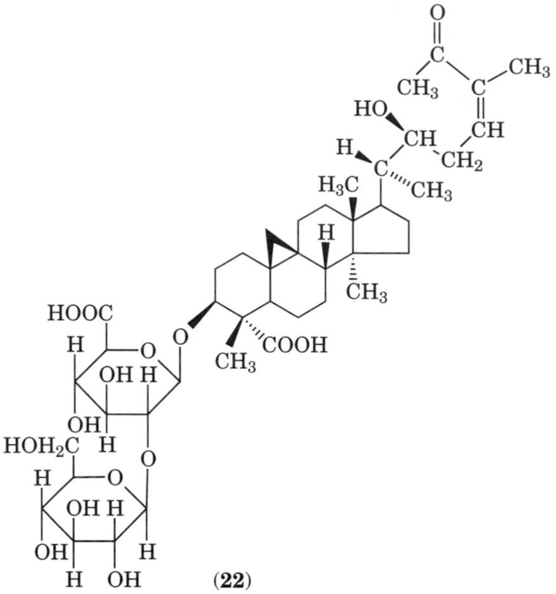

Fig. 8. Structures of other compounds reported to be sweet. See Table 2.

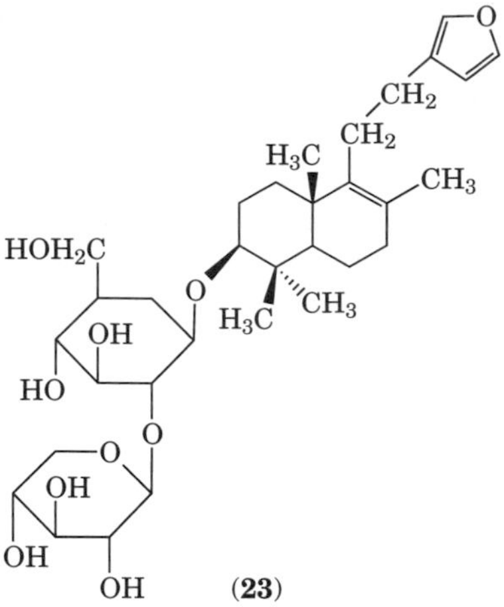

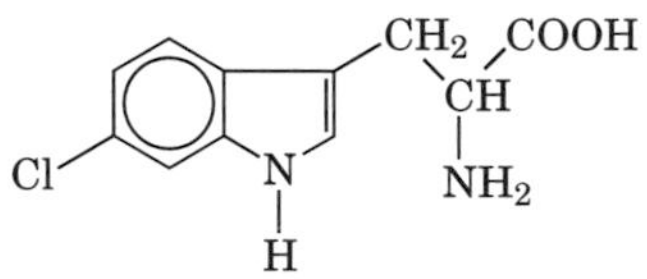

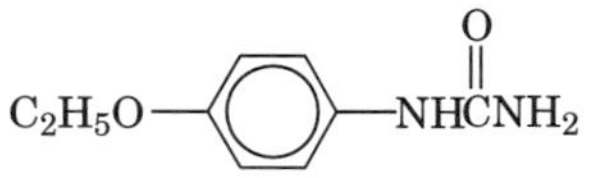

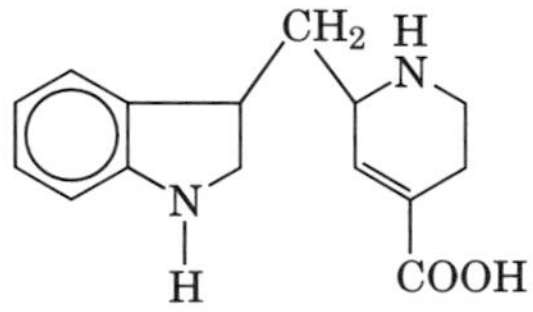

Computer technology has improved the traditional SAR methodology through the incorporation of molecular modeling (qv) techniques. The Tinti and Nofre sweetener model built upon theories proposed by several researchers (117–120) is depicted in Figure 9. Based on this model, sucrononic acid [*116869-55-7*] (**37**) (200,000X), one of the most potent sweeteners known, was discovered. NutraSweet is developing a next-generation sweetener, Sweetener 2000 (122), the structure of which is not known.

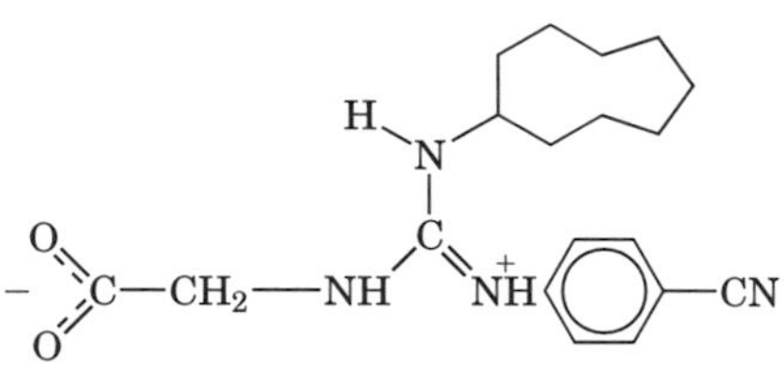

(**37**)

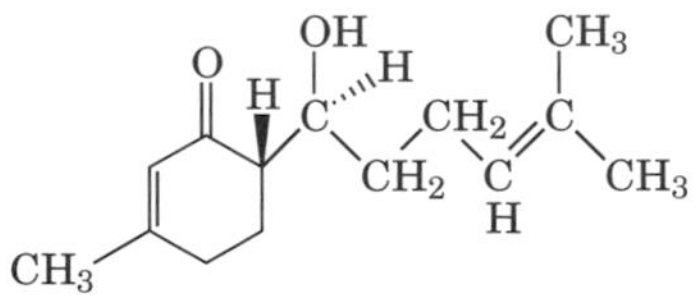

(27)

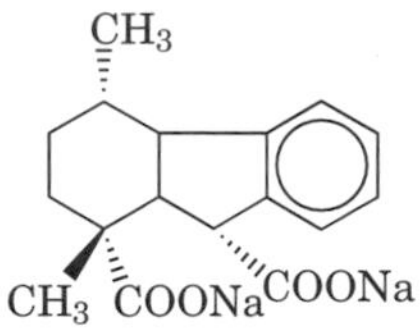

(28)

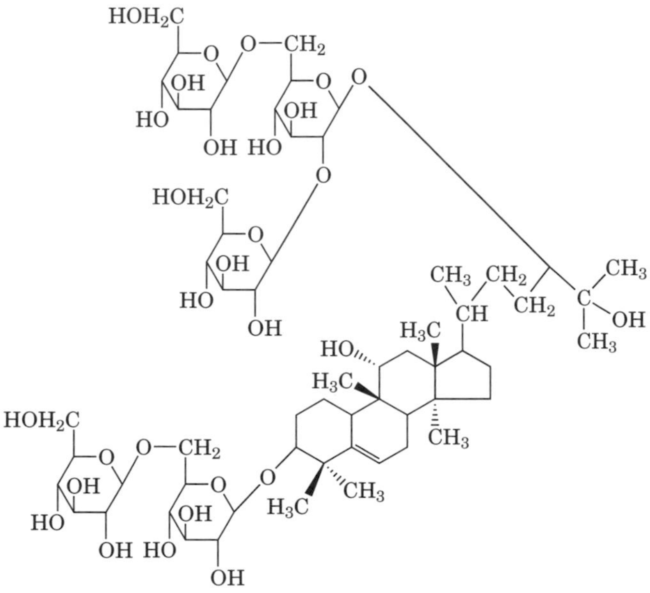

(29)

Fig. 8. (*Continued*)

An alternative view (123) is that no single model can adequately explain why any given compound is sweet. This hypothesis derives from several features. First, there is the observation that all carbohydrates having a critical ratio of OH to C are sweet tasting. In other words, there are no structural constraints to the sweetness of carbohydrates. Second, not all sweeteners can be fit to the same SAR model. Rather, some fit one, others fit another. Third, studies on the transduction mechanisms of sweetness suggest more than a single mechanism for sweet taste, implying multiple receptors for sweeteners.

Sweetness Enhancers, Inducers, and Inhibitors

Enhancers and Inducers.   A sweetness enhancer is defined as a compound that imparts no taste per se, but when combined with a sweetener in small quantities, increases sweetness intensity. A true sweetness enhancer has yet to be found. However, a good sweetness inducer, miraculin [*143403-94-5* or *125267-18-7*] (124), is known. Miraculin is a glycoprotein found in the fruit (called Miracle Fruit) of a West African shrub, *Richardella dulcifica*. By itself, miraculin

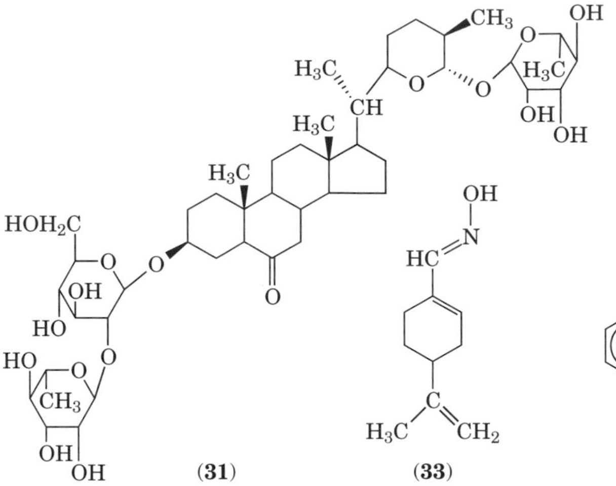

Fig. 8. *(Continued)*

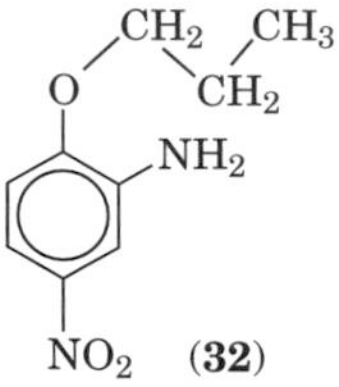

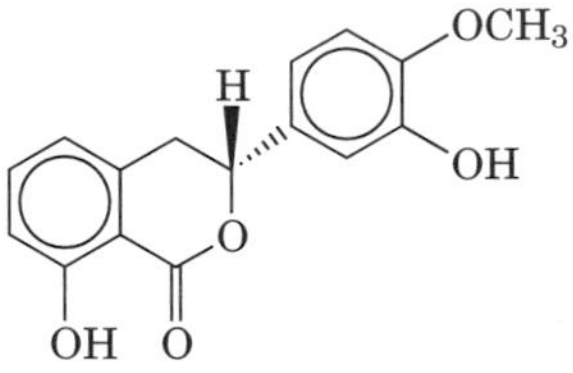

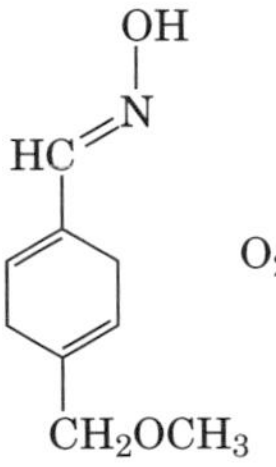

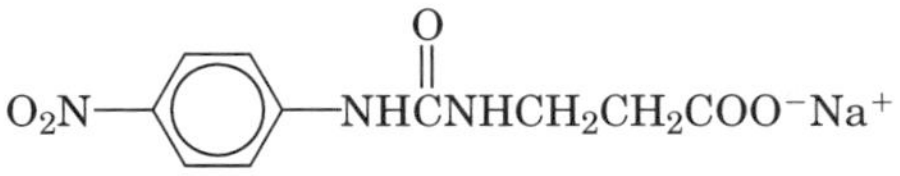

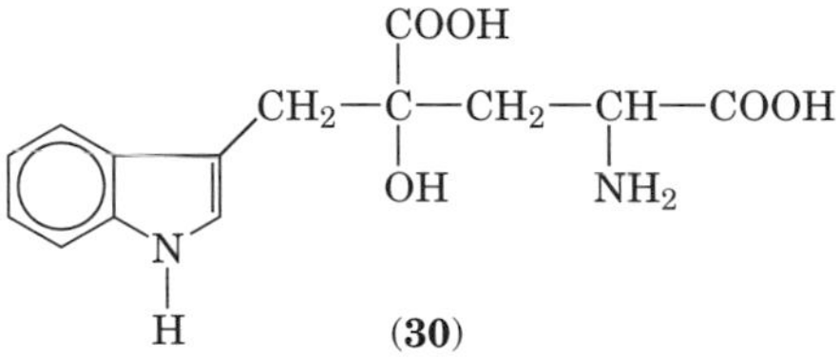

imparts no sweetness. When activated in the mouth by acidic substances, however, a sucrose-like sweetness is perceived. Thus, sour lemon, lime, grapefruit,
rhubarb, and strawberry taste sweet when combined with miraculin. The taste
conversion effect can last an hour or longer.

In 1974, a petition for affirmation of the GRAS status of miracle fruit
was submitted by the Miralin Company, mainly based on the fact that miracle
fruits have been consumed by humans since before 1958. In 1977, the petition
was denied by the FDA. However, miraculin remains a research curiosity. Its
structure was elucidated in 1989 (125). Another protein, curculin [151404-13-
6] (126), has also been reported to exert a sweet-inducing activity similar to
miraculin.

Inhibitors.   Sugar is used in large quantities in fruit jams as a preservative. The strong sweetness, however, prevents fruity flavors from being noticed.
For these and other foods that must use a large amount of sugar for purposes
other than sweet taste, there is need for a sweet-taste inhibitor.

Lactisole [13794-15-5], the sodium salt of racemic 2(4-methoxyphenoxy)-
propionic acid, is a sweet-taste inhibitor marketed by Domino Sugar. It was
affirmed as a GRAS flavor (FEMA no. 3773). At a concentration of 100 to
150 ppm, lactisole strongly reduces or eliminates the sweet taste of a 10% sugar
solution. This inhibition appears to be receptor-related because lactisole also

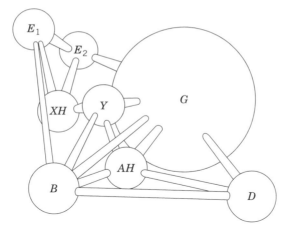

Fig. 9. Sweetener receptor binding sites postulated by Tinti and Nofre, where B is an anionic group, eg, CO_2^-, SO_3^-, or CN_4^-; AH, a hydrogen bond donor group, eg, NH or OH; G, a hydrophobic, hydrocarbon group; D, a hydrogen bond acceptor group, eg, CN, NO_2, or Cl; Y, E_1, and E_2 are hydrogen bond acceptors, eg, CO or halogen atoms; and XH is a hydrogen bond donor group, eg, NH or OH (121). Courtesy of American Chemical Society.

inhibits the sweet taste of aspartame. The S-$(-)$-enantiomer [4276-74-8] (**38**), isolated from roasted coffee beans, is the active isomer; the R-$(+)$-enantiomer is inert (127).

$$H_3CO-\bigcirc-O-\overset{\overset{\displaystyle CH_3}{|}}{\underset{\underset{\displaystyle COOH}{|}}{C}}\cdots H$$

(**38**)

Several natural products, eg, gymnemic acid [122168-40-5] and ziziphin [73667-51-3], have also shown sweet-inhibiting activities. These are not allowed for foods in the United States, however.

Sweet-Taste Transduction Mechanisms

Significant progress has been made since the late 1980s on understanding the transduction mechanisms for the basic tastes, ie, sweet, sour, salty, bitter, and possibly umami, eg, brothy. In general, transduction of taste starts at the binding of the stimulus compound with receptors or stimulus influx through ion channels. For sweet taste, however, this concept encounters difficulties because the existence of sweetener receptors is very much under debate. On the one hand, the stereochemical requirements for exerting potent sweetness and the inhibition of sweetness by lactisole strongly indicate the existence of specific sweetener receptors. On the other hand, L-sugar is just as sweet as D-sugar, and a high concentration of sugar is required for the development of a sweet sensation, implying that specific receptors do not exist. If sweeteners do bind to receptors, then the nature of these receptors is unclear.

There is evidence to suggest that the binding of sugars to receptors activates a secondary messenger cascade (128,129) involving a guanosine triphosphate (GTP) binding protein, called a G-protein (130,131), and adenylyl cyclase, which produces cyclic adenosine monophosphate (AMP). Cyclic AMP in turn may activate a cyclic AMP-dependent protein kinase, causing the closure of potassium channels and the influx of calcium ions from extracellular space. These events lead to the depolarization of the cell and the release of neurotransmitter (132). Neural signals encoding sweet-taste information are then carried to the central nervous system.

In addition to the mechanism involving cyclic AMP, nonsugar sweeteners, eg, saccharin and a guanidine-type sweetener, have been found to enhance the production of another second messenger, inositol 1,4,5-trisphosphate (IP_3), causing the closure of potassium channels and the release of calcium ions from intracellular storage depots (133). These diverse mechanisms support the hypothesis that multiple receptor systems exist for sweetness. Results from studies of amiloride [2016-88-8], a well-characterized suppressor of sodium ion transport, where sweet taste was partially inhibited, suggest the existence of a receptor–(sodium) ion channel complex for sweet taste (134,135).

Finally, some amphiphilic sweeteners, eg, aspartame, saccharin, and neohesperidin dihydrochalcone, have been shown to be capable of stimulating a purified G-protein directly in an *in vitro* assay (136). This suggests some sweeteners may be able to cross the plasma membrane and stimulate the G-protein without first binding to a receptor. This type of action could explain the relatively longer response times and the lingering of taste associated with many high potency sweeteners.

BIBLIOGRAPHY

"Sweeteners, Nonnutritive" in *ECT* 2nd ed., Vol. 19, pp. 593–607, by K. M. Beck, Abbott Laboratories; "Sweeteners" in *ECT* 3rd ed., Vol. 22, pp. 448–464, by R. Mazur, G. D. Searle and Co.

1. A. Bar, in L. O. Nabors and R. C. Gelardi, eds., *Alternative Sweeteners*, 2nd ed., Marcel Dekker, Inc., New York, 1991, p. 358.
2. P. Crapo and co-workers, *Diabetes Care*, **5**, 512 (1982).
3. G. E. DuBois, in Y. H. Hui, ed., *Encyclopedia of Food Science and Technology*, Vol. 4, John Wiley & Sons, Inc., New York, 1991, p. 2470.
4. SRI International, *Chem. Ind. News.* (Mar./Apr. 1994).
5. G. V. Levin, in L. O. Nabors and R. C. Gelardi, eds., *Alternative Sweeteners*, Marcel Dekker, Inc., New York, 1986, p. 155.
6. R. H. Mazur and co-workers, *J. Med. Chem.* **16**, 1284 (1973).
7. B. E. Homler, R. C. Deis, and W. H. Shazer, in Ref. 1., p. 46.
8. U.S. Pat. 3,868,472 (Feb. 25, 1975), G. H. Berg and J. F. Trumbetas (to General Foods).
9. U.S. Pat. 3,922,369 (Nov. 25, 1975), M. Glicksman and B. N. Wankier (to General Foods).
10. U.S. Pat. 3,939,289 (Feb. 17, 1976), J. Hornyak and H. D. Stahl (to General Foods).

11. U.S. Pat. 3,956,507 (May 11, 1976), M. D. Shoaf and L. D. Pischke (to General Foods).
12. U.S. Pat. 3,757,739 (Aug. 21, 1973), J. A. Cella and W. H. Schmitt (to Alberto Culver Co.).
13. U.S. Pat. 3,934,048 (Jan. 20, 1976), I. Furda and J. F. Trumbetas (to General Foods).
14. U.S. Pat. 4,059,706 (Nov. 22, 1977), M. D. Shoaf and L. D. Pischke (to General Foods).
15. U.S. Pat. 4,001,456 (Jan. 4, 1977), M. Glicksman and B. N. Wankier (to General Foods).
16. U.S. Pat. 4,007,288 (Feb. 8, 1977), M. Glicksman and B. N. Wankier (to General Foods).
17. U.S. Pat. 3,761,288 (Sept. 25, 1973), M. Glicksman and B. N. Wankier (to General Foods).
18. U.S. Pat. 4,517,214 (May 14, 1985), M. D. Shoaf and L. D. Pischke (to General Foods).
19. U.S. Pat. 4,448,716 (May 15, 1984), J. H. Tsau (to G. D. Searle).
20. U.S. Pat. 4,704,288 (Nov. 3, 1987), J. H. Tsau and J. G. Young (to NutraSweet).
21. R. Mazur and A. Ripper, in C. A. M. Hough, K. J. Parker, and A. J. Vlitos, eds., *Developments in Sweeteners—1*, Applied Science Publishers, London, U.K., 1979, p. 131.
22. B. E. Homler, R. C. Deis, and W. H. Shazer, in Ref. 1, p. 41.
23. L. N. Bell and M. J. Hageman, *J. Agric. Food Chem.* **42**, 2398 (1994).
24. U.S. Pat. Appl. 550,670 (July 10, 1990), M. J. Greenberg and S. Johnson (to Wm. Wrigley Co.).
25. Y. Gelman, *Chem. Abstr.* **115**, 280584m (1991).
26. C.-P. Yang and C.-S. Su, *J. Org. Chem.* **51**, 5186 (1986).
27. U.S. Pat. 3,475,403 (Oct. 28, 1969), R. H. Mazur, A. H. Goldkamp, and J. M. Schlatter (to G. D. Searle).
28. U.S. Pat. 3,786,039 (Jan. 15, 1974), Y. Ariyoshi and co-workers (to Ajinomoto).
29. U.S. Pat. 3,808,190 (Apr. 30, 1974), J. J. Dahlmans, B. H. N. Dassen, and W. H. J. Boesten (to Stamicarbon N.V.).
30. U.S. Pat. 3,833,553 (Sept. 30, 1974), Y. Ariyoshi and co-workers (to Ajinomoto).
31. U.S. Pat. 3,853,835 (Dec. 10, 1974), R. H. Mazur and J. M. Schlatter (to G. D. Searle).
32. U.S. Pat. 3,875,137 (Apr. 1, 1975), D. A. Jones and R. H. Mazur (to G. D. Searle).
33. U.S. Pat. 3,879,372 (Apr. 22, 1975), W. H. J. Boesten (to Stamicarbon, B.V.).
34. U.S. Pat. 3,901,871 (Aug. 26, 1975), G. W. Anderson (to American Cyanamid).
35. U.S. Pat. 3,907,766 (Sept. 23, 1975), M. Fujino and co-workers (to Takeda Chemical Industries).
36. U.S. Pat. 3,933,781 (Jan. 20, 1976), G. L. Bachman, M. L. Oftedahl, and B. D. Vineyard (to Monsanto).
37. U.S. Pat. 4,017,472 (Apr. 12, 1977), W. G. Farkas, W. M. Hoehn, and J. F. Zawadzki (to G. D. Searle).
38. U.S. Pat. 4,025,551 (May 24, 1977), A. H. Goldkamp, R. H. Mazur, and J. M. Schlatter (to G. D. Searle).
39. U.S. Pat. 4,256,836 (Mar. 17, 1981) and 4,436,925 (Mar. 13, 1984), Y. Isowa and co-workers (to Toyo Soda Mfg. Co., and Sagami Chem. Res. Ctr.).
40. U.S. Pat. 4,284,721 (Aug. 18, 1981), K. Oyama and co-workers (to Sagami Chem. Res. Ctr, Ajinomoto, and Toyo Soda Mfg. Co.).
41. U.S. Pat. 4,293,648 (Oct. 6, 1981), A. A. Davino (to G. D. Searle).
42. U.S. Pat. 4,309,341 (Jan. 5, 1982), M. Kubo, Y. Nonaka, and K. Kihara (to Sagami Chem. Res. Ctr., Ajinomoto, and Toyo Soda Mfg. Co.).
43. U.S. Pat. 4,332,718 (June 1, 1982), S. Takahashi and K. Toi (to Ajinomoto).
44. U.S. Pat. 4,333,872 (June 8, 1982), P. S. Sampathkumar and B. K. Dwivedi (to Chimicasa GmbH).
45. U.S. Pat. 4,480,112 (Oct. 30, 1984), H. L. Dryden and J. B. Hill (to G. D. Searle).

46. U.S. Pat. 4,506,011 (Mar. 19, 1985), T. Harada, H. Takemoto, and T. Igarashi (to Toyo Soda Mfg. Co.).
47. U.S. Pat. 4,507,231 (Mar. 26, 1985), M. Gourbault (to Lab. Fournier S.A.).
48. Y. Ariyoshi, N. Yasuda, and T. Yamatami, *Bull. Chem. Soc. Japan*, **47**, 326 (1974).
49. U.S. Pat. 4,654,219 (Mar. 31, 1987), G. M. Roy, R. E. Barnett, and P. R. Zanno (to General Foods).
50. U.S. Pat. 4,766,246 (Aug. 23, 1988), P. R. Zanno, R. E. Barnett, and G. M. Roy (to General Foods).
51. U.S. Pat. 4,774,028 (Sept. 27, 1988), T. D. Lee (to General Foods).
52. M. Fujino and co-workers, *Chem. Pharm. Bull.* **24**, 2112 (1976).
53. U.S. Pat. 4,692,512 (Sept. 8, 1987), J. M. Janusz (to Procter & Gamble).
54. G. A. King, III, J. G. Sweeny, and G. A. Iacobucci, *J. Agri. Food Chem.* **39**, 52 (1991).
55. C. Mapelli and co-workers, *Int. J. Peptide Prot. Res.* **30**, 498 (1987).
56. W. D. Fuller, M. Goodman, and M. S. Verlander, *J. Am. Chem. Soc.* **107**, 5821 (1985).
57. U.S. Pat. 5,286,509 (Feb. 15, 1994), L. L. D'Angelo, and J. G. Sweeny (to Coca-Cola).
58. U.S. Pat. 4,371,464 (Feb. 1, 1983), W. H. J. Boesten and L. A. C. Schiepers (to Stamicarbon, B.V.).
59. U.S. Pat. 4,645,678 (Feb. 24, 1987), C. Nofre and T. M. Tinti (to Universite Claude Bernard).
60. M. Lapidus and M. Sweeney, *J. Med. Chem.* **16**, 16 (1973).
61. Eur. Pat. Appl. 0,186,292 (July 2, 1986), T. Takemoto, T. Hijiya, and H. Yukawa (to Ajinomoto).
62. J. M. Janusz, in T. H. Grenby, ed., *Progress in Sweeteners*, Elsevier Science Publishing Co., Inc., New York, 1989, p. 42.
63. G. von Raymon Lipinski, in Ref. 1, p. 11.
64. U.S. Pat. 5,336,513 (Aug. 9, 1994), J. A. Riemer (to Kraft General Foods).
65. L. L. Hood and M. Schoor, *Cereal Foods World*, **35**, 1184 (1990).
66. U.S. Pat. 4,158,068 (June 12, 1979), G. von Raymon Lipinski and E. Luck (to Hoechst).
67. U.S. Pat. 3,689,486 (Sept. 5, 1972), K. Clauss and H. Jessen (to Hoechst).
68. K. Claus and H. Jensen, *Angew. Chem. Int. Ed.* **12**, 869 (1973).
69. U.S. Pat. 3,357,804 (Dec. 12, 1967), R. Appel (to Olin Mathieson Chem. Co.).
70. M. L. Mitchell and R. L. Pearson, in Ref. 1, p. 127.
71. L. F. Andrieth and M. Sveda, *J. Org. Chem.* **9**, 89 (1944).
72. U.S. Pat. 2,275,125 (Mar. 3, 1942), L. F. Andrieth and M. Sveda (to E. I. du Pont de Nemours & Co., Inc.).
73. B. A. Bopp and P. Price, in Ref. 1, p. 72.
74. U.S. Pat. 2,804,472 (Aug. 27, 1957), D. J. Loder (to E. I. du Pont de Nemours & Co., Inc.); U.S. Pat. 2,804,472 (Aug. 27, 1957), H. S. McQuaid (to E. I. du Pont de Nemours & Co., Inc.).
75. U.S. Pat. 3,060,231 (Dec. 31, 1962), D. Mueller and R. Trefzer (to CIBA).
76. U.S. Pat. 3,366,670 (Jan. 30, 1968), O. G. Birsten and J. Rosin (to Baldwin-Montrose Chem. Co.).
77. U.S. Pat. 3,194,833 (July 13, 1965), M. Freifelder and B. Meltsner (to Abbott Laboratories).
78. A. I. Bakal and L. O. Nabors, in Ref. 5, p. 295.
79. J. M. Pezzuto and co-workers, *Proc. Natl. Acad. Sci. USA*, **82**, 2478 (1985).
80. M. Bluementhal, *Herbalgram*, **35**, 17 (1995).
81. A. D. Kinghorn and D. D. Soejarto, in Ref. 1, p. 157.
82. K. C. Phillips, in T. H. Grenby, eds., *Developments in Sweeteners—3*, Elsevier Science Publishing Co., Inc., New York, 1979, p. 1.

83. J. D. Higginbotham, in T. H. Grenby, K. J. Parker, and M. G. Lindley, eds., *Recent Developments in Non-Nutritive Sweeteners—2*, Applied Science Publishers, London, U.K., 1983, p. 146.
84. L. O. Nabors and G. E. Inglet, in Ref. 5, p. 309.
85. A. D. Kinghorn and D. D. Soejarto, *CRC Crit. Rev. Plant Sci.* **4**, 79 (1986); *Med. Res. Rev.* **9**, 91 (1989).
86. H. van der Wel, A. van de Heijden, and H. G. Peer, *Food Rev. Internat.* **3**, 193 (1987).
87. U.S. Pat. 4,343,934 (Aug. 10, 1982), M. R. Jenner and D. Waite (to Talres Development).
88. U.S. Pat. 4,362,869 (Dec. 7, 1982), M. R. Jenner and co-workers (to Talres Development).
89. U.S. Pat. 4,435,440 (Mar. 6, 1984), L. Hough, S. P. Phadnis, and R. A. Khan (to Tate & Lyle).
90. G. A. Miller, in Ref. 1, p. 190.
91. U.S. Pat. 4,495,170 (Jan. 22, 1985), P. K. Beyts and Z. Latymer (to Tate & Lyle).
92. U.S. Pat. 5,380,541 (Jan. 10, 1995), P. K. Beyts, D. W. Lillard, and C. K. Batterman (to Tate & Lyle).
93. U.S. Pat. 4,927,646 (May 22, 1990), M. R. Jenner and G. Jackson (to Tate & Lyle).
94. G. A. Miller, in Ref. 1, p. 177.
95. U.S. Pat. 4,411,925 (Oct. 25, 1983), T. M. Brennan and M. E. Hendrick (to Pfizer).
96. M. E. Hendrick, in Ref. 1, p. 32.
97. Eur. Pat. Appl. 0,386,93A2 (Sept. 12, 1990), R. C. Glowaky, C. Sklavounos, and A. Torres (to Ammonia Casale SA).
98. R. M. Horowitz and B. Gentili, in Ref. 1, p. 107.
99. *Food Business*, **3**(16), 26 (1990).
100. J. D. Higginbotham, in Ref. 5, p. 103.
101. H. van der Wel, in B. J. F. Hudson, ed., *Developments in Food Proteins—4*, Elsevier Applied Science Publishers, Ltd., London, U.K., 1986, p. 219.
102. U.S. Pat. 5,198,427 (Mar. 30, 1993), A. D. Kinghorn, and Y.-H. Choi (to Res. Corp. Technologies).
103. A. D. Kinghorn and C. M. Compadre, in Ref. 1, p. 206.
104. U.S. Pat. 5,346,998 (Sept. 13, 1994), B. G. Hellekant and D. Ming (to Wisconsin Alumni Research Foundation).
105. M. Windholz, ed., *Merck Index*, 4th ed., Merck & Co., Inc., Rahway, N.J., 1976, p. 272.
106. T. Moriya, K. Hagio, and N. Yoneda, *Bull. Chem. Soc. Jap.* **48**, 2217 (1975).
107. A. D. Kinghorn and C. M. Compadre, in Ref. 1, p. 210.
108. A. Hofmann, *Helv. Chim. Acta.* **55**, 2934 (1972).
109. A. D. Kinghorn and C. M. Compadre, in Ref. 1, p. 204.
110. U.S. Pat. 3,734,966 (May 22, 1973), A. Tahara and co-workers (to Rikagaku Kenkyusho).
111. A. D. Kinghorn and C. M. Compadre, in Ref. 1, p. 199.
112. R. Vleggaar and co-workers, *J. Chem. Soc. Perkin Trans.* **1**, 3089 (1992).
113. A. D. Kinghorn and C. M. Compadre, in Ref. 1, p. 203.
114. P. E. Verkade, *Farmaco Ed. Sci.* **23**, 248 (1968).
115. A. D. Kinghorn and C. M. Compadre, in Ref. 1, p. 209.
116. A. D. Kinghorn and C. M. Compadre, in Ref. 1, p. 201.
117. R. S. Shallenberger, *J. Food Sci.* **28**, 584 (1963).
118. R. S. Shallenberger, *Taste Chemistry*, Blackie Academic & Professional, London, U.K., 1993, p. 292.
119. L. B. Kier, *J. Pharm. Sci.* **61**, 1394 (1972).
120. R. Rohse and H.-D. Belitz, in D. E. Walters, F. T. Orthoefer, and G. E. DuBois, eds., *Sweeteners: Discovery, Molecular Design and Chemoreception*, American Chemical Society, Washington, D.C., 1991, p. 176.

121. J.-M. Tinti and C. Nofre, in Ref. 119, p. 206.
122. D. Best, *Prepared Foods*, **161**(2), 39 (1992).
123. G. E. DuBois, *Proceedings of the Firmenich Jubilee Symposium*, Geneva, Switzerland, 1996.
124. L. O. Nabors and G. E. Inglett, in Ref. 5, p. 309.
125. S. Theerasilp and co-workers, *J. Biol. Chem.* **264**, 6655 (1989).
126. H. Yamashita and co-workers, *J. Biol. Chem.* **265**, 15770 (1990).
127. E. B. Rathbone and co-workers, *J. Org. Chem.* **37**, 54 and 58 (1989).
128. D. L. Kalinoski and co-workers, in J. G. Brand and co-workers, eds., *Chemical Senses*, Vol. 1, Marcel Dekker, Inc., New York, 1989, p. 85.
129. J. G. Brand and A. M. Feigin, *Food Chemistry* **56**, 199 (1996).
130. M. Rodbell, *Nature*, **284**, 17 (1980).
131. S. McLaughlin and R. F. Margolskee, *Am. Scientist*, **82**, 538 (1994).
132. J. G. Brand and B. P. Bryant, *Food Quality Preference*, **5**, 31 (1994).
133. S. J. Bernhardt and co-workers, *J. Physiol.* **490**, 325 (1996).
134. S. Mierson and co-workers, *J. Gen. Physiology*, **92**, 87 (1988).
135. S. S. Schiffman, E. Lockhead, and F. W. Maes, *Proc. Nat. Acad. Sci. USA*, **80**, 6136 (1983).
136. M. Naim and co-workers, *Biochem. J.* **297**, 451 (1994).

General References

References 1, 5, 21, 62, 82, 118, and 120 are all good sources of general references.
M. Mathlouthi, J. A. Kanters, and G. G. Birch, *Sweet-Taste Chemoreception*, Elsevier Science Publishing Co., Inc., New York, 1993.

THOMAS D. LEE
Kraft Foods

SYNTHETIC LUBRICANTS.   See LUBRICATION AND LUBRICANTS.

SYRUPS

Corn sweeteners, maple syrup, and molasses, all commercially available syrups, are concentrated solutions of carbohydrate. These products, produced for a variety of food and nonfood applications, are in some cases also available in a dry form. Corn sweeteners are prepared from hydrolyzed starch (qv) and include dextrose [50-99-7] (D-glucose), high fructose corn syrup (HFS), regular corn syrup, and maltodextrin (see SWEETENERS), which all have in common the raw material source, general methods of preparation, and many properties and applications. Dextrose, the common or commercial name for D-glucose, is available as a syrup or as a pure crystalline solid. HFS is produced by the partial enzymatic isomerization of dextrose. Corn syrups and maltodextrins are clear, colorless, viscous liquids prepared by hydrolysis of starch to solutions of dextrose, maltose, and higher molecular weight saccharides. Maple syrup, like corn syrup, is a nutritive sweetener produced as a concentrated carbohydrate (sucrose) solution. Molasses is a syrup produced as a by-product of sugar (qv) manufacture.

Dextrose

Dextrose (D-glucose) is by far the most abundant sugar in nature. It occurs either in the monosaccharide form (free state) or in a polymeric form of anhydrodextrose units. As a monosaccharide, dextrose is present in substantial quantities in honey, fruits, and berries. As a polymer, dextrose occurs in starch, cellulose (qv), and glycogen. Sucrose is a disaccharide of dextrose and fructose.

Dextrose was first prepared in pure form from grapes in the seventeenth century (1). The finding remained a laboratory curiosity until an attempt was made to prepare dextrose commercially from grapes in 1801 (2). In 1815 it was reported that acid conversion of starch to sugar was the result of hydrolysis of the starch rather than dehydration, and that the starch sugar was identical to grape sugar (3). It is generally conceded, however, that Kirchoff's work in 1811 was the forerunner of the starch hydrolyzate industry (4). A sweet substance was produced by cooking a mixture of potato starch and sulfuric acid. The process was improved to crystallize dextrose from a syrup, although attempts to commercialize the process were later abandoned. By 1842, however, a starch industry had been developed in the United States and crystalline dextrose became an important commercial product once crystallization methods were discovered in the 1920s (5).

Commercial dextrose products are produced in both dry and syrup forms. Dry products are prepared by crystallization (qv) to either an anhydrous, $C_6H_{12}O_6$, or hydrated, $C_6H_{12}O_6 \cdot H_2O$, form. These include dextrose hydrate [16824-90-1], anhydrous α-D-glucose [26655-34-5] (**1**), and anhydrous β-D-glucose [28905-12-6] (**2**). Syrup products are produced that contain from 95 to over 99% dextrose.

(**1**)   (**2**)

Properties.   Physical properties of the three crystalline forms of dextrose are listed in Table 1. In solution, dextrose exists in both the α- and β-forms. When α-dextrose dissolves in water, its optical rotation, $[\alpha]_D$, diminishes gradually as a result of mutarotation until, after a prolonged time, an equilibrium value is reached (see Table 1). At this point, about 62% of the dextrose is present in the β-form. This equilibrium value is not significantly changed over a wide range of temperatures and concentrations. The same equilibrium value exists in anhydrous melts as well as in glassy materials. Pure crystalline β-dextrose is quite sensitive to moisture, changing to the more stable α-form if exposed to

Table 1. Physical Properties of D-Glucose

Property	α-D-Glucose	α-D-Glucose hydrate	β-D-Glucose
molecular formula	$C_6H_{12}O_6$	$C_6H_{12}O_6 \cdot H_2O$	$C_6H_{12}O_6$
mp, °C	146	83	150
solubility at 25°C, g/100 g solution	$62 \rightarrow 30.2 \rightarrow 51.2^a$	$30.2 \rightarrow 51.2^{a,b}$	$72 \rightarrow 51.2^a$
$[\alpha]_D$	$112.2 \rightarrow 52.7^a$	$112.2 \rightarrow 52.7^{a,b}$	$18.7 \rightarrow 52.7^a$
heat of solution at 25°C, J/g^c	-59.4	-105.4	-25.9

[a]Initial value through solution equilibrium value. See text.
[b]Anhydrous basis.
[c]To convert J to cal, divide by 4.184.

high humidity. At 25°C, α-dextrose monohydrate dissolves fairly rapidly, yielding a solution containing ca 30 wt % dextrose. Very slowly thereafter, further
quantities of dextrose dissolve until a saturated solution containing ca 50 wt %
dextrose is obtained. The first phase of the dissolving process results from the
limited solubility of α-dextrose. The slow, subsequent dissolution is caused by the
transformation of part of the dissolved α-dextrose to the more soluble β-form.
When saturation is finally reached, a mixture of α- and β-dextrose in solution is
in equilibrium with solid α-dextrose hydrate. At 25°C, anhydrous α-dextrose dissolves rapidly and beyond the limit of solubility of α-dextrose hydrate. Because
the hydrate is the stable form at this temperature, crystallization of the hydrate
occurs to its limit of solubility and the pattern then follows that of the hydrate.
The rate of attainment of equilibrium is increased by heating or in the presence
of acids or bases. The solubility of the equilibrium mixtures is given in Table 2.
Additional data on the solubility of the crystalline forms are available (7).

Dextrose in solution or in solid form exists in the pyranose structural
conformation. In solution, a small amount of the open-chain aldehyde form exists
in equilibrium with the cyclic structures (**1**) and (**2**). The open-chain form is
responsible for the reducing properties of dextrose.

Table 2. Solubility of Dextrose in Water^a

Temperature, °C	Dextrose in solution, wt %
0	34.9
5	38.0
10	41.2
15	44.5
20	47.8
25	51.3
30	54.6
40	61.8
50	70.9
60	74.8
70	78.2
80	81.3

[a]Interpolated from data in Reference 6.

Dextrose shows the reactions of an aldehyde, a primary alcohol, a secondary alcohol, and a polyhydric alcohol. In acid solution, either after standing for a prolonged time or after heating, dextrose undergoes polycondensation, ie, dehydration, yielding a mixture of di- and oligosaccharides, most of which are the disaccharides gentiobiose [554-91-6] and isomaltose [499-40-1]. In acid solution and at high temperature, dehydration leads to formation of 5-hydroxymethylfurfural [67-47-0], which is a water-soluble, high boiling, and relatively unstable compound. Polymerization of 5-hydroxymethylfurfural yields dark-colored compounds that act as intermediates in the discoloration of sugar solutions. Further degradation of dextrose under these conditions also yields levulinic acid [123-76-2] and formic acid [64-18-6].

In mildly alkaline solution, the principal reaction of dextrose is partial transformation, ie, isomerization, to fructose and other ketoses. D-Mannose, saccharinic acids, and other decomposition products form to a lesser extent. In highly alkaline solution and particularly in the presence of atmospheric oxygen, a complex mixture of products of decomposition and rearrangement results. Mild oxidation in slightly alkaline solution results in a quantitative yield of D-gluconic acid. More vigorous oxidation with nitric acid yields glucaric acid, tartaric acid, oxalic acid, and other compounds resulting from fragmentation of the dextrose molecule. Alkaline Fehling's solution is reduced by dextrose. Roughly five atoms of copper are reduced per molecule of dextrose. Catalytic hydrogenation of dextrose is practiced commercially to manufacture sorbitol [50-70-4] (see SUGAR ALCOHOLS).

When dextrose is heated with methanol containing a small amount of anhydrous hydrogen chloride, α-methyl-D-glucoside is obtained in good yield and can be isolated by crystallization. Similar reactions occur with higher alcohols, but the reaction products are more difficult to isolate by crystallization. Dextrose reacts with acid anhydrides in the presence of basic catalysts, yielding esters. Complete reaction gives the pentaacylated derivative.

The reaction of dextrose with a nitrogen-containing compound, eg, amino acids or proteins, yields a series of intermediates which form pigments of varied molecular weight (Maillard reaction). The type of pigments produced is dependent on reaction conditions such as pH, temperature, and concentration of reactants.

Dextrose is the fundamental intermediary metabolite in carbohydrate metabolism. Other utilizable monosaccharides are largely converted to dextrose before being further metabolized. Starch, glycogen, and the common disaccharides are hydrolyzed enzymatically in the alimentary canal. The resulting monosaccharides are absorbed into the portal vein blood, by which they are transported first to the liver and then to all other parts of the body. In the liver, monosaccharides other than dextrose, eg, galactose and fructose, are largely converted to dextrose before they reach other tissues. When dextrose is taken up by body tissues, it is phosphorylated to glucose-6-phosphate, which can enter the glycolytic pathway or be stored as glycogen. Dextrose metabolism is discussed in more detail in Reference 8.

Manufacture. Dextrose is manufactured almost exclusively from corn (maize) starch in the United States. In other countries, starch from sorghum (milo), wheat, rice, potato, tapioca (yucca, cassava), arrowroot, and sago are used to varying degrees along with corn starch. Prior to the 1960s, commercial dextrose was produced using acid and acid–enzyme hydrolysis processes that yielded only about 86 and 92–94% dextrose, respectively. The development of thermostable bacterial α-amylase enzymes led to total enzyme processes that eliminated acid degradation products and increased dextrose yield to about 95–97% (see ENZYME APPLICATIONS).

In an enzymatic process, starch is hydrolyzed (thinned or liquefied) with a bacterial α-amylase. The resulting substrate is then hydrolyzed to dextrose (saccharified) using glucoamylase, a fungal enzyme that preferentially cleaves dextrose from the partially degraded starch. The initial extent of liquefaction is generally in the range of 10–20 dextrose equivalent (DE). DE is a measure of the reducing-sugar content calculated as dextrose and expressed as a percentage of the total dry substance. Several industrial enzyme liquefaction processes are used commercially (9) (Fig. 1). These processes are referred to as (1) enzyme–heat–enzyme, (2) low temperature, (3) dual enzyme/dual heating, (4) dual enzyme/single heating, and (5) thermal liquefaction.

Enzyme–Heat–Enzyme Process. The enzyme–heat–enzyme (EHE) process was the first industrial enzymatic liquefaction procedure developed and utilizes a *B. subtilis*, also referred to as *B. amyloliquefaciens*, α-amylase for hydrolysis. The enzyme can be used at temperatures up to about 90°C before a significant loss in activity occurs. After an initial hydrolysis step a high temperature heat treatment step is needed to solubilize residual starch present as a fatty acid/amylose complex. The heat treatment inactivates the α-amylase, thus a second addition of enzyme is required to complete the reaction.

In the EHE process, a starch slurry is prepared and calcium, as the chloride or hydroxide, is added as a cofactor to provide heat stability to the enzyme. The starch slurry is passed through a stream injection heater and held at temperature for about one hour. The resulting 4–8 DE hydrolyzate is then subjected to a heat treatment in a holding tube, redosed with enzyme, and allowed to react for one hour to a DE level of 10–15.

Low Temperature Process. The low temperature process was developed when *B. licheniformis* and *B. stearothermophilus* α-amylases became commercially available in the 1970s. These enzymes are more thermostable, more aciduric, and require less calcium for stability than the *B. subtilis* enzyme used in the EHE process. Consequently, the high temperature EHE heat treatment step was no longer required to attain efficient liquefaction.

In the low temperature process, the slurry is heated to 105–108°C and held at temperature for 5–10 minutes. The resulting 1–2 DE hydrolyzate is flashed to atmospheric pressure and held at 95–100°C for one to two hours in a batch or continuous reactor. Because the enzyme is not significantly deactivated at the first-stage temperature, a second enzyme addition is not needed. This process is used worldwide throughout the starch-based sweetener industry and has been judged the most efficient process for dextrose production.

Dual-Enzyme Processes. In some cases, especially in syrup production in Europe, a liquefaction process is used that incorporates both a thermostable enzyme and a high temperature heat treatment. This type of process provides better hydrolyzate filterability than that attained in an acid liquefaction process (9). Consequently, dual-enzyme processes were developed that utilized multiple

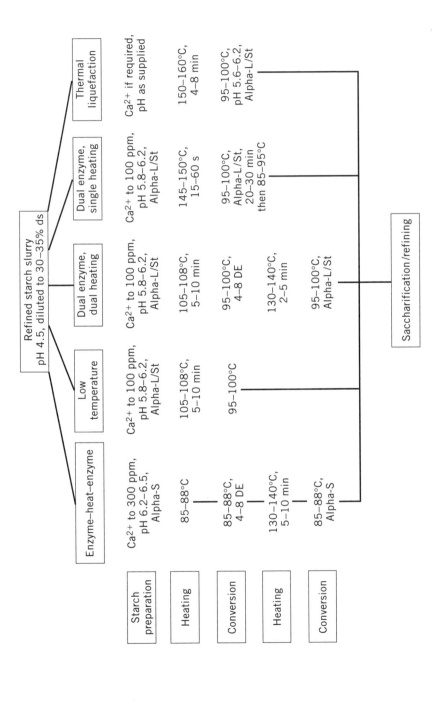

Fig. 1. Enzymatic liquefaction processes (9). Alpha-S is the *α*-amylase from *Bacillus subtilis*; alpha-L/ST are *α*-amylases from *B. licheniformis* or *B. stearothermophilus*, DE is dextrose equivalent, and ds is dry substance.

additions of either *B. licheniformis* or *B. stearothermophilus* α-amylase and a heat treatment step (see Fig. 1).

In these processes, the starch slurry is prepared in the same manner as in the low temperature process. In a dual-enzyme/dual-heating process, the steps are the same as the low temperature process until the completion of the second-stage reaction. Then, a 2–5-min heat treatment followed by a second enzyme addition and another reaction step is employed. In a dual-enzyme/single-heating process, the starch slurry is immediately heated to 145–150°C for one minute or less. Although the enzyme is rapidly inactivated, sufficient hydrolysis takes place to provide a partially thinned hydrolyzate that can be pumped to a second stage where additional enzyme is added and the reaction continued at 95–100°C for 20–30 minutes. The temperature is then lowered for the remainder of the reaction.

Thermal Liquefaction Process. In the thermal liquefaction process (see Fig. 1), a starch slurry containing no enzyme or added calcium is heated for several minutes. The slurry is slightly acidic and sufficient acid liquefaction is achieved to reduce viscosity. The hydrolyzate (at essentially zero DE) is flash-cooled to 95–100°C, α-amylase is added, and the pH is adjusted. The reaction then goes to completion.

Saccharification. Regardless of the starch liquefaction process, dextrose is produced by the action of glucoamylase during a subsequent saccharification step. Glucoamylase is produced in submerged fermentation from strains of *Aspergillus niger*. A broth is obtained containing two or more glucoamylase isozymes and an α-amylase. Other enzymes, such as transglucosidase and protease, as well as cellulase may be present in small quantities. The presence of α-amylase is beneficial in saccharification. Transglucosidase, however, is not desirable because this enzyme catalyzes the formation of isomaltose and reduces dextrose yield. Generally, *Aspergillus* mutants that do not produce transglucosidase are used.

Saccharification is generally conducted batchwise (although continuous systems are also used) in large agitated tanks containing several hundred thousand or million liters of hydrolyzate. The liquefied hydrolyzate is adjusted to 58–60°C and 4.0–4.5 pH and dosed with an amount of glucoamylase to provide a maximum dextrose level in 24–96 hours. At the normal saccharification solids of 30–35 wt %, ca 95–96% dextrose is produced. If the reaction is allowed to continue beyond the normal end point, dextrose level is lowered because glucoamylase catalyzes a reverse reaction that forms isomaltose and maltose by the condensation of a β-anomer of D-glucopyranose with either an α- or β-D-glucose molecule. Because of the reverse reaction, maltose level reaches an equilibrium early in the saccharification. Isomaltose formation, however, is slow. This sugar continues to accumulate. An equilibrium level of 12–14% disaccharides can eventually be reached. The production of reversion products via the reverse reaction is the principal deterrent to achieving a quantitative yield of dextrose in an industrial process.

Dextrose yield, however, can be increased by conducting saccharification at a lower solids level where the reverse reaction is minimized. For instance, dextrose yields of 98.8, 98.2, 97.5, and 96.9% dry basis can be achieved at solids levels of 10, 15, 20, and 25%, respectively (10). Low solids operation, however, is not used commercially owing to problems associated with microbial contamination and cost of water removal. Dextrose level can be increased by 0.5–1.5% at normal reaction solids by using an enzyme such as pullulanase (11) or a *B. megaterium* amylase (12) in conjunction with glucoamylase. Each enzyme works by a different mechanism (13). Both are effective in lowering levels of not only isomaltose but also higher saccharides and providing an increase in dextrose yield.

Refining. After saccharification, the hydrolyzate is clarified by precoat filtration, or possibly membrane filtration, to remove traces of insoluble fat, protein, and starch. Treatment with powdered carbon, granular carbon, and/or ion-exchange resins is then used to remove residual trace impurities, color, and inorganic constituents. The refined hydrolyzate can be dried to a solid product, evaporated to a high dextrose syrup, or processed to crystalline monohydrate or anhydrous dextrose. A typical process for production of crystalline dextrose is shown in Figure 2 (7).

For production of crystalline dextrose monohydrate, a refined 95–96% dextrose hydrolyzate is evaporated under reduced pressure to a syrup containing 75–78 wt % solids and then cooled and passed to crystallizers. The common form of crystallizer is a horizontal cylindrical tank fitted with a slowly turning agitator, cooling jacket, and cooling coils. A substantial bed of seed crystals comprising 20–25 wt % of the previous batch is left in the crystallizer and the refined syrup, at ca 46°C, is mixed with the seed crystals at an initial temperature of ca 43°C. Alternatively, a continuous precrystallizer may be used to form seed crystals (7). The agitated mass is slowly cooled to ca 20–40°C over a period of 3–5 days. By then, ca 60 wt % of the dextrose has crystallized as the monohydrate in a form suitable for separation and washing. The actual values for concentration, initial and final temperature, and rate of temperature drop vary depending primarily on the dextrose content of the hydrolyzate. The resulting magma in the crystallizers is passed into perforated-screen centrifuge baskets, spun to remove the mother liquor, and sprayed with water while spinning to wash out residual mother liquor. The wet sugar, at 99.5% or higher purity and containing ca 14 wt % total moisture, proceeds to rotary dryers, where it is dried in a stream of warm air to a final moisture of 8.5–8.9 wt %, slightly less than the theoretical 9.1 wt % for one molecule of water of crystallization. The lower final moisture content is a way of minimizing caking tendency by reducing the probability of free water present in the product. Crystals are screened to produce fine and course-grade products.

Mother liquor at 90% dextrose from the first crop of crystals can be concentrated and crystallized in a similar manner to recover an additional crystal crop. Depending on quality, some mother liquor can be partially recycled to the initial crystallization step to increase crystal-phase yield as a single crystal crop. The overall yield depends on the dextrose content of the original hydrolyzate and the extent to which hydrolyzates are refined. The mother liquor from the second crystallization contains less than 80% dextrose. The material is evaporated to 71% solids and sold as hydrol to the tanning and fermentation (qv) industries and for the manufacture of caramel color.

An exceptionally pure grade of dextrose monohydrate is obtained by dissolving crystalline monohydrate dextrose in water, carbon-treating to remove color and other impurities, ultrafiltering to remove pyrogens, and crystallizing

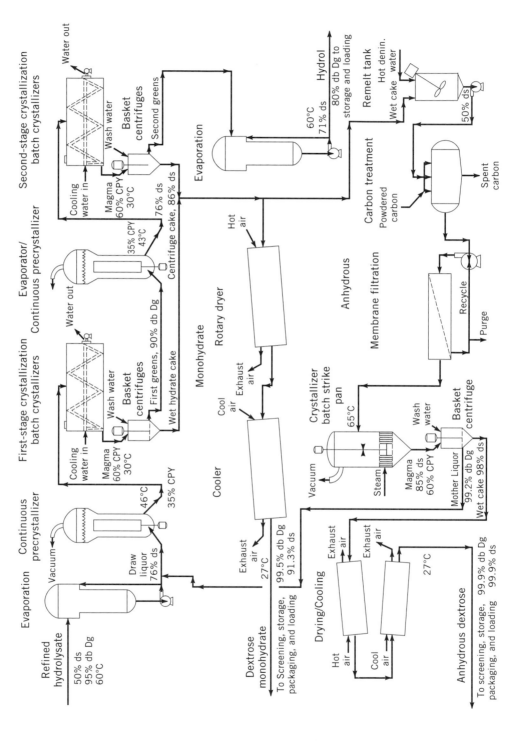

Fig. 2. Production of crystalline dextrose (7), where ds is dry substance; db, dry basis; Dg, D-glucose; and CPY, crystal-phase yield.

an α-anhydrous fraction containing 99.9% dextrose. This recrystallized dextrose is marketed as a USP-grade product for special therapeutic purposes, eg, intravenous injection.

Anhydrous α-dextrose is manufactured by dissolving dextrose monohydrate in water and crystallizing at 60–65°C in a vacuum pan. Evaporative crystallization is necessary to avoid color formation at high temperatures and hydrate formation at low temperatures. The product is separated by centrifugation, washed, dried to a moisture level of ca 0.1%, and marketed as a very pure grade of sugar for special applications.

Anhydrous β-dextrose can be crystallized from concentrated solutions or melts at 90% solids at a temperature above 100°C. By carefully controlling conditions, a product containing a high percentage of anhydrous β-dextrose can be obtained.

Total sugar products are also produced by dehydrating hydrolyzate to a mixture of crystals and amorphous glass. This product is not produced in significant quantities in the United States or Europe but is popular in Japan and Korea where it represents 40–50% of total crystalline dextrose sold (14).

Syrup products at 95–96% dextrose are prepared by evaporating refined hydrolyzate to 71 wt % solids. Liquid $\geq$99% dextrose products are prepared by dissolving crystalline monohydrate dextrose in water or by chromatographic separation of refined hydrolyzate. In the latter case, a 95% dextrose-refined hydrolyzate at 60% solids is fed to a simulated moving bed column containing a cation exchange resin in the sodium form. The separation process produces a product containing $\geq$99% dextrose and an oligosaccharide by-product containing 60–80% dextrose (7).

Dry dextrose is shipped in 45.4-kg bags or in bulk by trucks and railcars. These products are hygroscopic and need to be protected against high temperatures and humidity during storage. Best preventive storage conditions are 30°C and 55% relative humidity. If the temperature is allowed to reach 40°C, the monohydrate can be transformed to the anhydrous form. Liquid products are shipped in tank trucks, railcars, and barges. Occasionally, dextrose may crystallize from solution at low temperatures. In this case, steam is applied to solubilize the dextrose prior to unloading.

Economic Aspects. The average yearly price of dextrose monohydrate from 1975 through 1995 is given in Table 3. Factors contributing to dextrose price include corn costs, demand, processing capacity, production costs, and sucrose price. Dextrose production in the 1994–1995 fiscal year was estimated at 661,000 t (dry basis) (18). Production capacity has not changed significantly since the mid-1980s because of the expense of dextrose crystallizers. This factor, along with high production costs, has caused dextrose prices to remain above those of HFS and corn syrup. Of all the corn sweeteners, dextrose is the most expensive and has the most stable list price. Distribution of dextrose from 1970 through 1992 to various industries in the United States is shown in Table 4.

Analysis and Specifications. Typical product analyses include solids level, ash, color, conductivity, purity, and minor saccharide levels (19). Specifications for anhydrous and monohydrate crystalline dextrose are available (15). High quality anhydrous dextrose produced for the pharmaceutical industry is prepared in accordance with additional specifications (20).

Table 3. Yearly Price of Dextrose, HFS, and Corn Syrup in the United States[a], $/kg

Year	Dextrose[b]	HFS-42[c]	HFS-55[d]	Corn syrup[e]
1975	0.43	0.35		0.27
1980	0.59	0.37		0.25
1985	0.49	0.28	0.34	0.20
1990	0.50	0.31	0.37	0.25
1991	0.50	0.33	0.39	0.27
1992	0.50	0.32	0.39	0.27
1993	0.50	0.29	0.35	0.23
1994	0.52	0.32	0.38	0.27
1995		0.27	0.30	

[a]Refs. 15–17.
[b]In 45.4-kg (100-lb) bags, monohydrate basis (92 wt % solids).
[c]In tank cars, 71 wt % solids.
[d]In tank cars, 77 wt % solids.
[e]In tank cars, 80.3 wt % solids.

Health Factors. Dextrose products are substances that are presumed to be GRAS by the FDA (21). A study of the health aspects of dextrose, fructose, and corn syrups has indicated that these sweeteners are not hazardous at levels of normal human consumption with the exception of a small contribution to the formation of dental caries (22).

Uses. The main use of dextrose is in food processing (qv), where it is of value for its physical, chemical, and nutritive properties. Dextrose is also used in nonfood applications in the chemical, drug, and pharmaceutical industries. This latter segment of the market represents about 21% of dextrose usage (13). In the baking industry, dextrose is used as a fermentable sugar to provide crust color and, also, to supply strength, develop flavor, and optimize texture. In the beverage industry, it is used as a source of fermentables in low calorie beer (qv) and as a sweetener in beverage powders. In the canning industry, it supplies sweetness, body, and osmotic pressure. It also contributes to better natural color retention in certain products. In the confectionery industry, it is used to supply sweetness and softness control and to regulate crystallization. Dextrose is also used to make tableted products. Flavor is often enhanced by the cooling effect obtained when dextrose hydrate dissolves in the mouth. In frozen desserts, it prevents oversweetness and improves flavor. Dextrose is also used in seasoning formulations for sausages and for color in hams, in peanut butter for mouthfeel and chewability, in jams and jellies for controlling sweetness and providing stability by controlling osmotic pressure, as a substrate for vinegar (qv) fermentation, and in prepared mixes for biscuits, pancakes, waffles, doughnuts, and icings.

In many cases, dextrose is used in conjunction with sucrose. Although dextrose by itself is somewhat less sweet than sucrose, the combination of the two may be as sweet as pure sucrose at the same concentration (23). Dextrose is also used in the pharmaceutical industry for intravenous feeding as well as for tableting and other formulations. In fermentation, dextrose is a raw material for biochemical synthesis of organic acids, vitamins, antibiotics, enzymes, amino

Table 4. Distribution by Industry of Syrups in the United States[a], t × 10³

Year	Baking and cereal	Confectionery	Processed foods	Dairy products	Beverages	Total[b]
			Dextrose[c]			
1970	230	57	49	7	12	511
1975	195	65	28	8	25	501
1980	104	59	12	2	74	427
1985	122	72	17	1	104	490
1990	138	85	24	1	121	552
1992	148	85	18	4	105	572
			HFS-42[d]			
1970	18	4	15	9	6	61
1975	150	14	111	42	229	676
1980	389	14	427	141	476	1932
1985	442	33	508	201	656	2355
1990	412	36	698	235	1463	3272
1992	471	33	799	277	1495	3548
			HFS-55[e]			
1970	0	0	0	0	0	0
1975	0	0	0	0	0	0
1980	5	1	20	6	619	697
1985	11	6	73	15	3885	4141
1990	20	7	147	24	4045	4393
1992	26	8	191	24	4119	4510
			Corn syrup[f]			
1970	263	506	229	243	137	1621
1975	372	464	370	312	288	2299
1980	236	480	328	260	434	2157
1985	199	554	448	324	453	2439
1990	241	557	483	319	467	2839
1992	250	644	475	395	478	3052

[a]Ref. 16.
[b]Includes other applications not listed.
[c]Monohydrate basis.
[d]71 wt % solids basis.
[e]77 wt % solids basis.
[f]80.3 wt % solids basis, including syrup solids and maltodextrin.

acids, and polysaccharides (24). A principal industrial fermentation use of dextrose is for the production of fuel ethanol as an oxygenate or octane enhancer (25). Dextrose is hydrogenated to sorbitol for use in various food and nonfood applications. Reaction of dextrose with sorbitol and citric acid produces a low calorie dextrose polymer, which is used as a bulking agent in reduced-calorie foods (26). Other industrial applications of dextrose include use in wallboard as a humectant, in concrete as a setting retardant, in resin formulations as a plasticizer, in adhesives for flow control, and for the production of methyl glucoside.

High Fructose Corn Syrups

High fructose corn syrups (HFS, HFCS, isosyrup, isoglucose) are concentrated carbohydrate solutions containing primarily fructose and dextrose as well as lesser quantities of higher molecular weight saccharides. A 42 wt % fructose syrup is produced by partial enzymatic isomerization of dextrose hydrolyzate. A 55 wt % fructose syrup is produced by a combination of enrichment and blending. Liquid products containing 80–95 wt % fructose are also manufactured. Pure crystalline fructose is produced in a dry form.

In nature, fructose (levulose, fruit sugar) is the main sugar in many fruits and vegetables. Honey contains ca 50 wt % fructose on a dry basis. Sucrose is composed of one unit each of fructose and dextrose combined to form the disaccharide. Fructose exists in polymeric form as inulin in plants such as Jerusalem artichokes, chicory, dahlias, and dandelions, and is liberated by treatment with acid or enzyme.

Fructose was first isolated in 1847 (27). In 1874, it was recognized that fructose had advantages over sucrose as a sweetener for diabetics (28). Based on the discovery of the alkaline conversion of dextrose to fructose in 1895 (29), a considerable number of investigations were conducted in an attempt to develop a commercial process (30). However, because of problems associated with color, off-flavor, degradation products, and low fructose yield, an alkaline conversion process was never commercialized. Enzymatic conversion of dextrose to fructose using glucose isomerase was first reported in 1957 and patented in 1960 (31,32). Research in this area continued for several years in Japan, resulting in a U.S. patent (33) and commercial production in Japan in 1966 (34). Japanese technology was then licensed by a U.S. company and production initiated in 1967 by a batch process (35). A 15 wt % fructose syrup was made first, followed in 1968 by a 42 wt % fructose product. In 1972, a continuous system was initiated using an immobilized enzyme process (36,37).

Properties. Fructose, a ketohexose monosaccharide, crystallizes as β-D-fructopyranose [7660-25-5] (**3**) and has a molecular weight of 180 and a melting point of 102–104°C. At equilibrium (25°C) the solubility of fructose in water is 80 wt %. In solution at 36°C, fructose undergoes rapid mutarotation to an equilibrium mixture of 3% α-D-fructopyranose [10489-81-3] (**4**), 57% β-D-fructopyranose, 9% α-D-fructofuranose [10489-79-9] (**5**), and 31% β-D-fructofuranose [470-23-5] (**6**) (38). In solution, the concentration of (**3**) varies from 77% at 0.5°C to 48% at 61°C (39). The crystalline form of fructose is 10–80% sweeter than sucrose and 50–100% sweeter than dextrose (13), depending on temperature, pH, concentration, and the presence of other additives.

(**3**)   (**4**)

Because of the presence of fructose, HFS is sweeter than conventional corn syrups, although intensity of sweetness results from many factors. In general, 42% HFS is about 8% less sweet than sucrose, 55% HFS exhibits about the same sweetness as sucrose, and 90% HFS is about 6% sweeter than sucrose (40). Other properties of HFS include high solubility, which reduces the possibility of crystallization during shipment; humectant properties allowing for increased shelf life of bakery products; easier decomposition of fructose during baking resulting in improved color and flavor; and high osmotic pressure for containment of microbial growth.

Manufacture.   HFS containing 42% fructose is produced commercially by column isomerization of clarified and refined dextrose hydrolyzate using an immobilized glucose isomerase. Enriched syrup containing 90% fructose is prepared by chromatographic separation and blended with 42% HFS to obtain 55% HFS.

Glucose isomerase is produced for commercial use from a variety of bacterial organisms, including *Actinoplanes missouriensis, Bacillus coagulans, Microbacterium arborescens, Streptomyces olivochromogenes, S. griseofuscus, S. murinus, S. phaeochromogenes,* and *S. rubiginosus* (41). The enzyme is immobilized and used in a continuous column operation to produce fructose from a dextrose feed. The high concentration of enzyme allows a short reaction time of 0.5–4 hours that minimizes problems associated with long batch reactions such as formation of color, off-flavors, and by-products. In addition, continuous use of the enzyme reduces enzyme cost to an economical level.

Two types of immobilization are used for immobilizing glucose isomerase. The intracellular enzyme is either immobilized within the bacterial cells to produce a whole-cell product, or the enzyme is released from the cells, recovered, and immobilized onto an inert carrier. An example of the whole-cell process is one in which cells are disrupted by homogenization, cross-linked with glutaraldehyde, flocculated using a cationic flocculent, and extruded (42). In a second example, a cell–gelatin mixture is cross-linked with glutaraldehyde (43). When soluble enzyme is used for binding, the enzyme is first released from the cell, then recovered and concentrated. Examples of this type of immobilization include binding enzyme to a DEAE-cellulose–titanium dioxide–polystyrene carrier (44) or absorbing enzyme onto alumina followed by cross-linking with glutaraldehyde (45,46).

HFS is produced from 93–96% dextrose hydrolyzate that has been clarified, carbon-treated, ion-exchanged, and evaporated to 40–50% dry basis. Magnesium is added at a level of 0.5–5 mM as a cofactor to maintain isomerase stability and to prevent enzyme inhibition by trace amounts of residual calcium. The feed may also be deaerated or treated with sodium bisulfite at a level of 1–2-mM SO$_2$ to prevent oxidation of the enzyme and a resulting loss in activity.

Hydrolyzate at about 55–61°C and 7.5–8.2 pH is passed through a fixed bed of immobilized isomerase at a controlled flow rate. Maximum fructose content at equilibrium is 50–55% dry basis; however, residence time is adjusted to attain 42–45% fructose because a greatly increased reaction time is required to attain higher levels. The enzyme can be used for as long as 10 weeks, during which time loss in activity is compensated for by regulating the residence time, ie, the flow rate through the column. Enzyme reactors are operated in parallel or in series and are replaced individually as isomerase is inactivated. The isomerized hydrolyzate at about 42 wt % fructose is adjusted to pH 4–5, carbon-treated to remove color and off-flavors, ion-exchanged to remove salts and residual color, and concentrated by evaporation to ca 71 wt % solids. Dextrose content in the 42% HFS product is 50–52%, thus storage at 35–41°C is recommended. At a lower temperature dextrose crystallization is a possibility and at a higher temperature color development may occur. Product is shipped in tank trucks or by railcars. If crystallization occurs, the syrup is heated to as high as 54°C to dissolve crystals prior to unloading.

Products containing higher levels of fructose are produced by chromatographic separation of 42 wt % HFS through a column of absorbent containing calcium or other cation groups. Fructose is retained to a greater degree than dextrose or oligosaccharides, and therefore a nonfructose stream is collected first as raffinate, followed by elution of fructose extract with water. In commercial operation, separation is achieved in a batchwise, semicontinuous, or continuous fashion. The continuous procedure involves a simulated moving bed in which feed and desorbent enter the column at different points while fructose and raffinate streams are withdrawn. Points of entry and withdrawal are changed periodically to correspond to flow through the column, and hence separation efficiency is maximized. In a typical operation, 42% HFS at a dry substance content of 50% is separated in a column at 50–70°C to generate a high fructose fraction and a by-product raffinate stream. The enriched fructose stream contains 80–90% fructose and 7–19% dextrose; the raffinate is composed of 80–90% dextrose and 5–10% fructose. Raffinate is recycled to isomerization columns for production of 42% HFS. Enriched HFS is blended with 42% HFS to produce a product containing 55% fructose and evaporated to 77% solids for shipment. Crystallization is not a concern because of the low dextrose level and storage temperature is generally maintained at 24–29°C. Other HFS products containing 80, 90, and 95% fructose (18, 9, and 4% dextrose, respectively) are also produced and evaporated to 77% solids (40) for shipment as an essentially noncrystallizable syrup.

Crystalline fructose differs from HFS in that it is >99.5% pure, contains less than 0.5% dextrose, and is dried to 99.5% solids. Aqueous, solvent, and mixed solvent processes have been proposed for fructose crystallization. The aqueous process is the primary mode of manufacture in the United States and elsewhere, although some aqueous–ethanol systems are also used in Europe and the Far East (39). In a typical process, a 90% fructose feed is added to an aqueous, alcohol, or combined solvent at 60–85°C and a pH of 3.5–8. Crystallization is accomplished by lowering temperature to 25–35°C and/or reducing pressure during 2–180 hours (39). Crystals are separated by centrifugation or filtration, impurities removed by washing, and the mass dried and screened. Shipment is by bags, totes, or railcars. A crystalline fructose syrup can be prepared by dissolving crystalline fructose in water to 77% solids (47).

Economic Aspects. Prices of 42 and 55% HFS are listed in Table 3. HFS price is influenced by the price of sugar, cost of corn, corn quality, and amount of excess wet-milling capacity. In the United States, 42 and 55% HFS sell at a discount to wholesale refined beet sugar that has averaged about 16 and 11¢/kg, respectively. Prices exhibit a seasonal pattern having summer highs that are generally about 30% more than the yearly low price. The seasonal pattern results from increased consumption of soft drinks containing HFS during the summer months. The 1995 price of crystalline fructose (truckload quantities in 22.7-kg bags) was \$0.86/kg (48). World consumption of crystalline fructose is about 54×10^3 t/yr (49).

Analysis, Specifications, and Health Factors. Methods of analysis and health aspects are the same as those for dextrose. Specifications for HFS are the same as those for corn syrup. HFS is presumed to be GRAS by the FDA (50). Health effects of fructose are discussed elsewhere (51).

Uses. High fructose syrup is used as a partial or complete replacement for sucrose or invert sugar in food applications to provide sweetness, flavor enhancement, fermentables, or humectant properties. It is used in beverages, baking, confections, processed foods, dairy products, and other applications. Worldwide HFS production in the 1994–1995 fiscal year was estimated at about 8.6×10^6 t (dry basis) (18). About 75% of total world production is in the United States.

The primary application of HFS is in soft drinks as emphasized by the rapid increase in usage of 55% HFS in the beverage industry since the 1980s (Table 4). HFS containing 42% fructose was first used for this application in 1974. Because 55% HFS exhibits the same sweetness as sucrose, the use of a 50–50 blend of 55% HFS and sucrose was authorized by soft-drink producers in 1980 and 100% substitution was approved in 1984. HFS is also used in other applications such as baking where it acts as a source of fermentables and improves flavor, aroma, and texture. It is also used in ice cream to control crystallization and add body, in dairy products to provide body and improve mouthfeel and texture, in confectionery to control grain and humectancy, and in canned goods as a preservative and for adding sheen.

HFS containing 90% fructose is used in low calorie or specialty foods because of its high sweetness and, therefore, reduced usage level and lower caloric value. Crystalline fructose is essentially pure and used at a level that provides sweetness at a lower caloric level than other sweeteners (qv). Initial use was in diet and nutritious foods but application has now been extended to many other food areas, such as powdered beverages, dry mix desserts, dairy products, and confections.

Corn Syrups

Corn syrups [8029-43-4] (glucose syrup, starch syrup) are concentrated solutions of partially hydrolyzed starch containing dextrose, maltose, and higher molecular weight saccharides. In the United States, corn syrups are produced from corn starch by acid and enzyme processes. Other starch sources such as wheat, rice, potato, and tapioca are used elsewhere depending on availability. Syrups are generally sold in the form of viscous liquid products and vary in physical properties, eg, viscosity, humectancy, hygroscopicity, sweetness, and fermentability.

Properties. Corn syrups are defined as those starch hydrolysis products exhibiting a DE of 20–99.4. Lower DE products are classified as maltodextrins [9050-36-6] and higher DE products as dextrose. Syrups are often described in terms based on the type of production process, ie, acid conversion, acid–enzyme conversion, and enzyme–enzyme conversion; on the degree of hydrolysis (high conversion); or on a particular saccharide in the syrup (high maltose). Examples of some of these syrups are shown in Table 5. The most adequate characterization is with respect to the concentration of individual saccharides. In many cases, it is the individual saccharides or groups of saccharides that determine syrup characteristics. Consequently, corn syrups exhibit many functional properties, including fermentability, viscosity, humectancy–hygroscopicity, sweetness, colligative properties, and browning reactions, which differ from the properties of other syrups.

Fermentability of corn syrups by yeast is important in certain food applications, eg, baking and brewing. The fermentable sugars present in corn syrup are dextrose, maltose, and maltotriose. Fermentability of maltose or maltotriose depends on the specific fermentation process and organism. In general, greater fermentability is obtained at higher DE levels.

Viscosity of corn syrup is a function of DE value, temperature, and solids concentration. Viscosity decreases with increasing DE and temperature but increases with increasing concentration. For a 43-DE corn syrup, viscosity at 1.42 sp gr (43° Be′) is 56,000 mPa·s(=cP) at 27°C, 14,500 mPa·s at 38°C, and 4,900 mPa·s at 49°C. Corresponding values for a 55-DE syrup at the same density are 31,500, 8,500, and 2,900 mPa·s.

The hygroscopic and humectant properties of corn syrups are of great importance in many applications. Depending on the type of syrup and on the specific conditions of temperature and humidity, the products may either resist or facilitate moisture loss or moisture absorption. The ability to attract moisture or retard its loss increases with increasing DE value. Prevention of moisture pickup is more characteristic of syrups having low DE values.

Table 5. Composition of Corn Syrups and Maltodextrins

Type	Amount of constituent[a], wt %				
	DE	DP-1	DP-2	DP-3	DP-4+
acid-converted syrup	30	9	10	12	69
	36	14	12	10	64
	42	19	14	12	55
	55	31	18	13	48
high maltose syrup	42	6	44	13	37
	48	9	52	15	24
	50	3	75	13	9
high conversion syrup	64	37	31	11	21
	70	43	30	7	20
maltodextrin	10	0	3	4	93
	15	1	5	7	87
	20	1	8	9	82

[a]DP refers to degree of polymerization, where DP-1 is dextrose, DP-2 is a disaccharide, etc. DE = dextrose equivalent.

Sweetness is primarily a function of the levels of dextrose and maltose present and therefore is related to DE. Other properties that increase with increasing DE value are flavor enhancement, flavor transfer, freezing-point depression, and osmotic pressure. Properties that increase with decreasing DE value are bodying contribution, cohesiveness, foam stabilization, and prevention of sugar crystallization. Corn syrup functional properties have been described in detail (52).

Manufacture. Corn syrups are manufactured by acid, acid–enzyme, or enzyme–enzyme hydrolysis processes. Acid hydrolysis of starch is conducted by batch or continuous processes. Batchwise conversion is carried out in large cookers or converters, which are usually built of manganese–bronze and have capacities of ca 10 m^3 (10^4 L). A suspension of starch at 35–40 wt % dry solids is passed to the converter, hydrochloric acid is added to a concentration of 0.015–0.02 N, and the converter is steam-heated until a temperature of 140–160°C is reached. The mixture is held at temperature for a period of time, usually 15–20 minutes, to produce the desired degree of hydrolysis. Improved process control and therefore better product uniformity is achieved by a continuous process using indirect heating. In this type of process, acidified starch slurry is pumped at a constant rate through a series of heat exchangers at reaction conditions similar to those used in batch operations. In either process, hydrolyzate is neutralized to pH 4–5.5 by addition of soda ash, clarified by centrifugation or filtration, evaporated to ca 60 wt % solids, carbon-treated to remove color and acid degradation products, and concentrated to 77–85 wt % solids. Sulfur dioxide is added during evaporation to some grades of syrup to reduce color development. The standard acid-converted syrup is typically about 42 DE, although higher and lower DE syrups are also produced by this process. Syrup DE is limited to about 30–55 because at lower DE residual starch may cause a haze problem and at higher DE excess color and off-flavor are produced owing to acid-catalyzed side reactions. Composition of 30–55-DE acid-converted syrups are shown in Table 5.

Syrups are also produced by acid–enzyme or enzyme–enzyme processes. Starch is first hydrolyzed by acid or by enzyme. The latter is similar to the method for liquefaction processes (see Fig. 1). Either treatment is followed by saccharification with one or more enzymes to the desired composition. Maltose syrup [69-79-4], for example, can be prepared from a 10–20-DE partially hydrolyzed starch substrate by saccharification using a maltose-producing enzyme at 50–55°C and pH 5. Maltogenic enzymes, eg, β-amylase extracted from germinated barley or fungal α-amylase derived from *Aspergillus oryzae*, are used to produce a hydrolyzate containing about 40–50 wt % maltose (Table 5). Higher levels of maltose, ie, 60–80 wt %, are produced by saccharification with a combination of a maltogenic enzyme and a debranching enzyme such as pullulanase. The pullulanase hydrolyses α-1,6 linkages during saccharification and provides additional substrate for the action of the maltogenic enzyme.

High conversion syrups of 60–70 DE containing intermediate levels of dextrose and maltose are also produced (Table 5). Saccharification is conducted with combination of glucoamylase and maltogenic enzyme at pH 4.8–5.2 and 55–60°C for several hours or days until the desired DE is reached. Maltose and high conversion syrups are refined and shipped in rail cars and tank trucks. Syrups are generally heated to ca 38°C to facilitate unloading. Some syrups, particularly those having a limited extent of hydrolysis, are reduced to dry form by spray-drying or roll-drying. These products are commonly called corn syrup solids and are shipped in moistureproof bags. Products of <20 DE are referred to as maltodextrins or hydrolyzed cereal solids (Table 5). These products are bland-tasting, free-flowing, and nonhygroscopic. Maltodextrins are prepared from regular or waxy corn starch by enzyme or acid processes to products of 10–20 DE, then clarified, refined, spray-dried to a moisture content of 3–5 wt %, and finally shipped in 45.4-kg bags. In some cases, a syrup of ca 75 wt % solids is produced.

Economic Aspects.   Prices of corn syrup since 1975 are listed in Table 3. Production capacity, demand, and corn prices affect corn syrup price. U.S. syrup production for the 1994–1995 fiscal year was estimated at 2.9×10^6 t (dry basis) (15).

Analysis, Specifications, and Health Factors.   Corn syrups are usually sold with a specification of the Baumè measurement which is related to solids content. A typical value is 43.5° Be′, corresponding to 80.3 wt % solids for a 42-DE acid-converted syrup. Solids content for syrups at 41, 42, 43, 44, and 45° Be′ (sp gr 1.39, 1.41, 1.42, 1.43, and 1.45) are 75.0, 77.1, 79.1, 81.2, and 83.3 wt %, respectively. Higher DE syrups exhibit slightly higher solids levels at the same densities. DE is determined by copper reducing methods. Saccharide composition is determined by high performance liquid chromatography. Other analyses include color, iron, and pH. Specifications for glucose syrup and dried glucose syrup are available (53). Maltodextrin and corn syrup products are substances that are presumed to be GRAS by the FDA (54). Health and safety aspects are the same as those for dextrose.

Uses.   Principal uses of corn syrups are shown in Table 4. The specific type of syrup employed depends on the properties desired in the final product. Syrup properties are important to varying degrees in different products, and changes in formulation can affect the choice of syrup required to supply the most desirable properties. In many cases, corn syrups are used as supplementary sweeteners when sucrose is the primary one.

In the confectionery industry, corn syrups are used extensively in nearly every type of confection, ranging from hard candy to marshmallows. In hard candies, which are essentially solid solutions of nearly pure carbohydrates, corn syrup contributes resistance to heat discoloration, prevents sucrose crystallization, and controls hygroscopicity, viscosity, texture, and sweetness. Maltose syrups, high conversion syrups, and acid-converted syrups (36 and 42 DE) are used for this application.

In the canning and preserving industries, corn syrups are used to prevent crystallization of sucrose, provide body, accentuate true fruit flavors, and improve color and texture. In the beverage industry, the predominant use is in the beer and malt-liquor areas. High conversion syrups are used to replace dry cereal adjuncts, provide fermentable sugars, enhance flavor, and provide body. These syrups contain controlled amounts of dextrose and maltose for proper fermentation.

Corn syrups used in baking are generally of the high conversion type. They are incorporated into cakes, cookies, icings, and fillings to increase the amount of moisture retained, retard crystal growth of other sugars, enhance tenderness, and increase shelf life. In yeast-raised goods, fermentability is of importance, and therefore only high DE syrups are used.

Corn syrups used in ice cream and frozen desserts are generally 36- or 42-DE acid-converted syrups. The syrup serves primarily to provide maximum flexibility in adjusting flavor, texture, body, and smoothness. It also aids in grain control and in the modification of meltdown and shrinkage characteristics of the frozen product.

Syrups of 25–30 DE are used as spray-drying aids in products such as coffee. High conversion syrup, maltose syrup, and 42-DE syrup are used in jams and jellies. Corn syrup is also used in table syrups, baby food, meat packing, breakfast foods, salad dressing, pickles, dehydrated powdered foods, medicinal syrups, textile furnishings, adhesives, and numerous other products and processes.

Maple Syrup

Maple syrup is prepared by concentrating (evaporation or reverse osmosis) sap from the maple tree to a concentrated solution containing predominantly sucrose. Its characteristic flavor and color are formed during evaporation. Maple syrup is produced from the sap of several varieties of mature maple trees, eg, the sugar maple (*Acer saccharum*) and black maple (*Acer nigrum*).

Collection of sap is made sometime between late fall and mid-spring, depending on weather conditions. The best time is when the temperature is ca 7°C during the day and below freezing at night. Sap generally contains 2–3 wt % solids, of which ca 96% is sucrose; the remainder is composed of other carbohydrates, organic acids, ash, protein, and lignin-like materials. A taphole is drilled into the tree and a spout is driven in the opening. Sap is collected in a bucket, in a bag, or alternatively in plastic tubes directed to a centralized collection tank by gravity or vacuum. Evaporation is conducted at atmospheric pressure until a boiling point of 104°C is reached to produce a syrup meeting governmental specifications of at least 66 wt % solids (54). A syrup concentrated to only 65 wt % solids has a thin taste, and one concentrated to ≥67 wt % crystallizes when cooled. A refractometer or hydrometer is used to measure concentration. Final specific gravity should be 1.35 (37.75° Baumè) at 15.6°C. Flavor and color develop during evaporation as a result of loss of water and reactions occurring between sugar and other components. Syrup is clarified, graded as to color, flavor, and density, and finally packaged in small containers for retail sale as table syrup. Typically the product contains 88–99 wt % sucrose and 0–12 wt % invert sugar. Maple syrup is also used in candy manufacture by blending with sucrose. Other applications include addition to cookies, ice cream, baked beans, baked ham, and baked apples.

Maple sugar is prepared by concentrating sap to a high solids content, ie, a boiling temperature of 116–121°C, and then allowing the supersaturated solution to crystallize or solidify during cooling. Maple cream or maple butter is made by stirring a supersaturated solution while cooling rapidly to produce a product of creamy texture.

The main maple-syrup-producing areas are located in the northeastern and midwestern United States and eastern Canada. In the United States, leading producing states in 1994 were Vermont (accounting for 33% of total production), New York (19%), Maine (11%), and Wisconsin (10%) (55). U.S. production and imports as well as prices for maple syrup are shown in Table 6. Total production in the United States and Canada for 1995 is estimated to be about 25 million kg (57).

Table 6. Production, Imports, Exports, and Price of Maple Syrup in the United States[a]

Year	Production, $L \times 10^6$	Imports[b], $L \times 10^6$	Exports[b], $L \times 10^6$	Price[c], $/L
1980	3.7	3.7	0.5	4.31
1985	5.1	5.3	0.9	4.50
1990	4.1	7.2	0.8	6.63
1991	4.9	7.4	1.2	6.16
1992	6.2	8.7	1.7	6.29
1993	3.8	10.1	2.2	6.16
1994	5.0	12.5	2.2	6.45
1995	4.1			

[a]Refs. 55 and 56.
[b]Includes maple sugar.
[c]Northeastern United States.

Molasses

Molasses, another type of syrup, is a by-product of the sugar industry. It is the mother liquor remaining after crystallization and removal of sucrose from the juices of sugar cane or sugar beet and is used in a variety of food and nonfood applications. Molasses, first produced from sugarcane in China and India centuries ago and later in Europe and Africa, was introduced as the by-product of cane-sugar production into Santo Domingo by Columbus in 1493. During Colonial times, molasses was very important to the American colonies for the production of rum. In 1733, the British Parliament passed the Molasses Act to tax molasses imported from foreign countries. This attempt to restrict trade was ignored by the colonies and was, in part, responsible for the American Revolution.

Manufacture. Raw sugar (qv) is produced from sugarcane by a process that involves extraction of the sugar in water, treatment to remove impurities, concentration, and several crystallizations. After the first crystallization and removal of first sugar, the mother liquor is called first molasses. First molasses is recrystallized to obtain a second lower quality sucrose (second sugar) and a second molasses. After a third crystallization, the third molasses contains considerable nonsucrose material, and additional recovery of sucrose is not economically feasible. The third molasses is sold as blackstrap, final, or cane molasses. Raw sugar obtained from the above process is mixed with water to dissolve residual molasses and then separated by centrifugation. This process is called affination and the syrup is referred to as affination liquor. The sugar is dissolved in water, treated to remove color and impurities, and subjected to several crystallizations to obtain refined sugar. The mother liquor from the final crystallization is combined with affination liquor and crystallized to produce a dark sugar (remelts) which is recycled to raw sugar. The remaining mother liquor is called refiners molasses and is similar to final molasses but usually of better quality.

In beet sugar manufacture, the beet juice does not contain reducing sugars such as fructose and glucose, which are present in cane juice, but may contain raffinose. Because of the absence of reducing sugars, sucrose level in beet molasses is not reduced to the same extent as for cane. Final molasses from beet contains ca 60 wt % sucrose (dry basis) compared to 30 wt % sucrose (dry basis) in cane molasses. Treatment of diluted beet molasses with calcium oxide precipitates sucrose as tricalcium sucrate (Steffen process), which is recycled to the incoming hot beet juice. During recycling, raffinose accumulates in the final molasses and retards crystallization if not removed. Therefore, a portion of the final molasses, called discard molasses, is periodically removed.

Ion-exclusion chromatography is increasingly being used to remove sugar from beet molasses (58). As much as 90% sugar recovery is achieved by this technique, which involves passing diluted and clarified molasses through a column containing a strong acid cation exchange resin. Sucrose is absorbed on the resin and nonsucrose is recycled back to the sugar process. The resulting molasses contains 12–20% residual sugar (58) compared to traditional molasses that contains ≥50% sugar.

High test molasses (invert molasses) is produced from cane sugar when sucrose manufacture is restricted because of overproduction. The cane sugar at ca 55 wt % solids is enzymatically converted to invert syrup to prevent crystallization and evaporated to a syrup. The product is used in the same applications as blackstrap molasses.

Molasses from other sources include citrus and corn sugar (hydrol) molasses. Citrus molasses is produced from citrus waste and contains 60–75% sugars. Corn sugar molasses is the mother liquor remaining after dextrose crystallization and contains a minimum of 43% reducing sugars expressed as dextrose.

Molasses is shipped in drums, barrels, tank trucks, tank cars, barges, and sea vessels. Because of high viscosity, molasses must be heated in some situations to facilitate pumping. However, prolonged heating must be avoided to prevent caramelization.

Composition. Molasses composition depends on several factors, eg, locality, variety, soil, climate, and processing. Cane molasses is generally at pH 5.5–6.5 and contains 30–40 wt % sucrose and 15–20 wt % reducing sugars. Beet molasses is ca 7.5–8.6 pH, and contains ca 50–60 wt % sucrose, a trace of reducing sugars, and 0.5–2.0 wt % raffinose. Cane molasses contains less ash, less nitrogenous material, but considerably more vitamins than beet molasses. Composition of selected molasses products is listed in Table 7. Procedures for molasses analysis are available (59).

Uses. The primary use of molasses is in animal feed. Molasses, which provides a carbohydrate source, salts, protein, vitamins, and palatability, may be used directly or mixed with other feeds. The carbohydrate content of 24.6 L (6.5 gal) of blackstrap molasses is considered to be equal to 0.035 m^3 (one bushel) of corn as measured by the energy produced from 0.035 m^3 of corn and the amount of molasses required to produce the same amount of energy. When

Table 7. Composition of Various Types of Molasses, wt %

Molasses type	Solids	Total sugars as invert	Crude protein	Total ash
cane				
Louisiana	80.8	59.5	3.0	7.2
refiners	75.4	55.9	2.1	8.6
high test (Cuba)	80.4		0.7	1.4
beet (Wisconsin)	78.6	52.7	11.4	9.3
corn	74.9	50.3	0.4	8.9
citrus	71.4	42.4	4.7	4.8

molasses is less expensive than corn, sales increase; when the reverse is true, sales decrease.

Molasses is also used as an inexpensive source of carbohydrate in various fermentations for the production lactic acid, citric acid, monosodium glutamate, lysine, and yeast (60). Blackstrap molasses is used for the production of rum and other distilled spirits.

Food applications utilize first and second molasses in baking (bread, cakes, cookies) for the molasses flavor. Molasses is also used in curing of tobacco and meats, in confections such as toffees and caramels, and in baked beans and glazes.

Production and Economics.   Total U.S. molasses production, imports, and prices since 1980 are shown in Table 8. Sources of imports to the United States are Australia, Dominican Republic, Guatemala, Mexico, and Poland (58). U.S. molasses price is regulated by supply and demand and the world sugar price. Price of beet molasses is generally higher than that of cane molasses owing to higher levels of sucrose and nitrogen.

Table 8. Production, Imports, and Price of Molasses in the United States[a]

Year	Production, $t \times 10^6$	Imports, $t \times 10^6$	Price[b], $/t Blackstrap[c]	Price[b], $/t Beet[d]
1980	1.49	0.93	106.39	106.45
1985	1.84	1.65	55.43	74.42
1990	1.91	1.08	68.02	64.31
1991	2.18	1.26	73.89	72.94
1992	1.93	1.12	67.55	63.39
1993	1.74	1.04	61.17	71.05
1994		1.66	72.19	
1995			78.23	

[a]Refs. 55 and 56.
[b]Free-on-board tankcar or tank truck.
[c]New Orleans.
[d]Wyoming and Montana.

BIBLIOGRAPHY

"Dextrose and Starch Syrups" in *ECT* 1st ed., Vol. 4, pp. 961–969, by G. R. Dean, Corn Products Refining Co.; "Molasses" in *ECT* 1st ed., Vol. 9, pp. 167–180, G. T. Reich, Consultant; "Dextrose and Starch Syrups" in *ECT* 2nd ed., Vol. 6, pp. 919–932, by E. R. Kooi, Corn Products Co.; "Molasses" in *ECT* 2nd ed., Vol. 13, pp. 613–633, by A. G. Keller, Louisiana State University; "Syrups" in *ECT* 3rd ed., Vol. 22, pp. 499–522, by R. E. Hebeda, CPC International.

1. E. Martin, *Dextrose Therapy in Everyday Practice*, Harper, New York, 1937, p. 5.
2. H. Wichelhaus, *Der Starkezucker*, Akademische Verlagsgesellschaft, Leipzig, Germany, 1913.
3. T. de Saussere, *Ann. Phys.* **49**, 129 (1815).
4. G. S. C. Kirchoff, *Acad. Imp. Sci. St. Petersbourg, Mem.* **4**, 27 (1811).
5. U.S. Pat. 1,471,347 (1923), W. B. Newkirk (to Corn Products Refining Co.).
6. R. F. Jackson and C. G. Silsbee, *Natl. Bur. Stand. (U.S.) Sci. Paper*, **437**, 715 (1922).
7. P. J. Mulvihill, in F. W. Schenck and R. E. Hebeda, eds., *Starch Hydrolysis Products*, VCH Publishers, Inc., New York, 1992, pp. 121–176.
8. W. L. Dills, Jr., in Ref. 7, pp. 395–416.
9. A. Reeve, in Ref. 7, pp. 79–120.
10. R. E. Hebeda, in T. Nagodawithana and G. Reed, eds., *Enzymes in Food Processing*, 3rd ed., Academic Press, Inc., San Diego, Calif., 1993, p. 333.
11. B. F. Jensen and B. E. Norman, *Proc. Biochem.* **19**(4), 129–134 (1984).
12. R. E. Hebeda, C. R. Styrlund, and W. M. Teague, *Starch / Stärke*, **40**(1), 33–36 (1988).
13. R. E. Hebeda, in G. Reed and T. W. Nagodawithana, eds., *Enzyme, Biomass, Food and Feed: Biotechnology*, 2nd ed., VCH Verlagsgesellschaft mbH, Weinheim, Germany, 1995, p. 737.
14. F. W. Schenck, in *Ullmann's Encyclopedia of Industrial Chemistry*, Vol. A 12, VCH Publishers, Inc., New York, 1989, p. 468.
15. Code of Federal Regulations, 168.110 and 168.111, Title 21, Pts. 100–169, 1994.
16. F. Gray, P. Buzzanell, and W. Moore, *U.S. Corn Sweetener Statistical Compendium*, Bulletin 868, USDA, Washington, D.C., 1993.
17. Ref. 18, SSSV21N1, Mar. 1996.
18. *Sugar and Sweetener Situation and Outlook Report*, SSSV20N2, Commercial Agriculture Division, Economic Research Service, USDA, Washington, D.C., June 1995.
19. R. Bernetti, in Ref. 7, pp. 367–394.
20. *The United States Pharmacopeia, The National Formulary*, (USP 23-NF 18), The United States Pharmacopeial Convention, Inc., Rockville, Md., 1995, p. 483.
21. Code of Federal Regulations, 184.1857, Title 21, Pts. 170–199, 1994.
22. *Evaluation of the Health Aspects of Corn Sugar (Dextrose), Corn Syrup, and Invert Sugar as Food Ingredients*, SCOGS-50, DHEW contract no. FDA 223-75-2004, Life Sciences Research Office, Federation of American Societies for Experimental Biology, Bethesda, Md., 1976.
23. C. Nieman, *Manuf. Confect.* **40**(8), 19 (1960).
24. B. L. Dasinger and co-workers, in G. M. A. van Beynum and J. A. Roels, eds., *Starch Conversion Technology*, Marcel Dekker, Inc., New York, 1985, pp. 237–262.
25. *Corn Annual*, Corn Refiners Association, Inc., Washington, D.C., 1994.
26. A. Torres and R. D. Thomas, *Food Technol.* **35**(7), 44–49 (1981).
27. T. Doty, in P. Koivistoinen and L. Hyvonen, eds., *Carbohydrate Sweeteners in Foods and Nutrition*, Academic Press, London, 1980, p. 259.
28. C. Morris, *Food Eng.* 95 (May 1980).
29. C. A. Lobry DeBruyn and W. Alberda van Eckenstein, *Rec. Trav. Chim.* **14**, 203 (1895).

30. M. Seidman, in G. G. Birch and R. S. Shallenberger, eds., *Developments in Food Carbohydrate*, Vol. 1, Applied Science Publishers Ltd., London, 1977, p. 19.
31. R. O. Marshall and E. R. Kooi, *Science*, **125**(3249), 648 (1957).
32. U.S. Pat. 2,950,228 (Aug. 23, 1960), R. O. Marshall (to Corn Products Co.).
33. U.S. Pat. 3,616,221 (Oct. 26, 1971), Y. Takasaki and O. Tanabe (to Agency of Industrial Science and Technology, Tokyo).
34. Y. Takasaki, *Agric. Biol. Chem.* **30**, 1247 (1966).
35. N. H. Mermelstein, *Food Technol.* **29**(6), 20–26 (1975).
36. U.S. Pat. 3,694,314 (Sept. 26, 1972), N. E. Lloyd and co-workers (to Standard Brands, Inc.).
37. U.S. Pat. 3,788,945 (Jan. 29, 1974), K. N. Thompson, R. A. Johnson, and N. E. Lloyd (to Standard Brands, Inc.).
38. D. Doddrell and A. Allerhand, *J. Am. Chem. Soc.* **93**, 2779 (1971).
39. L. M. Hanover, in Ref. 7, 201–231.
40. J. S. White, in Ref. 7, 177–199.
41. W. M. Teague, in Ref. 7, 45–77.
42. O. B. Jorgensen and co-workers, *Starch/Stärke*, **40**(8), 307–313 (1988).
43. J. V. Hupkes, *Starch/Stärke*, **30**(1), 24–28 (1978).
44. R. L. Antrim and A. L. Auterinen, *Starch/Stärke*, **38**(4), 132–137 (1986).
45. U.S. Pat. 4,141,857 (Feb. 27, 1979), J. Levy and M. C. Fusee (to UOP Inc.).
46. U.S. Pat. 4,268,419 (May 19, 1981), R. P. Rohrbach (to UOP Inc.).
47. L. M. Hanover and J. S. White, *Am. J. Clin. Nutrition*, **58**(5), 724–732 (1993).
48. *Chem. Mark. Rep.* **247**(26), 39 (1995).
49. S. Vuilleumier, *Am. J. Clin. Nutrition*, **58**(5), 733–736 (1993).
50. Code of Federal Regulations, 182.1866, Title 21, Pts. 170–199, 1994.
51. A. L. Forbes and B. A. Bowman, *Am. J. Clin. Nutrition*, **58**(5), 721–823 (1993).
52. D. Howling, in Ref. 7, 277–317.
53. Code of Federal Regulations, 168.120 and 168.121, Title 21, Pts. 100–169, 1994.
54. Code of Federal Regulations, 184.1444, 184.1865, Title 21, Pts. 170–199, 1994.
55. Ref. 18, SSSV19N2, Commodity Economics Division, June 1994.
56. Ref. 18, SSSV20N4, Dec. 1995.
57. L. H. Reynolds, personal communication, International Maple Syrup Institute, Hortonville, Wis., June 2, 1995.
58. *Sugar y Azucar*, **89**(2), 16–38 (1994).
59. F. Schneider, ed., *Sugar Analysis: ICUMSA Methods*, International Commission for Uniform Methods of Sugar Analysis, Peterborough, U.K., 1979.
60. B. S. Purchase, *Int. Sugar J.* **97**(1154), 70–81 (1995).

RONALD E. HEBEDA
Corn Products, a Division of
CPC International Inc.

TACK. See RUBBER COMPOUNDING.

TACONITE. See IRON.

TALC

Talc [14807-96-6], a naturally occurring mineral of the general chemical composition $Mg_3Si_4O_{10}(OH)_2$, is a crystalline hydrous magnesium silicate belonging to the general mineral family of the layered silicates. Other layered silicates are kaolin, mica, and pyrophyllite (1).

Geology and Occurrence

Talc deposits are of four types and each has a different group of accessory minerals (2). The most common type is of ultramafic origin, where talc is formed by alteration of serpentinite [12108-92-2], $Mg_3Si_2O_5(OH)_4$, to talc–carbonate rock. This type of talc deposit is common in Vermont, Quebec, and Finland, and has magnesite [13717-00-5], $MgCO_3$, and chlorite [14998-27-7], $Mg_3Al_3Si_4O_{10}(OH)_8$, as accessory minerals. Talc content is typically in the range of 50 to 70%. Talc of mafic origin is common in Virginia, North Carolina, and Georgia. It is formed by the hydration of mafic rock to serpentinite, followed by alteration of the serpentinite to talc–carbonate. It is usually so badly contaminated by chlorite, quartz (SiO_2), and other deleterious minerals that it is rarely usable.

Talc of metasedimentary origin is formed by hydrothermal alteration of a dolomitic host rock by a silica-containing fluid. This type of deposit is typical of Montana and Australia. It is usually quite pure with talc content of 90 to 98% and often very white as well. Dolomite [17069-72-6], $CaMg(CO_3)_2$, is the most common accessory mineral. The fourth type is of metamorphic origin, where a siliceous dolostone is first converted to tremolite [14567-73-8] or actinolite [13768-00-8] and then partially converted to talc. The Balmat, New York, and Death Valley, California, deposits are of this type. Tremolite, dolomite, and serpentine are common accessory minerals. This type of talc deposit has a variable talc content (30–80%), but is usually white and often commercially exploited because of the properties of its accessory minerals rather than the talc.

Mining and Processing

The commercial value of a talc ore is based on its color, purity, accessibility, proximity to the market, and accessory minerals. Of these the most critical is color. Most talc is mined by open-pit methods, but there are also underground mines in the United States, Canada, Italy, India, and China. In open-pit mining, overburden is removed and the talc is mined via conventional benching techniques. All the mining is highly selective, using much smaller shovels and trucks than those for conventional base-metal mining. Because talc ore is very sensitive to contamination during mining, it is always better to remove better-quality ore selectively during mining (high grade) than to leave this in the greater portion of the ore and try to beneficiate later. In many cases mechanical or hand sorting is done at the mine or in a separate process building close to the mine. A typical mine can produce multiple grades based on color, purity, and accessory minerals (Fig. 1). The ore is generally stored or shipped in lump (−15 cm (−6 in.)) form.

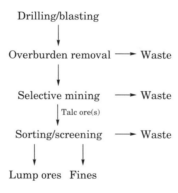

Fig. 1.   Mining of talc.

There is some beneficiation of talc by froth flotation (qv), practiced especially on ultramafic-type deposits. In this process (Fig. 2), talc is milled to its liberation size (−100 mesh (ca 0.15 mm)) using ball mills or ring-type roller mills and then slurried at 10–30% in water. Flotation is done in conventional multistage float cells using methyl amyl alcohol as a frother. Typically two to four stages are required to upgrade the ore from 50–70% talc to 90–98%. The product is filtered and then flash-dried and milled to a final product.

Dry ore processing of talc is summarized in Figure 3. Lump ore is usually crushed to −2 cm (−3/4 in.) and then reduced in ring-roller mills to −200- or

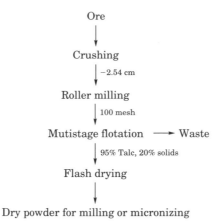

Fig. 2. Flotation of talc (100 mesh ≈0.15 mm).

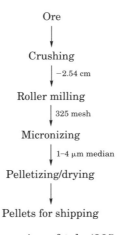

Fig. 3. Dry processing of talc (325 mesh ≈44 μm).

−325-mesh (ca −74−44 μm) products. This type of milling is preferred because it delaminates the talc and preserves its platy structure. Much product is sold in this size range. For finer products, the talc is further processed in micronizing mills to generate products with median particle sizes from 1 to 5 μm and top sizes down to 10 μm. Mills used included jet mills with integral air classifiers, fluid energy mills with external air classifiers, and hammer mills with integral air classifiers. Products are packaged and shipped in 50-lb (23-kg) bags or bulk sacks, but are also shipped as bulk powder. Because of its low bulk density when micronized, some talc is compacted and shipped in pelletized form. This is generally done by wetting the talc with water and processing through pellet mills. For paper applications, especially to control pitch deposits, the pellets are shipped with 2−10% moisture; for other applications such as rubber and plastics, however, the pellets are dried to <1% moisture.

Economic Aspects

According to statistics collected by the U.S. Geological Survey (3), U.S. production of crude talc in 1995 was 1,050,000 metric tons. Montana, Texas, Vermont, and New York were the principal producing U.S. states. Worldwide production was estimated to be 5,845,000 t. China, having 2,400,000 t, was the largest producer in the world; after China and the United States, Finland, India, Brazil, France, Italy, and Canada are the next principal producers. World production of talc in 1994 is listed in Table 1.

The value of crude ore produced in the United States was estimated to average $32.50/t in 1994. Product pricing was reported to vary from $99/t for New York State 200-mesh paint grade to $220/t for ultrafine Montana paint grade. Cosmetic grades were quoted at $263/t.

Table 1. 1994 World Production of Talc

Country and region	Production, 10^3 t/yr
North America	
United States	935
Canada	130
Mexico	15
Total North America	*1080*
Europe	
Austria	130
Finland	400
France	275
Germany	21
Italy	165
Norway	50
Spain	65
others	37
Total Europe	*1143*
Asia	
Australia	210
China	2400
India	360
Japan	61
South Korea	60
others	50
Total Asia	*3141*
Others	
Brazil	320
Africa	11
CIS	150
Total world	*5845*

Properties

The crystal structure of talc, illustrated in Figure 4, consists of repeating layers of a sandwich of brucite [1317-43-7], Mg(OH)$_2$, between sheets of silica [7631-86-9], SiO$_2$. The layers of silica are not strongly bonded to each other (except for van der Waals forces) and thus it is easy to fracture talc along this surface, which corresponds to delamination. This surface is covalent and hydrophobic. If talc is fractured across the brucite layer, the surfaces generated are ionic and hydrophilic in nature. Thus talc has a natural balance of hydrophilic and hydrophobic surfaces, giving it surfactant properties and consequently the name soapstone which is used in many parts of the world.

The mineral talc is extremely soft (Mohs' hardness = 1), has good slip, a density of 2.7 to 2.8 g/cm^3, and a refractive index of 1.58. It is relatively inert and nonreactive with conventional acids and bases. It is soluble in hydrofluoric acid. Although it has a pH in water of 9.0 to 9.5, talc has Lewis acid sites on its surface and at elevated temperatures is a mild catalyst for oxidation, depolymerization, and cross-linking of polymers.

Pure talc is thermally stable up to 930°C, and loses its crystalline bound water (4.8%) between 930 and 970°C, leaving an enstatite (dehydrated magnesium silicate) residue. Most commercial talc products have thermal loss below 930°C on account of the presence of carbonates, which lose carbon dioxide at 600°C, and chlorite, which loses water at 800°C. Talc is an insulator for both heat and electricity.

Talc products are also characterized by their crystallinity or relative platiness. Mycrocrystalline talc products typical of Montana and Australia tend to have very fine (~10 μm) natural grain sizes and thus are easily milled to very fine products of higher surface area (10–20 m^2/g). Macrocrystalline talc ores typical of Vermont and California have much larger grain sizes (50–200 μm). They tend to have a higher aspect ratio, lower surface area (<5 m^2/g), and are much more difficult to micronize. A micrograph illustrating the platiness of a macrocrystalline talc is shown in Figure 5.

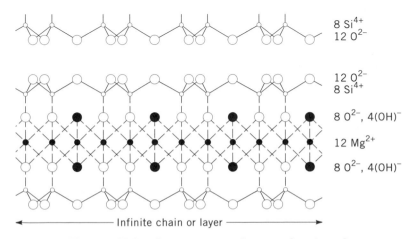

Fig. 4.   Molecular structure of pure talc mineral.

Fig. 5. Scanning electron microscope photograph of macrocrystalline talc at ×1000.

Uses

Talc is sold for use in a wide variety of applications, including paper (qv), ceramics (qv), roofing, paint (qv), plastics, rubber (qv), cosmetics (qv), pharmaceuticals (qv), adhesives (qv), sealants (qv), and animal feedstuffs (see FEEDS AND FEED ADDITIVES). In all of these applications it is a functional ingredient with specific beneficial properties. Talc is rarely used as a filler because it is much more expensive than alternative minerals such as limestone and clay.

The market for talc in the United States based on the 1995 U.S. Geological Survey Annual Review (3) is summarized in Table 2. Ceramics was the biggest market, having almost 35% of the total, followed by paint, paper, and plastics. Outside of the United States, especially in Asia, paper is the principal application. In that region talc is available locally at lower cost than competitive minerals such as kaolin.

In many markets talc competes not only against alternative minerals, but also against chemicals and other materials. In the paper pitch control market, dispersants are the main competition. In plastic, talc-reinforced polypropylene competes against engineering polymers such as ABS and polycarbonate in automotive parts. In cosmetics, talc and starch are both used as the basis of baby and body powder formulations. Thus the processing and marketing of talc has a much greater similarity to the specialty chemicals industry than to the traditional mining industry sector.

Paper. Paper is the principal application for talc throughout the world. It is used as a filler, a coating pigment, and a process aid in pulping and deinking for pitch and stickies control. As a paper filler, talc is used because of its platiness, softness, whiteness, ability to space TiO_2, and ink receptivity. It has

Table 2. Uses for Ground Talc in 1995

Application	Volume, 10^3 t/yr	% of Total
ceramics and refractories	265	34.7
paint	142	18.6
paper	121	15.8
plastics	45	5.9
roofing	38	5.0
cosmetics	31	4.1
rubber	19	2.5
other uses	103	13.5
Total	*764*	

better retention and bonding strength in the paper fiber than other minerals such as kaolin and $CaCO_3$, but it is also more expensive. As a coating pigment, talc is used in rotogravure printing papers to improve printing quality and in offset coatings to give a matte (low gloss) finish. Talc is much more difficult to wet out in coating formulations than kaolin or $CaCO_3$ and special formulations combining kaolin, talc, and dispersants are required for optimum performance.

In conventional chemical pulping systems, micronized talc is used for pitch control. The mechanism is not well understood, but talc does adsorb on the pitch and detackify it, preventing it from agglomerating and attaching on equipment. In de-inking talc is used both to enhance ink removal in conventional screening/centrifugation methods and as a carrier in flotation systems.

Ceramics. In ceramics, talc is widely used in wall tile and hobbyware bodies, in electrical porcelains, and in cordierite formulations. Wall tile and hobbyware are talc–clay bodies that are pressed and fast-fired to a high porosity (bisque) and then glazed and refired to produce the final product. Talc containing tremolite and carbonate is preferred to ensure good porosity.

In electrical porcelains (often called steatite bodies), high purity talc products with low levels of alkali metals are preferred. A typical steatite is made from 85% talc, 10% plastic kaolin, and 5% $BaCO_3$. Steatites are used as insulators on high voltage equipment such as automotive starters, microwave oven generators, and laser generators.

Cordierite [12182-53-5], $Mg_4Al_4Si_5O_{18}$, is a ceramic made from talc (25%), kaolin (65%), and Al_2O_3 (10%). It has the lowest thermal expansion coefficient of any commercial ceramic and thus tremendous thermal shock resistance. It has traditionally been used for kiln furniture and more recently for automotive exhaust catalyst substrates. In the latter, the cordierite raw materials are mixed as a wet paste, extruded into the honeycomb shape, then dried and fired. The finished part is coated with transition-metal catalysts in a separate process.

Roofing. Coarse talc products (-40 mesh) are widely used as both fillers and parting agents for asphalt-based roofing products. They are used as fillers in polymer-modified asphalt sheet products because they wet out easily and because they improve weathering and are inert to algal attack. Talc is used as a backcoating and parting agent on the backside of conventional fiber glass asphalt shingles. It coats the hot asphalt and prevents adjacent shingles from sticking to each other.

Paint.   In coatings, talc is used for flatting, corrosion resistance, TiO_2 spacing, and as a general extender. Fine talc, because of its platy, disordered structure, can roughen the surface of a conventional solvent-cast coating on a microscale and reduce the gloss. The platy structure of talc reduces the permeability of all coatings to water and the inert character of talc makes it immune to weathering or acid attack. Platy talc is an excellent spacer of TiO_2 and can improve the opacifying efficiency of this expensive pigment.

Plastics.   In plastics, talc is used as a reinforcing agent in polypropylene, an antiblock in polyethylene film, and a nucleating agent in nylon and injection-molded polypropylene (4,5). In polypropylene, −325-mesh (ca 44 μm) talc is used at loading levels of 15–40% to increase the stiffness, increase the heat stability, and reduce the shrinkage of homopolymer and copolymer injection-molding grades. In linear low density polyethylene film, a −500-mesh (ca 28 μm) talc is used at 0.5–1.5% to roughen the surface and reduce the tack of film so that it does not adhere to itself. Micronized talc (1–2 μm median particle size) is capable of promoting crystal growth in semicrystalline polymers such as nylon and polypropylene. At levels of 0.2–2% it is used to reduce cycle times in the injection molding and thermoforming of large parts.

Rubber.   In rubber, talc is used as a reinforcing agent and processing aid in mechanical rubber goods and as a parting agent in a variety of thermoset rubber processes. Talc, micronized by fluid energy milling, is capable of increasing the modulus of typical cured synthetic elastomers by 300% at loadings of 100 phr and is used in a wide range of belt, cable, and hose applications. It is also a rheology improver and process aid for extruded rubber goods at use levels of 10 to 50 phr. Coarser (−200-mesh (ca 74 μm)) talc is an excellent parting agent for all types of molding applications such as tires, elastomeric thread, and printing trays.

Cosmetics.   Talc is widely used in baby and body powders, pressed powders, creams, and antiperspirants. Its softness, slip, inertness (fragrance retention), and relative safety make this one of the oldest and most widely recognized applications for talc. Talc is also used in chewing gum as a detackifier and in tableting as a lubricating process aid.

Miscellaneous.   Talc is used in gypsumboard joint compounds as a high end filler to promote smoothness, sandability, and sag resistance. It is used in automotive primers and polyester body repair compounds to promote sandability. It is used in a wide variety of caulking compounds to improve rheology and sag resistance.

In many parts of the world, eg, China and northern Canada, block talc is used by the native artisans as a carving material and some of the sculptures produced are widely treasured. Block talc is also used for manufacture of fireplaces in Finland and Vermont. Machined talc pencils have been used for marking steel during processing since the 1890s.

Specifications, Standards, and Test Methods

Most talc sold to paper, ceramics, and other industrial customers is manufactured to specifications agreed to between the producer and consumer. In paper, properties such as color, abrasion, surface area, and tint are most important, whereas in ceramics, oxide chemistry, fired color, pressing characteristics, and alkali metal content are more important. There are some military specifications for talc used in corrosive coatings (6) and for cosmetic talc products used for cleaning of personnel in chemical warfare zones (7).

Talc sold to the cosmetics and baby powder markets must meet the Cosmetic, Toiletries and Fragrance Association (CTFA) specifications (8). For more stringent applications there are *United States Pharmacopeia* (USP) and *Food Codex* specifications.

Talc producers most commonly use screens for particle size analysis of coarser products (+325 mesh (ca 44 μm)) and the Micromeritics (Georgia) sedigraph for particle size analysis of finer products. The Hegman fineness of grind gauge is used for control of topsize.

Color is measured by a variety of machines using Tappi brightness, General Electric brightness, and Hunter L,a,b scales. Surface area is measured by nitrogen or argon adsorption using the BET method and loose and tapped bulk density using CTFA procedures. Mineralogy is characterized using x-ray diffraction and differential thermal analysis.

Health and Safety

Talc is considered a nuisance dust and subject to regulation in the workplace by both the Occupational Health and Safety Administration and the Mine Safety and Health Administration. Eight-hour exposure limits for talc dust are two milligrams of talc per cubic meter.

There has been a significant amount of litigation since the 1970s concerning the regulatory status of tremolitic talc products produced from the New York State and Madoc, Ontario deposits. In 1993 the U.S. federal government agreed that there were two forms of tremolite: so-called blocky tremolite and asbestiform tremolite. The first form is not regulated and the second is regulated as asbestos.

A thorough review of the health aspects of talc was presented in an FDA-sponsored seminar in Bethesda, Maryland, in January 1994 (9). The executive summary states that the probability of human risk is likely nonexistent under customary conditions of use. Used for decades in a wide variety of cosmetic and other applications, talc has proven to be among the safest of all consumer products.

BIBLIOGRAPHY

"Talc" in *ECT* 1st ed., Vol. 13, pp. 566–572, by H. Thurnauer, American Lava Corp.; in *ECT* 2nd ed., Vol. 19, pp. 608–614, by H. T. Mulryan, United Sierra Division, Cyprus Mines Corp.; in *ECT* 3rd ed., Vol. 22, pp. 523–531, by H. T. Mulryan, Cyprus Industrial Minerals.
1. F. H. Pough, *Rocks and Minerals*, Houghton Mifflin Co., Boston, Mass., 1988.
2. *Industrial Minerals and Rocks*, Society of Mining Engineers, Littleton, Colo., 1994, pp. 1049–1069.
3. *Mineral Industry Surveys: Talc and Pyrophyllite*, U.S. Geological Survey, Washington, D.C., July 1996.
4. E. F. McCarthy, *Talc/Polyolefin Composites for Exceptional Cost Performance in Automotive Applications*, Schotland Business Research, Princeton, N.J., 1992.

5. J. A. Radosta and N. C. Trivedi, *Handbook of Fillers for Plastics*, Van Nostrand Reinhold Co., Inc., New York, 1987, pp. 216–232.

6. Military Specification MIL-P-2441/IB(SH), U.S. Government Printing Office, Washington, D.C., May 15, 1986.

7. Federal Specification U-T-30C, U.S. Government Printing Office, Washington, D.C., Sept. 24, 1975.

8. *CTFA Compendium of Cosmetic Ingredient Composition–Specifications*, Cosmetic, Toiletry and Fragrance Association, Washington, D.C., 1990.

9. C. J. Carr, *Reg. Toxic. Pharmaco.* **21**, 211–215 (1995).

EDWARD F. MCCARTHY
Luzenac America

TALL OIL

Tall oil [*8002-26-4*] is a by-product of kraft pulping of pine wood. *Tall* is the Swedish word for pine, *kraft* the German word for strength. Crude tall oil (CTO), formed by acidifying black liquor soap skimmings with sulfuric acid, is a dark oily liquid with 26–42% resin acids or rosin, 36–48% fatty acids, and 10–38% neutrals. CTO is an excellent source of oleic/linoleic fatty acids and resin acids or rosin (1,2). Extensive fractional distillation is required, not only to separate these desirable products but also to remove the neutrals. The bulk of these neutrals, largely esters of fatty acids, sterols, resin and wax alcohols, and hydrocarbons, boil at either lower or higher temperatures than the boiling range of the fatty and resin acids. The wax alcohols and related components with boiling points in the rosin range are notable exceptions, but they are minor constituents.

Considering their heat sensitivity, the separation of fatty acids and rosin with minimal degradation by fractional distillation under vacuum and/or in the presence of steam is surprisingly good (3). Tall oil rosin (TOR) contains about 2% fatty acid and small amounts of neutrals. Tall oil fatty acid (TOFA) contains as little as 1.2% rosin and 1.7% neutrals. In typical U.S. TOFA, 49% of the fatty acids is oleic, 45% linoleic, and 3% palmitic, stearic, and eicosatrienoic acid. TOR and TOFA are upgraded to resins and chemicals for the manufacture of inks (qv), adhesives (qv), coatings (qv), and lubricants (see LUBRICATION AND LUBRICANTS).

The 1995 annual global CTO production was about 1.7 million metric tons. About half of that output was in the United States and one quarter in Europe outside the CIS. U.S. CTO production climbed 4.2% per year from 0.45 million metric tons in 1963 to 0.68 in 1973. After that the average annual increase slowed to 1%. The five U.S. CTO processors are listed in Table 1.

CTO prices are closely tied to the cycles of the U.S. economy and the paper industry. They vary between $120 and $220 per metric ton. In 1995 they were close to $200/t (4). With 50% of pine wood being converted to linerboard valued at $400–$600 per ton, pulp manufacturers do not focus on optimum black liquor

Table 1. U.S. Tall Oil Processors

Company	Capacity[a], 10^3 t/yr
Arizona Chemical	360
Georgia Pacific	70
Hercules	170
Union Camp	125
Westvaco	210
Total	*935*

[a]1995 estimated U.S. fractionation.

soap recovery, which only amounts of 60–70 kg/t of southern pine pulp. This soap is converted to 30–35 kg of CTO, worth \$6–\$7 or less than a little over 1% of the pulp value (5). This recovery is only 45% of the CTO available in the pine tree. With more care and higher CTO prices, 10–15 kg of additional CTO could be obtained per ton of pulp (6).

Composition and Properties

Variations found in CTO composition result primarily from the species of wood pulped and the location and climate where the trees are grown. Pulping process variations further affect CTO composition. The best CTO is produced from pine wood. However, many U.S. mills mix hardwood with pine to reduce fiber costs, or mix hardwood black liquor with pine black liquor. This lowers the rosin content. The composition of CTO produced in the southeastern United States and of typical Canadian and Scandinavian CTOs are shown in Table 2.

Table 2. Typical Tall Oil Compositions

Component, wt %	Southeastern United States	Canadian	Scandinavian
Principal fatty acids			
palmitic and lower	2.5	1.6	1.2
oleic	16.3	9.1	9.2
linoleic	13.6	10.9	14.4
conjugated linoleic and linolenic	6.9	13.1	15.9
stearic	0.7	0.3	0.3
eicosatrienoic and eicosadienoic	6.0	3.0	2.0
Total fatty acids	*46.0*	*38.0*	*43.0*
Principal rosin acids			
pimaric type	8.2	4.7	4.8
abietic type	19.2	11.1	12.3
dihydroabietic	4.4	4.3	3.6
neoabietic	5.4	1.9	3.7
other resin acids	2.8	6.0	4.6
Total rosin acids	*40.0*	*28.0*	*2.09*
Other			
neutral components[a]	14.0	34.0	28.0
acid number	167.0	128.0	138.0

[a]Unsaponifiables and esters.

Tall Oil Recovery and Processing

Process Sequence.   The process sequence consists of recovery of tall oil soap from the pulping black liquor, acidulation, ie, conversion of the soap into CTO with sulfuric acid, fractional distillation to separate rosin, and fatty acids and purification of the fatty acid fraction.

The kraft pulping process yields strong cellulose fibers by digesting pinewood chips for about two hours with an aqueous mixture of sodium hydroxide and sodium sulfide at 165–175°C under pressure. During pulping, the 2–3% resin and fatty acids that naturally occur in resinous wood are saponified. After filtration of the fibers, pulping black liquor is concentrated by multistage evaporation prior to feeding to a furnace for the recovery of the sodium salts and energy values.

Black Liquor Soap Recovery.   Black liquor soap consists of the sodium salts of the resin and fatty acids with small amounts of unsaponifiables. The soap is most easily separated from the black liquor by skimming at an intermediate stage, when the black liquor is evaporated to 25% solids (7). At this solids level, the soap rises in the skimmer at a rate of 0.76 m/h. At higher solids concentrations, the tall oil soap is less soluble, but higher viscosity lowers the soap rise rate and increases the necessary residence times in the soap skimmer beyond 3–4 hours. The time required for soap recovery can be reduced by installing baffles, by the use of chemical flocculants (8,9), and by air injection into the suction side of the soap skimmer feed pump. Soap density is controlled by the rate of air injection. Optimum results (70% skimmer efficiency) are obtained at a soap density of 0.84 kg/L (7 lb/gal). This soap has a minimum residual black liquor content of 15% (10–12).

Proper soap removal increases the pulp mill's evaporator capacity by 5%. Therefore, soap is often recovered as well at other places in the pulp mill, such as the weak black liquor storage tanks and the heavy black liquor oxidation system (13,14).

Only approximately 45% of CTO available in the pine tree is recovered. The rest is lost during woodyard operations (20%), pulping (15%), black liquor recovery (15%), and acidulation (5%). Several processing changes have been proposed to improve this yield. For instance, woodyard operations have become more efficient, with a turnover of one week, as opposed to two months, when these numbers were recorded. This has compensated for CTO losses that resulted from an increased use of hardwood. CTO losses due to soap adsorption on the pulp can be reduced, too. In a 1400 t/d pulp mill, about 25 t/d of soap is left on the pulp. Much of this soap can be recovered by adding *N,N*-dimethyl amides of tall oil fatty acids to the wash water of the rotary drum vacuum filter in the third and final pulp washer stage. Also, the addition of 6–7 grams of propyl stearic amide to the wash system per ton of pulp has been reported to increase tall oil soap yields significantly (15,16).

Black Liquor Soap Acidulation.   Only two-thirds of a typical black liquor soap consists of the sodium salts of fatty acids and resin acids (rosin). These acids are layered in a liquid crystal fashion. In between these layers is black liquor at the concentration of the soap skimmer, with various impurities, such as sodium carbonate, sodium sulfide, sodium sulfate, sodium hydroxide, sodium lignate, and calcium salts. This makes up the remaining one-third of the soap. Crude tall oil is generated by acidifying the black liquor soap with 30% sulfuric acid to a pH of 3. This is usually done in a vessel at 95°C with 20–30 minutes of vigorous agitation. Caution should be taken to scrub the hydrogen sulfide from the exhaust gas.

The long reaction time needed for this apparently simple neutralization is on account of the phase inversion that takes place, namely, upon dilution, the soap liquid crystals are dispersed as micelles. Neutralization of the sodium ions with sulfuric acid then reverses the micelles. The reverse micelles have a polar interior and a hydrophobic exterior. They coalesce into oil droplets.

When the acidified soap solution is allowed to stand, it separates into four different and fairly distinct phases. The uppermost layer is CTO. The second layer is a rag layer containing a mixture of CTO, lignates, and other solids. Next is a 15% sodium sulfate brine, commonly called spent acid. At the bottom precipitated calcium sulfate can be found. CTO can be recovered by gravity separation or centrifugation. The CTO phase separation rate can be accelerated 10-fold by the addition of a small amount (0.5% based on the gross weight of soap) of dispersants, such as certain commercial lignosulfonates (17). Dispersants also raise the acid number of the recovered CTO by as much as 12 points in the case of the addition of 0.42% Lignosol SFX. This indicates that the dispersant promotes the coalescence of the reverse micelles of fatty and resin acids, which are formed upon neutralization of the sodium salts.

The typical CTO yield from early batch acidulation systems was 75–80% of the total available in the skimmed soap. Direct steam injection was used for both soap heating and agitation. With improved reactor vessel design and mechanical agitation, CTO yield has been raised to 80–85%. High losses of CTO trapped in the lignin layer and the excessive amount of sulfuric acid required to speed up processing are the main disadvantages of the batch recovery system. Incorporation of self-cleaning centrifuges into a continuous acidulation process improved CTO recovery to 90%. The internal bowl of these centrifuges collects interfacial solids. These solids are ejected by a self-opening through a slot once every five minutes (18).

Even centrifuged CTO contains 2–3 wt % of water captured by the fatty and resin acids in reverse micelle structures. The water content can be reduced to the standard 1.5% by gravity settling at 80°C in a large tank. The remaining water is removed in a vacuum dehydrator before the CTO distillation. It is desirable to filter or centrifuge the CTO after dehydration. This removes residual lignin, estimated at 0.9%, which was finely dispersed in CTO, together with sodium sulfate crystals, estimated at 0.3–0.4% (19,20). This solids removal step prevents fouling of the distillation column and the power boiler in which tall oil pitch is burned. The residual sodium salts can also be removed by washing of CTO with pH-adjusted water.

Distillation. Separation of rosin from fatty acids is an essential step in utilizing CTO. The basic patent for tall oil distillation was granted in 1911 and the first commercial plant was constructed in Kotka in 1913 (21), making Finland the birth place of the tall oil industry. In the United States, CTO distillation started in the early 1930s (22). The first continuous large-scale fractionation plant began operations in 1949. It was built by the Badger Company of Boston, Massachusetts, adapting vacuum fractionation techniques from the petroleum industry to heat-sensitive CTO. Steam was injected to further reduce the partial pressure of the components. The plant was operated by Arizona Chemical Company. During the 1950s a number of large CTO fractionation plants were constructed (23).

The first step in CTO distillation is depitching. A relatively small distillation column is used as a pitch stripper. The vapor from the pitch stripper is fed directly into the rosin column, where rosin and fatty acids are separated. Rosin is taken from the bottoms of the column and fatty acids as a sidestream near the top. Palmitic acid and light neutrals are removed in the rosin column as heads. The operation is designed to minimize holdup and product decomposition. Care is taken to prevent carryover of some of the heavier neutrals, such as the sterols, from the depitcher to the rosin column (24).

The crude tall oil fatty acids obtained from the rosin column usually contain about 5% rosin because the boiling points of the heavier fatty acids and the lighter resin acids overlap. By adding the intermediate fraction to the fatty acid, rosin does not have to be redistilled.

The rosin column split is controlled by the fatty acid content specified for rosin. This is usually set at 2% fatty acids. At the high temperature near the bottom of the column and the reboiler, rosin dimerizes to some extent. By taking rosin from the column as a sidestream above the bottom, its rosin dimer content is minimized. Because of its high purity, sidestream rosin product is prone to crystallization.

Higher grade fatty acids with less than 2% rosin are obtained by further distillation. Union Camp uses two columns to achieve this. The first column is used to separate light ends and the second column to separate a mixture of rosin and higher boiling fatty acids. This mixture with about 40% rosin is sold as Distilled Tall Oil (DTO). Standard specifications for TOFA grades have been established by ASTM (25) as shown in Table 3. Also, the Pulp Chemicals Association provides specifications on TOFA having either more or less than 2% rosin.

The older tall oil distillation columns used bubble cap trays. In new columns, structured packing is preferred. Because of the low pressure drop of structured packing, steam injection is no longer necessary. The low liquid holdup of this packing minimizes the reactions of the fatty and resin acids. A specific distillation sequence for vacuum columns using structured packing of Sulzer has been described (26). Depitching is carried out at a vacuum of 0.26–1 kPa

Table 3. Grades I–III of Tall Oil Fatty Acids

Specification	ASTM method	Values		
		III	II	I
acid number	D1980	190–194	192–197	197–198
rosin acid, wt %	D1240	4.5–6.0	0.9–2.0	0.5–1.0
unsaponifiables, wt %	D1965	2.7–4.0	1.3–2.0	0.7–1.0
fatty acids, wt %	D1983	90.0–92.8	96.0–97.8	98.0–98.8
Gardner color	D1544	5–7	3–4	2–3
iodine value	D1959	131	130	130

(2–8 mm Hg) and a temperature of 220–270°C. The rosin column is operated at 0.4–1.2 kPa (3–9 mm Hg), a temperature of 204–260°C, and a reflux ratio of 2–3. These conditions give a bottoms rosin with a softening point of 80°C and an acid number of 178–180, as well as a crude fatty acid stream with 3% rosin.

Environmental Considerations. All CTO processing plants have environmental safeguards (27,28). CTO contains small amounts of volatile odor compounds that are exhausted by the vacuum systems of the distillation columns. They are captured, along with the more volatile fatty acids, in the cooling water of the barometric condensers. This oily water is recycled from the receivers or hot wells to the barometric condensers through heat exchangers, where it is cooled indirectly with water from the cooling towers. The oily water purge stream is skimmed and passed to the wastewater treatment system. The organic phase, consisting mostly of neutrals, palmitic acid, and organosulfur compounds, is usually sent to the heads storage tank.

All vapors, including hotwell odors, are captured in a header system linked with the incineration air of a steam boiler or hot oil vaporizer. Drain seals avoid escape of odors from the sewer lines. This completely eliminates total reduced sulfur (TRS) emissions. The SO_2 emissions are subject to local regulations.

Uses of Tall Oil Products

Tall Oil Rosin. U.S. production of tall oil rosin (TOR) in the 1990s is about 255,000 metric tons per year. It was introduced in the 1950s as a low cost substitute for gum and wood rosin, particularly for paper size. In 1960, 122,000 tons was used in this application in the United States. The introduction of the more economical dispersed rosin size (requiring only 0.25 and 0.75% alum based on printing paper weight) and the inroad of alkaline size have led to a decrease in this use to 50,000 t/yr (29). Substitution of rosin by alkaline size has been accelerated by the use of precipitated calcium carbonate filler in printing and writing papers. Waste-paper mills, finding large quantities of calcium carbonate in their furnish, are also switching to alkaline size. More than 80% of the printing and writing paper in the mid-1990s is alkaline-sized. Europe leads in this endeavor. The largest U.S. TOR uses for oligomeric resins in printing inks are 85,000 metric tons, and in adhesives, 60,000 metric tons (30). Another use of rosin is as an emulsifier in the manufacture of synthetic rubber, eg, styrene–butadiene rubber (SBR), by emulsion polymerization.

Tall Oil Fatty Acids. U.S. TOFA production in the mid-1990s is about 210,000 t/yr (see CARBOXYLIC ACIDS, FATTY ACIDS FROM TALL OIL). TOFA can replace fatty acid mixtures from vegetable oil sources in industrial applications, such as the manufacture of drying alkyd resins (qv) (31). At least one-third of TOFA is turned into dimer acids (qv) with a yield of about 60%, or 42,000 t/yr of dimer acids. They, in turn, by reaction with diamines, are converted to noncrystalline polyamides with low softening points and low transition temperatures. These polyamides find application in hot-melt adhesives, printing ink resins, and epoxy curing agents.

Distillation By-Products. Of the CTO distillation by-products, ie, pitch, heads, and Distilled Tall Oil (DTO), only the last, a unique mixture of rosin and fatty acids, has significant commercial value. Pitch and heads are used as fuel;

the former has a fuel value of 41,800 kJ/kg. Tall oil heads have outstanding solvent properties, but also have a bad odor, which is hard to remove. They contain a relatively high fraction of palmitic acid which can be recovered by crystallization.

BIBLIOGRAPHY

"Tall Oil" in *ECT* 1st ed., Vol. 13, pp. 572–577, by R. H. Stevens, Herty Foundation; in *ECT* 2nd ed., Vol. 19, pp. 614–629, by D. C. Tate, U.S. Plywood-Champion Papers Inc.; in *ECT* 3rd ed., Vol. 22, pp. 531–541, by H. G. Arlt, Jr., Arizona Chemical Co.

1. B. L. Browning, ed., *The Chemistry of Wood*, Wiley-Interscience, New York, 1963, p. 491.
2. E. Fritz and R. W. Johnson, in R. W. Johnson and E. Fritz, eds., *Fatty Acids in Industry*, Marcel Dekker, Inc., New York, 1989, p. 9.
3. L. A. Agnello and E. O. Barnes, *Ind. Eng. Chem.* **52**, 725 (1960).
4. A. Kimber, *Eur. Chem. News* **65**, 14–15 (July 31–Aug. 6, 1995).
5. R. L. Logan, *J. Am. Oil. Chem. Soc.* **56**, 777A (1979).
6. M. A. Lake, *Pulp Paper*, **54**(2), 130 (1980).
7. V. C. Uloth, A. Wong, and J. T. Wearing, *Naval Stores Rev.* **105**(2), 6 (1995).
8. U.S. Pat. 3,966,698 (June 29, 1976), R. V. Gossage (to Alfa Laval AB).
9. U.S. Pat. 4,085,000 (Apr. 18, 1978), J. V. Otrhalek, G. S. Gomes and G. H. Elfers (to BASF Wyandotte Corp.).
10. C. D. Foran, *Naval Stores Rev.* **94**(3), 14 (1984).
11. C. D. Foran, *Pulp Pap.* **58**(11), 104 (1984).
12. M. R. Dusenbury, *Pulp Pap.* **54**(5), 184 (1980).
13. R. W. Ellerbe, *Pap. Trade J.* 40 (June 25, 1973).
14. M. Ketcham, *Tappi J.* **73**(2), 107 (1990).
15. J. Drew, *Naval Stores Rev.* **102**(1), 9 (1992).
16. M. A. Lake, *Pulp Pap.* **54**(2), 130 (1980).
17. A. Wong, V. C. Uloth, and M. D. Ouchi, *Pulp Pap. Canada*, **82**(8), 97 (1981).
18. A. L. Goble, *Pulp Pap.* **54**(11), 147 (1980).
19. U.S. Pat. 3,948,874 (Apr. 6, 1976), F. T. E. Palmqvist (to Nalco Chemical Co.).
20. P. W. Sandermann, *Naturharze Terpentinol Tallol: Chemie und Technologie*, Springer-Verlag, Berlin, 1960, p. 320.
21. Anon., *Eur. Chem. News, Finland Suppl.* **65**, 28 (May 1995).
22. J. Drew and M. Probst, *Tall Oil*, Pulp Chemical Association, New York, 1981.
23. L. A. Agnello and E. O. Barnes, *Ind. Eng. Chem.* **52**, 726 (1960).
24. D. T. A. Huibers and E. Fritz, in Ref. 2, p. 96.
25. *ASTM Standards*, Vol. 06.03, American Society for Testing and Materials, Philadelphia, Pa., 1988, p. 304.
26. A. Ruetti, *Fette Seife Anstrichm.* **80**, 515 (1986).
27. D. T. A. Huibers and R. R. Rogers, *Proceedings of the 85th AOCS Annual Meeting*, Atlanta, Ga., 1995.
28. F. Prado and J. Drew, *Naval Stores Rev.* **100**(3), 12 (1990).
29. J. L. Latta, *Naval Stores Rev.* **100**(5), 12 (1990).
30. D. F. Stauffer, *Naval Stores Rev.* **100**(5), 2 (1990).
31. A. L. Stubbs, *J. Oil Col. Chem. Assoc.* **58**, 258 (1975).

DERK T. A. HUIBERS
Union Camp Corporation

TANKS AND PRESSURE VESSELS

Tanks are used in innumerable ways in the chemical process industry, not only to store every conceivable liquid, vapor, or solid, but also in a number of processing applications. For example, as well as reactors, tanks have served as the vessels for various unit operations such as settling, mixing, crystallization (qv), phase separation, and heat exchange. Herein the main focus is on the use of tanks as liquid storage vessels. The principles outlined, however, can generally be applied to tanks in other applications as well as to other pressure-containing equipment.

The most fundamental classification of storage tanks is based on whether they are above or below ground. The underside of aboveground tanks is usually placed directly on an earthen or a concrete foundation. Sometimes these tanks are placed on grillage, structural members, or heavy screen so that the bottom of the tank can be inspected on the underside and leaks can be easily detected. Alternatively, horizontal tanks are often placed on steel support saddles. The aboveground tank is usually easier to construct, costs less, and can be built in far larger capacities than underground storage tanks. It is also treated differently from an underground tank by the regulatory community.

Another type of aboveground tank is the elevated tank. These tanks, elevated by structural supports, are almost exclusively relegated to the domain of the municipal water supply companies. Because the municipal water supply is considered a vital public resource, tanks are often elevated for the reason that gravity as a source of pressure is considered more reliable for distribution to the market. Although the same effect can be accomplished by placing tanks on hills, where this is not possible the tanks are elevated by structural steel supports.

The other important type of tanks is the underground tank. These are usually limited to between 500–20,000 gal (2–75 m^3), although most are under 12,000 gal (45 m^3). Used to store fuels as well as a variety of chemicals, these tanks require special consideration for the earth pressure and settlement loads to which they are subjected. Because buoyancy must also be taken into account, underground tanks are often anchored into the ground so that they do not pop out when surrounded by groundwater. In addition, these tanks are subject to severe conditions of corrosion. Placement of special backfill, cathodic protection, and coatings and liners are some of the corrosion-prevention measures necessary to ensure good installations. Regulatory requirements are such that most underground tanks must have a means of being monitored for leakage. This may take the form of a double-wall tank where monitoring occurs in the interstitial space.

At retail refueling stations such as service stations, marinas, and convenience stores, fire protection codes generally restrict the use of aboveground tanks. Another reason for using underground tanks in chemical and processing plants is premium land values, where underground storage provides an answer to space needs. Moreover, temperatures underground are relatively constant, thus evaporation losses in the storage of fuels and hazardous chemicals are reduced in underground installations.

In the 1970s and 1980s the discovery of groundwater contamination in many areas of the United States resulted in the 1984 Subtitle 1 to the Resource Conservation and Recovery Act (RECRA), which required the Environmental

Protection Agency (EPA) to develop regulations for all underground tanks. These rules apply to any tank having 10% or more of its volume underground or covered by ground. Tanks in basements or cellars are not subject to these rules.

Because there is no uniform regulation requiring registration of tanks, the number of tanks being utilized is not known. An American Petroleum Institute (API) survey indicates there are about 700,000 petroleum storage tanks (1). The EPA estimated in 1990 that there were approximately 1.3×10^6 regulated underground storage tanks and an additional unknown number of exempt underground tanks used for home heating oil and farm fuel storage. The number of tanks in use in the chemical, petrochemical, pulp and paper, food, and pharmaceutical industries is unknown.

Basic Concepts

Although as of this writing (ca 1997), U.S. domestic regulations are generally using metric values to specify tank sizes and vapor pressures, American industry still uses gallon or barrel to designate tank capacities. For low pressures, the inch water column (in. wc) or the ounce per square inch (osi) are common in industry. For higher pressures, the units of pounds per square inch absolute (psia) or gauge (psig) are used. Additionally, in the petroleum and chemical industries, specialized units such as barrels (bbl), Reid vapor pressure, and degrees Baumé (Bé) are often used. A barrel (bbl) is 42 U.S. gallons.

Most tanks store liquid rather than gases or solids. Characteristics and properties such as corrosiveness, internal pressures of multicomponent solutions, tendency to scale or sublime, and formation of deposits and sludges are vital for the tank designer and the operator of the tank and are discussed herein. Excluded from the discussion are the unique properties and hazards of aerosols (qv), unstable liquids, and emulsions (qv). A good source of information for liquid properties for a wide range of compounds is available (2).

Density and Specific Gravity. Water has a density, mass per unit volume, of about 62.4 lb/ft^3 (1.000 g/cc) at 0°C, whereas mercury, also a liquid, has a density of about 842 lb/ft^3 (13.5 g/cc) at the same temperature. All things being equal, greater densities mean thicker required tank shell thicknesses.

Specific gravity (sp gr) is a measure of the relative weight of one liquid compared to a universally familiar liquid, generally water. More specifically, sp gr is a ratio of the density of a liquid divided by the density of liquid water at 16°C (60°F). Specific gravities of selected liquids are shown in Table 1.

In the petroleum industry, a common indicator of specific gravities, known as the API gravity or °API, is usually applied to crude oils. The formula for the API gravity is

$$°\text{API} = \frac{141.5}{\text{sp gr}} - 131.5$$

Thus, the higher the specific gravity, the lower the API gravity. Water, having a specific gravity of 1, has an API gravity of 10.

Table 1. Specific Gravity of Liquids

Liquid	Sp gr	Liquid	Sp gr	Liquid	Sp gr
acetic acid	1.06	gasoline	0.70	petroleum oil	0.82
alcohol, commercial	0.83	kerosene	0.80	phosphoric acid	1.78
ethanol	0.79	linseed oil	0.94	rape oil	0.92
ammonia	0.89	mineral oil	0.92	sulfuric acid	1.84
benzene	0.69	muriatic acid	1.20	tar	1.00
bromine	2.97	naphtha	0.76	turpentine oil	0.87
carbolic acid	0.96	nitric acid	1.50	vinegar	1.08
carbon disulfide	1.26	olive oil	0.92	water	1.00
cottonseed oil	0.93	palm oil	0.97	water, sea	1.03
				whale oil	0.92

Another common indicator of specific gravities used in the chemical industry is degrees Baumé (°Bé). For liquids heavier than water,

$$°Bé = \frac{140}{sp\ gr} - 145$$

For liquids lighter than water,

$$°Bé = \frac{140}{sp\ gr} - 130$$

When tank operators change a stored liquid, care must be exercised. If there is a significant increase in the specific gravity of the new liquid, the effective hydrostatic pressure acting on the tank walls is greater if the design liquid level is not reduced.

Temperature. Temperature may be measured on an absolute or relative scale. The two most common relative scales are the Celsius and the Fahrenheit scales. The Celsius scale is defined as 0°C at the freezing point (triple point) of water and 100°C at the boiling point. The Fahrenheit scale is arbitrarily defined by assigning it a temperature of 32 degrees at the freezing point of water and 212°F at the boiling point of water (see TEMPERATURE MEASUREMENTS).

The absolute temperature scale that corresponds to the Celsius scale is the Kelvin scale; for the Fahrenheit scale, the absolute scale is called the Rankine scale. The Celsius scale reads 0 when the Kelvin scale reads 273; the Fahrenheit scale reads 0 when the Rankine scale reads 460. These relationships are shown in Figure 1.

Tanks are used to store liquids over a wide temperature range. Cryogenic liquids, such as liquefied hydrocarbon gases, can be as low as −201°C (−330°F). Some hot liquids, such as asphalt (qv) tanks, can have a normal storage temperature as high as 260–316°C (500–600°F). However, most storage temperatures are either at or a little above or below ambient temperatures.

At very high and very low temperatures, material selection becomes an important design issue. At low temperatures, the material must have sufficient toughness to preclude transition of the tank material to a brittle state. At

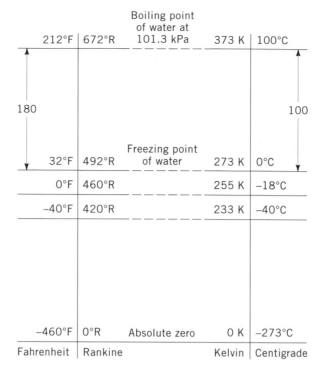

Fig. 1. Temperature scales (3).

high temperatures, corrosion is accelerated, and thermal expansion and thermal stresses of the material occur.

Vapor Pressure and Boiling Point.   Vapor pressure is important in liquid storage tank considerations. It affects the design and selection of the tank evaporation losses and is crucial for characterizing fire hazards of flammable and combustible liquids. The boiling point is also important because liquids should usually be stored at temperatures well below the boiling point. Flammable and combustible liquids are expressly prohibited by the fire codes for storage at temperatures above the boiling points. The National Fire Protection Association (NFPA) uses vapor pressure to classify the degree of fire hazardousness of liquids. The lower the boiling point, the lower the vapor pressure is for liquids stored at ambient temperatures. The boiling points of selected substances are given in Table 2.

The vapor pressure of a pure liquid is the pressure of the vapor space of a closed container. It is a specific function of temperature and always increases with increasing temperature (Fig. 2). If the temperature of a liquid in an open container is increased until its vapor pressure reaches atmospheric pressure, boiling occurs. The temperature of a pure liquid does not increase beyond its boiling point as heat is supplied. Rather, all the liquid evaporates at the boiling point. Because standard atmospheric pressure at sea level is 14.7 psia (101.3 kPa), this also presents the vapor pressure of a boiling liquid. Atmospheric pressure varies, however, with altitude. Water, for example, boils at 100°C (212°F) at sea level but

Table 2. Boiling Points of Liquids[a]

Substance	Boiling point, °C (°F)	Substance	Boiling point, °C (°F)	Substance	Boiling point, °C (°F)
aniline	184 (363)	chloroform	60 (140)	saturated brine	108 (226)
ethanol	78 (173)	linseed oil	313 (597)	sulfur	472 (833)
ammonia	−33 (−28)	mercury	358 (676)	sulfuric acid	310 (590)
benzene	80 (176)	naphthaline	220 (428)	water, pure	100 (212)
bromine	63 (145)	nitric acid	120 (248)	water, sea	100.6 (213.2)
carbon bisulfide	48 (118)	oil of turpentine	151 (315)	wood alcohol	66 (150)

[a]At atmospheric pressure, 101.3 kPa (760 mm Hg).

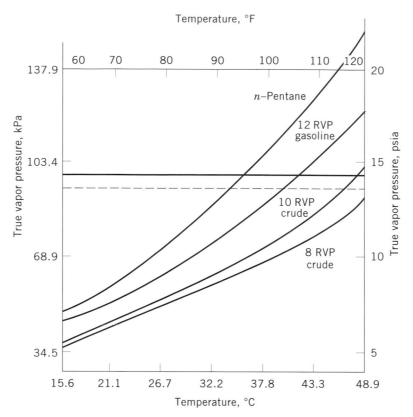

Fig. 2.   Vapor pressure and temperature (4), where the bold and dashed horizontal lines represent normal atmospheric pressure at sea level and at 609.6 m (2,000 ft), respectively. RVP = Reid vapor pressure.

at approximately 98°C (208°F) at 2000-ft (600-m) altitude. Barometric pressure changes slightly alter the boiling point of liquids in tanks as well.

In the fire codes, the atmospheric boiling point is an important physical property used to classify the degree of hazardousness of a liquid. If a mixture of liquids is heated, it starts to boil at some temperature but continues to rise in temperature over a boiling temperature range. Because the mixture does not have a definite boiling point, the NFPA fire codes define a comparable value of boiling point for the purposes of classifying liquids. For petroleum mixture, it is based on the 10% point of a distillation performed in accordance with ASTM D86, Standard Method of Test for Distillation of Petroleum Products.

Vapor pressure has also become a means of regulating storage tank design by the EPA. Because increasing vapor pressure tends to result in an increase in volatile emissions, the EPA has specific maximum values of vapor pressure for which various tank designs may be used.

Flash Point.   As a liquid is heated, its vapor pressure and, consequently, its evaporation rate increase. Although a liquid does not really burn, its vapor mixed with atmospheric oxygen does. The minimum temperature at which there is sufficient vapor generated to allow ignition of the air–vapor mixture near the surface of the liquid is called the flash point. Although evaporation occurs below the flash point, there is insufficient vapor generated to form an ignitable mixture below that point.

For flammable and combustible liquids, flash point is the primary basis for classifying the degree of fire hazardousness. NFPA Classifications 1, 2, and 3 designate the most to the least fire hazard liquids, respectively. In essence, low flash point liquids are high fire hazard liquids.

Pressure. Pressure, defined as force per unit area, can be expressed as an absolute or relative value. Although atmospheric pressure constantly fluctuates, a standard value of 101.3 kPa (14.7 psia) has been assigned as the accepted value at sea level. The "a" in the psia stands for absolute, ie, the pressure is 14.7 psi (101.3 kPa) above zero pressure or a vacuum. Most ordinary pressure-measuring instruments do not measure true pressure, but rather a pressure relative to the barometric or atmospheric pressure. This relative pressure is called gauge pressure. The atmospheric pressure is defined to be 1 psig, in which the "g" indicates that it is relative to atmospheric pressure. Vacuum is the pressure below atmospheric pressure and is, therefore, a relative pressure measurement as well. The relationship between absolute and relative pressure is shown in Figure 3 (see PRESSURE MEASUREMENT; VACUUM TECHNOLOGY).

For tank work, inches water column (in. wc) or ounces per square inch (osi) are commonly used to express the value of pressure or vacuum in the vapor space

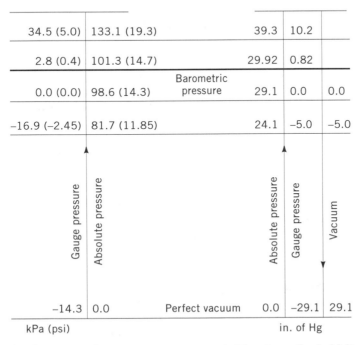

Fig. 3. Standard system of pressure measurement (3), where the bold line represents standard atmospheric pressure at sea level.

of a tank. These pressures are usually very low relative to atmospheric pressure. The common measures of pressure are compared as follows:

psi	in. wc	osi	Pa
1	27.68068	16	6894.7
0.03612628	1	0.5780205	249
0.0625	1.730042	1	431

Although both cylindrical shapes and spherical shells have simple theories to determine the strength and thus the thickness of tanks, the region of the tank that is most complex to design is the roof-to-shell junction. When there is internal pressure that exceeds the weight of the roof plates and framing the roof, the roof tends to separate from the shell. When tanks are subjected to pressures sufficient to cause damage, the roof-to-shell junction is usually the first area to show damage.

Internal and External Pressure. The difference in pressure between the inside of a tank or its vapor space and local barometric or atmospheric pressure is called internal pressure. When the internal pressure is negative it is simply called a vacuum. The pressure is measured at the top of the liquid in the tank because the liquid itself exerts hydrostatic pressure, thus increasing to a maximum value at the base of the tank.

Because tanks can be large structures, even small internal pressures can exert large forces which must be considered in design and operation. For example, a 100-ft (30.5-m) diameter tank having only 1-in. wc internal pressure exerts a force of almost 41,000 lb (9217 N) on the roof of the tank.

When the vapor space of a tank is open to the atmosphere or is freely vented, then the internal pressure is always zero or atmospheric. No pressure buildup can occur. This, of course, does not apply to dynamic conditions that occur in explosions or deflagrations. Most tanks, however, are not open to the atmosphere, but are provided with some form of venting device usually called a pressure-vacuum (PV) valve. A primary purpose of these valves is to reduce the free flow of air and vapors into and out of tanks, thereby reducing fire hazards and/or pollution. These valves are designed to open when the internal pressure builds up to some level in excess of atmospheric pressure, and keep the internal pressure from rising high enough to damage the tank. Typical flat-bottom tank design pressures range from 1 in. wc (0.25 kPa) to several psi. Conversely, the vacuum portion of the valve prevents the vacuum inside the tank from exceeding certain limits. Typical internal vacuum is 1 to 2 in. wc (249–498 Pa).

Internal pressure may be caused by several potential sources. One source is the vapor pressure of the liquid itself. All liquids exert a characteristic vapor pressure which varies with temperature. As the temperature increases, the vapor pressure increases. Liquids that have a vapor pressure equal to atmospheric pressure boil. Another source of internal pressure is the presence of an inert gas blanketing system. Inert gas blankets are used to pressurize the vapor space of a tank to perform specialized functions, such as to keep oxygen out of reactive liquids. The internal pressure is regulated by PV valves or regulators.

The most fundamental limitation on pressure is at 15 psig (101.4 kPa). Containers built to pressures exceeding this value are usually called pressure vessels and are covered by the American Society of Mechanical Engineers (ASME) Boiler and Pressure Vessel Code. For all practical purposes, tanks are defined to have internal pressures below this value.

External pressure implies that the pressure on the outside of the tank or vessel is greater than that in its interior. For atmospheric tanks, the development of an interior vacuum results in external pressure. External pressure can be extremely damaging to tanks because the surface area of tanks is usually large, generating high forces. The result of excessive external pressure is a buckling of the shell walls or total collapse. In some cases wind velocities during hurricanes have been sufficient to knock down and collapse tanks.

Miscellaneous Properties. Other properties such as viscosities, solidification temperature, pour point, and cubical rate of thermal expansion are all important for the tank designer or operator to consider and understand.

Tank Classification

There are many ways to classify a tank. Although there is no universal method, a classification commonly employed is based on the tank's internal pressure.

Atmospheric Tanks. By far, the most common type of tank is the atmospheric tank. These tanks are usually operated at internal pressures slightly above atmospheric pressure. Fire codes define an atmospheric tank as operating from atmospheric up to 0.5 psi (3448 Pa) above atmospheric pressure.

Low Pressure Tanks. Low pressure in the context of tanks means tanks designed for a higher pressure than atmospheric tanks. In other words, these are relatively high pressure tanks, designed to operate from atmospheric pressure up to 15 psig (101.4 kPa).

Pressure Vessels. High pressure tanks are vessels operating above 15 psig. These are really pressure vessels and the term high pressure tank is basically never used. Pressure vessels are a specialized form of container treated separately from tanks by all codes, standards, and regulations.

When the internal design pressure of a container exceeds 15 psig (101.3 kPa), it is called a pressure vessel. The ASME Boiler and Pressure Vessel Code is one of the primary standards used throughout the world to ensure safe storage vessels. Various substances, such as ammonia (qv) and many hydrocarbons (qv), are frequently stored in spherically shaped vessels that are often referred to as tanks. Most often the design pressure is 15 psig (101.3 kPa) or above. These are really spherical pressure vessels and fall under the rules of the ASME Boiler and Pressure Vessel Code. Discussion of pressure vessels are available (5,6); these are not covered in detail herein.

Tank Components

To a large extent, the vapor pressure of the substance stored determines the shape and, consequently, the type of tank used. The roof shape of a tank may be used to classify the type of tank. This classification is self-explanatory to tank fabricators and erectors. Also important is the tank bottom.

Fixed-Roof Tanks. The effect of internal pressure on plate structures, including tanks and pressure vessels, is important to tank design. If a flat plate is subjected to pressure on one side, it must be made quite thick to resist bending or deformation. A shallow cone-roof deck on a tank approximates a flat surface and is typically built of 3/16-in. (4.76-mm) thick steel (Fig. 4a). This is unable to withstand more than a few inches of water column pressure. The larger the tank, the more severe the effect of pressure on the structure. As pressure increases, the practicality of fabrication practice and costs force the tank builder to use shapes more suitable for internal pressure. The cylinder is an economic and easily fabricated shape for pressure containment. Indeed, almost all large tanks are cylindrical. The problem, however, is that the ends must be closed. The relatively flat roofs and bottoms or closures of tanks do not lend themselves to much internal pressure. As internal pressure increases, tank builders use roof domes or spheres. The spherical tank is the most economic shape for internal pressure storage in terms of required thickness, but it is generally more difficult to fabricate than a dome- or umbrella-roof tank because of its compound curvature.

Cone-Roof Tanks. Cone-roof tanks are cylindrical shells having a vertical axis of symmetry. The bottom is usually flat and the top made in the form of a shallow cone. These are the most widely used tanks for storage of relatively large quantities of fluid because they are economic to build and the market supports a number of contractors capable of building them. They can be shop-fabricated in small sizes but are most often field-erected. Cone-roof tanks typically have roof rafters and support columns except in very small-diameter tanks when they are self-supporting (see Fig. 4b and c; Table 3).

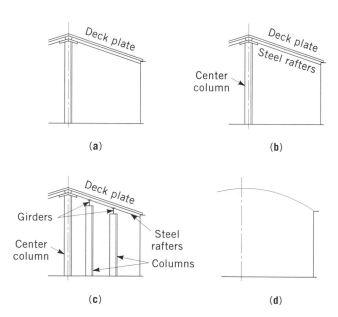

Fig. 4. Fixed-roof tanks: (**a**) self-supported cone roof; (**b**) center-supported cone roof; (**c**) column-supported cone roof; and (**d**) dome or umbrella roof.

Table 3. Comparisons of Tank Roof Types

Type	Advantages	Disadvantages
Fixed roofs[a]		
self-supported cone roof	minimum internal obstructions; relatively inexpensive; suitable for internal protective coating; makes cost-efficient conversion to internal floating roof	may require heavier roof-deck plate; only suitable for small tanks
center-supported cone roof	simple structural design; minimum internal obstructions; relatively inexpensive; makes cost-efficient conversion to internal floating roof	less ideal for internal protective coating; tank diameter limited by span of rafters
column-supported cone roof	simple structural design; relatively inexpensive; suitable for any diameter tank	poor for internal protective coating; many internal obstructions; difficult to inspect; makes costly conversion to internal floating roof
dome or umbrella roof	good design for internal coating; excellent design for high corrosion services such as sulfur; adequate for higher internal pressures	more expensive than cone roof; suitable for only small and medium tank (20 m); roof-deck-plate-only structural support, except for larger-diameter tanks; not frangible (API 650)
External floating roof[b]		
low deck floating pontoon	cheaper to construct than double deck for 5–50-m diameter; suitable for high vapor pressure stocks; capable of in-service repair of appurtenances; good buoyancy	poor design for roof insulation; structurally weaker than double deck; a leak could result in stock on the deck causing a fire hazard and oil in the roof drain system and an emissions violation

633

Table 3. (Continued)

Type	Advantages	Disadvantages
double-deck floating pontoon	can be easily insulated if structurally very strong; suitable for high vapor pressure stocks; capable of in-service repair of appurtenances; a leak will not put oil on the roof or in the roof drain system; better buoyancy	more expensive than pontoon; lose capacity because of the high amount of freeboard required
	Internal floating roof[c]	
vented at top of shell	good venting maximizes tank capacity	more expensive than air scoop roof; not suitable for retrofit
roof vents and shell overflow	suitable for retrofit; inexpensive installation	additional loss in tank capacity
noncontact aluminum	cheapest internal design; can be installed through shell manway; suitable for high vapor pressure stocks; rapid field installation	very weak structurally; aluminum limits services; not suitable for high turbulence; not suitable for high viscosity services; shorter life expectancy than steel; corrosion of aluminum
contact aluminum	can be installed through shell manway; suitable for high vapor pressure stocks; rapid field installation; easier repaired than noncontact; stronger than noncontact; less likely to sink; more fire resistant	aluminum limits services; shorter life expectancy
floating pan[d]	low cost	extremely vulnerable to sinking or capsizing; fire hazard

[a]See Figure 4.
[b]See Figure 5.
[c]See Figure 6.
[d]Larger roofs have truss system and external rafters on top of deck.

634

Umbrella- and Dome-Roof Tanks.   Umbrella-roof tanks are similar to cone-roof tanks, but have roofs that look like umbrellas. They are usually constructed to diameters not much larger than 60 ft (18 m). These tank roofs can be self-supporting, ie, having no column supports that must be run to the bottom of the tank (see Fig. 4**d**).

Dome-roof tanks are similar to umbrella-roof tanks except that the dome more nearly approximates a spherical surface than the segmented sections of an umbrella roof.

Aluminum geodesic dome roof tanks are becoming popular. These are often the economic choice. They offer superior corrosion resistance for a wide range of conditions, and are clear span structures not requiring internal supports. They can also be built to any required diameter. However, domes cannot handle more than a few inches of water column internal or external pressure.

Floating-Roof Tanks.   All floating-roof tanks have vertical, cylindrical shells just like a fixed-cone-roof tank. These common tanks have a cover that floats on the surface of the liquid. The floating cover or roof is a disk structure that has sufficient buoyancy to ensure that the roof floats during all expected conditions, even if leaks in the roof develop. The roof is built having approximately a gap of 8 to 12 in. (20–30 cm) between the roof and shell so that the roof does not bind as it moves up and down with the liquid level. The clearance between the floating roof and the shell is sealed by a device called a rim seal. The floating roof may be of any number of designs. The shell and bottom are similar to those of an ordinary vertical cylindrical fixed-roof tank. The two categories of floating roof tanks are external (Fig. 5) and internal (Fig. 6).

If the tank is open on top, it is called an external floating-roof (EFR) tank. If the floating roof is covered by a fixed roof on top of the tank, it is called an internal floating-roof (IFR) tank. The function of the cover is to reduce evaporation losses and air pollution by reducing the surface area of liquid that is exposed to the atmosphere. Fixed-roof tanks can easily be converted to internal floating-roof tanks by installing a floating roof inside the fixed-roof tank. Conversely, external floating-roof tanks can be easily converted to internal floating-roof tanks simply by covering the tank with a fixed roof.

EFR tanks have no vapor space pressure associated with them and operate strictly at atmospheric pressure. IFR tanks, like fixed-roof tanks, can operate at or above atmospheric pressure in the space between the floating roof and the fixed roof.

The fundamental requirements for floating roofs are dependent on whether the roof is for an internal or an external application. The design conditions of the external floating roof are more severe in that these must handle rainfall, wind, as well as dead-load and live-load conditions comparable to, and at least as severe as, building roofs.

External Floating Roofs.   Pontoon roofs are common for floating roofs from diameters of approximately 30–100 ft (10–30 m). The roof is simply a steel deck having an annular compartment that provides buoyancy (Fig. 5**a**). Double-deck roofs (Fig. 5**b** and 5**c**) are built for very small floating roofs up to about 30 ft (10 m) in diameter. These are also used on diameters that exceed about 100 ft (30 m). These roofs are strong and durable because of the double deck and are suitable for large-diameter tanks.

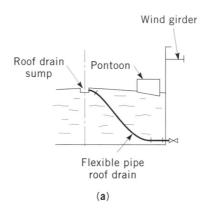

(a)

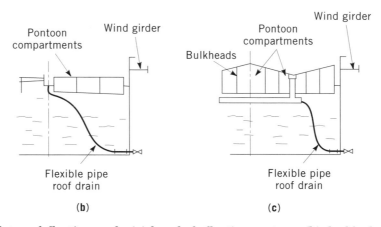

(b)   (c)

Fig. 5.   External floating roofs: (**a**) low deck floating pontoon; (**b**) double-deck floating pontoon for small tanks; and (**c**) double-deck floating pontoon for large tanks.

Internal Floating Roofs.   Pan roofs are simple sheet steel disks where the edge is turned up for buoyancy. These roofs are prone to capsizing and sinking owing to the fact that a small leak can cause them to sink (see Fig. 6).

The bulkhead pan roof has open annular compartments at the periphery to prevent the roof from sinking should a leak develop.

Skin and pontoon roofs are usually constructed of an aluminum skin supported on a series of tubular aluminum pontoons. These have a vapor space between the deck and the liquid surface.

The honeycomb roof is made from a hexagonal cell pattern similar to a beehive in appearance. The honeycomb is glued to a top and bottom aluminum skin that seals it. This roof rests directly on the liquid.

The plastic sandwich roof is made from rigid polyurethane foam panels sandwiched inside a plastic coating.

Tank Bottoms.   The shape of cylindrical tank closures, both top and bottom, is a strong function of the internal pressure. Because of the varying conditions to which a tank bottom may be subjected, several types of tank bottoms (Fig. 7; Table 4) have evolved. These may be broadly classified as flat bottom, conical, or domed or spherical. Flat-bottom tanks only appear flat. These usually

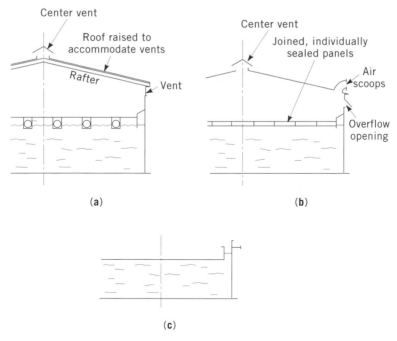

Fig. 6. Internal floating roofs: (**a**) vented at top of shell; (**b**) contact aluminum; and (**c**) floating pan.

have designed slope and shape and are subclassified according to the following: flat, cone up, cone down, or single slope.

Tank bottom slope is important because sediment, water, and heavy phases settle at the bottom. Corrosion is usually the most severe at the bottom, and the design of the bottom can have a significant effect on the life of the tank. In addition, if the liquid stock is changed, it is usually desirable to remove as much as the previous stock as possible. Therefore, designs that allow for the removal of water or stock and the ease of tank cleaning have evolved. In addition, specialized tank bottoms have resulted from the need to monitor and detect leaks. Tank bottoms in contact with the soil or foundations are one of the primary sources of leaks from aboveground tanks.

Flat.   For tanks less than 20–30 ft (6–10 m) in diameter, a flat bottom is used (Fig. 7**a**). The inclusion of a slope does not provide a substantial benefit, so they are fabricated as close to flat as practical.

Cone Up.   Cone-up tank bottoms are built to have a high point in the center of the tank (Fig. 7**b**). This is accomplished by crowning the foundation and constructing the tank on the crown. The slope is limited to about 1–2 in. (2.5–5 cm) for every 10 ft (3 m) of run. Therefore, the bottom may appear flat, but heavy stock or water tends to drain to the edge where it can be removed almost completely from the tank.

Cone Down.   The cone-down design slopes toward the center of the tank (Fig. 7**c**). Usually, there is a collection sump at the center (Fig. 7**d**). Piping under the tank is then drained to a wall or sump at the tank periphery. Although very effective for water removal from tanks, this design is inherently more

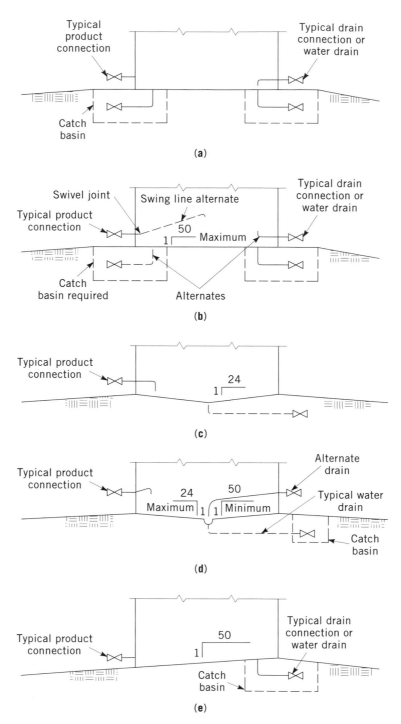

Fig. 7. Tank bottom designs: (**a**) flat bottom; (**b**) cone-up bottom; (**c**) cone-down bottom; (**d**) cone-down bottom with sump; and (**e**) single slope. In (**c**), buried lines are not used frequently because of the difficulty of inspection and possibility of accelerated corrosion.

Table 4. Comparisons of Tank Bottoms

Type	Uses	Advantages	Disadvantages
flat bottom[a]	mostly small tanks of ≤20-ft dia; suitable for filed-run tanks, gauge tanks, treating tanks, etc; widely used by chemical industry	simple and economical to fabricate and install in small sizes; bottom connections accessible for inspection and maintenance, similar to cone-up and single-slope bottom tanks	difficult to drain thoroughly on account of low spots (bird baths) caused by foundation settling and bottom plate warping; siphon does not drain completely because must clear bottom
cone-up bottom[b]	widely used by petroleum industry	less likely to collect water under bottom than flat horizontal or cone-down bottom tanks; better drainage than flat horizontal bottom tanks; shell and bottom connections accessible for inspection and maintenance; permits increased differential settlement of tank bottom; suitable for stock having specific gravity > water (>1.0 sp gr); easy to construct	less capacity than the conedown bottom tanks[c]; does not drain clean to low peripheral line; settlement reduces bottom slope and causes buckles, resulting in bird baths; drains to shell but does not drain well peripherally to water draw, at same elevation

Table 4. (*Continued*)

Type	Uses	Advantages	Disadvantages
cone-down bottom[d] and cone-down bottom with sump[e]	suitable for refined products where minimum contact of product with water is desired, eg, marketing bulk plant tanks >20-ft dia	good for tanks that undergo frequent product changes and in which complete drainage and/or water removal is required; complete drainage and drawdown; center sump decreases area of water in contact with product and with tank bottom	invites corrosion problem from collection of water under bottom plates; requires internal piping to tank center; reduced capacity for differential settlement; center drain should not be rigidly attached to bottom, merely guided; siphon drain only acceptable type
single slope[f]	suitable for tanks <100-ft dia; good for tanks that undergo frequent product change and in which complete drainage and/or water removal is required	improved drainage over cone-up and flat horizontal bottom tanks; bottom connections accessible for inspection and maintenance	installed cost more than cone-up or cone-down bottom tanks because of design and cost of foundation and erection shell; shallow slope problematic for sediment-containing tanks; sediment can form water pockets that do not drain

[a]See Figure 7**a**.
[b]See Figure 7**b**.
[c]1160 bbls for a 100-ft-dia tank and bottom slopes as shown in Figure 7**b**.
[d]See Figure 7**c**.
[e]See Figure 7**d**.
[f]See Figure 7**e**.

640

complex because of underground piping and the external sump. The design is also particularly prone to corrosion problems unless very meticulous attention is given to design and construction.

Single Slope. The single-slope design uses a planar bottom, but is tilted slightly to one side (Fig. 7e), allowing for drainage to be directed to the low point on the perimeter where it may be effectively collected. Because there is a constant rise across the diameter of the tank, the difference in elevation from one side to the other can be quite large. Therefore, this design is limited to about 100 ft (30 m) maximum.

Steep-Angle Conical Bottoms. Tanks often have a conical bottom where the slope exceeds 15–20° from the horizontal. This provides for complete drainage or even solids removal. Because these types of tanks are more costly, they are limited to smaller sizes. Such tanks are often found in the chemical industry or in processing plants.

Small Tanks

Numerous types of small tanks have been developed as a result of increasingly stringent regulations regarding leaks, spills, and containment.

Single Wall. Single-wall tanks are usually cylindrical and may have either vertical or horizontal orientation. Horizontal tanks are generally supported by two saddle supports and use more space than vertical tanks. Horizontal tanks have an advantage in that leaks can be seen as they occur. Water can also easily be drained from a drain valve located on the bottom.

Double Wall. Double-wall tanks have become more common for both above- and underground applications because the outer tank can contain a leak from the inner tank. This also serves as a means of detecting leaks. Such tanks are usually cylindrical tanks and may have either vertical or horizontal orientation.

Diked or Unitized Secondary Containment. Small tanks can have a secondary-containment dam built integrally into the tank. This is essentially within a steel box. These tanks may be either vertically or horizontally oriented in both cylindrical and rectangular shapes. The secondary-containment dikes may be open or closed. Closing the dikes makes access to the primary-containment tank more difficult, but keeps out rainwater.

Vaulted. Vaulted tanks are installed inside a concrete vault. The vault, itself a liquid-tight compartment, reduces the fire protection requirements as the NFPA and the International Fire Code Institute (IFCI) recognize these tanks as fire-resistant aboveground storage tanks. The vault provides a two-hour fire wall, thermal protection that minimizes tank breathing losses and pollution, secondary containment, and ballistic protection.

Engineering Considerations

Required Component Thicknesses. The tank design codes consider all strength calculations to be independent of temperature from ambient up to some upper limit. For example, when the temperature exceeds 93.3°C (200°F), the designer must reduce the allowable stresses. At high design temperatures, the various codes provide derating factors for steel, aluminum, and stainless steel. However, these codes provide little guidance for handling temperature-dependent effects, such as thermal expansion and creep. The tank designer must use good principles and practices to avoid problems such as fatigue or excessive distortions.

When the design temperatures are significantly below ambient temperature, the primary threat to tank integrity is failure of the material by brittle fracture. The tank design codes usually provide thorough treatment of this topic to prevent catastrophic failure. Additionally, there is the consideration of corrosion allowance, defined as extra thickness added beyond that required for strength. Corrosion allowance is not discussed herein.

Tank Bottom. In the fabrication of large steel structures, the minimum thickness is often governed by the minimum necessary for weldability and fabricability and not necessarily by strength requirements. A good example is the thickness requirements specified by the tank design codes for flat-bottom tanks: typically 0.25 in. (6 mm). The design codes treat the bottom as simply a spillproof membrane without any particular requirement for stresses. However, after settlement occurs, significant stresses may develop. API Standard 653 provides guidelines for the maximum degree of settlement while maintaining the tank bottom within allowable stresses.

Tank Shell. Another example of where thickness is set by minimums for fabricability but not for strength is in small-diameter tanks. For example, a water storage tank built using a steel of an allowable stress of 20,000 psi (138 mPa), 9 ft (3 m) in diameter by 21-ft (7-m) high, requires a shell thickness to resist hoop stress of only 0.023-in. (0.58-mm) thick. However, if built to API Standard 650, the shell would be fabricated at least 0.1875-in. (4.76-mm) thick. The code requires this thickness so that when fabrication, welding, and tolerances are considered, a tank of acceptable quality and appearance meeting the requirements of most services in most locations is provided.

In the large-diameter vertical cylindrical tanks, because hoop stress is proportional to diameter, the thickness is set by the hydrostatic hoop stresses. Although the hydrostatic forces increase proportionally with the depth of liquid in the tank, the thickness must be based on the hydrostatic pressure at the point of greatest depth in the tank. At the bottom, however, the expansion of the shell owing to internal hydrostatic pressure is limited so that the actual point of maximum stress is slightly above the bottom. Assuming this point to be about 1 ft (0.305 m) above the tank bottom provides tank shells of adequate strength. The basic equation modified for this anomaly is

$$t = \frac{\rho(H - H_0)D}{2\sigma_{\text{allow}}}$$

where t is the required wall thickness exclusive of corrosion allowance; ρ, the fluid density; H, the maximum shell design liquid level; H_0, 12 in. (30.48 cm); D, the tank diameter; and σ_{allow}, the shell material allowable stresses. This 1-ft (0.305-m) equation is slightly conservative. For tanks over 200 ft (61 m) in diameter, it is worthwhile using the variable point computation which takes into account the actual point of maximum hoop stress. This iterative procedure is illustrated in API Standard 650.

When tanks are built having an open top, the wind pressure may cause buckling of the shell. A wind girder of sufficient section modulus is used to stiffen the open top according to

$$Z = 5.8 \times 10^{-8} D^2 H$$

where Z is the minimum required section modulus; D, the tank diameter; and H, the tank height.

   Tank Roof.   The roof of a vertical cylindrical tank is treated like a building structure and uses the same basic rules as the building codes. For example, the API codes require a roof to be designed for the dead load plus a 122-kg/m^2 (25-lb/ft^2) live load. The minimum fabrication thickness of roof plates is 3/16 in. (4.8 mm).

   Live and dead loads generate hoop forces in the area of the roof-to-shell junction for a tank having a cone roof. For dead loads plus live loads, the roof-to-shell junction is assumed to carry most of the tensile forces generated. The minimum area required is computed assuming that the membrane force transmitted to the roof-to-shell junction varies with the sine of the angle of the roof:

$$\sigma = \frac{PD^2}{8A \sin \alpha}$$

where σ is the stress in the roof-to-shell region; P, the live load plus dead load repressed as pressure; D, the tank diameter; A, the roof-to-shell area which acts to resist the hoop forces; and α, the angle of conical roof.

   For tank internal pressures that do not exceed the weight of the roof plates, most tanks have conical roofs because these are the simplest and most cost-effective. When the pressure is increased beyond the weight of the roof plates, the roof-to-shell area goes into hoop compression. A small portion of the roof, the roof-to-shell angle, and the top few centimeters of the shell act as a compression ring to resist the unbalanced forces from internal pressure on the conical roof. The internal design pressure for this case may be

$$A = \frac{W}{2\pi\sigma_y \tan \theta}$$

where σ_y is the yield strength of steel, W the weight of shell and framing supported by the shell, and θ the angle the roof makes from the horizontal.

   Materials of Construction.   Tanks are constructed from a number of materials based on cost and availability of the material, ease of fabrication, resistance to corrosion, and compatibility with stored fluid. Sometimes specialized composites and techniques are used in tank construction. These are the exception.

   Carbon steel, or mild steel, is by far the most common material for tank construction. It is readily available and, because of the ease with which it is fabricated, machined, formed, and welded, results in low overall costs. Austenitic 300 series stainless steel is another important material used for storage of corrosive chemicals and liquids. Although the cost of the austenitic group of stainless steels is significantly more than steel itself, the stainless steels offer the same advantages of fabricability and availability as carbon steel. The API 650 Standard provides details on how to design stainless steel tanks. Fiber glass-reinforced plastic (FRP) tanks are noted for resistance to chemicals. Many times stainless steel or aluminum tanks are not acceptable. The fabrication and construction techniques for FRP are somewhat more specialized than for metals fabrication. Because of the lack of fire resistance, FRP tanks are not normally used to store flammable or combustible liquids. FRP tanks have been used to store water, water-treating chemicals, fire-fighting foam, wastes, lubricants, and nonflammable chemicals and corrosives. Aluminum tanks are suitable for a limited number of materials. Historically, FRP tanks were used for cryogenic applications owing to the fact that aluminum remains ductile at temperatures much lower than carbon steel. However, nickel steels and stainless steels have largely supplanted the market for aluminum tanks. Aluminum is still used for some acids, fertilizers, and demineralized water applications. However, in general, the use of aluminum for storage tanks has been low. Concrete tanks have been used in the water and sewage treatment business for a long time. However, because of the relatively high cost, these are not in common use in the 1990s.

Tank Selection Criteria. The selection of tanks is a complex process optimizing an array of information to yield a particular design. Figure 8 gives some guidelines. Once the specific liquid(s) to be stored is established, the liquid's physical properties determine the range of possible tank types. Although vapor pressure is a principal component in tank selection, other properties such as flash point, potential for explosion, temperature, and specific gravity all factor in the selection and design of tanks. A simplified example of tank selection by fluid stored is shown in Table 5. In addition to fundamental physical properties influencing tank selection, size, regulations, best practices, external loads such as wind, snow, and seismic loads, as well as numerous additional engineering issues, all play a role. Material selection, corrosion prevention systems, and environmental requirements and considerations may also influence the selection. Layout of the tank within existing or new facilities is always an important consideration as the limited plot space for the tank is a factor on the type of tank selected. The requirements of the fire authority having jurisdiction almost always have property line and public way setbacks as well as distance requirements between tanks and equipment or other tanks. The ultimate selection criteria are keenly dependent on such factors as the actual site-specific conditions, local regulations, cost considerations, required operating life, space availability, and potential for fires and explosion. These should all be evaluated and considered by the responsible engineering designer and documented for the protection of the plant owner and operator.

Special Engineering Considerations. Because tanks are used in so many different ways, some specialized applications have been developed that have become fairly commonplace.

Cryogenic Tanks. Low temperature tanks are used for liquefied hydrocarbon gases (LHG); liquefied natural gas (LNG); various liquefied gases such as air, nitrogen (qv), or oxygen (qv), and ammonia (qv); and other refrigerated

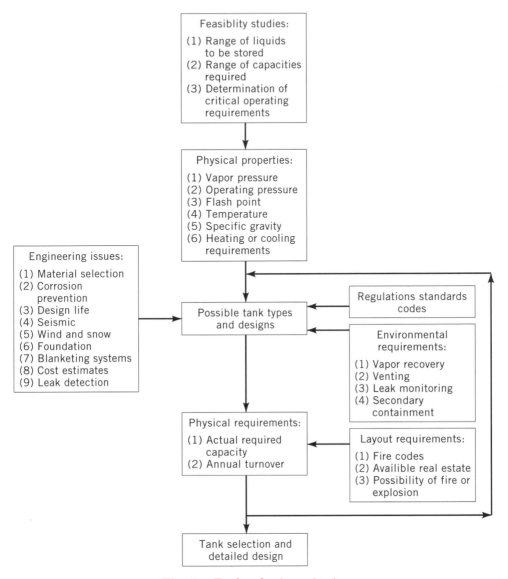

Fig. 8. Tank selection criteria.

liquids (see CRYOGENICS; LIQUEFIED PETROLEUM GAS). As a general rule, larger quantities of stored liquids that have high vapor pressures favor low temperature or cryogenic storage. Although the use of these products may be in the gaseous state, a higher quantity can be stored in the liquid state, so cryogenic storage is often the preferred method for storing large quantities of gases. Not only must these materials be cooled to temperatures substantially below ambient, they may need to be kept under pressure as well. Because of these requirements, the tanks often require accessory cooling systems. Some of these systems cool the vapor in the tank space and return it to the tank; others cool the liquid. In addition, to reduce the size of the cooling equipment and conserve energy consumption,

Table 5. Storage Tank Type for Liquids at 25°C (77°F)[a]

Chemical	Tank type[b]	Chemical	Tank type[b]
acetaldehyde	H	ethylenediamine	A
acetamide	A	ethylene dichloride	L
acetic acid	A	ethylene glycol	A
acetone	L	ethylene glycol monoethyl ether	A
acetonitrile	L	formic acid	L
acetophenone	A	freons	H
acrolein	L	furfural	A
acrylonitrile	L	gasoline	A
allyl alcohol	L	glycerine	A
ammonia	H	hydrocyanic acid	L
benzene	L	isoprene	L
benzoic acid	A	methyl acrylate	A
butane	L	methyl amine	A
carbon disulfide	L	methylchloride	A
carbon tetrachloride	A	methyl ethyl ketone	A
chlorobenzene	L	methyl formate	L
chloroethanol	A	naphtha	A
chloroform	L	nitrobenzene	A
chloropicrin	L	nitrophenol	A
dilorosulfonic acid	A	nitrotoluane	A
cumene	A	pentane	L
cyclohexane	L	petroleum oil	A
cyclohexanane	A	propane	H
dichloromethane	L	pyridine	A
diesel oil	A	styrene	A
diethyl ether	L	sulfuric acid	A
dimethylformanide	A	sulfur trioxide	L
dimethyl pthalate	A	tetrachloroethane	A
dioxane	L	tetrahydrofuran	L
epichlorohydrin	A	toluene	A
ethanol	L	trichloroethylene	L
ethyl acetate	L	xylene	A
ethylbenzene	A		

[a]Ref. 7.
[b]A = atmospheric pressure, <0.5 psig; L = low pressure, <15 psig but >0.3 psig; H = high pressure, >15 psig.

these tanks must be insulated. For flammable materials such as LNG or LHG, additional fire code requirements may stipulate that a secondary containment tank be provided which can contain the contents of the inner tank in case of failure.

Careful material selection is required to prevent brittle failure of tanks at low temperatures. In addition, for tanks where the service temperatures are reduced, it is essential that an engineering analysis be performed to ensure that the tanks are not subject to brittle failure at the house temperature. The tank and vessel codes usually specify allowable materials based on design temperature. Further information about selection of metals for low temperature is available (8).

Heated Tanks. Many compounds either freeze, solidify, or thicken to the point where they cannot be transferred through piping and equipment unless maintained at some minimum temperature. Examples are heavy oils, asphalts, sulfur, highly concentrated salt solutions, caustic soda solutions, or even molasses and foodstuffs. The storage tanks for these fluids must be heated and maintained to some minimum temperature. There are several ways to heat tanks, as shown in Figure 9. Heat transfer raises a rather complex engineering optimization problem to minimize the heat-transfer surfaces because these tend to be costly. This is done by mixing and/or pumping the fluids, varying the insulation requirements of the tank, picking the proper heat-transfer medium such as steam (qv), or optimizing the type of heat-transfer surface (extended or finned surface versus tubular). Of particular importance in establishing heating requirements is the rate of bringing a cold tank up to temperature versus the rate of heat needed to maintain a minimum temperature under design conditions. If the heatup rate is too high, a large heating system is required; if it is too small, the tank can take excessively long to heat up or it may not heat up to its design minimum under adverse conditions. The temperature of the heating fluid must be kept low where the product is heat-sensitive so that the product does not suffer degradation. Another case where temperature levels of the heat-transfer surface must be limited is where stress corrosion cracking can occur, as in some saline solutions or caustic solutions. Other design requirements involve the prevention of stratification of hot and cold layers and the ability to remelt or heat up the tank quickly and efficiently. These considerations usually involve use of mixers and eductors or of thermal circulation within the tank (see MIXING AND BLENDING).

The exposed surface area of a tank is relatively large, thus heated tanks are almost always insulated. Another reason for insulating is that the external corrosion rate of the steel owing to atmospheric conditions increases with increasing temperature. Insulation, if properly installed, reduces external corrosion. Many types of insulation systems are available. The difficulty with most is that if rainwater gets into them, the water becomes trapped inside and tends to accelerate corrosion. Much of the more recent improvements have been directed in keeping water out of the insulation.

An extremely important safety consideration for both heated and cryogenic tanks is that lower boiling liquids must not be introduced into the tank. These liquids can boil and cause a frothover or a violent evolution of vapor, followed by tank failure.

Design Considerations. Most of the design codes and standards for tanks provide checklists and concepts to prevent the designer from making gross mistakes. In particular, tank standards are issued by the American Petroleum Institute (API), which remove most of the risks of the catastrophic results that can occur without considering material selection, brittle fracture, insufficient welding or joining methods, fabrications methods, etc. In fact, these standards are recognized worldwide to be of the highest caliber. As a result, they have been used in industries such as chemicals, pulp and paper, food, and a host of others. Even using API standards many site-specific considerations still exist that can have a substantial impact on the design life of the tank as well as its safe operations. All of these considerations should be documented in records

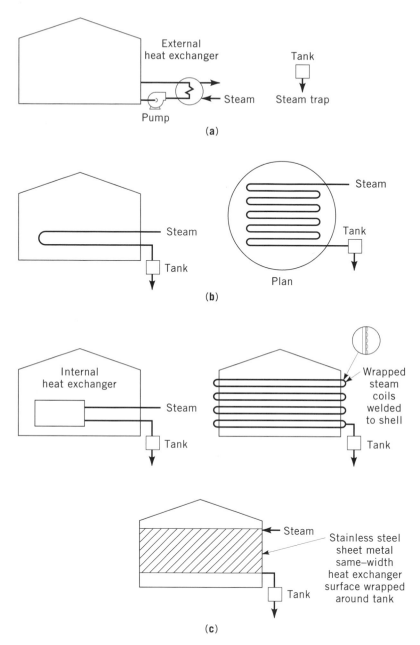

Fig. 9. Methods of heating tanks: (**a**) external heat exchanger; (**b**) serpentine steam coils; and (**c**) plate coils.

maintained by the owner or operator of the tank. Some of these elements are determination of appropriate standard and code to build the tank; compliance with fire codes; material selection; linings and coatings; cathodic protection; wall thickness selection; appurtenance design, eg, ladders, internal piping, and instrumentation; establishing and designing for external anticipated loads, eg, seismic or snow loads; venting and emergency venting; foundation and settling criteria; pollution prevention, eg, secondary containment, leak detection, and seal selection; fabrication, erection, inspection, and testing; and electrical area classification.

Regulations

Regulations and laws are mandatory requirements with which a tank owner or operator must comply. Most regulatory requirements are channeled through an agency whose general responsibility is the safety, well-being, and protection of the public or the environment. The authority having jurisdiction may be a federal, state, local, or regional agency, an individual such as a fire chief or marshal, a labor or health department, or a building official or inspector.

The general rule of thumb is that most tank facilities are subject to multiple authorities. When this is the case and the rules have overlapping or even conflicting provisions, the facility must comply with all the requirements of the multiple authorities. In short, one authority's requirements does not preempt or even satisfy the requirements of the other agency even if they accomplish exactly the same thing. There are many examples of this type of inefficiency in regulations.

For aboveground tanks, there is no comprehensive regulation or program as there is for underground tanks under the RECRA program. Most tanks are unregulated or regulated only if they contained flammable liquids under the jurisdiction of local fire authorities. This is likely to change.

Federal Regulations.   Federal regulations tend not to be aimed at spill and tank-bottom leak prevention but rather on spill response. These therefore address issues such as containment of spills, financial liability and responsibility, discharge of contaminated stormwater, reporting, and response requirements.

Statues.   The framework from which all regulations that affect the petroleum, chemical, and petrochemical industries are derived is essentially the result of nine statues shown in Figure 10. These statues address the issues of limiting exposure of substances that may be harmful to human health or the environment to acceptably low levels; assigning financial as well as criminal responsibility for damaging human health or the environment; and reporting of data, incidents, information that may affect the regulatory agencies for enforcing the regulations associated with the listed statues.

The relationship of codes and standards to the regulatory framework should be clearly understood. By themselves, industry codes and standards have no authority. However, an examination of both regulations and codes shows that the governmental jurisdictions that have authority over tanks usually rely on the codes and standards to form the technical basis of their requirements. In many cases, they simply refer to the code or standard by name, which then elevates the latter to a legal requirement. Frequently, the authority having jurisdiction often adds other requirements, such as registration fees, to their technical requirements to support the administrative costs of implementing the inspection and enforcement of their responsibilities. Identifying and using codes and standards is one of the first steps to take when considering a new tank installation, regulating it, or simply trying to understand it.

Clean Air Act (CAA)	Clean Water Act (CWA)	Resource Conservation and Recovery Act (RCRA)
regulates:	regulates:	regulates:
(1) emissions of pollutants to atmosphere	(1) discharges of waste waters to receiving waters	(1) generation, transportation, storage, treatment, and disposal of hazardous solid wastes
(2) emissions by treatment technology unless air quality requires tighter limits	(2) discharges by treatment technology unless water quality requires tighter limits	(2) storage of fuels in underground tanks for nonhazardous waste

Comprehensive Environmental Response, Compensation, and Liability Act (CERCLA)	Superfund Amendment and Reauthorization Act Title III (SARA Title III)	Occupational Safety and Health Act (OSHA)
regulates:	regulates:	regulates:
(1) cleanup of leaking landfills	(1) emergency response plans	(1) employee right-to-know
(2) reporting spills of certain chemicals	(2) right-to-know issues	(2) employment free of recognized hazards
(3) responsibility and liability for contaminated disposal cleanup	(3) chemical release reporting	(3) specific standards for job and industry safety

Toxic Substances Control Act (TSCA)	Hazardous Materials Transportation Act (HMTA)	Safe Drinking Water Act (SDWA)
regulates:	regulates:	regulates:
(1) commercial use of most chemicals	(1) hazardous materials when transported in commerce	(1) enforceable quality standards to drinking water
(2) use and disposal of asbestos, chlorinated biphenyls, and chlorinated fluorocarbons	(2) activities associated with identifying and classifying the material, marking, labeling and placarding, packaging, and documenting the hazardous material	(2) protection of groundwater sources
(3) reporting of all adverse health effects	(3) loading, unloading, and incidental storage	
(4) use, labeling, and documentation for chemicals that pose risk to health or the environment	(4) reporting of unintentional releases and injury or death to a person or property damage exceeding $50,000	

Fig. 10. Summary of principal environmental laws (9).

Spills, Leaks, and Prevention

Leaks and spills from aboveground storage tank (AST) facilities have had more impact on change in the way tanks are regulated and will be regulated, as well as on the design and operation of tanks, than any other single factor. Leaks and spills have had a substantial impact on public awareness as well. Leaks and spills are associated with groundwater, which in turn is associated with public water supplies and irrigation, thus the public has had little tolerance for environmental accidents.

Causes of Spills and Leaks.   There are numerous causes of tank leaks and spills. Some examples, as well as causes for bulk storage and handling facilities and piping, are noted in Table 6.

Corrosion.   Corrosion is one of the most prevalent and insidious causes of leaks. Because the large surface area of either aboveground or underground tanks cannot be easily inspected, leaks that develop tend to go on for long periods of time and large underground contamination pools can result. Corrosion can be mitigated by proper material selection, use of lining and coatings for both topside and bottomside corrosion, cathodic protection, and chemical inhibition. A good tank inspection program, such as API 653, which requires periodic internal inspections, is one of the best ways to ensure that leaks do not go on undetected for long periods of time.

Operations.   Overfill of tanks, owing to any number of reasons, is a common occurrence. This results from inoperative or failed equipment such as level alarms, instrumentation, and valves, as well as operator error or lack of training. A comprehensive tank management program addresses these kinds of problems.

Another occurrence is the leakage of product through the roof drains on external floating-roof tanks and out at the roof drain discharge nozzle. These spills are sometimes caused by equipment failures but also result from operator error or lack of training. Because it is easier to leave secondary-containment valves and roof drain valves open so that they do not need to be opened in periods of rainfall, the effectiveness of this equipment is reduced. This may be classified as operator error. A comprehensive tank management program addresses these kinds of problems.

Yet another significant cause of contamination is draining the water bottoms from tanks. This escapes through the secondary-containment system or stagnates in pools on the ground, resulting in contamination. A change in operation, as well as inclusion of the procedures for disposal of tank water bottoms in an overall tank management program, eliminates this form of leak and spill.

Tank Breakage.   Tank breakage, owing to either brittle fracture or ductile tearing during earthquakes or some uncorrected settlement problems, results in sudden and total loss of the tank contents. Such occurrences are relatively rare. Proper engineering and design, such as materials selection, application of the various codes and standards, and carefully detailed designs, can prevent these incidents.

Maintenance.   The lack of maintenance and investment in inspection programs results in not only poor housekeeping but unnoticed leaks due to corrosion, leaking flanges and valves, and inoperative instrumentation that could prevent spills.

Vandalism.   Vandalism is a surprisingly significant cause of spills. If a facility adopts a tank management program aimed at preventing leaks and spills, one of its elements addresses facility security.

Piping.   Piping is a principal cause of leaks and spills, having many of the same characteristics as tank leaks and spills. Indeed, piping is always connected to tankage. Perhaps the most significant leaks have resulted from pressurized underground piping.

Design.   Design deficiencies can result in almost any combination of the problems described above. It can also result in catastrophic and significant

Table 6. Causes of Leaks and Spills

Leak or spill source	Characteristics	Root causes	Preventive measures
corrosion	most common in tank bottoms and underground piping; low rate; lack of warning; may continue for years undetected; large volumes released over long periods; common	corrosion; materials selection; costs of corrosion prevention methods	careful design and engineering; inspection per API 653; tank management program
operations			
overfills or transfers	larger quantities released; quickly discovered; hazardous potential for fires; relatively common	operator error; instrumentation or equipment failure; lack of training; failure to maintain overfill systems	tank management program having written operating procedures; training and drills; periodic testing of instrumentation
roof drains	large volumes released; easily discovered; usually occur in stormy weather; relatively rare	equipment failure; failure to use secondary containment properly	tank management program having written operating procedures; training and drills; periodic testing of instrumentation
leaks	leaks relatively common in piping, valves, and fittings, pump seals, or in penetrations through secondary-containment areas		tank management program having written operating procedures; training and drills; periodic testing of instrumentation
tank breakage			
brittle fracture	occurs in cold weather; catastrophic failure mode; entire tank contents can empty; extremely rare	materials selection; poor fabrication details; failure to hydrotest	careful design and actual solution; assessment of brittle fracture and seismic after each significant charge of service; assessment of fabrication details per API 653; documentation of all work and engineering performed on all tasks

Table 6. (Continued)

Leak or spill source	Characteristics	Root causes	Preventive measures
seismic	damage to piping; tearing of repads and appurtenances; loss of tank contents; relatively rare	ground acceleration	careful design and actual solution; assessment of brittle fracture and seismic after each significant charge of service; assessment of fabrication details per API 653; documentation of all work and engineering performed on all tasks
maintenance	corrosion leaks; instrumentation malfunction	poor tank management programs	establish tank management program; periodically test all instrumentation; establish API 653 program; document all work on all tanks
vandalism	damage from opening valves; damage from gunfire; ignition of tank contents; bombs or explosions	poor security	improve security systems
piping	principal cause of leaks; failure to provide sufficient capacity to diked areas; leakage of product through second-containment penetration; poor or improper tank materials selection, corrosion prevention, design; charge of services	inadequate or no design engineering or periodic assessment	comply with API 570
fire or explosion	spills inside and outside secondary containment; fire likely to spread to all leaking tanks; rim seal fires relatively minor; spill fires can quickly become dangerous	improper design or operation	set up tank management program; ensure compliance with NFPA; conduct fire and safety audits; document results; establish emergency command system and resources; conduct process management safety review

653

spills. The best prevention is to use established codes, such as those provided by the API.

Fire and Explosion.   Most fires are associated with spills from overfills or leaking pump seals that form pools and clouds of flammable material. Periodic fire compliance and safety reviews can prevent many problems. Compliance with API and NFPA codes is good insurance. If fires do develop, then a program of emergency response and preparedness, as well as the establishment of an incident command system, can mitigate unforeseeable disasters. Careful design details and operational procedures are more effective than pumping resources into costly fire-fighting equipment in general.

Spill Anatomy and Remediation.   Contrary to past arguments that leaks or spills from aboveground tanks would stay near the surface, they go straight down into the aquifer and spread out. Various obstacles, such as clay lenses, rock, or impermeable layers of soil, simply divert the downward path. Slow leaks from tank bottoms tend to form a narrow plume, whereas larger spills cover much wider areas. When the contaminant reaches groundwater, it tends to be dispersed in the direction of the groundwater current and movement.

The properties of the material spilled, of course, play a key role in the dispersion of the spill. Not only are the various hazardous properties (toxicity, flammability, reactivity) important for affecting the levels at which these substances can damage the environment or health of life forms, but these properties also play a crucial role in dispersion and, consequently, potential for cleanup. Most petroleum products are relatively insoluble in water and therefore tend to remain in distinct and separate phases when they contact groundwater. Because petroleum is lighter than water, it tends to collect in pools above the groundwater and can be withdrawn by drilling wells into the high point of these pockets. However, petroleum also moves with the groundwater and often seeps into creeks, lakes, or rivers. Miscible products such as alcohols, oxygenates used for motor fuels, and various chemicals can mix with the groundwater. Once mixed, the entire groundwater current that passes through the spill region must be treated in order to remove the contaminant. Products having specific gravities greater than 1.0, such as halogenated hydrocarbons, tend to sink to the bottom of the aquifer. In spite of efforts to remove spilled chemicals by withdrawal from the ground through wells, a substantial amount remains embedded in the soil and groundwater in the form of coated particles or dissolved in the moisture of the soil and aquifers. Sometimes portions of the groundwater stream are pumped to the surface for treatment and reinjection, or neutralizing chemicals or oxidizers, such as air, are injected into the ground to reduce the effects of the spill and leak. None of these methods, however, is 100% effective.

Spill Prevention and Detection.   It is far better to prevent a leak or a spill than to clean one. The fundamental rule of leak and spill prevention is to reduce the possibility for contamination by directing resources as close to the source as possible (Fig. 11). In addition to increasing the effectiveness of a spill and leak prevention program, the costs are lower if the focus is placed on preventing the occurrence in the first place. Regulatory trend, however, is to require methods that respond to leaks after they occur. In addition to being more costly, this type of requirement is often a disincentive to prevent the leaks in the first place, because of the additional cost.

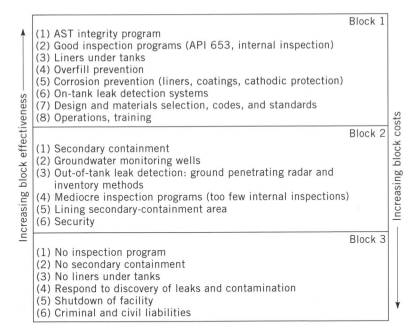

Fig. 11. Leak and spill prevention.

Leak and spill prevention comprises a system of management or a program embodying many facets which, when all working together, virtually eliminate the possibility of leaks and spills. In all cases, documentation and recordkeeping of all aspects of the tank management system are good practice.

Engineering Controls. The design, engineering, and maintenance of storage tanks strongly affect the potential for leaks and spills. For example, corrosion resistance can be increased substantially by use of coatings or cathodic protection. Instrumentation can be designed that reduces the likelihood of overfills, fires, or other accidents. Good engineering practice can virtually eliminate brittle fracture. Control of fabrication and inspection of new or repaired tanks can reduce the chances of all kinds of failures. Other parameters subject to engineering controls are foundation design and settlement, seismic capability, fire resistance, resistance to vandalism, etc.

Operation Controls. Standard written operating instructions go a long way to ensure that operators not only know what to do but have sufficient understanding to act effectively in the event of a leak or spill. These instructions should include information on the material stored and its properties, notification of the appropriate authorities in the event of spills, emergency shutdown procedures, and availability and use of emergency and protective equipment. Operational controls should have provisions for inspection. One of the most effective leak prevention methods available is a daily walk-through in the plant by an experienced operator. As a result, operation controls should include provisions for regular training, which should address contract personnel as well. Spill response and planning should be an integral part of the operational controls.

Secondary Containment. One of the most effective methods for mitigating large and catastrophic spills is the concept of secondary containment, where the entire tank field is surrounded with a dike wall or impound area from which the volume of the largest tank can be contained. This way, even if the tank were to fail from a sudden and total release, the contents would be captured in the secondary-containment area for immediate removal and disposal. In fact, any facility near a navigable waterway must use secondary containment according to the Spill Prevention Control and Countermeasures (SPCC) rules enacted by the EPA. However, if the drain valves that allow rainwater to escape are left open, the secondary-containment area can fail to operate as intended. Only operating procedure, knowledge, training, and practice can ensure that these systems work as intended.

Leak Detection. Leak detection methods may be subclassified according to whether or not they are on the tank. On-tank leak detection systems operate immediately upon leakage.

On-Tank Leak Detection Systems. Tanks having leak detection bottoms have a means of directing any leaks to the outside of the tank perimeter where these can be visually observed. Before any significant contamination can occur, the leaks are discovered and the tank taken out of service to address the leak.

Precision mass and volumetric methods use very precise measurements of pressure and/or level in the tank to detect leaks. The tank must be closed so that no liquid enters or leaves the tank. The threshold of detection and funnel required to perform a reliable test become greater as tank size increases.

Hydrocarbon sensors (qv) placed directly below the tank bottoms can be effective. However, old contamination or contamination from other tanks or piping can yield misleading results. In addition, the low permeability of some areas in the soil can prevent the migration of vapors to the sensing ports under the tank bottom.

Tracer methods involving chemical markers injected into the contents of the tank may be used. Instrumentation capable of picking up the chemical marker can then determine the presence of a leak caused by seepage of the tracer into the ground. This, like the hydrocarbon sensing method, is generically referred to as soil vapor monitoring. This method suffers the same weaknesses that have to do with undertank soil permeabilities.

By listening to the sound emitted from leaking tanks, it is possible to estimate not only the existence of, but also the location of, leaks in tank bottoms. Much work needs to be done in this area before it can be considered reliable.

Off-Tank Leak Detection Systems. Monitoring wells, drilled near the tank site, are effective only after large losses and resulting contamination have occurred. Inventory reconciliation, a method of detecting discrepancies in receipts and disbursements of product through metered piping, is sometimes employed. This method is relatively inaccurate, however, and a substantial leak can escape detection. The advantage to this method, which depends on a substantial amount of lost product and contamination, is that it requires little capital investment as most of the metering is usually already in place.

Use of Liners. The use of impermeable liners and membranes, often called release prevention barriers (RPBs) under tanks, may be the most effective leak detection and prevention method. On new tanks, it is relatively easy to install these systems, and large numbers of tanks are being built with this type of system in the 1990s. For existing tanks, however, it would be very costly if not impractical to install liners. For existing tanks, the combination of other methods as well as an effective inspection program can be more effective as a substitute for a release prevention barrier.

There has been much debate about using liners for the entire secondary-containment area in addition to the area under the tank. Lining the entire secondary-containment area is costly and probably ineffective for the following reasons. (*1*) Large spills and leaks into a secondary-containment area are not left for long periods of time where these can permeate into the ground, but are generally cleaned up immediately. Most secondary-containment areas are relatively impermeable for the short duration for which spills reside. (*2*) Lining the secondary-containment area is difficult to achieve completely. At walls, partitions, piping penetrations, and equipment foundations there are joints and cracks which permit fluids to migrate under the liner. Once under the liner, the fluids cannot be cleaned up. From this perspective, it is worse to use a liner. (*3*) According to the API, it may be cheaper to remediate than to provide lining. (*4*) A small fraction of the resources poured into liners would do far more good for the environment if used as prevention, eg, inspection, leak detection, operator training, and tank programs.

Inspection Programs.   One of the most effective ways to reduce leaks and spills resulting from mechanical failure or corrosion is to implement an inspection program. The American Petroleum Institute has issued API Standard 653, which provides a rational and reasonable approach to the problem of inspecting tanks. Because tanks are very costly to empty, remove from service, clean, and prepare for internal inspection, past practices requiring the entry of inspectors into the interior of tanks have been avoided. As a result, there have been numerous long-term leaks resulting from corrosion that have gone undetected. API Standard 653 provides a basis for scheduling internal inspection based on anticipated corrosion rates. For existing tanks that cannot easily be fitted with leak detection systems or liners, the use of this type of inspection program with appropriately spaced internal inspections is effective in reducing leaks.

BIBLIOGRAPHY

1. R. A. Christensen and R. F. Eibert, *Aboveground Storage Tank Survey*, Entropy, Ltd., for the American Petroleum Institute, Apr. 1989.
2. J. H. Perry, *Chemical Engineer's Handbook*, 4th ed., McGraw-Hill Book Co., Inc., New York, 1963.
3. C. G. Kirkbride, *Chemical Engineering Fundamentals*, McGraw-Hill Book Co., Inc., New York, 1947.
4. W. B. Young, *Floating Roofs: Their Design and Application*, #73-PET-44, American Society of Mechanical Engineers, New York, 1981.
5. E. F. Megyesy, *Pressure Vessel Handbook*, 5th ed., Pressure Vessel Handbook Publishing, Inc., Tulsa, Okla., 1985; H. H. Bednar, *Pressure Vessel Design Handbook*, Krieger Publishing Co., Melkar, Fla., 1991.
6. J. F. Harvey, *Theory and Design of Pressure Vessels*, 2nd ed., Von Nostrand Reinhold Co., Inc., New York, 1991.

7. Technical data, Ecology and Environment, Inc., Buffalo, New York, 1982.
8. J. E. Campbell, *Metals Handbook*, 9th ed., American Society of Metals.
9. *Chevron Environmental, Health, and Safety Regulatory Summary and Desk Reference*, Chevron Research and Technical Co., Orinda, Calif.

General References

API Standard 2610, "API Design, Construction, Operation, Maintenance and Inspection of Terminal and Tank Facilities," American Petroleum Institute.
R. P. Benedetti, *Flammable and Combustible Liquids Code Handbook*, Quincy, Mass., 5th ed., 1994.

PHILIP MYERS
Chevron Research and Technical Company

TANTALUM AND TANTALUM COMPOUNDS

Tantalum [7440-25-7], atomic number 73, is the heaviest element in Group 5 (VA) of the Periodic Table. This tough, ductile, silvery gray metal has an atomic weight of 180.948 amu. The element was discovered by A. K. Ekeberg in 1802 in minerals taken from Kimito, Finland, and Ytterby, Sweden (1). The element was named after the mythological Greek king Tantalus because the oxide was tantalizingly difficult to dissolve in acid. Most of the early efforts to isolate the metal focused on the reduction of the oxide with carbon and probably yielded oxide- and carbide-contaminated material. Berzelius produced oxide-contaminated metal in 1824. Tantalum was first isolated in the pure ductile state (2) by disassociation of the oxide at high temperature in a vacuum. Siemens & Halske AG, Berlin modified methods used by others (3,4) to reduce potassium or sodium heptafluorotantalate with sodium to produce tantalum metal on a commercial scale beginning in the early 1900s. Tantalum was first produced commercially in the United States by the Fansteel Metallurgical Corporation in the 1930s. The method involved electrolysis of tantalum pentoxide dissolved in a molten mixture of potassium heptafluorotantalate, potassium chloride, and potassium fluoride. By 1995, the bulk of the tantalum produced in the world was made by H. C. Starck and Cabot Corporation at facilities located in the United States, Germany, and Japan.

Physical and Chemical Properties

The physical properties of tantalum are presented in Table 1 and represent the best values from several compilations (5–8). Original sources of the data are given in the literature.

Table 1. Physical Properties of Tantalum

Property	Value	Reference
at no.	73	
at wt	180.9479	5
at vol, cm^3/mol	10.9	6
atomic radius, pm	147	6
crystal structure	bcc	
lattice constant, nm	0.33	6
coordination number	8	
density, at 20°C, g/cm^3	16.62	7
mp, °C	2996	7
bp, °C	5427 ± 100	7,8
enthalpy of fusion, J/mol^a	31,400	7
enthalpy of vaporization, J/mol^a	7.53 × 10^5	7
heat capacity, J/(K·mol)^a, t from 25 to 2000°C	24.2 + 3.0 t + 0.2 × 10^{-6} t^2	6
vapor pressure, log $P_{kPa}{}^b$, T from 290 K–mp	−40,800/T + 9.41	6
thermal conductivity, J/(cm·°C)^a		
20°C	0.540	6
568°C	0.680	6
828°C	0.720	6
coefficient of linear thermal expansion, °C^{-1}		
at 20°C	6.5 × 10^{-6}	8
for t = 20–500°C	$L_0(1 + (6.59 t + 0.00008 t^2) \times 10^6$	8

^{a}To convert J to cal, divide by 4.184.
^{b}Log $P_{mm\,Hg}$ = −40,800/T + 10.29.

Tantalum metal is easily oxidized to the +5 valence state although this reactivity is usually partly masked by the presence of a stable, adherent, passivating oxide layer on the surface of the metal. Because of the protective and chemically inert oxide, tantalum is not attacked by concentrated nitric, hydrochloric, or sulfuric acids at temperatures below 200°C and the metal is thus extremely useful in corrosive chemical environments. Hydrofluoric acid and acidic fluoride solutions react with the thermal oxide to destroy the protective layer and expose the true chemical reactivity of the metal especially in oxidizing environments. Mixtures such as HF–H$_2$O$_2$ and HF–HNO$_3$ violently react with tantalum if the reactions are not carefully controlled. Tantalum is also attacked by hot concentrated alkali solutions. The metal is, however, very resistant to attack by liquid metals such as sodium, potassium, and mercury, provided these do not contain oxygen. Tantalum reacts with the halogens at elevated temperatures to give the corresponding pentahalides. The Group 5 (VB) elements, nitrogen, phosphorus, and arsenic, also react with tantalum at temperatures above 500°C to give compounds where the compositions depend on the reaction conditions (9).

Occurrence

In nature, tantalum occurs not in the free state, but in several complex oxidic minerals, often in solid solution with a variety of other elements such as tin, titanium, thorium, and uranium. The main source of tantalum is an isomorphous series of minerals containing oxides of tantalum, niobium, iron, and manganese. This tantalite–columbite occurs as an accessory mineral distributed in granitic rocks or pegmatites. The microlite–pyrochlor minerals series is a second source of tantalum. These minerals consist essentially of complex oxides of tantalum, niobium, sodium, and calcium in combination with hydroxyl ions and fluorides. Struverite, a variation of the titanium containing mineral rutile, is a low grade source of tantalum. A considerable amount of tantalum slags from the tin smelting processes in Thailand, Malaysia, and Brazil makes up the rest of the Western world's tantalum supply (see TIN AND TIN ALLOYS). The tantalum-containing minerals fergusonite, samarskite, and euxenite are of minor importance. Table 2 summarizes the composition and the Ta_2O_5 and Nb_2O_5 content of the main tantalum-containing minerals (10). The world tantalum reserves that can be economically extracted are given in Table 3 (11).

Table 2. Composition of Tantalum Containing Minerals

Mineral	CAS Registry Number	Composition	Ta_2O_5, %	Nb_2O_5, %
tantalite	[1306-08-7]	$(Fe, Mn)Ta_2O_6$	42–84	1–40
microlite	[12173-96-5]	$(Na,Ca)(Ta,Nb)_2O_6F$	60–70	5–10
columbite	[1306-08-7]	$(Fe,Mn)(Nb,Ta)_2O_6$	1–40	40–75
wodginite	[12178-62-0]	$(Ta,Nb,Sn,Mn,Fe,Mn)_{16}O_{32}$	45–56	3–15
yttrotantalite	[12199-77-8]	$(Y,U,Ca)(Ta,Fe^{+3})_2O_6$	14–27	41–56
fergusonite		$(Re^{+3})(Nb,Ta)O_4$	4–43	14–46
strueverite	[12199-39-2]	$(Ti,Ta,Nb,Fe)_2O_4$	7–13	9–14
tapiolite	[1310-29-8]	$(Fe,Mn)(Nb,Ta,Ti)_2O_6$	40–85	8–15
euxenite	[1317-53-9]	$(Y,Ca,Ce,U,Th)(Nb,Ta,Ti)_2O_6$	1–6	22–30
samarskite	[1317-81-3]	$(Fe,Ca,U,Y,Ce)_2(Nb,Ta)_2O_6$	15–30	40–55

Table 3. World Tantalum Reserves and Reserve Base 1992, t^a

Country	Reserves	Reserve base
Australia	4,500	9,100
Brazil	900	1,400
Canada	1,800	2,300
Malaysia	900	1,800
Nigeria	3,200	4,500
Thailand	7,300	9,100
Zaire	1,800	4,500
other market economy countries	1,400	1,800
World total	*21,800*	*34,500*

^{a}Tantalum content.

Processing

Ore Dressing.   The mining of pegmatite deposits, either open-pit (Gwalia Consolidated Ltd., Greenbushes, Australia) or underground (Tantalum Mining Corporation of Canada Ltd., Tanco, Canada), is done by conventional techniques like blasting and crushing (see MINERALS RECOVERY AND PROCESSING). The materials are then dressed mainly by gravity concentration. Mineral processing of tantalum and niobium ores has been described in detail, including flow sheets of the main producing plants (12).

Upgrading of Tin Slags. The 0.2–17% Ta_2O_5-containing tin slags are upgraded in a sequence of three pyrometallurgical processing steps as outlined in Figure 1. In the first step (Fig. **1a**) the slags are intensively mixed with additives like Fe_2O_3 and CaO and the mixture is continuously fed into a three-stage electric arc furnace. By adding coke and charcoal, the oxides of tantalum, niobium, tungsten, titanium and trace metals are reduced to a high carbon-containing ferroalloy with the typical composition of 10–20% Ta, 10–20% Nb, 40–60% Fe, 5–10% Ti, and 3–10% C. The ferroalloy and the slag are separated by conventional techniques like tapping. The second step (Fig. **1b**) involves oxidation of the comminuted and crushed alloy by the addition of Fe_2O_3 (13) or by a roasting process. In the last processing step (Fig. **1c**), the oxidized material is reduced in an electric arc furnace under strictly controlled conditions in order to collect all the impurities like tungsten, tin, and phosphorus in an alloy while tantalum and niobium form a synthetic concentrate of oxides in the furnace slag. A typical analysis of this concentrate is given in Table 4 (14).

Separation of Tantalum and Niobium. *Solvent Extraction.* The industrial separation of tantalum from niobium was carried out historically by the Marignac process of fractional crystallization of potassium heptafluorotantalate and potassium heptafluoroniobate (15,16) or the long-established Fansteel process (17), which involved the decomposition of the ore by a caustic fusion procedure. Processors have replaced these expensive processes by procedures based on solvent extraction. This technique was developed in the United States at Ames Laboratory and the U.S. Bureau of Mines (18). Figure 2 shows the flow sheet of an industrial installation for the hydrometallurgical processing of tantalum–niobium raw materials.

Tantalite and columbite, either naturally occurring or synthetically produced as concentrates from tin slags, are digested with hydrofluoric and sulfuric acid at an elevated temperature. Along with the tantalum and niobium, which both form the complex heptafluorides H_2TaF_7 and H_2NbOF_5 or H_2NbF_7, other elements, eg, Fe, Ti, and Mg, are dissolved. After filtering out the insoluble residues to remove fluorides of alkaline earth and rare earth metals, the aqueous solution of Ta–Nb in hydrofluoric acid is extracted in several continuously operating mixer–settler systems or extraction columns with an organic solvent like methyl isobutyl ketone (MIBK) (19). The complex fluorides of niobium and tantalum are extracted into the organic phase; most of the impurities, such as iron, manganese, and titanium, remain in the aqueous phase. In practice, Nb_2O_5 + Ta_2O_5 concentrations of 150–200 g/L are achieved in the organic phase. The organic phase is then scrubbed with 6–15 N H_2SO_4 and extracted with water or dilute sulfuric acid to separate the niobium from the tantalum by selective stripping. The aqueous phase takes up the complex fluoroniobate and free hydrofluoric acid while the complex fluorotantalate remains dissolved in the organic phase. The aqueous niobium solution is contacted with a small amount of MIBK to remove traces of coextracted fluorotantalate. The tantalum salt is extracted from the organic phase with water (20) or dilute aqueous ammonium fluoride solution. The niobium and tantalum salts can be produced with yields

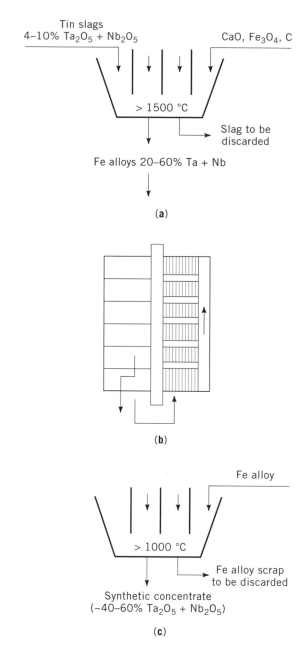

Fig. 1. The H. C. Starck pyrometallurgical upgrading process of Ta–Nb-containing tin slags where the solid vertical bars represent electrodes. The product of (**a**) electrothermic reduction of Sn slag goes to (**b**) oxidation of Fe alloy 20–60 Ta + Nb and then to (**c**) electrothermic reduction of oxidized Fe alloy. Courtesy of H. C. Starck GmbH & Co. KG.

in excess of 95% and greater than 99.9% purity by using sophisticated process optimization and control.

Modifications and improvements to the basic process have been made to reduce the quantity of waste products (21,22) in the wet chemical process, to recover

Table 4. Enrichment of Tantalum in the Pyrometallurgical Upgrading Process

Component	Composition, wt %	
	Slag[a] (Malaysia)	Synthetic concentrate
Ta_2O_5	2	20.8
Nb_2O_5	2.2	25.6
TiO_2	10.5	25.0
SnO_2	1.2	0.1
WO_3	2.6	0.1
P_2O_2	1.2	0.1
FeO	15	8

[a] Also contains CaO, MgO, Al_2O_3, and SiO_2.

HF, and to economically process low Ta, high Nb containing raw materials (23). Several alternative extraction media have been reported in the literature. Most, except for tributylphosphate (TBP) (24) and tri-*n*-octylphosphine oxide (TOPO) (25), have never been used in industry.

Chlorination.   The chlorination process is an alternative to solvent extraction. In reductive chlorination, the ore or concentrate is pelletized with coal or coke and pitch, dried, and then allowed to react in a stream of chlorine at 900°C. The nonvolatile alkaline earth metal chlorides remain in the bottom of the reactor while tungsten oxychloride, $WOCl_4$; the readily volatilized tetrachlorides of silicon, tin, titanium; and the pentachlorides $NbCl_5$ and $TaCl_5$ are distilled (26). The waste gas containing chlorine and phosgene must be rigorously controlled. The alternative chlorination of ferroalloys is much simpler and more economical (27). Ferroniobium–tantalum produced by the aluminothermic or electrothermic process is comminuted and fed together with sodium chloride into a $NaCl$–$FeCl_3$ melt. The chlorinating agent is $NaFeCl_4$. Chlorine is passed into the melt, continuously regenerating $NaFeCl_4$. The complete reaction is described by:

$$Fe(TaNb) + NaCl + 7\ NaFeCl_4 \longrightarrow (TaNb)Cl_5 + 8\ NaFeCl_3$$

$$8\ NaFeCl_3 + 4\ Cl_2 \longrightarrow 8\ NaFeCl_4$$

The reaction temperature of 500–600°C is much lower than that required for the reductive chlorination. The volatile chlorides evolve from the molten salt bath. The boiling points of $NbCl_5$, $TaCl_5$, and $WOCl_4$ lie between 228 and 248°C. These compounds must therefore be separated by means of a distillation column. The chlorination of ferroalloys produces very pure tantalum pentachloride in tonnage quantities. The $TaCl_5$ contains less than 5 μg Nb/g Ta, and other metallic impurities are only amount to 1–2 μg/g Ta.

Production

Tantalum Compounds.   Potassium heptafluorotantalate [*16924-00-8*], K_2TaF_7, is the most important tantalum compound produced at plant scale. This compound is used in large quantities for tantalum metal production. The fluorotantalate is prepared by adding potassium salts such as KCl and KF to

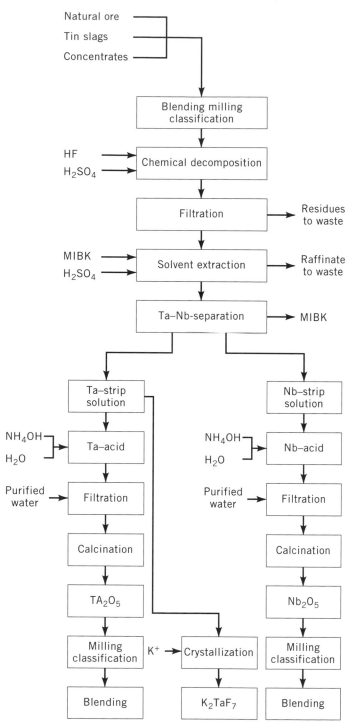

Fig. 2. The processing of Ta−Nb raw materials where MIBK = methyl isobutyl ketone.

the hot aqueous tantalum solution produced by the solvent extraction process. The mixture is then allowed to cool under strictly controlled conditions to get a crystalline mass having a reproducible particle size distribution. To prevent the formation of oxyfluorides, it is necessary to start with reaction mixtures having an excess of about 5% HF on a wt/wt basis. The acid is added directly to the reaction mixture or together with the aqueous solution of the potassium compound. Potassium heptafluorotantalate is produced either in a batch process where the quantity of output is about 300–500 kg K_2TaF_7, or by a continuously operated process (28).

Tantalum pentoxide [1314-61-0], Ta_2O_5, is prepared by calcining tantalic acid or hydrated tantalum oxide [75397-94-3], $Ta_2O_5 \cdot nH_2O$, at temperatures between 800 and 1100°C. This oxide hydrate is produced by adding gaseous or aqueous ammonia to the solvent extraction produced aqueous tantalum solution in a continuous (29) or batch process.

Tantalum. Numerous methods developed to extract tantalum metal from compounds included the reduction of the oxide with carbon or calcium; the reduction of the pentachloride with magnesium, sodium, or hydrogen; and the thermal dissociation of the pentachloride (30). The only processes that ever achieved commercial significance are the electrochemical reduction of tantalum pentoxide in molten K_2TaF_7/KF/KCl mixtures and the reduction of K_2TaF_7 with sodium.

Electrochemical Reduction. The electrochemical reduction of Ta_2O_5 in molten K_2TaF_7/KF/KCl mixtures was used for many years by Fansteel Metallurgical Corporation to manufacture tantalum metal in the United States. A mixture of K_2TaF_7, KF, and KCl was placed in an iron crucible that served as the cathode of an electrochemical cell, and the mixture was heated to above the melting point. A few percent of tantalum oxide was added to the melt, and the electrolysis was carried out using a carbon rod as the anode. Periodic additions of tantalum oxide were made to replace that consumed during the electrolysis process. The tantalum metal was recovered by crushing the cooled salt/metal mass and leaching out the salts using a mineral acid and water. The dendritic tantalum was pressed and sintered into bars and rods before conversion to sheet, wire, and powder for capacitors. The powder was quite coarse and thus is not suitable for use in the tantalum capacitor applications of the 1990s.

Chemical Reduction. The reduction of K_2TaF_7 with sodium, used almost exclusively in 1990s to manufacture tantalum metal, was first utilized on a commercial scale by Siemens & Halske AG, Berlin, in the early 1900s. Alternate layers of the salt and small cubes of sodium were packed in a steel tube. The covered tube was heated with a gas-fired ring burner near the top of the charge to initiate the reaction that quickly propagated down the tube. The excess sodium was neutralized by adding methanol to the cooled charge. The tantalum was separated from the salt residues by leaching with hot water. The inability to mix the reactants and to control the reduction temperature seriously limited the usefulness of the process.

The demands of the solid tantalum capacitor industry for high quality, high surface area tantalum powders have driven improvements in the sodium reduction of K_2TaF_7 process since the 1960s. Powder quality and performance improved rapidly following the invention of the stirred reactor in 1960 (31).

Several variations of this process have been described (32–36) but the underlying concepts have changed little. A typical reduction consists of placing the K_2TaF_7 and one or more of the diluent salts, eg, NaCl, KCl, or KF, into a reactor made of Inconel or nickel-clad Inconel. The reactor is usually 0.3–1 m in diameter and covered with a tight lid fitted with a stirrer as shown in Figure 3. The charged reactor is placed in a furnace and heated to a temperature sufficiently high to melt the salt mixture, but usually below 1000°C.

Molten sodium is injected into the retort at a prescribed rate and the temperature of the system is controlled by adjusting the furnace power or with external cooling. The variables that control the quality and physical properties of the powder are the reduction temperature and its uniformity, diluent type and concentration, sodium feed rate, and stirring efficiency. Optimizing a variable for one powder attribute can adversely affect another property. For example, a high reduction temperature tends to favor improved chemical quality but lowers the surface area of the powder.

The cooled reaction mass is extracted from the retort, crushed and leached first with dilute mineral acid, and then with water to separate the tantalum powder from the salts. After drying and classification, the primary powder is ready for processing to sheet, rod, wire, or capacitor-grade powder.

The dramatic improvements in the physical and chemical properties of tantalum powder produced by the sodium reduction process are evident in the lessening of chemical impurities (see Table 5). The much-improved chemistry reflects the many modifications to the process put in place after 1990. The dramatic reductions in alkalies and carbon concentrations have led to significant improve-

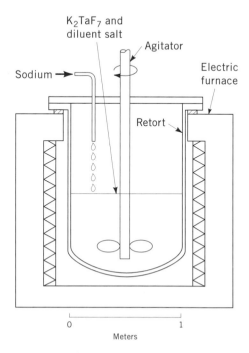

Fig. 3.   Schematic of a reactor used to produce tantalum powder by the sodium reduction process.

Table 5. Comparison of Impurity Levels in Tantalum Powders

Element	Concentration, ppm	
	Powder A[a]	Powder B[b]
carbon	38	14
sodium	12	2
chromium	21	12
iron	21	5
nickel	19	8

[a]Processing using 1960s technology.
[b]Post-1990 processing.

ments in the electrical quality as measured by the performance of tantalum capacitors.

Post-Reduction Processing.   The primary tantalum powder produced by the sodium reduction process is treated to convert the metal to a form suitable for use as capacitor-grade powder and feedstock for wire and sheet.

For use in capacitors, the primary powder is not suitable. Poor flow and pressing characteristics make it difficult to press this powder into pellets. Physical properties are improved by agglomerating the powder by vacuum sintering at a temperature in the range of 1300–1600°C. During agglomeration, the oxygen of the thermal oxide surface layer diffuses into the bulk tantalum and must be replaced to prevent uncontrolled oxidation when the powder is exposed to the atmosphere. This is achieved after the charge has cooled by the controlled addition of small amounts of oxygen, usually as air. The resulting passivated sinter cake is milled and classified to give a powder that flows into the press die cavity and yields a pressed pellet of sufficient crush strength to withstand subsequent handling before sintering.

The consequence of the dissolution of the surface oxygen during agglomeration and its subsequent replacement is a 500–1000-ppm increase in the oxygen content of the powder. This oxygen increase, coupled with an additional increase associated with pellet sintering, can seriously degrade the electrical quality of the powder (37). The excess oxygen is removed by a deoxidation process (38); the agglomerated powder is blended with magnesium and heated for several hours at a temperature in the 900–1100°C range. The magnesium reduces the tantalum oxide on the surface and, given sufficient time, reacts with the oxygen dissolved in the metal. The cooled charge is carefully passivated, and the magnesium oxide and excess magnesium are separated from the powder by leaching with dilute mineral acid. Finally, the powder is blended into lots. More and more frequently the powder is packaged under vacuum in a plastic pouch to prevent the pick-up of oxygen over time, which can amount to several hundred parts per million in 2–3 mo.

Some powder is pressed into bars and sintered at high temperature in a vacuum furnace. The sintered compacts are then arc-melted (39) to remove impurities. The resulting tantalum ingots are hydrided to embrittle the metal (40), which is then ground to a powder, dehydrided, and classified. As seen in Figure 4, these powders are composed of fairly regular solid grains, in contrast to the capacitor-grade powders produced directly by the sodium reduction

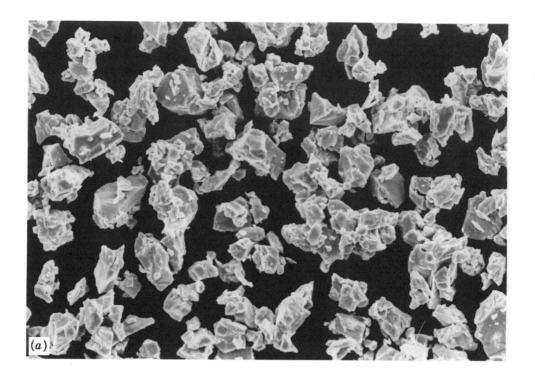

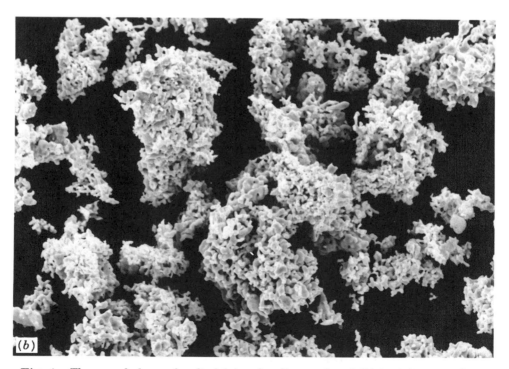

Fig. 4. The morphology of melted (**a**) and sodium-reduced (**b**) tantalum powders.

process, which consist of irregular, spongy particles. The melted powders have considerably lower surface area than the sodium-reduced powders, but higher chemical purity. They are used mostly in low specific capacitance, high voltage, high reliability applications.

Powder destined for wire is blended with high grade tantalum scrap and pressed into bars. The bars are sintered and then rolled into small-diameter rods suitable for drawing into wire with a diameter in the range of 0.2–0.8 mm. The rolling and drawing processes are interrupted periodically for annealing to reverse work hardening and the wire is carefully cleaned to remove traces of lubricant used during the drawing operation. Finally, the wire is straightened and wound on spools.

Economic Aspects

Tantalum metal is used as powder and wire in solid tantalum capacitors. The metal is fabricated into shields, trays, and heating elements in vacuum sintering furnaces. The chemical industry uses the metal for equipment such as retorts that are exposed to corrosive environments. Some biomedical implants are made from tantalum. Figure 5 shows the quantity of tantalum metal products produced annually over the time period 1984–1994. In 1995, about 820 metric tons of tantalum metal was produced. The bulk of this metal, 640 metric tons, was in the form of capacitor-grade powder and had an average value of $360/kg. The remaining 180 metric tons was split equally between capacitor-grade wire and fabricated parts. The average selling price of this material was $440/kg.

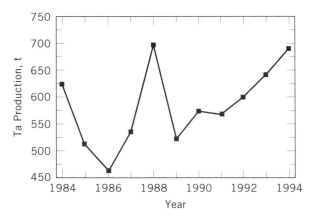

Fig. 5. Summary of tantalum metal production between 1984 and 1994. Courtesy of the Tantalum–Niobium International Study Center.

Uses of Tantalum Metal

The first commercial use of tantalum was as filaments in incandescent lamps but it was soon displaced by tungsten. Tantalum is used in chemical industry equipment for reaction vessels and heat exchangers in corrosive environments. It is usually the metal of choice for heating elements and shields in high temperature vacuum sintering furnaces. In 1994, over 72% of the tantalum produced in the world went into the manufacturing of over 10×10^9 solid tantalum capacitors for use in the most demanding electronic applications.

Capacitor-Grade Powder. Since 1984 about 72 percent of the tantalum metal produced annually has gone into the manufacturing of solid tantalum capacitors (STC). The principal steps in the STC manufacturing process are as follows. Tantalum powder is pressed into a pellet weighing from one to several hundred milligrams and then sintered in a vacuum at a temperature above 1300°C to give a porous compact. The pellet is formed or anodized by making it the anode of an electrochemical cell containing typically a dilute aqueous solution of phosphoric acid as the electrolyte. Upon application of a voltage, anodic tantalum oxide grows on the surface of the tantalum to a thickness proportional to the applied voltage. The resulting oxide film is an excellent electrical insulator and can serve as the dielectric in the conductor/insulator/conductor configuration of a capacitor. One conductor is the tantalum and the other conducting electrode is MnO_2, a semiconductor, which is deposited on the anodic oxide by the pyrolysis of manganese nitrate. The unit is finally encapsulated usually in a variation of the surface mount chip configuration shown in Figure 6.

Solid tantalum capacitors have a high volumetric capacitance which makes them attractive for use in miniaturized electronic systems like cellular telephones, hand-held video cameras, and personal computers. The insensitivity of their capacitance to temperature and their ability to operate at temperature extremes explains why these devices are used in such harsh environments as automobile engine compartments. Solid tantalum capacitors are extremely reliable and, therefore, are often the capacitor of choice in critical applications like spacecraft electronics, pacemakers, and safety equipment.

Wire. Tantalum wire is used primarily as the anode lead wire in solid tantalum capacitors. Since the 1970s, the average weight of tantalum in a solid tantalum capacitor has dropped from several hundred milligrams to less than 50 mg; but the consumption of tantalum powder for capacitors has remained

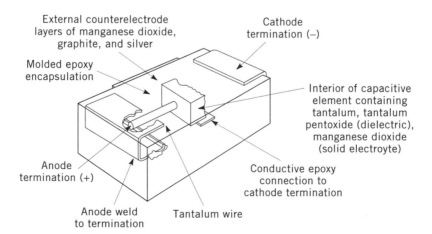

Fig. 6. A chip-type surface mount solid tantalum capacitor. Courtesy of AVX Corp.

relatively constant because of the dramatic increase in the number of capacitors manufactured. The weight of wire per capacitor has remained relatively constant and thus wire consumption has increased steadily.

Fabricated Parts. The starting point for manufacturing tantalum fabricated parts is mainly high grade tantalum scrap that is converted to ingots by a combination of electron beam and arc melting. The ingots are transformed into flat stock and rod by pressing and rolling. The fabrication of functional parts uses many standard metal working techniques including turning on a lathe, cutting, bending to shape, and welding under a helium atmosphere (see METAL TREATMENTS).

Typical equipment made from tantalum includes heat exchangers, reaction vessels liners, thermowells, and heating elements or heat shields for high temperature vacuum sintering furnaces. Tantalum fabricated parts are found in the manufacturing of pharmaceuticals, explosives, insecticides, dyes, acidic baskets for silver cyanide barrel platers, and in hydrochloric and hydrobromic acid condensers.

Biomedical Devices. Detailed reviews of the biomedical uses of tantalum are available (41,42). The biocompatibility of a metal is related to its corrosion resistance and toxicity of the metal ion (43). The very low corrosion rate and inertness of tantalum make it attractive for many biomedical applications. Surgeons have used the metal as sutures (qv), bone screws, and plates since the 1940s. Other uses include cartilage wire, nets to hold bone grafts in place, braided suture wire for skin closures and tendon repair, clips for ligature of vessels and bile ducts, mesh for abdominal wall reconstruction after hernias, plates for cranioplasty, and dental implants (see DENTAL MATERIALS). An undisturbed healing reaction and excellent long-term performance are found with tantalum. Mechanical performance is good and resistance to infection is high. Anodized, porous, sintered tantalum is used as nerve stimulators (44,45). The biocompatibility of tantalum coupled with the metal's high density has lead to widespread use in radiology; as powder it is used mainly for tracheography and bronchography of the respiratory tract. Tantalum for use in implants is sterilized by steam or γ-radiation (see PROSTHETIC AND BIOMEDICAL DEVICES).

Corrosion of Tantalum

The corrosion behavior of tantalum is well-documented (46). Technically, the excellent corrosion resistance of the metal reflects the chemical properties of the thermal oxide always present on the surface of the metal. This very adherent oxide layer makes tantalum one of the most corrosion-resistant metals to many chemicals at temperatures below 150°C. Tantalum is not attacked by most mineral acids, including aqua regia, perchloric acid, nitric acid, and concentrated sulfuric acid below 175°C. Tantalum is inert to most organic compounds; organic acids, alcohols, ketones, esters, and phenols do not attack tantalum.

Tantalum is not resistant to substances that can react with the protective oxide layer. The most aggressive chemicals are hydrofluoric acid and acidic solutions containing fluoride. Fuming sulfuric acid, concentrated sulfuric acid above 175°C, and hot concentrated alkali solutions destroy the oxide layer and, therefore, cause the metal to corrode. In these cases, the corrosion process occurs because the passivating oxide layer is destroyed and the underlying tantalum reacts with even mild oxidizing agents present in the system.

The excellent corrosion resistance means that tantalum is often the metal of choice for processes carried out in oxidizing environments or when freedom from reactor contamination of the product or side reactions are necessary, as in food and pharmaceutical processing. Frequently, the initial investment is relatively high, but this is offset by low replacement costs owing to the durability of the metal.

Health and Safety

The corrosion resistance imparted to tantalum by the passivating surface thermal oxide layer makes the metal inert to most hazards associated with metals. Tantalum is noncorrosive in biological systems and consequently has a no chronic health hazard MSDS rating.

Above 300°C in air, fine tantalum powder can ignite because of the rapid oxygen diffusion through the oxide layer into the substrate. Once burning starts, a dramatic temperature rise usually occurs and the oxidation reaction can accelerate rapidly. Tantalum powder can be ignited by the localized heating associated with a static electrical discharge or contact with a hot surface.

Tantalum metal powder does not explode in the normal sense of the word, because there is no release of gas when tantalum burns. The burning of suspended tantalum powder in a confined gaseous environment containing an oxidizing agent, for example air, only produces a pressure rise associated with the heating of the gas. The potentiality of a disastrous explosion occurring exists if hot tantalum comes in contact with an oxidizing agent that reacts with the metal to produce a combustible gas. For example in the presence of water, hot tantalum reacts to produce hydrogen gas, which will then likely detonate, with tragic consequences.

A tantalum fire cannot be extinguished with the usual firefighting chemicals. The only effective way to control a tantalum fire is to smother it with a nonoxidizing chemical like sodium chloride or argon. Fire extinguishers containing sodium chloride under pressure are available for this purpose but a container of salt and a noncombustible scoop will work. Under no circumstances should burning tantalum be allowed to come in contact with water. If tantalum powder is routinely handled in a closed environment, it is prudent to design the system so that it can be flooded with argon.

Good housekeeping practices to prevent the accumulation of tantalum dust and a proper passivation procedure will prevent most tantalum fires. All equipment used to handle the powder should be properly grounded and contact with hot surfaces or flames should be avoided.

Exposure to tantalum metal dust may cause eye injury and mucous-membrane irritation. The threshold limit value (TLV) in air is 5 mg/m^3, LD$_{50}$ is <400 mg/kg and the Occupational Safety and Health Administration (OSHA) time weighted average (TWA) exposure limit is 5 mg/m^3 (47). The immediate dangerous to life or health (IDLH) concentration is 2500 mg/m^3 (48). Whereas some skin injuries from tantalum have been reported, systemic industrial poisoning is apparently unknown (47).

Anodic Oxide Films on Tantalum

Anodic Oxidation. The ability of tantalum to support a stable, insulating anodic oxide film accounts for the majority of tantalum powder usage (see THIN FILMS). The film is produced or formed by making the metal, usually as a sintered porous pellet, the anode in an electrochemical cell. The electrolyte is most often a dilute aqueous solution of phosphoric acid, although high voltage applications often require substitution of some of the water with more aprotic solvents like ethylene glycol or Carbowax (49). The electrolyte temperature is between 60 and 90°C.

The thickness of the film depends on the applied voltage and is on the order of 1.6 nm/V. Film thickness is in the range of the wavelengths of visible light, and anodized tantalum has a characteristic interference color corresponding to the formation voltage. For example, a 70-V film produced at 85°C has a magenta color; a film formed to 100 V appears dark green.

Electronic Leakage. Many studies of anodic oxide films on tantalum have focused on the electrical properties. Any appreciable electronic leakage arising from a voltage applied across the oxide film degrades the quality of the dielectric and has a detrimental effect on the performance of tantalum capacitors. Considerable research has focused on work designed to explain the mechanism of electronic leakage through anodic oxide films on tantalum and similar systems. As of this writing (ca 1997), the pertinent literature remains contradictory and confusing.

Electronic leakage in anodic oxide films can be separated into intrinsic bulk conduction through the oxide and localized conduction at defect sites in the oxide film. Because the resistivity of stoichiometric anodic tantalum oxide is very high, on the order of 10^{15} Ω·cm, the bulk conduction electronic current density generated by applied voltages less than necessary for ionic conduction is very low and does not contribute significantly to poor dielectric performance. The localized electronic conduction associated with chemical impurities at the surface of the tantalum and structural flaws in the anodic oxide is usually orders of magnitude greater than the electronic conduction.

Many studies (50–56) have attempted to explain bulk conduction through anodic oxide films on tantalum foils or sputtered tantalum substrates. The results of several studies were interpreted by the Poole-Frenkel mechanism of field-assisted release of electrons from traps in the bulk of the oxide. In other studies, the Schottky mechanism of electron flow controlled by a thermionic emission over a field-lowered barrier at the counter electrode oxide interface was used to explain the conduction process. Some results suggested a space charge-limited conduction mechanism operates. The general lack of agreement between the results of various studies has been summarized (57).

Flaws in the anodic oxide film are usually the primary source of electronic conduction. These flaws are either structural or chemical in nature. The structural flaws include thermal crystalline oxide, nitrides, carbides, inclusion of foreign phases, and oxide recrystallized by an applied electric field. The roughness of the tantalum surface affects the electronic conduction and should be classified as a structural flaw (58); the correlation between electronic conduction and roughness, however, was not observed (59). Chemical impurities arise from metals alloyed with the tantalum, inclusions in the oxide of material from the formation electrolyte, and impurities on the surface of the tantalum substrate that are incorporated in the oxide during formation.

In one series of investigations (59,60) a redox printing technique was developed that allowed direct observation of the location of sites where electronic leakage through the oxide film occurred. The presence of Fe, Ni, or C on the surface of the tantalum was shown to cause an increase in electronic conduction. A related series of studies (61–63), showed that the presence of semiconducting oxides such as Fe_2O_3 or MnO_2 on the surface of the tantalum oxide caused the electronic conduction of the film to increase when a field slightly less than the anodization field was applied. Thus, a model for the system in which oxygen vacancies are generated in the tantalum oxide film at the interface with the semiconducting oxide was proposed. These oxygen vacancies provide a source of electrons which are released by the applied field, causing the conductivity of the tantalum oxide to increase.

Heat Treatment of Anodic Oxide Film on Tantalum.   One step in the manufacturing of solid tantalum capacitors is the deposition of the MnO_2 semiconductor by thermal decomposition of manganese nitrate (see SEMICONDUCTORS). This requires exposing the $Ta-Ta_2O_5$ system to a temperature in the range between 250 and 450°C for 15 to 30 min. This heat treatment can cause an increase in the frequency, temperature, and anodic bias dependence of the capacitance, as shown in Figure 7. The source of and explanation for these effects were the subject of a significant investigation (64–67). At temperatures above 300°C, the tantalum substrate can extract oxygen atoms from the anodic oxide film, leaving behind an exponential gradient of oxygen vacancies in the oxide. These vacancies introduce n-type semiconduction into the oxide and, if in sufficient concentration, will cause the film to behave as a conductor. The model is shown schematically in Figure 8. The critical conductivity, σ_0, separates the conducting from the insulating portion of the film. As the temperature increases, the level of conductivity across the film rises and the effective insulating layer becomes thinner, imparting an additional change in the capacitance besides that associated with the temperature dependence of the dielectric constant. Applying an anodic bias causes electrons trapped in the vacancies to move into the tantalum, increasing the slope of the conductivity gradient, and the thickness of the effective dielectric, thereby reducing the capacitance. The critical conductivity increases with frequency. Thus, as the circuit frequency goes up, σ_0 increases, and the inter-

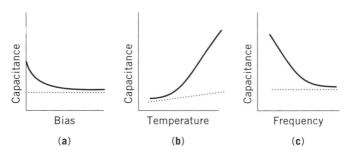

Fig. 7. Anodic oxide films on tantalum before ($\cdots$) and after (——) heat treatment. Dependences of capacitance on (**a**) bias, (**b**) temperature, and (**c**) frequency.

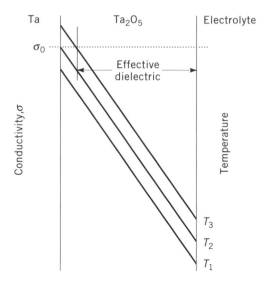

Fig. 8. Model of the conductivity profile in an anodic oxide film on tantalum after heat treatment, where $T_1 < T_2 < T_3$.

section between conducting and nonconducting oxide shifts toward the tantalum oxide interface, which results in a lowering of the capacitance. An understanding of the effect of heat treatment structure of anodic oxide films on tantalum led to changes in the process for manufacturing solid tantalum capacitors, such as lowering the manganese nitrate pyrolysis temperature, in order to avoid the creation of a conductivity profile in the finished capacitor.

Tantalum Compounds

Potassium Heptafluorotantalate. Potassium heptafluorotantalate [16924-00-8], K_2TaF_7, crystallizes in colorless, rhombic needles. It hydrolyzes in boiling water containing no excess of hydrofluoric acid. The solubility of potassium heptafluorotantalate in hydrofluoric acid decreases from 60 g/100 mL at 100°C to 0.5 g/100 mL at room temperature. The different solubility characteristics of K_2TaF_7 and K_2NbOF_5 are the fundamental basis of the Marignac process (16). A phase diagram exists for the system K_2TaF_7–NaCl–NaF–KCl (68). Potassium heptafluorotantalate has an LD_{50} value of 2500 mg/kg. The recommended TWA maximum work lace exposure for K_2TaF_7 in air is 2.5 mg/m^3 (fluoride base) (69).

Tantalum Oxides. Tantalum pentoxide [1314-61-0], Ta_2O_5, (mp = 1880°C, density = 8.73 g/cm^3) is a white powder existing in two thermodynamically stable modifications. The orthorombic β-phase changes at 1360°C into the tetragonal α-modification. The existence of an ϵ-modification has also been reported (70). Tantalum pentoxide reacts slowly with hot hydrofluoric acid but is insoluble in water and in most solutions of acids and alkalies. For analytical purposes, it can be dissolved by fusion with alkali hydroxides, alkali carbonates, and potassium pyrosulfate.

Tantalum(II) oxide [*12035-90-4*], TaO, is the only other oxide the existence of which has been confirmed. It can be prepared from Ta_2O_5 by reduction with carbon at 1900°C or with H_2 at 1100°C.

Tantalum Halides. *Fluorides.* Tantalum pentafluoride [*7783-71-3*], TaF_5, (mp = 96.8°C, bp = 229.5°C) is used in petrochemistry as an isomerization and alkalation catalyst. In addition, the fluoride can be utilized as a fluorination catalyst for the production of fluorinated hydrocarbons. The pentafluoride is produced by the direct fluorination of tantalum metal or by reacting anhydrous hydrogen fluoride with the corresponding pentoxide or oxychloride in the presence of a suitable dehydrating agent (71). The ability of TaF_5 to act as a fluoride ion acceptor in anhydrous HF has been used in the preparation of salts of the AsH_4, H_3S, and PH_4 ions (72). The oxyfluorides $TaOF_3$ [*20263-47-2*] and TaO_2F [*13597-27-8*] do not find any industrial application.

Chlorides. Tantalum pentachloride [*7721-01-9*], $TaCl_5$, (mp = 216°C, bp = 242°C) forms strongly hydroscopic, needle-shaped white crystals. It is produced on an industrial scale in large tonnage by the chlorination of tantalum scrap or ferrotantalum with $NaFeCl_4$ or by the reductive chlorination of natural ores or synthetic raw materials. High purity $TaCl_5$ is hydrolyzed with steam and thus converted to Ta_2O_5 of the highest purity (73). The reduction of $TaCl_5$ with Mg or Al was repeatedly attempted but this themite-type process never attained any industrial application. $TaCl_5$ is soluble in absolute alcohol, forming the corresponding alkoxides (74). The alkoxides as well as the chlorides are suitable products for the deposition of tantalum or tantalum pentoxide layers.

Other Halides. $TaBr_5$ [*13451-11-1*] and TaI_5 [*14693-81-3*] are well known but do not find industrial application. An excellent overview of the various halides, oxyhalides, and their reactions is available (75).

Tantalum Carbide. Tantalum monocarbide [*12070-06-3*], TaC, is a gold-colored powder produced industrially by direct reaction of carbon with either tantalum scrap or tantalum pentoxide at temperatures up to 1900°C. It is added in small amounts (0.2–2 wt %) in the form of TaC or mixed carbides like TaNbC and WTiTaC to tungsten carbide–cobalt-based cutting tools in order to reduce grain growth. Adding 2–15 wt % TaC to cemented carbides considerably increases their thermal shock resistance and their resistance to cratering and oxidation. The tantalum–carbon phase diagram shows the existence of several subcarbides (76).

Tantalum Nitrides. Tantalum nitride [*12033-62-4*], TaN, is produced by direct synthesis of the elements at 1100°C. Very pure TaN has been produced by spontaneous reaction of lithium amide, $LiNH_2$, and $TaCl_5$ (77). The compound is often added to cermets in 3–18 wt %. Ta_3N_5 [*12033-94-2*] is used as a red pigment in plastics and paints (78).

Tantalic Acid and Tantalates. Tantalic acid [*75397-94-3*], $Ta_2O_5 \cdot nH_2O$, is the name of the white insoluble precipitate formed by hydrolysis of alkali hydroxide or alkali carbonate fusions containing tantalum, or by adding ammonia to an acidic solution containing tantalum ions. Tantalic acid is characterized by a high surface acidity, affording it potential use as a catalyst.

Lithium tantalate [*12031-66-2*], $LiTaO_3$, is the most important tantalate. The crystal structure is related to that of perovskite. $LiTaO_3$ exhibits high spontaneous polarization and a high Curie temperature. For several applications in electronic devices such as surface acoustic wave (SAW) filters, second harmonic generators (SHG), and wave guides, crystals are grown by the conventional Czochralski technique from melts made of Li_2CO_3 and Ta_2O_5 (79). Similarly, other tantalates like $Ba_3MgTa_2O_9$ [*12231-81-1*] and $Ba_3ZnTa_2O_9$ [*12231-88-8*] also have the perovskite structure and are used in high frequency resonators in satellite communication systems.

Hazards of Tantalum Compounds. The toxicity of tantalum compounds depends on their solubility. Tantalum pentoxide is poorly absorbed and nontoxic perorally. The pentachloride, on the other hand, shows an LD_{50} of 985 mg/Kg administered perorally.

BIBLIOGRAPHY

"Tantalum and Tantalum Compounds" in *ECT* 1st ed., Vol. 13, pp. 600–613, by D. F. Taylor, Fansteel Metallurgical Corp.; in *ECT* 2nd ed., Vol. 19, pp. 630–652, by D. F. Taylor, Fansteel, Inc.; in *ECT* 3rd ed., Vol. 22, pp. 541–564, by R. E. Droegkamp, M. Schussler, J. B. Lambert, and D. F. Taylor, Fansteel, Inc.

1. A. G. Ekeberg, *Ann. Chem.* **43**, 276 (1802).
2. W. Von Bolton, *Z. Elektrochem.* **11**, 45 (1905).
3. J. J. Berzelius, *Ann. Phys. Lpy.* **4**, 10 (1825).
4. H. Rose, *Ann. Phys. Lpy.* **99**, 69 (1856).
5. D. R. Lide, *CRC Handbook of Chemistry and Physics*, 70th ed., Chemical Rubber Publishing Co., CRC Press, Inc., Boca Raton, Fla., p. B-138, 1989.
6. D. Brown, *Comprehension Inorganic Chemistry*, Vol. 5, Pergamon Press, Oxford, U.K., pp. 553–622, 1973.
7. F. Fairbrother, *The Chemistry of Niobium and Tantalum*, Elsevier Publishing Company, New York, p. 16, 1967.
8. G. L. Miller, *Tantalum and Niobium*, Academic Press, Inc., New York, Chapt. 8, 1959.
9. Ref. 6, Chapt. 7.
10. B. Elvers, S. Hawkins, and W. Russey, eds., *Ullmann's Encyclopedia of Industrial Chemistry*, Vol. A26, VCH Verlagsgesellschaft GmbH, Weinheim, Germany, p. 72, 1995.
11. *Columbium (Niobium) and Tantalum in 1994*, U.S. Bureau of Mines, Mineral Industry Survey, 1995.
12. R. Burt, *High Temp. Mater. Process.* **11**, 35 (1993).
13. U.S. Pat. 3,721,727 (Mar. 20, 1973), R. A. Gustison (to Kawecki Berylco Industries).
14. P. Moller, P. Cerny, and F. Saupe, *Lanthanides, Tantalum, and Niobium*, Springer Verlag, Berlin, p. 348, 1989.
15. J. C. De Marignac, *Ann. Chem. (Phys)*. **9**, 249 (1866).
16. J. C. De Marignac, *Arch. Sci. Phys. Nat.* **29**, 265 (1867).
17. C. W. Balke, *Ind. Eng. Chem.* **27**, 1166 (1935).
18. K. B. Higbie, J. R. Werning, *Separation of Tantalum–Columbium by Solvent Extraction*, Bulletin No. 5239, U.S. Bureau of Mines, Washington, D.C., 1956.
19. U.S. Pat. 2,962,372 (Nov. 29, 1960), R. A. Foos (to Union Carbide Corp.).
20. U.S. Pat. 3,117,883 (Jan. 14, 1964), J. A. Pierret (to Fansteel Metallurgical Corp.).
21. Ger. Pat. DE 4,021,207 (Sept. 2, 1993), W. Bludssus and J. Eckert (to H. C. Starck GmbH & Co. KG).
22. Ger. Pat. DE 4,404,374 (May 18, 1995), W. Bludssus and J. Eckert (to H. C. Starck GmbH & Co. KG).

23. Ger. Pat. DE 4,207,145 (Apr. 29, 1993), W. Bludssus and J. Eckert (to H. C. Starck GmbH & Co. KG).

24. J. M. Fletcher, F. C. Morris, and A. G. Wain, *Trans. Instn. Min. Metall. Lond.* **65**, 487 (1956).

25. Ger. Pat. DE 3,241,832 (June 19, 1987), J. Eckert and G. J. Bauer (to GFE Gesellschaft für Electrometallugie).

26. Ref. 10, Vol. A17, 1991.

27. W. Rockenbauer, in H. Stuart, ed., *Niobium–Proceedings of the International Symposium*, The Metallurgical Society of AIME, New York, 1984, pp. 133–153.

28. Ger. Pat. DE 2,551,028 (Sept. 11, 1986), E. Hogan and E. Enskep (to Mallinckrodt, Inc.).

29. Ger. Pat. 3,428,788 (July 10, 1986), G. J. Bauer and J. Eckert.

30. Ref. 8, Chapt. 5.

31. U.S. Pat. 2,950,185 (Aug. 23, 1960), E. G. Hellier and G. L. Martin (to National Research Corp.).

32. U.S. Pat. 4,684,399 (Aug. 4, 1987) R. M. Bergman and C. E. Moshein (to Cabot Corp.).

33. U.S. Pat. 4,231,790 (Nov. 4, 1980), R. Hahn and D. Behrens (to Hormann C. Starck Berlin).

34. U.S. Pat. 5,442,978 (Aug. 22, 1995), R. W. Hildreth and co-workers (to H. C. Starck, Inc.).

35. U.S. Pat. 4,645,533 (Feb. 24, 1987), T. Izumi (to Showa Cabot Supermetals KK).

36. U.S. Pat. 4,149,876 (Apr. 17, 1979), C. F. Rerat (to Fansteel Inc.).

37. K. Tierman and R. J. Millard, *1983 Proceedings of the Electronic Components Conference*, IEEE Cat. No. 83CH1904-2, New York, 1983, p. 157.

38. U.S. Pat. 4,537,641 (Aug. 27, 1985), W. W. Albrecht and co-workers (to Hermann C. Starck Berlin).

39. Ref. 8, Chapt. 6.

40. M. Stein and C. R. Bishop, in F. T. Sisco and E. Epremiam, eds., *Columbium and Tantalum*, John Wiley & Sons, Inc., New York, pp. 310–22, 1963.

41. H. Plank and S. Schider, in M. B. Berer, ed., *Encyclopedia of Material Sciences and Engineering*, Vol. 7, the MIT Press, Cambridge, Mass., pp. 48–9, 1986.

42. J. Black, *Clin. Mater.* **16**, 167 (1994).

43. D. F. Williams, ed., *Fundamental Aspects of Biocompatibility*, Vol. VI, CRC Press, Boca Raton, Fla., 1981, p. 3.

44. D. L. Guyton and F. T. Hambrecht, *Med. Biol. Eng. Comput.* **12**, 613 (1974).

45. B. R. Dworkin and S. Dworkin, *Behav. Neurosci.* **109**, 1119 (1995).

46. M. Schussler and C. Pokross, *Corrosion Data Survey On Tantalum*, 2nd ed., Fansteel Inc., North Chicago, Ill., 1985.

47. N. I. Sax, *Dangerous Properties of Industrial Materials*, 6th ed., Van Nostrand Reinhold Co., New York, 1984, p. 2498.

48. *NIOSH Pocket Guide to Chemical Hazards*, DHHS (NIOSH) Publication No. 94-116, U.S. Government Printing Office, Washington, D.C., 1994.

49. B. Melody and B. Chavey, *Proceedings of the 13th Capacitor and Resistor Technology Symposium*, Components Technology Institute, Inc., Huntsville, Ala., 1992, p. 40.

50. J. G. Simmons, *Phys. Rev.* **155**, 657 (1967).

51. F. C. Aris and T. J. Lewis, *J. Phys. D: Phys.* **6**, 1067 (1973).

52. V. Trifonova and A. Girginow, *J. Electroanal. Chem.* **107**, 105 (1980).

53. F. C. Aris and T. J. Lewis, *Proc. Brit. Ceram. Soc.* **23**, 115 (1972).

54. M. W. Jones and D. M. Hughes, *J. Phys. D: Appl. Phys.* **7**, 112 (1974).

55. A. K. Jonscher, *J. Electrochem. Soc.* **116**, 217C (1969).

56. A. K. Jonscher, *Thin Solids Films* **1**, 213 (1967).

57. C. J. Dell'Oca, D. L. Pulfrey, and L. Young, *Phys. Thin Films* **6**, 1 (1971).

58. D. A. Vermilyea, *J. Electrochem. Soc.* **110**, 250 (1963).
59. G. P. Klein, *J. Electrochem. Soc.* **113**, 348 (1966).
60. G. P. Klein, *J. Electrochem. Soc.* **117**, 1483 (1970).
61. G. P. Klein and N. I. Jaeger, *J. Electrochem. Soc.* **117**, 1483 (1970).
62. N. I. Jaeger, G. P. Klein, and B. Myrvaagness, *J. Electrochem. Soc.* **119**, 1531 (1972).
63. G. P. Klein and N. I. Jaeger, *Thin Solid Films* **43**, 103 (1977).
64. D. M. Smyth, G. A. Shirn, and T. B. Tripp, *J. Electrochem. Soc.* **110**, 1264 (1963).
65. D. M. Smyth and T. B. Tripp, *J. Electrochem. Soc.* **110**, 1271 (1963).
66. D. M. Smyth, G. A. Shirn, and T. B. Tripp, *J. Electrochem. Soc.* **111**, 1331 (1964).
67. D. M. Smyth, T. B. Tripp, and G. A. Shirn, *J. Electrochem. Soc.* **113**, 100 (1966).
68. V. E. Kartsev, F. V. Kovalev, B. G. Korshunov, *Zhur. Neorg. Khim.* **20**, 2204 (1975).
69. Ref. 10, p. 81.
70. F. Iyumi and H. Kodama, *J. Less-Common Met.* **63**, 305 (1979).
71. U.S. Pat. 5,091,168 (1992), J. Nappa.
72. R. Gut, *Inorg. Nucl. Chem. Lett.* **12**, 149 (1976).
73. Ref. 26, p. 254.
74. N. Sato and M. Nanjo, *High Temp. Mater. Proc.* **8**, 39 (1988).
75. S. A. Cotton and F. A. Hart, *The Heavy Transition Elements*, Halstead Press, a division of John Wiley & Sons, Inc., New York, 1975, pp. 16–22.
76. Ref. 10, Vol. A5, 1986.
77. P. I. Parkin and T. A. Rowley, *Adv. Mater.* **6**, 780 (1994).
78. Ger. Pat. DE 4,234,939 (Apr. 21, 1994), M. Jansen, H. P. Letschert, and D. Speer (to Cerdec AG).
79. S. Kimura, K. Kitamura, *J. Ceram. Soc. Jpn. Int. Ed.* **101**, 21 (1993).

TERRANCE B. TRIPP
J. ECKERT
H. C. Starck Inc.

TAR AND PITCH

Most organic substances, other than those of simple structure and low boiling point, when pyrolyzed, ie, heated in the absence of air, yield dark-colored, generally viscous liquids termed tar or pitch. The differentiation between these terms is not precise. When the by-product is a liquid of fairly low viscosity at ordinary temperature, it is regarded as a tar; if of very viscous, semisolid, or solid consistency, it is designated as a pitch. Thus, in preparative organic chemistry, a tar or pitch is frequently the distillation residue. Some thermal decomposition always accompanies the distillation of vegetable and animal fats and oils, resulting in the production of small amounts of vegetable-oil pitch, wool-grease pitch, stearin pitch, and so on. Such products, of poorly defined composition and properties, have little industrial importance and are mainly burned as fuel.

Large amounts of tar or pitch by-products are produced by industrial processes. The distillation of crude petroleum (qv) yields a pitch-like residue termed bitumen or asphalt (qv). In the United States, these terms are interchangeable, but in Europe the term asphalt is generally restricted to naturally occurring rock or lake asphalt, whereas the residual product of crude-oil distillation is termed bitumen. Although these are important industrial materials produced in millions of metric tons annually, they are not included herein (see ASPHALT; PETROLEUM, PRODUCTS).

With regard to coal-derived tar and pitch, the following definitions are appropriate to distinguish them from ostensibly similar materials from other sources and from crude oil in particular.

Coal tar is the condensation product obtained by cooling to approximately ambient temperature, the gas evolved in the destructive distillation of coal. It is a black viscous liquid denser than water and composed primarily of a complex mixture of condensed ring aromatic hydrocarbons. It may contain phenolic compounds, aromatic nitrogen bases and their alkyl derivatives, and paraffinic and olefinic hydrocarbons. Coal-tar pitch is the residue from the distillation of coal tar. It is a black solid having a softening point of 30–180°C (86–359°F).

In the processing of crude oil eventually bitumen or vacuum residues arise, ie, asphalt, obtained as a nonvolatile residue from the distillation of crude oil and contains a high proportion of hydrocarbons having carbon numbers predominantly greater than C_{25} and high carbon-to-hydrogen ratios; and residues petroleum vacuum, a complex residue from the vacuum distillation of the residuum from atmospheric distillation of crude oil, consisting of hydrocarbons having carbon numbers predominantly greater than C_{34} and boiling point of approximately 495°C (923°F). However, to complicate matters, Ashland (United States) also markets a material known as petroleum pitch, although technically it should be called bitumen.

Wood Tar

The pyrolysis or carbonization of hardwoods, eg, beech, birch, or ash, in the manufacture of charcoal yields, in addition to gaseous and lighter liquid products, a by-product tar in ca 10 wt % yield. Dry distillation of softwoods, eg, pine species, for the production of the so-called DD (destructively distilled) turpentine yields pine tar as a by-product in about the same amount. Pine tar, also called Stockholm tar or Archangel tar, was at one time imported from the Baltic by European maritime countries for the treatment of cordage and ship hulls; it was an important article of commerce from the seventeenth to the nineteenth century. The small amount produced in the late twentieth century is burned as a crude fuel. Charcoal production from hardwoods, on the other hand, has increased in the 1990s years.

Composition, Processing, and Uses.   There are no statistics available for the amount of wood tar processed, but almost all of it is burned. The commercial by-products from wood carbonization are limited to methanol, denatured methanol, methyl acetate, and acetic acid. These products are derived from the aqueous phase of the condensed products, the so-called pyroligneous acid. On distillation, pyroligneous acid yields wood spirit, acetic acid having small amounts of propionic and butyric acids, and soluble tar. The wood spirit, on refining, yields methanol as well as methyl acetate and acetone. The aqueous acid fraction is neutralized with milk of lime to give gray acetate of lime, 82–84% $(CH_3COO)_2Ca$, which is neutralized to give acetic acid or pyrolyzed to give acetone. The soluble tars are mainly condensation products of aldehydes and phenols; they are burned as fuel.

Small amounts of the sedimentation tar, ie, the separated organic layer from the condensed wood-carbonization vapors, are distilled, first at atmospheric pressure to give wood spirit, crude acetic acid, and light wood oils. The first two of these fractions are added to similar fractions from the pyroligneous acid for further processing. The light wood oil is treated with permanganate and sulfuric acid and used as a solvent or for giving a wood smoke or tar note to perfumes, soaps, and shampoos. Further distillation under reduced pressure yields wood creosote, which is used for the preservation of cordage, timber, and Hessian sacks, and as a disinfectant, component of wood stains, and froth-flotation agent.

Chemically, wood tar is a complex mixture that contains at least 200 individual compounds, among which the following have been isolated (1): 2-methoxyphenol, 2-methoxy-4-ethylphenol, 5-methyl-2-methoxyphenol, 2,6-xylenol, butyric acid, crotonic acid, 1-hydroxy-2-propanone, butyrolactone, 2-methyl-3-hydroxy-4*H*-pyran-4-one, 2-methyl-2-propenal, methyl ethyl ketone, methyl isopropyl ketone, methyl furyl ketone, and 2-hydroxy-3-methyl-2-cyclopenten-1-one.

Coal Tar

By far the largest source of tar and pitch is the pyrolysis or carbonization of coal (qv). Generally, the terms tar and pitch are synonymous with coal tar and the residue obtained by its distillation (see COAL-CONVERSION PROCESSES, CARBONIZATION). The importance of coal tar as an industrial raw material dates back to the first half of the eighteenth century, when the carbonization of coal and the by-product production of tar were expanding rapidly in the United Kingdom. Initially, the crude tar was subjected to a simple flash distillation in pot stills to yield a solvent naphtha, creosote for timber preservation, and a residue of pitch that found an outlet as a binder for coal briquettes. Later, coal tar was the main source of aromatic hydrocarbons, phenols, and pyridine bases needed by the rapidly expanding dyestuffs, pharmaceuticals, and explosives industries. The development of by-product coke ovens and crude-benzene recovery at both coke ovens and gasworks greatly increased the supplies of crude tar and tar distillates for the recovery of tar chemicals, ie, benzene, toluene, xylenes, phenol, cresols and cresylic acids, pyridine and methylpyridines, and naphthalene and anthracene, in addition to the so-called bulk products, eg, creosote, tar paints, road tars, and pitch binders (2–4).

Until the end of World War II, coal tar was the main source of these aromatic chemicals. However, the enormously increased demands by the rapidly expanding plastics and synthetic-fiber industries have greatly outstripped the potential supply from coal carbonization. This situation was exacerbated by the cessation of the manufacture in Europe of town gas from coal in the early 1970s, a process carried out preponderantly in the continuous vertical retorts (CVRs), which has led to production from petroleum. Over 90% of the world production of aromatic chemicals in the 1990s is derived from the petrochemical industry, whereas coal tar is chiefly a source of anticorrosion coatings, wood preservatives, feedstocks for carbon-black manufacture, and binders for road surfacings and electrodes.

Apart from the presence of a few percent (usually below 5%) of aqueous liquor containing inorganic salts and a percent or so of coal-char-coke dust arising from carryover of particles in the carbonization process, coal-tar consists essentially of two parts. The first, which at atmospheric pressure distills up to about 400°C, is primarily a complex mixture of mono- and polycyclic aromatic hydrocarbons, a proportion of which are substituted with alkyl, hydroxyl as well as amine and/or hydro sulfide groups, and to a lesser extent their sulfur-, nitrogen-, and oxygen-containing analogues and, for those tars produced at the lower coal carbonization temperatures, they contain in addition hydroaromatics, alkanes, and alkenes. The second part is the residue from the distillation, amounting to at least 50% of the coal-tar products by high temperature carbonization and consisting of a continuation of the sequence of polynuclear aromatic, aromatic, and heterocyclic compounds, but reaching molecules containing 20 to 30 rings.

Manufacture and Processing.   The largest volume of coal is carbonized in batch coke ovens to produce a hard coke suitable for blast furnaces for the reduction of iron ore. Oven temperatures, as measured in the flues, are between 1250 and 1350° and residence time varies between 17 and 30 h. The gas made in this process is mainly used as fuel and other applications in the steel works (see FUELS, SYNTHETIC).

Until 1960–1970, in countries where natural gas was not available, large amounts of coal were carbonized for the production of town gas, as well as a grade of coke which, although unsuitable for metallurgical use, was satisfactory as a domestic fuel in closed stoves. The early cast-iron and silica horizontal retorts used at gasworks were replaced by continuous vertical retorts. These operated at flue temperatures of 1000–1100°C. The volatile products were rapidly swept from the retort by the introduction of steam at 10–20% by weight of the coal carbonized.

The passing of the Clean Air Act in the United Kingdom in 1956 resulted in a revival of interest in low temperature carbonization to produce a very reactive coke suitable for open fires. In the Coalite process, the coal is heated at 600–650°C for 4 h in small retorts each holding 6–7 metric tons (5). The Rexco process employed large internally heated retorts in which charges of 34 metric tons were heated to 700–750°C for 6 h, but is no longer in operation in the United Kingdom (6).

In the future, crude, low temperature tar may be supplied as by-products of synthetic natural gas (SNG) and syncrude from coal substitute. A number of the more advanced SNG processes such as Lurgi, Bi-Gas, and Cogas employ low temperature pyrolysis of coal and yield a by-product tar. The tar obtained from the Lurgi process appears to be similar to Coalite tar. Lurgi gasifiers were chosen for five large U.S. plants for the conversion of western lignite and subbituminous coals to SGN. In full operation, these plants were expected to produce 2×10^6 metric tones of by-product tar per year. Lurgi gasifiers are also employed to produce synthesis gas for South Africa's three Fischer-Tropsch plants, which yield ca 250,000 metric tons of tar.

Primary Distillation. As produced, crude coal tar is of value only as a fuel. Although formerly large amounts were burned, the practice has largely been abandoned. In the 1990s, 99% of the tar produced in the United Kingdom and Germany and 75% of U.S. production is distilled. Most of the crude tar regarded as being burned in the United States is first topped in simple continuous stills to recover a chemical oil, ie, a fraction distilling to 235°C that contains the bulk of naphthalene and phenols.

Although 10–30-t mild-steel or wrought-iron pot stills, equipped with fractionating columns, are still in use at one tar works in Spain, continuous stills that have daily capacities of 100–700 t are preferred and used exclusively in the rest of the world.

The various designs of continuous tar stills are basically similar. The crude tar is filtered to remove large-sized solid particles, dehydrated by heat exchange and passage through a waste heat coil, then heated under pressure to ca 360°C, and flashed to separate volatile oils from the involatile pitch. The volatile oils are separated into a series of fractions of increasing boiling range by fractional condensation in a side-stream column or a series of columns. The various designs differ in the extent to which heat exchange is used, in the plan of the pipe-still furnace, in the distillation pressure, and in whether recycle of pitch or base tar is involved.

The Abderhalden design, employed at some French distilleries and used in the United Kingdom for distilling tar to a base-tar residue, is probably the simplest single-pass, atmospheric-pressure design (7). After straining and the addition of alkali as a corrosion inhibitor, the crude tar is heat-exchanged with the distillation side-streams and the hot pitch and then passed through an economizer coil in the convection section of the furnace. This furnace is essentially a rectangular chamber lined with refractory brick divided into two sections by a curtain wall, which has apertures to enable the hot flue gases to pass from one compartment to the other. The first section serves as the combustion and radiant-heating section in which coke-oven gas, fuel oil, or creosote-pitch fuel is burned at specially designed nozzles projecting into the chamber. The products of combustion are drawn by a fan from the combustion chamber into the second, or convection, chamber before exhaustion to a chimney stack via a waste-heat coil. The coils through which the tar is pumped are, in some designs, set in the walls of the convection chamber only. In other designs, the main heating coil is partly set in the combustion chamber where it is heated by direct radiation.

In the Abderhalden plant, the crude tar leaves the waste-heat coil at ca 150°C under a back pressure and is expanded into the dehydrator. The temperature in this cylindrical baffled vessel is maintained by circulating hot, dehydrated tar through a coil at the base. The moisture and light oils flash off and pass directly to the side-stream fractionating column. The dehydrated tar from the base of the dehydrator is pumped at 400–500 kPa (ca 4–5 amt) through the main heating coil in the radiant section of the furnace. The hot tar at 350–360°C is injected into the pitch flash chamber together with superheated steam. Steam and oil vapors join the dehydrator overhead distillate at the base of the fractionation column, from which an overhead stream of water and light oils, side-stream oil fractions, and a heavy oil residue are taken. The overhead stream is separated in a decantor and the oil layer partly returned to the bubble-cap fractionating column for reflux. A similar design differs only in the extent of heat exchange (8).

Several descriptions have been published of the continuous tar stills used in the CIS (9–11). These appear to be of the single-pass, atmospheric-pressure type, but are noteworthy in three respects: the stills do not employ heat exchange and they incorporate a column having a bubble-cap fractionating section and a baffled enrichment section instead of the simple baffled-pitch flash chamber used in other designs. Both this column and the fractionation column, from which light oil and water overhead distillates, carbolic and naphthalene oil side streams, and a wash oil-base product are taken, are equipped with reboilers.

The original Koppers design, still in use at some older plants in Europe (12), is a single-pass, atmospheric-pressure design, but differs from other designs by employing separate bubble-tray columns for each oil fraction. After straining, alkali doping, and dehydration, the tar is pumped through the radiantly heated main furnace coil under pressure and, at 360°C, injected into the baffled-pitch column where the pitch is separated from the oil vapors. Steam is added to the pitch column to assist vaporization. These vapors then pass through four columns operated at atmospheric pressure and progressively lower temperature to yield the various oil fractions as base products. The temperature at the top of each column is automatically controlled by recycling part of the bottoms from each column as reflux to the preceding column. The overhead distillate from the final column consists of the benzole fraction and steam. It is condensed, the two layers are separated in a decantor, and part of the benzole fraction product is recirculated as reflux.

The simplest unit employing vacuum fractionation is that designed by Canadian Badger for Dominion Tar and Chemical Company (now Rütgers VFT Inc.) at Hamilton, Ontario (13). In this plant, the tar is dehydrated in the usual manner by heat exchange and injection into a dehydrator. The dry tar is then heated under pressure in an oil-fired helical-tube heater and injected directly into the vacuum fractionating column from which a benzole fraction, overhead fraction, various oil fractions as side streams, and a pitch base product are taken. Some alterations were made to the plant in 1991, which allows some pitch properties to be controlled because pitch is the only product; the distillate oils are used as fuel.

A unique design is employed in the Clairton refinery of the Aristech Chemical Corporation (formerly United States Steel Corporation), where a sequence of fractionation stages is operated at increasing temperature (14). Each stage consists of a helical-coil heater, a vacuum flash drum, and a vacuum fractionating column. Dehydration and the removal of light oil are carried out at atmospheric pressure. The topped tar is pumped to the first stage which yields a phenolic oil overhead. The bottoms pass to the second stage which yields a close cut naphthalene oil as the overhead product. The third stage separates residual oils from the pitch, and in the fourth the residual oils are separated into fractions required for various grades of creosote.

The modern Teerverwertung-Koppers design, based on the experience at the Rütgerswerke refinery at Meiderich (which was closed in 1994), was a single-pass atmospheric- and reduced-pressure unit (Fig. 1) (15). The crude tar was

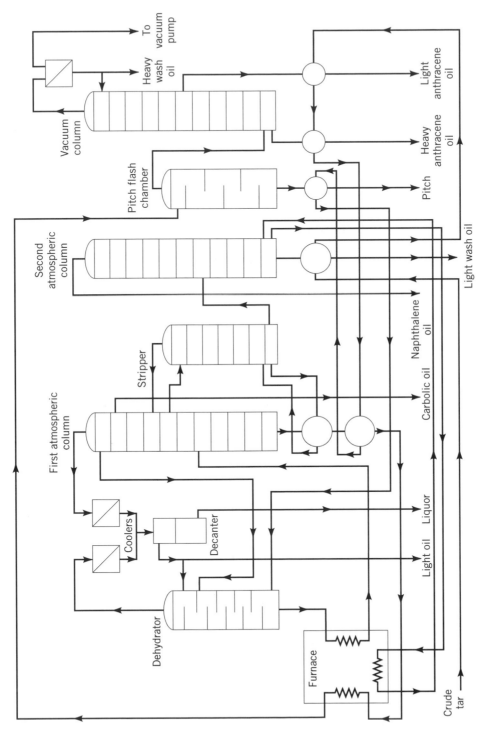

Fig. 1. Teerverwertung-Koppers tar distillation plant.

pumped through five heat exchangers to the dehydrator. The benzole fraction and steam from the top of this unit were cooled, condensed, and separated, and some of the oil recycled as reflux. The dehydrated tar was pumped through a heating coil in the tube furnace and, at 300–310°C, injected into the atmospheric-pressure fractionation stage, which consisted of two fractionating columns and a splitting column. The products taken from this first stage include residual light oils and water, which were taken overhead from the first column and passed back to the dehydrator. Naphthalene oil was taken as overhead distillate from the second column, whereas wash oil and methylnaphthalene oil were taken as the base product from the second column. The base tar from the base of the first column was picked up by a pump and passed through a separate coil in the furnace and, at 300°C, injected into the pitch flash chamber which, like the final fractionating column, was maintained under a vacuum of 13.3 kPa (100 mm Hg). The residual oil vapors, separated from the pitch residue, were split into a heavy wash oil and anthracene oils 1 and 2 in the vacuum column.

The former Rütgerswerke (now VFT, AG) plant at Castrop-Rauxel produces closely fractionated concentrates of a number of polynuclear hydrocarbons that are further separated and purified to provide the main source of these chemicals for western Europe (16).

The crude coke-oven tar is dehydrated by heat exchange with the dehydrator overhead distillate and passage through a low pressure steam heater. The dehydrated tar is heat-exchanged with the carbolic-oil stream and the pitch and enters the midpoint of the carbolic-oil column at 250°C. This column operates at atmospheric pressure and contains 40 bubble-cap trays; the reflux ratio is 16:1. Carbolic oil is taken overhead and naphthalene oil is taken as a side stream. The latter is upgraded in a 20-tray splitting column to give a fraction containing 84–88% naphthalene, equivalent to 95% of the naphthalene in the crude tar. The bottoms from the carbolic-oil column passes next to a vacuum fractionating stage consisting of a 40-tray column and a 20-tray splitting column to yield a methylnaphthalene concentrate, a light wash oil, and a fraction rich in fluorene and acenaphthene. The residual base tar is heated to 280°C and pumped to the anthracene column which is maintained at 9.3 kPa (70 mm Hg). Anthracene oil and heavy oil are separated from pitch and the overheads are further fractionated in a separate vacuum column into an anthracene–phenanthrene concentrate, a fraction containing 95% carbazole, and a heavy oil from which pyrene, fluoranthene, and chrysene can be isolated. In addition, separate base products from the splitting columns are reboiled by heat from the bottoms products of the main columns.

The most widely used design of continuous coal-tar is probably the Wilton design supplied by Bitwater Industrial Process Plant of Heywood (formerly Chemical and Thermal Engineering, Ltd.) in United Kingdom. Plants were installed in several refineries in the United Kingdom, Australia, New Zealand, India, Spain, Argentina, and Korea. The popularity of this design, which operates on the recycle principle, is the result of good thermal efficiency and flexibility in handling any type of tar over a wide range of throughputs (Fig. 2). The plant contains a helical-coil furnace, a dehydration column, two valve-tray fractionating columns, and a pitch column divided into an upper stripping section and a lower mixing section. The crude tar is pumped through coarse filters and, after

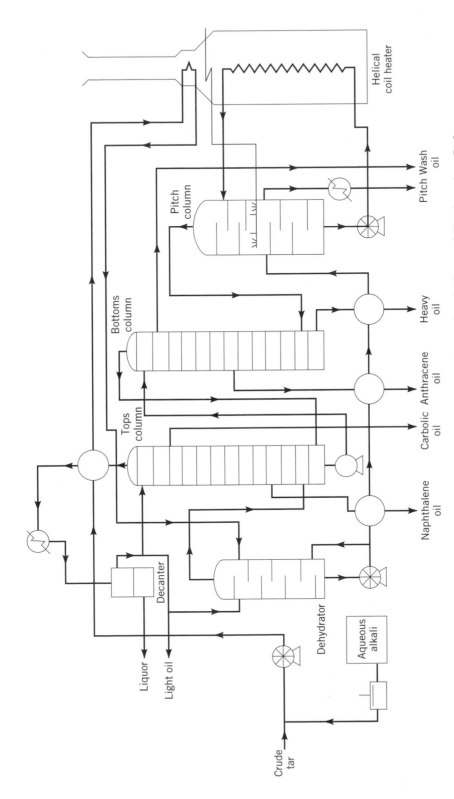

Fig. 2. The Wilton tar distillation plant. Courtesy of Chemical and Thermal Engineering, Ltd.

the addition of alkali solution, passed via a heat exchanger to the waste-heat coil in the convection part of the furnace. The heated tar, under a moderate back pressure, enters the baffled dehydration column where the entrained liquor and light oils vaporize, leaving the dehydrated tar which is pumped to the bottom or mixing section of the pitch column. Here, it is mixed with a greater volume of hot pitch overflowing from the stripping section. This pitch-dehydrated tar mixture is pumped through the radiantly heated main furnace coil, which is backpressured to prevent vaporization.

At 350–360°C, the mixture leaves the furnace and enters the upper part of the pitch column where the distillate oil vapors flash off, leaving the pitch residue. Vaporization is assisted by injecting superheated steam at the base of the stripping section. A level of some pitch is maintained at this point, where the main pitch product is drawn off. The remainder of the pitch overflows into the mixing section from which it is recycled together with the dehydrated tar. The total vapors from the pitch column, including those produced in the mixing section by the contact of hot pitch with the dehydrated tar, pass to the base of the bottoms fractionating column. Here, wash oil and anthracene oil are taken as side streams and a heavy oil is taken as base product. Uncondensed vapors from this column, together with the vapors from the dehydrator, are fed to the base of the tops column from which an overhead benzole fraction–water fraction, carbolic oil, and naphthalene-oil side-streams are taken, together with a base product which is returned to the bottoms column as reflux. The benzole fraction–water fraction is condensed and separated in a decantor: part of the light oil serves as reflux to the tops column. In South Africa the Wilton plant was decommissioned in the 1980s and replaced by a Koppers-design continuous distillation unit.

There are two serious problems associated with continuous tar distillation. Coal tar contains two types of components highly corrosive to ferrous metals. The ammonium salts, mainly ammonium chloride, associated with the entrained liquor remain in the tar after dehydration, tend to dissociate with the production of hydrochloric acid and cause rapid deterioration of any part of the plant in which these vapors and steam are present above 240°C. Condensers on the dehydration column and fractionation columns are also attacked. This form of corrosion is controlled by the addition of alkali (10% sodium carbonate solution or 40% caustic soda) to the crude tar in an amount equivalent to the fixed ammonia content.

The higher boiling phenols, present in considerable amounts in CVR and low temperature tars, are corrosive to mild steel, especially above 300°C. Cast iron, chrome steel, and stainless steel are more resistant. Furnace tubes, the insides of fractionating columns, and the rotors of pumps handling hot pitch and base tar are generally constructed of these metals. Nevertheless, to ensure satisfactory furnace tube life, particularly in plants processing CVR or low temperature tars, the tube temperature should be kept to a minimum.

A more important reason for operating furnaces at the lowest possible temperature is the fact that coal tar is thermally unstable. If heated above a certain temperature, coal tar decomposes, forming coke that rapidly blocks the furnace tubes. The critical temperature for coke formation is lower for vertical retort and low temperature tars than for coke-oven tars. To guard against overheating, different designs offer various options. In the Teerverwertung-Koppers design, in which vacuum is employed in the second stage, the tar need never be heated above 300–310°C, which is well below the critical temperature. If vaporization is allowed to proceed to any considerable extent in the furnace tubes combined with a high heat flux, tube wall temperatures might rise above the decomposition threshold at the zone where the tube contents change from foam to mist. Most plants operate the furnace tubes under a back pressure, which may be as high as 1 MPa (ca 10 atm) to suppress vaporization. The recycle of pitch in the Wilton design also minimizes vaporization and, by increasing the velocity through the tubes, improves the heat-transfer rate.

The main objectives in the primary distillation of crude tar are to obtain a pitch or refined tar residue of the desired softening point, to concentrate in certain fractions components that are subsequently to be recovered, and to yield distillate oils that are suitable for blending. A tar-distilling plant is designed to yield fractions that meet the specifications set by the purchaser, which depend on the product sales pattern. This pattern differs for different companies, and the result is that each plant is, to some extent, tailor-made and unique in its details and performance. Although at some refineries as many as seven distillate oil fractions are taken, in most plants the number of primary fractions is smaller. The situation is complicated by the fact that different names are given to identical fractions and, conversely, the same names are given by different distillers to fractions of different boiling ranges.

A typical primary distillation product pattern at a coke-oven tar-processing plant is given in Table 1. At some coke-oven distilleries, only one fraction, designated naphthalene oil, is taken between 180 and 240°C. Two fractions, light creosote or middle oil (230–300°C) and heavy creosote or heavy oil (above 300°C), are taken between the naphthalene oil and pitch.

When using a continuous vertical retort, phenols, cresols, and xylenols are collected in one fraction. A typical range of primary distillation fractions is given in Table 2.

The only valuable components in low temperature tar are the phenols and an oil fraction distilled over the range of 180–310°C, which is collected for tar-acid recovery is taken. A typical primary distillation is given in Table 3.

Table 1. Primary Coke Oven Tar Distillation

Product	CAS Registry Number	Distillation range, °C	Dry tar, %
distillates coal tar benzole fraction	[84650-02-2]	80–160	0.5–1.0
light oils or carbolic oil	[84650-03-3]	150–210	3–4
naphthalene oil	[84650-04-4]	200–250	10–12
wash oil or methyl-naphthalene oil	[90640-84-9]	240–280	6–8
anthracene oil 1	[90640-80-5]	300–400	20–25
anthracene oil 2	[90640-86-1]	300–400	20–25
pitch coal-tar[a]	[65996-93-2]		53–58

[a]High temperature.

Table 2. Primary CVR Tar Distillation

Product	Distillation range, °C	Dry tar, %
light oil	90–170	2–4
carbolic oil	180–240	12
light creosote oil	230–300	14
heavy creosote oil	275–360	11.5
residual oil	300–395	12.1
medium-soft pitch residue		44–46

Table 3. Primary Low Temperature Tar Distillation

Product	Distillation range, °C	Dry tar, %
light oil[a]	up to 180	3.5
middle oil	180–310	29
wax oil[b]	300–400	37.5
medium-soft pitch residue		26

[a]Also called fresh oil.
[b]Solidifies at 20°C because of high paraffin wax content.

Secondary Processing of Tar Distillate Oils. *Benzole (light oil) fraction, Refining of Benzene and Naphtha.* The only processing that light oils might receive at the refinery is a fractional distillation into crude benzene (formerly called benzol or benzole) distilling up to 150°C, a naphtha fraction distilling from 150 to 190°C, and a creosote residue (see BTX PROCESSING; BENZENE). Crude benzene from coke-oven tar is normally refined together with crude benzene separated from coke-oven gas. The naphtha is washed with alkali and acid to remove tar acids and bases, then treated with a small amount of sulfuric acid to remove sulfur compounds and olefins, and finally redistilled to give refined solvent naphtha. Alternatively, a special fraction rich in indene can be collected from which indene–coumarone resins are produced.

In the refining of the combined crude benzene, a defronting steam-stripping operation removes the lower boiling components. The defronted benzene may be fractionated in batch or continuous stills to yield a mixture of crude benzene, toluene, and mixed xylenes, and a naphtha residue. Each fraction is purified to meet grade specifications. Alternatively, the crude benzene, after defronting, may be refined and the refined product fractionated. In the refining of crude benzene or its distillate fractions, nonaromatic hydrocarbons and thiophene derivatives are removed. Formerly the oil was usually mixed with 2–4%-concentration sulfuric acid for a short time, washed, and separated from the spent acid. This treatment is followed by a dilute alkali and a water wash, and the product is refractionated. In the 1990s, however, a number of refineries in the United States, South Africa, and Europe employ hydrorefining, in which the vaporized hydrocarbon feedstock and hydrogen are passed over a cobalt molybdate-on-alumina catalyst at 400°C and 3550 kPa (500 psig). The olefins are converted to paraffins or cycloparaffins, and thiophene compounds to hydrogen sulfide. After washing with alkali to remove H_2S, small amounts of nonaromatic hydrocarbons are removed by the Udex process, the Shell Sulfolane process, or the Nmethylpyrrolidone extraction process of Metallgessellschaft AG (see EXTRACTION, LIQUID–LIQUID EXTRACTION).

Gasworks crude benzene and the lower boiling liquid products from low temperature carbonization, so-called fresh oil or low temperature spirit, are too low in aromatic hydrocarbon content to be refined economically to synthesis-grade benzene and toluene. Formerly, these fractions were used as a gasoline additive; in the 1990s, the lower boiling liquid products are used as fuel at the refinery.

Coumarone–Indene Resins. These should be called polyindene resins (17) (see HYDROCARBON RESINS). They are derived from a close-cut fraction of a coke-oven naphtha free of tar acids and bases. This feedstock, distilling between 178 and 190°C and containing a minimum of 30% indene, is warmed to 35°C and polymerized by adding 0.7–0.8% of the phenol or acetic acid complex of boron trifluoride as catalyst. With the phenol complex, tar acids need not be completely removed and the yield is better. The reaction is exothermic and the temperature is kept below 120°C. When the reaction is complete, the catalyst is decomposed by using a hot concentrated solution of sodium carbonate. Unreacted naphtha is removed, first with live steam and then by vacuum distillation to leave an amber-colored resin. It is poured into trays, allowed to cool, and broken up for sale.

Pyridine Bases. Formerly, pyridine bases were recovered from coal-tar light oils (18), but in more recent years synthetic pyridine and methylpyridine have mostly replaced the coal-tar products.

First, the tar acids were removed from the naphtha fractions of light oils and, in the case of CVR tars, carbolic oil. The oils were then mixed with 25–35% sulfuric acid. After separation of the sulfates, the aqueous solution was diluted with water and the resinous material skimmed off. The diluted sulfate solution was boiled to expel any neutral oils, dried by the addition of solid caustic soda or azeotropically with benzene, and fractionated to yield pyridine, 2-methylpyridine (α-picoline), and a fraction referred to as 90/140 bases, which consisted mainly of 3- and 4-methylpyridines and 2,6-dimethylpyridine (2,6-lutidine). Higher boiling fractions were termed 90/160 and 90/180 bases because 90% of the product distilled at 160 and 180°C, respectively.

Carbolic Oils and Low Temperature Tar Middle Oil, Tar Acids. The fractions of some coke-oven tars, distilling in the range of 180–240°C, and the middle oil fraction (180–310°C) from low temperature tars are treated for the recovery of tar acids (19).

The oils are mixed with a slight excess of 10% aqueous caustic soda in stirred vessels, continuous extraction columns, or, at some refineries, by circulating the contents of the mixing vessel by a gear pump. The extraction is carried out at ambient temperature or just above the crystallizing point of the oil. The extraction is best carried out in two stages, using 90% of the alkali to contact the fresh oil and the remainder to complete the removal of the phenols. The crude phenolate or cresolate solution is separated, and contains some neutral and basic material which must be removed by extracting with phenol-free light oil or crude benzole and/or treating with live steam.

The purified sodium phenolate solution is then decomposed by passing it down ring-packed or coke-filled towers at 80–85°C countercurrent to a gas containing 25–30% CO_2, generated in lime-coke kilns, or a flue gas containing

10–15% CO_2. This operation is called springing. Because the rate of CO_2 absorption depends on the CO_2 partial pressure, the richer gas reduces the number and size of the springing towers required. An upper layer of crude wet tar acids and a lower layer of sodium carbonate solution is obtained. The crude wet tar acids contain ca 20–25% of sodium carbonate solution. To reduce their moisture content to ca 10%, the crude wet tar acids are passed down an after-carbonation tower countercurrent to a stream of the CO_2-containing gas. Separation of the layers, into which the effluent from the various carbonation tower divides, yields crude tar acids containing ca 10% liquor and a solution of sodium carbonate and bicarbonate.

Caustic soda is removed from the carbonate–bicarbonate solution by treating with a slight excess of hard-burned quicklime (or slaked lime) at 85–90°C in a stirred reactor. The regenerated caustic soda is separated from the calcium carbonate precipitate (lime mud) by centrifuging or rotary vacuum filtration. The lime mud retains 30–35% liquid and, to avoid loss of caustic soda, must be well-washed on the filter or centrifuge. Finally, the recovered caustic solution is adjusted to the 10% level for recycle by the addition of 40% makeup caustic soda.

Disposal of the washed lime mud poses a problem. Drying and calcining to regenerate lime and CO_2 require an uneconomical expenditure of energy and, as of the mid-1990s, lime mud is dumped. The production of lime mud can be avoided by using sulfuric acid for springing and the partial electrolysis of the sodium sulfate to give a mixture of caustic soda and sodium sulfate as the recycled extracting medium. This alternative to caustic soda extraction and lime recausticization, although successful in laboratory and pilotscale tests, has not been applied in commercial practice as of 1996.

In the next stage in the recovery and refining of tar acids, water and pitch are removed from the crude tar acids in a continuous-vacuum still heated by superheated steam or circulating hot oil. The aqueous phenol overhead distillate is recycled, the stream of once-run tar acids is refined, and the phenolic pitch bottoms are burned.

The once-run tar acids are fractionated in three continuous-vacuum stills heated by superheated steam or circulating hot oil. These stills contain 40–50 bubble trays and operate at reflux ratios between 15 and 20:1. The overhead product from the first column is 90–95% phenol; from the second, 90% o-cresol; and from the third, a 40:60 m-cresol–p-cresol mixture. Further fractionation gives the pure products.

At other refineries, only two continuous stills in series are used, but these are of 80–100 plate efficiency and yield pure grades of phenol and o-cresol and a base mixture of cresols, xylenols, and higher boiling tar acids. The latter are fractionated batchwise to various saleable grades of cresylic acids.

Fractionation columns in tar-acid refineries are generally operated under vacuum and heated by high pressure steam or circulating hot oil. Calandria in the reboilers, condensers, rundown lines, and receiving tanks are constructed of stainless steel, or, in the case of the condensers, of tin or nickel. Cast-iron column shells are satisfactory, but stainless-steel bubble or valve trays are preferred. A flow sheet of a typical tar acid extraction and refining plant is shown in Figure 3.

In the case of low temperature tar, the aqueous liquor that accompanies the crude tar contains between 1 and 1.5% by weight of soluble tar acids, eg,

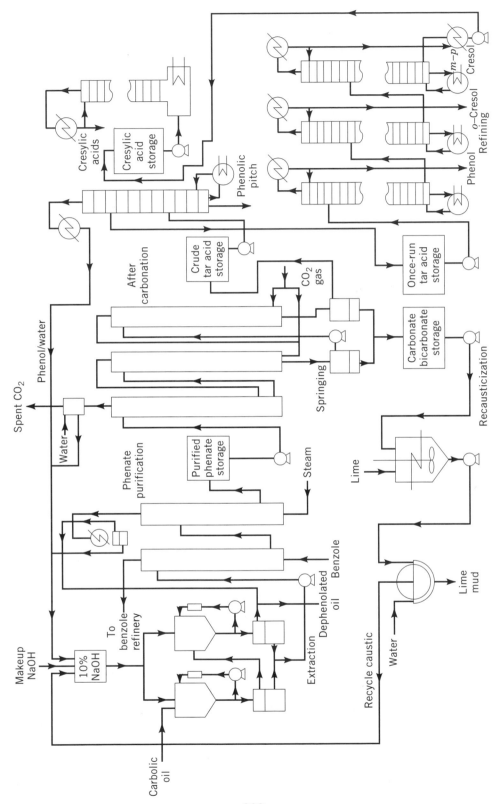

Fig. 3. Tar-acid extraction and refining.

phenol, cresols, and dihydroxybenzenes. Both for the sake of economics and effluent purification, it is necessary to recover these, usually by the Lurgi Phenosolvan process based on the selective extraction of the tar acids with butyl or isobutyl acetate. The recovered phenols are separated by fractional distillation into monohydroxybenzenes, mainly phenol and cresols, and dihydroxybenzenes, mainly o-dihydroxybenzene (catechol), methyl o-dihydroxybenzene, (methyl catechol), and m-dihydroxybenzene (resorcinol). The monohydric phenol fraction is added to the crude tar acids extracted from the tar for further refining, whereas the dihydric phenol fraction is incorporated in wood-preservation creosote or sold to adhesive manufacturers.

Naphthalene Oils.   Naphthalene is the principal component of coke-oven tars and the only component that can be concentrated to a reasonably high content on primary distillation. Naphthalene oils from coke-oven tars distilled in a modern pipe still generally contain 60–65% of naphthalene. They are further upgraded by a number of methods.

In the older method, still used in some CIS and East European tar refineries, the naphthalene oil is cooled to ambient temperatures in pans, the residual oil is separated from the crystals, and the crude drained naphthalene is macerated and centrifuged. The so-called whizzed naphthalene crystallizes at ca 72–76°C. This product is subjected to 35 MPa (350 atm) at 60–70°C for several minutes in a mechanical press. The lower melting layers of the crystals are expressed as liquid, giving a product crystallizing at 78–78.5°C (95.5–96.5% pure). This grade, satisfactory for oxidation to phthalic anhydride, is referred to as hot-pressed or phthalic-grade naphthalene.

The more modern processes adopted in the United Kingdom and some European plants (20) are also based on crystallization of the primary naphthalene oil, which is diluted with lower crystallizing material to give a feedstock crystallizing point at 55°C. This material is cooled in closed, stirred tanks to 30–35°C and the resultant slurry of naphthalene crystals and mother liquor is centrifuged, washed, and spun-dried. These operations are automatically timed and controlled. At other European and U.S. refineries, coke-oven naphthalene oil is upgraded by fractionation and further purification.

In a novel but no longer used naphthalene-purification method, oil containing ca 70% of naphthalene was charged to tall cylindrical tanks equipped with internal heating or cooling coils and cooled to 35°C. Mother liquor was allowed to drain from the base of the tank and, when drainage was complete, the tank was filled with methanol until the solvent levels was above the surface of the crystallized naphthalene. After standing for one hour, the methanol was drained. This procedure was repeated. Finally, the purified naphthalene was melted by passing steam through the internal coils and the liquid product discharged to storage. This product was pure enough for the manufacture of 2-naphthol.

Several refineries in Europe employ the Pro-Abd refiner to upgrade whizzed naphthalene to phthalic-grade quality or to convert the latter into the purer chemical grade. The device consists of a rectangular tank fitted with a nest of coils through which either steam or water circulates. The tank is filled with the feedstock, which is crystallized by circulating cold water in the coils. When the contents of the tank have solidified, a tap at the base is opened and hot water is circulated until the temperature is just below the desired crystallizing point of the product. This condition is maintained until no more oil drains from the base. The bottom tap is then closed and the contents of the tank are melted by steam circulation and drained.

In the Sulzer-MWB process the naphthalene fractions produced by the crystallization process are stored in tanks and fed alternately into the crystallizer. The crystallizer contains around 1100 cooling tubes of 25-mm diameter, through which the naphthalene fraction passes downward in turbulent flow and partly crystallizes out on the tube walls. The residual melt is recycled and pumped into a storage tank at the end of the crystallization process. The crystals that have been deposited on the tube walls are then partly melted for further purification. Following the removal of the drained liquid, the purified naphthalene is melted. Four to six crystallization stages are required to obtain refined naphthalene with a crystallization point of 80°C, depending on the quality of the feedstock. The yield is typically between 88 and 94%, depending on the concentration of the feedstock fraction.

In the late 1980s, Brodie crystallizers were installed in the United Kingdom and in France for upgrading phthalic-grade naphthalene to 99% purity or better. This apparatus, developed by Union Carbide Corporation, Australia, for separating o- and p-dichlorobenzene, was adapted for naphthalene refining. The one installed in the United Kingdom, however, has been closed (21) (see NAPHTHALENE).

Wash Oils and Methylnaphthalene Oils. No tar chemicals are extracted commercially from tar oils distilling in the range of 250–300°C. Although the wash-oil fraction of coke-oven tars, distilling mainly in the range of 250–280°C, is employed at coking installations to scrub benzene from coal gas, most oils in this boiling point range are used in creosote blends.

Anthracene Oils. In Europe but not in the United States, crude anthracene is isolated from coke-oven anthracene oils. In some cases where the anthracene oil gives only a small residue at 360°C, the oil is diluted using half its volume of drained naphthalene oil or a light wash oil and this blend is cooled to 35°C. The resulting solid–liquid slurry is filtered or centrifuged to give a crude anthracene containing 40–45% anthracene. With other samples of oil, it is necessary to recrystallize the first cake obtained by cooling the undiluted anthracene oil to 30–35°C and separating the crystals by filtration or centrifugation. This first crop of crystals generally contains 12–15% of anthracene and is recrystallized. The anthracene concentrate, termed 40-s anthracene, is sold as such.

Physical Properties. The physical properties of crude tars vary over a wide range. Investigation has been mainly concerned with establishing correlations between the more readily determined chemical and physical properties of the distillate oils and residual pitch, and other properties. Based on the correlations, other properties can be predicted with an accuracy sufficient for such purposes as plant design (Table 4).

Viscosity of Coal-Tar Pitch and Change with Temperature. Because pitch is mainly used as a hot-applied binder or adhesive, the viscosity and its change with temperature are important in industrial practice. Some useful correlations, by which the viscosity of pitch at any temperature can be predicted, have been developed. The data on which such correlations are based may be from one of the fixed equiviscous points that characterize a pitch (Table 5).

Table 4. Correlations for Predicting the Physical Properties of Tars and Tar Products

Property	Applicable to	Correlation expression	Comments and references
density, d at 20°C, g/cm^3	coke-oven dry tars and tar oils	$d_{20} = 1.877 \times 10^{-3} M^a$ $+ 0.808$	22
		$d_{20} = 7.337 \times 10^{-4} t_b^b$ $+ 0.890$	23
variation of sp gr with temperature, t, g/cm^3	dry tars	$\dfrac{\text{sp gr}}{t} = 0.001778$ $- 0.00098 \text{sp gr}$ at 15.6°C	nomograph available; 24
variation of density with temperature, t, g/cm^3	dry tars, tar oils, and pitches	$d_{t_1} = d_{t_2} - b(t_1 - t_2)^c$	25
	dry coke-oven tars and tar oils	$\beta = \dfrac{d_0 - d_t}{d_0 t}$	β has negative value; 25
		$\beta = \left(-10.5433 \right.$ $\left. + \dfrac{0.0122}{t_b} \right) \times 10^{-4},$ $\beta = (-12.6114$ $+ 0.0346 M^a) \times 10^{-4}$	
viscosity, η, mPa·s (=cP)	tar oils	$\log \eta_{20} = 0.0078 t_b$ $- 1.123^d$	nomograph available; 26
specific heat capacity, C, kJ/(kg·K)^e	tar oils	$C_t = ((0.7360 + 0.8951 d_{20}$ $+ 0.00360t)/d_{20})$ $+(0.00904$ $- 0.0000221 t_b)T_a^f$	27
	pitches	$C_t = \dfrac{3.665}{d_{20}} - 1.729$ $+ 0.00389t$	28
thermal conductivity, K, W/(m·K)	tar oils	$K = (1.34 - 0.00084t$ $\pm 0.084) \times 10^{-5}$	
	pitches	$K = (1.423 \pm 0.084)$ $\times 10^{-5}$	no significant variation over range 25–105°C; 29
surface tension, S, mN/m (=dyn/cm)	dry tars, tar oils, and pitches	$S = 93.8 d_{20} - 0.0496 t_b,$ $- 47.5, S_t = 18.4 d_t^4,$ $S_t = \dfrac{S_{20} d_t}{d_{20}^4}$	nomograph available; 30, 31
latent heat of vaporization, L, kJ/kg^b	tar oils	$L = d_{20}(486.1$ $- 0.599 t_b)^g$	32

^{a}Average molecular weight.
^{b}Average boiling point defined as the mean of the temperatures in °C at which 10%, 20%, . . . 90% by volume distills in a standard flask distillation.
$^c b = 0.00068 \pm 0.00005$ for dry tars and $(162.7 - 86.2 d_{20} \pm 8) \times 10^{-5}$ for tar oils and pitches.
^{d}At other temperatures, $\log \eta$ varies linearly with the absolute temperature. At t_b, the viscosity of any tar oil is approximately 0.25 mPa·s (=cP).
^{e}To convert J to cal, divide by 4.184.
$^f T_a$, % tar acids, is defined as the percentage by volume extracted by 10% aqueous caustic soda.
^{g}At the average boiling point.

Table 5. Viscosities of Fixed Points for Coal-Tar Pitch

Fixed point	Viscosity, Pa·s(=10^3 cP)	Difference between fixed point and EVT, °C	References
equiviscous temperature, EVT, °C	25		34
R-and-B softening point[a], °C	800	−19	37
K-and-S softening point, °C	5500	−27	38
penetration of 200[b]	2×10^4	−30	36
penetration of 10[b]	10^7	−51	39[c]
ductility point, °C[d]	10^7	−51	40
Fraas brittle point, °C	4×10^7	−65	41
transition to glassy state	10^{12}	−90	

[a]Ring-and-ball method (ASTM D36-26). In the United States, two other softening point methods are employed: cube-in-air (35) and cube-in-water (36). Cube-in-air softening point = R-and-B softening point +4°C. Cube-in-water softening point = R-and-B softening point +10°C.
[b]Penetration of 20 or 1 mm using a 100-g weight for 5 s at 25°C.
[c]Ref. 39 gives the relationship between viscosity and penetration (mm × 10): $\eta(Pa·s) = (1.58 \times 10^9)/penetration^{2.16}$.
[d]When a pitch is tested for ductility, the sample either suffers brittle fracture without elongation or elongates to the maximum distance without breaking. When tested at increased temperatures at a particular point, ie, the ductility point, the behaviour changes from the first type to the second.

The viscosity of a straight-run or fluxed-back pitch can be calculated from the R-and-B (ring-and-ball) softening point:

$$\log \eta_t = -4.175 + \frac{711.8}{86.1 - t_S + t} \qquad (1)$$

where η_t is the viscosity at temperature t °C in Pa·s × 10^{-1} (=P) and t_S is the R-and-B softening point in °C (42). Mettler softening point is widely used in the 1990s as the equivalent to R and B + 3°C (ASTM D3104). Equation 1 covers a temperature range of 180°C and viscosity of 10^8. It cannot be used to predict the viscosity accurately because the softening point itself is only an approximate measure of the viscosity and also because there are small differences in the temperature susceptibility of different types of pitch. However, equation 1 provides a useful means of calculating viscosities at any temperature above the softening point, exhibiting an accuracy sufficient for most design purposes.

The general equation for the temperature function of coal-tar pitch takes the following form (43):

$$\log \log(\nu + 0.8) = K + K_1 \log t \qquad (2)$$

where ν is the kinetic viscosity (mm^2/s (=cSt)), t is the temperature in K and K and K_1 are constants. K_1 depends on the toluene-insoluble (TI) content (TI in wt %).

$$K_1 = 5.8 - 0.03TI$$

This relationship applies to a wide range of straight-run and heat-treated pitches but not to fluxed pitches. If the K-and-S (Krämer and Sarnow) softening point is taken as a fixed point, equation 2 can be written as

$$\log \log(\nu + 0.8) = 0.08 + 0.04 \log \text{TI} + (5.8 - 0.03\text{TI})(\log(t_S + 273) - \log t)$$

where t_S is the K-and-S softening point in °C.

Another equation has the form

$$\log \eta = AT^{-5} + B \tag{3}$$

where η is the dynamic viscosity in mPa·s(=cP), t is the temperature in K, and A and B are constants derived from data relating to a range of pitches and refined tars varying in softening point from 54 to 104°C. The value of A in equation 3 is closely related to the softening point and can be expressed as

$$A = (0.358t_S + 3.74) \times 10^{12} \tag{4}$$

where t_S is the R-and-B softening point using glycerol as the heating medium or, using water as the heating medium, as follows:

$$A = (0.358t_S + 4.82) \times 10^{12} \tag{5}$$

B in equation 3 is related to the TI content:

$$B = 0.0206\text{TI} - 0.378 \tag{6}$$

Using equation 3, the viscosity of any pitch can be calculated from two measurements in the range of 10^4-10^6 mPa·s(=cP), exhibiting a precision similar to what may be expected of direct measurement. By employing equations 3, 4 or 5, and 6, the viscosity of pitch at any temperature can be calculated, with an accuracy adequate for most engineering purposes, from the R-and-B softening point and the TI content.

Chemical Composition.   The tars recovered from commercial carbonization plants are not the primary products of the thermal decomposition of coal. The initial products undergo a complex series of secondary reactions. Although pilot-plant tests carried out by the U.S. Bureau of Mines show that at any given carbonizing temperature the yield of tar is related approximately linearly to the content of volatile matter of the coal, no clear trends could be distinguished between the nature of the coal and the physical properties or chemical composition of the tar produced under the same carbonizing conditions (44). More recent studies have given a clearer picture. The application of statistical structural analysis based on ir spectrometry and proton nmr (45), a combination of uv and ir spectrometry with gas chromatography (gc) of coal pyrolysis products (45,46), and correlation of refractive index, molecular weights, and densities, have made important contributions (47,48). Evidence suggests that, when coal is heated in the absence of air, relatively small molecules are released that are absorbed in the micropore structure of the coal, including straight-chain and branched-chain alkanes and alkenes having terminal double bonds and aromatic hydrocarbons such as benzene, toluene, naphthalene, methyl- and dimethylnaphthalenes, acenaphthene, and fluorene. The principal decomposition occurs at 450–500°C with the volatilization of 15–20% of the coal. This reaction appears to be depolymerization, probably by breaking the methylene, polymethylene, or conjugated olefin bridges between the aromatic ring clusters to yield free radicals. These radicals are believed to have molecular weights of 300–800 and to consist of systems containing 4–10 fused aromatic or cycloaromatic rings having paraffinic side chains or hydroxy substituents. Some of these free radicals condense to high molecular weight resins which remain in the coke; others are stabilized by hydrogen transfer. The stabilized free radicals undergo secondary cracking reactions as they come in contact with the hot coke.

The nature of the secondary reactions is uncertain. Some believe that the primary tar components are broken down to small free radicals that recombine as they travel toward the retort exit; others suggest that some components remain relatively intact except for the removal of peripheral substituent groups and that the higher molecular weight components of coal tar are, in effect, slightly altered fragments of the original coal structure.

Although these speculations are of academic interest only, it is clear that even tars produced at the lowest commercial carbonization temperatures are very different from primary tars. In fact, low temperature tar, continuous vertical-retort (CVR) tar, and coke-oven tar form a series in which the yield of tar decreases, the aromaticity of the tar increases, the content of paraffins and phenols decreases, and the ratio of substituted aromatic and heterocyclic compounds to their unsubstituted parent molecules decreases. These differences are reflected in the densities and carbon/hydrogen ratios of the tars recorded in Table 6. Higher aromaticity correlates with higher density and C/H ratio. The reactions accounting for these changes, ie, cracking and cyclization of paraffins, dehydration of phenols, and dealkylation of aromatic and heterocyclic ring compounds, are those that would be expected, on thermodynamic grounds, to occur at the temperatures prevailing in carbonization retorts.

The application of modern methods of analyses such as gc, low ionization voltage mass spectrometry, high pressure liquid chromatography (hplc), and nmr has greatly increased knowledge of coal-tar composition without materially altering the qualitative picture.

Coke-oven tar is an extremely complex mixture, the main components of which are aromatic hydrocarbons ranging from the monocyclics benzene and alkylbenzenes to polycyclic compounds containing as many as twenty or more rings. Heterocyclic compounds containing oxygen, nitrogen, and sulfur, but usually only one heteroatom per ring system are present. Small amounts of paraffinic, olefinic, and partly saturated aromatic compounds also occur. The aromatic and heterocyclic structures occur in both unsubstituted and substituted forms. The main substitutes are methyl, ethyl, or hydroxyl. Although longer aliphatic side chains are encountered in small amounts in the lower boiling fractions, they tend, like hydroxyl groups, to disappear in the higher boiling fractions. Molecules containing up to four rings are generally fully condensed, but in the more complex

Table 6. Properties of Coal Tars

| Property | Coke-oven tars | | | | Fromer CVR tars | | Low temperature tars | Lurgi tars |
| | U.K | | Germany | U.S. | U.K. | | U.K. | U.K. |
	Average	Range	Average	Average	Average	Range	Average	Average
yield, L/t	33.6	16.8–43.2	26.8		70.9	62.7–75.0	95.5	12.7
density at 20°C, g/cm^3	1.169	1.138–1.180	1.175	1.180	1.074	1.066–1.083	1.029	1.070
water, wt %	4.9	3.2–6.5	2.5	2.2	4.0	2.1–6.3	2.2	2.8
carbon, wt %	90.3	86.8–93.5	91.4	91.3	86.0	85.3–87.2	84.0	84.2
hydrogen, wt %	5.5	5.1–6.0	5.25	5.1	7.5	7.2–7.8	8.3	7.7
nitrogen, wt %	0.95	0.73–1.07	0.86	0.67	1.21	0.84–1.65	1.08	1.09
sulfur, wt %	0.84	0.68–0.96	0.75	1.2	0.90	0.67–1.16	0.74	1.39
ash, wt %	0.24	0.03–0.67	0.15	0.03	0.09	0.03–0.24	0.10	0.02
TI[a], wt %	6.7	2.7–7.5	5.5	9.1	3.1	0.9–5.1	1.2	0.7

[a]Toluene-insoluble components.

molecules, the degree of condensation is less complete and branched ring–chain structures, in which most of the rings are fused to no more than three other rings, predominate.

The normally distillable part at atmospheric pressure of coke-oven tar boiling up to about 400°C and amounting to up to 50% of the whole, contains principally aromatic hydrocarbons. In particular, benzene, toluene, and the xylene isomers, tri- and tetra- methylbenzenes, indene, hydrindene (indane), and coumarone occur in the first fraction normally taken, which represents about 3.5% of the tar and boils up to about 200°C. This fraction also contains polar compounds including the tar acids, phenol and cresols, and the tar bases, pyridine, picolines (methylpyridines), and lutidines (dimethyl pyridines). The most abundant component of this type of tar is naphthalene, which is taken in the second fraction and represents about 10% of the tar. It is contaminated with small but significant amounts of thionaphthene, indene, etc. The next fraction contains the two methylnaphthalene isomers equivalent to 2% of the tar. Then follow biphenyl, acenaphthene, and fluorene, all in the range 0.7 to 1% of the tar in each case and then diphenylene oxide at about 1.5%. The components anthracene and phenanthrene are usually present at about 1 and 6%, respectively. The series continues with the components boiling up to 400°, which about represents the limit of the usual commercial distillation range, ie, pyrene and fluoranthene.

The by the 1990s far less abundant continuous vertical-retort tars differ from coke-oven tars. Whereas the latter contain relatively small amounts of nonaromatic hydrocarbons, CVR tars contain a relatively high proportion of normal straight-chain or slightly branched-chain paraffins, the content of which decreases from ca 20% in the lower boiling fractions to 5–10% in the higher distillate oils. They contain 20–30% of hydroxy-substituted hydrocarbons, of which some 15% (based on tar) can be extracted from the oils distilling up to 350°C. These phenols are mainly xylenols; methylethyl- and diethylphenols; dihydric phenols, eg, catechol, resorcinol, and their methyl derivatives; hydroxy hydrindenes and their methyl and dimethyl derivatives; naphthols and hydroxy-substituted acenaphthenes; anthracenes and phenanthrenes. Furthermore, the aromatic hydrocarbons and heterocyclic structures in CVR tar are predominantly substituted by one or more methyl groups, and the content of unsubstituted ring structures, which are the main components of coke-oven tars, is comparatively small.

Low temperature tars contain 30–35 wt % nonaromatic hydrocarbons, ca 30% of caustic-extractable phenols in the distillate oils, and 40–50% of aromatic hydrocarbons. The latter usually contain one or more alkyl substituent groups. On atmospheric distillation, coke-oven tars yield 55–60% pitch, whereas CVR tars give 40–50% pitch. The pitch yield from low temperature tars is in the 26–30% range.

The number of individual components in coal tar can only be guessed. As many as 461 compounds have been identified (49). As an example, the minor components in a neutral fraction of coke-oven naphtha and naphthalene oil were investigated by a combination of high efficiency fractionation, zone refining, and ir spectroscopy. Although the oil analyzed constituted only 13.5% by weight of the tar, 49 components were identified, including all possible C_6–C_9 aromatic hydrocarbons, thiophene and indan and four methylindans, benzofuran (coumarone) and four methylbenzfurans, benzonitrile and three isomeric tolunitriles, and styrene, α-methylstyrene, *trans-β*-methylstyrene, and indene. Table 6 gives the average properties of various types of tar and Table 7 lists the amounts of components that have either current or potential industrial possibilities.

Of the total tar bases in U.K. coke-oven and CVR tars, pyridine makes up about 2%, 2-methyl pyridine 1.5%, 3- and 4-methylpyridines about 2%, and ethylpyridine and dimethylpyridines 6%. Primary bases, anilines and methylanilines, account for about 2% of the bases in coke-oven and CVR tars and 3.5% of the bases in low temperature tars. The main basic components in coke-oven tars are quinoline (16–20% of the total), isoquinoline (4–5%), and methylquinolines. These dicyclic bases are less prominent in CVR and low temperature tars, in which only a minority of the basic constituents have been identified.

Although gas chromatography and high pressure liquid chromatography have greatly assisted in the elucidation of the structure of tar distillate oils, the former has restricted applicability to pitch. It has, however, been applied to solvent extracts and vacuum distillates of pitch (50). More precise information about the principal components of the lower molecular weight fractions of coke-oven pitch have been obtained by the application of high pressure liquid chromatography, thin-layer chromatography (51), and low voltage mass spectrometric examinations of pitches and pitch solvent fractions at the U.S. Bureau of Mines (52). The results of these studies indicate that pitch contains the following high molecular weight constituents: aromatic hydrocarbons having four rings, eg, chrysene, fluoranthene, pyrene, triphenylene, naphthacene, and

Table 7. Constituents of Coal Tar

Component, wt % dry tar	Coke-oven tars				Former CVR tars		Low temperature tars	Lurgi tars
	U.K.		Germany	U.S.	U.K.		U.K.	U.K.
	Average	Range	Average	Average	Average	Range	Average	Average
benzene	0.25	0.12–0.42	0.4	0.12	0.22	0.14–0.26	0.01	0.02
toluene	0.22	0.09–0.35	0.3	0.25	0.22	0.17–0.29	0.12	0.05
o-xylene	0.04	0.02–0.07		0.04	0.06	0.05–0.08	0.05	0.05
m-xylene	0.11	0.08–0.18	0.2	0.07	0.13	0.11–0.18	0.10	0.07
p-xylene	0.04	0.02–0.07		0.03	0.05	0.04–0.07	0.04	0.03
ethylbenzene	0.02	0.01–0.05		0.02	0.03	0.02–0.04	0.02	0.04
styrene	0.04	0.02–0.04		0.02	0.04	0.02–0.06	0.01	0.01
phenol	0.57	0.14–1.15	0.5	0.61	0.99	0.49–1.36	1.44	0.97
o-cresol	0.32	0.10–0.34	0.2	0.25	1.33	0.77–1.53	1.48	1.14
m-cresol	0.45	0.16–1.0	0.4	0.45	1.01	0.58–1.33	0.98	1.83
p-cresol	0.27	0.07–0.70	0.2	0.27	0.86	0.50–1.03	0.87	1.51
xylenols	0.48	0.13–1.30		0.36	3.08	2.70–3.58	6.36	5.55
high boiling tar acids	0.91	0.31–2.09		0.83	8.09	6.57–11.1	12.89	11.95
naphtha	1.18	0.52–2.66		0.97	3.21	2.86–3.84	3.63	3.02
naphthalene	8.94	7.29–11.31	10.0	8.80	3.18	2.15–3.84	0.65	2.01
α-methylnaphthalene	0.72	0.60–0.86	0.5	0.65	0.54	0.46–0.64	0.23	0.63
β-methylnaphthalene	1.32	1.15–1.63	1.5	1.23	0.68	0.62–0.74	0.19	1.05
biphenyl	0.73	0.34–1.04			0.42	0.30–0.49	0.4	0.9
acenaphthalene	0.96	0.42–1.28	0.3	1.05	0.66	0.50–0.80	0.19	0.57
fluorene	0.88	0.46–1.80	2.0	0.64	0.51	0.33–1.23	0.13	0.62
diphenylene oxide	1.50	1.40–2.00	1.4		0.68	0.62–0.74	0.19	0.57
anthracene	1.00	0.52–1.38	1.8	0.75	0.26	0.18–0.30	0.06	0.32
phenanthrene	6.30	2.30–9.80	5.7	2.66	1.75	0.40–2.90	1.60	0.28
carbazole	1.33	0.58–1.73	1.5	0.60	0.89	0.14–1.43	1.29	0.22
tar bases	1.77	1.25–2.60	0.73	2.08	2.09	1.81–2.47	2.09	2.50
medium-soft pitch	59.8	49.5–63.9	54.4	63.5	43.7	41.0–49.3	26.0	33.1

benzanthracene. Five-membered ring systems are represented by picene, benzopyrenes (BaP and BeP), benzofluoranthenes, and perylene. The next-highest fraction contains as the main components six-membered ring systems, such as dibenzopyrenes, dibenzofluoranthenes, and benzoperylenes. Seven-ring systems, eg, coronene, have also been identified. These basic hydrocarbon structures are accompanied by methyl and polymethyl derivatives and, in the case of the pitches from CVR and low temperature tars, by mono- and polyhydroxy derivatives. As in the case of the distillate oil range, heterocyclic compounds are also present.

Above this relatively low molecular weight range, which constitutes ca 40–50% of a medium soft coke-oven pitch, the information concerning the chemical structure of pitch is only qualitative and derived mainly from statistical structural analysis and mass spectra. The same sequence of polynuclear aromatic and heterocyclic compounds appears to continue, reaching molecules having 20–30 rings. As molecular weight increases, more heterocyclic atoms appear in the molecule, whereas the number and length of alkyl chains decreases and the hydrocarbon structures are not fully condensed. In the lower temperature pitches, some ring structures appear to be partly hydrogenated.

Shipment, Storage, and Handling. *Crude Tar and Tar Products.* Where the tar distillery is sited close to the carbonizing plant, the crude tar is transferred directly from the tar–liquor separating vessels on the by-product recovery unit to the storage tanks. Otherwise, it is shipped in rail or road tankers or by barge. Crude tar is stored in mild-steel tanks maintained at 40–50°C by steam coils.

Liquid tar products, eg, light oils, cresols, cresylic acids, creosote oil, and road tars, are generally transported in bulk in insulated mild-steel road or rail tankers. They are loaded at a temperature sufficiently high to ensure delivery at the desired viscosity. Small quantities are generally delivered in drums that may have to be steam-heated to ensure complete liquidity before discharge.

Pitch. Pitch used to be stored in solid form at the tar distillery in open bays, from which it was removed by small explosive charges. Loading of the lump pitch by mechanical shovel created a dust hazard both at the tar installation and at the customer's, where the lumps had to be ground before use. In the 1990s, pitch is stored in tanks heated by superheated steam or circulating hot-oil coils and transported in liquid form in insulated rail, road tankers, or ships. When transport as a hot liquid is not feasible, not acceptable by the customer, or for small amounts, the pitch is converted into a dust-free particulate form, ie, short rods termed pencils, pastilles, or flakes.

Several plants employ cooled-belt flakers. These consist of flexible steel belts, ca 1-m wide and up to 50-m long, that have short rubber skirting at the edges. Molten pitch flows from a thermostatically controlled tank over a weir to give a flat thin sheet on the belt, which is cooled from below by water sprays. At the end of the belt, the solid pitch is broken up by rotating tines. The pitch flakes are drained and transported to a covered storage silo by belt conveyor, during which time the surface moisture evaporates.

Direct water cooling is also employed in the pitch-pencilling plant at VFT, AG (formerly Rütgerswerke). Pitch at ca 50°C above its softening point is pumped through nozzles at a rate and pressure to give a series of almost horizontal jets. These jets project a short distance into horizontal tubes through which a cocurrent streamline flow of cooling water is maintained. The tubes extend for some 25 m and are bent into a semicircle at their exit end. The stream of pitch solidifies into solid rods about 10–20-mm thick, which break up into short lengths as they are forced round the bends. The pitch pencils are delivered into a cooling pond situated underneath the pipe assembly. A conveyor belt picks up the pitch from the pond and transports the pencils to a chute from which they fall to another conveyor belt which transports them to the storage silo. Surface drying is accomplished by warm-air jets.

Another type of pitch-pencilling plant, designed by Biwater Industrial Processes (formerly Chemical and Thermal Engineering, Ltd.), employs a vertical cooling tank. Liquid pitch at ca 150°C, held in a thermostatically controlled tank, is circulated round a ring main by a centrifugal pump. A horizontal section of the main contains an extrusion manifold consisting of a number (usually 6–10) of 9–12-mm-dia nozzles set 2000-mm apart. The circular streams of molten pitch from the nozzles fall by gravity into a deep rectangular tank containing water at 40–50°C where the pitch hardens into solid rods.

Economic Aspects.   *Crude Tar.*   Current world output of crude tar is estimated at between 11 and 12×10^6 t/yr. Table 8 gives such details as are available. The amount distilled is about 10×10^6 t/yr. Production of low temperature tar included in the U.K. total for 1994 was 35,000 t; at its zenith 203,000 t were produced in 1975. World production of all coal tar in 1975 was estimated at 17.3×10^6 t. The quantity of tar distilled at five yearly intervals since 1984 is shown in Table 9. Some comparative figures for the prices of coal-tar bulk products for 1982 and 1994 are given in Table 10.

Table 8. Crude Tar Production[a]

Country	Volume, t $\times 10^3$		
	1984	1989	1994
Belgium	195	258	~160
Brazil			318
Czech Republic	393	392	330
Canada	193	205	174
China			2800
Germany	890	739	390
France	354	250	206
Italy	267	232	182
Japan	2550	2200	1855
Netherlands	100	134	140
Poland			535[b]
CIS			2189[c]
Spain	141	129	125
South Africa			188
South Korea	246	361	445
United Kingdom	238	378	245
United States	1662	1794	1026
Total			*11,348*

[a]Ref. 53.
[b]40 imports.
[c]Production value is from 1993, consisting of Russia, 1175; Ukraine, 918; and Kagakhstan, 96.

Table 9. Distillation of Coal Tar[a]

Country	Volume, t × 10^3		
	1984	1989	1994
Belgium	189	183	~160
Brazil			223
Czech Republic	331	331	240
China			1250
Denmark	85	140	170
Germany	990	874	460
France	398	340	212
Italy	166	162	118
Japan	1840	2150	1758
Netherlands	100	105	115
Poland	228	277	290
CIS		3300^b	
Spain	140	170	210
South Africa			189
South Korea	248	320	403
United Kingdom	178	308	245
United States	2205^c	2422^c	1026

[a]Ref. 53.
[b]Production value is from 1990.
[c]Capacity.

Table 10. Tar Bulk Prices

Product	1982		1994	
	$/t	$/L	$/t	$/L
United States				
wood-preservation creosote				
U.S. bulk		0.31		
U.S. drums		0.33		0.21−0.24
electrode pitch	281−295			
aluminium	305−325		320−322	
graphitized steel	325−340		330−332	
impregnated pitch for refractories	340			
pitch for fiber pipes and coal briquettes	235			
roofing pitch	290		340−345	
United Kingdom				
electrode pitch	297		242	
carbon black feedstock and coal-tar fuels	171		104−106	
Western Europe				
indene-coumarone resins	810			
wood-preservation creosote	252−261	0.25−0.26	258−273	0.26−0.27

Specifications. Tar bulk products are covered by both national specifications and those formulated by the user. For instance, creosote for timber preservation is covered by the American Wood Preserving Association Standards (AWPA) and ASTM D350.

In the United States creosote specification AWPA P1/89 is intended for the treatment of timber for land and fresh-water use, and the heavier grade AWPA P13/89 for the preservation of marine piling and timber. In the United Kingdom a British Standard Specification, BS.144/90, Part 1, specifies three grades of creosote: two for pressure impregnation and one for brushing application. The standards of the West European Institute for Wood Preservation (WEI) are often used in Europe.

These specifications include specific gravity, maximum water content, maximum values for toluene- or benzene-insoluble material, and maximum amounts distilling at 230°C, 270°C, 315°C, and 355°C. In the case of the AWPA specifications, there are minimum limits to the specific gravities of each of the distillate fractions; in the case of the WEI specifications, limits for the contents of benzo[a]pyrene and water-soluble phenols (tar acids).

Other national specifications for wood preservation creosote are CAM 972 (Argentina), AS T505/1965 and AS 1143/1978 (Australia), NBN 439 (Belgium), DS Rec. M314/TP (Denmark), IS 218-1961 (India), DGN R21-1952 (Netherlands), SASS 17-1943 (South Africa), and GOST 2770-59 (CIS). These are generally similar to the U.S. and U.K. specifications.

Tar-based road binders are the subject of ASTM Specification D3515-77 (United States), BSS 76/1974 (United Kingdom), AS 63-1947 (Australia), DIN 51-551 (1961) (Germany), IS 215-1961 (India), JIS K2472 (Japan), and GOST 4641-49 (CIS). These cover a range of road tars of different viscosity ranges and generally include maximum values for moisture content, phenols, toluene-insoluble material, and, in some cases, naphthalene content. They also give the permissible ranges for specific gravity, for the amount of oils distilling between 200 and 270°C and between 270 and 300°C, and for the softening point of the residue above 300°C. No tar-based road binders are used in Germany, the Netherlands, or Denmark in the 1990s.

A few national specifications cover electrode-binder pitch, eg, GOST 10200-62 (CIS), but in general, this product is supplied to specifications set by the aluminium- or steel-manufacturing companies. The requirements for the binder for Söderberg electrodes differ from those for prebaked electrodes and include a minimum specific gravity or density, a narrow softening point range (95–100°C (R and B) for Söderberg electrodes and 105-115°C for prebaked electrode binder), a minimum C/H ratio, a minimum coking value, maximums for the permissible ash content, moisture content, and amount of volatile matter at 360°C, a range for quinoline-insoluble matter (8–13% for Söderberg electrode binder and 5–10% for prebaked electrode binder), and a minimum of ca 20% for the amount of toluene-insoluble/quinoline-soluble (β-resin) content. However, in addition to the published specification requirements, aluminum-smelting companies demand that the pitch binder should pass certain preacceptance tests involving flow measurement at various temperatures of the pitch-petroleum coke electrode paste and measurements of strength, porosity, electrical resistance, and reactivity of test electrodes made by slowly carbonizing the electrode paste under compression to 900°C (see CARBON AND ARTIFICIAL GRAPHITE).

Coal-tar pitch for other uses is similarly subject to a few national specifications but mainly sold to users' specifications. Pitch intended for roofing, dampproofing, and waterproofing is the subject of ASTM specification D450 and Federal specification R-P-381; hot-applied tar-based coatings (pipeline enamels) are the subject of BSS 4164/1987, amended in 1988, and, in the United States, of American Water Works Association (AWWA) specification C203.

Analytical Methods. *Tar.* Before the development of gas chromatography (gc) and high pressure liquid chromatography (hplc), the quantitative analyses of tar distillate oils involved tedious high efficiency fractionation and refractionation, followed by identification or estimation of individual components by ir or uv spectroscopy. In the 1990s, the main components of the distillate fractions of coal tars are determined by gc and hplc (54). The analytical procedures included in the specifications for tar bulk products are given in the relevant Standardization of Tar Products Tests Committee (STPTC) (33), ISO (55), and ASTM (35) standards.

Pitch. For the solvent analysis of pitch, a number of methods have been proposed. The solvents may be used sequentially or a fresh sample may be used with each solvent. Either the least or the most powerful solvent may be used first. The ratio of solvent to pitch or pitch fraction and the temperature and time of extraction vary.

In the Broche-Nedelmann procedure, the pitch is separated into material insoluble in benzene (α-component), material soluble in benzene but insoluble in petroleum ether (β-component), and material soluble in petroleum ether (γ-component). A modification of this procedure is now most widely used in the United States and the United Kingdom. Separate samples of the predried pitch are extracted using petroleum ether having 100–120°C boiling range (5 g of pitch is extracted using 100-mL portions of solvent until the extract is colorless), toluene (1 g of pitch is extracted using 100 mL of toluene under reflux for 20 min and the residue is washed using hot toluene until the washings are colorless), and quinoline (1 g of pitch is extracted using 25 mL of quinoline at 70–80°C for 20 min and the residue is washed with hot quinoline). The residues are dried and weighed and the analysis calculated to give the m/m percentages of the following four fractions: (*1*) Petroleum ether-soluble (crystalloids or γ-fraction), generally containing 4–50% of medium soft coke-oven pitch and having a molecular weight range of 175–300; (*2*) toluene-soluble–petroleum ether-insoluble (β-resins), generally containing 20–30% of a medium soft coke-oven pitch and having a molecular weight range of 300–700; (*3*) quinoline-soluble–toluene-insoluble (C-2 fraction or resins), generally containing 8–15% of a medium soft coke-oven pitch and having a molecular weight range of 1000–2000; and (*4*) quinoline-insoluble (C-1 fraction), the amount of which in coke-oven pitches varies from 5 to 15% but rarely exceeds 5% in CVR pitches or 1% in low temperature pitches. Quinoline-insoluble appears to be similar to carbon black. In an earlier U.K. procedure, the crystalloid fraction was recovered by evaporation of the petroleum ether extract, and, after drying and weighing, reextracted using boiling benzene until the solution was colorless. The β-resin fraction was recovered by evaporating the benzene extract. Finally, the benzene-insoluble residue was treated with boiling pyridine to give the pyridine-insoluble material. The pyridine extract was evaporated to obtain the C-2 fraction. The extractions can be carried out in a Soxhlet extraction apparatus.

The Mallison solvent analysis method is still used in Europe (56) in the 1990s. A sample of the dry pitch is extracted using a mixture of anthracene oil and pyridine to leave a residue of pyridine-insoluble material, the so-called H-resins. This fraction is roughly equivalent to the C-1 fraction obtained by the methods described earlier. A second sample of pitch is extracted on the water bath using benzene and the fraction-insoluble in this solvent is isolated, dried, and weighed. This residue is termed the H + M-resins (high and medium molecular weight resins). By subtraction, the M-resins (roughly equivalent to the C-2 fraction) are obtained. A third pitch sample is extracted using hot methanol to give an insoluble fraction regarded as the H + M + N-resins (high, medium, and low molecular weight resins). Again, by subtraction, the N-resins are obtained. The methanol extract is diluted using ammonium chloride solution which precipitates a fraction termed the m-oils, whereas the fraction soluble in the diluted methanol is termed the n-oils.

Pitch used as an anode binder for the aluminium industry may be analyzed and tested by the following International Organization for Standardization: ISO 6257, ISO 5939, ISO 5940, ISO 6376, ISO 6791, ISO 6998, ISO 6999, and ISO 8006 (55).

Health and Safety Factors. The volatile components of coal tar, ie, mononuclear aromatic hydrocarbons, phenols, and pyridine bases, are toxic when ingested, inhaled, or absorbed through the skin and the usual precautions must be taken when crude benzene or tar light oils are handled. Most polynuclear aromatic compounds are primary skin and eye irritants but are tolerated internally. Naphthalene was at one time prescribed as an internal antiseptic and antihelminthic in daily doses of 0.1–0.5 g. The probable lethal dose is between 5 and 15 g. Anthracene passes through the gastrointestinal tract mainly unchanged, as do most higher polynuclear hydrocarbons. The lowest lethal oral dose for coal-tar creosote is given as 140-mg/kg body weight in the 1974 Toxic Substances Control Act (TSCA) list.

In the European Union, coal-derived complex chemical substances, ie, those contained in the European Inventory of Existing Commercial Chemical Substances, have been classified for carcinogenicity in the twenty-first adaptation to technical progress of the European Commission (EC) Dangerous Substances Directive 1994 67/548/EEC (57). The EC Regulation 793/93 requires data sets to be submitted by producers or importers to the European Commission for these and other substances by nominated dates. The toxicological data and estimation of exposure will form the basis of risk analysis and determination of the appropriate restriction and control of substances in the work place (58). Restriction of the sales of dangerous substances and preparations to the general public is enforced under Directive 76/769 EC (59).

Carcinogenic Hazard of Tar and Tar Products. The main health hazard usually associated with coal tar and its products is carcinogenicity. Although this hazard undoubtedly exists, the risk is by no means as serious as some reports suggest and is constantly being reduced by improving working conditions. There is no evidence that the use of tar products in road surfacing, preservation of telegraph and transmission poles, pitch fiber or coal-tar enamels for water pipes constitutes any danger to the general public.

Cancer was first recognized as an occupational hazard in 1775 when the prevalence of scrotal cancer among London chimney sweeps was noticed. The chemical origin of this form of cancer was not universally accepted until 1922 when it was demonstrated that tumors could be induced on mouse skin using an etheral soot extract (60). Skin cancer was also noted to be an occupational hazard of workers exposed to pitch dust in the coal-briquetting industry (61) and to workers exposed to crude tar (62). In 1915, tumors were produced in rabbits' ears by prolonged application of crude coal tar (63). Some 38 cases per year have been reported of cutaneous epithelioma in the tar distilling industry over a 25-year period up to 1945 (64).

However, it was shown that only the higher boiling fractions were carcinogenic and that the spectra of these fractions resembled those of benz[a]anthracene (65). This work culminated in the isolation of a highly carcinogenic compound, benzo[a]pyrene (BaP) from coal-tar pitch (66). Cancerous skin lesions of workers exposed to pitch dust undoubtedly support the belief that these lesions are caused by polynuclear aromatic hydrocarbons, although it had not been possible to demonstrate their carcinogenic action in animals more closely related to humans, such as monkeys.

Although high temperature BaP-containing coal-tar creosote causes skin cancer in mice (67), evidence of skin cancer among workers in the timber-creosoting industry is limited. Moreover, modern creosotes contain <50 ppm of BaP. However, protective clothing should be worn to avoid skin irritation. Although epitheliomas can result from prolonged exposure to creosote (68), there is no evidence that the large volume of creosote used for wood preservation constitutes a hazard to the workers or the public. Similar large amounts of coal-tar pitch and tar oils were employed for the construction and maintenance of roads but are now being replaced by bitumen because of the latter's greater availability and cheaper price. Again, no evidence links any health hazard to either the road-working personnel or the general public.

The main risk is long-term, continual exposure of the skin to finely divided solid pitch (dust). In a relatively small, but statistically significant proportion of persons exposed in this manner, premalignant pitch warts appear, usually around the scrotum, hands, and face, particularly around the nostrils. These lesions, if untreated, develop into malignant epitheliomas. Such skin cancers are readily curable if diagnosed early. Premalignant warts usually heal spontaneously after excision, but malignant epitheliomas must be treated by radiation. A total surface treatment of 50 Gy (5000 rad) is generally enough to destroy the malignant cells. The condition clears up about one month after treatment, leaving some alteration in skin pigmentation. There are no reports of recurrence (69). Skin cancer caused by coal-tar pitch may take many years to manifest itself. Epidemiological studies at creosoting plants have proved inconclusive.

A more serious hazard might be thought to exist in the pitch-roofing and road-tar industries where some personnel are continually exposed to the inhalation of pitch fumes, which contain BaP and other carcinogens (70). However, epidemiological studies of such workers have not revealed any increased risk of cancer of the lungs, esophagus, or internal organs (71,72).

The risk of skin cancer by contact with pitch dust has in more recent years been reduced by the transport and handling of pitch as a liquid or as dust-free flakes or pencils. Nevertheless, in handling coal-tar products, certain precautions should be taken. These have become obligatory in tar distilleries and plants using pitch or creosote.

Where the lower boiling products are handled, effective ventilation must be ensured. Protective clothing, including eye protection and PVC gloves, must be worn, suitable respirators must be available, and regular medical checkups must be carried out on all personnel, including those who are at risk of exposure to benzene in significant concentrations. A high standard of personal hygiene must be maintained. Barrier creams, properly formulated against aromatic hydrocarbons, should be provided and used.

Uses. Coumarone-indene resins have outlets in paints, as tackifiers in rubber compounding, and as adhesives in the manufacturing of flooring tiles (see HYDROCARBON RESINS).

Cresylic Acids. The higher boiling cresylic acids are mixtures of cresols or xylenols with higher boiling phenols (see PHENOL; ALKYLPHENOLS). Their main uses are in phenol-formaldehyde resins, solvents for wire-coating enamels, as metal-degreasing agents, froth-flotation agents, and synthetic tanning agents. Statistics do not distinguish between the various grades or between cresylic acids derived from coal tar or petroleum or made synthetically (see PHENOLIC RESINS).

Naphthalene. Until the 1960s, the principal outlet for naphthalene was the production of phthalic anhydride; however, more recently, o-xylene has replaced naphthalene as the preferred feedstock (see PHTHALIC ACIDS). Nevertheless, of the 201,000 t produced in 1994 in Japan, 73.2% was used for phthalic anhydride production. The rest was consumed in dye stuffs manufacture and a wide variety of other uses. Naphthalene is also used to produce phthalic anhydride in the United Kingdom, Belgium, and the Czech Republic, and can be used by Koppers in the United States in time of o-xylene shortages. In Europe, the traditional uses for naphthalene have been for the manufacture of β-naphthol and for dye stuff intermediates (see DYES AND DYE INTERMEDIATES).

In more recent times, naphthalene has been used in condensation products from naphthalene sulfonic acids, utilizing formaldehyde as additives to improve the flow properties of concrete; these are referred to as superplasticizers. Another newer application is the production of diisopropylnaphthalenes. The mutual depression of the melting points in the mixture gives a liquid which is used as a solvent for dyes in the production of carbonless copy paper.

Creosote. In coal-tar refining, the recovery of tar chemicals leaves residual oils, including heavy naphtha, dephenolated carbolic oil, naphthalene drained oil, wash oil, strained anthracene oil, and heavy oil. These are blended to give creosotes conforming to particular specifications.

Creosote oils are by far the most widely used timber preservatives (see WOOD). This use dates back to 1850. For the treatment of railway ties and marine pilings, the Bethell or full-cell process is preferred. The timber to be treated is charged to a pressure cylinder, which is evacuated to extract the air from the wood cells. The cylinder is then filled with hot creosote and the pressure increased to 0.8–1 MPa (ca 8–10 atm) to force the oil into the cells. When uptake of the oil has ceased, residual creosote is drained from the cylinder and the treated timber is briefly subjected to a vacuum before discharge.

For the treatment of telegraph and transmission poles, fence posts, and farm buildings, where a clean outer surface is desired, the empty-cell process is generally used. The Rueping process employs an impregnation cycle in which the timber is first subjected to a moderate air pressure to compress the air in the cells before the cylinder is charged with hot creosote. The pressure in the cylinder is increased to force the oil into the wood cells. When no more oil is absorbed, the pressure is released, allowing the air inside the timber to expand and force the creosote out; a protective film remains on the cell wall. After discharging surplus creosote, the treated timber is evacuated for a short time to complete the removal of oil before the timber is removed from the cylinder. The Lowry process is similar but omits the initial pressurization.

Timber-preservation creosotes are mainly blends of wash oil, strained anthracene oil, and heavy oil having minor amounts of oils boiling in the 200–250°C range. Coal-tar creosote is also a feedstock for carbon black manufacture (see CARBON, CARBON BLACK). Almost any blend of tar oils is suitable for this purpose, but the heavier oils are preferred. Other smaller markets for creosote were for fluxing coal tar, pitch, and bitumen in the manufacture of road binders and for the production of horticultural winter wash oils and disinfectant emulsions.

Pitch. The principal outlet for coal-tar pitch is as the binder for the electrodes used in aluminum smelting. These are of two types. Older plants employ Söderberg furnaces, which incorporate paste electrodes consisting of a mixture of about 70% graded petroleum coke or pitch coke and 30% of a medium-hard coke-oven pitch. This paste is added periodically to the top of the monolithic electrode as it is consumed. The more modern smelters employ prebaked electrodes requiring less binder, about 18%.

The specification requirements for electrode binder pitch, eg, high C/H ratio, high coking value, and high β-resin content, effectively ruled out pitches from gasworks or low temperature tars. The crude tar is distilled to a medium-soft pitch residue and then hardened by heating for several hours at 385–400°C. This treatment increases the toluene-insoluble content and produces only a slight increase in the quinoline-insoluble (QI) material, the latter by the formation of mesophase.

Coke-oven tar distilleries are usually equipped for the heat treatment of pitch, either batch or continuous, but apparently not in either the United States or Australia. In some cases, pot stills, arranged in cascade, are still used. The more sophisticated plants employ one or more carbon steel or cast-iron vessels heated electrically and equipped with temperature controls for both the bulk liquid and the vessel walls. Contact time is usually 6–10 h. However, modern pitches are vacuum-distilled, producing no secondary quinoline insolubles, to improve the rheological properties.

The demand for electrode binder pitch has grown as aluminium output has expanded and the requirement for aluminium smelting is now between 1.5 and 2×10^6 t/yr. In Japan pitch is used for mixing with coal for carbonization in coke ovens to make metallurgical coke.

Coal-tar pitch of electrode binder quality but on a smaller scale is used as the binder for graphitized electrodes used in electric-arc steel-making convertors. Pitch of low quinoline-insoluble content is used for impregnation of such electrodes before graphitization. More recent uses are in the manufacture of carbon fibers and premium coke (73). These uses, which are not restricted to coal-tar pitch, depend on the formation of mesophase, an optically visible liquid crystal phase appearing as anisotropic spherulites, observable under the microscope using polarized light and size ranged upward from submicron to several microns in diameter. These are formed when coal-tar pitch, among others, is thermally treated at temperatures of between 300 and 500°C for protracted periods, usually several hours. Mesophase of this origin is insoluble in quinoline (74).

Mesophase formation in coal-tar pitch is encouraged by a reduction of the natural quinoline-insoluble matter content, which resembles carbon black but is not optically anisotropic and is characterized by an atomic carbon hydrogen ratio of 4:1. In contrast, the atomic carbon hydrogen ratio of mesophase is about 2:1.

The property of mesophase that makes it suitable for carbon fiber and premium coke manufacture is that it forms ordered structures under stress which persist following carbonization. However, most carbon fiber production in the 1990s is based on polyacrylonitrile (PAN).

In Europe, the production of coal briquettes by pressing a mixture of powdered bituminous coal using 8–10% medium-soft pitch (80°C R-and-B softening point), heated to 90°C in a pug mill, into blocks or ovoids was formerly a big outlet for pitch. Smokeless precarbonized briquettes using indigenous pitch and coal were made in the United Kingdom as a product called Phurnacite, produced from steam coal fines briquetted again using medium-soft coke-oven pitch and subjected to carbonization in a continuous vertical retort. This is now made in a different way using molasses as the binder in place of pitch.

In North America, coal-tar pitch is used as an adhesive in membrane roofs (see ADHESIVES; ROOFING MATERIALS). However, its uses for the same purpose in the Netherlands have been abandoned since the 1980s. The concrete or timber decking of flat-topped buildings is first covered with a layer of soft coke-oven pitch. While the pitch is still fluid, this base is covered with tar-impregnated paper on which a second layer of molten pitch is poured. This sandwich construction is repeated until the required number of pitch-paper layers (from 3 to 5) has been installed. The top paper membrane is coated with molten pitch and covered with slag chippings or gravel.

Pitch Coke. The manufacture of pitch coke provides a large tonnage outlet for coke-oven pitch in Japan, the CIS and, until more recently, Germany (75,76). Pitch coke is used either alone or mixed with petroleum coke as the carbon component of electrodes, carbon brushes, and shaped carbon and graphite articles.

In the CIS pitch coke is made by carbonizing a hard coke-oven pitch in modified coke ovens. The hard pitch has an R-and-B softening point of 140–150°C and is made by air-blowing a mixture of medium-soft pitch and recycled coking oils. This feedstock is charged in the molten state over a period of 5 h and coked for 17–18 h at 1250–1300°C. The coke yield is 70%. Oils, which are recycled, amount to 20% by weight of the pitch fed. The gas yield (80% hydrogen) is 10%.

A more modern process used in Japan and a unit in Germany employs the Lummus delayed-coking process (77). The feedstock is a soft coke-oven pitch from adjacent coke ovens. This flows into the bottom section of the so-called combination tower where it mixes with the vapors from the active coking drum after these have passed through the empty drum for preheating. The lighter components of the feedstock and the vapors are fractionated into fuel gas and light distillates in the upper part of the combination tower. The liquid from the base is heated to its incipient coking temperature and discharged into the preheated drum. The exothermic coking reaction proceeds for about 20 h. At the end of this period, the contents of the drum are steamed and cooled and the green coke discharged by high pressure water lances. The empty drum is steamed, then pressure-tested; subsequently it is ready to be preheated in preparation for the next charge. The green coke is produced in ca 61% yield and has an apparent density of 960–1088 kg/m^3 and a volatile matter content of 7.5–9.5%. It is calcined in a separate kiln to reduce the volatile matter to the 0.5% level required for electrode carbon. Figure 4 illustrates the Lummus delayed-coking plant installed at the refinery of the Nittetsu Chemical Industries at Tobata, Japan, which makes needle coke.

Other uses for coal-tar pitch include production as a binder for foundry cores, as a sealant for dry batteries, and in the manufacture of clay pigeons. Pelleted pitch used as the binder in foundry cores is a hard pitch supplied as spherical granules which are formed by a spray-cooling process. Clay pigeons consist of disks molded from a mixture of hard pitch and a mineral filler such as clay or limestone dust.

Fluxed Pitches and Refined Tars. *Road Tars.* In the United States, which has a large supply of bitumen, tar is little used in road construction or maintenance, but in Europe road binders still constitute an important, though declining, market for tar bulk products, mainly in France, Belgium, and Luxemburg. This is little used now in the United Kingdom and discontinued in Denmark, Netherlands, Germany or Spain. Road tars consist essentially of medium-soft pitch fluxed back using higher boiling tar oils. They are produced in a range of viscosities to suit the particular application and the climatic conditions under which they are to be used. They are employed as binders in road bases and base

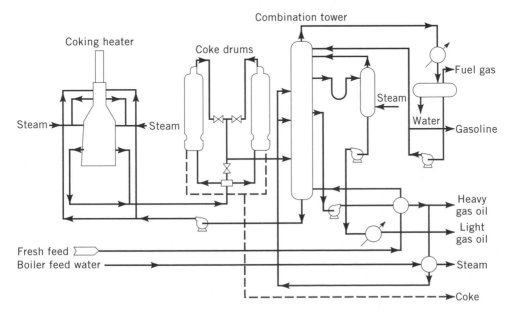

Fig. 4. Lummus delayed-pitch coking process. Courtesy of Lummus Co., Ltd.

courses and in tar macadam. For road maintenance, they are used as inexpensive surface dressing to restore the riding qualities of a road surface. The road surface is first swept free of dust, the potholes are filled, and a layer of hot tar sprayed on the surface. While the tar is still fluid, the surface is covered with an evenly applied layer of stone or slag chippings which are rolled into the new surface.

As a road binder, tar has advantages and disadvantages. Because the temperature interval between the R-and-B softening point and the brittle point is only 46°C (see Table 5), tar tends to flow in hot weather and suffer brittle fracture in cold weather. Its internal cohesive strength is low. On the other hand, it adheres well to stone and the bond is impervious to water or petroleum-based oils. The disadvantage of the high temperature coefficient of viscosity of road tars can be overcome without sacrificing the favorable properties by blending it with petroleum bitumen. Tar–bitumen blends are now preferred surface dressing binders. By adding 1–2% of PVC or certain types of synthetic rubber to road tar, a useful degree of elasticity is imparted. For surface dressing of heavily trafficked roads, polymer-modified road tars are popular, particularly in France.

The most widely used binder for the hot-rolled asphalt or asphaltic concrete-wearing courses of heavily trafficked roads is a penetration-grade petroleum bitumen. Such surfacings tend to polish under traffic and lose their resistance to skidding. Addition of 20–25% of coal-tar pitch improves and maintains the skid resistance. Asphalt road surfacings containing pitch–bitumen binder are used to some extent in France, Belgium, and Luxemburg.

The market for tar-based road binders has declined considerably for a variety of reasons. Less crude tar is available and the profits from the sales of electrode pitch and wood-preservation creosote or creosote as carbon-black feedstock are higher than those from road tar. In most industrial countries, road construction in more recent years has been concentrated on high speed motorways. Concrete, petroleum bitumen, or lake asphalt are used in the construction of these motorways. In the United Kingdom, for example, the use of tar products in road making and maintenance had fallen from 330,000 t in 1960 to 100,000 t in 1975 and is less than 100 t in 1994, mainly based on low temperature pitch which is not suitable for electrode or briquetting binders, but which is perfectly satisfactory as the basis for road binders.

Surface Coatings. Tar-based surface coatings range from the so-called black varnishes, which consist of a soft pitch fluxed back to brushing or spraying consistency using coal-tar naphtha, to pipe-coating enamels and pitch-polymer coatings. The black varnishes, occasionally mixed with small amounts of chlorinated rubber or powdered coal to improve film elasticity, are still used to some extent for the protection of industrial steel work and timber buildings and as antifouling marine paints.

The pipe-coating enamels are used for the corrosion protection of buried gas, water, or oil pipes. They are made from a ground coal dispersed in a coke-oven pitch fluxed back using strained anthracene oil as the basis and have a softening point of 105–125°C. Suitable mineral filler, eg, slate dust, is added to these enamels. To ensure good adhesion, the pipe is first primed with chlorinated rubber and chlorinated paraffin wax or chlorinated diphenyl dissolved in a mixture of toluene and trichloroethylene, and the coated pipe is wrapped with glass fiber impregnated with coal tar while the enamel coating is still hot and plastic. They are produced in the United States, Colombia, Australia, Denmark, and the United Kingdom.

The demand for pipe-coating enamels tends to fluctuate, depending on the success of oil and gas exploration. In the United Kingdom plastic-coated pipes are used on shore. Most production is exported for use in the Middle and Far Eastern regions.

Refined tars are used as extenders for epoxy and polyurethane resins to give surface coatings, which have developed into a sizeable and growing market (78). These formulations harden into tough abrasion-resistant, waterproof films. They are now widely used for coating underground storage tanks, structural steel items, water storage tanks, marine pilings, and bridge decks. The refined tar supplied for this purpose must meet stringent specifications requiring high aromaticity, low phenolic and base contents, and low quinoline-insoluble components. Such properties are secured by fluxing a pitch, made by distillation of a centrifuged coke-oven tar to a relatively low softening point residue, with coke-oven-strained anthracene or heavy oils that have been treated for the removal of phenols and bases. In the Netherlands this use will be forbidden probably at the end of the 1990s, on the grounds of pollution of their waterways.

The increasing use of carbon-bonded and carbon-impregnated refractories is providing another growing market for refined tars. These vary from a soft pitch to a refined tar of as low as 30°C equiviscous temperature (EVT) (see REFRACTORIES). Liquid fuels were formerly important outlets for coal-tar, pitch–oil blends, and topped tar which could not be disposed of more profitably (79). However, as a result of reduced tar supplies this usage has been phased out.

BIBLIOGRAPHY

"Tar and Pitch" in *ECT* 1st ed., Vol. 13, pp. 615–632, by E. O. Rhodes, Koppers Company, Inc.; in *ECT* 2nd ed., Vol. 19, pp. 653–682, by D. McNeil, The Coal Tar Research Association; in *ECT* 3rd ed., Vol. 22, pp. 564–600, by D. McNeil, Consultant.

1. A. W. Goos and A. A. Reiter, *Ind. Eng. Chem.* **38**, 132 (1946).
2. D. McNeil, *Coal Carbonization Products*, Pergamon Press, Oxford, U.K., 1966, Chapt. 3.
3. W. G. Adam, *J. Junior Inst. Eng.* **55**, 325 (1945).
4. D. Chandler and A. D. Lacey, *The Rise of the Gas Industry in Britain*, British Gas Council, London, U.K., 1949.
5. G. S. Pound, *J. Inst. Fuel*, **24**, 61 (1951).
6. *J. Gas*, **291**, 273 (1957).
7. P. Le Roux, *Goudron Routes*, **12**, (June 1958).
8. P. van Loyen and O. M. Stuhrmann, *Erdoel Kohle*, **13**, 758 (1960).
9. L. S. Elkind, *Coke Chem. USSR*, (6), 36 (1962).
10. P. Y. Kuleshov and co-workers, *Coke Chem. USSR*, (7), 38 (1960).
11. A. M. Kaznacheev, *Coke Chem. USSR*, (3), **44** (1969).
12. *Gas World*, 3 (Apr. 1959).
13. *Chem. Eng. Prog.* **55**(6), 84 (1959).
14. H. J. Turnbell, *Blast Furn. Steel Plant*, 412 (50) (May 1962).
15. H. G. Franck and C. Collin, *Steinkohlenteer: Chemie, Technologie, und Verwendung*, Springer Verlag, Berlin, Germany, 1968, p. 35.

16. *Ibid.*, p. 36.
17. R. C. Peter, *Ind. Chem.* **31**, 141 (1955).
18. Ref. 2, p. 109.
19. Ref. 2, p. 100.
20. Ref. 2, p. 90.
21. J. G. D. Molinari and B. V. Dodgson, *Chem. Eng.* **287**, 460 (1974).
22. B. S. Gurevich, S. C. Sabirova, and V. M. Rednov, *Coke Chem. USSR*, (2), 33 (1973).
23. *Standard Methods for Testing Tar and Its Products*, 7th ed., Serial No. COI 3-79, Standardization of Tar Products Tests Committee, Chesterfield, U.K., 1979, p. 477.
24. G. E. Mapstone, *J. Appl. Chem.* **5**, 582 (1955).
25. D. K. H. Briggs, Report No. 0126, Coal Tar Research Association, Gomersal, U.K., 1955.
26. D. K. H. Briggs, Report No. 0308, Coal Tar Research Association, Gomersal, U.K., 1963.
27. D. K. H. Briggs and F. Popper, *J. Appl. Chem.* **7**, 401 (1957).
28. D. K. H. Briggs and W. D. Drake, *Chem. Ind.* 666 (1957).
29. D. K. H. Briggs and F. Popper, *Fuel*, **33**, 222 (1954).
30. D. K. H. Briggs and K. E. Speak, *J. Appl. Chem.* **16**, 137 (1966).
31. D. K. H. Briggs, *Fuel*, **43**, 439 (1964).
32. D. K. H. Briggs and F. Popper, *Trans. Inst. Chem. Eng.* **35**, 369 (1957).
33. Ref. 23; Standardization of Tar Products Tests Committee (STPTC) is now at Tar Industries Services, Chesterfield, U.K.
34. Ref. 23, Serial No. R.T. 3-79, p. 194.
35. *American Society for Testing and Materials Annual Book of Standards*, D2319-76, Part 15, ASTM, Philadelphia, Pa., 1977.
36. *Ibid.*, D2319-61–75.
37. Ref. 23, Serial No. P.T. 3-79, p. 543.
38. *Standard Methods of Testing Tar and Its Products*, 6th ed., Serial No. PT. 2-67, Standardization of Tar Products Test Committee, Chesterfield, U.K., 1967.
39. R. M. Saal and G. Koens, *J. Inst. Pet. Tech.* **19**, 1976 (1933).
40. Ref. 35, Designation D113-44.
41. A. Fraas, *Asphalt Teer Strassenbautech.* **13**, 367 (1930).
42. E. H. Binns, Report No. 0167, Coal Tar Research Association, Gomersal, U.K., 1956.
43. H. G. Franck and O. Wegener, *Brennst. Chem.* **39**, 195 (1958).
44. J. G. Walters, C. Ortuglio, and J. Glaenzer, *U.S. Bureau of Mines Bulletin*, No. 643, 1967.
45. R. Chauvin and R. Deelder, *Bull. Soc. Chim. Fr.* (11), 3916 (1969).
46. C. Karr, Jr., and J. R. Comberiati, *U.S. Bureau of Mines Bulletin*, No. 636, 1966.
47. M. Vahrman, *Fuel*, **49**, 5 (1970).
48. M. Vahrman and R. H. Watts, *Fuel*, **51**, 131 (1972).
49. K. F. Lang and L. Eigen, *Forschr. Chem. Forsch.* **8**(1), 91 (1967).
50. R. Ferrand, *Goudron Routes*, (3), 5 (1964).
51. D. W. Grant, *Carbonization Research Report*, No. 23, British Carbonization Research Association, Chesterfield, U.K., 1976.
52. J. L. Schultz, R. A. Friedel, and A. G. Sharkey, Jr., *Fuel*, **44**, 55 (1965).
53. *International Tar Association Statistics*, Tar Industries Services, Chesterfield, U.K., 1994.
54. D. McNeil, in M. A. Elliot, ed., *High Temperature Coal Tar*, John Wiley & Sons, Inc., New York, 1981, p. 1013.
55. International Organization for Standardization.
56. H. Mallison, *Bitumén Teer Asphalte Peche*, **1**, 313 (1950).
57. EC 21st Adaptation to Technical Progress of the Dangerous Substances Directive 67/548/EEC (to be published).

58. Council Regulation (EEC) No. 793/93, *Official Journal of the EC*, L84 5.4.93, p. 1–75, 1993.

59. 14th Amendment to the Council Directive 76/769/EEC relating to the marketing and use of certain dangerous substances and preparations becomes Directive 94/60/EC, *Official Journal of the EC*, L356 31.12.94, p. 1, 1994.

60. R. D. Passey, *Brit. Med. J.* **2**, 1112 (1922).

61. A. Manouvirez, *Ann. Hyg. Publique*, **45**, 459 (1876).

62. R. Volkmann, *Beitrage zum Chirurgie*, Breitkopf and Hartel, Leipzig, Germany, 1875, p. 370.

63. K. Yamagiwa and K. Ichikawa, *Milt. Med. Fak. Tokio*, **15**, 295 (1915).

64. S. A. Henry, *Burt. Med. Bull. H*, 389–401 (1947).

65. E. L. Kennaway, *J. Ind. Hyg.* **5**, 462 (1924).

66. J. H. Cook, C. L. Hewitt, and I. Hieger, *Nature (London)*, **130**, 926 (1932).

67. W. E. Poel, *J. Natl. Cancer Inst.* **18**(1), 41 (1957).

68. H. I. Cookson, *Brit. Med. J.* **1**, 368 (1924).

69. F. Pierre and co-workers, *Arch. Mol. Prof. Med. Trav. Secur. Soc.* **T26**, 475 (Sept. 1965).

70. E. Sawicki and co-workers, *Am. Ind. Hyg. J.* **23**, 482 (1963).

71. E. C. Hammon, I. J. Selikoff, and P. J. Lawther, *N.Y. Acad. Sci.* **271**, 116 (1976).

72. V. P. Puzinawkas and L. W. Corbett, Research Report 78-1, The Asphalt Institute, Maryland, 1978.

73. H. G. Franck and J. W. Stadelhofer, *Industrial Aromatic Chemistry*, Springer Verlag, New York, 1988, pp. 368 and 382.

74. J. D. Brooks and G. H. Taylor, *Nature*, **206**(4985), 697–699 (1965).

75. S. Sarkar, *J. Mines Met. Fuels*, **10**, 16 (1962).

76. K. H. Osthaus, *Koppers-Mitt.* **3**, 115 (1960).

77. R. Ramirez, *Chem. Eng.* **76**, 74 (1969).

78. P. H. Pinchbeck, *Aust. Oil Colour Chem. Assoc. Proc. News*, **14**(6), 10 (1977); **14**(7), 6 (1977).

79. W. H. Huxtable ed. *Coal Tar Fuels*, The Association of Tar Distillers, London, U.K., 1960.

W. D. BETTS
Tar Industries Services

TAR SANDS

In addition to conventional petroleum (qv) and heavy crude oil, there remains another subclass of petroleum, one that offers to provide some relief to potential shortfalls in the future supply of liquid fuels and other products. This subclass is the bitumen found in tar sand deposits (1,2). Tar sands, also known as oil sands and bituminous sands, are sand deposits impregnated with dense, viscous petroleum. Tar sands are found throughout the world, often in the same geographical areas as conventional petroleum.

Petroleum, and the equivalent term crude oil, cover a vast assortment of materials consisting of gaseous, liquid, and solid hydrocarbon-type chemical compounds that occur in sedimentary deposits throughout the world (3). When petroleum occurs in a reservoir that allows the crude material to be recovered by pumping operations as a free-flowing dark- to light-colored liquid, it is often referred to as conventional petroleum.

Heavy oil is another type of petroleum, different from conventional petroleum insofar as the flow properties are reduced. A heavy oil is much more difficult to recover from the subsurface reservoir. These materials have a high viscosity and low API gravity relative to the viscosity and API gravity of conventional petroleum (Fig. 1) (3,4), and recovery of heavy oil usually requires thermal stimulation of the reservoir.

The definition of heavy oil is usually based on API gravity or viscosity, but the definition is quite arbitrary. Although there have been attempts to rationalize the definition based on viscosity, API gravity, and density (2,3), such definitions, based on physical properties, are inadequate, and a more precise definition would involve some reference to the recovery method.

In a general sense, however, the term heavy oil is often applied to a petroleum that has a gravity <20°API. The term heavy oil has also been arbitrarily used to describe both the heavy oil that requires thermal stimulation for recovery from the reservoir and the bitumen in bituminous sand (also known as tar sand or oil sand) formations, from which the heavy bituminous material is re-

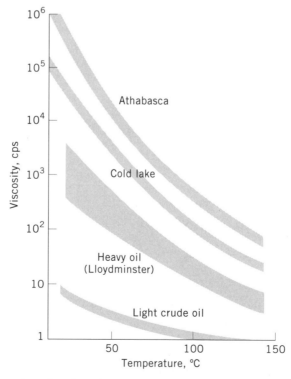

Fig. 1. Relative viscosity data for conventional petroleum, heavy oil, and bitumen.

covered by a mining operation. Extra heavy oil is the subcategory of petroleum that occurs in the near-solid state and is incapable of free flow under ambient conditions. The bitumen from tar sand deposits is often classified as an extra heavy oil.

Tar sand, also variously called oil sand (in Canada) or bituminous sand, is the term commonly used to describe a sandstone reservoir that is impregnated with a heavy, viscous black extra heavy crude oil, referred to as bitumen (or, incorrectly, as native asphalt). Tar sand is a mixture of sand, water, and bitumen, but many of the tar sand deposits in the United States lack the water layer that is believed to cover the Athabasca sand in Alberta, Canada, thereby facilitating the hot-water recovery process from the latter deposit. The heavy asphaltic organic material has a high viscosity under reservoir conditions and cannot be retrieved through a well by conventional production techniques.

It is incorrect to refer to bitumen as tar or pitch. Although the word tar is somewhat descriptive of the black bituminous material, it is best to avoid its use in referring to natural materials. More correctly, the name tar is usually applied to the heavy product remaining after the destructive distillation of coal (qv) or other organic matter. Pitch is the distillation residue of the various types of tar (see TAR AND PITCH).

Physical methods of fractionation of tar sand bitumen usually indicate high proportions of nonvolatile asphaltenes and resins, even in amounts up to 50% wt/wt (or higher) of the bitumen. In addition, the presence of ash-forming metallic constituents, including such organometallic compounds as those of vanadium and nickel, is also a distinguishing feature of bitumen.

Asphalt (qv) is prepared from petroleum and often resembles bitumen. When asphalt is produced simply by distillation of an asphaltic crude, the product can be referred to as residual asphalt or straight-run petroleum asphalt. If the asphalt is prepared by solvent extraction of residua or by light hydrocarbon (propane) precipitation, or if blown or otherwise treated, the term should be modified accordingly to qualify the product, eg, propane asphalt.

Origin of Bitumen

There are several general theories regarding the origin of the bitumen. One theory is that the oil was formed locally and has neither migrated a great distance nor been subjected to large overburden pressures. Because under these conditions the oil cannot have been subjected to any thermal effects with the resulting decomposition or molecular changes, it is geologically young and therefore dense and viscous.

Another theory promotes the concept of a remote origin for the bitumen, or, more likely, the bitumen precursor, both geographically and in geological time. The bitumen precursor, originally resembling a conventional crude oil, is assumed to have migrated into the sand deposit, which may originally have been filled with water. After the oil migrated, the overburden pressures were relieved, and the light portions of the crude evaporated, leaving behind a dense, viscous residue.

Included in the remote origin theory is the postulate that the light hydrocarbons were destroyed by bacteria carried into the petroleum reservoirs in oxygenated, meteoric waters. The remote origin theory would explain the water layer surrounding sand grains in the Athabasca deposit. However, because the metals and porphyrin contents of bitumen are similar to those of some conventional Alberta crude oils of Lower Cretaceous age and because Athabasca bitumen has a relatively low coking temperature, the bitumen may be of Lower Cretaceous age. This is the age of the McMurray formation (Canada), which is geologically young. This evidence supports the theory that the oil was formed *in situ* and is a precursor, rather than a residue of some other oil. The issue remains unresolved as of this writing (ca 1997).

Occurrence

Many of the reserves of bitumen in tar sand formations are available only with some difficulty, and optional refinery methods are necessary for future conversion of these materials to liquid products, because of the substantial differences in character between conventional petroleum (qv) and bitumen (Table 1).

Table 1. Bitumen vs Conventional Petroleum Properties

Property		Bitumen	Conventional
gravity, °API		8.6	25–37
distillation	Vol %	IBP[a], °C	
	5	221	
	10	293	
	30	437	
	50	543	
viscosity, suspension			
at 38°C		35,000	<30
at 99°C		513	
pour point, °C		10	≤0
elemental analysis, wt %			
carbon		83.1	86
hydrogen		10.6	13.5
sulfur		4.8	0.1–2.0
nitrogen		0.4	0.2
oxygen		1.1	
hydrocarbon type, wt %			
asphaltenes		19	≤5
resins		32	
oils		49	
metals, ppm			
vanadium		250	
nickel		100	
iron		75	≤100
copper		5	
ash, wt %		0.75	0
Conradson carbon, wt %		13.5	1–2
net heating value, kJ/g[b]		40.68	ca 45.33

[a]IBP = initial boiling point.
[b]To convert kJ/g to btu/lb, multiply by 430.2.

Because of the diversity of available information and the continuing attempts to delineate the various world oil sands deposits, it is virtually impossible to reflect the extent of the reserves in terms of barrel units with a great degree of accuracy. The potential reserves of hydrocarbon liquids that occur in tar sand deposits have, however, variously been estimated on a world basis to be in excess of 477×10^9 m^3 (3×10^{12} bbl). Reserves that have been estimated for the United States are believed to be in excess of 795×10^4 m^3 (50×10^6 bbl), although estimates vary. Bitumen reserves throughout the world can compare favorably with reserves of conventional crude oil.

Tar sand deposits are widely distributed throughout the world (Fig. 2) (5,6) and the various deposits have been described as belonging to two types: stratigraphic traps and structural traps (Table 2; Fig. 3) (7). However, there are the inevitable gradations and combinations of these two types of deposits, and thus a broad pattern of deposit entrapment is believed to exist. In general terms, the entrapment character of the very large tar sand deposits involves a combination of both stratigraphic and structural traps.

The largest tar sand deposits are in Alberta, Canada, and in Venezuela. Smaller tar sand deposits occur in the United States (mainly in Utah), Peru, Trinidad, Madagascar, the former Soviet Union, Balkan states, and the Philippines. Tar sand deposits in northwestern China (Xinjiang Autonomous Region) also are large; at some locations, the bitumen appears on the land surface around Karamay, China. The largest deposits are in the Athabasca area in the province of Alberta, Canada, and in the Orinoco region of east central Venezuela.

The Athabasca deposit, along with the neighboring Wabasca, Peace River, and Cold Lake heavy oil deposits, have together been estimated to contain 1.86 $\times 10^{11}$ m^3 ($>1.17 \times 10^{12}$ bbl) of bitumen. The Venezuelan deposits may at least

Fig. 2. Principal tar sand deposits of the world, where ● represents >2,385,000 m^3 (>15 $\times 10^6$ bbl) bitumen; ▲, probably <159,000 m^3 (<1 $\times 10^6$ bbl) bitumen; and ▼, reported occurrence information limited.

Table 2. Tar Sand Deposits and Mode of Entrapment[a]

Number	Deposit	Location
1.	stratigraphic trap: structure of little importance; short-distance migration assumed	Sunnyside, P.R. Springs, Santa Cruz
2.	structural/stratigraphic trap: folding/faulting and unconformity equally important	Oficina–Temblador tar, Bemolanga, Asphalt Ridge, Melville Island, Guanoco, Kentucky deposits
3.	structural trap: structure important; long-distance migration assumed; unconformity may be absent	Whiterocks, La Brea
4.	intermediate between 1 and 2	Athabasca, Edna, Sisquoc, Santa Rosa
5.	intermediate between 2 and 3	Selenizza, Derna

[a]See Fig. 3.

Fig. 3. Types of traps for tar sand deposits, where ○ represents a stratigraphic trap, ×, an intermediate between stratigraphic and structural/stratigraphic traps; ■, a structural/stratigraphic trap; ●, an intermediate between structural/stratigraphic and structural traps; and △, a structural trap.

contain $>1.60 \times 10^{11}$ m^3 (1.0×10^{12} bbl) bitumen (2). Deposits of tar sand, each containing $>3 \times 10^6$ m^3 (20×10^6 bbl) of bitumen, have also been located in the United States, Albania, Italy, Madagascar, Peru, Romania, Trinidad, Zaire, and the former Soviet Union, comprising a total of ca 450×10^9 m^3 (2.8×10^{12} bbl).

The Alberta (Athabasca) tar sand deposits are located in the northeast part of that Canadian province (Fig. 4). These are the only mineable tar sand deposits undergoing large-scale commercial exploitation as of this writing (ca 1997).

Fig. 4. Tar sand and heavy oil deposits in Alberta, Canada.

The Athabasca deposits have been known since the early 1800s. The first scientific interest in tar sands was taken by the Canadian government in 1890, and in 1897–1898, the sands were first drilled at Pelican Rapids on the Athabasca River. Up until 1960, many small-scale commercial enterprises were attempted but not sustained. Between 1957 and 1967, three extensive pilot-plant operations were conducted in the Athabasca region, each leading to a proposal for a commercial venture, eg, Suncor and Syncrude.

The Venezuelan tar sands are located in a 50–100-km belt extending east to west for >700 km, immediately north of the Orinoco River. The precise limits of the deposit are not well defined because exploration efforts in the past concentrated on light and medium crude accumulations.

The geological setting of the Orinoco deposit is complex, having evolved through three cycles of sedimentation. The oil is contained by both structural and stratigraphic traps, depending on location, age of sediment, and degree of faulting. The tar sands are located along the southern flanks of the eastern Venezuelan basin, where three distinct zones are apparent from north to south: a zone of tertiary sedimentation, a central platform with transgressive overlapping sediments, and a zone of erosional remnants covered by sediments. The deposit also contains three systems of faulting. All the faults are normal and many are concurrent with deposition.

Tar sands in the United States are contained in a variety of separate deposits in various states (Fig. 5) but because many of these deposits are small, information on most is limited (8). Attempts at development of the deposits have occurred primarily in Utah.

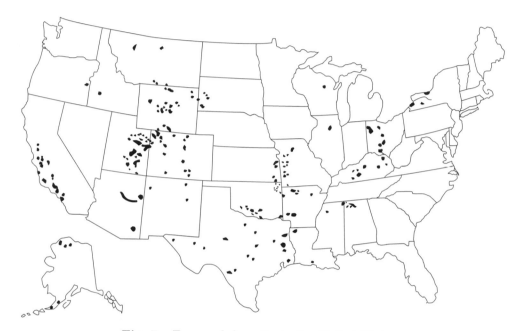

Fig. 5. Tar sand deposits in the United States.

Properties

Tar sand has been defined as sand saturated with a highly viscous crude hydrocarbon material not recoverable in its natural state through a well by ordinary production methods (2–8). Technically the material should perhaps be called bituminous sand rather than tar sand because the hydrocarbon is bitumen, ie, a carbon disulfide-soluble oil.

The data available are generally for the Athabasca materials, although workers at the University of Utah (Salt Lake City) have carried out an intensive program to determine the processibility of Utah bitumen and considerable data have become available. Bulk properties of samples from several locations (Table 3) (9) show that there is a wide range of properties. Substantial differences exist between the tar sands in Canada and those in the United States; a difference often cited is that the former is water-wet and the latter, oil-wet (10).

Table 3. Bulk Properties of Tar Sands

Property	Alberta	Asphalt Ridge[a]	P.R. Springs[a]	Sunnyside[a]	Tar Sand Triangle[a]	Texas	Alabama
bulk density, g/cm^3	1.75–2.19		1.83–2.50				
porosity, vol %	27–56	16–27	6–33	16–28	9–32	32	6–25
permeability, m$^2 \times 10^{-16b}$	99–5,900	4,905–5,950	553–14,902	5,265–7,402	2,043–7,777	3158	9.9–6,316
specific heat, J/(g·°C)^c	1.46–2.09						
thermal conductivity, J/(s·°C·cm)^c	0.0071–0.0015						

[a]Deposit in Utah.
[b]To convert m^2 to millidarcies, multiply by 1.013×10^{12}.
[c]To convert J to cal, divide by 4.184.

Canada	United States
sand is water-wet, thus disengagement of bitumen is efficient using hot-water process (caustic = sodium hydroxide; bitumen recovery >98%)	sand is oil-wet, thus efficient disengagement of bitumen requires high shear rates (caustic = sodium carbonate; bitumen recovery ~95%)
formations usually unconsolidated	formations usually consolidated to semiconsolidated by mineral cementation
few deposits have been identified (Alberta contains ca 0.4 m^3 bitumen)	numerous deposits identified (33 major deposits = 12 m^3 bitumen; 20 minor deposits = 12 m^3 bitumen); total resource = 6.5 m^3 bitumen (2.6 m^3 measured and 3.8 m^3 billion speculative)
problems exist in settling and removal of clay from tar sand deposits and process streams	little is known about the nature and effect on processing of clays
bitumen properties fairly uniform (sulfur = 4.5–5.5 wt %, nitrogen = 0.1–0.5 wt %; H/C ratio ~1.5; API gravity from 6 to 12°)	bitumen properties diverse (sulfur = 0.5–10 wt %, nitrogen 0.1–1.3 wt %; H/C ratio = 1.3–1.6; API gravity from −2 to 14°)
bitumen deposits large with uniform quality; recovery and upgrading plants on-stream since 1970s	bitumen deposits small and not of uniform quality; recovery and upgrading methods need to be site-specific

The sand component is predominantly quartz in the form of rounded or angular particles (11), each of which is wet with a film of water. Surrounding the wetted sand grains and somewhat filling the void among them is a film of bitumen. The balance of the void volume in the Canadian sands is filled with connate water plus, sometimes, a small volume of gas. Usually the gas is air but methane has been reported from some test borings in the Athabasca deposit. Some commercial gas deposits were developed in the late 1980s. The sand grains are packed to a void volume of ca 35%, corresponding to a mixture of ca 83 wt % sand; the remainder is bitumen and water which constitute ca 17 wt % of the tar sands.

Bitumen.   There are wide variations both in the bitumen saturation of tar sand (0–18 wt % bitumen), even within a particular deposit, and the viscosity. Of particular note is the variation of density of Athabasca bitumen with temperature, and the maximum density difference between bitumen and water (70–80°C (160–175°F)); hence the choice of the operating temperature of the hot-water bitumen-extraction process.

The API gravity of tar sand bitumen varies from 5 to ca 10°API, depending on the deposit, and the viscosity is very high. Whereas conventional crude oils may have a high (>100 MPa·s(=cP)) viscosity at 40°C, tar sand bitumen has a viscosity on the order of $10^5–100$ kPa·s ($10^5–10^6$ P) at formation temperature (ca 0–10°C), depending on the season. This offers a formidable obstacle to bitumen recovery and, as a result of the high viscosity, bitumen is relatively nonvolatile under conditions of standard distillation (Table 4) (12,13), which influences choice of the upgrading process.

Minerals.   Usually >99% of the tar sand mineral is composed of quartz sand and clays (qv). In the remaining 1%, more than 30 minerals have been identified, mostly calciferous or iron-based (14). Particle sizes range from large grains (99.9% finer than 1000 μm) to 44 μm (325 mesh), the smallest size that

Table 4. Distillation Data for Various Bitumens

Cut point, °C	Athabasca, wt % distilled[a]	NW Asphalt Ridge, wt % distilled[a]	P.R. Springs, wt % distilled[a]	Tar Sand Triangle, wt % distilled[a]
200	3.0	2.3	0.7	1.7
225	4.6	3.3	1.4	2.9
250	6.5	4.4	2.4	4.4
275	8.9	5.8	3.8	5.9
300	14.0	7.5	4.9	8.4
325	25.9	8.8	6.8	12.4
350	18.1	11.7	8.0	15.2
375	22.4	13.8	10.1	18.6
400	26.2	16.8	12.5	22.4
425	29.1	19.5	16.0	26.9
450	33.1	23.7	20.0	28.9
475	37.0	28.4	22.5	32.3
500	40.0	34.0	25.0	35.1
525	42.9	40.0	27.3	38.5
538	44.6	44.2	28.0	40.0
538+	55.4	55.8	72.0	60.9

[a]Cumulative.

can be determined by dry screening. The size between 44 and 2 μm is referred to as silt; sizes <2 μm (equivalent spherical diameter) are clay.

Clays (qv) are aluminosilicate minerals, some of which have definite chemical compositions. In regard to tar sands, however, clay is only a size classification and is usually determined by a sedimentation method. According to the previous definition of fines, the fines fraction equals the sum of the silt and clay fractions. The clay fraction over a wide range of fines contents is a relatively constant 30% of the fines.

The Canadian deposits are largely unconsolidated sands having a porosity ranging up to 45% and good intrinsic permeability. However, the deposits in Utah range from predominantly low porosity, low permeability consolidated sand to, in some instances, unconsolidated sands. In addition, the bitumen properties are not conducive to fluid flow under normal reservoir conditions in either Canadian or U.S. deposits. Nevertheless, where the general nature of the deposits prohibits the application of a mining technique, as in many of the U.S. deposits, a nonmining technique may be the only feasible bitumen recovery option (6).

Recovery

Oil prices and operating costs are the key to economic development of tar sand deposits. However, two technical conditions of vital concern for economic development are the concentration of the resource (percent bitumen saturation) and its accessibility, usually measured by the overburden thickness.

The remoteness of the U.S. tar sands is often cited as a deterrent to development but topography of the site, overburden-to-ore body ratio, and richness of the ore body are also important. In the 1990s context of mining tar sand deposits in the United States, the Utah deposits (Tar Sand Triangle, P.R. Springs, Sunnyside, and Hill Creek) generally have an overburden-to-net pay zone ratio above the 0.4–1.0 range, with a lean oil content. On the other hand, the Asphalt Ridge deposit is loosely consolidated and could be mined using a ripper/front-end loader (without drilling and blasting) at the near-surface location of the deposit.

Recovery methods are based either on mining combined with some further processing or operation on the oil sands *in situ* (Fig. 6). The mining methods are applicable to shallow deposits, characterized by an overburden ratio (ie, overburden depth-to-thickness of tar sand deposit) of ca 1.0. Because Athabasca tar sands have a maximum thickness of ca 90 m and average ca 45 m, there are indications that no more than 10% of the in-place deposit is mineable within 1990s concepts of the economics and technology of open-pit mining.

The bitumen in the Athabasca deposit, which has a gravity on the API scale of 8°, is heavier than water and very viscous. Tar sand is a dense, solid material, but it can be readily dug in the summer months; during the winter months when the temperatures plunge to −45°C, tar sand assumes the consistency of concrete. To maintain acceptable digging rates in winter, mining must proceed faster than the rate of frost penetration; if not, supplemental measures such as blasting are required.

Nonmining Methods. Nonmining (*in situ*) processes depend on injecting a heating-and-driver substance into the ground through injection wells and

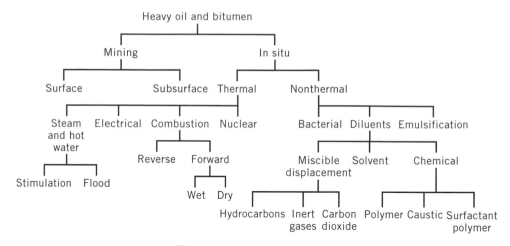

Fig. 6. Recovery processes.

recovering bitumen through production wells. Such processes need a relatively thick layer of overburden to contain the driver substance within the formation between injection and production wells (2).

In principle, the nonmining recovery of bitumen from tar sand deposits is an enhanced oil recovery technique and requires the injection of a fluid into the formation through an injection well. This leads to the *in situ* displacement of the bitumen from the reservoir and bitumen production at the surface through an egress (production) well. There are, however, several serious constraints that are particularly important and relate to the bulk properties of the tar sand and the bitumen. In fact, both recovery by fluid injection and the serious constraints on it must be considered *in toto* in the context of bitumen recovery by nonmining techniques (see PETROLEUM, ENHANCED OIL RECOVERY).

Another general constraint to bitumen recovery by nonmining methods is the relatively low injectivity of tar sand formations. It is usually necessary to inject displacement/recovery fluids at a pressure such that fracturing (parting) is achieved. Such a technique, therefore, changes the reservoir profile and introduces a series of channels through which fluids can flow from the injection well to the production well. On the other hand, the technique may be disadvantageous insofar as the fracture occurs along the path of least resistance, giving undesirable or inefficient flow characteristics within the reservoir between the injection and production wells, which leave a part of the reservoir relatively untouched by the displacement or recovery fluids.

In steam stimulation, heat and drive energy are supplied in the form of steam injected through wells into the tar sand formation. In most instances, the injection pressure must exceed the formation fracture pressure in order to force the steam into the tar sands and into contact with the oil. When sufficient heating has been achieved, the injection wells are closed for a soak period of variable length and then allowed to produce, first applying the pressure created by the injection and then using pumps as the wells cool and production declines.

Steam can also be injected into one or more wells, with production coming from other wells (steam drive). This technique is effective in heavy oil formations but has found little success during application to tar sand deposits because of the difficulty in connecting injection and production wells. However, once the flow path has been heated, the steam pressure is cycled, alternately moving steam up into the oil zone, then allowing oil to drain down into the heated flow channel to be swept to the production wells.

If the viscous bitumen in a tar sand formation can be made mobile by an admixture of either a hydrocarbon diluent or an emulsifying fluid, a relatively low temperature secondary recovery process is possible (emulsion steam drive). If the formation is impermeable, communication problems exist between injection and production wells. However, it is possible to apply a solution or dilution process along a narrow fracture plane between injection and production wells.

To date (ca 1997), steam methods have been applied almost exclusively in relatively thick reservoirs containing viscous crude oils. In the case of heavy oil fields and tar sand deposits, the cyclic steam injection technique has been employed with some success. The technique involves the injection of steam at greater than fracturing pressure, usually in the 10.3–11.0 MPa (1500–1600 psi) range, followed by a soak period, after which production is commenced (15).

Variations include the use of steam and the means of reducing interfacial tension by the use of various solvents. The solvent extraction approach has had some success when applied to bitumen recovery from mined tar sand but when applied to unmined material, losses of solvent and bitumen are always an obstacle. This approach should not be rejected out of hand because a novel concept may arise that guarantees minimal acceptable losses of bitumen and solvent.

Combustion has also been effective for recovery of viscous oils in moderately thick reservoirs where reservoir dip and continuity promote effective gravity drainage, or where several other operational factors permit close well spacing. During *in situ* combustion or fire flooding, energy is generated in the formation by igniting bitumen in the formation and sustaining it in a state of combustion or partial combustion. The high temperatures generated decrease the viscosity of the oil and make it more mobile. Some cracking of the bitumen also occurs, and the fluid recovered from the production wells is an upgraded product rather than bitumen itself.

The recovery processes using combustion of the bitumen are termed forward combustion or reverse combustion, depending on whether the combustion front moves with or counter to the direction of air flow. In either case, burning occurs at the interface where air contacts hot, unburned oil or, more likely, coke. Thus, if the flame front is ignited near the injection well, it propagates toward the production well (forward combustion). However, if the front is ignited near the production well, it moves in the opposite direction (reverse combustion). In forward combustion, the hydrocarbon products released from the zone of combustion move into a relatively cold portion of the formation. Thus, there is a definite upper limit of the viscosity of the liquids that can be recovered by a forward combustion process. On the other hand, because the air passes through the hot formation before reaching the combustion zone, burning is complete; the formation is left completely cleaned of hydrocarbons. In reverse combustion, some hydrocarbons are left in the formation. The theoretical advantage of reverse combustion is that the combustion products move into a heated portion of the formation and therefore are not subject to a strict viscosity limitation. However, most attempts to implement reverse combustion in field pilot installations have been unsuccessful. In many cases, the failure resulted from the onset of secondary combustion at the production well.

Using combustion to stimulate bitumen production is attractive for deep reservoirs and in contrast to steam injection usually involves no loss of heat. The duration of the combustion may be short (days) depending on requirements. In addition, backflow of oil through the hot zone must be prevented or excessive coking occurs (15,16). Another variation of the combustion process involves use of a heat-up phase, then a blow-down (production) phase, followed by a displacement phase using a fire−water flood (COFCAW process).

Mining Methods. The alternative to *in situ* processing is to mine the tar sands, transport them to a processing plant, extract the bitumen value, and dispose of the waste sand (17,18). Such a procedure is often referred to as oil mining. This is the term applied to the surface or subsurface excavation of petroleum-bearing formations for subsequent removal of the oil by washing, flotation, or retorting treatments. Oil mining also includes recovery of oil by drainage from reservoir beds to mine shafts or other openings driven into the oil rock, or by drainage from the reservoir rock into mine openings driven outside the oil sand but connected with it by bore holes or mine wells.

On a commercial basis, tar sand is recovered by mining, after which it is transported to a processing plant, where the bitumen is extracted and the sand discharged. For tar sands of 10% wt/wt bitumen saturation, 12.5 metric tons of tar sand must be processed to recover 1 m^3 (6.3 bbl) of bitumen. If the sand contains only 5% wt/wt bitumen, twice the amount of ore must be processed to recover this amount. Thus, it is clear that below a certain bitumen concentration, tar sands cannot be processed economically (19).

The Athabasca tar sands deposit in Canada is the site of the only commercial tar sands mining operations. The Suncor operation (near Fort McMurray, Alberta), started production in 1967. The Syncrude Canada project, located 8 km away, started production in 1978. In both projects, about half of the terrain is covered with muskeg, an organic soil resembling peat moss, which ranges from a few centimeters to 7 m in depth. The primary part of the overburden, however, consists of Pleistocene glacial drift and Clearwater Formation sand and shale. The total overburden varies from 7 to 40 m in thickness, and the underlying tar sand strata averages about 45 m, although typically 5−10 m must be discarded because of a bitumen content below the economic cut-off grade of ca 6% wt/wt.

Mining of the Athabasca tar sands presents two principal issues: in-place tar sand requires very large cutting forces and is extremely abrasive to cutting edges, and both the equipment and pit layouts must be designed to operate during the long Canadian winters at temperatures as low as −40°C.

There are two approaches to open-pit mining of tar sands. The first uses a few mining units of custom design, which are necessarily expensive, eg, bucket-wheel excavators and large drag lines in conjunction with belt conveyors. In the second approach, a multiplicity of smaller mining units of conventional design is employed at relatively much lower unit costs. Scrapers and truck-and-shovel operations have been considered. Each method has advantages and risks. The first approach was originally adopted by Suncor and Syncrude Canada, Ltd., with Suncor converting to large-scale truck and shovel technology in 1993.

In the Suncor pit design, the ore body is divided into two layers (benches), each nominally 23 m high. The pit floor and the dividing plane between the upper and lower bench are roughly horizontal, and 7300-t/h bucket-wheel excavators are employed as the primary mining equipment (Fig. 7). Tar sands loosened from the face of each bench by the bucket-wheels are discharged onto a series of conveyors. The overburden is stripped by an electric shovel that discharges to trucks for removal of the overburden material. Syncrude utilizes a single-bench design with four 60-m³ capacity draglines as the primary mining equipment (Fig. 8). The draglines pile tar sands in windrows along the edge of the pit; four 60,000-t/h bucket-wheels transfer the tar sands to a system of trunk conveyor belts that move the material to the extraction plant. The mining operations at the two plants differ by choice of the primary mining equipment; the bucket-wheel excavators sit on benches, whereas the draglines sit on the surface.

Bucket-wheel excavators use units having a 10 m dia digging wheel on the end of a long boom. Each wheel has a theoretical capacity of 8700 t/h, but the average output from digging is about 4500 t/h. At the rate of 122,000 t/d, tar sand can be transferred from mine to plant by a system of 152-cm wide conveyor belts and 183-cm trunk conveyors, operating at 333 m/min. The bucket-wheel excavators are supplemented by front-end loaders used to dig overburden and load it through twin chutes onto ca 135 t capacity trucks. Additional equipment is used for maintaining the haul roads and for spreading and compacting the

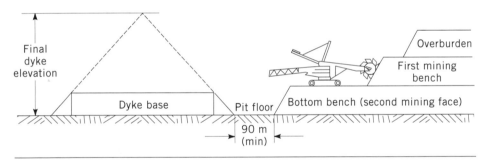

Fig. 7. Mining with a bucket-wheel excavator, where (– – –) represents the ultimate size of the dyke.

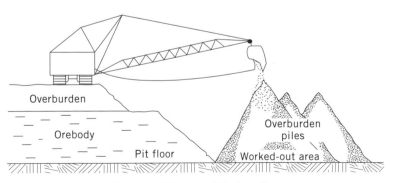

Fig. 8. Mining with dragline.

spoiled material. Overburden may be stripped with 14-m^3 hydraulically operated shovels and a fleet of ca 135 t trucks.

Draglines are equipped with a 71-m^3 bucket at the end of a 111-m boom and can be employed to dig both a portion of the overburden, which is free-cast into the mining pit, and the tar sand, which is piled in windrows behind the machine. Bucket-wheel reclaimers, similar to bucket-wheel excavators, load the tar sand from the windrows onto conveyor belts which transfer it to the plant.

Processing

Hot-Water Process. The hot-water process is the only successful commercial process to be applied to bitumen recovery from mined tar sands in North America as of 1997 (2). The process utilizes linear and nonlinear variations of bitumen density and water density, respectively, with temperature so that the bitumen that is heavier than water at room temperature becomes lighter than water at 80°C. Surface-active materials in tar sand also contribute to the process (2). The essentials of the hot-water process involve conditioning, separation, and scavenging (Fig. 9).

In the conditioning step (mixing or pulping), tar sand feed is heated and mixed with water to form a pulp of 60–85% solids at 80–90°C. First the lumps of tar sand as mined are reduced in size by ablation, ie, successive layers of lump are warmed and slough off, revealing cooler layers. The pulp is mechanically mixed, reacts with any chemicals added, is further heated to the process temperature, and is conditioned by open-steam heating in a horizontal rotating drum. With regard to equipment scale-up, conditioning is essentially a heat-transfer process. The effluent from the conditioning drum is screened to remove tramp material or lumps that were not sufficiently reduced in size. The screened pulp is mixed with any added water, adjusted to the proper consistency for pumping as described below, and sent to the separation cell.

The separation cell is an open vessel with straight sides and a cone bottom. Mechanical rakes on the bottom move the sand toward the center for discharge. Wiper arms rotating on the surface push the froth to the outside of the separation cell, where it overflows into launders for collection. The cell acts like two settlers, one on top of the other. In the lower settler sand settles down; in the upper settler, bitumen floats. The bulk of the sand in the feed is removed from the bottom of the separation cell as tailings. A large portion of the feed bitumen floats to the surface of the separation cell and is removed as froth. A middlings stream consists mostly of water, with some suspended fine minerals and bitumen particles. A portion of the middlings may be returned for mixing with the conditioning drum effluent in order to dilute the separation cell feed for pumping. The remainder of the middlings is called the drag stream, which is withdrawn from the separation cell to be rejected after processing in the scavenger cells.

Tar sand feed contains a certain portion of fine minerals that, if allowed to build up in concentration in the middlings, increases viscosity and eventually disrupts settling in the separation cell. The drag stream is required as a purge in order to control the fines concentration in the middlings. The amounts of water

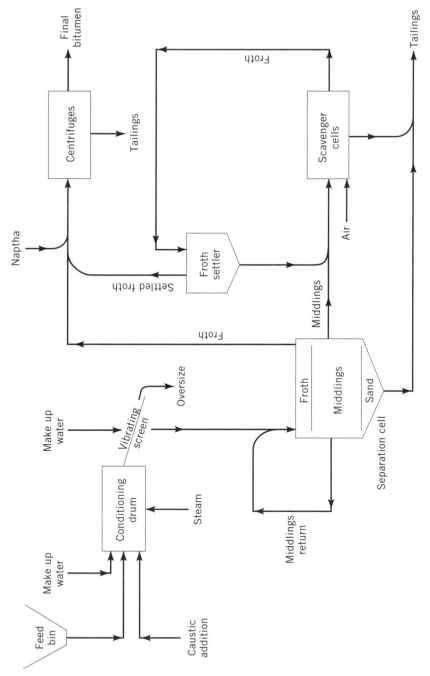

Fig. 9. Hot-water recovery process.

733

that can enter with the feed and leave with the separation cell tailings and froth are relatively fixed. Thus, the size of the drag stream determines the makeup water requirement for the separation cell.

The third step in the hot-water process is scavenging. Depending on the drag-stream size and composition, enough bitumen may leave the process in the drag stream to make another recovery step economical. Froth flotation with air is usually employed, and the scavenger froth is combined with the separation cell froth to be further treated and upgraded to synthetic crude oil. Tailings from the scavenger cell join the separation cell tailings stream and go to waste. Conventional froth-flotation cells are suitable for this step.

Froth from the hot-water process may be mixed with a hydrocarbon diluent, eg, coker naphtha, and centrifuged. The Suncor process employs a two-stage centrifuging operation, and each stage consists of multiple centrifuges of conventional design installed in parallel. The bitumen product contains 1–2 wt % mineral (dry bitumen basis) and 5–15 wt % water (wet diluted basis). Syncrude also utilizes a centrifuge system with naphtha diluent.

An attempt has been made to develop the hot-water process for the Utah sands (Fig. 10) (20). With oil-wet Utah sands, this process differs significantly from that used for the water-wet Canadian sands, necessitating disengagement by hot-water digestion in a high shear force field under appropriate conditions of pulp density and alkalinity. The dispersed bitumen droplets can also be recovered by aeration and froth flotation (21).

The hot-water separation process involves extremely complicated surface chemistry with interfaces among various combinations of solids (including both silica sand and aluminosilicate clays), water, bitumen, and air. The control of pH is critical. The preferred range is 8.0–8.5, achievable by use of any of the monovalent bases. Polyvalent cations must be excluded because they tend to flocculate clays and thus raise viscosity of the middlings in the separation cell.

One problem resulting from the hot-water process is disposal and control of the tailings. Each ton of oil sand in place has a volume of ca 0.45 m^3, which generates ca 0.6 m^3 of tailings and gives a substantial volume gain. If the mine produces 200,000 t/d of oil sand, volume expansion represents a considerable solids disposal problem.

Environmental regulations in Canada and the United States do not allow the discharge of tailings streams into the river, onto the surface, or onto any area where contamination of groundwater domains or the river may occur. The tailings stream is essentially high in clays and contains some bitumen; hence the need for tailings ponds, where some settling of the clay occurs (Fig. 11). In addition, an approach to acceptable reclamation of the tailings ponds must be accommodated at the time of site abandonment. Problems may be alleviated somewhat by the development of process options that require considerably less water in the sand–bitumen separation step. Such an option would allow a more gradual removal of the tailings ponds.

Cold-Water Process. The cold-water bitumen separation process has been developed to the point of small-scale continuous pilot plants. The process uses a combination of cold water and solvent. The first step usually involves disintegration of the tar sand charge, which is mixed with water, diluent, and reagents. The diluent may be a petroleum distillate fraction such as kerosene and is added

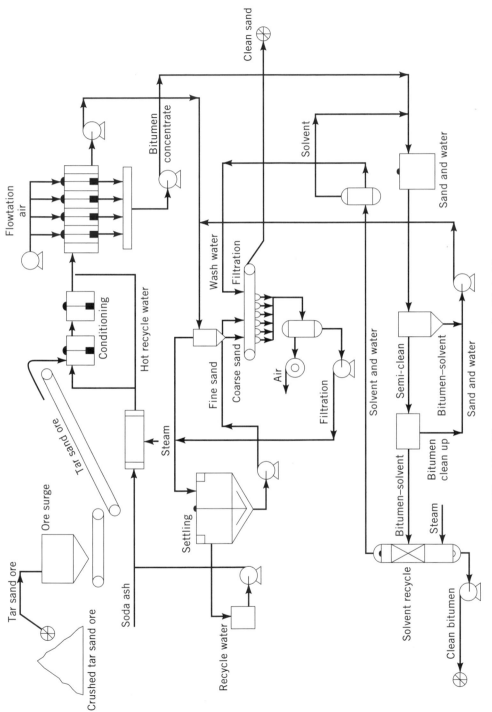

Fig. 10. Hot-water recovery for Utah bitumen.

735

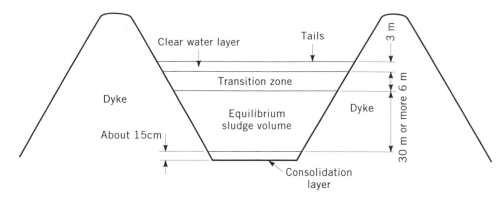

Fig. 11. Aqueous zones in tailings pond.

in a ca 1:1 weight ratio to the bitumen in the feed. The pH is maintained at 9–9.5 by addition of wetting agents and ca 0.77 kg of soda ash per ton of tar sand. The effluent is mixed with more water, and in a raked classifier the sand is settled from the bulk of the remaining mixture. The water and oil overflow the classifier and are passed to thickeners, where the oil is concentrated. Clay in the tar sand feed forms emulsions that are hard to break and are wasted with the underflow from the thickeners.

The sand reduction process is a cold-water process without solvent. The objective is removal of sand to provide a feed suitable for a fluid coking process. In the first step, the tar sand feed is mixed with water at ca 20°C in a screw conveyor at a ratio of 0.75–3 t/t tar sand (lower range preferred). The mixed pulp from the screw conveyor is discharged into a rotary-drum screen, which is submerged in a water-filled settling vessel. The bitumen forms agglomerates that are retained by an 840-μm (20-mesh) screen. These agglomerates settle and are withdrawn as oil product. The sand readily passes through the 840-μm (20-mesh) screen and is withdrawn as waste stream. Nominal composition of the oil product is 58 wt % oil (bitumen), 27 wt % mineral, and 15 wt % water.

A process called spherical agglomeration closely resembles the sand-reduction process. Water is added to tar sands and the mixture is ball-milled. The bitumen forms dense agglomerates of 75–87 wt % bitumen, 12–25 wt % sand, and 1–5 wt % water.

Solvent Extraction. An anhydrous solvent extraction process for bitumen recovery has been attempted and usually involves the use of a low boiling hydrocarbon. The process generally involves up to four steps. In the mixer step, fresh tar sand is mixed with recycle solvent that contains some bitumen and small amounts of water and mineral. Solvent-to-bitumen weight ratio is adjusted to ca 0.5. The drain step consists of a three-stage countercurrent wash. Settling and draining time is ca 30 min for each stage. After each extraction step, a bed of sand is formed and the extract drained through the bed until the interstitial pore volume of the bed is emptied. The last two steps of the process are devoted to solvent recovery from the solids. Although solvent extraction processes have been attempted and demonstrated for the Athabasca, Utah, and Kentucky tar sands, solvent losses influence economics of such processes and they have not yet been reduced to commercial practice.

Bitumen Conversion

Bitumen is a hydrogen-deficient oil that is upgraded by carbon removal (coking) or hydrogen addition (hydrocracking) (2,4). There are two methods by which bitumen conversion can be achieved: by direct heating of mined tar sand and by thermal decomposition of separated bitumen. The latter is the method used commercially, but the former has potential for commercialization (see FUELS, SYNTHETIC).

Direct Heating of Mined Tar Sand. An early process (Fig. 12) involved a coker for bitumen conversion and a burner to remove carbon from the sand (22). A later proposal suggested that the Lurgi process might have applicability to bitumen conversion (23). A more modern approach has also been developed which also cracks the bitumen constituents on the sand (24). The processor consists of a large horizontal rotating vessel which is arranged in a series of compartments, a preheating zone, and a reaction zone.

Direct coking of tar sand using a fluid-bed technique has also been tested. In this process, tar sand is fed to a coker or still, where the tar sand is heated to ca 480°C by contact with a fluid bed of clean sand from which the coke has been removed by burning. Volatile portions of the bitumen are distilled, whereas nonvolatile material is thermally cracked, resulting in the production of more liquid products and the deposition of a layer of coke around each sand grain. Coked solids are withdrawn down a standpipe, fluidized with air, and transferred to a burner or regenerator, operating at ca 800°C where most of the coke is burned off the sand grains. The clean, hot sand is withdrawn through a standpipe. Part (20–40%) is rejected and the remainder is recirculated to the coker to provide the heat for the coking reaction. The products leave the coker as a vapor, which is condensed in a receiver. Reaction off-gases from the receiver are recirculated to fluidize the clean, hot sand, which is returned to the coker.

Conversion of Separated Bitumen. The overall upgrading process by which bitumen is converted to liquid products is accomplished in two steps (Fig. 13). The first step is the primary upgrading process, which improves the hydrogen-to-carbon (H/C) ratio by either carbon removal or hydrogen addition,

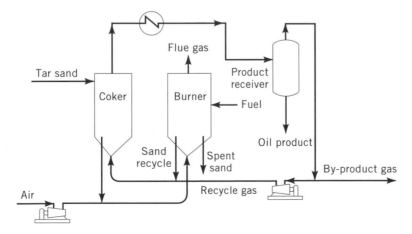

Fig. 12. Direct heating of tar sand for oil recovery.

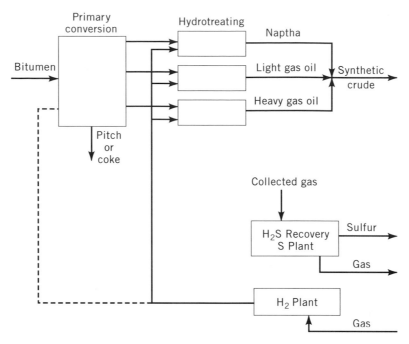

Fig. 13. Conversion of separated bitumen.

cracking the bitumen to lighter products which are more easily processed downstream. The secondary upgrading process involves hydrogenation of the primary products and is the means by which sulfur and nitrogen are removed from the primary products. The upgraded or synthetic crude can then be refined to consumer goods such as gasoline, jet fuel, and home heating oil by conventional means.

Conversion of petroleum feedstocks is accomplished using two basic process concepts: carbon rejection, of which the coking processes are examples, and hydrogen addition, of which the hydroprocesses are examples (3). The conversion of a feedstock such as tar sand bitumen has added another dimension to upgrading insofar as the feedstock is one of the most complex accepted by any refinery. Thus, coking has become the process of choice for bitumen conversion and bitumen is converted commercially by delayed coking (Suncor) and by fluid coking (Syncrude). In each case the charge is converted to distillate oils, coke, and light gases. The coke fraction and product gases can be used for plant fuel. The coker distillate is a partially upgraded material in itself and is a suitable feed for hydrodesulfurization to produce a low sulfur synthetic crude oil. In each case, bitumen conversion to liquids is on the order of 75%+. Fluid coking gives a generally higher (+1–5%) yield of liquids compared to delayed coking. The remainder appears as coke (ca 15% wt/wt) and gases.

Sulfur is distributed throughout the boiling range of the delayed coker distillate, as for distillates from direct coking. Nitrogen is more heavily concentrated in the higher boiling fractions but is present in most of the distillate fractions. Raw coker naphtha contains significant quantities of olefins and diolefins which must be saturated by downstream hydrotreating. The gas oil has a high aromatic content typical of coker gas oils.

Finishing and stabilization (hydrodesulfurization and saturation) of the liquid products are achieved by hydrotreating the liquid streams as two or three separate streams, because of the variation in conditions and catalysts necessary for treatment of a naphtha fraction relative to conditions necessary for treatment of a gas oil (13). It is more efficient to treat the liquid product streams separately and then to blend the finished liquids to a synthetic crude oil. In order to take advantage of optimum operating conditions for various distillate fractions, the Suncor coker distillate is treated as three separate fractions: naphtha, kerosene, and gas oil. In the operation used by Syncrude, the bitumen products are separated into two distinct fractions: naphtha and mixed gas oils. Each plant combines the hydrotreated fractions to form synthetic crude oil (Table 5), which is then shipped by pipeline to a refinery (see PIPELINES). Other processes which have received attention for bitumen upgrading include partial upgrading (a form of thermal deasphalting), flexicoking, the Eureka process, and various hydrocracking processes.

Partial coking or thermal deasphalting processes provide minimal upgrading of bitumen. In partial coking, the hot-water process froth is distilled at atmospheric pressure, and minerals and water are removed. In flexicoking a gasifier vessel is added to the system in order to gasify excess coke with a gas–air mixture to a low heating value gas, which can be desulfurized and used as a plant fuel. The Eureka process is a variant of delayed coking and uses steam stripping to enhance yield and produce a heavy pitch rather than coke by-product. Hydrocracking has also been proposed as a means of bitumen upgrading. The overall liquid yield of direct hydrogenation or hydrocracking of bitumen is substantially higher than that of coking, and significant amounts of sulfur and nitrogen are removed. Large quantities of external fuel or hydrogen plant feedstock are required, however. To prevent coking, the processes operate at high pressure, with direct contact between bitumen feed and circulating hydrogen. Hydrocracking processes include the H-Oil process, the LC-Fining process, the Vebe process, and the Chiyoda process.

The hydrocracker products have higher hydrogen and lower sulfur and nitrogen contents than those from the coking route and require less secondary

Table 5. Properties of Synthetic Crude Oil

Property	Value
gravity, °API	32
sulfur, wt %	0.15
nitrogen, wt %	0.06–0.10
viscosity, at 37°C, mm^2/s (=cSt)	<10
components, vol/vol %	
C$_4$	4
C$_5$/220°C	24
430/346°C	32
650/550°C	40
550°C+	0

upgrading. However, disadvantages of the hydrogen route include relatively high hydrogen consumption and high pressure operation. Processes that use conventional, eg, Co–Mo or Ni–Mo, catalysts are susceptible to metals poisoning, which may limit applicability to, or economics of, operation on feeds high in metals such as bitumen.

Health and Safety Factors

Health and safety factors associated with tar sand processing depend on the nature of the process and products (Table 6; Fig. 14). Issues arising from tar sand mining (Table 7) (25) are similar to other large-scale surface mining operations involving large equipment and the movement of huge quantities of material. The principal environmental consideration relating to the mining process is land reclamation following the completion of the mining and, in particular, those areas affected by the deposition of tailings. Both air and liquid effluents are subject to controls.

Health and safety factors in *in situ* operations are associated with high temperature, high pressure steam, or high pressure air. Environmental considerations relate to air and water quality and surface reclamation. In some environmentally sensitive areas such as the oil sands deposits in Utah, environmental considerations may make development unfeasible.

Table 6. Emissions from Tar Sand Plant

Source	Potential contaminant
process wastewater	suspended solids, dissolved solids, phenols, ammonia, oils, organics, sulfides, metals
sanitary wastewater	suspended solids, dissolved solids, biochemical oxygen demand, organics, nitrates, phosphate, residual chlorine, coliform organisms, metals
runoff from upgrader area: coke storage pile, sulfur storage pile, solid waste landfills	suspended solids, oils, organics, inorganics, sulfur, metals
power plant stacks	particulates, SO_2, NO_x, CO
sulfur plant stacks	sulfides, SO_2, H_2S, particulates
upgrader heaters	SO_2, NO_x, CO_2, hydrocarbons, particulates

Future Outlook

The government of the Canadian Province of Alberta has announced a standard royalty formula for the oil sands industry and embraced the principles and, to a large degree, put into practice the recommendations of the National Task Force on Oil Sands Strategies. The Canadian government plans to extend the mining tax regulation to include *in situ* operations. Over $3,400,000,000 (Canadian $) in new projects and expansions have been set aside to allow the industry to move forward with projects in the initial stages of development (Table 8) which should encourage further development of Canadian tar sand resources.

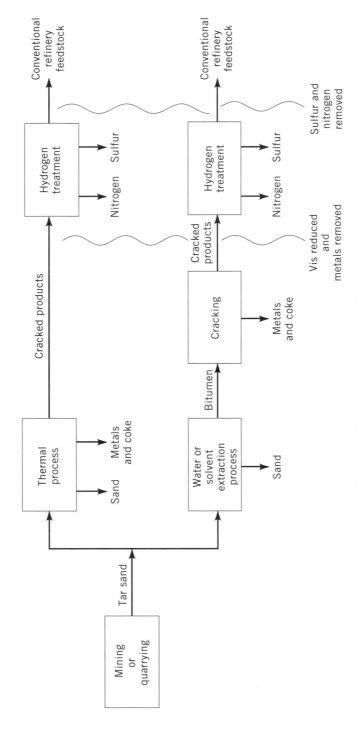

Fig. 14. Products from tar sand processing.

Table 7. Environmental Impact of Surface Mining Operations

Operation or source of impact	Surface changes			Topographic changes	Drainage diversion	Increased noise	Changes in ground water regime	
	Increased landslide risk	Destruction of existing vegetation	Alteration of habitats				Physical	Chemical
site preparation		+	+	+	+	+		
surface cleaning (cleared area)		+	+	+	+	+		
stripping (stripped area)		+	+	+			+	+
tar sand extracting (mined area)				+		+	+	+
haul road transportation (construction)		+	+	+	+	+		
tailings disposal						+		
bitumen in tailings or low grade tar sand waste		+				+		+
fines in tailings								
stripped waste			+					
solubles or water transportation particles in overburden							+	+
new surface								+
increases in surface slope from waste disposal	+							
rehandling of materials: back-filling, grading, and recontouring						+		

Table 8. Projects for Bitumen and Heavy Oil Recovery/Conversion

Project	Investment, $ \times 10^6$	Timing	Description[a]	Production, m³/d (bbl/d)[b]
Amoco				
Primrose	500	immediate	SAGD heavy oil production	6,000 (50,0000) heavy oil
Alberta Energy Co., Ltd.				
Cold Lake	13	immediate	SAGD heavy oil pilot plant	120 (1,000) heavy oil
Foster Creek	200	1997–1998	commercial heavy oil	3,600 (30,000) heavy oil
Elan Energy, Ltd.				
Lindbergh	60	immediate	SAGD heavy oil plant	increased 1,680–2,640 (14,000–22,000)
Elk Point				
Gibson Petroleum, Ltd.,	7	1996–1997	SAGD bitumen project	increased 240–540 (2,000–4,500)
Utf				
Gulf Canada Resources, Ltd.				
Surmont	30	1996	phase 1 SAGD heavy oil	360 (3,000)
		1998–1999	phase 2	increased 2,400 (20,000)
Imperial Oil Limited				
Cold Lake	250	immediate	phases 9–10 in situ bitumen production	increased 14,400; 3,000 (120,000; ~25,000+)
Cold Lake (proposed)	250	1997	phases 11–12 in situ bitumen production following completion of phases 9–10	add 2,400 (~20,000)
Imperial				
Cold Lake	40	immediate	heavy oil development below current Cold Lake lease	1,128 (9,400) heavy oil
Koch Oil Co., Ltd.				
Cold Lake		1997–1998	in situ bitumen project	6,000 (50,000)
Reita Lake				

743

Table 8. (Continued)

Project	Investment, $ × 10^6	Timing	Description[a]	Production, m^3/d (bbl/d)[b]
Shell Canada Ltd.				
Peace River				
Solvex	3	immediate	SAGD heavy oil project	increased
Bitumount	170	1997	bitumen mine/extraction complex, minerals extraction plant	168 (14,000) bitumen, 100,000 t alumina, 200,000 t synthetic silica (tailings)
Suncor Inc.				
Primrose	17	1996–1997	phase 1, SAGD pilot plant	300 (2,500) heavy oil
Burnt Lake	122	1998–1999	phase 2, commercial plant	increased 1,500 (12,500) heavy oil
Suncor Inc., Osg				
Steepbank Mine	300	2001	new bitumen mine	increased production 10,800 (90,000+) light sweet crude, custom blends
Fort McMurray	300	1998–2001	upgrader improvements, plant modifications	increased 9,600 (80,000) light sweet crude, custom blends
Fort McMurray	190	immediate	environmental measures	reduced SO$_2$ emissions
Syncrude Canada, Ltd.				
North Mine Mildred Lake	500	1998	new bitumen mine, debottlenecking plant	increased 13 × 10^6 m^3
Syncrude Canada, Ltd.				
Aurora Mine	500	2001	new bitumen, remote plant, debottleneck	first train, 2001; increased 11.3 (94 × 10^6 bbl/yr)
Total	*3,452 $ × 10^9*			

[a]SAGD = steam-assisted gravity draining.
[b]To convert m^3/d to bbl/d, multiply by 6.29.

BIBLIOGRAPHY

"Tar Sands" in *ECT* 1st ed., Vol. 13, pp. 633–645, by K. A. Clark, Research Council of Alberta, and G. B. Shea (U.S. Deposits), U.S. Department of the Interior; in *ECT* 2nd ed., Vol. 16, pp. 682–732, by F. W. Camp, Sun Oil; in *ECT* 3rd ed., Vol. 22, pp. 601–627, by D. Towson, Petro-Canada.

1. N. Berkowitz and J. G. Speight, *Fuel* **54**, 138 (1975).
2. J. G. Speight, in J. G. Speight, ed., *Fuel Science and Technology Handbook*, Part II, Marcel Dekker, Inc., New York, 1990.
3. J. G. Speight, *The Chemistry and Technology of Petroleum*, 2nd ed., Marcel Dekker, Inc., New York, 1991.
4. M. R. Gray, *Upgrading Petroleum Residues and Heavy Oils*, Marcel Dekker, Inc., New York, 1994.
5. P. H. Phizackerley and L. O. Scott, in G. V. Chilingarian and T. F. Yen, eds., *Bitumens, Asphalts, and Tar Sands*, Elsevier, Amsterdam, the Netherlands, 1978, p. 57.
6. R. F. Meyer and W. D. Dietzman, in R. F. Meyer and C. T. Steele, eds., *The Future of Heavy Crude Oil and Tar Sands*, McGraw-Hill Book Co., Inc., New York, 1981, p. 16.
7. E. J. Walters, in L. V. Hills, ed., *Oil Sands: The Fuel of the Future*, Canadian Society of Petroleum and Geology, Calgary, Alberta, Canada, 1974, p. 240.
8. L. C. Marchant and C. A. Koch, in R. F. Meyer, J. C. Wynn, and J. C. Olson, eds., *Proceedings of the 2nd International Conference on the Future of Heavy Crude and Tar Sands*, McGraw-Hill Book Co., Inc., New York, 1984, p. 1029.
9. D. Smith-Magowan, A. Skauge, and L. G. Hepler, *Can. J. Petrol. Technol.* **21**(3), 28 (1982).
10. *Heavy Oiler* (Jan. 1986).
11. M. C. Harris and J. C. Sobkowicz, in D. A. Redford and A. G. Winestock, eds., *The Oil Sands of Canada–Venezuela*, Special Publication No. 17, Canadian Institute of Mining and Metallurgy, Ottawa, Canada, 1977, p. 270.
12. J. W. Bunger, K. P. Thomas, and S. M. Dorrence, *Fuel* **58**, 183 (1979).
13. J. G. Speight, *The Desulfurization of Heavy Oils and residua*, Marcel Dekker, Inc., New York, 1981.
14. W. N. Hamilton and G. S. Mellon, in M. A. Carrigy and J. W. Kramers, eds., *Guide to the Athabasca Oil Sands Area*, Information Series No. 65, Alberta Research Council, Alberta, Canada, 1978.
15. J. Burger, in Ref. 5, p. 191.
16. R. Mungen and J. H. Nicholls, *Proc. 9th World Petrol. Cong.* **5**, 29 (1975).
17. R. A. Dick and S. P. Wimpfen, *Sci. Amer.* **243**(4), 182 (1980).
18. R. Houlihan, in Ref. 8, p. 1076.
19. F. K. Spragins, in Ref. 5, p. 92.
20. J. C. Miller and M. Misra, *Fuel Process. Technol.* **6**, 27 (1982).
21. K. E. Hatfield and A. G. Oblad, in Ref. 8, p. 1175.
22. P. E. Gishler, *Can. J. Res.* **27**, 104 (1949).
23. R. W. Rammler, *Can. J. Chem. Eng.* **48**, 552 (1970).
24. W. Taciuk, *Energy Proc. Canada* **74**(4), 27 (1981).
25. N. A. Frazier and co-workers, Report No. 68-02-1323, United States Environmental Protection Agency, Washington, D.C., 1976.

JAMES G. SPEIGHT
Western Research Institute

TARTARIC ACID. See HYDROXY DICARBOXYLIC ACIDS.

TEA

Tea, a "gift from God" according to the ancient Chinese, has been enjoyed for thousands of years (1). This fragrant brew is prepared from the leaves of the plant *Camellia sinensis*. At least two well-defined varieties of *Camellia sinensis* are recognized: *assamica*, a large-leaved (15–20-cm) plant, and *sinensis*, a smaller-leaved (5–12-cm) variety (2). Valued historically as the beverage of the social elite, as of the latter half of the twentieth century tea was a beverage of the masses and per capita consumption exceeded 40 liters annually (3). The processing, chemistry, and physiological functionality of tea beverages have intrigued scholars and tea drinkers over the millennia. These challenge researchers yet.

Tea is native to the East Asia region, especially the People's Republic of China, Burma, Laos, and Vietnam. Its first recorded use dates from the fourth century AD in China. A flourishing trade eventually developed, which led to the cultivation of tea. The modern tea industry has its origins in the spread of cultivated tea from China into Japan (ca 610 AD). Tea reached Europe in the sixteenth century and by the latter part of the seventeenth century had become a popular beverage, served in numerous tea houses in London. The cultivation of tea rapidly spread throughout the Indian subcontinent from 1878 to 1834. Development of tea plantations and migration of the technology of plantation operation from India to tropical areas in Africa (1850–1878), South America (ca 1900), Russia (Georgia) (1913), and Australia and the Pacific islands (1824–1909) led to a variety of localized practices and tea products (2,4–6).

Through cultivation tea has become an important agricultural product throughout the world, particularly in regions lying close to the equator. Geographical areas which receive annual rainfall of at least 19.7 cm/yr (50 in./yr) and have a mean average temperature of 30°C and slightly acidic soil are the most favorable for growth and agriculture of tea (2,6–8). Tea is generally pruned and maintained as a shrub-like bush of 1–1.5 m in height (7).

Traditionally, *Camellia sinensis* was propagated and bred through seeds; however, this practice led to genetic variability, loss of consistency of yield, and poorer beverage quality. Vegetative propagation has become a common practice. This helps to maintain genetic purity and aids in more rapid establishment of new productive stands of tea (Fig. 1). New clones of tea are generally selected based on criteria such as beverage quality, yield, ease of establishment, pest resistance, and frost resistance. Once established, new plantings of tea are economically viable for decades, barring diseases, pest infestation, or other destruction (8).

Tea estates range in size from small local holdings of 1 ha or less to large establishments of up to 800 ha. Harvesting worldwide is generally done by hand using a small knife. Shears, hand-held cutters powered by back-carried gasoline engines (9), and small self-propelled harvesters which straddle dome-contoured rows are used for harvesting in Japan. Mechanical harvesting methods have been developed and are popular only where labor is expensive and where tea is not grown on steep mountain-slopes. Large-scale mechanical harvesting equipment is most commonly used in Georgia, Australia, and Argentina. Tea is manufactured into a consumable product in tea factories, which are usually located near large plantations. Leaf harvested from small holdings is generally combined and processed at central factories.

Fig. 1. Clonal tea propagation: cuttings and seedlings.

New growth is harvested at intervals of 6–12 d, depending on the climatic conditions. Growth is most rapid during warm weather and heavy rainfall. In some areas, such as North India (Assam), Japan, and Russia, a period of dormancy occurs during the cold season. In South India, Sri Lanka, Indonesia, and Africa, production continues year-round, permitting more efficient use of labor and manufacturing facilities. The length of the growing season, however, has little effect on annual yield.

The flush of a tea shoot is defined as the apical bud and two new leaves below it (Fig. 2). This is the ideal target for harvesting fresh tea of optimum quality. Commonly, three or even four leaves are plucked in an attempt to increase crop yield.

Composition of Fresh Tea and Biosynthesis of Tea Polyphenols

The leaves of *Camellia sinensis* are similar to most plants in general morphology and contain all the standard enzymes and structures associated with plant cell growth and photosynthesis (10–12). Unique to tea plants are large quantities of flavonoids and methylxanthines, compounds which impart the unique flavor and functional properties of tea. The general composition of fresh tea leaves is presented in Table 1.

Flavonoids. Green tea leaves contain many types of flavonoids, the most important of which are the flavanols (catechins), the flavonols, and flavanol glycosides. Tea catechins are water-soluble, colorless substances which impart the bitter and astringent taste characteristic of green teas. Localized within the cytoplasmic region of leaf cells, the flavanols (catechins) generally make up

Fig. 2.   The flush: two leaves and a bud.

Table 1. General Composition of Fresh Tea Leaves[a]

Components	Quantity, wt %[b]
flavanols	25.0
flavonols and flavonol glycosides	3.0
polyphenolic acids and depsides	5.0
other polyphenols	3.0
caffeine	3.0
theobromine	0.2
amino acids	4.0
organic acids	0.5
monosaccharides	4.0
polysaccharides	13.0
cellulose	7.0
protein	15.0
lignin	6.0
lipids	3.0
chlorophyll and other pigments	0.5
ash	5.0
volatiles	0.1

[a]Ref. 11.
[b]On a dry weight basis.

25–40% of the water-soluble solids of tea (Table 2). Catechins are easily oxidized when catalyzed by enzymes of the general class called oxidases and autoxidize in an alkaline environment. The oxidation products of catechins are the red-brown pigments found in brewed and instant teas. They also form complexes with many

Table 2. The Principal Tea Flavanols (Catechins)

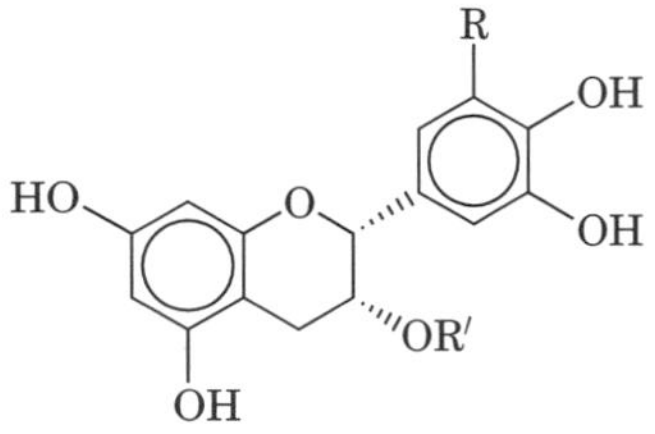

(**1**)

Name	Abbreviation	R	R'
epicatechin	EC	OH	H
epicatechin gallate	ECG	OH	gallate
epigallocatechin	EGC	OH	H
epigallocatechin gallate	EGCG	OH	gallate

other substances such as proteins and caffeine [58-08-2] (13). These polyphenolic constituents are the key reactants involved with the enzymatic fermentation of green tea to black tea. The quality of tea infusions correlates with the flavonol content of fresh green leaves (10,14). A number of flavonols including quercetin [117-39-5], kaempferol [520-18-3], myricetin [529-44-2], and their glycosides are also found in tea leaves (Table 3) (11,15,16). Flavonol glycosides generally make up 2–3% of the water-soluble solids of tea. The flavonol aglycones are not found in tea beverages owing to poor solubility in water.

Other Phenolic Compounds. There are several phenolic acids important to tea chemistry. Gallic acid (**3**) and its quinic acid ester, theogallin (**4**), have been identified in tea (17,18) and have been detected by hplc (19,20).

Table 3. The Flavonol Glycosides

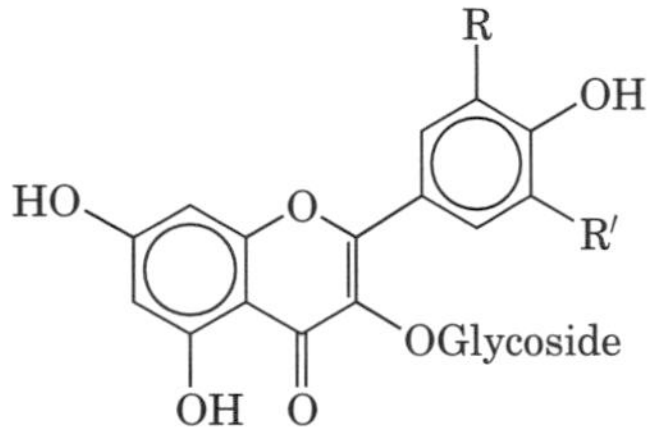

(**2**)

Name	Abbreviation	R	R'
kaempherol glycoside	KaG	H	H
quercetin glycoside	QuG	OH	H
myricetin glycoside	MyG	OH	OH

(3)   (4)

Caffeine and Other Xanthines. Tea flush contains 2.5–4.0% caffeine (**5**) on a dry weight basis and much smaller quantities of the related methylxanthine theobromine [*83-67-0*] (**6**). Whereas theophylline [*58-55-9*] has been reported to be a constituent of tea (21), there are no recent reports that theophylline can be found in tea and it is not possible to detect theophylline in tea beverages using modern analytical techniques (22). On average, a 6-ounce (180-cm^3) cup of tea contains 20–70 mg of caffeine, compared to 40–155 mg of caffeine in a 6-ounce (180-cm^3) cup of freshly brewed coffee (qv). Infusions of black, green, and oolong teas all contain about the same amounts of caffeine when prepared using similar amounts of leaves. The amount of caffeine in a tea beverage is largely determined by brewing conditions (time, temperature, leaf size, and amount of tea). Caffeine forms complexes with the polyphenolic constituents in tea and these complexes have poor solubility and often precipitate under cold storage. This precipitate is called cream because of its milky appearance. This physical and chemical property of tea affects the behavior of iced tea beverages as well as the technology of instant-tea manufacture.

(5)   (6)

Theanine and Other Amino Acids. Amino acids (qv) make up 4–8% of the soluble solids found in brewed tea. There is an amino acid unique to tea, γ-*N*-ethyl glutamine, called theanine [*3081-61-6*] (**7**) (23).

(7)

Minerals and Ash. The water-soluble extract solids which infuse from tea leaves contain 10–15% ash. The tea plant has been found to be rich in potassium (24) and contains significant quantities of calcium, magnesium (25), and aluminum (26). Tea beverages are also a significant source of fluoride (27), owing in part to the uptake of aluminum fluoride from soils (28,29).

Volatiles or Aroma. The essential oil, or aroma, of tea provides much of the pleasing flavor and scent of green and black tea beverages. Despite this, volatile components comprise only ~1% of the total mass of the tea leaves and tea infusions. Black tea aroma contains over 300 characterizing compounds, the most important of which are terpenes, terpene alcohols, lactones, ketones, esters, and spiro compounds (30). The mechanisms for the formation of these important tea compounds are not fully understood. The respective chemistries of the aroma constituents of tea have been reviewed (31).

Enzymes. The enzymes most important to the chemistry and manufacturing of tea are those responsible for the biosynthesis of tea flavonoids (Table 4) and those involved in the conversion of fresh leaf into manufactured commercial teas.

Alcohol dehydrogenase (5) and leucine α-ketoglutarate transaminase (33, 34) contribute to the development of aroma during black tea manufacturing. Polyphenol oxidase and peroxidase are essential to the formation of polyphenols unique to fermented teas.

Polyphenol oxidase (PPO) (EC 1.14.18.1; monophenol monooxygenase [tyrosinase] or EC 1.10.3.2; o-diphenol: O_2-oxidoreductase) is one of the more important enzymes involved in the formation of black tea polyphenols. The enzyme is a metallo-protein thought to contain a binuclear copper active site. The substance PPO is an oligomeric particulate protein thought to be bound to the plant membranes. The bound form of the enzyme is latent and activation is likely to be dependent upon solubilization of the protein (35). PPO is distributed throughout the plant (35) and is localized within in the mitochondria (36), the cholorplasts (37), and the peroxisomes (38). Using antibody techniques, polyphenol oxidase activity has also been localized in the epidermis palisade cells (39). Reviews on the subject of PPO are available (40–42).

Table 4. Enzymes Involved With Biosynthesis of Tea Polyphenols[a]

Enzymes	EC number
acetyl-CoA carboxylase	6.4.1.2
phenylalanine ammonia-lyase	4.3.1.5
cinnamate 4-hydroxylase	1.12.12.11
4-coumarate-CoA ligase	6.2.112
chalcone synthase	2.3.1.74
chalcone isomerase	5.5.1.6
2-hydroxyisoflavanone synthase	
flavone synthase	
(2S)-flavanone 3-hydroxylase	1.14.11.9
flavonol synthase	
dihydroflavonol 4-reductase	
flavan-3,4-cis-diol 4-reductase	

[a] Ref. 32.

PPO from tea reacts effectively with both 3'-4' and 3'-4'-5-hydroxylated catechins, with specificity for the o-diphenol (43,44). Studies defining the kinetics of PPO from tea in relation to substrate type are lacking as of this writing (ca 1997). Tea PPO has good functionality in the pH range 4.6–5.6 (43,45–48).

Peroxidase (POD) (EC 1.11.1.7) is thought to play in integral role in the fermentation process and is found in fresh green leaf (43,49,50). It is a heme-based enzyme which catalyzes the reductive decomposition of hydrogen peroxide to water, and organic peroxide species to the corresponding alcohol. PPO is thought to produce peroxide, which activates the POD system (50). However, catalase is quite active in tea and rapidly removes peroxides as they form. Along with PPO, POD plays a role in the oxidation processes involved with the formation of the black tea components.

Biosynthesis of Tea Flavonoids. The pathways for the *de novo* biosynthesis of flavonoids in both soft and woody plants (Figs. 3 and 4) have been generally elucidated and reviewed in detail (32,51). The regulation and control of these pathways in tea and the nature of the enzymes involved in synthesis in tea have not been studied exhaustively. The key enzymes thought to be involved in the biosynthesis of tea flavonoids are 5-dehydroshikimate reductase (52), phenylalanine ammonia lyase (53), and those associated with the shikimate/arogenate pathway (52). At least 13 enzymes catalyze the formation of plant flavonoids (Table 4).

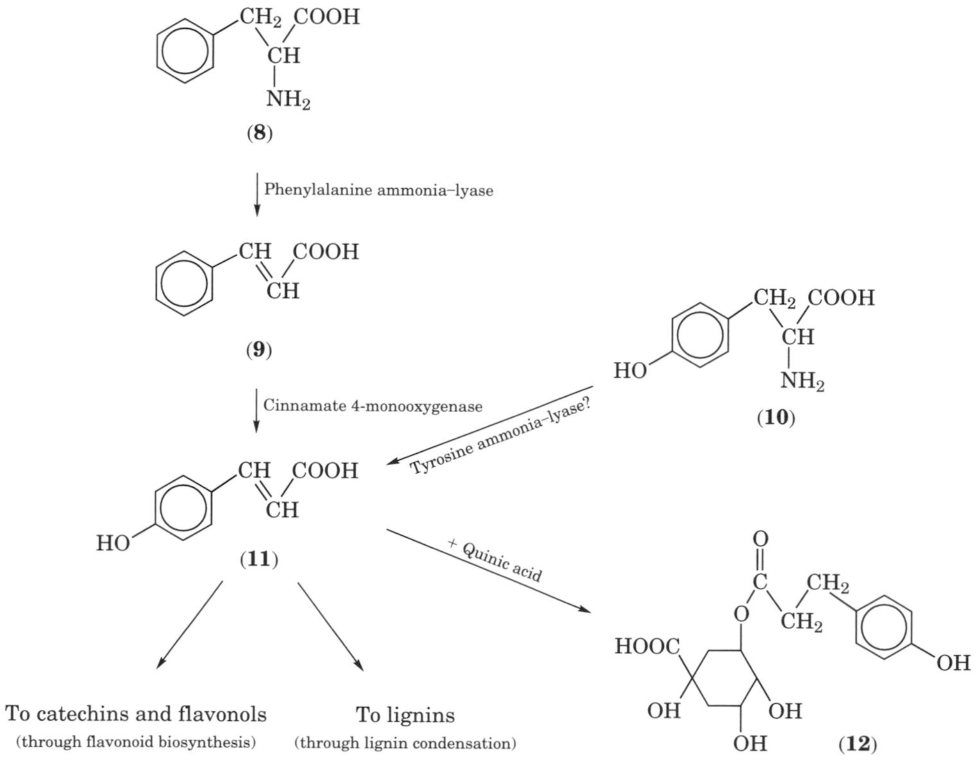

Fig. 3. Cinnamate biosynthesis from phenylalanine (**8**) to cinnamic acid (**9**) or from tyrosine (**10**) to coumaric acid (**11**). Coumarylquinic acid (**12**) is also formed.

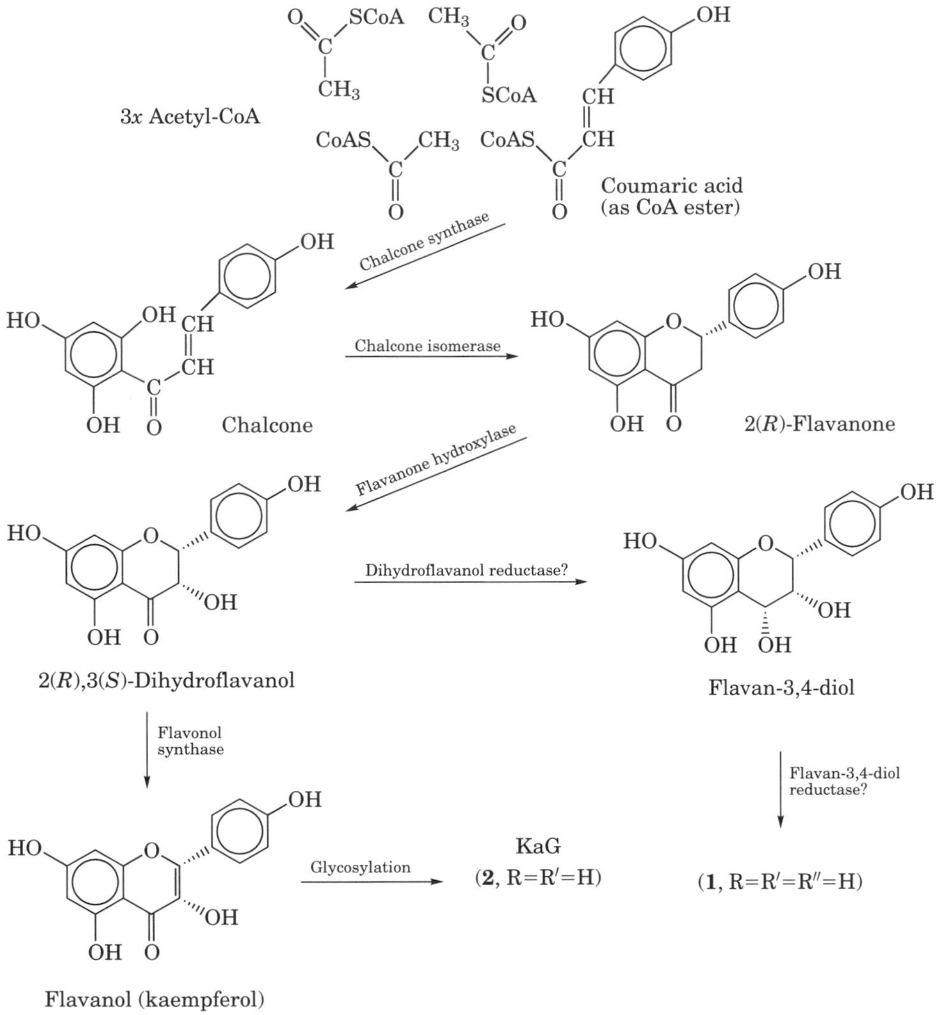

Fig. 4. Biosynthesis of flavonoids where CoA = coenzyme A.

Manufacturing

Freshly harvested tea leaves require manufacturing to be converted into green, oolong, and black teas (Fig. 5). Black tea, the dominantly manufactured tea product worldwide, is made through a polyphenol oxidase-catalyzed oxidation of fresh leaf catechins. Green tea is processed in a manner designed to prevent the enzymatic oxidation of catechins before drying. Green tea consumption is growing worldwide, but Japan, the People's Republic of China, North Africa, and the Middle East are traditionally the sites of greatest consumption. Oolong tea, a partially oxidized tea, is manufactured primarily in the People's Republic of China and Taiwan. Instant tea, usually a powder, is generally prepared by the aqueous extraction of tea leaves, followed by concentration and drying. In some cases, instant teas are prepared by removal of cold water-insoluble

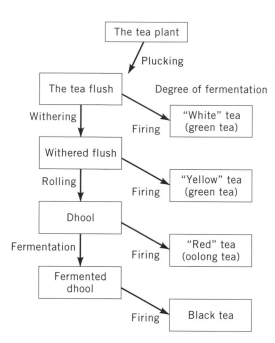

Fig. 5. Tea manufacturing processes.

constituents via filtration (qv) or centrifugation (see SEPARATION, CENTRIFUGAL). Technology, ie, tannase (galloyl esterase) treatment, exists to solubilize the usually cold-insoluble constituents of black tea. It is also possible to make instant tea from green tea or from oxidized leaf before the drying step of the process. Tea concentrates, often dried by spray drying, can also be dried by freeze drying or vacuum drying technologies.

Chemistry of Tea Fermentation Oxidation. The chemical changes which take place during the manufacture of green, oolong, and black teas are responsible for the unique color and flavors characteristic of each tea type. The most significant and well-understood of these reactions occurs during the manufacture of black tea. During the process called tea fermentation the colorless catechins (**1**) found in green tea (see Table 2) (54,55) proceed through a series of oxidative condensation reactions leading to the formation of a range of products of orange-yellow to red-brown color, plus the development of a large number of unique volatile constituents. These changes are reflected in the red amber color, reduced astringency, and more complex flavor found in black tea beverages.

Flavonol Oxidation. The fermentation process is initiated by the oxidation of catechins (**1**) to reactive catechin quinones (**13**), a process catalyzed by the enzyme polyphenol oxidase (PPO) (56). Whereas the gallocatechins, epigallocatechin, and epigallocatechin gallate, are preferred, polyphenol oxidase can use any catechin (Table 2) as a substrate. This reaction is energy-dependent and is the basis of the series of reactions between flavanoids that form the complex polyphenolic constituents found in black and oolong teas.

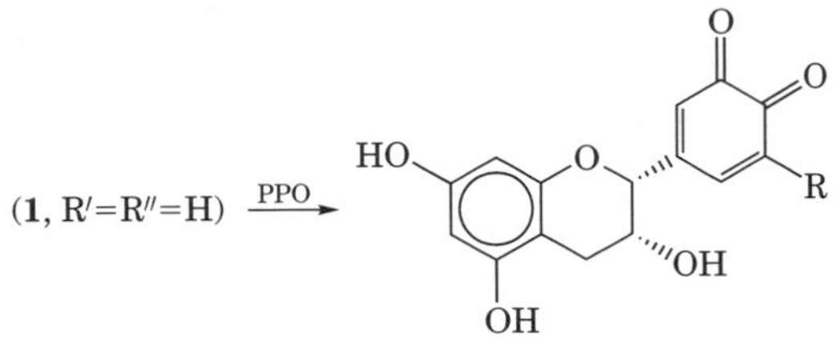

$$(1, R'=R''=H) \xrightarrow{\text{PPO}}$$

(13)

Theaflavins. One of the more well-defined groups of flavonoid polymers that forms during black tea manufacturing is that of the theaflavins (**14**). Exhibiting a bright orange-red color in solution, these are important contributors of brightness, a desirable visual attribute used by professional tasters to describe the appearance of tea infusions.

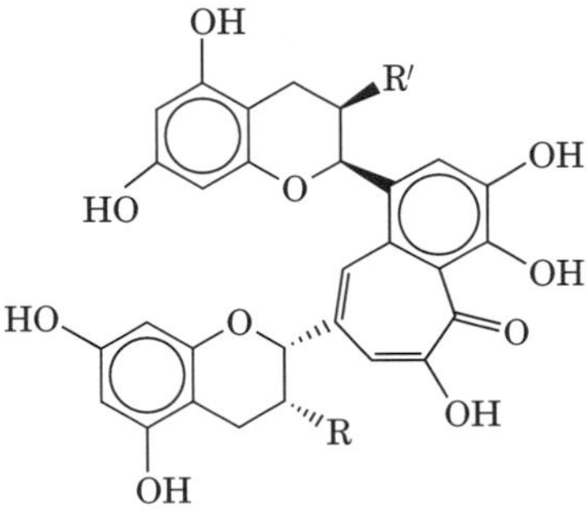

(14)

There are four main theaflavins common to black teas, and a second group of minor theaflavins, including the isotheaflavins (55) and neotheaflavins (57) (Table 5). The total theaflavin concentration in black tea leaves does not usually exceed 2% and can be as low as 0.3%. At most, only 10% of the catechins in tea flush can be accounted for as theaflavins in black tea and the fate of the remaining catechins is less clear. Theaflavins can be readily determined by direct hplc analysis of tea beverages (48,58,59).

Theaflavic Acids and Theaflagallins. Although gallic acid (**3**) is not directly oxidized by polyphenoloxidase, it can be converted to gallic acid quinone through other oxidation processes occurring during black tea fermentation (55,60). Gallic

Table 5. The Theaflavins (14)

Name	Abbreviation	R	R'
theaflavin	TF	H	H
theaflavin 3-gallate	TF3G	gallate	H
theaflavin 3'-gallate	TF3'G	H	gallate
theaflavin 3,3'-digallate	TFDG	gallate	gallate

acid reacts with catechin quinones (**13**) to form a group of compounds called theaflavic acids, eg, epitheaflavic acid (**15**), or a gallocatechin quinone to form theaflagallins such as epitheaflagallin (**16**) (61). The theaflavic acids are bright red, acidic substances present only in small quantities in black tea (11).

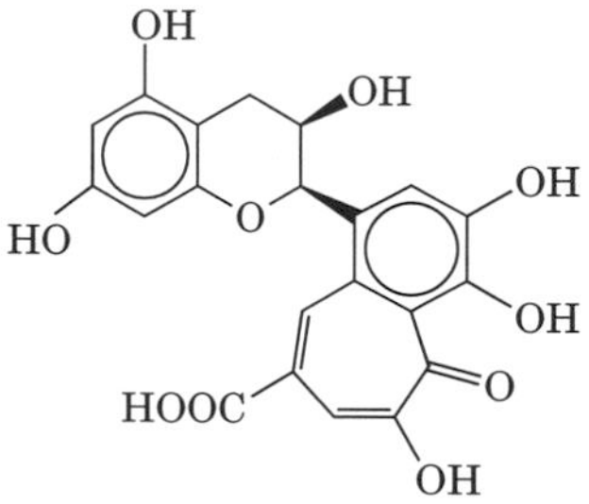

(**15**) 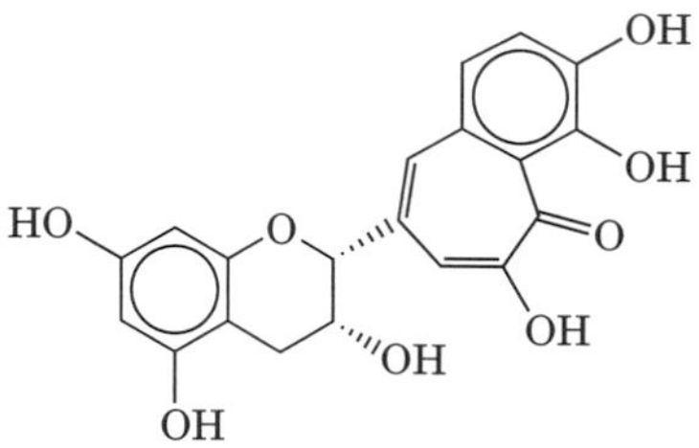 (**16**)

Bisflavanols (Theasinensins). The paired condensation of two gallocate-chins forms a group of colorless substances called bisflavanols (62,63). The bis-flavanols have been isolated and reclassified under the name of theasinensins (64) and have been found in both green and oolong teas (65). Bisflavanols such as theasinensin A (**17**, R = OH) and theasinensin F (**17**, R = H) are reactive com-pounds and can in part rearrange to form other black tea flavonoid compounds.

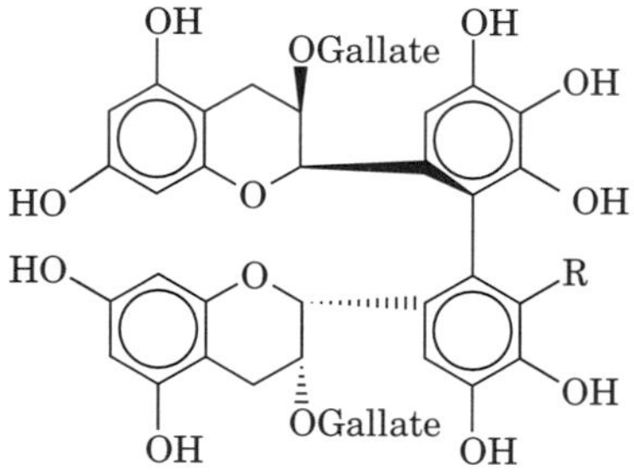

(**17**)

Complex Tea Polyphenols. The catechins (**1**) are reduced by ~85% dur-ing black tea manufacturing, yet only ~10% can be accounted for in the form of theaflavins (**14**) and theaflavic acids. The remaining ~75% of the catechins form undefined, water-soluble polyphenolic substances thought to be the pig-ments which provide tea its brown-to-black color. These compounds have histor-ically been termed thearubigens (66,67). One subgroup of this complex mixture of materials has been classified as proanthocyandin polymers (68). These form cyanidin [528-58-5] (**53**) and delphinidin [528-53-0] (**56**) upon acid hydrolysis. At-tempts to separate and purify the thearubigens of black tea via chromatography (qv) and reverse-phase hplc have not met with much success (59,69). Purification of a black tea fraction termed theafulvin (70) and oolongtheanin (71) has been accomplished. This class of polyphenolic compounds might make up a portion of the group of polyphenols of the general class historically termed thearubigens.

Black Tea. The black tea manufacturing process has evolved over hundreds of years, until the early part of the twentieth century, little was known about the chemical changes. The process consists of the unit operations of withering, rolling, fermentation, firing, and sorting (4).

Withering. Immediately after harvesting, tea leaves are brought to factories situated close to the tea gardens for manufacturing. Leaves are handled carefully to prevent bruising and to dissipate the heat generated by biochemical reactions associated with continued respiration. The leaves are distributed in thin layers on open-meshed fabric netting or, more efficiently, in troughs designed for effective air flow through deep (up to 30-cm) beds of leaf. This procedure allows more rapid loss of water without the heat buildup which is detrimental to quality.

Reduction of the leaf moisture converts the turgid leaf to a flaccid material which is easily handled without excessive fracture. The withering process takes a period of 6–18 h, depending on the weather conditions and the type of equipment employed. Biochemical changes to the tea leaves occur during withering. For example, cell membrane permeability is increased, allowing for easier disruption of the cell contents during the succeeding stage of the process; the concentration of amino acids, caffeine, organic acids, and polyphenols increases; and polyphenol oxidase is transformed from its latent to active form (72). These biochemical changes occur as a function of leaf senescence rather than owing to water loss and thus has been termed chemical withering (73).

Rolling. The next step is the initiation of the enzymic oxidation of the flavanols by molecular oxygen. This is accomplished by disruption of the cell structure to effect enzyme–substrate contact. The primary unit operation of this process, called rolling, is primarily a cutting and squeezing process but includes a twisting action to enhance leaf appearance. The traditional roller consists of a circular table, 1–1.3-m in diameter, equipped with battens. Above this, a smaller circular sleeve moves eccentrically over the surface of the table. Withered leaf is charged into the sleeve and a pressure cap is lowered into the sleeve. The result depends on the pressure and the rate of movement. Intermittently, leaf sufficiently rolled and reduced in size is separated by sieving and the remainder is rerolled. At the end of the rolling process, leaf juices are well-distributed on the surface of the cut, twisted leaf.

More modern equipment has come into use in most of the tea-growing areas. The McTear Rotorvane consists of a cylinder (20 or 37.5 cm dia) lined with vanes, through which a vaned rotorshaft propels withered leaf, thereby enhancing maceration. Rotating cutter blades are sometimes added to the shaft. Following Rotorvane treatment, leaves are frequently passed through a crush, tear, curl (CTC) machine, which consists of a pair of ridged cylindrical rollers revolving at different speeds and having a narrow clearance between rolls.

The Rotovane–CTC combination provides efficient and uniform maceration and is well suited to the establishment of a continuous manufacturing process. Macerated leaf is fluffed up by passage through a high speed Rotorvane without back pressure to eliminate matting and facilitate air diffusion. Leaf temperature should be kept below 35°C during maceration to prevent flavor deterioration.

The Legg cutter, originally designed for cutting tobacco, is widely used in North India for macerating unwithered leaf and is called an LTP process in the trade.

Fermentation.   The term fermentation arose from the misconception that black tea production is a microbial process (73). The conversion of green leaf to black tea was recognized as an oxidative process initiated by tea–enzyme catalysis circa 1901 (74). The process, which starts at the onset of maceration, is allowed to continue under ambient conditions. Leaf temperature is maintained at less than 25–30°C as lower (15–25°C) temperatures improve flavor (75). Temperature control and air diffusion are facilitated by distributing macerated leaf in layers 5–8 cm deep on the factory floor, but more often on racked trays in a fermentation room maintained at a high rh and at the lowest feasible temperature. Depending on the nature of the leaf, the maceration techniques, the ambient temperature, and the style of tea desired, the fermentation time can vary from 45 min to 3 h. More highly controlled systems depend on the timed conveyance of macerated leaf on mesh belts for forced-air circulation. If the system is enclosed, humidity and temperature control are improved (76).

During the oxidation process, leaf color changes from green to copper and a pleasant characteristic aroma develops. In most instances, the proper termination point is determined by the skill of the process supervisor (tea maker) on whose judgment the value of the final product is highly dependent. However, some attempts to control a suitable end point by instrumental techniques have been made. The fermentation step is terminated by firing (drying).

Firing.   A hot-air oven having forced circulation in a countercurrent mode is used to dry the fermented tea leaves and inactivates the key enzymes required for fermentation. The firing process generally occurs over an 18–20-min period, which is optimum for normal process efficiencies.

The firing process reduces the moisture of the fermented leaf mass from ~60–70% to 2.5–3.0%. The chemical and enzymatic reactions which occur during firing lead to important changes in leaf color owing to chlorophyll degradation (77) as well as development of many volatile constituents which characterize green, oolong, and black teas. Lower boiling aroma components are volatilized and alcohols and carbonyl compounds are oxidized during firing. Cyclic nitrogen compounds, such as pyrazines, pyridines, and quinolines, form owing to degradation of amino acids and proteins.

Grading.   Fired tea is fractionated into characterizing grades using a series of oscillating screens. The most common grades of tea in descending particle size are: orange pekoe (OP), pekoe fannings (PF), broken orange pekoe (BOP), broken orange pekoe fannings (BOPF), fannings, and dust. Dust, debris, and some fiber are removed by winnowing. Stalky materials are generally more moist than the fired leaf and may be removed using an electrostatic sorter although tea has been traditionally packed in foil-lined, plywood chests measuring $40 \times 50 \times 60$ cm^3 and holding ~60 kg, paper sacks have become very common. The use of chests is being phased out.

Process Variations.   The conventional techniques for tea manufacture have been replaced in part by newer processing methods adopted for a greater degree of automation and control. These newer methods include withering modification (78), different types of maceration equipment (79), closed systems for fermentation (80), and fluid-bed dryers (81). A thermal process has been described which utilizes decreased time periods for enzymatic reactions but depends on heat treatment at 50–65°C to develop black tea character (82). It is claimed that tannin–protein complex formation is decreased and, therefore, greater tannin extractability is achieved. Tea value is believed to be increased through use of this process.

Green Tea. Green tea is made by rolling and firing without enzymic oxidation. Fresh leaf is subjected to a rapid steaming process in a rotating cylindrical drum for ~20 s. Steamed leaf is cooled and rolled lightly in a hot-air environment. Partially dried leaf is then rolled more vigorously, dried further, and the process repeated until moisture is reduced to 3%. Drying temperatures are lower than for black tea, yielding further differences in both the volatile and nonvolatile fractions. The aroma components of green tea have been extensively studied and compared with those of fresh leaf and black tea (83). A high proportion of catechins (**1**) are retained in green tea, and the amino acid content is much higher. In the People's Republic of China, dry heat is commonly used to inactivate enzymes and can impart a smokey flavor to the tea.

Green or black tea can be prepared from any fresh leaf. In practice, however, different tea varieties and horticultural techniques are used, depending on the product desired. Plants grown for green tea production usually have smaller leaves, fewer catechins, and more amino acids (10). The green tea beverage is pale yellow to green, slightly more astringent than black tea, and has a brothy characteristic imparted by theanine and other amino acids.

Other Types. *Oolong Tea.* Oolong tea is produced in the People's Republic of China and on Taiwan. Withered leaf is lightly rolled and only partially fermented. Enzymatic action is terminated by a heating process. Vigorous rolling to achieve the desired appearance, additional heat treatment, and firing complete the process. In some respects, the beverage characteristics are between those of green tea and black tea, but oolong contains a unique array of volatile and polyphenolic compounds not found in either green or black tea.

Minor Varieties. Brick teas are prepared in the former Soviet Union and in parts of the People's Republic of China (82). These products are often cooked as a soup with butter or other fats. Flavorants such as jasmine flowers may be added during processing. Oil of bergamot is used to prepare Earl Grey tea.

Blending, Packaging, and Forms of Consumption

Blending. The tea taster plays an important role in purchasing and blending tea. The goal is usually the establishment and maintenance of a chosen standard of tea under constantly changing conditions. It is necessary to include teas from many countries and many gardens within a country in a blend to ensure constancy in flavor, color, and price over a long period of time. The tea taster's jargon includes terms such as bold, tippy, brisk, bright, dull, pungent, flavory, all of which need careful definition for use (84).

Packaging. In most countries, tea is sold in packets. In the United States, more than 90% of leaf tea is sold in tea bags, frequently packaged at 200 bags/lb (454 g). Larger, family-sized bags containing about 7 g are now very common and bags of up to 28 g are also used, especially in food service operations. An important consideration is tea bag paper quality which must be selected so as to avoid flavor contamination, allow efficient infusion, and retain tea fines. Papers must also have sufficient wet strength so as not to break open when gas is released as hot water is poured onto the bags.

Innovative developments have occurred in the form and functionality of tea bags or infusion packages. Round tea bags have been introduced into the consumer markets in both Europe and the United States (85). Pyramid or tetrahedral bags have also been developed and sold commercially in the United Kingdom and Japan (86). Functional tea bags are also being developed. Some of these bags have strings which when pulled, squeeze the bags to remove the retained liquid (70,87).

The outer packaging must protect the tea from light and moisture absorption. Polypropylene or coated cellophane outer wraps for paper board tea packages provide a barrier to loss of tea aroma and retard permeation of oxygen and foreign flavors. Low temperature improves storage stability. Properly packaged and stored teas retain acceptable flavor for about a year.

Forms of Consumption.   Hot black tea is the most common form of tea consumed worldwide. Tea in general is prepared from tea bags infused in hot water at a water-to-tea ratio of ~100:1. The beverage achieves a solids concentration of ~0.35% in ~3 min. The composition of typical green and black tea beverages as a percentage of extract solids is given in Table 6. A typical tea beverage contains 2500–3500 ppm solids. Hot black tea is generally consumed with milk and sugar in the United Kingdom and India, and neat in Europe, Asia, and the United States. In the last-named country, about 80% of all tea consumed is in the form of iced tea, a beverage which is becoming more common in the rest of the world, especially in the form of ready-to-drink canned and bottled teas. Flavoring agents such as lemon, herbs, spices, and fruit flavors are commonly added to hot or cold tea beverages.

Table 6. The Composition of Typical Green and Black Tea Beverages, % wt/wt Solids

Substance(s)	Green tea	Black tea
catechins	30	9
theaflavins		4
simple polyphenols	2	3
flavonols	2	1
other polyphenols	6	23
theanine	3	3
amino acids	3	3
peptides/protein	6	6
organic acids	2	2
sugars	7	7
other carbohydrates	4	4
lipids	3	3
caffeine	3	3
other methylxanthines	<1	<1
potassium	5	5
other minerals/ash	5	5
aroma	trace	trace

Green tea, consumed hot or cold, is most common in Japan and China and is growing rapidly in the United States owing to the numerous reports of health benefits. The market has doubled from about 110 metric tons in 1993 to just over 227 t in 1996. Green tea consumption in the United States remains a small segment (less than 2%) of the total tea market.

Instant Teas

Instant tea is manufactured in the United States, Japan, Kenya, Chile, Sri Lanka, India, and China. Production and consumption in the United States is greater than in the rest of the world. World production capacity of instant teas depends on market demand but is in the range of 8,000 to 11,000 t/yr (3). The basic process for manufacture of instant tea as a soluble powder from dry tea leaf includes extraction, concentration, and drying. In practice, the process is considerably more complicated because of the need to preserve the volatile aroma fraction, and produce a product which provides color yet is soluble in cold water, all of which are attributes important to iced tea products (88).

Extraction. Traditionally tea leaf is extracted with hot water either in columns or kettles (88,89), although continuous liquid solid-type extractors have also been employed. To maintain a relatively low water-to-leaf ratio and achieve full extraction (35–45%), a countercurrent system is commonly used. The volatile aroma components are vacuum-stripped from the extract (90) or steam-distilled from the leaf before extraction (91). The diluted aroma (volatile constituents) is typically concentrated by distillation and retained for flavoring products. Technology has been developed to employ enzymatic treatments prior to extraction to increase the yield of solids (92) and induce cold water solubility (93,94).

Decreaming. Extract is separated from leaf by centrifugation or filtration. The resulting extract is cooled and a milky precipitate, or cream, forms. Tea cream is a coacervate formed of a complex between caffeine, catechins, and the oxidized catechins of black tea. The conditions under which cream forms determine the chemical composition of the complex and its physical properties. This material is typically removed in order to obtain a product that is soluble in cold water. Several techniques have been described for the solubilization of this fraction to permit the reincorporation of the desirable substances into the product (88,95). The most common of these techniques is an air–oxygen-induced oxidation reaction, induced under alkaline conditions (96). Enzymatic treatment of tea cream using an enzyme called tannase has also been used to render cream cold water soluble (93). The addition of green teas or green tea polyphenols to black teas has been reported to block or reduce formation of cold water insoluble complexes in black tea (97). In practice, a significant portion of the cream solids are entrapped water-soluble tea constituents. Recovery of these solids through extraction using spent tea has been reported (98).

Concentration. Tea extracts are generally concentrated under vacuum to the solids content desired for drying. Freeze concentration has been described (99), as has reverse osmosis (qv) (100). Preserved aroma and the solubilized cream fraction may be added before drying.

Drying. Spray drying is the method most typically used for producing powdered teas. Freeze drying and vacuum tray drying have also been used.

Low (0.1 g/cm^3) density products are generally prepared to assure solubility and to provide for delivery of the proper weight (700 mg) of tea solids for a cup of beverage in a teaspoon (700 mg/200 mL). However, high density products are also prepared when the materials are to be reconstituted for preparation of ready-to-drink beverages, tea concentrates, or powdered drink mixes.

Other Methods.   Instant tea may also be prepared by directly extracting fermented leaf before the firing stage. These processes, only suited for practice in the tea-growing countries, have been carried out on a small scale in Sri Lanka, India, and Kenya (101).

Instant Tea-Based Products.   Powdered soft drinks and ready-to-drink teas are produced by formulating instant teas with acids, flavors, sugars, or noncaloric sweeteners. Lemon is by far the predominant flavor used but tropical, citrus, and berry flavors are also quite common.

Decaffeination

Decaffeinated teas are now commonly available, as caffeine can be removed from tea leaves or instant tea by a variety of processes. Solvent extraction of caffeine from tea leaf has been described in several patents (102). The typical solvents are ethyl acetate and methylene chloride. Ethyl acetate is an approved solvent in the United States; methylene chloride is approved for use in Canada and Europe. Supercritical carbon dioxide is also used to decaffeinate teas (see SUPERCRITICAL FLUIDS). The most recent decaffeination technologies have explored decaffeination of fresh green tea using either supercritical carbon dioxide (103) or methylene chloride, followed by fermentation of the fresh green tea to a final black tea product. This process leads to products of superior flavor because most of the characteristic flavors of black tea develop during fermentation and firing. One drawback of this approach is that it requires a source of fresh tea leaves and therefore is easily accomplished only in the growing regions. Instant teas can be decaffeinated by solvent (ethyl acetate or methylene chloride) extraction and solid-phase adsorption (qv) processes.

Physiological and Health Effects of Tea

There are numerous synthetic and natural compounds called antioxidants which regulate or block oxidative reactions by quenching free radicals or by preventing free-radical formation. Vitamins A, C, and E and the mineral selenium are common antioxidants occurring naturally in foods (104,105). A broad range of flavonoid or phenolic compounds have been found to be functional antioxidants in numerous test systems (106–108). The antioxidant properties of tea flavonoids have been characterized using models of chemical and biological oxidation reactions.

Chemical Antioxidant Systems.   The antioxidant activity of tea extracts and tea polyphenols have been determined using *in vitro* model systems which are based on hydroxyl-, peroxyl-, superoxide-, hydrogen peroxide-, and oxygen-induced oxidation reactions (109–113). The effectiveness of purified tea polyphenols and crude tea extracts as antioxidants against the autoxidation of fats has been studied using the standard Rancimat system, an assay based on air oxidation of fats or oils. A direct correlation between the antioxidant index of a tea extract and the concentration of epigallocatechin gallate in the extract was found (107).

A model system which determines the oxygen radical absorbance capacity (ORAC) was used to evaluate the antioxidant potency of extracts of teas or vegetables in reactions against hydroxyl or peroxyl radicals (109). Green and black tea extracts were found to be eight times more active in blocking peroxyl radical-induced oxidation reactions than any vegetable extract tested. Teas were also effective in blocking hydroxyl-induced oxidation reactions.

The total antioxidant activity of teas and tea polyphenols in aqueous phase oxidation reactions has been determined using an assay based on oxidation of 2,2'-azinobis-(3-ethyl benzothiazoline-sulfonate) (ABTS) by peroxyl radicals (114–117). Black and green tea extracts (2500 ppm) were found to be 8–12 times more effective antioxidants than a 1-mM solution of the water-soluble form of vitamin E, Trolox. The most potent antioxidants of the tea flavonoids were found to be epicatechin gallate and epigallocatechin gallate. A 1-mM solution of these flavanols were found respectively to be 4.9 and 4.8 times more potent than a 1-mM solution of Trolox in scavenging an ABTS$^+$ radical cation.

Biological Antioxidant Models. Tea extracts, tea polyphenol fractions, and purified catechins have all been shown to be effective antioxidants in biologically-based model systems. A balance between oxidants and antioxidants is critical for maintenance of homeostasis. Imbalances between free radicals and antioxidants may be caused by an increased production of free radicals or decreased effectiveness of the antioxidants within the reaction system. These imbalances can be caused by the radicals overwhelming the antioxidants within the system, or by an excess of antioxidants leading to a prooxidant functionality (105–118). When antioxidant defense systems are consistently overwhelmed by oxidative reactions, significant damage can occur, leading to the development of chronic diseases such as cancer and coronary heart disease (105,119–121).

Coronary Heart Disease. A theory for atherogenesis (120) has been developed whereby oxidation of low density lipoprotein (LDL) within the arterial wall is the critical first step. It has been hypothesized that sufficient intake of antioxidants would prevent oxidation of LDL and reduce development of coronary heart disease (122). Interest in determining the role of antioxidants in blocking LDL oxidation has led to the development of *in vitro* test systems.

Tea extracts and tea polyphenols inhibit copper- and peroxide-induced oxidation of LDL *in vitro* (116,123,124). The inhibitory concentration for 50% reduction (IC$_{50}$) values for inhibition of copper-induced oxidation of LDL by some phenolic antioxidants are listed in Table 7. The IC$_{50}$ for epigallocatechin gallate was found to be 0.075 μM, which was the most potent of all the phenolic antioxidants tested (123,124). Similar results have been reported elsewhere (115,116,125,126).

Cancer. Carcinogenesis may be inhibited or even prevented by numerous factors, including diet, cessation of smoking, or the use of blocking or suppressive agents (127–129). Mutation of protooncogenes and tumor suppressor genes are important steps in the development of cancer (130). Oxidative reactions are a possible cause of these mutational events. The role of dietary constituents (127–129) and tea (131) in prevention of cancers has been the subject of numerous scientific

Table 7. Antioxidant Potency of Vitamins and Phenolics Based on LDL Oxidation *In Vitro*[a]

Compound	IC$_{50}$, μM
trolox	1.26
vitamin E	1.45
beta-carotene	4.30
EGCG	0.075
EGC	0.097
ECG	0.142
catechin	0.187
genistein	14.3
naringenin	>16.0
quercetin	0.224
rutin	0.512
gallic acid	1.25

[a]Adapted from Ref. 123.

papers. The role of tea and phenolic antioxidants in control of carcinogenesis has been reviewed in detail (121,131–133).

Animal studies have shown that teas are effective in blocking or slowing carcinogenesis (121,131,133). Administration of teas or tea polyphenols to mice or rats have also been shown to decrease oxidative biomarkers, suggesting that tea polyphenols act as antioxidants (125,134).

Studies to determine the physiological effects of tea consumption associated with antioxidant activity and other relevant biomarkers of cancer risk have been conducted with human volunteers. Glucuronide and sulfate conjugates of tea catechins can be measured in human blood plasma at levels of up to 200 ng/mL after consumption of 1–2 cups of green tea (135), demonstrating that tea polyphenols are absorbed. Consumption of black and green teas (300 mL) resulted in an increase in the antioxidant status of blood plasma (136). Two human trials have found that tea drinking prevented oxidative damage to genetic material of blood cells, ie, white blood cell micronuclei formation (137) and sister chromatid exchange frequency (138) induced by tobacco smoking. Tea drinking was also found to block nitrosation reactions in human subjects (139). These studies demonstrate that tea polyphenols are absorbed into the body and appear to have physiological effects that are consistent with antioxidant activity.

BIBLIOGRAPHY

"Tea" in *ECT* 1st ed., Vol. 13, pp. 656–666, by G. F. Mitchell, General Foods Corporation; "Tea" in *ECT* 2nd ed., Vol. 19, pp. 743–755, by E. Hainsworth, Tea Research Institute of East Africa; in *ECT* 3rd ed., Vol. 22, pp. 628–644, by H. Graham, Thomas J. Lipton, Inc.

1. W. I. Kaufman, *The Tea Cookbook*, Doubleday Co., New York, 1966.
2. International Tea Committee, Ltd., *Annual Bulletin of Statistics*, London, 1995.
3. B. Banerjee, in K. C. Wilson and M. N. Clifford, eds., *Tea: Cultivation to Consumption*, Chapman and Hall, London, 1992, p. 25.

4. W. H. Ukers, *All About Tea*, The Tea and Coffee Trade & Journal Co., New York, 1935.

5. J. Weatherstone, in Ref. 3, pp. 1–23.

6. T. Eden, *Tea*, 3rd ed., Longman Group, London, 1976.

7. C. R. Harler, *The Culture and Marketing of Tea*, 3rd ed., Oxford University Press, London, 1964.

8. K. C. Wilson, in Ref. 3, pp. 201–263.

9. B. Banerjee, *Two Buds* **27**, 80 (1980).

10. R. L. Wickermasinghe, in C. O. Chichester, ed., *Advances in Food Research*, Vol. 24, Academic Press, Inc., New York, 1978, pp. 229–286.

11. G. W. Sanderson in V. C. Runeckles, ed., *Structural and Functional Aspects of Phytochemistry*, Academic Press, Inc., New York, 1972, p. 231.

12. D. J. Millin and D. W. Rustridge, *Process Biochem.* **2**, 9 (1967).

13. E. A. H. Roberts in T. A. Geissman, ed., *The Chemistry of Flavonoid Components*, The Macmillan Co., New York, 1962, pp. 471–479.

14. I. S. Bahtia and Ullah, M. R., *J. Sci. Food Agric.* **19**, 535 (1968).

15. M. G. L. Hertog, P. C. H. Hollman, and B. Van de Putte, *J. Agric. Food Chem.* **41**, 1241 (1993).

16. A. Finger, S. Kuhr, and U. Engelhardt, *J. Chromat.* **634**, 293 (1992).

17. R. A. Cartwright and E. A. H. Roberts, *J. Sci. Food Agric.* **5**, 593 (1954).

18. R. A. Cartwright and E. A. H. Roberts, *J. Sci. Food Agric. Chem. Ind.*, 230 (1955).

19. R. G. Bailey, I. McDowell, and H. E. Nursten, *J. Sci. Food Agric.* **52**, 509 (1990).

20. F. Hashmito, G. Nonoka, and I. Nishioka, *Parm. Bull.* **40**, 1983 (1992).

21. H. N. Graham, *Tea: The Plant and Its Manufacture: Chemistry and Consumption of the Beverage*, Alan R. Liss, Inc., New York, 1984, pp. 30–48.

22. M. B. Hicks, Y-H. P. Hsieh, and L. N. Bell, *Food Res. Int. 1996* **29**(3–4), 325–330 (1996).

23. Y. Sakato, *J. Agric. Chem. Soc. Japan*, **23**, 262 (1950).

24. G. W. Sanderson and co-workers, in G. Charalambous and I. Katz, eds., *Sulfur and Nitrogen Compounds in Food Flavors*, American Chemical Society, Washington, D.C., pp. 14–16, 1976.

25. J. M. Kalita and P. K. Mahante, *J. Sci. Food Agric.* **62**, 103 (1993).

26. E. M. Chenery, *Plant Soil* **6**, 174 (1955).

27. M. Elivin-Lewis, M. Vitali, and T. Kopjas, *J. Prev. Dentistry* **6**, 273 (1980).

28. H. Yamada and T. Hattori, *Jpn. J. Soil Sci. Plant Nutr.* **51**, 361 (1980).

29. T. K. Takeo, *Phytochemistry* **20**, 2145 and 2149 (1981); O. G. Vitzthum, P. Werkhoff, and P. Hubert, *J. Agric. Food Chem.* **23**, 999 (1975).

30. J. M. Robinson and P. O. Owuor, in Ref. 3, pp. 603–639.

31. G. W. Sanderson and H. N. Graham, *J. Agric. Food Chem.* **21**, 576 (1973).

32. W. Heller and G. Forkmann, in J. B. Harborne, ed., *The Flavonoids: Advances in Research Since 1986*, Chapman and Hall, London, 1994, pp. 499.

33. N. E. Tolbert, *Plant Physiol.* **51**, 234 (1973).

34. M. A. Bokuchava, T. K. Shalamberidze, and G. A. Soboleva, *Dokl. Akad. Nauk. SSSR* **192**, 1374 (1970).

35. S. V. Durmishidzern and G. N. Puridze, *Soviet Plant Physiol.* **27**, 1064 (1980).

36. E. A. H. Roberts, *Biochem. J.* **35**, 909 (1941).

37. C. Kata, I. Uritani, R. Saijo, and T. Takeo, *Plant Cell Physiol.* **17**, 1045 (1976).

38. R. L. Wickremashinghe, G. R. Roberts, and K. P. W. C. Perera, *Tea Q.* **38**, 309 (1967).

39. J. Zawistowski, C. G. Biladeris, and N. A. Eskin, in D. S. Robinson, ed., *Oxidative Enzymes in Foods*, Elsevier, London, 1991, pp. 217–273.

40. S. R. Whitaker, *Food Sci. Tech.* **61**, 543 (1994).

41. J. C. Steffens, E. Harel, and M. D. Hunt, *Recent Adv. Phytochem.* **28**, 275 (1994).

42. R. P. F. Gregory and D. S. Bendall, *Biochem. J.* **101**, 569 (1966).
43. L. Vamos-Vigyazo, *CRC Critic. Rev. Food, Sci. Nutr.* **15**, 49 (1981).
44. T. Takeo, *Agric. Biol. Chem.* **29**, 558 (1965).
45. T. Takeo and L. Uritani, *Agric. Biol. Chem.* **30**, 155 (1946).
46. K. P. W. C. Perera and R. L. Wickremashinghe, *Tea. Q.* **43**, 153 (1972).
47. A. Robertson and D. S. Bendall, *Phytochem.* **22**, 883 (1983).
48. M. A. Bokuchava and V. R. Popov, *Akad. Nauk. SSSR* **60**, 619 (1948).
49. Y. Jiang and P. W. Miles, *Phytochem.* **33**, 29 (1993).
50. J. C. Jain and T. Takeo, *J. Food Biochem.* **8**, 243 (1984).
51. H. A. Stafford, *Ann. Rev. Plant Physiol.* **15**, 459 (1974).
52. K. Iwasa, *Jpn. Agric. Res. Q.* **10**, 89 (1976).
53. E. A. H. Roberts, *Biochem. J.* **33**, 218 (1939). Other work by Roberts is reported in a series of papers through 1963.
54. M. A. Bokuchava and N. I. Skobeleva, in I. D. Morton and A. J. Macleod, eds., *Food Flavors, Part B.: The Flavors of Beverages*, Elsevier Science Publishers BV, 1986, Amsterdam, the Netherlands, p. 49.
55. J. M. Robinson and D. O. Owuor, in Ref. 3, p. 603.
56. D. T. Coxon, A. Holmes, and W. D. W. Ollis, *Tetrahedron Lett.* **60**, 5241 (1970).
57. B. Steinhaus and U. H. Englehardt, *Z. Lebensm. Unters. Forsch.* **188**, 509 (1989).
58. J. E. Berkowitz, P. Coggon, and G. W. Sanderson, *Phytochemistry* **16**, 2271 (1971).
59. A. Kiechne and U. H. Englhardt, *Z. Lebensm. Unters. Forsch.* **202**, 299 (1966).
60. G. Nonaka, F. Hashimuto, and I. Nishioka, *Chem. Pharm. Bull.* **34**, 61 (1986).
61. Y. Takino, *Jpn. Agric. Res. Q.* **12**, 94 (1978); D. J. Cattell and H. E. Nursten, *Phytochemistry* **16**, 1269 (1977).
62. L. Vuataz and H. Brandenberger, *J. Chromatogr.* **5**, 17 (1961).
63. G. Nonaka, O. Kawahara, and I. Nishuika, *Chem. Pharm. Bull.* **31**, 3906 (1983).
64. F. Hashimoto, G. Nonaka, and C. Nishioka, *Chem. Pharm. Bull.* **36**, 1676 (1988).
65. A. G. Brown, W. B. Eyton, and W. D. Ollis, *Phytochemistry* **8**, 2333 (1969).
66. E. A. H. Roberts, R. A. Cartwright, M. Oldschool, *J. Sci. Food Agric.* 8, 72−80, (1957).
67. E. A. H. Roberts, *J. Sci. Food Agric.* **9**, 381−390 (1958).
68. R. G. Bailey, H. E. Nursten, and C. McDowell, *J. Sci. Food Agric.* **59**, 365 (1992).
69. R. G. Bailey, H. E. Nursten, and I. McDowell, *J. Chrom. A.* **62**, 101−112 (1994).
70. U.S. Pat. 5,552,164 (Sept. 3, 1996), R. H. A. Haak, J. J. Kuipers, and C. S. McLean, (to Thomas J. Lipton Co.).
71. Hashimoto and co-workers, *Chem. Pharm. Bull.* **36**, 1076 (1988).
72. G. W. Sanderson, *Tea Q.* **35**, 146 (1964).
73. *The Tea Cyclopedia*, Whittingham, London, 1882, pp. 211−212.
74. J. B. Cloughley, *J. Sci. Food Agric.* **31**, 911 (1980).
75. Brit. Pat. 1,268,231 (Mar. 22, 1972) (to the Tea Research Association).
76. R. L. Wichremashinghe, *J. Natl. Sci. Counc. Sri Lanka* **1**, 111 (1973).
77. Brit. Pat. 2,026,668 (Feb. 6, 1980), D. W. Brooks (to Hambro Machinery Ltd.).
78. J. B. Cloughley and R. T. Ellis, *J. Sci. Food Agric.* **31**, 924 (1980).
79. Brit. Pat. 1,274,002 (May 10, 1972), A. C. K. Krishmaswami and C. Hariprased (to Walker & Greig).
80. Brit. Pat. 1,484,540 (Sept. 1, 1977), D. Kirtisimghe and D. P. Ranasinghe (to the Tea Research Institute).
81. M. A. Bokuchava and N. I. Skobeleva, in T. E. Furia, ed., *CRC Critical Reviews in Food Science and Nutrition*, CRC Press, Boca Raton, Fla., 1980, p. 303.
82. T. Yamanishi, *Nippon Nogei Kagaku Kaishi* **49**, 1 (1975).
83. C. R. Harier, *Tea Manufacture*, Oxford University Press, London, 1963, pp. 102−105.
84. *Annual Bulletin of Statistics*, International Tea Committee, London, 1981.
85. U.S. Pat. 5,233,813 (Aug. 10, 1993), A. G. Kenney and J. D. Wood (to AG Patents Ltd.).

86. Brit. Pat. 2,256,415 (Dec. 9, 1992), J. Kataoka (to Kataoka Bussan KK).

87. U.S. Pat. 5,552,164 (Sept. 3, 1996), J. Kuipers and C. S. McLean, (to Thomas J. Lipton Co.).

88. N. D. Pintaro, *Tea and Soluble Products Manufacture*, Noyes Data Corp., Park Ridge, N.J., 1971; M. Saltmarsh in Ref. 3, pp. 535–553.

89. U.S. Pat. 2,902,368 (Sept. 1, 1959), E. Seltzer and F. A. Saporito (to Thomas J. Lipton, Inc.); U.S. Pat. 3,451,823 (June 24, 1969), A. R. Mishkin, W. C. March, A. W. Fobes, and J. L. Ohler (to Agico SA).

90. U.S. Pat. 2,927,860 (Mar. 8, 1960), E. Seltzer and F. A. Saporito (to Thomas J. Lipton Co.).

91. Brit. Pat. 946,346 (Jan. 8, 1964), (to Afico, SA).

92. U.S. Pat. 4,483,876 (Nov. 20, 1984), B. R. Peterson (to Novo Industri A/S).

93. U.S. Pat. 3,959,497 (May 25, 1996), Y. Takino (to The Coca-Cola Co.).

94. U.S. Pat. 4,639,375 (Jan. 27, 1987), C. H. Tsal (to Procter & Gamble Co.).

95. U.S. Pat. 3,151,985 (Oct. 6, 1964), A. Forbe (to Afico, SA); U.S. Pat. 3,787,590 (Jan. 22, 1974), B. Borders, H. Rivkowich, and W. C. Rehman (to Tetley, Inc.); Brit. Pat. 1,294,932 (Oct. 12, 1971), (to Brooke Bond Ltd.).

96. U.S. Pat. 3,163,539 (Dec. 29, 1964), W. E. Barch (to Standard Brands, Inc.).

97. U.S. Pat. 4,680,193 (July 14, 1987), T. L. Lander, B. Hoffmann, and C. M. Nielsen, (to Nestec SA).

98. Eur. Pat. 699,393 A1 (Mar. 6, 1996), T. L. Lunder (to Societé Des Produits Nestlé SA).

99. U.S. Pat. 3,598,608 (Aug. 10, 1971), N. Geniaris (to Struthers Scientific and International Corp.).

100. DE Pat. 3,025,095 (1981), M. Buhler, M. Olofsson (to Nestle).

101. U.S. Pat. 3,649,297 (Mar. 14, 1970), D. J. Millin (to Tenco Brooke Bond, Ltd.); U.S. Pat. 3,392,028 (July 9, 1968), L. Vuataz (to Afico, SA).

102. U.S. Pat. 4,167,589 (Sept. 11, 1979), O. Vitzhum and P. Hubert.

103. Ger. Pat. 3,414,767 (Nov. 7, 1985), D. E. Wolnzach (to Hopfenextraktion HVG Barth, Raiser & Co.).

104. M. Namiki, *Antioxidants/Antimutagens In Food, Crit. Rev. Food Sci. Nutr.* **29**(4), 273 (1990).

105. B. Frei, *Am. J. Med.* **97**(Suppl. 3A), 5–13 (Sept. 25, 1994).

106. C. T. Ho, *Phenolic Compounds In Food and Their Effects On Health II—Antioxidants and Cancer Prevention*, ACS Symposium Series 507, American Chemical Society, Washington, D.C., pp. 2–7, 1992.

107. T. L. Lunder, *ibid.*, pp. 114–120.

108. T. Osawa, in Ref. 106, pp. 135–149.

109. G. Cao, E. Sofic, and R. L. Prior, *J. Agric. Food Chem.* **44**, 3426–3431 (1996).

110. B. Zhao, X. Li, R. He, S. Cheng, and X. Wenjuan, *Cell Biophysics*, Vol. 14, The Humana Press Inc., Clifton, N.J., pp. 175–185, 1989.

111. G. C. Yen and H. Y. Chen, *J. Agric. Food Chem.* **43**(1), 27–32 (1995).

112. J. P. Hu and co-workers, *Structure-Activity Relationship of Flavonoids with Superoxide Scavenging Activity*, Biological Trace Element Research, Vol. 47, The Humana Press Inc., Clifton, N.J., pp. 327–331, 1995.

113. S. V. Jovanovic, Y. Hara, S. Steenken, and M. Simic, *J. Am. Chem. Soc.* **117**, 9881–9888 (1995).

114. C. A. Rice-Evans, N. J. Miller, and G. Paganga, *Free Rad. Biol. Med.* **20**(7), 933–956 (1996).

115. C. A. Rice-Evans and N. J. Miller, *Bioactive Components of Food*, Biochemical Society Transactions, Vol. 24, Biochemical Society, London, pp. 790–794, 1996.

116. N. J. Miller, C. Castelluccio, L. Tijburg, and C. A. Rice-Evans, *FEBS Lett.* **392**, 40–44 (1996).

117. N. Salah and co-workers, *Arch. Biochem. Biophys.* **322**(2), 339–346 (Oct. 1, 1995).

118. O. I. Aruom, A. Murcia, J. Butler, and B. Halliwell, *J. Agric. Food Chem.* **41**, 1880–1885 (1993).

119. N. Ramarathnam, T. Osawa, H. Ochi, and S. Kawakishi, *Trends Food Sci. Technol.* **6**, 75–82 (Mar. 1995).

120. S. M. Grundy, *Clin. Cardiol.* **16**(Suppl. 1) (Apr. 1993).

121. J. H. Weisburger, *Handbook of Antioxidants*, Marcel Dekker, Inc., New York, Chapt. 15, pp. 469–482, 1996.

122. M. J. Stampfer and co-workers, *N. Engl. J. Med.* **328**(20), 1444–1449 (May 20, 1993).

123. J. A. Vinson, Y. A. Dabbagh, M. M. Serry, and J. Jang, *J. Agric. Food Chem.* **43**, 2800–2802 (1995).

124. J. A. Vinson and co-workers, *J. Agric. Food Chem.* **43**, 2798–2799 (1995).

125. I. Tomita and co-workers, in R. G. Cutler and co-workers, eds., *Oxidative Stress and Aging*, Birkhauser Verlag, Basel, Switzerland, pp. 355–365, 1995.

126. S. Miura, J. Watanabe, T. Tomita, M. Sano, and I. Tomita, *Bio. Pharm. Bull.* **17**(12), 1567–1572 (1994).

127. J. H. Weisburger and A. Rivenson, *Oncology* **17**(11) (Suppl.), 19–25 (1995).

128. P. Greenwald, *Sci. Am.*, 96–99 (Sept. 1996).

129. L. Wattenberg, M. Lipkin, C. W. Boone, and G. J. Kelloff, *Cancer Chemoprevention*, CRC Press, Inc., Boca Raton, Fla., 1992.

130. R. A. Weinberg, *Sci. Am.*, 62–70 (Sept. 1996).

131. C. S. Yang and Z. Y. Wang, *J. Natl. Cancer Inst.* **85**(13), 1038–1049 (July 17, 1993).

132. G. D. Stoner and H. Mukhtar, *J. Cellular Biochem.* (Suppl. 22), 169–180 (1995).

133. S. K. Katiyar and H. Mukhtar, *Exp. Stud. (Rev.), Int. J. Oncol.* **8**, 221–238 (1996).

134. H. Wei and K. Frenkel, *Carcinogenesis*, **14**(6), 1195–1201 (1993).

135. M. J. Lee and co-workers, *Biomarkers Prev.* **4**, 393–399 (June 1993).

136. M. Serafini, A. Ghiselli, and A. Ferro-Luzzi, *Eur. J. Clin. Nutr.* **50**, 28–32 (1996).

137. K. Xue and co-workers, *Int. J. Cancer* **50**, 702–705 (1992).

138. J. S. Shim and co-workers, *Cancer Epidemiol. Biomarkers Prev.* **4**(14), 387–391 (June 1995).

139. W. Y. Ning, W. H. Zhou, L. J. Sheng, and H. Chi, *Biomed. Envir. Sci.* **6**(3), 237–258 (1993).

Douglas A. Balentine
Lipton

TECHNETIUM.　See Radioactive Tracers.

TECHNICAL SERVICE

The principal objective of technical service in the chemical industry is to provide timely and professional information and support to downstream customers regarding chemical products and their uses. It is neither cost-effective nor necessary for a consumer of chemical products to develop a staff of specialists having detailed expertise in all aspects of chemical raw materials and their uses, particularly in a time of increasingly complex and rapidly technologically driven economies. Rather, this variety of expertise is provided in the chemical marketplace by technical service professionals whose knowledge and skills are made available by chemical products suppliers. As such, successful chemical companies provide technical service as a critical element of their offerings to the marketplace making use of this aspect of the value chain to enhance their competitiveness.

Technical service as a field of endeavor within the chemical industry is a relatively recent phenomenon, largely relegated to the period from the 1920s to present. The great European chemical companies began providing petroleum-based raw materials and dyes (qv) and pigments (qv) in the latter half of the nineteenth century (1). During the first half of the twentieth century, many breakthroughs, eg, nylon (see POLYAMIDES), rayon (see FIBERS, CELLULOSIC), polycarbonates (qv), penicillin (see ANTIBIOTICS), fluorochemicals, and antiknock additives, were made (2–4). The manufacture and supply of chemicals rapidly diversified and entirely new materials became items of commerce. The barrier for research and manufacturing personnel in embryonic industries, such as synthetic fibers, polymers, and large-scale organic syntheses of agricultural and pharmaceutical materials, to remain fully knowledgeable regarding the nature of raw materials and intermediates provided to their operations by other firms became effectively insurmountable. It therefore became incumbent on these suppliers to develop a means by which answers to questions regarding these materials could be readily obtained. This situation was a primary driver in the origins of what has become known as technical service.

As the twentieth century progressed, Dow, Monsanto, DuPont, Union Carbide, Bayer, BASF, and other large firms began training personnel to provide both direct technical service to customers and to participate in a wide variety of developmental programs aimed at improving financial performance by providing greater value to their customers (5). An example of the support of these efforts was the construction at DuPont in the 1950s of the Chestnut Run Technical Service Laboratory near Wilmington, Delaware. This complex was constructed to support personnel involved in direct technical service interactions and to provide laboratory and testing facilities (laboratory-scale through semiworks, in many cases) for a broad range of DuPont products. It was deliberately constructed near the home of DuPont's Central Research and Development Department allowing for ease of interaction between personnel at the two sites (6). A large level of support was also reflected in the actions of Union Carbide, as the firm designated its Tarrytown, New York, site as a technical service laboratory in the early 1960s. The presence of a formal technical service function in the U.S. chemical industry rapidly became ubiquitous.

The general focus of technical service in the chemical industry was, at the outset, largely tied to a firm's direct sales and marketing efforts. As applications became more complex, however, and customer requirements became more and more specific, a need evolved in many areas of the chemical industry to provide in-depth technical support having direct ties to the research and manufacturing arms, thus allowing rapid responses to increasingly demanding needs of customers. Some firms took the approach of placing the technical service function into the research and development organization. This could be viewed as placing the three critical aspects of what is generically looked upon as the research function, ie, research, development, and technical service, into a single function. An example of this was the Development and Technical Service function within Dow. Other firms created stand-alone technical service organizations. The stated mission of these organizations was originally restricted to direct customer support. Some firms trained their field sales personnel to act also as technical service professionals, a practice used by many companies into the mid-1990s. Time and experience have demonstrated that regardless of what the technical service function calls its home organization, the integrated organization generally surpasses the performance of one that is held as a physically separate function (7).

The Spectrum of Technical Service

Some firms have seen fit to blur the distinction between technical service per se and the research function. Others maintain a technical service organization as a stand-alone function while maintaining a high level of integration with other functions such as sales, marketing, research, and manufacturing. This integration is a critical structural element in ensuring the provision of up-to-date technical information to customers and in allowing the preservation of a two-way conduit between supplier and customer for information allowing, for example, product improvements and the rapid solution of customer problems.

The question "what uses are there for this product?" became chronic as the diversity of products available from the chemical industry began to increase in the mid-1900s. In order to properly address this question, a customer needed to know the possible uses and the value a new material might bring to their business. Again, such questions were typically answered by research personnel prior to the advent of specifically defined technical service organizations. A negative aspect of this approach was that the lack of specific training for research personnel in the processes utilized by customers sometimes led to fragmentary or incorrectly formulated answers. In similar fashion, sales representatives did not generally possess the training required to allow them to address technical issues raised by their customers.

One factor accelerating the development of the modern technical service organization was the dilution of effort resulting from the use of research personnel to answer questions regarding the nature of and possible applications for new products. Research organizations must provide support over a large manifold of responsibilities (6,8). Given a plethora of interests and responsibilities, a migration toward the formation of technical service organizations that could provide focused support for functions closer to the customer began and had its end point in technical service.

A well-integrated technical service function having strong ties to the sales organization at the customer interface and to the research and manufacturing functions at the production interface allows a company to provide rapid and accurate responses to customer needs as a singular function. The success level of such efforts using only sales personnel or only research personnel has not been considered to be particularly high. This is owing to the lack of linkage among persons needed to provide support rather than a fundamental fault on the part of the persons involved in the information chain.

The level of technical service support provided for a given product generally tracks in large part where the supplier considers their product to be located within the spectrum of commodity to specialty chemicals. Technical service support levels for pure chemicals usually provided in large quantities for specific synthetic or processing needs, eg, ammonia (qv), sulfuric acid (see SULFURIC ACID AND SULFUR TRIOXIDE), formaldehyde (qv), oxygen (qv), and so forth, are considerably less than for more complex materials or blends of materials provided for multistep downstream processes. Examples of the latter are many polymers, colorants, flocculants, impact modifiers, associative thickeners, etc. For the former materials, providing specifications of purity and physical properties often comprises the full extent of technical service required or expected by customers. These materials are termed undifferentiated chemicals (9), although the term commodity chemicals is a more common usage. For the latter materials, technical service support is considerably more complex.

Technical Service Functions

The largest number of technical service inquiries from customers involve questions regarding the performance of an existing product already in use by the customer. A typical question is whether product X would work in application Y. The answer may be quite straightforward, or it may require a substantial applications research effort. For example, a customer produces a rigid poly(vinyl chloride) (PVC) compound. In this instance the compound is a pre-mixed powdery material, containing PVC, an impact modifier, TiO_2, colorants, and other additives, that is used as a feedstock to an extruder and post-former to produce textured rigid vinyl siding, for use in the manufacture of woodgrain vinyl siding for home construction. If the customers wishes to investigate an alternative impact modifier to the one used in the formula, a common practice would be to call the supplier of the alternative material to determine specifications, availability, comparative performance data, and related information. The customer would then produce multiple samples of the vinyl compound. Some would contain the impact modifier under usage, others, the new material. At this point, the producer often works directly with a member of the supplier's technical service staff to carry out both in-house and external, ie, at the supplier's laboratories, evaluations. The intent is to obtain highly accurate performance data on vinyl siding produced using the various sample compounds. Test results can be relatively quick and easy to obtain, eg, impact strength, or lengthy and considerably more difficult, eg, exterior photodurability. Once the results have been obtained and discussed with the supplier's technical service personnel, the customer can make an informed decision regarding the use of the newer impact modifier in their vinyl compound. Whereas this may seem to be an exceptional amount of work, because much of the vinyl siding sold in the United States is guaranteed to last 30 years, most manufacturers are reluctant to make raw material changes. Similar issues exist for automotive and industrial paints (qv), as well as for certain polymers used in automotive applications (10).

Another typical customer question is "why did product X do Y in my process?" This is the troubleshooting, consultative part of the technical service function. The range of effort required to answer this question is as broad as for the first example. Consider the case of a converter firm that produces acrylonitrile–butadiene–styrene (ABS) compounds containing carbon black for use in the downstream manufacture of telephone housings. The customer reports that the L*, ie, a color measurement parameter, of a large batch of compound was too high for sale. Upon visiting the customer, the technical service person investigates the situation and finds that the mixers used to prepare the compound were not being operated in a manner to ensure a proper level of shear during mixing. As such, the level of dispersion of the carbon black in the ABS was not sufficient, leading to the drop in color as determined by a measurement of L*.

Both prototypal questions related illustrate the need for a successful technical service professional to have a strong understanding of the customer's applications and processes, within proper intellectual property considerations. This need for a thorough understanding is not always straightforward. A common example of the complications that can arise is provided from the paint (qv) industry (11). If, for instance, a calcium carbonate supplier would like a paint manufacturer to use their material versus a competitive one, the onus is on the supplier to show that the material can be successfully used in the paint formula of interest. However, many such formulas are held as proprietary. The technical service professional therefore does not know the components of the paint. This would lead to an unworkable situation from an evaluation standpoint save for the fact that the paint company may supply a millbase or other intermediate form of the paint to allow a proper comparison of carbonates to be carried out. Thus mutual benefits can result and no loss of proprietary information occur.

Simulation of a customer process at the laboratory scale is sometimes requested, usually to allow the ready surveying of a variety of raw materials or the evaluation of the impact of process changes at a scale more economical than full-plant capability. This is simple for certain requests, ie, the evaluation of a series of antioxidants (qv) in a given polymer formulation, yet nearly impossible for others, ie, studies of flocculation/dispersion phenomena at plant-scale shear rates in multicomponent systems. Generally, laboratory-scale efforts are of the survey variety. These are quite valuable, as they provide a means to identify any gross incompatibilities or other system problems prior to carrying out plant-scale changes. This is a typical approach for, as an example, the substitution of a pure compound from one supplier with that of another supplier in a batchwise chemical synthesis. Even in this simple case, however, it is important that such parameters as mixing and heat transfer are faithfully replicated in the laboratory-scale system.

A common requirement of the technical service professional is the support of either semiworks or full plant-scale trials of a material in a new application at a customer's site. This is often a very demanding responsibility. An example is the evaluation of the use of a new associative thickener in an established line of architectural paint. The technical service person must be familiar with the customer's processing equipment, operating practices, and raw materials if a smooth evaluation is to be ensured. As for any complex manufacturing process, experimental design is a crucial aspect of a successful line evaluation, both for the execution of the test and for the evaluation of data from the test (12). It is sometimes necessary for the technical service professional to spend an extended period of time at the customer's site. In fact, renting nearby lodgings for weeks at a time and working essentially full-time side-by-side with a customer's manufacturing and technical staff is not unknown.

Plant-scale trials are indeed quite common in the chemical and converting industries. This is largely because a true quantitative understanding is rarely known for a great many manufacturing processes. As such, it is common to run a full-scale plant or line trial in order to evaluate a new raw material head-to-head versus a raw material in use, while holding all typical operating parameters at identical control points. In so doing it is possible to detect relatively minor changes in downstream product performance or process performance upon substitution of the new raw material. An example is to change an optical brightener being used in a coated paperboard application and thence to look for changes in the behavior of recycle streams, runnability, printability, and appearance of the final board stock, and other process- and product-related variables that, to the uninitiated, might seem to be totally unconnected from the change in process feedstocks (13). A full-scale mill trial of this type is the only means by which seemingly unrelated variables can be tested in a manner that provides meaningful real-world results for many chemical and manufacturing processes. Semiworks and laboratory methods are progressing rapidly, however.

New product introductions are generally heavily supported by the technical service function. Many customers using chemical feedstocks to produce multicomponent products for the consumer market require extensive on-line evaluations of new raw materials prior to their acceptance for use. An example of this would be the use of a new engineering polymer for the fabrication of exterior automobile structural panels. Full-scale fabrication of the part followed by a detailed study of parameters, such as impact strength, colorant behavior, paint receptivity, exterior photodurability, mar resistance, and others, would be required prior to making a raw materials change of this nature.

Similar requirements exist for the evaluation of a new raw material versus the performance of a current competitive material. Support from technical service personnel can range from experimental design of the line trials through use evaluations. Coordination of evaluations and data analyses resulting from downstream customer testing is also often required. Acquisition and coordination of specialized use or other physical testing is almost always the responsibility of the technical service professionals. For instance, microstructural studies using atomic force microscopy (afm) is hardly commonplace. However, a supplier of raw materials to the microelectronics industry that can provide this powerful tool to a customer has a competitive advantage over a supplier who cannot do so.

An often-neglected aspect of the technical service function is the value technical service personnel can bring to the direct marketing and advertising functions. Accurate and detailed technical information are becoming more a part of print advertising as product cost/performance ratios become increasingly important and as the communication of the performance advantages of a given product becomes a more critical aspect of a successful sale. Such information is generally provided either by research and development personnel or by technical service personnel. In many companies the latter are more heavily involved in applications research, particularly that involving actual customer formulations. Thus these are often in a superior position to provide the sort of information needed. Even a brief perusal of trade magazines, eg, *Chemical Marketing Reporter* or *Plastics Engineering*, yields a substantial fraction of chemical advertisements that contain well-selected technical aspects of the product being advertised.

Many chemical companies also place into the technical service function the responsibility for the generation of a wide variety of printed materials containing customer-oriented information. These may range from a simple one-page technical specification handout to lengthy publications on application tips and recommendations for a broad range of uses (14). The sales value of such materials, when well prepared, is quite high. Most firms provide such information via mail, telephone, or fax order, by providing them to their field sales personnel, or at booths and hospitality suites at conventions and trade shows. The cost of physically preparing these documents has dropped precipitously owing to the advent of desktop publishing technology.

In-house training at a customer site is another valuable technical service function. Often this training is not restricted to the performance and uses of a given raw material, although this is the most common variety of in-house training. Almost every experienced technical service professional has a set of presentations on topics germane to the instruction of typical customers for a given line of products. Generally targeted to a broad audience at the customer's site, these are usually well received. In-depth presentations and follow-up discussions of presentations previously given at trade shows or technical meetings may also be provided as a service to the customer. Another extremely valuable area is operator training. This may involve product safety, receipt and unloading of unusual or new types of product packaging, operating parameter selection, and any other topics of value to the customer's operating staff.

More general training in a given field in which the technical service professional possesses expertise and is of value to the customer is often provided. Examples are training in flocculation phenomena for persons involved in water treatment, optics for companies using dyes and pigments, and overview courses in areas such as experimental design and process troubleshooting for a wide variety of downstream customers. Most customers find these types of presentations useful, both as a result of the information imparted to the organization and owing to the generally high cost of obtaining an outside expert to provide similar support.

It is important to recognize that even specialty chemicals have a product life cycle (Figs. 1 and 2) driven by changes in technology and other factors (15). As such, proper provision of detailed technical information on product performance and applications is critical to an extended life cycle for the product. The level of technical support for a material is generally highest at its introduction, drops somewhat in mid-cycle, and picks up again sharply in the latter stages of the product life cycle.

Technical Service Organization.   There are a tremendous number of possible structures for technical service organizations. This leads to tailoring of the structure within any given firm or product line. Requisite elements of a successful technical service organization generally include the following: (*1*) highly trained technical professionals having a high level of expertise; (*2*) appropriate

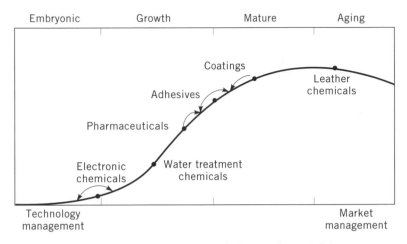

Fig. 1. Life cycle of selected chemical specialties.

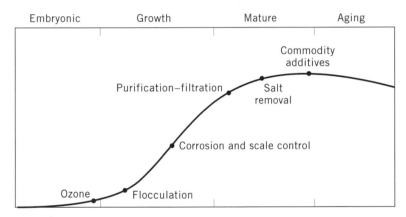

Fig. 2. Life cycle of selected water treatment chemical business areas.

support staffing, ie, laboratory technicians, clerical staff, etc; (*3*) sufficient physical facilities, such as a laboratory, instrumentation, computational support, etc; (*4*) strong integration with other company functions, such as manufacturing, marketing, sales, engineering, and research; and (*5*) the ability to leverage support from others.

Some organizations are constructed such that technical service reports to the research function, others to sales, and others to marketing. The line of reportage is not as important as ensuring that the requisite level of functioning capability is in place for the organization. A manager having prior experience in several business functions is best suited to providing balanced support to such an organization (*16*). A well-rounded perspective is crucial to the success of a technical service organization.

Training Requirements for Personnel.   It is the practice of most chemical firms to provide only personnel having a solid technical background for the provision of technical service regarding the use of their products. The most common academic training for such personnel is a four-year degree in chemical engineering, although other fields of engineering and the physical sciences, chemistry in particular, are also typical. Some firms also have the opportunity to assign PhD-level scientists and engineers to technical service roles.

It is generally of great value to assign personnel to technical service after they have had multiple experiences elsewhere in a business. As an example, a technical service person having a background including assignments in research and manufacturing can provide a balanced response to queries from customers regarding the possible creation and production of a new product. Another common request from customers is for a material with specifications tighter than those typically provided for product release. A technical service person with a varied background is in a better position to quickly ascertain the level of difficulty associated with such a customer request and to work both with the customer and internally to find a way to meet the customer needs in a commercially acceptable manner.

Specialized training is an absolute requirement for technical service personnel. A typical example is a person involved in supporting a polymer for which the use is the manufacture of rotationally molded consumer products. The technical service person is expected to be reasonably familiar with topics such as polymer rheology evaluations, gel-permeation chromatography, rotational molding, color science, regulatory requirements for use, mechanical and photochemical behavior of the pigmented polymer, optics, and so forth. Expertise of this variety cannot be expected to be obtained without careful planning and execution of tailored training for technical service personnel. Both internal and external resources are used to provide specialized training in appropriate areas.

The area of regulatory requirements is of particular importance. The extreme complexity and constant changes of regulations applying to chemicals used in commerce dictate that personnel providing technical service support be well-versed in regulations specific to the product(s) they support. They should have access to the *Code of Federal Regulations* as well as to either in-house or outside experts possessing in-depth expertise in regulatory matters. Many customers require written documentation from suppliers regarding suitability-for-use of a given material in a particular use, eg, use of a particular ionomeric polymer in a food packaging (qv) application. Although most companies make use of in-house or outside professionals to handle such requests, the technical service professional is often a part of the communication loop leading to the successful generation and delivery of such documentation. As such, familiarity with requirements, for this and many other reasons, is a necessity. It is not reasonable, however, to expect technical service personnel to also function as legal professionals in this or in any other sense. Similarly, it is essential that technical service personnel be well aware of other requirements for products, including disposal considerations, toxicity behavior, impact on effluents, special shipping and handling requirements, and other similar factors with regulatory aspects.

The value of mentoring, ie, the practice of providing a seasoned professional to assist in the training and development of less-experienced personnel, cannot be overstated (17). The typical technical service professional in the chemical industry must have at least a reasonable grasp of a broad variety of technical areas. Interaction with a senior member of the technical service staff to obtain guidance in the types and depth of understanding required to meet the needs of customers in a technical service role environment greatly facilitates the learning process. This also helps to ensure an accurate portfolio of "tools in the toolkit" for the less-experienced person, thus providing higher value to the customer via more rapid turnaround of information and higher quality of responses. Finally, the more senior member of the staff accrues a developmental benefit from the mentoring interaction.

Many firms tend to staff their technical service efforts with personnel that have experience solely in the particular business involved. It is often of value to consider individuals having experience in related industries, both to provide a different set of skills and to provide a different perspective than that of in-house experts. This is particularly important in the specialty chemicals arena, where formulations experience is generally of paramount importance. This variety of knowledge is often largely experiential, as much of the treatment and behaviors of formulations in the chemical industry are empirical or experience-based rather than quantitatively understood. It is therefore extremely difficult to train personnel in dealing with the often nebulous aspects of formulations technology.

Participation in Trade and Professional Organizations. Participation in appropriate trade and professional organizations by technical service personnel can provide a myriad of benefits to both the supplier and the customer. Participation can include a wide variety of activities. Examples are the presentation of technical papers, organizing and chairing technical sessions and workshops, serving as officers, attending local and national section meetings, and publishing in refereed professional journals or trade magazines. Examples of professional organizations typically considered to be appropriate for persons working in the chemical industry include the American Chemical Society, the American Institute of Chemical Engineers, the Materials Research Society, the International Union of Pure and Applied Chemistry (IUPAC), the American Crystallographic Association, and the American Physical Society. Examples of trade organizations include the Federation of Societies for Coating Technology, the Society of Polymer Engineers, and the Technical Association of the Pulp and Paper Industry (TAPPI). Many more examples of both types of organizations exist.

Many companies provide support for both their research personnel and technical service personnel to participate in these types of activities. Even a policy of one conference per person per year, if applied correctly, provides a great deal of value to the individual and to the organization. The returns on this investment include intangibles such as additional technical contacts and enhancement of the technical reputation of the company, as well as tangibles such as personnel possessing the latest information in their fields of endeavor, allowing them to better address customer concerns and needs, and developing ideas for process and product improvements.

The value of presenting original work, whether curiosity-driven or applied, should not be overlooked. A great many technical service professionals have extraordinary expertise in some areas of substantial specialization. It is not often straightforward to identify persons with such expertise in a given area. Presentation of a paper or conducting a seminar in one's field of expertise can provide a critical link in the search process customers use to seek assistance on technical issues.

Finally, there is great value in providing access to contemporary and archived technical journals and trade magazines to technical service professionals. This can be achieved in a number of ways, whether by generation of an in-house library, use of nearby university or governmental facilities, support of individual subscriptions, or some combination of the above. Internet access to on-line journals is also valuable.

Technical Service and the Development Process

The technical service function acts as the interface between the user and the research and manufacturing arms of the supplier for many firms. A consequence of this is that the technical service function can act as a conduit of information between the customer and personnel involved in the development process.

Many companies make use of multifunctional teams to drive the development process (18). This is true both for new products and for enhanced performance characteristics for existing offerings. The time required for either a significant enhancement of an existing product or for the launch of a new product can be greatly reduced by making intelligent use of the technical service organization as an information conduit. This generally results in both a greater financial reward for the supplier and in valuable product enhancements for the user. It is not unusual for the technical service function, or a subset of it, to act as the clearinghouse for new or improved product ideas and provide technical leadership for the evaluation and ultimate implementation of those ideas. Because technical service personnel generally possess the greatest familiarity with customer experiences and needs coupled with a strong understanding of internal company capabilities, they are in a unique position to provide guidance toward an accurate assessment of what is and is not feasible.

The troubleshooting aspect of technical service de facto results in the technical service professional having the best perspective on any faults that may be identified by a customer for a given application. Taking timely action on such observations enhances a supplier's credibility with a given customer and quite often provides an improvement to the product of general value to other members of the supplier's customer base. Technical service professional work with colleagues in research and manufacturing to develop process improvements, identify more appropriate specification ranges, identify necessary product enhancements, and provide support to other related efforts aimed at addressing the need identified as a result of technical interchanges at a customer's site. A completely accurate understanding of the true needs of the customer is an incontrovertible requirement of a new product development process that provides maximum value both to the customer and the supplier (19). The costs of an ill-defined set of product characteristics can be catastrophically high in capital investment, time, and supplier reputation.

A critical but often overlooked aspect of the technical service function is the value of relationships that develop between more senior members of the technical service staff and their colleagues at customer sites. Professional relationships provide value both to supplier and customer by virtue of the trust and respect generated in a successful technical service interaction, analogous to the types of mutually valuable relationships that can develop between supplier sales personnel and a customer's purchasing personnel. A more efficient process of gathering and dispersing new and useful technical information can result from these relationships, because the persons involved work in different environments and are exposed in the course of their work to different external sources of information.

Technology

Computers. It is difficult to envision how technical service in the chemical industry could be carried out without computer technology (qv). The increase in availability of low cost computing power has greatly enhanced the problem-solving capability of the technical service professional. The availability of highly portable notebook computers allows great freedom to transport substantial computational support anywhere in the world. The applications of this relatively new capability are myriad. Some examples include the ability to rapidly carry out complex fluid mechanical calculations in modeling a customers' mixing problem, using finite element analysis to determine the probable source of structural failures in large ceramic bodies and polymeric parts, or carrying out high level statistical analyses on-site using actual customer data in real time and directly interfacing to customer processing equipment to allow either an in-shop or after-the-fact analysis of data obtained during process disruptions. Some remote control aspects of computers include the ability to access, via telephone line connections, databases (qv) in the suppliers' computers, using powerful in-house computers to remotely execute molecular mechanics simulations of reaction pathways, or carrying out gas chromatographic analyses while away from the laboratory. The latter function has become increasingly routine as laboratory information management systems (LIMS) have evolved into very flexible installations allowing ready manipulation of multiple instruments while physically away from the laboratory.

The Internet. One of the most important changes in the manner in which technical service is provided to customers of the manufacturing and service industries is the advent of the Internet and its use in providing direct access to a wide range of technical service and related information. It is possible, either through the use of an in-house computer acting as a server or by renting space on a server at an internet service provider (ISP) site, to provide almost any sort of information that a customer might request. Many chemical companies have created extremely detailed web sites containing a vast array of information (20). As the costs associated with accessing the Internet have dropped, many firms have moved to provide full-time access to their technical staff members. This practice is cost-effective even for small companies having limited resources.

It has also become quite cost-effective to provide a substantial portion of typical technical service information via the Internet owing to the advent of inexpensive desktop computers possessing large amounts of storage capacity and the capability of acting as servers. At the time of this writing (ca 1997) it is possible to purchase a fully configured system capable of meeting the needs of companies from small to intermediate size for <$10,000. Desktop servers are projected to be able to supersede desktop personal computers as the technology of choice and of lowest cost by the end of 1998. Finally, there are several powerful operating systems available (21) that allow sophisticated use of this new tool without extensive formal computer training. Using authoring software, information, including audio and video, if desired, can be placed directly into a web site with minimal training.

In its simplest form, a technical service site on the Internet can provide a set of answers to queries that can generally be expected for the industry involved. Over 10,000 addresses for access to technical information may be yielded from a search of the term Technical Service/Chemicals. For the chemical industry it is common to provide such information as material safety data sheets (MSDS), compositional information, simple applications information, lists of available materials, returns policies, shipping information, purity, and regulatory status. It is also common to provide a query form that can be filled out electronically by the person accessing the site to obtain information not directly available from the site. These are then automatically forwarded to members of the technical staff for the formulation and communication of responses to the query. This can in a sense be looked at as a subset of electronic data interchange (EDI). This capability is becoming increasingly popular. Less common, but still growing, is the use of expert systems (qv) architectures to lead a customer through a series of questions to obtain answers to, for example, simple formulations issues associated with feedstock changes.

A large amount of information can be provided on a technical service Internet site by means of hyperlinks, ie, clickable terms or addresses that allow the reader to access multiple layers of information by a simple click of a control device. Some companies have provided their entire library of open literature on their products in the form of material stored within a hyperlinked page. The reader can either browse this information directly on the site or download it to their own computer to be read at their convenience.

The Next Stages of Technical Service

The need for well-trained technical service professionals is expected to continue as an essential aspect of the chemical industry, despite the phenomenal growth in electronic methods of information storage, retrieval, and transmission. Advanced troubleshooting of complex customer processes and accelerated accurate product development and market introductions should continue to be principal elements of technical service personnel duties. Increased levels of integration, perhaps blurring the lines between supplier and customer, may come to pass. There are already instances of personnel swapping between customers and suppliers for extended periods to allow cross-fertilization of ideas and provide more accurate perspectives for the companies involved in these efforts. Technical service and research personnel have been those persons most directly involved in such efforts.

BIBLIOGRAPHY

"Technical Service" in *ECT* 3rd ed., Vol. 22, pp. 645–657, by R. J. Ziegler and H. Lieberman, Betz Laboratories.

1. A. S. Travis, *The Rainbow Makers: The Origins of the Synthetic Dyestuffs Industry in Western Europe*, Lehigh University Press, Bethlehem, Pa., 1993.
2. *Our Story So Far: Notes on the First Seventy-Five Years of the 3M Company*, Minnesota Mining and Manufacturing Co., St. Paul, Minn., 1977.
3. D. J. Forrestal, *Faith, Hope and $5,000: The History of Monsanto*, Simon and Schuster, New York, 1977.
4. T. Mahoney, *The Merchants of Life: An Account of the Pharmaceutical Industry*, Harpers, New York, 1959.
5. R. Titleman, *Profits of Science*, Basic Books, New York, 1994.
6. D. Hounshell and J. Kenly Smith, Jr., *Science and Corporate Strategy: DuPont R&D 1902–1980*, Cambridge University Press, New York, 1988.
7. P. G. Smith, *Developing Products in Half the Time*, Van Nostrand Rheinhold, New York, 1991.
8. *Science: The Endless Frontier*, National Science Foundation Report, NSF 90-8, The National Science Foundation, Washington, D.C., 1990.
9. C. Kline, *CHEMTECH*, 110 (1976).
10. J. F. Rabek, *Photostabilization of Polymers*, Elsevier Science Publishers Ltd., Essex, U.K., 1990, Chapt. 8.
11. T. C. Dalton, *Paint Flow and Pigment Dispersion*, 2nd ed., John Wiley and Sons, Inc., New York, 1979, for information on complexities that can arise in such systems.
12. S. L. Meyer, *Data Analysis for Scientists and Engineers*, John Wiley and Sons, Inc., New York, 1975.
13. R. W. Hagemeyer, ed., *Pigments for Paper*, TAPPI Press, Altanta, Ga., 1991.
14. L. Leonard, ed., *Plastics Compounding 1995/96 Redbook*, Advanstar Communications, Cleveland, Ohio, 1995.
15. Ref. 7, p. 4.
16. P. F. Druker, *Frontiers of Management*, Part III, E. P. Dutton, New York, 1986.
17. G. Dalton and P. Thompson, *Novations: Strategies for Career Management*, Scott, Foresman and Co., Glenview, Ill., 1986, pp. 24–28.
18. A. Roussel, K. N. Saad, and T. J. Erikson, *Third-Generation R&D: Managing the Link to Corporate Strategy*, Harvard Business School Press, Boston, Mass., 1991, p. 160.
19. Ref. 7, p. 50.
20. http://www.dow.com or http://www.dupont.com, for example.
21. http://www.wcmh.com/uworld for information on UNIX; http://www.ziff.com/~macuser/ for the Macintosh operating system; http://www.ibm.com/ for information on OS/2; and the *Linux Journal*, Specialized System Consultants, Inc., Seattle, Wash., for information on the LINUX operating system.

AUSTIN H. REID, JR.
E. I. du Pont de Nemours & Co.

TELLURIUM AND TELLURIUM COMPOUNDS

Tellurium [13494-80-9], Te, at no. 52, at wt 127.61, is a member of the sixth main group, Group 16 (VIA) of the Periodic Table, located between selenium and polonium. Tellurium is in the fifth row of the Table, between antimony and iodine, and has an outer electron configuration of $5s^25p^4$. The four inner principal shells are completely filled. Tellurium is more metallic than oxygen, sulfur, and selenium, yet it resembles them closely in most of its chemical properties. Whereas oxygen and sulfur are nonmetals and electrical insulators, selenium and tellurium are semiconductors, and polonium is a metal. Tellurium forms inorganic and organic compounds superficially similar to the corresponding sulfur and selenium compounds, yet dissimilar in properties and behavior. The valence states assigned to the central atom in tellurium compounds are -2, 0, $+2$, $+4$, $+6$.

Tellurium was discovered in 1782 in Transylvanian gold ore. Its name is derived from the Latin *tellus*, meaning earth (1). At 0.047 atoms per 10,000 atoms Si, tellurium is about the fortieth element in the order of cosmic abundance. Along with platinum, palladium, and ruthenium, it ranks about seventy-first in the order of crustal abundance. The average amount in crustal rocks is 0.01 ppm. The chalcogens, sulfur, selenium, and tellurium, are primarily components of intrusive and extrusive magmas and volcanic gases, and hence of volcanic sulfur deposits. Nevertheless, selenium and tellurium are not essential components of the common igneous rock-forming minerals. Tellurium is widely distributed in the earth's crust in deposits of many different types, from magmatic and pegmatitic to hydrothermal, especially where these deposits are associated with epithermal gold and silver deposits (2–4).

Small concentrations but large quantities of tellurium are present in copper porphyries, massive pyritic copper sulfide and nickel sulfide deposits, and frequently in lead sulfide deposits. In pyritic deposits, tellurium is concentrated chiefly in pentlandite [53809-86-2], $(FeNi)_9S_8$; chalcopyrite [1308-56-1], $CuFeS_2$; and pyrite [1309-36-0], FeS_2 in decreasing order; and least in sphalerite [12169-28-7], ZnS, and pyrrhotite [12063-67-1], FeS. Other tellurium-bearing deposits are copper–molybdenum, lead–zinc, gold, tungsten–bismuth, uranium, and mercury–antimony. The S:Se:Te ratio varies widely among the deposits and within the same deposit. At times, tellurium concentration exceeds that of selenium, and often it increases with depth. Little is known about tellurium in sedimentary rocks, though some shales contain 0.1–2 ppm; manganese nodules from the Pacific and Indian Ocean floors contain 0.5–125 ppm (see OCEAN RAW MATERIALS).

Like selenium, tellurium usually forms binary minerals; but whereas selenium isomorphously replaces a part of the sulfur in sulfides, tellurium occurs only as discrete tellurium minerals of microscopic size. Tellurium is a main component of ca 40 mineral species, including 24 tellurides, two tellurates, native tellurium, and a selenium–tellurium alloy. It is a minor constituent of an undetermined number of minerals. It occurs in combination with oxygen, sulfur, and 10 other elements with high atomic numbers. At least ten tellurium minerals occur with gold and silver, ten with bismuth, and six with iron. Some of the better known minerals are hessite [12002-98-1], Ag_2Te; petzite [1317-73-

3], Ag$_3$AuTe$_2$; calaverite [*37043-71-3*], AuTe$_2$; sylvanite [*1301-81-1*], AuAgTe$_4$; altaite [*12037-86-4*], PbTe; tetradymite [*1304-78-5*], Bi$_2$Te$_2$S; rickardite [*12134-39-31*], Cu$_4$Te$_3$; nagyagite [*12174-01-5*], Au(Pb,Sb,Fe)$_8$(Te,S)$_{11}$; and tellurite [*14832-87-2*], TeO$_2$.

Like selenium, tellurium minerals, although widely disseminated, do not form ore bodies. Hence, there are no deposits that can be mined for tellurium alone, and there are no formally stated reserves. Large resources however, are present in the base-metal sulfide deposits mined for copper, nickel, gold, silver, and lead, where the recovery of tellurium, like that of selenium, is incidental.

Physical Properties

At least 21 tellurium isotopes are known, with mass numbers from 114 to 134. Of these, eight are stable, ie, 120, 122–126, 128, 130. The others are radioactive and have lifetimes from 2 min to 154 d; the heaviest six, 131m, 131, 132, 133m, 133, and 134, are fission products (see RADIOISOTOPES). Tellurium illustrates the rule that elements having even atomic numbers have more isotopes than elements having odd atomic numbers.

The physical properties of tellurium are given in Table 1. The vapor pressure between 511 and 835°C is given by the equation shown where temperature, T, is in K (5): Pressures in kPa may be converted to mm Hg by multiplying by 7.5.

$$\log p_{kPa} = 6.7249 - (5960.2/T)$$

At ordinary temperature and pressure, solid tellurium, unlike sulfur and selenium, has only one structural form. Tellurium crystallizes in a trigonal lattice with a space group P3$_1$21 (or P3$_2$21). Lattice parameters are $a = 44.6$ pm, $c = 593$ pm, and a bond angle of 103.2°. The crystal structure may be considered as a set of helical chains parallel to the c-axis, held together by relatively weak atomic forces. The helical screw directions can be either right-handed or left-handed, and a plane-polarized electromagnetic wave travelling along the c-axis can be rotated in a clockwise or anticlockwise direction. Light is strongly absorbed by tellurium for wavelengths, λ, smaller than about 3.5 micrometers, corresponding to the energy gap of this semiconductor. Above this absorption edge wavelength, the element is essentially transparent, having weak absorption owing to free carriers, characterized by the absorption coefficient α increasing as a function of λ^2. For plane-polarized infrared light where the electric vector E is parallel to the c-axis, there is a pronounced absorption peak at $\lambda = 11$ μm. This is absent for E perpendicular to c. Below room temperature, tellurium shows photoconductivity for wavelengths smaller than about 4 μm and has been considered as a cooled infrared detector for this wavelength range.

Single crystals of tellurium can readily be grown from the melt by the Czochrlaski method. Where the seed is parallel to the c-axis, the resulting crystal ingot is usually six-sided, with (1010) planes, reflecting the three-fold symmetry of the material. Crystals can also be grown perpendicular to the c-axis. The

Table 1. Physical Properties of Tellurium

Property	Value
specific gravity[a] at 18°C	
crystalline	6.24
amorphous	6.0–6.2
hardness, Mohs[b]	2.0–2.5
modulus of elasticity, MPa[c]	4140
Poisson ratio at 30°C	0.33
heat capacity at 25°C, kJ/mol[d]	25.70
entropy at 25°C, J/(K·mol)[d]	49.70
heat of fusion, kJ/mol[d]	17.87
mp, °C	450
viscosity at mp, mPa·s (=cP)	1.8–1.95
bp, °C[e]	990
heat of formation[f], kJ/mol[d]	171.5
heat of vaporization, kJ/g[d]	46.0
thermal conductivity at 20°C[g], W/(m·K)	0.060
Te–Te bond energy, kJ/mol[d]	138
covalent radius, pm	137
electronegativity[h]	ca 2.1
first ionization potential, eV	9.01
electron affinity, eV	
first	2.3
second	ca 3.0
volume shrinkage on solidification, %	5–7

[a]Increases under pressure.
[b]Anisotropic.
[c]To convert MPa to psi, multiply by 145.
[d]To convert J to cal, divide by 4.184.
[e]Extrapolated.
[f]Te (g) atom to Te_2 (g) molecule.
[g]Polycrystalline material; in single crystals, it is anisotropic and affected by impurities and lattice imperfections.
[h]Pauling scale.

prismatic (1010) planes can be etched to reveal characteristic four-sided etch pits; six-sided etch pits having a 3-fold symmetry can be obtained on the (0001) basal plane. Thermal etch pits of these symmetries can also be revealed by heating tellurium crystals under reduced pressure.

The physical properties of tellurium are generally anistropic. This is so for compressibility, thermal expansion, reflectivity, infrared absorption, and electronic transport. Owing to its weak lateral atomic bonds, crystal imperfections readily occur in single crystals as dislocations and point defects. Tellurium is diamagnetic below its melting point. Its intrinsic electrical resistivity at room temperature is about 0.25 ohm·cm, when the current is parallel to the c-axis, and decreases with increasing temperature and pressure. The element forms a continuous range of isomorphous solutions with selenium, consisting, in the solid state, of chains of randomly alternating Se and Te atoms.

Tellurium has a semiconducting thermal energy gap of 0.33 eV and, if sufficiently purified by zone refining in hydrogen, is intrinsic at room temperature with a carrier concentration of about 4×10^{15} cm^{-3}. At liquid nitrogen temperature (77 K), tellurium has always shown p-type extrinsic behavior; n-type extrinsic conduction has never been observed. Thus, whereas no donors have been found, the Group 15 (V) elements Bi, Sb, As, and P act as acceptors and can increase the extrinsic hole concentration. As the temperature of sufficiently pure tellurium is raised from 77 K, the Hall coefficient changes sign from positive to negative at an inversion temperature which increases with the extrinsic hole concentration. However, another sign reversal of the Hall coefficient takes place at 514 K from negative back to positive again. This time the reversal temperature is fixed and does not vary with the extrinsic hole concentration. The first sign reversal results from the larger mobility of the electrons compared to that of holes; the second reversal has been attributed to the existence of higher conduction band having a very low mobility.

An excellent review of the semiconducting properties of tellurium is available (6).

Chemical Properties

Although tellurium resembles sulfur and selenium chemically, it is more basic, more metallic, and more strongly amphoteric. Its behavior as an anion or a cation depends on the medium, eg:

$$TeO_2 + 2\ KOH \longrightarrow K_2TeO_3 + H_2O$$
$$TeO_2 + 4\ HCl \rightleftharpoons TeCl_4 + 2\ H_2O$$

In an acid medium, the following reactions take place:

$$Te^0 - 4e^- \longrightarrow Te^{4+} \qquad Te^{4+} + 4e^- \longrightarrow Te^0 \qquad Te^{4+} - 2e^- \longrightarrow Te^{6+}$$

In an alkaline medium:

$$Te^0 - 4e^- + 3\ H_2O \longrightarrow TeO_3^{2-} + 6\ H^+$$
$$TeO_3^{2-} + 4e^- + 6\ H^+ \longrightarrow Te^0 + 3\ H_2O$$
$$TeO_3^{2-} - 2e^- + H_2O \longrightarrow TeO_4^{2-} + 2\ H^+$$

Tellurium forms ionic tellurides with active metals, and covalent compounds with other elements. The valence states are -2, $+4$, and $+6$. Solid crystalline tellurium tarnishes slightly in the air, and more rapidly and to a greater degree in the powdered state. Moist, precipitated tellurium oxidizes on drying, especially above 100°C. Molten tellurium is readily oxidized to tellurium dioxide, which can be volatilized by blowing air through the melt. Tellurium reacts with halogens and halogenating agents, and mixes in all proportions with sulfur and selenium. Oxidation with nitric acid, and ignition of the resulting 2 TeO$_2$·HNO$_3$

[*23624-18-2*] yields very pure TeO_2. Tellurium dioxide can be used as an oxidant for acetoxymethylation reactions (7):

$$ArH \xrightarrow[\text{LiBr, 120 or 160°C}]{\text{TeO}_2,\ \text{CH}_3\text{COOH}} ArCH_2O\overset{\overset{\displaystyle O}{\|}}{C}CH_3$$

where Ar = aryl. Tellurium reacts with concentrated, but not with dilute, sulfuric acid to form tellurium sulfite [*84074-47-5*]:

$$Te^0 + H_2SO_4 \rightleftharpoons TeSO_3 + H_2O$$

Dilution with water reverses the reaction, and heating the solution liberates sulfur dioxide. Upon being added to a solution of tellurides, tellurium forms colored polytellurides. Unlike selenium, tellurium is not soluble in aqueous sodium sulfite. This difference offers a method of separating the two elements. Like selenium, tellurium is soluble in hot alkaline solutions except for ammonium hydroxide solutions. Cooling reverses the reaction. Because tellurium forms solutions of anions, Te^{2-}, and cations, Te^{4+}, tellurium films can be deposited on inert electrodes of either sign.

Elemental tellurium liberates chlorine from compounds such as $AsCl_3$ and $AuCl_3$; it reduces $FeCl_3$ partially to $FeCl_2$, and SO_2Cl_2 to SO_2 and Cl_2 gases. Oxidation of metals by tellurium gives metallic tellurides. Tellurium itself is oxidized by strong reagents such as $Na_2Cr_2O_7$, $KMnO_4$, $Ca(OCl_2)_2$, H_2O_2, and $HClO_3$. Tellurium dioxide and tellurous acid and its salts are readily reduced to the element, Te^0 with $SnCl_2$, H_2S, and $Na_2S_2O_4$. These compounds are oxidized to the Te(VI) state with PbO_2, and $KMnO_4$. Selenium, H_2S, and HCl reduce TeO_4^{2-} ions to TeO_3^{2-}. Solid tellurium oxides can be reduced by heating with hydrogen, carbon, and carbon monoxide.

The stability of organic chalcogen compounds decreases mostly in the order sulfur > selenium > tellurium.

Manufacture

Recovery.   No attention is paid to the recovery of tellurium in the mining, flotation and concentration of copper ores. In the further process steps of roasting, smelting, and refining, tellurium is found in detectable quantities but only on rare occasions are operations altered to enhance recovery or deliberately redistribute tellurium. Pyrometallurgical lead refineries do, however, actively monitor tellurium and recover tellurium-bearing intermediates when concentrations are high enough (>0.025%).

Most commercial tellurium is recovered from electrolytic copper refinery slimes (8–16). The tellurium content of slimes can range from a trace up to 10% (see SELENIUM AND SELENIUM COMPOUNDS). Most of the original processes developed for the recovery of metals of value from slimes resulted in tellurium being the last and least important metal produced. In recent years, many refineries have changed their slimes treatment processes for faster recovery of precious metals (17,18). The new processes have in common the need to remove the copper in slimes by autoclave leaching to low levels (<1%). In addition, this autoclave pretreatment dissolves a large amount of the tellurium, and the separation of the tellurium and copper from the solution which then follows places tellurium recovery at the beginning of the slimes treatment process.

Typically, the removal of copper in slimes is accomplished by the autoclaving of slimes at elevated temperature with sulfuric acid and oxygen. Use of temperatures of 120°C and oxygen pressures of 345 kPa (50 psig) allows almost complete copper extraction and tellurium extractions ranging from 50 to 80%. The range of tellurium extraction is wide because the mode of occurrence in slimes varies significantly (19). Selenium and intermetallic selenides are more resistant to oxidation and these remain essentially unchanged in the residue. The tellurium solubilized by autoclaving may be present in both the tetravalent and hexavalent forms, the proportion of hexavalent rising with increasing oxygen pressure, acidity, and temperature. The overall reactions are:

$$Cu_2Te + 2\ O_2 + 2\ H_2SO_4 \longrightarrow 2\ CuSO_4 + H_2TeO_3 + H_2O$$

$$H_2TeO_3 + 1/2\ O_2 \longrightarrow H_2TeO_4$$

Tellurium is recovered from solution by cementation with copper at elevated (>90°C) temperature.

$$H_2TeO_3 + 4\ Cu^0 + 2\ H_2SO_4 \longrightarrow Cu_2Te + 2\ CuSO_4 + 3\ H_2O$$

$$H_2TeO_4 + 5\ Cu^0 + 3\ H_2SO_4 \longrightarrow Cu_2Te + 3\ CuSO_4 + 4\ H_2O$$

Although this procedure yields tellurium as the same compound found in the original feedstock, the copper telluride is recovered in a comparatively pure state which is readily amenable to processing to commercial elemental tellurium or tellurium dioxide. The upgraded copper telluride is leached with caustic soda and air to produce a sodium tellurite solution. The sodium tellurite solution can be used as the feed for the production of commercial grade tellurium metal or tellurium dioxide.

If the final product desired is tellurium metal, excess free caustic soda is required in the sodium tellurite solution. The solution is electrolyzed in a cell using stainless steel anodes to produce tellurium metal (20). This technology is used at the CCR Division of Noranda Metallurgy Inc., Canada, and at Pacific Rare Metals Industries Inc., the Philippines. Typical electrolysis conditions are given in Table 2.

Alternatively, if tellurium dioxide is the product desired, the sodium tellurite solution can be neutralized in a controlled fashion with sulfuric acid. As the pH is lowered, precipitates containing impurities such as lead and silica that form are filtered off. At pH 5.6 the solubility of tellurous acid reaches a minimum and essentially all of the tellurium precipitates (>98%). After filtration and drying, commercial tellurium dioxide is obtained. A diagram for the process of detellurizing of slimes and recovering tellurium products is shown in Figure 1.

Tellurium is still recovered in some copper refineries by the smelting of slimes and the subsequent leaching of soda slags which contain both selenium

Table 2. Electrolysis Conditions for Tellurium Recovery[a]

Parameter	CCR Refinery	PRM Refinery
Te, g/L		
in fresh electrolyte	150–200	250
in spent electrolyte	90–140	80
NaOH, g/L		
in fresh electrolyte	40	85
in spent electrolyte	80	200
electrolyte temperature, °C	40–45	45
flow rate, L/min	60	40
number of anodes	15	9
number of cathodes	14	8
cathodic current density, A/m^2	160	160
cell voltage, V	2.0–2.5	2.5
deposition time, days	3	8
current efficiency, %	ca 90	90

[a]Ref. 20.

and tellurium. The caustic slags are leached in water and, using the controlled neutralization process, tellurium is recovered as tellurium dioxide.

To produce commercial (99.5%) tellurium, tellurium dioxide is dissolved in hydrochloric acid. The tellurium solution is saturated with sulfur dioxide gas to yield commercial tellurium powder, which is washed, dried, and melted.

$$H_2TeO_3 + H_2O + 2\,SO_2 \longrightarrow Te^0 + 2\,H_2SO_4$$

Purification. Tellurium can be purified by distillation at ambient pressure in a hydrogen atmosphere. However, because of its high boiling point, tellurium is also distilled at low pressures. Heavy metal (iron, tin, lead, antimony, and bismuth) impurities remain in the still residue, although selenium is effectively removed if hydrogen distillation is used (21).

Ultrahigh (99.999+%) purity tellurium is prepared by zone refining in a hydrogen or inert-gas atmosphere. Single crystals of tellurium, tellurium alloys, and metal tellurides are grown by the Bridgman and Czochralski methods (see SEMICONDUCTORS).

Commercial Products. Tellurium dioxide [7446-07-3], TeO$_2$ (79.9% Te theoretically), is made by heating an aqueous suspension of tellurous acid. The acid is purified, if necessary, by redissolving in caustic soda solution and neutralizing with sulfuric acid.

Sodium tellurate [10101-25-8], Na$_2$TeO$_4$, (53.7% Te theoretically), is made by oxidizing sodium tellurite solution with hydrogen peroxide. The reaction is exothermic.

Ferrotellurium or iron telluride [12125-63-2], FeTe, which usually has a Te content near the theoretical 69.5%, is made by the exothermic melting iron and tellurium powders in stoichiometric proportions.

Tellurium diethyldithiocarbamate [20941-65-5], [(C$_2$H$_5$)$_2$NC(S)S]$_4$Te, is made by the reaction of diethylamine, carbon disulfide, and tellurium dioxide in an alcoholic solution.

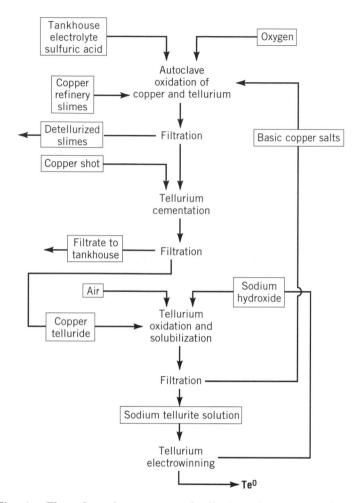

Fig. 1. Flow sheet for recovery of tellurium from copper slimes.

Metal tellurides for semiconductors are made by direct melting, melting with excess tellurium and volatilizing the excess under reduced pressure, passing tellurium vapor in an inert gas carrier over a heated metal, and high temperature reduction of oxy compounds with hydrogen or ammonia.

Economic Aspects

Tellurium has been produced commercially in the United States since 1918 and in Canada since 1934 (10). The principal world recoverers and producers are given in Table 3.

Production and trade figures are published by the U.S. Bureau of Mines (22), the American Bureau of Metal Statistics (23), the Canadian Department of Energy, Mines, and Resources (24), and the World Bureau of Metal Statistics (25). Production figures are often incomplete because of the need to avoid disclosure of proprietary data. Hence, marketing data is often estimated.

Table 3. Principal Tellurium Producers in 1995[a]

Country	Company
Belgium	Union Minière BU Hoboken, Hoboken
Canada	Inco Limited, Toronto
	Noranda Minerals Inc., CCR Division, Montreal East, Quebec
Peoples Republic of China	Guangzhan Smelter, Guangzhan
	Shanghal Smelter, Shanghai
	Shenyong Smelter, Shenyong
	Zhuzhan Smelter, Zhuzhuo
CIS	Almalysky Metallurgical Plant, Tashkent Region, Almalyk, Uzbekistan
	Balashmed Plant, Dzheskazgen Region, Russia
	Copper Refinery, Uralelecktromed, Yekaterinburg, Russia
	Krasny Chemik, St. Petersburg, Russia
	Krastvetmet, Krasnoyarsk, Russia
	Kyshtym Copper Electrolytic Plant, Chelyabimsk Region, Russia
	Norilsk Nickel Plant, Krasnoyarsk Region, Russia
	Shelkovskoje Plant, Agrochim, Shelkovo, Moscow Region, Russia
	Ust-Kamenogorsky Lead–Zinc Plant, East Kazakhstan, Russia
Germany	Norddeutsche Affinerie AG, Hamburg
Japan	Japan Energy Corporation, Tokyo
	Mitsubishi Materials Corporation, Tokyo
	Mitsui Smelting and Refining Co., Ltd., Tokyo
	Sumitomo Metal Mining Co., Ltd., Tokyo
Peru	Centromin-Peru, La Oroya
Philippines	Pacific Rare Metals Industries Inc., Quezon City (Manila)
U.K.	Mining and Chemical Products, Alperton (London)
United States	Asarco Inc., New York, N.Y.
	Kennecott Corporation (USA), Salt Lake City, Utah
	Phelps Dodge Refining Corporation, El Paso, Tex.

[a]Tellurium or tellurium-containing products.

The estimated world production of tellurium from 1991–1994 is shown in Table 4; the average annual production in that time period was 240 tonnes. The data for 1994 are considered to be the most accurate because of the direct reporting, for the first time, of production from plants in the former Soviet Union. The main uses for tellurium are as a free machining additive in ferrous metallurgy (55%), as a free machining additive in nonferrous metallurgy (10%), in chemicals

Table 4. World Production of Tellurium, t

Location	Year			
	1991	1992	1993	1994
Europe	66	81	61	112
North America	50	46	36	46
South America	0	0	4	4
Asia	24	50	57	60
others	59	74	58	72
Total	*199*	*251*	*216*	*294*

(30%), in electrical applications such as solar cells and thermoelectrics (3%), and miscellaneous (2%).

There were no significant market changes for tellurium from the 1970s through the mid-1990s. Technical efforts to develop new applications have been concentrated in the area of electronics. Steady improvements in performance for cadmium telluride [1306-25-8], CdTe, based solar cells have been noted and similarly for bismuth telluride [1304-82-1], Bi_2Te_3, thermoelectrics. Neither application has yet become a principal market for tellurium and, consequently, the price of tellurium has been related to variations in output from producers rather than to market demand. Prices for commercial grade (99.5+%) tellurium rose steadily from $24/kg in 1984 to a peak of $100/kg in 1992, before falling back to under $50/kg in 1995. Prices for high purity (99.999+%) tellurium fluctuated only slightly from $130/kg during the same time period.

Specifications and Grades

There are no official U.S. specifications for tellurium and producers publish their own standards. Typically the producer specifies the weight and shape of the pieces, a screen analysis of powders, and a maximum content of certain impurities.

The common grades marketed in the U.S. and Canada are given in Table 5. Commercial-grade tellurium is available in tablets (1 and 3 g), slabs (2.27 kg (5 lbs)), sticks (0.454 and 0.908 kg (1 and 2 lbs)) and powder (−100 and −200 mesh (<149 and <74 μm)). High purity tellurium is sold as cast cakes or chunks.

Table 5. Tellurium Grades

Element, wt %	Commercial	High purity
tellurium	99.7	99.999
copper	.01	
lead	.003	
selenium	.01	
silica	.04	
iron	.01	
other impurities		10[a]

[a]Value is maximum in units of ppm.

Analytical Methods

Comprehensive accounts of the analytical chemistry of tellurium have been published (5,26–30). The analytical methods for the determination of tellurium are to a considerable extent influenced by the element's resemblance, in many of its properties and in its limited terrestrial abundance, to selenium.

A solution of tetravalent or hexavalent tellurium is commonly detected by precipitation as the black elemental tellurium, Te^0, when sulfur dioxide is bubbled through dilute hydrochloric acid solutions. Tellurium(IV) or tellurium(VI) are also detected as elemental tellurium when treated with reducing agents such as tin(II) chloride, hypophosphorous acid, hydrazine hydrochloride, aluminum amalgam, zinc, or magnesium. A solution of tellurium(IV) reacts with hydrogen sulfide to form the reddish-brown tellurium sulfide [*16608-21-2*], TeS. Elemental tellurium and tellurides, Te^{2-}, but not the oxidized tellurium compounds such as tellurites, TeO_3^{2-}, or tellurates, TeO_3^{2-}, give a purplish-red color when warmed in concentrated sulfuric acid solutions.

Several common acid treatments for sample decomposition include the use of concentrated nitric acid, aqua regia, nitric–sulfuric acids, and nitric perchloric acids. Perchloric acid is an effective oxidant, but its use is hazardous and requires great care. Addition of potassium chlorate with nitric acid also assists in dissolving any carbonaceous matter.

Organic tellurium compounds and siliceous materials, ie, rock, ore, or concentrates, are fused with mixtures of sodium carbonate and alkaline oxidants, ie, sodium peroxide, potassium nitrate, or potassium persulfate. For volatile compounds, this fusion is performed in a bomb or a closed-system microwave digestion vessel. An oxidizing fusion usually converts tellurium into Te(VI) rather than Te(IV).

Many analytical methods depend on the conversion of the tellurium in the sample to tellurous acid, H_2TeO_3. Should tellurous acid precipitate on dilution, it can be redissolved with hydrochloric acid. Although tellurium is not as readily volatile as selenium, precautions should be taken to prevent the volatilization of tellurium when halogen or hydrohalide media are used during sample decomposition.

Depending on the tellurium content and the nature of the material being analyzed, tellurium can be determined by a number of gravimetric methods, notably precipitation as the element using either sulfur dioxide or hydrazine hydrochloride as reductants. This method suffers from coprecipitation of other elements and oxidation of the precipitate on drying. Alternatively, tellurium can be precipitated from Te(IV) solutions using ammonia, pyridine, and hexamethylenetetramine, but heavy metals and other elements are coprecipitated, unless complexed with ethylenediaminetetraacetic acid (EDTA) (31), citric acid, or tartaric acid. Not all of the many ions that can precipitate at the recommended pH can be complexed effectively, and the possibility of interferences must be carefully considered in each case when applying a gravimetric method.

Although gravimetric methods have been used traditionally for the determination of large amounts of tellurium, more accurate and convenient volumetric methods are favored. The oxidation of tellurium(IV) by ceric sulfate in hot sulfuric acid solution in the presence of chromic ion as catalyst affords a convenient volumetric method for the determination of tellurium (32). Selenium(IV) does not interfere if the sulfuric acid is less than 2 N in concentration. Excess ceric sulfate is added, the excess being titrated with ferrous ammonium sulfate using *o*-phenanthroline ferrous–sulfate as indicator. The ceric sulfate method is best applied in tellurium-rich materials such as refined tellurium or tellurium compounds.

The oxidation of tellurium(IV) by permanganate as an analytical method has been studied in some detail (26). The sample is dissolved in 1:1 nitric–sulfuric acid mixture; addition of potassium bisulfate and repeated fuming with sulfuric acid volatilizes the selenium. The tellurite is dissolved in 10 vol % sulfuric acid, followed by threefold dilution with water and titration with potassium permanganate:

$$4\,H_2TeO_3 + 2\,KMnO_4 + 5\,H_2SO_4 + 4\,H_2O \longrightarrow 4\,Te(OH)_6 + 2\,KHSO_4 + Mn_2(SO_4)_3$$

Any manganese(IV) which is produced can be sequestered by the addition of fluoride. Satisfactory results can be obtained by titration of a solution containing approximately 3 vol % H_2SO_4 and 1 % NaF and not more than 0.10 g of tellurium per 300 mL. The potassium permanganate solution containing 3.485 g/L of $KMnO_4$ (Te equivalence approximately 0.005 g/mL Te) is standardized against pure tellurium or telluric acid. Dissolution of the sample and its preparation for titration must be carried out so that tellurium is maintained in the tetravalent state while other elements are fully oxidized. There is a tendency for this method to give slightly high results.

From a toxicological and physiological point of view, the determination of very small amounts of tellurium is becoming increasingly important. Interest is environmental and human health has promoted development in analytical techniques and methods for the trace and ultra trace levels (see TRACE AND RESIDUE ANALYSIS).

Numerous methods have been published for the determination of trace amounts of tellurium (33–42). Instrumental analytical methods (qv) used to determine trace amounts of tellurium include: atomic absorption spectrometry, flame, graphite furnace, and hydride generation; inductively coupled argon plasma optical emission spectrometry; inductively coupled plasma mass spectrometry; neutron activation analysis; and spectrophotometry (see MASS SPECTROMETRY; SPECTROSCOPY, OPTICAL). Other instrumental methods include: polarography, potentiometry, emission spectroscopy, x-ray diffraction, and x-ray fluorescence.

A widely used procedure for determining trace amounts of tellurium involves separating tellurium in (1:1) hydrochloric acid solution by reduction to elemental tellurium using arsenic as a carrier and hypophosphorous acid as reductant. The arsenic, reduced from an addition of arsenite to the solution, acts as a carrier for the tellurium. The precipitated tellurium, together with the carrier, is collected by filtration and the filter examined directly in the wavelength-dispersive x-ray fluorescence spectrometer.

Health and Safety Factors

Elemental tellurium and the stable tellurides of heavy nonferrous metals are relatively inert and do not represent a significant health hazard (43–47). Other, more reactive tellurides, including soluble and volatile tellurium compounds such as hydrogen telluride [7783-09-7], tellurium hexafluoride [7783-80-4], and alkyl tellurides, should be handled with caution. Some of these materials can enter the body by absorption through the skin or by inhalation and ingestion of dust or fumes. No serious consequences or deaths have been reported in workers exposed to tellurium and its compounds in industry (48).

The soluble tellurites are more toxic than the selenites and arsenites. Hydrogen telluride, formed by the action of water on aluminum telluride [12043-29-7], Al_2Te_3, is a toxic gas (43).

The unusual physical complaints and findings in workers overexposed to tellurium include somnolence, anorexia, nausea, perspiration, a metallic taste in the mouth and garlic-like odor on the breath (48). The unpleasant odor, attributed to the formation of dimethyltelluride, has not been associated with any adverse health symptoms. Tellurium compounds and metabolic products have been identified in exhaled breath, sweat, urine, and feces. Elimination is relatively slow and continuous exposure may result in some accumulation. No definite pathological effects have been observed beyond the physical complaints outlined. Unlike selenium, tellurium has not been proved to be an essential biological trace element.

Industrial precautions for handling tellurium include the common-sense measures of good housekeeping, adequate ventilation, personal cleanliness, and frequent changes of clothing. Gloves and safety glasses should be worn at all times, and dust masks and chemical goggles should be used where needed. Oral administration of ascorbic acid (or chlorophyll) was recommended at one time to alleviate unpleasant breath, but such treatment may enhance toxic effects by reducing the tellurium compounds to more toxic constituents, and could also serve to give workers a false sense of security (49).

The threshold limit value (TLV) set by the American Conference of Industrial Hygienists (ACGIH) for tellurium and its compounds is 0.1 mg/m^3 which is about ten times the amount which has been known to produce the adverse garlic odor (45,50). The ACGIH TLV for tellurium hexafluoride is 0.1 mg/m^3 or 0.02 ppm of air. Likewise, the U.S. Occupational Safety and Health Administration (OSHA) has established its permissible exposure limit (PEL) for tellurium and its compounds at 0.1 mg/m^3; the PEL for tellurium hexafluoride is 0.2 mg/m^3 or 0.02 ppm of air (50).

Inorganic Compounds

Tellurium forms inorganic compounds very similar to those of sulfur and selenium. The most important tellurium compounds are the tellurides, halides, oxides, and oxyacids (5). Techniques and methods of preparation are given in the literature (51,52). The chemical relations of tellurium compounds are illustrated in Figure 2 (53).

Tellurides. Most elements form compounds with tellurium. Binary compounds of tellurium with 69 elements and alloys of two others have been described (5,54); 58 are metals, 3 are metalloids, and 10 are nonmetals. Most tellurides are prepared by direct reaction, varying from very vigorous with alkali and some alkaline-earth (Fig. 2) metals to sluggish and requiring a high temperature with hydrogen. The alkali and alkaline-earth tellurides are colorless ionic solids rapidly decomposed by air, especially in the presence of atmospheric moisture. In aqueous solutions, these hydrolyze less than sulfides and selenides but are very easily oxidized. The solutions dissolve excess tellurium, forming dark red tellurides. The alkali metal tellurides are strong reductants. Hydrotellurides, such as sodium hydrotelluride [23624-18-2], NaHTe, are also known. Some metals form more than one telluride, and some metal tellurides show nonstoichiometry; many of them exhibit semiconductor properties.

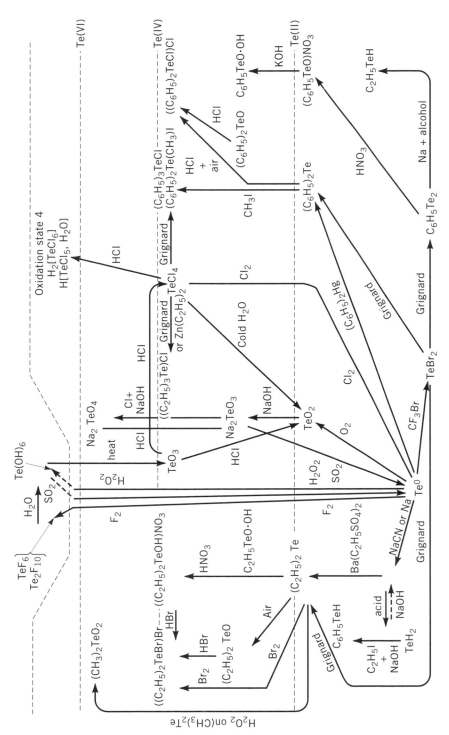

Fig. 2. Chemical relations of the compounds of tellurium (53).

795

Hydrogen Telluride.   Hydrogen telluride, H_2Te, is a colorless, toxic gas having an odor resembling that of arsine. It is also colorless in the liquid and solid states. Hydrogen telluride is prepared by the action of water on aluminum telluride in the absence of air, or by electrolysis of cold 15–20% sulfuric acid at a tellurium cathode. Hydrogen telluride and its aqueous solution decompose slowly in air. Unless perfectly dry, the compound also decomposes when exposed to light. Hydrogen telluride is a strong reductant and is rapidly oxidized. It reacts with alkalies and the solutions of many metal salts to form the corresponding tellurides. Hydrogen telluride solutions are weakly acidic and ionize to HTe^- and Te^{2-}.

Tellurium Sulfide.   In the liquid state, tellurium is completely miscible with sulfur. The Te–S phase diagram shows a eutectic at 105–110°C when the sulfur content is 98–99 atom % (94–98 wt %). Tellurium–sulfur alloys have semiconductor properties (see SEMICONDUCTORS). Bands attributed to tellurium sulfide [16608-21-2], TeS, molecules have been observed.

Tellurium Selenides.   Tellurium selenides or selenium tellurides are unknown. The molten elements are miscible in all proportions. The mixtures are not simple solid solutions but have a complex structure. Like the sulfides, the selenides exhibit semiconductor properties.

Carbon Sulfotelluride.   Carbon sulfotelluride [10340-06-4], CSTe, exists as a yellow-red liquid having a garlic-like odor. It is decomposed by light, even at −50°C, to carbon disulfide, carbon, and tellurium.

Carbonyl Telluride.   Little is known about carbonyl telluride [65312-92-7], COTe. It is formed in poor yield by passing carbon monoxide over tellurium at a high temperature. It is less stable than the selenide.

Tellurium Nitride.   Tellurium nitride [12164-01-0], Te_3N_4, is an unstable, citron-yellow solid that detonates easily when heated or struck, but it can be kept under dry chloroform. It is said to explode on contact with water, possibly because of the heat of wetting.

Tellurium Halides.   Tellurium forms the dihalides $TeCl_2$ and $TeBr_2$, but not TeI_2. However, it forms tetrahalides with all four halogens. Tellurium decafluoride [53214-07-6] and hexafluoride can also be prepared. No monohalide, Te_2X_2, is believed to exist. Tellurium does not form well-defined oxyhalides as do sulfur and selenium. The tellurium halides show varying tendencies to form complexes and addition compounds with nitrogen compounds such as ammonia, pyridine, simple and substituted thioureas and anilines, and ethylenediamine, as well as sulfur trioxide and the chlorides of other elements.

Tellurium Tetrafluoride.   Tellurium tetrafluoride [15192-26-4], TeF_4, forms white, hygroscopic needles melting at 129.6°C. It decomposes at 194°C to TeF_6 and is readily hydrolyzed. Tellurium tetrafluoride attacks glass, silica, and copper at 200°C, but it does not attack platinum below 300°C.

Tellurium Decafluoride.   Tellurium decafluoride [53214-07-6], Te_2F_{10}, is a stable, volatile, colorless liquid, melting at −33.7°C and boiling at 59°C.

Tellurium Hexafluoride.   Tellurium hexafluoride, TeF_6 melts at −38°C and sublimes at −39°C, forming a colorless gas. It hydrolyzes slowly to orthotelluric acid and is reduced by tellurium to TeF_4.

Tellurium Dichloride.   Tellurium dichloride [10025-71-5], $TeCl_2$, is a black hygroscopic solid, melting at 208°C to a black liquid; it boils at 328°C to a bright red vapor. The solid is stable when pure. It disproportionates in organic solvents and is decomposed by acids and alkalies. It is hydrolyzed to H_2TeO_3, Te, and HCl, and decomposed by HCl to Te and H_2TeCl_6 [17112-43-5]. Air oxidizes tellurium dichloride to TeO_2 and HCl.

Tellurium Tetrachloride. Tellurium tetrachloride [10026-07-0], $TeCl_4$, forms white, hygroscopic crystals which melt at 225°C to a dark red liquid. The vapor (bp ca 390°C) is also red. It is prepared by the action of chlorine on tellurium, or of SCl_2, CCl_4, or $AsCl_3$ on tellurium or TeO_2. The crystals are soluble in benzene, toluene, and ethyl alcohol, but not in ethyl ether. It hydrolyzes in cold water to TeO_2, and in hot water gives a clear solution. With dilute HCl, it yields H_2TeO_6 [56367-33-0], TeO_2, $HTeCl_5$ [22742-12-7], and H_2TeCl_6, and with concentrated HCl, $TeCl_3^{2-}$. On evaporation of the acid solution to dryness, $TeCl_4$ forms $H_2TeCl_6 \cdot 2H_2O$ [12192-31-3], which decomposes at 350°C to TeO_2. With phenylmagnesium chloride, it yields $Te(C_6H_5)_3Cl$ [1224-13-1]. Tellurium tetrachloride forms addition compounds with various ligands.

Tellurium Dibromide. Tellurium dibromide [7789-54-0], $TeBr_2$, must be prepared very carefully, because it is unstable and has a strong tendency to disproportionate to $TeBr_4$ and Te.

Tellurium Tetrabromide. Tellurium tetrabromide [10031-27-3], $TeBr_4$, forms yellow hygroscopic crystals which decompose above 280°C and melt at 363°C under bromine vapor. It boils at 414–427°C, dissociating into $TeBr_2$ and bromine. It is soluble in ether and chloroform but not in CCl_4, and is readily hydrolyzed in water.

Tellurium Tetraiodide. Tellurium tertaiodide [7790-48-9], TeI_4, forms gray-black volatile crystals which above 100°C decompose into the elements. They melt at 280°C in a sealed tube.

Tellurium Oxychlorides. Tellurium oxychlorides, $Te_6O_{11}Cl_2$ [12015-54-2], and $TeOCl_2$ [16981-34-2], have been reported.

Tellurium Oxydibromide. Tellurium oxydibromide [66461-30-1], $TeOBr_2$, has been prepared.

Tellurium Oxides, Oxyacids, and Salts. The white crystals of tellurium dioxide [7446-07-3], TeO_2, melt at 733°C to a clear, dark red liquid which vaporizes at 790–940°C. Tellurium dioxide is made by dissolving tellurium in strong HNO_3 solution and decomposing the resultant $2\ TeO_2 \cdot HNO_3$ at 400–430°C; by heating Te in air or oxygen; or by thermal dehydration of H_6TeO_6. It is slightly soluble in water with the formation of H_2TeO_3, and readily soluble in halogen acids, forming complex anions such as $TeCl_3^{2-}$. It dissolves in solutions of al-kali–metal carbonates only on boiling. The standard redox potential is between those of SO_2 and SeO_2. In solution, TeO_2 is a weaker oxidant for SO_2 than SeO_2. Tellurium dioxide is amphoteric, having minimum solubility at pH 3.8–4.2, its isoelectric point. It forms tellurites, $MTeO_3$, with alkali metals, but not with NH_3. With some acids it forms basic salts such as $TeO_2 \cdot 2\ HCl$ [83543-00-4], $TeO_2 \cdot 3\ HCl$, $TeO_2 \cdot 2\ HBr$ [83543-02-6], $2\ TeO_2 \cdot HNO_3$, $2\ TeO_2 \cdot SO_3$ [1206584-8], and $2\ TeO_2 \cdot HClO_4$. Heating with metal oxides yields tellurites, but with higher oxides, such as PbO_2, or with oxidants such as KNO_3 or $KClO_3$, tellurates are formed. Heating with metals such as Al, Cs, and Zn, with C (at red heat), or with S, ZnS, HgS, or PbS, reduces TeO_2 to Te. Hydrogen does not reduce it completely even at high temperatures, and H_2O_2 oxidizes aqueous TeO_2 suspension to H_4TeO_6 [41673-77-2].

Tellurium Trioxide. Tellurium trioxide [13451-18-8], TeO_3, exists in two modifications, the yellow-orange α-form and the grayish β-form. The α-form, sp gr 5.07, decomposes above 360°C to TeO_2. It is prepared by dehydrating H_6TeO_6 at 300–360°C, α-Tellurium trioxide is a strong oxidant, it reacts violently with metals such as Al or Sn, and with nonmetals such as C, P, and S. Although insoluble in water, dilute mineral acids, and dilute alkalies, it dissolves in hot concentrated alkalies with the formation of tellurates. α-Tellurium trioxide is reduced to TeO_2 and $TeCl_4$ by boiling with concentrated HCl (with the liberation of Cl_2). The β-form, a grayish solid, sp gr 6.21, is obtained by heating orthotelluric acid or α-TeO_3, alone or with H_2SO_4, for 10–12 h at 300–350°C. It is soluble in concentrated Na_2S solution. Less reactive than the α-form, it is reduced by hydrogen at 400°C. The β-form does not react with water at 150°C in the presence of acid or alkali. Most likely it is a polymer, but there are no structural data for either modification.

Tellurous Acid. Tellurous acid [10049-23-7], H_2TeO_3 is an unstable white solid that dehydrates readily to TeO_2. It is prepared by the acidification of a tellurite solution with HNO_3 or by cold hydrolysis of a tetrahalide. Tellurous acid is a much weaker acid than H_2SeO_3. Both normal and acid tellurites are known. Alkali–metal tellurites are obtained from alkali hydroxides and TeO_2; other tellurites are prepared by double decomposition with metal tellurites, or by fusing TeO_2 with metal oxides or carbonates. The alkali metal tellurites are water soluble, the alkaline-earth tellurites are less so. Other tellurites are insoluble. Tellurites are oxidized to tellurates by heating in air or by treating a solution with H_2O_2 and other oxidants. In acid solution, the salts are reduced to Te by SO_2, Sn, Zn, Cu, and some organic compounds.

Orthotelluric Acid. The white crystals of orthotelluric acid [7803-68-1], H_6TeO_6, are sparingly soluble in cold water and easily soluble in hot water and mineral acids, with the exception of HNO_3. It is made by oxidizing Te or TeO_2, for example by refluxing with H_2O_2 in concentrated H_2SO_4. It exists in a cubic form and a monoclinic β-form, which is stable at room temperature. The β-form is obtained from the α-form by heating. Orthotelluric acid tends to polymerize. It is weakly dibasic, forming salts such as KH_5TeO_6 [15855-68-2], $Li_2H_4TeO_6$ [20730-51-2], and $Na_2H_4TeO_6$ [20730-46-5]. An exception is Ag_6TeO_6. Heating dehydrates H_6TeO_6 to TeO_3, further heating gives TeO_2 and oxygen. The acid is fairly strong oxidant; it liberates iodine from KI and oxidizes HCl and HBr. It is reduced by H_2S, SO_2, Zn, Fe^{2+} and hydrazine. It forms complex acids and salts with other elements.

Polymetatelluric Acid. Polymetatelluric acid, H_2TeO_{4n} (where $n \simeq 10$), is a white, amorphous, hygroscopic solid obtained by partial dehydration of orthotelluric acid at 100–200°C in air. Polymetatelluric acid always contains some of the ortho acid and reverts to it in solution. It forms esters and salts. Orthotelluric and polymetatelluric acids are the only two telluric acids known. The so-called allotelluric acid [13520-55-3], a colorless syrup, is a mixture of orthotelluric and polymetatelluric acids. It is prepared by heating the ortho acid in a sealed tube at 305°C.

Tellurates. The water-soluble alkali metal and alkaline-earth tellurates are prepared by chlorinating alkaline solutions of the tellurites or by heating solid tellurites with KNO_3, $KClO_3$, or PbO_2 to form, for example, potassium tellurate

[*7790-58-1*], K_2TeO_3. The insoluble tellurates are made by double decomposition of alkali–metal salt solutions.

Other Inorganic Compounds. Alkali–metal and ammonium telluropentathionates $(Te(S_2O_3)_2)^{2-}$, have been prepared. The S_2O_3 group can be replaced by ethyl xanthate or diethyl dithiocyanate.

Tellurium pseudohalides, such as the dicyanide [*14453-24-8*] $Te(CN)_2$, the dithiocyanate [*83543-04-8*] $Te(SCN)_2$, and thiourea complexes with $Te(SCN)_2$, have been prepared. These are similar to the halides in properties.

Basic tellurium nitrate, TeO_2, is made by dissolving Te in HNO_3. Thermal decomposition begins at 190°C and is complete at 300°C.

Basic tellurium sulfate $2\,TeO_2{\cdot}SO_3$ [*12068-84-8*]; $2\,TeO_2{\cdot}SeO_3$ selenate; and tellurate, $2\,TeO_2{\cdot}TeO_3$, are known. The sulfate is made by the slow evaporation of a TeO_2 solution in H_2SO_4. It is stable up to 440–500°C and is hydrolyzed slowly by cold, and rapidly by hot water.

Tellurium perchlorate, iodate, and methylthiosulfate, as well as Te(IV) salts of aliphatic and aromatic acids, have been prepared.

Organic Compounds

The chemical properties of organosulfur, organoselenium, and organotellurium compounds are markedly similar. Because bond stability with carbon decreases with the increasing atomic number of the element, thermal stability decreases, whereas oxidation susceptibility increases to such an extent that alkyl tellurides are oxidized rapidly by air at room temperature. As a result, less has been written concerning the chemistry of organotellurium than organoselenium compounds. Nevertheless, a sizable literature exists (29,55–59).

Organotellurium compounds range from the simple carbon sulfotelluride to complex heterocyclic compounds and organotellurium ligands (60). Tellurium analogues of alcohols and mercaptans are prepared by reacting their vapors with aluminum telluride and protecting the products in an atmosphere of hydrogen. Ditellurides, R–Te–Te–R, are also sensitive to atmospheric oxygen. Dimethyltellurium dihalides, R_2TeX_2, have been prepared as *cis-trans* isomers (61), and organometallic compounds including both Group III and Group IV elements, involve a metal–tellurium bond. Various types of tellurium compounds and specific examples are listed in Table 6. Not included are the many organic complexes of inorganic tellurium compounds and their ions. For detailed descriptions of the laboratory preparation of organotellurium compounds, see References 56 and 62–64.

Uses

Free-Machining Steels. Tellurium has been shown to be the best additive for improving machinability in several types of ferritic steel (65–67). Its effectiveness in this area represents over 50% of tellurium consumption. The Inland Steel Company (East Chicago, Indiana) developed the largest single outlet for tellurium as an additive to plain-carbon, leaded, and leaded resulfurized steels (68). The presence of 0.023–0.057% tellurium in cold drawn steels, containing 0.06–0.09% carbon, 0.28–0.37% sulfur, and 0.15–0.22% lead, increases their

Table 6. Isolated Organic Tellurium Compounds[a]

Compounds	CAS Registry Number	Formula	Mp, °C	Bp, °C
tellurols or tallanes		RTeH		>90
ethanetellurol	[83270-38-6]	C_2H_5TeH		
benzenetellurol	[69577-06-6]	C_6H_5TeH		
alkyl, aryl, and cyclic tellurides				
dimethyltelluride (methyl telluride)	[593-80-6]	$(CH_3)_2Te$		82
diphenyl telluride	[1202-36-4]	$(C_6H_5)_2Te$	53.4	182
tetraphenyl telluride	[64109-07-5]	$(C_6H_5)_4Te$	104–106	
tetrahydrotellurophene	[3465-99-4]	$TeCH_2(CH_2)_2CH_2$		
ditellurides		R_2Te		
diphenyl ditelluride	[32294-60-3]	$C_6H_5Te-TeC_6H_5$	53–54	
alkyltellurium dihalides		$RTeX_3$		
methyltellurium tribromide	[20350-53-2]	CH_3TeBr_3	dec 140 C	
dialkyltellurium dihalides		R_2TeX_3		
dimethyltellurium dichloride	[24383-90-3]	$(CH_3)_2TeCl_2$	92(α), 134(β)	
tellurium salts		R_3TeX		
trimethyltellurium iodide	[18987-26-3]	$(CH_3)_3TeI$		
telluroxides		RTeO		
diphenyl telluroxide	[51786-98-2]	$(C_6H_5)_2TeO$	185	
tellurones		R_2TeO_2		
dimethyl tellurone	[83270-39-7]	$(CH_3)_2TeO_2$		
telluroketones		R_2CTe		
dimethyl telluroketone	[83270-40-0]	CH_3CTeCH_3	63–66	
tellurinic acid		RTeO·OH		
phenyltellurinic acid	[83270-41-1]	$C_6H_5Te·OH$	211	
heterocyclic compounds				
1,4-oxatellurane	[5974-87-8]	$OCH_2CH_2TeCH_2CH_2$		
3,5-telluranedione	[24572-07-4]	$CH_2COCH_2TeCH_2CO$		

[a]Ref. 39.

machinability by 30–50% and increases the feed rates. The machined surface finish is superior, the life of the cutting tool is extended, the metal chips are more uniform, and there is less buildup on the tool edges. Tellurium addition affects the mechanical properties very little, raises the recrystallization temperature only slightly without affecting the heat-treating properties, and has a minor effect on grain refining. In addition, it is less detrimental than sulfur in lowering the corrosion resistance. The addition of Te and Ca has been found to enhance the machining performance of a number of steels by controlling the morphology of sulfide and oxide inclusions. In the case of silicon-killed free-machining steels, the addition of Ca forms the plastic-type silicate inclusion $(Ca,Mn)O-Al_2O_3-$

2 SiO$_2$, which acts as an internal lubricant while the addition of Te promotes the globularization of MnS inclusions and increases their rigidity, preventing them from elongating in hot-rolled products which cause directional properties. The presence of well-distributed globular inclusions of MnS significantly improves the machinability of steel (69). Tellurium can also be added to low and high alloy steels that are normally difficult to machine (70) and tellurium improves the machinability of powder-metallurgy steels (71).

Chilled Castings. Since about 1940, tellurium has been used to control the chill depth of iron castings (see IRON). Hard, wear-resistant, chilled surfaces were originally produced on gray iron castings by pouring molten iron over a metal surface. A thin layer of white iron formed on rapid cooling. By the addition of less than 0.1% Te to the melt, chilled surfaces could be obtained, with higher yields and extended mold life.

Tellurium is 100–150 times as effective as chromium in producing a given depth of chill. It is a powerful carbide stabilizer and minute quantities convert gray iron to white. Because tellurium is highly volatile, it is added to the ladle just before pouring. It may be added as pellets, tablets, powder (in aluminum or copper containers), ferrotellurium or copper telluride [12181-15-6], Cu$_2$Te, granules, or any other convenient form. Alternatively, the mold is treated by coating the desired areas of the mold surface with a tellurium-bearing wash, surfacing the sand mold with a layer of tellurium-rich sand, or using tellurium-containing tape (72).

The result is a hard, abrasion-resistant surface, important in many applications of cast iron. The depth of the chill may be controlled by regulating the amount of tellurium added. The casting shows a sharp demarcation line between the chilled and unchilled regions; there is no intermediate or mottled zone. Yet, the chilled portion shows excellent resistance to spalling from thermal or mechanical shock. Tellurium-treated iron is more resistant to sulfuric and hydrochloric acids than is untreated, unchilled gray iron. The amount added ranges from 0.005 to 0.1%; ca 60% is lost by volatilization. Excessive addition causes porosity in the castings.

Tellurium-chilled iron has been used in mining, automotive, railroad, and other equipment. Tellurium used as a coating for molds and cores, so-called corewashes, eliminates troublesome localized shrinkage in castings. These corewashes may contain ca 25% tellurium; the remainder consists mainly of silica and some bentonite.

Other Ferrous Metal Uses. Tellurium shows an extremely high surface activity in liquid metals (66,73–75). As a result, addition of tellurium sharply decreases the rate of nitrogen absorption in liquid iron and steel (74). This allows the use of cheaper, lower purity oxygen in pneumatic or basic oxygen steelmaking, and decreases the nitrogen pickup during supplementary scrap melting.

In 0.1% and smaller additions, tellurium acts as a mild deoxidizer of liquid steel, improves the soundness of castings, refines the grain size, and counteracts the detrimental effect of sulfur on the ductility (75). The addition of 0.002–0.009% Te reduces impact and ductility anisotropy in plain carbon steels by controlling the Te addition so as not to exceed its solubility limit in the MnS inclusions and form tellurides. The globular shape of MnS inclusions is optimized with the presence of about 2% Te in solid solution, resulting in the enhancement of both the transverse and through thickness properties. In highly deoxidized steels, Te is also reported to improve the inclusion distribution and eliminates sulfide clustering at the grain boundaries (76).

The low temperature impact strength of structural steels was found to be significantly improved with the additions of 0.022–0.008% Te. The low (−40°C) temperature impact strength of a steel annealed at blue brittle temperature of 350°C did not show any embrittlement effect with the addition of .002% Te (76).

The presence of 0.004–0.0062% Te in a welding rod or wire produces a positive surface tension coefficient which enhances the heat transfer to the weld root, producing a weld pool fluid flow that results in solidified weld beads with high depth to width ratios. This characteristic is particularly desirable for stainless steel weldments (76).

The addition of 0.002–0.02% Te to a bearing steel is reported to improve the fatigue strength and antifriction characteristics by promoting the spheroidicity of the sulfide inclusions. It is specified that the ratio of Te to S be 0.1:1.0 (76).

Tellurium also improves the properties of electrical steels by aiding in the magnetic anisotropy, malleable cast iron (77), and spheroidal (graphitic) cast irons (see also METAL SURFACE TREATMENTS).

Copper Alloys. Tellurium is alloyed with copper for various purposes. Frequently the tellurium is added to molten copper as a copper telluride (46.3% Te) master alloy, taking advantage of the peritectic melting point of 1051°C.

A 99.5% Cu–0.5% Te alloy has been on the market for many years (78). The most widely used is alloy No. CA145 (number given by Copper Development Association, New York), nominally containing 0.5% tellurium and 0.008% phosphorous. The electrical conductivity of this alloy, in the annealed state, is 90–98%, and the thermal conductivity 91.5–94.5% that of the tough-pitch grade of copper. The machinability rating, 80–90, compares with 100 for free-cutting brass and 20 for pure copper.

Unlike lead and other additives, tellurium has no adverse effect on the hot-working properties of copper, and does not cause segregation and firecracking. Although the alloy is somewhat less ductile at room temperature than pure copper, it may be extensively hot- and cold-worked, and intermediate annealing is required when the cross section reduction is >40%. Machining methods are the same as for free-cutting brass. Because the copper telluride particles are harder than the lead particles in free-machining brass, carbide-tipped cutting tools are recommended.

A copper alloy, containing 0.02–0.04% tellurium, 0.002–0.015% phosphorous, and 0.002–0.05% oxygen is recommended for use in automobile radiators (79). A continuously cast Amtel copper alloy, containing 0.4–0.6% tellurium, 0.007–0.012% phosphorous, 0.02% sulfur max, and a copper–silver–tellurium–phosphorous alloy (99.90% max) was developed by AMAX Base Metals R&D, Inc. (80).

Tellurium–copper alloys are recommended for situations demanding a high production rate with no significant sacrifice in conductivity. These alloys can be soldered, brazed, or welded without incurring embrittlement. They are used in vacuum applications, forgings, screw-machine parts, welding-torch tips, transistor bases, semiconductor heat sinks, electrical connectors (qv), motor and switch parts, and nuts, bolts, and studs. Addition of tellurium significantly improves the surface of machined parts.

Copper–lead–tellurium alloys have high wear resistance in sliding contacts. In copper–zinc alloys, the benefits of tellurium decrease with increasing zinc content and almost disappear when the zinc content exceeds 35%.

Telnic bronze, Copper Development Association (CDA) No. 191, an alloy that hardens with age, contains 98.3% Cu, 1.1% Ni, 0.5% Te, and 0.2% P. It was developed by Chase Division, Kennecott Copper Corp., in 1950. The machinability rating is 80. The hot-working properties are comparable to those of a high copper commercial bronze and are superior to lead-containing free-machining copper alloys. The tensile strength is high, and the electrical and thermal conductivities are similar to those of nickel bronze CDA No. 191, and about one half those of copper (see COPPER ALLOYS).

Lead Alloys. A tellurium–lead alloy containing 0.02–0.1% tellurium, with or without antimony, was introduced in 1934 (81) as tellurium lead or Teledium. This alloy has higher recrystallization temperatures and corrosion resistance and takes a significantly longer time to soften at 25°C after cold work.

The addition of tellurium has a deoxidizing action and confers on lead useful work and precipitation-hardening properties. It also improves resistance to wear, vibration, and mechanical breakdown. These properties, along with improved corrosion resistance, are utilized in the sheathing of power communication and marine cables, in chemical equipment (especially those exposed to sulfuric acid), or where resistance to fatigue is important. Tellurium has successfully replaced tin (1–3% Sn) in the sheathing. Thinner sheaths can be extruded, and the extrusion pressure for 0.05% Te-bearing lead is about the same as for a 3% Sn alloy.

Adding tellurium to lead and to lead alloyed with silver and arsenic improves the creep strength and the charging capacity of storage battery electrodes (see BATTERIES). These alloys have also been suggested for use as insoluble anodes in electrowinning.

Other Metals. Tellurium has been added to copper-base, lead-base, and tin-base bearing alloys. In babbit-type alloys, tellurium controls the structure and improves uniformity and fatigue resistance by restraining the tendency to segregation (see BEARING MATERIALS).

Adding 0.05–1% tellurium to tin and tin-base bearing alloys improves the workhardening properties, tensile strength, and creep resistance.

Tellurium has been recommended as an additive to magnesium to increase corrosion resistance (see CORROSION AND CORROSION CONTROL). The addition is highly exothermic but can be controlled by adding one tellurium tablet at a time to a sufficiently large bath of liquid magnesium. The addition to tellurium and chromium improves the stress-corrosion resistance of aluminum–magnesium alloys.

In permanent-magnet alloys of high cobalt–titanium type, tellurium significantly increases the coercive force. Tellurium alloyed with gold produces a yellowish-green color useful for ornamental applications. A gold alloy with 43% Te, centrifically cast, appears light yellowish-green after polishing.

Metal Coatings. Tellurium chlorides, as well as tellurium dioxide in hydrochloric acid solution, impart permanent and attractive black antique finish to silverware, aluminum, and brass. Anodized aluminum is colored dark gold by tellurium electrodeposition. A solution containing sodium tellurate and copper ions forms a black or blue-black coating on ferrous and nonferrous metals and alloys. Addition of sodium tellurite improves the corrosion resistance of electroplated nickel. Tellurium diethyldithiocarbamate is an additive in bright copper electroplating (see ELECTROPLATING).

Pigments and Glass.   Tellurium has served as base for ultramarine-type cadmium sulfotelluride (82) and cadmium telluride pigments (83) (see PIGMENTS, INORGANIC). In addition, small amounts of tellurium have been used in glass and ceramics to produce blue to brown colors (see COLORANTS FOR CERAMICS).

Catalysts.   In industrial practice the composition of catalysts are usually very complex. Tellurium is used in catalysts as a promoter or structural component (84). The catalysts are used to promote such diverse reactions as oxidation, ammoxidation, hydrogenation, dehydrogenation, halogenation, dehalogenation, and phenol condensation (85–87). Tellurium is added as a passivation promoter to nickel, iron, and vanadium catalysts. A cerium tellurium molybdate catalyst has successfully been used in a commercial operation for the ammoxidation of propylene to acrylonitrile (88).

Lubricants.   Tellurides of titanium, zirconium, molybdenum, tungsten, and other refractory metals are heat- and vacuum-stable. This property makes them useful in solid self-lubricating composites in the electronics, instrumentation, and aerospace fields (see LUBRICATION AND LUBRICANTS). Organic tellurides are antioxidants in lubricating oils and greases.

Rubber.   At one time the largest single market for tellurium was the rubber industry. This has been replaced by the steel industry. Suspected health hazards of tellurium compounds used in rubber have hampered its applications (89). Traditionally, small additions of powdered tellurium have been used in the rubber industry as a secondary vulcanizing agent in hard natural rubber compositions to reduce curing time and increase flexibility and abrasion resistance. It is also used in the production of soft natural rubber and styrene–butadiene rubber vulcanizates of enhanced toughness and heat resistance. In many cases an addition of 0.5% Te increases the rate of vulcanization and improves the aging and mechanical properties of sulfurless and low sulfur stocks. Tellurium is particularly effective when used with tetramethylthiuram disulfide, (bis(dimethylthiocarbamoyl)disulfide) and with selenium diethyldithiocarbamate in sulfurless cures. It is frequently used to eliminate porosity in thick-molded sections. Tetrakis (dimethylthiocarbamato) tellurium(IV), when added to natural rubber, was found to offer an advantage over elemental tellurium in prevulcanizates because it reacts fast at low temperatures normally employed in natural rubber latex technology and, along with certain organic activators, results in a rubber with vastly enhanced high temperature and oxidation resistance and superior processing qualities in extrusion and calendering applications (90). Tellurium dioxide improves the heat stability and aromatic fuel resistance in cured polysulfide elastomers, such as Thiokol. Tellurium rubber is extremely resistant to heat and abrasion. It is used in all-rubber-jacketed portable cables in mining, dredging, welding, etc and conveyor belts for special applications (see RUBBER CHEMICALS).

Tellurium dimethylthiocarbamate in combination with mercaptobenzothiazole, with or without tetramethylthiuram disulfide, is the fastest known accelerator for butyl rubber. It is used extensively in butyl tubes for buses and similar vehicles and in other butyl applications (see ELASTOMERS, SYNTHETIC; RUBBER, NATURAL).

Tellurium is also useful in the coupling of rubber to metals.

Explosives. Sodium tellurite is used as a jelling promoter in explosive compositions that can be readily poured or pumped into drillholes (91) (see EXPLOSIVES AND PROPELLANTS, EXPLOSIVES).

Medical and Biological. A very small, yet very important application of tellurium is in organic derivatives and radioactive isotopes for use as biological tracers, x-ray-contrast agents, and diagnostic aids, and for the treatment of thyroid diseases (see MEDICAL IMAGING TECHNOLOGY; RADIOACTIVE TRACERS). Binary tellurides involving antimony, bismuth, cadmium, cobalt, or copper have found uses as fungicides (92). Some organic tellurobromides show strong bacterial activity (93). Terpene ether tellurocyanates are effective parasiticides, used either alone or with carriers (94).

Electronic and Optoelectronic Applications of Tellurides. Most metal tellurides are semiconductors with a large range of energy gaps and can be used in a variety of electrical and optoelectronic devices. Alloys of the form HgCdTe and PbSnTe have been used as infrared detectors and CdTe has been employed as a gamma ray detector and is also a promising candidate material for a thin-film solar cell.

A slow but growing area of use has been for thermoelectric applications. Here, cooling by the Peltier effect has been used to make refrigerators of relatively low efficiency for special purposes. Such devices, which have no moving parts (except for a fan), employ modules containing a $Bi_2Te_3-Sb_2Te_3$ alloy for the p-type region of each junction and a $Bi_2Te_3-Sb_2Se_3$ alloy for the n-region. From the inverse thermoelectric phenomenon of the Seebeck effect, heat can be used to generate electrical power if one junction is heated and the other is kept at a lower temperature. For such power conversion, the modules employed have been based on PbTe [1314-91-6], SnTe [12040-02-7], and MnTe [12032-88-1] as well as bismuth telluride alloys (95).

BIBLIOGRAPHY

"Tellurium and Tellurium Compounds" in *ECT* 1st ed., Vol. 13, pp. 666–676, by E. M. Elkin, Canadian Copper Refiners Ltd., and J. L. Margrave, University of Wisconsin; in *ECT* 2nd ed., Vol. 19, pp. 756–774, by E. M. Elkin, Canadian Copper Refiner, Ltd.; in *ECT* 3rd ed., Vol. 22, pp. 658–679, by E. M. Elkin, Noranda Mines, Ltd.

1. M. A. Weeks and H. M. Leicester, "Discovery of the Elements," 7th ed., *J. Chem. Educ.* (1968).

2. B. Mason, *Principles of Geochemistry*, 3rd ed., John Wiley & Sons, Inc., New York, 1971.

3. N. D. Sindeeva, *Mineralogy and Types of Deposits of Selenium and Tellurium*, Wiley-Interscience, New York, 1964.

4. A. M. Lansche, *Selenium and Tellurium—A Materials Survey*, U.S. Bureau of Mines Information Circular 8340, U.S. Government Printing Office, Washington, D.C., 1967.

5. W. A. Dutton in W. C. Cooper, ed., *Tellurium*, Van Nostrand Reinhold Co., New York, 1971, pp. 100–183.

6. P. Grosse, *Die Festkörpereigenschaften von Tellur*, Springer-Verlag, Berlin, Heidelberg, New York, 1969.

7. J. Bergman and L. Engman, *J. Org. Chem.* **47**, 5191 (1982).

8. J. H. Schloen and E. M. Elkin, *Trans. AIME* **188**, 764 (1950); A. Butts, ed., *Treatment of Electrolytic Copper Refinery Slimes*, Reinhold Publishing Corp., New York, 1954, Chapt. 11.

9. L. A. Soshnikova and M. M. Kupchenko, *Pererabotka Mednelektrolitnykh Shlamov (Treatment of Copper Electrolysis Slimes)*, *Metallurgiya*, Moscow, 1978.

10. S. C. Carapella, ed., *Proceedings of the 2nd Symposium on Industrial Uses of Selenium and Tellurium*, Toronto, Canada, Oct. 1980, Selenium–Tellurium Development Association, Grimbergen, Belgium, 1981.

11. D. H. Jennings, R. T. McAndrew, and E. S. Stratigakos, TMS paper no. A68-9 The Metallurgical Society, Warrendale, Pa., 1968.

12. R. K. Manahan and F. Loewen, *Treatment of Anode Slimes at the INCO Copper Refinery*, paper presented at 1972 CIM Meeting, Canadian Institute of Mining/Metallurgy.

13. U.S. Pat. 4,047,939 (Sept. 13, 1977), B. Morrison (to Noranda).

14. U.S. Pat. 2,990,248 (June 27, 1961), L. E. Vaaler (to Diamond Alkali Co.).

15. T. Yonagida and N. Hosoda, TMS paper no. A75-42 (1975); see Ref. 11.

16. U.S. Pat. 4,094,668 (June 13, 1978), J. C. Yannopoulos and B. M. Borham (to Newmont).

17. J. E. Hoffmann, *J. Metals*, 50–54 (Aug. 1990).

18. J. E. Hoffmann and B. Westrom, *Hydrometallurgy 1994*, Institution of Mining and Metallurgy, Chapman and Hall, London, pp. 69-105.

19. T. T. Chen and J. E. Dutrizac, *Journal of Metals 1990*, pp. 39–64.

20. R. Breese, D. Vleeschhouwer, and J. Thiriar, in Ref. 10, pp. 31–49.

21. P. H. Jennings and J. C. T. Farge, *Canad. Mining Metallurg. Bull.*, 193–200, (Feb. 1966).

22. *Minerals Yearbook*, U.S. Dept. of the Interior, Bureau of Mines, Government Printing Office, Washington, D.C. (published annually).

23. *American Bureau of Metal Statistics*, New York (published annually).

24. *Canadian Minerals Yearbook*, Canada Department of Energy, Mines, and Resources, Ottawa, Canada (published annually).

25. World Bureau of Metal Statistics, London (published annually).

26. W. C. Cooper, in N. H. Furman, ed., *Standard Methods of Chemical Analysis*, 7th ed., Vol. 1, D. Van Nostrand Co., Inc., Princeton, N.J., 1962, pp. 925–949; and in Ref. 5, pp. 281–312.

27. T. W. Green and M. Turley, in I. M. Kolthoff and P. J. Eavings, eds., *Treatise on Analytical Chemistry*, Vol. 7, Part 2, Interscience Publishers, a division of John Wiley & Sons, Inc., New York, 1961, pp. 137–205.

28. D. I. Ryabchikov, L. L. Nazarenko, and I. P. Alimarin, eds., *Analysis of High Purity Materials*, Israel Program for Scientific Information, Jerusalem, 1968.

29. K. J. Irgolic, ed., *The Organic Chemistry of Tellurium*, Gordon and Breach Science Publishers, Inc., New York, 1973.

30. W. C. Cooper, in Ref. 5, pp. 281–312.

31. K. L. Cheng, *Anal. Chem.* 33, 761 (1961).

32. H. H. Willard and P. Young, *J. Am. Chem. Soc.* **52**, 553 (1930).

33. V. K. Panday and A. K. Ganguly, *Atomic Absorp. Newsl.* **7**(3), 50–52 (May–June 1968).

34. B. V. Narayana and N. Appala Raju, *Analyst* **107**, 392–397 (Apr. 1982).
35. S. K. Aggarwal, M. Kinter, J. Nicholson, and D. A. Herold, *Anal. Chem.* **66**(8), 1316–1322 (Apr. 1994).
36. G. J. Desai and V. M. Shinde, *Talanta* **39**, 405–408 (1992).
37. E. M. Donaldson and M. E. Leaver, *Talanta* **37**(2) 173–183 (1990).
38. D. Bakhthr, G. R. Bradford, and L. J. Lund, *Analyst*, **114**, 901–909 (Aug. 1989).
39. J. Bozic, D. Maskery, S. Maggs, and H. Susil, *Analyst* **114**, 1401–1403 (Nov. 1989).
40. M. Tatro, *Spectroscopy* **5**(3) 14–17 (1990).
41. J. R. Clark, *Anal. Chem.* **53**(1) 61–64 (Jan. 1981).
42. B. A. T. Horler, *Analyst* **114**, 919–922 (Aug. 1989).
43. G. D. Clayton and F. E. Clayton, eds., *Patty's Industrial Hygiene and Toxicology*, 4th ed., Vol. 2, Pt. A, John Wiley & Sons, Inc., New York, 1993, pp. 801–805.
44. N. J. Sax, *Dangerous Properties of Industrial Materials*, 5th ed., Van Nostrand Reinhold Publishing Corp., New York, 1992, pp. 3186–3188.
45. T. Devlin, *A Review of the Toxicology of Tellurium and its Compounds*, Sandia Laboratories, Albuquerque, N. Mex., and Livermore, Calif., 1975.
46. V. B. Voulk in *Toxicology of Metals*, Vol. 2, Environmental Health Effects Research Series PB 268 324, U.S. Environmental Protection Agency, Research Triangle Park, N.C. reproduced by National Technical Information Service, May 1977, pp. 370–383.
47. J. R. Glover and V. Vouk, in L. Friberg and co-workers, eds., *Handbook on Toxicity of Metals*, Elsevier/North Holland Biomedical Press, Amsterdam, the Netherlands, 1979, Chapt. 355, pp. 587–598.
48. American Conference of Governmental Industrial Hygienists (ACGIH), *Guide to Occupational Exposure Values*, 1992, pp. 1489–1493.
49. L. Parmeggiani, ed., *Encyclopedia of Occupational Health and Safety*, 3rd rev. ed., Vol. 2, International Labour Office, Geneva, Switzerland, 1983, pp. 2156–2157.
50. *U.S. Fed. Reg.* **54**(12); 29 CFR Pt. 1910, p. 2953 (Jan. 19, 1989).
51. *Inorganic Synthesis*, McGraw-Hill Book Co., New York, published at irregular intervals.
52. G. Brauer, *Handbook of Preparative Inorganic Chemistry*, Vol. 1, Academic Press, New York, 1963.
53. P. J. Durrant and B. Durrant, *Introduction to Advanced Inorganic Chemistry*, Longmans, Green & Co., London, 1962, p. 881.
54. D. M. Chizhikov and V. P. Schastlivyl, *Tellurium and the Tellurides*, Collets Publishers, Ltd., London and Wellingborough, U.K., 1970.
55. K. W. Bagnall, *The Chemistry of Selenium, Tellurium, Polonium*, Elsevier Publishing Co., Amsterdam, 1966; *Comprehensive Inorganic Chemistry*, Vol. 2, Pergamon Press, Oxford, U.K., 1973, Chapts. 23–24.
56. J. Newton Friend, *A Text-Book of Inorganic Chemistry*, Vol. 11, Pt. 4, Charles Griffin & Co., London, 1937, Chapt. XI, pp. 166–260.
57. E. Krause and A. von Grosse, *Die Chemie der Metallorganischen Verbindungen*, Berlin, 1937.
58. R. A. Zingaro and K. J. Irgolic, in Ref. 29, pp. 184–280.
59. N. Petragnani and M. DeMoura Campos, *Organometall. Chem. Rev.* **2**, 61 (1967).
60. H. J. Gysling, *The Synthesis of Organotellurium Ligands*, Kodak Laboratory Chemical Bulletin, Vol. 53, No. 1, Eastman Kodak Co., Rochester, N.Y., 1982.
61. R. H. Vernon, *J. Chem. Soc.* **117**, 889 (1920).
62. *Organ. Synth.* (published annually since 1921, John Wiley & Sons, Inc., New York).
63. J. Houben, ed., *Die Methoden der Organischen Chemie (Weyls Methoden)*, Georg Thieme Verlag, Stuttgart, Germany, 1925–1941 (reproduced by Edwards Bros., Inc., Ann Arbor, Mich.).
64. V. Migrdichian, *Organic Syntheses*, Vols. 1 and 2, Reinhold Publishing Corp., New York, 1957.

65. E. S. Nachtman, in Ref. 5, pp. 373–388.
66. R. J. Raudebaugh, in Ref. 10, pp. 201–221.
67. D. Bhattacharya, in Ref. 10, pp. 222–243.
68. U.S. Pats. 3,152,889 and 3,152,890 (Oct. 13, 1964), M. O. Holowaty (to Inland Steel Co.); U.S. Pat. 3,1,69857 (Feb. 16, 1965), A. E. Rathke and A. T. Morgan (to Inland Steel Co.).
69. M. Confente and J. Bellot, in *Proceedings of the 3rd International Symposium on Industrial Uses of Selenium and Tellurium*, Saltsjöbaden, Sweden, Oct. 1984, Selenium–Tellurium Development Association, Grimbergen, Belgium, p. 129.
70. U.S. Pats. 2,009,713–2,009,716 (July 30, 1935), F. R. Palmer (to Carpenter Steel Corp.).
71. P. W. Taubenblat and W. E. Smith, in Ref. 10, pp. 254–266.
72. U.S. Pat. 2,979,793 (Apr. 18, 1961), R. L. Wilson, F. B. Hedihy, and T. J. Wood (to American Brake Shoe Co.).
73. R. H. Aborn, in Ref. 5, pp. 389–409.
74. U.S. Pat. 3,134,668 (May 26, 1966), R. D. Pehlke and M. Weinstein.
75. U.S. Pat. 2,258,604 (Oct. 14, 1961), A. P. Gagnebin (to General Metals Co.).
76. R. J. Raudebaugh, in Ref. 69, pp. 119–127.
77. U.S. Pats. 2,250,488 and 2,250,489 (July 29, 1941), C. H. Lorig and D. E. Krause (to Battelle Memorial Institute); U.S. Pats. 2,253,502 (Aug. 26, 1941) and 2,331,886 (Oct. 19, 1943), A. L. Boegehold (to General Motors Corp.); Brit. Pat. 552,390 (Apr. 3, 1943), E. Morgan and E. Hinchcliffe (to British Cast Iron Research Association).
78. U.S. Pat. 2,027,087 (Jan. 13, 1936), H. L. Brinkerhoff and D. E. Dawson (to Chase Co., Inc.).
79. R. J. Raudebaugh, in Ref. 10, p. 217.
80. P. W. Taubenblat, G. C. Van Tilburg, and F. Gruss, in Ref. 10, pp. 267–276.
81. W. Singleton and B. Jones, *J. Inst. Metals* **51**, 71 (1933); Brit. Pat. 441,524 (June 8, 1934), W. Singleton, W. Hulme, and B. Jones (to Goodlass Wall and Lead Industries, Ltd.).
82. U.S. Pat. 3,012,899 (Dec. 12, 1961), A. Giordano (to Harshaw Chemical Co.).
83. Brit. Pat. 2,033,418 (May 21, 1980), B. J. Chase and co-workers (to Johnson, Matthey & Co.).
84. Jpn. Pat. 02,141,549 (May 30, 1990), K. Nohara and co-workers (to Ishifuku).
85. V. Kollonitsch and C. H. Kline, *Hydrocarbon Process. Petrol. Refiners* **43**(6), 139 (1964).
86. *Proceedings of the 4th International Symposium on Uses of Selenium and Tellurium*, Banff, Alberta, Canada, May 1989, Selenium–Tellurium Development Association, Grimbergen, Belgium, 1989, pp. 609–672.
87. W. C. Cooper, in Ref. 5, pp. 414–416; A. B. Stiles in Ref. 10, pp. 286–290.
88. J. C. J. Bart and co-workers, in Ref. 69, p. 200.
89. W. R. McWhinnie and co-workers, in Ref. 69, p. 206.
90. K. G. Karnika de Silva and co-workers, in Ref. 86, p. 673.
91. U.S. Pat. 4,032,376 (June 28, 1977), K. R. Fossan and G. O. Ekman (to Nitrol-Nobel AB).
92. R. J. Brysson, B. J. Trask, and A. S. Cooper, Jr., *1968 Amer. Digest. Rep.* **57**, 12 (1968).
93. G. T. Morgan, E. A. Cooper, and A. C. Rawson, *J. Soc. Chem. Ind.* **45**, 106 (1926).
94. U.S. Pat. 2,263,716 (Nov. 25, 1941), J. N. Borglin (Hercules Powder Co.).
95. C. H. Champness in Ref. 5, pp. 322–372.

General References

J. Newton Friend, *A Text-Book of Inorganic Chemistry*, Vol. 7, Pt. 2, Charles Griffin & Co., London, 1931.
J. W. Mellor, *A Comprehensive Treatise on Inorganic and Theoretical Chemistry*, Vol. 11, Longmans, Green & Co., Inc., London, 1931.

Gmelins Handbuch der Anorganischen Chemie, 8th ed., System-Nummer 11, Verlag Chemie, Berlin, 1940.

N. V. Sidgwick, *The Chemical Elements and Their Compounds*, Vol. 2, Oxford University Press, London, 1950.

P. Pascal, *Nouveau Traité de Chemie Minérale*, 10th ed., Vol. 13, Masson & Cie., Editeurs, Paris, 1960.

A. F. Wells, *Structural Inorganic Chemistry*, 3rd ed., Oxford University Press, London, 1962.

A. A. Kudryavtsev, *The Chemistry and Technology of Selenium and Tellurium*, Collets Publishers, Ltd., London and Wellingborough, U.K., 1974.

F. A. Cotton and G. Wilkinson, *Advanced Inorganic Chemistry*, 5th ed., John Wiley and Sons, Inc., New York, 1988.

JAMES E. HOFFMANN
Jan Reimers and Associates USA Inc.

MICHAEL G. KING
Selenium–Tellurium Development Association

S. C. CARAPELLA
Consultant

J. E. OLDFIELD
Oregon State University

R. D. PUTNAM
Putnam Environmental Services

TEMPERATURE MEASUREMENT

The Kelvin Thermodynamic Temperature Scale

Temperature is a measure of the hotness of something. For a measure to be rational, there must be agreement on a scale of numerical values defining hotness and on devices for realizing and displaying these values. The single temperature scale having an absolute basis in nature is the Kelvin thermodynamic temperature scale (KTTS), or absolute scale, which is based on principles that can be deduced from the first and second laws of thermodynamics (qv). The most commonly used practical scale, however, is the International Temperature Scale (ITS), on which temperatures are designated as Celsius degrees (°C). Because the lower limit of the KTTS is absolute zero and the scale extends indefinitely upward and is by definition linear, only one nonzero reference point is required to stipulate its slope. The first such reference point was the equilibrium temperature of pure water in its liquid and solid phases at 101.3 kPa (1 atm) pressure, which was assigned the value 0°C, or 273.15 K. In 1954 the present KTTS was established by changing the reference point to the more reproducible equilibrium temperature of water in liquid–solid equilibrium under its own vapor pressure, ie, the triple point of water, assigned the value 273.16 K, or 0.01°C. (The unit of temperature on the KTTS is the kelvin, abbreviated K. The interval 1 K is identical to the interval 1°C. Thus these symbols may be used interchangeably to indicate an interval, but not to indicate a temperature. The symbol for a KTTS temperature is T; that for a temperature on any other scale is t).

Most thermometry using the KTTS directly requires a thermodynamic instrument for interpolation. The vapor pressure of an ideal gas is a thermodynamic function, and a common device for realizing the KTTS is the helium gas thermometer. The transfer function of this thermometer may be chosen as the change in pressure with change in temperature at constant volume, or the change in volume with change in temperature at constant pressure. It is easier to measure pressure accurately than volume; thus, constant volume gas thermometry is the usual choice (see PRESSURE MEASUREMENT).

A simplified gas thermometer is illustrated in Figure 1, showing that the temperature of the liquid–solid equilibrium of water (freezing point) under standard pressure is 273.15 K. A spherical bulb of constant volume is connected by constant-volume tubing to a U-tube manometer. The second connection to the manometer leads to a reservoir of mercury including a means for adjusting the height of mercury in the manometer, in this case a plunger. The bulb and tubing contain an ideal gas. The bulb is first surrounded by an equilibrium mixture of ice and water under a pressure of 1 standard atmosphere, or 101.325 Pa (C, in Fig. 1a), and when the gas A is in thermal equilibrium with the water and ice in the bath (at 0°C), the plunger is used to adjust the helium pressure so that both columns of mercury are at the same height, corresponding to the index mark 1.

The bulb is then surrounded with an equilibrium mixture of liquid and vapor water, ie, the boiling point of water under a pressure of 1 standard atmosphere (D, in Fig. 1b). As the gas A is heated to the temperature of the water, its pressure increases, displacing the mercury in both legs of the manometer. The

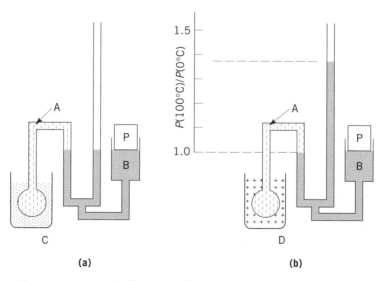

Fig. 1. Gas thermometer: A, helium gas; B, mercury; P, plunger for adjusting mercury column; and P, pressure, where in (**a**), bulb is surrounded by water and ice (0°C); (**b**) water and steam (100°C). See text.

mercury in the left leg of the U-tube can be readjusted to the index mark by operating the plunger, ie, restoring the condition of constant volume. However, the mercury in the right leg is now above the index mark. The difference in the heights of the mercury legs, which was zero at 0°C, is the measure of a new pressure which is 1.366099 times the initial, or 0°C, pressure. From the expression

$$(100°C - 0°C)/(1.366099 - 1) = 273.15 \tag{1}$$

the interval of the scale can be defined as 1/273.15 of the pressure ratio change between 0°C and 100°C. Thus, if the temperature is reduced below 0°C by 273.15 kelvins or Celsius units, an absolute zero is reached (in theory) that is the zero of the KTTS. The zero of the Celsius scale is therefore 273.15 K, and

$$T = t + 273.15 \tag{2}$$

The zero and the interval of the KTTS are defined without reference to properties of any specific substance. Real measurements with real gas thermometers are much more difficult than the example suggests, and all real gases condense before 0 K is reached.

The Fixed Points of Practical Temperature Scales

Accurate temperature measurements in real-life situations are difficult to make using the KTTS. Most easily used thermometers are not thermodynamic; that is, they do not operate on principles of the first and second laws. Most practicable thermometers depend upon some principle that is a repeatable and single-valued analogue of temperature, and they are used as interpolation devices of practical and utilitarian temperature scales which are themselves artifacts. Such principles include the expansion and contraction of liquids and solids, changes in the electrical properties of conductors and semiconductors, and the color and brilliance of light emitted from a very hot source. Any such principles may be used to make a thermometer. Because they are nonthermodynamic, they require construction of a consensus scale to relate the properties of a prescribed interpolation device to the KTTS.

The KTTS depends upon an absolute zero and one fixed point through which a straight line is projected. Because they are not ideally linear, practicable interpolation thermometers require additional fixed points to describe their individual characteristics. Thus a suitable number of fixed points, ie, temperatures at which pure substances in nature can exist in two- or three-phase equilibrium, together with specification of an interpolation instrument and appropriate algorithms, define a temperature scale. The temperature values of the fixed points are assigned values based on adjustments of data obtained by thermodynamic measurements such as gas thermometry.

Two-phase equilibria may be solid–liquid, liquid–vapor, or solid–vapor. As is evident from the phase rule of Gibbs, two-phase equilibria are pressure-dependent:

$$P + V = C + 2 \tag{3}$$

where P is an integer equal to the number of phases present, C is the number of components (for an ideally pure material $C = 1$), and V is an integer giving the number of degrees of freedom. Phase equilibria involving a vapor phase are much more pressure-dependent than liquid–solid equilibria. Triple points, ie, equilibria of all three phases, are independent of external pressure.

The triple point of water is the most important of the fixed points, because it is the one point that the KTTS and the ITS have in common, and because it can be realized with great accuracy. A water triple-point cell is shown in Figure 2. The cylindrical envelope is made of borosilicate glass and has a well for a thermometer. The cell is almost completely filled with very pure water of specified isotopic content. The small headspace contains water vapor, pressure of which at the triple-point temperature is about 600 Pa (4.5 torr). The integrity of the cell can be checked by inverting it carefully. If the cell has retained its integrity, a sharp click is heard as the water, uncushioned by air, strikes the opposite wall of the cell; also, as the bubble of water vapor is caught in the curved portion of the handle, it is compressed until it is almost invisible. These checks are all that is necessary to confirm the integrity of the cell, although they do not confirm the chemistry of the water. A technical guide and standard for the qualification and use of water triple point cells is found in Reference 1.

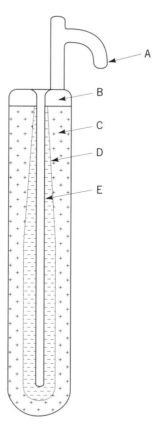

Fig. 2.   A water triple-point cell. A, handle; B, water vapor; C, liquid vapor, D, solid water (ice mantle); and E, thermometer well.

In use, a mantle of ice is frozen onto the outer surface of the thermometer well. A common way to do this is to fill the well with crushed dry ice until the mantle achieves a good thickness. Descriptions of the technique for doing this are given in several publications and in manufacturers' literature. The temperature of the water triple point is 0.01°C, or 273.16 K, by definition. In practice, that temperature can be realized in the cell within ~0.00015 K of the definition. In contrast, a bath of ice and water for producing the temperature 0°C is difficult to establish with an accuracy better than 0.002°C.

Any laboratory requiring precise temperature measurement should be equipped to realize the triple point of water, if no other fixed point. In addition to its status as a primary definition of a temperature, it is a valuable check on the laboratory's thermometers When thermometers change calibration, they tend to change at all temperatures within their ranges. If the calibration is correct at the water triple point, the probability is very high that it is also correct at any other temperature. The cost of recalibrating a standard platinum resistance thermometer has become extremely high, and such a check, which may avoid the need for recalibration, may be established for the price of one or two calibrations by the National Institute for Standards and Technology (NIST).

An important class of fixed points is the freezing points of high purity metals. As solid–liquid equilibria these are pressure-dependent, but this dependence is small (for tin, eg, 3.0×10^{-8} K/Pa (0.003 K/atm)) and can be corrected for completely. The metals used in fixed-point cells are better than 99.9999% (6 nines) pure. Figure 3 is one design of such a cell. The metal is contained in a

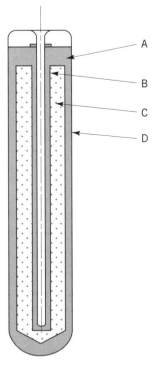

Fig. 3. Metal freezing-point cell: A, pure graphite crucible; B, thermometer well; C, highly pure metal; and D, quartz envelope.

crucible of purified graphite having a graphite cover and sleeve. The crucible is enclosed in an envelope of quartz, the quartz extending into the graphite sleeve to provide a well for the thermometer; thus the graphite assembly is completely enclosed. The cell is charged with pure metal, filled with argon or another inert gas at 101.325 kPa (1 atm) at the freezing temperature of the metal, and sealed. Thus it is protected against contamination and contains an atmosphere of the correct pressure at the freeze equilibrium temperature. Details of the construction of such cells have been published, and cells are available commercially, as are furnaces specifically designed for melting and freezing the metals (see *General References*). Sealed cells of this sort have shown no measurable change in freezing temperature after 20 years of continuous use. Table 1 shows a list of defining fixed points including the metal freezing points.

To perform calibrations using the freezing equilibria of these metals, the cells are heated in furnaces with relatively massive cores to about 10 K above the melting temperature. As the temperature of the metal increases, the onset of melting is indicated when the monitoring thermometer indicates a suddenly constant temperature, called the melt arrest, as the heat added is used in changing phase rather than changing temperature. When the melt is complete, the temperature increase continues to the furnace setpoint. The furnace control is

Table 1. Values of Defining Fixed Points of ITS-90 and Some Fixed Points of IPTS-68

Material[a]	State[b]	$T(90)$, K	$t(90)$, °C	$t(68)$, °C
He	VP	3–5	−270.15 to −268.15	
e-H$_2$	TP	13.8033	−259.3467	−259.34
e-H$_2$	VP, CVGT	~17	~−256.15	−256.108
e-H$_2$	VP, CVGT	~20.3	~−252.85	−252.87
Ne	TP		−248.5939	−248.595
Ne	BP	54.3584		−246.084
O$_2$	TP		−218.7916	−218.789
N$_2$	TP			−210.002
N$_2$	BP			−195.802
O$_2$	BP	83.8058		−182.962
Ar	TP		−189.3442	
Hg	FP	234.3156		−38.862[c]
Hg	TP	273.16	−38.8344	
H$_2$O	TP		0.01	0.01
H$_2$O	BP			100
Ga	MP	302.9146	29.7646	
In	FP	429.7485	156.5985	156.634[c]
Sn	FP	505.078	231.928	231.9681
Zn	FP	692.677	419.527	419.58
Al	FP	933.473	660.323	660.37[c]
Ag	FP	1234.93	961.78	961.93
Au	FP	1337.33	1064.18	1064.43
Cu	FP	1357.77	1084.62	1084.88[c]

[a] e-H$_2$ represents hydrogen with equal distribution of ortho and para states.
[b] VP = vapor pressure point; CVGT, constant volume gas thermometer point; TP, triple point; MP, melting point; FP, freezing point. Note: MP and FP at 101.325 Pa (1 atm) ambient pressure.
[c] In the IPTS-68 table designates a secondary reference point.

then reduced to slightly below the melt temperature until the metal is observed to be in freezing equilibrium. The technique for initiating nucleation, managing the supercool, etc, differs for each metal and is described in the literature.

The ITS and Interpolation Instruments

The ITS is an artifact scale, designed to relate temperature measurements made with practicable instruments as closely as possible to the thermodynamic scale. The scale is established and controlled by the International Committee of Weights and Measures (BIPM) through its Consultative Committee on Thermometry, which was established in 1937. The BIPM itself is established to maintain and implement the Treaty of the Meter, to which most nations of the world subscribe; thus the ITS has not only scientific but legal status in most nations. Within nations, the Temperature Scale is maintained by national standards establishments, eg, in the United States the National Institute for Standards and Technology (NIST), in England the National Physical Laboratory (NPL), and in Germany the Physikalisch-Technische Bundesanstalt (PTB).

Through the years the ITS has been changed a number of times, the better to fit its values to new information regarding the thermodynamic values of the fixed points. In recent history the International Practical Temperature Scale (IPTS) of 1927 has been replaced by the IPTS of 1948, by the IPTS of 1968 (IPTS-68) (2), and most recently, by the ITS of 1990 (ITS-90) (3), effective on January 1, 1990. These changes have included revisions of (1) the range of the scale, (2) the interpolation instruments and their algorithms, and (3) values of the defining fixed points. The changes have been, for most purposes, nontrivial, and it is essential in reading a temperature from the literature or in citing a temperature to specify which version of the IPTS or ITS is meant.

The ITS-90 has its lowest point at 0.65 K and extends upward without specified limit. A number of values assigned to fixed points differ from those of the immediately previous scale, IPTS-68. In addition, the standard platinum resistance thermometer (SPRC) is specified as the interpolation standard from 13.8033 K to 961.78°C, and the interpolation standard above 961.78°C is a radiation thermometer based on Planck's radiation law. Between 0.65 and 13.8033 K interpolation of the scale relies upon vapor pressure and constant-volume gas thermometry. The standard thermocouple, which in previous scales had a range between the upper end of the SPRT range and the lower end of the radiation thermometer range, has been deleted.

Platinum Resistance Thermometer Range: 13,8033–1234.93 K

Temperatures on the ITS-90 are expressed in terms of the ratio of the resistance of the SPRT at temperature to the resistance at the triple point of water. The resistance ratio $W(T_{90})$ is calculated as

$$W(T_{90}) = \frac{R(T_{90})}{R(273.16 \text{ K})} \qquad (4)$$

In all previous scales the denominator of the ratio W was the resistance at 0°C = 273.15 K. Restrictions on the SPRT, which are intended indirectly to specify the purity of the platinum from which it is made, are

$$W(302.9146 \text{ K}) \geq 1.11807 \tag{5}$$

$$W(234.3156 \text{ K}) \leq 0.844235 \tag{6}$$

where 302.9146 K is the melting temperature of gallium and 234.3156 K is the triple-point temperature of mercury. An SPRT for use to the silver freezing point has an additional requirement:

$$W(1234.93 \text{ K}) \geq 4.2844 \tag{7}$$

where 1234.93 K (961.93°C) is the freezing point of silver. The temperature range over which an SPRT can be used is a function of its design, but no single SPRT can be used over the full range from 13 K to 961°C. As a generality, capsule-type thermometers having a 0°C resistance of about 25.5 Ω may be used from 13 K to about 156°C, 25.5 Ω long-stem thermometers may be used from about 77 K to 660°C, and long-stem thermometers having 0°C resistance of 0.25 or 1 Ω from about 0°C to about 962°C.

The temperature T_{90} is calculated from the resistance-ratio relationship

$$W(T_{90}) - W_r(T_{90}) = \Delta W(T_{90}) \tag{8}$$

where $W(T_{90})$ is the observed value, $W_r(T_{90})$ is a value calculated from a reference function, and $\Delta W(T_{90})$ is the deviation of the observed $W(T_{90})$ of the individual SPRT from the reference function.

There are two reference functions $W_r(T_{90})$, one for the range from 13 K to 0.01°C, another for the range 0.01–962°C. The reference functions represent the calibration of a fictitious thermometer developed from experience in the calibration of many SPRTs over many years. Below 0.01°C, the reference function is

$$\ln[W_r(T_{90})] = A_0 + \sum_{i+1}^{12} A_i \left(\frac{\ln(T_{90})/273.16 \text{ K}) + 1.5}{1.5} \right)^i \tag{9}$$

Above 0.01°C the reference function is

$$W_r(T_{90}) = C_0 + \sum_{i=1}^{9} C_i \left(\frac{T_{90}/\text{K} - 754.15}{481} \right)^i \tag{10}$$

The ITS-90 provides for 11 SPRT ranges. Three of these are for ranges which extend below −189°C and can be considered to be of interest chiefly to specialists in very low temperatures. The other eight ranges are shown in Table 2. The range should be specified when ordering a thermometer calibration. When the thermometer design permits, a combination of ranges may be specified, eg, −189 to +420°C. The specification of so many ranges benefits the user of the SPRT over limited ranges, in that (1) accuracy of interpolation may be higher as the range is limited and (2) limited ranges may require fewer fixed-point determinations and consequently substantially lower calibration cost.

Table 2. Platinum Resistance Thermometer Ranges of ITS(90)

Range	Lower limit	Fixed points required[a]
		Upper Limit of 0.01°C (eq. 9)
1	13.8033K	e-H_2 (TP), e-H_2 (VP), e-H_2 (VP), Ne (TP), O_2 (TP), Ar (TP), Hg (TP), H_2O (TP)
2	24.5561	e-H_2 (TP), Ne (TP), O_2 (TP), Ar (TP), Hg (TP), H_2O (TP)
3	54.3584	O_2 (TP), Ar (TP), Hg (TP), H_2O (TP)
4	83.8058	Ar (TP), Hg (TP), H_2O (TP)
		Lower Limit of 0°C (eq. 10)
5	961.78°C	H_2O (TP), Sn (FP), Zn (FP), Al (FP), Ag (FP)
6	660.323°C	H_2O (TP), Sn (FP), Zn (FP), Al (FP)
7	419.527	H_2O (TP), Sn (FP), Zn (FP)
8	231.928	H_2O (TP), In (FP), Sn (FP)
9	156.5985	H_2O (TP), In (FP)
10	29.7646	H_2O (TP), Ga (MP)
		From −38.8344 to +29.7646°C[b]
11		Hg (TP), H_2O (TP), Ga (MP)

[a]Abbreviations as are given in Table 1.
[b]Reference function is equation 9 from −38.8344°C to 0.01°C; and equation 10 from 0.01°C to +29.7646°C.

Each range has a specific deviation function to be used in equation 8. For example, over the range 0–961.78°C, the deviation function is

$$\Delta W(T_{90}) = a[W(T_{90}) - 1] + b[W(T_{90}) - 1]^2 + c[W(T_{90}) - 1]^3$$
$$+ d[W(T_{90}) - W(933.473 \text{ K})]^2 \tag{11}$$

where a, b, c, and d are coefficients derived from calibration of the individual thermometer at specified fixed points, for this range, the water triple point and the freezing points of tin, zinc, aluminum, and silver. Deviation functions for other ranges may be found in Reference 3.

The Radiation Thermometry Range: Over 1234.93 K

Above the freezing point of silver, T_{90} is defined by the relationship

$$\frac{L\lambda(T_{90})}{L\lambda[T_{90}(X)]} = \frac{\exp\{c_2/\lambda[T_{90}(X)]\} - 1}{\exp[x_2/\lambda(T_{90})] - 1} \tag{12}$$

where $L\lambda(T_{90})$ and $L\lambda[T_{90}(X)]$ are the spectral concentrations of the radiation of a blackbody at wavelength (in vacuum) λ at T_{90} and at $T_{90}(X)$, respectively. Here $T_{90}(X)$ refers to the silver freezing point (T_{90}(Ag) = 1234.93 K), the gold freezing point (T_{90}(Au) = 1337.33 K), or the copper freezing point (T_{90}(Cu) = 1357.77 K), and C_2 is the second radiation constant of Planck's radiation formula with the value c_2 = 0.014388 m·K. The usual interpolation instrument is a radiation thermometer, eg, an optical pyrometer, which is in itself a thermodynamic device.

Resistance Thermometers

Some of the characteristics of the SPRT, as an interpolation device for the ITS-90, have been discussed. In order to meet these requirements, the thermometer must be made from almost ideally pure platinum wire mounted in a physical construction which will keep it in a strain-free condition. The conventional resistance 25.5 Ω or some convenient submultiple is historical; over limited ranges it permitted a rough but quick estimate of temperature, because a 1-K change is nearly equivalent to a 0.1-Ω change.

The SPRT is always provided with two current and two potential leads extending from the actual sensing element so that the element can be measured as a four-terminal resistor and the effects of lead resistance eliminated. Capsule thermometers, limited in range from cryogenic temperatures to about 156°C, are usually 6 mm in diameter by less than 6 cm long and are sealed into platinum tubes. They are usually mounted directly in the temperature zone of equipment, where the extension of a long tube from that zone to the external world would represent an unacceptable heat leak. Another common use of capsule thermometers is in calorimeters. The long-stem SPRT with quartz sheaths and an ice point resistance of 25.5 Ω is generally useful from −189 to +660°C. The sheath is commonly 40 cm long and must be immersed at least 15 cm into the medium where the temperature is to be measured. High temperature SPRTs (HTSPRTs) have been developed specifically to meet the requirements of the ITS-90 and are used from some temperature in the vicinity of 0°C, eg, −20°C, to an upper limit of the silver freezing point, 962°C. Because platinum has low mechanical strength at these upper temperatures, the winding must be made of large-diameter wire and is of low electrical resistance, usually 0.25 Ω at 0°C. It is this low resistance and the attendant measurement difficulty which limit the lower end of usefulness.

The SPRTs are devices of superb accuracy and resolution, but they are fragile and can easily be broken. They can also be put out of calibration by strain, induced by even slight mechanical shock or vibration. The principal use of SPRTs in science and industry is to maintain the calibrations of working thermometers.

Working-grade metallic resistance thermometers are made of metal wire, usually high-purity platinum, but also, more commonly in past years, copper, nickel, and alloys. The sensitive element is usually protected by a covering of ceramic tubing or ceramic cement, or the individual turns of wire are fixed in place by some means. Glass is occasionally used as a coating, but differences in temperature coefficient of expansion and imperfect insulating properties of glasses at high temperatures limit its use. These forms of protection stabilize the sensing element against shifts resulting from vibration and shock, while somewhat reducing the temperature coefficient of resistance; consequently they usually do not meet the criteria for interpolation devices of the ITS-90 mentioned in equations 5−7.

Working-grade thermometers, conventionally called industrial resistance thermometers, are generally smaller than the SPRT element and may be as small as 2.5 mm in diameter and 10 mm in length. These are available in various 0°C resistances, eg, 100, 200, and 500 Ω. They are available as unsheathed elements or in a wide variety of sheaths and enclosures, both standard and custom. They are relatively inexpensive. They are usually made to be interchangeable, without relying on individual calibration, within limits of 0.25 K or closer upon special order. A typical tolerance statement for a precision-class industrial resistance thermometer is

$$\text{deviation from nominal} \leq (0.1 \pm 0.002t)°\text{C} \tag{13}$$

In general, manufacturers do not report the calibrations of individual sensors to the purchaser, except upon request, but instead publish tables of resistance vs temperature and tolerance charts for each class. Deviation here means departure from a nominal set of values of resistance versus temperature given in a manufacturer's literature.

Industrial resistance thermometers are also the subject of a number of national and international standards, which describe both calibration constants and classes of accuracy and interchangeability. IEC publication 751 was revised in 1976 to conform to ITS-90, and national standards will be revised to conform to this document. IEC 751 uses the fixed-point values of ITS-90 with the simpler algorithm of IPTS-48:

$$R(t) = R(0°\text{C})[1 + A_{90}t + B_{90}t^2 + C_{90}t^3(t - 100°\text{C})] \tag{14}$$

where $C_{90} = 0$ for values of $t \geq 0°\text{C}$ (4).

The following values for the A, B, and C coefficients are to be recommended:

Coefficient	For ITS-90	Replacing IPTS-68
A_{90}	$3.9083 \times 10^{-3}°\text{C}^{-1}$	$3.90802 \times 10^{-3}°\text{C}^{-1}$
B_{90}	$-5.775 \times 10^{-7}°\text{C}^{-2}$	$-5.802 \times 10^{-7}°\text{C}^{-2}$
C_{90}	$-4.183 \times 10^{-12}°\text{C}^{-4}$	$-4.2375 \times 10^{-12}°\text{C}^{-4}$

These ITS-90 coefficients provide that the α value be $0.00385055°\text{C}^{-1}$, which is

$$\alpha = \frac{R(100°\text{C})}{R(0°\text{C})} - 1 \tag{15}$$

very close to the old standard of 0.00385. These are said to permit the calibration of a real thermometer having the same initial sensitivity as the table to within $\pm0.1°\text{C}$ from -200 to $+600°\text{C}$.

Since the early 1970s, development work has been reported on thick-film, thin-film, and other resistance elements which are made by deposition rather than wound from wire. Such thermometers might have certain advantages of size, response time, cost, etc. These do not have the stability or interchangeability of wrought-wire elements.

Among nonmetallic resistance thermometers, an important class is that of thermistors, or temperature-sensitive semiconducting ceramics (5). The variety of available sizes, shapes, and performance characteristics is very large. One manufacturer lists in the catalog a choice of characteristics ranging from 100 Ω at 25°C to 1 MΩ at 25°C.

The thermistor material is usually a metal oxide, eg, manganese oxide. Dopants, eg, nickel oxide or copper oxide, may be added to obtain a variety of resistance and slope characteristics. The material is usually sintered into a disk or bead with integral or attached connecting wires. Figure 4 shows a typical series of steps in the production of a disk thermistor.

Bead thermistors are formed by placing two wires, commonly of platinum, in close proximity and parallel to each other and bridging them with a drop of slurry, which is then sintered into a hard bead and encapsulated in protective glass. Such thermistors are quite stable, approaching, over narrow temperature limits, the stability of industrial metallic thermometers. However, the resistance tolerance may vary from unit to unit by as much as ±20%, and matching or interchangeability is usually achieved by selection. Beads can be made quite small, which may allow application in, eg, temperature probes mounted in intravenous needles.

Disk thermistors can be produced to close limits of interchangeability, eg, ±0.1 and ±0.05°C. Disks cannot be made as small as the smallest beads; 2 mm diameter seems an approximate practicable limit. Disks historically have been considered to be less stable than good beads. They are commonly protected with a coating of epoxy resin, which provides less compressive support than the glass coating of bead thermistors. More recent developments have resulted in interchangeable glass-encapsulated disk thermistors which have the stability characteristics of the best beads.

Two characteristics of thermistors are distinctly different from those of metallic resistance thermometers. For all but special types, the temperature-resistance characteristic is negative; that is, resistance decreases with increasing temperature (all metal resistance thermometers exhibit a positive characteristic). Also, the temperature-resistance characteristic is very nonlinear. One typical production thermistor changes resistance 376 Ω/K at 0°C and 13.5 Ω/K at 70°C. Over favorable portions of the nonlinear characteristic, the change in resistance can be more than an order of magnitude larger than that of metallic resistance thermometers. This high dR/dT is a distinct advantage where high sensitivity is required over narrow ranges, as it is in many biological, medical, and environmental situations.

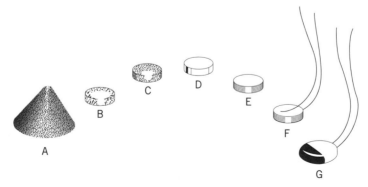

Fig. 4. Steps in the manufacture of a disk thermistor: A, ball-milled powder; B, pressed disk; C, sintered disk; D, soldered disk; E, edge-ground disk; F, lead wires attached; and G, epoxy-coated.

Many attempts have been made to develop an equation based on the semi-conducting transfer function of thermistors. These have usually been of the form

$$p^{-1} = f(T)\exp\left(-\frac{\Delta E}{2kT}\right) \tag{16}$$

where E is the energy gap between the valence band energies, T is the kelvin temperature, and k is Boltzmann's constant. None of these efforts has produced a simple law which fits the calibration data. The equation generally used is an empirically derived polynomial (4):

$$T^{-1} = A + B[\ln R(T)] + C[\ln R(T)]^3 \tag{17}$$

known as the Steinhart-Hart equation. Here A, B, and C are derived from calibration at suitable temperatures. It is often useful to have this equation explicit in R, and this may be written as follows:

$$R = \exp\left[\left(\delta - \frac{1}{2}\alpha\right)^{1/3} - \left(\delta + \frac{1}{2}\alpha\right)^{1/3}\right]^{1/2} \tag{18}$$

where $\alpha = (A - 1/T)/C$ and $\delta = (\frac{1}{4}\alpha^2 + \frac{1}{27}\beta^3)^{1/2}$, where $\beta = B/C$ and A, B, and C are the coefficients of equation 17.

The nonlinearity of thermistors has been a challenge to a number of designers. Using modern circuit components, easily the most flexible way to linearize a curve is using digital active circuitry, breaking the thermistor characteristic at a suitable number of inflection points and assigning an appropriate slope to each line segment between these. Other solutions, not requiring active circuitry but often involving networks of several thermistors combined with several fixed resistors, have also been found. These passive circuits achieve approximate linearity at the cost of sensitivity, the sensitivity being not much higher than that of platinum sensors, which are essentially linear over short temperature intervals.

Seebeck Effect Thermometers (Thermocouples)

Thermocouples are composed of two dissimilar materials, usually in the form of wires, that accomplish a net conversion of thermal energy into electrical energy with the occurrence of an electrical current. Unlike resistance thermometers, where the response is proportional to temperature, the response of thermocouples is proportional to the temperature difference between two junctions. Figure 5 illustrates such a circuit.

There are three fundamental rules regarding thermocouples: (*1*) no thermal electromotive force (emf) is produced if heat is applied to a circuit comprising

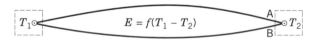

Fig. 5.　Basic thermocouple circuit. A and B are wires of different materials.

a single homogeneous conductor, (2) the thermal emf of a circuit comprising any number of conductive materials at uniform temperature is zero, and (3) the thermal emf developed by a pair of homogeneous but dissimilar conductors having junctions T_1 and T_3 is the same as the thermal emf developed by an arrangement of the same conductors having junctions at T_1 and T_2 and at T_2 and T_3. This latter principle is illustrated in Figure 6.

It is a common misconception that the generation of the thermal emf is often thought to be an effect which takes place at the junction between the two wires. Rather, it is a bulk effect of the material. Electrons in metals have an electrical charge and a kinetic energy. Along a thermal gradient, charge does not change but kinetic energy does; therefore a transfer of heat must occur. The thermal emf results from a slight rearrangement of electrons in the conductor to respond to thermal transfer. The function of the junction is to assure electrical conductivity between the conductors, and as a consequence, it does not matter how the junction is made, eg, by welding, brazing, clamping, etc, as long as electrical conductivity is preserved. The assumption of homogeneity of the first and third rules is a condition never actually met in practice and cannot be assumed to be preserved throughout the working life of the thermocouple because of strain, ion migration, etc. For precise work, it is important to verify the performance of the thermocouple by calibration at appropriate intervals.

The IPTS-68 stipulated a standard thermocouple where the negative leg was pure platinum and the positive leg was an alloy of 90% platinum and 10% rhodium as the interpolation standard instrument between the freezing point of antimony (630.74°C) and the freezing point of gold (1064.43°C). (These temperatures are given on the IPTS-68; different numbers have been assigned on the ITS-90.) Few sensing elements appear more simple than a thermocouple; few sensors (qv) are more subtly and deceptively complex. Errors of the type S Standard thermocouple can result from accidental alloying of the constituent metals owing to metallic vapors in the immediate atmosphere; nonmetallic vapors such as phosphorus, sulfur, arsenic; or easily reduced oxides. The positive element of 90Pt–10Rh can be affected by neutron flux, because rhodium gradually transmutes to palladium. The negative Pt leg forms stable isotopes and is relatively immune to transmutation but may be work-hardened by fast-neutron bombard-

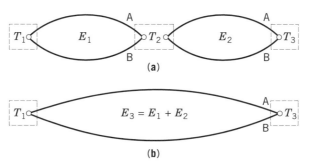

Fig. 6. Rule 3 of thermocouples, where (**a**) represents a four-junction and (**b**) a two-junction thermocouple. A and B, the two legs of the thermocouple, are wires of dissimilar materials. Junctions are at temperatures T_1, T_2, and T_3. E_1, E_2, and E_3 are thermal emfs. The thermal emf at T_3 is the same for both circuits.

ment. Calibrations are valid only in the fully annealed state. For these reasons, standard thermocouples increased the uncertainty of realization of the IPTS-68 by a factor of about 10, compared to the SPRT. In the ITS-90, the standard thermocouple has been eliminated by extending the range of the SPRT upward to 962°C and the range of the radiation thermometer downward to this same temperature.

Whereas it is no longer an interpolation standard of the scale, the thermoelectric principle is one of the most common ways to transduce temperature, although it is challenged in some disciplines by small industrial platinum resistance thermometers (PRTs) and thermistors. Thermocouple junctions can be made very small and in almost infinite variety, and for base metal thermocouples the component materials are very cheap. Properties of various types of working thermocouple are shown in Table 3; additional properties are given in Reference 5.

The wires of thermocouple pairs must be electrically insulated from each other and protected from the environment. Manufacturers' catalogs indicate the sort of insulation and protection available, including servings of high temperature ceramic and glass fibers, bare wire in double-bore ceramic beads or tubes, and metal tubing filled with compacted ceramics. Both common and exotic materials are included, and manufacturers' recommendations should be sought. Essentially, the enclosure should provide mechanical protection, but not, through differential expansion, contribute to mechanical strain; provide protection against contamination and not contribute to it via ion migration or evolution of contaminating vapors; and maintain the interconductor insulation. To most potential alterations of state, particularly progressive surface alterations, thick wires may be less susceptible than thin wires.

Another potential source of error, particularly in type K thermocouples between 200 and 500°C, is reversible change of state. Such changes are thought to be the result of lattice ordering–disordering. Reversible changes are particularly subtle, may not be readily discoverable by calibration, and may seriously compromise measurement results. Also, although letter designations by the American Society for Testing and Materials (ASTM) for thermocouple pairs are specific throughout the United States, the nominal calibration designated by the same

Table 3. Properties of Various Thermocouple Pairs

ASTM Type	Materials	Range °C	$\mu V/°C^a$
J	Iron-constantan	0–760	64.3
K	Chromel-alumel	0–1260	36.5
R	Pt/87Pt-13Rh[b]	0–1450	13.8
S	Pt/90Pt-10Rh[b]	0–1450	11.8
T	Copper-constantan	−183–375	53.0
E	Chromel-constantan	0–875	78.5
B	70Pt30Rh/94Pt6Rh[b]	870–1700	11.6

[a]Called the Seebeck coefficient. It is given here at the high-temperature end of the range; many thermocouples are nonlinear.
[b]Indicates, for example, one leg of pure platinum, the other of an alloy of platinum and rhodium. Alloy percentages are shown.

letters may be different in other countries and in international standards. It is important to determine the specific characteristics of a specific lot of wire, particularly where making replacements.

In industrial applications it is not uncommon that the thermocouple must be coupled to the readout instrument or controller by a long length of wire, perhaps hundreds of feet. It is obvious from the differential nature of the thermocouple that, to avoid unwanted junctions, extension wire be of the same type, eg, for a J thermocouple the extension must be type J. Where the thermocouple is of a noble or exotic material, the cost of identical lead wire may be prohibitive; manufacturers of extension wire may suggest compromises which are less costly. Junctions between the thermocouple leads and the extension wire should be made in an isothermal environment. The wire and junctions must have the same electrical integrity as the thermocouple junction. Because the emf is low, enclosure in a shield or grounded conduit should be considered.

The thermal emf of the thermocouple is a function of the difference between the hot end and the cold end, the latter usually located at the readout instrument; thus the measurement can be no more accurate than the isothermality of the leads at the cold end and the accuracy with which this temperature is known. Three common methods for addressing these problems follow.

1. The cold-junction temperature can be fixed by immersing the cold junctions into some known thermal environment: an ice bath or a properly maintained water triple-point cell. A temperature-controlled oven at a temperature above ambient may be used.
2. The cold junction may be attached to or contained within an isothermal mass, eg, a large block of copper, adequately insulated from ambient, whose temperature is measured by some independent means, eg, an embedded thermistor or PRT.
3. Cold-junction compensation can be provided by a network which includes a constant voltage source and a temperature-sensitive bridge to provide an offset voltage which is proportional to the temperature sensitivity of the thermocouple and of opposite sign.

Not all elements of the industrial thermocouple need to be wire. For example, if a copper pipe contains a flowing fluid whose temperature is to be measured, a constantan wire attached to the pipe will form a T, or copper–constantan, thermocouple. Such arrangements are difficult to calibrate and require full understanding of the possible inherent problems. For example, is the copper pipe fully annealed? Homogeneous? Pure, or an alloy? Many ingenious solutions to specific measurement problems are given in Reference 6.

Measurement of Emitted Radiation

Above 962°C, the freezing point of silver, temperatures on the ITS-90 are defined by a thermodynamic function and an interpolation instrument is not specified. The interpolation instrument universally used is an optical pyrometer, manual or automatic, which is itself a thermodynamic device.

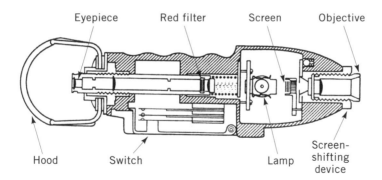

Fig. 7. Optical pyrometer.

The main elements of a disappearing-filament optical pyrometer are shown in Figure 7. There is an optical system for viewing the radiating target, a lamp filament, appropriate filters, an eyepiece, and a calibrated means for varying the lamp current and consequently the brightness of the filament. In the optical path, the filament is visible against the radiating source. The lamp filament current is varied until the image of the glowing filament is of such brilliance that its separate image is indistinguishable from the brilliance of the radiating source. The lamp current is then noted. If a second radiating source also causes the lamp filament to be indistinguishable at that lamp current, the temperature of the second source is said to be the same as that of the first.

Absolute calibrations on the ITS-90 are performed by sighting on a target which is a blackbody at the freezing temperature of silver, gold, or copper. The ideal blackbody is a closed cavity whose walls are at uniform temperature. At every point on the interior wall, radiation is emitted and absorbed, so that the emission and absorption are in equilibrium. If this condition is fulfilled, there is no net loss of energy, and the internal energy is a function of temperature only. For the sake of sighting into such a cavity, a small hole in the wall may be provided which, if it is of proper size and shape, does not disturb the equilibrium condition in a significant way. Under these circumstances the energy emerging from the viewing port has the same energy as the interior of the cavity, which corresponds to the temperature of the blackbody.

A simplified drawing of a blackbody and its furnace is shown in Figure 8. A graphite cavity is surrounded by a graphite crucible, so that the cavity can be surrounded by a shell of silver, gold, or copper. One end of the cavity is open and leads to a conical viewing port. The integral furnace is used to melt and freeze the metal. Such a system, idealized and properly proportioned, may have a temperature reproducibility of 0.02°C and an effective emissivity of 0.99999%; ie, it departs from the ideal by only 10 ppm. The simple blackbody and furnace shown is intended as a calibration device for use by industrial standardizing laboratories.

The total energy radiating from a blackbody of unit area is given by the Stefan-Boltzman law:

$$E = \sigma T^4 \tag{19}$$

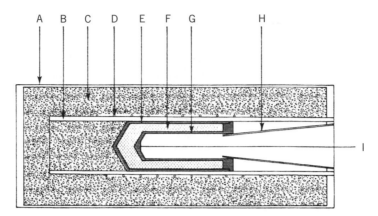

Fig. 8. Blackbody and furnace: A, furnace shell; B, furnace tube; C, thermal insulation; D, heater winding; E, outer wall of graphite crucible; F, pure metal (gold or copper); G, inner wall of graphite crucible; H, sighting cone; I, target.

where T is the absolute temperature in kelvins and σ is the Stefan-Boltzmann constant, 5.6697×10^{-12} W/cm^{-2} T^{-4}. This expression contains no term for frequency or wavelength, which is distributed over a large spectrum with most of the energy in the infrared. If the spectral radiance N_{bg} is defined as that portion of the blackbody radiance which is at wavelength λ, then

$$N_{b\lambda} = \frac{C_1 \lambda^{-5}}{\exp(C_2/\lambda T) - 1} \tag{20}$$

where C_1 and C_2 are the first and second radiation constants, respectively. The term -1 in the denominator was added by Planck when he observed that Wien's law did not fit experimental data at high temperatures. To explain this stroke of pure intuition, it was necessary for him to invent the quantum theory, for which he received the Nobel Prize in 1918.

This equation permits the evaluation of the energy emitted by a blackbody at a specific wavelength and comparison of it to the energy emitted by a second blackbody at the same wavelength but at a different temperature. Letting T_0 be a reference temperature,

$$N_b(\lambda, T)/N_b(\lambda, T_0) = \exp\left(\frac{C_2}{\lambda T_0}\right) - \frac{1}{\exp(C_2/\lambda T)} - 1 \tag{21}$$

A strip lamp is a convenient means for the calibration of secondary pyrometers (Fig. 9). The notched portion of the tungsten strip is the target. A pyrometer which has been calibrated against a radiating blackbody is sighted on the target, and the strip lamp current is adjusted to radiate at the intensity of the blackbody, as transferred by the primary pyrometer. The secondary pyrometer is then substituted for the primary, and the current required to raise the lamp filament to the brilliance of the target of the strip lamp is noted.

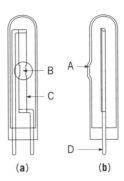

Fig. 9. (a) Front and (b) side views of a strip lamp. A, plane window; B, target area of strip; C, strip; D, base pins.

Measurement of Radiation From Real Objects

Real objects, when they are in an environment generally hotter or cooler than themselves, radiate, absorb, and reflect energy. The portion radiated is called the emissivity ϵ. If the portion reflected is r, then

$$\epsilon + r = 1 \qquad (22)$$

No object can radiate more energy than can a blackbody at the same temperature, because a blackbody in equilibrium with a radiation field at temperature T radiates exactly as much energy as it absorbs. Any object exhibiting surface reflection must have emissivity of less than 1. Pyrometers are usually calibrated with respect to blackbodies. This can cause a serious problem in use. The emissivities of some common materials are listed in Table 4.

The filter and screen of the pyrometer shown in Figure 9 require specific mention. From equation 21 it is evident that the observed radiation must be limited to a narrow bandwidth. Also, peak intensity does not occur at the same wavelength at different temperatures. The pyrometer is fitted with a filter (usually red) having a sharp cut-off, usually at 620 nm. The human eye is

Table 4. Emissivities of Common Materials, at 0.65 μm

Material	Emissivity	
	Solid	Liquid
Carbon	0.8–0.9	
Chromium	0.34	0.39
Copper	0.10	0.15
Gold	0.14	0.22
Iron	0.35	0.37
Manganese	0.59	0.59
Molybdenum	0.36	0.37
Platinum	0.30	0.38
Silver	0.07	0.07
Steel	0.35	0.37
Cast iron	0.37	0.40

insensitive to light of wavelength longer than 720 nm. The effective pyrometer wavelength is $\approx$655 nm.

It is also necessary to reduce the intensity of the radiation admitted into the pyrometer, because pyrometer lamp filaments should not be subjected to temperatures exceeding 1250°C. The reduction is accomplished by a screen or screens; in manually operated secondary pyrometers they are usually neutral-density filters.

Several types of secondary pyrometer are available. In addition to those that measure by varying lamp current, some pyrometers maintain the lamp at constant current but interpose a wedge of graduated neutral density, whose position is a measure of temperature. Also, automatic pyrometers are available in which the eye is replaced by a detector and the measuring element is operated by a servo. In general, the accuracy of the automatic pyrometer is somewhat less than that achieved manually by a skilled operator.

The problem of emissivity from real materials has stimulated the study of pyrometers that measure radiation at two different wavelengths. The principle of the two-color pyrometer is that the energy radiated from a source of one wavelength increases with temperature at a rate different from that radiated at another wavelength. Thus temperature can be deduced from the ratio of the intensities at the two wavelengths, regardless of emissivity. Two-color pyrometers are not widely used.

Other Electrical Thermometers

Many special-purpose electrical thermometers have been developed, either for use in practical temperature measurement, or as research devices for the study of temperature and temperature scales. Among the latter are thermometers which respond to thermal noise (Johnson noise) and thermometers based on the temperature dependence of the speed of sound.

A novel and useful thermometer, based on the change in resonant frequency of a quartz crystal, is produced by Hewlett-Packard. The sensor probe contains a precisely cut quartz crystal about 8 mm in diameter. Each quartz sensor has a unique temperature-vs-frequency characteristic, and each sensor is individually calibrated with its operating electronics. Sensitivities of 100 μK are obtained at -50 to 150°C, the effective limits of measurement, and digital readout is provided. The input signal to the conditioner is very low, and the long-term stability is such that frequent adjustment of zero against a reference standard, eg, the melting point of gallium, is required. However, direct readout in units of temperature and great sensitivity make this system a useful and practicable device in many environmental and energy situations, and it is particularly favored by oceanographers.

Yokagawa Electric Works has developed a thermometer based on the nuclear quadrupole resonance of potassium chlorate, usable over the range from -184 to 125°C. This thermometer makes use of the fundamental properties of the absorption frequency of the ^{35}Cl nucleus, and its calibration is itself a constant of nature.

Conventional electrical thermometry is difficult or impossible or involves unacceptable risks in some measurement situations. These include measurements where the process would interfere with the measurement, eg, in microwave and dielectric or inductive heating, high voltage, high frequency testing, and studies of the biological effects of electromagnetic radiation; measurements where the presence of an electrical potential is unacceptable, such as where flammable or explosive materials are present; or where the presence of any electrical field would perturb the quantity to be measured. An example is medical patient monitoring during electrosurgery or electrocautery, since the insulated leads of electrical thermometers at these frequencies may become capacitive conductors. A thermometer introduced by Luxtron uses rare-earth phosphors at the end of a thin fiber-optic line and resolves many of these problems. The phosphors are excited by ultraviolet illumination. The fluorescent emissions at different wavelengths have different temperature dependences, but the intensities have a linear dependence upon intensity of excitation. Thus the ratio of emissions is dependent upon temperature but independent of excitation.

Nonelectrical Thermometers (Liquid in Glass)

The thermal expansions of liquids are reliable analogues of temperature. Liquids are suitable for use between their freezing and boiling points; mercury and colored alcohols are common materials. The typical liquid-in-glass thermometer includes a thin-walled glass bulb attached to a capillary stem partially filled with a visible liquid and sealed against the environment. The portion of the capillary above the liquid is usually filled with a dry gas under some pressure to avoid column separation. A scale is etched or fused onto the stem. If the range of the thermometer does not include 0°C, a reference mark below the scale at 0°C is usually provided to give a calibration check at the ice point. A contraction chamber may be provided below the main capillary to avoid the need for a long stem before the low end of the calibrated portion of the stem.

There are problems to be considered and avoided when using liquid-in-glass thermometers. One type of these is pressure errors. The change in height of the mercury column is a function of the volume of the bulb compared to the volume of the capillary. An external pressure (positive or negative) which tends to alter the bulb volume causes an error of indication, which may be small for normal barometric pressure variations but large when, for example, using the thermometer in an autoclave or pressure vessel.

Immersion errors are another type of problem. Not only the liquid, but the glass as well, has a thermal expansion coefficient. Three types of thermometer, each having a different immersion requirement, are common. Complete-immersion thermometers are calibrated for use with the entire thermometer immersed. Total-immersion thermometers are calibrated for use with that portion of the thermometer immersed which contains the liquid column. Partial-immersion thermometers are calibrated with the thermometer immersed to a specific index mark on the stem. Most liquid-in-glass thermometers used for accurate measurement are of the total immersion type. However, it may be difficult to read a thermometer whose fluid column is at or below the surface of the medium. In practice, the thermometer is immersed to a depth that allows the meniscus to be seen. The emergent portion of the liquid is then subject to temperature gradients. In this case, and in the case of partial-immersion thermometers, it is necessary to make a correction for the temperature of the emergent stem which contains liquid by measuring the stem temperature with a second thermometer and calculating the correction by means of the equation

$$C_s = KN(t_1 - t_2) \tag{23}$$

where C_s is the correction for stem emergence to be added algebraically to the temperature indicated by the thermometer, K is a coefficient representing the difference in expansion coefficient between the glass and the thermometric liquid (for mercury and usual glass this is 0.00016 for Celsius thermometers), N is the length of liquid column extending above the medium expressed in degrees of the scale, t_1 is the temperature of the thermometer bulb, and t_2 is the average temperature of the liquid column extending above the medium.

Glass-creep errors are also encountered. The liquid-in-glass thermometer should always be used to measure temperatures in ascending order. If the thermometer is stored at room temperature, a temporary ice point depression results, which may be as much as 0.01 K for 10 K of temperature difference, when the thermometer is heated above room temperature. If the thermometer is used to measure a temperature and must then be used to measure a lower temperature, the thermometer should be stored at a still lower temperature for at least 3 days prior to use to assure recovery of bulb dimensions.

Parallax errors are occasioned by failure to read the thermometer from a position exactly normal to its tube axis. Parallax can be reduced by taking the mean of several observations by skilled observers but can be eliminated only by use of a filar telescope mounted so as to move along an axis parallel to the thermometer stem.

Mercury thermometers are subject to separation of the mercury column or to inclusion of bubbles of the fill gas. These may result from shipping and handling and cause a scale offset which can usually be seen upon visual examination, and they are always recognized by a 0°C verification check. Manufacturers will suggest means by which these temporary defects may be cured.

Frictional errors result from failure of the liquid to perfectly wet the small thermometer bore. They can usually be removed by lightly tapping the thermometer stem, for example with a wooden pencil, before reading.

Despite these problems, liquid-in-glass thermometers are a generally inexpensive means for making temperature measurements of modest accuracy. Many types of liquid-in-glass thermometer are available. Among them are thermometers intended for the performance of specific tests, such as ASTM tests; thermometers for very short ranges, eg, the NIST Standard Reference Materials 933 and 934 used in clinical chemistry; thermometers for measuring very small differentials, eg, the Beckman thermometer design; thermometers with fixed or movable electrical contacts for alarm and control functions; and maximum, minimum, or maximum–minimum reading thermometers, which show the extremes of temperatures reached during the period since last reset.

Other Nonelectrical Thermometers

Other nonelectrical thermometers are bimetal, filled-system, and pyrometric cone thermometers. In bimetal thermometers, two strips of metal of differing expansion characteristics are welded together face-to-face. If one end of such a strip is fixed, the strip bends in response to temperature change as the metal of higher coefficient produces a longitudinal dimensional change. If the strip is formed into a helix, the free end rotates and can actuate a pointer or switch. The filled-system thermometer comprises a rigid bulb, a connecting metal capillary, and a flexible bulb, eg, a metal bellows. The whole system is filled with a liquid or gas of known thermal expansion properties. As temperature change causes the gas to expand, the flexible bulb is displaced and the displacement is sensed as a mechanical motion. Pyrometric cones are geometric shapes, commonly conical, which slump or collapse at some (usually high) temperature. One principal use for them is in estimating peak temperature in kilns.

See Table 5 for a list of temperature-measuring devices.

Table 5. Temperature-Measuring Devices

Device	Manufacturer	Price[a]
water triple-point cells	Isothermal Technology Ltd. Southport, England, and New York	$1,500
water triple-point maintenance baths	Isothermal Technology Ltd. Southport, England, and New York	$4,000 to $18,000
cells, furnaces, etc, for freezing points of metals	Isothermal Technology Ltd. Southport, England, and New York	cells $8,000 to $18,000 furnaces $11,000 to $20,000
standard platinum resistance thermometers	Isothermal Technology Ltd. Southport, England, and New York YSI Inc., Yellow Springs, Ohio	$4,000–6,000, depending on type and calibration
industrial-grade resistance thermometers	Isothermal Technology Ltd. Southport, England, and New York	≥$50, depending on configuration, etc
	Minco Corp., Minneapolis, Minn.	≥$50, depending on configuration, etc
	DeGussa AG, Hanau, Germany	≥$50, depending on configuration
precision thermistors	YSI Inc., Yellow Springs, Ohio Thermometrics, Inc., Edison, N.J.	≤$40, varies with configuration
standard thermocouples	Isothermal Technology Ltd. Southport, England, and New York	$1,000–5,000, depending on configuration and calibration
industrial thermocouples	Isothermal Technology Ltd. Southport, England, and New York	varies with configuration
	Omega Engineering, Stamford, Conn.	varies with configuration
	Yamari Industries, Osaka and Tokyo, Japan	varies with configuration
optical pyrometers	Micron Instruments, Inc., Simi Valley, Calif.	varies with configuration
quartz thermometers	Hewlett-Packard Co., Mountain View, Calif.	upon inquiry
fiber-optic thermometers	Luxtron Corp., Mountain View, Calif.	upon inquiry
liquid-in-glass thermometers	Cole-Parmer Co., Chicago, Ill. Fischer Scientific Co., Chicago, Ill. Ever Ready Thermometer, Brooklyn, N.Y.	price varies with range, quality, application, certification, etc

[a]As of 1997.

BIBLIOGRAPHY

"Temperature Measurement" in *ECT* 1st ed., Vol. 13, pp. 677–699, by P. H. Dike, Leeds and Northrup Co.; in *ECT* 2nd ed., Vol. 19, pp. 774–802, by W. T. Gray, Leeds & Northrup Co.; in *ECT* 3rd ed., Vol. 22, pp. 679–708, by H. E. Sostmann, Yellow Springs Instrument Co., Inc.

1. H. E. Sostmann, *Isotech J. Thermomet.* **3**(2), 106–113 (1992).
2. H. Preston-Thomas, *Metrologia* **12**, 1 (1976).
3. H. Preston-Thomas, *Metrologia* **27**, 3–10 (1990).
4. International Electrotechnic Commission, *International Standard 751, Amendment 2 (1005)*, 1976.
5. H. B. Sachse, *Semi-Conducting Temperature Sensors and Their Applications*, John Wiley & Sons, Inc., New York, 1975.
6. H. D. Baker, E. A. Ryder, and N. H. Baker, *Temperature Measurement in Engineering*, John Wiley & Sons, Inc., New York, 1961.

General References

H. E. Sostmann, "Temperature Measurements," in A. Bisio and S. Boots, eds., *Encyclopedia of Energy Technology and the Environment*, Vol. 4, John Wiley & Sons, Inc., New York, 1995, pp. 2595–2607.
R. P. Benedict, *Fundamentals of Temperature, Pressure and Flow Measurement*, Wiley, New York, 1969.
F. G. Brickwedde, ed., *Temperature, Its Measurement and Control in Science and Industry*, Vols. 3–6, Reinhold, New York, 1962–1992.
Manual on the Use of Thermocouples, ASTM STP 470, American Society for Testing and Materials, Philadelphia, Pa., 1993.
T. J. Quinn, *Temperature*, Academic Press, Inc., New York, 1983.
J. F. Schooley, *Thermometry*, CRC Press, Boca Raton, Fla., 1986.
J. V. Nicholas and D. R. White, *Traceable Temperatures*, John Wiley & Sons, Inc., New York, 1994.

HENRY E. SOSTMANN
Consultant

TERBIUM.   See LANTHANIDES.

TEREPHTHALIC ACID.   See PHTHALIC ACID AND OTHER BENZENEPOLY-CARBOXYLIC ACIDS.

TERGITOL.   See SURFACTANTS; DETERGENCY.

TERNE PLATE.   See METALLIC COATINGS.

TERPENES AND TERPENOIDS.   See TERPENOIDS.

TERPENOIDS

Terpenes are found as constituents of essential oils and oleoresins of plants. Since antiquity they have been isolated and used in flavor and fragrance applications. Many important constituents of the essential oils have been identified and syntheses for them developed (see OILS, ESSENTIAL).

Terpenes are characterized as being made up of units of isoprene in a head-to-tail orientation. This isoprene concept, invented to aid in the structure determination of terpenes found in natural products, was especially useful for elucidation of structures of more complex sesquiterpenes, diterpenes, and polyterpenes. The hydrocarbon, myrcene, and the terpene alcohol, α-terpineol, can be considered as being made up of two isoprene units in such a head-to-tail orientation (**1**).

(**1**)

The terpenes are further classified by the number of isoprene units in their carbon skeletons, as shown in Table 1. Many natural products contain an isolated isoprene unit derived from mevalonic acid and these compounds are thus considered hemiterpenoid derivatives. Most of the terpenes covered in this article are monoterpenoids mainly as a result of the development of an important industry, beginning in the late 1950s, based on turpentine from the pine tree. Turpentine is clearly the largest-volume natural terpene source in commercial use in the 1990s. The term terpenoids designates the various derivatives of the terpene hydrocarbons; and many of the oxygenated derivatives such as alcohols, aldehydes, esters, and ketones are very important flavor and fragrance chemicals. Many are listed as GRAS (generally recognized as safe) as food additives and flavorings (1,2). Numerous monographs on terpenes and their derivatives as fragrance materials and on their biological aspects have been published (3,4).

Table 1. Classification of Terpenes

Isoprene units	Carbon atoms	Classification
1	5	hemiterpene
2	10	monoterpene
3	15	sesquiterpene
4	20	diterpene
5	25	sesterterpene
6	30	triterpene
8	40	tetraterpene
>8	>40	polyterpene

Terpene chemists use trivial names for most of the compounds because the systematic names are much more complex. Common or trivial names, CAS Registry Numbers, and properties of selected terpenes and terpenoids are listed in Tables 2 and 3. Compounds that exhibit chirality also have other Registry Numbers for specific optical isomers. For commercial products, a material safety data sheet (MSDS), which is required by OSHA, frequently lists multiple names such as a product name, trivial name, IUPAC name and the TSCA name. The MSDS is a good source of information about physical properties, potential health hazards, and other useful information for the safe handling of the materials. When the product is a mixture, the components and their amounts are usually listed along with their Registry Numbers.

Terpene chemists use mainly gas chromatography in dealing with terpene mixtures in research and development as well as in quality control. Capillary gas chromatography with stable bonded-phase columns, the primary analytical

Table 2. Properties of Selected Monoterpene Hydrocarbons and Ethers

Common name	CAS Registry Number	Bp, kPa^a, °C 13.33	Bp, kPa^a, °C 101.3	d^{20}, g/cm^3	n_D^{20}	$[\alpha]_D$
tricyclene	[508-32-7]	85	152			0
α-pinene	[80-56-8]	89	156	0.8595	1.4658	±51
α-fenchene	[471-84-1]	91	157	0.8697	1.4740	±44
camphene	[79-92-5]	91	158			±108
β-pinene	[127-91-3]	98	165	0.8722	1.4790	±22
myrcene	[123-35-3]	93^b	167	0.7880	1.4692	0
cis-pinane	[6876-13-7]	101	168	0.8575	1.4629	±23
cis/trans-p-8-menthene	[6252-33-1]		169	0.8142	1.4554	0
trans-2-p-menthene	[1124-26-1]		169	0.8100	1.4500	±132
p-3-menthene	[500-00-5]	102	169	0.8129	1.4519	±141
trans-p-menthane	[1678-82-6]	103	170	0.7928	1.4367	0
3-carene	[13466-78-9]	104	170	0.8617	1.4742	±16
cis-p-menthane	[6069-98-3]	105	172	0.8002	1.4431	0
1,4-cineole	[470-67-7]	105	172	0.8986	1.4446	0
1,8-cineole	[470-82-6]	108	174	0.9245	1.4574	0
α-terpinene	[99-86-5]	108	175	0.8315	1.4755	0
p-1-menthene	[499-94-5]	110	176	0.8246	1.4563	±118
p-4(8)-menthene	[34105-55-0]	110	174	0.819	1.4568	0
limonene	[5989-27-5]	110	176.5	0.8411	1.4730	±124
p-cymene	[25155-15-1]	110	177	0.8570	1.4905	0
γ-terpinene	[99-85-4]	116	182	0.8455	1.4715	0
p-3,8-menthadiene	[586-67-4]	117	183.5	0.8510	1.4876	±140
p-2,4(8)-menthadiene	[138-86-3]	120	185	1.5030	1.5030	±49
terpinolene	[586-62-9]	120	186	1.4861	1.4861	0
isobornyl methyl ether	[5331-32-8]	77^c	193	0.9235	1.4664	
α-terpinyl methyl ether	[14576-08-0]		212	0.9100	1.4630	

^{a}To convert kPa to mm Hg, multiply by 7.5.
^{b}At 9.33 kPa.
^{c}At 2.0 kPa.

Table 3. Properties of Selected Terpene Alcohols, Aldehydes, and Ketones

Common name	CAS Registry Number	Bp, kPa[a], °C /P	101.3	Mp, °C	d^{20}, g/cm³	n_D^{20}	$[\alpha]_D$
fenchone	[1195-79-5]	122/13.33	193	5	0.9452	1.4628	±70
linalool	[78-70-6]	79.8/1.33	199	48;39(±)	0.8607	1.4616	±22
α-fenchol	[1632-73-1]	133/13.33	200		0.9350	1.4734	±12.5
citronellal	[106-23-0]	107/2.0	203		0.8510	1.4467	±12
terpinen-1-ol	[586-82-3]	106/3.33	208		0.9171	1.4701	
camphor	[76-22-2]	182/53.3	209	179			±45
trans-β-terpineol	[7299-40-3]	139/13.33	209	33	0.9190	1.4712	0
trans-menthone	[89-8-5]	138/13.33	210	−6	0.8903	1.4500	±29
terpinene-4-ol	[562-74-3]	123/6.66	211		0.9259	1.4762	±29
neomenthol	[491-01-0]	108/2.67	212	−22;53(±)	0.8917	1.4604	20
borneol	[507-70-0]		212	209;210(±)	1.0110		±38
isoborneol	[124-76-5]		214	212(±)			±34
menthol	[1490-04-6]	111/2.67	216	43;38(±)	0.8911	1.4615	±50
γ-terpineol	[586-81-2]	132/6.66	218	70	0.9412	1.4912	0
α-terpineol	[98-55-5]	149/13.33	219	40;35(±)	0.9336	1.4831	±112
citronellol	[106-22-9]	105/1.47	224		0.8550	1.4559	±7
nerol	[106-25-2]	128/3.33	225		0.8735	1.4736	0
geranial	[141-27-5]	77/0.24	228		0.8972	1.4898	0
neral	[106-26-3]	76.5/0.31				1.4869	
geraniol	[106-24-1]	131/3.33	230		0.8770	1.4756	0
carvone	[99-49-0]	157/13.33	231		0.0955	1.4990	±62
hydroxy-citronellal	[107-75-5]	116/0.67			0.9220	1.4494	±10.5
1,8-terpin	[80-53-5]		258	105;117[b]			0
α-ionone	[6901-97-9]	123/1.33	258		0.9309	1.5020	±400
β-ionone	[14901-07-6]	132/1.33	271	−35	0.9461	1.5202	0
nerolidol	[40716-66-3]	145/1.60	276		0.8778	1.4795	±15

[a]To convert kPa to mm Hg, multiply by 7.5.
[b]Hydrate.

method, is also being used more frequently in the 1990s in product quality control because its greater resolution is helpful in producing consistent products.

Fractional vacuum distillation is the method used to separate terpene mixtures into their components. The terpene chemist usually has in the laboratory a range of columns with differing numbers of theoretical stages. Experimental distillation in the laboratory is useful in providing data for manufacturing plants that produce commercial quantities of terpene products.

Capillary gc/ms, hplc, nmr, ir, and uv are all analytical methods used by the terpene chemist; with a good library of reference spectra, capillary gc/ms is probably the most important method used in dealing with the more volatile terpenes used in the flavor and fragrance industry (see FLAVORS AND SPICES). The physical properties of density, refractive index, boiling point, melting point of derivatives, and specific rotation are used less frequently but are important in defining product specifications.

Monoterpenes

Production of Hydrocarbons from Turpentine. In 1993, U.S. production of crude turpentine was over 128 million liters at an average price of $0.21/kg and includes crude sulfate turpentine and turpentine from thermomechanical processes (5). In the same year, over 5.9 million liters of gum, wood, or sulfate turpentine was imported into the United States, with the majority coming from Canada; exports from the United States amounted to 6.16 million liters.

The majority of the turpentine comes from the southeastern United States, which consists of 60–70% α-pinene, 20–25% β-pinene, and 6–12% other components. Because there is variation in components from different species of the pine tree as well as variation from the many paper pulp mills, there is obviously variation in the analysis of sulfate turpentines. Some of the other components consist of p-menthadienes, alcohols, ethers such as anethole [104-46-1] and methylchavicol [104-67-0], and the sesquiterpene hydrocarbon, β-caryophyllene [87-44-5].

Turpentine from the western United States is different from that of the southern states in that it contains 3-carene ranging from 12–43%, depending on the species of pine tree. Indian turpentine also contains about 60% 3-carene and about 15% of the sesquiterpene longifolene. Turpentine from Sweden, Finland, CIS, and Austria all contain 3-carene; however, α- and β-pinene are commercially the most important components of the turpentines.

The crude sulfate turpentine coming from the mills of the Kraft sulfate process for paper is a commercially important raw material for obtaining α- and β-pinene for synthesizing aroma chemicals identical to natural products. The crude sulfate turpentine (CST) is shipped in tank cars or tank trucks and stored in large tanks. The turpentine is pumped into a continuous, usually about a 10-stage column to remove lower boiling sulfur compounds such as methyl mercaptan, dimethyl sulfide, and dimethyl disulfide. The material is then known as topped sulfate turpentine (TST) and is continuously distilled on highly efficient, reduced pressure columns to separate the principal components α-pinene (60–70%), β-pinene (20–25%), limonene/phellandrene (3–10%), and methyl chavicol and anethole/caryophyllene (1–2%). The α- and β-pinene are then used as the raw materials for producing higher valued flavor and fragrance compounds.

Although the turpentine is largely desulfurized in the stripping stage and again in the fractionation stages, many applications for α- and β-pinene require further desulfurization. Such methods involve adsorption on carbon, hypochlorite treatment, hydrogen peroxide treatment, treatment with metals, or a combination of techniques (6–15).

Synthesis of β-Methylheptenone from Petrochemical Sources. β-Methylheptenone (**1**) is an important intermediate in the total synthesis of terpenes. Continuous hydrochlorination of isoprene [78-79-5] produces prenyl chloride [503-60-6], which then reacts with acetone with a quaternary ammonium catalyst and sodium hydroxide to give β-methylheptenone (6-methylhept-5-en-2-one [*110-93-0*]) (eq. 1) (16–19).

$$(1)$$

Another process involves a one-step reaction of isobutylene with formaldehyde and acetone under high temperature and pressure (eq. 2) (20). α-Methylheptenone (**2**) (6-methylhept-6-en-2-one [*10408-15-8*]) is the product, but it is easily catalytically isomerized to β-methylheptenone (21,22). Unconverted isobutylene and acetone can be recycled to the process, thus making it commercially viable (23,24). Variations of this process have also been described in the literature (25–28).

$$(2)$$

The methylheptenones are intermediates for the synthesis of linalool (**3**). A continuous ethynylation with sodium hydroxide and *N*-methylpyrrolidinone has been developed (eq. 3) (29).

α-,β-linalool

$$(3)$$

Sodium acetylide addition to acetone gives the 2-methyl-3-butyne-2-ol, which is partially reduced to 2-methyl-3-butene-2-ol using a Lindlar catalyst (30–32). In a Carroll reaction with diketene in the presence of sodium methylate, 2-methyl-3-butene-2-ol reacts to give an acetoacetate intermediate, which undergoes a Claisen-Cope rearrangement at elevated temperatures to form β-methylheptenone (30,32–35). 2-Methyl-3-butene-2-ol will also undergo an acid-catalyzed transetherification with isopropenyl methyl ether; when the intermediate vinyl allyl ether is heated it undergoes a sigmatropic rearrangement to β-methylheptenone (36).

Rearrangement of dehydrolinalool (**4**) using vanadate catalysts produces citral (**5**), an intermediate for Vitamin A synthesis as well as an important flavor and fragrance material (37). Isomerization of the dehydrolinalyl acetate (**6**) in the presence of copper salts in acetic acid followed by saponification of the acetate also gives citral (38,39). Further improvement in the catalyst system has greatly improved the yield to 85–90% (40,41).

Dimerization of Isoprene.   Isoprene is becoming an increasingly important raw material for the production of terpenes. For example, myrcene (**7**) can be produced by the dimerization of isoprene (2-methyl-1,3-butadiene) (42–44); and myrcene is very useful for synthesizing a number of oxygenated terpenes important in the flavor and fragrance industry.

α-Pinene Manufacture.   Industrially, α-pinene produced from the fractionation of sulfate turpentine can be used directly for most of its applications. The bulk price of technical-grade α-pinene, min 92%, was $1.32/kg in 1995 (45). The commercial product is shipped in tank cars, tank trucks, or deck tanks.

Uses and Reactions. α-Pinene (**8**) is useful for synthesizing a wide variety of terpenoids. Hydration to pine oil, acid-catalyzed isomerization to camphene, thermal isomerization to ocimene and alloocimene, and polymerization to terpene resins are some of its direct uses. Manufacture of linalool, nerol, and geraniol has become an economically important use of α-pinene.

Synthetic pine oil is produced by the acid-catalyzed hydration of α-pinene (Fig. 1). Mineral acids, usually phosphoric acid, are used in concentrations of 20–40 wt % and at temperatures varying from 30–100°C. Depending on the conditions used, alcohols, chiefly α-terpineol (**9**), are produced along with *p*-menthadienes and cineoles, mainly limonene, terpinolene, and 1,4- and 1,8-cineole (46–48). Various grades of pine oil can be produced by fractionation of the crude products. Formation of terpin hydrate (**10**) from α-terpineol gives β-terpineol (**11**) and γ-terpineol (**12**) as a consequence of the reversible reactions.

Acid-catalyzed isomerization of α-pinene is carried out by heating with TiO_2 catalysts or other activated clays. Camphene (**13**) and tricyclene [*508-32-7*] (**14**) are the products obtained in a 4:1 equilibrium mixture in about an 80% yield (Fig. 2). Tricyclene undergoes reactions identical to those of camphene; therefore, the crude material is often used for producing other products.

Thermal isomerization of α-pinene, usually at about 450°C, gives a mixture of equal amounts of dipentene (**15**) and alloocimene (**16**) (49,50). Ocimene (**17**) is produced initially but is unstable and rearranges to alloocimene, which is subject to cyclization at higher temperatures to produce α- and β-pyronenes (**18** and **19**). The pyrolysis conditions are usually optimized to give the maximum amount of alloocimene. Ocimenes can be produced by a technique using shorter contact time and rapid quenching or steam dilution (51).

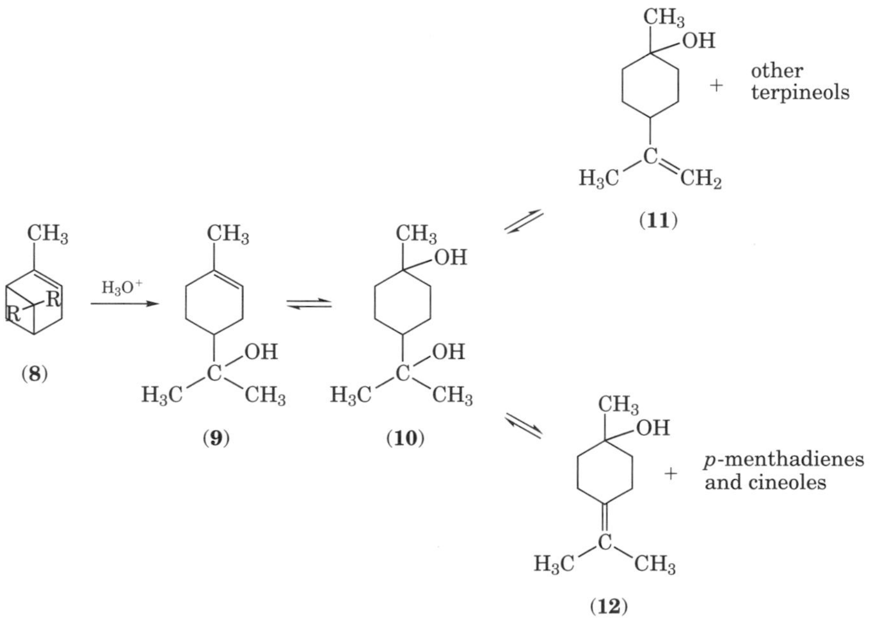

Fig. 1. Synthetic pine oil from α-pinene; R = CH_3.

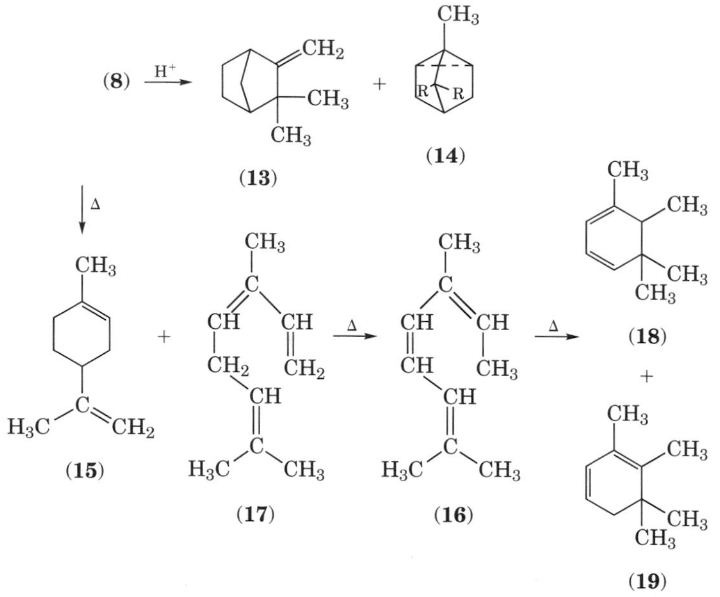

Fig. 2. Acid treatment and thermal isomerization of α-pinene where R = CH$_3$. The dipentene (**15**) is d,l-limonene; (**18**) is α-pyronene; (**19**) is β-pyronene.

α-Pinene can be isomerized to about 4% β-pinene (**20**) using a supported palladium catalyst (52). Although the amount of β-pinene in equilibrium with α-pinene is low, the use of an efficient fractional distillation column with continuous processing makes the process feasible (Fig. 3).

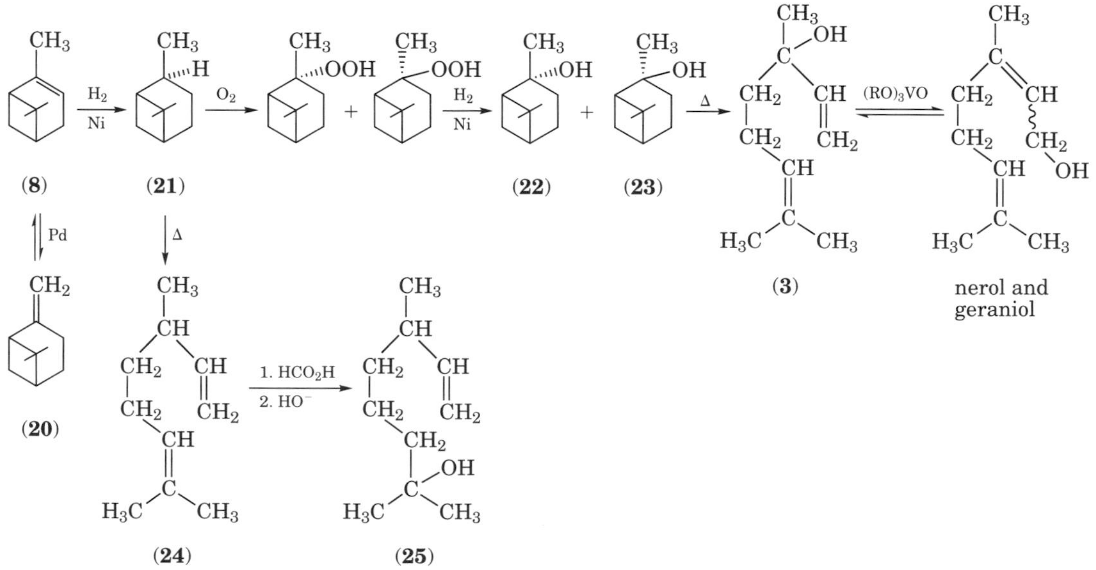

Fig. 3. Conversion of α-pinene (**8**) to β-pinene (**20**) and to *cis*-pinane (**21**) and subsequent reactions.

Another important use of α-pinene is the hydrogenation to *cis*-pinane (**21**). One use of the *cis*-pinane is based on oxidation to *cis*- and *trans*-pinane hydroperoxide and their subsequent catalytic reduction to *cis*- and *trans*-pinanol (**22** and **23**) in about an 80:20 ratio (53,54). Pyrolysis of the *cis*-pinanol is an important route to linalool; overall the yield of linalool (**3**) from α-pinene is about 30%. Linalool can be readily isomerized to nerol and geraniol using an orthovanadate catalyst (55). Because the isomerization is an equilibrium process, use of borate esters in the process improves the yield of nerol and geraniol to as high as 90% (56).

Pyrolysis of the *cis*-pinane produces dihydromyrcene (**24**) (citronellene) as the major product in 50–60% yield. Fractionation of the crude product then gives an 87 wt % dihydromyrcene (57). Dihydromyrcenol (**25**) produced from the dihydromyrcene is becoming increasingly important as a fragrance material. It has excellent stability and has a powerful, fresh, lime-like aroma. Hydration of citronellene using formic acid has become an important commercial method for producing dihydromyrcenol (58).

β-Pinene Manufacture. β-Pinene is obtained by fractionation of turpentine. The price of β-pinene, min 97%, was \$5.28/kg in 1995 and that quality is used mostly in flavor and perfumery applications (45). Most of the β-pinene produced by the turpentine fractionators is used captively for producing fragrance chemicals or for β-pinene resins. β-Pinene is shipped in tank cars, tank wagons, deck tanks, and lined drums. Prolonged storage requires conditions precluding autooxidation and polymerization.

Uses and Reactions.   Some of the principal uses for β-pinene are for manufacturing terpene resins and for thermal isomerization (pyrolysis) to myrcene. The resins are made by Lewis acid (usually $AlCl_3$) polymerization of β-pinene, either as a homopolymer or as a copolymer with other terpenes such as limonene. β-Pinene polymerizes much easier than α-pinene and the resins are useful in pressure-sensitive adhesives, hot-melt adhesives and coatings, and elastomeric sealants. One of the first syntheses of a new fragrance chemical from turpentine sources used formaldehyde with β-pinene in a Prins reaction to produce the alcohol, Nopol (**26**) (59).

     (**20**)   (**26**)

The acetate is also useful commercially and can be made by direct esterification of Nopol. The products are useful in soaps, detergents, polishes, and other household products.

The production of myrcene (**7**) from β-pinene is important commercially for the synthesis of a wide variety of flavor and fragrance materials. Some of those include nerol and geraniol, citronellol (**27**) and citral (**5**).

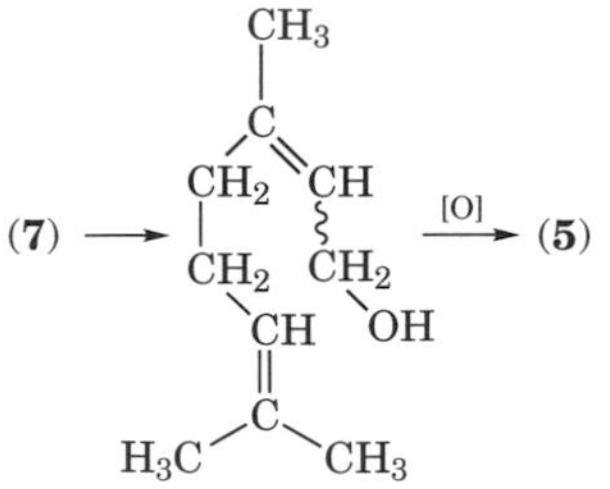

nerol and geraniol

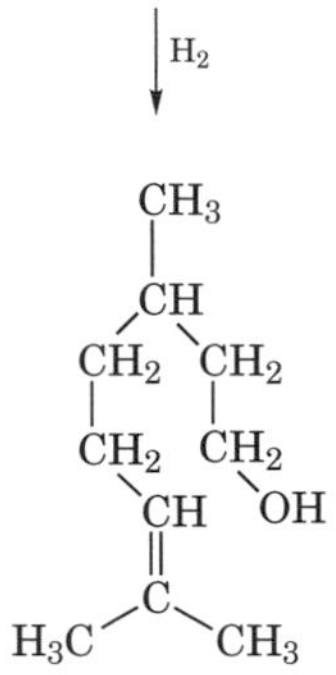

(27)

3-Carene Manufacture. 3-Carene is obtained by fractional distillation of turpentine. Turpentine from the western United States and Canada averages about 25% 3-carene; much of it is unutilized although it is obtained in high optical purity. Turpentines from the Scandinavian countries, the CIS, Pakistan, and India all contain significant quantities of 3-carene.

Uses and Reactions. The Prins reaction of 3-carene with formaldehyde in acetic acid gives mainly 2-carene-4-methanol acetate, which when saponified produces the 2-carene-4-methanol, both of which are commercial products of modest usage (60). 3-Carene (**28**) also reacts with acetic anhydride with a catalyst (ZnCl$_2$) to give 4-acetyl-2-carene (**29**) (61), which is also a commercial product. Although 3-carene does not polymerize to produce terpene resins, copolymerization with phenol has been successfully commercialized by DRT in France (62).

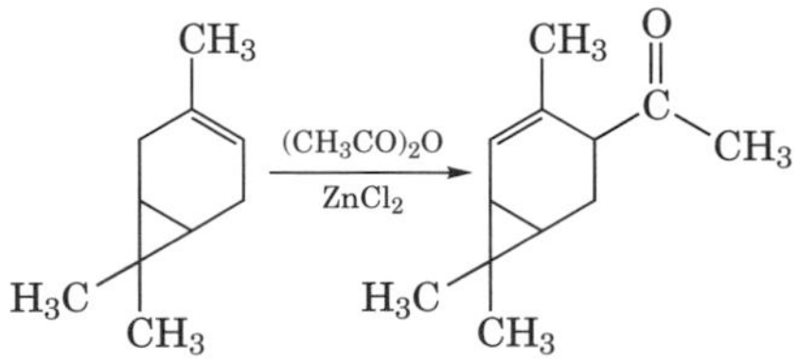

(28)   (29)

3-Carene has also been isomerized over an ϵ-alumina catalyst to a 50:50 mixture of dipentene (**15**) and carvestrene (**30**). The crude mixture can be readily polymerized to a terpene resin or copolymerized with piperylene (63,64).

(**28**)   (**15**)   (**30**)

Using strong bases such as t-C$_4$H$_9$OK/DMSO, sodium on Na$_2$CO$_3$ or sodium on alumina, (+)-3-carene can be isomerized to an equilibrium mixture containing 40% of (+)-2-carene [4497-92-1] (**31**) (65,66). The mixture can be thermally isomerized, even at fairly low (180°C) temperatures in the liquid phase, to produce (+)-*trans*-isolimonene [5133-87-1] (**32**). The isomerization reaction proceeds on (+)-2-carene via a 1,5-sigmatropic shift. The (+)-3-carene is unaffected and can be recovered by distillation and recycled to the isomerization reaction. The (+)-*trans*-isolimonene can be partially reduced to (+)-*trans*-2-menthene (**33**), which can be epoxidized, followed by hydrogenation of the epoxide to a mixture of (−)-menthol isomers and carvomenthol isomers.

(**31**)   (**32**)   (**33**)

Alternatively, (+)-*trans*-isolimonene can be isomerized to (+)-2,4(8)-p-menthadiene (**34**), which is partially reduced to (+)-3-menthene (**35**). Epoxidation of (+)-3-menthene gives the epoxide, which can be isomerized to (−)-menthone.

(**34**)   (**35**)

p-Menthadienes and p-Cymene Manufacture. The p-menthadienes are mainly produced as by-products from the manufacture of synthetic pine oil. The volume produced in the United States in 1993 was 27,509 t as monocyclic terpene hydrocarbons (Solvenol) (67). The mixtures of the p-menthadienes are also commonly referred to as dipentene and the price in 1995 was $1.17/kg for a solvent grade and $1.50/kg for a perfume grade (45). Important components of dipentene are limonene, terpinolene, α-terpinene, camphene, 1,4- and 1,8-cineole, α-pinene, p-cymene, 2,4(8)-p-menthadiene (**34**), α- and β-phellandrene, and γ-terpinene. Dipentene (**15**) is usually shipped in carbon-steel tank cars or in drums. (+)-Limonene [5989-27-5] is a large-volume p-menthadiene from the citrus industry in the United States and Brazil. The annual world production volume is large, estimated at over 50,000 t (68). Because of a growing demand for biodegradable solvents, the uses for this product have driven the sales price to $6.60/kg in 1995 (69). Subsequently, it has become almost impractical to consider using (+)-limonene as a synthetic raw material.

Dehydrogenation of p-menthadienes and α-pinene in the vapor phase over catalysts such as chromia–alumina produces p-cymene (70). p-Menthadienes can be disproportionated over a Cu–Ni catalyst to give a mixture of p-menthane and p-cymene (71).

Uses and Reactions. Dipentene is a good solvent for paints, varnishes, and enamels that contain synthetic resins, particularly phenolic resins. It is also used as an antiskinning agent and as a wetting agent in the dispersion of pigments. The solvency of dipentene for rubber and its swelling and softening properties make it useful in rubber reclaiming and in the processing of natural and synthetic rubbers. Dipentene is also formulated into a variety of cleaners similar to pine oil cleaners.

(+)- and (±)-limonenes are widely used in the manufacture of terpene resins. Additionally, a (−)-limonene and (+)-β-phellandrene mixture from sulfate turpentine has been used to produce terpene resins. (+)-Limonene from the citrus industry continually finds new uses as a solvent not only for its solvency properties but also for its orange oil fragrance.

(+)-Limonene is used for aqueous cleaning compositions using carefully selected and proportioned surface-active agents and a coupling agent, usually a glycol or glycol ether (72). They are meant to replace halogenated hydrocarbon solvents and are biodegradable. Another process has produced a multiuse cleansing agent from (+)-limonene that is basically harmless to human skin (73). An application for printed circuit board cleaners has also been developed using dipentene and limonene with emulsifying surfactants to facilitate removal by rinsing in water (74).

The p-menthadienes are isomerized to an equilibrium mixture by either strong acid or strong base. The equilibrium compositions have been determined; treating p-menthadienes with acid gives 53.3% α-terpinene (**36**), 13.8% γ-terpinene (**37**), 29.8% 2,4(8)-p-menthadiene (**34**), and 3.1% 3,8-p-menthadiene (**38**) (75).

p-menthadienes $\xrightarrow{H_3O^+}$

(**36**)　　　(**37**)　　　(**38**)

Limonene (**15**) can be isomerized to terpinolene (**39**) using liquid SO_2 and a hydroperoxide catalyst (*t*-butyl hydroperoxide (TBHP)) (76). Another method uses a specially prepared orthotitanic acid catalyst with a buffer such as sodium acetate (77). A selectivity of about 70% is claimed at about 50% conversion when run at 150°C for four hours.

$$(15) \qquad\qquad (39)$$

A mixture of the *p*-menthadienes can be hydrogenated to produce a mixture of *cis-p*-menthane [*6069-98-3*] and *trans-p*-menthane [*1678-82-6*]. Oxidation to a mixture of *p*-menthane hydroperoxides gives a useful polymerization initiator used in the rubber industry.

(+)-Limonene (+**15**) is an important raw material for producing (−)-carvone [*6485-40-1*]. The process uses nitrosyl chloride and proceeds via nitrosochloride and oxime (78,79). The (−)-carvone (**40**) is found as the main component of spearmint oil and the (+)-carvone produced from (−)-limonene has the characteristic odor of dill.

$$(15) \qquad\qquad\qquad\qquad\qquad\qquad\qquad (40)$$

Camphene Manufacture. Camphene (**13**) is produced by the reaction of α-pinene (**8**) with a TiO_2 catalyst (80). Preparation of the catalyst has a great influence on the product composition and yield. Tricyclene (**14**) is formed as a coproduct but it undergoes the same reactions as camphene; thus the product is generally used as a mixture. The *p*-menthadienes and dimers produced as byproducts are easily removed by fractional distillation and the camphene has a melting point range of 36–52°C, depending on its purity. Camphene is shipped in tank cars, deck tanks, and drums.

Uses and Reactions. Camphene is used for preparing a number of fragrance compounds. Condensation with acids such as acetic, propionic, isobutyric, and isovaleric produce useful isobornyl esters. Isobornyl acetate (**41**) has the greatest usage as a pine fragrance (81). The isobornyl esters of acrylic and methacrylic acids are also useful in preparing acrylic polymers.

Isobornyl acetate [125-12-2] is also useful in producing isoborneol (**42**) by saponification. Dehydrogenation or oxidation of isoborneol (*exo*-1,7,7-trimethylbicyclo[2.2.1]heptan-2-ol [124-76-5]) produces camphor (**43**), an important product used in religious ceremonies in Asian countries (82).

The Prins reaction with formaldehyde, acetic acid, acetic anhydride, and camphene gives the useful alcohol, 8-acetoxymethyl camphene, which has a patchouli-like odor (83). Oxidation of the alcohol to the corresponding aldehyde also gives a useful intermediate compound, which is used to synthesize the sandalwood compound dihydo-β-santalol.

Addition of alcohols and glycols to camphene catalyzed by strong acids such as Amberlyst 15 ion-exchange resin produces useful camphane ethers (84). Addition of 2-propanol to camphene gives the 2-isopropoxycamphane and addition of ethylene glycol gives 2-(β-hydroxyethoxy) camphane. The products are useful as woody, cedar-like fragrances.

When camphene reacts with guaiacol (2-methoxyphenol), a mixture of terpenyl phenols is formed. Hydrogenation of the mixture results in hydrogenolysis of the methoxy group and gives a complex mixture of terpenyl cyclohexanols (eg, 3-(2-isocamphyl) cyclohexanol [70955-71-4] (**45**)), which after fractional distillation produces a useful sandalwood fragrance product (85). A similar process has also been developed using catechol and camphene (86).

(45)

Myrcene Manufacture. An important commercial source for mycene is its manufacture by pyrolysis of β-pinene at 550–600°C (87). The thermal isomerization produces a mixture of about 75–77 wt % myrcene, 9% limonene, a small amount of Ψ-limonene [499-97-8] and some decomposition products and dimers. The crude mixture is usually used without purification for the production of the important alcohols nerol and geraniol. Myrcene may be purified by distillation but every precaution must be taken to prevent polymerization. The use of inhibitors and distillation at reduced pressures and moderate temperatures is recommended. Storage or shipment of myrcene in any purity should also include the addition of a polymerization inhibitor.

Purified myrcene has minimal use in flavor and fragrance applications. Production and cost figures for crude myrcene from β-pinene are not published to avoid disclosure of individual company operations, but the production volume is large (~30,000 t).

Uses and Reactions. The largest use of myrcene is for the production of the terpene alcohols nerol, geraniol, and linalool. The nerol and geraniol are further used as intermediates for the production of other large-volume flavor and fragrance chemicals such as citronellol, dimethyloctanol, citronellal, hydroxycitronellal, racemic menthol, citral, and the ionones and methylionones.

Nerol, geraniol, and linalool, known as the rose alcohols, are found widely in nature. Nerol and geraniol have mild, sweet odors reminiscent of rose flowers. They are manufactured by the hydrochlorination of mycene at the conjugated double bonds when a copper catalyst is used (88,89).

In the presence of the copper catalyst, myrcene (**7**) hydrochlorination proceeds initially to give predominately linalyl chloride (**46**), which then isomerizes to give about 40–50% neryl chloride, 50–55% geranyl chloride, and only 2–4% linalyl chloride. The crude mixture of chlorides is converted to the mixture of acetates (or formates) by the addition of sodium acetate or sodium formate with a phase-transfer catalyst (PTC) (90). Saponification of the acetates or formates gives the alcohols and the sodium acetate or formate for recycle. Fractionation of the crude alcohol mixture gives both nerol and geraniol products, usually as mixtures; high purity products are made by further distillation. There are usually many grades of nerol and geraniol produced, depending on the customer requirements.

(7) → (46) → neryl-/geranyl-chlorides

neryl-/geranyl-acetates → nerol(cis-) (47) geraniol(trans-) (48) + (3)

Myrcene with its conjugated diene system readily undergoes Diels-Alder reactions with a number of dienophiles. For example, reaction with 3-methyl-3-pentene-2-one with a catalytic amount of $AlCl_3$ gives an intermediate monocyclic ketone, which when cyclized with 85% phosphoric acid produces the bicyclic ketone known as Iso E Super [54464-57-2] (49). The product is useful in providing sandalwood-like and cedarwood-like fragrance ingredients (91).

(49)

Reaction of myrcene and sulfur dioxide under pressure produces myrcene sulfone. This adduct is stable under ordinary temperatures and provides a way to stabilize the conjugated diene system in order to hydrate it with sulfuric acid. The myrcene sulfone hydrate produced is pyrolyzed in the vapor phase in order to regenerate the diene system to produce myrcenol [543-39-5] (50).

myrcene sulfone   myrcene sulfone   (50)
                   hydrate

The sulfur dioxide can be recovered and recycled. The myrcenol, after purification by distillation, can condense with acrolein to produce Lyral [31906-04-4], an important aldehyde with a strong-lasting odor similar to lily-of-the-valley (92).

Another synthesis of Lyral (51) consists of the reaction of myrcene with acrolein to give the myrac aldehyde [37677-14-8] (52). The aldehyde group, which is sensitive to acid hydration conditions with strong acids, has to be protected by formation of the morpholine enamine. The enamine is then hydrolyzed on workup after the acid-catalyzed hydration to produce Lyral (93–95).

(52)   (51)

Alloocimene Manufacture. α-Pinene (8) is converted thermally first to cis-ocimene (17), which rearranges to give about 40–50 wt % of the two alloocimene isomers, ie, 4-trans-6-cis-alloocimene [7216-56-0] (53) and 4-trans-6-trans-alloocimene [3016-19-1] (54), along with dipentene (15), alloocimene dimers, and about 10–15% pyronenes (18 and 19). The mechanistic pathways for alloocimene formation have been determined (96,97). Alloocimene is shipped with an inhibitor in tank cars, deck tanks, and drums. Production data and costs are not available.

$$(8) \xrightarrow{\Delta} (17) \longrightarrow (53) + (54)$$

Uses and Reactions. Alloocimene isomers [637-84-7] can be hydrochlorinated cold with anhydrous hydrogen chloride diluted with nitrogen to about 50% concentration and run continuously coupled with hydrolysis of the mixture of chlorides formed (98). The hydrolysis must be done using sufficient aqueous base to neutralize all the liberated acid, and use of a surfactant is necessary for efficient hydrolysis. The mixture of alloocimenols [18479-54-4] consists of a mixture of three isomers, which is useful as a floral fragrance. Hydrogenation of the mixture gives essentially a 1:1 mixture (**55**) of tetrahydromyrcenol [18479-57-7] and tetrahydrolinalool [78-69-3] and has become a large-volume product. The same mixture can be made by hydrogenation of an equal mixture of dihydromyrcenol and linalool or a mixture of myrcenol and linalool.

Oxidation of alloocimene in the presence of a catalyst produces a polymeric peroxide, which can be thermally isomerized to produce alloocimene diepoxide [3765-28-4] (**56**) in 70–75% yield (99). The diepoxide has been used in the manufacture of resins and as an acid scavenger for halogenated solvents (100).

Dihydromyrcene Manufacture. 2,6-Dimethyl-2,7-octadiene, commonly known as dihydromyrcene (**24**) or citronellene, is produced by the pyrolysis of pinane, which can be made by hydrogenation of α- or β-pinene (101). If the pinene starting material is optically active, the product is also optically active (102). The typical temperature for pyrolysis is about 550–600°C and the crude product contains about 50–60% citronellene. Efficient fractional distillation is required to produce an 87–90% citronellene product.

Uses and Reactions. Dihydromyrcene is used primarily for manufacture of dihydromyrcenol (**25**), but there are no known uses for the pseudocitronellene. Dihydromyrcene can be catalytically hydrated to dihydromyrcenol by a variety of methods (103). Reaction takes place at the more reactive tri-substituted double bond. Reaction of dihydromyrcene with formic acid gives a mixture of the alcohol and the formate ester; and hydrolysis of the mixture with base yields dihydromyrcenol (104). The mixture of the alcohol and its formate ester is also a commercially available product known as Dimyrcetol. Sulfuric acid is reported to have advantages over formic acid and hydrogen chloride in that it is less complicated and gives a higher yield of dihydromyrcenol (105).

When dihydromyrcene is treated with formic acid at higher temperatures (50°C) than that required to produce dihydromyrcenol and its formate, an unexpected rearrangement occurs to produce α,3,3-trimethylcyclohexane methanol and its formate (106). The product is formed by cyclization of dihydromyrcene to the cycloheptyl carbonium ion, which rearranges to give the more stable cyclohexyl compound (107). The formate ester, α,3,3-trimethylcyclohexane methanol formate [*25225-08-5*], (**57**) is a commercially available product known as Aphermate.

(**57**)

Reaction of acetic acid and a catalytic amount of sulfuric acid at reflux temperatures for 6–8 hours with dihydromyrcene can cause rearrangement of the dihydromyrcenyl acetate to give a mixture of the cyclic acetates analogous to the cyclic formate esters (108). The stereochemistry has also been explained for this rearrangement, depending on whether (+)- or (−)-dihydromyrcene is used (109). The cyclic acetates are also commercially available products known as Rosamusk and Cyclocitronellene Acetate.

The addition of methanol to dihydromyrcene catalyzed by acid gives methoxycitronellene in good yield (110). Epoxidation of the remaining double bond with peracetic acid gives 2-methoxy-7,8-epoxy-2,6-dimethyloctane. Hydrogenation using nickel catalysts gives reduction of the epoxide to primarily methoxycitronellol (**58**). When the reduction is carried out with the nickel catalyst and a small amount of base such as sodium carbonate or triethylamine, the reduction gives 60% of the secondary alcohol, methoxyelgenol (**59**), and 40% of methoxycitronellol (**58**) (111). The methoxyelgenol possesses a woody, floral odor characteristic of sandalwood. The material has become an important component of synthetic sandalwood fragrances.

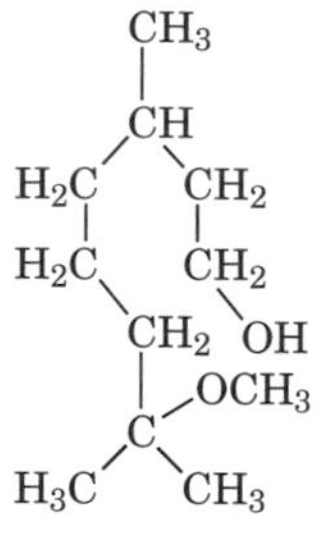

(**58**) 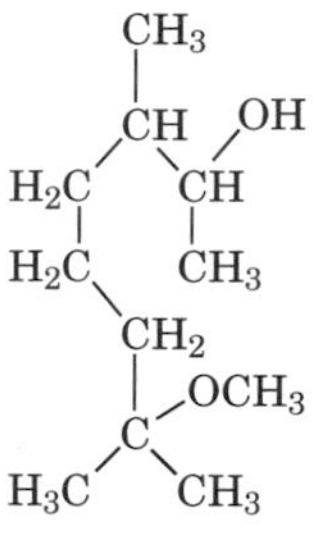   (**59**)

Epoxidation of dihydromyrcenol gives an intermediate epoxide that can be hydrogenated to hydroxycitronellol. Treatment of the diol with acid can also give the α-,β-citronellol and the α-citronellol can be isomerized to the β-citronellol (**26**) (110). The diol, hydroxycitronellol, is also useful for producing hydroxy-citronellal, a lily-of-the-valley fragrance material.

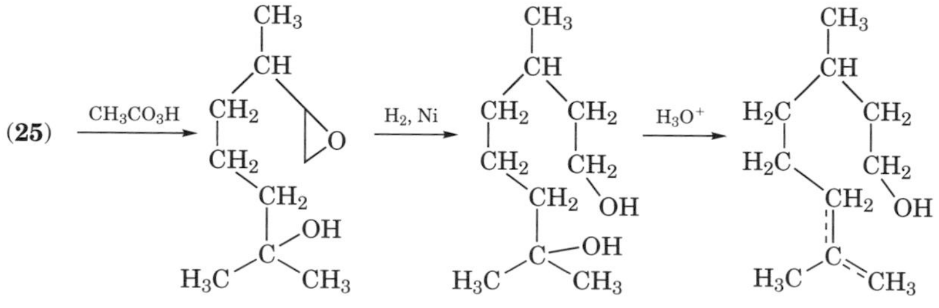

(**60**)   α,β-citronellol

Aluminum alkyls react by the Ziegler reaction with the least substituted double bond to give the tricitronellyl aluminum compound. Oxidation of the intermediate compound then produces the tricitronellyl aluminate, which is easily hydrolyzed with water to give citronellol (112,113). If the citronellene is optically active, optically active citronellol can be obtained (114). The (−)-citronellol is a more valuable fragrance compound than the (±)-citronellol.

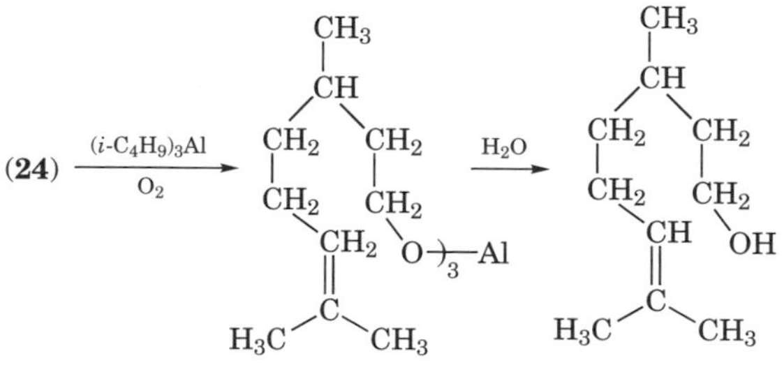

(**26**)

Although the Ziegler reaction provides a more direct method for producing primary alcohols, aluminum alkyl chemistry requires special handling and is fairly costly. The by-product aluminum salts usually require some treatment for disposal (115).

Monoterpene Alcohols

Pine Oil Manufacture. Synthetic pine oil manufacture is one of the principal uses of turpentine. U.S. production of synthetic pine oil in 1993 was 17,244 t at an average selling price of $1.10/kg (67). The amount of natural and sulfate pine oil was reported to be 1754 t. The world production of synthetic pine oil is estimated to be about twice the U.S. production figure. Natural pine oil is a product derived from the extraction of aged pine stumps, and sulfate pine oil is a product separated from crude sulfate turpentine in about 5% yield. The sulfate pine oil retains the sulfur odor of the sulfate turpentine, and its use is therefore limited to ore flotation and solvent applications.

Pine oil production in general has continued to decline. One of the principal factors is the decrease in the amount of pine oil used in cleaner and disinfectant products. The pine oil content of those products has dropped from 70–90% to 10–30% (116).

Synthetic pine oil is produced by the acid-catalyzed hydration of mainly α-pinene derived from sulfate turpentine, followed by distillation of the crude mixture of hydrocarbons and alcohols. The predominant alcohol obtained is α-terpineol, although under the usual conditions of the reaction, reversible and dehydration reactions lead to multiple hydrocarbon and alcohol components (Fig. 1).

Principal terpene alcohol components of pine oils are α-terpineol, γ-terpineol, β-terpineol, α-fenchol, borneol, terpinen-1-ol, and terpinen-4-ol. The ethers, 1,4- and 1,8-cineole, are also formed by cyclization of the p-menthane-1,4- and 1,8-diols. The bicyclic alcohols, α-fenchol [512-13-0] (**61**) and borneol (**62**), are also formed by the Wagner-Meerwein rearrangement of the pinanyl carbonium ion and subsequent hydration. Borneol is *endo*-1,7,7-trimethyl-bicyclo[2.2.1]heptan-2-ol [507-70-0]. Many other components of pine oils are also found, depending on the source of the turpentine used and the method of production.

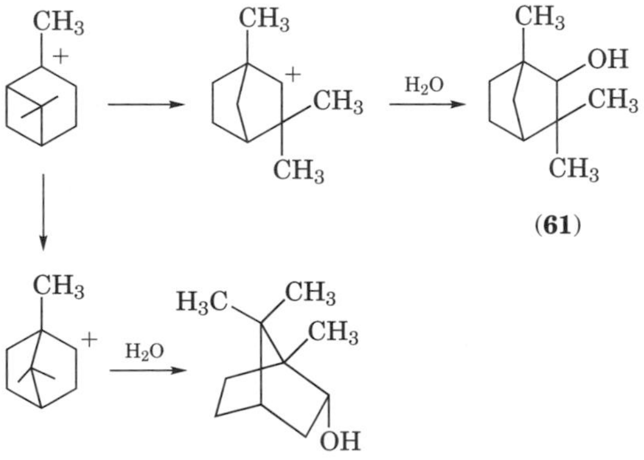

(**61**)

(**62**)

Mineral acids are used as catalysts, usually in a concentration of 20–40 wt % and temperatures of 30–60°C. An efficient surfactant, preferably one that is soluble in the acid-phase upon completion of the reaction, is needed to emulsify the α-pinene and acid. The surfactant can then be recycled with the acid. Phosphoric acid is the acid commonly used in the pine oil process. Its mild corrosion characteristics and its moderate strength make it more manageable, especially because the acid concentration is constantly changing in the process by the consumption of water. Phosphoric acid is also mild enough to prevent any significant dehydration of the alcohols formed in the process. Optimization of a process usually involves considerations of acid type and concentration, temperature, surfactant type and amount, and reaction time. The optimum process usually gives a maximum of alcohols with the minimum amount of hydrocarbons and cineoles.

Many commercial grades of pine oil are available and are specified by physical properties and total alcohol content. Some commercial pine oils and the typical physical properties are listed in Table 4. Other grades of pine oil may constitute a blend of synthetic and natural pine oil and give the product a different odor characteristic. The odor difference is caused by the presence of phenolic ethers anethole and methyl chavicol.

Pine oils can be fractionally distilled to produce a higher α-terpineol product, but usually contain borneol and γ-terpineol, along with small amounts of other components. High grade perfumery α-terpineol can be made by the partial dehydration of p-menthane-1,8-diol (terpin hydrate) under mildly acidic conditions (117,118).

Uses and Reactions. The largest use of pine oil is in the manufacture of cleaners and disinfectants. It is effective against gram-negative enterobacteria but not against gram-positive organisms. Lower grades such as sulfate pine oil are used for the flotation of metallic sulfide ores, including copper, zinc, nickel, iron, and lead. In textiles, the most important property of pine oil is its ability to reduce surface tension and interfacial tension between fiber and solution. Pine oil allows ingredients in wet-processing baths to get into fibers and to work immediately. Also, because of its bacteriocidal activity, it is used in almost all wet-processing of cotton, silk, rayon, and woolen goods.

Table 4. Typical Physical Properties of Commercial Synthetic Pine Oil[a]

Property	α-Terpineol	Unipine 90	Unipine 85	Unipine 60
terpene alcohols, %	99.0	95	86.3	63.9
specific gravity, 15.5/15.5°C	0.938	0.936	0.932	0.914
moisture, wt %	0.2	0.3	0.2	0.2
color (APHA)	10	20	35	25
flash point, °C	190	180	163	138
distillation range, °C				
5%	215	206	204	183
95%	228	222	218	220

[a]Unipine products are manufactured by Bush Boake Allen, Inc., a subsidiary of Union Camp Corp.

α-Terpineol (**9**) can be readily acetylated to its acetate using acetic anhydride (119). α-Terpinyl acetate [*80-26-2*] has a sweet, herbaceous, refreshing odor of the bergamot-lavender type and is used in scenting soaps and other household products such as cleansers and polishes. The selling price of terpinyl acetate in 1995 was $3.96/kg (45).

Hydrogenation of α-terpineol gives dihydroterpineol [*498-81-7*], a mixture of cis- and trans-compounds (120). Dihydroterpinyl acetate [*80-25-1*] produced from the mixture is also a useful fragrance compound.

Nerol and Geraniol Manufacture. Citronella oil is the principal natural source of geraniol. It occurs as a mixture with citronellol and citronellal and can be separated by fractional distillation. Nerol and geraniol are produced synthetically from turpentine sources, either from α-pinene or β-pinene in far greater quantities than are produced from natural sources. The equilibrium mixture is comprised of 60% geraniol and 40% nerol and the mixture of isomers is the predominant product manufactured. The isomers, however, can be separated by efficient fractional distillation, and products of many compositions are available. Shipment is generally made in tank trucks, deck tanks, or drums. Production figures are not available, but a considerable amount ($\sim$4000 t) of nerol and geraniol is used captively by the producers as intermediates for manufacture of other terpene products, namely citral, ionones, methyl ionones, citronellol, citronellal, hydroxycitronellal, ($\pm$)–menthol, vitamins A and E, and carotenoids. The price, which depends largely on the quality and volume purchased, for a 90–92% grade of geraniol in drums in 1995 was $14.30/kg (45).

Uses and Reactions. Nerol (**47**) and geraniol (**48**) can be converted to citronellol (**26**) by hydrogenation over a copper chromite catalyst (121). In the absence of hydrogen and under reduced pressure, citronellal is produced (122). If a nickel catalyst is used, a mixture of nerol, geraniol, and citronellol is obtained and such a mixture is also useful in perfumery. Hydrogenation of both double bonds gives dimethyloctanol, another useful product.

Nerol and geraniol can be converted to citral (**5**) in the vapor phase over a copper catalyst (123,124). Oppenauer oxidation of nerol and geraniol using catalytic amounts of aluminum isopropoxide and with aldehydes as the hydrogen acceptor also produces citral in good yields (125). The citral produced by this process gives a good intermediate product for producing high quality ionones and methyl ionones.

Acid-catalyzed esterification of nerol and geraniol with acid anhydrides produces the corresponding esters. The acetates and isobutyrates are also available commercial products. U.S. production of neryl acetate [*141-12-8*] in 1993 was 18 t at a price of $11.56/kg and that of geranyl acetate [*105-37-3*] was 132 t at a price of $9.86/kg (67).

Linalool Manufacture. The most important natural source of linalool is bois de rose oil from which it is separated by distillation. By far, more linalool (**3**) is produced synthetically than is obtained from the natural sources (126).

Isoprene (2-methyl-1,3-butadiene) can be telomerized in diethylamine with *n*-butyllithium as the catalyst to a mixture of *N*,*N*-diethylneryl- and geranyl-amines. Oxidation of the amines with hydrogen peroxide gives the amine oxides, which, by the Meisenheimer rearrangement and subsequent pyrolysis, produce linalool in an overall yield of about 70% (127–129).

One of the important processes for manufacturing linalool is from the β-methylheptenone intermediate produced by the methods from petrochemical sources discussed earlier. For example, addition of sodium acetylide to β-methylheptenone gives dehydrolinalool (**4**), which can be selectively hydrogenated, using a Lindlar catalyst, to produce linalool.

Another important process for linalool manufacture is the pyrolysis of *cis*-pinanol, which is produced from α-pinene. The α-pinene is hydrogenated to *cis*-pinane, which is then oxidized to *cis*- and *trans*-pinane hydroperoxide. Catalytic reduction of the hydroperoxides gives *cis*- and *trans*-pinanol, which are then fractionally distilled; subsequently the *cis*-pinanol is thermally isomerized to linalool. Overall, the yield of linalool from α-pinene is estimated to be about 30%.

Linalool can also be made from nerol and geraniol by the orthovanadate-catalyzed isomerization. Because linalool is lower boiling than nerol and geraniol, the isomerization can be run under distillation conditions to remove the linalool overhead while continually adding nerol and geraniol to the distillation kettle for further isomerization (56).

Linalool can also be made along with nerol and geraniol via the hydrochlorination of myrcene. After conversion of the chlorides to acetates followed by saponification of the acetates, the mixture of alcohols is obtained. Fractionation of the mixture gives linalool in about 95% purity, but the presence of close boiling impurities prohibits manufacture of a perfumery-quality product.

Linalool is shipped in cans, drums, and deck tanks. U.S. production in 1993 was not given to avoid disclosure of individual company operations, but the quantity is large (>13,000 t). The price of synthetic linalool in 1995 was $12.10/kg (45).

Uses and Reactions.   Linalool can be esterified to linalyl acetate by reaction with acetic anhydride. Linalyl acetate [*115-95-7*] has a floral-fruity odor, reminiscent of bergamot and lavender. The price of the acetate in 1995 was $14.30/kg (45). Linalool is subject to dehydration and to isomerization to nerol and geraniol during the esterification. However, if the acetic acid formed during the esterification is removed in a distillation column, the isomerization can be minimized and good yields of the acetate obtained (130).

Linalool can be hydrogenated to dihydrolinalool [*18479-49-7*] and tetrahydrolinalool [*78-69-3*], both of which are used in perfumery. The sales price of tetrahydrolinalool in 1995 was $13.20/kg (45).

Linalool can be converted to geranyl acetone (**63**) by the Carroll reaction (34). By transesterification with ethyl acetoacetate, the intermediate ester thermally rearranges with loss of carbon dioxide. Linalool can also be converted to geranyl acetone by reaction with methyl isopropenyl ether. The linalyl isopropenyl ether rearranges to give the geranyl acetone.

Geranyl acetone is an important intermediate in the synthesis of isophytol [505-32-8], farnesol [106-28-5], and nerolidol [40716-66-3]. Isophytol is used in the manufacture of Vitamin E.

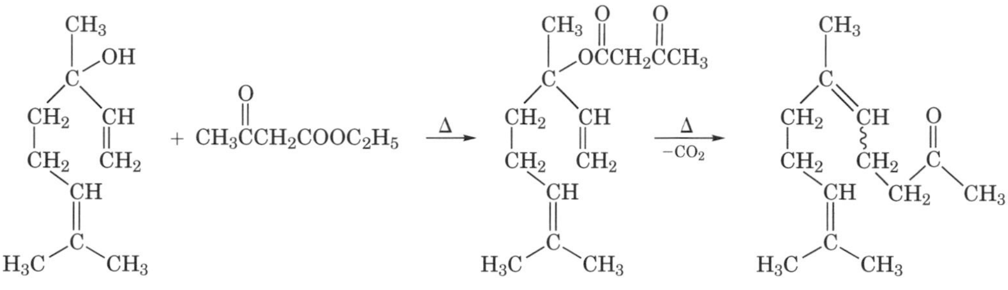

(3)   (63)

Linalool has been used to prepare a mixture of terpenes useful for enhancing the aroma or taste of foodstuffs, chewing gums, and perfume compositions. Aqueous citric acid reaction at 100°C converts the linalool (3) to a complex mixture. A few of the components include α-terpineol (34%) (9), Bois de Rose oxide (5.1%) (64), ocimene quintoxide (0.5%) (65), linalool oxide (0.3%) (66), *cis*-ocimenol (3.28%) (67), and many other alcohols and hydrocarbons (131).

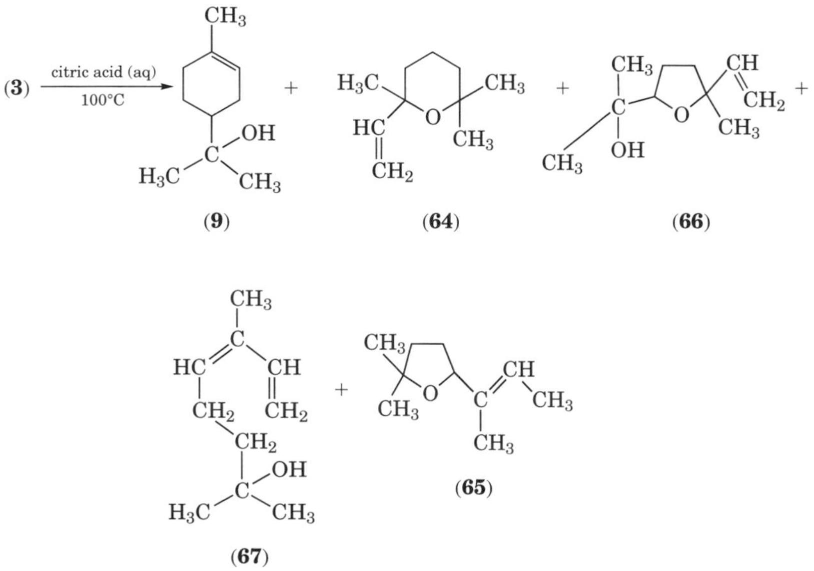

(9)   (64)   (66)

(67)   (65)

Citronellol Manufacture. Citronellol is found widely in nature and in both optically isomeric forms. Prior to the development of synthetic citronellol, this alcohol was obtained from certain oils of the *Rosaceae* family or by hydrogenation of citronellal isolated from citronella oil. Citronellol has a floral odor resembling that of roses.

Citronellol (**26**) is manufactured on a commercial scale by the hydrogenation of nerol (**47**) and geraniol (**48**) made either from α- or β-pinene. Hydrogenation of nerol and geraniol over a copper chromite catalyst gives a high yield of citronellol (121). Fractional distillation of the crude product produces a perfumery-quality citronellol. Partial hydrogenation using a nickel catalyst also gives some dimethyloctanol, but the products are nevertheless useful as perfumery materials (132). The products containing nerol, geraniol, and dimethyloctanol have odor properties closer to the products derived from citronella oil.

Citronellol can also be made by the selective hydrogenation of citral (**5**) and citronellal using a chromium-promoted Raney nickel catalyst (133,134). Geraniol can also be hydrogenated selectively to citronellol without forming dimethyloctanol using a Raney cobalt catalyst (135). A chiral ruthenium catalyst is used for the asymmetric hydrogenation of geraniol to produce optically active citronellol in high optical purity (136). Application of the Zeigler reaction on dihydromyrene (**24**) also produces citronellol; and if the dihydromyrcene is optically active, the citronellol will also be optically active.

Cyclodimerization of isoprene to 1,5-dimethylcycloocta-1,5-diene and disproportion with a rhenium oxide catalyst and isobutene produce 2,6-dimethylhepta-1,5-diene. The diene is hydroformylated to citronellal, which after hydrogenation produces citronellol (137).

The 1,5-dimethylcycloocta-1,5-diene can be partially hydrogenated and the monoene then pyrolyzed to 2,6-dimethylocta-1,7-diene, which can be converted subsequently to citronellol in several steps (138).

Citronellol is generally shipped in lined drums, deck tanks, or pails. U.S. production in 1993 was 1,714 t (67). The price of citronellol in drums in 1995 was $7.59/kg (45). The price varies according to quality and quantity purchased.

Uses and Reactions. The main use for citronellol is for use in soaps, detergents, and other household products. It is also important as an intermediate in the synthesis of other important fragrance compounds, such as citronellyl acetate and other esters, citronellal, hydroxycitronellal, and menthol.

Dehydrogenation of citronellol over a copper chromite catalyst produces citronellal [*106-23-0*] in good yield (110). If the dehydrogenation is done under distillation conditions in order to remove the lower boiling citronellal as it is formed, polymerization or cyclization of citronellal is prevented.

Citronellol is easily esterified with acid anhydrides or carboxylic acids, catalyzed by mineral acids. The price of citronellyl acetate [*150-84-5*] in 1995 was $10.45/kg (45). Other esters such as the formate and isobutyrate are also used.

The hydration of citronellol to hydroxycitronellol [*107-74-4*] (**60**) is important for the commercial synthesis of hydroxycitronellal (139). The use of solvents, such as 2-propanol (140) and acetone (141) with a strong acid ion-exchange resin are advantageous for continuous processing. They help in maintaining a homogeneous reaction solution, thus improving the rate of reaction. They also help in preventing fouling of the resin bed. Addition of alcohols such as methanol can be catalyzed by acid to give the alkoxycitronellols.

Menthol Manufacture.   Of the menthol isomers, only (−)-menthol [*2216-51-5*] and (±)-menthol [*15356-70-4*] are of commercial importance. The most important natural sources of (−)-menthol are the oils of *Mentha arvensis* (75–90%) and *Mentha piperita* (50–65%). The main suppliers are Japan, China, Brazil, and Taiwan for the former and the United States, CIS, Bulgaria, and Italy for the latter. (−)-Menthol is known for its refreshing, diffusive odor characteristic of peppermint. It also is known for its strong physiological cooling effect, which is useful in cigarettes, dentifrices, cosmetics, and pharmaceuticals.

Natural menthol is obtained by freezing the essential oil, eg, *Mentha arvensis*, and the menthol crystals are separated by centrifuging the supernatant liquid away from the crystals. The supernatant oil is then called dementholized cornmint oil. Impurities in the crystals come from the essential oil and usually give a slight peppermint aroma to the crystallized menthol. The cornmint oil, rich in (−)-menthone (~28%) and (−)-menthol (~32%), can be further processed to give additional natural menthol.

The 1995 price of natural menthol, which varies considerably according to source and availability, was $19.80/kg from China, $8.00/kg from India, and $19.25/kg from Singapore (69).

All (±)-menthol is made by synthetic methods. One method involves the cyclization of (+)-citronellal (**68**). Using a mild acid catalyst, (+)-citronellal [2385-77-5] undergoes an ene-reaction to produce a mixture of isopulegols (142). Catalytic hydrogenation of the isopulegol mixture gives a mixture of menthol and its isomers. The (±)-menthol is obtained after efficient fractional distillation and the remaining isomers can be equilibrated, usually with sodium mentholate or aluminum isopropoxide. An equilibrium mixture is obtained, comprised of 62 wt % (±)-menthol, 23 wt % (±)-neomenthol, 12 wt % (±)-isomenthol, and 3 wt % (±)-neoisomenthol. The equilibrium mixture can be distilled to recover additional (±)-menthol.

(±)-Menthol can also be made synthetically by hydrogenation of thymol [89-83-8], which can be produced by isopropylation of *m*-cresol with propylene (143,144).

The menthol isomers are equilibrated, then fractionally distilled, to recover the (±)-menthol. The other menthol isomers are recycled.

(−)-Menthol can also be synthesized from optically active terpenoids such as (+)-citronellal, (−)-β-phellandrene, and (+)-3-carene. The synthesis from (+)-3-carene has already been discussed in the section on hydrocarbons. Such methods must avoid any racemization during the course of a usually multiple-step synthesis. One disadvantage of such methods is that the other menthol diastereoisomers must be equilibrated and recycled.

The method for preparing (−)-menthol (**73**) from (+)-citronellal (**68**), which can be fractionally distilled from citronella oil, is cyclization by the ene-reaction.

The reaction can be done thermally or using alumina and silica catalysts (145–147).

High selectivities (95%) to (−)-isopulegol (**69**) are obtained using zinc halides as catalysts and hydrogenation of (−)-isopulegol gives (−)-menthol (Fig. 4) (148,149). A pyrolytic method of converting the undesired isopulegol isomers back to (+)-citronellal has also been developed (150).

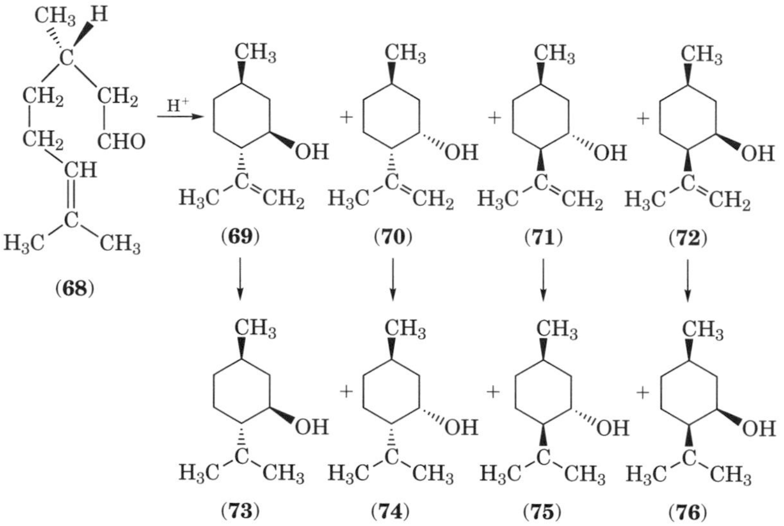

Fig. 4. Mild acid treatment of (+)-citronellal (**68**) produces a mixture, 70% (−)-isopulegol (**69**), 23% (+)-neoisopulegol (**70**), 8% (+)-isoisopulegol (**71**), and 2% (+)-neoisoisopulegol (**72**). Catalytic hydrogenation produces (−)-menthol (**73**) from (**69**); (+)-neomenthol (**74**) from (**70**); (+)-isomenthol (**75**) from (**71**); and (+)-neoisomenthol (**76**) from (**72**).

A synthesis of optically active citronellal uses myrcene (**7**), which is produced from β-pinene. Reaction of diethylamine with myrcene gives *N,N*-diethylgeranyl- and nerylamines. Treatment of the allylic amines with a homogeneous chiral rhodium catalyst causes isomerization and also induces asymmetry to give the chiral enamines, which can be readily hydrolyzed to (+)-citronellal (151).

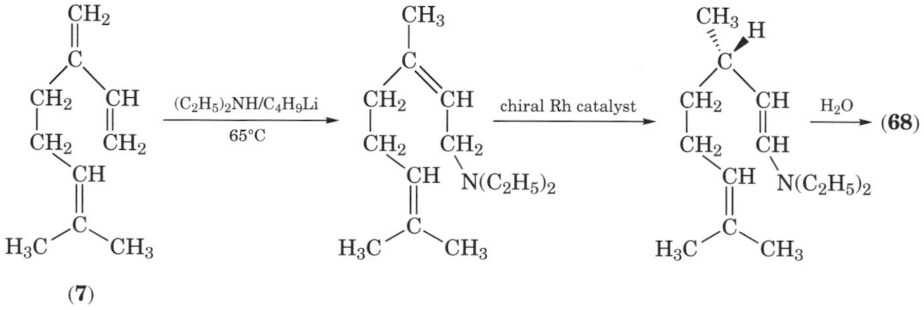

(−)-β-Phellandrene [*6153-17-9*] (**77**) is contained in about 2% in southeastern U.S. turpentine, and processing the turpentine gives a fraction containing about 28% (−)-β-phellandrene and 62% (−)-limonene. The (−)-β-phellandrene in the fraction can be selectively hydrochlorinated to piperityl chloride (Fig. 5) (152). Solvolysis of the resulting chloride under weakly alkaline conditions to control the pH produces a mixture of mainly cis- and trans-2-menthen-1-ol and a minor amount of piperitol. Treatment with dilute aqueous acid isomerizes the allylic alcohols to an equilibrium mixture in which (−)-cis-piperitol [65733-28-0] (78) and (+)-trans-piperitol [65733-27-9] (79) predominate. The two isomers can be separated by efficient fractional distillation (153). The (+)-trans-piperitol is the isomer that is hydrogenated to (+)-isomenthol (75) in high stereoselectivity (95%) (154). The (+)-isomenthol must be distilled away from the small amount (5%) of (+)-menthol also produced in the hydrogenation. The (+)-isomenthol is then epimerized using sodium mentholate or aluminum isopropoxide to obtain the equilibrium mixture containing about 56% (−)-menthol (73), 30% (+)-neomenthol (74), 13% (+)-isomenthol (75), and 1% (+)-neoisomenthol (76) (155). After fractional distillation to produce the USP (−)-menthol, the other menthol isomers are recycled to produce additional (−)-menthol.

Optical resolution is another method of producing (−)-menthol from racemic materials. (±)-Menthol is treated with optically active resolving agents to separate the (−)-menthol from the (+)-menthol, which is further processed by racemization over a nickel catalyst and recycled (156).

Resolution methods using nonoptically active agents are also used by taking advantage of the fact that certain benzoic acid derivatives of (±)-menthol can be inoculated with crystals of one enantiomer to induce immediate crystallization of that enantiomer. Although repeated crystallizations and separations must be done, the technique has been successful for (−)-menthol (157).

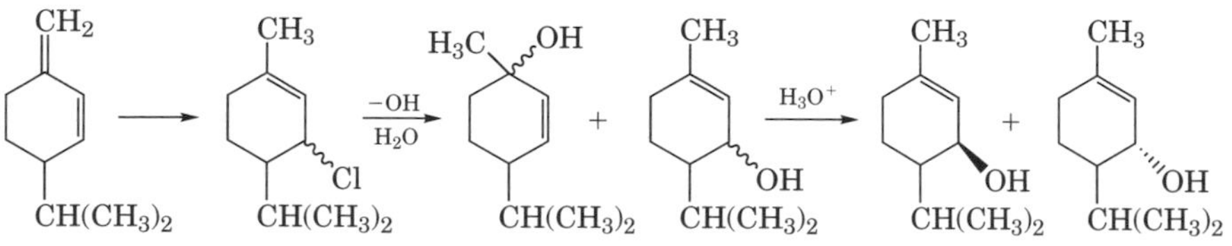

Fig. 5. Alkaline hydrolysis of piperityl chloride to cis- and trans-2-menthen-1-ol and acid-catalyzed rearrangement to piperitols (78) and (79).

Monoterpene Aldehydes and Ketones

Campholenic Aldehyde Manufacture. Campholenic aldehyde is readily obtained by the Lewis-acid-catalyzed rearrangement of α-pinene oxide. It has become an important intermediate for the synthesis of a wide range of sandalwood fragrance compounds. Epoxidation of (+)-α-pinene (8) also gives the (+)-cis-α-pinene epoxide [1686-14-2] (80) and rearrangement with zinc bromide is highly stereospecific and gives (−)-campholenic aldehyde [4501-58-0] (81) (158,159).

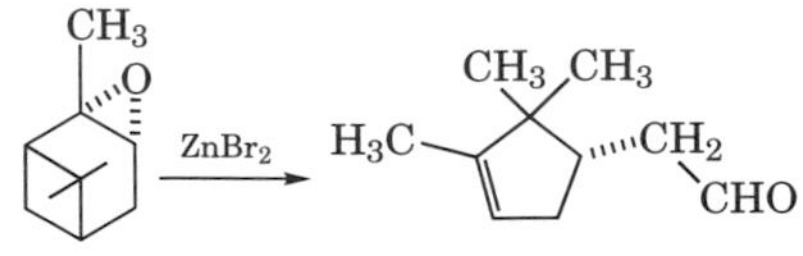

(80)   (81)

Aldol reaction of the campholenic aldehyde with 2-butanone gives the intermediate ketones from condensation at both the methyl group and methylene group of 2-butanone (Fig. 6). Hydrogenation results in only one of the two products formed as having a typical sandalwood odor (160).

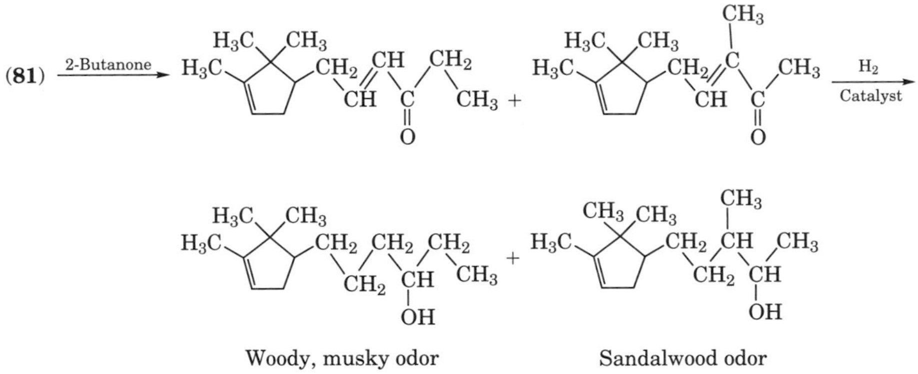

Woody, musky odor   Sandalwood odor

Fig. 6. Campholenic aldehyde (**81**) reacts with 2-butanone to produce ketones that are hydrogenated to alcohols having the odors indicated.

Aldol reaction of campholenic aldehyde with propionic aldehyde yields the intermediate conjugated aldehyde, which can be selectively reduced to the saturated alcohol with a sandalwood odor. If the double bond in the cyclopentene ring is also reduced, the resulting product does not have a sandalwood odor (161). Reaction of campholenic aldehyde with *n*-butyraldehyde followed by reduction of the aldehyde group gives the allylic alcohol known commercially by one manufacturer as Bacdanol [*28219-61-6*] (**82**).

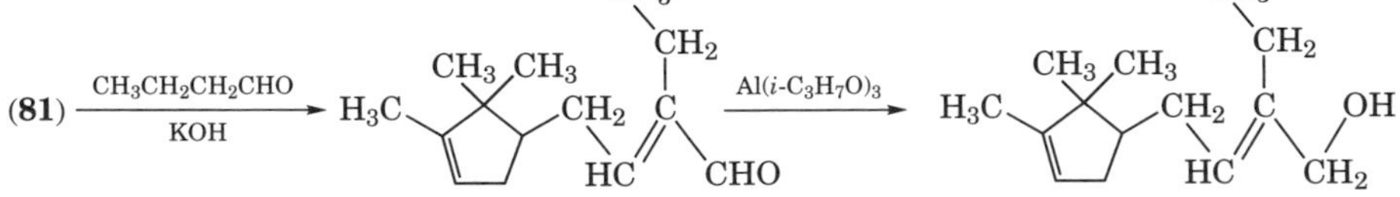

(82)

Condensation of campholenic aldehyde with ethyl acetoacetate with subsequent saponification and decarboxylation gives the intermediate unsaturated ketones.

$$(81) + CH_3\overset{O}{\overset{\|}{C}}CH_2COOC_2H_5 \longrightarrow$$

When the α,β-unsaturated ketone is hydrogenated to the alcohol, a product with an intense sandalwood odor is produced (162). Many other examples of useful products have been made by condensation of campholenic aldehyde with ketones such as cyclopentanone and cyclohexanone.

Citral Manufacture. Natural sources of citral are lemongrass oil and *Litsea cubeba*. Both oils contain 70–80 wt % citral. Synthetic citral is made from terpene sources such as nerol and geraniol and in multitonnage quantities from petrochemical sources.

The price of natural citral from *Litsea cubeba* in 1995 was $17.60–18.70/kg and the price of terpene-based synthetic citral was for $6.60–8.80/kg (69). Higher grades of synthetic citral are available for flavor and fragrance uses and price largely depends on the quality and quantity purchased. Shipment of citral is usually made in lined drums, pails, or aluminum cans.

Most terpene-based citral (**5**) produced is based on the catalytic oxidative dehydrogenation of nerol (**47**) and geraniol (**48**), or by the Oppenauer oxidation of nerol and geraniol (123–125).

Petrochemical-based methods of citral manufacture are very important for the large-scale manufacture of Vitamin A and carotenoids. Dehydrolinalool and its acetate are both made from the important intermediate, β-methylheptenone.

More recently, citral is produced in large quantity from prenal and prenol, which are made from isobutylene and formaldehyde. Prenal is 3-methyl-2-butenal and prenol is 3-methyl-2-butenol. The diprenol acetal of prenal is made and rearranged using a small amount of acid catalyst such as phosphoric acid. The rearrangement is combined with a distillation to remove prenol as it is formed (163). Cleavage of the acetal produces the allyl vinyl ether, which undergoes a Claisen rearrangement followed by a Cope rearrangement to produce citral.

Reactions. Besides the large quantity of citral used in the Vitamin A and carotenoid industry a large amount is used to produce perfume ionones and methyl ionones. An intermediate grade of citral of 70–80 wt % purity is generally used as the raw material. Citral is also useful as a lemon flavoring and fragrance, but it must first be purified. Distillation is the method most used commercially for purifying liquids, but citral is subject to thermal isomerization when distilled (164). The addition of mildly acidic materials such as ascorbic acid prevents isomerization of the citral to isocitrals and thus gives a higher purity (98 wt %) product (165).

Citral readily forms acetals by acid-catalyzed addition of alcohols or by the use of trialkoxyorthoformates. Citral dimethyl acetal [7549-37-3] is stable under alkaline conditions, whereas citral is not. Neryl and geranyl nitriles can be made by oximation of citral and dehydration of the intermediate oxime. For instance, geranonitrile [31983-27-4] is made as follows:

$$\text{(5)} \quad \xrightarrow[\text{2. H}_2\text{SO}_4]{\text{1. H}_2\text{NOH}}$$

(5)

The products have the characteristic lemon odor of citral and also have greater odor strength and chemical stability than citral. As the need for more stable citrus-like fragrances for use in bleach developed, other nitrile compounds have been made available commercially. Citronellyl nitrile is made from citronellal; dimethyloctanenitrile is produced from dimethyloctanal by the oximation method.

Ionones and Methyl Ionones Manufacture. The discovery of ionones and methylionones was an early example of the need to develop synthetic fragrance materials because of the high cost of natural materials. The aroma of violet flowers was important to perfumery and led to the development of ionones and methylionones at the end of the nineteenth century.

Citral reacts in an aldol condensation using excess acetone and a basic catalyst, usually sodium hydroxide. The excess acetone can be recovered for recycle. The resulting intermediate pseudoionone [141-10-6] (83) after cyclization with phosphoric acid gives predominantly α-ionone [127-41-3] (84), which is the isomer commercially important in flavors and fragrances. A hydrocarbon solvent is generally necessary in order to get high yields. β-Ionone [14901-07-6] (85) is the predominant isomer if sulfuric acid is used as the catalyst but lower temperature than that for cyclization to α-ionone is required. γ-Ionone [79-6-5] (86) is also produced.

$$\text{(5)} \xrightarrow[\text{NaOH}]{\text{acetone}} \quad \xrightarrow{\text{H}_3\text{PO}_4}$$

(84)   (85)   (86)

Processes are usually carried out at atmospheric pressure, but use of temperatures higher than 25–75°C above that of the lowest boiling component and at higher pressures can be carried out continuously at 0.5 MPa (5 bar) (166).

Use of 2-butanone in the aldol condensation with citral gives the methyl pseudoionones (Fig. 7). Condensation with citral occurs both on the methyl group of 2-butanone to give the normal methyl pseudoionones and on the methylene

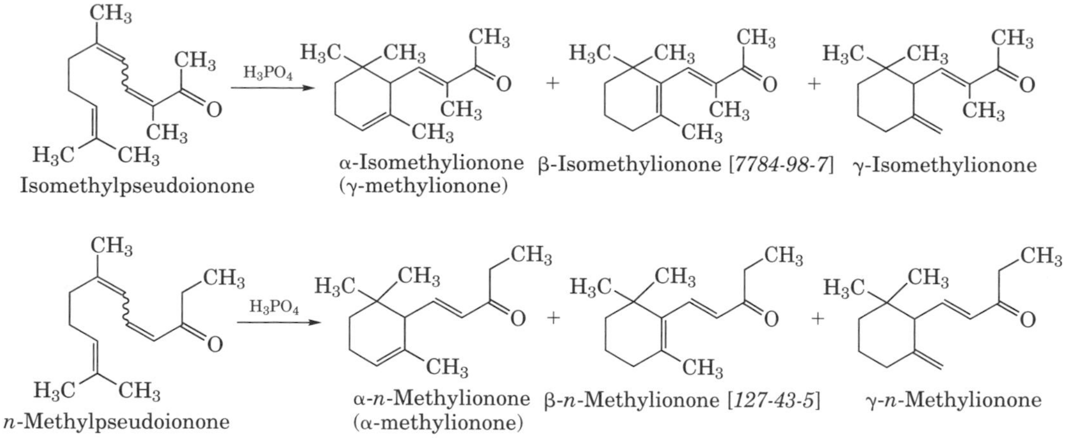

Fig. 7. Methyl pseudoionones formed from reaction of citral with 2-butanone. Reaction at the 3-position of 2-butanone yields isomethylpseudoionone [1117-41-5], whereas reaction at the 1-position gives normal-methylpseudoionone [26651-96-7].

group to give the isomethyl pseudoionones. Conditions have been optimized by methylionone manufacturers to give the methyl pseudoionone mixture that after cyclization either gives more α-n-methylionone [1322-70-9] or α-isomethylionone [7779-30-8] as the principal product. Mixtures of the two isomers are also used in perfumery, but the α-isomethylionone has the largest use. Commercially the α-isomethylionone is sold as γ-methylionone and the α-n-methylionone is sold as α-methylionone. Fractionation of the products are carefully carried out and fractions blended to meet customer specifications. The γ-methylionone [7779-30-8] has an orris-violet odor and its odor strength is said to have about two or three times that of the α-methylionone [1322-70-9] (167).

In the condensation of 2-butanone with citral, if the reaction temperature is kept at 0–10°C, higher yields of the isomethyl pseudoionones, which are the more thermodynamically stable isomers, are obtained. The aldol intermediates have more time to equilibrate to the more stable isomers at the lower temperature. The type of base used and a cosolvent such as methanol are also very important in getting a high yield of the isomethyl pseudoionones (168).

The production of α- and β-ionone in the United States in 1993 was about 120 t at an average price of \$26.48/kg (67). α-Ionone is more valuable than β-ionone in perfumery and sells for an average price of \$30.41/kg. β-Ionone is produced in large volumes and production figures are not available. It is used mainly as an intermediate for producing Vitamin A. Production figures for methylionones are not available but they are much larger than the α- and β-ionones used for perfumery.

Camphor Manufacture. Camphor is obtained both naturally and synthetically. Natural camphor is obtained from the wood of the camphor tree, *Cinnamormum camphora*, which grows in China and Japan. The camphor is isolated by combination of steam distillation, filtration, distillation, and sublimation (169). Natural camphor is the (+)-camphor, whereas synthetic camphor is racemic; both products are recognized by the USP. In 1995, the price of synthetic camphor was \$7.15/kg (45). In 1992–1993, the total production of synthetic camphor in India was 3800 t, which is estimated to be about 40% of the world consumption (170). The largest single use (80%) of camphor is for religious purposes in Asian countries.

Most synthetic camphor (**43**) is produced from camphene (**13**) made from α-pinene. The conversion to isobornyl acetate followed by saponification produces isoborneol (**42**) in good yield. Although chemical oxidations of isoborneol with sulfuric/nitric acid mixtures, chromic acid, and others have been developed, catalytic dehydrogenation methods are more suitable on an industrial scale. A copper chromite catalyst is usually used to dehydrogenate isoborneol to camphor (171). Dehydrogenation has also been performed over catalysts such as zinc, indium, gallium, and thallium (172).

(13)   (42)   (43)

Citronellal Manufacture. Natural sources of citronellal are citronella oil and *Eucalyptus citridora*. In 1995 the price of citronella oil from Java and China was \$18.90/kg and the price of citronellal in 25-lb cans was \$15.40–23.10/kg (69).

Synthetic methods for the production of citronellal include the catalytic dehydrogenation of citronellol (110), the telomerization of isoprene (151), and the lithium-catalyzed reaction of myrcene with secondary alkylamines (128).

Uses and Reactions. Citronellal undergoes an ene-reaction to form isopulegols, which are used mainly for the manufacture of menthol (148,149). If the citronellal is optically active, the isopulegol is also optically active. This is very important for the synthesis of (−)-menthol from (+)-citronellal. The aldehyde group of citronellal forms an adduct with sodium bisulfite and this protects the acid-sensitive aldehyde group so that it can be hydrated to hydroxycitronellal, an important lily-of-the-valley fragrance product (173). Hydroxycitronellal (**87**) is found in almost all floral fragrances. In 1995 its price was \$17.29/kg for synthetic and \$37.40/kg for an extra pure grade (69). The dimethylacetal of hydroxycitronellal can be readily made and its price in 1995 was \$36.50/kg (69).

(87)

Other methods of protecting the aldehyde group include formation of an enol acetate, an enamine, or an imine (174,175). In the enamine route, regeneration of the aldehyde is accomplished simply by the addition of water.

Citronellal can also be converted to the *cis-* [*92471-23-3*] (**88**) and *trans-p-*menthane-3,8-diol [*91739-72-9*] (**89**) by reaction with dilute acids (176,177). The glycol mixture can be readily purified by distillation and the two isomers easily separated. The glycols are useful as insect repellents (qv) and are especially effective against mosquitos (178). Derivatives of the glycols have been prepared and are useful as insecticides and plant growth regulators (179).

(**88**)   (**89**)

Sesquiterpenes

Sesquiterpenes are formed by the head-to-tail arrangement of three isoprene units (15 carbon atoms); there are, however, many exceptions to the rule. Because of the complexity and diversity of the substances produced in nature, it is not surprising that there are many examples of skeletal rearrangements, migrations of methyl groups, and even loss of carbon atoms to produce norsesquiterpenoids.

Important commercial sesquiterpenes mostly come from essential oils, for example, cedrene and cedrol from cedarwood oil. Many sesquiterpene hydrocarbons and alcohols are important in perfumery as well as being raw materials for synthesis of new fragrance materials. There are probably over 3000 sesquiterpenes that have been isolated and identified in nature.

The sesquiterpenes found in essential oils have low volatilities compared with monoterpenes and so are isolated mainly by steam distillation or extraction, but some are also isolated by distillation or crystallization. Most of the sesquiterpene alcohols are heavy viscous liquids and many crystallize when they are of high enough purity. Sesquiterpene alcohols are important in perfume bases for their odor value and their fixative properties as well. They are valuable as carriers of woody, balsamic, or heavy oriental perfume notes.

Caryophyllene. (−)-Caryophyllene can be isolated from Indian turpentine and has been used to prepare a number of woody aroma products. The epoxides are produced by reaction with peracids. Acetylation of caryophyllene also gives a useful methyl ketone (180) (Fig. 8). Acid-catalyzed rearrangement of caryophyllene in the presence of acetic acid gives a mixture of esters, which are related to caryolan-1-ol and clovan-2-ol (181).

Longifolene. There are at least four commercially important aroma chemicals made from (+)-longifolene and about thirteen products made from (−)-isolongifolene (**90**) (182). Acetoxymethyl longifolene or the formate are

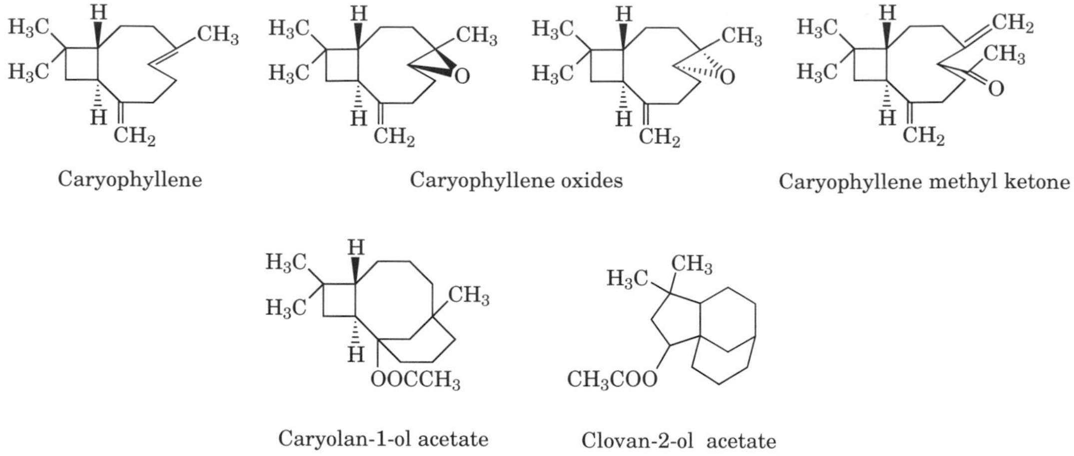

Caryophyllene   Caryophyllene oxides   Caryophyllene methyl ketone

Caryolan-1-ol acetate   Clovan-2-ol acetate

Fig. 8.   Caryophyllene and its derivatives.

formed during the Prins reaction on (+)-longifolene. Saponification of the esters gives the useful perfumery alcohol (183) (Fig. 9).

Longifolene has been found to rearrange easily to isolongifolene, which is of more interest commercially for that reason. Epoxidation of isolongifolene with peracids gives the epoxide, which on rearrangement produces the saturated ketone, 8-oxo-7β-H-isolongifolane, with a useful woody fragrance (184). The product is made by a number of manufacturers and is sold commercially under the names of Valenone B, Isolongifolanone, Timberone, and Piconia.

The Prins reaction on isolongifolene has also produced a number of useful products (185,186). All the products have amber, woody odors, and are known under the names of Amborol, Amboryl Formate, and Amboryl Acetate.

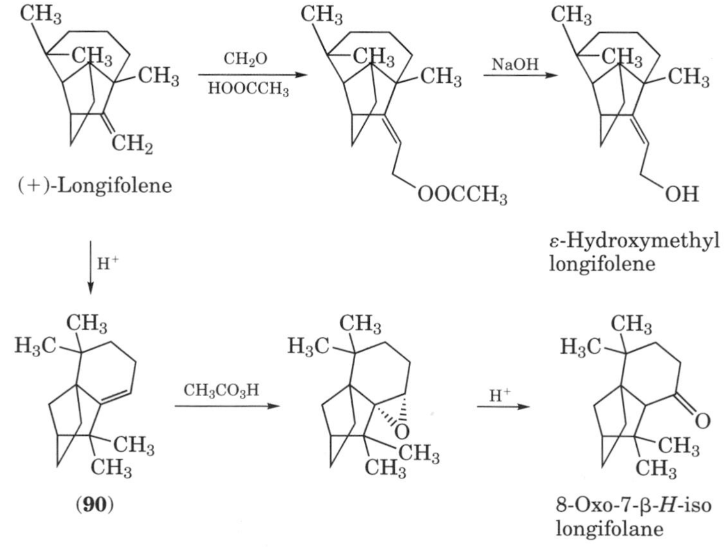

(+)-Longifolene

ε-Hydroxymethyl longifolene

(90)

8-Oxo-7-β-H-iso longifolane

Fig. 9.   Longifolene and isolongifolene (**90**).

amborol, amboryl formate, and amboryl acetate

Cedrene and Cedrol.   Cedarwood oil is one of the essential oils whose production is large and provides a source for synthesizing a number of derivatives. Cedrene (**91**) and thujopsene (**92**) are the two main sesquiterpene hydrocarbons found in the oil, along with a number of minor components (187). Cedrol [77-53-2] (**93**) is the main alcohol component of the oil.

(**91**)   (**92**)   (**93**)

The price of Cedarwood oil from Texas in 1995 was $7.70/kg and the price of the oil from Virginia was $15.18/kg (69). Distillation of the oil gives two main fractions, the cedrene or hydrocarbon fraction, and the alcohol fraction consisting of impure cedrol, which when purified by crystallization can be used directly in perfumery. Cedrenol is a product comprising a mixture of cedrol and its isomers widdrol and some ketone components.

Acetylation of the alcohol mixture yields cedryl acetate [61789-42-2] (**94**) and a range of qualities are available from manufacturers. Cedryl methyl ether [19870-74-7] is readily made and is useful in formulating woody-amber fragrances (188).

(**94**)

Acetyl cedrene (**95**) is obtained by the acetylation of the hydrocarbon fraction, α-cedrene [469-61-4] (**91**). The product has a woody, warm-ambergris, and musk odor, and many types of products are available (189).

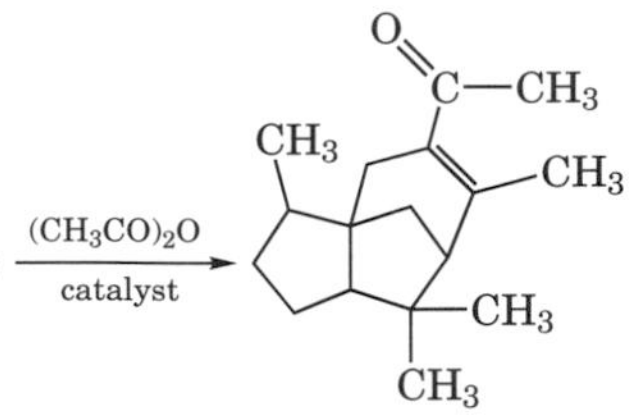

(95)

Nerolidol and Farnesol.   The alcohols nerolidol [7212-44-4] (**96**) and farnesol [4602-84-0] (**97**) are isomeric and are both important perfumery products.
Nerolidol has been isolated from neroli oil, jasmine, citronella oil, and pepper oil.
It has a mild and woody-floral, slightly green odor with excellent tenacity and
good blending and fixative properties (190).

Nerolidol can be manufactured synthetically from geranyl acetone using
either turpentine or petroleum sources. Geranyl acetone [396-70-1] can be produced from turpentine sources through linalool. Geranyl acetone from petroleum
sources is produced from acetylene and acetone. It then reacts with sodium
acetylide to give dehydronerolidol [2387-68-0] (**98**), which after hydrogenation
using a Lindlar catalyst gives nerolidol (191).

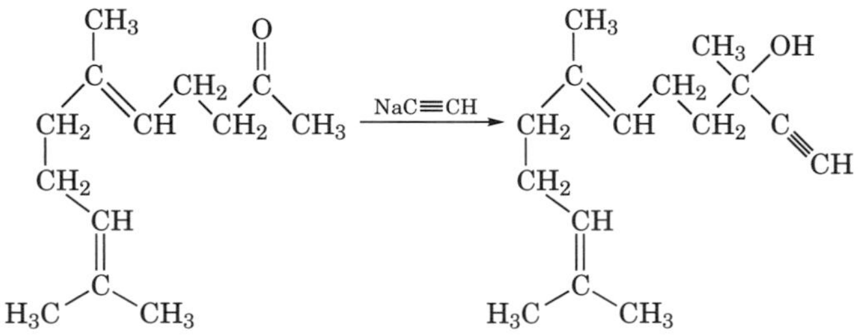

(98)

Geranyl acetones also reacts with vinyl Grignard reagent to produce nerolidol
directly (192).

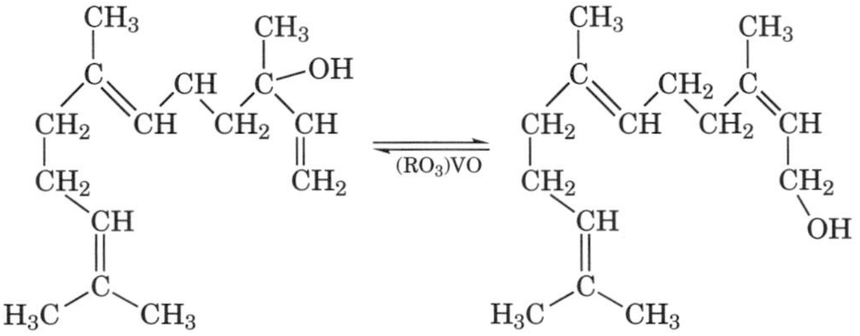

(96)   (97)

Farnesol is manufactured from nerolidol by isomerization over a vanadium catalyst (55). Farnesol occurs in several essential oils, such as ambrette

seed, neroli, rose, cyclamen, and jasmine; it is also used in floral and oriental fragrances.

α-Bisabolol. α-Bisabolol (**99**) occurs in camomile flowers and has been shown to be the main antiphlogistic and spasmolytic component of the European medicinal plant. The use of bisabolol is recommended mainly for its pharmaco-dynamic properties in cosmetic preparations for skin protection. In perfumery bisabolol functions as a fixative. The racemic form is manufactured by the acid-catalyzed cyclization of nerolidol (193).

(**99**)   (**100**)   (**101**)

α-Santalol. Sandalwood oil is comprised of mainly (90%) α- and β-san-talol, which gives the oil the woody, tenacious sandalwood odor. An impure α-santalol [115-71-9] (**100**) can be isolated by the distillation of the oil. β-Santalol [77-42-9] (**101**) is also isolated but in much smaller amounts. The price of sandalwood oil, E. Indian, in 1995 was $286/kg, and the oil from Indonesia was $187/kg (69). The high price of these oils has created the need to synthesize new materials with the sandalwood odor. The terpenophenols are manufactured by condensation of camphene with phenolic compounds followed by hydrogenation to the cycloaliphatic alcohols (194).

Valencene, Nootkatene, and Nootkatone. Valencene (**102**) has been iso-lated from orange juice and orange peel oil. Nootkatone is an important flavor material isolated from grapefruit oil (195). Nootkatene (**103**) is readily obtain-able from wood of *Chamaecyparis nootkatensis* by steam distillation and can be converted to nootkatone (**104**) by hydrochlorination followed by oxidation with Jones' reagent (196). The commercial importance of valencene and nootkatene lie in the fact that they are crucial raw materials for the synthesis of nootka-tone, which has a powerful, sweet, and citrusy odor. At concentrations greater than 7 ppm, (+)-nootkatone is the main ingredient responsible for the odor of grapefruit; it also contributes to the bitter flavor of grapefruit juice (182).

(**102**)   (**103**)   (**104**)

Patchouli alcohol. Patchouli oil comes from *Pogostemon patchouli* and the main constituent is patchouli alcohol [*5986-55-0*] (**105**) or patchoulol. Another component of the essential oil is norpatchoulenol (**106**), a norsesquiterpene derivative as a minor (3–5%) constituent, important in determining the overall odor of the essential oil (197). The price of patchouli oil in 1995 was $20.90/kg from Indonesia (69). A large proportion of the oil (40–60%) is comprised of sesquiterpene hydrocarbons that do not have much odor value. World production of the oil was at about 750 t in 1984. It is valuable in perfumery bases because of its characteristic woody, herbaceous odor (198).

(**105**)   (**106**)

Guaiol and Bulnesol. The main constituents of guaicwood oil are the sesquiterpene alcohols guaiol [*489-86-1*] (**107**) and bulnesol (**108**). When the alcohols are dehydrated with acid, many of the hydrocarbons formed are also found in patchouli oil.

Guaiol is valuable as a raw material for preparing guaiazulene (**109**), an antiinflammatory in pharmaceutical preparations for skin cosmetics. Guaiol is also used as a low cost fixative in perfumery. The acetate, guaiyl acetate [*134-28-1*], is also useful in perfumery. Bulnesol is not used as such but contributes to the fixative properties of guaicwood oil. The price of the oil in 1995 was $9.90/kg (69).

(**107**)   (**108**)   (**109**)

Diterpenes. Diterpenes contain 20 carbon atoms. The resin acids and Vitamin A are the most commercially important group of diterpenes. Gibberellic acid [*77-06-5*] (**110**), produced commercially by fermentation processes, is used as a growth promoter for plants, especially seedlings.

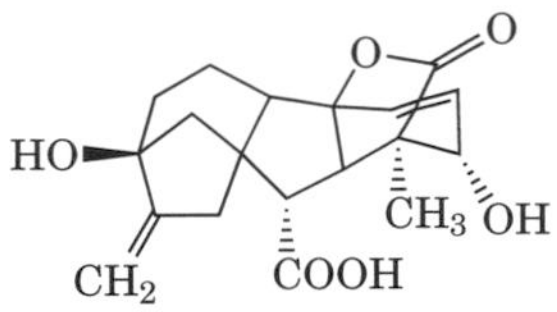

(**110**)

Phytol [*505-06-5*] (**111**) and isophytol [*150-86-7*] (**112**) are important inter-mediates used in commercial synthesis of Vitamins E and K. There is a variety of synthetic methods for their manufacture. Chlorophyll [*479-61-8*] is a phytyl ester.

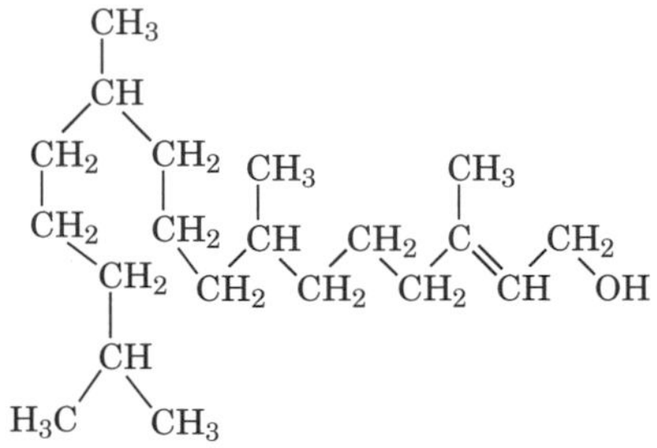

(**111**) 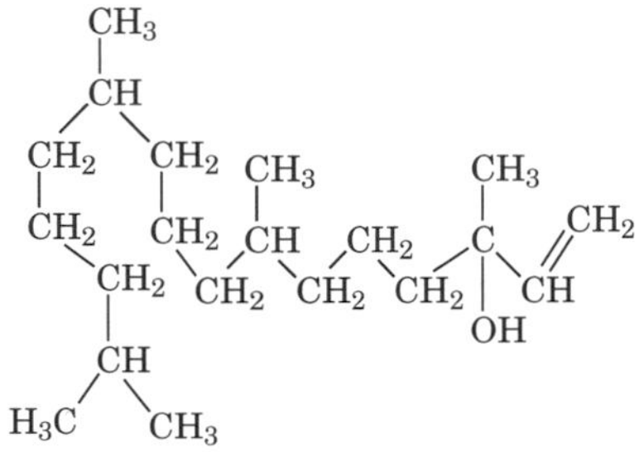   (**112**)

The tricyclic diterpenoid, taxol, is obtained by extraction from the bark of the Pacific Yew tree, *Taxus brevifolia*, which grows in forests of the western United States and Canada. Taxol (**113**) has shown promising results in fighting advanced stages of ovarian, breast, and other cancers. The yew trees, however, are slow growing and isolation of the taxol is difficult and expensive (199). The patent and chemical literature abound with efforts to synthesize taxol and analogues. One such analogue is Taxotere, which differs from taxol by having a *tert*-butoxy carbonyl group instead of a benzoyl group on the C-13 side chain and a hydroxyl group instead of an acetoxy group at C-10. These structural changes give Taxotere better water solubility and hence better bioavailability than taxol. The partial synthesis of Taxotere starts with 10-deacetylbaccatin III, which can be extracted in high yield from the leaves of *Taxus baccata*, a European yew bush.

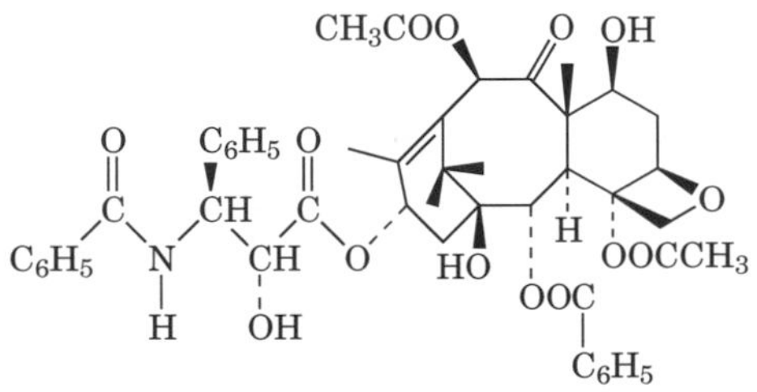

(**113**)

One method of synthesis of taxol analogues starts with α-pinene (**8**), the readily available and inexpensive monoterpene derived from the processing of turpentine from the pine tree (200). The α-pinene is oxidized to verbenone, which is then alkylated and converted to taxol analogues in a multistep process.

Another monoterpene used as a starting material for taxol analogues is camphor (**43**), which is readily available naturally or can be produced synthetically (201,202). Total synthesis of taxol analogues may be the answer toward finding new compounds for the treatment of many types of cancer.

Triterpenes.   The triterpenes (30 carbon atoms) are widely found in nature, especially plants, both in the free state and as esters or glycosides. A smaller but important group, including lanosterol [79-63-0] (**114**), occurs in animals. The triterpene hydrocarbon, squalene [111-02-4] (**115**), occurs in the liver oils of certain fish, especially those of sharks.

(**114**)   (**115**)

Squalene is also an intermediate in the synthesis of cholesterol. Structurally, chemically, and biogenetically, many of the triterpenes have much in common with steroids (203). It has been verified experimentally that *trans*-squalene is the precursor in the biosynthesis of all triterpenes through a series of cyclization and rearrangement reactions (203,204). Squalene is not used much in cosmetics and perfumery formulations because of its light, heat, and oxidative instability; however, its hydrogenated derivative, squalane, has a wide use as a fixative, a skin lubricant, and a carrier of lipid-soluble drugs.

Squalane [111-01-3] (fully saturated squalene) is produced synthetically by the coupling of two molecules of geranyl acetone with diacetylene, followed by dehydration and complete hydrogenation (205). Squalane can also be made by dimerization of dehydronerolidol, followed by dehydrogenation and hydrogenation (206).

Tetraterpenes.   Carotenoids make up the most important group of C_{40} terpenes and terpenoids, although not all carotenoids contain 40 carbon atoms. They are widely distributed in plant, marine, and animal life. It has been estimated that nature produces about 100 million t/yr of carotenoids; synthetic production amounts to several hundred tons per year (207,208).

Carotenoids may be acyclic, monocyclic, or bicyclic. The respective parent compounds for these categories are lycopene [502-65-8], γ-carotene [472-93-5], and β-carotene [7235-40-7]. The geometrical configuration of the double bonds is usually trans. The prefix neo is often used to designate isomers containing at least one cis-configuration. The prefix apo indicates carotenoids that are oxidative degradation products retaining more than half of the carotene structure. There are about 400 naturally occurring carotenoids of known structure (209).

An important function of certain carotenoids is their provitamin A activity. Vitamin A may be considered as having the structure of half of the β-carotene molecule with a molecule of water added at the end position. In general, all carotenoids containing a single unsubstituted β-carotene half have provitamin A activity, but only about half the activity of β-carotene. Provitamin A compounds are converted to Vitamin A by an oxidative enzyme system present in the intestinal mucosa of animals and humans. This conversion apparently does not occur in plants (see VITAMINS, VITAMIN A).

Because of the multiple conjugated olefinic structure in the molecule, pure crystalline carotenoids are very sensitive to light and air and must be stored in sealed containers under vacuum or inert gas to prevent degradation. Thus, commercial utilization as food colorings was initially limited; however, stable forms were developed and marketed as emulsions, oil solutions and suspensions, and spray-dried forms.

Carotenoids are also used as pigments and dietary supplements in animals and poultry feedstuffs. They are added to pharmaceutical products to provide a form of control during manufacturing and to distinguish one product from another. They also enhance the aesthetic aspects of the products (210).

β-Carotene is prescribed in the treatment of the inherited skin disorder erythropoietic protoporphyria (EPP) to reduce the severity of photosensitivity reactions in such patients. The essential theoretical background relevant to the role of carotenoids as photoconductors has been reviewed (211). β-Carotene has also been used as a photoconductor in recording-media film.

BIBLIOGRAPHY

"Terpene Resins" in *ECT* 1st ed., Vol. 13, pp. 700–704, by W. J. Roberts and A. L. Ward, Pennsylvania Industrial Chemical Corp.; "Terpenes and Terpenoids" in *ECT* 1st ed., Vol. 13, pp. 705–771, by R. S. Ropp, Hercules Powder Co.; J. E. Hawkins, University of Florida, E. G. Reitz, Chicago City Colleges (Wright Branch), P. de Mayo, Birkbeck College, University of London, and G. C. Harris, Hercules Powder Co.; in *ECT* 2nd ed., Vol. 19, pp. 803–838, by S. J. Autenrieth and A. B. Booth, Hercules Incorporated; "Camphor" in *ECT* 2nd ed., Vol. 4, pp. 54–58, by G. Etzel, Camphor & Allied Products Ltd.; "Terpenoids" in *ECT* 3rd ed., Vol. 22, pp. 709–762, by J. M. Derfer and M. M. Derfer, SCM Corp.

1. *Code of Federal Regulations*, Title 21, Section 172–182.
2. *Food Chemicals Codex*, 3rd ed., National Research Council, National Academy Press, Washington, D.C., 1981.
3. D. L. J. Opdyke, ed., *Monographs on Fragrance Raw Materials*, Research Institute for Fragrance Materials, Pergamon Press, Inc., New York, 1979.
4. R. J. Lewis, Sr. and R. L. Tatken, eds., *Registry of Toxic Effects of Chemical Substances*, 1979 ed., Vols. 1–2, U.S. Dept. of Health and Human Services, Public Health Services Center for Disease Control, National Institute for Occupational Safety and Health Service, Cincinnati, Ohio, Sept. 1980.
5. *Naval Stores Review International Yearbook*, Naval Stores Review, New Orleans, La., 1993, p. 4.
6. U.S. Pat. 3,655,803 (Mar. 11, 1972), F. L. Miller (to SCM Corp.).
7. U.S. Pat. 3,778,486 (Dec. 11, 1973), C. B. Hamby, Jr. (to SCM Corp.).
8. U.S. Pat. 3,325,553 (June 13, 1967), J. M. Derfer (to The Glidden Co.).
9. U.S. Pat. 3,660,512 (May 2, 1977), C. B. Hamby, Jr., C. W. Barrett, and J. M. Derfer (to SCM Corp.).
10. Can. Pat. 973,206 (Aug. 14, 1975), O. Prochazka (to Anglo Paper Products, Ltd.).

11. U.S. Pat. 3,359,342 (Dec. 19, 1967), J. M. Derfer (to The Glidden Co.).

12. U.S. Pat. 3,360,581 (Dec. 26, 1967), J. M. Derfer (to The Glidden Co.).

13. U.S. Pat. 3,420,910 (Jan. 7, 1969), C. Bordenca, J. M. Derfer, and C. B. Hamby, Jr. (to SCM Corp.).

14. Eur. Pat. Appl. EP 243,238A1 (Oct. 28, 1987), F. Casbas, D. Duprey, J. Ollivier, and R. Rolley (to Societe Nationale Elf Aquitaine).

15. L. F. Fieser and M. Fieser, *Reagents for Organic Synthesis*, Vol. 1, p. 475, John Wiley and Sons, Inc., New York, 1967.

16. Fr. Pat. 1,548,516 (Oct. 28, 1968), M. Pichou (to Societe des Usines Chimiques Rhone-Poulenc).

17. Can. Pat. 766,787 (Sept. 5, 1967), W. C. Meuley and P. Gradeff (to Rhodia, Inc.).

18. Fr. Pat. 1,384,137 (Jan. 4, 1965), W. C. Meuly and P. Gradeff (to Rhône-Poulenc SA).

19. U.S. Pat. 3,983,175 (Sept. 28, 1976), Y. Tamai and co-workers (to Kuraray Co., Ltd.).

20. Ger. Pat. 1,259,876 (Feb. 1, 1968), H. Pommer, H. Mueller, and H. Overwien (to Badische Anilin- & Soda-Fabrik A-G).

21. U.S. Pat. 3,670,028 (June 13, 1972), H. Mueller, H. Koehl, and H. Pommer (to Badische Anilin- & Soda-Fabrik A-G).

22. Ger. Pat. 1,643,668 (June 9, 1971), H. Pommer, H. Mueller, H. Koehl, and H. Overwien (to Badische Anilin- & Soda-Fabrik A-G).

23. H. Pommer and A. Nurrenbach, *Pure Appl. Chem.* **43**, (3–4), 527 (1975).

24. W. Reif and H. Grassner, *Chem. Ing. Tech.* **45**(10a), 646 (1973).

25. Ger. Pat. 973,089 (Dec. 3, 1959), W. Friedrichsen (to Badische Anilin- & Soda-Fabrik A-G).

26. Ger. Pat. 1,268,135 (May 16, 1968), H. Pommer, H. Mueller, and H. Overwien (to Badische Anilin- & Soda-Fabrik A-G).

27. Ger. Pat. 1,286,020 (Jan. 2, 1969), H. Pommer, H. Mueller, and H. Overwien (to Badische Anilin- & Soda-Fabrik A-G).

28. U.S. Pat. 3,686,321 (Aug. 22, 1972), H. Mueller and H. Overwien (to Badische Anilin- & Soda-Fabrik A-G).

29. Fr. Pat. 1,573,026 (July 4, 1969), E. Monrier (to Societe des Usines Chimiques Rhône-Poulenc).

30. W. Kimel, J. D. Surmatis, J. Weber, G. O. Chase, N. W. Sax, and A. Ofner, *J. Org. Chem.* **22**, 1611 (1967).

31. H. Lindlar, *Helv. Chim. Acta* **35**, 446 (1952); U.S. Pat. 2,681,938 (Dec. 31, 1954), H. Lindlar (to Hoffmann-LaRoche, Inc.); U.S. Pat. 3,674,888 (July 4, 1972), M. Derrien and J. F. Le Page (to Institut Francais du Petrole, des Carburants et Lubrifiants).

32. W. Kimel, N. W. Sam, S. Kaiser, G. G. Eichmann, G. O. Chase, and A. Ofner, *J. Org. Chem.* **23**, 153 (1958).

33. U.S. Pat. 3,023,246 (Feb. 27, 1962), H. Pasedach and M. Seefelder (to BASF A-G).

34. M. F. Carroll, *J. Chem. Soc.*, 704 (1940).

35. *Ibid.*, 507 (1941).

36. G. Saucy and R. Marbet, *Helv. Chim. Acta.* **50**, 2091 (1967).

37. Ger. Pat. 1,811,517 (July 17, 1969), P. Charbardes (to Rhône-Poulenc).

38. Brit. Pat. 1,204,754 (Sept. 9, 1970), P. Chabardes and Y. Querou (to Rhône-Poulenc SA).

39. U.S. Pat. 3,920,751 (Nov. 18, 1975), P. Chabardes and Y. Querou (to Rhône-Poulenc SA).

40. H. Pauling, D. A. Andrews, and N. C. Hindley, *Helv. Chim. Acta* **59**, 1233 (1976).

41. U.S. Pat. 3,981,896 (Sept. 21, 1976), H. Pauling (to Hoffmann-LaRoche Inc.).

42. K. Takabe, A. Agata, T. Katagiri, and J. Tanaka, *Synthesis*, 307 (1977).

43. U.S. Pat. 4,186,148 (Jan. 29, 1980), A. Murata, S. Tsuchiya, A. Konno, and J. Uchida (to Nissan Chemical Industries, Ltd.).

44. K. Takabe, T. Katagiri, and J. Tanaka, *Tetrahedron Lett.* (34), 3005 (1975).

45. *Chem. Mark. Rep.* (Apr. 10, 1995).

46. U.S. Pat. 2,898,380 (1959), R. Herrlinger (to Hercules, Inc.).

47. U.S. Pat. 2,178,349 (1939), D. H. Sheffield (to Hercules, Inc.).

48. G. W. Gladden and G. Watson, *Perf. Ess. Oil Rec.* **55**, 793 (1964).

49. L. A. Goldblatt and S. Palkin, *J. Am. Chem. Soc.* **63**, 3517 (1941).

50. R. E. Fuguitt and J. E. Hawkins, *J. Am. Chem. Soc.* **67**, 242 (1945).

51. U.S. Pat. 3,281,485 (1966), R. L. Blackmore.

52. U.S. Pat. 3,278,623 (Oct. 11, 1966), J. M. Derfer (to The Glidden Co.).

53. U.S. Pat. 3,723,542 (1974), R. R. Riso (to Stepan Chemical Co.).

54. G. Ohloff and E. Klein, *Tetrahedron*, **18**, 37 (1962).

55. U.S. Pat. 3,925,485 (Dec. 9, 1975), P. Chabardes, C. Grard, and C. Schneider (to Rhône-Poulenc).

56. U.S. Pat. 4,254,291 (1982), B. J. Kane (to SCM Corp.).

57. U.S. Pat. 3,277,206 (Oct. 4, 1966), J. P. Bain (to The Glidden Co.).

58. U.S. Pat. 3,487,118 (1969), J. H. Blumenthal (to International Flavors & Fragrances).

59. J. P. Bain, *J. Am. Chem. Soc.* **68**, 638 (1946).

60. G. Ohloff, H. Farnow, and W. Phillip, *J. Liebigs Am. Chem.* **43**, 613 (1958).

61. P. J. Kropp, D. C. Heckert, and T. J. Flautt, *Tetrahedron*, **24**, 1385 (1968).

62. *Resins for Adhesives*, product brochure, Derives Resiniques et Terpeniques (DRT).

63. U.S. Pat. 3,594,438 (1971), J. O. Bledsoe, Jr. (to SCM Corp.).

64. U.S. Pat. 3,466,267 (1969), J. M. Derfer (to SCM Corp.).

65. U.S. Pat. 3,407,241 (Oct. 22, 1968), A. B. Booth (to Hercules, Inc.).

66. U.S. Pat. 3,407,242 (Oct. 22, 1968), A. B. Booth (to Hercules, Inc.).

67. *Synthetic Organic Chemicals, U.S. Production and Sales*, USITC Publication 2810, U.S. International Trade Commission, Washington, D.C., 1993, p. 3.

68. A. F. Thomas and Y. Bessiere, *Nat. Prod. Rep.* **6**(3), 291 (1989).

69. *Chem. Mark. Rep.*, 34 (July 10, 1995).

70. A. Stanislaus and L. M. Yeddanapalli, *Can. J. Chem.* **50**, 113 (1972).

71. U.S. Pat. 2,211,432 (1941), R. C. Palmer and C. H. Bibb.

72. U.S. Pat. 4,511,488 (Apr. 16, 1985), G. B. Matta (to Penetone Corp.).

73. U.S. Pat. 4,533,487 (Aug. 6, 1985), C. L. Jones (to Pitre-Jones).

74. U.S. Pat. 4,640,719 (Feb. 3, 1987), M. E. Hayes, C. C. Hood, R. E. Miller, and R. Sharp (to Petroleum Fermentations, NV).

75. R. B. Bates, E. S. Caldwell, and H. P. Klein, *J. Org. Chem.* **34**, 2615 (1969).

76. U.S. Pat. 4,544,780 (Oct. 1, 1985), S. E. Wilson (to Shell Oil Co.).

77. U.S. Pat. 4,551,570 (Nov. 5, 1985), W. E. Johnson, Jr. (to SCM Corp.).

78. U.S. Pat. 3,293,301 (Dec. 20, 1966), J. M. Derfer, B. J. Kane, and D. G. Young (to The Glidden Co.).

79. U.S. Pat. 3,014,47 (Dec. 19, 1961), J. P. Bain, W. Y. Gary, and E. A. Klein (to The Glidden Co.).

80. British Intelligence Objectives Subcommittee Report No. 1240, British Intelligence Committee, London, 1946.

81. Jpn. Kokai 7,413,158 (Feb. 5, 1974), Y. Matoubara and H. Yada (to Yoshitomi Pharmaceutical Industries).

82. E. Klein and W. Rojahn, *6th International Congress on Essential Oils*, paper No. 163, San Francisco, Calif., 1974.

83. G. Buchhaner and G. Popp, *Chem. Ztg.* **107**, 327 (1983).

84. U.S. Pat. 3,354,225 (Nov. 21, 1967), B. J. Kane (to The Glidden Co.).

85. U.S. Pat. 3,499,937 (1970), J. Dorsky and W. M. Easter, Jr. (to Givaudan Corp.).

86. U.S. Pats. 4,104,203 (Aug. 1, 1978), 4,131,555 and 4,131,557 (Dec. 26, 1978) (to International Flavors & Fragrances, Inc.).

87. U.S. Pat. 2,420,131 (May 6, 1947), L. A. Goldblatt and S. Palkin.
88. U.S. Pat. 2,882,323 (1959), R. Weiss.
89. U.S. Pat. 3,016,408 (1962), R. L. Webb (to The Glidden Co.).
90. U.S. Pat. 3,076,839 (1963), R. L. Webb (to The Glidden Co.).
91. U.S. Pat. 3,907,321 (Sept. 23, 1975), 3,911,018 (Oct. 7, 1975), J. B. Hall and J. M. Sanders (to International Flavors & Fragrances, Inc.).
92. U.S. Pat. 3,176,022 (1965), J. H. Blumenthal (to International Flavors & Fragrances, Inc.).
93. U.S. Pat. 4,007,137 (1977), J. M. Sanders, W. L. Schreiber, and J. B. Hall (to International Flavors & Fragrances, Inc.).
94. U.S. Pat. 4,031,161 (1977), J. M. Sanders, W. I. Taylor, I. D. Hill, and J. J. Kryschuk (to International Flavors & Fragrances, Inc.).
95. K. Kogami, O. Takahashi, and J. Kumanothani, *Can. J. Chem.* **52**, 125 (1974).
96. K. J. Crowley, *J. Org. Chem.* **33**, 3679 (1968).
97. K. J. Crowley and S. G. Traynor, *Tetrahedron*, **34**, 2783 (1978).
98. U.S. Pat. 2,867,668 (1959), E. T. Theimer (to International Flavors & Fragrances, Inc.).
99. V. N. Kraseva, V. G. Cherkaev, and F. M. Raitses, *Katal. Reakts. Zhidk. Faze Tr. Vses. Konf., 2nd. Alma-Ata, Kaz. SSR*, 593 (1966).
100. U.S. Pat. 3,468,854 (Sept. 23, 1969), E. E. Royals (to Tenneco Chemicals, Inc.).
101. U.S. Pat. 4,018,842 (Apr. 19, 1977), L. A. Canova (to SCM Corp.).
102. U.S. Pat. 3,277,206 (Oct. 4, 1966), J. P. Bain (to The Glidden Co.).
103. U.S. Pat. 2,902,510 (Sept. 1, 1959), R. L. Webb (to The Glidden Co.).
104. U.S. Pat. 3,487,118 (Dec. 30, 1969), J. H. Blumenthal (to International Flavors and Fragrances, Inc.).
105. Fr. Demande FR 2,597,861 (Oct. 30, 1987), J. Ibareq and B. Lahourcade (to Derives Resiniques et Terpeniques).
106. U.S. Pat. 3,847,975 (Nov. 12, 1974), J. B. Hall (to International Flavors and Fragrances, Inc.).
107. J. B. Hall and L. K. Lala, *J. Org. Chem.* **37**, 920 (1972).
108. H. R. Ansari, *Tetrahedron*, **29**, 1559 (1973).
109. Brit. Pat. 1,254,198 (1970), H. R. Ansari and B. J. Jaggers (to Bush Boake Allen).
110. U.S. Pat. 3,038,431 (Apr. 3, 1962), R. L. Webb (to The Glidden Co.).
111. U.S. Pat. 3,963,648 (June 15, 1976), B. N. Jones, H. R. Ansari, B. N. Jaggers, and J. F. Janes (to Bush Boake Allen).
112. K. Ziegler, H. Martin, and F. Krupp, *Liebig Am. Chem.* **629**, 14 (1960).
113. K. Ziegler, F. Krupp, and K. Zosel, *Liebig. Ann. Chem.* **629**, 241 (1960).
114. Ger. Pat. 1,118,775 (Dec. 7, 1961), R. Rienacker and G. Ohloff.
115. U.S. Pat. 2,961,452 (Nov. 22, 1960), R. A. Raphael (to The Glidden Co.).
116. M. J. Kelley and A. E. Pohl, in D. F. Zinkel and J. Russel, eds., *Naval Stores*, Pulp Chemicals Association, Inc., New York, 1989, p. 561.
117. U.S. Pat. 2,521,399 (Sept. 5, 1950), S. G. Norton (to Hercules Powder Co.).
118. U.S. Pat. 2,628,258 (Feb. 10, 1953), J. E. Sapp, W. F. Gillespie, and P. A. McKim (to Gaylord Container Corp.).
119. H. Paillard and P. Tempia, *Helv. Chim. Acta*, **14**, 1314 (1931).
120. O. Zeitshel and H. Schmidt, *Ber.*, **60**, 1372 (1927).
121. U.S. Pat. 3,346,650 (Oct. 10, 1967), B. J. Kane (to The Glidden Co.).
122. USSR Pat. 118,498 (Mar. 10, 1959), V. N. Kraseva, A. A. Bag, L. L. Malkina, O. M. Khol'mer, and I. M. Lebedev.
123. Ger. Pat. 2,338,291 (Feb. 21, 1974), L. M. Polinski, I-Der Hang, and J. Dorsky (to Givaudan).
124. K. Kogami and J. Kumanotani, *Bull. Soc. Chem. Jpn.* **41**, 2508 (1968).

125. U.S. Pat. 4,055,601 (Oct. 25, 1977), W. J. Ehmann (to SCM Corp.).

126. Fr. Pat. 1,132,659 (Mar. 14, 1957), P. Teisseire (to Etabissements Roure-Bertrand Fils and J. Dupont).

127. K. Takabe, T. Katagiri, and J. Tanaka, *Tetrahedron Lett.* **34**, 3005 (1975).

128. K. Takabe, T. Katagiri, and J. Tanaka, *Chem. Lett.* **9**, 1025 (1977).

129. U.S. Pat. 4,247,480 (Jan. 27, 1981), A. Murata and co-workers (to Nissan Chemical Industries, Ltd.).

130. U.S. Pat. 3,661,978 (May 9, 1972), P. S. Gradeff and B. Finet (to Rhodia, Inc.).

131. U.S. Pat. 5,137,741 (Aug. 11, 1972), M. J. Zampino and B. D. Mookherjee (to International Flavors & Fragrances, Inc.).

132. A. F. Thomas, in J. Apsimon, ed., *The Total Synthesis of Natural Products*, Vol. 2, John Wiley & Sons, Inc., New York, 1973, pp. 1–195.

133. U.S. Pat. 4,029,709 (June 14, 1977), R. S. DeSimone and P. S. Gradeff (to Rhodia, Inc.).

134. P. S. Gradeff and G. Formica, *Tetrahedron Lett.*, 4681 (1976).

135. U.S. Pat. 3,275,696 (Sept. 27, 1966), E. Goldstein (to Universal Oil Products Co.).

136. U.S. Pats. 4,739,084 and 4,739,085 (Apr. 19, 1988), H. Takaya, J. Ohta, R. Noyori, N. Sayo, H. Kumobayashi, and S. Akutagawa (to Takasago Perfumery Co.).

137. Ger. Pat. 2,652,202 (1977) (to Shell International Research).

138. Ger. Pat. 2,704,547 (1977) (to Shell International Research).

139. Brit. Pat. 1,278,178 (1972) (to Givaudan).

140. U.S. Pat. 4,206,766 (Apr. 29, 1980), W. Hoffmann (to BASF).

141. U.S. Pat. 4,482,765 (Nov. 13, 1984), M. S. Pavlin (to Union Camp Corp.).

142. S. Kumura, *Bull. Chem. Soc. Jpn.* **10**, 330 (1935).

143. J. C. Davis, *Chem. Eng.* **85**, 62 (May 22, 1976).

144. M. Nitta, K. Aomura, and K. Yamaguchi, *Bull. Chem. Soc. Jpn.* **47**, 2360 (1974).

145. Brit. Pat. 942,054 (1961), P. S. Williams and B. T. D. Sully.

146. Fr. Pat. 1,374,732 (1964), A. Boake (to Roberts and Co., Ltd.).

147. G. Ohloff, *Tetrahedron Lett.* **10** (1960).

148. Y. Nakatani and K. Kawashima, *Synthesis*, 149 (1978).

149. Jpn. Kokai Tokkyo Koho, 116,348 (1978) (to Takasago Perfumery Co.).

150. C. A. Henrick, W. E. Willy, J. W. Baum, T. A. Baer, B. A. Garcia, T. A. Mastre, and S. M. Chang, *J. Org. Chem.* **40**, 1 (1975).

151. K. Tani, T. Yamamgata, S. Otsuka, H. Kumobayashi, and S. Akutagawa, *Org. Syn.* **67**, 33–43 (1989).

152. U.S. Pat. 2,827,499 (1958), J. P. Bain, A. B. Booth, and E. A. Klein (to The Glidden Co.).

153. U.S. Pat. 2,894,040 (1950), J. P. Bain, A. B. Booth, and W. Y. Gary (to The Glidden Co.).

154. U.S. Pat. 4,058,572 (1977), B. J. Kane, K. E. Irving, J. O. Bledsoe, Jr., and L. A. Canova (to SCM Corp.).

155. T. Yoshida, A. Komatsu, and M. Indo, *Agr. Biol. Chem.* **29**, 824 (1965).

156. R. Stroh, R. Seydel, and W. Hahn, *Angew. Chem.* **69**, 699 (1957).

157. Ger. Pat. 2,109,456 (1972), J. Fleicher, K. Bauer, and R. Hopp (to Haarman and Reimer).

158. B. Arbusow, *Ber.* **68**, 1430 (1935).

159. J. B. Lewis and G. W. Hedrick, *J. Org. Chem.* **30**, 4271 (1965).

160. U.S. Pat. 4,052,341 (1976), R. E. Naipawer and W. M. Easter (to Givaudan Corp.).

161. Dt. Offenleg. 2,827,957 (1978), E. J. Brunke and E. Klein.

162. U.S. Pat. 4,188,310 (1980), B. J. Willis and J. M. Yurecko, Jr.

163. U.S. Pat. 4,288,636 (Sept. 8, 1981), A. Nissen, W. Rebafka, and W. Aquila (to BASF).

164. G. Ohloff, *Tetrahedron Lett.* **11**, 10 (1960).

165. U.S. Pat. 5,094,720 (Mar. 10, 1992), D. E. Sasser (to Union Camp Corp.).

166. U.S. Pat. 4,431,844 (Feb. 14, 1984), L. Janitschke, W. Hoffman, L. Arnold, M. Stroezel, and H. J. Sheiper (to BASF).

167. P. Z. Bedoukian, in E. T. Theimer, ed., *Fragrance Chemistry, The Science of the Sense of Smell*, Academic Press, Inc., New York, 1982, p. 286.

168. U.S. Pat. 3,840,601 (Oct. 8, 1974), P. S. Gradeff (to Rhodia, Inc.).

169. E. Guenther, *The Essential Oils*, Vol. IV, D. Van Nostrand Co., Inc., New York, 1950, pp. 256–328.

170. R. Soman, *Chem. Eng. World*, **30**(2), 34 (1995).

171. Y. Matsubara, M. Kasano, K. Tanaka, A. Takita, and K. Itsuki, *Chem. Abstr.* **84**, 31257c (1976).

172. Jpn. Kokoi Tokkyo Koho, 77,48,645 (1977), Y. Ogino, Y. Saito, and K. Itoi.

173. U.S. Pat. 2,235,840 (Mar. 25, 1941), W. C. Meuly (to E. I. du Pont de Nemours & Co., Inc.).

174. U.S. Pat. 3,869,517 (Mar. 14, 1975), P. S. Gradeff and C. Bertrand (to Rhodia, Inc.).

175. U.S. Pat. 3,852,360 (Dec. 3, 1974), M. Vilkas and G. Senechal (to L'Air Liquide Societe Anonyme Pour Etude et Exploitation des Procedes Georges Claude).

176. H. E. Zimmerman and J. English, *J. Am. Chem. Soc.* **75**, 2367 (1953).

177. B. C. Clark, T. S. Chamblee, and G. A. Iacobucci, *J. Org. Chem.* **49**, 4557 (1984).

178. U.S. Pat. 5,130,136 (July 14, 1992), Y. Shono, K. Watanabe, H. Sekihachi, H. Kakimizu, Y. Suzuki, and N. Matsuo (to Sumitomo Chem. Co., Ltd.).

179. Jpn. Kokai Tokkyo JP 01,283,253 (Nov. 14, 1989), H. Nishimura, S. Akimoto, and T. Yasukochi (to Nippon Oils and Fats Co., Ltd.).

180. Ger. Pat. 2,440,025 (Sept. 11, 1975), K. H. Schulte-Elte, M. Joyeus, and G. Ohloff (to Firmenich, SA).

181. Ger. Pat. 2,440,024 (Sept. 11, 1975), K. H. Schulte-Elte, M. Joyeus, and G. Ohloff (to Firmenich, SA).

182. G. Ohloff, *Scent and Fragrances*, transl. by W. Pickenhagen and B. M. Lawrence, Springer-Verlag, Berlin, 1994, p. 136.

183. U. R. Nayak, T. Santhanakrishnan, and S. Dev, *Tetrahedron*, **19**, 2281 (1963).

184. T. S. Santhanakrishnau, R. Sohti, U. R. Nayak, and S. Dev, *Tetrahedron*, **26**, 65 (1970).

185. G. Ferber, *Perf. Cosmet.* **68**, 18 (1978).

186. U.S. Pat. 4,100,110 (July 11, 1978), H. R. Ansari, N. Unwin, and H. R. Wagner (to Bush Boake Allen).

187. G. C. Kitchens, J. Dorsky, and K. Kaiser, *Givaudanian*, **1**, 3 (1971).

188. U.S. Pat. 3,373,208 (Mar. 12, 1968), J. H. Blumenthal (to International Flavors & Fragrances).

189. T. F. Wood, *Givaudanian*, **1**, 3 (1970).

190. S. Arctander, *Perfume and Flavor Chemicals*, Vol. I and II, published by author, Montclair, N.J., 1969.

191. O. Isler, R. Ruegg, L. Chopard-dit-Jean, H. Wagner, and K. Bernhard, *Helv. Chim. Acta*, **39**, 897 (1956).

192. A. Ofner, W. Kimel, A. Holmgren, and F. Forrester, *Helv. Chim. Acta.* **42**, 2577 (1959).

193. C. D. Gutsche, J. R. Maycock, and C. T. Chang, *Tetrahedron*, **24**, 859 (1968).

194. E. Demole, *Helv. Chim. Acta.* **47**, 319 (1964).

195. G. L. K. Hunter and W. B. Brogden, *J. Food Sci.* **30**, 876 (1965).

196. Ger. Pat. 1,948,033 (Sept. 3, 1970), G. Ohloff (to Firmenich).

197. P. Bay and G. Ourisson, *Tetrahedron Lett.*, 2211 (1975).

198. B. M. Lawrence, *Perf. & Flavor*, **10**, 1 (1984).

199. S. Borman, *Chem. Eng. News*, 11 (Sept. 2, 1991).

200. P. A. Wender and T. P. Mucciaro, *J. Am. Chem. Soc.* **114**(14), 5878 (1992).

201. R. A. Holton and co-workers, *ACS Symp. Ser.*, **583**, 288–301 (1995).

202. L. A. Paquette, *ibid.*, pp. 313–325.

203. W. Templeton, *An Introduction to the Chemistry of Terpenoids and Steroids*, Butterworth Publishers & Co. Ltd., Inc., London, 1969.

204. A. A. Newman, ed., *Chemistry of Terpenes and Terpenoids*, Academic Press, Inc., New York, 1972.

205. Y. Fujita and T. Nishida, *Kukagaku*, **29**, 814 (1980).

206. Y. Fujita, Y. Ninagawa, T. Nishida, and K. Itoi, *Yuki Gosei Kagaku Kyokaishi*, **37**, 224 (1979).

207. O. Isler, R. Rüegg and U. Schwieter, *Pure Appl. Chem.* **14**, 245 (1976).

208. O. Isler, *Pure Appl. Chem.* **51**, 447 (1979).

209. T. W. Goodwin, ed., *Chemistry and Biochemistry of Plant Pigments*, 2nd ed., Academic Press, Inc., New York, 1976.

210. K. Münzel, in J. C. Bauernfeind, ed., *Carotenoids as Colorants and Vitamin A Precursors, Technological and Nutritional Applications*, Academic Press, Inc., New York, 1981, pp. 781–813.

211. J. O. Williams, in Ref. 200, pp. 787–813.

General References

S. Arctander, *Perfume and Flavor Chemicals (Aroma Chemicals)*, Vols. I–II, published by author, Montclair, N.J., 1969.

P. Z. Bedoukian, *Perfumery and Flavoring Synthetics*, 2nd rev. ed., Elsevier Publishing Co., New York, 1967.

S. Dev, *CRC Handbook of Terpenoids, Monoterpenoids*, Vols. I–II, CRC Press, Inc., Boca Raton, Fla., 1982.

T. K. Devon and A. I. Scott, *Terpenes*, Vol. II, *Handbook of Naturally Occurring Compounds*, Academic Press, Inc., New York, 1972.

W. F. Erman, *Studies in Organic Chemistry*, Vol. II, *Chemistry of the Monoterpenes: An Encyclopedic Handbook*, Marcel Decker, New York, 1985.

T. E. Furia and N. Bellanca, eds., *Fenaroli's Handbook of Flavor Ingredients*, 2nd ed., Vol. 2, CRC Press, Inc., Cleveland, Ohio, 1975.

E. Guenther, *The Essential Oils*, Vols. 1–6, Van Nostrand Co., Inc., Princeton, N.J., 1948–1952.

A. A. Newman, ed., *Chemistry of Terpenes and Terpenoids*, Academic Press, Ltd., London, 1972.

G. Ohloff, *Scent and Fragrances, The Fascination of Odors and Their Chemical Perspectives*, transl. by W. Pickenhagen and B. M. Lawrence, Springer-Verlag, Berlin, 1994.

J. L. Simonson and co-workers, *The Terpenes*, Vols. 1–3, 2nd ed. rev., Cambridge University Press, New York, 1953 and 1957.

W. Templeton, ed., *An Introduction to the Chemistry of Terpenoids and Steroids*, Butterworths & Co., Ltd., London, 1969.

E. T. Theimer, ed., *Fragrance Chemistry*, Academic Press, Inc., New York, 1982.

A. F. Thomas and Y. Bessiere, in J. ApSimon, ed., *The Total Synthesis of Natural Products*, Vol. 4, John Wiley & Sons, Inc., New York, 1981.

D. F. Zinkel and J. Russell, eds., *Naval Stores*, Pulp Chemicals Association, Inc., New York, 1989.

Physical Properties and Spectral Data of Terpenes and Terpenoids

J. G. Grasselli and W. M. Ritchey, *CRC Atlas of Spectral Data and Physical Constants of Organic Compounds*, 2nd ed., CRC Press Inc., Boca Raton, Fla., 1975.

A. A. Swigar and R. M. Silverstein, *Monoterpenes, Infrared, Mass, ^{1}H-NMR, and ^{13}C-NMR Spectra, and Kovats Indices*, Aldrich Chemical Co., Inc., Milwaukee, Wis., 1981.

W. Jennings and T. Shibamoto, *Qualitative Analysis of Flavor and Fragrance Volatiles by Glass Capillary Gas Chromatography*, Academic Press, Inc., New York, 1980; also includes retention indexes and mass spectral data.

C. J. Pouchert, *The Aldrich Library of Infrared Spectra*, Aldrich Chemical Co., Inc., Milwaukee, Wis., 1974.

C. J. Pouchert and J. R. Campbell, *The Aldrich Library of NMR*, Vols. 1–11, Aldrich Chemical Co., Inc., Milwaukee, Wis., 1974.

JAMES O. BLEDSOE, JR.
Bush Boake Allen, Inc.

TETRACYCLINES. See ANTIBIOTICS, TETRACYCLINES.

TEXTILES

SURVEY

Textiles are among the most ubiquitous materials in society. They provide shelter and protection from the environment in the form of apparel, as well as comfort and decoration in the form of household textiles such as sheets, upholstery, carpeting, drapery, and wall covering, and they serve a variety of industrial functions, eg, as tire reinforcement, tenting, filter media, conveyor belts, insulation, and reinforcement media in various composite materials.

Textiles are manufactured from staple fibers (qv), which have finite lengths, and filaments, which have continuous lengths, by a variety of processes to form woven, knitted, and nonwoven or felt-like fabrics. In woven and knitted fabrics, the fibers and filaments are formed into continuous-length yarns, which are then either interlaced by weaving or interlooped by knitting into planar, flexible, sheet-like structures known as fabrics. Nonwoven fabrics (qv) are formed directly from fibers and filaments by chemically or physically bonding or interlocking fibers that have been arranged in a planar configuration.

Classification

Textile fibers may be classified according to their origin, as follows.

Naturally occurring fibers

Vegetable: (based on cellulose) cotton (qv), linen, hemp, jute, ramie
Animal: (based on proteins) wool (qv), mohair, vicuna, other animal hairs, silk
Mineral: asbestos (qv)

Manufactured fibers

Based on natural organic polymers

rayon: regenerated cellulose (viscose and cuprammonium)
lyocell: regenerated cellulose (solvent process)
acetate: partially acetylated cellulose derivative
triacetate: fully acetylated cellulose derivative
azlon: regenerated protein

Based on synthetic organic polymers

acrylic: polyacrylonitrile (also modacrylic)
aramid: aromatic polyamides
nylon: aliphatic polyamides
olefin: polyolefins (polyethylene and polypropylene)
polyester: polyesters of aromatic dicarboxylic acids and dihydric alcohols
spandex: segmented polyurethane
vinyon: polyvinyl chloride
vinal (or vinylon): poly(vinyl alcohol)
carbon/graphite: derived from polyacrylonitrile, rayon or pitch
specialty fibers such as those based on poly(phenylene sulfide),
    polyetheretherketone, polyimides, and others

Based on inorganic substances

glass (qv)
metallic
ceramic

Natural fibers are those derived directly from the animal, vegetable, and mineral kingdoms. With the exception of silk (qv), which is extruded by the silkworm as a continuous filament, natural fibers are of finite length and are used directly in textile manufacturing operations after preliminary cleaning. These fibers are known as staple fibers, and an estimate of their average length is referred to as their staple length. Other quality factors which affect the utility of natural fibers for textile purposes are fineness, presence of foreign matter, color, and spinnability. The latter term denotes the ability of a fiber to be spun economically into yarns by conventional textile processing procedures. Chemical properties such as ability to absorb common dyes, thermal and environmental stability, and resistance to chemical degradation, are also important.

The other major category of textile fibers is that of the manufactured fibers, which are produced from natural organic polymers, synthetic organic polymers, and inorganic substances. Glass fiber is the only inorganic manufactured fiber in common use, although other ceramic and metallic fibers are being developed, particularly for use in high performance fiber-reinforced composites. Qualitative and quantitative analyses of textiles fibers can be performed according to AATCC Test Methods 20 and 20A (1).

Fiber Consumption Trends

For centuries the textile industry relied exclusively on the natural fibers, particularly cotton, wool, and silk. With the commercialization of manufactured fibers, beginning with those based on natural polymers in the 1930s, followed by those based on synthetic polymers in the 1940s, the textile industry had undergone a true revolution. The 1994 total world production of textile fibers amounted to approximately 42 million metric tons, distributed among the principal types of fibers as shown in Table 1. It is noteworthy that nearly all regions of the world are involved in fiber production. Polyester fibers alone accounted for nearly half of the synthetic fibers produced worldwide in 1994, as shown by the data in Table 2. The textile industry as constituted ca 1997 finds itself with an ever-increasing number of fibers from which to manufacture its products. Not only are there many more fiber types for the textile industry to use, but it is also recognized that many advantages are to be gained by blending the various fibers in any of an almost infinite number of combinations. Many modern textile products are blends of natural and manufactured fibers, incorporating the desirable attributes of both.

The average annual per capita fiber consumption in the decade of the 1990s is about 8 kg. As might be expected from variations in climate and socioeconomic conditions, there are large variations in the per capita fiber consumption for the different regions of the world, ranging from about 1.5 kg in Africa and Asia to about 25 kg in North America.

Table 1. Worldwide Fiber Production by Region, 1994,[a] 10^6 t

	Synthetic fiber[b]		Natural fibers	
Region	Synthetic polymers	Cellulosics	Cotton	Wool
United States	3.19	0.23	4.09	0.04
Europe (includes former Soviet Union)	3.88	0.79	2.50	0.52
Canada and other Americas	1.08	0.11	1.26	0.20
Japan	1.40	0.22		
People's Republic of China	2.03	0.34	4.51	0.25
Asia/Oceania	5.86	0.63	5.26	1.15
Middle East/Africa	0.28	0.02	1.40	0.64
Total	*17.7*	*2.32*	*19.0*	*2.80*

[a]Ref. 2.
[b]Not including olefin fiber and glass fiber.

Manufactured Fibers

Manufactured fibers produced from natural organic polymers are either regenerated or derivative. A regenerated fiber is one which is formed when a natural polymer or its chemical derivative is dissolved and extruded as a continuous

Table 2. Worldwide Synthetic Fiber Production by Fiber,[a] 10[6] t

Fiber	1990	1994
polyester	8.68	11.2
polyamide	3.74	3.92
acrylic	2.32	2.47
olefin	2.94	3.71
glass	1.84	2.28
other	0.16	0.17
Total	*19.4*	*23.8*

[a]Ref. 2.

filament, and the chemical nature of the natural polymer is either retained or regenerated after the fiber-formation process. A derivative fiber is one which is formed when a chemical derivative of the natural polymer is prepared, dissolved, and extruded as a continuous filament, and the chemical nature of the derivative is retained after the fiber-formation process.

Manufactured fibers based on synthetic organic polymers, generally referred to as the synthetic fibers, have revolutionized the textile industry since their initial commercialization by the chemical industry in 1940. The generic names of manufactured fibers are defined and controlled by the Federal Trade Commission. The production of manufactured fibers is based on three methods of fiber formation, or extrusion spinning. In this context, spinning refers to the overall process of polymer liquefaction (dissolution or melting), extrusion, and fiber formation. The three principal methods are melt spinning, dry spinning, and wet spinning, although there are many variations and combinations of these basic processes. (The term spinning is otherwise customarily reserved for that textile manufacturing operation wherein staple fibers are formed into continuous textile yarns by several consecutive attenuating and twisting steps. A yarn so formed from natural or manufactured staple fibers is referred to as a staple or spun yarn.)

In melt spinning, the polymer is heated above its melting point and the molten polymer is forced through a spinneret. A spinneret is a die with many small orifices which may be varied in their diameter and shape. The jet of molten polymer emerging from each orifice in the spinneret is guided to a cooling zone where the polymer solidifies to complete the fiber-formation process. In dry spinning, the polymer is dissolved in a suitable solvent and the resultant solution is extruded under pressure through a spinneret. The jet of polymer solution is guided to a heating zone where the solvent evaporates and the filament solidifies. In wet spinning, the polymer is also dissolved in a suitable solvent and the solution is forced through a spinneret which is submerged in a coagulation bath. As the polymer solution emerges from the spinneret orifices in the coagulating bath, the polymer is either precipitated or chemically regenerated.

In most instances the filaments formed by melt, dry, or wet spinning are not suitable textile fibers until they have been subjected to one or more successive drawing operations. Drawing is the hot or cold stretching and attenuation of manufactured filaments to induce molecular orientation with respect to the fiber axis, and to develop a fiber fine structure. This fine structure is generally characterized by a high degree of crystallinity and by an orientation of both the crystallites and the polymer chain segments in the noncrystalline domains. The fine structure and physical properties of manufactured textile fibers are frequently further modified by a variety of thermomechanical annealing treatments, notably heat setting, to provide dimensional stability. A widely used process is that of texturing, which provides crimp to the filaments, and thereby bulk and softness to textile products manufactured from them.

In the case of melt spinning, the two-step process of spinning (extrusion) and drawing (structure development) is increasingly being combined in a single, high speed spinning process. In high speed spinning, where windup speeds on the order of 6,000 to 10,000 m/min are used, orientation and crystallinity are developed directly on the spinline. Bicomponent manufactured fibers, where two different polymers are extruded simultaneously in either side-by-side or skin–core configurations, are also an important category of fibers. Microfibers, generally defined as fibers with linear densities of less than 1 denier per filament, are increasingly being used in apparel and sportswear.

In the production of manufactured fibers, the filaments are obtained in continuous form. When several such filaments are combined together and slightly twisted to maintain unity, the product is called a multifilament yarn. A typical yarn may contain 100 single filaments. Individual filaments, considerably larger in cross section than those used in multifilament yarns, may also be used in certain applications, and these are referred to as monofilaments. Frequently it is desired to obtain manufactured fibers in finite lengths for subsequent processing into spun yarns by conventional textile spinning operations. Thousands of continuous filaments are collected together into a continuous rope of parallelized filaments called a tow. The tow is converted into staple-length fiber by cutting it into specified lengths. The staple length which is produced in this conversion process depends on the system of yarn processing which is to be used. The cotton yarn processing system requires lengths of approximately 0.038 m (1.5 in.), whereas the woolen or worsted systems require lengths between 0.076–0.127 m (3–5 in.). When manufactured fibers are produced in staple form for blending with natural fibers, dimensional properties such as length and fineness are matched to the natural fiber that is to be used in the blend.

Types of Textile Materials

Textile yarns are produced from staple (finite-length) fibers by a combination of processing steps, referred to collectively as yarn spinning (3). After preliminary fiber alignment, the fibers are locked together by twisting the structure to form the spun yarn, which is continuous in length and remarkably strong and uniform. Depending on the specific processing conditions, the degree of fiber parallelization and surface hairiness can vary over a considerable range, which strongly influences yarn physical properties. Sorption properties in particular are strongly affected by the fiber organization in a spun yarn. The staple fibers may be either natural fibers, such as cotton or wool, or any of a number of manufactured fibers.

Textile yarns are also produced from continuous manufactured filaments. Such multifilament yarns are characterized by nearly complete filament alignment and parallelization with respect to the yarn axis. The degree of twist introduced into a multifilament yarn is usually quite low and just adequate to produce some level of interfilament cohesion. Such yarns are quite compact and smooth in appearance. A variety of processes have been developed to introduce bulk and texture into multifilament yarns. These processes are designed to disrupt the high degree of filament alignment and parallelization and to produce yarns with properties generally associated with spun yarns. Schematic representations of typical yarns are given in Figure 1.

Yarns are used principally in the formation of textile fabrics either by weaving or knitting processes (4). In a woven fabric, two systems of yarns, known as the warp and the filling, are interlaced at right angles to each other in various patterns. The woven fabric can be viewed as a planar, sheet-like material with pores or holes created by the yarn interlacing pattern. The dimensions of the fabric pores are determined by the yarn structure and dimensions, and by the weaving pattern. The physical properties of woven fabrics are strongly dependent on the test direction.

Knitted fabrics are produced from one set of yarns by looping and interlocking processes to form a planar structure. The pores in knitted fabrics are usually not uniform in size and shape, and again depend largely on yarn dimensions and on the numerous variables of the knitting process. Knitted fabrics are normally quite deformable, and again physical properties are strongly dependent on the test direction. Tufting and stitchbonding are additional processes in fabric manufacture.

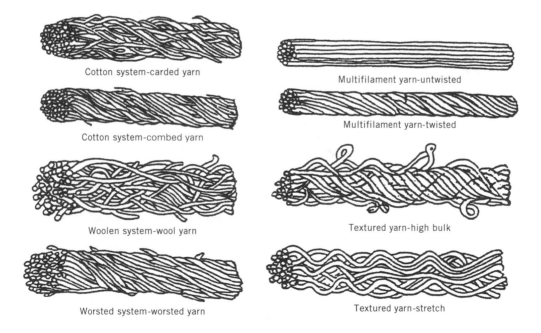

Cotton system-carded yarn

Cotton system-combed yarn

Woolen system-wool yarn

Worsted system-worsted yarn

Multifilament yarn-untwisted

Multifilament yarn-twisted

Textured yarn-high bulk

Textured yarn-stretch

Fig. 1. Schematic description of spun and multifilament yarns (3).

In the production of nonwovens, the fibers are processed directly into a planar, sheet-like fabric structure, bypassing the intermediate one-dimensional yarn state, and then are either bonded chemically or interlocked mechanically, or both, to achieve a cohesive fabric (5,6). Typically, staple-length fibers are dispersed in a fluid (liquid in the wet-laid process of manufacture or air in the dry-laid process of manufacture) and deposited in sheet-like planar form on a support base prior to bonding or interlocking. Carding of staple fibers is frequently the basis of dry-laid nonwoven manufacture. The paper-making process is a well-known example of a wet-laid nonwoven process which utilizes short paper (wood) fibers. The spunbond process differs from the dry- and wet-laid processes in that continuous length filaments are extruded, collected in a randomized planar network, and bonded together to form the final product. Spunbonded nonwovens are generally thin, strong, and almost film-like in appearance.

Within the plane of a nonwoven material, the fibers may be either completely isotropic or in a preferred orientation or alignment, usually with respect to a machine or processing direction. In the case of thicker, dry-laid nonwovens, fiber orientation may be developed in the third dimension, ie, that dimension which is perpendicular to the plane of the fabric, by a process known as needle-punching (7). This process serves to bind the fibers in the nonwoven by mechanical interlocking. Hydroentangling is an alternative to needle-punching (8).

Dyeing and Finishing

After the fabric formation process, textiles are generally subjected to either dyeing or printing and to a variety of mechanical and chemical finishing operations. The specific nature of the dyeing and finishing operations depends on the fiber type and on the intended use of the fabric.

Dyeing is normally conducted in an aqueous solution of dye (or a mixture of dyes) and involves partition or distribution of the dye from the aqueous phase to the solid fiber phase (9). Although dyeing is most commonly performed on fabrics, it may also be performed on raw stock (staple fiber) or on yarns. The dye is absorbed by the fiber and held by secondary intermolecular forces, but there are several reactive dyes that bond covalently to the fiber polymer. In order to achieve reasonable rates of dyeing, commercial dyeing is normally conducted at 100°C. In the case of some fibers, particularly polyester, dyeings are conducted under pressure to allow temperatures of approximately 130°C to be used, thereby increasing the dye diffusion rate. There are many different classes of dyes, including acid, direct, basic, sulfur, vat, disperse, and reactive, each of which has affinity for certain fiber types (10,11). Through the use of dye mixtures and blended yarns and fabrics, it is possible to achieve colorful patterns by dyeing. However, printing, utilizing insoluble pigments instead of dyes, is used more extensively to achieve multicolor patterns and designs. The printing process is much more rapid and less energy-intensive than dyeing, but generally less durable (see DYES AND DYE INTERMEDIATES; PIGMENTS).

Mechanical finishing is designed primarily to produce special fabric surface effects. Typical operations in this category are napping, shearing, embossing, and calendering. Chemical finishing may involve either the addition of functional chemicals or the chemical modification of some or all of the fibers comprising the fabric. It is not always possible to draw a clear distinction between additive chemical finishing and chemical modification. The primary purpose of chemical finishing is to confer special functional properties, for example, dimensional stability, wrinkle resistance, wash-and-wear or durable-press characteristics, soiling resistance, water repellency, and flame retardance (12).

Chemical finishing procedures, other than scouring and possibly bleaching, are performed principally on those fabrics that are composed either entirely or in part of cellulosic fibers (ie, cotton and rayon). These fibers have chemically reactive sites (OH groups) that lend themselves to chemical modification and additive chemical treatments. Of particular importance in this regard are treatments of rayon- and cotton-containing fabrics with difunctional reagents that are capable of chemically cross-linking cellulose chains (13). Such reactions improve fiber resilience, thereby enhancing fabric smoothness appearance, wrinkle resistance, and wrinkle recovery. The same reactions can be used to impart wash-and-wear and durable-press characteristics (see COATED FABRICS).

An important chemical finishing process for cotton fabrics is that of mercerization, which improves strength, luster, and dye receptivity. Mercerization involves brief exposure of the fabric under tension to concentrated (20–25 wt %) NaOH solution (14). In this treatment, the cotton fibers become more circular in cross-section and smoother in surface appearance, which increases their luster. At the molecular level, mercerization causes a decrease in the degree of crystallinity and a transformation of the cellulose crystal form. These fine structural changes increase the moisture and dye absorption properties of the fiber. Biopolishing is a relatively new treatment of cotton fabrics, involving cellulase enzymes, to produce special surface effects (15).

Wool-containing fabrics are also subjected to functional chemical finishing. The main purpose of these treatments is to control felting shrinkage which occurs when wool fabrics are subjected to mechanical agitation at elevated temperatures in the wet state, conditions typical of normal washing procedures. Felting shrinkage is caused by an inward migration of fibers in a yarn or fabric. This, in turn, is caused by a unique scalar structure of wool and other animal fibers that causes the frictional coefficient to be significantly higher in the against-scale direction than in the with-scale direction. The chemical finishing treatments are designed to minimize this differential friction by inactivating the scale edges. This is achieved by means of a variety of treatments, including chlorination, oxidation with permanganate solutions, polymer deposition, and combinations of these (16). Wool fabrics can also be chemically after-treated to impart wash-and-wear and durable-press properties.

Fabrics composed of synthetic polymer fibers are frequently subjected to heat-setting operations. Because of the thermoplastic nature of these fibers, eg, polyester, nylon, polyolefins, and triacetate, it is possible to set such fabrics into desired configurations. These heat treatments involve recrystallization mechanisms at the molecular level, and thus are permanent unless the fabrics are exposed to thermal conditions more severe than those used in the heat-setting process.

Although textiles are usually associated with apparel and household products, industrial or technical uses of fibers are of increasing importance. Of particular significance in this regard is the use of fibers to reinforce plastic materials, producing high performance, fiber-reinforced composites. Such materials are extensively used in construction and in the automotive, aerospace, and sports equipment industries. Their principal advantage lies in very high strength-to-weight ratios compared to typical metals and ceramics. The use of textiles in civil engineering applications, such as for road reinforcement, erosion control, and heat exchange devices, is becoming widespread throughout the world.

BIBLIOGRAPHY

"Textiles–Survey" in *ECT* 3rd ed., Vol. 22, pp. 762–768, by L. Rebenfeld, Textile Research Institute.

1. *Technical Manual*, American Association of Textile Chemists and Colorists, Research Triangle Park, N.C. (published annually).
2. *Fiber Organon*, Vol. 66, Fiber Economics Bureau, Inc., Roseland, N.J., 1995.
3. B. C. Goswami, J. G. Martindale, and F. L. Scardino, *Textile Yarns: Technology, Structure and Applications*, John Wiley & Sons, Inc., New York, 1977.
4. J. W. S. Hearle, P. Grosberg, and S. Backer, *Structural Mechanics of Fibers, Yarns, and Fabrics*, John Wiley & Sons, Inc., New York, 1969.
5. J. Lunenschloss and W. Albrecht, *Nonwoven Bonded Fabrics*, Halsted Press, a division of John Wiley & Sons, Inc., New York, 1985.
6. R. Krcma, *Manual of Nonwovens*, Textile Trade Press, Manchester, U.K. (and W. R. C. Smith Publishing Co., Atlanta, Ga.), 1971.
7. V. Mrstina and F. Fejgl, *Needle Punching Technology*, Elsevier, Amsterdam, the Netherlands, 1990.
8. D. Bertram, *INDA J. Nonwov. Res.* **5**(2), 34–41 (1993).
9. R. H. Peters, *Textile Chemistry*, Vol. III, *The Physical Chemistry of Dyeing*, Elsevier Science, Inc., New York, 1975.
10. *AATCC Dyeing Primer*, American Association of Textile Chemists and Colorists, Research Triangle Park, N.C., 1981.
11. H. Zollinger, *Color Chemistry: Syntheses, Properties and Applications of Organic Dyes and Pigments*, VCH Publishers, New York, 1987.
12. H. Goldstein, *Tex. Chem. Color.* **25**(2), 16 (1993).
13. G. C. Tesoro, in M. Lewin and S. B. Sello, eds., *Functional Finishes Part A, Handbook of Fiber Science and Technology*, Vol. II, Marcel Dekker, Inc., New York, 1983, pp.   .
14. R. Freytag and J. J. Donze, in *Fundamentals and Preparation Part A*, Ref. 13, Vol. I, pp.   .
15. G. Buschle-Diller and S. H. Zeronian, *Text. Chem. Color.* **26**(4), 17–24 (1994).
16. K. R. Makinson, *Shrinkproofing of Wool*, Marcel Dekker, Inc., New York, 1979.

LUDWIG REBENFELD
TRI/Princeton

FINISHING

Textile finishing includes various efforts to improve the properties of textile fabrics, whether for apparel, home, or other end uses. In particular, these processes are directed toward modifying either the fiber characteristics themselves or the gross textile end properties. Such modifications may be chemical or mechanical in nature. One modification that is not covered in this article relates to the dyeing of textiles and the dyestuffs employed for fibers; however, areas that involve chemical finishing designed to modify the normal dye receptivity and the growing use of enzyme treatments are included.

Fibers have been used by humans for thousands of years, but only in the twentieth century has there been such an explosion in fiber types available to the textile manufacturer. The advent of synthetic fibers possessing improved resiliency and dimensional stability has placed natural fibers, particularly cotton (qv), at an ostensible disadvantage. Before synthetics, various means to control the shrinkage, dimensional stability, and smooth-dry performance of cotton had been investigated, but the appearance of synthetics such as polyester has placed a greater sense of urgency on cotton interests to focus on the perceived deficiencies of natural fibers.

Textile finishing encompasses a broad range of approaches and may be directed toward needed properties such as shrinkage control or smooth-dry performance or toward developing properties for specific end uses such as flame retardance, soil release, smolder resistance, weather resistance, or control of static charges.

From a historical point of view, mechanical finishing processes have been directed toward improving shrink resistance via compressive shrinkage (Sanforized process) or calender finishing to give surface effects, which include shreinering, chintz finishing, and embossing. These latter effects are semidurable, and can be made more permanent by cross-linking with resins. However, such cross-linked fabrics seem to suffer greater loss in tearing strength when cross-linked in the calendered state. The need for calendered fabrics has decreased with the advent of cross-linked cotton, but compressive shrinkage to control garment shrinkage remains a significant component in textile mill practice, particularly for products that receive little or no resin treatments.

For fabrics of thermoplastic fibers, permanent effects are obtainable if heat and pressure are applied to soften the material. Processes dealing with carpets, nonwovens, and chemical modifications or additions that occur before the fiber is formed are not discussed herein (see NONWOVEN FABRICS).

Treatments with Chemicals or Resins. Resin treatments are divided into topical or chemical modifications of the fiber itself. Most chemical treatments of synthetic fibers are topical because of the inert character of the fiber itself and the general resistance of the fiber to penetration by reagents. By contrast, cellulosics and wool possess chemical functionality that makes them reactive with reagents containing groups designed for such purchases. Natural fibers also provide a better substrate for nonreactive topical treatments because they permit better penetration of the reagents.

Chemical Treatments

Chemical treatments of textiles encompass a variety of approaches. In one, the finisher may be attempting to form a chemical bond between the reactive group of the fiber (–OH group of cellulose) and the applied reagent. If the bond is resistant to hydrolysis, the property conferred to the garment, for example, smooth-dry performance, will be durable to laundering. If the bond is not durable, the change accomplished by finishing will gradually disappear.

Another approach in chemical finishing is to use reagent systems that are reactive with themselves but only to a limited extent or not at all with the fiber substrate. An example of such approaches are *in situ* polymer systems that form a condensed fiber system within the fiber matrix (1,2). A third type of approach may be the deposition of a polymer system on the fiber substrate. Once deposited, such systems may show a strong affinity to the fiber and may be quite durable to laundering. Polyacrylate and polyurethane are examples of durable deposits on cotton, which last through numerous launderings (3).

Methods of Application.   The predominant system used for finishing cotton and other cellulosics is the so-called pad, dry, and cure process. The padding is done normally by immersing the fabric in an aqueous solution, followed by squeezing it between two rollers, and finally drying and curing. This procedure places several requirements on the finishing components. Cross-linking agents and catalysts need to be water-soluble, and other bath components such as softeners are generally emulsified or dispersed in the bath. The amount of agent in the bath is controlled by the level of cross-linking desired in the fabric and by the percentage of wet pickup achieved in the padding operation. In conventional padding, wet pickup varies with fabric type, fiber content in blends, and pressure applied by pad rolls.

Following padding, the fabric is dried and cured. Removal of up to a kilogram of water per kilogram of fabric finished represents a substantial cost to the finisher. A reduction in the percentage of wet pickup is thus a desirable economic step, and several methods have been devised to accomplish this goal. In one system, the fabric is run against the bottom of a curved blade (4), the solution is applied by its flow down the surface of the blade into the fabric. The flow rate of the finishing agent is electronically controlled and meshed with the speed of the fabric to achieve the desired wet pickup, generally 25–30%. Another method to achieve reduced wet pickup in finishing is based on the use of a foamed finishing agent. As might be expected, this approach requires specialized finishing equipment (5–9). Other approaches utilized include specialized threading techniques in which padded fabric is run against dry incoming fabric to reduce wet pickup (10), and the use of a fabric loop in which the loop is immersed in the pad bath, squeezed to reduce wet pickup, then squeezed against the fabric being treated (11). This latter technique is very handy to small- and pilot-scale finishing for research investigation. Other so-called minimum application (MA) methods have also been developed. In the MA process developed in Europe, the fabric is passed over a lick roller, which supplies a controlled amount of solution to the fabric (12,13). A number of installations utilize this method; wet pickup is low, usually about 30%. Engraved kiss rolls are also being used (14).

A second general approach to achieve reduced wet pickup is based on the use of the vacuum slot technology (15,16). In this case the fabric may be padded conventionally, but is then run against a vacuum slot (17). This vacuum slot removes a certain amount of water solution from the fabric, so that a reduced level of wet pickup is achieved prior to drying. A lower level of wet pickup is achieved on blend fabrics than it is on 100% cellulosics using the vacuum slot technology.

Following fabric padding, the fabric in conventional processing is dried open-width in ranges at high temperatures, at processing speeds up to ~90–

183 m (100–200 yards) per minute. The fabric is then held in open-width in stentors for curing in a flat configuration. Temperatures ranging to 180°C require activated catalysts, which bring about cross-linking in a matter of seconds.

One concern in conventional processing is the achievement of uniform reagent application and uniform cross-linking (18). An area in which adequate treatment of all fibers is necessary is in flame-retardant finishing. One means of obtaining thorough treatment has been the use of vacuum impregnation, in which the fabric is first passed over a vacuum slot to remove air from the fabric interstices, followed by exposure to the phosphorus flame-retardant solution in the precondensate ammonia system (19).

The standard conventional finishing process has been modified to suit the purposes for different fabrics and garments. For example, tubular knits are frequently handled using specialized equipment to control tension and to get adequate padding. Some tubular knits are subjected to wet-on-wet padding, dried and cured in large drum dryers, and steam-treated to achieve a relaxed and nondistorted knit.

The general area of garment finishing has evolved over the years. Achievement of durable press (DP) garments in the 1960s involved padding fabric with resins, drying, and then processing fabric into garments that were subsequently cured. Although this process permitted the achievement of cured-in durable creases, it also brought along certain added problems. First, the uncured fabric generally exhibited a greater amount of formaldehyde release, which was a problem in sewing rooms. The presence of sharp creases and edges in these garments tended to exacerbate the abrasion problem of cellulosics. In a sense, this led to the marketing of blended cotton–polyester fabrics for these garments, particularly if good DP performance was desired. To a certain extent, post-curing of DP garments declined after this period, perhaps because heat setting could achieve a degree of creasing in blended garments. One specific area of garment treatment has endured, however, throughout the 1960s and 1970s. This is the vapor-phase formaldehyde process, in which a gaseous formaldehyde system is utilized to achieve garment application and curing of a cellulosic-containing fabric (20,21). The preferred fabric substrate is again cotton–polyester.

The reemergence of this concept in the 1990s has revived garment finishing with resins. In this case, the garments are usually 100% cellulosic and are normally manufactured in the undyed state, dyed in garment-dyeing machines, and then given an application of resin in a garment machine. Fabrics are then smoothed and subjected to curing in garment form. In another modification, garments such as denim may be manufactured conventionally, then given a treatment using stones, enzymes, or a combination of both to produce a stone-washed appearance. Such garments are then given a resin treatment in a garment-dying machine, then smoothed and processed in a curing oven. The finisher must be extremely careful with these all-cotton garments as the abrasion problem may recur to curtail these developments (22).

To many textile chemists, finishing refers primarily to chemical or resin finishing. The principal chemicals used in early finishing processes were aminoplasts capable of cross-linking cellulose as well as of homopolymerization. These agents have not only the ability to improve smooth-dry performance and dimensional stability of fabrics, but also other useful properties. These include improving durable color appearance in fabrics both by preventing fuzzing of fabric surface and by retention of dyestuff; providing a means of grafting other chemical agents to cellulose; and improving strength retention in outdoor fabrics.

Although in the early days of resin finishing there was continued debate over the mechanism by which these agents achieved smooth-dry performance, by the 1950s most textile chemists believed that cross-linking was primarily responsible for achieving smooth-dry performance.

Fabric Preparation.   Although for a standard woven fabric, desizing, scouring, and bleaching are usually performed to achieve a white fabric suitable for dyeing and finishing, for the purpose of this article, these processes are not considered finishing. Desizing, scouring, and bleaching are necessary so that a uniform fabric substrate is obtained. Nonuniformity and fabric impurities lead to problems that show up soon after the fabric is dyed. Resin treatments are generally acid-catalyzed and residual basicity can interfere with the desired cross-linking reaction.

Several other processes are also available. One fabric treatment is mercerization. This process is done to improve fabric strength, luster, and dye yield for those fabrics that are to be subsequently dyed. This improvement in strength, particularly tearing strength, is important because it leads to fabrics having better tearing strength and elongation after the fabric undergoes a cross-linking treatment (23). The impact of this pretreatment on the abrasion resistance of cross-linked fabrics, however, is not clear.

A second treatment that has gained a certain amount of use in modern textile practice is that of ammonia mercerization (24,25). In this process, the fabric is processed in equipment designed so that the fabric is immersed in liquid ammonia. Ammonia ranges are extremely expensive to build and operate. Fabrics so treated must be priced accordingly in the open market and represent sufficient volume to keep the ammonia line in operation. The most widespread application of this equipment was used to process denim fabrics. Fabrics treated with liquid ammonia are generally softer and exhibit better smooth-dry performance than fabrics without such treatment. In addition, ammonia treatment combined with a low wet pickup cross-linking treatment can lead to fabrics having better abrasion resistance than those fabrics given a straightforward cross-linking treatment (11,26).

Some research has been done to perform chemical treatments in the slashing or sizing operation, so that a permanent size could remain on the fabric subsequent to preparation. This permanent size would have utility in improving fabric resiliency and abrasion resistance (27,28). Another approach was based on the use of a reactive cross-linking system in the slashing operation (28). This would confer a permanent cationic character to the warp threads, thus permitting fabric dyeing to achieve a mock-denim in a range of colors.

Cross-Linking of Cellulosics Fabrics.   Essentially, any compound containing two reactive groups can be used to cross-link cotton. Exceptions are those that are too large to penetrate the fiber and perhaps those in which the reactive groups are widely separated. In the cross-linking reaction of equation 1, the fabric takes on a memory for its state at the moment of cross-linking:

$$2\,\text{cell} - \text{OH} + \text{HO} - \text{X} - \text{OH} \rightleftharpoons \text{cell} - \text{O} - \text{X} - \text{O} - \text{cell} + 2\,\text{H}_2\text{O} \qquad (1)$$

Thus, if the fabric is flat and smooth, it will tumble dry in that configuration. On the other hand, if the fabric is cross-linked in a creased condition, as in a pleated skirt, the original pleated skirt configuration should return on laundering and tumble drying.

Fabric can be cross-linked either in the dry or the wet state. If fabric is cross-linked in the dry state, smoothness returns on tumble drying. By contrast, if fabric is cross-linked in the wet state, smoothness is achieved by line-drying the fabric. This concept has been demonstrated using formaldehyde in pad–dry–cure or wet cure processes (eq. 2) (29).

$$2\,\text{cell}\!-\!\text{OH} + \text{CH}_2\text{O}\cdot\text{H}_2\text{O} \underset{}{\overset{\text{H}^+}{\rightleftharpoons}} \text{cell}\!-\!\text{OCH}_2\text{O}\!-\!\text{cell} + 2\,\text{H}_2\text{O} \qquad (2)$$

The wrinkle recovery angle provides a measure of the degree of chemical modification. This is calculated by blending a small sample and measuring the recovery to the flat configuration (180°). Whereas the untreated cotton recovers approximately 90°, the cross-linked cotton sample recovers 120–140°. If this is measured on dry fabric, it is termed conditional wrinkle recovery angle; if on wet fabric, it is termed wet wrinkle recovery. At one point, wet wrinkle recovery was important, particularly in Europe. In the United States, the widespread use of clothes dryers has made conditional wrinkle recovery important.

The effect of cross-linking a fabric or garment can last throughout the life of the garment, provided that the cross-link is retained. However, cross-linking reactions are reversible, so removing the cross-links through hydrolysis restores most of the original properties to fabric.

Whereas cross-linking imparts the desired resiliency and shrinkage control, it does have several undesirable effects. The most objectionable side effects are reductions in tearing strength, breaking strength, and abrasion resistance. The moisture regain of cross-linked cotton is less than that of untreated fabric. Moisture regain can be retained by the use of swelling agents or occasionally by retention of cross-links (30,31).

Another effect of cross-linking is reduction in accessibility of the fiber. Because of this, dyeing is normally performed before the fabric is processed. In a scheme of operation where fabric is dyed and cross-linked in the textile mill before being fabricated into garments, this approach is satisfactory. Since the rise of garment dyeing in the 1990s, which emphasizes just-in-time responses for color and style, an extensive finishing research effort to produce a dyeable cross-linked fabric was begun at the Southern Regional Research Center (New Orleans, Louisiana). Although these efforts have led to several chemical alternatives, the industry has opted for post-garment finishing using resins as the most reasonable alternative to this problem.

There is no question that the bane of textile chemists in the area of cross-linking for smooth-dry performance is the loss of abrasion resistance. This has been a continuing problem when durable press is pushed to high levels of performance. Numerous approaches to this problem have been explored (32). However, the simplest solution has been to blend cotton with synthetic fibers. A 50–50 cotton–polyester fabric can have excellent smooth-dry performance and yet be able to endure numerous launderings.

Early Cross-Linking Agents. Formaldehyde, urea–formaldehyde, and melamine–formaldehyde were among the earliest agents utilized for resin finishes. Concerns about the safety of formaldehyde, the need for lower formaldehyde release values, and the safety of exposure to melamine have reduced the use of these early cross-linking agents by industry substantially.

Formaldehyde, which produces a finish that resists hydrolysis and is inert, durable, and unaffected by heat or bleach, may be the cheapest and most economical cross-linking agent for cotton fabric. However, despite much research and promotion (33–36), formaldehyde has had only limited use in practical heat-cure and vapor-phase applications. One has only to stand by an oven or range using formaldehyde as the agent to understand the odor problem attached to the use of this agent. Vapor-phase treatments in garments are still in use in the 1990s, but containment and removal of unreacted formaldehyde is a must (20,21). One other problem concerning formaldehyde is that the lack of urea buffer tends to lead to greater damage to cotton in terms of strength retention.

Urea–formaldehyde and melamine–formaldehyde reagents are resin formers, which not only cross-link cotton but also copolymerize with themselves. These have been used both as simple cross-linkers or prepolymer systems. If too much of the polymerization is concentrated on the fiber surface, the fabric may be sufficiently stiffer that it takes on a boardy character. As such, the finisher must control the action of agent to give the desired crisp hand but prevent the development of boardiness. Melamines have been recommended for applications when complete shrink resistance is required. However, both finishes were rejected for the white-shirt market because of loss of strength when hypochlorite bleach is used on account of vulnerable NH groups and the ensuing discoloration (37).

Chemistry of *N*-Methylol Agents. The reaction of dimethylolurea and cellulose is illustrated in equation 3:

$$\underset{\displaystyle \text{HOCH}_2\text{NHCNHCH}_2\text{OH}}{\overset{\displaystyle \text{O}}{\overset{\|}{}}} + 2\ \text{cell}-\text{OH} \rightleftharpoons \text{cell}-\underset{\displaystyle \text{OCH}_2\text{NHCNHCH}_2\text{O}}{\overset{\displaystyle \text{O}}{\overset{\|}{}}}-\text{cell} + 2\ \text{H}_2\text{O} \qquad (3)$$

First, it should be noted that the *N*-methylol group is activated by the carbonyl group. This reactive group is present in almost all *N*-methylol systems. Second, the reaction is an equilibrium reaction so that both forward and reverse reactions can occur. Third, the agent is not simply a dimethylol agent, but is predominantly a mixture of mono- and di-substituted ureas.

Because of this, there are free NH groups bound to the fabric. Some of these NH groups are converted to N–Cl on bleaching. Subsequent exposure to ironing or hot drying can generate hydrogen chloride that leads to strength loss and discoloration. The chemical mechanisms involved in chlorine decomposition and retention have been studied in detail (37,38). The chloramide derived from the urea–formaldehyde finish forms and decomposes readily, leading to a heavily damaged fabric (39).

In order to overcome this problem, complete methylation of the NH groups is desired. Another approach is that certain molecular structures may yield cross-linkages where the presence of free NH groups is not a problem. It has been found that primary or unsubstituted amides that approach complete methylation are formamide, carbonates, and melamines. By contrast, the addition to secondary or mono-substituted amides as shown in equation 3 does not proceed to completion, so these type structures do not make good cross-linking agents. However, if these secondary amides are part of a ring system, such as dimethylodihydroxyethyleneurea (DMDHEU) (**1**), a good agent can be obtained. This agent is produced by reaction of urea, glyoxal, and formaldehyde (40).

(**1**)

Formation of a fully methylated cross-linking agent does not ensure that the finish is immune from damage to bleaching. The curing stage does not proceed to completion when even the best of finishes is employed. Furthermore, garments are subjected to alkaline washes in normal laundering, and acid sours in commercial laundering. A satisfactory finish thus needs to withstand both types of finishes.

Mechanisms for Formation and Hydrolysis of Finishes. The general mechanism for acid-catalyzed formation and hydrolysis of N-methylol cellulose cross-links has been shown to pass through a carbonium ion intermediate as in equations 4 and 5 (41):

$$\underset{\text{O}}{\overset{\text{O}}{\parallel}}\text{—CNHCH}_2\text{OH} + \text{H}^+ \rightleftharpoons \text{—CNHCH}_2\overset{\text{H}^+}{\overset{|}{\text{O}}}\text{H} \rightleftharpoons \text{—CNHCH}_2^+ + \text{H}_2\text{O} \tag{4}$$

$$\text{—CNHCH}_2^+ + \text{cell—OH} \rightleftharpoons \text{—CNHCH}_2\text{O—cell} + \text{H}^+ \tag{5}$$

Based on the principle of microscopic reversibility, it has been reasoned that a highly reactive agent would form an easily hydrolyzed product (42). This proved true in practice and the industry has adopted the phrase "easy on means easy off" (43).

Hydrolysis resistance of a finish is important because fabrics are generally not afterwashed. There may thus be a considerable time lapse before fabrics receive the initial wash to remove the acid catalyst needed for cross-linking. Furthermore, cotton and rayon fabrics regain moisture on standing (44,45). This combination of acid and moisture then has considerable time to act on the finish. Thus, the resistance of the finish to hydrolysis is of primary importance. On the other hand, finishes may need to be stripped or removed from the fabric to correct any errors in processing. Stripping, when necessary, must be performed in such a way so that excessive strength loss does not occur.

One such formulation for stripping utilizes treatment with 1.5% phosphoric acid and 5% urea for 30 minutes in an aqueous solution. Experiments performed using this agent have demonstrated that most N-methylol agents react with cotton but exhibit little or no modification of their structures during curing (46).

Alkaline conditions found in home laundering are generally not strong enough to cause significant hydrolysis of conventional finishes. However, the tendency of these finishes to release considerable amounts of formaldehyde after a strong alkaline home wash indicates that some breakdown of nonreactive pendent groups, if not the finish itself, may be occurring (47,48).

Curing Catalysts for N-Methylol Agents.   Many acid-type catalysts have been used in finishing formulations to produce a durable press finish. Catalyst selection must take into consideration not only achievement of the desired chemical reaction, but also such secondary effects as influence on dyes, effluent standards, formaldehyde release, discoloration of fabric, chlorine retention, and formation of odors. In much of the industry, the chemical supplier specifies a catalyst for the agent so the exact content of the catalyst may not be known by the finisher.

Types of catalysts used include mineral or organic acids and latent acids such as ammonurea salts, amine salts, and metal salts (49). One type of catalyst used in early commercial processing was amine hydrochloride. However, the fishy odor evolving from this material has caused this type of catalyst to be discarded. Metal salt catalysts such as magnesium chloride or zinc nitrate have been widely used over the years. However, because of effluent concerns, zinc salts have fallen into disuse. Magnesium nitrate has been suggested as an alternative catalyst (50).

Magnesium chloride is a very effective catalyst but has been deemed too mild for modern textile mill speeds. Catalysts that have come to the fore are mixed catalysts (51), which are usually a combination of a metal salt and an organic acid such as glycolic or citric acid. The predominant combination is probably magnesium chloride and citric acid. These catalysts are particularly suitable for the high speeds, high temperatures, and short (15 s) curing cycles utilized by many textile mills. Although organic acids, such as glycolic or citric acids, can lower pad bath pH below that observed when using metal salt catalysts, such acids are mild and tend to do less damage in terms of strength and loss of abrasion resistance (52). The exact nature of the complex by which metal salts act as catalysts has, however, eluded researchers (53,54).

Melamine and Other Amino-S-Triazines.   In the case of melamines, hypochlorite does not cause fabric degradation but does lead to yellow fabric (55). Commercial melamine derivatives used in finishing since the 1940s have generally been mixtures. A partially methylated di- or trimethylol melamine is an excellent polymer-former as well as an effective cross-linking agent. However, this polymer-forming characteristic can lead to poor shelf life and complications in finishing because the degree of self-polymerization before fabric treatment is a hard-to-control variable. Some highly methylated melamines have been shown to be effective cross-linking systems, showing only minor side deficiencies as a result of yellowness after bleaching (56).

Melamines have found utility as rotproofing and weatherproofing cellulosics (57,58), as binders for pigments and for transfer printing of cotton and cotton blends (59), as well as in numerous other applications. Guanamines have been suggested for many of the same applications as melamines, but the similar chemical structures of the two are likely to add to the same problems encountered with melamines.

Triazones, Urons, and Alkylene Ureas. Triazones, urons, and alkylene ureas were the first commercial N-methylol cross-linking agents for reducing or eliminating the hypochlorite bleach problem (60). These agents are still commercially available. The products are similar in structure and fully methylated products having a slight excess of formaldehyde are obtained. However, the release of formaldehyde by these agents has contributed to their decrease in popularity. The systemic names for these bis(hydroxy methyl) agents and their common names are tetrahydro-1,3,5-triazine-2(1H)-one (triazone), 2-imidazolidinone (ethyleneurea) (**2**), and tetrahydro 2-(1H)-pyrimidinone (propyleneurea) (**3**).

(**2**)   (**3**)

There was a tendency to use these resins mixed with urea–formaldehyde or melamine-type resins. Preparation of pure triazones or uron resins is difficult and expensive (61,62). Furthermore, the basic nature of the amine nitrogen in triazone permits the use of mixtures of triazones with other agents to yield finishes that retain strength in hypochlorite bleaching.

The evolution of triazones, from initial synthesis in 1874 to development as a generally used textile finishing agent in 1957, has been described (60). Five triazone formulations have shown improved resistance to acid hydrolysis and to damage from hypochlorite bleach. By using a special synthetic approach, 5-alkyl triazones containing C_8, C_{12}, and C_{18}-alkyl groups were prepared (63). Only the C_8-alkyl triazones produced a significant improvement in wrinkle recovery on reaction with cotton.

The synthesis of triazones by the methods of Burke (61) or Paquin (62) appears to be fairly general. This is a type of Mannich reaction in which the product may be formed either by the addition of dimethylol urea with a primary amine, or from the addition of an amine–formaldehyde condensate with urea. Triazones prepared from ethylamine have received the most commercial interest (64). Again the mixture of triazone and urea–formaldehyde is said to be cheaper than a pure triazone finish. However, for these types of finishes, odor problems arising from by-product formaldehyde or amines were a matter of concern. Although this problem could be overcome by an afterwash, such a solution is not feasible in modern textile mill practice (64). Triazones are used less often in the 1990s because of odor problems, high levels of formaldehyde release, and lower levels of resistance to acid hydrolysis.

The chemistry of the urons and their use in textile finishing have been studied (65,66). Interest in this system stems from the low cost of the starting materials: urea, formaldehyde, and methanol. Again, fabrics finished with the crude urons are vulnerable to strength losses during hypochlorite bleaching, but similarly finished fabrics finished with pure urons do not exhibit this problem (67). Because of the low cost of 1,3-bis(methoxymethyl) uron, this agent continues to be of some commercial interest, although no longer widely used.

Of the alkylene ureas, 1,3-dimethylolethyleneurea (DMEU) has had widespread usage in the United States as a replacement for earlier resins such as urea–formaldehyde (68), whereas 1,3-dimethylolpropyleneurea (DMPU) has been used more widely in Europe. These finishes yield fabrics that have excellent smooth-dry performance under mild curing conditions. However, they are extremely vulnerable to acid hydrolysis. In commercial laundry practice in the United States, an acid sour is an integral step in processing that leads to hydrolysis of much of the cross-links. For this reason as well as for formaldehyde release considerations, the industry has moved to cross-linking agents that produce finishes having a greater resistance to hydrolysis.

The ease of hydrolysis of a DMEU-treated fabric has been used to produce bicolored cotton fabrics. This was accomplished by applying a thickened DMEU solution in a print configuration to the pile of fabric, curing the resin, and dyeing the fabric. The DMEU-treated areas resisted dyeing because of the cross-links. Subsequently, the DMEU-crosslinks were removed via an acid hydrolysis and the entire fabric was overdyed to achieve the desired bicolored effect (69).

A general summary on the preparation, physical and chemical properties, and information relative to uses for ethylene urea is available (68). The widespread interest in ethyleneurea is reflected by the number of organizations holding patents in this area.

DMEU represented the first cross-linking agent that was a pure chemical rather than a mixture of components. As such, it provided research workers a tool to investigate the changes that take place in cellulose with cross-linking in a more exact manner (70–73).

Although cyclic ureas solved the problems associated with hypochlorite bleaching, a demand was arisen in the 1960s for garments that have higher levels of smooth-dry performance. This was termed durable press or permanent press. Such high level of performance brought on a higher level of chemical treatment for the individual garment or fabric. As a result of these fabrics, the formaldehyde release in the garment manufacturing plant and the necessity for finishes having a greater resistance to hydrolysis have led to new series of cross-linking agents.

Delayed Cure and Permanent Press.   The 1960s witnessed an explosive growth in finishing technology resulting from the debut of delayed cure systems. These systems were designed so that the chemical agent was applied in the textile mill but final curing was delayed until after the garment was fabricated and pressed into the final desired configuration. This approach has led to garments having a new, higher level of performance. Thus garments not only can be smooth-drying, but also have desired shapes fixed into them, eg, the permanent crease of trousers.

The initial application of this technique was performed on all-cotton fabric, but it soon became apparent that 100% cotton garments did not have a level of abrasion resistance to perform satisfactorily. As a reaction to this, cotton and commercial interests launched extensive research to find new approaches for enhancing the abrasion resistance of all-cotton fabrics. Examples of such approaches involved polymer deposition, surface polymer application, two-step cross-linking and fixing systems, wet cure, steam cures, grafting reactions, as well as various other approaches for controlling cross-link distribution (32). However, none of these systems has caught on to a substantial and enduring degree.

Two factors emerged to turn the focus of durable press: the discovery that incorporation of a level of nylon or polyester in the fabric can substantially increase the garments' abrasion resistance, and the realization that the marketplace preferred cotton–polyester blends in delayed cure operations, even though 85% cotton–15% nylon fabric yields a suitable product. The 50% cotton–50% polyester fabric seemed particularly appropriate because it contained sufficient cellulosic to benefit from a chemical finish and sufficient synthetic to provide strength and abrasion resistance.

Two types of approaches are available. In one, the fabric is padded with the cross-linker finish, dried, then sent to the garment cutter. The garments are then pressed and cured. In the second, the fabric is cured in fabric form, then fabricated into garments. It is then pressed and recured in hot-head presses. This double curing is particularly hard on the cellulosic fiber in terms of strength and abrasion resistance.

This change of putting a curing step after garment fabrication has put a new set of standards in place for the finishing industry. First, a change of resin type was dictated. An agent that released a minimum amount of formaldehyde in cutting and sewing rooms was required. Second, the agent needed to be unreactive in the time and conditions to which fabric is subjected before the final curing. Third, the matter of chlorine bleach resistance faded in importance because a truly permanent-press fabric required little or no ironing. This situation also required stronger interaction between various segments of the industry, such as chemical suppliers, finishers, dryers, garment makers, and garment component suppliers (74,75). By 1965, at least six permanent or durable processes were recognized and the patent situation on processing was confused. The principal chemical change brought on by this foray into durable press was the emergence of the urea–glyoxal formaldehyde adduct, 1,3-dimethyl-4,5-dihydroxyethyleneurea (DMeDHEU) (40), as the dominant cross-linking agent in the United States.

Although delayed cure cotton was the primary impetus in the rise to durable-press performance, the emergence of DP blends had the effect of reducing the importance of a true delayed cure. The industry tended to revert back to precure fabrics and the utilization of hot-head presses to set in creases, using the thermoplastic characteristics of the synthetic components as well as a touch of recure from the hot-head presses.

Sources of Formaldehydes in Textiles and Formaldehyde Analysis. Modern textile mill finishes need to be concerned with formaldehyde in the finishing garment plants and in the textile product itself. Formaldehyde can be present in the air, in cross-linking agent pad baths, and in the finished fabrics. The maximum concentration of formaldehyde in the air of workplaces should be dictated; in work areas, concentrations in air above 1 ppm are unusual. The free formaldehyde in finishing formulations has been substantially reduced and mill conditions in finishing areas and garment plants have been greatly improved.

Free formaldehyde is a mixture of formaldehyde, formaldehyde hydrates, and low molecular oligomers. It imparts a characteristic odor to padding bath or padded fabrics (76,77). Cellulosics fabrics are capable of retaining large quantities of free formaldehyde, which are gradually evolved. Because all finishes degrade to some extent, extractable formaldehyde and releasable formaldehyde must be considered with respect to user exposure.

Extraction tests are used primarily in Japan and Europe, a release test is used in the United States, and standard tests have been compared based on the sources of formaldehyde present in a finished fabric (76,78–80). Finished fabric may contain free formaldehyde, or formaldehyde released from unreacted N-methylol moieties.

The analytical methods (81–83) for the determination of free formaldehyde in the presence of N-methylol compounds are based on a low temperature (0–5°C) titration, which involves the reaction of sodium sulfite and formaldehyde (eq. 6).

$$Na_2SO_3 + HCHO + H_2O \longrightarrow NaOH + CH_2(NaSO_3)OH \qquad (6)$$

The analyses can be carried out in the presence of N-methylol groups. On fabric, the formaldehyde bisulfite compound is decomposed by excess sodium carbonate and the liberated sulfite is titrated with 0.1- or 0.01-N iodine solution (76). Commercial fabrics are seldom washed and dried before being used, and the free formaldehyde content may be between 50 and several hundred ppm, depending on finishing and storage conditions.

In the determination of free formaldehyde in solution, eg, commercial reagents and pad bath formulation, the conditions of analysis allow hydrolysis of the N-methylol groups, usually between <1% and several percent. The NaOH formed is titrated with hydrochloric acid (82). Because of an incomplete reaction of sulfite with free formaldehyde, these low temperature methods (83) detect only 80–90% of the free formaldehyde present. Skill is important for correct results.

Extraction methods at higher temperatures are used in some countries. These values are referred to as free formaldehyde contents even though the origin of part of the formaldehyde reported is the nitrogenous finish or unreacted cross-linking reagent on the treated fabric. These test methods have originated with the industry. In Japan, however, test methods were regulated by law in 1975. Guidelines were published in 1972. An extraction of treated fabrics at 25°C was involved. The mandatory standards involved an extraction at 40°C. An industrial method used in Europe (84) requires extraction at 22°C, followed by determination of total formaldehyde content of the extract. Use of this method identifies finished fabrics that have a high content of unreacted cross-linking agents.

Researchers had noted the release of formaldehyde by chemically treated fabric under prolonged hot, humid conditions (85,86). The American Association of Textile Chemists and Colorists (AATCC) Test Method 112 (87), or the sealed-jar test, developed in the United States and used extensively for 25 years, measures the formaldehyde release as a vapor from fabric stored over water in a sealed jar for 20 hours at 49°C. The method can also be carried out for 4 hours at 65°C. Results from this test have been used to eliminate less stable finishes.

Control of Formaldehyde Release. Once the sealed-jar test became a factor in measuring the formaldehyde release of fabrics supplied to garment cutters, limitations were placed on the allowable limits acceptable to the garment producers. These limits brought to the fore two classes of reagents: those based on DMDHEU, and those based on the N,N-dimethylolcarbamates (**4**) (88).

$$RO-\overset{\overset{\displaystyle O}{\|}}{C}-N\overset{\nearrow CH_2OH}{\searrow_{CH_2OH}}$$

(**4**)

Prior to 1965, it was not unusual for unwashed finished fabrics to release 3–5000 ppm of formaldehyde when tested by an AATCC test method. Formaldehyde release was reduced to the level of 2000 or less by application of DMDHEU or dimethylolcarbamates. This level was reduced to approximately 1000 in the mid-1970s. Modification of the DMDHEU system and use of additives demonstrated that release values below 100 ppm were achievable. As of this writing (1997), good commercial finishing ranges between 100 and 200 ppm of formaldehyde release.

Several factors were utilized in bringing formaldehyde release down. In particular, resin manufacturer executed more careful control of variables such as pH, formaldehyde content, and control of methylolation. There has also been a progressive decrease in the resin content of pad baths. The common practice of applying the same level of resin to a 50% cotton–50% polyester fabric as to a 100% cotton fabric was demonstrated to be unnecessary and counter productive (89). Smooth-dry performance can be enhanced by using additives such as polyacrylates, polyurethanes, or silicones without affecting formaldehyde release.

One technique that has been employed to lower formaldehyde release has been the alkylation of the N-methylol agent (90–93) with an alcohol (eq. 7):

$$(\mathbf{1}) + ROH \longrightarrow ROH_2C-\overset{HO\quad OH}{\underset{\underset{O}{\|}}{N\diagup\diagdown N}}-CH_2OH \qquad (7)$$

Alkylation involves reaction with an alcohol or polylol before formulation of a pad bath or by adding certain alcohols or polyols to the pad bath (94–99). In the latter case, the methylol agent is alkylated during the drying and curing steps. DMDHEU is used almost exclusively for this purpose. The degree of alkylation is only partial, in the order of 15–30%. However, even this low degree of alkylation suffices to lead to fabrics having formaldehyde release values of 100–400 ppm. Another chemical approved to lower formaldehyde release values involves the use of scavengers to tie up free formaldehyde or N-methylol groups (100). These low formaldehyde release values, however, did cause certain problems and some adjustments in test procedures were made for analyzing fabrics that have low formaldehyde release values (99,101,102).

Carbamates. Carbamates as finishing agents were an alternative to DMDHEU for a number of years and have gained wide commercial usage (88). Examples of carbamates used for commercial purposes were methyl, ethyl, isopropyl, isobutyl, hydroxyethyl, and methoxyethyl carbamates. Although the carbamates are vulnerable to forming *N*-chloro agents on bleaching, the *N*-chlorocarbamate is so stable that it does not lead to fabric damage. For many years carbamates gained substantial use because of price considerations and because they give an excellent white for fabrics such as linens or white shirting. However, their usage has declined substantially because it is much easier to control formaldehyde levels by using the DMDHEU system.

Other methylol agents have found use overseas besides the dimethylol-propyleneurea. There are other propylene ureas, including the 5-hydroxy- and 4-methoxy-5,5-dimethyl derivatives; the synthesis and finishing characteristics of these agents have been described (103,104). Research has also developed agents containing a tertiary amino group to provide a cross-linker having an internal buffering moiety (105).

Nonformaldehyde Finishing. The concern for formaldehyde release prompted interest in the development of cross-linking systems that did not contain formaldehyde. A number of systems were investigated but generally these systems seemed to fall short in performance (106,107). For example, 1,3-dimethyl-4,5-dihydroxyethyleneurea (DMeDHEU) (**5**) has been used in Japan since 1974. This same agent has been marketed in the United States and elsewhere, but generally the level of smooth-dry performance is substantially lower than the level achievable with DMDHEU. The cost of dimethylurea also raises the overall cost of DMeDHEU above that of DMDHEU.

(**5**)   (**6**)

Dihydroxyethyleneurea (DHEU) (**6**) is also a cross-linker, but is extremely vulnerable to hydrolysis in an alkaline home laundering. Another approach toward formaldehyde-free agents has been the use of glyoxal in the bridging group between cyclic DHEU and related ring systems (108).

Carboxylic Acids and Cross–Linkers. Perhaps the most intensely recorded research involving nonformaldehyde cross-linkers in more recent years is the work based on the use of polycarboxylic acids as cross-linking agents (109–115). Thus, in late 1988, it was shown that cotton fabric could be cross-linked with an agent such as 1,2,3,4-butanedicarboxylic (BTCA) to yield fabric quite comparable to DMDHEU in terms of smooth-dry performance. There has always been some question of the durability of such ester linkages to strong acid or basic conditions, but this type of cross-linked fabric has been found durable to repeated home launderings. The reaction has been postulated to go through an anhydride intermediate to form a linkage with cellulose. Anhydride formation and reaction with cellulose to form the second bond to cellulose complete the cross-link (eq. 8).

$$
\begin{array}{ccc}
\overset{\overset{O}{|}}{\underset{|}{\text{—C—C—OH}}} & & \\
\overset{\overset{O}{\parallel}}{\underset{|}{\text{—C—C—OH}}} & \xrightarrow[\text{catalyst}]{\text{heat}} & \\
\overset{\overset{O}{\parallel}}{\underset{|}{\text{—C—C—OH}}} & &
\end{array}
\qquad
\begin{array}{c}
\text{—C—C}\overset{O}{\diagdown}_{O} \\
\text{C—C}\diagup^{O} \\
\text{—C—COOH}
\end{array}
\xrightarrow{\text{cell—OH}}
\begin{array}{c}
\overset{\overset{O}{\parallel}}{\text{—C—C—O—cell}} \\
\overset{\overset{O}{\parallel}}{\text{—C—C—OH}} \\
\underset{\underset{O}{\parallel}}{\text{—C—C—OH}}
\end{array}
\tag{8}
$$

Several factors emerged to slow commercial acceptance of this finish. First, regarding DMDHEU, the polycarboxylic acid finish utilized much more expensive chemicals in both the agent and catalyst. Second, curing conditions tend to be more demanding in either time or temperature. Because BTCA is a commercially available material, an effort was made to make the price of this material more acceptable to the industry. Failing this, considerable effort has been expended for an acceptable product using a cheaper chemical such as citric acid. Although this material does lead to an evidently acceptable DP finish, yellowing during curing and a lower level of durability to laundering also result. However, the yellowing can be suppressed by incorporation of basic reactive additives in the finish (116). More recent research has focused on the use of simpler carboxylic acids such as maleic or malic acid. Another approach to the agent problem has been the use of *in situ* polymerization with materials such as maleic acid or combinations of related materials (117,118).

Hand in hand with this research on finding a suitable carboxylic acid chemical for cross-linker has been the search for an economical catalyst system. The catalyst found to be most effective for the esterification reaction was sodium hypophosphite (NaH_2PO_2). This material was also costly and out of range for the textile industry. Because weak bases function as catalyst, a range of bases has been explored, including the sodium salts of acids such as malic acid.

As of this writing (1997), researchers are exploring combinations of acids, additives, and catalysts to achieve a suitable economic finish. However, commercial application of these finishes would require costs akin to that of DMDHEU as well as compliance with formaldehyde release levels by consumers, regulators, and the textile industry. Another possible impetus could be marketing considerations. Nevertheless, this work has sparked intense effort in the use of cross-linkers containing ester cross-links and has broadened the scope of cross-linker research.

Special Finishing

Stone and Enzyme Treatments. In the 1980s, stone washing of heavyweight denim fabrics burst on the scene. The initial treatments consisted of various methods of tumbling garments using high ratios of pumic stones in garment-dyeing equipment. This treatment using stones was designed as a method of producing denim trousers having special effects in terms of softness, a worn look, and/or a modification of color appearance. This technique was hard on both fabric and equipment, but the premium price advantages of such garments in the marketplace seemed to justify the extra care and expense necessary for production (119–122). Care must be taken in processing as holes in garments or torn pockets, etc, may result. Although there was a period when such garment holes were fashionable, this trend seems to have abated.

The difficulties of maintaining a textile plant in which rocks were strewn through the plumbing system have led to the use of cellulose enzymes (123–130). These enzymes, which hydrolyze the 1,4-glycoside bond of the cellulose molecules, can also lead to a soft fabric having a different surface effect, eg, a denim fabric. The other effect produced by such enzymes has been termed biopolishing, which arises because the enzyme tends to hydrolyze cellulose in protruding fibers, thus giving the fabric a smooth or glossy appearance. This reduction of fuzzy appearance in certain fabrics or garments is a commercially valuable property.

With respect to the action of the enzyme itself, a loss of weight on account of cellulose hydrolysis, as well as loss in strength properties, occurs. Therefore, control of concentrations, temperature, and other processing conditions is important to achieve a product having the proper balance of properties.

In commercial practice, treatments may use stones only, enzymes only, or a combination of both. The desired end product determines the method employed. In certain stones-only treatments, stones may be soaked in bleach or oxidizing agents such as permanganate. Some techniques for differential dyeing use stones soaked in dyes, or soaked in finishes designed to lead to random dyed cotton garments (131,132). The technique of combining stones and enzymes offers the advantage of improving productivity from a garment machine. Much, but not all, of this type of finishing is applied to garments. A further step has been utilized in which enzyme-softened garments have been given a smooth-dry finish using the cross-linking agent added as part of a garment's treatment. The garment is subsequently cured in a garment-curing oven.

Modifying Dyeing. *Characteristics of Cotton.* The long-term approach in producing colored, smooth-dried garments has been to dye the fabrics first in a textile mill, then to apply the cross-linking finish, and, finally, to prepare the garments. In certain cases, the fabrics were only sensitized using cross-linker (dried only) before being converted to garments which were subsequently cured. This was termed delayed cure.

However, this procedure required color selection much before shipping of the garments to the retailer. It was considered better for just-in-time merchandising if the garment could be dyed just before product shipment. Therefore, an intense effort was made to produce a dyeable smooth-dye fabric. Although several functional moieties were grafted to cellulose and evaluated (133–141), the moiety that seemed to hold the most promise was the quaternary group. The addition of this group to cotton led to a cationic cotton, which was dyeable using anionic dyes. The reactive module first used for this purpose in conjunction with a cross-linking system was choline chloride (**7**) (133–140).

$$\text{(HOCH}_2\text{CH}_2\overset{\displaystyle \text{CH}_3}{\underset{\displaystyle \text{CH}_3}{\text{N}}}\text{—CH}_3)^+\text{Cl}^-$$

(**7**)

Reactive additive used for this purpose must contain a functional moiety, in this case alcohol group, by which the additives can be attached to cotton via the cross-linker. Other reactive additives have also been similarly employed. Although this technique has not gained commercial acceptance, use of cationic systems may yet find a place because of the high affinity of cationic cotton for anionic dyes. Because of concerns for uniformity, however, the industry has opted to employ garment dyeing as the first step, followed by garment application of resin and oven curing.

Outdoor Fabrics. The use of cotton in outdoor fabrics has a long history in terms of tents, tarps, and other applications. However, there has been an increasing use of synthetics in the 1990s, particularly in tents and awnings. Cotton in outdoor usage needs protection for mildew and algae, protection against sunlight, and a degree of waterproofing and water oil repellency (qv). Another property that is sometimes required of outdoor fabric is flame retardancy.

Early waterproofing treatments consisted of coatings of a continuous layer impenetrable by water. Later water-repellent fabrics permitted air and moisture passage to improve the comfort of the wearer. Aluminum and zirconium salts of fatty acids, silicone polymers, and perfluoro compounds are applied to synthetic as well as natural fibers. An increase in the contact angle of water on the surface of the fiber results in an increase in water repellency. Hydrophobic fibers exhibit higher contact angles than cellulosics but may still require a finish (142).

The microorganisms that grow on cotton have undesirable effects, particularly in terms of loss of strength and of aesthetic effects from staining of the fabric from mildew. Microorganisms are a more serious problem in hot, moist climates than in dry, arid areas. The familiar greenish color of cotton tents results from the so-called pearl grey finish of primarily chromic oxide. This finish provides a degree of protection against mildew and actinic degradation. A variety of phenols, metal salts, and complexes such as copper 8-quinolinolate have been used and demonstrated to be effective in resisting mildew and algae attack (143,144). For many years, mercury salts found wide usage for this purpose. Because of environmental concerns, however, such usage has been dramatically reduced (see FUNGICIDES, AGRICULTURAL).

For resistance to actinic degradation, the use of certain forms of titanium oxide is an alternative to chrome salts. Another approach has been the use of polymerized methylol melamine on cotton (145). In this case, the action of sunlight leads to gradual breakdown of the melamine polymer after several years. After this, actinic degradation of cotton proceeds as it does in unprotected cotton.

Flame Retardants. The amount of research expended to develop flame-retardant (FR) finishes for cotton and other fabrics has been extremely large in comparison to the total amount of fabrics finished to be flame retardant. The extent of this work can be seen in various reviews (146–148). In the early 1960s, a substantial market for FR children's sleepwear appeared to be developing, and substantial production of fabric occurred. In the case of cotton, the finish was based on tetrakis(hydroxymethyl)phosphonium chloride (THPC) or the corresponding sulfate (THPS). This chemical was partly neutralized to THPOH, padded on fabric, dried under controlled conditions, and ammoniated. The finish was subsequently oxidized, yielding a product that passed the test for FR performance. This process is widely preferred to the THPOH–NH$_3$ process.

A number of flame-retardant finishes have been developed for outdoor cotton fabrics. Various experimental and commercial finishes have been compared (149). Most noteworthy is that THPOH–NH$_3$ finishes do not perform as well outdoors as the THPOH–NH$_3$ precondensate finishes. Likewise, antimony oxide–halogen finishes perform exceptionally well on outdoor fabrics.

In the case of polyester, acetate, or triacetate fabrics, a tris(2,3-dibromopropyl)phosphate, generally referred to as Tris, was padded onto the fabric and forced into the fiber using a thermal treatment. However, in 1977, the Consumer Product Safety Commission labeled Tris as carcinogenic and banned its use on children's sleepwear. Although Tris was not used on cotton, the ban seemed to affect all sleepwear finishing, so the total volume of treated goods was substantially reduced. The other effect of this ban was the modification of the children's sleepwear standard so that thermoplastic fibers that melt at relatively low temperatures can pass the test without any chemical treatment. Because FR finishing is expensive, the chemical finishing of cotton is not economic for most of the children's sleepwear market.

The melt drip action of synthetics in certain industrial and combat situations has expanded the market considerably. Although there are many versions of THP-based products, the THPOH–NH$_3$ type predominated in the early 1960s. Subsequently, the precondensate ammonia finish has become dominant. In this case, the THP salt and urea are prereacted together to form the precondensate. Processing is analogous to the THPOH–NH$_3$ finish (150,151). The precondensate has the advantage of lowering the emission of phosphorus and formaldehyde odors during the drying step and the incorporation of a higher nitrogen content in the FR polymer. The latter substitutes a low cost nitrogen group for a part of the higher cost polymer component.

Another fire-related problem that has seen some research effort is that of smolder resistance of upholstery and bedding fabrics. Finishing techniques have been developed to make cotton smolder-resistant (152–156), but the use of synthetic barrier fabrics appears to provide a degree of protection. Work also has provided a means of producing cotton fabrics that have both smooth-dry and flame-retardant performance (150,151). In this case, the application of FR treatment should be performed first, and DP treatment should be modified to accommodate the presence of the FR polymer on the fabric.

In addition to FR treatments that are durable to laundering and weathering, work has also been done on a variety of treatments for the production of

FR fabrics using inorganic salt mixtures. These treatments have usually been used on drapes and related materials that are not exposed to laundering or washing.

A light-weight fabric generally requires a higher add-on of FR polymer or reactant than a heavy-weight fabric. A fabric that has caused finishers considerable grief is the 50% cotton–50% polyester blend. There is some indication that antimony–halogen finishes are effective for this purpose (157). There have also been efforts over the years to utilize blends of a FR fiber and cotton. Certain combinations utilizing modacrylic or aramid fibers appear to permit use of diminishing levels of treatment with increasing synthetic content (158) (see FLAME RETARDANTS).

Miscellaneous Finishing

Comfort has been a merchandising point for cotton and much of cotton's success has stemmed from this property (159–162). This has not, however, prevented cotton interests from trying to modify the moisture regain or content of finished fabrics (163–165). One unique finish imparts temperature-adaptable characteristics to cotton and particularly cotton blends (166–169). This technique uses a deposition of polymers of cross-linker (DMDHEU) and various polyethyleneglycols.

A variety of chemical products and fabrics are reputed to be antibacterial and to prevent odors and the spread of infection (170). One such finish is based on an organosilicon quaternary ammonium chloride compound (171). Chemical finishing of cotton has also been directed toward improving soil release (172,173), antistatic treatments (174), and rot resistance (175,176).

Finishing of Wool. Wool (qv) competes for markets where warmth, wrinkle recovery, and ability to set in creases are important. Wool problems relate to shrinkage, particularly to its tendency to felt. This is caused by scaly structure, which tends toward fiber entanglement when wet and subjected to mechanical action. In order to compensate for this tendency, wool needs to be set and also made shrinkproof if it is to be laundered.

Setting is an important step in the finishing of wool fabrics to impart dimensionally stable shape and hand. The mechanism of setting involves bond fusion and rebuilding of disulfide linkage. Chemicals used for this purpose are thiols, bisulfates, and thioglycolates (177). The problem of felting has been solved by altering the scaly structure of wool. Processors have used chlorination or some oxidation treatments. The chlorination step is followed by the application of a polymer to achieve desired shrinkage control.

In the CSIRO process, a reactive polyurethane prepolymer is applied to a garment from perchloroethylene. The garment is then pressed and subsequently steamed in an oven. A second polymer may sometimes be used in conjunction with the prepolymer. When this is employed, the process is termed the Serolan BAP Process (178). A number of alternative treatments are being investigated to achieve finishes that are more environmentally friendly (179).

Novel finishes have been developed from the traditional chlorination of wool (180). One, the IWS soft-handle process, gives an extremely soft hand to the wool fibers and reduces the prickly effect when wool is worn next to the skin (181). The other, the soft-handle luster treatment, improves the luster. This improvement is most apparent in knitted wool jersey fabrics.

Ozone is being investigated for shrinkage prevention (182). Wool and blends of wool, cotton, and polyester have been finished to provide improved flame-retardant, durable-press, and shrinkage properties (183,184). Fabrics of these types are often used for uniforms or protective clothing (185).

Blends of wool and cotton (80:20) are being used more and more. For durable-press properties, resins, catalysts, and polymeric additives in finishing systems must be adjusted (186).

Finishing of Synthetics. Although finishing is not as important to synthetics as it is to cotton and other cellulosics, there are still many opportunities for its use in synthetics. However, the finishing of synthetics, particularly polyester fabric, is not directed toward improving resiliency and smooth-dry performance by chemical treatments as in the case of cellulosics. In the case of polyester fabric, simple heat-setting treatment suffices to set the fabric for desired smoothness and resiliency. Most heat setting of synthetics is done before dyeing and printing, and fabrics are maintained in a smooth configuration in a hot-air oven using a stenter or a perforated drum dryer. The general conditions for heat-setting various synthetics have been described (187). The optimum setting temperature range for polyester is between 210 and 215°C. If the fabric is set after dyeing and finishing, all previous operations are to be carried out open-width and dyes must be fast to heat-setting temperatures. One synthetic normally set after dyeing is triacetate.

A concern for polyester fabrics in general is the hydrophobic character of the fiber. In order to improve the comfort characteristics of these fabrics, several approaches have been developed. In contrast to cotton, fiber engineering can be used to produce a fiber having better water absorption or wicking characteristics. Techniques to control the shape of the spun fiber and/or the incorporation of soluble removable components in the spinning mix are used to reduce hydrophobic character (188,191). However, the most common chemical treatment employed for improving the hydrophilic character of polyester is the use of aqueous sodium hydroxide solutions, which leads to a softer fabric having a less synthetic hand (192). Different effects are noted for increasingly vigorous treatments. These include lowering pilling tendency, increased water wettability and water wicking, decreased soiling problems, as well as increasing weight loss and production of silklike and fine denier fabrics (189,191–195).

The alkaline solutions can remove water-soluble polymers in the spinning mix and inert products such as titanium dioxide. Basic treatments can also hydrolyze a certain amount of the polyester itself. For some silk-like applications or for producing fine denier fabrics, this basic treatment can produce a 10–30% weight loss of polyester (190,196). Certain polyesters such as anionically modified polyester can undergo more rapid weight loss than regular polyester (189).

A number of after-treatments with polyester copolymers carried out after sodium hydroxide processing are reported to produce a more hydrophilic polyester fabric (197). Likewise, the addition of a modified cellulose ether has improved water absorbency (198). Other treatments used on cotton and blends are also effective on 100% polyester fabrics (166–169). In this case, polymerization is used between an agent such as DMDHEU and a polyol to produce a hydrophilic network in the synthetic matrix (166–169).

Synthetic fabrics can also be finished to achieve a number of specific characteristics (199). For example, increased electrical conductivity can improve the antistatic character of polyester. Similarly, finishes that improve hydrophilic character also improve properties related to soil release and soil redeposition (199,200).

Other characteristics for which finishes have been designed include water repellancy (201), antipilling (202), and flame retardancy. For synthetics, an inherently flame-retardant fabric may be produced either by changing the polymer structure or by incorporating an agent into the fiber melt. Other agents, usually based on halogens, phosphorus, or both, may be applied after weaving (203).

BIBLIOGRAPHY

"Fire-Resistant Textiles" under "Textile Technology" in *ECT* 2nd ed., Suppl. Vol., pp. 944–964, by G. L. Drake, Jr., United States Department of Agriculture; "Soil-Release Finishes" under "Textile Technology" in *ECT* 2nd ed., Suppl. Vol., pp. 964–973, by S. Smith, Minnesota Mining and Manufacturing Co.; "Textiles (Finishing)" in *ECT* 3rd ed., Vol. 22, pp. 769–802, by S. L. Vail, United States Department of Agriculture.

1. W. Brenner, B. Rugg, and W. Liu, *Text. Res. J.* **40**, 318 (1970).
2. W. K. Walsh, C. R. Jin, and A. A. Armstrong, Jr., *Text. Res. J.* **39**, 560 (1969).
3. R. J. Harper, Jr., E. J. Blanchard, and J. D. Reid, *Tex. Ind.* **131**, 172 (1967).
4. G. Davis, *Text. Chem. Color.* **17**(1), 16 (1985).
5. R. S. Gregorian, *Text. Chem. Color.* **19**(4), 13 (1987).
6. U.S. Pat. 4,099,913 (July 11, 1978), A. T. Walter, G. M. Bryant, and R. L. Readshaw (to Union Carbide Corp.).
7. U.S. Pat. 4,118,526 (Oct. 3, 1978), R. S. Gregorian and C. G. Nambodri (to United Merchants Corp.).
8. H. B. Goldstein and H. W. Smith, *Text. Chem. Color.* **12**, 49 (1980).
9. South Central Section Committee, *Text. Chem. Color.* **15**(12), 25 (1983).
10. E. Rossler and G. Pusch, *Chemie Fasern/Textilind.* **24**, 763 (1974).
11. R. J. Harper, Jr., *Text. Chem. Color.* **11**(6), 21 (1979).
12. M. Schwemmer, H. Bors, and A. Gotz, *Textilvered.* **10**, 15 (1975).
13. U.S. Pat. 3,811,834 (May 21, 1974), M. Schwemmer, H. Bors, and A. Gotz (to Triatex International).
14. B. W. Jones, J. D. Turner, and L. G. Snyder, *Text. Ind.* **148**, 25 (Oct. 1984).
15. M. R. Fox and B. N. Parsons, *J. Soc. Dyers Colour.* **89**, 474 (1983).
16. J. A. Ostervold, *Int. Dyer Text. Printer*, 17 (Oct. 1984).
17. R. A. Holser, R. J. Harper, and A. H. Lambert, *Text. Chem. Color.* **18**, 29 (1986).
18. N. R. Bertoniere, W. D. King, and S. P. Rowland, *Text. Res. J.* **51**, 242 (1981).
19. N. Cashen, *Text. Chem. Color.* **6**(3), 21 (1974).
20. R. S. Swidler, J. P. Gamarra, and B. W. Jones, *Text. Chem. Color.* **3**(2), 41 (1971).
21. G. L. Payet, *Text. Res. J.* **43**, 194 (1973).
22. J. Turner, *Int. Text. Bull. Dye Pract. Finish.* **40**(2), 50 (1994).
23. J. D. Reid, R. M. H. Kullman, and E. J. Blanchard, *Amer. Dyest. Rep.* **52**, 946 (1963).
24. F. H. Burkitt, *Text. Month*, 63 (Mar. 1974).
25. S. A. Heap, *Colourage.* **24**(7), 15 (1977).
26. B. W. Jones, J. D. Turner, and D. O. Luparello, *Text. Res. J.* **50**, 165 (1980).
27. J. T. Lofton and co-workers, *Text. Ind.* **134**, 56 (1970).
28. Gulf Coast Section Committee, *Text. Chem. Color.* **22**(3), 33 (1990).

29. R. J. Harper and co-workers, *Text. Chem. Color.* **2**(1), 38 (1970).

30. A. G. Pierce, Jr., J. G. Frick, Jr., and J. D. Reid, *I&EC Prod. Res. Dev.* **5**, 23 (Mar. 1966).

31. R. W. Liggett and co-workers, *Text. Res. J.* **38**, 375 (1968).

32. R. J. Harper, in J. G. Cook, ed., *Merrow Monographs*, Merrow Publishing Co., Ltd., Waterford, U.K., 1971.

33. W. A. Reeves, R. M. Perkins, and L. H. Chance, *Text. Res. J.* **30**, 179 (1960).

34. D. D. Gagliardi and A. C. Nuessle, *Am. Dyest. Rep.* **39**, 12 (1950).

35. H. Tovey, *Text. Res. J.* **31**, 185 (1961).

36. B. W. Jones and co-workers, *Text. Res. J.* **52**, 157 (1982).

37. W. F. Herbes, S. J. O'Brien, and R. G. Weyker, in H. Mark, N. Woodling, and S. M. Atlas, eds., *Chemical Aftertreatment of Textiles*, Wiley-Interscience, New York, 1971.

38. L. B. Arnold and co-workers, *Am. Dyest. Rep.* **49**, 843 (1960).

39. S. J. O'Brien, *Am. Dyest. Rep.* **54**, 477 (1965).

40. U.S. Pat. 2,764,573 (Sept. 25, 1956), B. V. Reibnitz, A. Woerner, and H. Scheuermann (to Badische Anilin-and Soda-Fabrik Akt.).

41. W. A. Reeves, S. L. Vail, and J. G. Frick, *Text. Res. J.* **32**, 305 (1962).

42. S. L. Vail, *Text. Res. J.* **39**, 774 (1969).

43. H. Petersen, *Text. Res. J.* **41**, 239 (1971).

44. R. A. Gill and R. Steele, *Text. Res. J.* **32**, 338 (1962).

45. R. J. Harper and co-workers, *Text. Res. J.* **46**, 82 (1976).

46. S. L. Vail, *Text. Res. J.* **42**, 360 (1972).

47. S. L. Vail and R. M. Reinhardt, *Text. Chem. Color.* **13**, 131 (1981).

48. R. M. Reinhardt, B. A. K. Andrews, and R. J. Harper, *Text. Res. J.* **51**, 263 (1981).

49. S. Buckholz, *Am. Dyest. Rep.* **56**, 1025 (1967).

50. W. A. Reeves, A. A. Cory, and K. Phillips, *Amer. Dyest. Rep.* **74**(7), 48 (1985).

51. A. G. Pierce, Jr., and J. G. Frick, Jr., *Amer. Dyest. Rep.* **57**(22), 47 (1968).

52. R. J. Harper, Jr., and co-workers, *Text. Chem. Color.* **3**(5), 65 (1971).

53. H. Petersen, *Textilveredlung*, **8**, 412 (1973).

54. A. G. Pierce, S. L. Vail, and E. A. Boudreaux, *Text. Res. J.* **41**, 1006 (1971).

55. W. F. Herbes, S. J. O'Brien, and R. G. Weyker in Ref. 37, p. 319.

56. S. L. Vail, J. G. Frick, and J. D. Reid, *Am. Dyest. Rep.* **51**, 622 (1962).

57. H. H. St. Mard, C. Hamalainen, and A. S. Cooper, Jr., *Amer. Dyest. Rep.* **55**, 1046 (1966).

58. U.S. Pat. 4,236,890 (Dec. 2, 1980), E. J. Blanchard, G. A. Gautreaux, and R. J. Harper, Jr.

59. U.S. Pat. 4,304,565 (Dec. 8, 1981), E. J. Blanchard, G. A. Gautreaux, and R. J. Harper, Jr.

60. J. D. Reid and co-workers, *Amer. Dyest. Rep.* **48**, 81 (1959).

61. W. J. Burke, *J. Amer. Chem. Soc.* **69**, 2136 (1947).

62. A. M. Paquin, *J. Org. Chem.* **14**, 189 (1949).

63. S. L. Vail and co-workers, *Amer. Dyest. Rep.* **50**, 200 (1961).

64. R. L. Wayland, *Text. Res. J.* **29**, 170 (1959).

65. M. T. Beachem and co-workers, *J. Org. Chem.* **28**, 1876 (1963).

66. C. D. Egginton and C. P. Vale, *Text. Res. J.* **39**, 140 (1969).

67. R. L. Arceneaux and J. D. Reid, *Ind. Eng. Chem. Prod. Res. Dev.* **1**, 181 (1962).

68. P. K. Shenoy and J. W. Pearce, *Amer. Dyest. Rep.* **57**, 352 (1968).

69. E. J. Blanchard and co-workers, *Text. Chem. Color.* **8**, 92 (1976).

70. J. G. Frick, B. A. Kottes, and J. D. Reid, *Text. Res. J.* **29**, 314 (1959).

71. P. C. Mehta and J. R. Mody, *Text. Res. J.* **31**, 951 (1961).

72. J. D. Reid and co-workers, *Text. Res. J.* **27**, 252 (1957).

73. J. G. Roberts, *J. Text. Inst.* **58**, 418 (1964).

74. H. B. Goldstein and J. M. May, *Amer. Dyest. Rep.* **54**, 738 (1965); R. L. Stultz, p. 744; W. L. Beaumont, p. 746; A. S. Cooper and co-workers, p. 749; R. Nirenberg, p. 755; J. Midholland, p. 757; R. W. Malburg, p. 759.

75. R. M. Reinhardt, *Text. Bull.* **91**(4), 64 (1965).

76. S. L. Vail and R. M. Reinhardt, *Text. Chem. Color.* **13**, 131 (1981).

77. P. J. Jaco and J. E. Hendrix, *Text. Chem. Color.* **14**, 194 (1982).

78. R. M. H. Kullman, A. B. Pepperman, and S. L. Vail, *Text. Chem. Color.* **9**, 195 (1978).

79. H. Bille and H. Petersen, *Melliand Textilber.* **57**, 155 (1976).

80. K. Schliefer and U. Beines, *Melliand Textilber.* **60**, 960 (1979).

81. S. L. Vail and A. G. Pearce, *J. Org. Chem.* **37**, 391 (1972).

82. J. C. Morath and J. T. Woods, *Anal. Chem.* **30**, 1437 (1958).

83. C. M. Moran and S. L. Vail, *Am. Dyest. Rep.* **54**, 185 (1965).

84. G. Lund, *Shirley Inst. Bull.* **48**, 17 (1975).

85. A. C. Nuessle, *Amer. Dyest. Rep.* **55**, 646 (1966).

86. P. X. Riccobono, R. N. Ring, and A. Roth, *Text. Chem. Color.* **8**, 108 (1976).

87. *Method 112-1975, Technical Manual*, American Association of Textile Chemists and Colorists (AATCC), Research Triangle Park, N.C., 1975.

88. R. L. Arceneaux and co-workers, *Amer. Dyest. Rep.* **50**, 849 (1961).

89. R. J. Harper and J. S. Bruno, *Text. Res. J.* **42**, 433 (1972).

90. S. L. Vail and W. C. Arney, *Text. Res. J.* **44**, 400 (1974).

91. S. L. Vail and G. B. Verburg, *Text. Res. J.* **42**, 367 (1972).

92. S. L. Vail, F. W. Snowden, and E. R. McCall, *Amer. Dyest. Rep.* **56**, 856 (1967).

93. U.S. Pat. 3,622,261 (Nov. 23, 1971), J. F. Cotton, J. W. Reed, and W. C. Monk (to West Point-Pepperell, Inc.).

94. P. N. Abhyankar and co-workers, *Text. Res. J.* **57**, 395 (1987).

95. B. A. K. Andrews and R. M. Reinhardt, *Text. Res. J.* **52**, 123 (1982).

96. B. A. K. Andrews and R. J. Harper, *Text. Res. J.* **50**, 177 (1980).

97. B. A. K. Andrews, R. J. Harper, and S. L. Vail, *Text. Res. J.* **50**, 315 (1980).

98. B. A. K. Andrews and co-workers, *Text. Chem. Color.* **12**, 287 (1980).

99. H. Petersen and P. S. Pai, *Text. Res. J.* **51**, 282 (1981).

100. J. D. Turner and N. A. Cashen, *Text. Res. J.* **51**, 271 (1981).

101. R. S. Perry, C. Tsou, and C. S. Lee, *Text. Chem. Color.* **12**, 311 (1980).

102. C. Tomasino and M. B. Taylor, *Text. Chem. Color.* **16**, 259 (1984).

103. H. Petersen, *Text. Res. J.* **38**, 156 (1968).

104. H. Petersen, *Synthesis*, 143 (1973).

105. S. P. Rowland, C. P. Wade, and W. E. Franklin, *Text. Res. J.* **44**, 869 (1974).

106. S. L. Vail and co-workers, *Amer. Dyest. Rep.* **50**, 27 (1961).

107. K. Yamamoto, *Text. Res. J.* **52**, 357 (1982).

108. S. L. Vail, *Textilver Edlung.* **14**, 436 (1979).

109. C. M. Welch, *Text. Res. J.* **58**, 480 (1988).

110. C. M. Welch and B. A. K. Andrews, *Text. Chem. Color.* **21**(2), 13 (1989).

111. C. M. Welch, *Text. Chem. Color.* **22**(5) (1990).

112. U.S. Pat. 4,820,307 (Apr. 11, 1989), C. M. Welch and B. A. K. Andrews.

113. U.S. Pat. 4,936,865 (Jun. 26, 1990), C. M. Welch and B. A. K. Andrews.

114. U.S. Pat. 4,975,209 (Dec. 4, 1990), C. M. Welch and B. A. K. Andrews.

115. C. M. Welch, *Amer. Dyest. Rep.* **83**(9), 19 (1994).

116. B. A. K. Andrews, E. J. Blanchard, and R. M. Reinhardt, *Text. Chem. Color.* **25**(3), 52 (1993).

117. H. Choi, *Text. Res. J.* **62**, 614 (1992).

118. H. Choi and C. M. Welch, *Proceedings of the AATCC National Meeting*, 1992, p. 287.

119. L. Olson, *Text. Rental*, **9**, 26 (1987).

120. D. Milora, *Ind. Launderer*, **1**, 24 (1989).

121. R. M. Tyndall, *Amer. Dyest. Rep.* **79**(5), 22 (1990).

122. J. H. Hoffer, *Text. Chem. Color.* **25**(2), 13 (1993).

123. L. Olson, *Amer. Dyest. Rep.* **77**(5), 19 (1988).

124. AATCC Technology Committee on Garment Wet Processing, *Text. Chem. Color.* **23**(1), 23 (1991).

125. R. M. Tyndall, *Text. Chem. Color.* **24**(6), 23 (1992).

126. D. Kochavi, T. Videback, and D. Cedroni, *Amer. Dyest. Rep.* **79**(9), 24 (1990).

127. U.S. Pat. 5,006,126 (Apr. 9, 1991), L. A. Olson and P. M. Stanley.

128. S. Klahorst, A. Kumar, and M. M. Mullins, in Ref. 118, p. 243.

129. G. Screws and D. Cedroni, in Ref. 118, p. 250.

130. H. Koo and co-workers, *Text. Res. J.* **64**, 70 (1994).

131. R. J. Harper and A. H. Lambert, *Amer. Dyest. Rep.* **79**(5), 17 (1990).

132. R. J. Haprer, Jr., and A. H. Lambert, *Text. Chem. Color.* **24**(2), 13 (1992).

133. R. J. Harper, Jr., and R. L. Stone, *Text. Chem. Color.* **18**(11), 32 (1986).

134. E. J. Blanchard and R. M. Reinhardt, *Text. Chem. Color.* **21**(3), 19 (1989).

135. U.S. Pat. 3,807,946 (Apr. 30, 1974), R. J. Harper, Jr., G. A. Gautreaux, and E. J. Blanchard.

136. R. M. Reinhardt, E. J. Blanchard, and E. E. Graves, *Amer. Dyest. Rep.* **81**(6), 27 (1992).

137. E. J. Blanchard, R. M. Reinhardt, and B. A. K. Andrews, *Text. Chem. Color.* **23**(5), 25 (1991).

138. R. M. Reinhardt and co-workers, *Polym. Preprints*, **33**(2), 254 (1992).

139. J. G. Frick, Jr., *Text. Res. J.* **56**, 124 (1986).

140. U.S. Pat. 4,629,470 (Dec. 16, 1986), R. J. Harper, Jr.

141. R. J. Harper, Jr., and A. H. Lambert, *J. Coated Fabrics*, **17**, 197 (1988).

142. E. I. Valko, in H. Mark, S. M. Atlas, and E. Cernia, eds., *Man-Made Fibers*, Vol. 3, Wiley-Interscience, New York, 1968, p. 499.

143. C. J. Conner and co-workers, *Text. Chem. Color.* **10**(4), 70 (1978).

144. C. J. Conner and R. J. Harper, Jr., *Text. Chem. Color.* **11**(3), 62 (1979).

145. H. H. St. Mard, R. J. Harper, Jr., and W. A. Reeves, *J. Fire Retard. Chem.* **5**, 174 (Nov. 1978).

146. V. M. Bhatnagar, ed., *Advances in Fire Retardant Textiles*, Technomic, Westport, Conn., 1975.

147. J. W. Lyons, *The Chemistry and Uses of Fire Retardants*, Wiley-Interscience, New York, 1970.

148. W. A. Reeves, G. L. Drake, and R. M. Perkins, *Fire-Resistant Textiles Handbook*, Technomic, Westport, Conn., 1974.

149. D. A. Yeadon and R. J. Harper, *J. Fire Retard. Chem.* **7**, 228 (1980).

150. U.S. Pat. 3,096,201 (July 2, 1963), H. Coates and B. Chalkley (to Albright and Wilson).

151. T. A. Clamari, R. J. Harper, and S. P. Schreiber, *Text. Chem. Color.* **7**, 146 (1979).

152. D. J. Donaldson, H. H. St. Mard, and R. J. Harper, Jr., *Text. Res. J.* **49**, 185 (1979).

153. D. J. Donaldson and R. J. Harper, Jr., *Text. Res. J.* **50**, 205 (1980).

154. D. J. Donaldson and R. J. Harper, Jr., *J. Consum. Prod. Flam.* **7**, 40 (1980).

155. D. J. Donaldson, D. A. Yeadon, and R. J. Harper, Jr., *Text. Res. J.* **51**, 196 (1981).

156. J. T. Gill and A. Shaw, *Text. Chem. Color.* **17**(11), 25 (1985).

157. U.S. Pat. 4,618,512 (Oct. 21, 1986), R. J. Harper, Jr.

158. G. F. Ruppenicker and co-workers, *Proceedings of the Fifth Fire Safety Conference and Exhibition*, Orlando, Fla., 1993, p. 117.

159. D. G. Mehrtens and K. C. McAlister, *Text. Res. J.* **32**, 658 (1962).

160. N. R. S. Hollies and co-workers, *Text. Res. J.* **49**, 557 (1979); N. R. S. Hollies and R. F. Goldman, eds., *Clothing Comfort*, Ann Arbor Science Publishers, Inc., Ann Arbor, Mich., 1977.

161. L. Fourt and N. R. S. Hollies, *Clothing Comfort and Function*, Marcel Dekker, Inc., New York, 1970.
162. E. T. Renbourn, in J. G. Cook, ed., *Merrow Monographs*, Merrow Publishing Co., Ltd., Waterford, U.K., 1971.
163. A. G. Pierce, Jr. and J. G. Frick, Jr., *J. Appl. Poly. Sci.* **2**, 2577 (1967).
164. A. G. Pierce, Jr., J. G. Frick, Jr., and J. D. Reid, *Text. Res. J.* **34**, 552 (1964).
165. A. G. Pierce, *Text. Bull.* **94**(5), 41 (1968).
166. T. L. Vigo and J. S. Bruno, *J. Appl. Poly. Sci.* **37**, 371 (1989).
167. U.S. Pat. 4,851,291 (July 25, 1989), T. L. Vigo and co-workers.
168. T. L. Vigo and J. S. Bruno, *Text. Res. J.* **57**, 427 (1987).
169. J. S. Bruno and T. L. Vigo, *J. Coated Fabrics*, **16**, 264 (1987).
170. T. L. Vigo, *Chem. Tech.* **6**, 455 (1976).
171. P. A. Walters, E. A. Abbott, and A. J. Isquith, *App. Microbiol.* **25**, 253 (1973).
172. A. Hebeish and co-workers, *Amer. Dyest. Rep.* **72**(9), 48 (1983).
173. B. M. Latta and S. B. Sello, *Text. Res. J.* **51**, 579 (1981).
174. R. J. Harper, Jr., J. S. Bruno, and G. A. Gautreaux, *Text. Res. J.* **47**, 340 (1977).
175. W. N. Berard, G. A. Gautreaux, and W. A. Reeves, *Text. Res. J.* **29**, 126 (1959).
176. M. H. St. Mard, C. Hamalainen, and A. S. Cooper, Jr., *Text. Chem. Color.* **2**(8), 27 (1970).
177. R. H. Mehra, A. R. Mehra, and A. R. Mehra, *Colourage*, **38**(7), 83 (1991).
178. K. W. Fincher and M. A. White, CSIRO Division of Textile Industry Report G-30, 1977.
179. I. Holme, *J. Text. Inst.* **84**(4), 520 (1993).
180. K. M. Byrne, *Conference Papers*, Haddersfield Polytechnic, 1991, p. 7.
181. R. K. Grnsworthy and co-workers, *Australasian Textiles*, **8**(4), 26 (1988).
182. W. J. Thorsen, D. L. Sharp, and V. G. Randall, *Text. Res. J.* **49**, 190 (1979).
183. J. V. Beninate, B. J. Trask, and G. L. Drake, *Text. Res. J.* **51**, 217 (1981).
184. P. G. Gordon, R. I. Logan, and M. A. White, *Text. Res. J.* **54**, 559 (1984).
185. P. N. Mehta, *Text. Res. J.* **50**, 185 (1980).
186. R. J. Harper and P. Mehta, *Text. Res. J.* **55**, 761 (1985).
187. D. Behv, *Knitting Tech.* **14**(6), 409 (Nov. 1992).
188. U.S. Pat. 4,371,485, (Feb. 1, 1983), N. W. Matles, W. Lange, and K. Gerbach.
189. S. H. Zeronean and M. J. Collins, *Textile Chem. Color.* **20**(4), 25 (1988).
190. S. Davies, *Textile Horizons*, 27 (Feb. 1994).
191. M. Fukuhara, *Text. Res. J.* **63**, 387 (1993).
192. U.S. Pat. 2,590,402 (Mar. 25, 1952) J. D. H. Hall, B. P. Ridge, and J. R. Whinfield.
193. N. T. Lijemark and H. Asres, *Text. Res. J.* **41**, 732 (1971).
194. G. Elefante and F. Giammanco, *Chemiofasern/Textileindustrie*, **36**(88), 892 (1986).
195. C. G. G. Namboodri, *Text. Chem. Color.* **1**(2), 24 (Jan. 15, 1969).
196. B. M. Latta, *Text. Res. J.* **54**, 766 (1984).
197. U.S. Pat. 4,370,143 (Jan. 25, 1983), J. Bauer.
198. U.S. Pat. 4,136,218 (Jan. 23, 1979), E. Nischwitz and co-workers.
199. R. V. Jaiswal, *Text. Dyer Printer*, **25**(2), 23 (Jan. 22, 1992).
200. N. C. Maity, K. P. Kartha, and H. C. Srivastava, *Colourage*, **31**(24), 11 (Nov. 29, 1984).
201. J. R. Caldwell and C. C. Dannely, *Amer. Dyest. Reptr.* **56**, 77 (1967).
202. A. A. Vaidya and J. K. Nigam, *Man-Made Text. India.* **22**, 363 (1979).
203. K. Masuda, *Japan Text. News.* **268**, 106 (1977).

TIMOTHY A. CALAMARI, JR.
ROBERT J. HARPER
United States Department of Agriculture

TESTING

Knowledge of fiber and yarn properties, including mechanical, physical, and chemical behaviors, is fundamental in understanding textile structures, whether the fabric structure is woven, knitted, or nonwoven. Because fabric performance is a function of the application of chemical, thermal, or mechanical finishes, the effects of these treatments must be studied. Textile testing is the use of engineering principles in the measurement of properties of textile fibers, yarns, and fabrics. Tests performed on textile structures relate to, but do not necessarily define, textile's use performance. These tests can be categorized as either objective or subjective. Objective testing relates to physical performances, including strength, elastic behavior, shrinkage, color fastness, as well as tear, abrasion, pilling, and degradation resistance. Subjective testing relates to aesthetics and includes fabric hand, appearance following laundering or dry cleaning, luster, and comfort.

Properties of finished textile structures or fabrics are an accumulation of properties of the fiber (or fibers if it is a blend), yarn configuration, construction, and selected finish. Fiber properties include length, crimp, transverse dimensions, density, cross-sectional shape, shrinkage, friction, moisture absorption, electrostatic and thermal properties, as well as optical, tensile, and elastic properties. Yarn properties include size and number, twist, strength, evenness, friction, and texture. Fabric properties include construction, thickness, air permeability, strength and elongation, snag, pilling and abrasion resistance, shrinkage, thermal and moisture transmission, color and wash fastness, flammability, hand, drape, wrinkle resistance, luster, and comfort. The properties of textile structures are dictated by interactions of all of these parameters.

In general, textile materials are moisture- and temperature-sensitive. As a result, all tests should be performed at ambient conditions of 21°C and 65% relative humidity unless otherwise stated. Most of the testing procedures discussed for textile materials are available in the literature (1,2).

Fiber Properties

Fibers (qv) have been defined by the Textile Institute as units of matter characterized by flexibility, fineness, and a high ratio of length to thickness (3). For use in textile applications, fibers should have adequate temperature stability, strength, and extensibility. Other important qualities include cohesiveness or spinability and uniformity. There are also several secondary characteristics that improve customer satisfaction and therefore may be desirable. These include cross-sectional shape, specific gravity or density, moisture regain, resiliency, luster, elastic recovery, and resistance to chemicals, environmental conditions, and biological organisms.

Length.   Fiber lengths are classified as either staple or continuous filament. In filament yarns each filament is considered to be infinitely long, where the length is generally determined to be the total length of yarn on the package. Fibers included in this category are silk and synthetic fibers. Staple fibers must have a high (>1000) length-to-diameter ratio to facilitate efficient conversion to yarn. Staple fibers include all natural fibers other than silk and synthetic fibers which are cut into staple lengths.

During processing, many machine settings are established based on staple length. Thus uniformity of staple length is extremely important. Consequently, test methods used to measure fiber length also include measurement of distribution of fiber length. ASTM test methods D1575 (wool), D519 (wool), D1234 (greased wool and other animal hair fibers), D1440 (cotton), D3660 (synthetic staple), and D5103 (natural and synthetic fibers) include techniques for laying out a uniform array of fiber lengths progressing from the longest to the shortest. In these methods, both the average or mean staple length and the distribution of fiber lengths are recorded. ASTM D3661 gives the procedures for measuring average staple length and distribution of staple length for synthetic fibers ranging in length from 25 to 250 mm. Because synthetic filaments are cut or broken into staple fibers, the possibility of multiple lengths of fiber exists. ASTM D3513 gives the procedures for determining the percentage of multiple-lengths (over length) fibers using visual inspection of a combed sample of fibers.

There are also ASTM standards for determining length and length distribution of natural fibers. These include ASTM D1447 (Fibrograph), D5332 (Fiber Length Measuring Unit, AL-101), D4604 (Motion Control High Volume Instrument Fiber Information System), and D4605 (Spinlab System High Volume Instrument). The test instruments in these methods use either photoelectric, capacitance or pneumatic scanning devices to determine fiber length. The results from these methods, however, do not agree with those obtained from ASTM D1440 (Manual Array) because of the differences both in the treatment of fiber crimp and in the definition of the length being measured.

Other automatic and semiautomatic systems exist for determining length and length distribution for wool and synthetic staple. These systems are not described in ASTM testing procedures but can be found in the literature (4–8). An apparatus using electrooptical sensors to measure length and length distribution also exists. The use of photoelectric scanning has also been described (3).

Crimp. Practically all staple fibers have crimp, which is defined as waviness of the fiber. Fibers such as cotton and wool have a natural spiraling or helical crimp. In synthetic fibers crimp is imposed by mechanically deforming the fiber. The elasticity of the crimp may vary, especially in synthetic fibers where the method of deformation and heat setting are important factors. Essential in the conversion of fiber to yarn, crimp determines the capacity of fibers to entangle during processing and thus determines the cohesiveness of card webs as well as the hairiness of the resultant yarn (9). Crimp is the principal feature governing bulking power of textile materials and, generally, the specific volume of yarns and fabrics (3). Whereas ASTM D3937 describes the method for determining crimp frequency in synthetic staple fibers, no specific method is available for measuring crimp in cotton fibers because of the low frequency and amplitude of the convolutions. However, tensile test as described in ASTM D1774, Standard Test Method For Elastomeric Properties of Textile Fibers, may be used to determine the amount of crimp in textile fibers. As of this writing (1997), no test exists for measuring the amplitude of crimp, which is needed in determining the energy or work required to remove crimp from fibers. A discussion of problems in measuring fiber crimp energy is available (10), as is that of an experimental method for determining fiber crimp (11).

Transverse Dimensions or Fineness. Historically, the quantity used to describe the fineness or coarseness of a fiber was the diameter. For fibers that have irregular cross-sections or that taper along their lengths, the term diameter has no useful meaning. For cylindrical fibers, however, diameter is an accurate measurement of the transverse dimension. Though textile fibers can be purchased in a variety of cross-sectional shapes, diameter is still a useful descriptor of the transverse dimension. Fiber diameter is important in determining not only the ease with which fibers can be twisted in converting them to yarns, but also fiber stiffness, ie, fabric stiffness, and, alternatively, fabric softness and drapeability.

Fiber weight per unit length, called linear density, provides the most general way of describing fiber size. Linear density is useful because it describes yarn size and, for a given size yarn, allows the determination of the average number of fibers in the yarn cross-section. The linear density of natural staple fibers is normally given in micrograms per inch. For silk, synthetic continuous filament, and synthetic staple, the term tex or decitex (dtex) is used. Tex is the weight in grams of 1000 meters of material; dtex is the weight in grams of 10,000 meters of material. When weight per unit length is used to define fiber size, the smaller the numerical value is, the finer the fiber.

Several methods for determining fiber fineness or linear density are based on the resistance principle of air flowing through a plug of fibers, which can be directly related to fiber fineness. These methods include ASTM D1448 (cotton), which describes the use of an apparatus called the micronaire; D4604 and D4605, which describe the use of high volume instruments (HVI); D3818 (cotton), which describes the use of the IIC-Shirley fineness or maturity tester; and D1282 (wool), which describes the use of the Port-Ar and the wira fiber fineness meter. Overview of micronaire testing procedures, including the use of near-infrared spectroscopic image analysis, is available (12).

ASTM D1577 describes the method for determining individual fiber linear density of textile fibers using a vibroscope. This, the most practical method for measuring fiber weight per unit length, uses the principle of a vibrating string, and is most applicable to staple fibers having linear densities below 1 tex. The linear density is calculated from the fundamental resonant frequency of transverse vibration of the fiber measured under known conditions of length and tension. The method not only describes procedures for both crimped and uncrimped fibers, but also gives procedures for determining linear density using the direct weighing method. In this method, the average linear density of single fibers in a bundle is calculated after determining the bundle mass, individual lengths of fibers in the bundle, and the number of fibers in the bundle. ASTM D2130 covers the procedures for determining average diameter and fiber diameter variations of wool and other animal fibers using microprojection. This standard also describes a method for estimating the linear density of wool fibers using the diameter measured by microprojection.

Information on measurement of fiber diameter using optics is available (13). ASTM D629, Sections 23 through 28, describes procedures for determining fiber diameter using microscopic analysis. Characterization of cotton fibers by cross-section, fineness, and maturity has been discussed (3,14), as have one simple method for determining fiber cross-sectional area and length (15) and general descriptions of methods for measuring and calculating fiber transverse dimensions (3).

Density. Density, the mass per unit volume, is useful when comparing substances having different volumes. Several methods have been used in the past to determine fiber density; three of which are summarized in ASTM D276. Determining fiber density by displacement of liquids has also been discussed (3,16,17). The density of fibers can be determined by classical physical means, ie, precise microscopic measurement of length, cross-sectional area, and weight. For fibers, the relative mass per unit volume, called the specific gravity, is often given. The specific gravities of most common textile fibers have been listed (3,18–21).

Cross-Sectional Shape. Fibers vary in cross-sectional shape both naturally and by design (1,2,19). Whereas wool fibers are essentially round, cotton fibers are elliptical or kidney-shaped. In synthetic fibers, the cross-sectional shape is determined by the method of spinning and the shape of the spinnerette hole through which the fiber is extruded (3,22).

Cross-sectional shape influences the stiffness (flexural rigidity) of fibers (3,23), the tendency of fibers to pack together in yarns, as well as fiber, yarn, and consequently fabric luster. In addition, fibers having specific cross-sectional shapes are often engineered to provide particular fiber properties. Circular hollow fibers can be used to improve fluid transport and insulation properties. Du Pont manufactures a fiber that has a square cross-section containing four longitudinal, continuous holes (Antron III), used to hide soil. Eastman manufactures an irregular cross-sectioned, deep-groove fiber (4DG) used for fluid movement and absorbency.

ASTM D629 describes procedures for determining cross-sectional shapes for natural fibers using microscopic analysis. Cross-sectional shape of synthetic fibers also can be verified by using microscopic analysis.

Shrinkage. Shrinkage of a bundle of crimped or uncrimped fibers that contract at least 10% in boiling water or hot air can be determined by ASTM D2102. A more reliable technique is to measure the fiber linear density by a vibroscope both before and after subjecting the fiber to the contracting medium, eg, hot air, hot water, or solvent. Another method, ASTM D5104, explains the procedure for measuring shrinkage of crimped and uncrimped single fibers exposed to hot air or hot water. Mechanisms giving rise to fiber shrinkage have been discussed (24–27).

Friction. Friction is the force that holds together fibers in spun yarns and interlacing yarns in fabrics. High fiber friction can be both an advantage and a disadvantage. If the fiber-to-fiber friction is too low, yarn strength will be reduced. If the yarn-to-yarn friction is too low in woven fabrics, the fabric dimensional stability will be reduced. Friction also plays an important role in the processing of fibers into yarns. To avoid excessive fiber breakage, the tension buildup in the fiber during processing must be less than the average fiber strength.

In sewing, high yarn friction may result in needle heating, which in turn may cause yarn fusing. Also, high yarn-to-yarn friction may hinder the needle from passing between yarns during stitching, causing poor stitch quality. Fabric properties that are influenced by frictional effects are fabric hand, strength, elongation, abrasion resistance, dimensional stability, and seam slippage. In polymeric materials, the friction coefficient can vary with surface speed (28). During high speed applications, the buildup of frictional forces can cause fusing in some thermoplastic fibers. It is therefore important that frictional properties be measured under simulated or use conditions.

Fiber friction can be determined by physical methods. For rapid evaluation of fiber or yarn friction, the capstan method is used, where a yarn or fiber is pulled over a cylindrical surface. The frictional coefficient, μ, can be determined according to the formula

$$\mu = \ln \frac{\left(\frac{t_2}{t_1}\right)}{\theta}$$

where t_1 is the entry tension; t_2, the exit tension; and θ, the contact angle.

Methods for determining fiber-to-fiber friction have been developed (29–31). The friction coefficient can also be measured in terms of the force required to pull entwined fibers apart (32–34) or the force necessary to remove a single fiber from a mass of fibers under pressure (35). Another test involves an apparatus wherein one or a series of parallel fibers are mounted across a small bridge similar to a violin bridge. This is then pressed against a surface that may be another fiber or some other material, and the fibers alternately slip and stick as they slide across each other (36,37).

Because fiber frictional properties are so important in the conversion of staple yarns to spun yarns, ASTM D2612 has been designed to measure the cohesive force encountered in the drafting or fiber alignment of sliver and top under static conditions. This frictional force is affected by surface lubrication, linear density, surface configuration, fiber length, and fiber crimp.

Optical Properties. When light falls on an object, it is either partially absorbed, reflected, or transmitted. The behavior of the object as it relates to each of these three possibilities determines visual appearance. Optical properties of fibers give useful information about the fiber structure; refractive indexes correlate well with fiber crystalline and molecular orientation; and birefringence gives a measure of the degree of anisotropy of the fiber.

Optical properties of fibers are measured by light microscopy methods. ASTM D276 describes the procedure for fiber identification using refractive indexes and birefringence. Other methods for determining fiber optical properties have been discussed (3,38–44). However, different methods of determining optical properties may give different results (42).

Tensile Properties. Probably the most important properties of textile fibers are the mechanical properties. A knowledge of fiber tensile properties is essential in understanding fiber behavior during processing. Also, the final properties of yarns and ultimately fabrics are dependent on both fiber properties and fiber arrangement within the yarn or fabric. Because many important processing decisions and final product properties depend on the results of fiber tensile tests, much precision is desired when running the test. Accurate tensile testing of fibers can be extremely difficult, especially when testing single fibers. This difficulty arises from the need to manipulate single fibers and the large number of tests that must be performed to get a representative average.

Although there are a variety of tensile testers for yarns and fabrics, which work by such principles as constant rate of traverse, constant rate of loading, or constant rate of extension, the field is significantly reduced with respect to single-fiber testing. The main reason for this is the high level of sensitivity required to detect the low loads associated with the breaking strain of a single fiber. Variations in load of as little as 0.05 g must be easily detectable. Most natural fibers break at loads of 2–60 g. The load required to break a fiber depends also on the speed at which the load is applied and the length of the specimen. Crimp in a fiber can also affect reported result (45–47).

The essential features of any tensile-testing method consist of the jaws in which the ends of the specimen are held, the type of specimen used, the method of varying the load and elongation, and the means of recording their values to give the load–elongation curve (2). For single-fiber testing, a constant rate of extension machine is used, rather than a constant rate of loading machine (incline-plane technique) or a constant rate of traverse machine (pendulum technique).

Methods used for the tensile testing of single fibers and fibers taken from yarns and tows are discussed in ASTM D3822 and D2101. Measurement equipment used in fiber tensile testing is described in ASTM D76. An overview of test procedures and their significance is also available (3,10).

Because testing of single fibers is tedious and time-consuming, parallel bundles of fibers may be tested also. ASTM D1445 describes the measurement of bundle strength using incline-plane and pendulum-type machines. Bundle strength can also be measured using ASTM D4604 and D4605. The results of such tests show lower mean strengths than those derived in tests for single fibers (3).

Stress–Strain Curve. Other than the necessity for adequate tensile strength to allow processibility and adequate finished fabric strength, the performance characteristics of many textile items are governed by properties of fibers measured at relatively low strains (up to 5% extension) and by the change in these properties as a function of varying environmental conditions (48). Thus, the whole stress–strain behavior of fibers from zero to ultimate extension should be studied, and various parameters should be selected to identify characteristics that can be related to performance.

Typical patterns of stress–strain behavior and the relationship of molecular motion on stress–strain behavior have been discussed (10,18,19,21,49–51). At times, it becomes desirable to characterize stress–strain behavior numerically so that a large amount of information can be condensed and many fibers exhibiting different behaviors can be compared. Procedures for measurement of stress–strain parameters are described in ASTM D3822 and D2101 (10).

Effect of Fiber Variability on Stress–Strain Behavior. Textile fibers are not uniform, varying in both composition and fineness within a fiber and between fibers. The length of the fibers is also different from fiber to fiber. Because the fibers vary, their tensile properties also vary. The variation in properties from one fiber to another influences the distribution of loads on fibers in a textile structure (3). The mean measured strength of a sample decreases as the sample length increases. The probability of finding more weak links along the fiber increases as the fiber length increases. Because weak links can not share equally in the distribution of the load, the maximum load bearing capacity of the fiber is thus reduced. The weak link effect (52–55) also affects breaking extension. If a fiber breaks under a light load owing to the presence of a weak place, the rest of the specimen will have a comparatively small extension and the breaking extension will be low. The mean breaking extension decreases as the specimen length increases (3). The strain of a fiber is also affected by nonuniform diameter along the fiber length, ie, the thin places extend more than the thick places. Variations in fiber morphology may cause the modulus to vary from place to place in a fiber or between fibers in a given sample (3).

Effect of Strain Rate on Stress–Strain Behavior.   One of the most important factors in testing is the duration of an individual test or the time scale of the experiment. Only by standardization of strain rate can reliable data be obtained. The factors governing the standardization of strain rate largely concern test instruments to be used, rather than the particular deformation rates that fibers may encounter during processing and in commercial usage. Exceptions to this rule, ie, where some attempt has been made to simulate actual use behavior, include several tests involving stress-and-strain cycles of several minutes. On the other extreme, there are tests of extremely short duration for simulation of high speed impact, eg, ballistic testing for military body armor (56,57,58). Between the two extremes of testing time scales is a wide region of deformation rates that are particularly significant to the fiber producer and mill representative. These deformation rates approximate the strain history encountered by a fiber in processing from fiber to finished fabric. Testing speeds for fibers depend on the estimated breaking elongation. Normal tensile-testing speed for fibers range from approximately 0.01 to 10 cm/min, whereas processing velocities range from 1.5 to 50 m/s. For example, carding involves a speed of motion of approximately 40–50 m/s, normal weft insertion during weaving involves speeds of 13–33 m/s, and drafting may involve speeds of 4–13 m/s.

The particular fiber gauge length, ie, the effective fiber length that is accelerated, must also be considered. For example, a velocity of 1.5 m/s on an effective fiber length of 2.5 cm results in a strain rate of 360,000%/min. If the effective gauge length is 1.25 cm, the strain rate doubles at the same velocity. Because the effective fiber length may be extremely short in certain processing procedures, it is theoretically possible to achieve strain rates of up to several million percent per minute in the deformation of fibers in processing. It is important to consider carefully each material and application for the range of strain rates over which reliable stress–strain data are needed. Conclusions about performance based solely on data obtained at one testing speed can be misleading.

Loop and Knot Strength.   The loop strength of textile fibers has been described (59). There is a direct relationship between tensile elongation to rupture and loop-strength efficiency (18). If a filament is loaded in a bent state, it will break more easily than when it is straight. This is a result of the initiation of breakage by the high extension of the outside layers. The reduction in strength is greatest in fibers that have the lowest elongation at break (3). A similar effect is observed when there is a knot in the filament (3). Knot strength values have been given in the literature (60,61). Test methods for determining the loop and knot strengths of single fibers are given in ASTM D3217 (1). Both loop-breaking and knot-breaking tenacities are fundamental properties used to establish limitations on fiber processing and end-use applications. Physical properties, such as brittleness, that are not well-defined by tests for breaking load and elongation can be estimated from the ratio of breaking tenacity, measured by loop and knot tests, and normal tenacity, measured by ASTM D2101 (1).

Elastic Properties. The ability of a fiber to deform under below-rupture loads and to return to its original configuration or dimension upon load removal is an important performance criterion. Permanent deformation may be as detrimental as actual breakage, rendering a product inadequate for further use. Thus, the repeated stress or strain characteristics are of significance in predicting or evaluating functional properties.

A perfectly elastic material is one that obeys Hooke's law, ie, the stress–strain diagram of the material is linear and the unloading curve superposes the loading curve. A completely elastic material has a stress–strain diagram that is nonlinear and a recovery curve that does not necessarily superimpose its loading curve but whose recovery, ie, restoration to its original dimension, is complete. An imperfectly elastic material is one whose stress–strain diagram is nonlinear, whose recovery curve does not superimpose its loading curve, and whose recovery is incomplete. On a molecular scale, recoverable or elastic deformation results from a stretching of interatomic and intermolecular bonds, whereas nonrecoverable or plastic deformation results from a breaking of bonds and their reforming in new positions or to the stabilization of new chain conformations (3).

When a fiber is stressed, the instantaneous elongation that occurs is defined as instantaneous elastic deformation. The subsequent delayed additional elongation that occurs with increasing time is creep deformation. Upon stress removal, the instantaneous recovery that occurs is called instantaneous elastic recovery and is approximately equal to the instantaneous elastic deformation. If the subsequent creep recovery is 100%, ie, equal to the creep deformation, the specimen exhibits primary creep only and is thus completely elastic. In such a case, the specimen has probably not been extended beyond its yield point. If after loading and load removal, the specimen fails to recover to its original length, the portion of creep deformation that is recoverable is still called primary creep; the portion that is nonrecoverable is called secondary creep. This nonrecoverable elongation is typically called permanent set.

It is generally accepted that, all other things being equal, the lower the secondary creep, the better the fiber is in terms of wear, shape retention, and crease resistance. This does not mean that glass, which has no secondary creep, is better in abrasion resistance than high tenacity viscose rayon, which has secondary creep, because the respective energy absorption capacities of these two materials, exclusive of secondary creep, are not equal. Nor does it mean that fibers that exhibit secondary creep are of no value. For fabrics to meet the requirements of wear, crease resistance, and shape retention, the load and extension yield points should not be exceeded during use.

Elastic performance coefficient (EPC) is used as a criterion of repeated stress performance. This value reflects the effect of immediate elastic, primary, and secondary creep deformation, and the ability of the material repeatedly to absorb and return energy during repeated stressing. The importance of energy-absorbing ability is evident, because to resist destruction the specimen must be capable both of absorbing energy imparted to it upon stress application and of releasing this energy upon stress removal. The EPC is a normalized index that expresses degree of perfect elasticity. A material that exhibits identical elastic properties, both before and after repeated stressing, has an EPC of 1.0. A perfectly viscous material, having no recovery, has an EPC of zero.

The more quickly and completely a fiber recovers from an imposed strain, the more nearly perfectly elastic it is. The ratio of the instantaneous elastic deformation to the total deformation may be used as a criterion of elasticity (62). The integrated divergences from a theoretical graph of perfect elasticity versus elongation is also used as a criterion for determination of the elasticity index.

A method of testing immediate elastic recovery, delayed recovery, and permanent set of fibers is given in ASTM D1774. British Standard 4029:1966 (63) also gives a procedure for calculating elastic recovery. The effects of strain rate, temperature, length of time strained, etc, have been discussed (9,64–66). Elastic recovery, which are very sensitive to test conditions, may be plotted against stress or strain. The first shows the extent to which a given force will cause permanent damage to a fiber; the second what proportion of a given extension will be recovered and the amount of the permanent deformation (3).

Other Fiber Deformations.   Deformations such as bending, torsion, shear, and compression are of practical importance in textile applications. Bending and twisting of yarns, both influential in the development of bulk and stretch in filament yarns, are also important in the production of staple yarns. Bending characteristics are important in crush resistance in carpets. Bending and shear are factors that influence the hand and drape of apparel fabrics, whereas compression influences the recovery of fabrics after such processes as winding.

Textile fibers must be flexible to be useful. The flexural rigidity or stiffness of a fiber is defined as the couple required to bend the fiber to unit curvature (3). The stiffness of an ideal cylindrical rod is proportional to the square of the linear density. Because the linear density is proportional to the square of the diameter, stiffness increases in proportion to the fourth power of the filament diameter. In addition, the shape of the filament cross-section must be considered also. For textile purposes and when flexibility is requisite, shear and torsional stresses are relatively minor factors compared to tensile stresses. Techniques for measuring flexural rigidity of fibers have been given in the literature (67–73).

The torsional rigidity of a fiber, or resistance to twisting, is defined as the couple needed to insert unit twist, that is, angular deflection between the ends of a specimen of unit length (3). Torsional rigidity for a cylindrical rod, such as bending rigidity, is proportional to the square of the linear density, and thus the fourth power of diameter. Again, fiber cross-sectional shape must be considered. Methods for characterizing torsional behavior of fibers are available (74–81).

The problem of shear is usually of much greater significance in brittle, high modulus materials, where, although absolute shear strengths are higher than for low modulus materials, inability to deform can cause the development of extremely high stress concentrations at certain points, leading eventually to failure. Shear modulus, defined as the ratio of shear stress to shear strain, is difficult to measure directly. Discussion of techniques for characterizing shear properties of fibers is available (74,82), as is that of the effect of compressive stresses on a mass of fibers (83). Fibers that show good tensile recovery generally also show high recovery after compression (3).

Moisture Absorption. Absorption of moisture changes the properties of fibers (3). It may cause swelling to occur, which alters dimensions of the fiber. This in turn causes changes in fiber size, shape, and stiffness. The mechanical and frictional properties are altered also, thus affecting the behavior of the fibers during processing. Wetting and drying may lead to permanent set or creasing. Moisture conditions are also one of the most important factors in determining electrical properties (3).

The amount of moisture present in a mass of fibers, a yarn, or a fabric is calculated as moisture regain or moisture content. Moisture regain is the difference in wet and dry fiber weight relative to the dry fiber weight. Moisture content is water weight loss as a percentage of the combined fiber and water weight. The normal methods for determining these values involve weighing, bone-drying, and reweighing. A discussion of other methods and problems associated with determining moisture properties is available (3). Moisture regain values of fibers at standard testing conditions of 65% relative humidity and 21°C and at higher relative humidities can be found in the literature (19,84–95). ASTM D2654-89 is a test method for measuring moisture in textiles and ASTM D2462 gives a method for determining moisture specifically in wool. Fibers that are capable of absorbing water, when exposed to progressively increasing and then decreasing relative humidities, show regain–relative humidity curves that do not coincide (3). The area between the two curves or amount of hysteresis generally correlates directly with the moisture regain determined at standard conditions. Thus, a hydroscopic fiber brought to equilibrium from the wet side always has a higher moisture regain than the same fiber brought to equilibrium from the dry side.

Because the mechanical properties of hydrophilic fibers are critically dependent on moisture regain, it is vital that such fibers be tested under constant conditions of temperature and humidity. Standard conditions used in the textile industry are 65% relative humidity and 21°C (1,2,21,96). ASTM D1909, D2118, and D2720 list accepted commercial moisture regain values used in the buying and selling of fibers.

Because textiles are sized, bleached, dyed, and finished, they must have tension, torsion, and bending properties to withstand the required wet-finishing operations. For example, the yield point must not be reduced by wetting to a value such that the resulting wet fiber or fabric cannot be processed without becoming permanently and irrevocably distorted. Fabric wetting, however, is necessary in many finishing operations. The use of water to generate viscous behavior in textile materials, so that these materials can be molded into a required geometric configuration and then set in that configuration by redrying, is the basis of many finishing operations. If a fiber is to be used in any environment where it is apt to become wet or subjected to high humidities, the fiber must be able to withstand such conditions, eg, in laundering at elevated temperatures.

There is little difference between the wet and the dry stress–strain diagrams of hydrophobic fibers, eg, nylon, acrylic, and polyester. Hydrophilic protein fibers and regenerated cellulose exhibit lower tensile moduli on wetting out, that is, the elongations increase and the strengths diminish. Hydrophilic natural cellulosic fibers, ie, cotton, linen, and ramie, are stronger when wet than when dry.

As a general qualitative rule, the greater the ability of the fiber to swell, the greater is the moisture regain. The amount of moisture regain and swelling is a function of the ability of the fiber to absorb water and depends on the amorphous regions of the fibers. Little water enters the crystalline regions. Swelling mechanics, methods for determining swelling, and effects of swelling on fiber properties have been discussed (3,97–102).

Electrostatic Properties.   In the past it was believed that static electricity was generated when two different materials were rubbed together. Though the separation of two unlike surfaces does result in a separation of charge, it has been shown that the asymmetric rubbing of two identical surfaces also results in generation of charge (103). If the resulting charges are retained and accumulated, a measurable potential is generated. If the material under consideration is a conductor and is grounded, the charge is removed as fast as it is deposited and there is no static electricity. However, if a material is a dielectric (3), charge builds up to the point where it may interfere with textile processing such as carding and spinning, fabric spreading, and ply separation in the cut-and-sew industry. Buildup of static charge may also be objectionable to a person wearing the material because of its clinging and sparking.

Four criteria are used to study static electricity (3): surface resistivity, volume resistivity, rate of charge buildup and release, and maximum charge capable of being retained. The American Association of Textile Chemists and Colorists (AATCC) lists the following methods pertaining to the assessment of static in textiles: AATCC 84-1960 for electrical resistivity of yarns, AATCC 76-1964 for electrical resistivity of fabrics, and AATCC 115-1965 for electrostatic clinging of fabrics. ASTM D4238 gives the method for determining electrostatic propensity of textiles. A specimen is inductively charged, and both the maximum voltage developed in the specimen and the rate of decay of the charge are measured. Another test method for static has also been described (104).

Synthetic hydrophobic fibers have high dielectric strengths and are thus excellent electrostatic generators. The natural and synthetic hydrophilic fibers, because of their higher moisture regains, present less of a processing-and-use problem, particularly at high relative humidities. However, at low relative humidities these fibers do present electrostatic problems. Although electrostatic properties are influenced by intrinsic fiber properties and moisture regain, these two parameters do not completely explain the electrostatic ranking of various fibers.

Thermal Properties.   Textile fibers may be classified into two main groups according to their reaction to heat: thermoplastic and nonthermoplastic. The natural fibers, eg, wool, silk, cotton, linen, and jute, and the synthetic regenerated rayons, proteins, and aramids, are nonthermoplastic, ie, they do not have a fixed melting point. Upon heating these fibers char and decompose. Most of the synthetic fibers, eg, polyamides, polyesters, acrylics, olefins, and acetates, soften and melt with increasing temperature. Many thermoplastic fibers have no precise melting points because they have two-phase structures, ie, crystalline and amorphous. The amorphous region melts at a lower temperature than the crystalline region. It is customary to report, as the melting temperature, the crystalline melting point as determined by x-ray observation at elevated temperatures. ASTM D276 gives procedures for identifying fibers using fiber melting point.

A group of fibers that are resistant to high temperature have been prepared from highly aromatic compounds, eg, Kevlar (Du Pont), Nomex (Du Pont), and PBI (Hoechst-Celanese) (3). These materials exhibit no clear melting point, degrade only at very high temperatures, and maintain a high percentage of their original strength even at elevated temperatures. These fibers are primarily being used in body armor (Kevlar) and protective clothing, eg, firefighter, race car driver, and astronaut apparel. Their great thermal stability is responsible for their resistance to burning.

Fiber degradation from heat is affected by temperature, time of exposure, relative humidity, and air circulation. Air circulation may be important because many aging and heat degradation processes involve oxidation, and the greater the air circulation, the more oxygen there is available to the fiber. With respect to ironing and pressing, pressure and heat transfer also affect the amount of degradation. The effect of heat on fibers has been described (3,19,105), as have contraction temperatures and melting points of thermoplastic fibers (3,106).

The criterion usually used to measure degradation is breaking strength, but, depending on the performance requirements, other criteria are also used. General practice for the evaluation of sunlight aging of textile materials is given in ASTM G23–25. These methods involve accelerated testing as well as direct exposure to sunlight and exposure under window glass. The direct sunlight methods, although reliable if exposures of test samples take place simultaneously, fail to take into account the shifting concentration of the short-wavelength ultraviolet light from season to season. A procedure exists that corrects this seasonal shift in the most damaging element of sunlight, thereby allowing test results obtained at different times of the year to be compared directly (107,108).

Thermal Conductivity and Heat Capacity. Most fibers have similar thermal conductivities and heat capacities. The insulating characteristics of textiles are more related to fabric geometry than they are dependent on fiber thermal characteristics.

Yarn Properties

Size or Number. Contrary to the linear density method for measuring fibers in which weight for a specified length is recorded, yarns are usually characterized by count or the units of length for a specific weight. Though the linear density in tex units is sometimes used for yarns, most spun yarns are sized by the count method. The count of a yarn of a given cross-sectional diameter varies according to the particular spinning process used. Table 1 lists the most common forms of yarn count, which apply for any fiber or blend of fibers processed by a particular system.

Determination of yarn count or number is made by weighing a specific length of yarn as described in ASTM D1907 (skein lengths) and ASTM D1059

Table 1. Common Forms of Yarn Count

System	System units	Length of hank, yd	Count
English cotton	cotton hanks per pound	840	1 cotton hank/pound (1695 m/kg or 840 yd/lb)
English worsted	worsted hanks per pound	560	1 worsted hank/pound (1130 m/kg or 560 yd/lb)
U.S. woolen	woolen hanks per pound	1600	1 woolen hank/pound (3225 m/kg or 1600 yd/lb)

(short lengths). Like fibers, continuous-filament yarns are normally sized using the direct system, ie, mass per unit length. Methods for determining the linear density of yarns is described in ASTM D861. ASTM D2260 provides a conversion table for equivalent yarn numbers measured in various numbering systems.

Twist. In staple yarns, twist is required to force the fibers into contact with one another, thus increasing cohesion and thereby developing yarn strength. Without twist fibers in staple yarns, fibers would slip past one another easily. In filament yarns, each filament is as long as the yarn itself, thus eliminating the concern of yarn breakage resulting from fiber slippage. Thus, for continuous-filament yarns, twist has a secondary effect on yarn properties. Without twist in continuous-filament yarns, the individual filaments would spread out, separate from one another, and cause the bundle to lack unity. Twist is also useful in continuous-filament yarns to help prevent snagging by tucking loose filaments into the yarn bundle.

Two methods are used for determining the twist in conventional ring spun yarns. ASTM D1423 and D1422 describe the direct-counting method and the untwist–retwist method, respectively. In the latter case, a yarn is untwisted and then retwisted again. The number of turns recorded divided by two is an estimate of the actual twist level in the yarn. Care must be taken during twisting to avoid ballooning of the yarn, which causes drafting of the fibers, thus increasing the length of the yarn beyond that associated with a loss in contraction. This drafting of fibers and increased length in the yarn can result in a higher quantitative value of twist than actually exist in the yarn. Twist results have been discussed (109).

Measurement of twist in open-end spun yarns, a newer and faster method of yarn spinning as compared to conventional ring spinning, is more difficult. The difficulty arises because of wrapper fibers on the yarn surface. Because these fibers in part lay loosely on the yarn surface, during the untwisting of the yarn they tend to become twisted and lead to an erroneous measurement of yarn twist. Optical observation of the surface fibers has been suggested as the most reliable method for measuring twist in open-end spun yarns (110). As of this writing (1997), no test method exists for determining twist level in open-end spun yarns.

The selection of twist level is important not only in establishing the surface characteristics of the yarn, eg, low twist for soft, fuzzy yarns and high twist for compact, smooth yarns, but also in determining yarn strength. Yarn strength as a function of yarn twist level is shown in Figure 1.

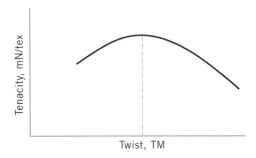

Fig. 1.   Yarn strength vs twist level (111).

There is an optimum twist level in terms of strength development in spun yarns. As the twist increases, the frictional forces holding the fibers together increases, thus reducing the chances of yarn breakage as a result of fiber slippage. However, twist in excess of the optimal development of frictional forces can cause a decrease in fiber orientation with respect to the yarn axis. This decrease in orientation causes a decrease in strength because only in the most oriented position do fibers bear the maximum stress in a tensile test. Therefore, optimization of twist level, ie, desired strength for a given application, and desired yarn surface characteristics must occur.

Because a given number of turns per length generates a greater helix angle in a thicker, ie, low count, yarn than in a thinner one, it is customary to express twist in the form of twist multiplier or twist factor. These terms take into account the turns per unit length required to give a constant helix angle in yarns of differing count.

$$\text{twist multiplier} = \frac{\text{turns per inch}}{(\text{yarn number})^{1/2}}$$

$$\text{twist factor} = \text{turns per centimeter (yarn linear density (in tex))}^{1/2}$$

If this unit of twist measurement is substituted for the horizontal axis in Figure 1, then it is possible to determine the optimum twist levels for maximum yarn strength for any size yarn of a given fiber composition.

Filament yarns require much lower twist factors for optimum strength than spun yarns. These yarns are often used as supplied by the manufacturer and contain producer twist only, which can be as low as 0.25 turns per centimeter.

Single twisted yarns are often twisted together to form plied yarns. Several plies are twisted together to form cables or ropes. A method for analyzing yarn construction and the nomenclature used to define yarn construction is found in ASTM D1244.

Fiber Migration.   Of importance in determining yarn properties is the desirability for individual fibers in spun yarns to alter their radial positions along their lengths in the yarn bundle. This enhances the interlocking of the fibers in the yarn structure and increases the number of contact points between each fiber. An in-depth discussion of fiber migration in the yarn is available (112), as is a comprehensive discussion of the effects of fiber migration or lack thereof on fabric drape, hand, pill resistance, abrasion resistance, and various methods for determining fiber migration (113).

Strength. Yarn strengths are important not only in governing the strengths of the resultant textile structures, but also in terms of the ease of conversion of the yarns into these structures. Knitting and weaving processes impart significant stresses on the component yarns. Weak yarns cause difficulties in the fabrication process, resulting in fabric defects and loss of manufacturing efficiency.

The translation of fiber strength to yarn strength is complex even in the case of filament yarns. The yarn strength may be expected to be equal to the fiber strength multiplied by the number of filaments in the bundle. In reality, however, the yarn strength is usually significantly lower than the sum of the individual fiber strengths. This is explained by the weak-link theory, which argues that there is a distribution of individual fiber strengths in the yarn bundle. If all filaments are stressed uniformly, the weakest fiber will break at a combined stress level less than $1/n$ of the average fiber strength, where n is the number of filaments in the bundle. The total stress in the yarn then must be shared by the remaining $n - 1$ fibers, which causes a stress surge in the remaining filaments. This increase in stress exposure thus causes the filaments to fail at some combined stress level below that calculated from the average fiber strength multiplied by the number of filaments in the yarn bundle. Consequently, the larger the bundle size, the lower the yarn strength is relative to the predicted value obtained from the average fiber strength, because the opportunity for the presence of a weak filament is increased by the larger number of fibers present.

In spun staple yarns, the yarn strength is only a fraction of that predicted by the individual fiber strengths and the number of fibers in the yarn cross-section. In addition to fiber strength, twist, fiber migration, and frictional forces all influence yarn strength. If a yarn contains two or more fibers of differing stress–strain behavior, it would fail at a stress governed by a combination of the breaking stress of the low elongation fiber and the stress development in the high elongation fiber, at the breaking extension of the low elongation fiber. This phenomenon has encouraged the synthetic fiber industry to develop special fiber variants for blending with cotton and wool to ensure compatibility of the stress–strain curves for both blended fibers.

Test methods for determining yarn strength are detailed in ASTM D2256 for single yarns and D1578 for yarn skeins. A method for testing tire cords and industrial filament yarns is given in ASTM D885 (1).

Evenness and Grade. Unevenness of diameter along the yarn length, which is more prevalent in spun yarns than in filament yarns, gives rise to poor strength and a nonuniform appearance in fabric form, ie, thick and thin spots. Yarn evenness is usually measured by passing a length of yarn through an instrument that measures capacitance on a continuous basis. Fluctuations in capacitance readings are proportional to yarn unevenness. This method is discussed in ASTM D1425. Though no test methods exist as of this writing, testing equipment is available using conventional optics and lasers to determine yarn evenness and hairiness (114).

Yarn grade is assessed using ASTM D2255, which is a subjective test requiring the evaluation of yarn wound on a board. More recently a method has been introduced to assign a yarn grade quantitatively (115).

Friction. Frictional properties of yarns are important in considering the performance life of some machine components, eg, yarn guides and knitting and sewing needles. Frictional properties also affect the quality and performance properties of yarns. In high speed sewing, yarn friction plays an important part in the fabric to thread interaction. Sewing threads must be highly lubricated to ensure good stitch quality. In sewing, high yarn friction may result in needle heating, which in turn may cause yarn fusing. Also, high yarn-to-yarn friction may hinder the needling from passing between yarns during stitching, thus causing poor stitch quality. In determining frictional properties, it must be remembered that no coefficient of friction exists for a single body. The coefficient of friction measures the frictional interaction between two bodies or elements. Test methods exist for determining yarn-to-yarn friction (ASTM D3412) and yarn-to-solid friction surfaces (ASTM D3108).

Textured Yarns. Filament yarns are textured to give stretch and bulk to the yarns, thereby increasing their cover factor. Texturing also alters filament hand and luster and improves insulating properties. In evaluating textured yarns, along with typical yarn properties such as linear density, strength, and elongation, yarn shrinkage and bulk are of great importance. ASTM offers two test methods for evaluating textured yarns: D2259 for determining shrinkage in yarns and D4031 for determining bulk properties. Other yarn properties are determined using methods for regular continuous-filament yarns (116–119).

Fabric Testing

OBJECTIVE TESTS

Fabric properties are dependent on the geometry and form of the fabric structure, as well as the properties of the yarns or fibers from which the fabric is constructed. The three principal types of fabric structures are woven, knitted, and nonwoven. In many cases, the types of tests required to characterize fabric properties are defined by the fabric structure itself. Those properties that are measurable in quantitative terms using instrumentation are referred to as objective properties, eg, tensile strength or color measurement. Other fabric characteristics must be evaluated, at least in part, by human observation and judgment. These are referred to as subjective properties. Fabric hand, drape, luster, and comfort are of this second category (18,19), although in more recent years, instrumentation has been developed to allow quantification of some of these properties. Many of the test methods for determining physical properties of fabrics are ASTM standards (1). Test methods requiring chemical procedures are primarily based on AATCC standards (2). Because of the moisture sensitivity of most textile materials, particularly hydrophilic fibers such as cotton, and the dependence of physical properties on the moisture content of the material, most test procedures should be carried out at standard conditions of temperature and humidity. For a textile testing laboratory, these conditions are generally 21 ± 1°C and 65 ± 2% relative humidity, as specified in ASTM D1776.

Fabric Construction.   Woven fabrics are characterized in terms of length (ASTM D3773); width (D3774); weight per unit area (D3776); yarn count, ie, number of warp and filling yarns per unit length (D3775); and yarn crimp (D3883). Fabric analysis may also include raveling warp and filling yarns to determine twist and yarn number (ASTM D1059). In addition, the weave of the fabric is usually specified, eg, plain, twill, or satin weave.

Knitted fabric construction characterization is discussed in ASTM D3887. Characterization of knitted fabrics includes yield (area per weight), width, length, and yarn count (number of courses and wales per unit length). The type of knit is also specified, eg, warp knit or weft knit. Yarn analysis may be difficult on account of problems in raveling individual yarns from the knitted structure, particularly warp knits.

Regardless of whether the fabric structure is woven or knitted, accurate characterization of construction parameters is necessary to evaluate test results for other properties and to explain differences noted when comparing test data from fabrics of different constructions.

Nonwoven fabrics are generally described by the method of production, eg, needle-punched or spun-bonded (120). ASTM D1117 discusses various physical properties to be determined for nonwoven fabrics and the standard procedures used.

Thickness.   Because two fabrics that have identical weight per unit area values may have widely varying bulks, the specification of thickness is essential for properly characterizing a fabric. Fabric thickness has been shown to be directly proportional to thermal insulation, or warmth (121). Fabric warmth is the result of the entrapment of air between fibers and yarns. A thicker fabric in general allows an increased amount of entrapped air and thus is warmer.

The proper methods for measuring fabric thickness are described in ASTM D1777. Because fabric thickness is dependent on the applied pressure, any measurement of thickness should also report the pressure at which the measurement was made. Thus, an apparatus capable of applying variable pressure to the sample while determining thickness would be desirable. Many instruments, however, allow only incremental increases in pressure, depending on the weight used. Regardless of the instrument used, it is always necessary to state the pressure under which the thickness was determined.

Air Permeability.   Air permeability is an important parameter for certain fabric end uses, eg, parachute fabrics, boat sails, warm clothing, rainwear, and industrial air filters. Air permeability of a fabric is related to its cover, or opacity. Both of these properties are related to the amount of space between yarns (or fibers in the case of nonwovens). The most common method for specifying air permeability of a fabric involves measuring the air flow per unit area at a constant pressure differential between the two surfaces of the fabric. This method, suitable for measuring permeability of woven, knitted, and nonwoven fabrics, is described in ASTM D737. Units for air permeability measured by this method are generally abbreviated as CFM, or cubic feet per square foot per minute.

An alternative method for measuring air permeability is based on measuring differential pressure when a constant rate of air flow passes through the fabric. In this case, air resistance is reported in typical units of kPa·s/m.

Breaking Strength and Elongation. The breaking strength and elongation (extensibility) of woven fabrics are often used as quality-control parameters. These properties are especially important for industrial uses where tensile strength is a principal consideration. For many apparel uses, however, breaking strength is of little consequence, although minimum standards are generally reported in most ASTM performance specifications. Most fabrics far exceed the reported minimum tensile strength standards. Nonetheless, tensile strength tests, generally easy to conduct, are relied upon as popular quality-control tests. In some cases, the maximum or minimum elongation produced at a selected stress level below rupture is specified as a criterion, eg, percent stretch in stretch fabrics.

Two types of fabric tensile strength tests are commonly used: the grab test (ASTM D5034) and the raveled strip method (D5035). Several modifications of these methods are described in the test standards. These tests are designed to be carried out on either constant rate of extension (CRE), constant rate of load (CRL), or constant rate of traverse (CRT) testing machines, although CRE testers are becoming the most commonly used. In this type of testing, the rate of increase of the specimen length is uniform over time.

For the raveled strip method, specimens are cut in both the warp and filling directions of the fabric. Initially samples are cut 3.8 cm × 15.2 cm, and the longer dimension is placed parallel to the direction being tested (warp or filling). The edge yarns in the longer direction are raveled and removed from the fabric until a width of 2.5 cm results, thus leaving a fringe of 0.65 cm on each side of the fabric. The test specimen is clamped lengthwise in the flat jaws of the testing machine. Each jaw is at least 5 cm wide so that the jaws extend at least 1.2 cm on each side of the test strip. The initial separation between the two sets of jaws is referred to as the gauge length and is specified by the test method to be 7.6 cm. The jaws are then separated at a rate of 300 mm/min until rupture of the test specimen occurs. Most modern tensile testing machines can be controlled by or interfaced with a computer, which provides load–elongation diagrams as well as a multitude of parameters derived from the load–elongation curve, such as breaking strength and elongation, energy required to rupture the specimen, and modulus (or stiffness) of the specimen.

If the fringe were not present during the raveled strip test, the edge yarns would pop out of the fabric during the progress of the test. The presence of the fringe keeps the edge yarns in place and allows testing of a consistent width of fabric. However, the raveled strip test requires careful, tedious preparation of the specimen and is therefore considered uneconomical by many mills. Specimens for the grab test are easier to prepare, and this method is considered by some to simulate more closely the way fabrics are actually used. The specimens are 10.2 × 15.2 cm. The front jaws of each set are 2.5 cm in width; the rear jaws are at least 5 cm wide. In this way, a 2.5-cm width of fabric is actually clamped. All other test parameters, eg, gauge length and rate of extension, are identical to the strip test. For the grab test, however, the interaction of warp and filling yarns adjacent to those being pulled, and the attendant frictional forces, results in a fabric assistance effect, thus increasing the load required to elongate and rupture the fabric. For this reason, grab tests generally produce higher strength values than strip tests for the same fabric.

Empirical attempts have been made to relate strip and grab test results, particularly for cotton fabrics, so that if one strength is known, the other can be calculated. The relationship is complex, depending on fiber strength and modulus, yarn size and crimp, yarn-to-yarn friction, fabric cover factor, weave, weight, and other factors (19).

Bursting Strength.   The tensile tests discussed above are not suitable for knitted fabrics because of the distribution of applied forces in all directions in a knitted structure. The strengths of knitted fabrics are measured by determining bursting strength. ASTM D3786 describes the measurement of bursting strength of knits of low-to-intermediate extensibility using a hydraulic burst tester. The fabric is held over an expanding diaphragm, and the hydraulic pressure in the diaphragm at the instant of fabric rupture is reported.

For knitted fabrics exhibiting high extensibility to failure, the ball burst test is recommended (ASTM D3787). A polished steel ball is pressed onto a rigidly held circle of knitted fabric. The force on the ball required to rupture the fabric is reported as breaking strength.

Tear Strength and Energy.   For flat, sheet-like materials such as woven fabric, films, paper, and leather, the breaking strength of a material in a tensile test is generally stronger than its tear resistance. Although it may be difficult to initiate a tear in any of these materials, once started the tear can usually be propagated using relatively low force. Three basic methods for determining tear strength have historically been used for testing textile fabrics: the tongue tear (ASTM D2261 and D2262), the Elmendorf tear (D1424), and the trapezoid tear (D1117). The tongue and Elmendorf (falling pendulum) methods are generally used for woven fabrics, and the trapezoid method is recommended for nonwoven fabrics.

The tongue tear test can be performed on either a CRE-type tester (ASTM D2261) or a CRT-type tester (D2262), although the former is preferred. A specimen 20 × 7.5 cm is cut to produce a 7.5-cm slit along the longer center line of the fabric. The two 7.5 × 3.8-cm tongues are placed in the upper and lower jaws of the testing machine. As the jaws separate, the yarns running along the shorter dimension of the specimen are ruptured one by one as the tear is propagated. A load–elongation curve showing a series of progressively increasing and then sharply decreasing loads is generated. The tear strength of the sample is then determined by one of several methods for determining average load over a specified distance of tear. Average load can easily be determined using the computer software available in most up-to-date tensile testers.

A specially designed falling-pendulum tester is required to perform the Elmendorf test. The apparatus makes it possible to measure quickly the average force required to propagate a tongue tear through a fixed distance in a woven fabric. The tester consists of a sector-shaped pendulum carrying a clamp that is in alignment with a fixed clamp when the pendulum is in the raised, starting position and contains maximum potential energy. The specimen is fastened in the clamps and the tear is started by a slit cut in the specimen. The pendulum is then released, and the specimen is torn as the clamps are separated by the pendulum motion. A scale attached to the pendulum is graduated to allow reading of the maximum tearing force in grams as a percentage of the maximum potential energy of the pendulum.

For both the tongue and Elmendorf test methods, it is important to observe the behavior of the specimen as the tear is propagated. In cases where the yarns in the test direction are much stronger than the perpendicular yarns, it is sometimes difficult or impossible to propagate the tear in the desired direction. In this case, a crosswise tear results. Tear resistance is primarily a function of fabric construction. Loose, open weaves such as cheesecloth tend to resist tear, whereas tight weaves tend to tear easily. In the open weave, the concentrated force field at the point of tear is dissipated by the compliance of the fabric structure to accommodate the stress field, thereby distributing the force over a greater number of yarns.

The trapezoid test method (ASTM D1117) is recommended for determining the tear resistance of nonwoven fabrics. An outline of a trapezoid is marked on a 7.5 × 15-cm specimen, and the nonparallel sides are clamped in the jaws of the tensile-testing machine. The load is applied to the specimen in such a way that the tear propagates across the specimen width. The value of the breaking load is obtained from the load–elongation curve and is determined primarily by the bonding or interlocking of the fibers of the composite structure.

Snag Resistance. Although knits do not tear easily, they are prone to snagging on sharp, pointed objects. ASTM D3939 describes a method for quantifying the tendency of a fabric to snag. The method applies to woven or knitted fabrics made from textured or untextured yarns containing staple or continuous filaments. Fabric specimens in tubular form are placed on a rotating cylindrical drum, and a mace is allowed to bounce randomly against the rotating specimen. The degree of snagging is evaluated by comparison of the tested specimen with visual standards of fabrics or photographs of fabrics. Resistance to snagging is reported on a scale from 5 (no snagging) to 1 (very severe snagging).

A second snag test method, described by ASTM D5362, is the bean bag snag test. Each fabric specimen is made into a cover for a bean bag, which is randomly tumbled for 100 revolutions in a cylindrical test chamber fitted on its inner surface with rows of pins. Evaluation is similar to that for the mace snag test.

For tests performed in simulated situations such as the snag tests, the results are meaningful only if an established correlation exists between performance in the tests and in wear situations. Thus, the acceptable level of performance in the test should not be selected arbitrarily but should be established in actual wear studies (122).

Pilling Resistance. Fabrics containing high strength synthetic fibers, especially in blends with weaker natural fibers, exhibit a tendency for pill formation in varying degrees. The mechanism of pill formation involves breaking of the weaker fibers and retention of the broken lengths by the stronger fibers, resulting in small balls of fiber adhering to the fabric surface. Pilling resistance testing involves rubbing the sample against a mildly abrasive surface, followed by visual comparison of the fabric sample to a series of photographic standards representing no pilling (No. 5) to very severe pilling (No. 1).

The most common method of testing for pilling resistance is the random tumble method (ASTM D3512), in which pill formation is caused by tumbling specimens in a cylindrical test chamber lined with a mildly abrasive material. A small amount of short-length cotton fiber is included in the chamber to simulate pills produced in actual wear. In the brush pilling test (ASTM D3511), fabrics are subjected to simulated wear conditions by first brushing them to form free fiber ends and then rubbing two specimens together in a circular manner to form pills. The elastomeric pad method (ASTM D3514) calls for laundering of the sample, followed by rubbing against an elastomeric pad.

Synthetic fiber producers have attempted to minimize the tendency for pilling by several methods, one of which is to reduce polymer molecular weight. The resulting lower strength fiber would break away from the fabric surface more readily. Another method for reducing pilling is to notch or etch the fiber surface either before or after incorporation in fabric form.

Abrasion and Wear Resistance. Abrasion resistance is generally measured by subjecting the fabric to some type of rubbing action under known conditions of pressure, tension, and abrasive action. The term wear is broader in scope and includes the combined effects of additional factors such as laundering, dry cleaning, ironing, and wearing of apparel. Correlation of laboratory abrasion resistance with general wear resistance is very difficult. More often than not, laboratory testing is useful for predicting the relative abrasion resistance of a series of samples that vary greatly in resistance properties. Resistance to abrasion is affected by such factors as fiber properties, yarn structure, fabric construction, dyes and finishes, as well as test factors such as the nature of the abradant, tension of the specimen, pressure between specimen and abradant, and dimensional changes in the specimen. Depending on the test method, the abradant itself may change over the course of the test. Seven common abrasion test methods are as follows.

Rotary Platform, Double-Head Method (ASTM D3884). A specimen is abraded by rotary rubbing action under controlled conditions of pressure and abrasive action. The specimen is mounted on a platform and rotates on a vertical axis against the sliding rotation of two abrading wheels. The wheels can be of a variety of materials and coarseness. One abrading wheel rubs the specimen outwardly and the other inwardly, resulting in a pattern of crossed arcs on the specimen. Abrasion resistance can be evaluated either by determining the loss in breaking load of the abraded area after a specified number of cycles or by determining the number of cycles required to give specified destruction, eg, color change as based on the AATCC Gray Scale.

Flexing and Abrasion Method (ASTM D3885). This method tests the resistance of woven fabrics to flexing and abrasion. The specimen is subjected to unidirectional reciprocal folding and rubbing over a bar under known conditions of pressure and tension. Resistance to flexing and abrasion is evaluated by determining the number of cycles to rupture the specimen, by comparing the breaking load of the abraded fabric to the breaking load for nonabraded fabric, or by examining the abraded specimen for visual changes to luster, color, napping, pilling, etc.

Inflated Diaphragm Method (ASTM D3886). This method is applicable both to woven and knitted fabrics. The specimen is abraded by rubbing either unidirectionally or multidirectionally against an abradant having specified surface characteristics. The specimen is supported by an inflated rubber diaphragm under a constant pressure. Evaluation of abrasion resistance can be either by determination of the number of cycles required to wear through the center of the fabric completely or by visual examination of the specimens after a specified number of cycles.

Oscillatory Cylinder Method (ASTM D4157).   This test is applicable to woven fabrics and measures abrasion resistance by subjecting the specimen to unidirectional rubbing action under known conditions of pressure, tension, and abrasive action. Abrasion resistance may be evaluated either by determining the number of cycles required to rupture the specimen, by comparing the breaking load of the abraded fabric to the breaking load for nonabraded fabric, or by examining the abraded specimen for visual changes to luster, color, napping, pilling, etc.

Uniform Abrasion Method (ASTM D4158).   This test is applicable to a wide range of textile fabrics and materials, including floor coverings. Abrasive action is applied uniformly in all directions in the plane of the specimen surface. The test may be run dry or wet. Evaluation is made by comparing initial and final value for various properties, which may be thickness, weight, electrical capacitance, or absorption of beta emission from a radioactive surface. An abrasion curve may be constructed by plotting values for the measured quantity against the number of rotations of the tester.

Martindale Abrasion Tester Method (ASTM D4966).   The Martindale tester is used to determine the abrasion resistance of woven or knitted textile fabrics by subjecting the specimen to straight-line rubbing motion, which becomes a gradually widening ellipse, until it forms another straight line in the opposite direction under known conditions of pressure and abrasive action. Evaluation of abrasion resistance is either by determining the number of cycles to break two or more threads on a woven fabric or by causing a hole in a knitted fabric. Change in shade, as evaluated using the AATCC Gray Scale for Color Change, can also be used to determine the end point.

Impeller Tumble or Accelerotor Method (AATCC 93).   A fabric specimen is driven by a rotor in a random path so that it repeatedly impinges the walls and abradant liner of the test chamber. The specimen is subjected to flexing, rubbing, shock, compression, stretching, and other mechanical forces during the test. Evaluation of abrasion resistance is based on weight loss of the specimen, loss in breaking load, or changes in various other properties, such as air permeability, light transmission appearance, or hand.

The variety of evaluation methods for abrasion resistance testing requires the tester to choose a method to align most closely with the desired performance of the fabric under end-use conditions. Although abrasion to rupture is an easy parameter to measure, unsightliness of a textile garment or other consumer product would precede this stage. Alternative techniques are the measurement of weight loss or the measure of remaining fabric strength after abrasion. However, even these results generally exceed the degree of fabric surface damage objectionable to the consumer. For example, in blended fabrics containing two fibers of slightly varying color shades, preferential wear of one fiber can cause a shade change in the abraded area known as frosting. Therefore, visual observations should be made frequently over the course of testing, and the first detectable appearance changes should be noted. The determination of the point where unacceptable damage to the fabric resulting from abrasion begins to occur is subjective and depends on the judgment of the observer. Thus, abrasion-testing results often show a great deal of variability. Acceptance standards and controls for evaluation of abrasion results have, in many cases, not been estalished. Therefore, acceptability criteria must be determined based on requirements for the specific material and end use.

Fiber requirements for high abrasion resistance are low modulus of elasticity, large immediate elastic deflection, high ratio of primary to secondary creep, high magnitude of primary creep, and high rate of creep recovery. Three methods of fiber abrasion are frictional wear, surface cutting, and fiber rupture or slippage. Surface cutting occurs when a fiber is subjected to metal-cutting or grinding, as when a fine abrasive or emery surface is rubbed across a fabric, and is applicable where surface projections of the abrading surface are small relative to fiber diameter. Fiber rupture, or plucking, may develop when the surface protuberances of the abradant are large compared to fiber diameter, and the normal forces between the abradant and cloth planes are high (62).

Abrasion resistance is also greatly dependent on fabric geometry. Factors that affect abrasion resistance are weave, area of contact between fabric and abradant, local pressures or stress concentrations that develop on specific yarn points, fabric density (ends and picks per inch), crown height (the extent of deformation out of the plane of the fabric resulting from the intersection of warp and filling yarns), yarn number, fabric thickness, yarn crimp, float length, yarn cohesiveness, compressible compliance, fabric tightness, cover factor, direction of abrasion, and magnitude and direction of tensions developed during the abrasive action (123).

In addition to fiber and fabric influences on abrasion resistance, chemical finishes must also be considered. Many thermosetting resins used to impart durable press characteristics to cellulosic fabrics reduce their resistance to abrasion as a result of fiber embrittlement.

Laundering Shrinkage. Shrinkage, often a result of laundering textile fabrics, can be of three types: relaxation, swelling, or felting. Relaxation shrinkage occurs when a fabric that has been finished in a stretched state is exposed to heat and/or moisture and allowed to relax. Swelling shrinkage may occur when a fabric that is composed of hydrophilic fibers is soaked in water, resulting in the increase of fiber diameter relative to the fiber length. Because hydrophobic fibers do not swell in water, they do not shrink by this mechanism when laundered. Hydrophobic fibers are much more susceptible to relaxation shrinkage. Felting shrinkage is associated exclusively with wool and other animal hair fibers and is characterized by a continued reduction in fabric area and increase in fabric density. Felting is caused by a combination of heat, moisture, and mechanical motion during aqueous washing.

Five AATCC methods are described for determining the dimensional stability of fabrics and garments. These methods are applicable to different types of laundering methods and to different substrates, depending on whether the material is in fabric or garment form and on its fiber composition. AATCC 135 and 150 describe determination of shrinkage under home laundering conditions for fabrics and garments, respectively. AATCC 96 describes a method for measuring stability of all fabrics except wool under commercial laundering conditions. Shrinkage during dry cleaning can be determined according to AATCC 158. Finally, AATCC 99 describes a test for determining the shrinkage specifically for wool fabrics.

Thermal Transmission. Thermal transmission of a fabric is an important property affecting the comfort of a garment made from that fabric. Dry heat transfer can be by either conduction, convection, or radiation. As related to comfort, heat transfer can also occur by evaporative heat loss from sweating skin. Thermal transmittance is defined as the overall heat transfer through a fabric resulting from a combination of the three mechanisms for dry heat transfer (124). Because the convection mechanism is based on the transport of heat by moving air or liquids, the insulating properties of a textile fabric are therefore improved by creating dead air spaces between and within fabric layers to reduce heat transfer by convection (121). Increasing fabric thickness can also reduce heat transfer. Although fabric thickness correlates well with thermal insulation, fabric weight per unit area shows almost no correlation to this property. Thus, the fabric's ability to maintain its thickness under conditions of compression, tension, bending, laundering, dry cleaning, and wear can affect its insulative properties under conditions of use. One additional factor is fabric density. Two fabrics may have the same thickness and, hence, the same thermal conductivity, but to obtain such equal thickness, different weights of fiber may be needed. Thus, aggregates of different fibers may have different bulk densities, and on this weight basis one fiber may exhibit a thermal insulating advantage over another.

Thermal conductivity of a fabric is related to its air permeability, or movement of air between the interstices of the yarn and fabric. For fabrics of a given thickness, the one that has greater air permeability allows greater heat dissipation by convection. Thus thermal insulation falls as air velocity rises.

A guarded hot-plate method, ASTM D1518, is used to measure the rate of heat transfer over time from a warm metal plate. The fabric is placed on the constant temperature plate and covered by a second metal plate. After the temperature of the second plate has been allowed to equilibrate, the thermal transmittance is calculated based on the temperature difference between the two plates and the energy required to maintain the temperature of the bottom plate. The units for thermal transmittance are $W/m^2 \cdot K$. Thermal resistance is the reciprocal of thermal conductivity (or transmittance). Thermal resistance is often reported as a clo value, defined as the insulation required to keep a resting person comfortable at 21°C with air movement of 0.1 m/s. Thermal resistance in $m^2 \cdot K/W$ can be converted to clo by multiplying by 0.1548 (121).

The guarded hot-plate method can be modified to perform dry and wet heat transfer testing (sweating skin model). Some plates contain simulated sweat glands and use a pumping mechanism to deliver water to the plate surface. Thermal comfort properties that can be determined from this test are clo, permeability index (i_m), and comfort limits. Permeability index indicates moisture–heat permeability through the fabric on a scale of 0 (completely impermeable) to 1 (completely permeable). This parameter indicates the effect of skin moisture on heat loss. Comfort limits are the predicted metabolic activity levels that may be sustained while maintaining body thermal comfort in the test environment.

Moisture Transmission.   Water vapor permeability is related to the suitability of fabrics for apparel purposes. Two methods for measuring the diffusion of moisture through a fabric barrier include the absorption-cup method and the evaporation method. The former involves ascertaining the amount of water that diffuses through the test fabric and into a cup of drying agent. The test fabric acts as the cover of the vessel containing the desiccant. The other side of the fabric is exposed to an atmosphere of constant humidity, temperature, and air velocity. The amount of water absorbed per unit area per unit time can be calculated by determining the increase in weight of the desiccant at selected time intervals.

The evaporation method consists of placing the test fabric over a carefully calibrated cup filled with water to a prescribed depth, and maintaining a specified distance between the water surface and the fabric. Again, the cup and fabric assembly is exposed to an atmosphere of constant temperature, humidity, and air velocity, and the rate of loss in water weight in the cup is determined. This method is similar to ASTM E96. Although the above methods give valid results for impermeable materials, eg, plastic films, and may be useful for comparing fabrics, these methods do not give precise results for open fabrics of high moisture transmission. As the water is absorbed into the desiccant or evaporates, the relative humidity of the air adjacent to the fabric may not remain constant. A modified evaporation method involves placing a permeable cover fabric 0.5 cm above the test fabric specimen (125). This causes a constant humidity to develop above and below the specimen.

If it is assumed that a moisture differential exists between two sides of a fabric, and then water vapor diffuses through the fabric at a rate depending primarily on the relative amount of air contained within the structure (126). The resistance of the fiber to the passage of water vapor is as much as several orders of magnitude higher than that of air. Because normal fabrics contain 50–90 vol % of air, a negligible amount of water vapor passes through the fibers, and differences between fibers are much less important than differences between fabric geometries. Very tightly woven fabrics have higher resistance to moisture vapor transfer than more open fabrics and may, for this reason, be less comfortable to wear.

The wettability of a fabric is measured by the Gravimetric Absorbency Testing System (GATS). This test is a measure of the ability of the fabric to take up liquid spontaneously in the direction perpendicular to its plane (lateral absorbency). The amount of water driven from a reservoir beneath the sample is determined.

Water Repellency and Water Resistance.   Water repellency is defined as the ability of a textile fiber, yarn, or fabric to resist wetting, whereas water resistance is a general term applied to a fabric's ability to resist wetting and penetration by water (2). A third term, waterproof, is applied to those fabrics that do not allow any water penetration at all. Waterproof fabrics are generally coated with an impermeable surface layer that does not allow air permeability. Water-repellent finishes are hydrophobic compounds that are applied to fabrics to inhibit water penetration while still allowing air permeability.

AATCC methods for determining water repellency are AATCC 22 (spray test) and AATCC 70 (tumble jar dynamic absorption test). In the spray test, water is sprayed against the taut surface of the test specimen to produce a wetted pattern the size of which depends on the repellency of the fabric. Evaluation is by comparing the pattern with a series of patterns on a standard chart. The latter method evaluates the percentage by weight of water absorbed by a sample after dynamic exposure to water for a specified period of time.

Water resistance test methods include AATCC 127 (hydrostatic pressure test), AATCC 42 (impact penetration test), and AATCC 35 (rain test). In the hydrostatic pressure test, a sample is subjected to a column of increasing water pressure until leakage occurs. The impact penetration test requires water to be sprayed on the taut surface of a fabric sample from a height of two feet. The fabric is backed by a blotter of predetermined weight, which is reweighed after water penetration. The rain test is similar in principle to the impact penetration test.

Stretch and Compressional Resilience. The growth and stretch properties of knitted fabrics can be determined according to ASTM D2594. This test is for fabrics intended for applications requiring low power stretch properties. To determine growth (or stretch recovery), the specimen is extended to a specified percentage stretch and held for a prescribed length of time. The specimen is then allowed to recover under zero load. During the recovery period, the specimen length is measured at various time intervals. Fabric stretch is measured by applying a load to a fabric specimen of known length and determining the length of the fabric before and after loading.

A similar method, ASTM D3107, has been developed for measuring stretch and stretch recovery of woven fabrics made in whole or in part from stretch yarns. The term stretch yarns refers to thermoplastic filament or spun yarns having a high degree of potential elastic stretch and rapid recovery. These yarns are characterized by a high degree of yarn curl.

Resilience of textile fabrics when compressed in the bent state is related to wrinkle resistance and retention of shape, drape, and hand. Resilience is an important parameter for evaluating blankets, wearing apparel in which warmth is a factor, pile fabrics including carpets, and bulk fiber utilization in mattresses, cushions, etc. The general method for determining compressional resilience is to compress and unload the material cyclically, creating a plot of compressive force versus fabric thickness.

Color. Spectrophotometers are commonly used to quantify color shade and intensity. White light impinged on the surface of a dyed fabric causes light of a specific wavelength to be absorbed or reflected, depending on the properties of the chromopores in the dye. The reflected light passes through a prism and is measured by a photoelectric cell. A plot of reflectance (or absorbance) versus wavelength results. Although this curve is adequate to define quantitatively any color, the transmission of these plots from customer to dyehouse is difficult. However, it is possible to define a color by comparing the intensities of the three reflected primary colors, ie, red, green, and blue. Thus, a system of tristimulus values allows the storage and transmission of color data. Dyestuff manufacturers list these tristimulus values for each of their dyes as measured for fabrics of specific fiber content, allowing the dyer to approximate closely the desired color. Final adjustment should still be made by a human eye skilled in shade matching.

The science of color measurement has been explored by various authors (127,128). AATCC evaluation procedure no. 6 describes a method for instrumental measurement of color of a textile fabric. AATCC evaluation procedure no. 7 may be used to determine the color difference between two fabrics of a similar shade. Instrumentation may be either a spectrophotometer for measuring reflectance versus wavelength, or a colorimeter for measuring tristimulus values under specified illumination. If a spectrophotometer is used, however, the instrument must be equipped with tristimulus integrators capable of producing data in terms of CIE X, Y, and Z tristimulus values.

Another test method applicable to textiles is ASTM E313, Indexes of Whiteness and Yellowness of Near-White, Opaque Materials. The method is based on obtaining G, ie, green reflectance, and B, ie, blue reflectance, from X, Y, and Z tristimulus values. Whiteness and yellowness index are then calculated from the G and B values. This method has particular applicability to measurement of whiteness of bleached textiles. AATCC test method 110 also addresses measurement of the whiteness of textiles.

Colorfastness.   A variety of test methods exist for determining the fastness, or color retention, properties of dyed fabric exposed to various conditions of weathering, laundering, or general exposure associated with the end use of the product. The AATCC Technical Manual should be consulted for the applicable test method for the expected exposure of the sample. Fastness to acids and alkalies, bleaching, gas fumes, crocking, dry cleaning, heat, light, perspiration, washing (laundering), seawater, and numerous other conditions can be determined according to AATCC methods. Four of the most common types of colorfastness tests are described below. Evaluation of the results is generally performed by visual comparison of the exposed samples to the unexposed against the AATCC Gray Scale for Color Change.

Lightfastness.   AATCC test method no. 16 describes various techniques for measuring lightfastness of textiles. Exposure can be directly to the sun through glass or by an accelerated method using a carbon-arc or xenon light source. The xenon light source is the most commonly used accelerated exposure type, has a spectral distribution similar to sunlight, and gives better correlation with direct sunlight than the carbon-arc lamp.

Laundering Fastness.   Colorfastness to laundering (washing) under various conditions of temperature, bleaching, and abrasive action is determined according to AATCC test method no. 61. A Launder-Ometer or other apparatus for rotating closed containers in a thermostatically controlled water bath is required. Abrasive action is accomplished by the use of a low liquor ratio and an appropriate number of steel balls enclosed with the sample in a cylindrical container. Samples are sewn to a multifiber test fabric, and the color change of both fabrics after laundering is evaluated using an appropriate color change scale.

Gas Fading.   Many dyes used in textiles, particularly disperse dyes on acetate, triacetate, and polyester, will fade on exposure to oxides of nitrogen derived from the combustion of illuminating or heating gas. Gas fading can also occur in some sensitive resin-bonded pigments applied to cotton fabrics. AATCC test method no. 23 describes a method for exposure of test specimens and a control fabric to oxides of nitrogen until the control shows a change in color corresponding to that of the standard of fading. The test specimen is then examined for fading. If none is observed, the cycle is repeated until fading is obvious. The specimen is then ranked according to the number of exposures required to produce a shade change and the change in color according to the AATCC gray scale.

Fastness to Crocking. Crocking is defined as the transfer of color from the surface of a dyed fabric to another surface by rubbing. AATCC test method no. 8 is a method by which a colored test fabric swatch is fastened to the base of a Crockmeter and rubbed against a white crock test cloth under controlled conditions. Color transfer to the white cloth is evaluated by comparison with the AATCC Chromatic Transference Scale. A similar method, AATCC 116, uses a Rotary Vertical Crockmeter, which requires a smaller area of test fabric than the Crockmeter.

Soil Redeposition and Soil Release. Hydrophobic synthetic fibers have affinity for oily materials and therefore attract oily soils to a greater extent than natural fibers and hydrophilic synthetic fibers. To promote release during laundering of oily stains, a number of soil-release and whiteness-retention finishes have been developed. Two factors are involved in evaluating the soiling characteristics of fabrics: the amount of soil that can be deposited on a fabric during laundering, and the ease with which soil can be removed during laundering. Several AATCC methods exist to evaluate soil release of fabrics. AATCC test method no. 130 measures a fabric's ability to release oily stains during laundering by using a weight to force a stain into a fabric and then rating the residual stain after laundering by comparison with a standard replica. AATCC test methods no. 151 measure a fabric's resistance to soil redeposition during laundering using a launder-ometer and a terg-o-tometer, respectively; the latter is a more accelerated test. This method evaluates soil redeposition by measuring the change in reflectance of the samples.

Carpet soiling can be evaluated by AATCC 121, 122, and 123. Method no. 121 describes a procedure for visually rating degrees of cleanness in floor coverings; 122 involves subjecting carpets to foot traffic; and 123 is an accelerated laboratory procedure for tumbling carpet samples and a prepared synthetic soil in a ball mill, followed by evaluation by the visual rating method. The cross-sections of many carpet fibers, eg, trilobal or internal voids, are designed to hide soil, so carpet appearance does not always correlate with the amount of dirt present in the carpet.

Flammability. The terminology relating to the testing of flammability of textile fabrics and the resulting classifications of materials may be somewhat confusing. ASTM D4391 defines standard terminology relating to the burning behavior of textiles. A material may be classified as combustible, flame-resistant, flammable, noncombustible, or nonflammable. Various organizations and government agencies have established test methods for determining flammability; these methods are indexed and described in ASTM D4723. The following methods are described for flammability of textile products: ASTM D1230 for apparel textiles, D2859 for finished textile floor-covering materials, D3411 for textile materials, D4151 for blankets, and D4372 for camping tentage. The Consumer Product Safety Commission (CPSC) has defined specifications and test methods for clothing and textiles, including 16 CFR 1610 for general wearing apparel; 16 CFR 1615 for children's sleepwear, sizes 0 through 6X; 16 CFR 1616 for children's sleepwear, sizes 7 through 14; and 16 CFR 1630 for carpets and rugs.

In addition to the conventional methods for determining flammability of the fabric itself, some laboratories are using techniques for studying the thermal protection afforded by a garment. A few research laboratories are equipped with mannequins containing heat sensors to provide a temperature profile when the garment covering the mannequin is exposed directly to flames representing flash fire conditions. Garments tested in this manner are generally made of high performance fibers such as Nomex (DuPont) and are designed for use by the military, race car drivers, or firefighters.

SUBJECTIVE TESTS

Fabric Hand or Handle.   Fabric handle is somewhat difficult to define, although the term generally refers to properties such as draping quality, fullness, or stiffness (120). ASTM D123 defines a list of eight terms relating to fabric properties that make up the components of the general term, hand. These properties are flexibility, compressibility, extensibility, resilience, density, surface contour, surface friction, and thermal character (121). AATCC Evaluation Procedure no. 5 also provides guidelines for subjective evaluation of hand properties.

Although the description of fabric handle may rely on subjective terms, several methods exist to quantify the stiffness of a fabric. ASTM D1388 is used to measure the flexural rigidity and bending length of a woven fabric by calculating its resistance to bending under its own weight. The cantilever test is the preferred method for this ASTM procedure, in which a strip of fabric is slid in a direction parallel to its long dimension so that the fabric end projects from the end of a horizontal surface. The length of overhang is measured when the tip of the specimen is depressed under its own weight to the point where the line joining the tip to the edge of the platform makes an angle of 41.5° with the horizontal. The bending length is defined as one-half the length of overhang. Bending length is related to the drape stiffness of the fabric. Flexural rigidity is defined as the cube of the bending length multiplied by the fabric weight per unit area. A fabric having a higher flexural rigidity is stiffer.

A second method, ASTM D4032, is a circular bend stiffness test, in which a fabric specimen is forced through a circular opening 1.5 in. in diameter by a plunger that is 1 in. in diameter. The maximum force to push the fabric through the opening is determined. This method is a good indicator of the three-dimensional bending stiffness of a woven, knitted, or nonwoven fabric (121).

Years of development have led to a standardized system for objective evaluation of fabric hand (129). This, the Kawabata evaluation system (KES), consists of four basic testing machines: a tensile and shear tester, a bending tester, a compression tester, and a surface tester for measuring friction and surface roughness. To complete the evaluation, fabric weight and thickness are determined. The measurements result in 16 different hand parameters or characteristic values, which have been correlated to appraisals of fabric hand by panels of experts (121). Translation formulas have also been developed based on required levels of each hand property for specific end uses (129). The properties include stiffness, smoothness, and fullness levels as well as the total hand value. In more recent years, abundant research has been documented concerning hand assessment (130–133).

Drape. Drape is closely related to fabric hand. Whereas hand is based on tactile criteria, drape refers more to a fabric's appearance by its tendency to fall into graceful, three-dimensional folds. Fabric drape depends to a large degree on the same properties that influence hand, ie, flexural rigidity, thickness, and compressibility. Although stiffness is normally measured in a single direction, drape implies bending in all directions.

Drape can be measured by placing a circular fabric specimen over a round table or pedestal and viewing from directly overhead. A drape coefficient is defined as the ratio of the area of the fabric's actual shadow to the area of the shadow if the fabric were rigid. Drape is closely related to stiffness: the drape coefficient for a stiff fabric approaches a value of 1; a limp fabric has a drape coefficient near 0. The Cusick drape tester is an example of this type of measurement. For this method, the relative weights of paper rings representing tracings of the fabric's shadows are used to calculate drape coefficient.

Crease Retention, Wrinkle Resistance, and Durable Press. On bending or creasing of a textile material, the external portion of each filament in the yarn is placed under tension, and the internal portion is placed in compression. Thus, the wrinkle-recovery properties must be governed in part by the inherent, tensional elastic deformation and recovery properties of the fibers. In addition to the inherent fiber properties, the yarn and fabric geometry must be considered.

AATCC Test Method no. 66 describes measurement of recovery angle after placing a crease in a specimen. The specimen is creased by subjecting is to a prescribed load for a length of time. The recovery angle is then measured after a controlled recovery period. Recovery angles of greater than 120° are considered to indicate good wrinkle resistance (121). Because wrinkle resistance is particularly important for summer clothing, the AATCC stipulates that recovery angles be measured at 90% as well as 65% relative humidity. The former is considered to represent the most humid condition to which apparel textiles may be subjected in summer.

Fabrics that have been given durable-press or permanent-press finishes are generally treated with an uncured resin. In the post-cure process, curing is done by the garment maker to set the garment in the desired configuration. In the precure process, the fabric is cured prior to garment manufacture. Although the latter process does not result in permanent pleats, the smoothness of the cured fabric is retained during wear and laundering. Thus, durable-press goods are tested for recovery angle under both wet and dry conditions.

A second wrinkle-recovery test, AATCC test method no. 128, describes the determination of the appearance of textile fabrics after intentional wrinkling followed by evaluation of appearance in comparison to standard replicas. A visual rating from 1 (wrinkled) to 5 (smooth) is assigned. This method may be used for both woven and knitted fabrics, whereas the recovery angle method is applicable only to woven fabrics.

Appearance after Laundering. The extent to which permanent-press fabrics retain smoothness after laundering is determined by conducting a series of tests devised by AATCC: 143 for appearance of apparel and other textile end products after repeated home laundering; 124 for appearance of durable press fabrics after repeated home laundering; as well as 88C and 88B for appearance of wash-and-wear items, including creases after home laundering and seams after home laundering.

In each of these tests, the specimens are subjected to standard procedures simulating home-laundering conditions. A choice is provided for alternative washing temperatures and drying procedures to conform with care instructions recommended for the fabric or final garment. Evaluation is performed by visual comparison with standards prepared by AATCC under prescribed lighting; five grades from very poor to excellent can be assigned.

Luster. Luster is defined as the amount of light reflected from a fabric at different angles. It depends on fiber, yarn, and fabric geometries which control the regularity of the fabric surface. The greater the degree of parallelization of fibers in yarns and yarns in fabrics is, the greater the luster. Luster is also affected by fiber cross-sectional shape and, in synthetic fibers, the addition of delustrants, eg, titanium dioxide. Luster can be evaluated quantitatively by measuring light reflectance and the angle of the reflected light, but visual observation is necessary for a complete appraisal of luster. In general, a fabric having a smooth flat surface shows higher luster than a fabric having a rough surface and loose fibers. The more irregular the surface of the fabric is, the greater the degree of scatter of reflected light.

Comfort. In the past, the evaluation of fabric or garment comfort has been a subjective process influenced by such variables as temperature, insulating efficiency, moisture absorption, drying speed, softness, bulk, fabric construction, and air permeability. Human factors must also be considered. To predict the comfort of a material, a combination of hand evaluation, eg, using the Kawabata system, as well as determination of the heat and moisture transport properties, is necessary. Often, these values are correlated with a sensory evaluation of the tactile qualities of the material by a human subject panel. A thorough discussion of the many physical and psychological factors affecting comfort is available (134,135).

FAST System. Fabric Assurance by Simple Testing (FAST) is a system developed in the 1980s for objective measurement of those properties important to the appearance, handle, and performance of fabrics (136). The system is designed for use by garment makers and worsted wool finishers. The critical fabric properties for predicting garment performance have been defined as extensibility, shear rigidity, bending rigidity, and dimensional stability. Extensibility and shear rigidity (looseness) both affect the sewability of the fabric. Stiffness affects the handle of the fabric. Dimensional stability consists of the relaxation shrinkage, hygral expansion, and the stability of the fabric's surface layer, which in turn affects the subjective property of smoothness. Thus, the FAST system provides a means for establishing the tailorability of fabric based on subjective properties of looseness, stiffness, and stability by simple objective measurements. The system consists of a compression meter, a bending meter, an extension meter, and a dimensional stability test. Test results for a fabric are compared to a control chart showing defined tolerance limits of each measured property.

Geotextile Testing

Geotextiles (qv) are defined as any permeable technical material used with foundation, soil, rock, earth, or any geotechnical-related material as an integral part of an artificial project, structure, or system (137). Geotextiles function mainly in separation, reinforcement, filtration, drainage and/or moisture barrier applications. In separation, geotextiles are used between different layers of earth. In reinforcing, geotextiles have the ability to distribute a concentrated load over a large area of subgrade, thus avoiding local overloading (138). Filtration in fabric-to-soil systems means water can move across the plane of the geotextile, whereas soil is retained indefinitely on the upstream side without clogging the fabric. This is especially important under road and railroad beds and airfields. The aim of drainage is to improve workability, accessibility, and stability of treated land or sites (139). Filtration (140,141) is important in chimney and dam drainage, and in water drainage beneath athletic fields. Moisture barriers use geotextiles called geomembranes for the purpose of preventing or slowing down water passage in either liquid or vapor state. Geomembranes are used as pond liners and covers, underneath landfills, etc.

ASTM has test methods specifically for testing of geotextiles (142). ASTM D4873 discusses proper identification, storage, and handling of geotextiles. Measuring of mass per unit area of a geotextile is described in ASTM D5261, and measuring of nominal thickness of geotextiles and geomembranes is discussed in D5199. Degradation of geotextiles is measured in D4355 and in ASTM D4594. Additional information on degradation of geotextiles is also available (143–145).

In applications for filtration and drainage, fabric permeability, pore size, and permittivity are important. These parameters can be measured using ASTM D4491, D5493, and D4751, respectively. Permittivity is defined as the volumetric flow rate of water per unit cross-sectional area per unit head under laminar flow conditions. Clogging of geotextiles can be measured by ASTM D1987 and D5101. The long-term behavior of geotextile fabrics used as filter media has been discussed (146), as have permeability constants for many geotextile fabrics (147).

In the past, tensile properties of geotextiles were measured using ASTM D1682, which has since been discontinued. In the 1990s, there are a number of methods for tensile testing, including ASTM D4595, Tensile Properties of Geotextiles by Wide Width Strip Method. This method is often used for fabrics that have a tendency to contract (or neck) during tensile testing (148). The greater width of specimen minimizes the contraction effect and provides results closer to those expected from geotextiles in the field. The grab breaking load and elongation of geotextiles are measured using ASTM D4632. ASTM D4533 describes the method for trapezoid tear testing of fabrics used in geotextiles. Discussions on tensile properties of geotextiles are available (149–153). In addition, ASTM D4884 describes seam strength of sewn geotextiles; D5262, creep behavior of geotextiles; and D5617, multiaxial tension test for geotextiles.

Abrasion testing for geotextiles is found in ASTM D4886 (154,155). ASTM D5321 covers the determination of the friction coefficient between soil and geotextiles (156,157). Other standards on testing of geotextiles include INDA (158) standards 180.2 (breaking load and elongation), 180.3 (trapezoid tear), 180.4 (puncture strength), 180.5 (bursting strength), 180.6 (pore size), and 180.7

(permittivity). Finally, a review of ASTM discontinued methods for testing of geotextiles is also available (148).

BIBLIOGRAPHY

"Textile Testing" in *ECT* 1st ed., Vol. 13, pp. 908–927, by E. R. Kaswell, Fabric Research Laboratories, Inc.; in *ECT* 2nd ed., Vol. 20, pp. 33–62, by E. R. Kaswell, G. A. M. Butterworth, and N. J. Abbott, Fabric Research Laboratories, Inc.; "Textile (Testing)" in *ECT* 3rd ed., Vol. 22, pp. 802–835, by R. W. Singleton, University of Connecticut.

1. *Book of ASTM Standards*, Vols. 7.01 and 7.02, ASTM, Philadelphia, Pa., 1995.
2. *AATCC Technical Manual*, Vol. 71, American Association of Textile Chemists and Colorists, Research Triangle Park, N.C., 1996.
3. W. E. Morton and J. W. S. Hearle, *Physical Properties of Textile Fibers*, The Textile Institute, Bath, U.K., 1993.
4. S. L. Anderson and R. C. Palmer, *J. Text. Inst.* **44**, T95 (1953).
5. *Wool Research*, Vol. 3, *Testing and Control*, Wira, Leeds, U.K., 1955, p. 41.
6. *Wool Sci. Rev.* **9**, 23 (1952).
7. R. Guse and co-workers, *Melliand Textilber./Intern. Text. Rep.* **76**, 25 (1995).
8. J. Yankey and G. Deaton, *Proceedings of the Beltwide Cotton Production Conference*, 1995.
9. T. Madeley and R. Postle, *Text. Hor.* **15**, 43 (1995).
10. R. W. Singleton, in H. F. Mark, S. M. Atlas, and E. Cernia, eds., *Man-made Fibers-Science and Technology*, Vol. 3, Wiley-Interscience, New York, 1968.
11. Y. Muraoka and co-workers, *Text. Res. J.* **65**, 454 (1995).
12. *ICAC Recorder*, **13**, 6 (1995).
13. M. Glass, T. Dabbs, and P. Chudleigh, *Text. Res. J.* **65**, 85 (1995).
14. K. Gilhaus and Luenenschloss, *Int. Text Bull. Spinning*, **2**, 117 (1980).
15. J. C. Abbott, L. B. Jaycox, and G. M. Ault, *TAPPI*, 119 (1979).
16. P. H. Hermans, *Physics and Chemistry of Cellulose Fibres*, Elsevier, Amsterdam, Netherlands, 1949, p. 197.
17. N. J. Abbott and A. C. Goodings, *J. Text. Inst.* **40**, T232 (1949).
18. E. R. Kaswell, *Textile Fibers, Yarns, and Fabrics*, Reinhold Publishing Corp., New York, 1953.
19. E. R. Kaswell, *Wellington Sears Handbook of Industrial Textiles*, Pepperell Co., Inc., West Point, N.Y., 1963.
20. M. A. Sieminski, *Rayon Text. Month.* **24**, 585 (1943).
21. J. H. MacGregor, *J. Text. Inst.* **42**, 525 (1951).
22. M. A. Sieminski, R. W. Singleton, and B. S. Sprague, *Text. Res. J.* **31**, 917 (1961).
23. H. DeW. Smith, *Am. Soc. Text. Mater. Proc.* **44**, 543 (1944).
24. J. F. Clark and J. M. Preston, *J. Text. Inst.* **44**, T596 (1953).
25. M. V. Forward and H. J. Palmer, *J. Text. Inst.* **45**, T510 (1954).
26. P. F. Dismore and W. O. Statton, *J. Polymer Sci. C.* **13**, 133 (1966).
27. W. S. Hearle, P. K. Sern Gupta, and A. Matthews, *Fibre Sci. Technol.* **3**, 167 (1971).
28. R. S. Merkel, *Text. Res. J.* **33**, 84 (1963).
29. E. H. Mercer and K. R. Makinson, *J. Text. Inst.* **38**, T227 (1947).
30. J. C. Guthrie and P. H. Oiver, *J. Text. Inst.* **43**, T579 (1952).
31. M. W. Pascoe and D. Tabor, *Proc. Roy. Soc.* **A235**, 210 (1956).
32. J. Lindberg and N. Gralen, *Text. Res. J.* **18**, 287 (1948).
33. E. Lord, *J. Text. Inst.* **46**, P41 (1955).
34. W. S. Hearle and A. K. M. M. Husain, *J. Text. Inst.* **62**, 83 (1971).
35. D. S. Taylor, *J. Text. Inst.* **46**, P59 (1955).

36. B. Speakman and E. Scott, *J. Text. Inst.* **22**, T339 (1931).
37. G. Howell and J. Mazur, *J. Test. Inst.* **44**, T59 (1953).
38. A. N. J. Heyn, *Text. Res. J.* **22**, 513 (1952).
39. A. N. J. Heyn, *Text. Res. J.* **23**, 246 (1953).
40. A. F. Wyssling, *Helv. Chim. Acta.* **19**, 900 (1936).
41. J. M. Preston and K. Freeman, *J. Text. Inst.* **34**, T19 (1943).
42. R. C. Faust, *Proc. Phys. Soc.* **B68**, 1081 (1951).
43. R. D. Andrews, *J. Appl. Phys.* **25**, 1223 (1954).
44. R. D. Andrews and J. F. Rudd, *J. Appl. Phys.* **27**, 990 (1956).
45. E. Alexander and co-workers, *Text.. Res. J.* **26**, 606 (1956).
46. M. Shiloh and co-workers, *Text. Res. J.* **31**, 999 (1961).
47. E. Alexander and co-workers, *Text. Res. J.* **32**, 898 (1963).
48. L. Beste and R. Hoffman, *Text. Res. J.* **32**, 721 (1961).
49. R. Meredith and J. W. S. Hearle, eds., *Physical Methods of Investigating Textiles*, Interscience Publishers, New York, 1956.
50. W. E. Morton and W. S. Hearle, *Physical Properties of Textile Fibers*, The Textile Institute, Butterworths, London, 1961.
51. P. F. Dismore and W. Statton, *J. Polym. Sci. Polym. Symp.* 133 (1966).
52. F. T. Peirce, *J. Text. Inst.* **17**, T355 (1926).
53. R. Meredith, *J. Text. Inst.* **37**, T205 (1946).
54. J. L. Spencer-Smith, *J. Text. Inst.* **38**, P257 (1947).
55. C. Nanjundayya, Ph.D. diss., University of Manchester, 1949.
56. J. C. Smith and co-workers, *Text. Res. J.* **32**, 721 (1961).
57. D. R. Petterson and co-workers, *Text. Res. J.* **30**, 411 (1960).
58. R. J. Coskren, H. M. Morgan, and C. C. Chu, *High Speed Testing*, Vol. 3, Wiley-Interscience, New York, 1961.
59. M. J. Coplan, *The Effect of Temperature on Textile Materials*, Documents 53-21, Wrighton Development Center, Dayton, Ohio, 1953.
60. F. Maillard, *J. Text. Inst.* **40**, 379 (1949).
61. H. Shiefer, L. Fourt, and R. Kropf, *Text. Res. J.* **18**, 18 (1948).
62. S. Backer, *Text. Res. J.* **21**, 453 (1951).
63. B.S. 4029:1966, British Standards Institution, London, 1966.
64. J. C. Guthrie and S. Norman, *J. Text. Inst.* **52**, T503 (1961).
65. J. C. Guthrie and J. Wibberley, *J. Text. Inst.* **56**, T97 (1965).
66. D. W. Hadley, *J. Text. Inst.* **60**, 301, 312 (1969).
67. P. W. Carlene, *J. Text Inst.* **38**, T38 (1947).
68. F. T. Peirce, *J. Text Inst.* **21**, T377 (1930).
69. P. W. Carlene, *J. Text Inst.* **41**, T159 (1950).
70. J. C. Guthrie, D. H. Morton, and P. H. Oliver, *J. Text Inst.* **45**, T192 (1954).
71. R. Khayatt and N. H. Chamberlain, *J. Text Inst.* **39**, T185 (1948).
72. J. W. Ballou and J. C. Smith, *J. Appl. Phys.* **20**, 493 (1949).
73. R. Meredith and B. S. Hsu, *J. Polymer Sci.* **61**, 271 (1962).
74. G. A. M. Butteworth and N. J. Abbott, *ASTM J. Materials*, **2**, 487 (1967).
75. S. F. Calil, B. C. Goswami, and J. W. S. Hearle, *J. Phys.* **13**, 725 (1980).
76. R. Meredith, *J. Text Inst.* **45**, T489 (1954).
77. W. E. Morton and F. Permanyer, *J. Text Inst.* **38**, T54 (1947).
78. W. E. Morton and F. Permanyer, *J. Text Inst.* **40**, T371 (1949).
79. F. Permanyer, Ph.D. diss., University of Manchester, 1947.
80. R. Schwab, *Klepzig's Textil-Z.* **42**, 397 (1939).
81. P. A. Koch, *Textil-Rdsch.* **4**, 199 (1949); **6**, 111 (1951).
82. D. Finlayson, *J. Text. Inst.* **38**, T50 (1947).
83. H. J. Kolb and co-workers, *Text Res. J.* **23**, 84 (1953).

84. P. W. Carlene, *J. Soc. Dyers Colour.* **60**, 232 (1944).
85. M. Harris, *Handbook of Textile Fibers*, Textile Book Publishers, Inc., New York, 1954.
86. *Textile World*, McGraw-Hill Book Co., Inc., New York, 1962.
87. A. R. Urquhart and A. M. Williams, *J. Text. Inst.* **15**, T138 (1924).
88. A. R. Urquhart and N. Eckersall, *J. Text. Inst.* **23**, T163 (1932).
89. E. A. Hutton and J. Gartside, *J. Text. Inst.* **40**, T161 (1949).
90. J. B. Speakman, *J. Soc. Chem. Industr.* **49**, 209T (1930).
91. A. Hutton and J. Gartside, *J. Text. Inst.* **40**, T170 (1949).
92. R. Hill, ed., *Fibres from Synthetic Polymers*, Elsevier, Amsterdam, Netherlands, 1953.
93. J. E. Ford, *Fibre Data Summaries*, Shirley Institute, Manchester, U.K., 1966.
94. A. R. Urquhart and A. M. Williams, *J. Text. Inst.* **16**, T155 (1925).
95. *Textile World*, Synthetic Fiber Table, 1953.
96. Textile Test Methods, Federal Test Methods Standard No. 191, Dec. 14, 1968.
97. I. H. Welch, *Am. Dyes. Rep.* **50**, 25 (1961).
98. J. M. Preston, *Trans. Faraday Soc.* **42B**, 131 (1946).
99. H. J. White and P. B. Stam, *Text. Res. J.* **19**, 136 (1949).
100. F. F. Morehead, *Text Res. J.* **17**, 96 (1947).
101. F. F. Morehead, *Text Res. J.* **22**, 535 (1952).
102. P. Denton, *J. Sci. Instrum.* **29**, 55 (1952).
103. P. S. H. Henry, *Brit. J. Appl. Phys.* **2**, S31 (1953).
104. O. Kovalcik and R. Simo, *Makrotest*, 175 (1980).
105. W. G. Wolfgang, in J. J. Press, ed., *Man-Made Textile Encyclopedia*, Interscience Publishers, New York, 1959.
106. J. M. Preston, *J. Text. Inst.* **40**, T767 (1949).
107. R. Kunkel, R. W. Singleton, and B. S. Sprague, *Text. Res. J.* **35**, 228 (1965).
108. R. W. Singleton and P. Cook, *Text. Res. J.* **39**, 43 (1969).
109. H. J. Vogt and R. Punthe, *Textilbetrieb (Wuerzberg, Ger.)*, **39**, 6 (1981).
110. P. R. Lord, *Text. Res. J.* **41**, 778 (1971).
111. P. R. Lord, *Economics Science & Technology of Yarn Production*, North Carolina State University, Raleigh, N.C., 1981.
112. J. W. S. Hearle, P. Grosberg, and S. Backer, *Structural Mechanics of Fibers, Yarns and Fabrics*, John Wiley & Sons, Inc., New York, 1969.
113. R. W. Singleton, *Text. Res. J.* **50**, 457 (1980).
114. S. Milosavljevic, T. Tadic, and S. Stankovic, *Text. Month*, 25 (Jan. 1995).
115. A. Nevel, F. Avsar, and L. Rosales, *JSN Intern.* **95**(8), 30 (Aug. 1995).
116. F. Frank, *J. Text. Inst.* **51**, T83 (1960).
117. P. Hempel, *Man-Made Text.* **38**, 36 (1961).
118. *Man-Made Text.* **39**, 48 (1962).
119. K. Baldwin, *Modern Text. Manage.* **7**, 48 (1971).
120. A. M. Collier, *A Handbook of Textiles*, 3rd ed., Wheaton, Exeter, 1980.
121. R. S. Merkel, *Textile Product Serviceability*, Macmillan, New York, 1991.
122. R. T. Cary, *Text. Res. J.* **51**, 61 (1981).
123. S. Backer and S. Tannenhaus, *Text. Res. J.* **21**, 635 (1951).
124. D. S. Lyle, *Performance of Textiles*, John Wiley & Sons, New York, 1977.
125. M. E. Whelan and L. E. MacHattie, *Text. Res. J.* **25**, 197 (1955).
126. P. Nordon and J. G. Downes, *J. Text. Inst.* **56**, T8 (Aug. 1965).
127. D. B. Judd and G. Wyszecki, *Color in Business*, 3rd ed., Wiley-Interscience, New York, 1975.
128. F. W. Billmeyer and M. Saltzman, *Principles of Color Technology*, Wiley-Interscience, New York, 1966.
129. S. Kawabata, *The Standardization and Analysis of Hand Evaluation*, 2nd ed., The Textile Machinery Society of Japan, Tokyo, 1980.

130. H. M. Behery, *Text. Res. J.* **56**, 227 (1986).
131. N. Pan and co-workers, *Text. Res. J.* **58**, 531 (1988).
132. N. G. Ly, *Text. Res. J.* **59**, 17 (1989).
133. T. J. Mahar and R. Postle, *Text. Res. J.* **59**, 721 (1989).
134. T. Wallenberger, *Text. Res. J.* **10**, 5 (May 1980).
135. N. R. S. Hollies and L. Fourt, *Clothing: Comfort and Function*, Marcel Dekker Inc., New York, 1970.
136. *Fabric Assurance by Simple Testing: Instruction Manual*, CSIRO Division of Wool Technology, Sydney, Australia, 1989.
137. T. F. Cooke, *Text. Month*, 25 (Aug. 1988).
138. R. V. van Zanten, ed., *Geotextiles and Geomembranes in Civil Engineering*, A. A. Balkema Publishing, Rotterdam, The Netherlands, 1986.
139. T. F. Cooke, *Text. Month*, 63 (Sept. 1988).
140. S. DeBerardino, *Geotech. Fabrics Rep.* **9**, 12 (1993).
141. S. DeBerardino, *Geotech. Fabrics Rep.* **8**, 4 (1993).
142. *Book of ASTM Standards*, Vol. 4.08, ASTM, Philadelphia, Pa., 1995.
143. Y. G. Hsuan, R. M. Koerner, *Geotech. Fabrics Rep.* **8**, 12 (1993).
144. M. Sotton, *Index 81 Congress Papers, Nonwovens For Technical Applications*, 1–15 (May 1981).
145. E. Martin, *Proceedings of the 2nd International Conference on Geotextiles*, 1982, pp. 751–756.
146. R. V. van Zanten and R. A. H. Thabet, in Ref. 145, p. 259.
147. G. Raumann, in Ref. 145, pp. 55–60.
148. T. F. Cooke, *Text. Month*, 42 (Oct. 1988).
149. K. Moritz and H. Murray, in Ref. 145, pp. 757–762.
150. J. Lafleur, S. Z. Akber, and Y. Hammamji, *Proceedings of the 3rd International Conference on Geotextiles*, 1986, pp. 935–940.
151. E. Leflaive, J. L. Paute, and M. Segouin, in Ref. 145, pp. 733–738.
152. E. A. Nowatzki and S. R. Pageau, *Geotech. Fabrics Rep.* **4**, 12–15 (1984).
153. J. Puig and co-workers, in Ref. 145, pp. 763–768.
154. C. G. Gray, in Ref. 145, pp. 817–821.
155. D. Van Dine, G. Raymond, and S. E. Williams, in Ref. 145, pp. 811–816.
156. H. Kabeya and co-workers, *Text. Res. J.* **63**, 604–610 (1993).
157. E. Dembicki and J. Alenowicz, *Geotext. Geomembranes*, **4**, 307–314 (1987).
158. INDA Association of the Nonwoven Fabrics Industries, Cary, North Carolina.

PAMELA BANKS-LEE
JAN PEGRAM
North Carolina State University

THALLIUM AND THALLIUM COMPOUNDS

Thallium Metal

Occurrence.   Thallium [7440-28-0], discovered by Sir William Crookes in 1861 (1), belongs to Group 13 (IIIA) of the Periodic Table along with boron, aluminum, gallium, and indium. ^{203}Tl [14280-48-9] (29.5%) and ^{205}Tl [14280-49-0] (70.5%) are the two stable naturally occurring isotopes, whereas $^{191-202}$Tl [32148-21-3; 18235-45-5; 32148-22-4; 18235-46-6; 26683-69-2; 18724-77-1; 14107-52-9; 15743-50-7; 15064-66-1; 15720-55-5; 15064-65-0; 15720-57-7], respectively; ^{204}Tl [13968-51-9]; and $^{206-210}$Tl [15035-09-3; 14133-67-6; 14913-50-9; 15690-73-0; 13966-01-3], respectively; are the decomposition products of natural radioactive series (see RADIOISOTOPES).

Thallium is not a particularly rare metal and its abundance in the earth's crust is ca 0.3 ppm. It occurs not only in oxide minerals but also as a chalcophilic element. The metal cation commonly occurs in potash minerals and in a number of thallium-containing minerals, eg, crookesite [12414-86-7], $(Cu,Tl,Ag)_2Se$ (17% Tl), in Sweden; lorandite [15501-93-6], $TlAgS_2$ (59% Tl), in Greece; hutchinsonite [12198-34-4], $(Tl,Pb)S \cdot (Ag,Cu)_2S \cdot 5As_2S_3$, in Switzerland; vrbaite [12006-31-4], $TlSbAs_2S_5$ (30% Tl); and avicennite [12022-82-1], $7Tl_2O_3 \cdot Fe_2O_3$. Of these, crookesite and lorandite are the most important.

Properties.   Thallium is grayish white, heavy, and soft. When freshly cut, it has a metallic luster that quickly dulls to a bluish gray tinge like that of lead. A heavy oxide crust forms on the metal surface when in contact with air for several days. The metal has a close-packed hexagonal lattice below 230°C, at which point it is transformed to a body-centered cubic lattice. At high pressures, thallium transforms to a face-centered cubic form. The triple point between the three phases is at 110°C and 3000 MPa (30 kbar). The physical properties of thallium are summarized in Table 1.

In moist air, thallium slowly oxidizes to thallium(I) oxide [1314-12-1]. Steam and air or oxygen react readily with thallium forming thallium(I) hydroxide [12026-06-1]. Thallium dissolves only slowly in sulfuric or hydrochloric acid and the resultant salts have low solubilities. It is not soluble in alkaline solutions.

Production and Economic Aspects.   Thallium is obtained commercially as a by-product in the roasting of zinc, copper, and lead ores. The thallium is collected in the flue dust in the form of oxide or sulfate with other by-product metals, eg, cadmium, indium, germanium, selenium, and tellurium. The thallium content of the flue dust is low and further enrichment steps are required. If the thallium compounds present are soluble, ie, as oxides or sulfates, direct leaching with water or dilute acid separates them from the other insoluble metals. Otherwise, the thallium compound is solubilized with oxidizing roasts, by sulfatization, or by treatment with alkali. The thallium precipitates from these solutions as thallium(I) chloride [7791-12-0]. Electrolysis of the thallium(I) sulfate [7446-18-6] solution affords thallium metal in high purity (5,6). The sulfate solution must be acidified with sulfuric acid to avoid

Table 1. Physical Properties of Thallium

Property	Value
atomic weight	204.37
melting point, °C	303
boiling point, °C	1457
density, g/cm^3	11.85
thermal conductance, W/(cm·K)^a	0.39
specific heat, 20°C, J/g^b	0.13
heat of fusion, J/g^b	21.1
heat of vaporization, J/g^b	795
first ionization energy, kJ/mol^b	590
electronegativity, Pauling's scale	1.8
covalent radius, pm	148
Brinell hardness	2
linear coefficient of expansion	28×10^{-6}
electrical resistivity, $\mu\Omega$·cm	18
tensile strength, MPa^c	9.0
crystal structure	
below 230°C	close-packed hexagonal
above 230°C	body-centered cubic
nuclear spin, $\hbar$	
^{203}Tl	0.5
^{205}Tl	0.5
nuclear magnetic moment, J/T^d	
^{203}Tl	1.495×10^{-23}
^{205}Tl	1.509×10^{-23}

^{a}To convert W/(cm·K) to (cal·cm)/(s·cm^2·°C), divide by 4.184.
^{b}To convert J to cal, divide by 4.184.
^{c}To convert MPa to psi, multiply by 145.
^{d}To convert J/T to Bohr magneton, divide by 9.274×10^{-24}.

cathodic separation of zinc and anodic deposition of thallium(III) oxide [1314-32-5]. The metal deposited on the cathode is removed, kneaded into lumps, and dried. It is then compressed into blocks, melted under hydrogen, and cast into sticks.

Demand for thallium has decreased since the early 1980s because, wherever possible, substitute materials are used due to its hazardous nature. Annual production is very small, estimated to be 10,000 kg/yr. The principal domestic producer is Noah Chemical Division, Noah Industrial Corporation. Thallium is available in three grades: 99.9%, 99.99%, and 99.999% pure. The 99.9% material is priced at $360/kg ($17/kg 1969).

Uses. Thallium has limited commercial applications because of its toxic nature. Thallium forms alloys readily with many metals and some of these alloys have unique properties. A number of binary, ternary, and quaternary eutectic alloys are known and have very low coefficients of friction and good resistance to acids (7). These alloys can be used in bearings, eg, Ag–Tl [83542-95-4] and Au–Al–Tl [83547-99-3]. Alloys of silver and thallium are used in contact points, and an alloy of lead, tin, and thallium is used in the production of anodes. The most important alloy of thallium is the mercury–thallium alloy [83542-96-5], which forms a eutectic at 8.7 wt % Tl and has a melting point of $-60°C$. It can be used as a substitute for mercury in switches and seals for equipment used in the polar region, stratosphere, or space program.

Thallium Compounds

Unlike boron, aluminum, gallium, and indium, thallium exists in both stable univalent (thallous) and trivalent (thallic) forms. There are numerous thallous compounds, which are usually more stable than the corresponding thallic compounds. The thallium(I) ion resembles the alkali metal ions and the silver ion in properties. In this respect, it forms a soluble, strongly basic hydroxide and a soluble carbonate, oxide, and cyanide like the alkali metal ions. However, like the silver ion, it forms a very soluble fluoride, but the other halides are insoluble. Thallium(III) ion resembles aluminum, gallium, and indium ions in properties.

Thallic salts are readily reduced to the thallous salts by common reducing agents. The standard potentials for the Tl(III)/Tl(I) and Tl(I)/Tl(0) systems in aqueous solutions have been measured (8):

$$Tl^+ \longrightarrow Tl^{3+} + 2\,e^- \qquad E^0 = -1.252 - 0.0295 \log \frac{[Tl^{3+}]}{[Tl^+]}$$

$$Tl^0 \longrightarrow Tl^+ + e^- \qquad E^0 = +0.336 - 0.0591 \log [Tl^+]$$

Thallium(I) salts are oxidized only by very powerful oxidizing agents, eg, MnO_4^- or Cl_2, in acid solutions. The properties of the more important thallium compounds are listed in Table 2.

Thallous sulfate, thallous nitrate, and thallous and thallic oxide are the main compounds produced in bulk quantities by Noah Chemical. Approximately 20 other thallium compounds are also available commercially from Noah Chemical, Cooper Chemical, and Alfa Products, Ventron Division, Thiokol Corporation, in research and production quantities. However, demand for thallium compounds is small and limited to such applications as synthetic or analytical reagents.

Carbonates.   Thallium forms only thallous carbonate and bicarbonate (9). Thallous carbonate, Tl_2CO_3, is prepared by saturating a hot aqueous solution of thallium hydroxide with carbon dioxide. The carbonate separates upon reduction of the volume and cooling of the concentrated solution. It dissolves in water, yielding a strongly basic solution because of hydrolysis. It is stable to 175°C in air but loses CO_2 on further heating. Thallous bicarbonate [65975-01-1], $TlHCO_3$, forms by the action of carbon dioxide on a dilute aqueous solution of Tl_2CO_3.

Carboxylates.   Thallium forms a large number of carboxylate derivatives in both Tl(I) and Tl(III) forms. The Tl(I) salts of carboxylic acids can be prepared from the acid and thallium(I) hydroxide or carbonate. Thus, thallous formate, $TlOOCH$, is prepared by treating Tl_2CO_3 or $TlOH$ with formic acid, followed by

Table 2. Properties of Thallium Compounds

Compound	CAS Registry Number	Formula	Color and crystal form	Mp, °C	Density, g/cm³	Solubility, g/100 g solvent[a] Water	Other
thallous carbonate	[29809-42-5]	Tl_2CO_3	colorless, monoclinic	272	7.16	5.2 (18), 22.4 (100)	methanol
thallous formate	[992-98-3]	TlO_2CH	colorless, needles	101	4.967	500 (10)	
thallous acetate	[563-68-8]	TlO_2CCH_3	silky white, needles	131	3.765	very soluble	alcohols
thallous fluoride	[7789-27-7]	TlF	colorless, cubic	327	8.23	78.6 (15)	alcohols,HF
thallic fluoride	[7783-57-5]	TlF_3	olive green, orthorhombic	550 dec	8.65	insoluble	
thallous chloride	[7791-12-0]	$TlCl$	colorless, cubic	430	7.0	0.32 (20), 2.38 (100)	
thallic chloride	[13453-32-2]	$TlCl_3$	hexagonal plate	155		very soluble	alcohols, ether
thallous bromide	[7789-40-4]	$TlBr$	green–yellow, cubic	456	7.5	0.05 (20), 0.25 (60)	
thallic bromide	[13701-90-1]	$TlBr_3$	yellow			soluble	alcohols
thallous iodide	[7790-30-9]	TlI	yellow, rhombic	440	7.29	0.0006 (20)	
thallous oxide	[1314-12-1]	Tl_2O	black	300	9.52	dec	
thallic oxide	[1314-32-5]	Tl_2O_3	black	7.7	10.11	insoluble	
thallous hydroxide	[12026-06-1]	$TlOH$	yellow, needles	139 dec	7.44	25.9 (0), 52 (40)	alcohols
thallous nitrate	[10102-45-1]	$TlNO_3$	colorless	206	5.55	8 (15), 594 (104.5)	
thallic nitrate	[13746-98-0]	$Tl(NO_3)_3$	colorless			dec	
thallous sulfate	[7446-18-6]	Tl_2SO_4	colorless, prisms	632	6.77	4.87 (20), 18.45 (100)	

[a]Value in parentheses is temperature in °C of solubility.

evaporation of the solution in vacuum (9). Thallium(I) acetate is prepared similarly. Alternatively, it can be obtained by adding acetic acid to a solution of thallium(I) ethoxide [20398-06-5] in ethanol (10). These carboxylates are colorless, light-insensitive, hygroscopic, sharp-melting, and stable solids. A 1:1 thallium formate–thallium malonate [2757-18-8] saturated aqueous solution, called Clerici's solution [61971-47-9], which has a density of 4.324 g/mL at 20°C, has been used as a heavy liquid in sink–float separation of minerals on a laboratory scale (see FLOTATION; MINERAL RECOVERY AND PROCESSING).

Thallium(III) salts of formic, acetic, and trifluoroacetic acids are prepared from the corresponding acid and thallium(III) oxide (11). Other thallium(III) carboxylates can be obtained from the metathesis of thallium(III) acetate and the carboxylic acid (12). They are colorless, hygroscopic solids.

Halides. Thallous halides are air-stable, light-sensitive, anhydrous salts (9). The water-soluble thallous fluoride is made from Tl_2CO_3 and aqueous hydrogen fluoride. The salt is obtained upon concentrating and cooling the solution. Thallous chloride bromide, and iodide are prepared by precipitation from aqueous solution of Tl_2SO_4 with dilute solutions containing the corresponding anions. The chloride can be purified by recrystallization from hot water, whereas the bromide is purified by digesting with boiling water. Thallium(I) halides form many double salts and complexes with nitrogen, oxygen, and sulfur ligands.

Thallium(III) fluoride has been prepared by the action of fluorine or bromine trifluoride on thallium(III) oxide at 300°C. It is stable to ca 500°C but is extremely sensitive to moisture. Thallium(III) chloride can be obtained readily as the tetrahydrate [13453-33-3] by passing chlorine through a boiling suspension of TlCl in water. It can be dehydrated with thionyl chloride. Thallium(III) bromide tetrahydrate [13453-29-7] is prepared similarly, whereas the iodide prepared in this manner is thallium(I) triiodide [13453-37-7], $Tl^+I_3^-$.

A large number of halide complexes of thallium(III) have been prepared by precipitation of the complexes from solution with a suitable cation, eg, Tl^+, $(C_2H_5)_4N^+$, $(C_6H_5)_4As^+$, and K^+. Both four-coordinated $[TlX_4]^-$ and six-coordinated $[TlX_6]^{3-}$ ions exist in solutions and in solid states.

Hydrides. Unlike boron and aluminum, which form many stable and useful hydrides, there is very little evidence for the existence of TlH and TlH_3 (13). Lithium tetrahydridothallate(III) [82374-47-8], $LiTlH_4$, has been obtained from $TlCl_3$ and LiH in ether at $-15°C$ (14). It decomposes rapidly above 30°C to hydrogen, thallium, and lithium hydride (9).

Thallium(I) salts of tetrahydridoborate and aluminate are obtained from a Tl(I) compound, eg, ethoxide, perchlorate, or nitrate, and $LiBH_4$ or $LiAlH_4$ in ether. Thallium(I) tetrahydridoborate [61204-71-5], $TlBH_4$, is unstable at 40°C, evolving diborane. Thallium(I) tetrahydridoaluminate [82391-12-6], $TlAlH_4$, is extremely unstable, decomposing to H_2 and Tl at $-80°C$.

Oxides and Hydroxides. Thallous oxide is most readily obtained by dehydration of TlOH in high vacuum at 50°C. It is black, crystalline, and hygroscopic. It reacts with water to form the hydroxide and dissolves in ethanol to yield the ethoxide (9).

Thallic oxide can be prepared by reaction of thallium with oxygen or hydrogen peroxide and an alkaline thallium(I) solution. However, it is more easily made from the oxidation of thallous nitrate by chlorine in aqueous potassium hydroxide solution. It is insoluble in water but dissolves in carboxylic acids to give carboxylates.

There is no evidence for the existence of thallic hydroxide; addition of hydroxide to an aqueous solution of a Tl(III) salt gives Tl_2O_3 instead. Thallous hydroxide can be isolated as yellow needles by the hydrolysis of thallous ethoxide [20398-06-5], which is conveniently prepared as a heavy oil by the oxygen oxidation of thallium metal in ethanol vapor. Thallous hydroxide darkens at room temperature and decomposes to Tl_2O and H_2O on warming.

Nitrates. Thallous nitrate, a convenient source of other thallium(I) derivatives, eg, halides, is prepared from the reaction of the pure metal with dilute nitric acid (9). The solid is stable to 300°C and decomposes at 800°C to Tl_2O, NO, and NO_2. Thallic nitrate is obtained as a trihydrate upon dissolving Tl_2O_3 in cold nitric acid. It decomposes to Tl_2O_3 on heating to 100°C or upon hydrolysis. Thallic nitrate, soluble in alcohols and ethers of polyethylene glycols, is often used as an oxidizing agent in organic syntheses.

Sulfates. Thallous sulfate is a commercial product produced by reaction of the metal with sulfuric acid and concentration of the solution until crystallization begins (9). It reacts with thallic sulfate, yielding $Tl^ITl^{III}(SO_4)_2$ [37475-01-7], and with SO_3, forming the pyrosulfate, $Tl_2S_2O_7$ [82391-11-5]. Thallic sulfate is extremely unstable and therefore cannot be isolated. Reaction of thallic oxide with dilute sulfuric acid results in $HTl(SO_4)_2$ [15478-75-8] or $Tl(OH)SO_4$ [37205-71-3], depending on the concentration of sulfuric acid used.

Organometallics. Organothallium compounds have attracted intense interest mostly because of the applications of these compounds in organic synthesis. Stable compounds occur in both Tl(I) and Tl(III) oxidation states. Only the cyclopentadienyl ligand is known to stabilize the Tl(I) ion, and the parent compound, cyclopentadienylthallium(I) [34822-90-7], is prepared from freshly distilled cyclopentadiene and thallium(I) hydroxide (15). It is a yellow, air-stable solid which is insoluble in most common organic solvents. X-ray and microwave spectroscopic studies in the gas phase indicate C_{5v} symmetry consistent with the half-sandwich structure. The crystal has a polymeric structure with zigzag chains of alternating C_5H_5 unit and Tl atoms. A number of alkylated derivatives of cyclopentadienylthallium(I), $(RC_5H_4)Tl$ where R = CH_3 [34034-67-8], C_2H_5 [28553-52-8], and C_4H_9 [28553-53-9], can also be prepared by the same procedure.

Organothallium(III) derivatives can be classified into three types: R_3Tl, R_2TlX, and $RTlX_2$. The trialkyl derivatives are reactive, unstable compounds, whereas the dialkyl derivatives are among the most stable and least reactive organometallic compounds. The monoalkyl compounds are rather unstable and often cannot be isolated. They are important intermediates in some Tl(III)-promoted organic reactions.

R_3Tl compounds, eg, methyl [3003-15-4], ethyl [687-82-1], isobutyl [3016-08-8], and phenyl [3003-04-1] thallium(III), are usually prepared by the reaction between a dialkyl or diarylthallium halide and an organolithium reagent in ether (16):

$$R_2TlX + LiR \longrightarrow R_3Tl + LiX$$

or a Grignard reagent in tetrahydrofuran (17):

$$R_2TlX + RMgX \longrightarrow R_3Tl + MgX_2$$

However, the most convenient method of preparation is from the reaction of thallium(I) iodide and organolithium:

$$TlI + 2\ CH_3Li + CH_3I \longrightarrow (CH_3)_3Tl + 2\ LiI$$

The intermediate involved in this reaction is thought to be CH_3Tl.

R_3Tl compounds react readily with acids, halocarbons, or sulfur dioxide to form R_2TlX. They also form neutral complexes ($R_3Tl{\cdot}L$) with Lewis bases (L), eg, amines and phosphines, in a similar manner as the gallium and indium analogues.

R_2TlX compounds are best prepared from thallium(III) halides and an excess of Grignard reagent (18)

$$TlX_3 + 2\ RMgX \longrightarrow R_2TlX + 2\ MgX_2$$

However, the yields are usually poor because of the oxidation of the Grignard reagent by thallium(III). Reaction of R_3Tl with acids also affords a convenient route to this type of compound.

Dialkyl and diarylthallium(III) derivatives are stable, crystalline solids that melt at 180–300°C. The dimethylthallium derivatives of CN^-, ClO_4^-, BF_4^-, and NO_3^- contain linear $(CH_3)_2Tl$ cations and the free anions (19). In aqueous solutions, they ionize to the $(CH_3)_2Tl(H_2O)_x^+$ ions, except those derivatives containing alkoxide, mercaptide, or amide anions, which yield dimeric structures (20,21).

$RTlX_2$ compounds, eg, dichlorophenylthallium(III) [19628-33-2], can be obtained from the reaction of thallic halide or carboxylate with an organoboron compound (22):

$$TlCl_3 + C_6H_5B(OH)_2 + H_2O \longrightarrow C_6H_5TlCl_2 + B(OH)_3 + HCl$$

or with an organomercury compound (23):

$$\underset{\text{Tl(OCR)}_3}{\overset{\text{O}}{\overset{\|}{}}} + (C_6H_5)_2Hg \longrightarrow \underset{C_6H_5Tl(OCR)_2}{\overset{\text{O}}{\overset{\|}{}}} + \underset{C_6H_5HgOCR}{\overset{\text{O}}{\overset{\|}{}}}$$

They can also be prepared by thallation (24), eg, tallium(III) trifluoroacetate [23586-53-1], which yields di(trifluoroacetato)phenylthallium(III) [23586-54-1]:

$$C_6H_6 + \underset{Tl(OCCF_3)_3}{\overset{\text{O}}{\overset{\|}{}}} \longrightarrow C_6H_5Tl(OOCCF_3)_2 + \underset{CF_3COH}{\overset{\text{O}}{\overset{\|}{}}}$$

or oxythallation (25), eg, tallic acetate [2570-63-0] and vinylbenzene, which yield diacetato 2-methoxy-2-phenylethane tallium(III) [37011-27-1]:

$$C_6H_5CH{=}CH_2 + Tl(O\overset{O}{\overset{\|}{C}}CH_3)_3 + CH_3OH \longrightarrow C_6H_5\underset{\underset{OCH_3}{|}}{CH}CH_2Tl(O\overset{O}{\overset{\|}{C}}CH_3)_2$$

RTlX$_2$ derivatives are covalent compounds, generally soluble in organic solvents. The aryl and vinyl derivatives are more stable than the corresponding alkyl compounds. This type of compound has been postulated to be an intermediate in many organic synthetic reactions involving thallium(III) species.

Uses. Thallium compounds have limited use in industrial applications. The use of thallous sulfate in rodenticides and insecticides has been replaced by other compounds less harmful to animals (see INSECT CONTROL TECHNOLOGY; PESTICIDES). Thallium sulfide has been used in photoelectric cells (see PHOTOVOLTAIC CELLS). A thallium bromide–thallium iodide mixture is used to transmit infrared radiation for signal systems. Thallous oxide is used in the manufacture of glass (qv) that has a high coefficient of refraction. Thallium formate–malonate aqueous solutions (Clerici's solution) have been used in mineral separations. Many thallium compounds have been used as reagents in organic synthesis in research laboratories.

Thallium(I) Compounds. Carboxylic anhydrides can be prepared by the reaction of acyl or aroyl halides in ether with thallium(I) carboxylates (26):

$$R'\overset{O}{\overset{\|}{C}}Cl + TlO\overset{O}{\overset{\|}{C}}R \longrightarrow R'\overset{O}{\overset{\|}{C}}O\overset{O}{\overset{\|}{C}}R + TlCl$$

Thallium(I) carboxylates also react with alkyl iodides to form the esters (27):

$$R'I + TlO\overset{O}{\overset{\|}{C}}R \longrightarrow R\overset{O}{\overset{\|}{C}}OR' + TlI$$

Thallium acetate reacts with iodine and an alkene to give the *trans*-iodoacetate derivatives in 90% yield (28):

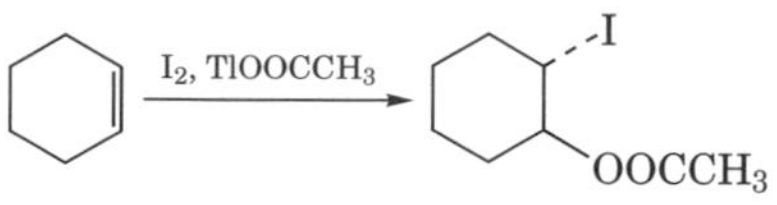

Cyclopentadienylthallium and its alkylated derivatives are used in the synthesis of metallocenes (qv) and other transition-metal cyclopentadienyl complexes (29).

Thallium(III) Compounds. Thallium(III) derivatives have been used extensively as oxidants in organic synthesis. In particular, thallic acetate and trifluoroacetate are extremely effective as electrophiles in oxythallation and thallation reactions. For example, ketones can be prepared from terminal acetylenes by means of $Tl(OOCCH_3)_3$ in acetic acid (oxythallation) (30):

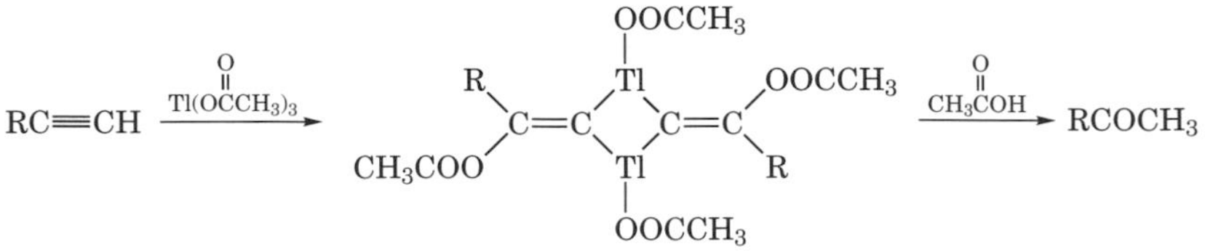

Oxythallation reactions of olefins provide useful synthetic routes to products, eg, glycols, aldehydes, and ketones (31):

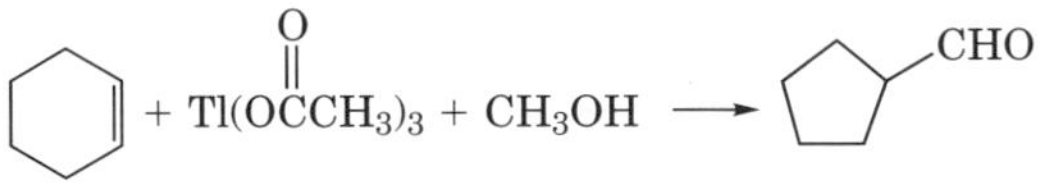

Thallium(III) nitrate in nitric acid and thallium(III) sulfate in sulfuric acid are also effective.

The enhancement of the electrophilic properties of thallium(III) trifluoroacetate makes it a very important thallation reagent. The products of thallation, eg, arylthallium bis(trifluoracetate), undergo a variety of substitution reactions, yielding iodides, fluorides, nitriles, thiophenols, phenols, and biaryls.

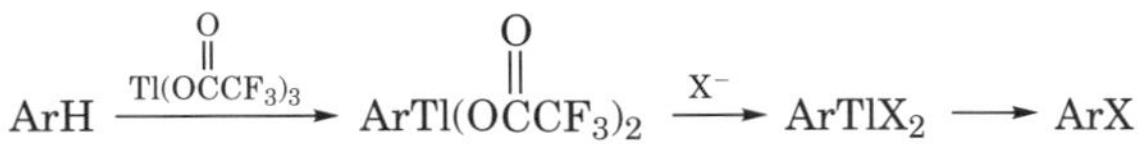

Thallium(III) trifluoroacetate promotes olefin cyclization reactions and intramolecular coupling reactions (32,33).

The highly ionic thallic nitrate, which is soluble in alcohols, ethers, and carboxylic acids, is also a very useful synthetic reagent. Oxidation of olefins, α,β-unsaturated carbonyl compounds, β-carbonyl sulfides, and α-nitrato ketones can all be conveniently carried out in good yields (31,34–36).

Toxicology

The relative toxicities of thallium compounds depend on their solubilities and valence states. Soluble univalent thallium compounds, eg, thallous sulfate, acetate, and carbonate, are especially toxic. They are rapidly and completely absorbed from the gastrointestinal tract, skin peritoneal cavity, and sites of subcutaneous and intramuscular injection. Thallium is also rapidly absorbed from the mucous membranes of the respiratory tract, mouth, and lungs following inhalation of soluble thallium salts. Insoluble compounds, eg, thallous sulfide and iodide, are poorly absorbed by any route and are less toxic.

Thallium, which does not occur naturally in normal tissue, is not essential to mammals but does accumulate in the human body. Levels as low as

0.5 mg/100 g of tissue suggest thallium intoxication. Based on industrial experience, 0.10 mg/m^3 of thallium in air is considered safe for a 40-h work week (37). The lethal dose for humans is not definitely known, but 1 g of absorbed thallium is considered sufficient to kill an adult and 10 mg/kg body weight has been fatal to children. In severe cases of poisoning, death does not occur earlier than 8–10 d but most frequently in 10–12 d. Thallium excretion is slow and prolonged. For example, thallium is present in the feces 35 d after exposure and persists in the urine for up to three months.

The usual symptoms in human thallotoxicosis resulting from acute, subacute, or chronic intoxication are generally the same. Common symptoms include nausea, vomiting, abdominal colic, pain in legs, nervousness and irritability, chest pain, gingivitis or stomatitis, and anorexia. Alopecia (hair loss) does not always occur, especially in cases of mass intake of thallium and low resistance where the patient may die before the occurrence of hair loss. Thallium intoxication during the first trimester of pregnancy can cause skeletal deformities, alopecia, rash, low birth weight, and premature birth (38).

A variety of therapies for thallium poisoning have been suggested by neutralizing thallium in the intestinal tract, hastening excretion after resorption, or decreasing absorption. Berlin-Blue (ferrihexacyanate) and sodium iodide in a 1 wt % solution have been recommended. Forced diuresis hemoperfusion and hemodialysis in combination results in the elimination of up to 40% of the resorbed thallous sulfate (39).

BIBLIOGRAPHY

"Thallium and Thallium Compounds" in *ECT* 1st ed., Vol. 13, pp. 927–932, by K. M. Herstein, Herstein Laboratories, Inc.; in *ECT* 2nd ed., Vol. 20, pp. 63–69, by H. E. Howe, American Smelting and Refining Co.; in *ECT* 3rd ed., Vol. 22, pp. 835–845, by B. C. Hui, Alfa Products.

1. W. Crookes, *Chem. News*, **3**, 193, 303 (1861).
2. U.S. Pat. 2,060,453 (Nov. 10, 1936), and 2,011,882 (Aug. 20, 1935), R. Teats (to American Smelting and Refining Co.).
3. R. Kleinert, *Z. Erzbergbau Metallhuettenwes.* **16**, 67 (1963).
4. J. D. Prater, D. Schlain, and S. F. Ravitz, *Recovery of Thallium from Smelter Products*, Rep. Inv. 4900, U.S. Bureau of Mines, Washington, D.C., 1963.
5. T. W. Richards and C. P. Smyth, *J. Am. Chem. Soc.* **44**, 525 (1922).
6. T. W. Richards and J. D. White, *J. Am. Chem. Soc.* **50**, 3290 (1928).
7. M. A. Filyand and E. I. Semenova, *Handbook of the Rare Elements*, MacDonald, London, 1969.
8. W. M. Latimer, *The Oxidation States of the Elements and Their Potentials in Aqueous Solutions*, 2nd ed., Prentice-Hall, Englewood Cliffs, N.J., 1952.
9. G. Brauer, *Handbook of Preparative Inorganic Chemistry*, 2nd ed., Academic Press, New York, 1963.
10. C. M. Fear and R. C. Menzies, *J. Chem. Soc.* **129**, 937 (1926).
11. A. McKillop and co-workers, *J. Am. Chem. Soc.* **93**, 4841 (1971).
12. J. K. Kochi and T. W. Bethea, *J. Org. Chem.* **33**, 75 (1968).
13. E. Wiberg, O. Dittmann, H. Noth, and M. Schmidt, *Z. Naturforsch. Teil B*, **12**, 61, 62 (1957).
14. E. Wiberg, O. Dittmann, and M. Schmidt, *Z. Naturforsch. Teil B*, **12**, 60 (1957).

15. H. Meister, *Angew. Chem.* **69**, 533 (1957).
16. H. Gilman and R. G. Jones, *J. Am. Chem. Soc.* **62**, 2357 (1940).
17. O. Yu, Okhlobystin, K. A. Bilevitch, and L. I. Zakharkin, *J. Organomet. Chem.* **2**, 281 (1964).
18. H. Kurosawa and R. Okawara, *Organomet. Chem. Rev.* **2**, 255 (1967).
19. G. B. Deacon, J. H. S. Green, and R. S. Nyholm, *J. Chem. Soc.* 3411 (1965).
20. M. G. Miles and co-workers, *Inorg. Chem.* **7**, 1721 (1968).
21. R. C. Menzies and A. R. P. Walker, *J. Chem. Soc.* 1678 (1936).
22. F. Challenger and O. V. Richards, *J. Chem. Soc.* 405 (1934).
23. H. J. Kabbe, *Ann. Chem.* **656**, 204 (1962).
24. A. McKillop, J. D. Hunt, and E. C. Taylor, *J. Organomet. Chem.* **24**, 77 (1970).
25. A. McKillop and E. C. Taylor, *Chem. Br.* **9**, 4 (1973).
26. E. C. Taylor, G. W. McLay, and A. McKillop, *J. Am. Chem. Soc.* **90**, 2422 (1968).
27. R. C. Menzies, *J. Chem. Soc.* 1378 (1947).
28. C. C. Hunt and J. R. Doyle, *Inorg. Nucl. Chem. Lett.* **2**, 283 (1966).
29. R. C. Cambie, B. G. Lindsay, P. S. Rutledge, and P. D. Woodgate, *Chem. Commun.* 919 (1978).
30. S. Uemura and co-workers, *Chem. Commun.* 218 (1976).
31. A. McKillop and co-workers, *Tetrahedron Lett.* 5275 (1970).
32. Y. Yamada and co-workers, *Tetrahedron Lett.* **22**, 1355 (1981).
33. E. C. Taylor, J. G. Andrade, and A. McKillop, *Chem. Commun.* 538 (1977).
34. A. McKillop, B. P. Swann, and E. C. Taylor, *Tetrahedron Lett.* 5281 (1970); A. McKillop, B. P. Swann, and E. C. Taylor, *Tetrahedron*, **26**, 4031 (1970).
35. Y. Nagao, K. Kaneko, and E. Fujita, *Tetrahedron Lett.* 4115 (1978).
36. A. McKillop and co-workers, *J. Org. Chem.* **43**, 3773 (1978).
37. A. Hamilton and H. L. Hardy, *Industrial Toxicology*, 3rd ed., Publishing Science Group Inc., 1974.
38. W. J. Stevens and co-workers, *Acta Clin. Belg.* **31**, 188 (1976).
39. S. Moeschlin, *Clin. Toxicol.* **17**, 133 (1980).

General References

N. V. Sidgwick, *The Chemical Elements and Their Compounds*, Vol. I, Oxford University Press, London, 1950.
A. G. Lee, *The Chemistry of Thallium*, Elsevier, Amsterdam, the Netherlands, 1971.
K. Wade and A. J. Banister, in J. C. Bailar, Jr., and co-workers, eds., *Comprehensive Inorganic Chemistry*, Vol. 1, Pergamon Press, Oxford, 1973.
F. A. Cotton and G. Wilkinson, *Advanced Inorganic Chemistry: A Comprehensive Text*, 4th ed., John Wiley & Sons, Inc., New York, 1980.
A. N. Nesmeyanov and R. A. Sokolik, *The Organic Compounds of Boron, Aluminum, Gallium, Indium and Thallium*, North-Holland Publishing Co., Amsterdam, the Netherlands, 1967.
A. McKillop and E. C. Taylor, in F. G. A. Stone and R. West, eds., *Advances in Organometallic Chemistry*, Vol. 2, Academic Press, New York, 1973, p. 147.
B. Venugopal and T. D. Luckey, *Metal Toxicity in Mammals*, Vol. 2, *Toxicity of Metals and Metalloids*, Plenum Press, New York, 1978.
J. F. Reith, in J. J. G. Prick, W. G. S. Smith, and L. Muller, eds., *Thallium Poisoning*, Elsevier, Amsterdam, the Netherlands, 1955.
Charles Pfizer and Co., Inc., *Thallium Poisoning, Spectrum* (*N.Y.*) **6**, 558 (1958).

Bob Blumenthal
Noah Technologies Corporation

THERMAL POLLUTION

An important by-product of most energy technologies is heat. Few energy conversion processes are carried out without heat being rejected at some point in the process stream. Historically, it has been more convenient as well as less costly to reject waste heat to the environment rather than to attempt significant recovery. The low temperatures of waste heat in relation to process requirements often make reuse impractical and disposal the only attractive alternative (see PROCESS ENERGY CONSERVATION).

Heat rejected to the environment by most industries is of little consequence. Cooling flows of air or water are deployed over equipment or through heat exchangers and the relatively small quantities of heat are dissipated to the surrounding air. Small cooling towers, often of the evaporative type, have become ubiquitous in an industrial facility.

Concern over heat rejection arose when quantities at localized sites rose dramatically as the electric utility industry shifted to water-cooled, thermal-electric generating stations of high unit capacity in the 1950s. The term thermal pollution took on fearsome portents among aquatic scientists, fishery managers, and eventually water pollution control agencies (1). Directly lethal effects of high temperatures on aquatic life were predicted and, where sublethal temperatures were maintained, effects on reproductive cycles, growth rates, migration patterns, and interspecies competition were hypothesized (2).

Much research involving monitoring of thermal effects at power stations was conducted in the 1970s and 1980s. The increased knowledge gained and more stringent regulations led to approaches for using biological requirements of aquatic organisms plus local environmental characteristics of the rivers, lakes, and estuaries used for cooling to design nondamaging cooling systems specific for a site. Because the rate of increase in demand for electricity and development of new generating stations also diminished, new power plants were evaluated more thoroughly and located in less susceptible environments. Approaches for regulatory thermal power stations that evolved by the mid-1980s were still in practice as of 1997 (see POWER GENERATION).

Cooling Techniques

Power station cooling is fairly straightforward. Generation of electricity by the steam (qv) cycle, the most common method regardless of fuel type, ie, coal, oil, gas, nuclear, solid waste, entails production of waste heat (Fig. 1). Although there are some atmospheric losses in the steam cycle, most of the heat is rejected to flowing water through the heat of condensing steam. For a modern 1200 MW (electrical) lightwater nuclear power plant, this release is ca 7100 kJ/(kW·h) (1700 kcal/(kW·h)), ie, approximately twice the thermal equivalent of the generated electric energy. The amount of reject heat is proportionately less in fossil-fueled plants, in which up to 40% of the fuel energy can be converted to electricity. The flowing water for the steam condenser was traditionally pumped from a nearby lake or river, then returned to this water body at an elevated temperature in the once-through or open-cycle system. A 1000-MW nuclear power plant having once-through cooling requires ca 49 m^3/s of water (ca 13,000 gal/s).

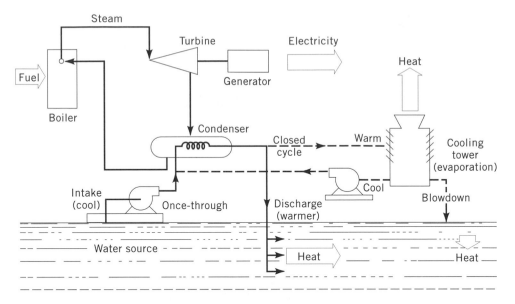

Fig. 1. The energy cycle of a thermal electric generating station having two alternative cooling systems: (—) the open-circuit or once-through system and (– – –) a representative closed-cycle, cooling-tower system. Reproduced by permission (3).

The water is elevated 10°C. Screens at the water intake prevent entry of objects larger than ca 1 cm. A biocide, usually chlorine, is used to clean heat-exchange surfaces and water conduits.

Risks

A risk is considered herein to be the biological or ecological damage that could be done by a human alteration of the environment with the likelihood or probability that the specific damaging alteration will actually occur. For purposes of clarifying risks to aquatic life, a clear distinction must be made between heat, a quantitative measure of energy that depends on the mass of an object (in this case the volume of cooling water), and temperature, a measure of energy intensity. An amount of heat distributed in two unit volumes of water yields half the elevation of temperature as the same amount of heat distributed in one unit. Although heat is the waste product of electricity generation, temperature is the environmental characteristic to which organisms respond. Thus, the quantity of water used to carry away the load of reject heat is crucial to determining the temperatures created in the environment.

Initially, the source of environmental risk from cooling water was assumed to be the pollutant discharged, ie, heat, in the form of the elevated temperature of the water released from the condensers. Heat is now recognized as being only one of several potential risks of power station cooling (Fig. 2).

Generally unheralded until the early 1970s was the physical entrapment and impingement of fish on cooling-water intake screens. Always occurring at chronic low levels at small fossil-fueled plants, the losses of fish and some large invertebrates rose markedly with the startup of new nuclear stations.

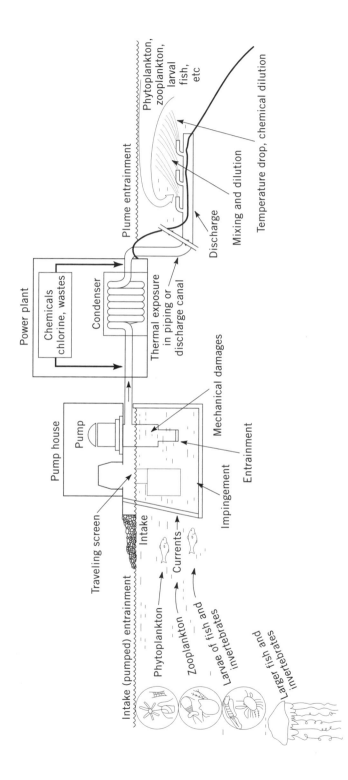

Fig. 2. Sources of potential biological damage in the immediate vicinity of a power station cooling system where the condenser may sustain mechanical damage or thermal shock. Reproduced with permission (4).

In particular, plants on the Hudson River began to impinge large numbers of juvenile striped bass, a species of considerable importance locally.

Impingement risks for affected species seem more direct and identifiable than the indirect aspects of temperature change because these are more visible. The effects of impingement include long periods of futile swimming in screen wells, being held by water currents against the screen mesh (often resulting in suffocation through inability to ventilate the gills), and physical injury from the rotating screen or high pressure water spray used to wash the screen. Beyond risks to individual fish are risks for populations of important species that could be seriously depleted by introducing a new form of chronic mortality into the life cycle. Impingement acts selectively, affecting some species more than others, ie, generally schooling, pelagic fish.

Serious analysis of the risks associated with power plant impingement began when environmental impact statements for nuclear power plants were initiated following the U.S. National Environmental Policy Act of 1969 (NEPA) and the ensuing Calvert Cliffs court decision which assigned responsibility for all cooling-water impacts of nuclear stations to the Atomic Energy Commission (now the Nuclear Regulatory Commission). The analogous problem of impingement of downstream migrating salmon on screens of irrigation diversion dams had been addressed for many years on the U.S. West Coast, but it was not until 1973 that the commonality of both of these problems and of possible engineering solutions was widely recognized (4). Emphasis on solutions was stimulated by the U.S. Federal Water Pollution Control Act Amendment of 1972, which allowed once-through cooling only where intakes minimized losses.

Biocides, principally chlorine, used periodically (0.5 h/d per condenser) for condenser cleaning were identified as toxic risks for any organisms in the cooling circuit at the time and for those in the vicinity of the discharge where the biocide dissipates. Concern over residual toxic effects of power-plant chlorination was also largely an outgrowth of NEPA reviews of nuclear stations in the early 1970s, when it coincided with renewed concern over ecological effects of chlorination of treated sewage effluents and evidence of chlorinated organic materials formed in chlorinated water (5,6).

A fourth risk is from combined damages, ie, thermal, physical, and chemical, sustained by small organisms that are pumped through the cooling system, ie, entrainment. These damages were among the earliest to be recognized, but the early emphasis on thermal stresses alone retarded examination of the physical components of the damage, which are appreciated as the principal risks at many installations (7). During entrainment, any organisms, including phytoplankton, zooplankton, larval fish, invertebrates, and many small fish that cannot swim against the induced current at the cooling-water intake, are drawn into the cooling circuit unless they are large enough to be screened out initially. In the cooling circuit, they receive in rapid sequence (usually <1 min) a series of stresses. These include a pressure drop in front of the pump impeller, risk of physical impact with the impeller or shear stress of a near miss, rapid pressurization downstream of the pump, shear stress as the cooling water is divided among hundreds of condenser tubes ca 2.5 cm in diameter, rapid temperature elevation as heat is transferred to the water through condenser tubes, maintenance of high temperature (usually 8–10°C above ambient) through the discharge system, decreasing pressure in discharge piping (sometimes below atmospheric), followed by turbulent mixing and cooling as the condenser water rejoins the source water body (8). Many entrained organisms do not survive. During periods of biocide treatment to remove heat-transfer-retarding biological slimes from condenser tubes, entrained organisms are also exposed to lethal concentrations of a toxicant, usually chlorine.

Many organisms are exposed to some of the thermal, chemical, and physical stresses of entrainment by being mixed at the discharge with the heated water; this is plume entrainment. The exact number exposed depends on the percentage of temperature decline at the discharge that is attributed to turbulent mixing rather than to radiative or evaporative cooling to the atmosphere.

Entrainment was in the forefront of concerns over condenser cooling in the early 1970s as two trends in the planning of power plants developed: gradual decrease in the temperature rise (ΔT) across condensers to reduce thermal effects, which was necessarily accompanied by an increase in volume of water pumped, and increased exploitation by large nuclear stations of estuarine waters, which generally contain far more planktonic eggs and larvae of important fishes and invertebrates than inland rivers or lakes. The long-term risk with entrainment lies in chronic losses of vulnerable stages to physical stresses which can contribute to instability or decline in populations of valued species. Complex scientific debates over the magnitudes and probabilities of population effects of entrainment mortalities have punctuated power plant licensing proceedings since 1971, particularly those for the Indian Point Nuclear Power Plant on the Hudson River (9). Computer models have been developed to estimate numbers of organisms, eg, fish larvae, that can be killed in a population by both natural and human-caused mechanism, without affecting the continued success of the population as a whole. The Electric Power Research Institute (Palo Alto, California) and Oak Ridge National Laboratory (Tennessee) have been leaders in developing such models.

Questions also arose over the risks to aquatic life from changes in gas content of the water as a by-product of temperature change. Warmer water holds less gas in solution. Dissolved oxygen was hypothesized to decrease below levels necessary for fish by a combination of physical solubility relationships and increased biological demand for oxygen at elevated temperatures. Supersaturation of dissolved gases in water was also viewed, particularly in the northwest United States, as a significant risk arising from heating already saturated river water. Fishery scientists on the Columbia River were especially sensitive to this potential problem, because high gas saturation levels from dam spillways were demonstrably harmful to experimental salmon and trout (10).

Physical changes in habitats near the cooling water intake and discharge structures of power stations were also identified in NEPA analyses as posing some risk or at least potential for change to segments of aquatic communities. Concrete structures, rock jetties, and altered current patterns contribute to habitat modifications that influence suitability of the area for desirable species and can be linked to the overall problem of dissipating rejected heat.

Risks from cooling towers were identified by comparative ecological analyses of cooling-system alternatives and the risks became part of the assessment process. Throughout the 1960s, evaporative cooling towers, whether natural or mechanical draft (see Fig. 1), were seen as the panacea: a closed cooling cycle to replace the transgression of the once-through system on natural waters.

The closed cycle is, however, not really closed. Among the first environmental risks recognized from cooling towers were increases in fogging and icing in northern climates. Wherever cooling towers are used for coastal waters, sea salts are dispersed from the towers to adjacent lands with gradual salt accumulation and detrimental effects on agriculture and corrosion of structural materials. Continuous evaporation gradually accumulates unwanted salts if there is not some flow through in addition to replacement of cooling water for evaporative losses. This flow through or blowdown carries with it numerous chemicals added to the cooling circuit to prevent corrosion, eg, chromates, zinc, and organophosphorus complexes, to slow mineral scaling of heat-exchange surfaces by, for example, acids or to eliminate biological fouling, eg, by chlorine. Blowdown was traditionally discharged untreated into water bodies that supplied the tower intake, and organisms susceptible to the chemicals were placed at risk. The chemicals also escape the towers in drift in the form of small water droplets and aerosols that are dispersed with air flows to terrestrial surroundings. Cooling-tower sludge, the solid material that accumulates in tower basins, must also be discarded. This sludge includes precipitated chemical elements from the water supply, corrosion inhibitors, and airborne dust, pollen, and tree leaves washed from the tower air flow. At the intake, entrainment and impingement are not eliminated by the addition of cooling towers, although water flows can be reduced to 5–10% of that needed for once-through cooling. The presumed advantage of reduction in water volume for minimizing entrainment risks may be illusory, as organisms entrained into cooling-tower circuits are exposed to lethal conditions, whereas many often survive a rapid once-through passage at well-designed open-cycle plants.

An alternative closed-cycle system that is used particularly in the Midwest, Southeast, and Texas is the artificial cooling lake. Such lakes exist in a gradient of designs from large multipurpose public reservoirs that provide essentially once-through cooling to privately controlled, diked ponds or canals that rely on evaporative cooling and are often augmented by spray modules. The risks to public resources also vary greatly. Despite the obvious multipurpose benefits of cooling reservoirs where water resources are scarce, these cooling systems were judged in violation of the 1972 Water Pollution Control Act and for several years were not allowed for new power stations. Thus, the number of cooling lakes constructed diminished greatly (11). There has been much debate over cooling lakes. Numerous water and fish management opportunities from their use has brought them back in favor. Many of the most productive fisheries are fresh water species in lakes also used for power plants.

Risk Minimization

Selection of the best cooling system in terms of minimal environmental damages involves matching engineering options to the local aquatic system potentially at risk. General principles of aquatic ecology and of the life histories and environmental requirements of species represented locally can be adapted to local water resource goals using detailed understanding of the local aquatic setting to achieve site-specific risk prevention.

Thermal Effects. Temperature is the most all-pervasive environmental factor that influences aquatic organisms. There is always an environmental temperature, whereas other factors may or may not be present. Nearly all aquatic organisms, with the exception of marine mammals, are for all practical purposes thermal conformers. As such, they are not able to exert significant influence on maintaining a certain body temperature by physiological means. Their body temperatures fluctuate in close accord with the temperature of the immediate surrounding water. Especially large, active-moving fish, such as tunas, do maintain deep-muscle temperatures slightly higher than the surrounding water. Intimate contact between body fluids and water at the gills and the high specific heat of water assure a near-identity of internal and external temperatures. Behavioral thermoregulation or the control of body temperature by selection of water temperature in natural gradients is, however, a common feature of many fish. Behavioral thermoregulation serves an important ecological role in partitioning aquatic habitats among species. Its understanding is a powerful feature for estimating and ameliorating the impacts of thermal discharges.

Thermal effects on aquatic organisms have been given critical scientific review. Annual reviews of the thermal effects literature have been published beginning in 1968 (12). Water temperature criteria for protection of aquatic life were prepared by the NAS in 1972, and these criteria have formed the basis of the EPA recommendations for establishing water temperature standards for specific water bodies (13,14).

The importance of temperature for organisms lies partly in its fundamental physical–chemical influences. The biochemistry of life is, in general, subject to the basic dictum that a rise of temperature by 10°C results in about a doubling of the rate of chemical reactions. Life also has an upper thermal limit, set partly by the chemical stability of organic molecules, eg, proteins or enzyme systems, and by the balance of food energy inputs and expenditures, that is characteristic of each species and, in some cases, different life stages. Although aquatic organisms must conform to water temperature, many have evolved internal mechanisms other than body-temperature regulation to allow continued functioning as temperature changes occur geographically, seasonally, and daily. Through gradual biochemical adjustments termed acclimation, physiological processes are maintained relatively constant over a prescribed thermal range which is species- or life-stage-specific. Within this range, each species has a zone of optimal physiological performance which, because it tends to coincide with temperatures preferred in gradients, has recently been termed its thermal niche in the environment.

Organisms evolving under annual temperature cycles and in environments with varying temperatures spatially have incorporated thermal cues in reproductive behavior, habitat selection, and certain other features which act at the population level. Thus, the balance of births and mortalities, which determines whether a species survives, is akin to the metabolic balance at the physiological level in being dependent upon the match, within certain limits, to prescribed temperatures at different times of year. At the ecosystem level, relationships among species, eg, predators, competitors, prey animals, and plant foods, are related to environmental temperatures in complex ways. Many of these interactions are poorly understood.

Despite the immensity and complexity of known and suspected roles of temperature in aquatic ecosystems, certain thermal criteria have been especially useful in minimizing risks from thermal discharges. More data have been organized at the physiological level than at higher levels of organization.

Preventing Mortality.   Upper and lower temperature tolerances of aquatic organisms have been well conceptualized and standardized methods are available for determining a species' tolerance ranges under different conditions of thermal history (Fig. 3a). On a short time scale, mortality is highly dependent on duration of exposure (Fig. 3b), so that brief exposures to potentially lethal temperatures are not actually lethal. Temperature elevations and duration of exposure can be tailored in a plant's piping or in the effluent mixing zone to maximize organism survival (13). This is usually accomplished using detailed mathematical models of cooling-water-effluent dispersion and heat dissipation in the near field where temperatures are highest (16,17). One option used extensively for this purpose in the 1970s was increasing the cooling-water flow and thus decreasing the temperature elevation to below lethal levels. This tactic had the undesirable result of increasing the numbers of organisms damaged by the physical effects of entrainment and impingement. As of the 1990s, a quantitative balance between minimizing direct thermal mortalities and increasing those risks related more to the volume of cooling-water flow is employed.

Preventing Stressful High Temperatures Over Long Periods.   For the long term (>1–2 d), simply preventing mortality is insufficient for protecting aquatic species. All of the physiological functions normally performed must be carried out to maintain healthy individuals that are capable of competing in the natural ecosystem. An aggregate measure, growth rate, has proved useful as an integrator of all physiological functions and some behavioral ones, eg, feeding rate. Growth occurs only if all other metabolic demands are being met and when sufficient food energy is left over for adding biomass. Typically, many physiological functions of well-fed, cold-blooded organisms proceed optimally over a temperature range in which growth rate is maximal (Fig. 4). Above the temperatures of maximum growth rate, the rate typically declines steeply to a temperature of zero growth, which often occurs 1–2°C below the temperature at which direct mortalities begin. Intuitively, the healthy fish becomes unhealthy as the long-term temperatures it experiences rise from those that yield maximum growth to that which stops growth. Alternative methods for calculating the upper danger level have been proposed, but each suggests that a long-term decline of growth rate below ca 75% of maximum at high temperatures is unduly risky (13,15).

Standards for upper limits on water temperatures for particular water bodies over periods of about one week or more can be based on species-specific growth rates. Using an inventory of important species and life stages in the area during the warm season, the upper temperature limit which does not stress those in the desired aquatic assemblage can be ascertained. Hydrothermal models of heat dissipation in the far field beyond the zone of effluent mixing are important for estimating the zones that may present a long-term risk from elevated temperatures.

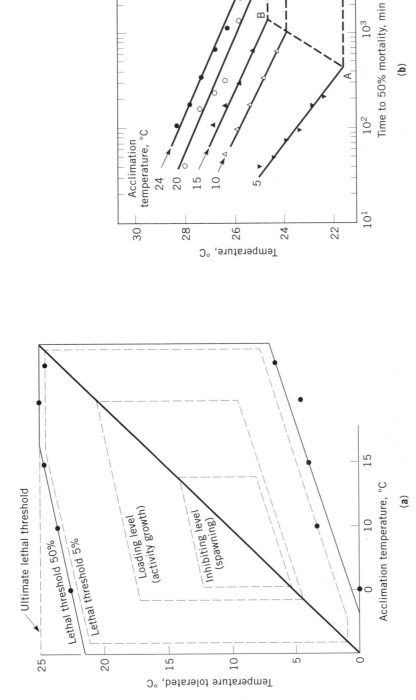

Fig. 3. Lethal temperature thresholds for aquatic species. Patterns are general for all species, but exact temperatures are species-specific. (**a**) Tolerance polygon of upper and lower lethal (50%) temperatures for one-week exposures of an example species (juvenile sockeye salmon) which has been held at the acclimation temperature, with more restrictive thresholds indicated as dashed lines; (**b**) time-dependent mortality (50%) of an example species (juvenile chinook salmon) at temperatures above the one-week lethal threshold after holding at different acclimation temperatures. The dashed line ABC indicates transition to less than 50% mortality at lower temperatures and coincides with the upper lethal threshold of this species' tolerance polygon. Reproduced by permission (15).

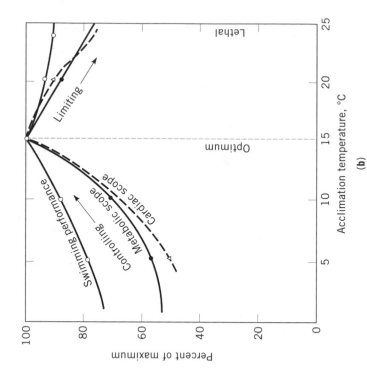

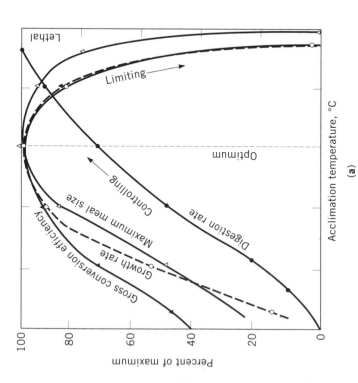

Fig. 4. Convergence of numerous physiological functions, (**a**) and (**b**), including overall growth rate, at an optimum acclimation temperature which is specific for the species (juvenile sockeye salmon). Reproduced by permission (15).

The preferred and avoided temperatures in a gradient have been used as surrogates for optimum growth and upper danger levels for fish. One practical drawback to using growth rate as an index of optimal temperatures is the experimental cost of determining it. As more and more data link temperatures of optimal growth rate with preferred temperatures for a species, there is an increasing tendency to conduct only thermal preference tests (18). Long-term abundance of species at power plant sites appears to be generally correlated with preferred and avoided temperatures (19).

Preserving Reproduction Cycles. Organisms in the temperate zone have evolved in concert with seasonal temperature variations. Reproduction success depends in part on the preservation of an annual temperature pattern, although the precise timing is usually not critical. Thermal discharges can be designed, usually with the help of far-field mathematical models, to assure the necessary thermal periodicity. Unless the cooling system is a heavily heat-loaded stream or cooling pond, the thermal output of a power station complex is rarely sufficient to offset the large natural cooling rates of winter. Thus, annual thermal cycles are generally maintained despite anthropogenic heat rejection.

Maintaining Ecosystem Structure and Function. Thermal heterogeneity of water bodies is an important structural feature of the environment and plays a large role in determining the composition and functioning of most aquatic systems. Vertical thermal stratification of lakes, reservoirs, and many estuaries in summer (Fig. 5) segregates an available habitat into discrete zones with differing thermal and water quality characteristics. In the evolution of aquatic species, these discrete zones have been the basis for partitioning the environment, such that different species or life stages within some species occupy discrete portions based, in part, on temperature selection behavior. The disruption of thermal structure can be more damaging to ecosystem composition and function than the direct killing of many individual organisms.

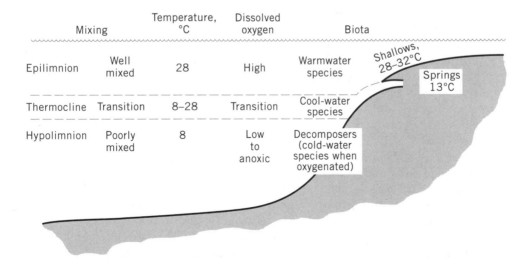

Fig. 5. Typical thermal stratification of a lake, reservoir, or poorly mixed estuary in summer which, because of density differences, establishes discrete zones with differing thermal, water quality, and biotic characteristics.

Particularly significant are thermal refuges during periods of extreme high or low temperatures. Aberrant weather or artificial thermal changes can cause large portions of a water body to exceed the physiological limits of some aquatic species for brief periods. The species usually affected are those at the northern or southern limits of their geographic ranges. These species often survive in isolated zones or refuges that retain suitable temperatures, eg, springs where trout can remain cool in the warmest summer months or water bodies where cold-sensitive species, eg, threadfin shad, find relatively warm waters in mid-winter. The demise or survival of such geographically marginal species, which is determined by the availability and sizes of such thermal refuges, can be paramount in establishing the ecological interactions of a water body (20). Power station discharges can prematurely force desirable cool-water species into refuges in summer and can provide essential warm refuges for desirable warm-water species in winter.

Thermal heterogeneity of the environment, thermal optima for organism growth rates, and behavioral temperature selection are being evaluated in ways that are useful for predicting ecosystem effects of thermal alterations (18). Thermal-niche concepts can be used to identify critical habitats for important species in ways that focus the potential influences of power-station discharges (20,21). For example, striped bass (*Morone saxatilis*), an East Cost species intro-duced on the West Coast and in freshwater reservoirs, partitions a water body in summer along thermal gradients among its age or size classes. Young bass prefer and grow optimally at high temperatures near 24–26°C, subadults prefer ca 22°C, and mature adults select temperatures of 20°C or less. Thermal dis-charges may benefit growth rates and survival of juveniles, but the adults face a different prospect. Forced to cool water in summer by their genetically based temperature preferences, they may find this habitat severely restricted by ther-mal additions or compromised by simultaneous depletion of dissolved oxygen as a result, in part, of decomposition of thermally stimulated plankton production. Overcrowding in limited thermal refuges that have sufficient dissolved oxygen has led to starvation, high disease incidence, and abnormally high fishing sus-ceptibility, all of which cause high numbers of deaths.

Altered predator–prey interactions, previously viewed as intractable long-term influences of changing water temperatures, can be estimated at least qualitatively through use of thermal-niche concepts. Adult striped bass that are restricted from access to surface waters by high temperatures seek food among the cool-water species, eg, trout in deep water, rather than among the surface-dwelling shad species which are their normal prey. Trout populations can be decimated, as they were in Lake Mead after introduced striped bass grew to the size and thermal preference of adults.

Careful planning of thermal additions, including creation of new cooling reservoirs, can yield thermal structures which enhance rather than damage desirable aquatic species. Knowledge of the thermal niches of these species and their potential competitors or predators can permit special provisions for thermal refuges, eg, cool summer zones in a heavily loaded cooling pond. Cooling-water circulation can be designed so that the thermal stratification patterns that are essential for some desired species are maintained. From a different perspective, aquatic species introduced to waters used for power-station cooling can be selected so that their thermal niche matches the thermal structure that the facility creates. Multiple species introductions can be evaluated beforehand to estimate whether available thermal patterns will abnormally aggregate or separate potential predators and prey.

Impingement.   Most of the techniques to avoid impingement have been reviewed (4,22,23). Repellents, eg, sound, electricity, and light, were used to keep susceptible fish away from intakes with only moderate success. More useful were orientations of the inlet structure such that screens flush with the shoreline allowed lateral escape (Fig. **6a**). Where this screen orientation was impossible, guidance systems were developed for intake bays to direct fish from the incoming water flows to alternative escape routes (Fig. **6b**). Guidance devices were also installed at the entrances to unscreened, offshore inlets, where a horizontal velocity cap placed a few feet over a vertical opening proved especially effective in preventing fish entrance (Fig. **6c**). Small-scale modifications to existing rotating trash screens allow many fish to survive as they are raised from the water and

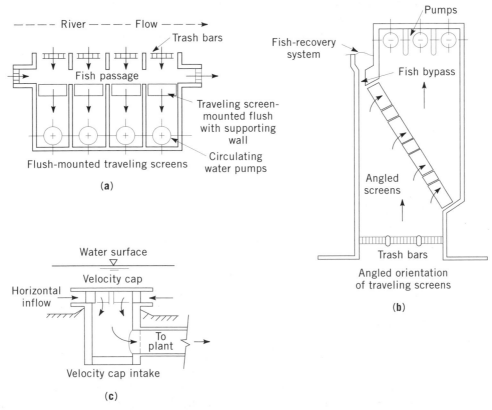

Fig. 6.   Approaches to minimizing entrapment and impingement of fish and large aquatic invertebrates, eg, blue crabs, on trash screens at intakes. (**a**) An inlet pump house with vertical traveling screens mounted flush with a river shoreline to minimize obstructions to animal movements; (**b**) parallel flow to direct fish to a recovery chamber that returns to the water body; (**c**) a velocity cap atop a vertical, offshore inlet; induces a horizontal flow which fish avoid instead of a vertical flow which they do not. Reproduced by permission (23).

deposited in sluiceways for return to the water. Expansion of hydroelectric power development has led to renewed interest in these guidance techniques.

Experience has also shown that deaths of fish on intake screens, although spectacular, often have minor consequences for local populations of many species (24). Most commonly impinged are open-water, schooling species that have high reproductive potential and a propensity for large natural variations in population numbers. The most common season for impingement mortalities, ie, winter, is a time of natural population reduction because of cold stress for several species. Cold-stressed fish may be impinged in large enough numbers to shut down the facility, but evidence suggests that most are moribund and destined not to survive regardless of cooling-water-intervention.

Minimization of impingement risks focuses on site-specific analyses of potentially vulnerable species and selection of engineering designs which, within acceptable cost limits, keep impingement deaths low.

Biocides.   Chlorine and other biocides are used occasionally in cooling water to kill and dispose of organic growths on heat-exchange surfaces and on piping where water flow could be hampered by such growth. Of necessity, organisms passing through the cooling circuit or residing in the effluent area during periodic chlorine injections experience the potentially lethal exposures. The objective is to maximize the intended kill and minimize extraneous damages, particularly in the receiving water (25).

Chlorine is a toxicant with a typical dose-response pattern for biota (Fig. 7). There is a time-dependent mortality at high concentrations and a low concentration above which long-term chronic effects are shown. Early methods for using the time-dependent effects of high temperature for predicting safe temperatures and exposure times during cooling-water exposures led to a similar approach for chlorine (3). Chlorine toxicity data were scant and different aggregates had to be developed for freshwater and marine assemblages because of the markedly different chlorine chemistry in fresh and salt water.

Entrainment.   Stresses to small nonscreenable organisms, eg, fish larvae, during passage through the cooling circuit come from a combination of thermal shock, physical abuses, and periodic injections of biocide. The physical abuses are less well-understood (8). Pressure changes, shear forces, and physical contact with pump impellers or tubing walls are sources of potential damage that vary from system to system. Minor differences in configuration of piping can mean the presence or absence of such devastating features as microcavitation cells. The physical configuration does not lend itself easily to laboratory experimentation, although three such physical simulators have been tried (26–28).

A principal frustration in attempts to minimize entrainment damages has been the contradictory demands of thermal and physical stresses. Thermal stresses can be quantitatively predicted based on dose-response data and minimized by increasing water-flow volumes, which dilute the fairly constant supply of rejected heat. The added volume of cooling water, however, includes proportionately more planktonic organisms, which are subjected to physical stresses. Attempts to minimize thermal damage in new power stations had the demonstrable effect of increasing physical damage. One proposed solution, which is based on the assumption of a high percentage of mortality resulting from physical stresses, is to return to high condenser temperatures at which all of the

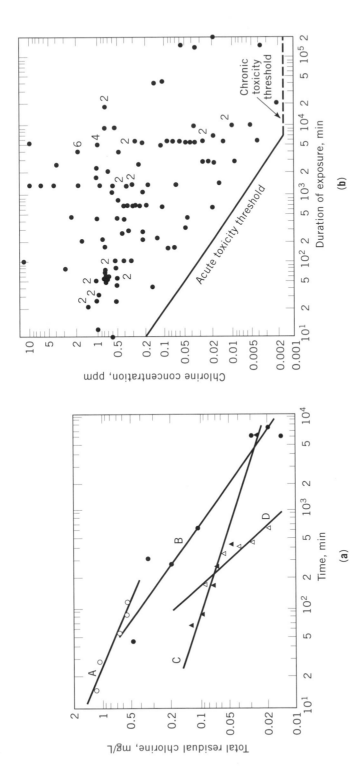

Fig. 7. Toxicity of chlorine to aquatic organisms. (**a**) Time-dependent mortality (50%) of four example species in various levels of total residual chlorine in the laboratory, where for A, *Alosa aestivalis*, and B, *Salmo gairdnerii*, r (correlation coefficient of the curve) = −0.96; and for C, *Pleuronectes platessa*, and D, *Salmo trutta*, r = −0.98. (**b**) A summary of chlorine toxicity to freshwater species, indicating overall no-effect thresholds for acute and chronic exposures. Numbers indicate where more than one test yielded the same result. A different summary figure applies to marine organisms because of differences in the chemistry of chlorine in seawater. Reproduced by permission (25).

relatively few organisms entrained would be killed, but those remaining in the waterbody would be assured survival (8).

The assumption of high percentage mortality resulting from physical stresses has been criticized. Newer methods of sampling entrained biota, which involve extensive precautions against damaging stresses during the sampling process, indicate a much higher survival of even delicate fish larvae than had been realized with earlier and more crude sampling methods. It appears that ameliorations of thermal stress with flow increases generally are not cancelled by additional physical mortalities except in cases of exceedingly high flows. An optimization procedure, such as that suggested by the Committee on Entrainment, appears fruitful for identifying on a site-specific and seasonally varying basis the most appropriate cooling-water flow regime (8).

There is also a compromise between entrainment and impingement. One of the engineering methods suggested for minimizing entrainment was replacement of 1-cm intake screens with screens of much smaller aperture. Although fewer organisms are entrained, most of those saved are, in fact, impinged on the smaller mesh of the screen, and the probable new impingement mortality may exceed the risks during entrainment.

Gas Balance.   Oxygen in solution does not generally change in a cooling circuit unless there is strong aeration of originally saturated water heated maximally, in which case some dissolved gas is lost. In no case does the loss reduce levels to below that required by aquatic life. When intake water is undersaturated, the agitation of the cooling circuit generally yields an increase. Abundant data support these generalities. Polluted receiving water does, however, have an accelerated microbial oxygen demand because of raised temperature, and this microbial deoxygenation can yield exaggerated dissolved oxygen sag zones downstream. The interaction of thermal effluent and waste stabilization in the river must be considered whenever polluted water is used for cooling, and engineering models for predicting oxygen sag zones are generally capable of incorporating temperature changes.

Damaging supersaturation of dissolved gases has occurred in some cooling-water discharges. Damage has been isolated to circumstances, usually in winter, when temperature elevations above the cool ambient are highest based on the assumption that even a large increase in winter results in temperatures below lethal level or prevalent water quality standards. The practice of winter increases in temperature rise across condensers by cutting back on pumping capacity has either ceased in general practice or the immediate discharge areas have been engineered to prevent long-term residence by susceptible biota. These remedial measures can be completely effective.

Cooling-Tower Chemicals.   The risk posed by changing power-station systems from traditional once-through or open-cycle cooling to cooling towers is from chemicals added to the recirculating water. Blowdown to aquatic systems and drift to the terrestrial landscape carry these chemicals to locations where natural biota can be damaged through direct poisoning or where toxicants can accumulate to potentially detrimental levels by food-chain transfer. Such risks can be minimized using fairly straightforward engineering approaches. Airborne drift has been reduced significantly by installation of physical baffles at the air outlets which intercept droplets, coalesce them, and return the water to the cooling-water flow. Blowdown can be treated for removal of chemical constituents, eg, chromates, with the additional benefit of chemical recycling and, thus, cost recovery. Chemical-laden sludges become a more long-term disposal problem. Practices include ponding and landfills. Chemical-recovery processes are also available for sludge treatment. Of significance for all of these processes is the cost which, when added to the initial capital cost of the cooling towers and the operating costs of pumps and fans, can reduce the attractiveness of closed-cycle cooling, especially when less costly mitigative measures exist for the open-cycle system at many locations.

Human Pathogens. Proper designs and good maintenance practices can reduce the potential risks to humans from pathogens stimulated by environmental conditions in cooling systems, eg, amoebae *Legionella*, to very low levels. Thermophilic protozoans which thrive in water near human body temperature that is also rich in organic matter find animal tissues similarly favorable, particularly the nasal passages which allow easy penetration to the brain. Field evidence indicates that these organisms are most abundant in stagnant pools and roadside ditches; fatal infections have generally been traced to exposures in such environments. The rapidly flowing water in power station discharges, although quite warm in summer, generally does not provide either stable substrates or high concentrations of organic material. Additionally, plant security measures generally ensure that bathers do not use the warmest zones. Allowing water contact by the public in discharges, particularly wherever the outlet is a slowly moving canal with abundant vegetation, is risky. The Legionnaires' disease microorganism, identified as a common inhabitant of many cooling-tower systems, especially small ones used for building-temperature control, can be held in check by systematic use of biocides in recirculating cooling water. Contact with the infectious organisms can be minimized if care is taken to isolate inlets for ventilating air from the drift aerosols which are emitted from cooling towers.

Maximizing the Benefits

During occasional episodes of energy scarcity, particularly in the 1970s, attention was given to finding productive uses for power plant waste heat. Potential physical applications of power plant rejected heat include industrial heating and biological applications, such as fish culture, soil warming, heating greenhouses, and livestock shelters.

The most important factor determining the possible uses of such heat is the temperature of the heat source. There is a threshold between low and high grade heat for engineering uses at ca 100°C. Most waste heat from electric power stations is in the form of low grade heat. This is so low in temperature that disposal to the environment has traditionally been considered the only practical alternative. Discharges from once-through cooling systems of thermal power plants have outlet temperatures of 10–40°C, depending on the season. Circulating water in closed-cycle cooling systems is only slightly warmer (20–50°C). These low temperatures are the result of careful engineering of power stations to extract the maximum amount of electrical energy from the fuel before the rejected heat is dumped to circulating water in steam condensers. Such engineering has produced high efficiencies compared to most other mechanical systems; 33% in nuclear plants to 40% in fossil-fueled plants is the fuel energy converted to electricity.

A more useful, higher grade heat is in a form that has not been degraded to the low temperatures of most power plant thermal discharges but which remains at 40–200 °C following production of initial amounts of electricity. Low pressure steam can be extracted from the turbine system of a power plant before condensation to waste-heat temperature, thus permitting its use for functions requiring higher temperatures. Although this gives better utilization of the energy remaining in the steam, the efficiency of the low pressure turbine for producing electricity is diminished. Condensers can also be operated with smaller cooling-water flows to yield higher discharge temperatures but with some penalty to generation efficiency because of higher turbine back pressures. The overall efficiency of fuel energy use in such multiple-purpose systems can exceed that of electricity production alone, however.

Ideally, a power station could be a potential supplier of electricity, high grade heat, and low grade heat, even though electricity is the main product and most new power stations are optimized accordingly. Depending on the desired uses at particular sites, future power stations could be designed to alter the ratios of electricity and different grades of heat to produce the most efficient total energy use. Such multiple use was common at early generating stations in the United States and is still common in Europe. Applications in North America are becoming more common.

Aquaculture. Culture of some aquatic species in essentially unmodified thermal effluents of power stations has been attempted both experimentally and commercially. The principle behind such culture is the temperature dependence of growth rates. Use of rejected heat to prolong optimal growth temperatures in cool months can significantly increase the sizes attained in a year. Even though power stations rarely provide high enough temperatures in winter for optimal rates, temperatures can usually be maintained at levels that increase growth rates. Maximum use of available temperatures has been made by culturing warm-water shrimp in summer and cool-water trout in winter, each of which can utilize the added temperature imparted by power station cooling (29). Most thermal aquaculture facilities at power stations have not been economical and have been closed. There has been interest in small-scale thermal aquaculture cogeneration facilities in which small fossil-fueled electricity generators are operated, with temperature control of fish tanks as the primary objective. Any electricity remaining after on-site use by the facility is sold to the local electrical utility.

Open-Field Agriculture. Use of warmed water for field crops and orchards has been tested in several studies. Buried pipes can convey thermal-discharge heat to soils where warming aids plant growth and extends the growing season. Soil warming with open irrigation water has also been studied. There are advantages of spraying thermal-effluent water on crops and orchards to prevent freezing of buds and blooms. Whether or not the heat is used, power plant cooling water can conveniently be combined with irrigation water systems to provide multiple uses of the water. An extensive demonstration project has been established by Electricité de France at their St. Laurent des Eaux power station.

Greenhouse Agriculture. Greenhouse agriculture is well known for its many advantages over open-field agriculture for certain crops. Yields are greater per land area and year-around culture is possible, which allows matching of crop harvest with high demand and price. One drawback is high expense for heating in winter, which can be the most costly part of greenhouse operation. The use of waste heat from steam-electric power plants therefore appears promising as a source of low cost heat for use in greenhouses, especially if the power stations have cooling towers with wintertime operating temperatures of 16°C or higher. The economic analysis is favorable (30).

Experimental greenhouses based on waste heat for temperature control have been operated at several sites and have demonstrated the principle that cooling water can be used to heat a greenhouse in winter. Some have also been based on power plant effluent for cooling in summer; inexpensive porous packing through which the heated water drips and air circulates is used (Fig. 8). Evaporation provides cooling in summer and sensible heat transfer warms the air in winter. Additional heat and humidity control can be provided with a finned-tube heat exchanger. The French have used pipes and plastic tubes laid in the floor of the greenhouse; warm thermal effluent or water warmed further with heat pumps operating from thermal discharge waters circulates through the pipes and tubes. As with aquaculture systems, greenhouses should affiliate with multiple-unit power stations to ensure continuous warm water supplies.

Animal Shelters. The advantages of temperature control for maximizing weight gain and avoiding animal losses in livestock and poultry are well known. Low grade heat from power station cooling offers the possibility of low cost heating of animal shelters, although few demonstration projects have been developed. Heating animal shelters is a special case of space heating, although capital costs generally must be lower for the system to be economical. The greatest potential for heating animal shelters appears to be in continental climates where exceedingly cold winters can lead to deaths of livestock.

Space Heating. A large percentage of the energy requirements of most countries in temperate zones is for heating and cooling of living and working

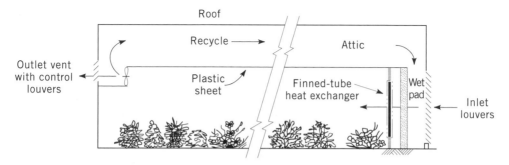

Fig. 8. Longitudinal section of an experimental waste-heat greenhouse in which temperature control in all seasons is provided by evaporation and heat transfer as air passes through a fiber pad soaked with power station cooling water or by heat transfer as air passes through a finned-tube heat exchanger that carries cooling water. A false ceiling provides for recycle of air through the heat-transfer medium. Reproduced by permission (31).

spaces and for hot water. The historical use of dual-purpose power plants for electricity generation and central district heating in the United States and their extensive use in such countries as the Russia, Sweden, and Germany suggests that expansion of this form of waste-heat utilization can contribute significantly to energy conservation and control of concentrated thermal discharges worldwide.

The most economical form of long-distance transport of thermal energy for space heating and cooling by adsorption methods is by heated water rather than steam. Steam had been used from early dual-purpose power plants prior to development of modern water-cooled condensers. It is still used in areas of dense loads, but its range of effective distribution is small because of large pressure drops in distribution systems. Modern experience with dual-purpose power plants (mostly in Europe) and new district heating systems in the United States, eg, in colleges, institutions, and shopping centers, has shown hot water to be superior for dispersed loads, including single-family residences. These dual-purpose stations operate at high thermal efficiency; the Swedish Malmo Plant is researching new technologies for distribution pipes in order to expand economical district heating to dispersed single-family residences.

Economic analyses for the United States indicate that supplying thermal energy to the commercial–residential sector from dual-purpose power stations is more economically competitive in new applications than in cases where existing buildings are to be serviced (32). There is little difference for supplying industrial heating. Such a thermal grid is most competitive with new fossil fuels where there is a high heat-load density and expensive fuel costs.

Industrial Process Heat. Many industries use process heat at 77–110°C. Much of this heat is supplied by combustion of oil and natural gas (qv). Equipment manufacturers are developing industrial heat pumps to capture free industrial plant waste heat and regenerate it to the desired process heat temperature, thereby greatly reducing energy costs associated with direct heating. Operations that use heat in the 88–110°C range that can be supplied by new high efficiency heat pumps based on rejected heat sources are washing, blanching, sterilizing, and cleaning operations in food processing; grain drying; metal cleaning and treating processes; distilling operations in the food and petrochemical industries; and industrial space heating. Power plant rejected heat could be valuable for developing industries that perform these processes (see PROCESS ENERGY CONSERVATION).

Cooling Reservoirs. The most extensively developed productive use for power plant cooling is in multiple-purpose cooling reservoirs. Small impoundments built especially for heat dissipation have been managed for extensive recreational uses as well. Highly productive fisheries for warm-water species, eg, largemouth bass (*Micropterus salmoides*) and channel catfish (*Ictalurus punctatus*), have made cooling reservoirs highly popular. Such reservoirs have been extensively developed in lake-free areas, eg, Texas and Illinois. Broad-scale ecological research on some of these cooling reservoirs has documented the valuable synergistic relationship between cooling-lake fisheries and power station heat dissipation (33).

Thermal pollution from energy generation is not viewed today as the threat to the environment that it appeared to be in the 1960s and 1970s. This is not because the potential environmental hazards of power station cooling are necessarily any less. The change in perspective has arisen because the hazards have been recognized, biological and other environmental constraints (and benefits, in some cases) have become understood, and good engineering practice has devised methods to minimize risks. Where location-specific controversies remain, multidisciplinary teams of biologists and engineers can usually design appropriate, site-specific studies to develop the most suitable solutions.

BIBLIOGRAPHY

"Thermal Pollution in Power Plants" in *ECT* 3rd ed., Vol. 22, pp. 846–868, by C. C. Coutant, Oak Ridge National Laboratory.

1. E. W. Moore, *Ind. Eng. Chem.* **50**, 87A (1958).
2. Pennsylvania Dept. of Health, *Heated Discharges—Their Effects on Streams*, Publication No. 3, Division of Sanitary Engineering, Pennsylvania Dept. of Health, Harrisburg, Pa., 1962.
3. C. C. Coutant, *Crit. Rev. Environ. Control.* **1**, 341 (1970).
4. L. D. Jensen, ed., *Entrainment and Intake Screening*, Publication No. 74-049-00-5, Electric Power Research Institute, Palo Alto, Calif., 1974.
5. J. A. Zillich, *J. Water Pollut. Control Fed.* **44**, 212 (1972).
6. R. L. Jolley, *J. Water Pollut. Control Fed.* **47**, 601 (1975).
7. S. Markowski, *J. Anim. Ecol.* **28**, 243 (1959).
8. J. R. Schubel and B. C. Marcy, eds., *Power Plant Entrainment—A Biological Assessment*, Academic Press, Inc., New York, 1978.
9. S. W. Christensen, W. Van Winkle, L. W. Barnthouse, and D. S. Vaughan, *Environ. Impact Assess. Rev.* **2**, 63 (1981).
10. W. J. Ebel, *Fishery Bull.* **68**, 1 (1969).
11. J. Z. Reynolds, *Science* **207**, 367 (1980).
12. *J. Water Pollut. Control Fed.*, annual review issues.
13. *Water Quality Criteria, 1972*, Report No. R-73-033, National Academy of Sciences–National Academy of Engineering, U.S. Environmental Protection Agency, Washington, D.C., 1973, pp. 151–171 and appendix.
14. *Quality Criteria for Water*, U.S. Environmental Protection Agency, Washington, D.C., 1976.
15. C. C. Coutant, *Crit. Rev. Environ. Control* **3**, 1 (1972).
16. D. R. F. Harleman, *Report of a Workshop on the Impact of Thermal Power Cooling Systems on Aquatic Environments*, CONF-750.980, EPRI SR-38, Asildmar, Pacific Grove, Calif., Sept. 28–Oct. 2, 1975, pp. 128–135.
17. H. H. Carter, J. R. Schubel, R. E. Wilson, and P. M. J. Woodhead, *Environ. Manage.* **3**, 353 (1979).
18. J. J. Magnuson, L. B. Crowder, and P. A. Medvick, *Am. Zool.* **19**, 331 (1979).
19. J. R. Gammon, unpublished results, Depauw University, Greencastle, Ind., 1981.
20. C. C. Coutant, *Trans. Am. Fish. Soc.* **114**, 31 (1985).
21. C. C. Coutant, *Striped Bass and the Management of Cooling Lakes, Waste Heat Utilization and Management*, Hemisphere Publishing Corp., Wash., 1983, pp. 389–396.
22. L. D. Jensen, ed., *Third National Workshop on Entrainment and Impingement*, Ecological Analysts, Inc., Melville, N.Y., 1977.
23. L. D. Jensen, ed., *Fourth National Workshop on Entrainment and Impingement*, Ecological Analysts, Inc., Melville, N.Y., 1978.
24. R. B. McLean, P. T. Singley, J. S. Griffith, and M. V. McGee, *Threadfin Shad Impingement: Effect of Cold Stress*, NUREG/CR-1044, ORNL/NUREG/TM-340, Oak Ridge National Laboratory, Oak Ridge, Tennessee, and National Technical Information Service, Springfield, Va., 1980.

25. J. S. Mattice, *Environmental Effects of Cooling Systems*, Technical Report Series 202, International Atomic Energy Agency, Vienna, 1980, pp. 12–26, 148–167.

26. R. J. Kedl and C. C. Coutant, in G. W. Esch and R. W. McFarlane, eds., *Thermal Ecology II*, Conf-750425, National Technical Information Service, Springfield, Va., 1976, pp. 394–400.

27. J. M. O'Connor, T. C. Ginn, and G. V. Pase, *The Evaluation and Description of a Power Plant Condenser Tube Simulator*, Report No. 8-0248-956, New York University Medical Center, New York, 1977.

28. J. S. Suffern, *The Physical Effects of Entrainment—Current Research at ORNL*, ORNL/TM-5948, Oak Ridge National Laboratory, Oak Ridge, Tennessee, and National Technical Information Service, Springfield, Va., 1977.

29. W. Majewski and D. C. Miller, eds., *Predicting Effects of Power Plant Once-Through Cooling on Aquatic Systems*, Technical Papers in Hydrology 20, United Nations Educational, Scientific and Cultural Organization (Unesco), Paris, 1979.

30. M. Olszewski, S. J. Hillenbrand, and S. A. Reed, *Waste Heat vs. Conventional Systems for Greenhouse Environment Control: An Economic Assessment*, ORNL/TM-5069, Oak Ridge National Laboratory, Oak Ridge, Tennessee, and National Technical Information Service, Springfield, Va., 1976.

31. S. E. Beall and G. Samuels, *The Use of Warm Water for Heating and Cooling Plant and Animal Enclosures*, ORNL-TM-3381, Oak Ridge National Laboratory, Oak Ridge, Tennessee, and National Technical Information Service, Springfield, Va., 1971.

32. M. Olszewski, *Preliminary Investigations of the Thermal Energy Grid Concept*, ORNL/TM-5786, Oak Ridge National Laboratory, Oak Ridge, Tennessee, and National Technical Information Service, Springfield, Va., 1977.

33. W. Larimore and co-workers, *Evaluation of a Cooling Lake Fishery*, EA-1148, 3 Vols., Electric Power Research Institute, Palo Alto, Calif., 1979.

General References

Environmental Effects of Cooling Systems, Technical Report Series No. 202, International Atomic Energy Agency, Vienna, 1980.

J. R. Schubel and B. C. Marcy, Jr., eds., *Power Plant Entrainment—A Biological Assessment*, Academic Press, New York, 1978.

G. F. Lee and C. Stratton, *Ind. Water Eng.* **9**, 12–16 (1972).

H. Precht, J. Christophensen, H. Hensel, and W. Larcher, *Temperature and Life*, Springer-Verlag, New York, 1973.

C. C. Coutant, *Crit. Rev. Environ. Control* **3**, 1 (1972).

S. E. Beall, C. C. Coutant, M. Olszewski, and J. S. Suffern, *Ind. Water Eng.* **14**, 8 (1977).

L. B. Goss and L. Scott, *Factors Affecting Power Plant Waste Heat Utilization*, Pergamon Press, New York, 1980.

C. C. Coutant, *Elec. Perspec.* **16**(4), 32 (1992).

C. C. Coutant, *Sci. Am.* **254**(8), 98 (1986).

M. C. Bell, *Fisheries Handbook of Engineering Requirements and Biological Criteria*, U.S. Army Corps of Engineers, North Pacific Division, Portland, Ore., 1991.

L. W. Barnthouse, R. J. Klauda, D. S. Vaughan, and R. L. Kendall, eds., *Science, Law, and Hudson River Power Plants, A Case Study in Environmental Impact Assessment*, American Fisheries Society Monograph 4, Bethesda, Md., 1988.

Trans. Am. Fish. Soc. **122**(3) (May 1993).

CHARLES C. COUTANT
Oak Ridge National Laboratory

THERMODYNAMICS

Thermodynamics is a deductive science built on the foundation of two fundamental laws that circumscribe the behavior of macroscopic systems: the first law of thermodynamics affirms the principle of energy conservation; the second law states the principle of entropy increase. In-depth treatments of thermodynamics may be found in References 1–7.

In the formal application of these laws, attention is focused on a specific object, a particular quantity of matter, or a bounded region of space, specified as the system. All else then constitutes the surroundings. The coordinates which characterize the systems of interest in chemical technology are temperature, T, pressure, P, molar volume, V, and composition. Such a PVT system may be open or closed to the exchange of matter with its surroundings, or it may be isolated from its surroundings, in which case it can exchange neither matter nor energy. Once isolated, a system is independent of its surroundings and can only progress toward an equilibrium state the thermodynamic coordinates of which have no further tendency to change. Systems not in equilibrium states undergo processes by which their coordinates alter. During such processes the system and its surroundings may exchange energy in the forms of heat, Q, and work, W.

The coordinates of thermodynamics do not include time, ie, thermodynamics does not predict rates at which processes take place. It is concerned with equilibrium states and with the effects of temperature, pressure, and composition changes on such states. For example, the equilibrium yield of a chemical reaction can be calculated for given T and P, but not the time required to approach the equilibrium state. It is however true that the rate at which a system approaches equilibrium depends directly on its displacement from equilibrium. One can therefore imagine a limiting kind of process that occurs at an infinitesimal rate by virtue of never being displaced more than differentially from its equilibrium state. Such a process may be reversed in direction at any time by an infinitesimal change in external conditions, and is therefore said to be reversible. A system undergoing a reversible process traverses equilibrium states characterized by the thermodynamic coordinates.

The Laws of Thermodynamics

First Law of Thermodynamics. The energy change of any system together with its surroundings is zero. Implicit in this declaration is the affirmation that there exists a form of energy, known as internal energy, which for systems in equilibrium states is an intrinsic property of the system. Internal energy is separate from the external energy forms, ie, the kinetic and potential energies of macroscopic bodies. Although a macroscopic property, internal energy originates in the kinetic and potential energies of molecules and submolecular particles. In applications of the first law of thermodynamics, energy in all its forms, both internal and external, must be considered.

Applied to a closed system which undergoes only an internal energy change, the first law of thermodynamics is given by equation 1:

$$dU^t = dQ + dW \tag{1}$$

where U^t is the total internal energy of the system and dU^t represents a differential change in this property. On the other hand, dQ and dW are differential quantities of heat and work representing energy in transit across the boundary of the system, and serving to account for the energy change of the surroundings. Integration of equation 1 gives for a finite process (eq. 2):

$$\Delta U^t = Q + W \tag{2}$$

where ΔU^t is the finite change from the initial to the final value of U^t. The heat, Q, and work, W, are finite quantities, neither properties of the system nor functions of the thermodynamic coordinates that characterize the system. By convention, the numerical values of Q and W are positive for the transfer of heat and work to the system and negative for transfer from the system.

Second Law of Thermodynamics. The entropy change of any system together with its surroundings is positive for a real process, approaching zero as the process approaches reversibility:

$$\Delta S_{total} \geq 0 \tag{3}$$

where S_{total} is the entropy of the system and its surroundings. Implicit in this declaration is the affirmation that there exists a thermodynamic property, known as entropy, which for systems in equilibrium states is an intrinsic property of the system. For reversible processes, changes in the entropy of the system are given by equation 4:

$$dS^t = dQ_{rev}/T \tag{4}$$

where S^t is the total entropy of the system and T is the absolute temperature of the system.

Each of the two laws of thermodynamics asserts the existence of a primitive thermodynamic property, and each provides an equation connecting the property with measurable quantities. These are not defining equations; they merely provide a means to calculate changes in each property.

Heat Engines and Heat Pumps

A heat engine is a device operating in cycles that takes in heat, Q_H, from a heat reservoir at temperature T_H, discards heat, Q_C, to another heat reservoir at a lower temperature T_C, and produces work. A heat reservoir is a body that can absorb or reject unlimited amounts of heat without change in temperature. Entropy changes of a heat reservoir depend only on the absolute temperature and on the quantity of heat transferred, and are always given by the integrated form of equation 4:

$$\Delta S^t = \frac{Q}{T} \tag{5}$$

Here, Q_{rev} is replaced by Q because the effect of heat transfer on a heat reservoir does not depend on its reversibility. Thus the entropy changes of the two heat reservoirs associated with a heat engine are given by equations 6 and 7:

$$\Delta S_H^t = \frac{Q_H}{T_H} \tag{6}$$

$$\Delta S_C^t = \frac{Q_C}{T_C} \tag{7}$$

In these equations Q_H and Q_C refer to the respective heat reservoirs, and numerical values are positive when heat flows into the reservoir and negative when heat flows out.

Because the engine operates in cycles, it experiences no change in its own properties; therefore the total entropy change of the engine and its associated heat reservoirs is given by equation 8:

$$\Delta S_{total} = \frac{Q_H}{T_H} + \frac{Q_C}{T_C} \tag{8}$$

The first law (eq. 2) is applied to the engine taken as the system:

$$\Delta U_{engine} = Q + W \tag{9}$$

The heat quantities Q_H and Q_C are the same whether considered with respect to the engine or with respect to the heat reservoirs, except that they have opposite signs. Thus for the engine,

$$Q = -Q_H - Q_C \tag{10}$$

Because $\Delta U_{engine} = 0$,

$$W = Q_H + Q_C \tag{11}$$

Elimination of Q_H between equations 8 and 11 gives equation 12:

$$W = T_H \Delta S_{total} - Q_C \left(\frac{T_H}{T_C} - 1 \right) \tag{12}$$

The work, W, can range from zero, if the engine is completely ineffective, to the limiting negative value attained for reversible operation. If $W = 0$, then the process degenerates to one of simple heat transfer, for which

$$\Delta S_{total} = Q_C \left(\frac{T_H - T_C}{T_H T_C} \right) \tag{13}$$

At the other limit, $\Delta S_{\text{total}} = 0$, and

$$W = -Q_{\text{C}}\left(\frac{T_{\text{H}}}{T_{\text{C}}} - 1\right) \tag{14}$$

This shows that if W is to be negative (work produced) and finite, then Q_{C} must be positive and finite. Thus even for reversible operation, a finite amount of heat must be rejected from the engine to the reservoir at T_{C}.

Equations 11 and 14 may be combined to yield Carnot's equations.

$$\frac{Q_{\text{C}}}{T_{\text{C}}} = \frac{-Q_{\text{H}}}{T_{\text{H}}} \tag{15}$$

$$\frac{W}{-Q_{\text{H}}} = \frac{T_{\text{C}}}{T_{\text{H}}} - 1 \tag{16}$$

These apply to all reversible heat engines operating between a heat source and a heat sink at fixed temperature levels, ie, to Carnot engines. In these equations Q_{C} and Q_{H} refer to the heat reservoirs, and this is the origin of the minus sign with Q_{H}. Common practice is to delete the minus signs, adding absolute value signs:

$$\frac{|Q_{\text{C}}|}{|Q_{\text{H}}|} = \frac{T_{\text{C}}}{T_{\text{H}}} \tag{17}$$

$$\frac{|W|}{|Q_{\text{H}}|} = 1 - \frac{T_{\text{C}}}{T_{\text{H}}} \tag{18}$$

Equation 18 expresses the thermal efficiency of a Carnot heat engine, ie, the fraction of the heat taken in that is converted into work. Because a Carnot engine is reversible, it can be run as a heat pump or refrigerator, as indicated in Figure 1. Equations 17 and 18 apply to the Carnot heat pump or refrigerator as well as to the Carnot engine, because all quantities are the same for the two cases. The measure of performance of the Carnot refrigerator is given by its coefficient of performance, obtained by dividing equation 17 by equation 18:

$$\frac{|Q_{\text{C}}|}{|W|} = \frac{T_{\text{C}}}{T_{\text{H}} - T_{\text{C}}} \tag{19}$$

For given temperatures T_{H} and T_{C}, no heat engine operating in a cycle can have a thermal efficiency greater than that of a Carnot engine, and no refrigerator can have a coefficient of performance greater than that of a Carnot refrigerator. For example, a power plant operating between a boiler temperature of 560 K and a condenser temperature of 310 K typically has a thermal efficiency (work out/heat in) of ~0.3, whereas a Carnot engine operating between these same temperatures has an efficiency (eq. 18) of 0.446.

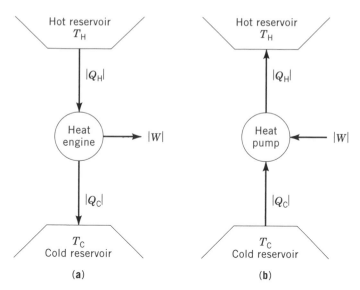

Fig. 1. Schematic representation of (**a**) Carnot heat engine and (**b**) Carnot refrigerator used as a heat pump.

PVT and Phase Behavior of Pure Substances

Although thermodynamics arose from the study of steam engines (early heat engines), its universality has long been recognized. Application to a wide variety of systems requires data on the volumetric and phase behavior of many substances. This behavior for pure substances is most readily described by reference to schematic *PT* and *PV* diagrams. Figure 2, a *PT* diagram, displays regions of single-phase equilibrium as areas.

In Figure 3, a *PV* diagram, regions of two-phase equilibrium appear as areas. Thus, the region of vapor–liquid equilibrium is delineated by curves representing saturated liquid and saturated vapor. These curves meet at the critical point c, and together form a dome below which lies the region in which vapor and liquid phases coexist in equilibrium. The critical point is also shown in Figure 2, where it is seen to mark the upper end of the liquid–vapor saturation curve. The critical point properties, temperature T_c, pressure P_c, and molar volume V_c, are characteristic of a fluid, and are the basis for generalized correlations of fluid properties. Tabulations of critical constants are given in References 8 and 9.

For temperatures below the vapor–liquid critical temperature, T_c, isotherms to the left of the liquid saturation curve (see Fig. 3) represent states of subcooled liquid; isotherms to the right of the vapor saturation curve are for superheated vapor. For sufficiently large molar volumes, V, all isotherms are approximated by the ideal gas equation, $P = RT/V$. Isotherms in the two-phase liquid–vapor region are horizontal. The critical isotherm at temperature T_c exhibits a horizontal inflection at the critical state, for which

$$\left(\frac{\partial P}{\partial V}\right)_{T;cr} = \left(\frac{\partial^2 P}{\partial V^2}\right)_{T;cr} = 0 \qquad (20)$$

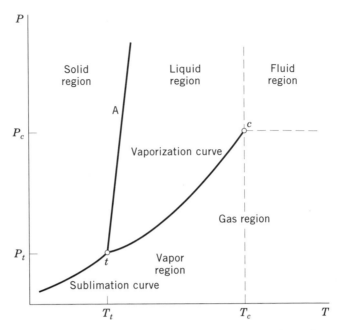

Fig. 2. *PT* diagram for a pure substance that expands on melting (not to scale). For a substance that contracts on melting, eg, water, the fusion curve, A, has a negative slope; point *t* is a triple state; point *c* is the gas–liquid critical state; (—) are phase boundaries representing states of two-phase equilibrium; and pressures, represented by curves, are saturation pressures, P^{sat}.

Many empirical expressions have been proposed to represent the temperature dependence of the vapor–liquid saturation pressure, as shown by the vaporization curve of Figure 2. The most popular is the Antoine equation:

$$\ln P^{\text{sat}} = A - \frac{B}{T + C} \tag{21}$$

The accurate representation of vapor pressure data over a wide temperature range requires an equation of greater complexity. The Wagner equation (eq. 22) expresses the reduced vapor pressure P^{sat}/P_c as a function of reduced temperature T/T_c:

$$\ln \frac{P^{\text{sat}}}{P_c} = \frac{A\tau + B\tau^{1.5} + C\tau^3 + D\tau^6}{1 - \tau} \tag{22}$$

where $\tau \equiv 1 - (T/T_c)$ and A, B, C, and D are constants. Values of the constants for this equation and for equation 21 are available (9).

The volumetric properties of fluids are conveniently represented by *PVT* equations of state. The most popular are virial, cubic, and extended virial equations. Virial equations are infinite series representations of the compressibility

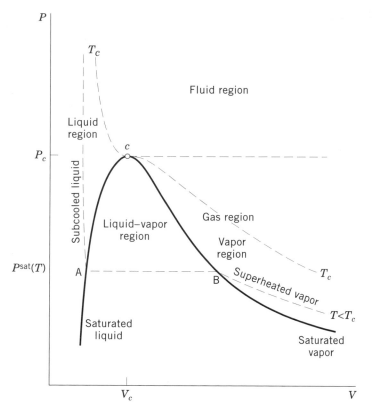

Fig. 3. *PV* diagram for a pure fluid (not to scale): point c is the gas–liquid critical state, $P^{\text{sat}}(T)$ is the constant pressure at which phase transition occurs at temperature T, and A and B are saturation states.

factor Z, defined as $Z \equiv PV/RT$, having either molar density, $\rho(\equiv V^{-1})$, or pressure, P, as the independent variable of expansion:

$$Z = 1 + B\rho + C\rho^2 + D\rho^3 + \cdots \tag{23}$$

$$Z = 1 + B'P + C'P^2 + D'P^3 + \cdots \tag{24}$$

Parameters $B, C, D, \ldots$, are density series virial coefficients, and $B', C', D', \ldots$, are pressure series virial coefficients. The second virial coefficients are defined by the following recipes:

$$B \equiv \frac{1}{1!}\left(\frac{\partial Z}{\partial \rho}\right)_{T,y;\rho=0} \qquad B' \equiv \frac{1}{1!}\left(\frac{\partial Z}{\partial P}\right)_{T,y;P=0} \tag{25}$$

Similarly, the third virial coefficients are defined by equation 26:

$$C \equiv \frac{1}{2!}\left(\frac{\partial^2 Z}{\partial \rho^2}\right)_{T,y;\rho=0} \qquad C' \equiv \frac{1}{2!}\left(\frac{\partial^2 Z}{\partial P^2}\right)_{T,y;P=0} \tag{26}$$

Higher virial coefficients are defined analogously. All virial coefficients depend on temperature and composition only.

The pressure series and density series coefficients are related to one another:

$$B' = \frac{B}{RT} \tag{27}$$

$$C' = \frac{C - B^2}{(RT)^2} \tag{28}$$

$$D' = \frac{D - 3\,BC + 2\,B^3}{(RT)^3} \tag{29}$$

Equation 24 for the pressure series expansion can therefore be written as follows:

$$Z = 1 + \frac{BP}{RT} + \left[\frac{C - B^2}{(RT)^2} \right] P^2 + \cdots \tag{30}$$

in which the density series coefficients are used. This form is preferred to equation 24 because density series coefficients are normally reported by experimentalists. Correlations for B and C are available (4,6).

Because virial coefficients beyond the third are rarely available, equations 23 and 30 are usually truncated after two or three terms. For low pressures, the two-term truncation of equation 30 is preferred:

$$Z = 1 + \frac{BP}{RT} \tag{31}$$

For higher pressures, equation 23 truncated to three terms is more suitable:

$$Z = 1 + B\rho + C\rho^2 \tag{32}$$

Equation 31 should not be used for densities greater than about half the critical value, nor equation 32 for densities exceeding ~85% of the critical value. Substantially lower limits may apply for polar gases, particularly those that associate.

The virial equations are unsuitable for liquids and dense gases. The simplest expressions appropriate (in principle) for such fluids are equations cubic in molar volume. These equations, inspired by the van der Waals equation of state, may be represented by the following general formula, where parameters $b, \theta, \delta, \epsilon$, and η each can depend on temperature and composition:

$$P = \frac{RT}{V - b} - \frac{\theta(V - \eta)}{(V - b)(V^2 + \delta V + \epsilon)} \tag{33}$$

Special cases of equation 33 are obtained on specification of values or expressions for various parameters. Because of their generality, two-parameter cubic equations are the most popular. The following equations are modern examples (10–12):

$$P = \frac{RT}{V - b} - \frac{\theta_{\mathrm{SRK}}}{V(V + b)} \tag{34}$$

$$P = \frac{RT}{V - b} - \frac{\theta_{\mathrm{PR}}}{V^2 + 2\,bV - b^2} \tag{35}$$

$$P = \frac{RT}{V - b} - \frac{\theta_{\mathrm{H}}}{V^2 + 3\,bV - 2\,b^2} \tag{36}$$

In each of these expressions, ie, the Soave-Redlich-Kwong, θ_{SRK} (eq. 34), Peng-Robinson, θ_{PR} (eq. 35), and Harmens, θ_{H} (eq. 36), parameter θ, different for each equation, depends on temperature. Numerical values for b and $\theta(T)$ are determined for a given substance by subjecting the equation of state to the critical derivative constraints of equation 20 and by requiring the equation to reproduce values of the vapor–liquid saturation pressure, P^{sat}.

Cubic equations, although simple and able to provide semiquantitative descriptions of real fluid behavior, are not generally useful for accurate representation of volumetric data over wide ranges of T and P. For such applications, more comprehensive expressions with large numbers of adjustable parameters are needed. The simplest of these are the extended virial equations, exemplified by the eight-constant Benedict-Webb-Rubin (BWR) equation of state (13):

$$Z = 1 + \left(B_0 - \frac{A_0}{RT} - \frac{C_0}{RT^3}\right)\rho + \left(b - \frac{a}{RT}\right)\rho^2 + \frac{a\alpha}{RT}\rho^5$$
$$+ \ \frac{c\rho^2}{RT^3}\,(1 + \gamma\rho^2)\exp(-\gamma\rho^2) \tag{37}$$

Although PVT equations of state are based on data for pure fluids, they are frequently applied to mixtures. The virial equations are unique in that rigorous expressions are known for the composition dependence of the virial coefficients. Statistical mechanics provide exact mixing rules which show that the nth virial coefficient of a mixture is nth degree in the mole fractions:

$$B = \sum_i \sum_j y_i y_j B_{ij} \tag{38}$$

$$C = \sum_i \sum_j \sum_k y_i y_j y_k C_{ijk} \tag{39}$$

$$\vdots$$

The subscripted coefficients, $B_{ij}, C_{ijk}, \ldots$, are functions of T only, and their numerical values are unchanged on permutation of the subscripts. Coefficients having identical subscripts, eg, B_{11} and C_{222}, are properties of pure species; those having mixed subscripts, eg, $B_{12} = B_{21}$, $C_{122} = C_{212} = C_{221}$, are mixture properties and are called interaction or cross-coefficients.

Mixing rules for the parameters in an empirical equation of state, eg, a cubic equation, are necessarily empirical. With cubic equations, linear or quadratic expressions are normally used, and in equations 34–36, parameters b and θ for mixtures are usually given by the following, where, as for the second virial coefficient, $\theta_{ij} = \theta_{ji}$.

$$b = \sum_i x_i b_i \tag{40}$$

$$\theta = \sum_i \sum_j x_i x_j \theta_{ij} \tag{41}$$

Fundamental Property Relations

The systems of interest in chemical technology are usually comprised of fluids not appreciably influenced by surface, gravitational, electrical, or magnetic effects. For such homogeneous fluids, molar or specific volume, V, is observed to be a function of temperature, T, pressure, P, and composition. This observation leads to the basic postulate that macroscopic properties of homogeneous PVT systems at internal equilibrium can be expressed as functions of temperature, pressure, and composition only. Thus the internal energy and the entropy are functions of temperature, pressure, and composition. These molar or unit mass properties, represented by the symbols V, U, and S, are independent of system size and are intensive. Total system properties, V^t, U^t, and S^t, do depend on system size and are extensive. Thus, if the system contains n moles of fluid, $M^t = nM$, where M is a molar property. Temperature and pressure are also intensive, but have no extensive counterparts.

Thermodynamics provides a set of differential equations that interrelate the properties of PVT systems. Most of these properties are abstract, but the equations provide a limited number of connections with measurable quantities. One of the functions of thermodynamics is to maximize the return in useful information for any investment in experiment.

Consider a closed, nonreacting PVT system containing n moles of a homogeneous fluid mixture. The mole numbers n_i of the individual chemical species sum to

$$n = n_1 + n_2 + \ldots + n_N \tag{42}$$

Equation 1, applied to this system as it undergoes a differential change of state in a reversible (rev) process, may be written as follows:

$$d(nU) = dQ_{\mathrm{rev}} + dW_{\mathrm{rev}} \tag{43}$$

According to equation 4,

$$dQ_{\mathrm{rev}} = Td(nS) \tag{44}$$

and by the definition of work,

$$dW_{rev} = -Pd(nV) \tag{45}$$

These three equations combine to give equation 46:

$$d(nU) = Td(nS) - Pd(nV) \tag{46}$$

Although derived for a reversible process, equation 46 relates properties only, irrespective of the process, and therefore applies to any change in the equilibrium state of a homogeneous, closed, nonreacting system.

As a result of equation 46 the functional relation (eq. 47) may be written as follows:

$$nU = f(nS, nV) \tag{47}$$

It follows that

$$d(nU) = \left[\frac{\partial(nU)}{\partial(nS)}\right]_{nV, n} d(nS) + \left[\frac{\partial(nU)}{\partial(nV)}\right]_{nS, n} d(nV) \tag{48}$$

The subscript n indicates that all mole numbers n_i are held constant. Comparison with equation 46 shows that

$$\left[\frac{\partial(nU)}{\partial(nS)}\right]_{nV, n} = T \tag{49}$$

$$\left[\frac{\partial(nU)}{\partial(nV)}\right]_{nS, n} = -P \tag{50}$$

For the more general case of a homogeneous fluid in which the n_i vary for whatever reason, it is assumed that nU is a function, F, of the n_i as well as of nS and nV:

$$nU = F(nS, nV, n_1, n_2, \dots, n_N) \tag{51}$$

Then

$$d(nU) = \left[\frac{\partial(nU)}{\partial(nS)}\right]_{nV, n} d(nS) + \left[\frac{\partial(nU)}{\partial(nV)}\right]_{nS, n} d(nV) + \sum_i \left[\frac{\partial(nU)}{\partial n_i}\right]_{nS, nV, n_j} dn_i$$

$$\tag{52}$$

where the summation is over all species i, and subscript n_j signifies that all mole numbers are held constant except the ith. The definition of the chemical potential,

$$\mu_i \equiv \left[\frac{\partial(nU)}{\partial n_i} \right]_{nS,nV,n_j} \tag{53}$$

along with equations 49 and 50 reduce equation 52 to equation 54:

$$d(nU) = Td(nS) - Pd(nV) + \sum_i \mu_i dn_i \tag{54}$$

This fundamental property relation is the basis for development of all other equations relating the properties of PVT systems.

Equation 54 implies that U is a function of S and V, a choice of variables that is not always convenient. Alternative fundamental property relations may be formulated in which other pairs of variables appear. They are found systematically through Legendre transformations (1,2), which lead to the following definitions for the enthalpy, H, Helmholtz energy, A, and Gibbs energy, G:

$$nH \equiv nU + P(nV) \tag{55}$$

$$nA \equiv nU - T(nS) \tag{56}$$

$$nG \equiv nH - T(nS) \tag{57}$$

The total differentials of these three equations in combination with equation 54 yield the following:

$$d(nH) = Td(nS) + nVdP + \sum_i \mu_i dn_i \tag{58}$$

$$d(nA) = -nSdT - Pd(nV) + \sum_i \mu_i dn_i \tag{59}$$

$$d(nG) = -nSdT + nVdP + \sum_i \mu_i dn_i \tag{60}$$

Equations 54 and 58 through 60 are equivalent forms of the fundamental property relation applicable to changes between equilibrium states in any homogeneous fluid system, either open or closed. Equation 58 shows that H is a function of S and P. Similarly, A is a function of T and V, and G is a function of T and P. The choice of which equation to use in a particular application is dictated by convenience. However, the Gibbs energy, G, is of particular importance because of its unique functional relation to T, P, and the n_i, the variables of primary interest in chemical technology. Thus, by equation 60,

$$\left[\frac{\partial(nG)}{\partial T} \right]_{P,n} = -nS \tag{61}$$

where n as a subscript indicates that all n_i are held constant. This equation may therefore be written

$$\left(\frac{\partial G}{\partial T}\right)_{P,x} = -S \tag{62}$$

where subscript x denotes constant composition. Similarly,

$$\left(\frac{\partial G}{\partial P}\right)_{T,x} = V \tag{63}$$

and

$$\left[\frac{\partial(nG)}{\partial n_i}\right]_{T,P,n_j} = \mu_i \tag{64}$$

When equations 54 and 58 through 60 are applied to the special case of one mole of a homogeneous fluid of constant composition, $n = 1$, $dn_i = 0$, and these equations reduce to the following:

$$dU = TdS - PdV \tag{65}$$

$$dH = TdS + VdP \tag{66}$$

$$dA = -SdT - PdV \tag{67}$$

$$dG = -SdT + VdP \tag{68}$$

Because these are exact differential expressions, Maxwell equations can be written by inspection. The two most useful ones are derived from equations 67 and 68:

$$\left(\frac{\partial S}{\partial V}\right)_T = \left(\frac{\partial P}{\partial T}\right)_V \tag{69}$$

$$\left(\frac{\partial S}{\partial P}\right)_T = -\left(\frac{\partial V}{\partial T}\right)_P \tag{70}$$

Enthalpy and Entropy as Functions of *T* and *P*

For a homogeneous fluid of constant composition,

$$H = H(T,P) \tag{71}$$

$$S = S(T,P) \tag{72}$$

Therefore,

$$dH = \left(\frac{\partial H}{\partial T}\right)_P dT + \left(\frac{\partial H}{\partial P}\right)_T dP \tag{73}$$

$$dS = \left(\frac{\partial S}{\partial T}\right)_P dT + \left(\frac{\partial S}{\partial P}\right)_T dP \tag{74}$$

By definition,

$$C_P \equiv \left(\frac{\partial H}{\partial T}\right)_P \tag{75}$$

is the heat capacity at constant pressure. If equation 66 is divided by dT and restricted to constant P,

$$\left(\frac{\partial H}{\partial T}\right)_P = T\left(\frac{\partial S}{\partial T}\right)_P \tag{76}$$

With equation 75 this becomes equation 77:

$$\left(\frac{\partial S}{\partial T}\right)_P = \frac{C_P}{T} \tag{77}$$

When equation 66 is divided by dP and restricted to constant T,

$$\left(\frac{\partial H}{\partial P}\right)_T = T\left(\frac{\partial S}{\partial P}\right)_T + V \tag{78}$$

In view of equation 70, this becomes equation 79:

$$\left(\frac{\partial H}{\partial P}\right)_T = V - T\left(\frac{\partial V}{\partial T}\right)_P \tag{79}$$

A combination of equations 73, 75, and 79 gives equation 80:

$$dH = C_P dT + \left[V - T\left(\frac{\partial V}{\partial T}\right)_P\right]dP \tag{80}$$

Similarly, combination of equations 74, 77, and 70 gives equation 81:

$$dS = \frac{C_P}{T}dT - \left(\frac{\partial V}{\partial T}\right)_P dP \tag{81}$$

For an ideal gas (*ig*),

$$PV^{ig} = RT \tag{82}$$

and

$$\left(\frac{\partial V^{ig}}{\partial T}\right)_P = \frac{R}{P} = \frac{V^{ig}}{T} \tag{83}$$

Thus, for an ideal gas, equations 80 and 81 reduce to the following:

$$dH^{ig} = C_P^{ig} dT \tag{84}$$

$$dS^{ig} = \frac{C_P^{ig}}{T} dT - \frac{R}{P} dP \tag{85}$$

Internal Energy and Entropy as Functions of *T* and *V*

Because $V = V(T, P)$ for a constant composition fluid, T and V can be considered independent variables, therefore

$$U = U(T, V) \tag{86}$$

$$S = S(T, V) \tag{87}$$

Then

$$dU = \left(\frac{\partial U}{\partial T}\right)_V dT + \left(\frac{\partial U}{\partial V}\right)_T dV \tag{88}$$

$$dS = \left(\frac{\partial S}{\partial T}\right)_V dT + \left(\frac{\partial S}{\partial V}\right)_T dV \tag{89}$$

By definition, equation 90 is the heat capacity at constant volume:

$$C_V \equiv \left(\frac{\partial U}{\partial T}\right)_V \tag{90}$$

Two relations follow from equation 65:

$$\left(\frac{\partial U}{\partial T}\right)_V = T\left(\frac{\partial S}{\partial T}\right)_V \tag{91}$$

$$\left(\frac{\partial U}{\partial V}\right)_T = T\left(\frac{\partial S}{\partial V}\right)_T - P \tag{92}$$

With equations 90 and 69, these become the following:

$$\left(\frac{\partial S}{\partial T}\right)_V = \frac{C_V}{T} \tag{93}$$

$$\left(\frac{\partial U}{\partial V}\right)_T = T\left(\frac{\partial P}{\partial T}\right)_V - P \tag{94}$$

Combining equation 88 with equations 90 and 92 yields equation 95:

$$dU = C_V dT + \left[T\left(\frac{\partial P}{\partial T}\right)_V - P\right]dV \tag{95}$$

Similarly, combination of equations 89, 93, and 69 gives equation 96:

$$dS = \frac{C_V}{T}dT + \left(\frac{\partial P}{\partial T}\right)_V dV \tag{96}$$

For an ideal gas,

$$\left(\frac{\partial P}{\partial T}\right)_V = \frac{R}{V^{ig}} = \frac{P}{T} \tag{97}$$

and equations 95 and 96 reduce to the following:

$$dU^{ig} = C_V^{ig}dT \tag{98}$$

$$dS^{ig} = \frac{C_V^{ig}}{T}dT + \frac{R}{V^{ig}}dV^{ig} \tag{99}$$

Equations 80, 81, 95, and 96 are basic to the calculation of numerical values for the thermodynamic properties U, H, and S from experimental heat capacity and PVT data.

Energy Equations for Steady-State, Steady-Flow Processes

Industrial production of chemicals is largely by continuous processes in which rates of inflow and outflow of mass are constant. Moreover, conditions at all points in the process are maintained constant with time. A simple steady-state, steady-flow process is represented in Figure 4. A process occurs within the fixed control volume between points 1 and 2 that changes the properties of a fluid element as it flows from point 1 to point 2. The system is the fluid in the control volume plus the fluid element of mass δm_1 (Fig. 4a) that enters the control volume at point 1 during time $\delta\tau$. After time interval $\delta\tau$, the system appears as in Figure 4b and consists of fluid in the control volume plus the fluid element of mass δm_2 that has left the control volume. Because the process is one of steady-state steady flow, $\delta m_2 = \delta m_1 = \delta m$, and the properties of the fluid in the control volume are unchanged. The fluid elements δm_1 and δm_2 have properties as measured at points 1 and 2, and these include a velocity, u, and an elevation above a datum level, z. Thus, these masses have kinetic and potential energy as well as internal energy, and the general energy balance for a closed (constant mass) system, as given by equation 2, must include terms to account for changes in these forms of energy:

$$\Delta U^t + \Delta E_K + \Delta E_P = Q + W \tag{100}$$

where Δ denotes the change from entrance to exit of the control volume. As applied to the change shown in Figure 4 during time interval $\delta\tau$, this equation becomes

$$\Delta(U\delta m) + \Delta E_K + \Delta E_P = \delta Q + \delta W \tag{101}$$

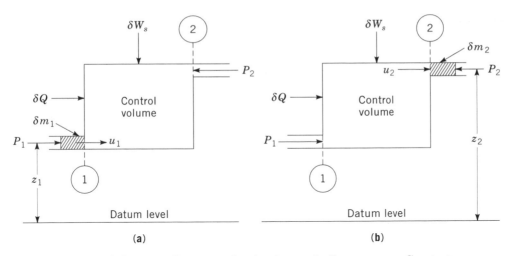

Fig. 4. Schematic diagram of a simple steady-flow process. See text.

where U is internal energy per unit mass, and

$$\Delta(U\delta m) \equiv (U_2 - U_1)\delta m \tag{102}$$

The kinetic and potential energies of the fluid at points 1 and 2 are given by equation 103:

$$E_K = \frac{1}{2}(\delta m)u^2 \tag{103}$$

and equation 104, where g is the local acceleration of gravity:

$$E_P = (\delta m)zg \tag{104}$$

The energy equation is therefore written as follows:

$$\Delta(U\delta m) + \Delta\left(\frac{u^2\delta m}{2}\right) + \Delta(zg\delta m) = \delta Q + \delta W \tag{105}$$

The quantity δW includes two kinds of work: shaft work, δW_s, shown in Figure 4, and work done by pressures P_1 and P_2 moving through volumes occupied by fluid masses δm_1 and δm_2. Thus,

$$\delta W = \delta W_s - P_2(V_2\delta m_2) + P_1(V_1\delta m_1) = \delta W_s - \Delta(PV\delta m) \tag{106}$$

and the energy equation becomes

$$\Delta(U\delta m) + \Delta\left(\frac{u^2\delta m}{2}\right) + \Delta(zg\delta m) + \Delta(PV\delta m) = \delta Q + \delta W_s \tag{107}$$

Factoring δm yields equation 108:

$$\Delta\left[\left(U + PV + \frac{u^2}{2} + zg\right)\delta m\right] = \delta Q + \delta W_s \qquad (108)$$

Because by definition $H = U + PV$,

$$\Delta\left[\left(H + \frac{u^2}{2} + zg\right)\delta m\right] = \delta Q + \delta W_s \qquad (109)$$

Division of equation 109 by the time interval, $\delta\tau$, transforms it into a rate expression:

$$\Delta\left[\left(H + \frac{u^2}{2} + zg\right)\dot m\right] = \dot Q + \dot W_s \qquad (110)$$

where $\dot m$ is mass flow rate, $\dot Q$ is heat-transfer rate, and $\dot W_s$ is power. Equations 109 and 110, developed for a process with one entering and one leaving stream, are applicable for any number of such streams. The Δ indicates the difference between the sum over all leaving streams and the sum over all entering streams. However, in the common case of a single entering and single leaving stream, $\dot m$ (eq. 110) must be the same for both, and this equation is then written as follows:

$$\Delta\left(H + \frac{u^2}{2} + zg\right)\dot m = \dot Q + \dot W_s \qquad (111)$$

Multiplication by the time required for a unit mass of fluid to enter or leave the control volume gives equation 112, where each term is for a unit mass of fluid.

$$\Delta H + \frac{\Delta u^2}{2} + g\Delta z = Q + W_s \qquad (112)$$

Partial Molar Properties

Because the macroscopic-intensive properties of homogeneous fluids in equilibrium states are functions of T, P, and composition, it follows that the total property of a phase nM can be expressed functionally as in equation 113:

$$nM = m(T, P, n_1, n_2, n_3, \ldots) \qquad (113)$$

The total differential of nM is therefore

$$d(nM) = \left[\frac{\partial(nM)}{\partial T}\right]_{P,n} dT + \left[\frac{\partial(nM)}{\partial P}\right]_{T,n} dP + \sum_i \left[\frac{\partial(nM)}{\partial n_i}\right]_{T,P,n_j} dn_i \qquad (114)$$

or

$$d(nM) = n\left(\frac{\partial M}{\partial T}\right)_{P,x} dT + n\left(\frac{\partial M}{\partial P}\right)_{T,x} dP + \sum_i \left[\frac{\partial(nM)}{\partial n_i}\right]_{T,P,n_j} dn_i \quad (115)$$

The derivatives in the summation are partial molar properties, denoted by $\overline{M}_i$; thus,

$$\overline{M}_i \equiv \left[\frac{\partial(nM)}{\partial n_i}\right]_{T,P,n_j} \quad (116)$$

This definition is the means by which partial properties are calculated from solution properties. Equation 115 can now be written as equation 117:

$$d(nM) = n\left(\frac{\partial M}{\partial T}\right)_{P,x} dT + n\left(\frac{\partial M}{\partial P}\right)_{T,x} dP + \sum_i \overline{M}_i dn_i \quad (117)$$

Important equations follow from this result through the following identities:

$$d(nM) \equiv ndM + Mdn \quad (118)$$

$$dn_i \equiv d(nx_i) = ndx_i + x_idn \quad (119)$$

Combining these expressions with equation 117 and collecting like terms give equation 120:

$$\left[dM - \left(\frac{\partial M}{\partial T}\right)_{P,x} dT - \left(\frac{\partial M}{\partial P}\right)_{T,x} dP - \sum_i \overline{M}_i dx_i\right]n + \left[M - \sum_i x_i\overline{M}_i\right]dn = 0 \quad (120)$$

Because n and dn are independent and arbitrary, the terms in brackets must each be zero, whence

$$dM = \left(\frac{\partial M}{\partial T}\right)_{P,x} dT + \left(\frac{\partial M}{\partial P}\right)_{T,x} dP + \sum_i \overline{M}_i dx_i \quad (121)$$

and

$$M = \sum_i x_i\overline{M}_i \quad (122)$$

This summability equation, the counterpart of equation 116, provides for the calculation of solution properties from partial properties.

Differentiation of equation 122 yields equation 123:

$$dM = \sum_i x_i d\overline{M}_i + \sum_i \overline{M}_i dx_i \tag{123}$$

Because equations 121 and 123 are both valid in general, their right-hand sides can be equated, yielding the following:

$$\left(\frac{\partial M}{\partial T}\right)_{P,x} dT + \left(\frac{\partial M}{\partial P}\right)_{T,x} dP - \sum_i x_i d\overline{M}_i = 0 \tag{124}$$

This result, known as the Gibbs-Duhem equation, imposes a constraint on how the partial molar properties of any phase may vary with temperature, pressure, and composition. In particular, at constant T and P it represents a simple relation among the $\overline{M}_i$ to which measured values of partial properties must conform.

Equations 116, 117, 121, 122, and 124 are the general property relations between partial molar properties and solution properties. The symbol M may represent the molar value of any extensive thermodynamic property, for example, V, U, H, S, A, or G. When $M = H$, the derivatives $(\partial H/\partial T)_{P,x}$ and $(\partial H/\partial P)_{T,x}$ are given by equations 75 and 79. Equations 121, 122, and 124 then become the following:

$$dH = C_P dT + \left[V - T\left(\frac{\partial V}{\partial T}\right)_{P,x}\right] dP + \sum_i \overline{H}_i dx_i \tag{125}$$

$$H = \sum_i x_i \overline{H}_i \tag{126}$$

and

$$C_P dT + \left[V - T\left(\frac{\partial V}{\partial T}\right)_{P,x}\right] dP - \sum_i x_i d\overline{H}_i = 0 \tag{127}$$

When $M = G$, the required derivatives are given by equations 62 and 63. Moreover, the derivative on the left side of equation 64 defines the partial molar Gibbs energy, $\overline{G}_i$. Therefore,

$$\mu_i = \overline{G}_i \tag{128}$$

and equation 117 reduces to equation 60. Equations 122 and 124 become the following:

$$G = \sum_i x_i \overline{G}_i = \sum_i x_i \mu_i \tag{129}$$

and

$$-S dT + V dP - \sum_i x_i d\overline{G}_i = 0 \tag{130}$$

If temperature and pressure are constant, equation 130 reduces to equation 131 (constant T, P), which is a common form of the Gibbs-Duhem equation.

$$\sum_i x_i d\overline{G}_i = \sum_i x_i d\mu_i = 0 \tag{131}$$

Residual Properties

A class of thermodynamic functions called residual properties is given generic definition by equation 132:

$$M^R \equiv M - M^{ig} \tag{132}$$

where M is the molar value of any extensive thermodynamic property of a fluid in its actual state and M^{ig} is the corresponding value for the fluid in its ideal gas state, ie, as an ideal gas at the same temperature, pressure, and composition. Residual property M^R reflects the contributions made to property M by intermolecular forces. For H^R and S^R,

$$\lim_{P \to 0} M^R = 0 \tag{133}$$

It follows from equations 132 and 133 (constant T, x) that

$$M^R = \int_0^P \left[\left(\frac{\partial M}{\partial P} \right)_T - \left(\frac{\partial M^{ig}}{\partial P} \right)_T \right] dP \tag{134}$$

Values and expressions for H^R and S^R are found from equation 134 through use of PVT data and equations of state.

Property changes are readily determined for fluids in the ideal gas state, and these in combination with residual properties are used to compute property changes of real fluids. The computational scheme is suggested in Figure 5, and is based on the following identity:

$$\Delta M \equiv -M_1^R + (M_{3'}^{ig} - M_{1'}^{ig}) + (M_{2'}^{ig} - M_{3'}^{ig}) + M_2^R \tag{135}$$

Thus, for $M = H$ (eq. 84),

$$\Delta H = -H_1^R + \int_{T_1}^{T_2} C_P^{ig} dT + H_2^R \tag{136}$$

and for $M = S$ (eq. 85),

$$\Delta S = -S_1^R + \int_{T_1}^{T_2} \frac{C_P^{ig}}{T} dT - R \ln \frac{P_2}{P_1} + S_2^R \tag{137}$$

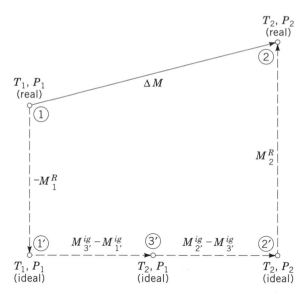

Fig. 5. Calculation path for determination of property change ΔM by residual properties (eq. 135).

Evaluation of the integrals requires an empirical expression for the temperature dependence of the ideal gas heat capacity, C_P^{ig} (8). The residual Gibbs energy is related to H^R and S^R by equation 138:

$$G^R = H^R - TS^R \tag{138}$$

Vapor–Liquid Phase Transition

The isothermal vaporization of pure liquid i represents its transition from saturated liquid to saturated vapor at temperature T and at saturation vapor pressure P_i^{sat}. The treatment of this transition is facilitated through use of property changes of vaporization ΔM_i^{lv}, defined by equation 139:

$$\Delta M_i^{lv} \equiv M_i^v - M_i^l \tag{139}$$

where M_i^l and M_i^v are molar properties of pure i for states of saturated liquid and vapor. The heat of vaporization, ΔH_i^{lv}, is directly related to the entropy change of vaporization:

$$\Delta H_i^{lv} = T\Delta S_i^{lv} \tag{140}$$

This equation follows from equation 66, because vaporization occurs at the constant pressure P_i^{sat}. Moreover, the heat of vaporization is related to the slope of the vapor–liquid saturation curve through the Clapeyron equation:

$$\frac{dP_i^{\text{sat}}}{dT} = \frac{\Delta H_i^{lv}}{T\Delta V_i^{lv}} \tag{141}$$

Ideal Gas Mixtures

An ideal gas is a model gas comprised of imaginary molecules of zero volume that do not interact. Each chemical species in an ideal gas mixture therefore has its own private properties, uninfluenced by the presence of other species. The partial pressure of species i $(i = 1, 2, \ldots, N)$ in an ideal gas mixture is defined by equation 142:

$$p_i \equiv x_i P \qquad (142)$$

where x_i is the mole fraction of species i. The sum of the partial pressures clearly equals the total pressure. Gibbs's theorem for a mixture of ideal gases may be stated as follows: a partial molar property (other than the volume) of a constituent species in an ideal gas mixture is equal to the corresponding molar property of the species as a pure ideal gas at the mixture temperature but at a pressure equal to its partial pressure in the mixture. This is expressed mathematically for generic partial property $\overline{M}_i^{ig}$ by equation 143:

$$\overline{M}_i^{ig}(T, P) = M_i^{ig}(T, p_i) \qquad (143)$$

where superscript ig denotes an ideal gas property, and $\overline{M}_i^{ig} \neq \overline{V}_i^{ig}$. For those properties of an ideal gas that are independent of P, eg, U, H, and C_P, this becomes simply

$$\overline{M}_i^{ig} = M_i^{ig} \qquad (144)$$

where M_i^{ig} is evaluated at the mixture T and P.

The entropy and Gibbs energy of an ideal gas do depend on pressure. By equation 85 (constant T),

$$dS_i^{ig} = -R d \ln P \qquad (145)$$

Integration from p_i to P gives equation 146:

$$S_i^{ig}(T, P) - S_i^{ig}(T, p_i) = -R \ln \frac{P}{p_i} = -R \ln \frac{P}{x_i P} = R \ln x_i \qquad (146)$$

Whence

$$S_i^{ig}(T, p_i) = S_i^{ig}(T, P) - R \ln x_i \qquad (147)$$

Substituting this result into equation 143, written for entropy, gives equation 148, where S_i^{ig} is evaluated at the mixture T and P:

$$\overline{S}_i^{ig} = S_i^{ig} - R \ln x_i \qquad (148)$$

For the Gibbs energy of an ideal gas mixture, $G^{ig} = H^{ig} - TS^{ig}$; the parallel relation for partial properties is equation 149:

$$\overline{G}_i^{ig} = \overline{H}_i^{ig} - T\overline{S}_i^{ig} \tag{149}$$

In combination with equation 144, written for the enthalpy, and equation 148, this becomes equation 150:

$$\overline{G}_i^{ig} = H_i^{ig} - TS_i^{ig} + RT \ln x_i \tag{150}$$

or equation 151:

$$\mu_i^{ig} \equiv \overline{G}_i^{ig} = G_i^{ig} + RT \ln x_i \tag{151}$$

Elimination of G_i^{ig} from this equation is accomplished by equation 68, written at constant T for pure species i:

$$dG_i^{ig} = V_i^{ig}dP = \frac{RT}{P} dP = RTd \ln P \tag{152}$$

Integration gives

$$G_i^{ig} = \Gamma_i(T) + RT \ln P \tag{153}$$

where $\Gamma_i(T)$, the integration constant for a given temperature, is a function of temperature only. Equation 151 becomes equation 154:

$$\mu_i^{ig} = \Gamma_i(T) + RT \ln x_i P \tag{154}$$

Fugacity and Fugacity Coefficient

The chemical potential, μ_i, plays a vital role in both phase and chemical reaction equilibria. However, the chemical potential exhibits certain unfortunate characteristics which discourage its use in the solution of practical problems. The Gibbs energy, and hence μ_i, is defined in relation to the internal energy and entropy, both primitive quantities for which absolute values are unknown. Moreover, μ_i approaches negative infinity when either P or x_i approaches zero. While these characteristics do not preclude the use of chemical potentials, the application of equilibrium criteria is facilitated by the introduction of a new quantity to take the place of μ_i but which does not exhibit its less desirable characteristics.

Equation 153 is valid only for pure species i in the ideal gas state. For a real fluid, an analogous equation is as follows:

$$G_i \equiv \Gamma_i(T) + RT \ln f_i \tag{155}$$

in which the new property f_i replaces the pressure, P. This equation serves as a partial definition of f_i, which is called the fugacity.

Subtraction of equation 153 from equation 155, both written for the same temperature, pressure, and composition, gives equation 156:

$$G_i - G_i^{ig} = RT \ln \frac{f_i}{P} \tag{156}$$

According to equation 132, $G_i - G_i^{ig}$ is the residual Gibbs energy, G_i^R. The dimensionless ratio f_i/P is a new property called the fugacity coefficient, ϕ_i. Thus,

$$G_i^R = RT \ln \phi_i \tag{157}$$

where

$$\phi_i \equiv \frac{f_i}{P} \tag{158}$$

The definition of fugacity is completed by setting the ideal gas state fugacity of pure species i equal to its pressure:

$$f_i^{ig} = P \tag{159}$$

Thus for the special case of an ideal gas, $G_i^R = 0$, $\phi_i^{ig} = 1$, and equation 153 is recovered from equation 155.

The definition of fugacity of a species in solution is parallel to the definition of pure species fugacity. Equation 154 is analogous to the ideal gas expression:

$$\mu_i \equiv \Gamma_i(T) + RT \ln \hat{f}_i \tag{160}$$

where $\hat{f}_i$ is the fugacity of species i in solution. Because it is not a partial property, it is identified by a circumflex rather than an overbar.

Analogous to the defining equation for the residual Gibbs energy is the definition of a partial molar residual Gibbs energy (eq. 161):

$$\overline{G}_i^R \equiv \overline{G}_i - \overline{G}_i^{ig} \tag{161}$$

Subtracting equation 154 from equation 160, both written for the same temperature and pressure, yields equation 162:

$$\mu_i - \mu_i^{ig} = RT \ln \frac{\hat{f}_i}{x_i P} \tag{162}$$

This result combined with equation 161 and the identity $\mu_i \equiv \overline{G}_i$ gives equation 163:

$$\overline{G}_i^R = RT \ln \hat{\phi}_i \tag{163}$$

where by definition

$$\hat{\phi}_i \equiv \frac{\hat{f}_i}{x_i P} \tag{164}$$

The dimensionless ratio $\hat{\phi}_i$ is called the fugacity coefficient of species i in solution.

Equation 163 is the analogue of equation 157, which relates ϕ_i to G_i^R. For an ideal gas, $\overline{G}_i^R$ is necessarily zero; therefore $\hat{\phi}_i^{ig} = 1$, and

$$\hat{f}_i^{ig} = x_i P \tag{165}$$

Thus the fugacity of species i in an ideal gas mixture is equal to its partial pressure.

Evaluation of Residual Properties

An alternative form of the fundamental property relation given by equation 60 is provided by the mathematical identity of equation 166:

$$d\left(\frac{nG}{RT}\right) \equiv \frac{1}{RT} d(nG) - \frac{nG}{RT^2} dT \tag{166}$$

Substitution for $d(nG)$ by equation 60 and for G by its definition, $G \equiv H - TS$, gives, after algebraic reduction, equation 167:

$$d\left(\frac{nG}{RT}\right) = \frac{nV}{RT} dP - \frac{nH}{RT^2} dT + \sum_i \frac{\overline{G}_i}{RT} dn_i \tag{167}$$

All terms in this equation have the units of moles; moreover, in contrast to equation 60, the enthalpy rather than the entropy appears on the right-hand side. Equation 167 is a general relation expressing G/RT as a function of all of its coordinates, T, P, and mole numbers. Because of its generality, equation 167 may be written for the special case of an ideal gas:

$$d\left(\frac{nG^{ig}}{RT}\right) = \frac{nV^{ig}}{RT} dP - \frac{nH^{ig}}{RT^2} dT + \sum_i \frac{\overline{G}_i^{ig}}{RT} dn_i \tag{168}$$

In view of equation 132, this equation may be subtracted from equation 167 to yield equation 169:

$$d\left(\frac{nG^R}{RT}\right) = \frac{nV^R}{RT} dP - \frac{nH^R}{RT^2} dT + \sum_i \frac{\overline{G}_i^R}{RT} dn_i \tag{169}$$

Equation 169, the fundamental residual property relation, is so general as to be useful for practical application only in restricted forms. Division by dP and restriction to constant T and composition leads to equation 170:

$$\frac{V^R}{RT} = \left[\frac{\partial (G^R/RT)}{\partial P} \right]_{T,x} \tag{170}$$

Similarly, division by dT and restriction to constant P and composition gives equation 171:

$$\frac{H^R}{RT} = -T \left[\frac{\partial (G^R/RT)}{\partial T} \right]_{P,x} \tag{171}$$

Equation 170 may be written as follows (constant T, x):

$$d\frac{G^R}{RT} = \frac{V^R}{RT} dP \tag{172}$$

Integration from zero pressure to arbitrary pressure P gives equation 173 (constant T, x):

$$\frac{G^R}{RT} = \int_0^P \frac{V^R}{RT} dP \tag{173}$$

Because $V = ZRT/P$, where Z is the compressibility factor, and $V^{ig} = RT/P$,

$$V^R \equiv V - V^{ig} = \frac{RT}{P} (Z - 1) \tag{174}$$

Equation 173 can now be written as follows (constant T, x):

$$\frac{G^R}{RT} = \int_0^P (Z - 1) \frac{dP}{P} \tag{175}$$

Differentiating this equation with respect to T in accord with equation 171 gives equation 176 (constant T, x):

$$\frac{H^R}{RT} = -T \int_0^P \left(\frac{\partial Z}{\partial T} \right)_{P,x} \frac{dP}{P} \tag{176}$$

Equations 175 and 176 combine with equation 138 to yield equation 177 (constant T, x):

$$\frac{S^R}{R} = -T \int_0^P \left(\frac{\partial Z}{\partial T} \right)_{P,x} \frac{dP}{P} - \int_0^P (Z - 1) \frac{dP}{P} \tag{177}$$

Equation 175 may be written for pure species i and combined with equation 157 (constant T):

$$\ln \phi_i = \int_0^P (Z_i - 1) \frac{dP}{P} \tag{178}$$

Equation 163, written as $\ln \hat{\phi}_i = \overline{G}_i^R/RT$, clearly shows that $\ln \hat{\phi}_i$ is a partial molar property with respect to G^R/RT. Multiplication of equation 175 by n and differentiation with respect to n_i at constant T, P, and n_j in accord with equation 116 yields, after reduction, equation 179 (constant T,x), where $\overline{Z}_i$ is the partial molar compressibility factor. This equation is the partial-property analogue of equation 178.

$$\ln \hat{\phi}_i = \int_0^P (\overline{Z}_i - 1) \frac{dP}{P} \tag{179}$$

Equations 175 through 179 allow calculation of thermodynamic properties from volume-explicit equations of state, ie, equations explicitly solvable for volume. If an equation of state is solvable explicitly for pressure but not for volume, then alternative formulas must be used, where ρ is molar density and subscript $\rho/n = 1/V^t$ indicates constancy of total volume. For equations 180, 181, and 183, T and x are constant; for equation 182, T is constant.

$$\frac{H^R}{RT} = -T \int_0^\rho \left(\frac{\partial Z}{\partial T}\right)_\rho \frac{d\rho}{\rho} + (Z - 1) \tag{180}$$

$$\frac{S^R}{R} = -\int_0^\rho \left[T\left(\frac{\partial Z}{\partial T}\right)_\rho + Z - 1 \right] \frac{d\rho}{\rho} + \ln Z \tag{181}$$

$$\ln \phi_i = \int_0^{\rho_i} (Z_i - 1) \frac{d\rho_i}{\rho_i} + Z_i - 1 - \ln Z_i \tag{182}$$

$$\ln \hat{\phi}_i = \int_0^\rho \left\{ \left[\frac{\partial(nZ)}{\partial n_i} \right]_{T,\rho/n,n_j} - 1 \right\} \frac{d\rho}{\rho} - \ln Z \tag{183}$$

The volumetric properties of fluids are represented not only by equations of state but also by generalized correlations. The most popular generalized correlations are based on a three-parameter theorem of corresponding states which asserts that the compressibility factor is a universal function of reduced temperature, reduced pressure, and a parameter ω, called the acentric factor:

$$Z = Z(T_r, P_r, \omega) \tag{184}$$

According to equation 184, all fluids having the same value of ω have identical values of Z when compared at the same T_r and P_r. This principle of corresponding states is presumed valid for all T_r and P_r, and therefore provides generalized correlations for properties derived from Z, ie, for residual properties and fugacity coefficients, which depend on T and P through Z and its derivatives.

The acentric factor is by definition (3):

$$\omega \equiv -\log_{10}(P_r^{\mathrm{sat}})_{T_r=0.7} - 1 \tag{185}$$

where P_r^{sat} is the reduced vapor pressure. Correlations for Z are written as follows:

$$Z = Z^0(T_r, P_r) + \omega Z^1(T_r, P_r) \tag{186}$$

which is a special form of equation 184. The linearity of this equation in ω implies similar linear correlations for the residual properties and fugacity coefficients. A comprehensive set of such correlations is presented by the Lee-Kesler tables (6,14). Values from these correlations are subject to errors of several percent for nonpolar fluids, but larger errors can be expected for polar fluids and fluids that associate.

Although generalized correlations are based on data for pure fluids, they are frequently applied to mixtures. The mole fraction is introduced as a variable through empirical recipes for the composition dependence of parameters upon which the correlation is based. The simplest such recipes provide pseudoparameters that are linear in mole fraction:

$$T_c = \sum_i x_i T_{c_i} \tag{187}$$

$$P_c = \sum_i x_i P_{c_i} \tag{188}$$

$$\omega = \sum_i x_i \omega_i \tag{189}$$

However, more complex expressions have been proposed (9).

The Ideal Solution

The ideal solution is a model fluid which serves as a standard to which real solution behavior can be compared. Equation 151, which characterizes the behavior of a constituent species in an ideal gas mixture, takes on a new dimension if G_i^{ig}, the Gibbs energy of pure species i in the ideal gas state, is replaced by G_i, the Gibbs energy of pure species i as it actually exists at the mixture T and P and in the same physical state (real gas, liquid, or solid) as the mixture. It can then be applied to species in real solutions, indeed to liquids and solids as well as gases. An ideal solution is therefore defined as one for which the following is true, where superscript id denotes an ideal solution property.

$$\overline{G}_i^{id} \equiv G_i + RT \ln x_i \tag{190}$$

All other thermodynamic properties for an ideal solution follow from this equation. In particular, differentiation with respect to temperature and pressure, followed by application of equations for partial properties analogous to equations 62 and 63, leads to equations 191 and 192:

$$\overline{S}_i^{id} = S_i - R \ln x_i \tag{191}$$

$$\overline{V}_i^{id} = V_i \tag{192}$$

Because $\overline{H}_i^{id} = \overline{G}_i^{id} + T\overline{S}_i^{id}$, substitutions by equation 190 and 191 yield equation 193:

$$\overline{H}_i^{id} = H_i \tag{193}$$

The summability relation (eq. 122) applied to the special case of an ideal solution is written as equation 194:

$$M^{id} = \sum_i x_i \overline{M}_i^{id} \tag{194}$$

Equations for the mixture properties of an ideal solution follow immediately.

A simple equation for the fugacity of a species in an ideal solution follows from equation 190. Written for the special case of species i in an ideal solution, equation 160 becomes equation 195:

$$\mu_i^{id} = \overline{G}_i^{id} = \Gamma_i(T) + RT \ln \hat{f}_i^{id} \tag{195}$$

When equations 195 and 155 are combined with equation 190, $\Gamma_i(T)$ is eliminated, and the resulting expression reduces to equation 196:

$$\hat{f}_i^{id} = x_i f_i \tag{196}$$

This equation, known as the Lewis-Randall rule, applies to each species in an ideal solution at all conditions of temperature, pressure, and composition. It shows that the fugacity of each species in an ideal solution is proportional to its mole fraction; the proportionality constant is the fugacity of pure species i in the same physical state as the solution and at the same T and P. Ideal solution behavior is often approximated by solutions comprised of molecules similar in size and of the same chemical nature.

Excess Properties

Liquid solutions are often most easily dealt with through properties that measure their deviations, not from ideal gas behavior, but from ideal solution behavior. Thus the mathematical formalism of excess properties is analogous to that of the residual properties.

If M represents the molar value of any extensive thermodynamic property, an excess property M^E is defined as the difference between the actual property value of a solution and the value it would have as an ideal solution at the same temperature, pressure, and composition. Thus,

$$M^E \equiv M - M^{id} \tag{197}$$

Excess properties have no meaning for pure species, but for species in solution,

$$\overline{M}_i^E = \overline{M}_i - \overline{M}_i^{id} \tag{198}$$

where $\overline{M}_i^E$ is a partial molar excess property.

Equation 167, written for the special case of an ideal solution, may be subtracted from equation 167 itself, yielding equation 199:

$$d\left(\frac{nG^E}{RT}\right) = \frac{nV^E}{RT}\,dP - \frac{nH^E}{RT^2}\,dT + \sum_i \frac{\overline{G}_i^E}{RT}\,dn_i \tag{199}$$

This is the fundamental excess property relation, analogous to the fundamental residual property relation (eq. 169).

The excess Gibbs energy is of particular interest. Equation 160 may be written for the special case of species i in an ideal solution, with $\hat{f}_i^{id}$ replaced by $x_i f_i$ in accord with the Lewis-Randall rule:

$$\overline{G}_i^{id} = \Gamma_i(T) + RT \ln x_i f_i \tag{200}$$

Subtraction from equation 160 yields equation 201:

$$\overline{G}_i - \overline{G}_i^{id} = RT \ln \frac{\hat{f}_i}{x_i f_i} \tag{201}$$

The difference on the left is the partial excess Gibbs energy $\overline{G}_i^E$; the dimensionless ratio $\hat{f}_i/x_i f_i$ on the right is called the activity coefficient of species i in solution, γ_i. Thus, by definition,

$$\gamma_i \equiv \frac{\hat{f}_i}{x_i f_i} \tag{202}$$

and

$$\overline{G}_i^E = RT \ln \gamma_i \tag{203}$$

For an ideal solution, $\overline{G}_i^E = 0$, and therefore $\gamma_i = 1$. Comparison shows that equation 203 relates γ_i to $\overline{G}_i^E$ exactly as equation 163 relates $\hat{\phi}_i$ to $\overline{G}_i^R$. Moreover, just as $\ln \hat{\phi}_i$ is a partial property with respect to G^R/RT, so $\ln \gamma_i$ is a partial property with respect to G^E/RT. Equation 116, the defining equation for a partial molar property, in this case becomes equation 204:

$$\ln \gamma_i = \left[\frac{\partial(nG^E/RT)}{\partial n_i} \right]_{T,P,n_j} \tag{204}$$

Moreover, the summability relation (eq. 122) is written as follows:

$$\frac{G^E}{RT} = \sum_i x_i \ln \gamma_i \tag{205}$$

and the Gibbs-Duhem equation (eq. 131) implies the following, where T and P are constant:

$$\sum_i x_i d \ln \gamma_i = 0 \tag{206}$$

Again, the generality of the fundamental excess property relation (eq. 199) precludes its direct practical application. However, its restricted forms find use:

$$\frac{V^E}{RT} = \left[\frac{\partial(G^E/RT)}{\partial P} \right]_{T,x} \tag{207}$$

$$\frac{H^E}{RT} = -T \left[\frac{\partial(G^E/RT)}{\partial T} \right]_{P,x} \tag{208}$$

The partial property analogues of equations 207 and 208 are as follows:

$$\left(\frac{\partial \ln \gamma_i}{\partial P} \right)_{T,x} = \frac{\overline{V}_i^E}{RT} \tag{209}$$

$$\left(\frac{\partial \ln \gamma_i}{\partial T} \right)_{P,x} = -\frac{\overline{H}_i^E}{RT^2} \tag{210}$$

Whereas the fundamental residual property relation derives its usefulness from its direct relation to experimental PVT data and equations of state, the excess property formulation is useful because V^E, H^E, and γ_i are all experimentally accessible. Activity coefficients are found from vapor–liquid equilibrium data, and V^E and H^E values come from mixing experiments. Property changes of mixing are defined by the equation,

$$\Delta M \equiv M - \sum_i x_i M_i \tag{211}$$

For the volume change of mixing and the enthalpy change (heat) of mixing, which are directly measurable, $\Delta V = V^E$ and $\Delta H = H^E$.

Criteria for Phase Equilibria

The following criterion of phase equilibrium can be developed from the first and second laws of thermodynamics: the equilibrium state for a closed multiphase system of constant, uniform temperature and pressure is the state for which the total Gibbs energy is a minimum, whence

$$G^t \text{ (constant } T, P) = \text{minimum} \tag{212}$$

It is presumed in this statement that equilibrium in a multiphase system implies uniformity of T and P throughout all phases. In certain situations, eg, osmotic equilibrium, pressure uniformity is not required, and equation 212 is then not a useful criterion. Here, however, it suffices.

If the T and P of a multiphase system are constant, then the quantities capable of change are the individual mole numbers n_i^p of the various chemical species i in the various phases p. In the absence of chemical reactions, which is assumed here, the n_i^p may change only by interphase mass transfer, and not (because the system is closed) by the transfer of matter across the boundaries of the system. Hence, for phase equilibrium in a π-phase system, equation 212 is subject to a set of material balance constraints:

$$\sum_p n_i^p = n_i^t \tag{213}$$

where n_i^t is the constant, total number of moles of species i in the system.

For the application of equation 212 to the case of multicomponent, two-phase equilibrium, the phases are denoted by α and β, and the following general expression is written for G^t:

$$G^t = n^t G = \sum_i n_i^\alpha \mu_i^\alpha + \sum_i n_i^\beta \mu_i^\beta \tag{214}$$

This expression is minimized subject to the material balance constraints of equation 213, which may be written as follows, where $i = 1, 2, \ldots, N$:

$$n_i^\alpha + n_i^\beta = n_i^t \tag{215}$$

Equation 215 asserts that only N of the $2N$ mole numbers are independently variable. If the independent mole numbers are chosen as n_i^α, it follows from equation 212 that a set of N conditions on G^t at the equilibrium state can be written as follows, where $i = 1, 2, \ldots, N$:

$$\left(\frac{\partial G^t}{\partial n_i^\alpha} \right)_{T, P, n_j^\alpha} = 0 \tag{216}$$

If equations 216 and 131 are applied to equation 214, with $\partial n_i^\beta / \partial n_i^\alpha = -1$, then,

$$\mu_i^\alpha = \mu_i^\beta \tag{217}$$

which is the familiar algebraic criterion for two-phase equilibrium.

This approach is easily extended to an arbitrary number of phases, π, and produces the following generalization of equation 217:

$$\mu_i^\alpha = \mu_i^\beta = \cdots = \mu_i^\pi \tag{218}$$

Both convention and convenience suggest use of the fugacity $\hat{f}_i$ in practical calculations in place of the chemical potential μ_i. Equation 218 is then replaced by the equal fugacity criterion which follows directly from equation 160:

$$\hat{f}_i^\alpha = \hat{f}_i^\beta = \cdots = \hat{f}_i^\pi \tag{219}$$

Vapor–Liquid Equilibria. By equation 219, a criterion for vapor–liquid equilibrium (VLE) is as follows:

$$\hat{f}_i^l = \hat{f}_i^v \tag{220}$$

Effective use of this general equation requires explicit introduction of the compositions of the phases. This is done either through the activity coefficient, γ_i, or the fugacity coefficient, $\hat{\phi}_i$. Two procedures are in common use. By the gamma–phi approach, activity coefficients for the liquid phase enter by equation 202 and fugacity coefficients for the vapor phase by equation 164; equation 220 then becomes equation 221:

$$x_i \gamma_i f_i = y_i \hat{\phi}_i P \tag{221}$$

Equation 221 is commonly used for low pressure VLE, and is rewritten as follows:

$$x_i \gamma_i^+ P_i^{\text{sat}} = y_i \Phi_i P \tag{222}$$

where P_i^{sat} is the vapor pressure of pure i at the equilibrium temperature. Quantity Φ_i is given rigorously by the following formula (7):

$$\Phi_i = \frac{\hat{\phi}_i}{\phi_i^{\text{sat}}} \exp\left[-\int_T^{T^+} \frac{\overline{H}_i^E}{RT^2} \, dT + \int_P^{P^+} \frac{\overline{V}_i^E}{RT} \, dP + \int_P^{P_i^{\text{sat}}} \frac{V_i}{RT} \, dP \right] \tag{223}$$

where all quantities within brackets refer to the liquid phase and T^+ and P^+ denote reference conditions at which the γ_i^+ are evaluated, presumably from an expression for G^E. For isothermal VLE, $T^+ = T$, and the first integral within brackets is zero. Moreover, the effect of pressure on G^E and activity coefficients is usually ignored; in this event $\gamma_i^+ = \gamma_i$ and the second integral within brackets is zero.

If the liquid phase is an ideal solution, the vapor phase an ideal gas mixture, and the liquid-phase properties independent of pressure, then $\gamma_i^+ = 1$, $\Phi_i = 1$, and equation 222 reduces to Raoult's law:

$$x_i P_i^{\text{sat}} = y_i P \tag{224}$$

Of the assumptions inherent in this approximate relation, that of liquid-phase ideality is usually least realistic. Therefore, the simplest generally useful special case of equation 222 is the modified Raoult's law expression:

$$x_i \gamma_i P_i^{\text{sat}} = y_i P \tag{225}$$

where the γ_i are functions of temperature and composition, but are assumed independent of pressure. Equation 225, unlike equation 224, can represent vapor–liquid azeotropy.

The second common procedure for VLE calculations is the equation-of-state approach. Here, fugacity coefficients replace the fugacities for both liquid and vapor phases, and equation 220 becomes equation 226:

$$x_i \hat{\phi}_i^l = y_i \hat{\phi}_i^v \tag{226}$$

Application of equation 226 requires the availability of a single equation of state suitable for both vapor and liquid mixtures. Cubic equations of state are widely used for VLE calculations.

The proper design of distillation and absorption columns depends on knowledge of vapor–liquid equilibrium, as do flash calculations used to determine the physical state of streams at given conditions of temperature, pressure, and composition. Detailed treatments of vapor–liquid equilibria are available (6,7).

Other Kinds of Phase Equilibria. Equation 219 is the basis for other kinds of phase-equilibrium formulations. Thus, for liquid–liquid equilibria (LLE):

$$\hat{f}_i^\alpha = \hat{f}_i^\beta \tag{227}$$

where α and β designate the two liquid phases. Eliminating fugacities in favor of activity coefficients gives the gamma–gamma formulation for LLE, where $i = 1, 2, \ldots, N$:

$$x_i^\alpha \gamma_i^\alpha = x_i^\beta \gamma_i^\beta \tag{228}$$

For most LLE applications, the effect of pressure on γ_i can be ignored, and thus equation 228 constitutes a set of N equations relating equilibrium compositions to each other and to temperature. Solution of these equations for a particular temperature requires a single expression for the composition dependence of G^E

suitable for both liquid phases. Not all expressions for G^E suffice, even in principle, because some cannot represent liquid–liquid phase splitting.

For vapor–liquid–liquid equilibria (VLLE), equation 219 gives the following:

$$\hat{f}_i^\alpha = \hat{f}_i^\beta = \hat{f}_i^v \tag{229}$$

where α and β designate the two liquid phases. With activity coefficients applied for the liquid phases and fugacity coefficients for the vapor phase, the $2N$ equilibrium equations for subcritical VLLE are as follows:

$$\left.\begin{aligned} x_i^\alpha \gamma_i^\alpha f_i^\alpha = y_i \hat{\phi}_i P \\ x_i^\beta \gamma_i^\beta f_i^\beta = y_i \hat{\phi}_i P \end{aligned}\right\} \tag{230}$$

As for LLE, an expression for G^E capable of representing liquid–liquid-phase splitting is required; as for VLE, a vapor-phase equation of state for computing $\hat{\phi}_i$ is also needed. Moreover, VLLE calculations can in principle and sometimes in practice be carried out with an equation of state valid for all coexisting phases.

Chemical Reaction Stoichiometry

In a system containing N chemical species, any or all of which can participate in r chemical reactions, the r reactions can be represented schematically by the following algebraic equation, where $j = \mathrm{I}, \mathrm{II}, \ldots, r$:

$$0 = \sum_i \nu_{i,j} A_i \tag{231}$$

The A_i represent formulas for the chemical species and $\nu_{i,j}$ is the stoichiometric number for species i in reaction j. Each $\nu_{i,j}$ has a magnitude and a sign:

$$\mathrm{sign}(\nu_{i,j}) = \begin{cases} - \text{ for a reactant species} \\ + \text{ for a product species} \end{cases}$$

If species i does not participate in reaction j, then $\nu_{i,j} = 0$.

The stoichiometric numbers provide relations among the changes in mole numbers of chemical species which occur as the result of chemical reaction. Thus, for reaction j:

$$\frac{\Delta n_{1,j}}{\nu_{1,j}} = \frac{\Delta n_{2,j}}{\nu_{2,j}} = \cdots = \frac{\Delta n_{N,j}}{\nu_{N,j}} \tag{232}$$

Because all of these terms are equal, they can be equated to the change in a single quantity ϵ_j, called the reaction coordinate for reaction j, thereby giving the following:

$$\Delta n_{i,j} = \nu_{i,j} \Delta \epsilon_j \quad \begin{cases} i = 1, 2, \ldots, N \\ j = \mathrm{I}, \mathrm{II}, \ldots, r \end{cases} \tag{233}$$

Now the total change in mole number n_i is just the sum of the changes $\Delta n_{i,j}$ resulting from the various reactions. Thus, by equation 233,

$$\Delta n_i = \sum_j \Delta n_{i,j} = \sum_j \nu_{i,j} \Delta \epsilon_j \qquad (234)$$

where $i = 1, 2, \ldots, N$. If the initial number of moles of species i is n_{i_0} and if the convention is adopted that $\epsilon_j = 0$ for each reaction in this initial state, then, for $i = 1, 2, \ldots, N$,

$$n_i = n_{i_0} + \sum_j \nu_{i,j} \epsilon_j \qquad (235)$$

Equation 235 is the basic expression of material balance for a closed system in which r chemical reactions occur. It asserts that in such a system there are at most r mole-number-related quantities, ϵ_j, capable of independent variation. Note the absence of implied restrictions with respect to chemical reaction equilibria; the reaction coordinate formalism is merely an accounting scheme, valid for tracking the progress of each reaction to any arbitrary level of conversion.

Standard Property Changes of Reaction

For a specific single reaction, equation 231 may be written as follows:

$$a\,A + b\,B \longrightarrow l\,L + m\,M$$

where lower case letters are stoichiometric coefficients and upper case letters stand for chemical formulas. A standard property change for this reaction is by definition the property change occurring when a moles of A and b moles of B in their standard states at temperature T react to form l moles of L and m moles of M in their standard states also at temperature T. A standard state of species i is its real or hypothetical state as a pure species at temperature T and at a standard-state pressure P°. The standard property change of reaction j, ΔM_j°, is given its general mathematical definition by equation 236:

$$\Delta M_j^\circ \equiv \sum_i \nu_{i,j} M_i^\circ \qquad (236)$$

For species present as gases in the actual reactive system, the standard state is the pure ideal gas at pressure P°. For liquids and solids, it is usually the state of pure real liquid or solid at P°. The standard-state pressure P° is fixed at 100 kPa. Note that the standard states may represent different physical states for different species; any or all of the species may be gases, liquids, or solids.

The most commonly used standard property changes of reaction are as follows:

$$\Delta G_j^{\circ} \equiv \sum_i \nu_{i,j} G_i^{\circ} = \sum_i \nu_{i,j} \mu_i^{\circ} \tag{237}$$

$$\Delta H_j^{\circ} \equiv \sum_i \nu_{i,j} H_i^{\circ} \tag{238}$$

$$\Delta C_{P_j}^{\circ} \equiv \sum_i \nu_{i,j} C_{P_i}^{\circ} \tag{239}$$

The standard Gibbs energy change of reaction, ΔG_j°, is used in the calculation of equilibrium compositions; the standard heat of reaction, ΔH_j°, is used in the calculation of heat effects of chemical reaction; and the standard heat capacity change of reaction is used for extrapolating ΔH_j° and ΔG_j° with T. Numerical values for ΔH_j° and ΔG_j° are computed from tabulated formation data (15,16), and $\Delta C_{P_j}^{\circ}$ is determined from empirical expressions for the T dependence of the $C_{P_i}^{\circ}$ (8).

Criteria for Chemical Reaction Equilibria

The general criterion of chemical reaction equilibria is the same as that for phase equilibria, namely that the total Gibbs energy of a closed system be a minimum at constant, uniform T and P (eq. 212). If the T and P of a single-phase, chemically reactive system are constant, then the quantities capable of change are the mole numbers, n_i. The independently variable quantities are just the r reaction coordinates, and thus the equilibrium state is characterized by the r necessary derivative conditions (and subject to the material balance constraints of equation 235) where $j = \mathrm{I, II}, \ldots, r$:

$$\left(\frac{\partial G^t}{\partial \epsilon_j} \right)_{T, P, \epsilon_k} = 0 \tag{240}$$

In the case of a single-phase, multicomponent system undergoing just a single reaction, the total Gibbs energy is as follows:

$$G^t = n^t G = \sum_i n_i \mu_i \tag{241}$$

and is minimized subject to the following constraints:

$$n_i = n_{i_0} + \nu_i \epsilon \tag{242}$$

Equation 240 requires that

$$\left(\frac{\partial G^t}{\partial \epsilon} \right)_{T, P} = 0 \tag{243}$$

If equations 243 and 131 are applied to equation 241, with $\partial n_i/\partial \epsilon = \nu_i$, then

$$\sum_i \nu_i \mu_i = 0 \tag{244}$$

which is the familiar algebraic criterion for single-reaction equilibria.

    This is easily extended to an arbitrary number of independent reactions r and produces the expected generalization of equation 244 where $j = \text{I}, \text{II}, \ldots, r$:

$$\sum_i \nu_{i,j} \mu_i = 0 \tag{245}$$

Equilibrium Constants

Convenience suggests elimination of the μ_i in equation 245 in favor of fugacities. Equation 155 for species i in its standard state is subtracted from equation 160 for species i in the equilibrium mixture, giving equation 246:

$$\mu_i = G_i^\circ + RT \, \ln \, (\hat{f}_i/f_i^\circ) \tag{246}$$

Substitution of equation 246 into equation 245 yields, upon rearrangement (all j):

$$\prod_i (\hat{f}_i/f_i^\circ)^{\nu_{i,j}} = K_j \tag{247}$$

where

$$K_j \equiv \exp(-\Delta G_j^\circ/RT) \tag{248}$$

Quantity K_j is the chemical reaction equilibrium constant for reaction j, and ΔG_j° is the corresponding standard Gibbs energy change of reaction (eq. 237). Although called a constant, K_j is a function of T, but only of T.

    Use of equation 247 for actual calculations requires explicit introduction of composition variables. As in phase-equilibrium calculations, this is normally done for gas phases through the fugacity coefficient and for liquid phases through the activity coefficient. Thus, either

$$\hat{f}_i/f_i^\circ = y_i \hat{\phi}_i P/f_i^\circ(P^\circ) \tag{249}$$

or

$$\hat{f}_i/f_i^\circ = x_i \gamma_i f_i(P)/f_i^\circ(P^\circ) \tag{250}$$

where the notations make explicit the pressures at which the fugacities are evaluated. For gases the standard state is the ideal gas state at P°, and in this case $f_i^\circ = P^\circ$. Equations 249 and 250, or variations on them, are the bases for the various forms into which equation 247 is usually cast. An important special case of this equation is obtained for gas-phase reactions when the phase can be assumed ideal. In this event $\hat{\phi}_i = 1$, and equations 247 and 249 combine to give equation 251 (all j):

$$\prod_i (y_i)^{\nu_{i,j}} (P/P^\circ)^{\nu_j} = K_j \tag{251}$$

where $\nu_j \equiv \sum_i \nu_{i,j}$. For liquid-phase reactions, the assumption of ideal solution behavior makes $\gamma_i = 1$; moreover, except for high pressures, $f_i(P) = f_i^\circ(P^\circ)$. For this case equations 247 and 250 combine to give equation 252 (all j):

$$\prod_i (x_i)^{\nu_{i,j}} = K_j \tag{252}$$

the so-called law of mass action. The significant feature of the equations for these two special cases is that the temperature-, pressure-, and composition-dependent terms are distinct and separate.

Gibbs's Phase Rule

For a PVT system of uniform T and P containing N species and π phases at thermodynamic equilibrium, the intensive state of the system is fully determined by the values of T, P, and the $(N - 1)$ independent mole fractions for each of the equilibrium phases. The total number of these variables is then $2 + \pi(N - 1)$. The independent equations defining or constraining the equilibrium state are of three types: equations 218 or 219 of phase-equilibrium, $N(\pi - 1)$ in number; equation 245 of chemical reaction equilibrium, r in number; and equations of special constraint, s in number. The total number of these equations is $N(\pi - 1) + r + s$. The number of equations of reaction equilibrium r is the number of independent chemical reactions, and may be determined by a systematic procedure (6). Special constraints arise when conditions are imposed, such as forming the system from particular species, which allow one or more additional equations to be written connecting the phase-rule variables (6).

The degrees of freedom, F, is by definition the difference between the number of variables and the number of independent equations:

$$F = [2 + \pi(N - 1)] - [N(\pi - 1) + r + s]$$

or

$$F = 2 - \pi + N - r - s \tag{253}$$

This is Gibbs's phase rule. It specifies the number of independent intensive variables that can and must be fixed in order to establish the intensive equilibrium state of a system and to render an equilibrium problem solvable.

NOMENCLATURE

Symbol	Definition	Units
A	Helmholtz energy	J
B	second virial coefficient, density expansion	cm^3/mol
C	third virial coefficient, density expansion	cm^6/mol^2
D	fourth virial coefficient, density expansion	cm^9/mol^3
B'	second virial coefficient, pressure expansion	kPa^{-1}
C'	third virial coefficient, pressure expansion	kPa^{-2}
D'	fourth virial coefficient, pressure expansion	kPa^{-3}
B_{ij}	interaction second virial coefficient	cm^3/mol
C_{ijk}	interaction third virial coefficient	cm^6/mol^2
C_P	heat capacity at constant pressure	$J/(mol \cdot K)$
C_V	heat capacity at constant volume	$J/(mol \cdot K)$
E_K	kinetic energy	J
E_P	gravitational potential energy	J
F	degrees of freedom as given by the phase rule (eq. 253)	
f_i	fugacity of pure species i	kPa
$\hat{f}_i$	fugacity of species i in solution	kPa
G	molar or unit-mass Gibbs energy	J/mol or J/kg
g	acceleration of gravity	m/s^2
H	molar or unit-mass enthalpy	J/mol or J/kg
K_j	equilibrium constant for chemical reaction j	
M	molar or unit-mass value of any extensive thermodynamic property of a solution	
M_i	molar or unit-mass value of any extensive property of pure species i	
$\overline{M}_i$	partial molar property of species i in solution	
ΔM	property change of mixing	
ΔM_j°	standard property change of reaction j	
m	mass	kg
$\dot{m}$	mass flow rate	kg/s
n	number of moles	
n_i	number of moles of species i	
P	absolute pressure	kPa
P_c	critical pressure	kPa
P_i^{sat}	saturation or vapor pressure of species i	kPa
Q	heat	J
$\dot{Q}$	rate of heat transfer	J/s
R	universal gas constant	$J/(mol \cdot K)$
r	number of independent chemical reactions (eq. 253)	
S	molar or unit-mass entropy	$J/(mol \cdot K)$ or $J/(kg \cdot K)$
s	number of special constraints (eq. 253)	
T	absolute temperature	K
T_c	critical temperature	K
U	molar or unit-mass internal energy	J/mol or J/kg
u	velocity	m/s
V	molar or unit-mass volume	m^3/mol or m^3/kg
W	work	J
W_s	shaft work for flow process	J

Symbol	Definition	Units
$\dot{W}_s$	shaft power for flow process	J/s
x_i	liquid-phase mole fraction of species i in solution	
y_i	vapor-phase mole fraction of species i in solution	
Z	compressibility factor	
z	elevation above a datum level	m

Superscripts

E	excess thermodynamic property	
id	value for ideal solution	
ig	value for ideal gas	
l	liquid phase	
lv	phase transition from liquid to vapor	
R	residual thermodynamic property	
t	total value of thermodynamic property	
v	vapor phase	

Subscripts

C	value for colder heat reservoir	
c	value for critical state	
H	value for hotter heat reservoir	
r	reduced value	

Greek letters

α, β	identify phases	
γ_i	activity coefficient of species i in solution	
μ_i	chemical potential of species i	
$\nu_{i,j}$	stoichiometric number of species i in reaction j	
π	number of phases (eq. 253)	
ρ	molar density	
τ	time	
Φ_i	see eq. 223	
ϕ_i	fugacity coefficient of pure species i	
$\hat{\phi}_i$	fugacity coefficient of species i in solution	
ω	acentric factor	

BIBLIOGRAPHY

"Thermodynamics" in *ECT* 1st ed., Vol. 14, pp. 1–37, by J. C. Chu, Polytechnic Institute of Brooklyn; in *ECT* 2nd ed., Vol. 20, pp. 118–146, by H. T. Chen, Newark College of Engineering; and D. F. Othmer, Polytechnic Institute of Brooklyn; in *ECT* 3rd ed., Vol. 22, pp. 868–899, by H. C. Van Ness and M. M. Abbott, Rensselaer Polytechnic Institute.

1. M. M. Abbott and H. C. Van Ness, *Schaum's Outline of Theory and Problems of Thermodynamics*, 2nd ed., McGraw-Hill Book Co., New York, 1989.
2. J. W. Tester and M. Modell, *Thermodynamics and its Applications*, 3rd ed., Prentice-Hall, Englewood Cliffs, N.J., 1996.
3. K. S. Pitzer, *Thermodynamics*, 3rd ed., McGraw-Hill Book Co., New York, 1995.
4. J. M. Prausnitz, R. N. Lichtenthaler, and E. G. de Azevedo, *Molecular Thermodynamics of Fluid-Phase Equilibria*, 2nd ed., Prentice-Hall, Englewood Cliffs, N.J., 1986.

5. S. I. Sandler, *Chemical and Engineering Thermodynamics*, 2nd ed., John Wiley & Sons, Inc., New York, 1989.

6. J. M. Smith, H. C. Van Ness, and M. M. Abbott, *Introduction to Chemical Engineering Thermodynamics*, 5th ed., McGraw-Hill Book Co., New York, 1996.

7. H. C. Van Ness and M. M. Abbott, *Classical Thermodynamics of Nonelectrolyte Solutions: With Applications to Phase Equilibria*, McGraw-Hill Book Co., New York, 1982.

8. T. E. Daubert, R. P. Danner, H. M. Sibul, and C. C. Stebbins, *Physical and Thermodynamic Properties of Pure Chemicals: Data Compilation*, Taylor & Francis, Bristol, Pa., 1995.

9. R. C. Reid, J. M. Prausnitz, and B. E. Poling, *The Properties of Gases and Liquids*, 4th ed., McGraw-Hill Book Co., New York, 1987.

10. G. Soave, *Chem. Eng. Sci.* **27**, 1197 (1972).

11. D.-Y. Peng and D. B. Robinson, *Ind. Eng. Chem. Fundam.* **15**, 59 (1976).

12. A. Harmens, *Cryogenics*, **17**, 519 (1977).

13. M. Benedict, G. B. Webb, and L. C. Rubin, *J. Chem. Phys.* **8**, 334 (1940); **10**, 747 (1942).

14. B.-I. Lee and M. G. Kesler, *AIChE J.* **21**, 510–527, 1040 (1975).

15. *TRC Thermodynamic Tables—Hydrocarbons* and *TRC Thermodynamic Tables—Nonhydrocarbons*, serial publications of the Thermodynamic Research Center, Texas A&M University System, College Station, Tex.

16. "The NBS Tables of Chemical Thermodynamic Properties," *J. Phys. Chem. Ref. Data*, **11**(suppl. 2) (1982).

HENDRICK C. VAN NESS
MICHAEL M. ABBOTT
Rensselaer Polytechnic Institute

THERMOELECTRIC ENERGY CONVERSION

Thermoelectric energy conversion is the science of the interchange of thermal and electrical energy in simple solid-state devices, generally heavily doped semiconductors (qv). Figure 1 shows two thermocouples connected in series. If thermal energy is applied at the top of the device from some heat source, and removed at the bottom by a heat sink, an electrical potential appears across the device as indicated by the polarity signs. The size of the ΔT and the basic material properties determine the voltage across the couple. Potential per couple is usually on the order of tenths of a volt. Practical working voltages are obtained by connecting a large number of these thermocouples in series. The amount of current is dependent on the cross-sectional area and length of the legs.

If an electric current were to be passed through the device from an outside source, heat would be absorbed at one junction and released at the other. The device then operates as a solid-state heat pump.

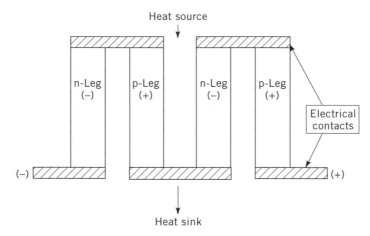

Fig. 1. Schematic of a simple thermoelectric device having two thermocouples in series.

Thermoelectric devices can be used to generate electrical power, to refrigerate, or to transfer large amounts of heat without the use of freons or compressors or any rotating machinery. However, the overall efficiency of these devices is somewhat lower than other systems, and the initial cost is often higher. Mass production, improved design and manufacturing techniques, and an extremely long (decades or more) operating life for many of these solid-state devices should both decrease initial cost and serve to aid in the recovery of the costs over time.

Segmenting of legs (Fig. 2) to take advantage of the higher performance of certain materials in a given temperature range has been used to increase conversion efficiency. However, this technique is limited because all of the thermal and electrical currents must flow through each segment. Every material has its own optimum current-to-charge (I/q) ratio. Thus in a given segmented couple it is normally impossible to have each segment operating at its own optimum efficiency unless the couple is used at precisely the design hot-and-cold junction temperatures.

In thermoelectric cooling applications, extensive use has been made of cascaded systems to attain very low temperatures, but because the final stage is so small compared to the others, the thermal flux is limited (Fig. 3). The

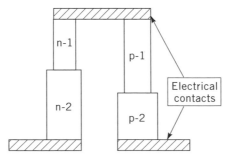

Fig. 2. Schematic of a segmented thermocouple. Each segment of each leg must be optimized in size for the design temperatures.

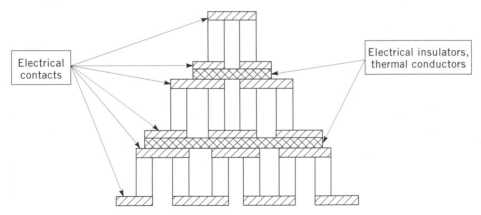

Fig. 3. Three-stage cascaded thermopile. The relative sizes of the stages must be adjusted to obtain the maximum ΔT.

relative sizes of the stages are adjusted to obtain the maximum ΔT. Thus, for higher cooling capacity, the size of each stage is increased while the area ratios are maintained.

A number of good references have been published on thermoelectric energy conversion, the first in 1957 (1). A thorough survey of the field, together with information on materials properties, is also available (2), as are works on the various approaches to the analysis, design, and construction of thermoelectric devices (3–6). The most current materials research and device design work is presented annually in two conferences: the International Conference on Thermoelectrics, sponsored by the International Thermoelectric Society, and the Intersociety Energy Conversion Engineering Conference. Work on thermoelectricity takes only a few sessions of this latter conference, which is hosted on a rotating basis by seven cooperating technical societies: ACS, American Institute of Aeronautics and Astronautics (AIAA), AIChE, American Nuclear Society (ANS), ASME, Institute of Electrical and Electronics Engineers (IEEE), and SAE.

Historical Background

The basic ideas of thermoelectricity have been known for nearly two centuries, but until well after the Second World War the primary use was for temperature measurement (qv) using metallic wires. Then, upon improvements in semiconductor technology, thermoelectric power generation and refrigeration came under serious consideration.

In 1822 Seebeck began to study the magnetic polarization of metals and ores produced by a temperature difference. His goal was to prove that the earth's magnetic field arose from temperature differences between the equator and the poles. The extensive data collected laid the foundation for later development, even though Seebeck himself had steadfastly refused to believe that the magnetic fields observed in the experiments came from the electric current flowing as a result of the electrical potentials generated by the temperature differences in the circuits of dissimilar materials.

The second principal thermoelectric effect was discovered by Peltier in 1834, who reported the production or absorption of heat at the interface of two dissimilar conductors when an electric current was passed through them. The significance of this discovery was misinterpreted, however. Peltier thought that Joule heating (I^2R) only occurred with strong electrical currents. It was not until 1838 that Lentz finally gave the correct interpretation of the Peltier effect. He provided a phenomenological demonstration by placing a drop of water at the junction of bismuth and antimony rods and caused it to freeze or melt, depending on which direction an electrical current was made to flow through the rods.

In 1857, Thomson (Lord Kelvin) placed the whole field on firmer footing by using the newly developing field of thermodynamics (qv) to clarify the relationship between the Seebeck and the Peltier effects. He also discovered what is subsequently known as the Thomson effect, a much weaker thermoelectric phenomenon that causes the generation or absorption of heat, other than Joule heat, along a current-carrying conductor in a temperature gradient.

In 1919, electrical power generators using these effects were attempted. Metallic components were used, which caused the efficiencies to fall well below one percent. These generators were not economically justifiable. However, the rapid growth of semiconductor theory brought a renewed interest in these devices. Practical power generators having from 6 to 11% efficiency were developed, which were especially useful for remote terrestrial and space applications. Indeed, radioisotope thermoelectric generators proved to be an enabling technology for deep-space exploration programs to the outer planets and beyond.

Thermoelectric Effects

The primary thermoelectric phenomena considered in practical devices are the reversible Seebeck, Peltier, and, to a lesser extent, Thomson effects, and the irreversible Fourier conduction and Joule heating. The Seebeck effect causes a voltage to appear between the ends of a conductor in a temperature gradient. The Seebeck coefficient, S, is given by

$$S = \frac{\delta V}{\delta T}$$

For metals, S generally varies between about 0 and 40 μV/K; low conductivity semiconductors and insulators have values of S up to 1000 μV/K and higher.

The Seebeck voltage is often referred to as being generated only at the junction of dissimilar conductors in an electrical circuit. Consideration of the Thomson effect, however, leads to the conclusion that the Seebeck voltage is generated along the entire thermoelectric element in a temperature gradient. The Thomson effect is the generation or absorption of heat, q_τ, other than I^2R heat, in a current-carrying conductor subjected to a thermal gradient:

$$q_\tau = \tau I$$

The Thomson coefficient, τ, is given by

$$\tau = -T\frac{\delta S}{\delta T}$$

Because the third law of thermodynamics requires $S = 0$ at absolute zero, the following equation is derived, which enables the determination of the absolute value of the Seebeck coefficient for a material without the added complication of a second conductor:

$$S = -\int_0^T \frac{\tau}{T} \, dT$$

Voltage measurement have been made at very low temperatures using a superconductor as one leg of a thermocouple. For a superconductor, S is zero, so the output of the couple is entirely from the active leg. The Thomson heat is then measured at higher temperatures to extend the absolute values of the Seebeck coefficients (7,8). The Thomson heat is generally an order of magnitude less than the Peltier heat and is often neglected in device design calculations.

The other primary thermoelectric phenomenon is the Peltier effect, which is the generation or absorption of heat at the junction of two different conductors when a current flows in the circuit. Whether the heat is evolved or absorbed is determined by the direction of the current flow. The amount of heat involved is determined by the magnitude of the current, I, and the Peltier coefficients, π, of the materials:

$$q_p = \pi_{A,B} I$$

Kelvin showed the interdependence of these phenomena by thermodynamic analysis, assuming that the irreversible processes were independent of the reversible ones. This approach was later proved theoretically sound using Onsager's concepts of irreversible thermodynamics (9).

$$\pi_{A,B} = S_{A,B} T$$

$$\tau_A - \tau_B = S_{A,B} - \frac{2\pi_{AB}}{2T}$$

The irreversible phenomena represent entropy gain through irrecoverable heat losses as follows, where λ is the thermal conductivity and l is the length:

electrical (Joule heating) $\qquad\qquad\qquad\qquad$ $q = I^2 R$

heat flow (Fourier conduction) $\qquad\qquad\qquad$ $q = \lambda \frac{A}{l} \Delta T$

Taking all of the above into account, it can be shown that the efficiency, η, of a thermoelectric power generator, neglecting Thomson heat, is given by

$$\eta = \frac{P_{out}}{Q_{in}} = \frac{I^2 R_0}{K\Delta T + S T_H I - \frac{1}{2} I^2 R}$$

Optimizing for both leg geometry and load resistance and using average values for S, ρ, and λ, the following equation is derived:

$$\eta = \frac{\Delta T}{T_H} \left(\frac{(1 + ZT_{\mathrm{av}})^{1/2} - 1}{(1 + ZT_{\mathrm{av}})^{1/2} + \frac{T_c}{T_H}} \right)$$

where

$$Z = \frac{(|S_A| + |S_B|)^2}{((\lambda_A \rho_A)^{1/2} + (\lambda_B \rho_B)^{1/2})^2} \ \deg^{-1}$$

The first term in the efficiency is simply the Carnot efficiency for a reversible heat engine and the second term is the reduction in efficiency owing to the irreversible effects. The important factors can easily be seen to be Z (the figure of merit) and ΔT. Increasing Z or T_H or decreasing T_c can increase the efficiency in general. However, because S, ρ, and λ are all temperature-dependent, changing T also changes the value of these properties, thus making the prediction of the overall effect less easy to estimate without detailed calculations using actual material properties.

For insulators, Z is very small because ρ is very high, ie, there is little electrical conduction; for metals, Z is very small because S is very low. Z peaks for semiconductors at $\sim 10^{19}$ cm^{-3} charge carrier concentration, which is about three orders of magnitude less than for free electrons in metals. Thus for electrical power production or heat pump operation the optimum materials are heavily doped semiconductors.

As indicated in Figure 4, the basic thermoelectric parameters are all functions of carrier concentration. Thus adjusting the dopant level to increase the output voltage generally also increases the electrical resistance. In addition, it affects the electronic component of the thermal conductivity. However, there are limitations on what can be accomplished by simply varying the carrier concentration in any given material.

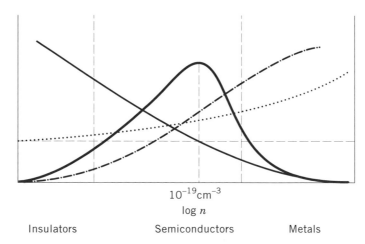

10^{-19}cm^{-3}

log n

Insulators   Semiconductors   Metals

Fig. 4. Dependence of thermoelectric parameters on carrier concentration, where (—) is the Seebek coefficient (S); (▬), the figure of merit (Z); (–·–), reciprocal of electrical resistivity, $\sigma = 1/\rho$; (····), thermal conductivity (λ); and (–––), (λ_1).

The lattice contribution to the thermal conductivity can be varied by other means, such as differences in the atoms of the crystal. Alloy disorder, resonance (impurity) scattering, charge carrier scattering, and grain boundary scattering are all aimed at reducing the phonon mean free path. Unfortunately, modifying the crystal structure to reduce the thermal conductivity also causes changes in the carrier mobilities and effective masses, which also influences S and P.

Theoretical calculations can only point to promising types of materials. Because the interactions between the various parameters are too complex for direct calculation, it is necessary to rely on experimental data to confirm that proposed materials really perform as predicted.

Operating in the opposite mode, ie, causing an electric current to flow through the device from an outside source rather than impressing a thermal gradient on it, can result in a cooling or heating system. In thermoelectric cooling applications, the effectiveness of a device is measured in terms of the coefficient of performance, β, rather than efficiency. This is the ratio of the heat removed from the cold reservoir to the electrical power input.

$$\beta = \frac{q_c}{P_e} = \frac{ST_c I - \frac{1}{2}I^2 R - \lambda \Delta T}{SI\Delta T + I^2 R}$$

When optimized for geometry and current this becomes

$$\beta = \frac{T_c}{\Delta T} = \left(\frac{(1 + ZT_{\text{av}})^{1/2} - \frac{T_H}{T_c}}{(1 + ZT_{\text{av}})^{1/2} + 1} \right)$$

Under these conditions the maximum heat pumping rate, $q_{c_{\text{max}}}$; the maximum temperature difference across the module, ΔT_{max}; and the minimum temperature attainable on the cold side, $T_{c_{\text{min}}}$, are given by the following relations:

$$q_{c_{\text{max}}} = \frac{S^2 T_c^2}{2R} - \lambda \Delta T$$

$$\Delta T_{\text{max}} = \frac{S^2 T^2}{((\rho_n \lambda_n)^{1/2} + (\rho_p \lambda_p)^{1/2})^2} = \frac{1}{2} T_c^2 Z$$

$$T_{c_{\text{min}}} = \frac{(1 + 2ZT_H)^{1/2}}{Z}$$

As of the mid-1990s, devices are generally capable of cooling rates of $2-3$ W/cm^2 using electrical current densities of 1 A/mm^2. Typical temperature drops are about 30°C across a single stage, but much larger ΔT_s can be obtained by cascading.

Technology

Materials. Because the Seebeck coefficient, the electrical resistivity, and the thermal conductivity of thermoelectric materials all vary with temperature, no single material can function well over the entire temperature range in which thermoelectric devices are used. Table 1 indicates the generally accepted operating range categories and the primary thermoelectric materials used in each range. Many other materials have been or are being developed to obtain either higher conversion efficiencies or less expensive manufacturing (2).

More recently the idea of using materials having various properties along the legs have been revived. Adjustment of dopant levels changes the carrier concentration along the thermoelement and allows each small segment of the leg to operate at optimum power conversion levels along the entire thermal gradient (10). However, this results in only incremental improvement over existing thermocouples.

In order to obtain a step function increase in conversion efficiency, new materials are needed. One promising class of compounds being investigated is the skutterudites, which have very good electrical properties, but rather high thermal conductivities in their simplest form. Work is proceeding on ways to reduce the thermal conductivity through the use of ternary alloys, and also by filling vacancies in the skutterudite structure with rare earth elements. As of this writing (1997), the most promising filled structure found is $CeFe_4Sb_{12}$, which has a figure of merit of 1.6×10^{-3} K^{-1} at 600°C (11).

An interesting and promising method of obtaining significant improvement in thermoelectric properties is by fabricating materials of good thermoelectricity in two-dimensional quantum well structures of very thin (<2.5 nm) layers, separated by large (>40 nm) band gap barrier layers in one sample. This multiple quantum well structure, fabricated by molecular beam epitaxy, gave an increase of a factor of three in S, but showing little effect on the other thermoelectric properties. Although this would be expected to lead to a five-fold increase in Z, the heat loss through the relatively thick barrier layers negates much of that increase. It is possible that barrier materials having a higher band gap would result in thinner barrier layers and real gains in Z (12).

The physical form of the thermocouples varies significantly according to applications. Most spacecraft power supplies utilize separate thermocouples that can be checked for performance at successive stages of manufacturing and be replaced if necessary. This approach fits in very well with the extremely high reliability requirements imposed on such systems. In terrestrial systems where such individualized attention is not economically feasible, modular assemblies are generally used, which can contain tens to hundreds of couples in a single unit.

One approach to improving the performance of existing materials through novel manufacturing techniques is the use of the mechanical alloying of ex-

Table 1. Operating Modes and Temperatures of Thermoelectric Materials

Temperature, °C		Predominant materials	Operating mode
Hot	Cold		
200	−130	bismuth telluride; bismuth antimony telluride	cooling or power
600	100	lead telluride; silver antimony germanium telluride	power
1000	300	silicon germanium	power

tremely fine powders of germanium silicide and its dopants through high energy ball milling. This method has resulted in an increase of the integrated figure of merit from 0.7×10^{-3} to 0.9×10^{-3} K^{-1} for n-type SiGe alloys (13).

Because most thermoelectric materials are very strong in compression but rather weak in tension and shear, most low and moderate temperature devices are either spring-loaded or made integral with the heat exchangers to avoid stress problems and to ensure good thermal contact for conductive coupling with the heat source and sink. The high operating temperature of silicon germanium devices and the physical ruggedness of these materials have allowed cantilever mounting of the thermocouples from the cold surface of the generator housing and the use of radiant heating to supply heat to the hot shoes.

Thermoelectric Power Generation

Thermoelectric devices represent niche markets, but as economic and environmental conditions continue to change, they appear poised to advance into more common use. Thermoelectric power generators are in use in many areas, including satellites, deep-space probes, remote-area weather stations, undersea navigational devices, military and remote-area communications, and cathodic protection.

Undoubtedly, the most exciting application of thermoelectric power has been in the radioisotope thermoelectric generators (RTG) of the U.S. space programs. These began very modestly. Two SNAP-3B three-watt demonstration units were flown as proof-of-principle tests on the mainly solar-powered Transit 4A and 4B navigational satellites in 1961. These were followed by the 27-watt SNAP-9A design on the Transit 5BN-1 and 5BN-2 satellites in 1963, the first spacecraft to be fully powered by radioisotope power sources. The U.S. RTG-powered missions are summarized in Table 2. The power levels have steadily increased up to the 850 watts scheduled for use on the Cassini mission, which is to send a probe into the atmosphere of Saturn's moon Titan and then spend four years orbiting Saturn, making multiple encounters with its other moons. The launch for this spacecraft is scheduled for October 1997.

Thermoelectric generators have made extended spacecraft missions possible for exploration of the giant outer planets and the regions beyond the far reaches of the solar system, eg, the two Pioneer and the two Voyager missions. As of this writing (1997), the Pioneer spacecrafts have been operating over twenty-five years in the frigid reaches of outer space, surviving extremely high radiation fields on very close flybys of Jupiter and sending back data from $>9.6 \times 10^9$ km away from the sun. None of these missions has encountered any failures in the thermoelectric power systems (14,15).

For terrestrial applications, 77 radioisotope-powered thermoelectric generators were put into operation between the years 1961 and 1995. Their electrical power outputs range from one to 500 watts. They have been used in a wide range of applications, mostly involving remote locations where supply and servicing is difficult and long-term unattended operation is of primary importance. Among the uses of these generators have been arctic weather stations, navigational light buoys, undersea acoustic beacons, oceanographic data collection, Federal

Table 2. Summary of Radioisotope Thermoelectric Generators Successfully Launched by the United States from 1961 to 1990

Power source	Initial average power, W	Spacecraft	Mission type[a]	Launch date	Orbital lifetime, yr
SNAP-3B7	2.7	Transit 4A	N	June 29, 1961	500
SNAP-3B8	2.7	Transit 4B	N	Nov. 15, 1961	1200
SNAP-9A	>25.2	Transit 5BN-1	N	Sept. 28, 1963	1900
SNAP-9A	26.8	Transit 5BN-2	N	Dec. 5, 1963	1800
SNAP-19B3	56.4	Nimbus III	M	Apr. 14, 1969	3600
SNAP-27	73.6	Apollo 12	L	Nov. 14, 1969	on lunar surface
SNAP-27	72.5	Apollo 14	L	Jan. 31, 1971	on lunar surface
SNAP-27	74.7	Apollo 15	L	July 26, 1971	on lunar surface
SNAP-19	162.8	Pioneer 10	P	Mar. 2, 1972	beyond solar system
SNAP-27	70.9	Apollo 16	L	Apr. 16, 1972	on lunar surface
Transit-RTG	35.6	Triad	N	Sept. 2, 1972	137
SNAP-27	75.4	Apollo 17	L	Dec. 7, 1972	on lunar surface
SNAP-19	159.6	Pioneer 11	P	Apr. 5, 1973	out of solar system
SNAP-19	84.6	Viking 1	ML	Aug. 20, 1973	on Martian surface
SNAP-19	84.2	Viking 2	ML	Sept. 9, 1975	on Martian surface
MHW-RTG	307.4	Les-8	C	Mar. 14, 1976	1,000,000
MHW-RTG	308.4	Les-9	C	Mar. 14, 1976	1,000,000
MHW-RTG	477.6	Voyager 2	P	Aug. 20, 1977	out of solar system
MHW-RTG	470.1	Voyager 1	P	Sept. 5, 1977	out of solar system
GPHS-RTG	576.8	Galileo	P	Oct. 18, 1989	second earth flyby
GPHS-RTG	~288	Ulysses	P	Oct. 6, 1990	solar–polar orbit

[a]C = communications; L = lunar; M = meteorological; ML = Mars landers; N = navigational; P = planetary.

Aviation Administration (FAA) communications relay stations, seismic sensors, and offshore oil well beacon lights and foghorn (16).

Because normal radioisotopic decay lowers the thermal output by about 2.5%/yr in these units, they are purposefully overdesigned for beginning of life conditions. Several of these generators have successfully operated for as long as 28 years. This is approximately equal to the half-life of the strontium-90 isotope used in the heat sources. The original SNAP-7 series immobilized the strontium-90 as the titanate, but the more recent ones have used it in the form of the fluoride, which is also very stable. A number of tiny nuclear-powered cardiac pacemaker batteries were developed, which have electrical power outputs of 33–600 μW and have been proven in use (17).

More mundane but, in the long run, much more important economically and environmentally are the terrestrial systems for power generation and cooling. The primary focus of the power-generating devices is to recapture some of the waste heat dumped by commercial processes and vehicle engines, or to make use of the heat from geothermal sources (see GEOTHERMAL ENERGY). In cooling devices the emphasis is on eliminating the use of refrigerant gases that could be environmentally harmful, and in providing very small systems for specialized uses.

The California Energy Commission, Energy Technology Advancement Program, and the U.S. Departmant of Energy jointly sponsored a project for reducing diesel engine fuel use and NO_x particulates (18). An engine-driven electrical generator uses up to 2200–3700 W (3–5 hp) from the engine drive shaft, whereas a thermoelectric generator uses only exhaust heat and coolant from the radiator. In the latter instance there is very little degradation of the engine power output. The unit is about 1-meter long and 25-cm in diameter, and uses bismuth telluride thermoelectric modules. Initially more expensive than the engine driver alternator, at 1997 fuel costs, however, the device could recover this cost differential in two years in the United States or 6–8 months in Europe. It is believed that on-board electrical usage in trucks for powering computer NO_x reduction systems, particulate trapping, etc, is only to increase in the future, making the gains from the thermoelectric system even better.

The U.S. Coast Guard has been experiencing very high fuel and maintenance costs on their diesel motor-generator-powered major aids to navigation's systems, and are therefore sponsoring work on such systems as photovoltaics, wind, or wave power generators to float-charge large battery packs. Because each of these new power supplies is subject to significant down time owing to adverse weather conditions, a very reliable low maintenance backup power supply is required. A 1.5-W thermoelectric generator burning diesel fuel is being investigated, which uses a segmented lead telluride–bismuth telluride module.

As an offshoot of the Saudi–German HYSOLAR program that uses photovoltaics to produce hydrogen by electrolysis of water, giving in effect storable solar energy, the hydrogen can later be burned at 200–700°C using catalytic combustion (19). This provides the heat to a modified Teledyne Energy Systems thermoelectric converter that was originally designed for natural gas, propane, or butane fuels. The H_2 is both storable and transportable and the combustion product is water.

Osaka University and the University of Wales are collaborating on a research project aimed at the recovery of electrical power from waste heat from commercial processes (20). This is a low (80°C) temperature low (~60°C) ΔT process that makes use of very thin thermoelements and very large area heat exchange devices. The initial costs could be quite high, but the low temperature operation should result in extremely long life times to depreciate the costs effectively. Because this system is using only waste heat from another process, electrical power is gained without additional pollution of any sort. The efficiency of these systems is rather low compared to other devices operating over wider temperature ranges, but because it is piggybacking on other processes that take care of the economic investment, it is a promising approach to alleviating any energy crunch in the future.

Cold climate electrical heaters for trucks in far northern areas of the United States, Canada, Northern Europe, and the CIS are very important. These are used when the trucks are idle and provide heat to the engine coolant and the cab without idling the engine for extended periods (21). This not only minimizes engine wear and reduces pollution but also provides significant economic savings. Many jurisdictions now prohibit extended idling of engines. In extremely cold environments, engines can quickly become difficult, sometimes nearly impossible, to start. If ordinary gasoline- or diesel-oil-fired heaters are used, the coolant circulation pump, air fan, etc, must be powered from the vehicle's batteries, thus curtailing the time the system can be used, especially at very low temperatures when it is needed the most. By adding PbTe thermoelectrics to such heater systems, about 2% of their thermal output can be turned into electricity to run the heater's electronics, fuel pump, combustion fan, and coolant circulation pump, with still sufficient power left over to keep the vehicle's battery fully charged. The market for such units is in the hundreds of thousands if manufacturing costs can be reduced.

Thermoelectric Cooling

In the 1960s there were a number of attempts to make thermoelectric cooling a significant industry. Although all of the systems worked very well, they could not compete economically with the low fuel costs of those years. Nevertheless, the groundwork was laid. A corporate headquarters building in Wisconsin was fitted with a thermoelectric air conditioning system by Carrier Corporation. It consisted of 30 decentralized units that were still operating well over 10 years later (22). However, Carrier has since withdrawn from the business.

Small refrigerators were developed by several companies and some were even installed in hotel rooms in Chicago. Borg-Warner and other companies produced many compact systems for laboratory uses (23). Air-Industry in France built an air conditioning system for a passenger railway coach that was still in daily use after 10 years of operations without a single thermoelectric failure (24).

With the international agreements concluded in the 1990s on banning chlorofluorocarbons by the year 2000, the efficiency of refrigeration systems using substitute working fluids has moved much closer to that of the thermoelectric cooling systems. Considering the economics of scale, it now seems possible to market thermoelectric refrigerators at only 10 to 20% above the cost of similar sized compressor units. The quieter operation and much longer life of thermoelectric units could make these competitive in the near future (25).

The U.S. Army is sponsoring work on a modular device to be used in protecting personnel in severe conditions. This is a thermoelectric cooling device, designed for mounting in a two-person army vehicle, which is to provide a controlled working atmosphere for two personnel wearing nuclear, biological, and chemical (NBC) protective clothing (26). The device consists of a thermoelectric cooling unit, a multiple-filter pack, and a blower assembly. The unit can supply 0.67 m³/min (24 ft³/min) of cooled, filtered air with 250 watts of cooling to each person. The unit operates off of the vehicle's 28-volt electrical system and can easily maintain an effective working climate inside the suits over an external ambient range from 35°C/85% rh to 51.5°C/3% rh.

By far the most important applications for thermoelectric devices as of the mid-1990s are the fields of small laboratory and medical devices and recreational coolers. These range in cooling capacity from fractions of a watt to 600 watts and minimum temperatures as low as −100°C up to reverse-mode heating temperatures of +100°C (27). Such devices range from simple laboratory measurement devices, uses in biotechnology and medical and surgical devices, to rugged coolers for photodetectors and lasers and portable coolers for boats and autos. The market is in the hundreds of thousands of thermoelectric modules per year.

NOMENCLATURE OF THERMOELECTRIC ENERGY CONVERSION

Symbols	Definition	Units
A	area	cm^2
I	electric current	A
K	device thermal conductivity	
l	length	cm
η	efficiency	
P	electrical power out	W
q	heat	W
Q	heat in	W
R	electrical resistance	Ω
S	Seebeck coefficient	V/K
T	temperature	K
V	electrical voltage	V
Z	figure of merit	K^{-1}
β	coefficient of performance	
λ	thermal conductivity	W/(cm·K)
π	Peltier coefficient	W/A
ρ	electrical resistivity	ohm·cm
τ	Thomson coefficient	V/K

BIBLIOGRAPHY

"Thermoelectric Power Conversion" in *ECT* 1st ed., Suppl. Vol., pp. 840–866, by W. Arbiter and R. C. Ross, Nuclear Development Corp. of America; "Thermoelectric Energy Conversion" in *ECT* 2nd ed., Vol. 20, pp. 147–172, by A. C. Glatz, National Aeronautics and Space Administration; in *ECT* 3rd ed., Vol. 22, pp. 900–917, by A. C. Glatz, General Foods Corp.

1. A. F. Joffe, *Semiconductor Thermoelements and Thermoelectric Cooling*, Infosearch Limited, London, 1957.
2. D. M. Rowe, ed., *CRC Handbook of Thermoelectrics*, CRC Press, London, 1993.
3. R. R. Heikes and R. W. Ure, Jr., *Thermoelectricity: Science and Engineering*, Interscience Publishers, New York, 1961.
4. S. W. Angrist, *Direct Energy Conversion*, Allyn and Bacon, Inc., Boston, Mass., 1965.
5. F. J. Blatt and co-workers, *Thermoelectric Power of Metals and Alloys*, Plenum Press, New York, 1976.
6. T. C. Harmon and J. M. Honig, *Thermoelectric and Thermomagnetic Effects and Applications*, McGraw-Hill Book Co., Inc., New York, 1967.
7. R. B. Roberts, *Phil. Mag.* **36**, 1:91 (1977).
8. N. Cusack and P. Kendall, *Proc. Phil. Soc.* **72**, 898 (1958).
9. C. A. Domenicalli, *Rev. Mod. Phys.* **26**, 237 (1954).
10. Y. Noda and co-workers, *Proceedings of the 15th International Conference on Thermoelectrics*, Kansas City, Mo., 1996, p. 146.
11. T. Caillat, J.-P. Fleurial, and A. Borshchevsky, in Ref. 10, p. 100.
12. L. D. Hicks and co-workers, in Ref. 10, p. 450.
13. B. A. Cook and co-workers, *J. Appl. Phys.* **78**, 5474 (1995).
14. E. A. Skrabek, *Proceedings of the 7th Symposium on Space Nuclear Power Systems*, Albuquerque, N.M., 1990.
15. G. L. Bennett and E. A. Skrabek, in Ref. 10, p. 357.

16. W. C. Hall, in Ref. 2, Chapt. 40.
17. D. M. Rowe, in Ref. 2, Chapt. 40.
18. J. C. Bass and N. B. Elsner, *Proceedings of the 11th International Conference on Thermoelectrics*, Arlington, Tex., 1992, p. 1.
19. M. Al-Garni and M. Bajunaaid, in Ref. 18, p. 4.
20. K. Mastsuura and co-workers, in Ref. 18, p. 10.
21. A. G. McNaughton and G. J. McBride, in Ref. 18, p. 17.
22. J. G. Stockholm, *Proceedings of the 9th International Conference on Thermoelectrics*, Pasadena, Ca., 1990, p. 90.
23. A. B. Newton, *ASHRAE Trans.*, 133 (July 1965).
24. L. P.-J. Stockholm and P. Sternat, *Proceedings of the 4th International Conference on Thermoelectrics*, Arlington, Tex., 1982.
25. P. M. Schliklin, in Ref. 22, p. 381.
26. P. Hanan and B. Mathiprakasam, in Ref. 18, p. 181.
27. K. Uemura, in Ref. 2, Chapt. 52.

E. A. SKRABEK
Orbital Sciences Corporation

THERMOGRAPHIC DYES. See ELECTROPHOTOGRAPHY; THERMOGRAPHY.

THIAZOLE DYES. See AZO DYES.

THIN FILMS

FILM FORMATION TECHNIQUES

A deposited thin film is a layer on a surface having properties that differ from those of the bulk material (substrate) that has been formed by the addition of solid material to the surface. Generally, the substrate material cannot be detected in the film, which can be an organic or inorganic material. This surface layer differs from surface conversion where the surface is chemically converted to another material, eg, anodization of aluminum (see METAL SURFACE TREATMENTS; SURFACE MODIFICATION TECHNIQUES (SUPPLE-MENT)). The term thin film is generally applied to layers that have thicknesses on the order of several micrometers or less. These films may be as thin as a few atomic layers. In many cases, thin films are formed by adding atoms or molecules to a substrate surface one at a time. Thicker layers are generally called coatings (qv). Although coatings can often be formed by the same processes that are used to form thin films, there are some coating processes (qv) that are not applicable to forming thin films. For example, thermal spray coating processes, which melt small particles, accelerate them to high velocities, and splat-cool them on surfaces, are not applicable to forming thin films.

The properties of thin films generally differ from the values for the material in bulk form (see THIN FILMS, FILM PROPERTIES). In many cases, the growth and properties of thin films are affected by the properties of the underlying substrate material. The properties of the film can also be affected by the high surface-to-volume ratio of the film.

Physical Deposition Techniques

Physical vapor deposition (PVD) processes are film deposition processes in which atoms or molecules of a material are vaporized from a solid or liquid source, transported in the form of a vapor through a vacuum or low pressure gaseous environment, and condense on the substrate. PVD processes, by far the largest-volume techniques for forming thin films, use vacuum to control the gaseous contamination in the deposition system and to provide a long mean-free path for collision for the vaporized material as it passes from the source to the substrate. Generally, vacuums having gas pressures of 10.3 Pa or less ($<10^{-4}$ atm) are employed (see VACUUM TECHNOLOGY). PVD processes can be used to deposit films of elemental, alloy, and compound materials as well as atomically dispersed mixtures and some polymeric materials. Typical PVD deposition rates are from 1 to 10 nm/s, although the rates for easily vaporized materials, such as zinc and cadmium, can be much higher. The films can be of single materials or multilayers of different materials, have a graded composition, or be very thick deposits (coatings). The deposited material can be amorphous, fine or coarse grained, or single-crystal, depending on the material and deposition conditions. Compounds can be deposited by vaporizing the compound material, eg, SiO or MgF_2, directly, or by allowing the depositing material, such as Ti, Zr, Si, or C, to react with an ambient gaseous environment or a codeposited species to form films of compound materials, such as TiN, ZrN, SiO_2, and TiC. Deposition with reaction is called reactive deposition and the term is often used as an adjective, ie, reactive sputter deposition or reactive ion plating.

Figure 1 shows the components of a typical PVD system. The system consists of a processing chamber, vacuum pumping system, an exhaust system, and a gas-handling system for admitting processing gases if used, or gases to return the system pressure to the ambient pressure. The processing chamber must contain the substrates being processed and the fixtures that hold them. The system and its cycle time must be designed to give a reproducible processing environment and provide the processing capacity, ie, the throughput, necessary for the intended application.

There are several types of deposition chamber configurations (Fig. 2). The batch-type system is the most commonly used, but the requirement that the system be returned or let-up to ambient pressure on each cycle can pose problems

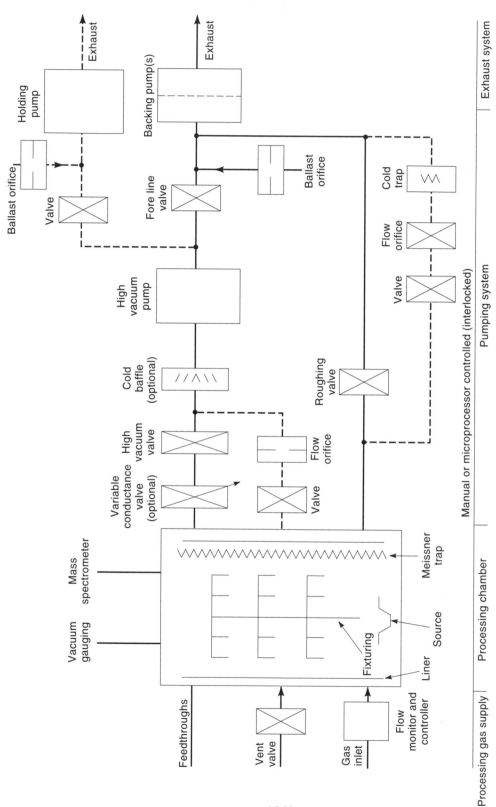

Fig. 1. Vacuum deposition system having a plasma processing capability, where the dashed lines represent optional additions to a system.

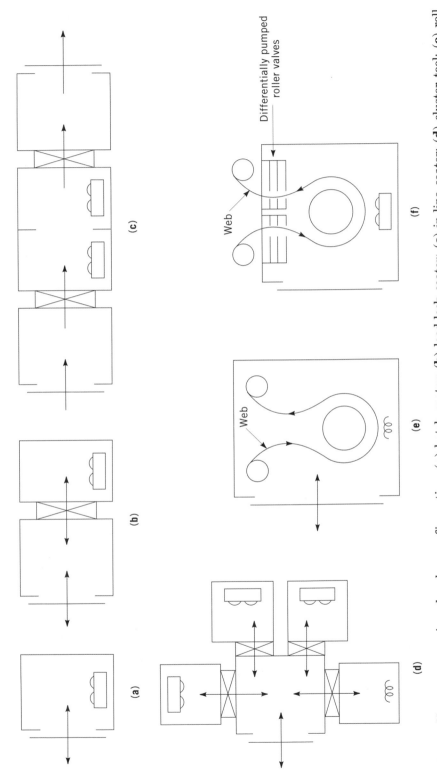

Fig. 2. Vacuum processing chamber configuration: (**a**) batch coater; (**b**) load-lock coater; (**c**) in-line coater; (**d**) cluster tool; (**e**) roll coater (batch); and (**f**) roll coater (air-to-air). ⊠ represents the isolation valve with transfer tooling; ↔, the motion of fixturing; and ⊣⊢, the access door.

Web

Differentially pumped
roller valves

Web

(a)

(b)

(c)

(d)

(e)

(f)

in obtaining a reproducible processing environment. The load-lock system and the in-line system allows the deposition chamber to be kept under vacuum at all times and the substrates introduced and removed through load-locks. The in-line systems generally have a much higher production volume than the batch-type systems. The cluster-tool chamber allows random access to processing chambers from the load-lock chamber and is widely used in semiconductor processing. The roll or web coating system is a specialized system for metallization of flexible polymer film for packaging and thermal control applications. The load-lock systems present the problem of requiring a long-lived vaporization source or a source where the vaporization material can be renewed remotely because the deposition chamber is not opened very often.

Vacuum Deposition. Vacuum deposition, sometimes called vacuum evaporation, is a PVD process in which the material is thermally vaporized from a source and reaches the substrate without collision with gas molecules in the space between the source and substrate (1–3). The trajectory of the vaporized material is therefore line-of-sight. Typically, vacuum deposition takes place in the pressure range of 10^{-3}–10^{-2} Pa (10^{-5}–10^{-9} torr), depending on the level of contamination that can be tolerated in the resulting deposited film. Figure 3 depicts a simple vacuum deposition chamber using a resistively heated filament vaporization source.

The substrate fixturing is an important component in any PVD system. The fixturing allows the substrates to be held in a desirable orientation, provides a means for heating or cooling the substrate, and allows movement during deposition that is important for obtaining a uniform deposition over the substrate surface(s). The shutter is important in that it allows the vaporization source to be heated without exposing the substrate to volatile contaminant material that may vaporize from the source. It also minimizes radiant heating from the vaporization source by allowing the source material to be premelted and thus wet the hot surface. By opening and closing the shutter, the deposition time can be accurately controlled. The glow bar allows the formation of a plasma in the system for *in situ* cleaning of substrate surfaces in the deposition chamber.

The vapor pressure is an important property of the material to be thermally vaporized. In a closed container at equilibrium, the number of atoms returning to the surface is the same as those leaving the surface, and the pressure above the surface is the equilibrium vapor pressure. Vapor pressures are strongly dependent on the material and the temperature (Fig. 4).

Materials having a higher vapor pressure at low temperatures are typically vaporized from resistively heated sources such as those shown in Figure 5a. Refractory materials require a high temperature to be vaporized. A focused high energy electron-beam heating is necessary for vaporization (Fig. 5b).

The vaporization rate from a hot surface into a vacuum, ie, free surface vaporization rate, depends on the temperature and the equilibrium vapor pressure of the material at that temperature. For thermal evaporation, a reasonable deposition rate can be obtained only if the free surface vaporization rate is fairly high and an equilibrium vapor pressure of ca 1 Pa (10^{-2} torr) is arbitrarily considered as the value necessary to provide a useful deposition rate. Materials having vapor pressure above the solid are described as subliming materials. Examples are Cr and C. Materials having vapor pressure above the liquid are

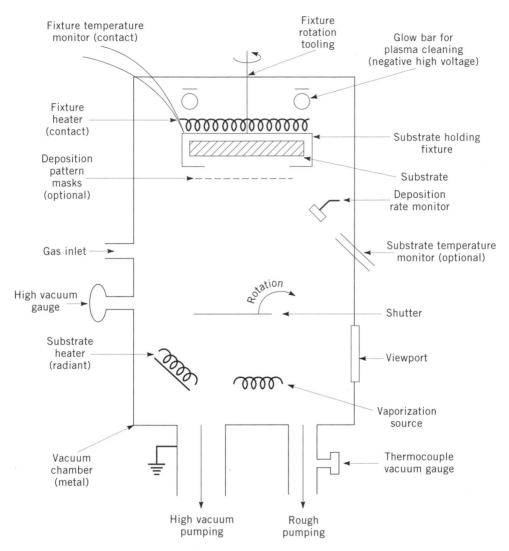

Fig. 3. Principal components of a vacuum deposition chamber.

described as evaporating materials. Many materials, such as titanium, can be deposited by either sublimation or evaporation, depending on the temperature of the source. For some materials, such as aluminum and tin, the temperature of the molten material must be significantly greater than the melting point in order to have a significant vaporization rate. For alloys, the thermal vaporization rate of each constituent is proportional to the relative vapor pressures (Raoult's Law). Therefore during vaporization, the higher vapor pressure material vaporizes more rapidly and the vaporization source is progressively enriched in the lower vapor pressure material as evaporation progresses.

Most elements thermally vaporize as atoms but some, such as Sb, C, and Se, have a portion of their vapor as clusters of atoms. For these materials, special vaporization sources, called baffle sources, can be used to ensure that the

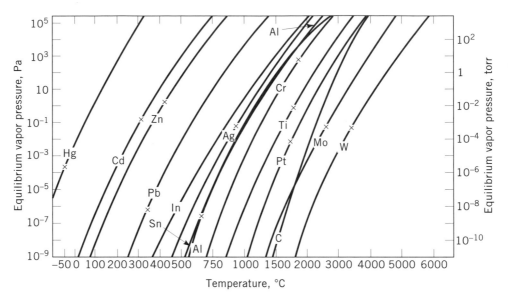

Fig. 4. Equilibrium vapor pressure of materials, where × indicates the melting point of the metal. The melting points for In and Sn are off the graph at 156 and 232°C, respectively. There is no melting point for C.

depositing vapor is in the form of atoms by causing the material to be vaporized from multiple hot surfaces before it leaves the source.

Some compounds sublime as molecules. However, the molecules of many compound materials partially dissociate on vaporization. Notable among the materials that vaporize without much dissociation are silicon monoxide, SiO, and magnesium fluoride, MgF_2, which are widely used in optical coating technology as high (SiO has $n_D \approx 2$) and low (MgF_2 has $n_D = 1.38$) index of refraction materials. The degree of dissociation of a compound depends strongly on the vaporization temperature. When depositing a compound that dissociates, the depositing film is generally deficient in the gaseous constituent. For example, in the evaporation of SiO_2, the resulting film is deficient in oxygen, giving a film of composition SiO_{2-x}, which has a brownish color. This loss of gaseous constituents during vaporization can be partially compensated for by using reactive or activated reactive evaporation, where there is a low pressure reactive gas or a low pressure plasma of the reactive gas in the deposition environment. At higher gas pressures, multibody gas-phase collisions can cause gas-phase nucleation (gas evaporation) of the vaporized material, which then deposits as a very low density film (1).

The material vaporized from a small area (point source) leaves the surface with a cosine distribution and thermal energies of a few tenths of an eV. The vaporized material arrives at a substrate surface having a mass per unit area (dm/dA) given by

$$dm/dA = M(1/r^2) \cos \phi \cdot \cos \theta \qquad (1)$$

where M is the mass vaporized, r is the source-to-substrate distance, and the angles are as shown in Figure 6. This gives a distribution on a planar surface above the source shown in Figure 6.

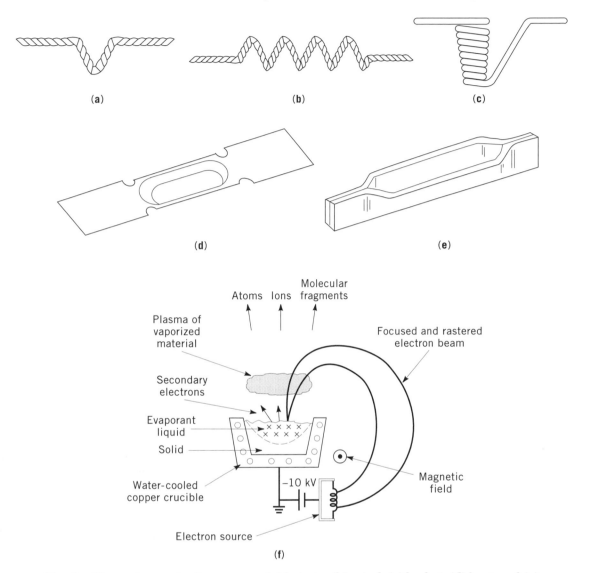

Fig. 5. Thermal vaporization sources: (**a**) hairpin; (**b**) spiral; (**c**) basket; (**d**) boat; and (**e**) canoe, which are all resistively heated sources; and (**f**) focused electron-beam evaporation from a high energy electron-emitting source.

In actuality, the flux distribution from the source may not be cosine-dependent. It can be modified by source geometry, collisions in the vapor above the vaporizing surface when there is a high vaporization rate, changes of vaporization source geometry with time, etc. The strong dependence of deposition rate on geometry and time often requires that fixturing be used to randomize the substrate position(s) during deposition in order to increase the film thickness uniformity. This fixturing also randomizes the angle of incidence and deposition rate of the depositing vapor flux, thereby increasing the uniformity of the film properties over the substrate surface(s) (4).

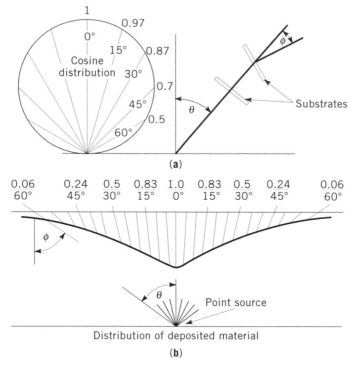

Fig. 6. Cosine distribution of vaporized material from a point source: (**a**) distribution of deposited material from a point source, and (**b**) distribution of film thickness on a planar surface above a point source. All thickness values normalized to 1.

The principal processing variables in vacuum deposition are deposition geometry, deposition rate, substrate temperature during deposition and the level of gaseous and vapor (eg, water vapor) contamination in the deposition environment. Deposition rates and amounts can be monitored *in situ* and in real-time by collecting the vapor on the surface of a quartz crystal oscillator, thus causing the oscillation frequency to change. Calibration allows the change in frequency to be related to deposited film mass and, by assuming a film density, the film thickness. In many applications the amount of material deposited is controlled by the evaporation-to-completion of a specific amount of material and using a specific deposition geometry. In some cases a property of the film, such as optical transmittance, is monitored during deposition and is used to control the amount of material deposited.

There are several advantages of vacuum deposition: (*1*) high purity films can be deposited from high purity source material; (*2*) vacuum premelting (with shutter closed) can be used to remove volatile contaminants from the source material; (*3*) the source of vaporized material can be a solid in any form and purity; (*4*) high vaporization (deposition) rates can be attained; (*5*) the line-of-sight trajectory and limited-area sources allow the use of shutters and masks to define deposition patterns; (*6*) deposition monitoring and control are relatively easy; (*7*) deposition systems can be pumped at a high rate (high throughput) during the deposition to remove gaseous contaminants; and (*8*) residual gases and vapors in the vacuum environment are easily monitored. Moreover, these are probably the least expensive of the PVD processes.

There are also disadvantages of vacuum deposition: (*1*) many alloy compositions and compounds can only be deposited with difficulty; (*2*) line-of-sight and limited-area sources result in not only poor surface coverage on complex surfaces, but also poor film thickness uniformity over large areas, without proper fixturing and fixture movement; (*3*) the source geometry can change during the deposition, thus changing the deposition rate; (*4*) for vaporizing large amounts of material, material often must be added to the source during the deposition run; and (*5*) few processing variables are available for film property control. In addition, the source material utilization may be poor, and high radiant heat loads can exist in the deposition chamber.

Vacuum deposition is used to form optical interference coatings, reflecting coatings, decorative coatings, permeation barrier films on flexible packaging materials, electrically conducting films, and corrosion protective coatings. The sophisticated vacuum deposition technique of molecular beam epitaxy (MBE) is used in the semiconductor industry to form single-crystal films of elemental and compound semiconductors.

Sputter Deposition. Sputter deposition, often called sputtering, is the deposition of atoms vaporized from a surface (sputtering target) by physical sputtering (5,6). Physical sputtering is a nonthermal vaporization process where surface atoms are physically ejected by momentum transfer from an atomic-sized energetic bombarding particle. This particle is usually a gaseous ion accelerated from a plasma or an ion gun (see PLASMA TECHNOLOGY). In the collision of atomic-sized particles, the energy transferred from the incident particle, E_i, to the target particle, E_t, is given by

$$E_t/E_i = \left[(4M_i M_t)/(M_i + M_t)^2 \right] \cos^2 \theta \qquad (2)$$

where M_i is mass of incident energetic particle; M_t, mass of target atom; and θ, the angle of impact. The maximum energy is transferred when the masses are equal and the collision is along a line joining their centers.

The surface atom that is struck can strike other atoms in the near-surface region, resulting in a collision cascade. Multiple collisions in the near-surface region can result in some momentum being directed back toward the surface. If the energy attained by a surface atom that is struck from below is sufficient, this atom can be physically ejected from the surface, ie, sputtered. Figure 7 depicts the processes that occur in the near-surface region of a bombarded surface. Most of the energy transferred by the bombarding particle appears as heat in the near-surface region. Moreover, many of the bombarding particles can be reflected from the surface as high energy reflected neutrals.

At all but the lowest bombarding energies, the flux of atoms that are sputtered from the surface leaves the surface with a cosine distribution (Fig. 6). The sputtered atoms have kinetic energies higher than those of thermally vaporized atoms, as well as a high energy tail in the energy distribution that can be several tens of eV.

The sputtering yield is the number of surface atoms that are sputtered for each incident energetic bombarding particle. The sputtering yield depends on

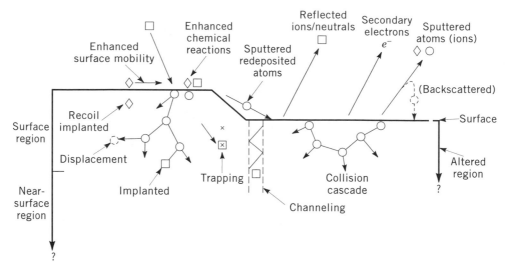

Fig. 7. Bombardment processes at the surface and in the near-surface region of a sputtering target, where □ represents the energetic particle used for bombarding the surface; ◇, an adsorbed surface species; ○, atoms; and ×, lattice defects.

the bombarding particle energy, relative masses of the bombarding and target species, the angle of incidence of the bombarding species, and the chemical bonding of the surface atoms. The most common inert gas used for sputter deposition is argon, $M_{Ar} = 40$. Figure 8 shows the sputtering yields from several materials sputtered with argon ions at various energies at normal incidence. As the angle of incidence of the bombarding particles becomes off-normal, the sputtering yield can increase two to three times, up to a point where the bombarding particle transfers little momentum because it is reflected from the surface and the sputtering yield drops off rapidly. The apparent sputtering yield can be affected by the surface topography because, in sputtering a rough surface, some of the sputtered particle are forward-sputtered and redeposited on the surface. At pressures above about 6.65 Pa (50 mtorr), the apparent sputtering yield is decreased by backscattering of some of the sputtered material to the target surface.

In the sputtering process, each surface atomic layer is removed consecutively. If there is no diffusion in the target, the composition of the vapor flux leaving the surface is the same as the composition of the bulk of the material being sputtered, even though the composition of the surface may be different from the bulk. This allows the sputter deposition of alloy compositions, which can not be thermally vaporized as the alloy because of the greatly differing vapor pressures of the alloy constituents.

Often the surface to be sputtered has a surface layer composed of a reacted material such as an oxide. Because the chemical bonding in compounds is generally stronger than that in elements, the sputtering yield may be low until the surface layer is removed. Also, if reactive gases are present, these can poison the target surface by continuously forming compounds on the target surface, thus giving low sputtering yields.

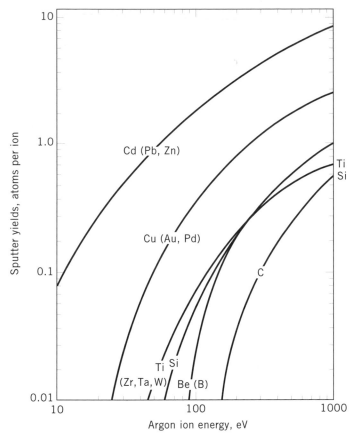

Fig. 8. Calculated sputtering yield of several materials bombarded with argon ions at various energy levels. The materials listed parenthetically also have similar sputtering yield curves.

When there are no gas-phase collisions, ie, at low gas pressure, particles that are sputtered or reflected from the sputtering target surface travel in a line-of-sight path to condense on or bombard the substrate. If the gas pressure is higher, eg, 0.6–4 Pa (5–30 mtorr), collisions in the gas phase take place, which reduce the energy of the reflected neutrals and sputtered atoms, scattering them from a line-of-sight path. If there are enough collisions, the energetic particles are thermalized to the energy (temperature) of the ambient gas. Energetic gaseous species bombarding the surface of the growing film and high energy depositing particles can affect the film formation process and the properties of the deposited film material (4). Thus it makes a difference in film properties whether the sputter deposition is done at a low, <0.6 Pa (<5 mtorr), or at a higher, >0.6 Pa (>5 mtorr), gas pressure.

The most simple plasma configuration for sputtering is d-c diode sputtering, where a high negative d-c voltage is applied to the surface of an electrically conductive material to be deposited in a low pressure inert gas. A plasma is formed, which fills the container, and positive ions are accelerated to the surface. In the d-c diode glow discharge, the electric field strength is high near the cathode and most of the applied voltage is dropped across a region called the cathode dark space near the cathode surface. This region is the primary region of ionization by electron atom collision. In the cathode dark space region, ions are accelerated from the plasma to impinge on the target surface with a high kinetic energy. This ion bombardment causes the ejection of secondary electrons, which are accelerated away from the cathode and cause ionization and atomic excitation by electron-atom collision. At equilibrium, enough electrons are created to cause sufficient ionization to sustain the discharge. The rest of the space between the cathode and the anode is filled with a plasma where there is little potential gradient. This is called the plasma region. The d-c diode plasma is generally weakly ionized, consisting perhaps of one ion to 10^4 neutrals, and has an average particle temperature of several electron volts in the plasma region.

Typically a d-c diode argon-sputtering plasma is operated at 1–3 Pa (10–20 mtorr) argon gas pressure. At this pressure, the cathode dark space has a width of about 1 cm. If a ground surface comes within a cathode dark space distance of the cathode, the plasma is extinguished between the two surfaces. This means that ground shields can be used to localize the plasma discharge to regions of the cathode where sputtering is desired.

The equipotential field lines around the cathode surface are conformal over flat or gently curved surfaces, but curve sharply around edges and underneath ground shields. Ions are accelerated normal to the field lines, which gives uniform bombardment and sputter-erosion over most of a planar target surface. Where there is appreciable field curvature, however, focusing of the ions can give higher bombardment fluxes.

In the d-c diode configuration, the ions that impinge on the target surface do not have the full cathode fall potential, because the gas pressure is high enough to give charge exchange collisions and momentum transfer collisions between the accelerating ions and the residual gas neutrals. This creates a flux of ions and high energy neutrals having a spectrum of energies that impinge on the cathode (target) surface. The higher the gas pressure, the less is the mean energy of particles that impinge on the target. Because the mean energy is much less than the applied voltage, 2000–5000-volts d-c is typically used for d-c sputtering, even though 500-eV particles are quite effective for sputtering. At typical d-c sputtering pressures, energetic particles reflected or sputtered from the target surface are thermalized to lower energies before striking the substrate surface. However, the electrons that are accelerated away from the cathode can attain high energies and bombard surfaces, thus giving electron heating that may be undesirable. Typically, d-c diode sputtering discharges are controlled by regulating the sputtering gas pressure, target voltage, and the target power (W/cm^2). Because most of the bombarding energy is given up as heat, the sputtering target must be actively cooled.

Reactive gases and vapors in a d-c plasma become activated by excitation and dissociation or fragmentation of molecules to form atoms, new chemical species, radicals, and excited species. These reactive species can poison the sputtering target surface. The formation of an electrically conductive compound, such as TiN, on the target surface reduces the sputtering yield significantly, whereas the formation of an electrically insulating layer, such as TiO_2, causes surface charge buildup and arcing. Activation of contaminant species in the plasma also increases the contamination rate of the deposited film material over what would occur if the contaminants were not activated.

The simple d-c diode sputtering configuration has the advantage that (1) large areas can be sputtered rather uniformly over long periods of time; (2) the target can be made conformal with the substrate; (3) the target-to-substrate distance can be made small compared to thermal vaporization; and (4) the target material is utilized effectively. Disadvantages of d-c diode sputter deposition include: (1) the rather low sputtering rate; (2) target poisoning by reactive gases; (3) surface heating from the electrons accelerated away from the target; and (4) limited choice of sputtering targets to only electrically conductive materials.

Simple d-c diode sputtering can best be used to deposit relatively nonreactive metals such as gold, copper, and silver. Most of these materials can also be deposited by thermal evaporation. However, this type of d-c sputter deposition has not been widely used. The advent of the use of magnetic fields to confine the plasma near the target surface, ie, magnetron configuration, to increase the plasma density as well as the sputtering rate has allowed d-c diode magnetron sputtering to be used to deposit elements and compound materials at high rates. In the reactive sputter deposition of compound materials such as TiN, the target is eroded so rapidly that the target is not significantly poisoned by the reactive gas used to form the compound film material.

Magnetron sputtering configurations use a magnetic field, $\vec{B}$, of a few hundredths of tesla having a component parallel to the sputtering target surface to confine the electrons near the surface. When an electron is ejected from the target surface, it is accelerated away from the surface by the electric field, $\vec{E}$, but is forced to spiral around the magnetic field lines. The $\vec{E} \times \vec{B}$ force causes the electron also to move normal to the $\vec{E} \times \vec{B}$ plane. If the magnetic field is arranged appropriately, the electrons form a closed-path circulating current on and near the target surface. This closed path can be easily produced on a planar surface or any surface of revolution such as the inside or outside of a cylinder, a cone, a hemisphere, etc. Usually permanent magnets are used to form the magnetic fields, but electromagnets can also be used.

By confining the electrons near the surface and by having a high electron density, the d-c gas discharge can be sustained at low pressures where collision and charge exchange collision probabilities are low. This allows the applied cathode potential to be low (<1000 V), having high energy, high current ion flux that gives a high sputtering yield and high sputtering rates. The various magnetron sputtering configurations are the most widely used sputter deposition configurations. Figure 9 shows a commonly used configuration, the planar magnetron configuration.

In the magnetron configurations, the sputter erosion is nonuniform. In the case of the planar magnetron, the erosion pattern looks like an elongated racetrack which, for a long narrow target, has an erosion pattern that gives two parallel line-sources of vaporized material. This means that the deposition pattern is nonuniform and film thickness uniformity must be accomplished by substrate (or target) movement. In the case of the planar magnetron, this is often done by passing flat substrates on a pallet fixture over the two line sources. The nonuniform erosion also means that the target material utilization is poor.

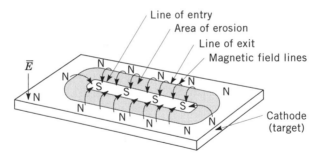

Fig. 9.   Direct-current diode planar magnetron sputtering configuration.

Utilization is often improved by shaping the magnetic field, moving the magnets during sputtering, or moving the target material over the magnets.

Confining the plasma to near the target surface also means that the plasma does not fill the space between the source and the substrate and is not available for plasma activation of reactive species for reactive sputter deposition. A plasma can be formed near the substrate by utilizing a radio frequency (r-f) potential on the sputtering target along with the d-c potential, by establishing an auxiliary discharge near the substrate, or by using an unbalanced magnetron configuration where some of the electrons ejected from the magnetron sputtering target surface are allowed to escape by having a portion of the magnetic field normal to the target surface as shown in Figure 10. This can be done either by using differing strengths for the north and south poles of the magnets in the target, placing magnetic materials in the target region, employing auxiliary

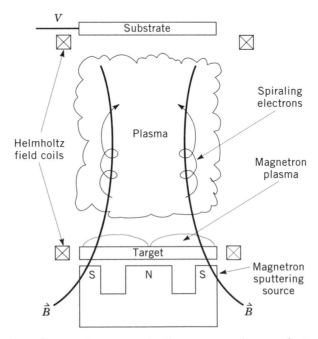

Fig. 10.   Unbalanced magnetron vaporization source using an electromagnetic field.

magnetic fields, or by using dual targets facing each other that have their north and south poles opposite one another.

In reactive sputter deposition (6,7) where reactive gases or gas mixtures are used, the gas pressure and composition are important. During film growth, new film material is continuously being deposited and the deposited film material must react rapidly with the ambient to form the desired compound (4). Too high a reactive gas pressure can result in poisoning the target surface with an associated reduction in sputtering rate. Too low a gas pressure can result in not enough reaction with the depositing film, thus resulting in an undesirable film composition. Unbalanced magnetron sputtering is particularly useful in reactive sputter deposition because it activates the reactive species in the vicinity of the substrate. However, the plasma generated in various regions near the target by the unbalanced configurations is generally nonuniform in plasma properties because of the nonuniform escape of the electrons from the target region. Often, when using reactive sputter deposition, mixtures of inert and reactive gases are employed because the mass of the reactive species is rather low, eg, $M_N = 14$, thus giving low sputtering yields.

When coating three-dimensional parts, the magnetron and unbalanced magnetron targets are often arranged so that the parts to be coated are passed between opposing targets. This allows deposition on all sides of the part even though the average angle of incidence of the depositing material varies over the surface of the part. This variation in angle of incidence can cause a variation in the properties of the deposited film (4). To minimize this variation, the part can be rotated in front of the magnetron target(s) using appropriate fixturing. When using facing unbalanced magnetron targets, it is best to have the north magnetic pole of one target be opposite the south magnetic pole of the other target.

By using an r-f potential on the sputtering target, electrically insulating surfaces, such as oxides, can be sputtered (r-f-sputtering) by periodically neutralizing any charge buildup on the surface. A typical commercial frequency used is 13.56 MHz. Because the thermal conductivity of electrical insulators is generally low, uneven heating of the target can cause thermal stresses and fracture the target. Quartz, SiO_2, which has a low thermal expansion coefficient and thus does not generate high thermal stresses on uneven heating, is one of the few materials routinely deposited by r-f-sputter deposition.

Principal sputter deposition processing variables for a specific material for nonreactive sputter deposition include sputtering gas; sputtering pressure; substrate temperature; deposition rate; target voltage; target power; deposition geometry, ie, angle of incidence of depositing flux; and level of gaseous contamination in the system. For reactive sputter deposition, the variables include those listed, plus partial pressures of the gases and vapors, availability of reactive species over the surface of the depositing material, and chemical reactivity (activation) of the reactive species.

Advantages of sputter deposition are as follows. Elements, alloys, and compounds can be deposited. The sputtering target provides a long-lived vaporization source having a stable geometry. The sputtering target can be positioned close to the substrate. In some configurations, eg, simple d-c diode, the sputtering target provides a large-area vaporization source that can be of any shape; in other configurations, the sputtering source can be a defined shape, such as a line (planar magnetron); and in still others, the reactive deposition can be easily accomplished using reactive gaseous species that are activated in the plasma, ie, reactive sputter deposition.

Disadvantages of sputter deposition include the following. Sputter vaporization rates are low compared to those that can be attained in thermal vaporization. In many configurations, the deposition flux distribution is nonuniform and requires fixturing to randomize the position of the substrates in order to obtain films of uniform thickness and isotropic properties. Sputtering targets are often expensive and material utilization can be poor. Most of the kinetic energy incident on the target appears as heat, which must be removed. Generally, the pumping speed (throughput) of the system is reduced during sputtering, and gaseous contaminants, released during processing, are not easily removed from the system. Furthermore, these contaminants are activated in the plasma. Finally, in some configurations, radiation and bombardment from the plasma or sputtering target can degrade the substrate material.

Sputter deposition is widely used to deposit thin-film metallization on semiconductor material, energy-conserving coatings on architectural glass, transparent conductive coatings on glass, reflective coatings on compact disks, magnetic films, dry film lubricants, wear-resistant coatings, and decorative coatings.

Ion Plating. In the early 1960s it was shown that controlled concurrent energetic bombardment of the depositing film material by particles of atomic or molecular size (atomic peening) could be used to modify and tailor the properties of the deposited film material. This form of PVD is called ion plating (8,9). In ion plating, the source of material to be deposited can be evaporation, sputtering, arc erosion, laser ablation, or other vaporization source. The energetic particles used for bombardment are usually ions of an inert or reactive gas. However, when using an arc erosion source, a high percentage of the vaporized materials is ionized and ions of the film material (film-ions) can be accelerated and used to bombard the growing film. Ion plating can be produced in a plasma environment where ions for bombardment are extracted from the plasma, or it may be produced in a vacuum environment where ions for bombardment are formed in a separate ion gun. The latter ion plating configuration is often called ion–beam–assisted deposition (IBAD). Figure 11 shows examples of plasma-based and vacuum-based ion plating configurations.

The most common form of ion plating is the plasma-based process, in which the substrate and/or its fixture is an electrode used to generate a d-c or an r-f plasma in contact with the surface being coated. If an elemental or alloy material is being deposited, the plasma can be of an inert gas, usually argon. In reactive ion plating, the plasma, generally a mixture of inert and reactive gases, provides ions of reactive species such as nitrogen or oxygen that are accelerated to the surface to form compounds such as oxides, nitrides, carbides, or carbonitrides. Typically, in plasma-based ion plating, the substrate fixture is the cathode of the d-c circuit. However, the plasma can also be formed independent of the substrate and ions accelerated from the plasma to the surface of the growing film.

Concurrent or periodic bombardment during film growth can modify the film properties by affecting the nucleation and surface mobility of the depositing atoms, densifying the film by compaction or atomic peening, introducing significant thermal energy directly into the substrate surface region, and sputtering

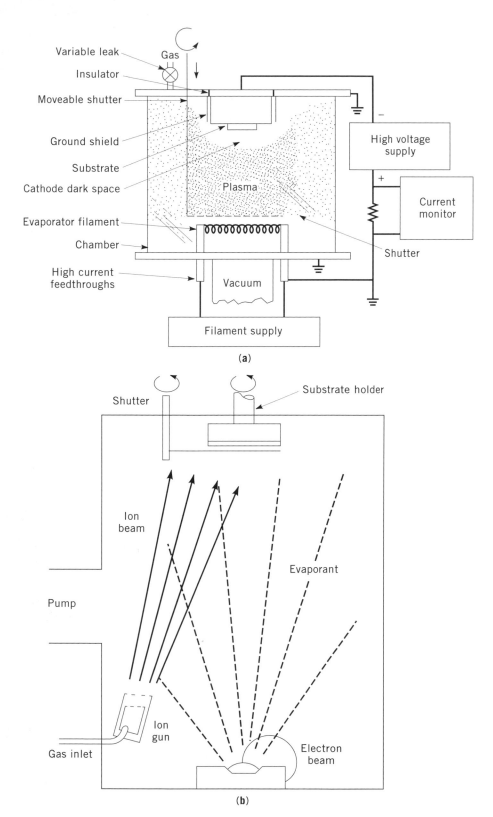

Fig. 11. Ion plating configurations: (**a**) plasma-based, where the substrate fixture is the cathode of the d-c circuit, and (**b**) vacuum-based (IBAD).

and redeposition of the film material (4). Energetic particle bombardment can also introduce compressive film stress by recoil implantation of the near-surface atoms into the lattice structure. In the case of reactive deposition in plasma-based ion plating or when using ion sources that employ a plasma of reactive species, the plasma also activates the reactive species, which enhances the kinetics of chemical reactions, by bombardment-enhanced chemical reactions and preferential sputtering, which can desorbing unreacted species (4,10).

For low (<50 eV) energy ion bombardment, momentum transfer is not sufficient to displace the film atoms and produce atomic peening. However, it does increase the surface mobility of the depositing atom and affect nucleation and film growth on the substrate surface. If the energy of the bombarding species is from 50 to about 300 eV, atomic peening occurs in the near-surface region and the film is densified. For higher energies, the bombarding species are incorporated into the film unless the substrate temperature is high. This gas incorporation can result in void formation and microporosity in the film. In cases of low (<50 eV) energy bombardment, the ratio of the bombarding flux rate to the deposition rate may be greater than 5 to 1. For energies greater than 50 eV, the flux ratio may be less. For example, at a deposition rate of 3 nm/s, the ion flux of 200-eV ions should be at least 10^{15} ions/(cm^2·s) or an ion current (singly charged ions) of about 0.1 mA/cm^2 to give significant densification. At these ion energies and fluxes, an appreciable portion (10–30%) of the depositing atoms is sputtered from the growing film surface, giving a growth rate less than the deposition rate.

In plasma-based ion plating, the negative potential on the substrate surface can be generated by applying either a continuous d-c potential to an electrically conductive surface, an r-f or a pulsed d-c (10–50 kHz) potential to an insulating surface, a combination of d-c and r-f bias, or by inducing a self-bias on an electrically insulating or electrically floating surface. When using an ion gun, the high energy ions are extracted from the source at a high energy and can be injected into a fieldfree region so that a negative potential does not have to be applied to the substrate to achieve high energy bombardment of the surface.

Concurrent bombardment during film growth affects film properties such as the film–substrate adhesion, density, surface area, porosity, surface coverage, residual film stress, index of refraction, and electrical resistivity. In reactive ion plating, the use of concurrent bombardment allows the deposition of stoichiometric, high density films of compounds such as TiN, ZrN, and ZrO$_2$ at low substrate temperatures.

A special case of ion plating uses arc vaporization of materials as the source of depositing material. In an arc, the electrode material is vaporized by the passage of a high current, low voltage current. In a vacuum, using closely spaced electrodes, the vaporized material is almost completely ionized, often by using a multiple charge. In the space between the electrodes, a positive space charge is developed as a result of the low mobility of the positive ions with respect to the electrons. The positive ions of electrode material are accelerated away from this space charge region and subsequently attain high kinetic energies. In a gaseous arc discharge, both the vaporized electrode material and the gas species are ionized. In the gas, the ions of the electrode material are thermalized to the ambient gas temperature, and the gas ions and film ions are accelerated to the substrate by a negative potential on the substrate.

A disadvantage of the arc vaporization source is the formation of liquid droplets, called macros, during the arcing process. These droplets deposit on the substrate, giving undesirable bumps on the surface. Such bumps can be dislodged, but will leave pinholes in the film. These droplets can be avoided by bending the plasma to a surface out of the line of sight of the source, but at a loss of deposition rate. Figure 12 depicts the vacuum arc source and the gaseous cathodic arc source.

The principal processing variables for ion plating for nonreactive deposition include substrate temperature; deposition rate; species, flux, and energy distribution of bombarding species; bombardment uniformity over the substrate surface; and gaseous contamination level. For reactive ion plating, the following are also necessary: partial pressures of the gases and vapors, availability of reactive species, chemical reactivity of the reactive species, and adsorption of reactive species on the surface.

The advantages of ion plating are as follows. Controlled bombardment can be used to modify film properties such as adhesion, film density, residual film stress, and optical properties. Surface coverage can be improved over vacuum evaporation and sputter deposition owing to gas scattering and sputtering/redeposition effects. Film properties are less dependent on the angle of incidence of the flux of depositing material than with sputter deposition and vacuum deposition owing to gas scattering, sputtering/redeposition, and atomic peening effects. In reactive ion plating, the plasma can be used to activate reactive species, and create new chemical species that are more readily adsorbed so as to aid in the reactive deposition process. Moreover, in reactive ion

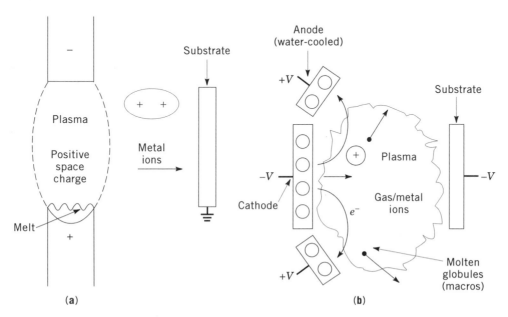

Fig. 12. Arc vaporization source configurations: (**a**) vacuum arc/molten anode, and (**b**) cathodic arc.

plating, bombardment can be used to improve the chemical composition of the film material by bombardment-enhanced chemical reactions, which lead to increased reaction probability, and preferential sputtering of unreacted species from the growing surface.

There are also disadvantages to ion plating. For instance, it is often difficult to obtain uniform ion bombardment over the substrate surface. This leads to film property variations over the surface. Substrate heating from bombardment can be excessive, and, under certain conditions, the bombarding gas may be incorporated into the growing film. Under other conditions, however, excessive residual compressive film stress can be generated by the bombardment. Furthermore, in plasma-based ion plating, the system pumping speed is limited, thus increasing film contamination problems.

Ion plating is used to deposit hard coatings of compound materials, adherent metal coatings, optical coatings having high densities, and conformal coatings on complex surfaces.

Thin Films from Vapors

Chemical Vapor Deposition.   Chemical vapor deposition (CVD) or thermal CVD is the deposition of atoms or molecules from a chemical vapor precursor, which contains the film material to be deposited (11). Chemical vapor precursors include chlorides such as titanium tetrachloride, $TiCl_4$; fluorides such as tungsten hexafluoride, WF_6; hydrides such as silane, SiH_4; carbonyls such as nickel carbonyl, $Ni(CO)_4$; and many others. Decomposition of the vapor is by chemical reduction or thermal decomposition. The reduction is normally accomplished by hydrogen at an elevated temperature. Some vapors, such as the carbonyls, can be thermally decomposed at relatively low temperatures. The deposited material may react with gaseous species such as oxygen, with a hydrocarbon gas such as methane or ammonia, or with a codeposited species to give compounds such as oxides, nitrides, carbides, and borides. These reactions are called synthesis reactions. CVD has numerous other names and adjectives. Examples include vapor–phase epitaxy (VPE), which takes place when CVD is used to deposit single-crystal films; metalorganic CVD (MOCVD), when the precursor gas is a metalorganic species; plasma-enhanced CVD (PECVD), when a plasma is used to induce or enhance decomposition and reaction; and low pressure CVD (LPCVD), when the reaction chamber pressure is less than ambient.

Some CVD chemical reactions are

Decomposition reactions	$Ni(CO)_4$ (g) $\rightarrow$ Ni (s) + 4 CO (g)
	SiH_4 (g) $\rightarrow$ Si (s) + 2 H_2 (g)
	$Al(CH_3)_3$ (g) $\rightarrow$ Al (s) + C_xH_y (g)
	B_2H_6 (g) $\rightarrow$ 2 B (s) + 3 H_2 (g)
Reduction reactions	WF_6 (g) + 3 H_2 (g) $\rightarrow$ W (s) + 6 HF (g)
	$SiCl_4$ (g) + 2 H_2 (g) $\rightarrow$ Si (s) + 4 HCl (g)
	2 WF_6 (g) + Si (substrate) $\rightarrow$ 2 W (s) + 3 SiF_4
Synthesis reactions	$TiCl_4$ (g) + CH_4 (g) $\rightarrow$ TiC (s) + 4 HCl (g)

$$4 \text{ BCl}_3 \text{ (g)} + \text{CCl}_4 \text{ (g)} + 8 \text{ H}_2 \text{ (g)} \rightarrow \text{B}_4\text{C (s)} + 16 \text{ HCl (g)}$$

$$3 \text{ SiH}_4 \text{ (g)} + 4 \text{ NH}_3 \text{ (g)} \rightarrow \text{Si}_3\text{N}_4 \text{ (s)} + 12 \text{ H}_2$$

$$\text{SiH}_4 \text{ (g)} + \text{O}_2 \text{ (g)} \rightarrow \text{SiO}_2 + 2 \text{ H}_2 \text{ (g)}$$

$$\text{SiCl}_4 \text{ (g)} + \text{CH}_4 \text{ (g)} \rightarrow \text{SiC (s)} + 4 \text{ HCl (g)}$$

$$2 \text{ Ga (g)} + 2 \text{ AsCl}_3 \text{ (g)} + 3 \text{ H}_2 \text{ (g)} \rightarrow 2 \text{ GaAs (s)} + 6 \text{ HCl (g)}$$

$$12 \text{ WF}_6 \text{ (g)} + \text{C}_6\text{H}_6 \text{ (g)} + 33 \text{ H}_2 \text{ (g)} \rightarrow 6 \text{ W}_2\text{C (s)} + 72 \text{ HF (g)}$$

$$12 \text{ TiCl}_4 \text{ (g)} + 12 \text{ NH}_3 \text{ (g)} + 6 \text{ H}_2 \rightarrow 12 \text{ TiN (s)} + 48 \text{ HCl (g)}$$

$$2 \text{ BCl}_3 \text{ (g)} + \text{TiCl}_4 \text{ (g)} + 5 \text{ H}_2 \rightarrow \text{TiB}_2 \text{ (s)} + 10 \text{ HCl (g)}$$

$$\text{Al}_2\text{Cl}_6 \text{ (g)} + 3 \text{ CO}_2 \text{ (g)} + 3 \text{ H}_2 \text{ (g)} \rightarrow \text{Al}_2\text{O}_3 \text{ (s)} + 3 \text{ CO (g)} + 6 \text{ HCl (g)}$$

CVD reactions are most often produced at ambient pressure in a freely flowing system. The gas flow, mixing, and stratification in the reactor chamber can be important to the deposition process. CVD can also be performed at low pressures (LPCVD) and in ultrahigh vacuum (UHVCVD) where the gas flow is molecular. The gas flow in a CVD reactor is very sensitive to reactor design, fixturing, substrate geometry, and the number of substrates in the reactor, ie, reactor loading. Flow uniformity is a particularly important deposition parameter in VPE and MOCVD.

The CVD process is accomplished using either a hot-wall or a cold-wall reactor (Fig. 13). In the former, the whole chamber is heated and thus a large volume of processing gases is heated as well as the substrate. In the latter, the substrate or substrate fixture is heated, often by inductive heating. This heats the gas locally.

The gas flow over the substrate surface establishes a boundary layer across which precursor species must diffuse in order to reach the surface and deposit. In the cold-wall reactor configuration, the boundary layer defines the temperature gradient in the vapor in the vicinity of the substrate. This boundary layer can

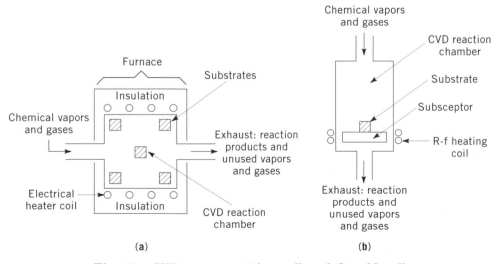

Fig. 13. CVD reactors: (**a**) hot-wall, and (**b**) cold-wall.

vary in thickness and turbulence, depending on the direction of gas flow. Direct impingement of the gas on the surface reduces the boundary layer thickness and increases the temperature gradient, whereas stagnant flow regions give much thicker boundary layers.

Each heating technique has its advantages and disadvantages, and changing from one technique to another may involve significant changes in the process variables. The cold-wall reactor is most often used in small-size systems. The hot-wall reactor, by contrast, is most often used in large-volume production reactors.

CVD processing can be accompanied by volatile hot-reaction by-products such as HCl or HF, which, along with unused precursor gases, must be removed from the exhaust gas stream. This is done by scrubbing the chemicals from the gas using water to dissolve soluble products or by burning the precursor gases to form oxides.

In many cases, the deposited material can retain some of the original chemical constituents, such as hydrogen in silicon from the deposition from silane, or chlorine in tungsten from the deposition from WCl_6. This can be beneficial or detrimental. For example, the retention of hydrogen in silicon allows the deposition of amorphous silicon, a-Si:H, which is used in solar cells, but the retention of chlorine in tungsten is detrimental to subsequent fusion welding of the tungsten.

The gases used in the CVD reactor may be either commercially available gases in tanks, such as Ar, N_2, WF_6, SiH_4, B_2H_6, H_2, and NH_3; liquids such as chlorides and carbonyls; or solids such as Mo carbonyl, which has a vapor pressure of 10 Pa (75 mtorr) at 20°C and decomposes at >150°C. Vapor may also come from reactive-bed sources where a flowing halide, such as chlorine, reacts with a hot-bed material, such as chromium or tantalum, to give a gaseous species.

Vapors from liquids can be put into the gas stream by bubbling the hydrogen or a carrier gas through the liquid or by using a hot surface to vaporize the liquid into the gas stream. Liquid precursors are generally metered onto a hot surface using a peristaltic pump and the gas handling system is kept hot to keep the material vaporized. In some cases, the vapor above a liquid may be used as the gas source.

Reactive-bed sources use heated solid materials, eg, chips and shavings, over which a reactive gas flows. The reaction produces a volatile gaseous species that can then be used as the precursor gas, eg,

$$Cr \ (s) + 2 \ Cl \ (g) \longrightarrow CrCl_2 \ (g)$$

By controlling the reaction-bed parameters, the stoichiometry of the resulting gas can be controlled. For example, $TiCl_3$ or $TiCl_4$ can be generated in the reaction bed. The $TiCl_3$, along with NH_3, can then be used to deposit TiN at 450–600°C at low (0.3–2.0 kPa (2–15 torr)) pressures to form hard coatings on tool steel without degrading the hardness of the steel. The $TiCl_4$ + NH_3 reaction requires several hundred degrees higher temperature for the same quality film.

The morphology, composition, crystalline structure, defect concentration, and properties of CVD-deposited material depends on a number of factors. An important variable in the CVD reaction is the effect of vapor supersaturation over the substrate surface and the substrate temperature. At low supersaturations, which also give a low deposition rate, nuclei initiate on isolated sites and grow over the surface, giving a high density film. At high temperatures and low supersaturations of the vapor, epitaxial growth (oriented overgrowth) can be obtained on appropriate substrates. This vapor-phase epitaxial (VPE) growth is used to grow doped layers of semiconductors, eg, Si doped with B on Si. At intermediate concentrations, a nodular growth structure may form. Figure 14 shows the nodular and columnar growth of CVD-TiC on graphite. At high supersaturations, the decomposition gives whiskers and dendritic structures having a low film density. In the extreme case, decomposition can occur in the gas phase and gas-phase nucleation produces ultrafine particles or soot. These particles are then swept through the system and no deposit occurs or the particles may deposit on a surface, giving a very low density deposit. Carrier gases are used to dilute the gas and thus change the concentration of the precursor gases over the substrate surface.

Many materials deposited by CVD have a high elastic modulus and a low fracture toughness and are therefore affected by residual film stresses. Film stress arises owing to the manner of growth and the coefficient of expansion mismatch between the substrate and film material (4). In many CVD processes, high temperatures are used. This restricts the substrate-coating material combinations to ones where the coefficient of thermal expansions can be matched.

Fig. 14. Titanium carbide coating on graphite showing the nodular growth with a columnar morphology and the fracturing resulting from a coefficient of expansion mismatch between the coating and the substrate.

High temperatures often lead to significant reaction between the deposited material and the substrate, which can also introduce stresses.

CVD processing can be used to provide selective deposition on certain areas of a surface. Selective tungsten CVD is used to fill vias or holes selectively through silicon oxide layers in silicon-device technology. In this case, the silicon from the substrate catalyzes the reduction of tungsten hexafluoride, whereas the silicon oxide does not. Selective CVD deposition can also be accomplished using lasers or focused electron beams for local heating.

Plasmas can be used in CVD reactors to activate and partially decompose the precursor species and perhaps form new chemical species. This allows deposition at a temperature lower than thermal CVD. The process is called plasma-enhanced CVD (PECVD) (12). The plasmas are generated by direct-current, radio-frequency (r-f), or electron-cyclotron-resonance (ECR) techniques. Figure 15 shows a parallel-plate CVD reactor that uses r-f power to generate the plasma. This type of PECVD reactor is in common use in the semiconductor industry to deposit silicon nitride, Si_3N_4, and glass (PSG) encapsulating layers a few micrometers-thick at deposition rates of 5–100 nm/min.

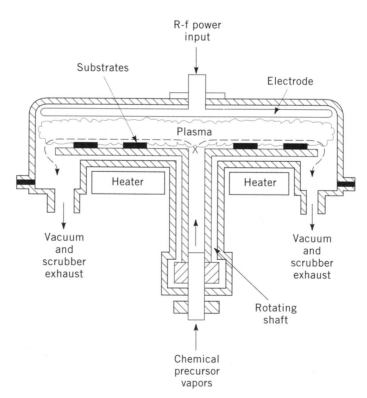

Fig. 15. Parallel-plate PECVD reactor, where typical parameters are radio frequency at 50 kHz to 13.56 MHz; temperature, 25–700°C; pressure, 13–270 Pa (100 mtorr to 2 torr); and gas flow rate, 200 cm^2/min.

Some typical PECVD reactions are

$$SiH_4 \xrightarrow{<300°C} a\text{-}Si\text{:}H$$

$$SiH_4 \xrightarrow{>500°C} Si_x$$

$$SiH_4 + O_2 \xrightarrow{>300°C} SiO_2$$

$$SiH_4 + PH_3 + N_2O \xrightarrow{>380°C} glass\ (PSG\ glass)$$

$$3\ SiH_4 + 4\ NH_3 \xrightarrow{>400°C} Si_3N_4 + 12\ H_2$$

Addition of nitric oxide, NO, to the plasma during SiO_2 deposition gives silicon oxynitride films.

The CVD mechanism under plasma conditions is complicated, being a combination of plasma processes and surface processes. The radicals, unique species, and excited species formed in the plasma can play an important role in adsorption and deposition from a gaseous precursor (10). For example, in the deposition of silicon from silane by plasma CVD, it has been proposed that disilane and trisilane are formed in the plasma, and that their adsorption on the surface, along with low energy particle bombardment, is important to the low temperature, high rate deposition of the amorphous silicon. Ion bombardment during PECVD can be an important factor and in the properties of oxides formed by PECVD. For example, it is found that low frequencies and low gas pressure give the best CVD-oxide films. This can be attributed to the increased ion bombardment during deposition (13).

In PECVD, the plasma generation region may be in the deposition chamber or precede the deposition chamber in the gas flow system. The latter configuration is called remote plasma-enhanced CVD (RPECVD). In either case, the purpose of the plasma is to give activation and partial reaction/reduction of the chemical precursor vapors so that the substrate temperature can be lowered and still obtain deposit of the same quality.

Many materials have been deposited by PECVD. Typically, the use of a plasma allows equivalent-quality films to be deposited at temperatures several hundred degrees centigrade lower than those needed for thermal CVD techniques. Often, the plasma-enhanced techniques give amorphous films and films containing incompletely decomposed precursor species such as amorphous silicon (a-Si:H) and amorphous boron (a-B:H).

Processing variables that affect the properties of the thermal CVD material include the precursor vapors being used, substrate temperature, precursor vapor temperature gradient above substrate, gas flow pattern and velocity, gas composition and pressure, vapor saturation above substrate, diffusion rate through the boundary layer, substrate material, and impurities in the gases. For PECVD, plasma uniformity, plasma properties such as ion and electron temperature and densities, and concurrent energetic particle bombardment during deposition are also important.

Safety is often a primary concern in CVD processing because of the hazardous nature of some of the gases and vapors that are used and the hot reaction products generated.

The density of CVD deposits is generally high (>99%), but dendritic or columnar growth can decrease the density. High purity films can be attained but incomplete decomposition of the precursor gases can leave residuals in the deposits particularly at low deposition temperatures. The CVD deposition process has good throwing power but the properties of the deposit often depend on deposition rate, gas flow impingement rate and direction. CVD deposition rates can range from <10 to >300 μm/h and generally have no restriction on thickness. The thickness of a CVD deposit is determined by the processing parameters and can range from very thin films to thick coatings to free-standing shapes.

Applications of CVD thin films exist in the semiconductor industry for semiconductor materials, eg, Si and Ge epitaxy, 3-5 compounds (GaAs), doped (As, P, B) epitaxial silicon (epi-silicon), polycrystalline silicon (polysilicon), and amorphous silicon for solar cells. Semiconductor metallization of tungsten and semiconductor insulators and encapsulation of Si_3N_4, phosphate silicate glass, SiO_2, and silicon oxynitrides are carried out by CVD. Wear- and erosion-resistant coatings of TiC, TiN, Al_2O_3, $MoSi_2$, SiC, and TiB_2 are also formed using CVD. The inclusion of dopants into the CVD-deposited films in semiconductor-device fabrication is an important technology. Arsine, phosgene, and diborane are common dopant precursors.

A unique CVD technique is used to deposit polycrystalline diamond films by passing a precursor gas mixture (1.3–13 kPa) (10–100 torr) of H_2 plus ~5% CH_4 over a hot (~2200°C) tungsten or tantalum filament (14). The hot filament is carburized and then dissociates the CH_4 to carbon plus other species and the H_2 to hydrogen radicals. The activated hydrogen preferentially reacts with and etches the deposited carbon that has the sp^2 bonding (graphitic-type bonding), as opposed to the sp^3-bonded carbon (diamond-type bonding), leaving predominately sp^3-bonded carbon in the film. At substrate temperatures above 600°C, the atoms arrange into the tetrahedral diamond structure, giving a polycrystalline diamond film. At lower temperatures, where the atomic mobility is less, the film may be composed of sp^3-bonded carbon but without the diamond crystal structure. The latter film is called a diamond-like carbon (DLC). In the DLC, films have many of the desirable properties of diamond. There is usually some portion of sp^2-bonded carbon in the DLC films and a high hydrogen concentration, which has a significant effect on its properties.

Plasma Polymerization. Many organic and inorganic monomers can be cross-linked in a plasma environment. Plasmas can be used to polymerize organic monomers to form thin polymer films (15). For example, a plasma is used to polymerize organosilicone thin films for protective coatings on aluminum reflector films for the automotive headlight industry. The plasma-deposited organosilicone films can be further oxidized by using an oxygen-containing plasma. Plasma-polymerized and -oxidized films have been formed using tetramethyldisiloxane (TMDSO) to form quartz-like transparent films that show good gas and moisture permeation barrier properties. The plasma-polymerized organosilicone films, which have excellent surface coverage ability, are hydrophobic, hard, and relatively pinhole-free when compared to films deposited by atomistic deposition techniques. The films are being used as clear protective top coats on optical reflective films.

Inorganic monomers can be used to plasma-deposit polymer-type films (16). At high plasma energies, the monomers are largely decomposed and can be used to form materials such as amorphous hydrogen-containing silicon films from SiH_4 for thin-film solar-cell materials.

Thin Films from Chemical Solutions

When a metal is dipped into a solution containing its own ions, some of the surface atoms dissolve and some of the ions in solution deposit. The difference in rates establishes a potential specific to the material. To measure this potential, a second electrode is needed. All electrode potentials, reported with respect to hydrogen, on platinum give the relative tendency of the material to gain or lose electrons. Table 1 gives the electrode potentials of some materials ranked in order of tendency to lose electrons or to become ionized. Metals such as lithium and zinc have positive oxidation potentials. These metals release electrons and go into solution more easily than does hydrogen. Gold dissolves less readily than does hydrogen and thus by definition has a negative potential.

Displacement Plating. Displacement or immersion plating results from the differences arising in electromotive potential between a surface and the ions in solution. In displacement plating, ions from the surface go into solution to be replaced by ion from the solution. For example, if a copper rod is placed in an acidic gold solution, the gold plates out as a porous coating on the copper. The resulting deposit is porous and must be buffed to give a dense and bright coating. Immersion plating can be used to plate zinc or tin on oxide-free aluminum, uranium, or other active metals. This is called the zincate or stannate process.

Electroplating. When ionically bonded molecules are dissolved in a solvent, some of the molecules dissociate into ions, whether the solvent is water, organic solvent, or a fused salt. A simple example is that of sulfuric acid or

Table 1. The Electromotive Series

Material	Potential, V	Material	Potential, V
$Li^0 \rightleftharpoons Li^+$	−3.045	$Co^0 \rightleftharpoons Co^{2+}$	−0.277
$Rb^0 \rightleftharpoons Rb^+$	−2.93	$Ni^0 \rightleftharpoons Ni^{2+}$	−0.250
$K^0 \rightleftharpoons K^+$	−2.924	$Sn^0 \rightleftharpoons Sn^{2+}$	−0.136
$Ba^0 \rightleftharpoons Ba^{2+}$	−2.90	$Pb^0 \rightleftharpoons Pb^{2+}$	−0.126
$Sr^0 \rightleftharpoons Sr^{2+}$	−2.90	$Fe^0 \rightleftharpoons Fe^{3+}$	−0.04
$Ca^0 \rightleftharpoons Ca^{2+}$	−2.87	$Pt/H_2 \rightleftharpoons H^+$	0.0000
$Na^0 \rightleftharpoons Na^+$	−2.715	$Sb^0 \rightleftharpoons Sb^{3+}$	0.15
$Mg^0 \rightleftharpoons Mg^{2+}$	−2.37	$Bi^0 \rightleftharpoons Bi^{3+}$	0.2
$Al^0 \rightleftharpoons Al^{3+}$	−1.67	$As^0 \rightleftharpoons As^{3+}$	0.3
$Mn^0 \rightleftharpoons Mn^{2+}$	−1.18	$Cu^0 \rightleftharpoons Cu^{2+}$	0.34
$Zn^0 \rightleftharpoons Zn^+$	−0.762	$Pt/OH \rightleftharpoons O_2$	0.40
$Cr^0 \rightleftharpoons Cr^{3+}$	−0.74	$Cu^0 \rightleftharpoons Cu^+$	0.52
$Cr^0 \rightleftharpoons Cr^{2+}$	−0.56	$Hg^0 \rightleftharpoons Hg_2^{2+}$	0.789
$Fe^0 \rightleftharpoons Fe^{2+}$	−0.441	$Ag^0 \rightleftharpoons Ag^+$	0.799
$Cd^0 \rightleftharpoons Cd^{2+}$	−0.402	$Pd^0 \rightleftharpoons Pd^{2+}$	0.987
$In^0 \rightleftharpoons In^{3+}$	−0.34	$Au^0 \rightleftharpoons Au^{3+}$	1.50
$Tl^0 \rightleftharpoons Tl^+$	−0.336	$Au^0 \rightleftharpoons Au^+$	1.68

copper sulfate in water, giving

$$H_2SO_4 \longrightarrow 2\,H^+ + SO_4^{2-} \qquad CuSO_4 \longrightarrow Cu^{2+} + SO_4^{2-}$$

The electrolyte thus formed can conduct electric current by the movement of ions under the influence of an electric field. A cell using an electrolyte as a conductor and a positive and a negative electrode is called an electrolysis cell. If a direct-current voltage is applied to a cell having inert electrode material such as platinum, the hydrogen ions (cations) migrate to the cathode where they first accept an electron and then form molecular hydrogen. The SO_4^- ions (anions) migrate to the anode, release two electrons, react with the water, and go back into solution and release molecular oxygen. The result is electrolytic decomposition of water to give hydrogen and oxygen.

If the cations in solution are condensable as a solid, such as copper, they can plate out on the cathode of the cell. As the same time, perhaps some hydrogen is also produced at the cathode. The SO_4^- can react with a copper anode material by taking it into solution to replace the lost copper ions. Thus the anode is a consumable electrode in the process.

Electroplating (qv), the deposition of metallic ions on a cathode in an electrolysis cell, is a way of depositing a limited number of materials on electrically conductive surfaces (17,18). Electroplating is often used to form coatings many micrometers thick. A thin film formed by electroplating is called a flash or strike. Figure 16 shows an elctrolytic-cell configuration for electroplating.

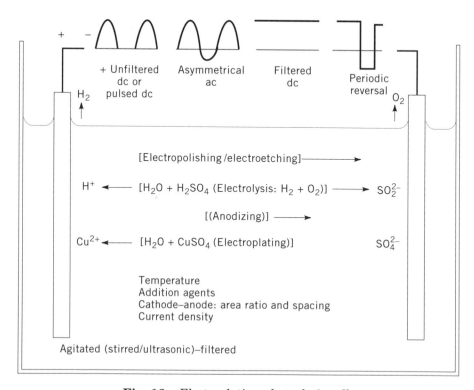

Fig. 16.   Electroplating electrolytic cell.

There are several potentials associated with the electrodes of an electrolysis cell. The principal ones are those given in Table 1. Other potentials exist in the presence of complexing agents such as cyanides, or for multivalent species such as chromium, which can exist as Cr^{2+}, Cr^{3+}, or Cr^{6+}. Another potential is generated by electrode polarization, also called the overpotential electrical double layer or Nernst diffusion layer, owing to changes in ion concentrations near the surface. This polarization varies with ion concentration, ion mobility, temperature, agitation, etc. The polarization layer is an important factor in electrode effects and determines the current densities through the electrolyte. In electroplating, an external voltage is applied to the electrolysis cell, which is greater than the sum of the electrode potentials of the two materials and the polarization voltage.

Faraday's Law of electrolysis states that the amount of chemical change, ie, amount dissolved or deposited, produced by an electric current is proportional to the quantity of electricity passed, as measured in coulombs; and that the amounts of different materials deposited or dissolved by the same quantity of electricity are proportional to their gram-equivalent weights (GEW) defined as the atomic weight divided by the valence. The weight in grams of material deposited, W, is given by

$$W = Itm/F$$

where I is the current in amperes, t the time in seconds, m the gram-equivalent weight, and F the Faraday constant, ie, the quantity of electricity that must be passed to dissolve or deposit one gram-equivalent weight of any substance, which is equal to 9.64×10^4 coulombs.

In the electrolytic cell, the release of hydrogen can compete with the deposition of metallic ions on the cathode. Current efficiency is the ratio of the metal deposited to that which would be deposited if no other reactions occur. For example, a 96% efficiency for plating means that 4% of the current passed has been used to release hydrogen or for some other reaction. In the deposition of some metals such as chromium, large amounts of hydrogen can be incorporated into the deposit and in the substrate. Hydrogen charging of high strength steel during cadmium electroplating leads to hydrogen embrittlement of the steel. A physical vapor deposition process, ie, vacuum cadmium plating, is often used to avoid this.

The deposition of ions at the cathode creates a depletion layer across which the ions must migrate in order to deposit. This layer can vary in thickness according to surface morphology. The depletion layer is more or less defined as the region where the ion concentration differs from that of the bulk solution by >1%. The layer thickness can be decreased by agitation.

In electroplating, the reactions at the cathode can be very complex. One example is the electroplating of chromium, which has an electronegativity close to that of hydrogen. There are three distinct cathodic reactions. First, when the potential is low, chromic acid is reduced from its hexavalent state to a trivalent state. Second, as the potential is made higher, hydrogen gas is formed. Third, at higher potentials, the hexavalent chromium is reduced to the metallic state through several intermediate species. At the anode, two oxidation reactions take place. One is the decomposition of water to give oxygen; the other is oxidation of the trivalent chromium ion to the hexavalent ion. Chromium plating takes place with the addition of a sulfate anion as a catalyst for the reduction of the hexavalent chromium ion. Using only the sulfate ion, the current efficiency is only about 12% and the rest goes into generating hydrogen and trivalent chromium, which is oxidized at the anode. If a mixture of sulfates and fluorides are used, the chromium deposition efficiency can be as high as 25%.

Current can be passed through the electrolyte with various waveforms. These include dc, asymmetrical ac, pulsed (on–off dc), and periodic reversal of the polarity of the electrodes, ie, periodic off-plating. A version of pulse plating uses an initial pulse of high voltage to increase nucleation density followed by lower voltage pulses. This procedure is used to give a fine grained deposit. Some of the advantages claimed for pulse plating compared to dc plating, are faster plating rates, less film porosity, high purity, smoother fine-grained deposits, reduced need for addition agents, less hydrogen evolution, and decreased stress. Pulse-periodic-reversal plating where the voltage is periodically made anodic to off-plate and remove roughness and decrease the columnar morphology often found in electrodeposits may also be used (see ELECTROPLATING).

Agents are added to the bath to modify the deposition, growth, and properties of the depositing atoms and allow the formation of deposits having a wide range of mechanical and microstructural properties. Often the purpose of the addition agents is to poison the nucleation sites and make the depositing material continuously renucleate during deposition to give a fine-grained structure and to disrupt columnar growth. Wetting agents are used to prevent hydrogen bubbles from adhering to the cathode surface and becoming incorporated into the electrodeposit.

Agitation plays an important role in the elctrodeposition process by increasing ion transport in the solution and allowing higher current densities to be obtained. It also reduces electrode polarization effects and prevents stratification of the chemicals. Hydrosonic and ultrasonic agitation are reported to yield higher current densities and thus high rate deposition; suppression of hydrogen evolution; reduced stress in the deposit; reduced porosity; increased brightness; improved adhesion; and increased hardness, especially in chromium deposits. Increasing the bath temperature of the electrolyte has similar effects.

Processing for electroplating involves surface preparation to ensure good adhesion and desirable coating properties. Etching by chemical solution or electrolytic stripping is often used to remove surface barrier layers such as oxides. After etching, the part is transferred through the rinse baths (to prevent carryover) into the plating bath while wet. Surfaces that form coherent oxides, eg, Cr, Ti, Zr, U, and Al, are difficult to plate. These surfaces can be electroplated by removing the oxide layer, displacement-plating a thin layer of tin (stannate) or zinc (zincate) on the surface, and then electrodepositing the desired material.

The ability to cover a surface and to reduce pinhole density in the film is affected by addition agents and deposition waveform. Throwing power, the ability of the depositing material to plate inside a deep, narrow recess, depends on a number of factors, including complexing of the ions, electrode polarization, current density, etc. Additives to increase throwing power are usually organic materials.

In an electrodeposit, the deposit thickness uniformity depends in part on electric field distribution, temperature uniformity, and solution availability. High field regions such as corners or points receive more deposit, thus causing a buildup of these areas. Shaped anodes may be used to control the current density to the cathode surface by controlling the electrode separation. Cathode guards (thieves, robbers) may be used on the cathode to prevent edge effects and control field distribution at the cathode. The shape and positioning of the anodes, shields, and robbers to give the desired local current density can be monitored using a current density probe in the bath.

Selective deposition may be used to minimize the use of expensive coating materials. Applications include stripe-on-strip followed by punching/stamping to form electrical contacts. Selective plating is most often achieved using masking techniques. Masking material may either be a physical shield or be painted on or applied as a tape. Masking patterns may also be generated using photolithographic techniques. In brush plating, a swab of absorbent material acts as the anode with the electrolyte in the swab. The swab is used like a brush to paint a thin coating on an area or repair a defect without having to place the part in an electrolyte. Barrel plating is used for complete coating of small parts by tumbling them in a barrel-shaped cathode having a grid structure.

Properties of electrodeposits can vary widely with plating solution and processing parameters. Plating variables are often interdependent. They include, for geometry, such variables as spacing, shape, area ratio, and field distribution; for substrate, cleaning, activation, and nucleation; for bath, addition agents, pH, temperature, agitation, impurities, composition, concentration, valence of species, and filtration; and for applied potential, voltage, waveform, current density, and ripple.

Only about 10 elements, ie, Cr, Ni, Zn, Sn, In, Ag, Cd, Au, Pb, and Rh, are commercially deposited from aqueous solutions, though alloy deposition such as Cu–Zn (brass), Cu–Sn (bronze), Pb–Sn (solder), Au–Co, Sn–Ni, and Ni–Fe (permalloy) raise this number somewhat. In addition, 10–15 other elements are electrodeposited in small-scale specialty applications. Typically, electrodeposited materials are crystalline, but amorphous metal alloys may also be deposited. One such amorphous alloy is Ni–Cr–P. In some cases, chemical compounds can be electrodeposited at the cathode. For example, black chrome and black molybdenum electrodeposits, both metal oxide particles in a metallic matrix, are used for decorative purposes and as selective solar thermal absorbers (19).

Aqueous electroplating has the advantages of low capital cost, except for pollution control, which has become a significant portion of the capital cost; low unit production cost; excellent adhesion; and a chemical-etch surface preparation step that is easily incorporated into the processing. Disadvantages are that many materials can not be electrodeposited; hydrogen can be incorporated into the substrate surface, giving hydrogen embrittlement; hydrogen can be incorporated into the deposit; bath composition is often an important process parameter and must be carefully controlled; and the costliness of pollution control. Applications of thin-film electroplates include deposition of electrical conductors, corrosion-protective films, eg, a gold flash, and decorative films.

Nonaqueous solvents can be used for electroplating. Nonaqueous electrolytes have the advantage that hydrogen is not given off during the deposition.

Moreover, the deposition takes place in a nonoxidizing environment, which allows the electrodeposition of oxygen-active materials such as aluminum from $AlCl_3$ and $LiAlH_4$ dissolved in tetrahydrofuran. Organic-based electrolytes are, however, highly flammable.

Molten or fused chloride and fluoride salts can be used to dissolve ionic materials. Electroplating from a molten salt is called fused salt electroplating or metalliding (20). The process allows the combining of electrodeposition and diffusion to give chemical conversion of surfaces to form boride, aluminide, and chromide coatings. Fluoride salts, such as 46.5 mol % LiF–11.5 mol % NaF and 42.0 mol % KF, are used at high temperatures of 700–900°C, whereas chloride salts such as the LiCl–KCl eutectic are used at lower temperatures of about 500°C. The elements that can be deposited/diffused include Ta, Nb, Cr, B, W–Co, Al, Co, Cr, Pt, Ti, Hf, Zr, Mo, Y, V, and some alloys. Aluminum can be deposited in a very pure form from a fused salt bath at 200°C. Tungsten carbide coatings can be deposited at 500°C from the ternary fluoride eutectic (Li, Na, K)F containing Na_2WO_4 and K_2CO_3. Coatings of Cr are deposited at 450°C from (Li–K)Cl plus $CrCl_2$(16 wt %).

Electroless Electrolytic Plating. In electroless or autocatalytic plating, no external voltage/current source is required (21). The voltage/current is supplied by the chemical reduction of an agent at the deposit surface. The reduction reaction must be catalyzed, and often boron or phosphorus is used as the catalyst. Materials that are commonly deposited by electroless plating (qv) are Ni, Cu, Au, Pd, Pt, Ag, Co, and Ni–Fe (permalloy). In order to initiate the electroless deposition process, a catalyst must be present on the surface. A common catalyst for electroless nickel is tin. Often an accelerator is needed to remove the protective coat on the catalysis and start the reaction.

Important bath constituents in electroless plating are metal ion concentration, catalyst, reducing agent(s), complexing agent(s), and bath stabilizer(s), along with pH adjusters. Important deposition parameters are temperature, pH, metal ion and reducer concentrations, stabilizer concentration, and trace impurities that can catalyze the decomposition of the solution. Typical reducing agents are sodium hypophosphite (for Ni and Co), formaldehyde, hydrazine and sodium borohydride (for Ni and Au).

Complexing agents, which act as buffers to help control the pH and maintain control over the free metal–salt ions available to the solution and hence the ion concentration, include citric acid, sodium citrate, and sodium acetate potassium tartrate ammonium chloride. Stabilizers, which act as catalytic inhibitors that retard the spontaneous decomposition of the bath, include fluoride compounds thiourea, sodium cyanide, and urea. Stabilizers are typically not present in amounts exceeding 10 ppm. The pH of the bath is adjusted.

In Ni–P electroless deposits, there can be as much as 10% by weight of phosphorus. The amount depends on the added complexing agents and the pH. The Ni–P deposits are fine-grained supersaturated solid solutions, which may be precipitation hardened by heat treatment to form dispersed Ni_3P particles in a nickel matrix.

Electroless deposition has the following advantages: very uniform deposits over complex surfaces in contact with the solution; generally the electroless deposits have low porosity; nonconductive substrates can be plated using suitable sensitizers and catalyzing agents; no electrical contacts are needed on the surface to be plated; and deposit contains P or B, which act as fluxing agents in subsequent brazing or can harden material by precipitation. The disadvantages of electroless deposition are that the solutions are expensive and unstable; deposition process is slow; process is sensitive to solution condition and processing; impure deposits are formed; and surfaces must be catalyzed.

The cost of the complexing and reducing agents, as well as the short shelf life of the chemicals used in electroless plating, makes this techniques generally noncompetitive with electroplating costwise. However, electroless plating is often used to form the electrically conductive surface necessary for subsequent electroplating. Its ability to deposit films on nonconductive surfaces and its good coverage over complex surfaces makes it particularly useful in some applications. The thickness of the deposit is limited only by the time that the surface is exposed to the plating solution having the correct composition.

Spray Pyrolysis. In spray pyrolysis, a chemical solution is sprayed on a hot surface where it is pyrolyzed (decomposed) to give thin films of either elements or, more commonly, compounds (22). For example, to deposit CdS, a solution of $CdCl_2$ plus NH_2CSNH_2 (thiourea) is sprayed on a hot surface. To deposit In_2O_3, $InCl_3$ is dissolved in a solvent and sprayed on a hot surface in air. Materials that can be deposited by spray pyrolysis include electrically conductive tin–oxide and indium/tin oxide (ITO), CdS, Cu–InSe$_2$, and CdSe. Spray pyrolysis is an inexpensive deposition process and can be used on large-area substrates.

Wetting. Films can be formed on surfaces by wetting the surface with a fluid containing the desired material usually dissolved in a solvent. When the solvent evaporates, the film material is left behind. The amount of material left behind depends on the thickness of the fluid layer and the solids content of the fluid. Thin fluid films can be formed by dip-coating or spin-coating and the resulting film thickness depends on the viscosity of the fluid. This is a common way to form polymer films. After the polymer is deposited, it can be cured by heat, uv-radiation, or electron-beam irradiation. This technology is important in forming the thin polymer films on semiconductor materials for photolithography. A thin organic film can be modified by post-deposition treatments. For example, a hydrocarbon film such as furfarane can be vacuum-pyrolyzed to give a carbon film on the surface.

Inorganic coatings can also be formed by wetting techniques. The sol–gel coating process can be used to form thin films of carbides, nitrides, oxides, and mixtures thereof by beginning with the correct precursor liquids (23,24) (see SOL–GEL TECHNOLOGY). For example, in forming an oxide film, the steps consists of hydrolysis of a metal alkoxide to form a fluid gel which is applied to the surface; drying the wet gel to obtain a porous dry gel, ie, a low density glass; and sintering (heating) the dry gel to get a dense glass film. An example of a sol–gel SiO_2 precursor solution is tetramethoxysilane, $Si(OCH_3)_4$, with water for hydrolysis and methanol as a solvent in a molar ratio of 1:4:4.5.

The drying and sintering of the gels is sensitive to heating rates. Fully dense sintering requires high temperatures. Combination of liquids can give glass layers that have varying compositions, such as silicon-oxy-carbide, and a wide range of coefficients of thermal expansions. Sol–gel coating techniques are used to apply antireflecting coatings on glass for optical applications. The coatings are left porous to reduce the index of refraction. Sol–gel coatings are also used in the coating of optical fibers. Typically sol–gel coatings are very thin (20–100 nm), but multilayers can be deposited and treated to build up the film thickness.

Spin-on-glass (SOG) is a technique for applying a fluid coating that is converted into a glass film and is used to planarize surfaces in semiconductor device technology. A solution, eg, polysilicates, polysiloxanes, or polysilsesquioxanes, is applied by spin-coating. The solution is then oxidized to SiO_2 by baking at temperatures of around 450°C. Unwanted crystallization, stress, cracking, adhesion, corrosion, coverage, outgassing, and particulate formation are the principal problems in SOG technology.

Reduction Reactions. Some thin films can be deposited from chemical solutions at low temperatures by immersion in a two-part solution that gives a reduction reaction. Chemical silvering of mirrors and dewar flasks is a common example (25,26). The glass surface to be silvered is cleaned thoroughly, then nucleated using a hot acidic stannous chloride solution, or by vigorous swabbing using a saturated solution of $SnCl_2$. The surface is then immediately immersed in the silvering solution, where a catalyzed chemical reduction causes silver to be deposited on the glass surface. Copper oxide, Cu_2O, films can be deposited from mixing solutions of $CuSO_4$ and sodium thiosulfate and NaOH.

Properties of Thin Films

Thin films formed by atomistic deposition techniques are unique materials that seldom have handbook properties. Properties of these thin films depend on several factors (4), including substrate surface condition, the deposition process used, details of the deposition process and system geometry, details of film growth on the substrate surface, and post-deposition processing and reactions. For some applications, such as wear resistance, the mechanical properties of the substrate is important to the functionality of the thin film. In order to have reproducible film properties, each of these factors must be controlled.

Film Adhesion. The adhesion of an inorganic thin film to a surface depends on the deformation and fracture modes associated with the failure (4). The strength of the adhesion depends on the mechanical properties of the substrate surface, fracture toughness of the interfacial material, and the applied stress. Adhesion failure can occur owing to mechanical stressing, corrosion, or diffusion of interfacial species away from the interface. The failure can be exacerbated by residual stresses in the film, a low fracture toughness of the interfacial material, or the chemical and thermal environment or species in the substrate, such as gases, that can diffuse to the interface.

Film Morphology and Density. Under many conditions, less-than-fully dense films are deposited by atomistic film deposition processes. This is usually a result of growth mode on the surface, which is controlled by geometric factors (4). For example, in the physical vapor deposition of films, submicrometer-size or larger features on a surface can cause geometrical shadowing, leading to growth of the peaks and shadowing of the valleys. This in turn leads to a columnar growth morphology, which is porous and less than fully dense, and has a high surface area.

Residual Film Stress. Invariably, atomistically deposited films have a residual stress that can be tensile or compressive in nature and that can approach the yield or fracture strength of the materials involved (27). The residual stresses can be a result of both stresses that arise from differences in coefficient of expansion of the film and substrate materials when using high temperature processing, and the growth or quenched-in stresses from the atoms not being in their lowest energy configurations when they become immobile on the substrate surface or from the incorporation of foreign species in the deposited material.

In physical vapor deposition processing, vacuum deposited films and sputter-deposited films prepared at higher (>0.6 Pa (5 mtorr)) pressures generally have tensile stresses that may be anisotropic with off-normal angle of incidence depositions. In low pressure sputter deposition and ion plating, energetic particle bombardment can give rise to high compressive film stresses resulting from the recoil displacement (atomic peening) of atoms in the near-surface region of the growing film. The residual film stress anisotropy may be very sensitive to the sputtering target configuration and gas pressure during sputter deposition.

In chemical vapor deposition processing, the principal source of residual stress is from a coefficient of expansion mismatch. One of the principal criteria for CVD processing is the matching of the coefficient of expansions of the film and substrate, which limits the possible film–substrate combinations that can be used.

In electrodeposition, the film stresses can vary with deposition parameters and be regulated by controlling the deposition parameters and the use of addition agents in the bath.

The lattice strain associated with the residual film stress represents stored energy, and this energy, along with a high concentration of lattice defects, can lead to: (1) lowering of the recrystallization temperature in crystalline materials; (2) a lowered strain point in glassy materials; (3) a high chemical etch rate; and (4) room temperature void growth in ductile materials and other such mass transport effects.

Surface Coverage. The surface-covering ability of deposition techniques is best when the materials are deposited from a vapor or from a fluid having no need for an applied voltage. The macroscopic and microscopic surface coverage of a thin film deposited by PVD techniques on a substrate surface may be improved by the use of gas scattering and concurrent bombardment during film deposition.

Electrical Properties. Generally, deposited thin films have an electrical resistivity that is higher than that of the bulk material. This is often the result of the lower density and high surface-to-volume ratio in the film. In semiconductor films, the electron mobility and lifetime can be affected by the point defect concentration, which also affects electromigration. These effects are eliminated by depositing the film at low rates, high temperatures, and under very controlled conditions, such as are found in molecular beam epitaxy and vapor-phase epitaxy.

Optical Properties. The index of refraction of a deposited material is sensitive to the film density. A lower index of refraction is found at less than bulk densities. The reflectance of a metallic surface is affected by the growth morphology of the film.

Mechanical Properties. The mechanical properties of thin films is affected by the film density and composition. In many applications, the properties of the substrate material is important to the functionality of the thin film in mechanical applications.

BIBLIOGRAPHY

"Film Deposition Techniques" in *ECT* 2nd ed., Vol. 9, pp. 186–220, by K. H. Behrndt, Bell Telephone Laboratories; in *ECT* 3rd ed., Vol. 10, pp. 247–283, by S. M. Lee, Ford Aerospace & Communications Corp.

1. D. M. Mattox, in *Surface Engineering, ASM Handbook*, Vol. 5, ASM International, 1994, p. 556.
2. R. Glang, in L. I. Maissel and R. Glang, eds., *Handbook of Thin Film Technology*, McGraw-Hill Book Co., Inc., New York, 1970, Chapt. 1.
3. H. K. Pulker, *Coatings on Glass*, Elsevier Science Publishing Co., Inc., New York, 1984, p. 170.
4. D. M. Mattox, in Ref. 1, p. 538.
5. J. H. Thornton, in R. F. Bunshah and co-workers, eds., *Deposition Technologies for Films and Coatings*, Noyes Data Corp., Park Ridge, N.J., 1982, Chapt. 5.
6. S. L. Rhode, in Ref. 1, p. 573.
7. W. D. Westwood, in S. M. Rossnagel, J. J. Cuomo, and W. D. Westwood, eds., *Handbook of Plasma Processing Technology*, Noyes Data Corp., Park Ridge, N.J., 1989, Chapt. 9.
8. D. M. Mattox, in Ref. 1, p. 582.
9. N. A. G. Ahmed, *Ion Plating Technology*, John Wiley & Sons, Inc., New York, 1987.
10. D. M. Mattox, *Appl. Surf. Sci.* **48/49**, 540 (1991).
11. H. O. Pierson, *Handbook of Chemical Vapor Deposition*, Noyes Data Corp., Park Ridge, N.J., 1993.
12. P. K. Tedlow and R. Reif, in *Surface Engineering*, Vol. 5, ASM International, Materials Park, Ohio, 1994, p. 532.
13. H. P. W. Hey, B. G. Sluijk, and D. G. Hemmes, *Solid State Tech.* **33**(4), 139 (1990).
14. R. J. Nemanich, *Ann. Rev. Mater. Sci.* **21**, 535 (1991).
15. H. Yasuda, *Plasma Polymerization*, Academic Press, Inc., New York, 1985.
16. R. D'Agnostino, ed., *Plasma Deposition, Treatment and Etching of Polymers*, Academic Press, Inc., New York, 1991.
17. J. W. Dini, *Electrodeposition*, Noyes Data Corp., Park Ridge, N.J., 1993.
18. M. E. Browning, in Ref. 12, p. 165.
19. R. R. Sowell and D. M. Mattox, *Plat. Surf. Finish*, **65**, 50 (1978).
20. N. Godshall, *J. Electrochem. Soc.* **123**, 137c (1976).
21. M. E. Browning, in Ref. 12, p. 164.
22. K. L. Chopra and co-workers, in G. Hass, M. Francombe, and J. L. Vossen, eds., *Physics of Thin Films*, Vol. 12, Academic Press, Inc., New York, 1982, p. 168.
23. C. J. Brinker and G. W. Scherer, *Sol-Gel Science*, Academic Press, Inc., New York, 1990.
24. B. J. J. Zelinski and D. R. Uhlmann, *J. Phys. Chem. Solids*, **45**, 1069–1090 (1984).
25. *National Bureau of Standards Circular No. 389*, National Bureau of Standards, Washington, D.C., 1931.
26. F. A. Lowenheim, in J. L. Vossen and W. Kern, *Thin Film Processes*, Academic Press, Inc., New York, 1975, p. 209.
27. D. W. Hoffman, *J. Vac. Sci. Tech.* **A12**(4), 953 (1994).

Donald M. Mattox
Management Plus, Inc.

MONOMOLECULAR LAYERS

Molecularly engineered materials can be fabricated from the molecular level up, and their physical properties can be both predicted and designed. Surface analytical tools enable investigations of monomolecular layers in previously unprecedented detail, leading to understanding of molecular packing and ordering. These tools also provide information to aid in understanding the relationships between the structure and properties of the individual molecule as well as of the material it forms (see SURFACE AND INTERFACE ANALYSIS).

Supramolecular assemblies are fabricated by assembling molecules that interlock in a planned, hierarchical manner, forming structures having specific desired functions. There are two principal methods. Using the Langmuir-Blodgett technique, molecular layers are formed at air–water interfaces under programmed external influence. The different kinds of monolayers are superimposed in an intelligently planned sequence, forming increasingly more complex supramolecular structures. In the self-assembly technique, layers are formed spontaneously by molecules self-organizing at a solid–liquid interface, and multilayer structures are formed only after the monolayer surface has been chemically modified.

Langmuir-Blodgett Films

The discovery that monolayer films can be transferred from the air–water interface onto a solid substrate by a simple dipping technique (1) and the subsequent report that multilayers can be built up by sequential monolayer transfer (2) opened a significant new area in science to further investigation (3,4). Studies of these multilayers led to a patent describing the use of Langmuir-Blodgett (LB) films for preparing nonreflecting glass (5).

Langmuir-Blodgett was the first technique to provide a practical route for the construction of ordered molecular assemblies. These monolayers, which provide design flexibility both at the individual molecular and at the material levels, are prepared at the water–air interface using a fully computerized trough (Fig. 1). Detailed discussions of troughs (4) and of surface pressure, π, and methods of surface pressure measurements are available (3,6).

Monolayers at the Air–Water Interface. Molecules that form monolayers at the water–air interface are called amphiphiles or surfactants (qv). Such molecules are insoluble in water. One end is hydrophilic, and therefore is preferentially immersed in the water; the other end is hydrophobic, and preferentially resides in the air, or in a nonpolar solvent. A classic example of an amphiphile is stearic acid, $C_{17}H_{35}COOH$, wherein the long hydrocarbon tail, $C_{17}H_{35}-$, is hydrophobic, and the carboxylic acid group, $-COOH$, is hydrophilic. This carboxylic group can dissociate in water to give a negatively charged ion, COO^-. Complex organic amphiphiles containing chromophores, various donor or acceptor groups, etc, can be designed and synthesized. Understanding the structure of such monolayers can be assisted by computer modeling (4) (see COMPUTER TECHNOLOGY; MOLECULAR MODELLING).

The monolayer resulting when amphiphilic molecules are introduced to the water–air interface was traditionally called a two-dimensional gas owing

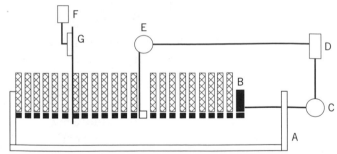

Fig. 1. A trough for deposition of monolayers on solid substrates: A, bath; B, a moving barrier; C, a motor; D, a pressure-control device; E, a surface pressure balance; F, a motor with a gearbox that lowers and raises the substrate; and G, a solid substrate. The film material (⊠) has a hydrophobic tail and (■) hydrophilic head.

to what were the expected large distances between the molecules. However, it has become quite clear that amphiphiles self-organize at the air–water interface even at relatively low surface pressures (7–10). For example, x-ray diffraction data from a monolayer of heneicosanoic acid spread on a 0.5-mM $CaCl_2$ solution at zero pressure (11) showed that once the barrier starts moving and compresses the molecules, the surface pressure, π, increases and the area per molecule, A, decreases. The surface pressure, ie, the force per unit length of the barrier (in N/m) is the difference between σ_0, the surface tension of pure water, and σ, that of the water covered with a monolayer. Where the total number of molecules and the total area that the monolayer occupies is known, the area per molecules can be calculated and a π-A isotherm constructed. This isotherm (Fig. 2), which describes surface pressure as a function of the area per molecule (3,4), is rich in information on stability of the monolayer at the water–air interface, the reorientation of molecules in the two-dimensional system, phase transitions, and conformational transformations.

As the barrier moves, the molecules are compressed, the intermolecular distance decreases, the surface pressure increases, and a phase transition may be observed in the isotherm. These phase transitions, characterized by a break in the isotherm, may vary with the subphase pH, and temperature. The first-phase transition, π_{LE} in Figure 2, is assigned to a transition from the gas to the liquid state, also known as the liquid-expanded, LE, state. In the liquid phase, the monolayer is more coherent and the molecules occupy a smaller area

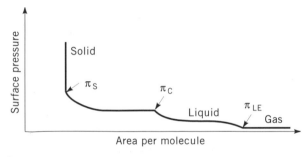

Fig. 2. A schematic π-A isotherm for a phospholipid of fatty acid (see text).

than in the gas phase, but have neither positional nor orientational order. The molecules have more degrees of freedom and gauche conformations can be found in the alkyl chains. When the barrier compresses the film further, a second phase transition, π_C, can be observed. This is from the liquid to the liquid-condensed (LC) state, where the molecules essentially are in a liquid crystalline state, and thus have some orientational order (12). The system may also undergo phase transitions between different liquid crystalline phases. The plateaus in the isotherm, where pressure does not change and area per molecule decreases, indicates an increasing orientational order.

The last phase transition is to the solid state, where molecules have both positional and orientational order. If further pressure is applied on the mono-layer, it collapses, owing to mechanical instability and a sharp decrease in the pressure is observed. This collapse-pressure depends on the temperature, the pH of the subphase, and the speed with which the barrier is moved.

It was established in 1945 that monolayers of saturated fatty acids have quite complicated phase diagrams (13). However, the observation of the differ-ent phases has become possible only much more recently owing to improvements in experimental optical techniques such as fluorescence, polarized fluorescence, and Brewster angle microscopies, and x-ray methods using synchrotron radia-tion, etc. Thus, it has become well accepted that lipid monolayer structures are not merely solid, liquid expanded, liquid condensed, etc, but that a fairly large number of phases and mesophases exist, as a variety of phase transitions be-tween them (14,15).

The Transfer of Monolayers to a Solid Substrate. Two methods of trans-fer of monolayers from the water–air interface onto a solid substrate are impor-tant. The first, and more conventional, method is the vertical deposition (16). A monolayer of amphiphiles at the water–air interface can be deposited by the displacement of a vertical plate (Fig. 3). When such a plate is moved through the monolayer at the water–air interface, the monolayer can be transferred dur-ing immersion (retraction or upstroke) or immersion (dipping or downstroke). A monolayer is usually transferred during retraction when the substrate surface is hydrophilic, and the hydrophilic head groups interact with the surface. On the other hand, if the substrate surface is hydrophobic, the monolayer is transferred in the immersion, and the hydrophobic alkyl chains interact with the surface. If the deposition process starts with a hydrophilic substrate, the surface becomes

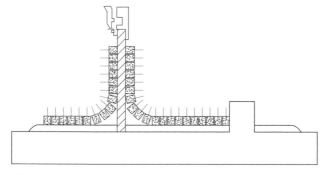

Fig. 3. Deposition of a monolayer from the water–air interface to a vertical plate.

hydrophobic after the first monolayer transfer. Thus the second monolayer is transferred in the immersion. This is the most usual mode of multilayer formation for amphiphilic molecules in which the head group is very hydrophilic and the tail is an alkyl chain. This mode is called the Y-type deposition (Fig. 4a). For very hydrophilic head groups, eg, $-COOH$, $-PO_3H_2$, etc, this is the most stable deposition mode, because the interactions between adjacent monolayers are then either hydrophobic–hydrophobic, or hydrophilic–hydrophilic. This mode produces centrosymmetric films comprised of bilayers.

Films may be formed only in downstroke (X-type, Fig. 4b). The deposition speed may affect the deposition mode (16,17). If deposition occurs only when films are formed in upstroke Z-type films result (Fig. 4c). These are cases where the head group is not as hydrophilic, eg, $COOCH_3$ (18), or where the alkyl chain is terminated by a weak polar group, eg, NO_2 (19). In both cases the interactions between adjacent monolayers are hydrophilic–hydrophobic. These multilayers are therefore less stable than the Y-type systems. Both X- and Z-type depositions are noncentrosymmetric.

The amount of amphiphile that can be deposited on a glass slide depends on several factors (2). The deposition ratio is defined as A_I/A_S, where A_S is the area of the substrate coated with a monolayer, and A_I is the decrease of area occupied by that monolayer at the water–air interface, at constant pressure. An ideal Y-type film is a multilayer system having a constant transfer ratio of one for both upstroke and downstroke. An ideal X-type film can be defined accordingly as a layer system where the transfer ratio is always one for the downstroke and

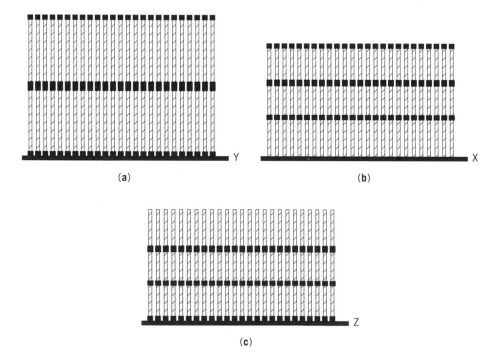

(a)   **(b)**

(c)

Fig. 4.   Multilayer films where ($\square$) represent a hydrophobic group and ($\blacksquare$) a hydrophilic one: (**a**) Y-type, (**b**) X-type, and (**c**) Z-type.

zero for the upstroke. In practice, there are deviations from the ideal, ie, the transfer ratio is ≤1, or not equal for upstroke and downstroke depositions in the case of Y-type films, or not zero in upstroke depositions for X-type films, giving a mixed X–Y-type film. This is a clear manifestation of the inherent instability of X films, because it suggests that the molecules in the X film flip over. For such X–Y films, a deposition can be defined by the ratio θ,

$$\theta = \frac{A_1^u}{A_1^d}$$

where A_1^u is the transfer ratio for the upstroke deposition, and A_1^d for the downstroke deposition. For an ideal Y film, $\theta = 1$; for X film, $\theta = 0$, and for Z film, $\theta = \infty$. Once the decrease of area of the monolayer at the water–air interface is plotted as a function of time during the deposition process, the deposition ratio for successive layers can be measured, and information on the deposition nature obtained.

Another way of building LB multilayer structures is the horizontal lifting or Schaefer's method introduced in 1938 (20). Schaefer's method is useful for the deposition of very rigid films, which are at the two-dimensional-solid region in the π-A diagram (Fig. 3). In this method, a compressed monolayer is formed at the water–air interface and then a flat substrate is placed horizontally on the monolayer film. When this substrate is lifted and separated from the water surface, the monolayer is transferred onto the substrate, in theory, keeping the same molecular direction (X-type, Fig. 4b). The deposition of high quality X-type LB films of ethyl stearate and octadecyl acrylate using a fully automated horizontal lifting has been reported (21,22). However, cadmium and lanthanum arachidate give LB films having x-ray diffraction corresponding to Y-type deposition, thus indicating that those molecules turn over either during, or after deposition. Monolayers of polymeric amphiphiles may be good candidates for horizontal lifting because of their high viscosity.

LB Films of Long-Chain Fatty Acids. LB films of saturated long-chain fatty acids have been studied since the inception of the LB technique. The most stable films of long-chain fatty acids are formed by cadmium arachidate deposited from a buffered CdCl$_2$ subphase. These films, considered to be standards, have been widely used as spacer layers (23) and for examining new analytical techniques. Whereas the chains are tilted ~25° from the surface normal in the arachidic acid, CH$_3$(CH$_2$)$_{18}$COOH, films (24), it is nearly perpendicular to the surface in the cadmium arachidate films (25).

LB films of ω-tricosenoic acid, CH$_2$=CH–(CH$_2$)$_{20}$COOH, have been studied as electron photoresists (26–28). A resolution better than 50 nm could be achieved. Diacetylenic fatty acids have been polymerized to yield the corresponding poly(diacetylene) derivatives that have interesting third-order nonlinear optical properties (29).

LB Films of Liquid-Crystalline Amphiphiles. Liquid-crystal (LC) phases are materials that have inherently ordered-layer structures, formed by self-organization of mesogenic compounds (30) (see LIQUID CRYSTALLINE MATERIALS). Therefore, by having a liquid-crystalline group in an amphiphile, enhanced order, thermal stability, and interesting physical properties can result. Furthermore, the study of liquid crystals at the water–air interface in a systematic way should add to the understanding of the two-dimensional organization, and the effect of the director on the relative orientation of molecules in the layers of a multilayer film. There have been a large number of studies on LB films of LCs (4).

Investigations of a terphenyl LC compound (**1**) showed that the hydrophilicity of the substrate and the ratio of the LC and CA determined the mode of transfer (31). The material has a short (C_5) alkyl chain and a weak hydrophilic headgroup (CN). The pure LC gives Z-type deposition on hydrophilic substrates. The head group is not highly hydrophilic. On the other hand, on hydrophobic surfaces, the deposition starts as Y-type, but gradually changes to Z-type. The inherent order of the LC material may be high enough so that a very short alkyl chain can be used. Furthermore, the order is given by the LC unit and there is no need for a strong hydrophilic head group. Therefore, Z-type deposition becomes more stable, and even preferred.

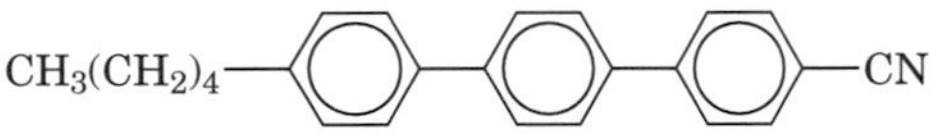

(1)

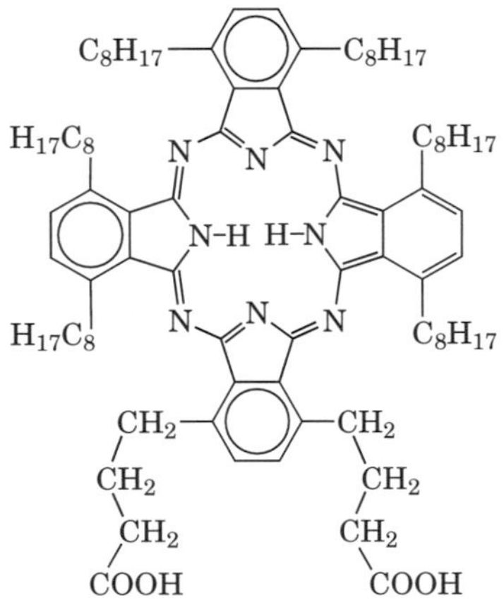

(2)

LB films of 1,4,8,11,15,18-hexaoctyl-22,25-bis-(carboxypropyl)-phthalocyanine (**2**), an asymmetrically substituted phthalocyanine, were stable monolayers formed at the water–air interface that could be transferred onto hydrophilic silica substrates (32–34). When a monolayer film of the phthalocyanine derivative was heated, there was a remarkable change in the optical spectrum. This, by comparison to the spectrum of the bulk material, indicated a phase transition from the low temperature herringbone packing, to a high temperature hexagonal packing.

LB Films of Porphyrins and Phthalocyanines. The porphyrin is one of the most important among biomolecules. The most stable synthetic porphyrin is 5,10,15,20-tetraphenylporphyrin (TPP). Many porphyrin and phthalocyanine (PC) derivatives form good LB films. Both these molecules are important for applications such as hole-burning that may allow information storage using multiple frequency devices. In 1937 multilayers were built from chlorophyll (35).

The first synthesis of amphiphilic porphyrin molecules involved replacement of the phenyl rings in TPP with pyridine rings, quaternized with $C_{20}H_{41}Br$

to produce tetra(3-eicosylpyridinium)porphyrin bromide (**3**) (36). The pyridinium nitrogen is highly hydrophilic: the long C_{20} hydrocarbon serves as the hydrophobic part. Tetra[4-oxy(2-docosanoic acid)]phenyl-porphyrin (**4**) has also been used for films (37).

Tetrakis(cumylphenoxy)phthalocyanine (**5**) a PC derivative, having liquid–crystalline-like substituents (38–43) was studied because the cross-section area of the substituents is much larger than that of a normal alkyl chain, and therefore, the requirement of minimized free volume in the assembly may be easier to accomplish.

An LB film from a new amphiphilic, a two-ring phthalocyanine, $(HO)GePc-O-SiPc(OSi(n\text{-}C_6H_{13})_3$ (**6**) (44), gave monolayers at the water–air interface where the rings were parallel to the water surface. The hydrophilic OH head and the hydrophobic $Si(C_6H_{13})_3$ tail help to ensure the desired molecular orientation, as well as provide high solubility. The monolayers thus formed are stable and robust, and can be deposited to form good multilayer films. The main interest in these monolayers is their apparent high anisotropy.

LB Films of Polymerizable Amphiphiles. Studies of LB films of polymerizable amphiphiles include simple olefinic amphiphiles, conjugated double bonds, dienes, and diacetylenes (4). In general, a monomeric amphiphile can be spread and polymerization can be induced either at the air–water interface or after transfer to a solid substrate. The former polymerization results in a rigid layer that is difficult to transfer.

In the first attempt to prepare a two-dimensional crystalline polymer (45), ^{60}Co γ-radiation was used to initiate polymerization in monolayers of vinyl stearate (**7**). Polymerization at the air–water interface was possible but gave a rigid film. The monomeric monolayer was deposited to give X-type layers that could be polymerized *in situ*. This polymerization reaction, quenched by oxygen, proceeds via a free-radical mechanism.

$$\text{CH}_3\text{—}(\text{CH}_2)_{15}\text{CH}_2\underset{O}{\overset{\overset{\displaystyle O}{\overset{\|}{C}}}{\diagup}}\overset{CH}{\underset{\displaystyle CH_2}{\diagdown}}$$

(**7**)

The pursuit of further miniaturization of electronic circuits has made submicrometer resolution lithography a crucial element in future computer engineering. LB films have long been considered potential candidates for resist applications, because conventional spin-coated photoresist materials have large pinhole densities and variations of thickness. In contrast, LB films are two-dimensional, layered, crystalline solids that provide high control of film thickness and are impermeable to plasma down to a thickness of 40 nm (46). The electron beam polymerization of ω-tricosenoic acid monolayers has been mentioned. Another monomeric amphiphile used in an attempt to develop electron-beam-resist materials is α-octadecylacrylic acid (**8**).

$$\begin{array}{c}\text{CH}_3\\|\\(\text{CH}_2)_{17}\\|\\C\\\diagup\diagdown\\CH_2\quad C\\|\\OH\end{array}\qquad\text{O}$$

(**8**)

$$\text{CH}_3\text{—}(\text{CH}_2)_m\text{—}\text{C}{\equiv}\text{C–C}{\equiv}\text{C–}(\text{CH}_2)_8\text{—COOH}$$

(**9**)

Diacetylenic amphiphiles have been studied in great detail because of the potential application of poly(diacetylene) films in nonlinear optical waveguide devices (47–51) (see NONLINEAR OPTICAL MATERIALS). The early systematic studies on diacetylenic amphiphiles (52–56) showed that ultrathin polyacetylenic films of (**9**) where $m \geq 8$, are very stable and have interesting physical properties, eg, photoconductivity (57). Diacetylene polymerization requires specific arrangement of the triple bonds, because the reaction proceeds via a 1,4-addition to the conjugate triple bond (58,59). In general, diacetylenic amphiphiles for rigid films aggregate even at zero surface pressure. Thus, they form domains at the air–water interface which make them rather useless in integrated optics application. A horizontal electric field of 104 V/m above the water surface can increase the domain size from 1–300 μm to 1 mm (60).

LB Films of Polymeric Amphiphile. Since the first successful deposition of a polymeric LB film (61), there have been a large number of studies examining different structural parameters on the transferability and stability of the polymeric LB films (4). One interesting idea for polymers for LB films is the use of a spacer group (mostly hydrophilic) to decouple the motion of the polymer from that of the lipid membrane (62,63). Monolayers from a polymer (**10**) having

$$CH_3-\underset{\underset{\underset{\xi}{|}}{\overset{|}{CH_2}}}{\overset{|}{C}}-\overset{\overset{O}{\|}}{C}-(O\text{-}CH_2CH_2)_4-O-\overset{\overset{O}{\|}}{\underset{\underset{O^-Na^+}{|}}{P}}-O-\overset{\overset{CH_2-O-(CH_2)_{15}CH_3}{|}}{\underset{\underset{CH_2}{|}}{CH-O-(CH_2)_{15}CH_3}}$$

(**10**)

hydrophilic phosphate groups and a tetraethylene oxide spacer were used to link a glycerol diether to the polymer chain (63).

Hydrophilic spacer groups may be introduced into a polymer through the side chain, the main chain, or both. Films can be prepared using different values of monomer feed (62).

Potential Applications of LB Films. LB films have long been expected to provide new technologies and novel materials, designed at the molecular level. Commercialization of any device would, however, require much faster deposition rates than those available as of this writing (ca 1997) when there is very little activity in U.S. Industrial laboratories.

Serious attempts to use LB films in commercial applications include the use of lead stearate as a diffraction grating for soft x-rays (64). Detailed discussion on applications of LB films are available (4,65). From the materials point of view, the ability to build noncentrosymmetric films having a precise control on film thickness, suggests that one of the first applications of LB films may be in the area of second-order nonlinear optics. Whereas a waveguide based on LB films of fatty acid salts was reported in 1977, a waveguide based on polymeric LB films has not yet been commercialized.

In 1983 the first paper on SHG from LB multilayers (66), using 4-octadecylamino-4'-nitroazobenzene (**11**) as the amphiphile, was reported.

$$O_2N-\bigcirc-N\underset{\diagdown}{\diagup}N-\bigcirc-NH\text{-}C_{18}H_{37}$$

(**11**)

A monolayer of the pyridine-substituted alkyl merocyanine (**12**) was prepared in the 1970s (67), and a noncentrosymmetric multilayer structure of merocyanine amphiphiles was later prepared (68) using derivatives, but introducing long-chain amines as the counter layer in an ABABAB system (69,70).

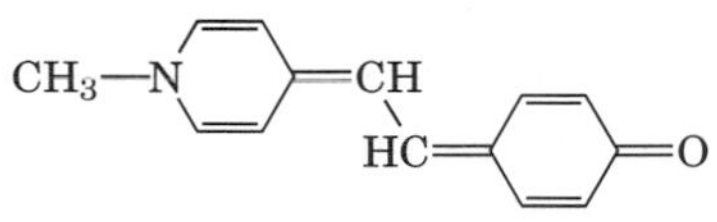

$$CH_3-N \quad =CH$$
$$HC= \quad =O$$

(12)

In the mid-1980s, SHG was measured from a merocyanine LB film (71) giving a value of 2.42×10^{-27} cm^5/esu for β_z of the dye. This is a very high number and may be resonance-enhanced at 2ω (533 nm).

An amphiphile having amide groups (**13**) in the alkyl chain, thus introducing H-bonding in addition to the van der Waals interaction, was prepared (20). A p-nitroaniline chromophore was used at the end of the alkyl group, allowing for hydrogen bonding to stabilize a Z-type multilayer.

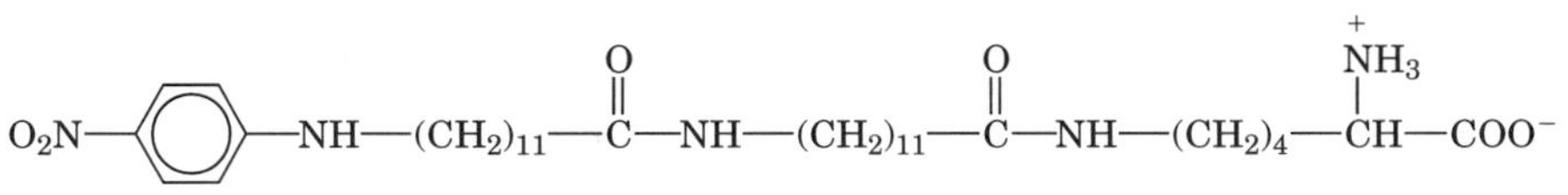

$$O_2N-\!\!\!\bigcirc\!\!\!-NH-(CH_2)_{11}-\overset{O}{\overset{\|}{C}}-NH-(CH_2)_{11}-\overset{O}{\overset{\|}{C}}-NH-(CH_2)_4-\overset{\overset{+}{NH_3}}{\overset{|}{CH}}-COO^-$$

(13)

Alternate-layer LB films (Y-type, ABAB) of long-chain amines and fatty acids may be used for pyroelectric applications (Fig. 5). Stearylamine, $C_{18}H_{37}NH_2$, and a series of straight-chain fatty acids, yield a thick film (several hundreds of monolayers) which gave a pyroelectric coefficient of ~0.05 nC/(cm^2·K) (72). A coefficient of 0.3 nC/(cm^2·K) for an 11-monolayer sample of ω-tricosenoic acid and docosylamine $C_{22}H_{45}NH_2$ has been reported (73).

In pyroelectric devices, a charge is developed across the film in response to heating and such devices may serve as ir-detectors (see INFRARED TECHNOLOGY AND RAMAN SPECTROSCOPY). Piezoelectric applications are promising as sound detectors, because for these, a charge is developed across the film in response to pressure. A review is available (74).

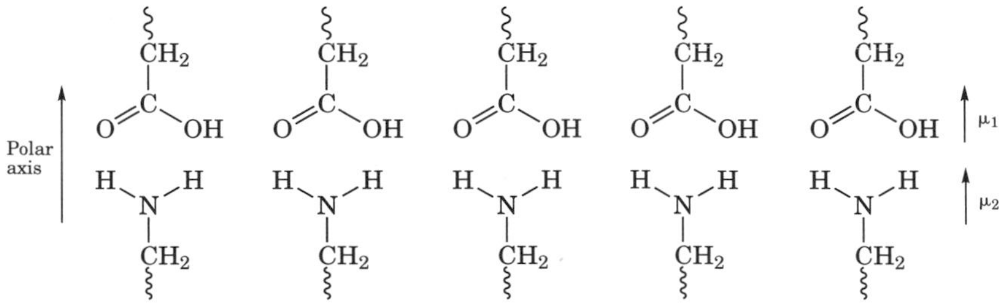

Fig. 5. Interaction between fatty acids and amines to produce an ABAB film having a polar axis.

The area of photoinduced electron transfer in LB films has been established (75). The ability to place electron donor and electron acceptor moieties in precise distances allowed the detailed studies of electron-transfer mechanism and provided experimental support for theories (76). This research has been driven by the goal of understanding the elemental processes of photosynthesis. Electron transfer is, however, an elementary process in applications such as photoconductivity (77–79), molecular rectification (79–84), etc.

Chemical and biological sensors (qv) are important applications of LB films. In field-effect devices, the tunneling current is a function of the dielectric constant of the organic film (85–90). For example, NO_2, an electron acceptor, has been detected by a phthalocyanine (or a porphyrin) LB film. The mechanism of the reaction is a partial oxidation that introduces charge carriers into the film, thus changing its band gap and as a result, its dc-conductivity. Field-effect devices are very sensitive, but not selective.

One of the most promising optical devices is that of the surface plasmon resonance (spr) (91). Surface plasmons are collective oscillations of the free electrons at the boundary of a metal and a dielectric. The surface plasmons are guided waves (92), and their resonance conditions are very sensitive to changes in the thickness and refractive index of the medium adjacent to the metal. Spr has been used to investigate the interaction between NO_x and LB films of tetra-4-*tert*-butylphthalocyanine-containing silicon (93). Another optical detector of toxic gases was demonstrated using fluorescent porphyrin LB films (94). Changes in fluorescence for NO_2, HCl, and Cl_2 gases were reported. The fluorescence could be quenched quantitatively in the cases of both NO_2 and HCl using NH_3 vapor. Oxygen could be detected using phosphorescent LB films of tetraphenylporphyrin palladium (95). A surface acoustic wave (SAW) oscillator incorporating LB films had a detection limit of 40 ppb NO_2 in dry air (84). The subject of chemical sensors has been reviewed (96).

The search for microbiosensors has brought the need for highly selective and highly sensitive organic layers, with tailored biological properties that can be incorporated into electronic, optical, or electrochemical devices (97). From the materials point of view, LB films are an excellent choice because of the high control on their chemical structure at the molecular level, and the ability to incorporate into them large biomolecules molecules. The limitation, however, is that these can be transferred only onto a flat surface and cannot, for example, be coated on an optical fiber. Also, it is difficult to imagine the transfer of an LB film onto a substrate having a very small surface area. Examples of biosensors using LB films appear in References (98–105) (see BIOSENSORS).

Self-Assembled Monolayers

The formation of monolayers by self-assembly of surfactant molecules at surfaces is one example of this general phenomenon. In nature, self-assembly results in supermolecular hierarchical organizations of interlocking components providing very complex systems (106). In 1946 as account was published of the preparation of monomolecular layers by adsorption (self-assembly) of a surfactant onto a clean metal surface (107). In the 1980s it was shown that self-assembled monolayers (SAMs) of alkanethiolates on gold can be prepared by adsorption of di-*n*-alkyl disulfides from dilute solutions (108). Many self-assembly systems have since been investigated, but monolayers of alkanethiolates on gold are probably the most studied.

The ability to tailor both head and tail groups of the constituent molecules makes SAMs excellent systems for a more fundamental understanding of phenomena affected by competing intermolecular, molecular–substrate and molecule–solvent interactions, such as ordering and growth, wetting, adhesion, lubrication, and corrosion. Because SAMs are well-defined and accessible, they are good model systems for studies of physical chemistry and statistical physics in two dimensions, and the crossover to three dimensions.

SAMs provide the needed design flexibility, both at the individual molecular and at the material levels, and offer a vehicle for investigation of specific interactions at interfaces, and of the effect of increasing molecular complexity on the structure and stability of two-dimensional assemblies. These studies may eventually produce the design capabilities needed for assemblies of three-dimensional structures (109).

The interest in the general area of self-assembly, and specifically in SAMs, stems partially from the perceived relevance to science and technology. In contrast to ultrathin films made by, for example, molecular beam epitaxy (MBE) and chemical vapor deposition (CVD), SAMs are highly ordered and oriented and can incorporate a wide range of groups both in the alkyl chain and at the chain termina. Therefore, a variety of surfaces having specific interactions can be produced with fine structural control (110). Owing to their dense and stable structure, SAMs have potential applications in corrosion prevention, wear protection, and other areas. In addition, the biomimetic and biocompatible nature of SAMs makes their applications in chemical and biochemical sensing promising. Their high molecular order parameter in SAMs makes them ideal as components in electrooptic devices. Work on nanopatterning of SAMs suggests that these systems may have applications in patterning of GaAs, and in the preparation of sensor arrays (111).

SAMs are ordered molecular assemblies formed by the adsorption (qv) of an active surfactant on a solid surface (Fig. 6). This simple process makes SAMs inherently manufacturable and thus technologically attractive for building superlattices and for surface engineering. The order in these two-dimensional systems is produced by a spontaneous chemical synthesis at the interface, as the system approaches equilibrium. Although the area is not limited to long-chain molecules (112), SAMs of functionalized long-chain hydrocarbons are most frequently used as building blocks of supermolecular structures.

Herein the focus is on SAMs of trichlorosilanes and thiols. SAMs of carboxylic acids are important as a connection between the LB and self-assembly techniques, but studies of their formation and structure have been relatively limited. SAMs of carboxylic acids on Al_2O_3, AgO, and CuO have also been carried out (113–124).

Monolayers of Organosilicon Derivatives. SAMs of alkylchlorosilanes, alkylalkoxysilanes, and alkylaminosilanes require hydroxylated surfaces as substrates for their formation. The driving force for this self-assembly is the *in situ* formation of polysiloxane, which is connected to surface silanol groups (–Si–OH) via Si–O–Si bonds. Substrates on which these monolayers have been success-

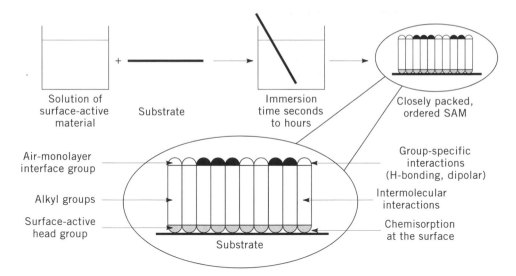

Fig. 6. Self-assembled monolayers are formed by immersing a substrate into a solution of the surface-active material. Necessary conditions for the spontaneous formation of the 2-D assembly include chemical bond formation of molecules with the surface, and intermolecular interactions.

fully prepared include silicon oxide (125–130), aluminum oxide (131,132), quartz (133–135), glass (130), mica (136–138), zinc selenide (131,132), germanium oxide (130), and gold (139–141). OTS monolayers on silicon oxide and on gold activated by uv-ozone exposure have been compared by ir spectroscopy, ellipsometry, and wetting measurements showing identical average film structures (142).

High quality SAMs of alkyltrichlorosilane derivatives are not simple to produce, mainly because of the need to carefully control the amount of water in solution (126,143,144). Whereas incomplete monolayers are formed in the absence of water (127,128), excess water results in facile polymerization in solution and polysiloxane deposition of the surface (133). Extraction of surface moisture, followed by OTS hydrolysis and subsequent surface adsorption, may be the mechanism of SAM formation (145). A moisture quantity of 0.15 mg/100 mL solvent has been suggested as the optimum condition for the formation of closely packed monolayers. X-ray photoelectron spectroscopy (xps) studies confirm the complete surface reaction of the $-SiCl_3$ groups, upon the formation of a complete SAM (146). Infrared spectroscopy has been used to provide direct evidence for the full hydrolysis of methylchlorosilanes to methylsilanoles at the solid/gas interface, by surface water on a hydrated silica (147).

Temperature has been found to play an important role in monolayer formation. The threshold temperature below which an ordered monolayer is formed is a function of the chain length, being higher (18°C) for an octadecyl than for a tetradecyl chain (10°C) (126). The issue is the competition between the reaction of hydrolyzed (or partially hydrolyzed) trichlorosilyl groups and other such groups in solution to form a polymer, and the reaction of such groups with surface Si–OH moieties to form a SAM. As temperature decreases, the preference for surface reaction increases. Moreover, as temperature decreases, reaction kinetics decrease as well, resulting in the diminution of thermal disorder in the forming monolayer, the formation of an ordered assembly, and the gain of van der Waals (VDW) energy. Solid-state ^{13}C nmr studies of OTS monolayers deposited on fumed silica particles have confirmed these results (148).

Substrates used in the formation of silane SAMs are amorphous; thus the packing and ordering of alkyl chains in SAMs of alkyl silanes are determined by the underlying structure of the surface polysiloxane chain. A schematic description of a polysiloxane at the monolayer substrate interface is shown in Figure 7. In this trimer, siloxane oxygen atoms occupy the equatorial positions and the alkyl chains are connected to the axial positions. The interchain distance is ca 0.44 nm, leaving very little free volume. This should require very little or no chain tilt. The connection between free volume and tilt is a general one, because the driving force for tilt is the reestablishment of van der Waals (VDW) contact among chains.

Alkyl chains in OTS monolayers of SiO_2 and oxidized gold are tilted at $10 \pm 2°$ from the normal and there is a significant gauche defect content at the chain termina (142). Based on both ellipsometry, and the concentration of gauche defects, it was concluded that the monolayer is $\sim96 \pm 4\%$ of the theoretical maximum coverage, which explains the observed average tilt. An important conclusion of this study is that surface hydration is responsible for decoupling of film formation from surface chemistry and the observed high film quality. Increasing the surface attachment of the forming siloxane chain through surface Si—OH groups introduces disorder and film defects.

Near edge x-ray absorption fine structure spectroscopy (nexafs) and x-ray photoelectron spectroscopy (xps) have been used to study SAMs of OTS, octade-cyltrimethoxysilane (OTMS), $CH_3(CH_2)_{17}Si(OCH_3)_3$, and (17-aminoheptadecyl)-trimethoxysilane (AHTMS), $H_2N(CH_2)_{17}Si(OCH_3)_3$ (149). A number of important observations have been reported. First, the chains in OTS SAMs are practically perpendicular to the substrate surface (tilt angle $0 \pm 5°$). Second, the adsorption mechanisms of trichlorosilane and trimethoxysilane groups are different, result-ing in a higher ($20 \pm 5°$) tilt angle of the chains in OTMS SAMs. Third, the

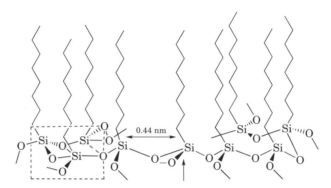

Fig. 7. Schematic description of a polysiloxane at the monolayer—substrate surface (4). The arrow points to an equatorial Si–O bond that can be connected either to another polysiloxane chain or to the surface. The dashed line on the left is a bond in a possible precursor trimer where the alkyl chains can occupy either axial or equatorial positions.

introduction of a polar amino group at the chain termina results in a more disordered monolayer, probably as a result of acid−base interactions with surface silanol groups. This last observation suggests that when such interactions exist, a preferred route may be to create surface functionalities by chemical reactions.

The reproducibility of alkyltrichlorosilane monolayers remains a problem. The quality of the monolayer formed is very sensitive to reaction conditions (126,127,150). Hexadecane may or may not be incorporated in OTS monolayers (129,143). It has been suggested that partial OTS monolayers have heterogeneous island structure (129,151). These incomplete monolayers may, however, be homogeneous and disordered (134,135,146,152,153). More recently atomic force microscopy (afm) studies have confirmed the island structure of partial monolayers. The adsorption of OTS onto glass and silicon oxide surfaces (150) and on mica (138) results in monolayers on mica by nucleating isolated domains, where the fractal dimensions increase with increased surface coverage. Other afm images of OTS SAMs on mica and on silica and silicon have also been produced (154−161). Pin holes in the OTS films from several nanometers to 100 nm in diameter, in monolayers on mica formed by self-assembly from a solvent mixture have been observed (154). Studies of OTS SAMs on silicon indicate that in order to obtain reproducible, good quality films, samples must be prepared under class 100 clean room conditions (158). OTS SAMs form on silicon, first by growth of large islands and then by filling-in with smaller islands until the film is complete. This growth mechanism has been utilized to form binary SAMs of OTS and 11-(2-naphthyl)undecyltrichlorosilane (162). Other researchers have suggested that in partial (25−30%) monolayers OTS molecules lie flat on the silicon surface, producing a water contact angle of 90° (159).

Differences in reported results also exist for other alkyltrichlorosilane systems. Surface coverages of vinyl-terminated alkyltrichlorosilane have been reported to be only ~63% (163), and well-packed monolayers (126). Surface coverage for monolayers of methyl-23-trichlorosilyltricosanoate (MTST), $H_3COOC−(CH_2)_{22}−SiCl_3$, have been reported to be ~93% (164) and full monolayers (165).

Patterns of ordered molecular islands surrounded by disordered molecules are common in Langmuir layers, where even in zero surface pressure molecules self-organize at the air−water interface. The difference between the two systems is that in SAMs of trichlorosilanes the island is comprised of polymerized surfactants, and therefore the mobility of individual molecules is restricted. This lack of mobility is probably the principal reason why SAMs of alkyltrichlorosilanes are less ordered than, for example, fatty acids on AgO, or thiols on gold. The coupling of polymerization and surface anchoring is a primary source of the reproducibility problems. Small differences in water content and in surface Si−OH group concentration may result in a significant difference in monolayer quality. Alkyl silanes remain, however, ideal materials for surface modification and functionalization applications, eg, as adhesion promoters (166−168) and boundary lubricants (169−171).

Surface modification can be achieved either by using ω-substituted alkyl silanes, or by surface chemical reactions. SAMs have been reported from alkyltrichlorosilanes having terminal functional groups of halogen (172−176), cyanide (173), thiocyanide (173), methyl ether (172), acetate (172), thioacetate (172,177),

α-haloacetate (174), vinyl (126,127,163,178–184), trimethylsilylethynyl (185), methyl ester (164,165), and p-chloromethylphenyl (174,186–189). Monolayers having low surface free energy have been prepared using partially fluorinated alkylsilanes (153,172,190,191). Surface modification can also be performed using various nucleophilic substitutions on SAMs of 16-bromohexadecylsilane (173). These SAMs were converted to the 16-thiocyanatohexadecylsilane monolayers by simply treating with a 0.1 M KSCN solution in DMF for 20 h. Similarly, NaN_3, Na_2S, and Na_2S_2 gave complete conversions of the bromo-terminated monolayers, as was evident from x-ray photoelectron spectroscopy (xps) (173). Reduction of the thiocyanato, cyanide, and azide surfaces by $LiAlH_4$ gave the mercapto-, and amino-terminated monolayers in complete conversions (173). Oxidation of the ω-thiol group gave sulfonic acid surfaces (173). XPS investigations of nucleophilic substitution at chain termina of alkyltrichlorosilane monolayers, using p-nitrothiophenolate as the nucleophile, have been carried out (174). The reaction rates obey the following order of leaving groups $I > Br > Cl$, and $X-CH_2-CO > C_6H_5-CH_2-X > CH_2-CH_2-X$. Competition reactions using thiolates and amines as nucleophiles show a clear thiolate preference. Reactions using small peptide fragments having cysteine moieties as the nucleophiles resulted in grafting of the monolayer surface with these peptides, which may be important for the development of biosensors. Patterned SAMs formed by microcontact printing of alkyltrichlorosilane on Al_2O_3/Al, SiO_2/Si, and TiO_2/Ti open new opportunities for preparation of sensors and electrooptical devices (192,193).

Surface modification reactions are important not only for engineering surface energy and interfacial properties such as wetting, adhesion, and friction, but also for providing active surfaces for the attachment of molecules having different properties. One example is the reaction of bromo-terminated alkylsilane monolayers with the lithium salt of 4-methylpyridine to provide pyridine surfaces (175,176). Such surfaces react with palladium (194), rhenium (176), and osmium complexes (175), and provide immobilization of organometallic moieties. Immobilized OsO_4 reacts with C_{60}-buckyballs, resulting in the formation of a C_{60}-monolayer (175) (see CARBON–C_{60} (see SUPPLEMENT)). Similar monolayers can be formed by the reaction of buckyballs with amino or azido surface groups (195–197). A cysteine-specific surface was prepared for the fabrication of metalloprotein nanostructures (198). These examples show the opportunities SAMs provide in the construction of layers and of new materials by combinations thereof.

Mixed monolayers provide an excellent route for surface engineering at the molecular level. Hence, by coadsorption of alkyltrichlorosilane with different ω-functionalities, surface free energy and chemical reactivity can be designed via the control of surface chemical functionalities. However, there are few reports on mixed monolayers of alkyltrichlorosilane, and most investigations were carried out on alkanethiolate monolayers on gold. When mixed monolayers of alkyltrichlorosilane and ω-vinyl or ω-2-naphthyl alkyltrichlorosilane were prepared by competitive adsorption, it was found that the composition of the monolayer is equal to the composition of the immersion solution (127,128,134). The gradual increase of the amount of excimers observed with the gradual increase of the naphthyl concentration supports the ideal mixing of the two silanes in the monolayer. When the preparation of mixed monolayers of alkyltrichlorosilanes having different chain lengths was investigated, ideal mixing was observed. The composition was determined by the relative rates of adsorption of the components (199).

Construction of multilayers requires that the monolayer surface be modified to a hydroxylated one. Such surfaces can be prepared by a chemical reaction and the conversion of a nonpolar terminal group to a hydroxyl group. Examples of such reactions are the $LiAlH_4$ reduction of a surface ester group (165), the hydroboration–oxidation of a terminal vinyl group (127,163), and the conversion of a surface bromide using silver chemistry (200). Once a subsequent monolayer is adsorbed on the "activated" monolayer, multilayer films may be built by repetition of this process (Fig. 8).

Using this strategy, construction of multilayer films of ~0.1 μm thickness by self-assembly of methyl 23-trichlorosilyltricosanoate (MTST) on silicon substrates has been demonstrated (Fig. 9) (165). The linear relationship between the film thickness and the layer number showed a slope of 3.5 nm/layer. Ellipsometry data, absorbance intensities, and dichroic ratios for the multilayers all suggest that the samples were composed of distinct monolayers. However, ir data indicated that there may be more tilting or disordering of the alkyl chains in the seven-layer sample than for the monolayer samples.

Despite the increasing level of monolayer disorder, the preparation of a multilayer film having thickness of ~0.1 mm was possible, indicating that the *in situ* formation of a polysiloxane backbone at the substrate–solution interface allows the monolayer to bridge over defects, such as pinholes and unreduced carbonyl groups. This repair mechanism may be very significant, because the construction of very thick films (1–2 μm, 250–500 layers) by self-assembly can be considered unlikely if defects inevitably propagate and grow.

Synchrotron x-ray diffraction studies were performed on a fifteen-layer thin film of MTST (201). The specular profile suggested a compression of the outermost layers from an average spacing of 3.190 ± 0.002 nm, which is interpreted

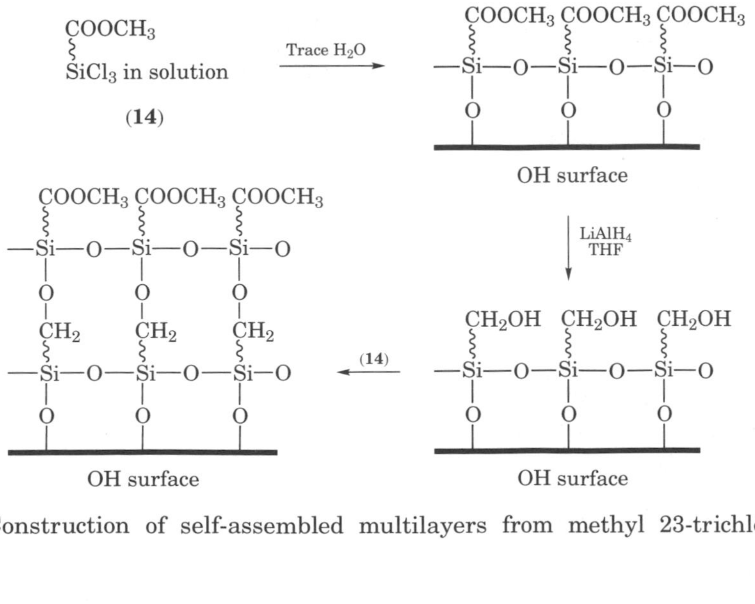

Fig. 8. Construction of self-assembled multilayers from methyl 23-trichlorosilyltricosanoate.

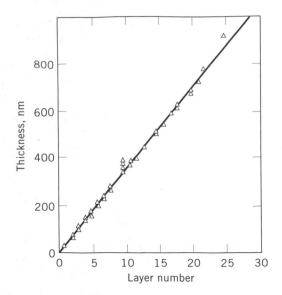

Fig. 9. Film thickness vs layer number (165).

as an increase in disordering near the film–air interface. Rocking curves of the specular profile suggest extremely rigid –SiO_2– layers. In-plane results also depict rigid –SiO_2– layers having spacing of 0.135 ± 0.003 nm and an increase in disorder at below critical angle measurements. The alkyl chains were shown to be hexagonally packed between these rigid layers. There was no observance of a chain tilt. A self-assembly strategy, where a –$SiCl_3$ group is attached to a small molecule, and an S_N2 reaction of the SAM introduced to a monolayer of nonlinear optically (NLO) active dyes has been developed. These were used in the construction of SAMs having second-harmonic nonlinear optical (NLO) properties (185–189). A significant improvement in the synthesis of multilayer structure reported recently is summarized in Figure 10. Using the process described in Figure 10, a three-layer system was prepared in one hour (202).

Hydrogen-bonded multilayers of self-assembling silanes have been reported (203,204). Using a combination of ftir spectroscopy and x-ray scattering a multilayer structure was observed as having district monolayers, coupled to each other in a flexible, nonepitaxial manner, via interlayer multiple hydrogen bonds. The hydrocarbon chains are perpendicular to the layer plane: a lateral packing density is 2.1 nm/molecule; and a positional coherence length is of ca 7.0 nm.

19-Trimethylsilyl-18-nonadecynylsilane monolayers can be polymerized to the corresponding polyacetylene systems (185). The treatment of the nonpolymerized monolayers with electron-beam radiation is dependent on ambient conditions. When irradiation was carried out under helium, the result was crosslinked monolayers; however, irradiation under nitrogen yielded cross-linking accompanied by the formation of amino terminal groups; and when irradiation was carried out under oxygen, cross-linked monolayers having hydroxyl, aldehyde, and carboxylic acid terminal groups were obtained. Using this technique, it was possible to fabricate well-ordered multilayer films (184) but, having a nonlinear relationship between film thickness and the number of layers (205). Attempts to

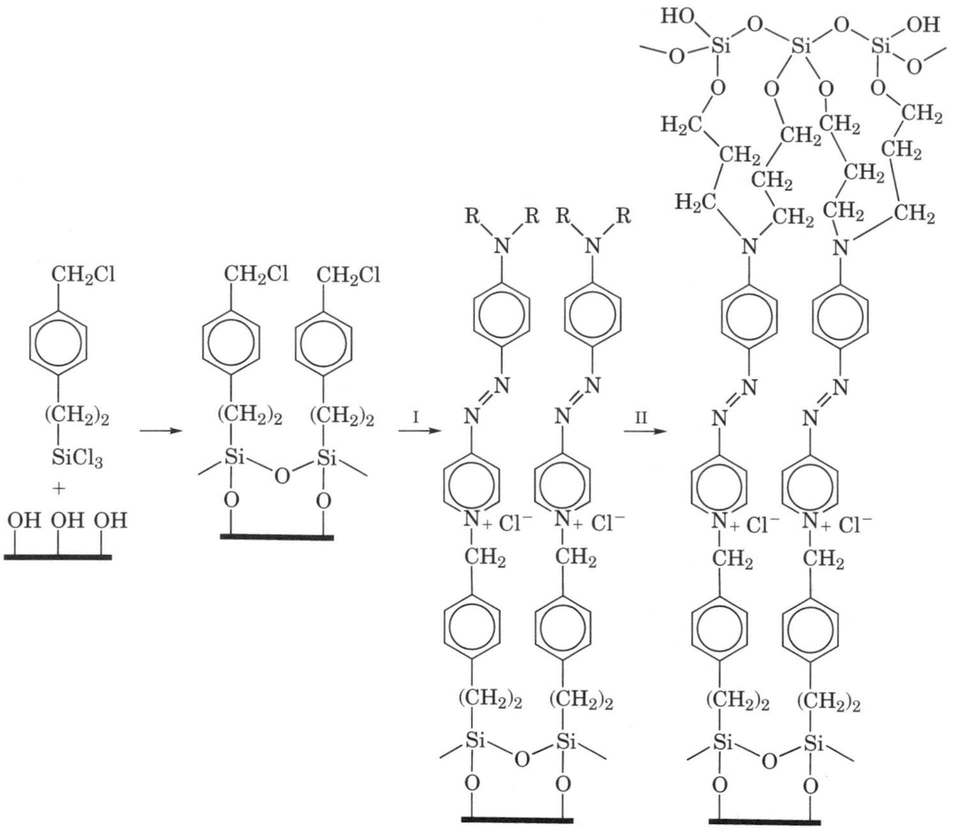

Fig. 10. Formation of noncentrosymmetric multilayer film by combining self-assembly and a surface S_N2 reaction, where $R = (CH_2)_3OH$; procedure I = spin-coating followed by annealing at 110°C; and procedure II = reaction of $Cl_3SiOSiCl_2OSiCl_3$, ie, a dilute solution of 4-[*N,N*,-bis-(3-hydroxypropyl)-aminophenylazo]-4'-pyridine on a benzyl chloride SAM surface was used, resulting in facile formation of SAMs having high order parameters.

prepare thicker films failed owing to increased disorder. A competition between irradiation damage and the formation of a new layer (205) appears to exist.

Trichlorosilane derivatives of large dye molecules are difficult to purify and owing to moisture sensitivity are hard to handle. Their organic solutions tend to become turbid rather quickly owing to the formation of insoluble polymers. Thus, solutions must be replaced frequently. An exception may be the combination of self-assembly and surface chemical reaction (186–189,202). On the other hand, ω-substituted alkyltrichlorosilane derivatives are easy to synthesize the purify. These could be used for the engineering of surface free energy through the control of chemical functionalities in their SAMs, or as active layers for attachment of biomolecules in biosensors.

Organosulfur Adsorbates on Metal and Semiconductor Surfaces. Sulfur compounds (qv) and selenium compounds (qv) have a strong affinity for transition metal surfaces (206–211). The number of reported surface-active organosulfur compounds that form monolayers on gold includes di-*n*-alkyl sulfide (212,213), di-*n*-alkyl disulfides (108), thiophenols (214,215), mercaptopyridines (216), mercaptoanilines (217), thiophenes (217), cysteines (218,219), xanthates (220), thiocarbaminates (220), thiocarbamates (221), thioureas (222), mercaptoimidazoles (223–225), and alkaneselenoles (226) (Fig. 11). However, the most studied, and probably most understood, SAM is that of alkanethiolates on Au(111) surfaces.

It has been suggested that gold does not have a stable surface oxide (227), and therefore, its surface can be cleaned simply by removing the physically and chemically adsorbed contaminants. However, more recently it has been shown that oxidation of gold by uv and ozone at 25°C gives a 1.7 ± 0.4-nm thick Au_2O_3 layer (228), stable to extended exposure to ultra high vacuum (UHV) and water and ethanol rinses.

Organosulfur compounds coordinate very strongly also to silver (229–233), copper (231–234), platinum (235), mercury (236,237), iron (238,239), nanosize γ-Fe_2O_3 particles (240), colloidal gold particles (241), GaAs (242), and InP surfaces (243). Octadecanethiol monolayers provide an excellent protection of the metal surface against oxidation (234). For example, silver surfaces having octadecanethiolate monolayers could be kept in the ambient without tarnishing for many months; copper surfaces coated with the same monolayer sustain dilute nitric acid (244).

Kinetic studies of alkanethiol adsorption onto Au(111) surfaces have shown that at relatively dilute ($10^{-3}M$) solutions, two distinct adsorption kinetics can be observed: a very fast step, which takes a few minutes, by the end of which the contact angles are close to their limiting values and the thickness about 80–90% of its maximum; and a slow step, which lasts several hours, at the end of which the thickness and contact angles reach their final values (245). The initial step is described well by diffusion-controlled Langmuir adsorption, and strongly depends on thiol concentration. At 1 mM solution the first step was over after ~1 minute, while it required over 100 minutes at 1 mM concentration (245). The second step can be described as a surface crystallization process, where alkyl chains get out of the disordered state and into unit cells, thus forming a two-dimensional crystal. Therefore, the kinetics of the first step is governed by the surface-head-group reaction, and the activation energy may

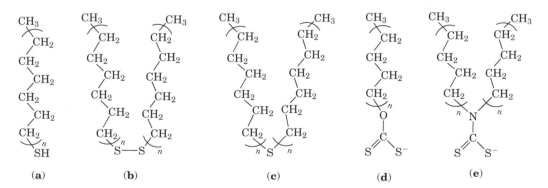

Fig. 11. Surface-active organosulfur compounds that form monolayers on gold: (**a**) alkanethiol; (**b**) dialkyl disulfide; (**c**) dialkyl sulfide; (**d**) alkyl xanthate; and (**e**) dialkylthiocarbamate.

depend on the electron density of the adsorbing sulfur. On the other hand, the kinetics of the second step is related to chain disorder, eg, gauche defects; the different components of chain–chain interaction, eg, VDW, dipole–dipole, etc; and the surface mobility of chains. The kinetics are faster for longer alkyl chains, probably owing to the increased VDW interactions (245).

Second-harmonic generation, and xps measurements (246,247), as well as near edge X-ray absorption fine structure spectroscopy (nexafs) studies confirm the two-step mechanism (248). Studies also showed pronounced differences between the short ($n < 9$) and long ($n > 9$) alkanethiolates, probably owing to the decreased rate of the second step which results from the diminution of the interchain VDW attraction energy. In the case of simple alkyl chains, the masking of adsorption sites by disordered chains is not a serious problem. However, if the chain contains a bulky group, the two steps are coupled, and the chemisorption kinetics is greatly impeded by the chain disorder (249). A direct competition between *tert*-butylmercaptan and *n*-octadecylmercaptan reveals that the latter adsorbed onto gold at greater efficiency than the former by a factor of 290–710 from ethanol (250). The additive effects of the stabilizing van der Waals interactions in the *n*-alkyl mercaptan monolayer and the sterically hindered *tert*-butylmercaptan explain the clear preference of the linear molecules.

Chemisorption of alkanethiols as well as of di-*n*-alkyl disulfides on clean gold gives indistinguishable monolayers (251) probably forming the Au(I) thiolate species. A simple oxidative addition of the S–S bond to the gold surface is possibly the mechanism in the formation of SAMs from disulfides:

$$R—S—S—R + Au_n^0 \longrightarrow R—S^-Au^+ \cdot Au_x^0$$

The rates of formation of SAMs from dialkyl disulfides or alkanethiols were indistinguishable, but the rate of replacement of molecules from SAMs by thiols was much faster than that by disulfides (251). Reaction of an unsymmetrical disulfide, $HO(CH_2)_{10}SS(CH_2)_{10}CF_3$, and a gold surface gave SAMs containing equal proportions of the two thiolate groups (252). Replacement experiments showed that the $S(CH_2)_{10}CF_3$ group in the mixed SAMs is replaced by $S(CH_2)_{10}CN$, on exposure to the $HS(CH_2)_{10}CN$ solution in ethanol, about 10^3 times faster than the $HS(CH_2)_{10}OH$ group. This is strong support for the disulfide bond cleavage mechanism and the subsequent formation of gold thiolate species. 4-Aminobenzenethiol has been reported to be spontaneously oxidized to 4,4'-diaminodiphenyl disulfide in the presence of gold powder (253). This, the first observation of its kind, hints that the stability of thiolate SAMs on gold may be related to the electron density on the thiolate sulfur. However, except for the report on the dimerization of alkanethiolates of Au(111) surface to form the dialkyl disulfides (254) there has been no other direct evidence supporting such a reaction.

In the alkanethiol case, the reaction may be considered formally as an oxidative addition of the S–H bond to the gold surface, followed by a reductive elimination of the hydrogen. When a clean gold surface is used, the proton probably ends as a H_2 molecule. Monolayers can be formed from the gas phase (241,255,256), in the complete absence of oxygen:

$$R—S—H + Au_n^0 \longrightarrow R—S^-Au^+ \cdot Au_x^0 + \frac{1}{2} H_2$$

The combination of hydrogen atoms at the metal surface to yield H_2 may be an important exothermic step in the overall chemisorption energetics. That the adsorbing species is the thiolate, RS–, has been shown by xps (231,257–259), Fourier transform infrared (ftir) spectroscopy (260), Fourier transform mass spectrometry (261), electrochemistry (262), and Raman spectroscopy (263–265). The bonding of the thiolate group to the gold surface is very strong. The homolytic bond strength is approximately 167 kJ/mol (40 kcal/mol) (206).

Based on the bond energies of RS–H, 364 kJ/mol (87 kcal/mol); H_2, 435 kJ/mol (104 kcal/mol); and RS–Au, 167 kJ/mol (40 kcal/mol), the net energy for adsorption of alkanethiolates on gold would be ca −20.9 kJ/mol (−5 kcal/mol) (exothermic). A value of −23.0 kJ/mol (−5.5 kcal/mol) has been calculated using electrochemical data (266), suggesting that the estimate of 167 kJ/mol for the S–Au bond strength is a good one. Based on similar calculations the value of ca −100 kJ/mol (−24 kcal/mol) was estimated for the adsorption energy of dialkyl disulfide, or −50 kJ/mol (−12 kcal/mol) per RS⁻. This is about twice as favorable as the adsorption energy calculated for the thiol mechanism involving molecular hydrogen (266). In view of the disulfide picture (254), desorption data applied to first order kinetics, gave a better correlation than for second order kinetics (266). This, however, cannot be considered as direct evidence for thiolate dimerization. It is not clear why a dialkyl disulfide molecule remains adsorbed as such, having gauche defects at the S–C bonds to allow the hydrocarbon chains to assume hexagonal close-packing, if it can simply adsorb as two all-trans alkanethiolates.

The incomplete stability of alkanethiolate SAMs can be concluded from a number of papers. Some loss in electroactivity of ferrocenyl alkathiolate SAMs upon soaking in hexane has been reported (267), although such loss was not observed when the same SAM was immersed in ethanol (268). Exposure of other electroactive SAMs to nonaqueous electrolytes also gave clues of instability (269–271). Alkanethiolates bearing radiolabeled (^{35}S) head groups have been incorporated into SAMs on a variety of substrates (266). The question of S–C bond cleavage during adsorption to yield adsorbed sulfide, S^{2-}, and thiolate, SH^-, has been raised after S–C cleavage was reported in organosulfides, R–S–R, on adsorption to gold, producing SAMs identical to those resulting from S–H breaking in the corresponding thiol or from S–S breaking in the corresponding disulfides (272). Based on coverage measurements it was concluded that if any C–S bond cleavage occurs it is minimal (266).

Thermal stability of alkanethiolate SAMs has also been addressed. Loss of sulfur from hexadecanethiolate monolayer on gold has been reported over the range of 170–230°C (267). Temperature-programmed desorption of methanethiolate SAMs on gold yielded a desorption maximum at ca 220°C (260). Detailed mass spectroscopic studies of *tert*-butanethiolate monolayers on gold showed maximum desorption at ca 200°C (272). Using radiolabeled hexadecanethiolate monolayers, a complete loss of surface sulfur at 210°C was observed with some loss occurring at 100°C (266).

Early electron diffraction studies, both high (273,274) and low energy (275) of monolayers of alkanethiolates on Au(111) surfaces show that the symmetry of sulfur atoms is hexagonal with an S–S spacing of 0.497 nm, and calculated area per molecule of 0.214 nm^2. Helium diffraction (276) and atomic force microscopy (afm) (277) studies confirmed that the structure formed by docosanethiol on Au(111) is commensurate with the underlying gold lattice (Fig. 12). Ultra high vacuum scanning transmission microscopy (stm) studies have added important information on the mechanism of SAM formation, revealing the coexistence of a two-dimensional (2D) liquid phase at room temperature of butane-, $CH_3(CH_2)_3S^-$, and hexanethiolate, $CH_3(CH_2)_5S^-$ monolayers on Au(111) (278). No 2D liquid was observed for octane-, $CH_3(CH_2)_7S^-$, and decanethiolate, $CH_3(CH_2)_9S^-$, monolayers. The short-chain homologues exhibited slow desorption of surface thiolate that led to the nucleation and growth of ordered domains. On the other hand, both octane- and decanethiolate form densely packed SAMs (279–281).

The above stm study also discovered a facile transport of surface gold atoms in the presence of the liquid phase, suggesting that the two-step mechanism does not provide a complete picture of the surface reactions, and that adsorption/desorption processes may have an important role in the formation of the final equilibrium structure of the monolayer. Support for the importance of a desorption process comes from atomic absorption studies showing the existence of gold in the alkanethiol solution. The stm studies suggest that this gold comes from terraces, where single-atomic deep pits are formed (281–283).

Ab initio calculations show that at the hollow site of Au(111), the sulfur charge is ca $-0.4\ e$ (211), whereas at the on-top site, this charge is ca $-0.7\ e$ (211). Because S–H bond cleavage occurs at the on-top site (284), if this cleavage is the rate-determining step, the adsorption rate should be faster in polar solvents, owing to the stabilization of the forming dipole. However, if the migration of a thiolate from the on-top to the hollow site is the slow step, the reaction should be faster in nonpolar solvent, owing to the diminished charge separation. Most recent second-harmonic generation studies have showed that whereas the rate constant in ethanol is $1.3 \times 106\ cm^3/(mol{\cdot}sec)$, it is $4.7 \times 106\ cm^3/(mol{\cdot}cm^3)$ in hexane (284), suggesting that S–H bond cleavage is not the rate-determining step.

Migration of thiolates between neighboring hollow sites is essential for healing of defects. Such migration should occur either through the on-top or the

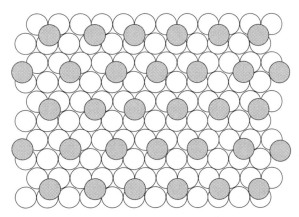

Fig. 12.   Hexagonal coverage scheme for alkanethiolates on Au(111), where (○) are gold atoms and (◉) are sulfur atoms.

bridge sites. In both cases, the transition state is more polar than the ground state, and hence should be sensitive to dielectric constant. Indeed, ethanol has been found to yield consistently highly ordered monolayers (245). The thiolate is chemically bonded to one gold atom at the on-top site, forming a neutral gold thiolate molecule, RS–Au. This may desorb before the thiolate moves to the hollow site, thus leaving a defect. More recent stm studies suggest that some of the pinholes observed in monolayers on Au(111) may be a result of such an etching process (283). However, it is also possible that alkanethiolates desorb from the surface as $RS^-Au_3^|$. These pinholes disappeared after annealing the monolayers at 77 or 100°C (285,286). If alkanethiolates increase, the mobility of gold atoms at the surface is not clear, however, the data so far indicate that surface migration of gold thiolate molecules, RS–Au, may be considered as a possible mechanism for healing monolayer defects.

Alkanethiolates have two binding modes at the Au(111) hollow site, one with a bend angle around the sulfur of 180° (sp) and the other of 104° (sp^3). The latter is being more stable by 1.7 kJ/mol (0.41 kcal/mol) (211). Thus, packing requirements may dictate the final surface–S–C angle. Many studies have suggested that this angle in monolayers on Au(111) surfaces must be tetrahedral (261). Modeling of terphenylthiolate, $C_6H_5–C_6H_4–C_6H_4–S–$, monolayers on Au(111) suggest a tilt angle of ~6° from the surface normal (214), and preliminary x-ray diffraction studies of 4-methyl-4'-mercaptobiphenyl monolayers on Au(111) single-crystal surfaces confirm this suggestion (239), thus providing the first evidence that a second chemisorption mode is possible.

The energy barrier between the two chemisorption modes on Au(111) is very small, 10.5 kJ/mol, (2.5 kcal/mol), (211), suggesting that the thiolate may easily cross from one of these minima to the other, enabling a facile annealing mechanism. This predicts that changing tilt direction may occur well below the melting point of the monolayer, and should be chain-length-dependent.

X-ray data show narrowing of the diffraction peak when monolayers of alkanethiolates on Au(111) were annealed (279). A development of larger domain size was the apparent result of the heating and cooling. Thus, close packing and high ordering of alkanethiolates on Au(111) may result from the relatively easy 2-D recrystallization process, as well as from the migration of gold thiolate molecules.

Molecular mechanics (MM) energy minimization indicates that the two modes lead to monolayers exhibiting different types of packing arrangements, but comparable in their ground state energies. (The monolayer resulting from the sp^3 mode is more stable by 2.5 kJ/mol (0.6 kcal/mol)) (211). Therefore, monolayers may consist of two different chemisorption modes ordered in different domains, simultaneously coexisting homogeneous clusters, each characterized by a different conformer in their unit cell. This may explain the observation of 2D liquid in butane- and hexanethiolate monolayers on gold (278), where VDW interactions do not provide enough cohesive energy to allow for small domains to coexist as a 2D solid.

The chemisorption of S atoms (287), SH (288), and SCH$_3$ groups (289,290) on Ag(111) can be described as ($\sqrt{7} \times \sqrt{7}$)R10.9° (Fig. 13), having an S–S distance of 0.441 nm slightly smaller than the interchain repeat distance in crystalline paraffins of 0.465 nm (291).

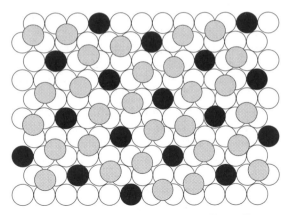

Fig. 13. A diagram showing one of the two possible $\sqrt{7} \times \sqrt{7}$ structures of alkanethiolates on Ag(111), where (○) represent the silver atoms, (●) represent thiolates at the on-top and (◎) at the hollow sites.

For octadecanethiolate, $CH_3(CH_2)_{17}S^-$, monolayers, grazing incident x-ray diffraction (gixd) shows a lattice constant of 0.46–0.47 nm, alkyl chains that are hardly tilted, and an overlayer very similar to $(\sqrt{7} \times \sqrt{7})R10.9°$, but with 12° rotation, and an outermost Ag(111) layer slightly expanded (292). The $(\sqrt{7} \times \sqrt{7})R10.9°$ requires that the thiolates at the on-top site be ca 0.05 nm higher than those residing at the hollow site. SAMs of decanethiolate, $CH_3(CH_2)_9S^-$, on Ag(111) using ultrahigh impedance STM have been studied (293). The average nearest-neighbor distance within a domain was shown to be 0.461 ± 0.015 nm; there are two domain types corresponding to two orientations of a six-fold symmetric lattice separated by $20.7 \pm 2.3°$; and fluctuations of heights of nearest neighbors far from domain boundaries are less than 0.01 nm.

Some thiophenolate monolayers also have been investigated. Thiophenolate, C_6H_5S-, forms ordered monolayers on Ag(111) with a $(\sqrt{7} \times \sqrt{3},88°)R40.9°$, and benzene rings closely packed in face-to-face stacked columns (294). Benzylthiolate (295), p-pyridinethiolate (296), and o-pyridinethiolate (296), also form ordered monolayers on Ag(111), but with fewer close-packed aromatic rings.

The ftir studies reveal that the alkyl chains in SAMs of thiolates on Au(111) usually are tilted ~26–28° from the surface normal, and display ~52–55° rotation about the molecular axis. This tilt is a result of the chains reestablishing VDW contact in an assembly with ~0.5 nm S–S distance, larger than the distance of ~0.46 nm, usually quoted for perpendicular alkyl chains in a close-packed layer. On the other hand, thiolate monolayers on Ag(111) are more densely packed owing to the shorter S–S distance. There were a number of different reports on chain tilt in SAMs on Ag(111), probably owing to different amounts of oxide, formed on the clean metallic surface (229,230,296,297). In carefully prepared SAMs of alkanethiolates on a clean Ag(111) surface, the alkyl chains are practically perpendicular to the surface.

Functionalized alkanethiolate SAMs are important both for engineering of surface properties and for further chemical reactions. Simple, eg, CH_3, CF_3, $CH=CH_2$, $C\equiv CH$, Cl, Br, CN, OH, OCH_3, NH_2, $N(CH_3)_2$, SO_3H, and $Si(OCH_3)_3$, COOH, $COOCH_3$, $CONH_2$ (206–211),219,260,298,299), as well as more complex functionalities, eg, ferrocenyl (268,300–307), biotinyl (308–312), 2,2-bipyridyl (313), tetrathiafulvalenecarboxylate (314), tetraphenylporphyrin (315,316), and ferrocenylazobenzene (317), were attached to the chain termina of alkanethiolate monolayers. These monolayers are thus becoming the system of choice for studies of surface phenomena, electron transfer, molecular recognition, etc.

Surface OH and COOH are very useful groups for chemical transformations. Monolayers having terminal COOH functionality react with alkanoic acids (318), and decylamine (319) to form bilayer H-bonding-stabilized structures, which lack long-term stability owing to the strong electrostatic repulsion in the newly formed charged interface. The caboxylate group can be transformed to the corresponding acid chlorides by using $SOCl_2$ (320). Further reactions with amines and alcohols yield bilayer structures with amide and ester linkages, respectively. Reaction of the acid chloride with a carboxylic acid-terminated thiol provides the corresponding thioester. This reaction has been used to form polymeric self-assembled monolayers and multilayers from the diacetylene $HS(CH_2)_{10}C \equiv C-C \equiv C(CH_2)_{10}COOH$ (321).

SAMs of OH-terminated alkanethiols have been used in many surface modification reactions (Fig. 14). These reacted with OTS to yield a well-ordered bilayer (322), with octadecyldimethylchlorosilane (323,324), with $POCl_3$ (325–327),

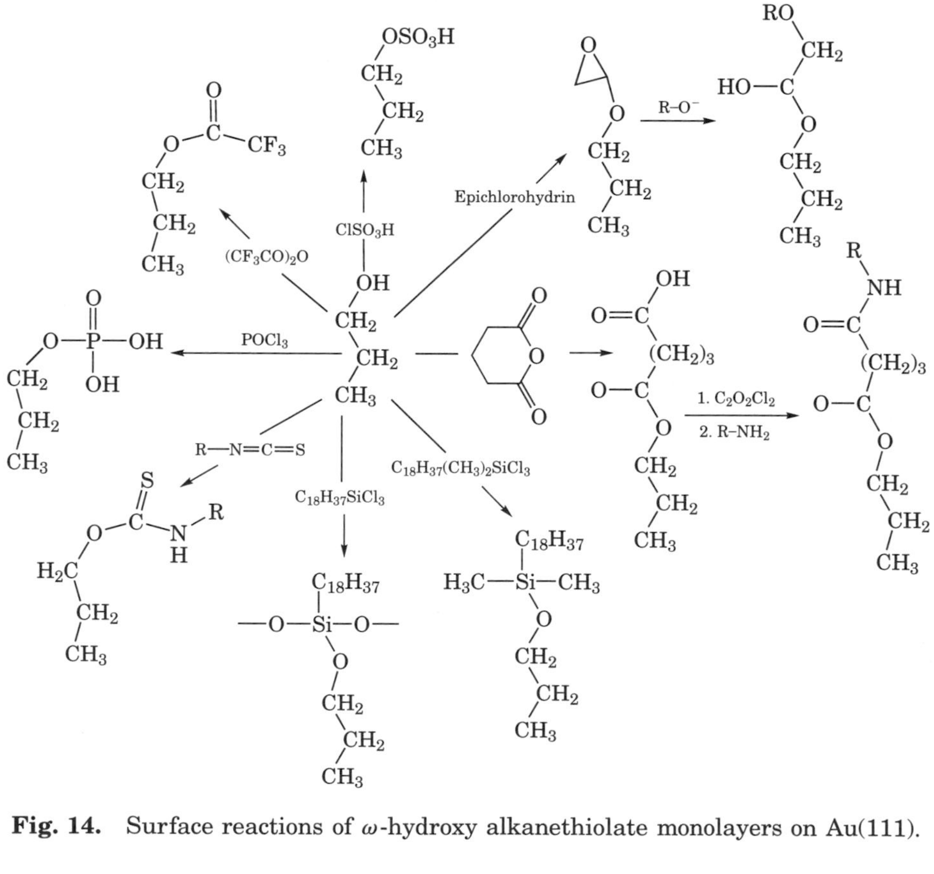

Fig. 14. Surface reactions of ω-hydroxy alkanethiolate monolayers on Au(111).

with trifluoroacetic anhydride (328), epichlorohydrin (329), with alkylisothio-cyanate (330), with glutaric anhydride (331), and with chlorosulfonic acid (327).

Alkyl Monolayers on Silicon. Robust monolayers can be prepared where the alkyl chains are covalently bound to a silicon substrate mainly by C–Si bonds (332,333). In the first experiments hydrogen-terminated silicon, H–Si(111) and H–Si(100), were used with diacetylperoxide (332). These monolayers, although exhibiting thickness, wettability, and methylene-stretching frequencies indica-tive of highly packed chains, lost ~30% of the chains when exposed to boiling water. The apparent conclusion was that hydrolyzable acyloxy groups are re-moved, leaving the robust alkyl chains bound to the surface by the C–Si bonds. In an attempt to reduce the fraction of surface acyloxy groups, a mixture of alkene and diacetylperoxide was used (333). Reaction of alkynes also yielded robust, closely packed monolayers, and chlorine-terminated olefins gave mono-layers having wettability indicative of Cl-terminated alkyl chains. The resulting monolayers are ~90% olefin-based, as shown by deuterium labeling experiments. The introduction of olefin molecules can be explained by a radical reaction. The surface radical (dangling bond ($\cdot$)) reacts with the double-bond to yield a sec-ondary carbon radical:

$$R'CH = CH_2 + \cdot Si(111) \longrightarrow R'CH(\cdot)-CH_2-Si(111)$$

This radical can either abstract another hydrogen from the allylic position of an olefin molecule:

$$R'CH(\cdot)-CH_2-Si(111) + R'CH_2CH = CH_2 \longrightarrow R'(CH_2)_2Si(111) + R'CH(\cdot)CH = CH_2$$

or with a surface Si–H group:

$$R'CH(\cdot)-CH_2-Si(111) + H-Si(111) \longrightarrow R'(CH_2)_2Si(111) + \cdot Si(111)$$

The monolayer density, as measured by x-ray reflectivity, is only ~90% of the value of a crystalline paraffin such as $n\text{-}C_{33}H_{68}$, suggesting a significant number of defects. Ellipsometry and infrared spectroscopy suggest that the chains are tilted ~45° from the surface normal, and that a twist angle of ~53° exists between the plane that bisects the methylene groups and the plane of the tilt. This tilt angle is not surprising because the interchain distance is 0.665 nm (334).

Monolayers of alkyl chains on silicon are a significant addition to the family of SAMs. An ability to directly connect organic materials to silicon allows a direct coupling between organic materials and semiconductors. The fine control of superlattice structures provided by the self-assembly technique offers a route for building organic thin films with, for example, electrooptic properties on silicon.

Multilayers of Diphosphates. One way to find surface reactions that may lead to the formation of SAMs is to look for reactions that result in an insoluble salt. This is the case for phosphate monolayers, based on their highly insoluble salts with tetravalent transition metal ions. In these salts, the phosphates form layer structures, one OH group sticking to either side. Thus, replacing the OH with an alkyl chain to form the alkyl phosphonic acid was expected to result in a bilayer structure with alkyl chains extending from both sides of the metal phosphate sheet (335). When zirconium(IV) is used the distance between next neighbor alkyl chains is ~0.53 nm, which forces either chain disorder or chain tilt so that VDW attractive interactions can be reestablished.

Self-assembled multilayers can be prepared simply by alternating adsorption of Zr^{4+} ions and α,ω-alkylidenediphosphate (**15**) on a phosphorylated surface (336,337). Other diphosphates have also been investigated (237,325, 326,338–340) (Fig. 15). These are all centrosymmetric multilayers.

For second-order NLO applications, the films need to be noncentrosymmetric. 4-Di(2-hydroxyethyl)amino-4′-azobenzenephosphonate was used to form SAMs on zirconium-treated phosphorylated surfaces. Further reaction with $POCl_3$ and hydrolysis created a new phosphorylated surface that could be treated with zirconium salt (341–343). The principal advantage of the phosphate systems is high thermal stability, simple preparation, and the variety of substrates that can be used. The latter is especially important if transparent substrates are required. Thiolate monolayers are not transparent, and alkyltrichlorosilanes have a serious stability disadvantage.

Surface Engineering Using SAMs. Independent control of surface structure and chemical properties and the resulting structure property relationships are scientifically interesting and technologically important. For many applications, controlling the properties of interfaces is very important. However, in real-life circumstances, interfaces that contain at least one polymer surface are typically irregular. Surface properties of polymers depend critically upon the chemical and physical details of molecular structure at the surface of the polymer. To control surface properties by manipulating surface structure, it is necessary to have an extensive database of detailed correlations between properties and structure for the polymer surface of interest.

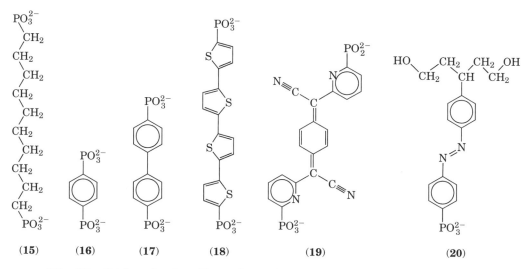

Fig. 15. Diphosphonic acids used in self-assembled multilayer preparation.

Surface properties are generally considered to be controlled by the outermost 0.5–1.0 nm at a polymer film (344). A logical solution, therefore, is to use self-assembled monolayers (SAMs) as model polymer surfaces. To understand fully the breadth of surface interactions, a portfolio of chemical functionalities is needed. SAMs are especially suited for the studies of interfacial phenomena owing to the fine control of surface functional group concentration.

In choosing a SAM system for surface engineering, there are several options. Silane monolayers on hydroxylated surfaces are an option where transparent or nonconductive systems are needed. However, trichlorosilane compounds are moisture-sensitive and polymerize in solution. The resulting polymers contaminate the monolayer surface, which occasionally has to be cleaned mechanically. Carboxylic acids adsorb on metal oxide, eg, Al_2O_3, AgO through acid–base interactions. These are not specific; therefore, it would be impossible to adsorb a carboxylic acid selectively in the presence of, for example, a terminal phosphonic acid group. In many studies SAMs of thiolates on Au(111) are the system of choice.

The structure of SAMs is affected by the size and chemical properties of surface functionalities. Indeed, the introduction of any surface functionality reduces monolayer order. The impetus toward disorder may result from sterically demanding terminal groups, eg, $-O-Si(CH_3)_2(C(CH_3)_3)$ (245) and $-C_5H_5N{:}Ru(NH_3)_5$ (345,346), or from very polar surface groups, eg, OH, COOH, etc. In both cases, the disorder introduced may be significant and not confined only to the surface.

The sensitivity of wetting to surface chemistry is evident from an OH-concentration-driven wetting transition of hexadecane (Fig. 16) (110), observed in the study of mixed SAMs containing varying proportions of hydrophobic, CH_3, and hydrophilic, OH, components. A mechanism based on the influence of surface-adsorbed water layers was supported by calculations based on a mean-field Cahn-type wetting analysis. These calculations also predicted the correct trend in the transition-onset position as a function of relative humidity (347). As relative humidity decreases, the transition-onset shifts to higher surface

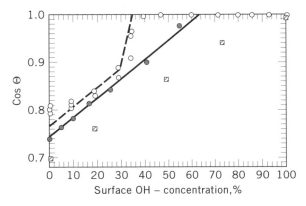

Fig. 16. Cos θ of hexadecane for $HO(CH_2)_{11}SH$ and $CH_3(CH_2)_{11}SH$ in 30% RH (○) and in ≤2% RH (◉), and for $HO(CH_2)_{11}SH$ and $CH_3(CH_2)_{13}SH$ (▨) mixed alkanethiolate SAMs on gold, as a function of surface OH-concentration. The lines represent theoretical calculations.

OH-concentration. This prediction was confirmed experimentally. In an experiment demonstrating the sensitivity of the wetting process to surface roughness at the molecular level, two CH_2 groups (together, 0.25 nm-long) were added to the hydrophobic component. The wetting transition disappeared, demonstrating the potential of surface engineering using SAMs, where changes at the molecular level made possible by utilizing mixed SAMs may result in control of macroscopic surface properties. The success of surface engineering at the molecular level requires surface stability, ie, that surface functional groups not initiate or promote surface reorganization. Moreover, since it can be expected that structural changes at the surface penetrate into the monolayer bulk, surface stability may have a significant effect on the equilibrium structure of the monolayer. Surface reorganization is a complex phenomenon. It is not clear a priori to what depth conformational changes that start at the surface can penetrate.

Instability in the wettability behavior of OH surfaces was noticed when OH-terminated silane monolayers were exposed to hydrophobic solvents, such as CCl_4 (175). Similarly, monolayers of 11-hydroxyundecane-thiol (HUT), $HO-(CH_2)_{11}-SH$, on Au(111) surfaces have been found to undergo surface reorganization by exposure to ambient atmosphere for a few hours (328). After that, the water contact angle reached a value of ca 60°, and only ca 25% of the OH groups could be esterified by trifluoroacetic anhydride. Molecular dynamics simulations verified that the driving force for the surface reorganization is the formation of surface-correlated H-bonds (348,349). Surface instability was also observed for mixed monolayers. The decrease of surface-free energy with time increases with the increasing number of surface OH groups, ie, with the increase of surface-free energy. These observations would support the assumption of a mechanism in which surface-free energy decreases owing to the decrease of surface OH groups, resulting from conformational changes at chain termina. As surface-free energy increases, the tendency toward reorganization, which results in exposure of CH_2 groups and a surface energy decrease, increases. This tendency can be offset by strong intermolecular interactions. Stability studies of monolayers made of a longer-chain derivative, $HO-(CH_2)_{21}-SH$, as a function of temperature showed that surface reorganization is indeed a function of monolayer melting point.

Every monolayer surface, even that made of CH_3 groups, is disordered at room temperature because of gauche defects at the chain termina. However, whereas the concentration of surface gauche defects is a function of free volume, the latter is a function of the adsorption scheme and of molecular cross-sectional area. Furthermore, surface reorganization may be augmented by the formation of H-bonds, as in the case of surface OH groups, or be restricted by the size and shape of the functional group, OH, vs COOH, or SO_3H. Temperature, relative humidity, and adsorption at the monolayer surface are other factors that affect surface stability. The equilibrium structure of a surface is the result of balancing all these factors, and is very hard to predict. However, the stability of a monolayer against reorganization may be increased by intermolecular interactions, as described; however, studies confirming this hypothesis have not yet been carried out.

Conclusions. Future strategies for building supramolecular devices may be based on molecular biology principles. The assembly of modules of increasing hierarchic order and the testing of modules before each assembly step, rejecting incorrect samples has been suggested (Fig. 17). Having such a sequence is a basic requirement for avoiding accumulating more defects when increasing the complexity of the assembly. The naturally occurring mechanism for such self-repair in biosystems is the aggregation of modules. Thus, incorrect samples that do not match are rejected and exchanged for the correct ones.

Beyond the self-organization of two-dimensional assemblies at interfaces, the next level of complexity requires controlling the third dimension. The different fabrication methods for organized molecular films offer a mechanism for building multilayer films, each having its own advantages and disadvantages. Amphiphiles used in the LB technique are reasonably stable, but the resultant films are unstable thermally, with the possible exception of those made of polymeric amphiphiles. In the latter case, however, the viscosity of the layer prevents fast deposition rates and may limit large-scale fabrication of useful devices. Understanding how the structure of a polymeric amphiphile, its molecular weight and molecular weight distribution, relate to the viscosity of its monolayer at the air–water interface is of crucial importance. Relating deposition rates to parameters such as surface viscosity and temperature is not a straightforward matter, and requires a large matrix of experiments. Analyzing the resulting films for defects and relating order parameters to deposition rates is also a complete task. Nevertheless, without such efforts, it is difficult to envision actual utilization of the LB technique in manufacturing.

The advantage of the LB technique is that it allows systematic studies of 2-D organization, both before and after transfer from the air–water interface onto a solid substrate. However, the coupling of 3-D self-organization of macromolecules in solution with organization at a solid surface may best be achieved using the self-assembly technique.

Whereas research in SAMs was originally motivated by a potential for application as building blocks for superlattices having engineered physical

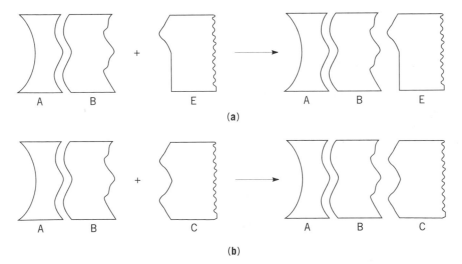

Fig. 17. Testing modules before each assembly step and rejecting incorrect samples: (**a**) E, an erroneous copy, is rejected; whereas (**b**) C, a correct copy, is accepted.

28. A. Barraud, *Thin Solid Films* **99**, 317 (1983).

29. F. Kajzar and J. Messier, *Thin Solid Films* **132**, 11 (1985).

30. H. Kelker and R. Hatz, *Handbook of Liquid Crystals*, Verlag Chemie, Weinheim, Germany, 1980.

31. T. Sakuhara, H. Nakahara, and K. Fukuda, *Thin Solid Films* **159**, 345 (1988).

32. M. J. Cook, M. F. Daniel, K. J. Harrison, N. B. McKeown, and A. J. Thomson, *J. Chem. Soc. Chem. Commun.*, 1148 (1987).

33. N. B. McKeown and co-workers, *Thin Solid Films* **159**, 469 (1988).

34. M. J. Cook, N. B. McKeown, and A. J. Thomson, *Chem. Mater.* **1**, 287 (1989).

35. I. Langmuir and V. J. Shaefer, *J. Am. Chem. Soc.* **59**, 2075 (1937).

36. A. Ruaudel-Teixier, A. Barraud, B. Belbeoch, and M. Roulliay, *Thin Solid Films* **99**, 33 (1983).

37. P. Lesieur, M. Vandevyver, A. Ruaudel-Teixier, and A. Barraud, *Thin Solid Films* **159**, 315 (1988).

38. A. W. Snow and N. L. Jarvis, *J. Am. Chem. Soc.* **106**, 4706 (1984).

39. W. R. Barger, A. W. Snow, H. Wohltjen, and N. L. Jarvis, *Thin Solid Films* **133**, 197 (1985).

40. D. P. Dilella, W. R. Barger, A. W. Snow, and R. R. Smardzewski, *Thin Solid Films* **133**, 207 (1985).

41. A. W. Snow, W. R. Barger, M. Klusky, H. Wohltjen, and N. L. Jarvis, *Langmuir* **2**, 513 (1986).

42. M. D. Pace, W. R. Barger, and A. W. Snow, *J. Mag. Res.* **75**, 73 (1987).

43. W. R. Barger, J. Dote, M. Klusty, R. Mowery, R. Price, and A. W. Snow, *Thin Solid Films* **159**, 369 (1985).

44. J. D. Shutt, D. A. Batzel, R. V. Sudiwala, S. E. Rickert, and M. E. Kenney, *Langmuir* **4**, 1240 (1988).

45. A. Cemel, T. Fort Jr., and J. B. Lando, *J. Polym. Sci. A-1* **10**, 2061 (1972).

46. A. Barraud, C. Rosilio, and A. Ruaudel-Teixier, *Microcircuit Engineering 79*, Institut of Semiconductors and Electronics, Aachen, Germany, 1979, p. 127.

47. H. G. Winful, J. H. Harburger, and E. Garmire, *Appl. Phys. Lett.* **35**, 379 (1979).

48. C. T. Seaton, X. Mau, G. I. Stegeman, and H. G. Winful, *Opt. Eng.* **24**, 593 (1985).

49. D. Sarid, *Opt. Lett.* **6**, 552 (1981).

50. A. Lattes, H. A. Haus, F. J. Leonberger, and E. P. Ipen, *IEEE J. Quant. Elect.* **QE-19**, 1718 (1983).

51. R. J. Leymour and co-workers, *Proc. SPIE* **578**, 137 (1985).

52. B. Tieke, G. Wegner, D. Naegele, and H. Ringsdorf, *Angew. Chem. Int. Ed. Engl.* **15**, 764 (1976).

53. B. Tieke and G. Wegner, in E. Kay and P. S. Bagus, eds., *Topics in Surface Chemistry*, Plenum Press, New York, 1978, p. 121.

54. D. Day and H. Ringsdorf, *J. Polym. Sci., Polym. Lett. Ed.* **16**, 205 (1978).

55. D. Day and H. Ringsdorf, *Makromol. Chem.* **180**, 1059 (1979).

56. D. Day, J. B. Lando, and H. Ringsdorf, *Polym. Prepr., Am. Chem. Soc., Div. Polym. Chem.* **19**, 176 (1978).

57. K. Lochner, H. Bässler, B. Tieke, and G. Wegner, *Phys. Stat. Solidi B* **82**, 633 (1978).

58. G. Wegner, *Z Naturfosch., Teil B* **24**, 824 (1969).

59. G. Wegner, *Pure Appl. Chem.* **49**, 443 (1977).

60. F. Grunfeld and C. W. Pitt, *Thin Solid Films* **99**, 249 (1983).

61. R. H. Tredgold and C. S. Winter, *J. Phys. D* **15**, L55 (1982).

62. H. Ringsdorf, G. Schmidt, and J. Schneider, *Thin Solid Films* **152**, 207 (1987).

63. R. Elbert, A. Laschewsky, and H. Ringsdorf, *J. Am. Chem. Soc.* **107**, 4134 (1985).

64. B. L. Henke, *Adv. X-Ray Anal.* **7**, 460 (1947).

65. G. Roberts, ed., *Langmuir-Blodgett Films*, Plenum Press, New York, 1990.

66. O. A. Aktsipetrov and E. D. Mishina, *JEPT Lett.* **37**, 207 (1983).

67. G. L. Gaines Jr., *Anal. Chem.* **48**, 450 (1976).

68. M. F. Daniel and G. W. Smith, *Mol. Cryst. Liq. Cryst.* **102**, 193 (1984).

69. J. E. Kuder and D. Wychick, *Chem. Phys. Lett.* **102**, 193 (1974).

70. G. L. Gaines Jr., *Nature* **298**, 544 (1982).

71. I. R. Girling, N. A. Cade, P. V. Kolinsky, and C. M. Montgomery, *Elect. Lett.* **21**, 169 (1985).

72. G. W. Smith, M. F. Daniel, J. W. Barton, and N. Ratcliffe, *Thin Solid Films* **132**, 125 (1985).

73. P. Christie, G. G. Roberts, and M. C. Petty, *Appl. Phys. Lett.* **48**, 1101 (1986).

74. M. B. Biddle and S. E. Rickert, *Ferroelectrics* **76**, 133 (1987).

75. H. Kuhn, *J. Photochem.* **10**, 111 (1979).

76. R. A. Marcus, *J. Chem. Phys.* **63**, 2654 (1965).

77. E. E. Polymeropoulos, D. Möbius, and H. Kuhn, *J. Chem. Phys.* **68**, 3918 (1978).

78. *Idem., Thin Solid Films* **68**, 173 (1980).

79. T. Tran-Thi, S. Palacin, and B. Clergeot, *Chem. Phys. Lett.* **157**, 92 (1989).

80. A. Aviram and M. A. Ratner, *Chem. Phys. Lett.* **29**, 277 (1974).

81. A. S. Martin, J. R. Sambles, and G. J. Ashwell, *Thin Solid Films*, **210–211**, 313 (1992).

82. D. J. Sandman and co-workers, *Synth. Met.* **42**, 1415 (1991).

83. N. J. Geddes, J. R. Sambles, D. J. Jarvis, W. G. Parker, and D. Sandman, *J. Appl. Phys. Lett.* **56**, 1916 (1990).

84. M. Fujihira, in A. Ulman, ed., *Thin Films*, Vol. 20, Academic Press, Boston, 1995.

85. G. G. Roberts, K. P. Pande, and W. A. Barlow, *Proc. IEE Part 1* **2**, 169 (1978).

86. M. C. Petty and G. G. Roberts, in G. G. Roberts, ed., *Proc. INFOS 79*, Institute of Physics, London, 1980, p. 186.

87. K. K. Kan, M. C. Petty, and G. G. Roberts, *Thin Solid Films* **99**, 291 (1983).

88. J. P. Lloyd, M. C. Petty, G. G. Roberts, P. G. LeComber, and W. E. Spear, *Thin Solid Films* **89**, 4 (1982).

89. J. P. Lloyd, M. C. Petty, G. G. Roberts, P. G. LeComber, and W. E. Spear, *Thin Solid Films* **89**, 2974 (1982).

90. G. G. Roberts, K. P. Pande, and W. A. Barlow, *Electron. Lett.* **13**, 581 (1977).

91. H. Raether, *Phys. Thin Films* **9**, 145 (1977).

92. J. P. Lloyd, C. Pearson, and M. C. Petty, *Thin Solid Films* **160**, 431 (1988).

93. R. B. Beswick and C. W. Pitt, *J. Colloid Interface Sci.* **124**, 146 (1988).

94. R. B. Beswick and C. W. Pitt, *Chem. Phys. Lett.* **143**, 589 (1988).

95. B. Holcroft and G. G. Roberts, *Thin Solid Films* **160**, 445 (1988).

96. T. Moriizumi, *Thin Solid Films* **160**, 413 (1988).

97. W. M. Reichert, C. J. Bruckner, and J. Joseph, *Thin Solid Films* **152**, 345 (1987).

98. A. Arya, U. J. Krull, M. Thompson, and H. E. Wong, *Anal. Chim. Acta* **173**, 331 (1985).

99. J. Anzai, K. Furuya, C. Chen, T. Osa, and T. Matsuo, *Anal. Sci.* **3**, 271 (1987).

100. J. Anzai, J. Hashinoto, T. Osa, and T. Matsuo, *Anal. Sci.* **4**, 247 (1988).

101. H. Tsuzuki and co-workers, *Chem. Lett.*, 1265 (1988).

102. M. Sriyudthsak, H. Yamagishi, and T. Moriizumi, *Thin Solid Films* **160**, 463 (1988).

103. Y. Okahata, T. Tsuruta, K. Ijiro, and K. Ariga, *Langmuir* **4**, 1373 (1988).

104. M. Aizawa, M. Matsuzawa, and H. Shinohara, *Thin Solid Films* **160**, 477 (1988).

105. J. Anzai, S. Lee, T. Osa, and T. Makromol. *Chem., Rapid Commun.* **10**, 167 (1989).

106. H. Kuhn and A. Ulman, in Ref. 86, pp.   .

107. W. C. Bigelow, D. L. Pickett, and W. A. Zisman, *J. Colloid Interface Sci.*, **1**, 513 (1946).

108. R. G. Nuzzo and D. L. Allara, *J. Am. Chem. Soc.* **105**, 4481 (1983).

109. P. Ball, *Designing the Molecular World*, Princeton University Press, Princeton, N.J., 1994.
110. A. Ulman, S. D. Evans, Y. Shnidman, R. Sharma, J. E. Eilers, and J. C. Chang, *J. Am. Chem. Soc.* **113**, 1499 (1991).
111. A. Kumar, H. A. Biebuyck, and G. M. Whitesides, *Langmuir* **10**, 1498 (1994).
112. D. A. Tirrell, *MRS Bulletin*, (July 23–28, 1991).
113. D. L. Allara and R. G. Nuzzo, *Langmuir* **1**, 45 (1985).
114. *Ibid.*, 52.
115. H. Ogawa, T. Chihera, and K. Taya, *J. Am. Chem. Soc.* **107**, 1365 (1985).
116. N. E. Schlotter, M. D. Porter, T. B. Bright, and D. L. Allara, *Chem. Phys. Lett.* **132**, 93 (1986).
117. D. Y. Huang and Y.-T. Tao, *Bull. Inst. Chem., Acad. Sin.* **33**, 73 (1986).
118. Y. Shnidman, A. Ulman, and J. E. Eilers, *Langmuir* **9**, 1071 (1993).
119. M. G. Samart, C. A. Brown, and J. G. Gordon, *Langmuir* **9**, 1082 (1993).
120. Y.-T. Tao, *J. Am. Chem. Soc.* **115**, 4350 (1993).
121. W. R. Thompson and J. E. Pemberton, *Langmuir* **11**, 1720 (1995).
122. E. L. Smith and M. D. Porter, *J. Phys. Chem.* **97**, 4421 (1993).
123. A. H. M. Soundag, A. J. W. Tol, and F. J. Touwslager, *Langmuir* **8**, 1127 (1992).
124. Y.-T. Tao, M.-T. Lee, and S.-C. Chang, *J. Am. Chem. Soc.* **115**, 9547 (1993).
125. J. Sagiv, *J. Am. Chem. Soc.* **102**, 92 (1980).
126. P. Silberzan, L. Léger, D. Ausserré, and J. J. Benattar, *Langmuir* **7**, 1647 (1991).
127. S. R. Wasserman, Y.-T. Tao, and G. M. Whitesides, *Langmuir* **5**, 1074 (1989).
128. J. D. Le Grange, J. L. Markham, and C. R. Kurjian, *Langmuir* **9**, 1749 (1993).
129. R. Maoz and J. Sagiv, *J. Colloid and Interf. Sci.* **100**, 465 (1984).
130. J. Gun and J. Sagiv, *J. Colloid and Interf. Sci.* **112**, 457 (1986).
131. J. Gun, R. Iscovici, and J. Sagiv, *J. Colloid and Interf. Sci.* **101**, 201 (1984).
132. N. Tillman, A. Ulman, and J. S. Schildkraut, and T. L. Penner, *J. Am. Chem. Soc.* **110**, 6136 (1988).
133. S. Brandriss and S. Margel, *Langmuir* **9**, 1232 (1993).
134. K. Mathauser and C. W. Frank, *Langmuir* **9**, 3002 (1993).
135. K. Mathauser and C. W. Frank, *Langmuir* **9**, 3446 (1993).
136. G. Carson and S. Granick, *J. Appl. Polym. Sci.* **37**, 2767 (1989).
137. C. R. Kessel and S. Granick, *Langmuir* **7**, 532 (1991).
138. D. K. Schwartz, S. Steinberg, J. Israelachvili, and Z. A. N. Zasadzinski, *Phys. Rev. Lett.* **69**, 3354 (1992).
139. H. O. Finklea and co-workers, *Langmuir* **2**, 239 (1986).
140. I. Rubinstein, E. Sabatani, R. Maoz, and J. Sagiv, *Proc. Electrochem. Soc.* **86**, 175 (1986).
141. I. Rubinstein, E. Sabatani, R. Maoz, and J. Sagiv, *Electroanal. Chem.* **219**, 365 (1987).
142. D. L. Allara, A. N. Parikh, and F. Rondelez, *Langmuir* **11**, 2357 (1995).
143. C. P. Tripp and M. L. Hair, *Langmuir* **8**, 1120 (1992).
144. D. L. Angst and G. W. Simmons, *Langmuir* **7**, 2236 (1991).
145. M. E. McGovern, K. M. R. Kallury, and M. Thompson, *Langmuir* **10**, 3607 (1994).
146. S. R. Wasserman and co-workers, *J. Am. Chem. Soc.* **111**, 5852 (1989).
147. C. P. Tripp and M. L. Hair, *Langmuir* **11**, 149 (1995).
148. W. Gao and L. Reven, *Langmuir* **11**, 1860 (1995).
149. K. Bierbaum and co-workers, *Langmuir* **11**, 512 (1995).
150. R. Banga, J. Yarwood, A. M. Morgan, B. Evans, and J. Kells, *Langmuir* **11**, 4393 (1995).
151. S. R. Cohen, R. Naaman, and J. Sagiv, *J. Chem. Phys.* **90**, 3054 (1986).
152. T. Ohtake, N. Mino, and K. Ogawa, *Langmuir* **8**, 2081 (1992).

153. I. M. Tidswell and co-workers, *Phys. Rev.* **B41**, 1111 (1990).
154. T. Nakagawa and K. Ogawa, *Langmuir* **10**, 367 (1994).
155. H. Okusa, K. Kurihara, and T. Kunitake, *Langmuir* **10**, 8 (1994).
156. M. Fujii, S. Sugisawa, K. Fukada, T. Kato, T. Shirakawa, and T. Seimiya, *Langmuir* **10**, 984 (1994).
157. N. Yoshino, *Chem. Lett.*, 735 (1994).
158. K. Bierbaum and M. Grunze, *Adhesion Soc.*, 213 (1994).
159. D. H. Flinn, D. A. Guzonas, and R.-H. Yoon, *Colloids Surf. A.* **87**, 163 (1994).
160. Y.-I. Rabinovich and R.-H. Yoon, *Langmuir* **10**, 1903 (1994).
161. C. A. Siedlecki, S. L. Eppell, and R. E. Marchant, *J. Biomed. Mater. Res.* **28**, 271 (1994).
162. K. Mathauer and C. W. Frank, *Langmuir* **9**, 3446 (1993).
163. L. Netzer, R. Iscovichi, and J. Sagiv, *Thin Solid Films* **100**, 67 (1983).
164. M. Pomerantz, A. Segmüller, L. Netzer, and J. Sagiv, *Thin Solid Films*, **132**, 153 (1985).
165. N. Tillman, A. Ulman, and T. L. Penner, *Langmuir* **5**, 101 (1989).
166. D. G. Kurth and T. Bein, *Langmuir* **11**, 2965 (1995).
167. *Ibid.*, 3061.
168. C. N. Durfor, D. C. Turner, J. H. Georger, B. M. Peek, and D. A. Stenger, *Langmuir* **10**, 148 (1994).
169. J. Lühe, V. J. Novotny, K. K. Kanazawa, T. Clarke, and G. B. Street, *Langmuir* **9**, 2383 (1993).
170. X.-D. Xiao, G.-Y. Liu, D. H. Charych, and M. Salmeron, *Langmuir* **11**, 1600 (1995).
171. X.-D. Xiao, J. Hue, D. H. Charych, and M. Salmeron, *Langmuir* **12**, 235 (1996).
172. M. K. Chaudhury and G. M. Whitesides, *Science* **255**, 1230 (1992).
173. N. Balachander and C. N. Sukenik, *Langmuir* **6**, 1621 (1990).
174. Y. W. Lee, J. Reed-Mundell, C. N. Sukenik, and J. E. Zull, *Langmuir* **9**, 3009 (1993).
175. J. A. Chupa and co-workers, *J. Am. Chem. Soc.* **115**, 4383 (1993).
176. S. Paulson, K. Morris, and B. P. Sullivan, *J. Chem. Soc. Chem. Commun.*, 1615 (1992).
177. S. R. Wasserman, H. Biebuyck, and G. M. Whitesides, *J. Mater. Res.* **4**, 886 (1989).
178. M. Maoz and J. Sagiv, *Langmuir* **3**, 1034 (1987).
179. M. Maoz and J. Sagiv, *Langmuir* **3**, 1045 (1987).
180. R. Maoz, L. Netzer, J. Gun, and J. Sagive, *J. Chim. Phys. (Paris)* **85**, 1059 (1988).
181. L. Netzer and R. Iscovici, J. Sagiv, *Thin Solid Films* **99**, 235 (1983).
182. R. Netzer and J. Sagiv, *J. Am. Chem. Soc.* **105**, 674 (1983).
183. R. Maoz and J. Sagiv, *Thin Solid Films* **132**, 135 (1985)K.
184. Ogawa, N. Mino, H. Tamura, and M. Hatada, *Langmuir* **6**, 851 (1990).
185. *Ibid.*, 1807.
186. D. Q. Li, M. A. Ratner, T. J. Marks, C. H. Zhang, J. Yang, and G. K. Wong, *J. Am. Chem. Soc.* **112**, 7389 (1990).
187. A. K. Kakkar and co-workers, *Langmuir* **9**, 388 (1993).
188. S. Yitzchaik and co-workers, *J. Phys. Chem.* **97**, 6958 (1993).
189. S. B. Roscoe, S. Yitzchaik, A. K. Kakkar, T. J. Marks, W. L. Lin, and G. K. Wong, *Langmuir* **10**, 1337 (1994).
190. C. P. Tripp, R. P. N. Veregin, and M. L. Hair, *Langmuir* **9**, 3518 (1993).
191. (to be supplied)
192. L. J. Jeon, R. G. Nuzzo, Y. Xia, M. Mrksich, and G. M. Whitesides, *Langmuir* **11**, 3024 (1995).
193. Y. Xia, M. Mrksich, E. Kim, and G. M. Whitesides, *J. Am. Chem. Soc.* **117**, 9576 (1995).
194. W. J. Dressick, C. S. Dulcey, J. H. Georger, and J. M. Calvert, *Chem. Mater.* **5**, 148 (1993).

195. K. Chen, W. B. Caldwell, and C. A. Mirkin, *J. Am. Chem. Soc.* **115**, 1193 (1993).
196. D. Q. Li and B. I. Swanson, *Langmuir* **9**, 3341 (1993).
197. V. V. Tsukruk, L. M. Lander, and W. L. Brittain, *Langmuir* **10**, 996 (1994).
198. H.-H. Hong, M. Jiang, S. G. Slinger, and P. Bohn, *Langmuir* **10**, 153 (1994).
199. D. A. Offord and J. H. Griffin, *Langmuir* **9**, 3015 (1993).
200. C. N. Sukenik, personal communication.
201. K. Robinson, A. Ulman, J. Lando, and A. J. Mann, personal communication.
202. T. J. Marks, preprint.
203. R. Maoz, R. Yam, G. Berkovic, and J. Sagiv, in Ref. 86, pp.
204. R. Maoz, J. Sagiv, D. Degenhardt, H. Möhwald, and P. Quint, *Supramol. Sci.* **2**, 9 (1995).
205. N. Mino, K. Ogawa, M. Hatada, M. Takastuka, S. Sha, and T. Moriizumi, *Langmuir* **9**, 1280 (1993).
206. L. H. Dubois and R. G. Nuzzo, *Ann. Phys. Chem.* **43**, 437 (1992).
207. C. D. Bain and G. M. Whitesides, *Adv. Mater.* **1**, 506 (1989).
208. J. P. Folkers, J. A. Zerkowski, P. E. Laibinis, C. T. Seto, and G. M. Whitesides, in T. Bain, ed., *Supramolecular Architecture, ACS Symposium Series 499*, American Chemical Society, Washington, D.C., 1992, pp. 10–23.
209. T. R. Lee, P. E. Laibinis, J. P. Folkers, and G. M. Whitesides, *Pure & Appl. Chem.* **63**, 821 (1991).
210. G. M. Whitesides and G. S. Ferguson, *Chemtracts-Organic Chemistry* **1**, 171 (1988).
211. H. Sellers, A. Ulman, Y. Shnidman, and J. E. Eilers, *J. Am. Chem. Soc.* **115**, 9389 (1993).
212. E. B. Troughton, C. D. Bain, G. M. Whitesides, D. L. Allara, and M. D. Porter, *Langmuir* **4**, 365 (1988).
213. E. Katz, N. Itzhak, and I. Willner, *J. Electroanal. Chem.* **336**, 357 (1992).
214. E. Sabatani, J. Cohen-Boulakia, M. Bruening, and I. Rubinstein, *Langmuir* **9**, 2974 (1993).
215. M. A. Bryant, S. L. Joa, and J. E. Pemberton, *Langmuir* **9**, 753 (1992).
216. W. Hill and B. Wehling, *J. Phys. Chem.* **97**, 9451 (1993).
217. T. T.-T. Li, H. Y. Liu, and M. J. Weaver, *J. Am. Chem. Soc.* **106**, 1233 (1984).
218. J. M. Cooper, K. R. Greenough, and C. J. McNeil, *J. Electroanal. Chem.* **347**, 267 (1993).
219. A. Ihs, K. Uvdal, and B. Liedberg, *Langmuir* **9**, 733 (1993).
220. Th. Arndt, H. Schupp, and W. Schepp, *Thin Solid Films* **178**, 319 (1989).
221. J. A. Mielczarski and R. H. Yoon, *Langmuir* **7**, 101 (1991).
222. T. R. G. Edwards, V. J. Cunnane, R. Parsons, and D. Gani, *J. Chem. Soc. Chem. Commun.*, 1041 (1989).
223. A. J. Arduengo, J. R. Moran, J. Rodriguez-Paradu, and M. D. Ward, *J. Am. Chem. Soc.* **112**, 6153 (1990).
224. G. Xue, X.-Y. Huang, J. Dong, and J. Zhang, *J. Electroanal. Chem.* **310**, 139 (1991).
225. S. Bharathi, V. Yegnaraman, and G. P. Rao, *Langmuir* **9**, 1614 (1993).
226. M. G. Samanat, C. A. Broen, and J. G. Gordon, *Langmuir* **8**, 1615 (1992).
227. G. A. Somorjai, *Chemistry in Two Dimensions—Surfaces*, Cornell University Press, Ithaca, N.Y., 1982.
228. D. E. King, *J. Vac. Sci. Technol.* 0000 (1995).
229. A. Ulman, *J. Mater. Educ.* **11**, 205 (1989).
230. P. E. Laibinis and co-workers, *J. Am. Chem. Soc.* **113**, 7152 (1991).
231. M. W. Walczak, C. Chung, S. M. Stole, C. A. Widrig, and M. D. Porter, *J. Am. Chem. Soc.* **113**, 2370 (1991).
232. P. E. Laibinis and G. M. Whitesides, *J. Am. Chem. Soc.* **112**, 1990 (1992).
233. A. Ihs and B. Liedberg, *Langmuir* **10**, 734 (1994).

234. P. E. Laibinis and G. M. Whitesides, *J. Am. Chem. Soc.* **114**, 9022 (1992).

235. K. Shimazu, Y. Sato, I. Yagi, and K. Uosaki, *Bull. Chem. Soc. Jpn.* **67**, 863 (1994).

236. A. Demoz and D. J. Harrison, *Langmuir* **9**, 1046 (1993).

237. N. Muskal, I. Turyan, A. Shurky, and D. Mandler, *J. Am. Chem. Soc.* **117**, 1147 (1995).

238. M. Stratmann, *Adv. Mater.* **2**, 191 (1990).

239. M. Volmer, M. Stratmann, and H. Viefhaus, *Surf. and Interf. Anal.* **16**, 278 (1990).

240. Q. Liu and Z. Xu, *Langmuir* **11**, 4617 (1995).

241. M. Brust, M. Walker, D. Bethell, D. J. Schiffrin, and R. Whyman, *J. Chem. Soc. Chem. Commun.*, 801 (1994).

242. C. W. Sheen, J. X. Shi, J. Martensson, A. N. Parikh, and D. L. Allara, *J. Am. Chem. Soc.* **114**, 1514 (1992).

243. Y. Gu, B. Lin, V. S. Smentkowski, and D. H. Waldeck, *Langmuir* **11**, 1849 (1995).

244. A. Ulman, personal communication.

245. C. D. Bain and co-workers, *J. Am. Chem. Soc.* **111**, 321 (1989).

246. M. Buck, F. Eisert, J. Fischer, M. Grunze, and F. Träger, *Appl. Phys.* **A53**, 552 (1991).

247. M. Buck, F. Eisert, and M. Grunze, *Ber. Bunsenges. Phys. Chem.* **97**, 399 (1993).

248. G. Hähner, Ch. Wöll, M. Buck, and M. Grunze, *Langmuir* **9**, 1955 (1993).

249. S. D. Evans, E. Urankar, A. Ulman, and N. J. Ferris, *J. Am. Chem. Soc.* **113**, 4121 (1991).

250. D. A. Offord, C. M. John, M. R. Linford, and J. H. Griffin, *Langmuir* **10**, 883 (1994).

251. H. A. Biebuyck, C. D. Bain, and G. M. Whitesides, *Langmuir* **10**, 1825 (1994).

252. H. A. Biebuyck and G. M. Whitesides, *Langmuir* **9**, 1766 (1993).

253. N. Mohri, M. Inoue, Y. Arai, and K. Yoshikawa, *Langmuir* **11**, 1612 (1995).

254. P. Fenter, A. Eberhardt, and P. Eisenberger, *Science* **266**, 1216 (1994).

255. R. C. Thomas, L. Sun, and M. Crooks, *Langmuir* **7**, 620 (1991).

256. O. Chailapakul, L. Sun, C. Xu, and M. Crooks, *J. Am. Chem. Soc.* **115**, 12459 (1993).

257. M. D. Porter, T. B. Bright, D. L. Allara, and C. E. D. Chidsey, *J. Am. Chem. Soc.* **109**, 3559 (1987).

258. R. G. Nuzzo, F. A. Fusco, and D. L. Allara, *J. Am. Chem. Soc.* **109**, 2358 (1987).

259. C. D. Bain, H. A. Biebuyck, and G. M. Whitesides, *Langmuir* **5**, 723 (1989).

260. R. G. Nuzzo, B. R. Zegarski, and L. H. Dubois, *J. Am. Chem. Soc.* **109**, 733 (1987).

261. R. G. Nuzzo, L. H. Dubois, and D. L. Allara, *J. Am. Chem. Soc.* **112**, 558 (1990).

262. Y. Li, J. Huang, R. T. McIver Jr., and J. C. Hemminger, *J. Am. Chem. Soc.* **114**, 2428 (1992).

263. C. A. Widrig, C. Chung, and M. D. Porter, *J. Electroanal. Chem.*, (1991).

264. M. A. Bryant and J. E. Pemberton, *J. Am. Chem. Soc.* **113**, 3630 (1991).

265. *Ibid.*, 8284.

266. J. B. Schlenoff, M. Li, and H. Ly, *J. Am. Chem. Soc.* **117**, 12528 (1995).

267. J. J. Hickman, D. Ofer, C. Zou, M. S. Wrighton, P. E. Laibinis, and G. M. Whitesides, *J. Am. Chem. Soc.* **113**, 1128 (1991).

268. D. M. Collard and M. A. Fox, *Langmuir* **7**, 1192 (1991).

269. K. A. Groat and S. E. Creager, *Langmuir* **9**, 3668 (1993).

270. M. S. Ravenscroft and H. O. Finklea, *J. Phys. Chem.* **98**, 3843 (1994).

271. L. S. Curtin and co-workers, *Anal. Chem.* **65**, 368 (1993).

272. D. M. Jaffey and R. J. Madix, *J. Am. Chem. Soc.* **116**, 3012 (1994).

273. L. Strong and G. M. Whitesides, *Langmuir* **4**, 546 (1988).

274. C. E. D. Chidsey and D. N. Loiacono, *Langmuir* **6**, 709 (1990).

275. L. H. Dubois, B. R. Zegarski, and R. G. Nuzzo, *J. Chem. Phys.* **98**, 678 (1993).

276. C. E. D. Chidsey, G.-Y. Liu, Y. P. Rowntree, and G. Scoles, *J. Chem. Phys.* **91**, 4421 (1989).

277. C. A. Alves, E. L. Smith, and M. D. Porter, *J. Am. Chem. Soc.* **114**, 1222 (1992).

278. G. E. Poirier, M. J. Tarlov, and H. E. Rushneier, *Langmuir* **10**, 3383 (1994).

279. P. Fenter, P. Eisenberger, and K. S. Liang, *Phys. Rev. Lett.* **70**, 2447 (1993).

280. N. Camillone, C. E. D. Chidsey, G.-Y. Liu, and G. Scoles, *J. Phys. Chem.* **98**, 3503 (1993).

281. G. E. Poirier and M. J. Tarlov, *Langmuir* **10**, 2859 (1994).

282. G. Edinger, A. Gölzhäuser, K. Demota, Ch. Wöll, and M. Grunze, *Langmuir* **9**, 4 (1993).

283. C. Schönenberger, J. A. M. Sondag-Huethorst, J. Jorritsma, and L. G. J. Fokkink, *Langmuir* **10**, 611 (1994).

284. M. Grunze, *Phys. Scripta*, 0000, (1993).

285. R. L. McCarley, D. J. Dunaway, and R. J. Willicut, *Langmuir* **9**, 2775 (1993).

286. J.-P. Bucher, L. Santesson, and K. Kern, *Langmuir* **10**, 979 (1994).

287. K. Schwaha, N. D. Spencer, and R. M. Lambert, *Surf. Sci.* **81**, 273 (1979).

288. G. Rovida and F. Pratesi, *Surf. Sci.* **104**, 609 (1981).

289. A. L. Harris, C. E. D. Chidsey, N. J. Levinos, and D. N. Loiacono, *Chem. Phys. Lett.* **141**, 350 (1987).

290. A. L. Harris, L. Rothberg, L. Dhar, N. J. Levinos, and L. H. Dubois, *J. Phys. Chem.* **94**, 2438 (1991).

291. T. Seto, T. Hara, and K. Tanaka, *Jpn. J. Appl. Phys.* **7**, 31 (1968).

292. P. Fenter and co-workers, *Langmuir* **7**, 2013 (1991).

293. A. Dhirani, M. A. Hines, A. J. Fisher, O. Ismail, and P. Guyot-Sionnest, *Langmuir* **11**, 2609 (1995).

294. J. Y. Gui, D. A. Stern, D. G. Frank, F. Lu, D. C. Zapien, and A. T. Hubbard, *Langmuir* **7**, 955 (1991).

295. J. Y. Gui, F. Lu, D. A. Stern, and A. T. Hubbard, *J. Electroanal. Chem.* **292**, 245 (1990).

296. A. Nemetz, T. Fischer, A. Ulman, and W. Knoll, *J. Chem. Phys.* **98**, 5912 (1993).

297. P. Fenter and P. Eisenberger, personal communication.

298. C. E. D. Chidsey and D. N. Loiacono, *Langmuir* **6**, 682 (1990).

299. K. Dohlhofer, J. Figura, and J.-H. Fuhrhop, *Langmuir* **8**, 1811 (1992).

300. C. E. D. Chidsey, C. R. Bertozzi, T. M. Putvinski, and A. M. Mujsce, *J. Am. Chem. Soc.* **112**, 4301 (1990).

301. K. Uosaki, Y. Sata, and H. Kita, *Langmuir* **7**, 1510 (1991).

302. C. E. D. Chidsey, *Science* **251**, 919 (1991).

303. D. D. Popenoe, R. S. Deinhammer, and M. D. Porter, *Langmuir* **8**, 2521 (1992).

304. Y. Sato, B. L. Frey, R. M. Corn, and K. Uosaki, *Bull. Chem. Soc. Jpn.* **67**, 21 (1994).

305. S. E. Creager and G. K. Rowe, *Anal. Chim.* **246**, 233 (1991).

306. G. K. Rowe and S. E. Creager, *Langmuir* **7**, 2307 (1991).

307. G. K. Rowe and S. E. Creager, *Langmuir* **10**, 1186 (1994).

308. L. Häussling, H. Ringsdorf, F.-J. Schmitt, and W. Knoll, *Langmuir* **7**, 1837 (1991).

309. L. Häussling, B. Michel, H. Ringsdorf, and H. Rohrer, *Angew. Chem. Int. Ed. Engl.* **30**, 679 (1991).

310. F.-J. Schmitt, L. Häussling, H. Ringsdorf, and W. Knoll, *Thin Solid Films* **210/211**, 815 (1992).

311. J. Spinke, J. Liley, H.-J. Guder, L. Angermaier, and W. Knoll, *Langmuir* **9**, 1821 (1993).

312. J. Spinke, M. Liley, F.-J. Schmitt, H.-J. Guder, L. Angemaier, and W. Knoll, *J. Chem. Phys.* **99**, 7012 (1993).

313. Y. S. Obeng and A. J. Bard, *Langmuir* **7**, 195 (1991).

314. C. M. Yip and M. D. Ward, *Langmuir* **10**, 549 (1994).

315. J. Zak, H. Yuan, K. Woo, and M. D. Porter, *Langmuir* **9**, 2772 (1993).

316. J. E. Hutchinson, T. A. Postlethwaite, and R. W. Murray, *Langmuir* **9**, 3277 (1993).

317. B. R. Herr and C. A. Mirkin, *J. Am. Chem. Soc.* **116**, 1157 (1994).
318. L. Sun, J. Kepley, and R. M. Crooks, *Langmuir* **8**, 2101 (1992).
319. L. Sun, R. M. Crooks, and A. J. Ricco, *Langmuir* **9**, 1775 (1993).
320. R. V. Duevel and R. M. Corn, *Anal. Chem.* **64**, 337 (1992).
321. T. Kim, R. M. Crooks, M. Tsen, and L. Sun, preprint.
322. A. Ulman and N. Tillman, *Langmuir* **5**, 1418 (1989).
323. L. Sun, R. C. Thomas, R. M. Crooks, and A. J. Ricco, *J. Am. Chem. Soc.* **113**, 8550 (1991).
324. C. Xu, L. Sun, L. J. Kepley, and R. M. Crooks, *Anal. Chem.* **65**, 2102 (1993).
325. M. L. Schilling and co-workers, *Langmuir* **9**, 2156 (1993).
326. S. F. Bent, M. L. Schilling, W. L. Wilson, H. E. Katz, and A. L. Harris, *Chem. Mater.* **6**, 122 (1994).
327. L. Bertilsson and B. Liedberg, *Langmuir* **9**, 141 (1993).
328. S. D. Evans, R. Sharma, and A. Ulman, *Langmuir* **7**, 156 (1991).
329. S. Löfås and B. Johnsson, *J. Chem. Soc. Chem. Commun.*, 1526 (1990).
330. I. Wilner, A. Riklin, B. Shoham, D. Rivenson, and E. Katz, *Adv. Mater.* **5**, 912 (1993).
331. H. Keller, W. Schrepp, and H. Fuchs, *Thin Solid Films* **210/211**, 799 (1992).
332. M. R. Linford and C. E. D. Chidsey, *J. Am. Chem. Soc.* **115**, 12631 (1993).
333. M. R. Linford, C. E. D. Chidsey, P. Fenter, and P. M. Eisenberger, *J. Am. Chem. Soc.* **116**, 0000 (1995).
334. A. Ulman, J. E. Eilers, and N. Tillman, *Langmuir* **5**, 1147 (1989).
335. G. Cao, H.-G. Hong, and T. E. Mallouk, *Acc. Chem. Res.* **25**, 420 (1992).
336. H. Lee, L. J. Kepley, H.-G. Hong, and T. E. Mallouk, *J. Am. Chem. Soc.* **110**, 618 (1988).
337. H. Lee, L. J. Kepley, H.-G. Hong, S. Akhter, and T. E. Mallouk, *J. Phys. Chem.* **92**, 2597 (1988).
338. H. E. Katz, M. L. Schilling, S. B. Ungahse, T. M. Putvinski, and C. E. D. Chidsey, in T. Bein, ed., *Supramolecular Architecture, ACS Symposium Series 499*, American Chemical Society: Washington, D.C., 1992, pp. 24–32.
339. H. E. Katz, M. L. Schilling, C. E. D. Chidsey, T. M. Putvinski, and R. S. Hutton, *Chem. Mater.* **3**, 699 (1991).
340. S. B. Ungahse, W. L. Wilson, H. E. Katz, R. G. Scheller, and T. M. Putvinski, *J. Am. Chem. Soc.* **114**, 8717 (1992).
341. T. M. Putvinski, M. L. Schilling, H. E. Katz, C. E. D. Chidsey, A. M. Mujsce, and A. B. Emerson, *Langmuir* **6**, 1567 (1990).
342. H. E. Katz, R. G. Scheller, T. M. Putvinski, M. L. Schilling, W. L. Wilson, and C. E. D. Chidsey, *Science* **254**, 1485 (1991).
343. H. E. Katz and M. L. Schilling, *Chem. Mater.* **5**, 1162 (1993).
344. D. L. Allara, in W. J. Feast, H. S. Munro, and R. W. Richards, eds., *Polymer Surfaces and Interfaces*, Vol. II, John Wiley & Sons, Ltd., Chichester, U.K., 1993, p. 27.
345. H. O. Finklea and D. D. Hanshew, *J. Am. Chem. Soc.* **114**, 3173 (1992).
346. H. O. Finklea and D. D. Hanshew, *J. Electroanal. Chem.* **347**, 327 (1993).
347. D. J. Olbris, A. Ulman, and Y. Shnidman, *J. Chem. Phys.* **102**, 6865 (1995).
348. J. Hautman and M. L. Klein, *Phys. Rev. Lett.* **67**, 1763 (1991).
349. J. Hautman, J. P. Bareman, W. Mar, and M. L. Klein, *J. Chem. Soc. Faraday Trans.* **87**, 2031 (1991).
350. S. J. Potochnik, P. E. Pehrsson, D. S. Y. Hsu, and J. M. Calvert, *Langmuir* **11**, 1842 (1995), and references therein.
351. A. Kumar, N. L. Abbott, E. Kim, H. A. Biebuyck, and G. M. Whitesides, *Acc. Chem. Res.* **28**, 219 (1995).
352. T. E. Mallouk and D. J. Harrison, eds., *Interfacial Design and Chemical Sensing, ACS Symposium Series 561*, American Chemical Society, Washington, D.C., 1994.

353. K. T. Carron, L. Pelterson, and M. Lewis, *Environ. Sci. Technol.* **26**, 1950 (1992).

354. L. J. Kepley, R. M. Crooks, and A. J. Ricco, *Anal. Chem.* **64**, 3191 (1992).

355. K. D. Schierbaum, T. Weiss, J. F. J. Thoden van Velzen, D. N. Reinhoudt, and W. Goepel, *Science* **265**, 1413 (1994).

356. C. M. Duan and M. E. Meyerhoff, *Anal. Chem.* **66**, 1369 (1994).

357. Y. Kajiya, T. Okamoto, and H. Yoneyama, *Chem. Lett.* **12**, 2107 (1993).

358. C. Chung and M. D. Porter, *Chem. Eng. News* 32 (May 1, 1989).

359. M. T. Rojas, R. Koniger, J. F. Stoddart, and A. E. Kaifer, *J. Am. Chem. Soc.* **117**, 336 (1995).

360. S. Steinberg, Y. Tor, E. Sabatani, and I. Rubinstein, *J. Am. Chem. Soc.* **113**, 5176 (1991).

361. P. G. deGennes, in J. Charvolin, J. F. Joanny, and J. Zinn-Justin, eds., *Conference Proceedings, Liquids at Interfaces*, Les Houches 1990, North Holland, Amsterdam, 1990.

362. E. Manias, G. Hadziioannou, I. Bitsanis, and G. ten Brinke, *Europhys. Lett.*, 1996, in press.

363. P. Silberzan and L. Leger, *Macromolecules* **25**, 1267 (1992).

364. V. J. Novotny, *J. Chem. Phys.*, 3189 (1990).

365. X. Zheng and co-workers, *Phys. Rev. Lett.* **74**, 407 (1995).

ABRAHAM ULMAN
Polytechnic University

THIOCYANATES.   See SULFUR COMPOUNDS.